Y0-EEH-211

Pure and Applied Chemistry Vol. 75, Nos. 11–12 (2003)

CONTENTS

Special Topic Issue on the Implications of Endocrine Active Substances for Humans and Wildlife

SCIENTIFIC COMMITTEE ON PROBLEMS OF THE ENVIRONMENT
AND
INTERNATIONAL UNION OF PURE AND APPLIED CHEMISTRY

IMPLICATIONS OF ENDOCRINE ACTIVE SUBSTANCES FOR HUMANS AND WILDLIFE

J. MIYAMOTO AND J. BURGER (EDITORS)

Associate Editors
John Ashby, William Kelce, Werner Klein, Kenneth Korach,
James Lamb, and Peter Matthiessen

SCOPE/IUPAC Project on Endocrine Active Substances Scientific Advisory Committee

Junshi Miyamoto, Chair, IUPAC/Chemicals Evaluation and Research Institute, Tokyo, Japan
Joanna Burger, Co-Chair, SCOPE/Rutgers University, New Jersey, USA

John Ashby, Syngenta Ltd., Cheshire, UK
William Kelce, Pharmacia Corporation, Michigan, USA
Werner Klein, IUPAC/Fraunhöfer Institute for Environmental Chemistry and Ecotoxicology, Schmallenberg, Germany
Kenneth Korach, National Institute of Environmental Health Sciences, North Carolina, USA
James Lamb, BBL Sciences, Virginia, USA
Tadao Matsumoto, Professor, Department of Biology, Graduate School of Arts and Sciences, The University of Tokyo, Japan
Peter Matthiessen, Center for Ecology and Hydrology, Cumbria, UK
Yasuyuki Oshima, President, Japan Wildlife Research Center, Japan
Tatsuo Urabe, Professor, Computation Center, Nagoya University, Japan

FUNDED BY:

BBL Sciences
Centre for Ecology and Hydrology, UK Natural Environment Research Council
Chemicals Evaluation and Research Institute, Japan (CERI)
Consortium for Risk Evaluation with Stakeholder Participation
Division of Life Science, Rutgers University
Environmental and Occupational Health Sciences Institute (EOHSI)
International Council of Chemical Industry Associations (ICCA) (European Chemical Industry Council, CEFIC; American Chemistry Council, ACC; Japan Chemical Industry Association, JCIA)
International Council for Science (ICSU)
Ministry of Economy, Trade and Industry, Japanese Government (METI)
Ministry of Agriculture, Forestry and Fisheries, Japanese Government
Ministry of Education, Culture, Sports, Science and Technology, Japanese Government
Ministry of Health, Labor and Welfare, Japanese Government
Ministry of Land, Infrastructure and Transport and Japanese Government
Organization for Economic Cooperation and Development (OECD)
Pharmacia/Pfizer
Society of Toxicology (SOT)
Syngenta Central Toxicology Laboratory
United Nations Educational, Scientific and Cultural Organization (UNESCO)
U.S. National Institute of Environmental Health Sciences (US NIEHS)
U.S. Environmental Protection Agency (USEPA)

Preface

Disruption of endocrine systems by anthropogenic as well as natural compounds has become an important global issue during the last decade, because it may encompass not only humans, but also a wide range of other organisms, and not only the present generation, but also future ones. Massive scientific research efforts are currently underway to assess the significance of reported adverse effects of exposure to xenobiotics on endocrine systems.

In 1996, the U.S. Congress mandated the EPA to establish the Endocrine Disruptor Screening and Testing Advisory Committee, which issued its final report in October 1998. The Committee recommended testing programs composed of tiered systems, starting with high-throughput pre-screening, followed by tier 1 screening using relatively simple in vivo screening, and tier 2 testing involving laborious and time-consuming tests such as rodent two-generation reproduction studies. Concomitantly, the Organization for Economic Cooperation and Development (OECD) initiated similar programs.

It was clear that there are still many scientific uncertainties to be resolved before acceptable testing procedures can be established. As pointed out in the recent "White Book" on Endocrine Disruptors by IUPAC, International Union of Pharmacology (IUPHAR), and International Union of Toxicology (IUTOX) ["Natural and Anthropogenic Environmental Estrogens: the Scientific Basis for Risk Assessment", *Pure and Applied Chemistry*, **70** (9), 1617–1865 (1998)], these uncertainties can only be resolved by conducting high-quality scientific investigations and a thorough peer review of the results. In this report, the Unions highlighted the need for a better understanding of the mechanisms by which the chemicals produce their effects, and for further examination of the relation between exposure and adverse effects both on humans and the environment. They called for the development of better methods of screening and testing chemicals. Finally, they pointed out the need for a review of existing and new risk assessment methods.

Most of the reviews undertaken about endocrine active substances had considered these issues at a national (e.g., USA, Japan) or European Union level, while the only truly international study, by IUPAC, IUPHAR, and IUTOX, took place at an early stage in the development of research in this field.

The present SCOPE/IUPAC project on endocrine active substance is the only one looking at endocrine active substances (EASs) on a world-wide basis, with emphasis on the specific situation in each region. Thus, a comparative approach was applied to look at the various aspects of the issues from different regional perspectives, and this gives the project significant added value.

Moreover, the project was also designed to take account of the significant advances in the current scientific understanding over recent years. It thus has yielded policy-relevant information and advice that could not be addressed by previous initiatives, due to the early stage of scientific information.

The issue surrounding EASs addressed by the project are a high priority on the decision-making and practitioner agendas regarding environmental chemicals, both in individual countries and globally. Scientific gaps and uncertainties remain high, and will continue for some time. However, the in-depth, comprehensive, authoritative review of EASs and their environmental and health effects by the SCOPE/IUPAC project on endocrine active substances will facilitate risk assessment and assist governmental and intergovernmental authorities, industry, and the wider public, in framing policies to address these issues.

The project culminated in a Symposium held 17–21 November 2002 in Yokohama, Japan. Scientists, managers, and public policy-makers presented papers in (1) human effects, (2) wildlife effects, (3) exposure assessment, and (4) testing for EASs and endocrine disruption (ED) effects, as well as in six workshops dealing with the effectiveness of QSAR, toxicogenomics, integrated monitoring systems, rapid assays, precautionary principle/weight of evidence approaches, and risk management options for endocrine disruptors. Overall, 408 scientists gathered from 31 countries, giving 84 talks and an additional 84 posters. The papers presented form the basis for this special issue of *Pure and Applied Chemistry*.

J. Burger
SCOPE/Rutgers University, New Jersey, USA

Dedication

On 14 April 2003, Dr. Junshi Miyamoto passed away unexpectedly after a short illness. All the members of the Scientific Advisory Committee of the SCOPE/IUPAC project on endocrine active substances, as well as all the contributors to the project find it hard to accept, given that he was so lively at the Yokohama Symposium in November 2002. With his death, we have lost the spiritus rector of the project and a very good friend.

Dr. Junshi Miyamoto, trained as a chemist, provided impressive scientific achievements during his career in the Japanese crop protection industry, where he had been Associate Director of the board of Sumitomo Chemical for ten years until retirement, and general manager of the Takarazuka Research Center for seven years. His major fields of research were toxicology, metabolism, and evaluation of the mode of action of pesticides and other xenobiotics. He also worked on biochemistry and molecular biology, environmental chemistry, and risk assessment of chemicals. In all, he published over 350 scientific publications.

The major objective of his professional activities was to contribute a science base to issues of safety and public concern and to promote the understanding of chemistry. Consequently, he was active in many international organizations, served IUPAC for more than 25 years in different positions, and was a major contributor to nine Environmental Health Criteria documents of the International Program on Chemical Safety (IPCS). He was also very active in SCOPE and the Intergovernment Forum on Chemical Safety (IFCS). His achievements were acknowledged by many foreign and Japanese awards.

From the beginning of the evolving concerns on endocrine disruption, he engaged himself in the sound scientific judgment of the potential risks for humans and the environment. This involved him in the IUPAC/IUPHAR/IUTOX "White Book" on the basis for scientific risk assessment of natural and anthropogenic estrogens. With continuous public concern and advances in scientific knowledge, and his tremendous efforts, scientific skills, and management capabilities, he initiated and led this current SCOPE/IUPAC project. It is the result of his outstanding leadership that over 80 leading international scientists in this area contributed to this unique and comprehensive project. Thus, it is our honor and deeply felt obligation to dedicate this publication to Dr. Junshi Miyamoto. The members of the Scientific Advisory Committee miss him deeply, as we have all lost a friend and a valued colleague.

Joanna Burger
Co-Chair of the SCOPE/IUPAC Project

Implications of endocrine active substances for humans and wildlife: Executive summary*

J. Miyamoto and J. Burger

INTRODUCTION

Understanding the scientific issues surrounding endocrine active substances (EASs) is an international priority. The present SCOPE/IUPAC project is a natural extension of the first project by IUPAC/IUTOX/IUPHAR (1998) [1], conducted in 1997. Many recommendations were made at that time, the chief of which was that progress made in understanding and responding to the global problem of EASs and endocrine disruption (ED) should be reviewed as appropriate. As defined during the 1996 Weybridge Workshop, "an endocrine disrupter is an exogenous substance that causes adverse health effects in an intact organism, or its progeny, secondary to changes in endocrine function." Advances in our understanding over the past five years have made it possible to review all of the major aspects of this problem and to refine goals and research needs. The project concentrated on four broad areas: (1) nuclear receptor mechanisms, (2) fate and metabolism of EASs, (3) effects in rodents and humans, and (4) effects in wildlife species. Workshops were also conducted on a variety of technical, regulatory, management, and policy issues.

Each of the four sections included presentations and chapters from experts in the area, and each section resulted in the production of a set of key recommendations. There was a degree of overlap in the recommendations between the sections, with some topics receiving unanimous support from all groups. As might be expected, more detailed laboratory and field experiments and observations are required to move the field forward and to provide sufficient data for all aspects of human and ecological risk assessment of EASs and risk management for human and ecological receptors. In the following summary, the broad sweep of these recommendations is presented. The complete report is available in this special issue of *Pure and Applied Chemistry*.

THE PROBLEM: SCIENTIFIC KNOWLEDGE AND OUR CURRENT IGNORANCE

The serious and concerted study of ED is barely a decade old, although toxicity tests in some laboratory animals for the types of adverse effects seen with ED date back more than three decades. Despite this, some effects in wildlife were noted much earlier, stimulating great concern. In the last decade, significant advances in our understanding of the underlying biology of endocrine control and the processes of reproduction and sexual development have occurred. Nonetheless, this recent progress has served mainly to highlight our current lack of knowledge regarding not only the underlying biological systems, but also whether low levels of EASs in the environment pose any appreciable degree of risk to humans or to wildlife. The endeavor before us, which commenced as the potential of exposures to relatively high levels of estrogens to induce ED effects in a few species, has grown to include all forms of life on Earth, all natural and anthropogenic sources of exposure to EASs, and a variety of mechanisms of endocrine control. Concomitant with this have been efforts, often uncoordinated, to devise and validate ED assays with which to tackle the task of hazard assessment.

*Report from a SCOPE/IUPAC project: Implication of Endocrine Active Substances for Human and Wildlife (J. Miyamoto and J. Burger, editors). Other reports are published in this issue, *Pure Appl. Chem.* **75**, 1617–2615 (2003).

One of the main questions facing scientists and policy-makers is: When is there enough scientific understanding to proceed with actions? We are, therefore, caught between many urgent calls for action, and the realization that the means and knowledge to achieve these actions are only inadequately understood. This has inevitably led to the adoption of simplified models with which to devise assays and hazard definition/risk assessment methodologies. While accepting that this involves necessary compromises, it is important not to forget that these compromises have been made, and to remain open to the impacts that new insights and understanding will have on these simplified, yet enabling models. For example, it is known that at the molecular level, the processes that regulate cell growth, differentiation, and/or function can be ligand-dependent or -independent, and that the consequent biological effects can be induced through both genomic and nongenomic regulatory pathways. Nonetheless, this complexity has to be expediently reduced to the enabling model in which xenobiotic ligands bind to nuclear receptors, leading to the induction or inhibition of downstream gene expression. The value of such simplified models applies to all aspects of this endeavor, and they remain valuable only so long as they are not allowed to outlive their usefulness.

There is a need for an international forum capable of assessing the impact of major advances in understanding, or of methodological advances, and relating these to current testing and risk assessment strategies. While the Organization for Economic and Cooperative Development (OECD) has contributed admirably in the test methods validation efforts, the pace has been slow. This international effort should involve basic and applied researchers from academic and research institutions, government agencies and laboratories, scientists, industrial and contract research facilities, and government regulators. Such a forum should consider new data on target tissues, critical life phases of exposure, refined endpoints, dose response, and improved methods of data interpretation. At present, this is done only at the national level, and often less than comprehensively.

In the following sections, human effects, wildlife effects, exposure, testing for EASs, and the importance of assays conducted in vivo are discussed. Each section contains an overview, recommendations, and management considerations. These sections are followed by a section of generic issues germane to the entire field of EASs and ED.

HUMAN EFFECTS

It is important to confirm the postulated range of human effects and their possible reversibility currently ascribed to EASs. It is necessary to establish whether exposures to EASs are causative, contributing, or unrelated to effects. Some early reports, for example, a decline in sperm counts, have not been confirmed in all locations by subsequent more extensive investigations. Other reported effects include acceleration or delay in the timing of human sexual maturation and increases in the incidence of human sexual developmental disorders. The role of EASs in these observations remains to be determined. Nevertheless, it is important to continue to research what role, if any, exposures to EASs play in established human epidemiologic trends, such as increases in incidence of cancer in hormone-sensitive tissues (e.g., testis, breast, prostate). The potential for effects occurring in utero following maternal exposure to EASs also requires further study. Some diseases may be evident only after chronic exposure or after a long latency. Therefore, even though toxicity testing methods, such as the developmental toxicity study and the mammalian multigeneration reproduction study, are capable of detecting ED disease conditions, including those unique effects that may occur from exposures in utero, as our knowledge increases, there is a need to consider revising such methods and perhaps developing newer, more sensitive and specific techniques. It is too early to reach firm conclusions about whether human populations are seriously at risk from potential exposures to EASs, and further vigilance is clearly required. However, it is somewhat reassuring that after substantial research in the past decade, there have been no conclusive findings of low-level environmental exposures to EASs causing human disease.

Chemical interferences with steroid biosynthesis and metabolism can produce adverse health effects, even though the inducing agent would not be detected as an EAS using receptor-based test systems. This is an important area of study because some examples of ED occurring in animals derive from exposure to inhibitors of steroidogenic enzymes such as 5α-reductase and aromatase. Some such agents are known to be active in humans and are used successfully in the treatment of a range of human hormonal conditions. Evaluation of such effects requires integrated screening that brings together in silico, in vitro, and in vivo technologies.

The following are research priorities and recommendations:

- Focus epidemiologic studies on testing hypotheses to understand risk factors, including exposure to low levels of EASs on acceleration or delay in puberty and its consequences to health and well-being.
- Study beneficial and hazardous effects of phytohormone and synthetic hormone exposure, including relative dose and exposure levels.
- Evaluate dose response and consequences of maternal exposure to EASs on development of offspring, including elucidation of mechanisms of effects.
- Identify sensitive developmental windows of exposure to natural and synthetic EASs and describe dose– and time–response relationships to advance our understanding of the potential for long-term effects that may arise well after exposure.
- Examine putative atypical dose–response curves with regard to overlapping mechanisms of action.
- Conduct experimental studies on the impact of mixtures, as well as the interaction of xenobiotics with endogenous hormones.
- Conduct epidemiologic studies of the consequences of exposures to defined EAS mixtures.
- Address the need for epidemiologic studies of putative endocrine effects that carefully take into account potential confounders and alternative risk factors.
- Use molecular epidemiology approaches to identify susceptible populations.

WILDLIFE EFFECTS

Field studies have shown that many individual organisms and populations have experienced some degree of exposure to EASs, and in some cases, both individuals and populations have been adversely affected. In some cases, these effects have influenced population stability and/or the integrity of relevant animal communities. Over 200 species are either known or are suspected to have been affected by EASs, including examples from at least two invertebrate phyla and all five major vertebrate classes. Although scientific knowledge of EAS interactions with wildlife species is accumulating, we are still struggling to answer the question of whether low levels of exposures to environmentally relevant concentrations of EAS in the environment pose an appreciable risk to many species. Field observations indicate there are sometimes large differences between closely related species and between individuals of the same species, but the factors that are responsible for these differences, such as differential exposures or differential metabolism, are poorly understood.

Most examples of ED in wildlife have been reported from Europe, North America, Japan, and Australasia, but this may simply reflect the current global distribution of research effort in ED. In general, the effects reported are confined to the more contaminated areas of the planet, although studies have shown that lower levels of exposures to certain substances can occur in areas far from the vicinity of large human populations (for example, exposures have been documented in the Arctic—an area previously considered pristine), presumably caused by the atmospheric distribution of persistent and bioaccumulative EASs. Most examples of ED in wildlife are associated with aquatic species and with consumers of aquatic species; these examples have been attributed to high or continuous exposure to EASs experienced by water-breathers and their predators. However, this conclusion may have been influenced by the fact that the majority of wildlife ED research has been focused on aquatic life. The large majority of current investigations have concentrated on the individual, but of greater potential concern may be effects on the population. Population effects, however, are difficult to discern without long-term population data sets.

The EASs associated with impacts on wildlife include representatives of natural and synthetic steroids; synthetic alkylphenols; natural phytoestrogens and phenolics; natural and synthetic polycyclic aromatic hydrocarbons (PAHs); and synthetic organohalogens and triorganotins. Some of these substances are now subject to government regulation, and for some of these substances, regulatory action has been accelerated because of ED concerns. Other actions are expected as the newest scientific knowledge of exposures and hazards is evaluated in the near term by regulatory agencies. However, those EASs that we currently recognize probably form only a small proportion of the total EAS burden in the environment. Perhaps the most intractable of current problems is that posed by natural steroids excreted by humans and livestock and discharged in a reactivated form in sewage effluents and in runoff from agricultural operations. In these cases, upgrading of sewage treatment plants is the appropriate solution.

The major recommendations for future wildlife studies are as follows:

- Vertebrates (terrestrial and aquatic) possess similar endocrine systems and can act to some extent as surrogates for each other in testing or monitoring programs. However, the >30 invertebrate phyla (comprising about 95 % of all known animal species) have diverse and often poorly understood endocrine systems. Apart from the long-term need to study these different systems, there is a current need to define sentinel species to reduce the impossible workload of studying all species to the same degree. Although it is probable that a relatively small number of screening tests will be sufficient to identify the large majority of potential EASs, a full range of apical (comprehensive, multi-endpoint) assays such as the rodent multigeneration assay and the fish life cycle test, will be required to clarify the environmental hazards posed by EASs to at least six invertebrate phyla and all classes of vertebrates. These apical assays must be practical and properly validated, and should be capable of measuring mixture and low-dose effects and anticipating population effects.
- The potential differential sensitivity to EASs of different life history stages and reproductive strategies in the many species subject to ED requires study. For example, in certain fish, processes such as hermaphroditism, reproductive behavior, larval-to-adult metamorphosis, smoltification, and osmoregulation are under endocrine control, but the response of these to EASs has yet to be systematically considered.
- There is evidence from studies conducted in fish that chronic exposure to a wide range of contaminants can inhibit normal responses to stress. The wide implications of this finding require further study. There is also surprisingly limited information on the existence of xenobiotics that mimic or antagonize the activity of corticosteroid hormones or interfere with the catecholaminergic system.
- Consideration should be given to potential interactions between the estrogen/androgen systems and the immune, corticosteroid, catecholamine, thyroid, and retinoid systems. At present, such interactions are either not considered or are not adequately illuminated due to experimental use of only a limited range of reference substances and because dose–response studies are often not integrated into study designs.
- It is critical to identify factors that could confound interpretation of suspected ED effects in wildlife. This will involve collection of appropriate background data from relatively uncontaminated environments and appreciation of natural variability. The factors that could cause natural variability include changes in the levels of steroid binding proteins, and differences in intrauterine positions, temperature, and food/water availability.
- There is a need for laboratory and field studies to separate the effects of different EASs that may be acting together, leading to greater understanding of the impact that mixtures have on wild populations.

- At present, only a small fraction of the many wildlife species and different ecosystems have been investigated with respect to ED. This increases the need to assess the causes and risks of ED in endangered species, many of which are found in tropical regions that so far have not been studied in this respect. This will require the development of noninvasive monitoring techniques, such as nondestructive biopsy.
- There is a need for development of robust population and ecosystem models that can be used to predict higher-order effects from knowledge of responses observed in individuals. Validation of these models is important.

EXPOSURE ASSESSMENT

Understanding and quantifying exposure is an essential feature of risk assessment and management. Exposure assessment includes the sources, fate, and transport of EASs in environmental media, contact with organisms, bioavailability and absorption, and distribution to target tissues or receptors. There is a substantial existing literature on exposure assessment to chemicals in general, which should be applied to EASs.

Monitoring programs for EAS occurrence and effects are important. Systematic sampling of indicator species and measurement of biomarkers should inform environmental management and provide early warning of problems. In coordination with these biological studies, reliable and valid information on the fate and transport of substances in the environment and humans is a prerequisite for the design of monitoring programs. Monitoring should identify and document changes in the spatial or temporal occurrence of hazards or risks from environmental media and foods. Indicators should include sensitive species or population subgroups. Sustained monitoring programs should include high-priority EASs, which may pose a risk for humans or the environment. Such programs require sustained investment and public support.

Retrospective studies using appropriately archived environmental and human samples can provide valuable baseline data, and appropriate archiving of biological and environmental samples for future studies is essential. This will allow further and future analysis for substances not recognized at the present or the use of more sensitive analytical technologies.

- Existing monitoring programs (e.g., the U.S. EPA's EMAP program) should be expanded to include EASs, and data should be compared to models to enhance exposure assessment and achieve comprehensive risk assessment for EASs.
- Such monitoring programs should be implemented in countries with economies in transition.
- Identification of the relative contribution of different EAS components of an environmental medium can be accomplished by toxicologically guided fractionation and analysis of complex environmental media, known as toxicity identification and evaluation (TIE). This approach can help set priorities for management.

Measurements of contaminants in the environment are important, but do not necessarily reflect internal exposures or dose of EASs to target organs. Dose to target is influenced by variation in bioavailability and absorption, and metabolism and transport. Care is, therefore, necessary when relating environmental measurements to predicted exposure or effects.

Specific research priorities to improve exposure assessment include:

- There is a need to conduct more field monitoring of exposures to highly potent EASs.
- Research is needed to investigate chemical activation (e.g., by hydroxylation of nonactive substances), which may influence exposure to or uptake of EASs.
- Research is needed on hormonally active pharmaceuticals and dietary supplements, particularly those or their metabolic products that are excreted into sewage. Such studies should include reactivation of conjugated metabolites in the environment.

- Improved exposure modeling and parameterization require more extensive data from field studies. This includes data on individuals and sensitive life stages, and seasonal or other variables. Field validation of models is particularly necessary.
- Processing of plant materials and consequent recycling and redistribution of phytohormones into the environment requires careful observation and monitoring. Examples include paper-pulp mills, food processing, and sewage effluents.
- Maternal–fetal exposure from dietary intake of phytohormones warrants additional research and monitoring.
- Chemical analytical methods have been developed, providing adequate detection limits and precision for the analysis of the most important groups of known EASs in food and the environment at levels of biological significance. Technical improvements are needed to simplify sample preparation, diminish confounders, enhance analytical sensitivity, and reduce cost.

TESTING FOR EASs AND ED EFFECTS

There is a growing range of assays suitable for defining the potential of chemicals to interact with several hormone receptors in vitro, perhaps also with concomitant expression of hormone receptor-regulated genes, or with steroidogenic enzymes in vitro. However, these assays cannot define ED activity, as there is no endocrine system being monitored. To gain information regarding whether an EAS will show ED activities, it is necessary to use assays based in a whole organism—such as the fish vitellogenin assay or the rodent uterotrophic and Hershberger assays. Above this class of assay are the apical test systems, such as the rodent multigeneration assay and the fish life cycle assay. The primary distinctions between the different classes of assay will be important to recognize when devising any form of coherent or tiered testing strategy.

The following points were considered important for future development of more efficient testing schemes:

- Individual (mechanism-dependent) structure–activity relationships (SARs) derived in silico, may be of value in prioritizing the evaluation of functional and structural congeneric classes of chemicals. However, SARs may be of limited value when screening all classes of chemicals for all possible mechanisms of action and for interactions with numerous molecular components of the endocrine system.
- Attempts should continue to enhance higher-level testing by adding more sensitive endpoints and maximizing use of the animals being studied. However, care must be taken not to overburden experimental protocols leading to reduced overall efficiency.
- Agreement should be sought on a base set of comprehensive test systems that will be capable of confirming or over-ruling initial indications of endocrine activity provided by lower-tier tests.
- There is a need to improve dose–response analyses and to understand how individual system characteristics can create different dose–response relationships. This is important for moving risk assessments away from default assumptions to more scientifically based approaches.
- Studies in amphibians and humans have established the developmental importance of the thyroid gland, and inclusion of evaluations of TSH, T3, and T4 assays in multigeneration, developmental, and neurotoxicity test protocols has recently been considered. Further investigations are required to establish the optimum way to monitor dose- and time-dependent changes in thyroid function in relation to exposures to postulated EDs and EASs.
- Further work is required regarding extrapolation of benchmark doses, derived from experimental animal studies, to predicted safe human or ecological exposure levels.
- Chemical analytical methods exist with adequate detection limits and precision for the analysis of the most important groups of known EASs in food and the environment. Refinement of these methods could include provision of more specific and sensitive biomarkers, better integration for

diverse classes of EASs, and more automated and selective clean-up procedures. Increased use of LCMS and LCMSMS and stable ELISA tests could contribute to improved quality assurance.

IMPORTANCE OF ASSAYS CONDUCTED IN VIVO

Although discrete mechanisms of ED action can be studied using assays in vitro or in silico, interactions within the endocrine system cannot be assessed in a comprehensive way by simple in vitro assays. Consequently, data from in vitro assays that assess specific aspects of ED activity should be combined with more apical short-term in vivo screens to capture substances interacting with the endocrine system at more complex levels. Such in vivo assays are uniquely able to take account of factors such as the following:

- Metabolic activation and/or deactivation of the EAS, including induction by nuclear receptors (pregnane x receptor and/or constitutively activated receptor) to increase cytochrome-P450 enzyme activity (Phase I metabolism), increase conjugating enzyme activity (Phase II metabolism), or increase drug transport proteins (*p*-glycoprotein), all of which affect exposure.
- Integration of the biological half-life of EASs and their relative binding affinities to chaperone proteins and receptors, including coactivator/corepressor interactions and growth factor interactions and signaling cross-talk.
- Cell- and tissue-specific expression of nuclear hormone receptors and their isoforms and phosphorylation states, including compensation through functional redundancy (receptor isoforms and/or convergent signaling pathways) and repair processes.

OVERARCHING ISSUES IN EAS RESEARCH

A range of needs were identified that apply to all aspects of the study of ED and EASs, as follows:

- The field of ED is rich in unexpected observations—consistent with evolving methodologies and the state of our understanding of the underlying biology of the endocrine systems. However, the science will be aided by a renewed commitment of researchers to follow the scientific method, by testing hypotheses, confirming unexpected findings (wherever possible, before publication), communicating data and results clearly, and including in analyses several alternative, biologically plausible and reasonable explanations for observations.
- Uncertainty regarding whether some EASs may possess the ability to induce effects at doses below those considered safe using current testing methodologies should be evaluated urgently and resolved. Central to this will be agreement on an EAS in a named system that shows such a nonmonotonic dose response. This will enable progress on establishing the range of dose–response relationships that may exist for EASs (monotonic/nonmonotonic, threshold/nonthreshold). It will also enable the molecular mechanism of nonmonotonic responses to be studied.
- Pharmacodynamic and pharmacokinetic factors, and the half-life and bioaccumulation potential of chemicals, should be incorporated into all risk assessment strategies. This requirement means that the route and method of chemical administration adopted in animal tests should be considered and justified, because such decisions may have a profound influence on the quality of the data generated and on their ability to be extrapolated to humans and wildlife species.
- In all areas of study, there may be sensitive subgroups of exposed individuals that may show much greater responses than the majority. This could lead to a loss of information if this potential is not recognized. We need better statistical methods to detect such effects, which may be obscured by population-based, parametric statistical analysis alone.
- Research should clearly identify statistical issues related to underlying variability in exposure or response, sample size, and power. Graphical representations should clarify this variability and

identify individual outliers with unusual exposure or responses, which may be of biological or ecological importance.

- Field studies are expensive to mount. Therefore, in all such cases, the possibility of cryo-preserved archiving of samples should be considered pending advances in assay techniques.
- Methodologies need to be developed to improve quantitation and understanding of both the certainties and the uncertainties associated with extrapolating experimental animal data, derived from multiple studies using different endpoints, to effects expected at ambient levels of environmental exposure.
- Consensus on definitions and applications of the *precautionary principle* and the *weight of evidence approach* should be sought. At present, these two concepts compete for attention and are subject to a range of definitions. For example, some regard the former as taking action in the absence of complete information in case a hazard should exist, but others regard it as taking action when an overwhelming case has been made, but which falls short of absolute proof. The precautionary approach also allows action when the probability of an adverse effect may be low, but consequences are considered large and/or irreversible, and the cost of preventive action is acceptable to society. Likewise, some regard the weight of evidence approach as a numerical averaging of positive and negative data sets, while others regard it as an expert integration of all available data, that may include the explicit consideration of some clear, but isolated positive or negative findings. The integration step includes consideration of the adequacy, strength, and consistency of the overall data set as well as the coherence of results with respect to toxicological relationships between affected endpoints and commonality of underlying mechanisms. Because screening assays provide qualitatively different information than definitive tests, the results from these dissimilar assays are used in a manner that is consistent with the scientific basis and purpose of each. To advance our understanding of the relative merits and disadvantages of these different approaches to risk management, it is essential to examine some examples of actions that have been taken on EASs, compare the different outcomes, and decide which are preferable.

CONCLUSIONS

That natural substances as well as synthetic chemicals have the potential to interact with the endocrine system of organisms in complex and often subtle ways is no longer surprising. It is well established that certain substances can interact with components of the endocrine system and produce adverse health effects. One important question is whether low levels of exposure pose any appreciable risk. Although low levels of some EASs are already known to have caused adverse effects in some wildlife species, the reported pervasiveness of effects attributable to low doses of some synthetic EASs invites validation. Failures in the past to provide full and complete data sets for scientific expert panel analysis cannot be perpetuated, because such actions impact the integrity and credibility of the research. Furthermore, the risk management implications of low-dose effects are potentially substantial and warrant careful scientific replication. Although the human and ecological consequences of ED may not be as universal as some have feared, there are sufficient examples and biological plausibility to leave little basis for complacency in the research community. Future well-designed research, encompassing temporal, spatial, and taxonomic trends, exploring multiple mechanisms of action, and clarifying interactions between endocrine and other (nervous, immune) systems, will elucidate the magnitude of the problem, identify target substances of concern, and advance our knowledge of human and wildlife health. In cases where there is documented scientific evidence based upon valid studies of serious and irreversible damage, but some degree of scientific doubt, it may be important to consider implementing interim precautionary measures or risk management actions that may avert harm, while ongoing research fills the knowledge gap. Risk assessment techniques that apply additional safety factors to make up for the lack of information and uncertainty of the quality of the database or suspected greater sensitivity of the subpopulation is an example of such a precautionary approach. Where biological systems appear well adapted to perturbations, a graded intervention may be acceptable. To tackle these problems, the Intergovernment Forum on Chemical Safety (IFCS) should, therefore, initiate integrated global management of the ED issue. At present, there is little, if any, coordination between research findings and national and international societal responses. Such cooperation and coordination is essential to further research and wise management of the EAS issue.

In conclusion, we have learned that the global effects attributed to EASs are not as all-pervading or fearsome as some have asserted, nor as trivial as others would wish. The beauty of science is that "more research is always needed", and our quest for understanding the world around us is boundless. However, the most important question regarding ED is: What are the significant effects of EASs in terms of health, well-being, and population stability of humans and wildlife?

REFERENCES

1. "Natural and Anthropogenic Environmental Estrogens: The Scientific Basis for Risk Assessment", *Pure Appl. Chem.* **70** (9), 1617–1865 (1998).

TOPIC 1

MOLECULAR MODE OF ACTION OF NUCLEAR RECEPTORS: FUNDAMENTALS FOR UNDERSTANDING THE ACTION OF ENDOCRINE ACTIVE SUBSTANCES

Pure Appl. Chem., Vol. 75, Nos. 11–12, pp. 1619–1664, 2003.

Topic 1.1

Nuclear receptor superfamily: Principles of signaling*

Pierre Germain[1], Lucia Altucci[2], William Bourguet[3], Cecile Rochette-Egly[1], and Hinrich Gronemeyer[1,‡]

[1]IGBMC - B.P. 10142, F-67404 Illkirch Cedex, C.U. de Strasbourg, France; [2]Dipartimento di Patologia Generale, Seconda Università di Napoli, Vico Luigi De Crecchio 7, I-80138 Napoli, Italia; [3]CBS, CNRS U5048-INSERM U554, 15 av. C. Flahault, F-34039 Montpellier, France

Abstract: Nuclear receptors (NRs) comprise a family of 49 members that share a common structural organization and act as ligand-inducible transcription factors with major (patho)physiological impact. For some NRs ("orphan receptors"), cognate ligands have not yet been identified or may not exist. The principles of DNA recognition and ligand binding are well understood from both biochemical and crystal structure analyses. The 3D structures of several DNA-binding domains (DBDs), in complexes with a variety of cognate response elements, and multiple ligand-binding domains (LBDs), in the absence (apoLBD) and presence (holoLBD) of agonist, have been established and reveal canonical structural organization. Agonist binding induces a structural transition in the LBD whose most striking feature is the relocation of helix H12, which is required for establishing a coactivator complex, through interaction with members of the p160 family (SRC1, TIF2, AIB1) and/or the TRAP/DRIP complex. The p160-dependent coactivator complex is a multiprotein complex that comprises histone acetyltransferases (HATs), such as CBP, methyltransferases, such as CARM1, and other enzymes (SUMO ligase, etc.). The agonist-dependent recruitment of the HAT complex results in chromatin modification in the environment of the target gene promoters, which is requisite to, or may in some cases be sufficient for, transcription activation. In the absence of ligands, or in the presence of some antagonists, certain NRs are bound to distinct multiprotein complexes through the interaction with corepressors, such as NCoR and SMRT. Corepressor complexes comprise histone deacetylases (HDACs) that have the capacity to condense chromatin over target gene promoters. Ligands have been designed that selectively modulate the interaction between NRs and their coregulators. Both HATs and HDACs can also modify the acetylation status of nonhistone proteins, but the significance in the context of NR signaling is unclear. NRs communicate with other intracellular signaling pathways on a mutual basis, and their functionality may be altered, positively or negatively, by post-translational modification. The majority of NRs act as retinoid X receptor (RXR) heterodimers in which RXR cannot a priori respond autonomously to its cognate ligand to activate target gene transcription. This RXR subordination allows signaling pathway identity for the RXR partner. The corresponding mechanism is understood and reveals cell and NR selectivity, indicating that RXR can, under certain conditions, act autonomously. NRs are

*Report from a SCOPE/IUPAC project: Implication of Endocrine Active Substances for Human and Wildlife (J. Miyamoto and J. Burger, editors). Other reports are published in this issue, *Pure Appl. Chem.* **75**, 1617–2615 (2003).

‡Corresponding author

regulators of cell life and death, and NR malfunction can be at the basis of both disease and therapy, as is impressively documented in the case of acute promyelocytic leukemia. Recently, several pathways have been uncovered that link NR action with cell proliferation and apoptosis.

GENERAL ORGANIZATION OF NUCLEAR RECEPTORS

Nuclear receptors regulate intra- and intercellular communication in metazoans

Multicellular organisms require specific intercellular communication to properly organize the complex body plan during embryogenesis and maintain its properties and functions during the entire life span. While growth factors, neurotransmitters, and peptide hormones bind to membrane receptors, thereby inducing the activity of intracellular kinase cascades or the JAK-STAT/Smad signaling pathways, other small, hydrophobic signaling molecules such as steroid hormones, certain vitamins, and metabolic intermediates enter, or are generated within, the target cells and bind to cognate members of a large family of nuclear receptors (NRs) (Fig. 1). NRs are of major importance for metazoan intercellular signaling, as they converge different intra- and extracellular signals to initiate and regulate gene expression programs. They act as transcription factors that (i) respond directly through physical association with a large variety of hormonal and other regulatory, as well as metabolic signals, (ii) integrate diverse signaling pathways as they correspond themselves to targets of posttranslational modifications, and (iii) regulate the activities of other signaling cascades (commonly referred to as "signal transduction cross-talk"). The genetic programs that they modulate affect virtually all aspects of the life of a multicellular organism, covering such diverse aspects as, for example, embryogenesis, homeostasis and reproduction, or cell growth and death. Their gene regulatory power and selectivity has prompted intense research on these key factors, which is now starting to decipher a complex network of molecular events accounting for their transcription regulatory capacity. The study of these molecular processes has also shed light on multiple general principles underlying transcription regulation, and it will be a future challenge to uncover the molecular rules that define selective NR-dependent spatial and temporal control of gene expression.

The family and its ligands

To date, 49 NRs [2,3] have been identified throughout the animal kingdom, ranking from nematodes to humans (Fig. 1). They constitute a family of transcription factors that share a modular structure of five to six conserved domains encoding specific functions [4,5]. The most prominent distinction to other transcription factors is their capacity to specifically bind small hydrophobic molecules. These ligands constitute regulatory signals, which, after binding, change the transcriptional activity of the corresponding NR. For some time, a distinction was made between classic NRs with known ligands and so-called "orphan" receptors, hence receptors without or with unknown ligand. However, recent years have seen the identification of ligands for many of these orphan receptors, making this distinction rather superficial (see, e.g., refs. [6,7–10]). Moreover, the classification of NRs into six to seven phylogenetic subfamilies with groups that comprise both orphan and nonorphan receptors further dismisses such discrimination [1]. The classification of NRs is done by virtue of the homology to other family members, with the DNA- and ligand-binding domains (DBDs and LBDs) having the highest evolutionary conservation.

Interestingly, some recently identified ligands for "orphan" NRs turned out to be metabolic intermediates. It appears therefore that in certain systems the control of built-up, break-down and storage of metabolic active substances is regulated at the level of gene expression, and that in many cases this "intracrine" signaling is brought about by NRs. Furthermore, gene knock-out experiments suggest that metabolic intermediates such as SF1 (NR5A1) or PPARγ (NR1C3) ligands may have regulatory func-

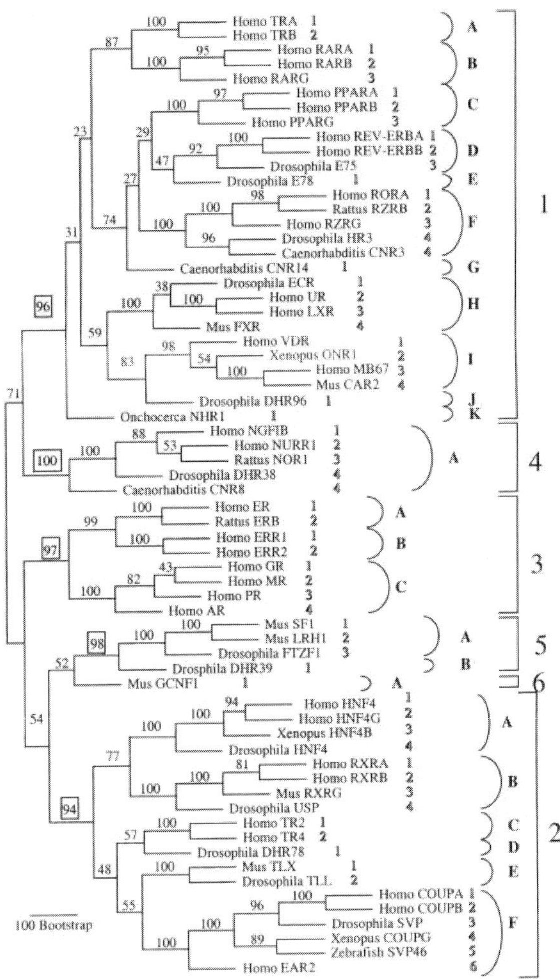

Fig. 1 Phylogenetic tree of 65 nuclear receptor genes in vertebrates, arthropoids and nematodes. For a detailed description, see Nuclear Receptors Nomenclature Committee (1999) [1] and the regular updates at <http://www.ens-lyon.fr/LBMC/LAUDET/nomenc.html>.

tion in specifying organ development [11,12]. Prominent metabolic ligands are bile and fatty acids, eicosanoides, and oxysterols. The group of steroid hormones encompasses estrogens, progestins, mineralocorticoids, glucocorticoids, androgens, and ecdysterones. Examples for vitamin–derived NR ligands are vitamin D_3 (VDR; NR1I1) or the vitamin A derivative, retinoic acid (RARs and RXRs; NR1B and NR2B, respectively). Thus, NRs function in endocrine (steroid hormone receptors), auto/paracrine (retinoid receptors), and intracrine [metabolic receptors such as LXRα (NR1H3), SF-1 (NR5A1), FXR (NR1H4), PXR (NR1I2), PPARs (NR1C), CARβ (NR1I4)] signaling pathways.

Genetics of nuclear receptors

Genetic programs consist typically of several hundred to thousand genes that are expressed in a spatially and temporally controlled fashion. Nuclear receptors act as master "switches" to initiate specific genetic programs that, for example, lead to cell differentiation, proliferation, or apoptosis or regulate homeostasis. In the context of other programs, these genetic activities support or initiate complex physiological phenomena, such as reproduction and organ function. Once activated by the cognate ligand, NRs regulate the primary and secondary target gene expressions that make up the corresponding physiological event. Throughout the life cycle of a multicellular organism, the coordinate interplay between programs defining cell fates in different tissues, organs, and finally the entire body is at the foundation of the organisms' development and subsistence. This is fully supported by the analysis of mice bearing mutations or deletions of one or several receptors (a searchable mouse knock-out and mutation database with PubMed can be found at <http://www.biomednet.com/db/mkmd>). Several NR knock-out animals (in particular, compound knock-out animals) die in utero or around birth, displaying severe malformations of organs that render them inviable (for examples, see ref. [13] and refs. therein; for reviews on earlier work, see refs. [14,15]). Others are viable under laboratory conditions, but display a reduced life span and are often infertile (see, e.g., ref. [11]). These knock-out animal models have helped in deciphering the physiology of NR action. Often they provided initial or additional evidence for new, yet undiscovered functions, exerted by the receptor, and thus initiate further research on previously unknown signaling pathways. One example is the involvement of retinoic acid receptor, RARβ (NR1B3), and retinoid X receptor, RXRγ (NR2B3), in long-term memory potentiation and depression [16]. Furthermore, knockouts have also provided insight into the distinct modes of transcriptional regulation by NRs. An elegant example is the mouse NR3C1 gene encoding the glucocorticoid receptor (GR). $GR^{-/-}$ mice die at birth due to respiratory failure, but replacement of the GR gene by the GR_{A458T} mutant that impairs binding to consensus GR response elements generates mice (termed $GR^{dim/dim}$ with "dim" indicating DBD-mediated dimerization deficiency) that were fully viable [17]. These mice could be used to define direct (cognate response element-mediated) and indirect (e.g., signaling cross-talk-dependent) actions of GR (see Topic 1.5). The indirect actions of the glucocorticoid receptor concern also the transrepression of the activating protein AP1, a phenomenon further discussed below.

An interesting result of studies with NR gene deletion models has been the discovery of redundancy and adaptivity among family members of the same group. In this respect, the interpretation of retinoic acid signaling, which is of remarkable complexity and displays a high degree of apparent retinoic acid receptor redundancy, may serve as an example (Topic 1.6).

Modular structure and function

N-terminal region A/B harbors cell-specific activation function(s) AF-1 of unknown structure

As schematically depicted in Fig. 2, NRs are composed of five to six regions (A to F; originally defined by [18]) that have modular character. The N-terminal A/B region harbors one (or more) autonomous transcriptional activation function (AF-1), which, when linked to a heterologous DNA-binding domain, can activate transcription in a constitutive manner. Note, however, that in the context of the full-length receptor, AF-1 is silent in the absence of agonist and certain antagonists. When comparing NRs from different subfamilies and groups, the A/B region displays the weakest evolutionary conservation, and the distinction between A and B regions is not always evident. A/B regions differ significantly in their length, ranging from 23 (vitamin D receptor, NR1I1) to 550 (androgen, NR3C4, mineralocorticoid, NR3C2, and glucocorticoid receptors, NR3C1) amino acids. No 3D structure of a nuclear receptor A/B region has been solved up to now, and structure prediction is not straightforward. A/B regions are subject to alternative splicing and differential promoter usage, and the majority of known NR isoforms differ in their N-terminal region. Through alternative splicing and differential promoter usage (PR forms A + B), the absence or presence of different activation functions found in the AB regions can be regulated [for details, see ref. 5]. Moreover, the N-terminus of NRs has reportedly been found as subject of posttranslational events such as phosphorylation (discussed further below). Finally, the activation function(s) AF-1 display cell, DBD and promoter specificity [19–21], the origin of which is still elusive but may be related to the posttranslational modification [22] and cell-specific action and/or expression of AF-1 coactivators [23].

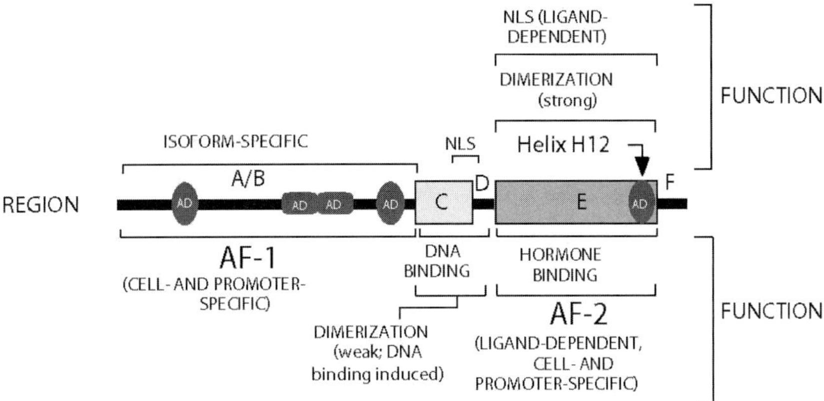

Fig. 2 Schematic illustration of the structural and functional organization of NRs. The evolutionary conserved regions C and E are indicated as boxes and a black bar represents the divergent regions A/B, D, and F. Note that region F may be absent in some receptors. Domain functions are depicted below and above the scheme. Two transcription activation functions (AFs) have been described in several nuclear receptors, a constitutively active AF-1 in region A/B and a ligand-inducible AF-2 in region E. Within these activation functions, autonomous transactivation domains (ADs) have been defined in the estrogen (ER) and progesterone receptor (PR) N-terminal regions. In the case of the estrogen, retinoid and thyroid hormone receptors an autonomous activation domain (AF-2 AD) encompassing helix H12 has been detected at the C-terminal end of the ligand binding domain E.

DNA-binding domain encompasses region C

The highly conserved domain C harbors the DBD of NRs, which confers sequence-specific DNA recognition. This domain has been extensively investigated, especially with respect to its selective response element recognition and dimerization properties (for details, see below). Several X-ray and NMR data sets are available for different NR C domains in their DNA complexed and uncomplexed forms (see Table 1 for PDB file names). The DBD is mainly composed of two zinc-finger motifs, the N-terminal motif Cys-X2-Cys-X13-Cys-X2-Cys (CI) and the C-terminal motif Cys-X5-Cys-X9-Cys-X2-Cys (CII); in each motif, two cysteine residues chelate one Zn^{2+} ion. Within the C domain, several sequence elements (termed P-, D-, T- and A-boxes) have been characterized that define or contribute to (i) response element specificity, (ii) a dimerization interface within the DBD, and (iii) contacts with the DNA backbone and residues flanking the DNA core recognition sequence. Figure 3 illustrates the three prototypic DNA binding modes of NRs: the estrogen receptor DBD as an example of a homodimer that binds to a palindromic response element (Fig. 3a), the RXR-TR as an example of an anisotropic (5′-RXR-TR-3′) heterodimeric complex on a DR1 direct repeat (Fig 3b), and NGFI-B as an example of a monomer that binds to an extended hexameric motif, the so-called NBRE (Fig. 3c). Note that the so-called A- and T-boxes of TR and NGFI-B are involved in additional minor groove DNA contacts (see below for details).

Table 1 List of presently reported 3D structures of NR DBDs together with their PDB assignments.

PDB ID	Receptors	Response element
Homodimers		
1GLU	GR	GRE
1LAT	GR mutant	non-cognate
1HCQ	ER	ERE
1A6Y	RevErb	DR2
1GA5	RevErb	DR2
1HLZ	RevErb	DR2
1BY4	RXR	DR1
1KB2	VDR	Osteopontin
1KB4	VDR	DR3
1KB6	VDR	Osteocalcin
Heterodimers		
2NLL	5′-RXR-TR-3′	DR4
1DSZ	5′-RAR-RXR-3′	DR1
Monomers		
1CIT	NGFI-B	NBRE
NMR structures		
2GDA	GR	–
1GDC	GR	–
1RGD	GR	–
1HCP	ER	–
1RXR	RXR	–
1HRA	RARβ	–
1LO1	ERR2	–

Region D, a hinge with compartmentalization functions

The D region of NRs is less conserved than the surrounding regions C and E. This domain appears to correspond to a "hinge" between the highly structured C and E domains. It might allow the DBDs and LBDs to adopt several different conformations without creating steric hindrance problems. Region D contains a nuclear localization signal (NLS), or at least some elements of a functional NLS (see, e.g., ref. [24]). The intracellular localization of NRs is a result of a dynamic equilibrium between nuclear-cytoplasmatic and cytoplasmatic-nuclear shuttling [25]. At equilibrium, the large majority of NRs is nuclear, while some steroid receptors (androgen, glucocorticoid, and mineralocorticoid receptors) apparently reside at cytoplasmic locations in the absence of their cognate ligands and translocate to the nucleus in a ligand-induced fashion [26,27].

Region E encompasses the ligand-binding domain and activation function 2

The hallmark of an NR is its LBD in the E region. This domain is highly structured and encodes a wealth of distinct functions, most of which operate in a ligand-dependent manner. The LBD harbors the ligand-dependant activation function AF-2, a major dimerization interface and often a repression function (for review and references, see ref. [5]). Detailed molecular insights into the structure-function relation of signal integration by NRs have been gained by the elucidation of the crystal structures of the E region alone or in presence of agonists, antagonists and coregulator peptides (see below and Table 2).

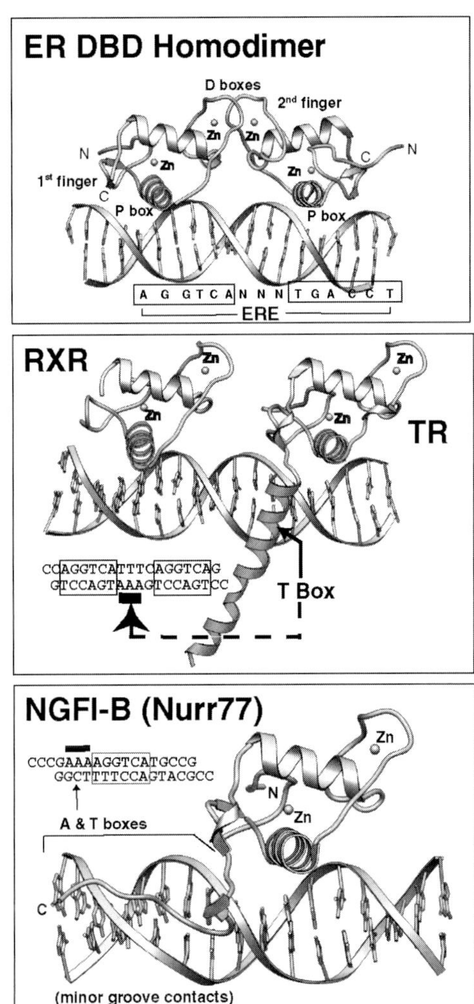

Fig. 3 3D structures of the three prototype DBDs of nuclear receptors obtained from X-ray crystal structure analyses of the DBD-DNA response element co-crystals. (a) Structure of the estrogen receptor α (ER) DBD on an ERE; one strand of the canonical ERE DNA sequence is given at the bottom. The various structural elements (Zn^{++} fingers, D and P boxes) are indicated. Note that the D-boxes form a DNA-induced DBD dimerization interface, while the P-box α-helices establish the selective base contacts in the major groove. (b) Crystal structure of the 5′-RXR-TR-3′ heterodimer on a cognate direct repeat response element spaced by four base pairs (DR4), depicted as a double-stranded DNA sequence at the bottom left. Note that the T-box makes minor groove contacts, thus specifying to some degree the DR4 spacer nucleotide sequence (arrow). (c) Crystal structure of the monomeric NGFI-B on its response element (NBRE). The double-stranded NBRE sequence is given at the top left. Note that the A and T-boxes define the 5′ AAA sequence that contacts the minor groove .

Table 2 List of all presently reported 3D structures of NR LBDs.

Receptors	Ligands	Remarks	PDB ID
Monomers			
AR	R1881, DHT	agonists	1e3g, 1i37
AR (T877A)	DHT	agonist	1i38
AR (« ccr » mutant)	9α-fluorocortisol	agonist	1gs4
DHR38	no	holo conformation	1pdu
FXR	6-ethyl- and 3-deoxy-CDCA	NR box complexes	1osv
	Fexaramine	agonist	1osh
LRH-1	no	holo conformation	1pk5
Nurr1	no	holo conformation	1ovl
PPARα	GW6471	CoRNR box complex	1kkq
	GW409544	NR box complex	1k7l
	AZ242	agonist	1i7g
PPARγ	no	–	3prg
	YPA	agonist	1knu
	AZ242	agonist	1i7i
	DRF	agonist	1nyx
PPARδ	GW2433, EPA	agonists	1gwx, 3gwx
	no	–	2gwx
PR	Progesterone, R1881	agonist	1a28, 1e3k
PXR	apo	–	1ilg
	SR12813, hyperforin	agonist	1ilh, 1m13
RARγ	T-RA	agonist	2lbd
	9C-RA, BMS961	agonists	3lbd, 4lbd
	BMS270394, BMS270395	agonists	1exa, 1exx
	BMS184394, Cd564	agonists	1fcx, 1cfy
	BMS181156	agonist	1cfz
	SR11254	agonist	1fd0
RXRα	9C-RA	agonist	1fby
RORα	Cholesterol	agonist	1n83
RORβ	Stearic acid	NR box complex	1k4w
	T-RA, ALRT1550	part. antagonist; NR box complex	1n4h, 1nq7
TRα	T3, Dimit	agonists	–
	IH5	TRβ agonist	1nav
TRβ	T3	NR box complex	1bsx
	IH5	agonist	1nax
	PFA	agonist	1n46
TRβ (A234T); (R243Q)	4HY	agonist	1nq0, 1nq1
USP	Phospholipid	antagonist	1hg4, 1g2n
VDR	Vitamin D3	agonist	1db1
	KH1060, MC1288	super-agonist	1ie8, 1ie9
Homodimers			
ERα	Estradiol	agonist	1ere, 1qku
	RAL	antagonist	1err
	DES	NR box complex	3erd
	OHT	antagonist	3ert
	Estradiol	agonist	1a52
	THC	NR box complex	1l2i
	Estradiol	NR box complex	1gwr
	RAL « core »	NR box complex	1gwq

(*continues on next page*)

Table 2 (*Continued*).

Receptors	Ligands	Remarks	PDB ID
Homodimers			
ERα (Cys mut.)	Estradiol	agonist	1qkt
ERβ	RAL, Genistein	antagonist, part agonist	1qkn, 1qkm
	ICI164384	antagonist	1hj1
	THC	antagonist	1l2j
ERR3	no	-	1kv6
HNF4	Fatty acid	agonist	1lv2, 1m7w
GR	Dexamethasone	NR box complex	1m2z, 1p93
	RU486	antagonist	1nhz
LXRβ	T0901317, Epoxycholesterol	agonists	1pqc, 1pq9, 1p8d
	GW3965	agonist	1pq6
PPARγ	no	–	1prg
	Rosiglitazone	NR box complex	2prg
	GW0072	partial agonist	4prg
RXRα	no	–	1lbd
	no, undefined ATRA metabolite	tetramer	1g1u, 1g5y
	DHA, BMS649	agonists	1mv9, 1mzn
RXRβ	LG100268	agonist	1h9u
Heterodimers			
LXRα/RXRβ	T1317/methoprene acid	NR box complex	1uhl
PPARγ/RXRα	Rosiglitazone/9C-RA	NR box complex	1fm6
	GI262570/9C-RA	NR box complex	1fm9
	GW409544/9C-RA	NR box complex	1k74
RARα/RXRα	BMS614/Oleic acid	antagonist/part. agonist	1dkf

Role of the C-terminal region F is unknown

Some receptors possess C-terminal of the LBD, a region F, which displays little evolutionary conservation. Note that the LBD is structurally defined as the domain generated by the elements between the beginning of helix H1 and the end of helix H12; this sequence is not necessarily identical to what is commonly referred to as region E from sequence alignments, and also receptors like the progesterone receptor possess some kind of E region. This region is, however, much longer in the cases of, for example, estrogen (NR3A) and retinoic acid (NR1B) receptors. There are no clues as to the function of the C-terminal sequence. Recent literature suggests that the F region might play a role in coactivator recruitment to the E domain and in determining the specificity of the LBD coactivator interface [28,29]. It seems clear that this domain also inherits little structural features. It is tempting to speculate that it sort of fine-tunes the molecular events associated with the transcriptional properties of the E domain, or the entire receptor, as it may affect antagonist action [30,31].

DNA RECOGNITION BY NUCLEAR RECEPTORS

Response elements of nuclear receptors

Common principle

All NRs recognize derivatives of the same hexameric DNA core motif, 5′-PuGGTCA (Pu = A or G). However, mutation, extension, and duplication, and, moreover, distinct relative orientations of repeats of this motifs generate response elements that are selective for a given (class of) receptors. Apparently coevolutionarily, NRs devised mechanisms to optimally interact with these sequences—they either modified residues, which establish contacts to the nucleotides that specify a given response element or they generated response element-adapted homo- or heterodimerization interfaces.

Spacer "rules" derived from synthetic response elements

To describe the preference of the various DR-recognizing receptors for elements with a certain spacer length, a simplified rule has been proposed [32,33] that is easily memorized.

Albeit a reasonable approximation, this rule should be used with care, since there are numerous exceptions, such as receptors binding to complex, unusual, or noncognate response elements, and different receptors may bind to common elements that are not predicted by the rule.

Table 3 "Spacer rules" for DNA binding response elements of nuclear receptors.

Spacer NTs	Systematic name	Acronym	Receptor complex
1	DR1	RXRE, PPARE,..	RXR-RXR, PPAR-RXR, RAR-RXR, ...
2	DR2	RARE	RAR-RXR
3	DR3	VDRE	RXR-VDR
4	DR4	TRE	RXR-TR
5	DR5	RARE	RXR-RAR

Variability of the binding motif, spacer sequence, and flanking nucleotides

It is important to point out that there is considerable degeneration in the sequence of half-site motifs of a given type of natural retinoid response element and that there is a distinct preference of the various receptors for a certain motif. For example, the preference for the half-site motif 5′-PuGGTCA over 5′-GTTCA follows the order TR > RXR > RAR [34].

In addition to a distinct preference for certain nucleotides in the half-site motif, there is also a receptor-specific preference for certain nucleotides in the DR spacer, which is easily rationalized in view of the crystallograhic data (see Table 1 and ref. [35]). See, in this respect, also the NGFI-B DNA complex, which illustrates the binding of A- and T- box residues to the 5′ minor groove of the NBRE (see Table 2 and ref. [36]).

Steroid hormone receptor response elements

Steroid hormone receptors bind to 3 bp-spaced palindromic arrangements (3 bp-spaced inverted repeats; generally termed IR3) of the prototypic recognition motif (see ref. [5] and refs. therein). The mutation of a single nucleotide at position 4 in each motif from T to A (5′-PuGGTCA to 5′-PuGGACA) will convert an estrogen (ERE) into a glucocorticoid response element (GRE). Note, however, that the classic GRE is rather a 5′-PuGAACA, which corresponds to the mutation of two nucleotides.

Although progesterone (PR), androgen (AR), and mineralcorticoid (MR) receptors bind to GREs, differences in their DNA-binding specificities have also been observed. A systematic mutational analysis concluded that GRs and PRs may not distinguish individual target sites, but may use the whole of the response element context differentially [37]. Notably, the chicken PR was shown to replace GR for the activation of the endogenous tyrosine-amino transferase gene, a cognate GR target gene [38]. This could suggest that in some cases, only the distinct expression profiles of some NRs and their ligands suffice to generate specificity.

Response elements for retinoid, thyroid, vitamin D, and peroxisome proliferator-activated receptors

For detailed discussion of this topic, the reader is referred to a number of extensive reviews [39–43]; please compare also the published crystal structure data on complexes between various NR DNA-binding domains and the cognate DNA-response elements (Table 1 and Fig. 3). The characteristics of the major response elements to which retinoid can potentially contribute, either as RAR-RXR heterodimer or through RXR as the heterodimeric partner, are reviewed below.

Retinoid response elements

The classic retinoic acid response element (RARE), which was found in the P2 promoter of the RARβ gene and gives rise to the RARβ2 mRNA, is a 5 bp-spaced direct repeat (generally referred to as DR5) of the motif 5′-PuGTTCA. In addition, response elements with a DR5 containing the motif 5′-PuGGTCA (also termed DR5G to distinguish it from the DR5T of the RARβ2 promoter) act as perfect RAREs [44–46] as well as direct 5′-PuGGTCA repeats spaced by 1 bp (DR1) or 2 bp (DR2). RAR-RXR heterodimers bind to, and activate transcription from, these three types of RAREs, provided target cells express both RARs and RXRs. DR1 elements bind, in addition to RAR-RXR heterodimers, also RXR homodimers in vitro, and RXRs can transactivate in response to an RXR ligand target genes containing DR1 elements. That DR1 elements can act as functional retinoid X receptor response elements (RXREs) in vivo is supported by their activity in yeast cells [47], in which any contribution of endogenous RAR via heterodimerization with RXR can be excluded. However, no natural RXRE has been found up to now. The only reported natural RXRE is a DR1-related element found in the rat CRBPII promoter [48].

Thyroid hormone receptor response elements

The thyroid response element (TRE) consensus sequence is 5′-AGGTCA, as for RARs, RXRs, PPARs, or VDRs. However, there is evidence for some differences in the natural response element repertoires of these receptors. It has been shown, for example, that TRα is able to bind to both 5′-AGGTCA and 5′-AGGACA motifs [49–51]. Such differences may lead to further specification by cooperative DNA binding with other promoter-bound factors and could contribute to the ability of a given target gene to preferentially respond to a particular signaling pathway.

Like other receptors, TRs are able to bind to a palindromic element (TREpal) [51–53]; but such an element has not yet been found in cellular genes. Furthermore, this element confers no hormonal specificity, since it can be recognized by a large number of other receptors. The most commonly found TREs are either direct repeats (DRs) or everted repeats (ERs). Examples of direct repeat TREs are discussed in ref. [5].

On direct repeats, TRs have a strong preference for DR4, i.e., direct repeats spaced by four nucleotides (for the corresponding crystal structure, see Table 1 and ref. [35]). Nevertheless, TRs are able to bind to direct repeats with a spacing other than four, such as DR5 [54], DR2, or DR0 [55]. TRs can also bind to inverted palindromes with a preferred spacing of six nucleotides [56].

Vitamin D response elements

Only a few natural vitamin D response elements (VDREs) are known; several of them contain DR3 elements. Studies with "optimized" synthetic response elements assembled from 5′-PuGGTCA motifs have confirmed that DR3 elements bind VDR-RXR heterodimers, and that the cognate ligands, vitamin D, and 9C-RA, activate the corresponding promoters. The promoter of the human/rat osteocalcin gene contains a complex VDRE with several possible combinations of the recognition motifs, including that of a DR6. For more extensive discussion on vitamin D response elements and action see recent reviews [57,58].

Peroxisome proliferator-activated receptor response elements

Natural peroxisome proliferator-activated receptor response elements (PPAREs), which have been found in enzymes that catalyze the peroxisomal β-oxidation and microsomal ω-hydroxylation in response to peroxisome proliferators, usually contain (degenerate) DR1 elements, but more complex PPAREs have been reported also (for reviews, see refs. [42,43,59,60]).

Principles of DNA recognition by nuclear receptor DNA-binding domains

Homo- and heterodimerization

Nuclear receptors can bind their cognate response elements as monomers, homodimers, or heterodimers with another family member (for reviews, see refs. [39,61]). Dimerization is a general mechanism to increase binding site affinity, specificity, and diversity due to (i) cooperative DNA binding (an extreme case of cooperative binding is the existence, in solution, of stable dimers), (ii) the lower frequency of two hexamer binding motifs separated by a defined spacer compared to that of single hexamers, and (iii) heterodimers that may have recognition sites distinct from those of homodimers.

Steroid hormone receptors bind generally as homodimers to their response elements, while RARs, RXRs, TRs, and VDRs can homo- and/or heterodimerize. RXRs play a central role in these various signal transduction pathways, since they can both homodimerize and act as promiscuous heterodimerization partner for RARs, TRs, VDRs and orphan receptors. Heterodimerization has a three-fold effect: it leads to a novel response element repertoire, increases the efficiency of DNA binding relative to the corresponding homodimers, and allows two signaling inputs, that of the ligands of RXR and its partner. Crystal structures of DBD homo- and heterodimers have defined the surfaces involved in dimerization (see ref. [61 and refs. therein, [62]). It is important to point out that the response element repertoire described above for receptor homo- and heterodimers is dictated by the DBD while LBDs stabilizes the dimers, but do not contribute to response element selection.

Specificity of DNA recognition (P-box, D-box, T-box, A-box)

The DNA response element specificity (half-site sequence, spacing, and orientation) is generated by (i) the recognition of the actual "core" or "half-site" motif and (ii) the dimerization characteristics (mono-, homo- or heterodimerization; structure of the actual dimerization interface) of the receptor(s).

Identification of the residues involved in distinguishing the hexameric half-site motives of EREs (5′-AGGTCA) and GREs (5′-AGAACA) was done by a series of refined swapping experiments. Initially, DBD swaps showed that specific half-site recognition depends on DBD identity [63], subsequently the N-terminal finger was found to differentiate between ERE and GRE recognition [64]. Finally, three studies identified two to three residues at the C-terminal "knuckle" of the N-terminal finger; commonly referred to as the P-box (proximal box; Fig. 3a), to be responsible for ERE vs. GRE recognition [65–67].

A second region, the D-box (distal box; N-terminal "knuckle" of the C-terminal finger; see Fig. 3b), was found to be involved in differentiating between the binding to a 3bp- (characteristic for steroid receptor REs) and a 0 bp-spaced (one type of TRE) palindrome [67]. As was later confirmed by the crystal structures of GR and ER DBDs, this region does indeed contribute to the DBD dimerization interface.

Two other boxes have been described within the DBDs of heterodimerizing receptors, the A- and the T-box (Fig. 3c). The A-box was originally described for NGFIB as the sequence responsible for the recognition of two or three additional A nucleotides in the minor groove 5′ of the hexameric core motif, thus generating an NGFIB response element (NBRE; 5′-(A)AAAGGTCA) [68,69]. This A-box was later found to play a similar role in heterodimers such are 5′-RXR-TR on DR4 elements, where it specifies to some extent the spacer 5′ of TR and sets a minimal spacing by steric hindrance phenomena [70,71]. Interestingly, in the 3D structure, the A-box presents as a helix contacting the minor groove and modeling is in keeping with its role in setting a minimal distance between the half-sites [35].

The T-box was originally defined in RXRβ (then H-2RIIBP) as a sequence required for dimerization on a DR1 element [72]. Its role as a RXR homo- and heterodimerization surface has been subsequently confirmed [70,71,73].

Three-dimensional structure of nuclear receptor DNA-binding domains

A significant amount of structural information has been accumulated during the past years, providing information about the solution structure of the GR, ER, RAR and RXR DBDs. Moreover, the 3D crystal structures have been solved of the GR DBD homodimer bound to noncognate DNAs, the crystal structure of ER DBD homodimer bound to consensus and nonconsensus natural EREs, the crystal structure of the RXR homodimer on a DR1 element and the TR-RXR DBD heterodimer bound to its cognate DR4 element, the structures of the NGFIB-NBRE, RevErb-DR2, and RAR-RXR-DR1 complexes, as well as the VDR homodimer on various response elements (for references to the original studies, see Table 1 (PDB accession numbers) and refs. [5,61]).

The 3D structure of the ER DBD-ERE co-crystal is shown in Fig. 3a [74]. The structure consists of a pair of amphipathic α helices packed at right angles and crossing near their midpoints. A zinc-binding pocket lies near the N-terminus of each of the two helices. Hydrophobic side chains form an extensive hydrophobic core between the two helices. The residues' N-terminal to the first helix are folded to form two loops. Hydrophobic residues at the tips of the two loops pack with hydrophobic residues in the core between the two helices.

Two ER DBD molecules bind to adjacent major grooves from one side of the DNA double helix. The protein makes extensive contacts to the phosphate backbone on one side, orienting the DBD such that the recognition helix enters the major groove, allowing surface side chains to make sequence-specific contacts to the base pairs. Although ER or GR DBDs are monomers in solution, they bind cooperatively to the cognate response elements owing to the DNA-induced formation of a dimerization interface which comprises also D-box residues. For further details, see the original publications [5].

STRUCTURE OF NUCLEAR RECEPTOR LIGAND-BINDING DOMAINS

Canonical fold of nuclear receptor ligand-binding domains

To date, the crystal structures of monomeric, homodimeric and heterodimeric NR LBDs in the presence of agonists (holo form), antagonists, partial agonists, fragments of coactivators and corepressors, or in the absence of bound hormone (apo form) have been described (Table 3). All these NR LBDs display a common fold, as originally predicted [75], with 12 α-helices (H1 to H12) and one β-turn arranged as an antiparallel α-helical "sandwich" in a three-layer structure (Fig. 4). Note that some variability exists; for example, no helix H2 was found in RARγ [76], while an additional short helix H2′ is present in PPARγ [77].

Mouse trap model

A comparison of the apo- and holo-LBD structures (Fig. 4) suggested a common mechanism by which the activation function AF-2 becomes transcriptional competent: upon ligand binding, H11 is repositioned in the continuity of H10, and the concomitant swinging of H12 unleashes the Ω-loop which flips over underneath H6, carrying along the N-terminal part of H3. In its final position, H12 seals as a "lid" the ligand-binding cavity and further stabilizes ligand binding (in some, but not all NRs) by contributing additional ligand–protein interactions. It is a general and essential feature of the ligand "activation" of NRs that the transconformation of H12, together with additional structural changes (such as bending of helix H3), creates distinct surface(s) on the apo- and holo-LBD. The novel surfaces generated upon agonist binding allow bona fide coactivators, such as the members of the SRC-1/TIF2 family, to bind and recruit additional transcription factors (see below). Concomitantly, corepressor proteins, which bind to a surface topologically related to that involved in coactivator interaction of the apo-LBD, dissociate upon agonist, but not necessarily antagonist, binding (see below). Notably, as is discussed further below, certain antagonists "force" H12 in a third position, distinct from the holo position whereby it impairs coactivator binding.

For a given receptor, the equilibrium between the apo and holo (or apo and antagonist) conformational states of a NR LBD can be affected through intramolecular interactions of H12, such as a salt bridge (holo LBD of RARγ [76]) or hydrophobic contacts (as suggested for apo-ER; [78]). This implies

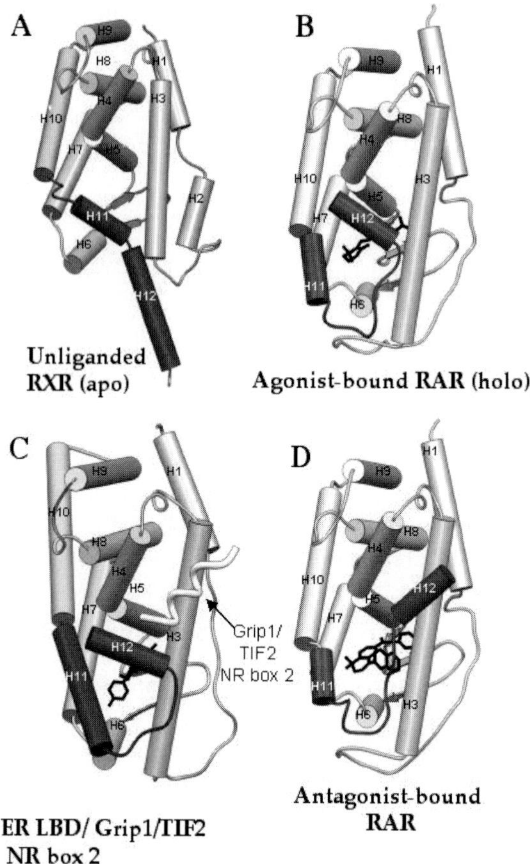

Fig. 4 Schematic illustration of the canonical 3D structures of NR LBDs. The canonical apo and holo structures of NR LBDs are shown in (A) and (B), respectively. The holo structure allows interaction with the NR box of nuclear receptor coactivators, such as TIF2/Grip1, SRC1, or RAC3/AIB1/ACTR (see text and reviews cited), as depicted in (C). This NR box binding site is occupied by helix H12 in the antagonist-bound LBD illustrated in (D).

that the apo conformation is not necessarily the default state, so that some NRs may be constitutive activators or repressors without possessing a cognate ligand. Moreover, an increase in coactivator concentration can generate a transcriptionally competent RAR under certain conditions [79] and the apo-ER conformation may be destabilized by phosphorylation [78,80]. Thus, overexpression of coactivators or receptor modification may generate ligand-independent receptors. Such scenarios could have significant implications for endocrine cancer therapies.

Dimer interface

Recently, the first structures of heterodimeric LBDs have been solved. One comprises the RARα LBD bound to the α-selective antagonist BMS614 and the constitutively active [81] RXR mutant F318A, which turned out to harbor an oleic acid-like ligand [82]. The other is the LBD of PPARγ bound to rosiglitazone or a synthetic ligand in a heterodimer with 9-*cis* RA-bound RXRα [83]. These structures allowed for the first time to compare the homo- and heterodimerization interfaces of several NR ligand-binding domains. In principle, the overall heterodimeric arrangement closely resembled that of homodimers of RXRα, ERα, and PPARγ [77,84,85]. However, in contrast to the nearly perfect symmetric organization of homodimer interfaces, the heterodimer interfaces are slightly asymmetric. Subsequently, the heterodimeric interface is described based on the results obtained by Bourguet and colleagues [82]: The interfaces comprises residues from helices H7, H9, H10, and H11, as well as loops L8-9 and L9-10. Within the heterodimer, the two protomers do not equally contribute to the heterodimerization interface. For example, RAR and RXR exhibit different patterns of contact areas, as helix H7 of RXR contributes to the interface four times more surface area than its RAR counterpart. Inversely, the contribution of RARα loop L8-9 to the interface is three times higher than the one of the corresponding loop in RXRαF318A. Recently, RXR homo- and heterodimerization has been separated by mutational analysis (V. Vivat et al., 2003, *Mol. Cell. Biol.*, in press).

Except for side chain rearrangements, the part of the RXR LBD involved in the heterodimeric interface does not differ significantly from that of the RXRα apo-LBD homodimer. In both dimers, helices H9 and H10 contribute to more than 75 % of the total surface and constitute the core of the dimer interfaces. However, some differences in the relative involvement of some RXR structural elements to the homo- or heterodimer interfaces are observed and originate from the nature of the interacting protomer. For example, in the heterodimer, helix H7 contributes twice as much surface area to the interface than in the homodimer. This is due to the different structure of RAR loop L8-9, which makes more extensive contacts with H7 of RXR than its RXR counterpart in the homodimer. The buried surfaces are larger for the RAR-RXR heterodimer (967 $Å^2$ per monomer). The smaller interface in RXR homodimers (915 $Å^2$ per monomer) suggests a weaker link between the promoters that could be related to the promiscuous character of this NR. The moderate stability of RXR homodimeric association may facilitate heterodimerization.

When compared to RAR and RXR, the ERα LBD structural elements generating the dimerization interface are identical. However, helices H8, H9, H10, and the loop L8-9 in ERα are longer and make additional contacts. As a consequence, the buried surfaces are larger for the ERα homodimer (around 1700 $Å^2$), suggesting its higher dimerization affinity when compared to RXR dimers.

In contrast, the recent crystal structure of the GR LBD suggests a novel mode of receptor dimerization [86]. This unique dimer configuration involves residues from the β-turn of strands 3 and 4, the extended strand between helices H1 and H3 as well as the last residue of helix H5. Compared to the dimerization surfaces observed in other NRs, formation of the GR homodimer buries only 623 $Å^2$ of solvent accessible surface, probably reflecting its weaker dimerization affinity (Kd = 1.5 μM).

Ligand-binding pocket

In all crystal structures presently available, the ligand is embedded within the protein with no clear accessible entry or exit site. PPARγ seems to be the only exception to that rule, since a potential access cleft to the LBP was observed between helix 3 and the β-turn, which could be of sufficient size to allow entry of small ligands without major adaptation. For all other receptors of known structure, significant conformational changes are necessary to generate potential entry sites. The mouse trap model provides an easy solution to the problem: The mobility of H12 opens a channel by removing the "lid" from the ligand pocket.

The ligand-binding pockets are lined with mostly hydrophobic residues. Few polar residues at the deep end of the pocket near the β-turn act as anchoring points for the ligand or play an essential role in the correct positioning and enforce the selectivity of the pocket. Most NRs contain a conserved arginine attached to helix H5, which points into this part of the cavity. These anchoring residues, conserved within a given subfamily, are indicative of the polar group characteristics of each family of ligands (i.e., carboxylate for retinoids and ketones for steroids).

In the case of retinoid receptors, it is the ligand that adapts to a fairly rigid ligand-binding pocket [87]. The ligand-binding pockets of some other NRs are significantly larger and use precise anchoring points for their ligands; in such cases, it is possible to generate ligands, which owing to additional contacts, have higher binding affinities that the natural ligands and may even act as "superagonists". In this context, the recent structure of the human nuclear xenobiotic receptor PXR (NR1I2) is interesting, because it helps to understand how the ligand-binding pocket of this receptor can accommodate such diverse ligands as dexamethasone, RU486, rifampicin, taxol, and others [88]. In contrast to other NRs, PXR contains a small number of polar residues distributed throughout its large hydrophobic ligand-binding pocket. These unique features allows PXR to bind a diverse set of chemicals containing polar groups capable of hydrogen bonding and to act as chemical sensor.

MOLECULAR MECHANISMS OF TRANSCRIPTIONAL REGULATION

The expression of a given gene can be regulated at several different levels (transcription, translation or RNA processing, posttranslation) and also transcription itself has multiple levels at which regulation might occur. The transcriptional activity of a gene can be controlled epigenetically via methylation, at the level of its chromatin structure and at the level of the assembly and activity of the initiating and elongating polymerase-complexes. The reported effects of NRs on transcription are so far restricted to the initiation of transcription by RNA polymerase II. Below, we will summarize our current knowledge about the molecular mechanisms/interactions through which NRs can positively or negatively regulate the expression of cognate genes. For further details and references, see recent reviews [89–100].

Most nuclear receptors contain at least two independent activation functions, AF-1 and AF-2

Nuclear receptors contain at least two distinct regions, termed activation functions AF-1 and AF-2 (Fig. 2), which, when tethered to a (heterologous) DBD, will transactivate transcription from response elements recognized by this DBD. In the context of the wild-type receptor, both AF-1 and AF-2 become active in response to the ligand, but AF-1 can act constitutively in fusion proteins with heterologous DBDs. AF-2 remains ligand-dependent even in such fusion proteins. Within AF-2, at the C-terminal boundary of the LBD, an autonomous constitutively active transactivation function (AF-2 AD) has been identified whose integrity is crucial for AF-2, as mutations in AF-2 AD abolish AF-2 activity structure. This AF-2 AD encompasses in helix H12 in the 3D structure of NR LBDs (see above). Note, however, that the constitutive activity of AF-2 AD is very weak compared with the full ligand-induced activity of AF-2.

Within the AF-1-containing A/B region, shorter regions have been described to display constitutive activation function. In the case of the human PR isoforms, the additional N-terminal sequence, which is unique to the larger ("form B") isoform, was found to be able to squelch and to display an autonomous transactivation activity on its own together with the homologous, but not with a heterologous, DNA-binding domain (for more information and original refs. see ref. [5]).

Several NRs exist as isoforms. As was originally shown for the PR forms A and B, both isoforms exhibit different promoter specificities. For a more detailed discussion on this topic, see the recent review by Conneely and colleagues [101], and for a discussion of the differential antagonist action of the two isoforms and the relevance to endocrine cancer therapy see the corresponding reviews from the Horwitz laboratory [102,103]. Note that the additional N-terminal sequence of PR B may have a peculiar structure [104]. A later study showed that isoform-specific transcription activation is not confined to PR, but can be observed also with the RARs [105].

It is important to point out that the activation functions of NRs act in both promoter context- and cell-specific fashion, as was best documented for ERs and RARs [20,105]. This selectivity may originate from cooperative/synergism with other promoter-bound transcription factors and/or the cell-specific expression of TIFs/coactivators (see below and Topic 1.3).

Chromatin-modifying nuclear receptor coregulators

Coactivators and HATs

So-called squelching [106] experiments paved the way to predict the existence of factors that would transmit the signal generated by the holo-receptor to the transcription machineries. Squelching occurs if a receptor inhibits the activity "off the DNA" of the same ("autosquelching", [19]) or a different ("heterosquelching", [107,108]) receptor in an agonist- and AF-2 integrity-dependent manner. These squelching data were interpreted as the result of sequestering, by either excess of the same, or addition of another ligand-activated receptor, of so-called transcription intermediary factors (TIFs) that mediate the action of the activation/repression functions of NRs and are limiting constituents of the machineries required for transcription initiation. This concept predicted the existence of TIFs that are shared between, and are critically involved in, the action of different receptors. Indeed, the subsequent cloning and characterization of TIFs, also known as coactivators and corepressors, has fully justified this concept. According to the squelching studies, bona fide coactivators were predicted to fullfill the following criteria: (i) interact directly with NR LBDs in an agonist- and activation function-dependent manner, (ii) enhance NR-dependent reporter gene activity when transiently expressed in mammalian cells, (iii) activate transcription autonomously when tethered to DNA via a heterologous DBD, and (iv) relieve squelching. The development of yeast "two hybrid" and direct cDNA expression library screening approaches has allowed to identify a great number of putative and bona fide of coactivators, corepressors, and other coregulators that are believed to transmit the NR signal to its molecular targets.

The cloning of coregulators was followed by the definition of the coactivator signature LxxLL NR box (where x is any amino acid) motifs embedded in a short α-helical peptide [109–111]. These NR boxes are necessary and sufficient for ligand-dependent direct interaction with a cognate surface in the NR LBD that constitutes the transcriptional activation function AF-2.

Recently, a second contact site for coactivators has been identified in NRs. Proteins from the TIF2 family are able to interact specifically with the A/B domains of estrogen and androgen receptors [114–118]. These interactions result in a stimulation of the transcriptional activity originating form AF-1. Moreover, it appears that simultaneous interaction of coactivators with both the AF-1 and the AF-2 of an NR accounts for the synergy between both transcriptional activation functions [119,120]. The structural features of the A/B domains that are recognized by the coactivator have not yet been defined.

The identification of specific NR coactivators has prompted the question of how they function on a molecular level in transcription. To this end, several observations have been made. It is now generally accepted that NR coactivators possess or recruit enzymatic activities, and that they form large coactivator complexes. CBP, p300, P/CAF, SRC-1, P/CIP, and GCN5 are reported to act as histone acetyltransferases. They are capable of acetylating specific residues in the N-terminal tails of different histones, a process that is believed to play an important role in the opening of chromatin during transcription activation, and also nonhistone targets. Note, however, that the HAT activity of SRC-1 and P/CIP, if real, is negligible compared to that of p300 or CBP. Thus, p160 coactivators may rather recruit such activities by physical association with histone acetyl transferases, such as CBP or p300, or with complexes containing such activities. Specifically, the activation domain AD1 of TIF2 has been demonstrated to function via the recruitment of cointegrator CBP [79], which apparently in turn acetylates TIF2 [121]. Besides HAT activities, also other enzymatic activities have been attributed to NR coactivator complexes. TIF2 proteins are able of interacting functionally via their activation domain AD-2 with a protein methyltransferase [122], via N-terminal region(s) with PIAS proteins that act as SUMO-ligases [123,124], and other factors [125,126]. The roles of these various interactions in epigentic gene regulation are being actively investigated.

In conclusion, bona fide coactivators (i.e., members of the TIF2/SRC-1/RAC3 family), together with the CBP/p300 cointegrators function by rendering the chromatin environment of an NR target gene prone to transcription. This opening of the chromatin environment is achieved by intrinsic or recruited

HAT activity. The HAT activities of different coactivators/coactivator complexes targets (i) the N-termini of histones, which have reduced DNA-binding activity upon acetylation, (ii) certain basal transcription factors, and (iii) at least some bona fide coactivators themselves. The chromatin modification step represents the first of at least two distinct steps in transcription activation by NRs (see Fig. 5 for an illustration of the "derepression" and further below for the second step).

Structural basis of coactivator binding to nuclear receptors

Biochemical [127,128] and structural [77,129,130] studies clearly demonstrated that the holo conformation of NR LBDs harbors an "active" surface that is able to recognize and dock a short α-helix present in the NR interacting domain (NID) of coactivators. This α-helix contains a LxxLL motif (with L being a leucine residue and x any amino acid), called nuclear receptor box (NR box) by Le Douarin and colleagues [109], and is present in several copies in the NID of coactivators [109,110,131]. The NR interaction surface is composed of a static part involving the C-terminal half of helix H3, helix H4, and the loop L3-4 connecting them, and of a mobile part corresponding to the activation helix H12. All together, these structural elements define a hydrophobic groove that can accommodate the coactivator NR boxes. The NR–coactivator interaction is mediated by the leucine residues of the NR box motif, which pack on the hydrophobic cleft but also by two conserved residues of the receptor (a lysine and a glutamic acid residues in helices H3 and H12, respectively) which are hydrogen-bonded to two main-chain carbonyl groups of the helical NR box. This "charge clamp" further stabilizes the interaction and strictly defines the length of the helix that can be docked into the NR recognition groove. Furthermore, biochemical experiments suggest that nonconserved residues adjacent to the LxxLL motif of coactivators make additional contacts with the NR LBD and may determine the specificity of NR–coactivator interaction [129,131].

Nuclear receptor corepressors, silencing, and HDACs

The second class of NR coregulators comprises the corepressors. Early on, it was shown that some NRs do actively repress transcription when in the apo form. This phenomenon had been particularly well established for retinoic acid and thyroid hormone receptors. Soon after the identification of ligand-recruited coactivators, similar approaches identified proteins that recognize the ligand-free NR. To date, several different NR corepressors have been identified; by far, most studies have been performed with NCoR and SMRT. For these two, it was shown recently that a conserved CoRNR box motif interacts with a surface on the ligand-binding domain which is topologically very similar to that recognized by coactivator LXXLL motifs, but does not involve helix H12 [132–135]. Corepressors are believed to reside in, or recruit, high-molecular-weight complexes that display the opposite activity of coactivator complexes. While coactivator complexes acetylate histones, thereby weakening the interaction of the N-terminal histone tails with the nucleosomal DNA, corepressors recruit histone deacetylase activities that reverse this process (illustrated in Fig. 5 as the "repression" step). Deacetylated histones are associated with silent regions of the genome, and it is generally accepted that histone acetylation and deacetylation shuffle nucleosomal targets between condensed and relaxed chromatin, the latter being requisite for transcriptional activation. An unresolved issue is whether all NRs are able of active repression. In concert with this observation, recent evidence has been presented that some steroid hormone receptors also bind to corepresssors in presence of certain antagonists [136–138].

Given this high number of different coregulators for NRs, two principal questions emerge: (i) what defines coactivator selectivity, and (ii) how is the assembly of different coactivator complexes with different intrinsic transcription activities regulated? Future research will have to address such questions, especially also in view of the therapeutic perspectives in disease.

Structural basis of corepressor binding to nuclear receptors

It has been proposed that corepressors interact with RAR and TR LBDs in a region encompassing helix H1 since mutation of residues Ala223, His224, and Thr227 in the so-called CoR box of TRβ abrogated the recruitment of N-CoR [139,140]. However, examination of this portion of NR LBD structures re-

A. REPRESSION

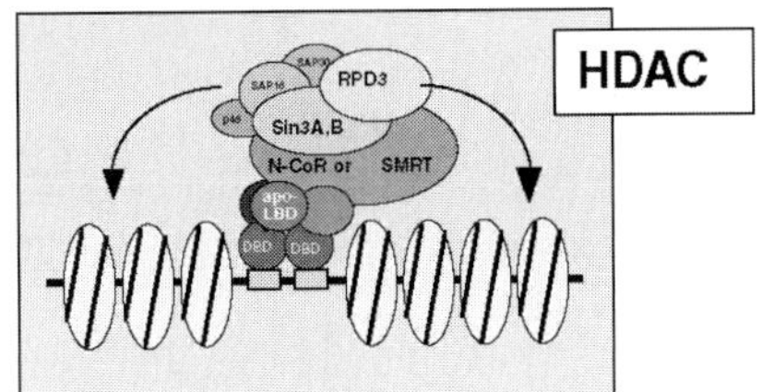

B. DEREPRESSION

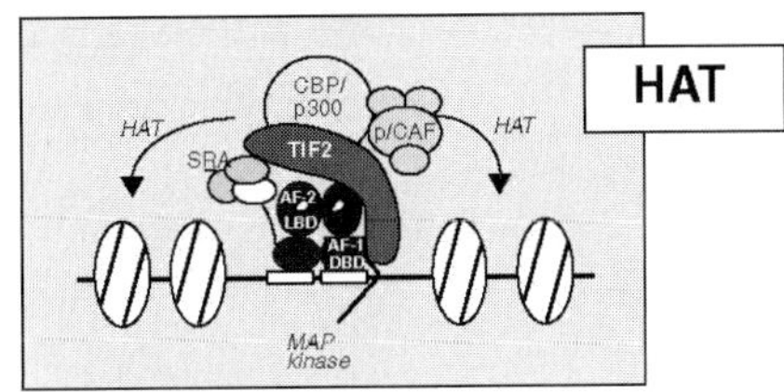

C.TRANSCRIPTION ACTIVATION

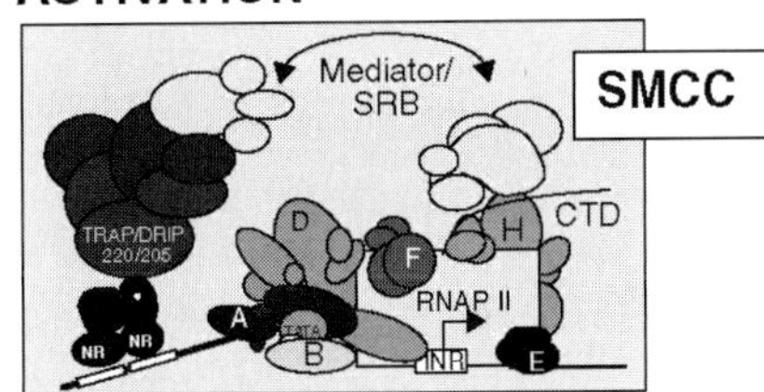

SMCC/SRB/MED-COMPLEX

Fig. 5 Hypothetical 3-step mechanism leading to transcriptional activation by NRs. Some NRs act as silencers of target gene transcription in the absence of ligand (or in the presence of certain antagonists). This "repression" step is due to the recruitment by the apo-NR of a corepressor complex that exerts HDAC activity. Ligand binding dissociates this complex and recruits the coactivator complex that displays HAT activity. The subsequent chromatin decondensation ("derepression") is believed to be necessary, but not sufficient, for target gene activation. At the third step, the HAT complex dissociates, possibly due to acetylation of the coactivator which decreases its ability to interact with the receptor, and the SMCC/DRIP/TRAP complex is assembled through its TRAP220/DRIP205 subunit. The SMCC complex is able to establish contacts with the basal transcription machinery, resulting in transcription. Note however, that it has been proposed that for some receptors p160 coactivator-mediated transactivation can occur without contribution of SMCC factors [112] initiation. This illustration is modified from ref. [113]. Note that the involvement of ATP-dependent chromatin remodeling machineries in NR actions is not considered in this cartoon.

veals that these residues cannot be part of a corepressor interacting surface. Indeed, they are engaged in intramolecular interactions with other structural elements (H3, H8,...) and are not solvent exposed. Most probably, these mutations destabilize the positioning of helix H1, which in turn can affect the structural stability of other parts of the LBD. Another possibility is that the opposite face of helix H1 contains residues important for corepressor binding. However, residues at the surface of helix H1 are not conserved among NRs that were shown to interact with SMRT and N-CoR. Therefore, helix H1 is probably not the major anchoring point of corepressors on NRs. Recent correlative analyses of biochemical data and sequences provided evidence that coactivator and corepressor recruitment share similar molecular basis. Evaluating corepressor binding to mutants in the coactivator binding site of TRα, Hu and Lazar (1999) demonstrated that mutations that impared activation and coactivator recruitment, also decreased repression and corepressor binding, indicating that corepressors N-CoR and SMRT bind to a NR surface topologically related to that involved in coactivator interaction. Mapping of NR determinants of N-CoR binding using a series of mutations introduced into TRβ indicated that the groove formed by residues from helices H3 and H4 corresponds to the primary corepressor binding site, but that H11 or H12 are not involved [133,134]. Using a similar approach, a parallel study was reported on the interaction between N-CoR and the orphan NRs Rev-erbA/RVR that are known to funtion as transcriptional silencers. In that particular case, the LBD interacting surface with corepressors would encompass H3–H4 and H11, which was correlated to the fact that Rev-erbA/RVR lack the C-terminal activation helix H12 [141]. The C-terminus of SMRT and N-CoR contain a region interacting with NRs composed of two independent interacting subdomains ID1 and ID2 [142,143]. Examination of the two interaction domains ID1 and ID2 in SMRT and N-CoR revealed sequences (CoRNR box 1 and CoRNR box 2 by analogy with NR boxes of coactivators) similar but not identical to the LxxLL motif of coactivators, which were also predicted to adopt an amphipathic helical conformation [132–134]. Chimaeric CoRNR box peptides were designed to show that although the consensus hydrophobic core ΦxxΦΦ is necessary and sufficient for corepressor binding, sequences flanking the CoRNR box strengthen the interaction and determine NR specificity. In addition, within a corepressor, the two CoRNR boxes are not equivalent. CoRNR box 1 interacts almost exclusively with RAR/TR, and CoRNR box 2 binds to both heterodimer partners. Perissi and coworkers further extended the analogy with the helical coactivator LxxLL motifs. They proposed a model in which the CoRNR box motif would fold into a significantly longer LxxI/HIxxxI/L helix when compared to the coactivator LxxLL NR box motif [134]. In the absence of agonists, the CoRNR box helix could interact with the H3-H4 hydrophobic groove displayed by apo-receptors. However, in the presence of an agonist, the repositing of the activation helix H12 in its active (holo) conformation would prevent this interaction. Indeed, in contrast to the unliganded NR case, the length of the helix that can be accommodated by the H12-containing groove is strictly defined by the presence of the charge clamp that specifically recognizes helices of the NR box type [77]. In this respect, the recent crystal structure of a ternary complex containing the PPARα LBD bound to the antagonist GW6471 and a SMRT corepressor motif [135] confirms that the corepressor fragment adopts a three-turn helix that binds into the groove which also serves as the coactivator-binding site.

Recruitment of the RNA polymerase II holoenzyme

The initial chromatin-modifying step carried out by NR coactivators (see above) has to be followed by the actual recruitment of the RNA polymerase II holoenzyme and initiation of transcription (illustrated in Fig. 5 as the "transcription activation" step). Comprehension of the recruitment of the polymerase II holoenzyme by NRs has only become at reach through the identification and cloning of the mammalian mediator complex as a thyroid hormone and vitamin D receptor coactivator [144–146]. The mammalian mediator came in several flavors. It was identified as the so-called SMCC, the Srb and Mediator protein containing complex [147], the TRAP complex, a thyroid hormone receptor associated protein complex [144], or the DRIP complex, a vitamin D receptor interacting protein complex [145]. Furthermore, common subunits are shared with PC2, the so-called positive coactivator 2 [148], the ARC [149], CRSP [150] and NAT [151] complexes. Hereafter, this complex is referred to as "SMCC". It is a large multisubunit protein complex that contains several homologs of the yeast mediator complex (RGR1, MED6, MED7, SRB7, SRB10, SRB11, NUT2, SOH1 [146,147]) as well as additional proteins of unknown function. As expected for a mediator complex, SMCC associates with the RNA polymerase II to form RNA pol II holoenzymes [146]. On the other hand, the SMCC complex is able to interact functionally with different transcription factors such as p53 and VP16 [146]. Furthermore, due to its identification as a thyroid hormone and vitamin D receptor-interacting complex it is believed to function as NR coactivator. This notion is supported by the demonstration that SMCC can enhance thyroid hormone and vitamin D transcription activation in in vitro transcription systems [145,152,153]. The subunit of the complex that is responsible for interaction with the agonist-bound LBD of NRs was identified as DRIP205 [154], which is identical to TRAP220 and contains a functional LxxLL NR box motif [155].

Interestingly, another subunit of the SMCC complex, DRIP150, interacts with the N-terminal region of the glucocorticoid receptor, which harbors the activation function AF-1 [156]. The current working hypothesis is that once the chromatin environment at target gene promoters has been decondensed by coactivator complexes containing members of the TIF2 and CBP families, the NR recruits RNA pol II holoenzymes via its association with the TRAP220/DRIP205 subunit of the SMCC. This switch between coactivators and the SMCC complex might be regulated by the acetylation of coactivators within the HAT complex [121], resulting in their dissociation from the NR, thus allowing the recruitment of factors such as SMCC via the LxxLL motif of the TRAP220/DRIP205 subunit.

The recruitment of the RNA polymerase II holoenzyme might also be enhanced by interactions of NRs with components of the SWI/SNF complex, which is part of RNA pol II holoenzymes.

Whether these various complexes are all necessary for a particular receptor-regulated gene program and, if so in which order, and with which stoichiometry they operate, if they act tissue, receptor, and ligand-selective, if they preexist, or get assembled by signaling events, are all questions that are being actively investigated.

STRUCTURAL BASIS OF LIGAND ACTION

Ligand selectivity

As shown in the cases of RARγ and TRβ, the shape of the ligand-binding pocket matches that of the ligand. The accordance of shape and volume maximize the number of mostly hydrophobic contacts, thus contributing to the stability of the complex and the selectivity of the pocket for the cognate ligand.

RAR possesses an interesting LBD, since it can bind equally well two chemically different ligands: all-*trans* retinoic acid and its 9-*cis* isomer. Crystallographic analysis [87] of the two ligands in the RARγ LBD showed that both adapt conformationally to the LBP that acts as matrix. Moreover, the conformation of a RARγ-selective agonist was also shown to closely match that of the natural ligands in their bound state [87]. The adaptation of ligands to the protein leads to an optimal number of interactions for binding and selectivity, and justifies modeling approaches for ligand design.

For steroid receptors, the LBP volume is significantly larger than that of the corresponding ligands, and the rigidity of the ligand does not allow adaptability. Therefore, selectivity cannot be driven by multiple hydrophobic contacts, which could anyway not suffice to discriminate between small structurally similar ligands. In this case, specific key interactions are more important. Note that very large LBP volumes allow for the binding multiple ligands of different stereochemistry such as in the case of PPAR [77], often at the expense of lower binding affinities.

A structure-based sequence alignment revealed that only three residues diverged in the LBPs of RARα, β, and γ, leading to the prediction that these divergent residues were critically involved in differentiating between isotype-selective retinoids [76]. Indeed, swapping of these residues confirmed this hypothesis [157]. Moreover, swapping of these residues not only mediated isotype-selective binding, but also the agonistic/antagonistic response of a cognate ligand onto any other RAR isotype, thus emphasizing the importance of these three residues in triggering the ligand-induced transcriptional response.

Antagonist action

To define or understand the agonistic/antagonistic features of a ligand, the following aspects have to be considered/elucidated: Ligands may positively or negatively interfere with receptor activities at various levels. In the case of steroid hormone receptors, they may affect: (i) the stability of the so-called hetero-oligomeric 8S complex, which is proposed to exist in hormonally naive cells and comprises, in addition to steroid receptors, the Hsp90 heat-shock protein and additional associated factors (note that certain receptors, such as TR and RAR, do not for complexes with Hsp90); (ii) the homo- or heterodimerization ability of the receptor; and (iii) its interaction with the cognate DNA response element.

Nuclear receptors harbor two activation functions (AF-1 and AF-2). These activation functions operate in a cell-type and promoter environment-dependent fashion. Thus, a given antagonist may antagonize only one or both AFs, and an AF-2 antagonist can act as an AF-1 agonist.

While the structural basis of AF-1 activity is still unknown, AF-2s correspond to agonist-induced surfaces that can interact with coactivators. Conversely, nonliganded receptors express a surface(s) that can accommodate corepressors. A given ligand may more or less precisely generate these surfaces and lead to different coregulator recruitment efficiencies. Thus, "superagonists" may enhance coactivator binding more efficiently than the natural ligand, while "inverse agonists" may stabilize the receptor–corepressor complex.

Ligands may act at various levels in the sequence of events that leads to transcriptional activation or silencing. Theoretically, the same ligand may stimulate the recruitment of SMCC, but not of the HAT complex. Indeed, initial evidence for the existence of ligands that differently affect HAT and SMCC subunit recruitment has been provided [154].

Several types of antagonists

Multiple aspects that have to be considered when analyzing the mechanism of action of an antagonist or when the aim is to design an antagonist with certain characteristics. Below we will discuss first some general principles originating from structural studies and then discuss particular mechanisms and individual antagonists. Note that some analyses have to be considered in the context of the experimental setting because some receptor activities, such as DNA-binding, can be ligand-independent in vitro, but are ligand-dependent in vivo (for an example, see refs. [158,159]). Therefore, some of the results/categories described below may have to be reconsidered as more in vivo data accumulate.

Structural basis of AF-2 antagonists: Steric hindrance by the ligand precludes holo position of H12

Agonists are ligands that lock the receptor in the active conformation. In contrast, antagonists should be viewed as molecules that prevent NRs to adopt this conformation. Helix H12 is a crucial component of the NR LBDs, because its ligand-induced repositioning in the holo NR contributes in a critical manner to the surface recognized by the LxxLL NR boxes of coactivators and thereby generates a transcriptional active AF-2 domain. The original structures of apo- and holo-LBDs (Fig. 4) revealed the ligand-induced conformational changes and suggested that the interactions between H12 or residues in its proximity and the ligand was critical for the control of agonist-antagonist properties of NRs [75,76,84]. The crystal structures of the ERα LBD complexes with raloxifen and tamoxifen or the RARα LBD with BMS614 confirmed this hypothesis and showed an alternative stable antagonist "position" for H12 (compare Fig. 4b), nested between H4 and H3 [82,85,130]. Apparently, steric hindrance upon binding of the bulkier ligands (compared with the agonists) prevents the proper positioning of H12 in its "agonistic" site.

In the antagonist conformation, a most important feature is the lengthening of the loop L11-12, resulting from the unwinding of the C-terminus of helix H11. This enables helix H12 to adopt a second low-energy position by binding to the coactivator LxxLL recognition cleft. In contrast to agonists, which stabilize a long H11 helical conformation, different ligand–receptor interactions at the level of H11 and of the surrounding regions (loop L6-7 and H3) most likely explain the antagonist-induced unwinding of the C-terminal part of this helix. Note that these structural features are found in all antagonist-bound LBD complexes crystallized so far. Hence, it appears that the action described above of pure AF-2 antagonists originates from at least two structural principles. The main feature is the presence of a large "antagonistic" ligand extension that sterically prevents helix H12 to adopt the holo position. Without a holo-H12, no LBD-coactivator interface can be formed. The second structural principle is the unwinding of helix H11, which allows H12 to bind to the binding groove of coactivator NR box LxxLL motifs. Thus, the second feature of antagonism is the competition between H12 and the NR boxes of coactivators for a common LBD surface.

However, there are additional structural principles of antagonism: In a recent report, Pike and colleagues [160] determined the structure of ERβ LBD in complex with the pure antagonist ICI 164,384. In this case, the ligand completely abolishes the association between the transactivation helix H12 and any part of the LBD. The lack of a stabilized interacting surface containing a holo H12 prevents the binding of a coactivator to the NR LBD, and thus accounts for the "complete" antagonism.

Structural basis accounting for full and partial AF-2 agonistic activity of a ligand

In addition to complete antagonists of the AF-2 function (e.g., raloxifen or tamoxifen for ER; BMS614 for RAR; see above), AF-2 partial agonists/antagonists (Table 3) have been crystallized with the corresponding receptors. In contrast to complete AF-2 antagonists, where there was a clear correlation between H12 positioning in the coactivator cleft (also referred to as "antagonist groove"), two crystal structures have been described in which there is an apparent discrepancy between the orientation of the AF-2 helix and the biological activity of the corresponding ligand. The ERβ/genistein and RXRαF318A/oleic acid LBD structures (for refs., see Table 3) show that H12 can adopt the antagonist conformation even though the corresponding ligand elicits a weak but clear transcriptional AF-2 activity. A likely explanation for the discrepancy between the antagonist location of H12 and the transcriptional activity of these complexes is that these ligands display some, but not all, features of pure AF-2 agonists or antagonists. They can thus be classified as partial AF-2 agonists/antagonists. A major difference between pure and partial antagonists lies in their steric properties. In contrast to full antagonists, genistein and oleic acid do not bear a bulky extension. Thus, they do not sterically preclude the agonist position of H12 and are in this respect similar to agonists. However, they induce unwinding of helix H11, which permits the positioning of helix H12 in the antagonist groove; in this respect, these ligands are similar to antagonists. The structure of PPARγ bound to the mixed agonist-antagonist GW0072 suggests that an additional mechanism might account for the particular biological properties of such ligands [161]. In this case, the partial activity of the ligand is attributed to a poor stabilization of the holo position of H12 as a result of a lack of contact between the ligand and the AF-2 helix. In the presence of such mixed ligands, the equilibrium between the agonist holo position of H12 and its antagonist position in the coactivator binding groove is likely to depend on the intracellular concentration of coactivators and corepressors, and these ligands may act as either AF-2 agonists or antagonists depending on the cellular context.

A novel mode of antagonism was recently suggested by the resolution of the crystal structures of ERα and β in complex with THC (5, 11-*cis*-diethyl-5, 6, 11, 12-tetrahydrochrysene-2, 8-diol). Interestingly, THC acts as an ERα agonist and as an ERβ antagonist. Structure comparison of the two ligand-receptor complexes reveals that THC, which lacks the bulky side chain of pure antagonists, antagonizes ERβ by stabilizing the conformation of several residue side chain from helix H11 and loop L11-12 in such a way that they do not create the proper hydrophobic binding surface for the holo helix H12 [162].

Inverse agonists: Stabilization of corepressor interaction

Whereas unliganded steroid hormone receptors do not appear to interact strongly with SMRT and N-CoR, it has been proposed that some antagonists enhance significantly this interaction [136,138,163]. Moreover, it appears that corepressors interact with steroid receptors occupied by partial antagonists, but not with receptors bound to pure antisteroids. Also, some RAR antagonists reinforce corepressor binding to RAR and enhance silencing [164,165]. The structural basis of this stabilization remains to be established. However, such ligands have to induce an alternative position for H12, which does not occlude the hydrophobic groove formed by H3 and H4. Thus, the AF-2 helix may be inhibitory for full corepressor binding, and its deletion or displacement by some antagonists can potentiate the interaction.

In this respect, the antagonist GW6471 binding to PPARα reinforces the corepressor interaction [135]. In contrast to other antagonist-bound NR structures, the AF-2 helix undergoes a rigid body shift toward the N-terminus of helix H3 and is loosely packed against this helix. The third helical turn in the corepressor motif occupies the space that is left by the repositioning of helix H12 and prevents this helix from adopting its agonist-bound conformation.

RXR SUBORDINATION IN HETERODIMERS

Many NRs use RXR promiscuously as heterodimerization partner. In contrast to homodimerization, heterodimerization allows, in principle, fine-tuning of NR action by using combinatorial sets of ligands. However, whereas RAR agonists can autonomously activate transcription through RAR-RXR heterodimers, RXR is unable to respond to RXR-selective agonists in the absence of a RAR ligand. Consequently, RXR-selective ligands on their own could not trigger RXR-RAR heterodimer-mediated RA-induced events in various cell systems [159,166,167]. Similarly, RXR cannot autonomously respond to its ligand in the corresponding TR and VDR heterodimers, unless those heterodimeric partners are liganded. This phenomenon, referred to as RXR "subordination" or "silencing", may be of utmost biological importance because it avoids confusion between retinoic acid, thyroid hormone, and vitamin D3 signaling pathways. RXR subordination is, however, not due to an inability of the RXR partner to bind its cognate ligand in DNA-bound heterodimers, as has been suggested [168], as RXR ligand binding has been demonstrated to occur in such complexes [158,169–172]. Several studies investigated whether coregulator interactions could account for the inability of RXR to respond to its agonists. Recently, it has been reported that RXR can bind its ligand and recruit coactivators in heterodimer with apoRAR. However, in the usual cellular environment, corepressors do not dissociate and thus compete with coactivators for binding [164]. Corepressor binding to RAR prevents liganded RXR in the context of the holoRXR/apoRAR heterodimer from assembling a transcription-inducing competent complex. Consequently, the only way for RXR to modulate transactivation in response to its ligand in RXR-RAR heterodimers is through synergy with RAR ligands. This synergy results from increased interaction efficiency of a single coactivator molecule with the heterodimer and requires two intact receptor-binding surfaces on the coactivator. Cooperative TIF2 binding was also observed with some RAR antagonists [164]. Together with such RAR antagonists, pure rexinoid agonists can activate transcription of endogenous target genes [159,164]. All these observations do not support the concept that RXR is a priori a transcriptionally "silent" partner in RAR-RXR heterodimers [168,173].

RXR subordination may not apply to all NR partners, as the ligand-induced RXR activity was permissive in heterodimers with FXR, LXR, PPAR, or NGFI-B [173–175]. However, neither the existence of an endogenous NGFI-B ligand nor a weak constitutive activity of the NGFI-B AF-2 can be excluded; both these scenarios would readily explain RXR activity and NGFI-B-RXR synergy due to the absence of RXR silencing.

POSTTRANSLATIONAL MODIFICATION BY PHOSPHORYLATION

NRs can be subjected to various posttraductional modifications such as phosphorylation, ubiquitylation, or acetylation [176,177], which act in concert to coordinate NR-mediated transcription [178]. Among these modifications, phosphorylation has been more extensively studied than the others and is increasingly recognized as a signaling cross-talk that affects most if not all NRs. Phosphorylation can modify all major domains of NRs, the N-terminal activation function (AF-1), the LBDs and DBDs (Fig. 2). Phosphorylation of NRs by kinases that are associated with general transcription factors (e.g., cdk7 within TFIIH), or are activated in response to a variety of signals (MAPKs, Akt, PKA, PKC), often facilitates the recruitment of coactivators, or of components of the transcription machinery, and therefore, cooperates with the ligand to enhance transcription activation. But phosphorylation can also contribute to termination of the ligand response through inducing DNA dissociation or NR degradation or through decreasing ligand affinity. These different modes of regulation reveal an unexpected complexity of the dynamics of NR-mediated transcription. Finally, as phosphorylation can occur in the absence of the ligand, deregulation of NR phosphorylation in certain diseases or cancers may lead to apparently ligand-independent activities.

Phosphorylation targets mainly two nuclear receptor activation functions and the DNA-binding domain

The majority of the NR-phosphorylated residues lie within the N-terminal A/B region. For PRs, phosphorylation of this region is rather complex, with more than 13 phosphorylation sites. For RARs or PPARs, region A/B phosphorylation concerns only one or two residues. In VDRs, this region is apparently not phosphorylated, probably due to its short size. Most of the modified residues are serines surrounded by prolines and therefore correspond to consensus sites for proline-dependent kinases, which include cyclin-dependent kinases (CDKs) [179,180] and MAP kinases [181–183]. Among these sites, many are phosphorylated by CDKs, either "constitutively" (i.e., in the absence of ligand) or in response to the hormone [184]. MAPKs can hormone-independently phosphorylate other sites in response to a variety of signals, such as growth factors, stress, or cytokines. Several serine kinase cascade pathways lead to activation of MAP kinases (Erks, JNKs, p38MAPK) that can enter the nucleus and phosphorylate NRs. Accordingly, the AF-1 function of PR [185], ERα [22,186], ERβ [187,188], AR [189], PPARs [190,191], RARγ [192,193] have been reported to be substrates for p42/p44 or p38 MAPKs, while that of RXRα is targeted by JNKs [194,195].

The N-terminal A/B region also contains consensus phosphorylation sites for the so-called kinase Akt or PKB [196], which plays an important role in cell survival and proliferation. Akt is regulated positively by the phosphoinositol 3-kinase (PI3K) pathway and negatively by a variety of downstream kinases. Upon translocation into the nucleus, it can phosphorylate NRs, such as ERα [197] and AR [198] in their N-terminal A/B region.

In addition to the N-terminal domain, the LBD of NRs is also a target for ligand-independent phosphorylation, involving the same proline-dependent kinases as above. Thus, RXRα can be targeted by stress kinases (JNKs) [194,195]. However, other kinases, such as tyrosine kinases, may phosphorylate ERα [199] and RXRα [195] or PKA for RARs [200,201].

Finally, NRs can be phosphorylated in their DBD. Phosphorylation of this domain involves either PKA in the case of ERα [202] or PKC for RARα [203] and VDR [204].

Nuclear receptor phosphorylation by cdks

When bound at their response elements, liganded NRs recruit the transcription machinery, including the general transcription factors. One of them is TFIIH, a multiprotein complex mediating transcription activation and nucleotide excision repair [205]. Several NRs, such as ERα [206,207], RARα [208], RARγ [209], and AR [210], have been reported to interact with TFIIH. Notably, the interaction of ERα with TFIIH involves the same surface as that required for the recruitment of coactivators [206] and, therefore, depends on the conformational changes induced by the ligand. In contrast, RARs interact with TFIIH through an alternative surface(s) that is not modified by the ligand [209].

TFIIH is composed of nine subunits, one of them, cdk7, having a cyclin-dependent kinase activity. As a consequence of their interaction with TFIIH, ERα and RARs have been shown to be phosphorylated in their N-terminal A/B region by cdk7 within TFIIH. This phosphorylation process plays a critical role in receptor-dependent transcription activation, because cells from patients bearing mutations in the XPD subunit of TFIIH, which results in an incorrect positioning of the cdk7 kinase relative to its substrate, have hypophosphorylated RARs that diplay a decreased ligand response [211]. As the serine residues targeted by cdk7 lie within surfaces that interact with transcription factors, their phosphorylation would help the recruiment of coactivators and thus would increase the efficiency of chromatin derepression. It could also facilitate the recruitment of components of the transcription machinery and therefore stabilize the formation of the NR transcription complex.

The critical role of NR phosphorylation by cdk7 has been further dissected in the case of RARγ, by using F9 cells which represent a cell-autonomous system for analyzing retinoid signaling (for review, see ref. [212]). In these cells, the retinoid-induced events (primitive endodermal differentiation, growth arrest, and the activation of expression of a number of genes) are transduced by RARγ/RXR heterodimers. Consequently, the various RA responses are abolished in RARγ null cells. By reexpressing in these RARγ null cells, RARγ mutated at the cdk7 phosphorylation sites located in the A/B region, it has been demonstrated that the integrity of these phosphorylation sites is indispensible to the activation of certain RA target genes and for RA-induced F9 cell differentiation [213].

Regulation of nuclear receptor-mediated transcription through phosphorylation

Positive regulation of nuclear receptor transactivation

NRs are substrates for a multitude of kinases activated by a variety of signals, independently of the ligand. For several NRs, phosphorylation of the N-terminal A/B region by MAPKs (Erks, p38MAPK, JNKs) or Akt facilitates the recruitment of coactivators and, thus, transcription activation.

Supportive evidence for a link between phosphorylation by MAPKs and NR-mediated transcription is the recent finding that the N-terminal AF-1 domain of RARγ can be phosphorylated by p38MAPK [192]. Importantly, this event, which is induced by RA, was found to be also a signal for RARγ ubiquitylation [214]. Moreover, mutations that block phosphorylation and/or ubiquitylation result in defects in the activation of RA target genes controlled by RARγ [192]. It is tempting to speculate that phosphorylation by p38MAPK positively modulates the transcriptional properties of RARγ through the recruitment of the ubiquitylation machinery. However, phosphorylation-dependent ubiquitylation also targets RARγ for degradation by the 26S proteasome [192]. Thus, the modulable equilibrium between transactivation and degradation may represent an efficient mechanism to simultaneously activate RARγ to initiate transcription and tag it for subsequent degradation by the proteasome in a step that attenuates transcription.

Also, phosphorylation of the AF-2 domain can modulate the transcription factor properties of NRs. Phosphorylation by Src kinases of ERα at tyrosine 537 which is close to helix 12, enhances ERα function [215]. Phosphorylation of RARα by PKA at serine 369 also modulates positively the transcriptional activity of the receptor [201]. Phosphorylation by PKA also modulates the parietal endoderm differentiation of F9 cells which occurs subsequently to primitive endodermal differentiation when RA is combined with cAMP and which involves RARα/RXR heterodimers. Accordingly, upon reexpressing in RARα null F9 cells the same receptor mutated at the PKA phosphorylation site, parietal differentiation is delayed [216].

Phosphorylation of nuclear receptor coregulators

The cross-talk between NRs and signal transduction pathways involves not only the phosphorylation of NRs, but also that of their coactivators and corepressors; SRC-1 [217], TIF2 [218], PGC-1 [219], AIB1 [220], and p300/CBP [221,222] are themselves targets for a variety of kinases. Phosphorylation may enhance their interaction with NRs, efficiency to recruit HAT complexes and/or enzymatic activity. In contrast, phosphorylation of corepressors such as SMRT subsequently to the activation of MAPKs cascades, correlates with an inhibition of their interaction with NRs and their redistribution from the nucleus to the cytoplasm [223].

Negative regulation of nuclear receptor transactivation by phosphorylation

Phosphorylation events can also inactivate NRs, possibly to switch off their activity. PKC-dependent phosphorylation of VDRs at the DBD inhibits transcription activation most likely by facilitating promoter escape [204]. Inhibition of the transcriptional activity of other NRs such as ERα and RARα, also occurs subsequently to phosphorylation of residues located within the DBD dimerization surface, by PKA [202] or PKC [203], respectively.

NUCLEAR RECEPTORS: PLATFORMS FOR MULTIPLE SIGNAL INTEGRATION

It has become increasingly well documented in the past few years that NR action is not confined to the positive and negative regulation of the expression of cognate target genes. Indeed, these receptors, and most likely also their "downstream" mediators, are targets of other signaling pathways and reciprocally, can modify the activity of such pathways. The best known examples of such a signal transduction "cross-talk" is the mutual repression of NR and AP1 (c-Fos/c-Jun) activities. A distinct type of cross-talk is the modification of NR AF activity by phosphorylation, e.g., by the MAP kinase pathway. The existence of signal transduction "cross-talks" is likely to reflect the integration of NR action in the context of the functional state of the cell in which it is expressed. The importance of signal transduction "cross-talk" in "real life" was recently impressively demonstrated by the observation that GR null mice die at birth, whereas mice harboring a GR mutant ($GR^{dim/dim}$) that can still cross-talk with AP1 but not activate target genes with consensus GR response elements are viable [17]. A particular interesting issue is the possibility to generate NR ligands that can "dissociate" such response element-dependent and cross-talk-dependent gene programming, thus giving rise to the hope of generating ligands with reduced side effects [224].

Signal transduction cross-talk between nuclear receptors and AP1

In addition to transactivation of their own target genes, certain NRs were shown to cross-talk with other signal transduction pathways. The original observation was made in 1990 when it was observed that GR could inhibit, in a ligand-dependent manner, the ability of AP1 (the heterodimer composed of the proto-oncogene products c-Fos and c-Jun) to transactivate its target gene promoters [225–227]. This transrepression is mutual and requires an unknown state of the receptor, which can be induced by both agonists and certain, but not all, antagonists. It is important to point out that the nuclear receptor-AP1 cross-talk does not per se imply negative regulation of transcription; several reports show that under certain conditions this cross-talk can lead to positive transcriptional effects [228–230].

The mechanism(s) on nuclear receptor-AP1 cross-talk has remained elusive despite, or because of, several contradictory reports and discrepancies between studies using in vivo and in vitro approaches. For example, the original proposal that AP1 and GR form DNA-abortive complexes (derived from in vitro evidence obtained in gel retardation experiments) appears to represent an in vitro artefact, as in vivo footprinting did not support this concept. Controversially discussed are studies proposing that sequestration of the coactivator CBP (or its homolog p300), which function as mediator of the transcriptional activities of both AP1 and NRs, accounts for the cross-talk phenomenon as transrepression could be relieved when CBP was overexpressed in cells with limiting endogenous levels of this protein [231]. Several other studies, however, have shown that this mechanism can at best only partially account for the observed phenomena [232]. In particular, the use of synthetic ligands that dissociate transrepression from coactivator recruitment to NRs argue against the involvement of coactivators, since it was shown that antagonists can still transrepress AP1 activity [233]. A second mechanism that has been proposed is based on the observation that estrogen receptors are capable of down-regulating the activity of Jun-kinase, leading to reduced AP1 activity [234]. Thirdly, NR-mediated effects on the dimerization of the AP1 subunits have been observed [235]. Still, further analysis is required to understand the contribution of these various mechanisms to the receptor-AP1 cross-talk. A fourth mechanism proposed to be involved in AP1 nuclear receptor cross-talk is direct physical contact between both factors when bound to so-called composite elements on a promoter [230]. Composite elements are thought to recruit both, AP1 proteins and NRs, bringing them into close physical contact. Depending on the nature of such response elements, and the activity of the participating proteins, steric effects might lead to differential regulation. It is currently not clear whether this mechanism applies only to promoters that carry composite elements or is of more general significance.

In addition to GR, mutual interference observed between the transcriptional activities of AP1 and RARs and RXRs, ER, TR/v-ErbA, PML-RARα, while MR appears to be insensitive to AP1. In addition, interference between the transactivation abilities of NRs and other transcription factors has been reported. For reviews and references to original work, see refs. [224,236,237].

NFκB and nuclear receptor cross-talk

The second-best-studied example of transcription factor cross-talk is the mutual interference between GRs and NFκB proteins. This pathway is again of significant importance, since it may also contribute to the anti-inflammative, as well as osteoporotic action of glucocorticoids. Albeit some have suggested that glucocorticoid action can be attributed to the increased production of the NFκB inhibitory molecule IκB, which in turn would sequester active NFκB in the cell nucleus [238], studies with mutant receptors and "dissociated" glucocorticoids showing IκB-independent repression of NFκB activity are incompatible with a simple IκB-mediated mechanism [239]. The underlying molecular events are still elusive, but may be related to those discussed above for the AP1 cross-talk. Note that as in the case of AP1 [233], squelching of limiting amounts of CBP, which also coactivates NFκB, is unlikely to be involved [240] and GR ligands that dissociate transactivation from transrepression still induce transcriptional interference with NFκB signaling [239]. For another NR, PPAR, a positive cross-talk with NFκB signaling pathways due to PPAR response element-independent IκB induction has been observed recently [241].

Other transcription factors that are cross-regulated in their activities by NRs are Oct 2A, RelA (another NFκB family member), STAT5, and Spi-1/PU.1 (for references to original work, see ref. [237]).

DEREGULATION IN DISEASE AND NOVEL THERAPEUTIC TARGETS

Given the major impact of NR signaling on animal physiology, it is no surprise to find aberrant NR function at the basis of multiple pathologies. Indeed, synthetic agonists and antagonists have been developed and are in clinical use for endocrine therapies of cancer as well as hormone replacement therapies in osteoporosis. Thiazolidinediones (TZDs), known as insulin "sensitizers" in the treatment of noninsulin-dependent diabetes, have been recognized as PPARγ agonists some time ago, and the recent establishment of a link between human type 2 diabetes and PPARγ mutation has proven that PPARγ malfunction can lead to severe insulin resistance, diabetes mellitus, and hypertension [242]. Considering in addition the success story of retinoids in the therapy of acute promyelocytic leukemia [243,244] and the promise of cancer therapy and prevention by NR-derived drugs [245–247], it becomes clear that the NR family is of outstanding importance for both diagnosis and drug design. But this is likely to represent only the top of the iceberg, and novel types of NR-based drugs are expected to be developed based on our increasing knowledge on the structural and molecular details of NR and ligand function, and the elucidation of the signaling pathways involved in (patho)physiological events.

Nuclear receptor-associated diseases

Nuclear receptors have a major role on human health and disease. Indeed, if their alteration causes pathological syndroms their activity contributes to therapy via agonist or antagonist drugs. Several observations have shown that deregulation of the NR genes leads to specific human diseases. For example, mutations of TRβ have been associated with the syndrome of resistance to thyroid hormone characterized by reduced thyroid hormone action in the presence of high levels of TSH, T3, and T4. Most commonly, these mutations—located in the LBD of the TRβ—reduce its affinity for thyroid hormones, interfere with the function of the wild-type TR, and impair interaction with cofactors involved in TR action. Affected patients present delayed bone maturation, heart abnormalities, hearing defects, and mental retardation [248–254].

Also, RARs have been associated with several diseases, among which cancer is one of the most important. The prototype of cancer that always involves RARα gene translocation is acute promyelocytic leukemia (APL) [244,245,255–257]. In the vast majority of cases, the origin of APL, according to the French-American-British classification a M3 type of acute myeloid leukemia (AML FAB-M3 or AML3), is a t(15;17)(22; q11.2-12) chromosomal translocation that fuses the PML (promyelocytic leukemia gene) and RARα genes. In rare cases, alternate chromosomal translocations generate RARα fusion proteins in which PML is replaced with PLZF [t(11; 17)(q23; q21)], NUMA PLZF [t(11; 17)(q13; q21)], NPM PLZF [t(5; 17)(q32; q21)], or STAT5b [258]. In contrast to wild-type RARα, only pharmacological doses of ATRA can dissociate the HDAC-containing corepressor complex from PML-RARα. PLZF-RARα binds corepressors through both the apo-RARα and PLZF moieties and ATRA cannot release HDACs. Consequently, PLZF-RARα remains a transcriptional repressor in the presence of ATRA. However, high concentrations of HDAC inhibitor (HDACi) convert also PLZF-RARα into an activator of the retinoic acid signaling pathway [259–261]. PML-RARα exhibits different oligomerization characteristics than RARα and essentially acts as a dominant silencing transcription factor that represses transcription activation mediated by the RAR-RXR heterodimer, which can still originate from the intact RARα allele. PML-RARα interacts with PML and causes nuclear body (NB) disintegration and aberrant localization of PML(-RARα) and other NB constituents. The consequence of the formation of PML-RARα is a block of differentiation at the promyelocytic stage. It is reasonable to assume that the altered functionality of PML in the fusion protein, such as its pro-apoptogenic activity, adds to the growth potential of APL blasts, while HDAC-dependent silencing of "normal" retinoid signaling during myelopoiesis causes the differentiation block.

The molecular analysis of NR action and APL explain also the basis of the retinoid therapy: ATRA, RARα agonists, or RARα agonists/antagonists in synergy with rexinoid agonists [159], bind to the PML-RARα LBD, resulting in allosteric transconformation [262] that dissociates the corepressor complex from the LBD. This event relieves the HDAC-dependent block of differentiation and through association of coactivator complexes, triggers the transcriptional regulation of cognate gene programs normally controlled by the RARα-RXR heterodimer. An exciting observation is originated from the recent analysis of the gene programs induced by ATRA in APL cells. In addition to the induction of anti-apoptotic and survival programs, ATRA induces postmaturation apoptosis through the induction of TRAIL [263], a tumor-selective death ligand [264], as well as caspase 8/10 that mediate TRAIL action through the cognate DR5 receptor.

NR deregulation is also linked with human metabolic disease. The PPARs have been implicated in dyslipidemia, diabetes, obesity, atherosclerosis, and inflammation control, and more recently their role in cancer has been suggested [59]. A mutation of the human PPARγ in the MAP kinase target sequence of the A/B region has been described in obese patients, while two mutations that destabilize helix H12 have been found in the LBD of PPARγ in patients with diabetes mellitus, insulin resistance, and hypertension [242,265,266]. Furthermore, a common polymorphism of PPARγ, Pro12 changed to Ala, has been associated with a decreased risk of type 2 diabetes [267]. The observation that the LBD of the human PPARγ gene was inactivated by point mutations or frame shifts in cases of sporadic colon cancer leads to the proposal that PPARγ is a tumor suppressor gene, and this is in accordance with its antiproliferative effect [268]. These results suggest that PPARγ ligands may have a potential as anticancer agents.

Also, VDR mutations have been linked to human pathology because they were found in hypocalcemic patients [269,270]. Interestingly, a link between VDR gene polymorphism and hyperparathyroidism has been reported [271].

Several studies have addressed the impact of ER mutations on cancer and osteoporosis. A patient that exhibited severe osteoporosis, cardiovascular alterations, with normal genitalia and sperm density, was found to have estrogen insensitivity due to homozygous mutation in the exon 2 of the ERα gene, resulting in a premature stop codon [272]. A large number of studies have shown that ERs can regulate genes implicated in the control of normal and tumoral cell growth. Epidemiological studies suggest that estrogens and ERs are critical in breast cancer etiology. Mutated, truncated, alternatively spliced versions of ERα have been detected in hormone-resistant breast cancer samples, but the precise role of these events in pathogenesis is still unclear [273]. As ER target gene, PR expression is a routine marker for endocrine therapy responsivity and prognosis of breast cancer.

Up to 200 different naturally occurring mutations of the AR gene have been described (see <http://ww2.mcgill.ca/androgendb/>). Mutations found in androgen insensitivity syndrome are located in DBDs and LBDs of the receptor causing androgen insensitivity due to alterations of DNA-binding or impaired ligand-binding activity. Both prostate cancer and spinal/bulbar muscular atrophy have been linked to variation in the number of Gln residues found in the polyglutamine repeat which normally contains 16–39 residues. Expansion of the repeat to 40–65 residues leads to spinal/bulbar muscular atrophy, whereas reduction in the number of repeats confers a higher risk for prostate cancer [274,275]. Furthermore, the amplification of AR is implicated in the androgen resistance of prostate tumors, although other mechanisms have been suggested. Recently, it was shown that two mutations in the LBD of AR allow this gene product to function as a high-affinity GR and reduce its ability to bind androgens [276,277].

Finally, the generalized inherited glucocorticoid resistance or familial glucocorticoid resistance (FGR) is associated with alterations of GR. These patients have mutations or deletions in the GR and exhibit high levels of circulating corticoids. For example, a single amino acid substitution in the GR LBD resulted in reduced binding affinity for glucocorticoids [278], while altered splicing of exon 6 led to FGR in a female patient presenting with hirsutism, menstrual abnormalities, and acne due to ACTH-induced hyper-secretion of androgens [279]. Further information can be found at the glucocorticoid receptor resource (<http://biochem1.basic-sci.georgetown.edu/grr/grr.html>).

Novel perspectives for nuclear receptor-based therapies

In addition to the well-established endocrine therapies of breast and prostate cancers, and to the more recent differentiation therapy of acute promyelocytic leukemia by retinoids, novel synthetic NR ligands are of considerable interest for the therapy and prevention of different types of cancers [246,247]. The development of novel types of NR ligands is facilitated by recent pharmacological and chemical developments, such as (i) combinatorial chemistry, computer-assisted ligand docking based on LBD crystal structure and ultra-high-throughput screening with NR-based reporter systems, (ii) the possibility to dissociate NR-associated functions such as transactivation and cross-talk with other signaling pathways, and (iii) the possibility to generate receptor and receptor isotype-selective ligands.

Nuclear receptor coactivators and cancer

It is tempting to speculate that coactivators are not entirely promiscuous in their choice of NRs. For example, only AIB1/RAC3 is found to be amplified in breast cancer cells while the expression level of the other two family members remains constant, reflecting estrogen receptor specificity [280]. Furthermore, in some types of acute myeloid leukemia (AML), a chromosomal translocation specifically fuses a monocytic zinc finger protein of unknown function (MOZ) to the C-terminus of TIF2 and not any of the other two TIF2 family members [281–283], again reflecting a bias toward one specific coactivator. Interestingly, the observation has been made that overexpression of coactivators of the TIF2/AIB1 family can lead in some systems to ligand-independent activity under certain conditions [79]. This suggests that transcriptional mediators may possibly be involved in the origin and/or progression of proliferative diseases and may become novel pharmacological targets. Indeed, mutations of the CBP gene have been implicated in the cause of Rubinstein-Taybi syndrome, and alterations of the p300 gene were found associated with gastric and colorectal carcinomas. Based on these various data, it is tempting to speculate that alterations in the cellular abundance of coregulators, or altered substrate specificity of the associated enzymatic functions may lead to pathological states.

Nuclear receptor phosphorylation and cancer

Several lines of evidence indicate that NR phosphorylation plays a crucial role in the development of certain cancers such as breast, ovarian, and prostate cancers. In most of these tumors, the MAPK and Akt kinase pathways exert increased activity, due to amplification of receptor protein-tyrosine kinases (RPTKs) [284] such as HER-2/neu [285] or deregulated activity of cytoplasmic protein-tyrosine kinases (c-Src, c-Abl, or bcr-Abl). This has been correlated with a ligand-independent transactivation of estrogen and androgen receptors [286]. Evidence is accumulating that this effect results from ligand-independent phosphorylation of the AF-1 domain of AR and ER by the disregulated MAPKs or Akt (see above). Phosphorylation would create receptors that activate transcription independently of the ligand. Such "outlaw" receptors would account for estrogen- or androgen-independent growth of prostate and breast cancer cells and for the failure of androgen ablation or tamoxifen therapy.

Toward novel types of synthetic nuclear receptor ligands for therapy

Synthetic ligands of NRs are classified as agonist and antagonist with respect to a particular receptor-associated function [224]. This discrimination is not always obvious, since a particular ligand might antagonize some activities while functioning as agonist for other activities. Examples are the ERα antagonists hydroxytamoxifen and ICI164.384. Hydroxytamoxifen antagonizes the activation function AF-2, but it acts as agonist for AF-1, whereas ICI164.384 in turn antagonizes both AF-1 and AF-2 [20]. Similarly, certain retinoid receptor antagonist are agonists for AP1 repression [233]. It is, therefore, important to consider, where possible, the molecular basis of the anticipated action of an NR to increase drug efficacy and limit side effects. If the molecular mechanism is unknown, it may be wise to use screening paradigms that consider the multiple dimensions of receptor activities.

Another twist to the classification of synthetic NR ligands results from the availability of isotype-specific ligands. These compounds affect one isotype of NR, but not another. The interspecies conservation of retinoid receptor isoforms, together with results obtained with isotype selective retinoids and gene ablation studies, have established that each of the three retinoic acid receptor genes has a cognate spectrum of functions [15]. Given the pharmacological potential of retinoids, the development of isotype-specific ligands has attracted much attention. Today, a wealth of synthetic retinoids exist, which display either isotype specificity or act as mixed agonists/antagonists for the three retinoic acid receptors [157]. Some of these retinoids were found to display cell specificity, and their pharmacological potential is currently investigated. RXR-specific ligands are also being developed, which is of particular interest in view of the role of retinoid X receptor as the promiscuous heterodimerization partner in a number of signaling pathways (see also "RXR subordination"). A recent report suggests that RXR ligands may stimulate insulin action in noninsulin-dependent diabetes [287] through a PPARγ-RXR heterodimer that is responsive to thiazolidinediones. It is thus conceivable that pathway-specific RXR ligands can be generated.

CONCLUSIONS

Nuclear receptors are ligand-regulated transcription factors that have evolved from an ancestral orphan receptor into a highly diverse family present throughout the entire animal kingdom and encompassing receptors for steroid and nonsteroid hormones, vitamins, corticoids, and metabolic intermediates. These receptors signal through endocrine, paracrine, autocrine, and intracrine modes of action to regulate multiple aspects of animal physiology, such as homeostasis, development, and reproduction. They regulate target genes that they either bind directly as mono-, homo- or heterodimers at cognate response elements, and have the ability to indirectly modulate other gene expression programs ("signal transduction cross-talks"). Through the coordinated expression of genetic programs, NRs contribute to cell fate-determining processes, thereby shaping and sustaining the organism. The inducing signal- binding of the ligand induces a major allosteric change in the LBD, which is transformed into cascades of protein-protein recognition paradigms inducing coregulator and cointegrator proteins.

Direct transcriptional repression in the absence of ligand or the presence of certain antagonists by some NRs is mediated by corepressor complexes that are associated with the unliganded receptor and condense the chromatin environment at the promoter region through histone deacetylation. Corepressors interact by virtue of their CoRNR boxes with nonliganded NRs. Upon ligand binding, the allosteric change in the LBD induces corepressor dissociation, and coactivator complexes are recruited. Bona fide coactivators recognize the active NR LBD (AF-2) via conserved LxxLL NR boxes and often the N-terminal activation function AF-1. The NR boxes of coactivators and CoRNR boxes of corepressor bind to topologically similar sites in the LBD, but the surfaces are entirely distinct due to the agonist-induced conformational changes. In particular, the holo-H12 is required for coactivator, but incompatible with corepressor binding. Coactivator complexes reverse the repressive effects of chromatin by specific histone acetylation, and allow access of the basal transcription machinery. In a subsequent step, the mammalian SMCC mediator is recruited to the NR and possibly stabilizes the formation of the preinitiation complex at target gene promoters. SMCC recruitment might be regulated by the acetylation and subsequent dissociations of TIF2 family members allowing thus SMCC-receptor association.

Despite their direct actions on the chromatin environment and the transcription machinery, NRs also regulate transcription by positive and negative interference with other signaling pathways. Different mechanisms for such transcription factor cross-talk have been described, but none of them is fully accepted and can explain all aspects of the particular cross-talk. The activity of NRs is regulated by phosphorylation that may serve to fine-tune the signaling and/or to establish a link to other signaling pathway. Finally, the promoter context, and the temporal order of incoming signals on a particular promoter have the likelihood of adjusting the transcriptional potential of NRs to particular situations. Taken together, NRs serve as platforms to coordinate cognate signals with those emanating from other signaling pathways, thereby integrating the NR signal into the functional context of cellular state and activity.

Nuclear receptors and their coregulators have been implicated in several diseases. Their role as key regulatory molecules in a wide variety of signaling pathways qualifies them as novel pharmacological targets. The ongoing improvement of synthetic NR ligands with altered specificity is likely to improve therapy and reduce side effects.

Future research on NRs still has to answer important questions. What are the constituents of the genetic programs that are governed by a given NR? How are the NR signals matched and complemented with other signaling cascades? What are the precise molecular events leading to the variety of transcriptional effects exerted by NRs? Once these questions have been addressed adequately, specific interference into these immensely complex systems might lead to the successful control and reprogramming of an organism's physiology and pathology. Understanding of NR-controlled transcription will shed light on the general and signaling pathway-selective control of gene expression. In this respect, the use of gene arrays together with the information derived from the genome sequencing will certainly have enormous impact.

ACKNOWLEDGMENTS

Work in the authors' laboratories has been supported by the Association for International Cancer Research, the Association pour le Recherche sur le Cancer, the Institut National de la Santé et de la Recherche Médicale, the Centre National de la Recherche Scientifique, the Hôpital Universitaire de Strasbourg, la Regione Campania, L.41/94, annualità '00, the Ministero della Salute R.F. 02/184, the French-Italian GALILEO program, the European Community (QLG3-CT2000-O844, QLG1-CT2000-01935, QLK3-CT2002-02029), and Bristol-Myers Squibb.

REFERENCES

1. A unified nomenclature system for the nuclear receptor superfamily [letter]. *Cell* **97**, 161–163 (1999).
2. M. Robinson-Rechavi, A. S. Carpentier, M. Duffraisse, V. Laudet. "How many nuclear hormone receptors are there in the human genome?", *Trends Genet.* **17**, 554–556 (2001).
3. J. M. Maglich et al. "Comparison of complete nuclear receptor sets from the human, Caenorhabditis elegans and Drosophila genomes", *Genome Biol.* **2** (2001).
4. P. A. Chambon. "Decade of molecular biology of retinoic acid receptors", *FASEB J.* **10**, 940–954 (1996).
5. V. Laudet and H. Gronemeyer. *The Nuclear Receptor Facts Book,* Academic Press, San Diego (2002).
6. T. T. Lu, J. J. Repa, D. J. Mangelsdorf. "Orphan nuclear receptors as eLiXiRs and FiXeRs of sterol metabolism", *J. Biol. Chem.* **276**, 37735–37738 (2001).
7. W. Xie and R. M. Evans. "Orphan nuclear receptors: the exotics of xenobiotics", *J. Biol. Chem.* **276**, 37739–37742 (2001).
8. G. B. Tremblay, D. Bergeron, V. Giguere. "4-Hydroxytamoxifen is an isoform-specific inhibitor of orphan estrogen-receptor-related (ERR) nuclear receptors beta and gamma", *Endocrinology* **142**, 4572–4575 (2001).
9. G. B. Tremblay et al. "Diethylstilbestrol regulates trophoblast stem cell differentiation as a ligand of orphan nuclear receptor ERR beta", *Genes Dev.* **15**, 833–838 (2001).
10. T. M. Willson and S. A. Kliewer. "PXR, CAR and drug metabolism", *Nat. Rev. Drug Discov.* **1**, 259–266 (2002).
11. X. Luo, Y. Ikeda, K. L. Parker. "A cell-specific nuclear receptor is essential for adrenal and gonadal development and sexual differentiation", *Cell* **77**, 481–490 (1994).
12. Y. Barak et al. "PPAR gamma is required for placental, cardiac, and adipose tissue development. *Mol. Cell* **4**, 585–595 (1999).
13. B. Mascrez et al. "The RXRalpha ligand-dependent activation function 2 (AF-2) is important for mouse development", *Development* **125**, 4691–4707 (1998).
14. M. Beato, P. Herrlich, G. Schutz. "Steroid hormone receptors: many actors in search of a plot", *Cell* **83**, 851–857 (1995).
15. P. Kastner, M. Mark, P. Chambon. "Nonsteroid nuclear receptors: what are genetic studies telling us about their role in real life?", *Cell* **83**, 859–869 (1995).
16. M. Y. Chiang et al. "An essential role for retinoid receptors RARbeta and RXRgamma in long-term potentiation and depression", *Neuron* **21**, 1353–1361 (1998).
17. H. M. Reichardt et al. "DNA binding of the glucocorticoid receptor is not essential for survival", *Cell* **93**, 531–541 (1998).
18. A. Krust et al. "The chicken oestrogen receptor sequence: homology with v-erbA and the human oestrogen and glucocorticoid receptors", *EMBO J.* **5**, 891–897 (1986).
19. M. T. Bocquel, V. Kumar, C. Stricker, P. Chambon, H. Gronemeyer. "The contribution of the N- and C-terminal regions of steroid receptors to activation of transcription is both receptor and cell-specific", *Nucleic Acids Res.* **17**, 2581–2595 (1989).
20. M. Berry, D. Metzger, P. Chambon. "Role of the two activating domains of the oestrogen receptor in the cell-type and promoter-context dependent agonistic activity of the anti-oestrogen 4-hydroxytamoxifen", *EMBO J.* **9**, 2811–2818 (1990).
21. C. A. Sartorius et al. "A third transactivation function (AF3) of human progesterone receptors located in the unique N-terminal segment of the B-isoform", *Mol. Endocrinol.* **8**, 1347–1360 (1994).
22. S. Kato et al. "Activation of the estrogen receptor through phosphorylation by mitogen- activated protein kinase", *Science* **270**, 1491–1494 (1995).

23. M. Watanabe et al. "A subfamily of RNA-binding DEAD-box proteins acts as an estrogen receptor alpha coactivator through the N-terminal activation domain (AF-1) with an RNA coactivator, SRA", *EMBO J.* **20**, 1341–1352 (2001).
24. T. Ylikomi, M. T. Bocquel, M. Berry, H. Gronemeyer, P. Chambon. "Cooperation of proto-signals for nuclear accumulation of estrogen and progesterone receptors", *EMBO J.* **11**, 3681–3694 (1992).
25. A. Guiochon-Mantel, K. Delabre, P. Lescop, E. Milgrom. "Nuclear localization signals also mediate the outward movement of proteins from the nucleus", *Proc. Natl. Acad. Sci. USA* **91**, 7179–7183 (1994).
26. D. Walker, H. Htun, G. L. Hager. "Using inducible vectors to study intracellular trafficking of GFP- tagged steroid/nuclear receptors in living cells", *Methods* **19**, 386–393 (1999).
27. G. L. Hager, C. S. Lim, C. Elbi, C. T. Baumann. "Trafficking of nuclear receptors in living cells", *J. Steroid Biochem. Mol. Biol.* **74**, 249–254 (2000).
28. G. A. Peters and S. A. Khan. "Estrogen receptor domains E and F: role in dimerization and interaction with coactivator RIP-140", *Mol. Endocrinol.* **13**, 286–296 (1999).
29. F. M. Sladek, M. D. Ruse, Jr., L. Nepomuceno, S. M. Huang, M. R. Stallcup. "Modulation of transcriptional activation and coactivator interaction by a splicing variation in the F domain of nuclear receptor hepatocyte nuclear factor 4alpha1", *Mol. Cell Biol.* **19**, 6509–6522 (1999).
30. M. M. Montano, V. Muller, A. Trobaugh, B. S. Katzenellenbogen. "The carboxy-terminal F domain of the human estrogen receptor: role in the transcriptional activity of the receptor and the effectiveness of antiestrogens as estrogen antagonists", *Mol. Endocrinol.* **9**, 814–825 (1995).
31. M. Nichols, J. M. Rientjes, A. F. Stewart. "Different positioning of the ligand-binding domain helix 12 and the F domain of the estrogen receptor accounts for functional differences between agonists and antagonists", *EMBO J.* **17**, 765–773 (1998).
32. K. Umesono, K. K. Murakami, C. C. Thompson, R. M. Evans. "Direct repeats as selective response elements for the thyroid hormone, retinoic acid, and vitamin D3 receptors", *Cell* **65**, 1255–1266 (1991).
33. S. A. Kliewer, K. Umesono, D. J. Noonan, R. A. Heyman, R. M. Evans. "Convergence of 9-cis retinoic acid and peroxisome proliferator signalling pathways through heterodimer formation of their receptors", *Nature* **358**, 771–774 (1992).
34. S. Mader et al. "The patterns of binding of RAR, RXR and TR homo- and heterodimers to direct repeats are dictated by the binding specificites of the DNA binding domains", *EMBO J.* **12**, 5029–5041 (1993).
35. F. Rastinejad, T. Perlmann, R. M. Evans, P. B. Sigler. "Structural determinants of nuclear receptor assembly on DNA direct repeats", *Nature* **375**, 203–11 (1995).
36. G. Meinke and P. B. Sigler. "DNA-binding mechanism of the monomeric orphan nuclear receptor NGFI-B", *Nat. Struct. Biol.* **6**, 471–477 (1999).
37. V. G. Thackray, B. A. Lieberman, S. K. Nordeen. "Differential gene induction by glucocorticoid and progesterone receptors", *J. Steroid Biochem. Mol. Biol.* **66**, 171–178 (1998).
38. G. Klock, U. Strahle, G. Schutz. "Oestrogen and glucocorticoid responsive elements are closely related but distinct", *Nature* **329**, 734–736 (1987).
39. C. K. Glass. "Differential recognition of target genes by nuclear receptor monomers, dimers, and heterodimers", *Endocr. Rev.* **15**, 391–407 (1994).
40. M. Tsai, and B. O'Malley. "Molecular mechanisms of action of steroid/thyroid receptor superfamily", *Ann. Rev. Biochem.* **63**, 451–486 (1994).
41. H. Gronemeyer and V. Laudet. "Transcription factors 3: Nuclear receptors", *Protein Profile* **2**, 1173–1308 (1995).
42. B. Desvergne and W. Wahli. "Peroxisome proliferator-activated receptors: nuclear control of metabolism", *Endocr. Rev.* **20**, 649–688 (1999).

43. S. A. Kliewer, J. M. Lehmann, M. V. Milburn, T. M. Willson. "The PPARs and PXRs: nuclear xenobiotic receptors that define novel hormone signaling pathways", *Recent Prog. Horm. Res.* **54**, 345–367 (1999).
44. H. de The, M. M. Vivanco-Ruiz, P. Tiollais, H. Stunnenberg, A. Dejean. "Identification of a retinoic acid responsive element in the retinoic acid receptor beta gene", *Nature* **343**, 177–180 (1990).
45. B. Hoffmann et al. "A retinoic acid receptor-specific element controls the retinoic acid receptor-beta promoter", *Mol. Endocrinol.* **4**, 1727–1736 (1990).
46. H. M. Sucov, K. K. Murakami, R. M. Evans. "Characterization of an autoregulated response element in the mouse retinoic acid receptor type beta gene", *Proc. Natl. Acad. Sci. USA* **87**, 5392–5396 (1990).
47. D. M. Heery, B. Pierrat, H. Gronemeyer, P. Chambon, R. Losson. "Homo- and heterodimers of the retinoid X receptor (RXR) activated transcription in yeast", *Nucleic Acids Res.* **22**, 726–731 (1994).
48. D. J. Mangelsdorf et al. "A direct repeat in the cellular retinol-binding protein type II gene confers differential regulation by RXR and RAR", *Cell* **66**, 555–561 (1991).
49. H. Chen, Z. Smit-McBride, S. Lewis, M. Sharif, M. L. Privalsky. "Nuclear hormone receptors involved in neoplasia: erb A exhibits a novel DNA sequence specificity determined by amino acids outside of the zinc- finger domain", *Mol. Cell Biol.* **13**, 2366–2376 (1993).
50. H. W. Chen and M. L. Privalsky. "The erbA oncogene represses the actions of both retinoid X and retinoid A receptors but does so by distinct mechanisms", *Mol. Cell Biol.* **13**, 5970–5980 (1993).
51. B. M. Forman, J. Casanova, B. M. Raaka, J. Ghysdael, H. H. Samuels. "Half-site spacing and orientation determines whether thyroid hormone and retinoic acid receptors and related factors bind to DNA response elements as monomers, homodimers, or heterodimers. *Mol. Endocrinol.* **6**, 429–442 (1992).
52. C. K. Glass et al. "A c-erb-A binding site in rat growth hormone gene mediates trans- activation by thyroid hormone", *Nature* **329**, 738–741 (1987).
53. C. K. Glass, J. M. Holloway, O. V. Devary, M. G. Rosenfeld. "The thyroid hormone receptor binds with opposite transcriptional effects to a common sequence motif in thyroid hormone and estrogen response elements", *Cell* **54**, 313–323 (1988).
54. P. P. Harding and G. Duester. "Retinoic acid activation and thyroid hormone repression of the human alcohol dehydrogenase gene ADH3", *J. Biol. Chem.* **267**, 14145–14150 (1992).
55. J. Burnside, D. S. Darling, F. E. Carr, W. W. Chin. "Thyroid hormone regulation of the rat glycoprotein hormone alpha- subunit gene promoter activity", *J. Biol. Chem.* **264**, 6886–6891 (1989).
56. A. Farsetti, B. Desvergne, P. Hallenbeck, J. Robbins, V. M. Nikodem. "Characterization of myelin basic protein thyroid hormone response element and its function in the context of native and heterologous promoter", *J. Biol. Chem.* **267**, 15784–15788 (1992).
57. M. R. Haussler et al. "The vitamin D hormone and its nuclear receptor: molecular actions and disease states. *J. Endocrinol.* **154 Suppl**, S57–73 (1997).
58. H. F. DeLuca and C. Zierold. "Mechanisms and functions of vitamin D", *Nutr. Rev.* **56**, S4–10; discussion S54–75 (1998).
59. S. Kersten, B. Desvergne, W. Wahli. "Roles of PPARs in health and disease", *Nature* **405**, 421–424 (2000).
60. T. M. Willson, P. J. Brown, D. D. Sternbach, B. R. Henke. "The PPARs: from orphan receptors to drug discovery", *J. Med. Chem.* **43**, 527–550 (2000).
61. S. Khorasanizadeh and F. Rastinejad. "Nuclear-receptor interactions on DNA-response elements", *Trends BioChem. Sci.* **26**, 384–390 (2001).
62. F. Rastinejad, T. Wagner, Q. Zhao, S. Khorasanizadeh. "Structure of the RXR-RAR DNA-binding complex on the retinoic acid response element DR1", *EMBO J.* **19**, 1045–1054 (2000).

63. S. Green and P. Chambon. "Oestradiol induction of a glucocorticoid-responsive gene by a chimaeric receptor", *Nature* **325**, 75–78 (1987).
64. S. Green, V. Kumar, I. Theulaz, W. Wahli, P. Chambon. "The N-terminal DNA-binding 'zinc finger' of the oestrogen and glucocorticoid receptors determines target gene specificity", *EMBO J.* **7**, 3037–3044 (1988).
65. M. Danielsen, L. Hinck, G. M. Ringold. "Two amino acids within the knuckle of the first zinc finger specify DNA response element activation by the glucocorticoid receptor", *Cell* **57**, 1131–1138 (1989).
66. S. Mader, V. Kumar, H. de Verneuil, P. Chambon. "Three amino acids of the oestrogen receptor are essential to its ability to distinguish an oestrogen from a glucocorticoid-responsive element", *Nature* **338**, 271–274 (1989).
67. K. Umesono and R. M. Evans. "Determinants of target gene specificity for steroid/thyroid hormone receptors", *Cell* **57**, 1139–1146 (1989).
68. T. E. Wilson, T. J. Fahrner, M. Johnston, J. Milbrandt. "Identification of the DNA binding site for NGFI-B by genetic selection in yeast", *Science* **252**, 1296–1300 (1991).
69. T. E. Wilson, T. J. Fahrner, J. Milbrandt. "The orphan receptors NGFI-B and steroidogenic factor 1 establish monomer binding as a third paradigm of nuclear receptor-DNA interaction", *Mol. Cell Biol.* **13**, 5794–5804 (1993).
70. C. Zechel et al. "The dimerization interfaces formed between the DNA binding domains of RXR, RAR and TR determine the binding specificity and polarity of the full-length receptors to direct repeats", *EMBO J.* **13**, 1425–1433 (1994).
71. C. Zechel, X. Q. Shen, P. Chambon, H. Gronemeyer. "Dimerization interfaces formed between the DNA binding domains determine the cooperative binding of RXR/RAR and RXR/TR heterodimers to DR5 and DR4 elements", *EMBO J.* **13**, 1414–1424 (1994).
72. T. E. Wilson, R. E. Paulsen, K. A. Padgett, J. Milbrandt. "Participation of non-zinc finger residues in DNA binding by two nuclear orphan receptors", *Science* **256**, 107–110 (1992).
73. M. S. Lee, S. A. Kliewer, J. Provencal, P. E. Wright, R. M. Evans. "Structure of the retinoid X receptor alpha DNA binding domain: a helix required for homodimeric DNA binding", *Science* **260**, 1117–11121 (1993).
74. J. W. Schwabe, L. Chapman, J. T. Finch, D. Rhodes. "The crystal structure of the estrogen receptor DNA-binding domain bound to DNA: how receptors discriminate between their response elements", *Cell* **75**, 567–578 (1993).
75. J. M. Wurtz et al. "A canonical structure for the ligand-binding domain of nuclear receptors", [published erratum appears in *Nat Struct Biol.* Feb. 3 (2), 206 (1996)]. *Nat. Struct. Biol.* **3**, 87–94 (1996).
76. J. P. Renaud et al. "Crystal structure of the RAR-gamma ligand-binding domain bound to all-trans retinoic acid. *Nature* **378**, 681–689 (1995).
77. R. T. Nolte et al. "Ligand binding and co-activator assembly of the peroxisome proliferator- activated receptor-gamma", *Nature* **395**, 137–143 (1998).
78. R. White, M. Sjoberg, E. Kalkhoven, M. G. Parker. "Ligand-independent activation of the oestrogen receptor by mutation of a conserved tyrosine", *EMBO J.* **16**, 1427–1435 (1997).
79. J. J. Voegel et al. "The coactivator TIF2 contains three nuclear receptor-binding motifs and mediates transactivation through CBP binding-dependent and -independent pathways", *EMBO J.* **17**, 507–519 (1998).
80. K. E. Weis, K. Ekena, J. A. Thomas, G. Lazennec, B. S. Katzenellenbogen, "Constitutively active human estrogen receptors containing amino acid substitutions for tyrosine 537 in the receptor protein. *Mol. Endocrinol.* **10**, 1388–1398 (1996).
81. V. Vivat et al. "A mutation mimicking ligand-induced conformational change yields a constitutive RXR that senses allosteric effects in heterodimers", *EMBO J.* **16**, 5697–5709 (1997).

82. W. Bourguet et al. Crystal structure of a heterodimeric complex of RAR and RXR ligand- binding domains", *Mol. Cell* **5**, 289–298 (2000).
83. R. T. Gampe, Jr. et al. "Asymmetry in the PPARgamma/RXRalpha crystal structure reveals the molecular basis of heterodimerization among nuclear receptors", *Mol. Cell* **5**, 545–555 (2000).
84. W. Bourguet, M. Ruff, P. Chambon, H. Gronemeyer, D. Moras. "Crystal structure of the ligand-binding domain of the human nuclear receptor RXR-alpha", *Nature* **375**, 377–382 (1995).
85. A. M. Brzozowski et al. "Molecular basis of agonism and antagonism in the oestrogen receptor", *Nature* **389**, 753–758 (1997).
86. R. K. Bledsoe et al. "Crystal structure of the glucocorticoid receptor ligand binding domain reveals a novel mode of receptor dimerization and coactivator recognition", *Cell* **110**, 93–105 (2002).
87. B. P. Klaholz et al. "Conformational adaptation of agonists to the human nuclear receptor RAR gamma", *Nat. Struct. Biol.* **5**, 199–202 (1998).
88. R. E. Watkins et al. "The human nuclear xenobiotic receptor PXR: structural determinants of directed promiscuity", *Science* **292**, 2329–2333 (2001).
89. J. D. Chen. "Steroid/nuclear receptor coactivators", *Vitam. Horm.* **58**, 391–448 (2000).
90. C. K. Glass and M. G. Rosenfeld. "The coregulator exchange in transcriptional functions of nuclear receptors", *Genes Dev.* **14**, 121–141 (2000).
91. H. Chen, M. Tini, R. M. Evans. "HATs on and beyond chromatin", *Curr. Opin. Cell Biol.* **13**, 218–224 (2001).
92. O. Hermanson, C. K. Glass, M. G. Rosenfeld. "Nuclear receptor coregulators: multiple modes of modification", *Trends Endocrinol. Metab.* **13**, 55–60 (2002).
93. N. J. McKenna and B. W. O'Malley. "Combinatorial control of gene expression by nuclear receptors and coregulators", *Cell* **108**, 465–474 (2002).
94. K. Jepsen and M. G. Rosenfeld. "Biological roles and mechanistic actions of co-repressor complexes", *J. Cell Sci.* **115**, 689–698 (2002).
95. A. Aranda and A. Pascual. "Nuclear hormone receptors and gene expression", *Physiol. Rev.* **81**, 1269–1304 (2001).
96. K. C. Lee and W. Lee Kraus. "Nuclear receptors, coactivators and chromatin: new approaches, new insights", *Trends Endocrinol. Metab.* **12**, 191–197 (2001).
97. M. Ito and R. G. Roeder. The TRAP/SMCC/Mediator complex and thyroid hormone receptor function", *Trends Endocrinol. Metab.* **12**, 127–134 (2001).
98. J. W. Lee, Y. C. Lee, S. Y. Na, D. J. Jung, S. K. Lee. "Transcriptional coregulators of the nuclear receptor superfamily: coactivators and corepressors", *Cell Mol. Life Sci.* **58**, 289–297 (2001).
99. P. Ordentlich, M. Downes, R. M. Evans. "Corepressors and nuclear hormone receptor function", *Curr. Top. Microbiol. Immunol.* **254**, 101–116 (2001).
100. M. G. Rosenfeld and C. K. Glass. Coregulator codes of transcriptional regulation by nuclear receptors", *J. Biol. Chem.* **276**, 36865–36868 (2001).
101. O. M. Conneely, J. P. Lydon, F. De Mayo, B. W. O'Malley. "Reproductive functions of the progesterone receptor", *J. Soc. Gynecol. Investig.* **7**, S25–32 (2000).
102. K. B. Horwitz, L. Tung, G. S. Takimoto. "Novel mechanisms of antiprogestin action", *J. Steroid BioChem. Mol. Biol.* **53**, 9–17 (1995).
103. C. A. Lange, J. K. Richer, K. B. Horwitz. "Hypothesis: Progesterone primes breast cancer cells for cross-talk with proliferative or antiproliferative signals", *Mol. Endocrinol.* **13**, 829–836 (1999).
104. D. L. Bain, M. A. Franden, J. L. McManaman, G. S. Takimoto, K. B. Horwitz. "The N-terminal region of the human progesterone A-receptor. Structural analysis and the influence of the DNA binding domain", *J. Biol. Chem.* **275**, 7313–7320 (2000).
105. S. Nagpal et al. "Promoter context- and response element-dependent specificity of the transcriptional activation and modulating functions of retinoic acid receptors", *Cell* **70**, 1007–1019 (1992).

106. G. Gill and M. Ptashne. "Negative effect of the transcriptional activator GAL4", *Nature* **334**, 721–724 (1988).
107. M. E. Meyer et al. "Steroid hormone receptors compete for factors that mediate their enhancer function", *Cell* **57**, 433–442 (1989).
108. D. Tasset, L. Tora, C. Fromental, E. Scheer, P. Chambon. "Distinct classes of transcriptional activating domains function by different mechanisms", *Cell* **62**, 1177–1187 (1990).
109. B. Le Douarin et al. "A possible involvement of TIF1 alpha and TIF1 beta in the epigenetic control of transcription by nuclear receptors", *EMBO J.* **15**, 6701–6715 (1996).
110. D. M. Heery, E. Kalkhoven, S. Hoare, M. G. Parker. "A signature motif in transcriptional co-activators mediates binding to nuclear receptor", *Nature* **387**, 733–736 (1997).
111. J. Torchia et al. "The transcriptional co-activator p/CIP binds CBP and mediates nuclear- receptor function", *Nature* **387**, 677–684 (1997).
112. Y. Shang, X. Hu, J. DiRenzo, M. A. Lazar, M. Brown. "Cofactor dynamics and sufficiency in estrogen receptor-regulated transcription", *Cell* **103**, 843–852 (2000).
113. L. P. Freedman. "Increasing the complexity of coactivation in nuclear receptor signaling", *Cell* **97**, 5–8 (1999).
114. C. A. Berrevoets, P. Doesburg, K. Steketee, J. Trapman, A. O. Brinkmann. "Functional interactions of the AF-2 activation domain core region of the human androgen receptor with the amino-terminal domain and with the transcriptional coactivator TIF2 (transcriptional intermediary factor2) ", *Mol. Endocrinol.* **12**, 1172–1183 (1998).
115. P. Alen, F. Claessens, G. Verhoeven, W. Rombauts, B. Peeters. "The androgen receptor amino-terminal domain plays a key role in p160 coactivator-stimulated gene transcription", *Mol. Cell Biol.* **19**, 6085–6097 (1999).
116. C. L. Bevan, S. Hoare, F. Claessens, D. M. Heery, M. G. Parker. "The AF1 and AF2 domains of the androgen receptor interact with distinct regions of SRC1", *Mol. Cell Biol.* **19**, 8383–8392 (1999).
117. B. He, J. A. Kemppainen, J. J. Voegel, H. Gronemeyer, E. M. Wilson. "Activation function 2 in the human androgen receptor ligand binding domain mediates interdomain communication with the NH(2)-terminal domain", *J. Biol. Chem.* **274**, 37219–37225 (1999).
118. H. Ma et al. "Multiple signal input and output domains of the 160-kilodalton nuclear receptor coactivator proteins", *Mol. Cell Biol.* **19**, 6164–6173 (1999).
119. E. M. Mcinerney, M. J. Tsai, B. W. Omalley, B. S. Katzenellenbogen. "Analysis of estrogen receptor transcriptional enhancement by a nuclear hormone receptor coactivator", *Proc. Natl. Acad. Sci. USA* **93**, 10069–10073 (1996).
120. A. Benecke, P. Chambon, H. Gronemeyer. "Synergy between estrogen receptor a activation functions AF1 and AF2 mediated by transcription intermediary factor TIF2", *EMBO Reports* **1**, 151–157 (2000).
121. H. Chen, R. J. Lin, W. Xie, D. Wilpitz, R. M. Evans. "Regulation of hormone-induced histone hyperacetylation and gene activation via acetylation of an acetylase", *Cell* **98**, 675–686 (1999).
122. D. Chen et al. "Regulation of transcription by a protein methyltransferase", *Science* **284**, 2174–2177 (1999).
123. N. Kotaja, U. Karvonen, O. A. Janne, J. J. Palvimo. "The nuclear receptor interaction domain of GRIP1 is modulated by covalent attachment of SUMO-1", *J. Biol. Chem.* **277**, 30283–30288 (2002).
124. A. M. Jimenez-Lara, M. J. Heine, H. Gronemeyer. "PIAS3 (protein inhibitor of activated STAT-3) modulates the transcriptional activation mediated by the nuclear receptor Coactivator TIF2", *FEBS Lett.* **526**, 142–146 (2002).
125. A. M. Jimenez-Lara, M. J. Heine, H. Gronemeyer. "Cloning of a mouse glucocorticoid modulatory element binding protein, a new member of the KDWK family", *FEBS Lett.* **468**, 203–210 (2000).

126. S. M. Huang and M. R. Stallcup. "Mouse Zac1, a transcriptional coactivator and repressor for nuclear receptors", *Mol. Cell Biol.* **20**, 1855–1867 (2000).
127. P. M. A. Henttu, E. Kalkhoven, M. G. Parker. "AF-2 activity and recruitment of steroid receptor coactivator 1 to the estrogen receptor depend on a lysine residue conserved in nuclear receptors", *Mol. Cell Biol.* **17**, 1832–1839 (1997).
128. W. Feng et al. "Hormone-dependent coactivator binding to a hydrophobic cleft on nuclear receptors", *Science* **280**, 1747–1749 (1998).
129. B. D. Darimont et al. "Structure and specificity of nuclear receptor-coactivator interactions", *Genes Dev.* **12**, 3343–3356 (1998).
130. A. K. Shiau et al. "The structural basis of estrogen receptor/coactivator recognition and the antagonism of this interaction by tamoxifen. *Cell* **95**, 927–937 (1998).
131. E. M. McInerney et al. "Determinants of coactivator LXXLL motif specificity in nuclear receptor transcriptional activation", *Genes Dev.* **12**, 3357–3368 (1998).
132. X. Hu and M. A. Lazar. "The CoRNR motif controls the recruitment of corepressors by nuclear hormone receptors", *Nature* **402**, 93–96 (1999).
133. L. Nagy et al. "Mechanism of corepressor binding and release from nuclear hormone receptors", *Genes Dev.* **13**, 3209–3216 (1999).
134. V. Perissi et al. "Molecular determinants of nuclear receptor-corepressor interaction", *Genes Dev.* **13**, 3198–3208 (1999).
135. H. E. Xu et al. "Structural basis for antagonist-mediated recruitment of nuclear co- repressors by PPARalpha", *Nature* **415**, 813–817 (2002).
136. T. A. Jackson et al. "The partial agonist activity of antagonist-occupied steroid receptors is controlled by a novel hinge domain-binding coactivator L7/SPA and the corepressors N-CoR or SMRT", *Mol. Endocrinol.* **11**, 693–705 (1997).
137. B. L. Wagner, J. D. Norris, T. A. Knotts, N. L. Weigel, D. P. McDonnell. "The nuclear corepressors NCoR and SMRT are key regulators of both ligand- and 8-bromo-cyclic AMP-dependent transcriptional activity of the human progesterone receptor", *Mol. Cell Biol.* **18**, 1369–1378 (1998).
138. X. Zhang, M. Jeyakumar, S. Petukhov, M. K. Bagchi. "A nuclear receptor corepressor modulates transcriptional activity of antagonist-occupied steroid hormone receptor", *Mol. Endocrinol.* **12**, 513–524 (1998).
139. A. J. Horlein et al. "Ligand-independent repression by the thyroid hormone receptor mediated by a nuclear receptor co-repressor", *Nature* **377**, 397–404 (1995).
140. R. Kurokawa et al. "Differential use of CREB binding protein-coactivator complexes", *Science* **279**, 700–703 (1998).
141. J. P. Renaud, J. M. Harris, M. Downes, L. J. Burke, G. E. Muscat. "Structure–function analysis of the Rev-erbA and RVR ligand-binding domains reveals a large hydrophobic surface that mediates corepressor binding and a ligand cavity occupied by side chains", *Mol. Endocrinol.* **14**, 700–717 (2000).
142. W. Seol, M. J. Mahon, Y. K. Lee, D. D. Moore. "Two receptor interacting domains in the nuclear hormone receptor corepressor RIP13/N-CoR", *Mol. Endocrinol.* **10**, 1646–1655 (1996).
143. I. Zamir et al. "A nuclear hormone receptor corepressor mediates transcriptional silencing by receptors with distinct repression domains", *Mol. Cell Biol.* **16**, 5458–5465 (1996).
144. J. D. Fondell, H. Ge, R. G. Roeder. "Ligand induction of a transcriptionally active thyroid hormone receptor coactivator complex", *Proc. Natl. Acad. Sci. USA* **93**, 8329–8333 (1996).
145. C. Rachez et al. "A novel protein complex that interacts with the vitamin D3 receptor in a ligand-dependent manner and enhances VDR transactivation in a cell- free system", *Genes Dev.* **12**, 1787–1800 (1998).
146. M. Ito et al. "Identity between TRAP and SMCC complexes indicates novel pathways for the function of nuclear receptors and diverse mammalian activators", *Mol. Cell* **3**, 361–370 (1999).

147. W. Gu et al. "A novel human SRB/MED-containing cofactor complex, SMCC, involved in transcription regulation", [published erratum appears in *Mol. Cell* April 3 (4), following 541 (1999)]. *Mol. Cell* **3**, 97–108 (1999).
148. M. Kretzschmar, G. Stelzer, R. G. Roeder, M. Meisterernst. "RNA polymerase II cofactor PC2 facilitates activation of transcription by GAL4-AH in vitro", *Mol. Cell Biol.* **14**, 3927–3937 (1994).
149. A. M. Naar et al. "Composite co-activator ARC mediates chromatin-directed transcriptional activation", *Nature* **398**, 828–832 (1999).
150. S. Ryu and R. Tjian. Purification of transcription cofactor complex CRSP. *Proc. Natl. Acad. Sci. USA* **96**, 7137–7142 (1999).
151. X. Sun et al. "NAT, a human complex containing Srb polypeptides that functions as a negative regulator of activated transcription", *Mol. Cell* **2**, 213–222 (1998).
152. J. D. Fondell, M. Guermah, S. Malik, R. G. Roeder. "Thyroid hormone receptor-associated proteins and general positive cofactors mediate thyroid hormone receptor function in the absence of the TATA box-binding protein-associated factors of TFIID", *Proc. Natl. Acad. Sci. USA* **96**, 1959–1964 (1999).
153. C. Rachez et al. "Ligand-dependent transcription activation by nuclear receptors requires the DRIP complex", *Nature* **398**, 824–828 (1999).
154. W. Yang and L. P. Freedman. "20-Epi analogues of 1,25-dihydroxyvitamin D3 are highly potent inducers of DRIP coactivator complex binding to the vitamin D3 receptor", *J. Biol. Chem.* **274**, 16838–16845 (1999).
155. C. X. Yuan, M. Ito, J. D. Fondell, Z. Y. Fu, R. G. Roeder. "The TRAP220 component of a thyroid hormone receptor- associated protein (TRAP) coactivator complex interacts directly with nuclear receptors in a ligand-dependent fashion", [published erratum appears in *Proc. Natl. Acad. Sci. USA* Nov 24; 95 (24), 14584 (1998)]. *Proc. Natl. Acad. Sci. USA* **95**, 7939–7944 (1998).
156. A. B. Hittelman, D. Burakov, J. A. Iniguez-Lluhi, L. P. Freedman, M. J. Garabedian. "Differential regulation of glucocorticoid receptor transcriptional activation via AF-1-associated proteins", *EMBO J.* **18**, 5380–5388 (1999).
157. M. Gehin et al. "Structural basis for engineering of retinoic acid receptor isotype- selective agonists and antagonists", *Chem. Biol.* **6**, 519–529 (1999).
158. S. Minucci et al. "Retinoid X receptor (RXR) within the RXR-retinoic acid receptor heterodimer binds its ligand and enhances retinoid-dependent gene expression", *Mol. Cell Biol.* **17**, 644–655 (1997).
159. J. Y. Chen et al. "Two distinct actions of retinoid-receptor ligands", *Nature* **382**, 819–822 (1996).
160. A. C. Pike et al. "Structural insights into the mode of action of a pure antiestrogen", *Structure (Camb.)* **9**, 145–153 (2001).
161. J. L. Oberfield et al. "A peroxisome proliferator-activated receptor gamma ligand inhibits adipocyte differentiation", *Proc. Natl. Acad. Sci. USA* **96**, 6102–6106 (1999).
162. A. K. Shiau et al. "Structural characterization of a subtype-selective ligand reveals a novel mode of estrogen receptor antagonism", *Nat. Struct. Biol.* **9**, 359–364 (2002).
163. R. M. Lavinsky et al. "Diverse signaling pathways modulate nuclear receptor recruitment of N-CoR and SMRT complexes", *Proc. Natl. Acad. Sci. USA* **95**, 2920–2925 (1998).
164. P. Germain, J. Iyer, C. Zechel, H. Gronemeyer. "Coregulator recruitment and the mechanism of retinoic acid receptor synergy", *Nature* **415**, 187–192 (2002).
165. E. S. Klein et al. "Identification and functional separation of retinoic acid receptor neutral antagonists and inverse agonists", *J. Biol. Chem.* **271**, 22692–22696 (1996).
166. J. Clifford, H. Chiba, D. Sobieszczuk, D. Metzger, P. Chambon. "RXRalpha-null F9 embryonal carcinoma cells are resistant to the differentiation, anti-proliferative and apoptotic effects of retinoids", *EMBO J.* **15**, 4142–4155 (1996).

167. R. Taneja et al. "Cell-type and promoter-context dependent retinoic acid receptor (RAR) redundancies for RAR beta 2 and Hoxa-1 activation in F9 and P19 cells can be artefactually generated by gene knockouts", *Proc. Natl. Acad. Sci. USA* **93**, 6197–6202 (1996).
168. R. Kurokawa et al. "Regulation of retinoid signalling by receptor polarity and allosteric control of ligand binding", *Nature* **371**, 528–531 (1994).
169. C. M. Apfel et al. "Enhancement of HL-60 differentiation by a new class of retinoids with selective activity on retinoid X receptor", *J. Biol. Chem.* **270**, 30765–30772 (1995).
170. Z. Chen et al. "Ligand- And DNA-induced Dissociation of RXR Tetramers", *J. Mol. Biol.* **275**, 55–65 (1998).
171. S. Kersten, M. I. Dawson, B. A. Lewis, N. Noy. "Individual subunits of heterodimers comprised of retinoic acid and retinoid X receptors interact with their ligands independently", *Biochemistry* **35**, 3816–3824 (1996).
172. H. Li, P. J. Gomes, J. D. Chen. "RAC3, a steroid/nuclear receptor-associated coactivator that is related to SRC-1 and TIF2", *Proc. Natl. Acad. Sci. USA* **94**, 8479–8484 (1997).
173. B. M. Forman, K. Umesono, J. Chen, R. M. Evans. "Unique response pathways are established by allosteric interactions among nuclear hormone receptors", *Cell* **81**, 541–550 (1995).
174. T. Perlmann and L. Jansson. "A novel pathway for vitamin A signaling mediated by RXR heterodimerization with NGFI-B and NURR1", *Genes Dev.* **9**, 769–782 (1995).
175. B. Blumberg and R. M. Evans. "Orphan nuclear receptors—new ligands and new possibilities", *Genes Dev.* **12**, 3149–3155 (1998).
176. M. Fu et al. "Androgen receptor acetylation governs trans activation and MEKK1- induced apoptosis without affecting in vitro sumoylation and trans- repression function", *Mol. Cell Biol.* **22**, 3373–3388 (2002).
177. C. Wang et al. "Direct acetylation of the estrogen receptor alpha hinge region by p300 regulates transactivation and hormone sensitivity", *J. Biol. Chem.* **276**, 18375–18383 (2001).
178. T. Kouzarides. "Acetylation: a regulatory modification to rival phosphorylation?", *EMBO J.* **19**, 1176–1179 (2000).
179. D. O. Morgan. "Principles of CDK regulation", *Nature* **374**, 131–134 (1995).
180. D. O. Morgan. "Cyclin-dependent kinases: Engines, clocks, and microprocessors", *Ann. Rev. Cell Dev. Biol.* **13**, 261–291 (1997).
181. Z. Chen et al. "MAP kinases", *Chem. Rev.* **101**, 2449–2476 (2001).
182. G. Pearson et al. "Mitogen-activated protein (MAP) kinase pathways: regulation and physiological functions", *Endocr. Rev.* **22**, 153–183 (2001).
183. L. Chang and M. Karin. "Mammalian MAP kinase signalling cascades", *Nature* **410**, 37–40 (2001).
184. N. L. Weigel. "Steroid hormone receptors and their regulation by phosphorylation", *BioChem. J.* **319**, 657–667 (1996).
185. T. Shen, K. B. Horwitz, C. A. Lange. "Transcriptional hyperactivity of human progesterone receptors is coupled to their ligand-dependent down-regulation by mitogen-activated protein kinase-dependent phosphorylation of serine 294", *Mol. Cell Biol.* **21**, 6122–31 (2001).
186. G. Bunone, P. A. Briand, R. J. Miksicek, D. Picard. "Activation of the unliganded estrogen receptor by EGF involves the MAP kinase pathway and direct phosphorylation", *EMBO J.* **15**, 2174–2183 (1996).
187. P. H. Driggers, J. H. Segars, D. M. Rubino. "The proto-oncoprotein Brx activates estrogen receptor beta by a p38 mitogen-activated protein kinase pathway", *J. Biol. Chem.* **276**, 46792–46797 (2001).
188. A. Tremblay, G. B. Tremblay, F. Labrie, V. Giguere. "Ligand-independent recruitment of SRC-1 to estrogen receptor beta through phosphorylation of activation function AF-1", *Mol. Cell* **3**, 513–519 (1999).

189. S. Yeh et al. "From HER2/Neu signal cascade to androgen receptor and its coactivators: a novel pathway by induction of androgen target genes through MAP kinase in prostate cancer cells", *Proc. Natl. Acad. Sci. USA* **96**, 5458–5463 (1999).
190. E. Hu, J. B. Kim, P. Sarraf, B. M. Spiegelman. "Inhibition of adipogenesis through MAP kinase-mediated phosphorylation of PPARgamma", *Science* **274**, 2100–2103 (1996).
191. C. E. Juge-Aubry et al. "Regulation of the transcriptional activity of the peroxisome proliferator-activated receptor alpha by phosphorylation of a ligand- independent trans-activating domain", *J. Biol. Chem.* **274**, 10505–10510 (1999).
192. M. Gianni, A. Bauer, E. Garattini, P. Chambon, C. Rochette-Egly. "Phosphorylation by p38MAPK and recruitment of SUG-1 are required for RA- induced RARgamma degradation and transactivation", *EMBO J.* **21**, 3760–3769 (2002).
193. M. Gianni et al. "Down-regulation of the phosphatidylinositol 3-kinase/Akt pathway is involved in retinoic acid-induced phosphorylation, degradation, and transcriptional activity of retinoic acid receptor gamma 2", *J. Biol. Chem.* **277**, 24859–24862 (2002).
194. S. Adam-Stitah, L. Penna, P. Chambon, C. Rochette-Egly. "Hyperphosphorylation of the retinoid X receptor alpha by activated c- Jun NH2-terminal kinases", *J. Biol. Chem.* **274**, 18932–18941 (1999).
195. H. Y. Lee et al. "Stress pathway activation induces phosphorylation of retinoid X receptor", *J. Biol. Chem.* **275**, 32193–32199 (2000).
196. M. P. Scheid and J. R. Woodgett. "PKB/AKT: functional insights from genetic models", *Nat. Rev. Mol. Cell Biol.* **2**, 760–768 (2001).
197. R. A. Campbell et al. "Phosphatidylinositol 3-kinase/AKT-mediated activation of estrogen receptor alpha: a new model for anti-estrogen resistance", *J. Biol. Chem.* **276**, 9817–9824 (2001).
198. H. K. Lin, S. Yeh, H. Y. Kang, C. Chang. "Akt suppresses androgen-induced apoptosis by phosphorylating and inhibiting androgen receptor", *Proc. Natl. Acad. Sci. USA* **98**, 7200–7205 (2001).
199. A. Migliaccio, M. Pagano, F. Auricchio. "Immediate and transient stimulation of protein tyrosine phosphorylation by estradiol in MCF-7 cells", *Oncogene* **8**, 2183–2191 (1993).
200. C. Rochette-Egly et al. Phosphorylation of the retinoic acid receptor-alpha by protein kinase A. *Mol. Endocrinol.* **9**, 860–871 (1995).
201. J. I. Huggenvik, M. W. Collard, Y. W. Kim, R. P. Sharma. "Modification of the retinoic acid signaling pathway by the catalytic subunit of protein kinase-A", *Mol. Endocrinol.* **7**, 543–550 (1993).
202. D. Chen, P. E. Pace, R. C. Coombes, S. Ali. "Phosphorylation of human estrogen receptor alpha by protein kinase A regulates dimerization", *Mol. Cell Biol.* **19**, 1002–1015 (1999).
203. M. H. Delmotte, A. Tahayato, P. Formstecher, P. Lefebvre. "Serine 157, a retinoic acid receptor alpha residue phosphorylated by protein kinase C in vitro, is involved in RXR.RARalpha heterodimerization and transcriptional activity", *J. Biol. Chem.* **274**, 38225–38231 (1999).
204. J. C. Hsieh et al. "Phosphorylation of the human vitamin D receptor by protein kinase C. Biochemical and functional evaluation of the serine 51 recognition site. *J. Biol. Chem.* **268**, 15118–15126 (1993).
205. J. M. Egly. "The 14th Datta Lecture. TFIIH: from transcription to clinic", *FEBS Lett.* **498**, 124–128 (2001).
206. D. Chen et al. "Activation of estrogen receptor alpha by S118 phosphorylation involves a ligand-dependent interaction with TFIIH and participation of CDK7", *Mol. Cell* **6**, 127–137 (2000).
207. J. M. Trowbridge, I. Rogatsky, M. J. Garabedian. "Regulation of estrogen receptor transcriptional enhancement by the cyclin A/Cdk2 complex", *Proc. Natl. Acad. Sci. USA* **94**, 10132–10137 (1997).
208. C. Rochette-Egly, S. Adam, M. Rossignol, J. M. Egly, P. Chambon. "Stimulation of RAR alpha activation function AF-1 through binding to the general transcription factor TFIIH and phosphorylation by CDK7", *Cell* **90**, 97–107 (1997).

209. J. Bastien et al. "TFIIH interacts with the retinoic acid receptor gamma and phosphorylates its AF-1-activating domain through cdk7", *J. Biol. Chem.* **275**, 21896–21904 (2000).
210. D. K. Lee, H. O. Duan, C. Chang. "From androgen receptor to the general transcription factor TFIIH. Identification of cdk activating kinase (CAK) as an androgen receptor NH(2)-terminal associated coactivator", *J. Biol. Chem.* **275**, 9308–9313 (2000).
211. A. Keriel, A. Stary, A. Sarasin, C. Rochette-Egly, J. M. Egly. "XPD mutations prevent TFIIH-dependent transactivation by nuclear receptors and phosphorylation of RARalpha", *Cell* **109**, 125–135 (2002).
212. C. Rochette-Egly and P. Chambon. "F9 embryocarcinoma cells: a cell autonomous model to study the functional selectivity of RARs and RXRs in retinoid signaling", *Histol. Histopathol.* **16**, 909–922 (2001).
213. R. Taneja et al. "Phosphorylation of activation functions AF-1 and AF-2 of RAR alpha and RAR gamma is indispensable for differentiation of F9 cells upon retinoic acid and cAMP treatment", *EMBO J.* **16**, 6452–6465 (1997).
214. E. Kopf et al. "Dimerization with retinoid X receptors and phosphorylation modulate the retinoic acid-induced degradation of retinoic acid receptors alpha and gamma through the ubiquitin-proteasome pathway", *J. Biol. Chem.* **275**, 33280–33288 (2000).
215. S. F. Arnold, M. Melamed, D. P. Vorojeikina, A. C. Notides, S. Sasson. "Estradiol-binding mechanism and binding capacity of the human estrogen receptor is regulated by tyrosine phosphorylation", *Mol. Endocrinol.* **11**, 48–53 (1997).
216. C. Rochette-Egly, J. L. Plassat, R. Taneja, P. Chambon. "The AF-1 and AF-2 activating domains of retinoic acid receptor-alpha (RARalpha) and their phosphorylation are differentially involved in parietal endodermal differentiation of F9 cells and retinoid-induced expression of target genes", *Mol. Endocrinol.* **14**, 1398–1410 (2000).
217. B. G. Rowan, N. L. Weigel, B. W. O'Malley. "Phosphorylation of steroid receptor coactivator-1. Identification of the phosphorylation sites and phosphorylation through the mitogen- activated protein kinase pathway", *J. Biol. Chem.* **275**, 4475–4483 (2000).
218. G. N. Lopez, C. W. Turck, F. Schaufele, M. R. Stallcup, P. J. Kushner. "Growth factors signal to steroid receptors through mitogen-activated protein kinase regulation of p160 coactivator activity", *J. Biol. Chem.* **276**, 22177–22182 (2001).
219. D. Knutti, D. Kressler, A. Kralli. "Regulation of the transcriptional coactivator PGC-1 via MAPK-sensitive interaction with a repressor", *Proc. Natl. Acad. Sci. USA* **98**, 9713–9718 (2001).
220. J. Font de Mora and M. Brown. "AIB1 is a conduit for kinase-mediated growth factor signaling to the estrogen receptor", *Mol. Cell Biol.* **20**, 5041–5047 (2000).
221. N. Vo and R. H. Goodman. "CREB-binding protein and p300 in transcriptional regulation", *J. Biol. Chem.* **276**, 13505–13508 (2001).
222. L. W. Yuan and J. E. Gambee. "Phosphorylation of p300 at serine 89 by protein kinase C", *J. Biol. Chem.* **275**, 40946–40951 (2000).
223. S. H. Hong and M. L. Privalsky. "The SMRT corepressor is regulated by a MEK-1 kinase pathway: inhibition of corepressor function is associated with SMRT phosphorylation and nuclear export", *Mol. Cell Biol.* **20**, 6612–6625 (2000).
224. M. Resche-Rigon and H. Gronemeyer. "Therapeutic potential of selective modulators of nuclear receptor action", *Curr. Opin. Chem. Biol.* **2**, 501–507 (1998).
225. M. I. Diamond, J. N. Miner, S. K. Yoshinaga, K. R. Yamamoto. "Transcription factor interactions: selectors of positive or negative regulation from a single DNA element", *Science* **249**, 1266–1272 (1990).
226. C. Jonat et al. "Antitumor promotion and antiinflammation: down-modulation of AP-1 (Fos/Jun) activity by glucocorticoid hormone", *Cell* **62**, 1189–1204 (1990).

227. H. F. Yang-Yen et al. "Transcriptional interference between c-Jun and the glucocorticoid receptor: mutual inhibition of DNA binding due to direct protein-protein interaction", *Cell* **62**, 1205–1215 (1990).
228. L. Shemshedini, R. Knauthe, P. Sassone-Corsi, A. Pornon, H. Gronemeyer. "Cell-specific inhibitory and stimulatory effects of Fos and Jun on transcription activation by nuclear receptors", *EMBO J.* **10**, 3839–3849 (1991).
229. A. Bubulya, S. C. Wise, X. Q. Shen, L. A. Burmeister, L. Shemshedini. "c-Jun can mediate androgen receptor-induced transactivation", *J. Biol. Chem.* **271**, 24583–24589 (1996).
230. D. Pearce, W. Matsui, J. N. Miner, K. R. Yamamoto. "Glucocorticoid receptor transcriptional activity determined by spacing of receptor and nonreceptor DNA sites", *J. Biol. Chem.* **273**, 30081–30085 (1998).
231. Y. Kamei et al. "A CBP integrator complex mediates transcriptional activation and AP-1 inhibition by nuclear receptors", *Cell* **85**, 403–414 (1996).
232. K. De Bosscher, W. Vanden Berghe, G. Haegeman. "Glucocorticoid repression of AP-1 is not mediated by competition for nuclear coactivators", *Mol. Endocrinol.* **15**, 219–227 (2001).
233. J. Y. Chen et al. "RAR-specific agonist/antagonists which dissociate transactivation and AP1 transrepression inhibit anchorage-independent cell proliferation", *EMBO J.* **14**, 1187–97 (1995).
234. C. Caelles, J. M. Gonzalez-Sancho, A. Munoz. "Nuclear hormone receptor antagonism with AP-1 by inhibition of the JNK pathway", *Genes Dev.* **11**, 3351–3364 (1997).
235. X. F. Zhou, X. Q. Shen, L. Shemshedini. Ligand-activated retinoic acid receptor inhibits AP-1 transactivation by disrupting c-Jun/c-Fos dimerization", *Mol. Endocrinol.* **13**, 276–285 (1999).
236. P. Webb, G. N. Lopez, R. M. Uht, P. J. Kushner. "Tamoxifen activation of the estrogen receptor/AP-1 pathway: potential origin for the cell-specific estrogen-like effects of antiestrogens", *Mol. Endocrinol.* **9**, 443–456 (1995).
237. M. Gottlicher, S. Heck, P. Herrlich. "Transcriptional cross-talk, the second mode of steroid hormone receptor action", *J. Mol. Med.* **76**, 480–489 (1998).
238. N. Auphan, J. A. DiDonato, C. Rosette, A. Helmberg, M. Karin. "Immunosuppression by glucocorticoids: inhibition of NF-kappa B activity through induction of I kappa B synthesis", *Science* **270**, 286–290 (1995).
239. S. Heck et al. I kappaB alpha-independent downregulation of NF-kappaB activity by glucocorticoid receptor", *EMBO J.* **16**, 4698–4707 (1997).
240. K. De Bosscher et al. "Glucocorticoids repress NF-kappaB-driven genes by disturbing the interaction of p65 with the basal transcription machinery, irrespective of coactivator levels in the cell", *Proc. Natl. Acad. Sci. USA* **97**, 3919–3924 (2000).
241. P. Delerive et al. "DNA Binding-Independent Induction of IkappaBalpha Gene Transcription by PPARalpha", *Mol. Endocrinol.* **16**, 1029–1039 (2002).
242. I. Barroso et al. "Dominant negative mutations in human PPARgamma associated with severe insulin resistance, diabetes mellitus and hypertension", *Nature* **402**, 880–883 (1999).
243. Z. Wang, G. Sun, Z. Shen, S. Chen, Z. Chen. "Differentiation therapy for acute promyelocytic leukemia with all-trans retinoic acid: 10-year experience of its clinical application", *Chin. Med. J. (Engl.)* **112**, 963–967 (1999).
244. L. Degos and Z. Y. Wang. "All trans retinoic acid in acute promyelocytic leukemia. *Oncogene* **20**, 7140–7145 (2001).
245. L. Altucci and H. Gronemeyer. "The promise of retinoids to fight against cancer", *Nat. Rev. Cancer* **1**, 181–193 (2001).
246. L. Altucci and H. Gronemeyer. "Nuclear receptors in cell life and death", *Trends Endocrinol. Metab.* **12**, 460–468 (2001).
247. M. B. Sporn and N. Suh. "Opinion: Chemoprevention: an essential approach to controlling cancer. *Nat. Rev. Cancer* **2**, 537–543 (2002).

248. S. Refetoff, R. E. Weiss, S. J. Usala. "The syndromes of resistance to thyroid hormone", *Endocr. Rev.* **14**, 348–399 (1993).
249. S. J. Usala et al. "A homozygous deletion in the c-erbA beta thyroid hormone receptor gene in a patient with generalized thyroid hormone resistance: isolation and characterization of the mutant receptor", *Mol. Endocrinol.* **5**, 327–335 (1991).
250. S. J. Usala et al. "A base mutation of the C-erbA beta thyroid hormone receptor in a kindred with generalized thyroid hormone resistance. Molecular heterogeneity in two other kindreds", *J. Clin. Invest.* **85**, 93–100 (1990).
251. T. N. Collingwood et al. "A role for helix 3 of the TRbeta ligand-binding domain in coactivator recruitment identified by characterization of a third cluster of mutations in resistance to thyroid hormone", *EMBO J.* **17**, 4760–4770 (1998).
252. T. N. Collingwood, M. Adams, Y. Tone, V. K. Chatterjee. "Spectrum of transcriptional, dimerization, and dominant negative properties of twenty different mutant thyroid hormone beta-receptors in thyroid hormone resistance syndrome", *Mol. Endocrinol.* **8**, 1262–1277 (1994).
253. R. E. Weiss et al. "Dominant inheritance of resistance to thyroid hormone not linked to defects in the thyroid hormone receptor alpha or beta genes may be due to a defective cofactor", *J. Clin. Endocrinol. Metab.* **81**, 4196–4203 (1996).
254. J. D. Safer, R. N. Cohen, A. N. Hollenberg, F. E. Wondisford. "Defective release of corepressor by hinge mutants of the thyroid hormone receptor found in patients with resistance to thyroid hormone", *J. Biol. Chem.* **273**, 30175–30182 (1998).
255. M. Alcalay et al. "Common themes in the pathogenesis of acute myeloid leukemia", *Oncogene* **20**, 5680–5694 (2001).
256. H. de The and M. K. Chelbi-Alix. "APL, a model disease for cancer therapies?", *Oncogene* **20**, 7136–7139 (2001).
257. F. Piazza, C. Gurrieri, P. P. Pandolfi. "The theory of APL. *Oncogene* **20**, 7216–7122 (2001).
258. E. M. Rego and P. P. Pandolfi. "Reciprocal products of chromosomal translocations in human cancer pathogenesis: key players or innocent bystanders?", *Trends. Mol. Med.* **8**, 396–405 (2002).
259. F. Grignani et al. "Fusion proteins of the retinoic acid receptor-alpha recruit histone deacetylase in promyelocytic leukaemia", *Nature* **391**, 815–818 (1998).
260. L. Z. He et al. "Distinct interactions of PML-RARalpha and PLZF-RARalpha with co- repressors determine differential responses to RA in APL", *Nat. Genet.* **18**, 126–135 (1998).
261. R. J. Lin et al. "Role of the histone deacetylase complex in acute promyelocytic leukaemia", *Nature* **391**, 811–814 (1998).
262. W. Bourguet, P. Germain, H. Gronemeyer. "Nuclear receptor ligand-binding domains: three-dimensional structures, molecular interactions and pharmacological implications", *Trends Pharmacol. Sci.* **21**, 381–388 (2000).
263. L. Altucci et al. "Retinoic acid-induced apoptosis in leukemia cells is mediated by paracrine action of tumor-selective death ligand TRAIL", *Nat. Med.* **7**, 680–686 (2001).
264. L. E. French and J. Tschopp. The TRAIL to selective tumor death [news; comment]. *Nat. Med.* **5**, 146–147 (1999).
265. M. Ristow, D. Muller-Wieland, A. Pfeiffer, W. Krone, C. R. Kahn. "Obesity associated with a mutation in a genetic regulator of adipocyte differentiation", *N. Engl. J. Med.* **339**, 953–959 (1998).
266. M. W. Schwartz and S. E. Kahn. "Insulin resistance and obesity", *Nature* **402**, 860–861 (1999).
267. D. Altshuler et al. "The common PPARgamma Pro12Ala polymorphism is associated with decreased risk of type 2 diabetes", *Nat. Genet.* **26**, 76–80 (2000).
268. P. Sarraf et al. "Loss-of-function mutations in PPAR gamma associated with human colon cancer", *Mol. Cell* **3**, 799–804 (1999).
269. M. R. Hughes et al. "Point mutations in the human vitamin D receptor gene associated with hypocalcemic rickets", *Science* **242**, 1702–1705 (1988).

270. H. H. Ritchie et al. "An ochre mutation in the vitamin D receptor gene causes hereditary 1,25- dihydroxyvitamin D3-resistant rickets in three families", *Proc. Natl. Acad. Sci. USA* **86**, 9783–9787 (1989).
271. T. Carling et al. "Vitamin D receptor genotypes in primary hyperparathyroidism", *Nat. Med.* **1**, 1309–1311 (1995).
272. E. P. Smith et al. "Estrogen resistance caused by a mutation in the estrogen-receptor gene in a man", *N. Engl. J. Med* **331**, 1056–1061 (1994).
273. T. A. Hopp and S. A. Fuqua. "Estrogen receptor variants. *J. Mammary Gland Biol. Neoplasia* **3**, 73–83 (1998).
274. A. R. La Spada, E. M. Wilson, D. B. Lubahn, A. E. Harding, K. H. Fischbeck. "Androgen receptor gene mutations in X-linked spinal and bulbar muscular atrophy", *Nature* **352**, 77–79 (1991).
275. A. N. Mhatre et al. "Reduced transcriptional regulatory competence of the androgen receptor in X-linked spinal and bulbar muscular atrophy", *Nat. Genet.* **5**, 184–188 (1993).
276. X. Y. Zhao et al. "Glucocorticoids can promote androgen-independent growth of prostate cancer cells through a mutated androgen receptor", *Nat. Med.* **6**, 703–706 (2000).
277. A. O. Brinkmann and J. Trapman. "Prostate cancer schemes for androgen escape", *Nat. Med.* **6**, 628–629 (2000).
278. D. M. Hurley et al. "Point mutation causing a single amino acid substitution in the hormone binding domain of the glucocorticoid receptor in familial glucocorticoid resistance", *J. Clin. Invest.* **87**, 680–686 (1991).
279. M. Karl et al. "Familial glucocorticoid resistance caused by a splice site deletion in the human glucocorticoid receptor gene", *J. Clin. Endocrinol. Metab.* **76**, 683–689 (1993).
280. S. L. Anzick et al. "AIB1, a steroid receptor coactivator amplified in breast and ovarian cancer", *Science* **277**, 965–968 (1997).
281. M. Carapeti, R. C. Aguiar, A. E. Watmore, J. M. Goldman, N. C. Cross. "Consistent fusion of MOZ and TIF2 in AML with inv(8)(p11q13) ", *Cancer Genet. Cytogenet.* **113**, 70–72 (1999).
282. J. Liang, L. Prouty, B. J. Williams, M. A. Dayton, K. L. Blanchard. "Acute mixed lineage leukemia with an inv(8)(p11q13) resulting in fusion of the genes for MOZ and TIF2", *Blood* **92**, 2118–2122 (1998).
283. M. Carapeti, R. C. Aguiar, J. M. Goldman, N. C. Cross. "A novel fusion between MOZ and the nuclear receptor coactivator TIF2 in acute myeloid leukemia", *Blood* **91**, 3127–3133 (1998).
284. P. Blume-Jensen and T. Hunter. "Oncogenic kinase signalling", *Nature* **411**, 355–365 (2001).
285. B. P. Zhou et al. "Cytoplasmic localization of p21Cip1/WAF1 by Akt-induced phosphorylation in HER-2/neu-overexpressing cells", *Nat. Cell Biol.* **3**, 245–252 (2001).
286. B. J. Feldman and D. Feldman. "The development of androgen-independent prostate cancer", *Nat. Rev. Cancer* **1**, 34–45 (2001).
287. R. Mukherjee et al. "Sensitization of diabetic and obese mice to insulin by retinoid X receptor agonists", *Nature* **386**, 407–410 (1997).

Pure Appl. Chem., Vol. 75, Nos. 11–12, pp. 1665–1669, 2003.

Topic 1.2
Nuclear receptor coregulators*

Neil J. McKenna and Bert W. O'Malley‡

Department of Molecular and Cellular Biology, Baylor College of Medicine, Houston, TX 77030, USA

Abstract: It has been postulated that nuclear receptors (NRs) regulate transcription via interactions with chromatin and the basal transcription machinery at the promoters of genes. Coregulators (coactivators or corepressors) are important in mediating these interactions and thereby modulating positive or negative receptor activity. A large number of putative coactivators have been isolated, several of which will be reviewed with respect to certain "criteria" initially proposed for coactivators. We will discuss, with reference to in vitro and in vivo experiments, the main steps in initiation that are influenced by coactivators: (1) initiation (e.g., SRC-1 family, CBP); (2) repetitive transcription (e.g., TRAPs/DRIPs); (3) RNA processing (PGC-1, etc); and (4) termination/turnover (E6-AP, etc). A variety of enzyme functions have been implicated in the coactivator complex including acetylase, methylase, ubiquitin ligase, kinase, and phosphatase activities. Moreover, coactivators and corepressors appear to exist in the steady-state cell as a series of multiprotein complexes referred to collectively as the "coregulatorsome". Different subcomplexes within the coregulatorsome may have different levels of preference for individual receptors or promoters, likely contributing to context-specific functions of NRs in target tissues.

INTRODUCTION

Nuclear receptors (NRs) are transcription factors that respond to modulation by lipophilic ligands and other signaling pathways to regulate the expression of genes in a tissue- and context-selective manner [1,2]. Recent evidence suggests that they achieve this effect by interacting with a group of molecules collectively referred to as coregulators. NR coregulators are defined as cellular factors recruited by NRs that complement the activity of NRs. They are generally divisible into coregulators that promote positive receptor activity when recruited (coactivators) and those that mediate negative receptor activity (corepressors). Receptor activity is generally thought to involve the interaction of the receptor with specific promoter sequences (hormone response elements) to activate or repress target genes, although receptors can influence events outside the nucleus ("nongenomic" action). In this report, we summarize selected advances in the coregulator field to date. While recognizing the importance of corepressor function in receptor action, we will focus our discussion on coactivators.

Using a variety of experimental techniques, from in vitro experiments to cultured cells, to null deletions in living animals, our laboratory and others initially established a number of criteria for designation of a molecule as a coactivator. These included: enhancement of the transcriptional activity of a receptor, demonstrated by addition and subtraction in cultured cells; relief of "squelching", or transcriptional interference between receptors competing for a limited, common pool of transcriptional me-

---

*Report from a SCOPE/IUPAC project: Implication of Endocrine Active Substances for Human and Wildlife (J. Miyamoto and J. Burger, editors). Other reports are published in this issue, *Pure Appl. Chem.* **75**, 1617–2615 (2003).

‡Corresponding author: Tel: (713) 798-6205; Fax: (713) 798-5599; E-mail: berto@bcm.tmc.edu

diators; and nuclear localization, reflecting the fact that many of these molecules appeared to be nuclear entities in most cells. Since then, an appreciation has grown of tissue-specific variations in coactivators' mode of action, their cellular localization, and their receptor/signaling pathway response preferences, all of which appear to be influenced, at least in part, by differential programming of these molecules by phosphorylation and other posttranslational modifications. The net result of this has been to broaden the compass within which NR coactivator functions are interpreted, and to define them simply as molecules that interact with and mediate NR functions. Due to space constraints, the reader is referred to more extensive reviews [3,4]

COACTIVATORS: A TENTATIVE CLASSIFICATION

Coactivators can be divided into a series of functional groups which, while structurally distinct, are functionally coordinated during transcriptional initiation by NRs. Over the past several years, rapid progress has been made toward the goal of establishing the mechanistic principles that govern the interaction between receptor, coactivator, and promoter in the activation of target genes. A composite model of ligand-mediated receptor activation envisages at least four main steps, each appearing to involve the mediation of distinct categories of coactivators: initiation, repetitive transcription (or reinitiation), RNA processing, and termination/turnover.

In NR-mediated transcriptional initiation, coactivators considered to be among the first recruited by activated receptor are the SRC/p160s, CREB-binding protein (CBP), and p300 [3,4 and refs. therein]. (Note that this does not preclude the initial involvement of ATP-coupled chromatin domain-remodeling machines such as the SWI/SNF complex, components of which have been shown to be required by individual receptors for efficient transcriptional initiation in certain experimental systems—see also refs. [3,4].) It was in the SRC/p160 family that an amphipathic helix conserved on the surface of most coactivators was demonstrated (the LXXLL motif or NR box) [5], which substantially determines the interaction between ligand-activated receptors and these molecules. The interface between ligand-dependent receptor motifs and the LXXLL motif has been the subject of intense study as a possible target for manipulation of NR pharmacology, and as a flexible, informative basis for ligand screening assays.

The physiological importance of SRC/p160s has been implied by knock-out studies of genes encoding these coactivators. Although the phenotypes of these knock-outs are largely subtle in nature, they provide clues as to their functions. SRC-1 knock-out mice show a partial resistance to hormones and a reduced growth and development of various steroid target organs [6]. SRC-3 knock-out mice show reduced growth and female reproduction, and lack of mammary gland development. In addition, mouse embryonic fibroblasts or liver cells derived from these SRC-$3^{-/-}$ mice are reported to be insensitive to growth stimulation by IGF-1 or growth hormone [7,8]. The participation of SRC-3 in cell growth is further supported by its role in various cancers—indeed, the correlation between SRC-3 expression and cancers is striking. It has been demonstrated that SRC-3 is amplified in 5–10 % of breast tumors and 7–8 % of ovarian cancer samples [9].

SRC/p160s and CBP/p300 contain acetyltransferase activity that targets specific lysines in nucleosomal histones to generate a transcriptionally permissive environment at NR-regulated promoters. The histone targeting specificity and the relative contribution of SRC/p160s and CBP or p300 to histone modification at various promoters may to some extent determine differential patterns of gene expression in various tissues, and in the same tissue in response to different signals. More recently, it has been shown that this acetyltransferase activity is used to choreograph specific protein–protein interactions during receptor-mediated assembly of the preinitiation complex [10].

The sequential model of NR-mediated transcriptional initiation suggests that following initial recruitment of SRC/p160s and CBP to effect modification of local histones and other proteins, members of the TRAP/DRIP complex directly contact components of the basal transcription machinery. The evolutionary conservation of various guises of this complex, from the yeast Mediator through to human

SMCC and others, is reflected in the fact that targeted deletion of the receptor-interacting subunit (TRAP220) results in embryonic lethality [11]. This phenotype stands in contrast to the less severe phenotypes of the SRC-1 and p/CIP/SRC-3 null mice, which exhibit a variety of partial aberrations in specific NR- and non-NR-mediated signaling pathways. PGC-1, a coactivator originally thought to be a receptor-specific coactivator for PPARγ, has been shown to interact with a variety of receptors, as well as with components of the RNA splicing machinery [12]. These studies raise the intriguing possibility that, through PGC-1 and similar coregulators, NRs might specify tissue- and ligand-specific edits of individual primary RNAs, further contributing to the functional diversity of NR signaling.

To continue the theme of employing posttranslational modifications to achieve efficient transcriptional initiation, the role of targeted ubiquitination in tagging components of the various complexes for removal and recycling is attractive from the point of view of refurnishing the promoter for subsequent rounds of transcriptional reinitiation. The E3 ubiquitin ligase E6-AP was the first ubiquitin ligase to be shown to coactivate NRs [13], and though the enzyme and transactivation functions of E6-AP are separable, the fact that the 26S proteasome is required for efficient ER-mediated transcriptional activation points to a fundamental role for protein turnover in receptor-mediated transactivation.

ROLE OF KINASE-BASED SIGNALING PATHWAYS

Studies on characterized NRs indicate that ligand-independent modulation through posttranslational modification is a common currency in control of protein–protein interactions between NR and "non-NR"-regulated pathways. While intricate patterns of reciprocal auto- and heterophosphorylation have long been known to mediate the functions of membrane-associated receptors, recent studies suggest that such modifications also directly influence protein–protein interactions in NR-regulated pathways. In addition, enzyme functions of coregulators themselves appear to target receptors, histones, and coregulators themselves, to modulate their molecular interactions [reviewed in 3,4]. It seems that coactivators can, depending upon the phosphorylating kinase, be commandeered by any one of a number of such pathways, with their ultimate promoter specificity being determined by the specific "phosphosignature" they bear. Evidence suggests that these modifications may be able to effect changes in coregulator concentration in individual intracellular compartments, alter functional specificity, or determine the final transcriptional complexes into which they are recruited. The term "coregulatorsome", coined to account for the myriad protein–protein interactions in which coactivators participate, likely represents a heterogeneous continuum of complexes, whose composition and promoter specificity are under constant scrutiny and revision according to a variety of parameters (Fig. 1). Discrimination by receptors between these complexes likely occurs on a trial-and-error basis according to the specific requirements of the promoter at a defined point in time (see refs. [3,4] for review).

ROLE OF LIGANDS

Ligands have been shown to be capable of influencing the pharmacokinetics of the interaction between the receptor AF-2 and the coregulator NR box (references), hinting at a possible basis for the endocrine activity of many exogenous, nonphysiological ligands. To illustrate this, it is becoming apparent that the type of agonist bound to a specific receptor is an important determinant of its affinity for a particular subset of coactivator complexes, thereby ultimately influencing the biological response to the ligand. In the case of the estrogen receptor (ER) and vitamin D receptor (VDR), ligands and ligand derivatives may elicit their distinct biological responses through effecting differential interactions of ligand-bound VDR with coactivators (reviewed in ref. [14]). Similar findings have been observed with PPAR ligands that specify distinct patterns of SRC recruitment by members of the PPAR family. The potential role of selective estrogen receptor modulators (SERMs) in influencing ER function in a tissue-specific manner has been the subject of much recent interest, and it appears that selective coregulator recruitment may contribute in part to the custom pharmacology of many of these compounds [15–17].

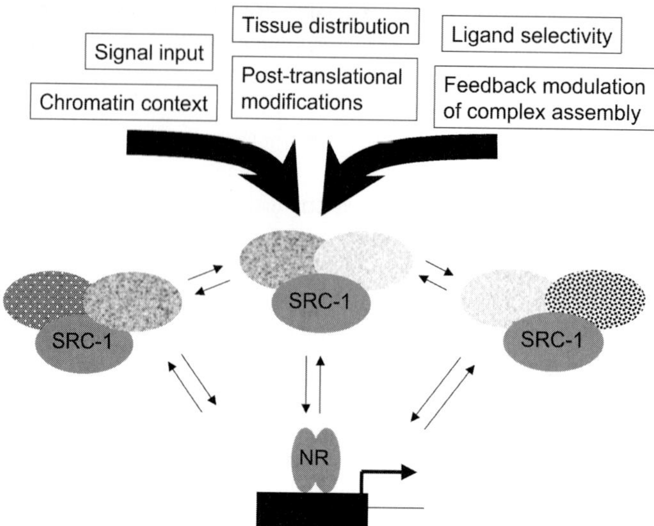

Fig. 1 Order and disorder in transcriptional complex assembly by nuclear receptors. A variety of factors influence the composition of coregulator complexes recruited by NRs (boxes). The "coregulatorsome" is likely defined by a spectrum of complexes in continuous flux, which are incorporated into the final transcriptional complex according to the specific requirements at a given time point (heterogeneity in SRC-1 complexes is shown as an example).

COREPRESSORS

In addition to potentiation of gene expression, many NRs possess a silencing, or repression function [18]. As was anticipated, this silencing function requires the participation of corepressors, of which SMRT and NCoR have been the most intensively studied (for a review, see ref. [3]). Corepressors were initially identified as receptor-interacting proteins in the absence of ligand, with this interaction being uncoupled in the presence of ligand. Just like their coactivator counterparts, corepressors play important roles in both health and disease. For example, both SMRT- and NcoR-mediated repression have been shown to play a role in promyelocytic and myelogenous leukemias as well as thyroid hormone resistance. Mutation of NCoR in mice results in defects in CNS, erythrocyte, and thymocyte development [19]. The association of corepressors with histone deacetylases [20] suggests that one of the primary functions of corepressors is to transform local chromatin structure to an inactive conformation by deacetylating histones. Remarkably, corepressors utilize an amphipathic helix related to the NR box (referred to as the "CoRNR" motif) that recognizes unliganded NRs in a manner that is inhibited by the activation helix of AF2 (H12) [21]. The mechanisms underlying coactivator and corepressor interactions with liganded and unliganded receptors, respectively, are therefore surprisingly similar, given their radically opposed transcriptional consequences.

FUTURE RESEARCH DIRECTIONS

It is likely that animal models will soon be developed in which the contribution of coactivators to the organization of genes along select metabolic pathways can be more closely scrutinized. These animal models may hold the key to examining in a more "physiological" context the functional relationships between SRMs and their ability to influence gene expression in a tissue-specific manner. Moreover, the potential for combination SRM/NR box peptide therapeutics holds much promise for the future, given its potential for further fine-tuning the tissue-specificity of SRM function. For example, the administration of a specific modulator along with a peptide that might inhibit a specific receptor conformation might abolish an unwanted effect while retaining a desirable effect of the SRM. In conclusion, a full appreciation of coregulator biology will involve characterization of both ligand-dependent and -independent activation pathways, and their role in determining the interactions that contribute to the regulation by NRs of complex spatiotemporal patterns of gene expression.

ACKNOWLEDGMENTS

We apologize to the many authors whose publications were, due to space constraints, left unreferenced in this review.

REFERENCES

1. R. M. Evans. *Science* **240**, 889–895 (1988).
2. M. J. Tsai and B. W. O'Malley. *Annu. Rev. Biochem.* **63**, 451–486 (1994).
3. M. G. Rosenfeld and C. K. Glass. *J. Biol. Chem.* **276**, 36865–36868 (2001).
4. N. J. McKenna and B. W. O'Malley. *Cell* **108**, 465–474 (2002).
5. D. M. Heery, E. Kalkhoven, S. Hoare, M. G. Parker. *Nature* **387**, 733–736 (1997).
6. J. Xu, Y. Qiu, F. J. DeMayo, S. Y. Tsai, M. J. Tsai, B. W. O'Malley. *Science* **279**, 1922–1925 (1998).
7. J. Xu, L. Liao, G. Ning, H. Yoshida-Komiya, C. Deng, B. W. O'Malley. *Proc. Natl. Acad. Sci. USA* **97**, 6379–6384 (2000).
8. Z. Wang, D. W. Rose, O. Hermanson, F. Liu, T. Herman, W. Wu, D. Szeto, A. Gleiberman, A. Krones, K. Pratt, M. G. Rosenfeld. *Proc. Natl. Acad. Sci. USA* **97**, 13549–13554 (2000).
9. S. L. Anzick, J. Kononen, R. L. Walker, D. O. Azorsa, M. M. Tanner, X. Y. Guan, G. Sauter, O. P. Kallioniemi, J. M. Trent, P. S. Meltzer. *Science* **277**, 965–968 (1997).
10. H. Chen, R. J. Lin, W. Xie, D. Wilpitz, R. M. Evans. *Cell* **98**, 675–686 (1999).
11. M. Ito, C. X. Yuan, H. J. Okano, R. B. Darnell, R. G. Roeder. *Mol. Cell* **5**, 683–693 (2000).
12. M. Monsalve, Z. Wu, G. Adelmant, P. Puigserver, M. Fan, B. M. Spiegelman. *Mol. Cell* **6**, 307–316 (2000).
13. Z. Nawaz, D. M. Lonard, C. L. Smith, E. Lev-Lehman, S. Y. Tsai, M. J. Tsai, B. W. O'Malley. *Mol. Cell. Biol.* **19**, 1182–1189 (1999).
14. N. J. McKenna and B. W. O'Malley. *J. Steroid Biochem. Mol. Biol.* **74**, 351–356 (2000).
15. C. L. Smith, Z. Nawaz, B. W. O'Malley. *Mol. Endocrinol.* **11**, 657–666 (1997).
16. Z. Liu, D. Auboeuf, J. Wong, J. D. Chen, S. Y. Tsai, M. J. Tsai, B. W. O'Malley. *Proc. Natl. Acad. Sci. USA* **99**, 7940–7944 (2002).
17. Y. Shang and M. Brown. *Science* **295**, 2465–2468 (2002).
18. A. Baniahmad, X. Leng, T. P. Burris, S.Y. Tsai, M. J. Tsai, B. W. O'Malley. *Mol. Cell Biol.* **15**, 76–86 (1995).
19. K. Jepsen, O. Hermanson, T. M. Onami, A. S. Gleiberman, V. Lunyak, R. J. McEvilly, R. Kurokawa, V. Kumar, F. Liu, E. Seto, M. G. Rosenfeld. *Cell* **102**, 753–763 (2000).
20. L. Nagy, H. Y. Kao, D. Chakravarti, R. J. Lin, C. A. Hassig, D. E. Ayer, S. L. Schreiber, R. M. Evans. *Cell* **89**, 373–380 (1997).
21. X. Hu and M. A. Lazar. *Nature* **402**, 93–96 (1997).

Pure Appl. Chem., Vol. 75, Nos. 11–12, pp. 1671–1683, 2003.

Topic 1.3

Function and mode of action of nuclear receptors: Estrogen, progesterone, and vitamin D*

Erin A. Kimbrel and Donald P. McDonnell‡

Duke University Medical Center, Durham, NC 27710, USA

Abstract: Estrogens and progestins play important roles in regulating growth and differentiation of a wide range of cell types, in both reproductive and nonreproductive tissues. Not surprisingly, therefore, endocrine active substances that mimic the actions of these steroid hormones have been demonstrated to have untoward effects on the reproductive function of a variety of animals. Although direct links between exposure to endocrine active substances of this class and reproductive abnormalities in humans have been difficult to establish, the potential for harm by this class of chemicals warrants further investigation. This chapter summarizes our current understanding of the molecular pharmacology of the estrogen and progesterone receptors, ER and PR, beginning with a historical perspective of ER pharmacology and ending with a comparison of these receptors to the vitamin D receptor (VDR), a nonsteroid hormone nuclear receptor involved in regulating growth and development in nonreproductive tissues. Major topics discussed include receptor subtypes (or isoforms), receptor functional domains, ligand-binding characteristics, receptor structure, cofactor binding, effects of phosphorylation, and nonclassical modes of action. This discussion will demonstrate the need for developing novel screens for potential endocrine disruptors that incorporate our current understanding of nuclear receptor pharmacology.

INTRODUCTION: HISTORICAL MODELS OF ENDOCRINE RECEPTOR ACTION

The modern era of ovarian steroid hormone pharmacology began nearly 40 years ago when Elwood Jensen and colleagues demonstrated that radiolabeled estradiol was specifically retained in cells that exhibit a phenotypic response to this hormone [1]. This initial finding led to the discovery that the biological activity of these hormones is mediated by specific high-affinity receptors for estrogens and progestins, ER and PR respectively, which are expressed in target cells. It is now well established that these receptors are ligand-regulated transcription factors which, when activated, can facilitate alterations in gene expression and effect a phenotypic change in target cells.

Being a lipophilic molecule, estrogen passively diffuses across cell membranes and binds to the estrogen receptor (ER), located in the nucleus. When a ligand such as estrogen binds to ER, it allows the receptor to undergo a conformational change, bind to its response element within the promoter of a target gene and regulate transcription of that gene. Historically, it was thought that ligands control nuclear receptor activity much like a switch; binding to the receptor was thought to convert it from an "off"

*Report from a SCOPE/IUPAC project: Implication of Endocrine Active Substances for Human and Wildlife (J. Miyamoto and J. Burger, editors). Other reports are published in this issue, *Pure Appl. Chem.* **75**, 1617–2615 (2003).

‡Corresponding author

to an "on" state. With the identification of SERMs or "selective estrogen receptor modulators", compounds that display tissue-specific agonist/antagonist activities, it was realized that this on/off model of nuclear receptor action did not adequately describe the pharmacology of known ligands. Rather, it now appears that different ligands acting through the same receptor can manifest different activities in different cells; thus explaining why it has been difficult to establish a single predictive screen for potential endocrine disruptors.

ER subtypes

There are two subtypes of ER—ERα, which was first described in 1962, and the more recently discovered ERβ, which was first cloned in 1996 from rat prostate [1,2]. As members of the nuclear receptor superfamily, ERα and β contain the hallmark features of this family of transcription factors. These include a two zinc-finger motif DNA-binding domain (DBD) and C-terminal ligand-binding domain (LBD), which is composed of 12 alpha helical structures [3,4]. ERα and ERβ each also contain two activation functions, AF-1 in the N-terminus and AF-2 in the C-terminus, within the LBD (Fig. 1). Overall, the receptors are only 47 % identical; the most conserved region between them is the DBD, while the most divergent region is the N-terminal AF-1 [5]. ERα and β interact with endogenous estrogens and many antiestrogens with similar affinity, however, some phytoestrogens, such as genistein and coumestrol, appear to have stronger affinity for ERβ [6]. Both receptors bind estrogen response elements (EREs) in the promoters of target genes equally well, yet, due to their sequence differences, especially those in the N-terminus, they are not functionally equivalent [7]. When occupied by estradiol, the transcriptional activity of ERβ only reaches 20–60 % of the activity of ERα in a given tissue [8]. Molecular dissection of the divergent N-terminus of ERβ reveals that it may contain a repressor domain that dampens its activity; removal of the N-terminus renders the receptor more transcriptionally active than its wild-type counterpart ERβ [8]. Additionally, it has been found that the SERMs tamoxifen, raloxifene, GW7604, and idoxifene do not display any partial agonism on ERβ when analyzed on canonical ERE-containing genes as they do on ERα. Coexpression studies of ERα and β reveal that heterodimerization between the two receptor subtypes can occur, possibly explaining how ERβ can dampen ERα's transcriptional activity in response to physiological levels of estradiol [8]. It may also explain how the partial agonist activity of tamoxifen on ERα can be abolished in cells where both subtypes are expressed. Superimposed on their distinct molecular pharmacology, differences in the tissue distribution and expression levels of the two ER subtypes add complexity to the biological roles of es-

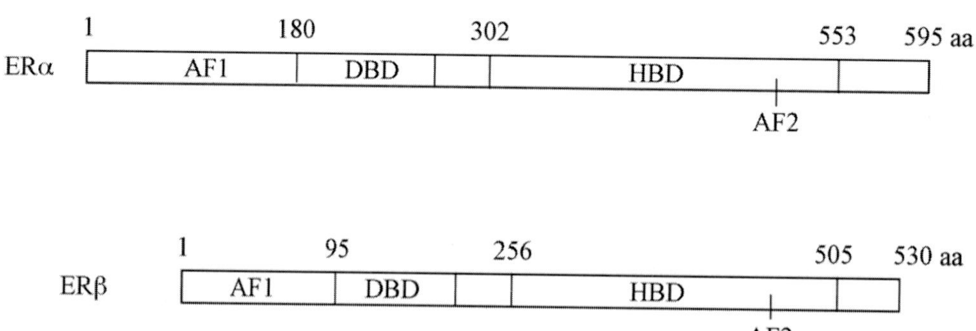

Fig. 1 Diagram of ERα and ERβ functional domains. These two subtypes of ER are 47 % identical overall, with ERα being slightly longer than ERβ (595 vs. 530 amino acids). The most conserved region between the two receptors is the DNA-binding domain (DBD), while the most divergent region is the amino-terminal activation function 1 (AF-1). Also shown are the hormone-binding domain (HBD) and the activation function 2 (AF-2), which resides within the HBD.

trogen. For instance, ERα is generally expressed at higher levels than ERβ and their expression patterns, while somewhat different, overlap in many target tissues such as the ovary, uterus, and testes [2,9]. The standardization of receptor quantitation techniques and their application to the evaluation of ERα and ERβ distribution would significantly help in studying estrogen action.

Molecular actions of ERα and β

In most tissue types, AF-2 is the dominant activation function for both ERα and β, and it operates in a ligand-dependent manner [10]. Upon ligand-binding, the p160 coactivators, SRC1, GRIP1, and ACTR are the principal coactivators that bind to AF-2 and each of these contains multiple copies of the binding motif, L*xx*LL (L = leucine, *x* = any residue) which is critical for their interaction with the AF-2 coactivator binding pocket. The p160 coactivators also serve as docking sites for the assembly of a large coactivator network, the function of which is to transmit information from the ER-ligand complex to the general transcription apparatus (GTA) and to promote transcriptional initiation [11–13]. A more detailed discussion of how coactivators coordinately regulate ER activity can be found elsewhere [14]. The AF-1 domain of ERα or β operates independently of ligand but contains sequences that allow the receptor to be controlled by other means, such as growth factor signaling. Studies with ERα show that EGF, IGF, and TGFα signaling can activate the receptor through phosphorylation of key residues in AF-1, such as serine 118 (in human ERα). Phosphorylation of these residues may then allow coactivators like SRC1 and CBP to be recruited to the AF-1 domain [15–17]. The AF-1 has also been found to bind to a variety of unique cofactors that do not interact with AF-2. These include the RNA coactivator, SRA (steroid receptor RNA coactivator), and p68 RNA helicase, a protein that preferentially binds to AF-1 when Ser 118 is phosphorylated [18,19]. Another AF-1-binding cofactor is the RNA-binding protein RTA (repressor of tamoxifen action), which represses the partial agonist activity of ER when bound by tamoxifen or other antiestrogens [20]. Differences in the expression levels of these non-AF-2 interacting cofactors suggest that communication between AF-1 and AF-2 may not occur in the same manner in all cells. When antiestrogens bind ER, corepressor proteins that impede transcriptional activation are recruited to the receptor. Corepressors like N-CoR and SMRT are thought to associate with a region encompassing the AF-2 domain while repressors like RTA, mentioned above, interact with the N-terminus [20,21]. These proteins facilitate the formation of a large corepressor complex on the receptor that opposes transcriptional activation, in part, through the enzymatic activity of histone deacetylases (HDACs) and chromatin remodeling factors [22–24]. Cofactors that bind agonist or antagonist-occupied ER have been found to cycle on and off ER target gene promoters in a highly regulated and dynamic pattern [25]. The availability of different coactivators and corepressors will help determine if a SERM acts as an ER agonist or antagonist in a given cell.

Molecular pharmacology of ER

In addition to the factors mentioned above, the differential activities of ER ligands are influenced by the specific ligand-induced conformation of the receptor [26]. Analysis of the crystal structures of several different ER-ligand complexes has been informative with regard to their agonist/antagonist actions. These structures have revealed that although narrow, the ligand-binding pocket of ER is flexible enough to accommodate a variety of ligands, some with bulky side chains and/or spatial requirements different than that of 17β estradiol (Fig. 2) [4,7]. Differences in ligand-binding characteristics will impact the structure of the coactivator-binding pocket, which is the primary region of contact between ER and the transcriptional apparatus. In an "agonist conformation", helix 12 of the LBD resides along the edge of the coactivator-binding pocket, allowing coactivators like GRIP1 to bind within the pocket using their L*xx*LL motifs [4] and activate transcription. However, when tamoxifen is bound, a conformational difference in the receptor causes helix 12 to wedge itself into the coactivator pocket through its own L*xx*LL-like region (aa540-544; LLEML). In this manner, helix 12 acts as a decoy coactivator to block the transcriptional coactivation of tamoxifen-bound ER [4]. By exposing different surfaces on the receptor, small ligand-induced conformational differences can translate into significant pharmacological differences by offering new surfaces for cofactors to bind or by covering up surfaces that otherwise would be used in binding cofactors. Phage display screening has been used successfully to identify peptides that interact with unique surfaces on ER that are exposed in the presence of different ligands. By binding to sites that are important for protein–protein interactions, many of the identified peptides have been found to act as antagonists, modulating the pharmacology of the receptor when bound by a particular SERM [27,28]. These peptide antagonist studies have demonstrated how ligand-induced conformational differences in ER help determine the activity of SERMs. Cumulatively, the differential activity of SERMs depend on (a) which ER subtype(s) is present in a given cell type, (b) cofactor expression levels in that cell type, and (c) the specific ligand-induced conformation of the receptor.

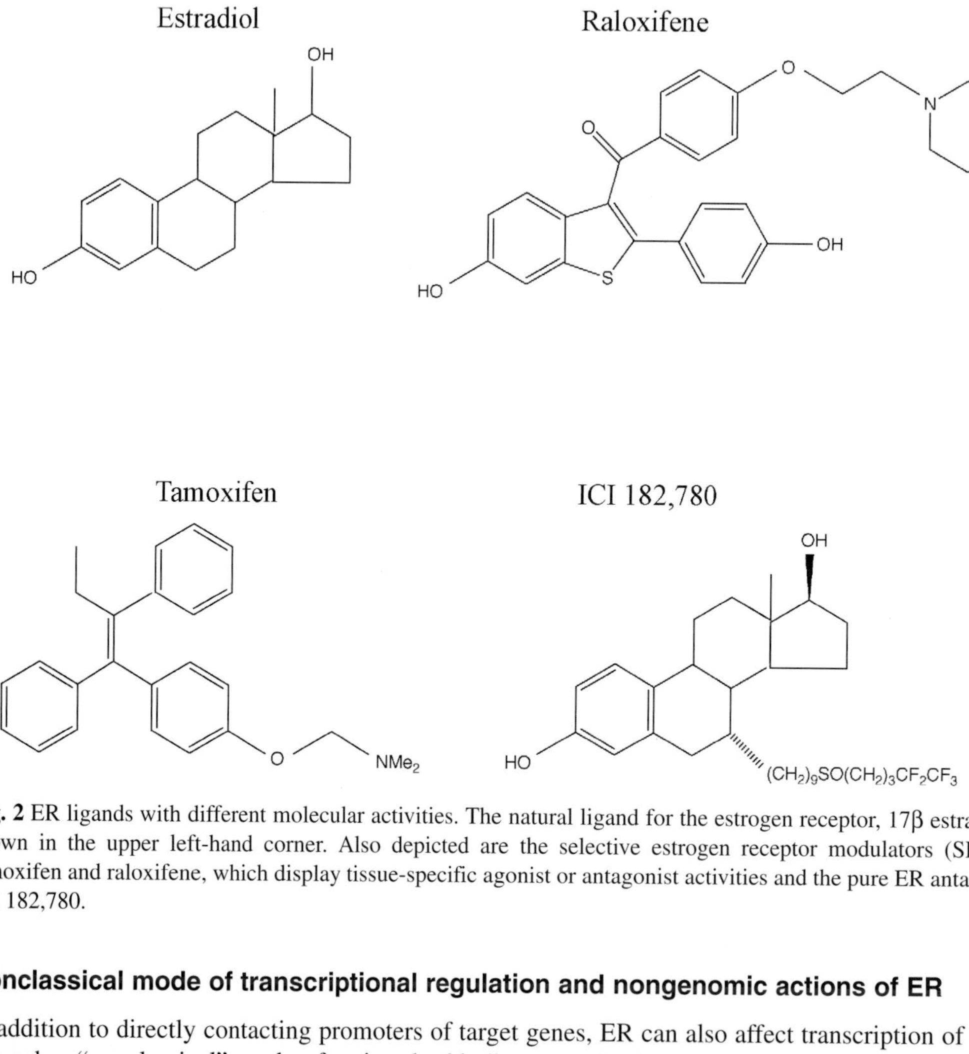

Estradiol

Raloxifene

Tamoxifen

ICI 182,780

Fig. 2 ER ligands with different molecular activities. The natural ligand for the estrogen receptor, 17β estradiol is shown in the upper left-hand corner. Also depicted are the selective estrogen receptor modulators (SERMs) tamoxifen and raloxifene, which display tissue-specific agonist or antagonist activities and the pure ER antagonist, ICI 182,780.

Nonclassical mode of transcriptional regulation and nongenomic actions of ER

In addition to directly contacting promoters of target genes, ER can also affect transcription of genes through a "nonclassical" mode of action, by binding to and influencing the activities of other transcription factors, such as NFκB and AP-1 (Fig. 3). For example, by binding to AP-1, ERα and ERβ can enhance transcription of the collagenase gene, which contains an AP-1 responsive promoter [29]. All ligands tested, including the SERMs tamoxifen, raloxifene, and GW7604 allow ER to regulate

AP-1 activity in this manner. Therefore, nonclassical modes of transcriptional regulation may need to be considered when evaluating the actions of ER ligands. To add further to the complexity of ER action, there is increasing evidence that ERα may have roles that do not require its transcriptional activation, such as modifying the activity of enzymes and ion channels. For example, liganded ERα but not ERβ has been found to increase nitric oxide (NO) release from endothelial cells, which may prevent leukocyte accumulation in the vasculature [30,31]. Additionally, ERα has also been shown to interact with and modulate the activity of the tyrosine kinase, c-src, which is involved in signal transduction pathways of cell cycle progression, cell proliferation, differentiation, adhesion, and migration [32]. However, more studies are needed to understand the potential biological significance of nongenomic ER actions.

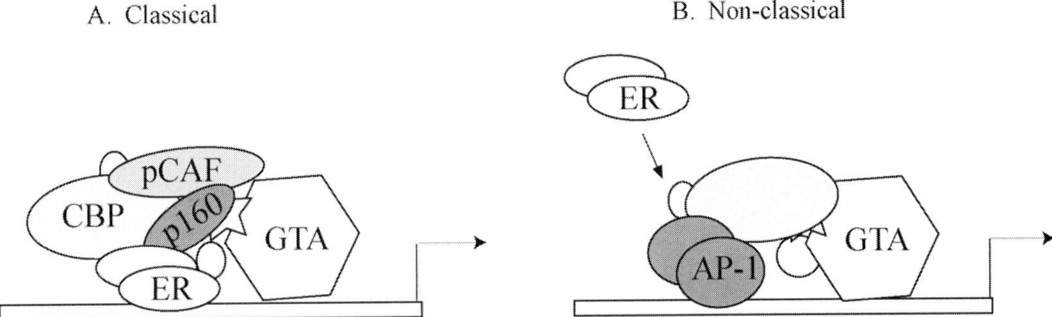

Fig. 3 Modes of ER action. (A) "Classical" mode of ER action occurs when ER directly binding to estrogen response elements (EREs) within the promoters of target genes. The coactivators pCAF and CBP, and a member of the p160 family of coactivators (SRC1, GRIP1, or ACTR) are tethered to ER and to the general transcription apparatus (GTA) to promote transcription. (B) "Nonclassical" mode of ER action occurs when ER regulates transcription in an indirect manner. ER binds to another transcription factors, such as AP-1, and either positively or negatively affects the activity of AP-1 on its target gene promoter.

PROGESTERONE RECEPTORS

Similar to ER, the human progesterone receptor (hPR) also exists in two forms, hPR-A and hPR-B (Fig. 4). Unlike ER, these two PR isoforms are derived from alternative start sites on the same gene. hPR-A and hPR-B are identical in amino acid sequence with the exception of an additional 164 aa on the N-terminus of hPR-B [33,34]. Despite their similarity, the two PR receptors have opposite transcriptional activities. hPR-B acts as a strong transcriptional activator, while hPR-A acts as only a weak activator or in many cases, a repressor of transcription [35–37]. Major target tissues of progesterone include those in the reproductive tract such as the ovary, uterus, and vagina, as well as the mammary gland, pituitary, and hypothalamus [38–42]. Progesterone is involved in maintaining female reproductive function and establishment and maintenance of pregnancy, making it a useful target for therapeutic intervention [43].

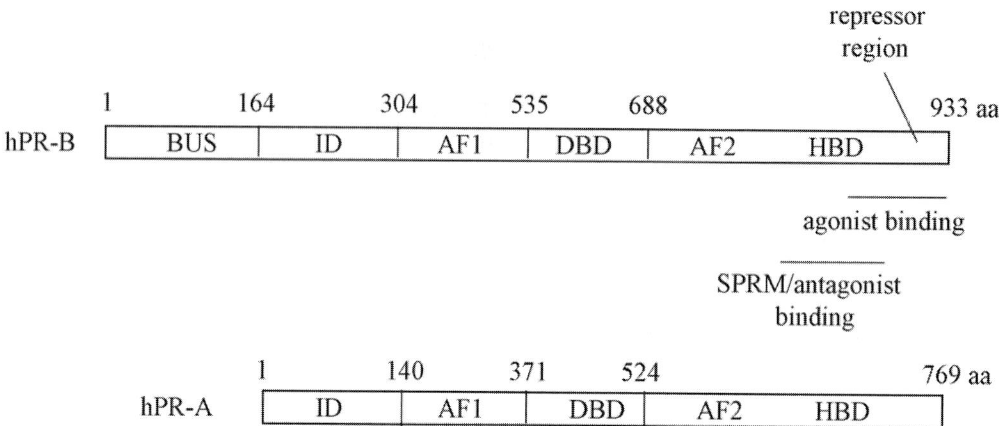

Fig. 4 Diagram of the two isoforms of human PR, hPR-B and hPR-A. These receptors are identical except for the 164 amino acid B-upstream sequence (BUS) on the N-terminus of hPR-B. The inhibitory domain (ID) confers a repressive function on hPR-A, causing this receptor to function as a repressor or weak activator of transcription. In hPR-B, the presence of the BUS overrides the function of the ID and allows hPR-B to function as a strong transcriptional activator. Also depicted on each receptor are two activation functions, AF-1 and AF-2, a DNA-binding domain (DBD), and hormone-binding domain (HBD). Residues located in the extreme C-terminus of each receptor (within the repressor region) are required for agonists to bind, whereas selective progesterone receptor modulators (SPRMs) and antagonists require residues more centrally located within the LBD of hPR-A or hPR-B.

PR domains

The primary sequences of PR-A and B contain the classic features of the nuclear receptor superfamily, but three additional features distinguish PR from other nuclear receptors (Fig. 4). These are (1) an amino-terminal inhibitory domain (ID), (2) a B upstream sequence or "BUS", which is only present in hPR-B, and (3) a small carboxy-terminal repressor region [36,44]. The ID, although present in both isoforms, functions as a repressor only within hPR-A. If the ID is removed, hPR-A turns into a strong transcriptional activator, similar to PR-B [36]. In the context of full-length hPR-B, however, the repressive influence of the ID is overcome by the BUS region, the N-terminal extension on PR-B that distinguishes it from PR-A. The BUS region allows hPR-B to function as an activator of transcription, most likely by imposing conformational changes that render the ID inactive [45]. Additionally, the BUS region may contain an additional AF between aa 54-154, called AF3, yet the relative contribution of this AF3 to the transcriptional activity of full-length hPR-B remains unclear [46]. The third feature of PR that distinguishes it from other nuclear receptors is its C-terminal repressor region. In an unliganded state, corepressor proteins are thought to bind to this region of the LBD and are only displaced by the binding of PR agonists. The binding of PR antagonists, however, does not displace these corepressors, and therefore they are able to contribute to the repression of the antagonist-bound receptor [47]. Like ER, PR activity is not only regulated by ligand binding, but it is also regulated by phosphorylation. To date, eight phosphorylation sites have been found on PR. Four of these sites, Ser 20, 81, 102, and 162 reside in the BUS region of hPR-B and the other four sites, Ser 190, 294, 345, and 400 are common to both hPR-A and B. The significance of phosphorylation of these residues is not entirely clear since only some of these residues appear to be required for PR activity [48]. Interestingly, some isoform preference for specific phosphorylation sites has been observed, and these sites may contribute to or reflect the different activities of hPR-A and B [49].

PR ligands

The physiological effects of progestins and antiprogestins can be difficult to interpret due to their potential to cross-react with other nuclear receptors, such as glucocorticoid, mineralocorticoid, and androgen receptors (GR, MR, and AR, respectively). The DBDs and LBDs of these receptors share a high degree of homology with PR, allowing ligands for one receptor to bind to and effect the activity of others. The noted cross-reactivity of ligands in the PR, GR, MR, and AR subfamily of receptors could confound the interpretation of results from whole animal studies set up to evaluate endocrine disruptors. In addition to cross-reactivity with other nuclear receptors, many PR ligands display agonist or antagonist effects in a cell- or tissue-specific manner. Analogous to SERMS, PR ligands with tissue-specific agonist/antagonist activities are now referred to as SPRMs for "selective progesterone receptor modulators". As with SERMs, SPRM action relies on the specific ligand-induced conformation of PR and the profile of cofactors available to modulate its activity (Fig. 5) [50–52]. In sum, the actions of PR ligands may be even more complex than those for ER because the actions of PR ligands depend on both their tissue-specific PR agonist or antagonist effects and their potential to cross-react with other nuclear receptors.

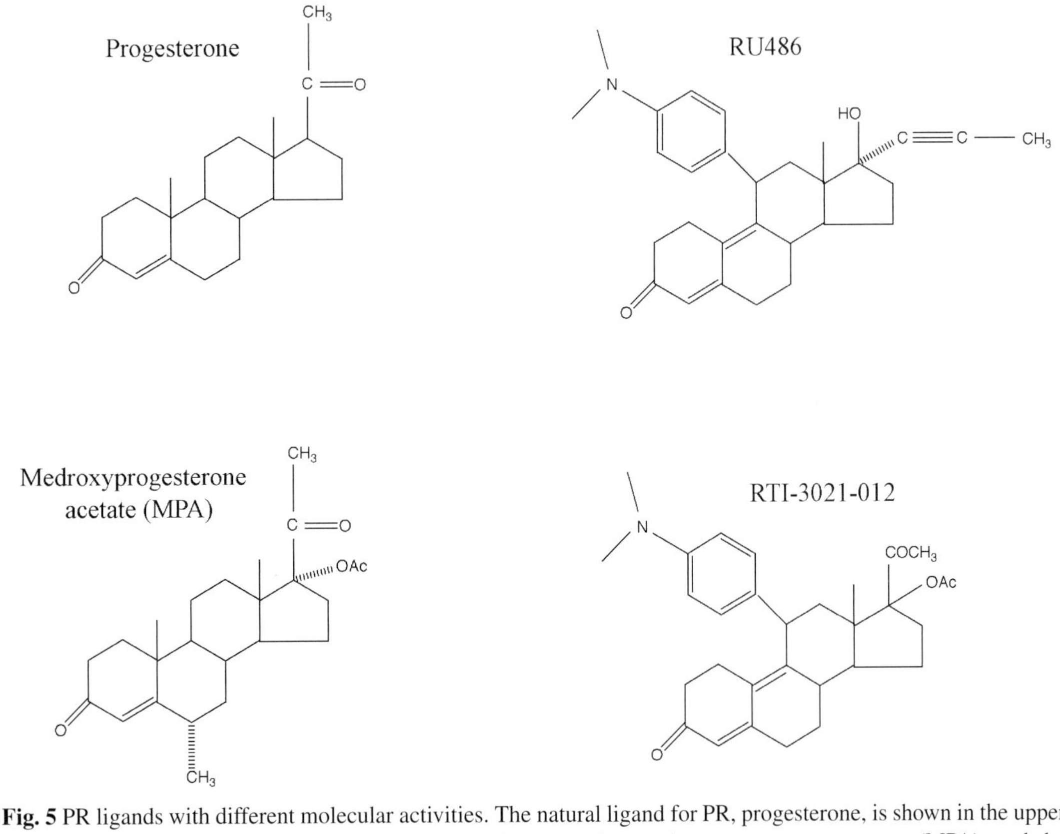

Fig. 5 PR ligands with different molecular activities. The natural ligand for PR, progesterone, is shown in the upper left-hand corner. Also shown are the clinically used PR agonist, medroxyprogesterone acetate (MPA), and the selective progesterone receptor modulators (SPRMs), RU486 and RTI-3021-012.

Antiestrogenic activities of PRs

The clinical uses of progestins rely primarily upon the ability of progestins to oppose ER activity [53,54]. The antiestrogenic effects of PR ligands can be accomplished through distinct processes regulated by either PR-A or B, the relative importance of which varies from cell to cell. For example, the estradiol-stimulated transcription of the lactoferrin gene is downregulated in mice when progesterone is administered. Studies with PR knock-out mice suggest that PR-B is responsible for this downregulation [38]. However, these studies also show that, in the lumen of the uterus, PR-A is responsible for opposing the estradiol-induced hyperproliferation of epithelial cells [38]. The antiestrogenic effects of PR may be accomplished by several different mechanisms. For example, PR-A or B may directly regulate the promoters of genes that are also regulated by ER. Alternatively, PR-A and B may regulate different genes whose protein products oppose the actions of those regulated by ER. Another possible mechanism by which PR could elicit its antiestrogenic effects is by interfering with the ability of ER to activate transcription. This nonclassical mode of PR action, termed "transrepression", has been observed in cell culture experiments for hPR-A but not hPR-B. When bound by either agonist or antagonist, hPR-A can inhibit or "transrepress" ER activity on ERE-containing reporter genes [35,55]. This PR-mediated transrepression occurs with both endogenously and exogenously expressed ERα and β in a variety of cell types, and the ID region within hPR-A seems to be important for this repressive function [36]. More studies are currently being performed in order to determine the importance of transrepression in the antiestrogenic activity of PR.

Nongenomic PR actions

Compelling evidence in support of nongenomic actions of PR has emerged recently. Specifically, it has been shown that both hPR A and B, in a ligand-dependent manner, can bind to the SH3 domain of the tyrosine kinase, c-src. The PR amino acid sequence contains an SH3 binding site in its AF-1 (aa 421-428, PPPPLPPR) [32]. Other nuclear receptors such as ER, GR, AR, and TR do not contain this SH3 binding site, although ER has been reported to bind the SH2 domain of the same protein, c-src. The interaction between PR and c-src appears to be rapid and transient and results in activation of c-src. However, the relative contributions of nongenomic actions to the biological function of PR remain to be determined.

VITAMIN D RECEPTORS

In contrast to ER and PR, the vitamin D receptor (VDR), a nonsteroid hormone nuclear receptor, does not form homodimers with itself, but by dimerizing with the retinoid X receptor (RXR) forms obligate heterodimers, enabling it to regulate transcription of target genes [56]. VDR's natural ligand, 1,25 dihydroxyvitamin D [1,25 $(OH)_2D_3$] is converted from its precursor, vitamin D, by two successive hydroxylations in the liver and kidney. When bound to VDR, the active ligand regulates transcription of genes involved in calcium and phosphate absorption and homeostasis. The main target tissues of VDR include the bone, where it functions in bone formation and remodeling, intestine, kidney, and skin [57]. In the intestine, however, a ligand for VDR other than 1,25 $(OH)_2D_3$ may also regulate its activity. Recently, it was discovered that VDR can bind lithocholic acid (LCA), a toxic bile acid that is produced in excess from continuous high-fat diets [58]. In the intestine, binding of LCA to VDR activates the receptor and enables it to upregulate expression of the p450 enzyme, cyp3A. By catabolizing LCA, cyp3A eliminates this toxic compound before it can accumulate in the colon. Adding to the biological function of VDR, recent research also indicates that VDR may also play a role in the immune, central nervous, muscle, and endocrine systems. On a molecular level, the 427 amino acid long VDR is the only one of its kind; no subtypes or isoforms of the receptor are known to exist. The DBD of VDR is located at the far amino terminus of the receptor and unlike ER and PR, VDR only contains one activation function,

AF-2, located within the LBD (Fig. 6). Several synthetic VDR ligands have been developed, and, similar to ER and PR ligands, many of these compounds display tissue-specific agonist/antagonist properties, making them selective vitamin D receptor modulators. Studies with deletion and mutant VDR proteins have revealed that the binding characteristics of agonists like 1,25 $(OH)_2D_3$ are different from those of mixed agonists/antagonists like ZK159222 (Fig. 6) and yet also different from the binding characteristics of antagonists, like ZK168281 [59]. Each of these three classes of VDR ligands can induce different conformational changes in the receptor, which alter the structure of the coactivator binding pocket and the positioning of helix 12 in the LBD. Similar to ER and PR, positioning of helix 12 regulates how well the liganded receptor will bind coactivator proteins. Additionally, the tissue-specific expression levels of coactivators and corepressors will help determine if a given VDR ligand will act as an agonist or antagonist. Regulation of VDR activity also depends on the phosphorylation state of the receptor. At least two serine residues within VDR are thought to be targets of phosphorylation. Ser 208 within the LBD represents the major phosphorylation site and contributes to the receptor's transactivation, while Ser 51 within the DBD may play a role in response element binding, nuclear localization, and transactivation [57]. Given VDR's important roles in growth, development, and differentiation, any chemicals that disrupt the receptor's normal activities could certainly impact the physiological well-being of an organism. However, the impact of interfering with VDR activity on endocrine function is not as clear as it is for ER or PR.

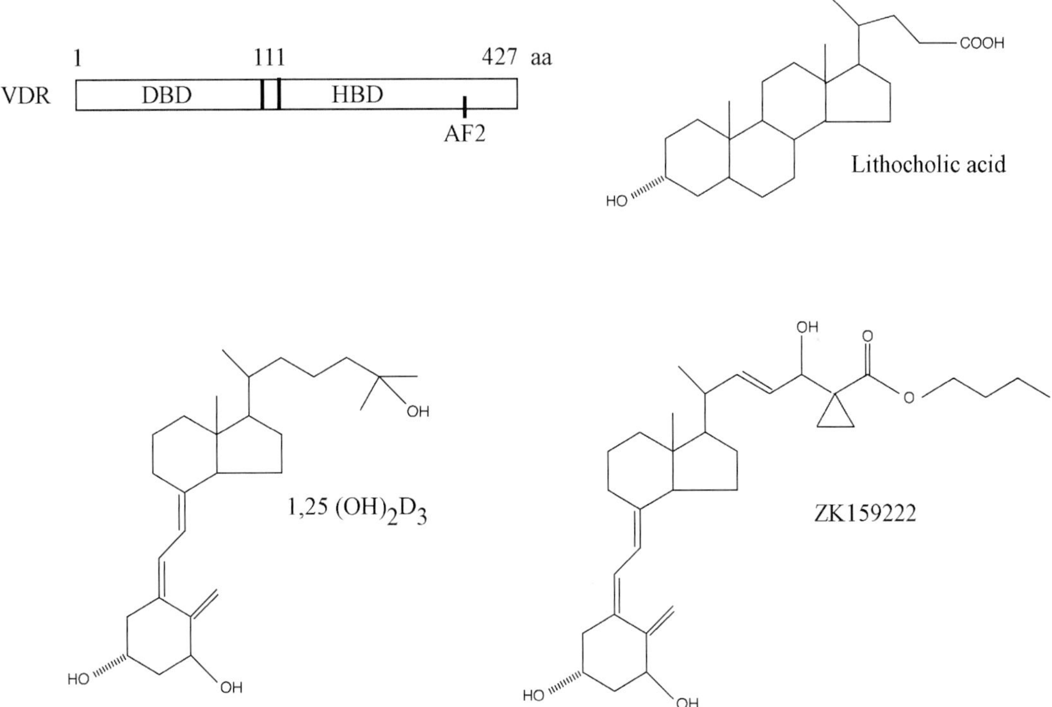

Fig. 6 (A) Diagram of the human VDR is shown in the upper left-hand corner. The natural VDR ligands, 1,25 $(OH)_2D_3$ and lithocholic acid are also shown. The synthetic VDR ligand, ZK159222, is shown in the bottom right-hand corner.

SCREENS FOR DETERMINING THE ROLE OF ER, PR, AND VDR IN ENDOCRINE DISRUPTION

The complexity of nuclear receptor pharmacology described in this topic illustrates the need for the development of screens that can accurately predict the effects of potentially hazardous compounds in different cell types. As we have discussed in this chapter, the ability of a compound to simply bind ER, PR, or VDR does not necessarily translate into its ability to affect the normal activities of the receptor. More accurately, the specific conformation of the receptor induced by binding of a particular ligand can help predict its effects. Screens that distinguish these various conformations can help determine the endocrine-disrupting potential of a new chemical. One such screen is the peptide-binding assay, which was recently developed through the isolation of ER-binding peptides from phage display analyses [27,28]. The binding of a SERM to ER changes the conformation of the receptor in such a way that each SERM/ER complex has a different profile of binding to various classes of peptides. The profile of peptide binding for a new or potentially hazardous compound could help categorize the compound as "tamoxifen-like" or "estradiol-like" depending on the similarities and differences in its peptide binding to those of known ER ligands. Another useful screen for predicting the effects of a potential endocrine active substance on ER, PR, or VDR activity is microarray analysis, which is covered in more detail in other topics. Given the sequencing of the entire human genome and advances in microarray chip design, thousands of genes from multiple tissue types can be classified according to their ability to be regulated by estrogens, progestins, or vitamin D. Comparison of these gene expression profiles to those induced by potentially hazardous chemicals can help predict their endocrine-disrupting abilities and should be a powerful tool for studying endocrine disruption in the future. However, these sorts of cell-based screening methods must be used in combination with other whole organism screens in order to take into account more global factors contributing to their endocrine-disrupting potential.

SUMMARY

This topic summarizes the molecular actions of the ligand-inducible transcription factors, estrogen, progesterone, and vitamin D receptors (ER, PR, and VDR). To begin with, two subtypes of ER, α and β, and two isoforms of PR, A and B, are responsible for responding to potential ligands of ER and PR. Their transcriptional regulatory function can occur through a direct mode, with the receptor binding to response elements within the promoters of target genes, or indirectly, through their effects on other transcription factors. The subtypes of ER and PR have different molecular activities, different expression patterns, and different expression levels throughout the body, which make it difficult to predict the effects of a given substance on ER or PR activity without taking into account which tissue is being considered. Although VDR only has one form, factors that determine its activity parallel many of those for ER and PR. To further complicate the actions of ER, PR, and VDR, the ligands that bind the receptors each do so in a slightly different manner. This imposes different conformational shapes on the receptor that ultimately affect the ability of coactivator and corepressor proteins to bind to and modulate the receptor's activity. The expression patterns and levels of coactivator and corepressor proteins as well as phosphorylation status of the receptor and its associated cofactors are also important regulators of the activities of ER, PR, and VDR. Therefore, the combination of many factors determines the potential endocrine-disrupting ability of a substance that acts through one of these receptors.

REFERENCES

1. E. V. Jensen and H. I. Jacobson. *Rec. Prog. Horm. Res.* **18**, 387–414 (1962).
2. G. G. J. M. Kuiper, E. Enmark, M. Pelto-Huikko, S. Nilsson, J.-A. Gustafsson. *Proc. Natl. Acad. Sci. USA* **93**, 5925–5930 (1996).
3. D. J. Mangelsdorf, C. Thummel, M. Beato, P. Herrlich, G. Schutz, K. Umesono, B. Blumberg, P. Kastner, M. Mark, P. Chambon, R. M. Evans. *Cell* **83**, 835–839 (1995).
4. A. K. Shiau, D. Barstad, P. M. Loria, L. Cheng, P. J. Kushner, D. A. Agard, G. L. Greene. *Cell* **95**, 927–937 (1998).
5. G. G. J. M. Kuiper, B. Carlsson, K. Grandien, E. Enmark, J. Häggblad, S. Nilsson, J.-A. Gustafsson. *Endocrinology* **138**, 863–870 (1997).
6. G. G. Kuiper, J. G. Lemmen, B. Carlsson, J. C. Corton, S. H. Safe, P. T. van der Saag, B. van der Burg, J. A. Gustafsson. *Endocrinology* **139**, 4252–4263 (1998).
7. A. C. W. Pike, A. M. Brzozowski, R. E. Hubbard, T. Bonn, A.-G. Thorsell, W. Engström, J. Ljunggren, J.-Å. Gustafsson, M. Carlquist. *EMBO J.* **18**, 4608–4618 (1999).
8. J. M. Hall and D. P. McDonnell. *Endocrinology* **140**, 5566–5578 (1999).
9. I. Choi, C. Ko, O.-K. Park-Sarge, R. Nie, R. A. Hess, C. Graves, B. A. Katzenellenbogen. *Mol. Cell Endocrinol.* **181**, 139–150 (2001).
10. J. D. Norris, D. Fan, M. R. Stallcup, D. P. McDonnell. *J. Biol. Chem.* **273**, 6679–6688 (1998).
11. J. DiRenzo, Y. Shang, M. Phelan, S. Sif, M. Myers, R. Kingston, M. Brown, *Mol. Cell Biol.* **20**, 7541–7549 (2000).
12. J. C. G. Blanco, S. Minucci, J. Lu, X.-J. Yang, K. K. Walker, H. Chen, R. M. Evans, Y. Nakatani, K. Ozato. *Genes Dev.* **12**, 1638–1651 (1998).
13. H. Ichinose, J. M. Garnier, P. Chambon, R. Losson. *Gene* **188**, 95–100 (1997).
14. N. J. McKenna, R. B. Lanz, B. W. O'Malley. *Endocrine Rev.* **20**, 321–344 (1999).
15. W. Feng, P. Webb, P. Nguyen, X. Liu, J. Li, M. Karin, P. J. Kushner. *Mol. Endocrinol.* **15**, 32–45 (2001).
16. A. Tremblay and V. Giguere. *J. Steroid Biochem. Mol. Biol.* **77**, 19–27 (2001).
17. S. Kato, H. Endoh, Y. Masuhiro, T. Kitamoto, S. Uchiyama, H. Sasaki, S. Masushige, Y. Gotoh, E. Nishida, H. Kawashima. *Science* **270**, 1491–1494 (1995).
18. R. B. Lanz, N. J. McKenna, S. A. Onate, U. Albrecht, S. Y. Tsai, M.-J. Tsai, B. W. O'Malley. *Cell* **97**, 17–27 (1999).
19. H. Endoh, K. Maruyama, Y. Masuhiro, Y. Kobayashi, M. Goto, H. Tai, J. Yanagisawa, D. Metzger, S. Hashimoto, S. Kato. *Mol. Cell Biol.* **19**, 5363–5372 (1999).
20. J. D. Norris, D. Fan, A. Sherk, D. P. McDonnell. *Mol. Endocrinol.* **16**, 459–468 (2002).
21. A. Marimuthu, W. Feng, T. Tagami, H. Nguyen, J. L. Jameson, R. J. Fletterick, J. D. Baxter, B. L. West. *Mol. Endocrinol.* **16**, 271–286 (2002).
22. X. Hu and M. A. Lazar. *TEM* **11**, 6–10 (2000).
23. R. M. Lavinsky, K. Jepsen, T. Heinzel, J. Torchia, T.-M. Mullen, R. Schiff, A. L. Del-Rio, M. Ricote, S. Ngo, J. Gemsch, S. G. Hilsenbeck, C. K. Osborne, C. K. Glass, M. G. Rosenfeld, D. W. Rose. *Proc. Natl. Acad. Sci. USA* **95**, 2920–2925 (1998).
24. I. Zamir, J. Dawson, R. M. Lavinsky, C. K. Glass, M. G. Rosenfeld, M. A. Lazar. *Proc. Natl. Acad. Sci. USA* **94**, 14400–14405 (1997).
25. Y. Shang and M. Brown. *Science* **295**, 2465–2468 (2002).
26. D. P. McDonnell, D. L. Clemm, T. Hermann, M. E. Goldman, J. W. Pike. *Mol. Endocrinol.* **9**, 659–668 (1995).
27. J. D. Norris, L. A. Paige, D. J. Christensen, C.-Y. Chang, M. R. Huacani, D. Fan, P. T. Hamilton, D. M. Fowlkes, D. P. McDonnell. *Science* **285**, 744–746 (1999).
28. L. A. Paige, D. J. Christensen, H. Grøn, J. D. Norris, E. B. Gottlin, K. M. Padilla, C.-Y. Chang, L. M. Ballas, P. T. Hamilton, D. P. McDonnell. *Proc. Natl. Acad. Sci. USA* **96**, 3999–4004 (1999).

29. P. Webb, G. N. Lopez, R. M. Uht, P. J. Kushner. *Mol. Endocrinol.* **9**, 443–456 (1995).
30. T. Simoncini, A. Hafezi-Moghadam, D. P. Brazil, K. Ley, W. W. Chin, J. K. Liao. *Nature* **407**, 538–541 (2000).
31. C. Pendaries, B. Darblade, P. Rochaix, A. Krust, P. Chambon, K. S. Korach, F. Bayard, J.-F. Arnal. *Proc. Natl. Acad. Sci.* **99**, 2205–2210 (2002).
32. V. Boonyaratanakornkit, M. P. Scott, V. Ribon, L. Sherman, S. M. Anderson, J. L. Maller, W. T. Miller, D. P. Edwards. *Mol. Cell* **8**, 269–280 (2001).
33. P. Kastner, A. Krust, B. Turcotte, U. Stropp, L. Tora, H. Gronemeyer, P. Chambon. *EMBO J.* **9**, 1603–1614 (1990).
34. K. B. Horwitz and P. S. Alexander. *Endocrinology* **113**, 2195–2201 (1983).
35. D. X. Wen, Y.-F. Xu, D. E. Mais, M. E. Goldman, D. P. McDonnell. *Mol. Cell Biol.* **14**, 8356–8364 (1994).
36. P. H. Giangrande, G. Pollio, D. P. McDonnell. *J. Biol. Chem.* **272**, 32889–32900 (1997).
37. P. H. Giangrande, E. A. Kimbrel, D. P. Edwards, D. P. McDonnell. *Mol. Cell. Biol.* **20**, 3102–3115 (2000).
38. B. Mulac-Jericevic, R. A. Mullinax, F. J. DeMayo, J. P. Lydon, O. M. Conneely. *Science* **289**, 1751–1754 (2000).
39. D. P. McDonnell. *Trends Endocrinol. Metab.* **6**, 133–138 (1995).
40. J. D. Graham, C. Yeates, R. L. Balleine, S. S. Harvey, J. S. Milliken, A. M. Bilous, C. L. Clarke. *Cancer Res.* **55**, 5063–5068 (1995).
41. R. S. Carroll, M. Brown, J. Zhang, J. Direnzo, J. F. D. Mora, P. M. Black. *Clin. Cancer Res.* **6**, 3570–3575 (2000).
42. K. W. Cheng, C.-K. Cheng, P. C. K. Leung. *Mol. Endocrinol.* **15**, 2078–2092 (2001).
43. R. L. Robker, D. L. Russell, S. Yoshioka, S. C. Sharma, J. P. Lydon, B. W. O'Malley, L. L. Espey, J. S. Richards. *Steroids* **65**, 559–570 (2000).
44. J. Xu, Z. Nawaz, S. Y. Tsai, M.-J. Tsai, B. W. O'Malley. *Proc. Natl. Acad. Sci. USA* **93**, 12195–12199 (1996).
45. M. J. Tetel, P. H. Giangrande, S. A. Leonhardt, D. P. McDonnell, D. P. Edwards. *Mol. Endocrinol.* **13**, 910–924 (1999).
46. L. Tung, T. Shen, M. G. Abel, R. L. Powell, G. S. Takimoto, C. A. Sartorius, K. B. Horwitz. *J. Biol. Chem.* **276**, 39843–39851 (2001).
47. B. L. Wagner, G. Pollio, P. Giangrande, J. C. Webster, M. Breslin, D. E. Mais, C. E. Cook, W. V. Vedeckis, J. A. Cidlowski, D. P. McDonnell. *Endocrinology* **140**, 1449–1458 (1999).
48. D. L. Clemm, L. Sherman, V. Boonyaratanakornkit, W. T. Schrader, N. L. Weigel, D. P. Edwards. *Mol. Endocrinol.* **14**, 52–65 (2000).
49. S. Kazmi, V. Visconti, R. Plante, A. Ishaque, C. Lau. *Endocrinology* **133**, 1230–1238 (1993).
50. B. L. Wagner, G. Pollio, S. Leonhardt, M. C. Wani, D. Y.-W. Lee, M. O. Imhof, D. P. Edwards, C. E. Cook, D. P. McDonnell. *Proc. Natl. Acad. Sci. USA* **93**, 8739–8744 (1996).
51. B. L. Wagner, J. D. Norris, T. A. Knotts, N. L. Weigel, D. P. McDonnell. *Mol. Cell. Biol.* **18**, 1369–1378 (1998).
52. G. Giannoukos, D. Szapary, C. L. Smith, J. E. Meeker, S. Stoney Simons, Jr. *Mol. Endocrinol.* **15**, 255–270 (2001).
53. J. G. M. Klign, B. Setyono-Han, J. A. Foekens. *Steroids* **65**, 825–830 (2000).
54. S. G. Lundeen, Z. Zhang, Y. Zhu, J. M. Carver, R. C. Winneker. *J. Steroid Biochem. Mol. Biol.* **78**, 137–143 (2001).
55. D. P. McDonnell, M. S. Shahbaz, E. Vegeto, M. E. Goldman. *J. Steroid Biochem. Mol. Biol.* **48**, 425–432 (1994).
56. D. P. McDonnell, D. J. Mangelsdorf, J. W. Pike, M. R. Haussler, B. W. O'Malley. *Science* **235**, 1214–1217 (1987).

57. P. W. Jurutka, G. K. Whitfield, J.-C. Hsieh, P. D. Thompson, C. A. Haussler, M. R. Haussler. *Rev. Endocr. Metab. Disorders* **2**, 203–216 (2001).
58. M. Makishima, T. T. Lu, W. Xie, G. K. Whitfield, H. Domoto, R. M. Evans, M. R. Haussler, D. J. Mangelsdorf. *Science* **296**, 1313–1316 (2002).
59. S. Vaisanen, M. Perakyla, J. I. Karkkainen, A. Steinmeyer, C. Carlberg. *J. Mol. Biol.* **315**, 229–238 (2002).

Pure Appl. Chem., Vol. 75, Nos. 11–12, pp. 1685–1697, 2003.

Topic 1.4

Biological function and mode of action of the androgen receptor*

Elizabeth M. Wilson

Laboratories for Reproductive Biology, and the Department of Pediatrics and the Department of Biochemistry and Biophysics, University of North Carolina, Chapel Hill, NC 27599, USA

Abstract: Chemical contaminants with antiandrogen or androgen activity have been identified in the environment. The mode of action of these endocrine disruptors derives from their ability to bind to the androgen receptor (AR), a member of the steroid receptor family of nuclear receptors. The AR is a ligand-activated transcription factor with properties unique among members of the steroid receptor family. Binding of endocrine disruptors to the AR impacts its ability to activate or inhibit AR-regulated genes. Most notable of these are the agonist-induced and antagonist-inhibited interdomain interactions that influence AR stability and function. Environmental antiandrogens identified thus far are metabolites of pesticides and herbicides, exerting androgen antagonist effects by blocking AR-induced gene transcription required for male sexual development. Environmental androgens can be precursors of the naturally occurring biologically active androgens testosterone and dihydrotestosterone or result from anabolic steroid use in the livestock industry. They have agonist activity by their ability to bind the AR and mimic the natural hormone, increasing AR-mediated transcription of androgen-responsive genes. The presence of masculinized female fish in polluted rivers indicates the presence of androgen-like pollutants in the environment.

STRUCTURAL AND FUNCTIONAL PROPERTIES OF THE ANDROGEN RECEPTOR

Domain structure of the androgen receptor

The androgen receptor (AR) is encoded by a single gene on the X chromosome at Xq11-12 [1,2]. The AR consists of 919 amino acid residues [3] and occurs as a single form in mammalian cells. The precise amino residue length of the AR can vary, however, due to the presence of polymorphic amino acid repeats. In contrast, most other steroid receptors are coded by more than one gene and are expressed in multiple forms. Previous reports of A and B forms of the AR appear to result from proteolytic breakdown during tissue preparation [4]. AR is sensitive to degradation during its isolation, and in some cases is not protected from degradation by the addition of protease inhibitors [5]. AR degradation is minimized by the binding of high-affinity androgens [6]. The overall structural arrangement of the AR is similar to other steroid receptors with several relatively autonomous domains contributing a specific function. These include the NH_2-terminal region, the DNA-binding domain, and the carboxyl-terminal ligand-binding domain.

*Report from a SCOPE/IUPAC project: Implication of Endocrine Active Substances for Human and Wildlife (J. Miyamoto and J. Burger, editors). Other reports are published in this issue, *Pure Appl. Chem.* **75**, 1617–2615 (2003).

NH_2-terminal region

The AR NH_2-terminal region is comprised of 558 amino acid residues and contains the major transactivation domain [7] referred to as activation function 1 (AF1). This is a poorly defined region that lies within amino acid residues 142–337 [8]. AF1 appears to be the predominant activation region in AR and depends on androgen binding for activation [7].

The NH_2-terminal region contains in addition several amino acid repeat sequences, the most notable being the CAG trinucleotide repeat that codes for a stretch of glutamine residues. The NH_2-terminal region also contains a glycine repeat that has a more constant length of about 23 residues. The function of the glutamine and glycine repeat in AR activity is not known. The CAG repeat region coding for polyglutamine is polymorphic in the normal population. Shorter CAG repeat lengths have an increased association with prostate cancer [9]. The AR CAG repeat length is relatively invariant even in its expanded form [10] and has been used as a marker in human population genetics. CAG repeat expansion to more than 40 glutamine residues results in Kennedy's disease or adult onset spinal bulbar muscular atrophy [11]. While the direct relationship between glutamine repeat length and Kennedy's disease is not understood, the expanded CAG repeat is associated with decreased expression of the AR messenger RNA, resulting in reduced AR protein levels in transfection assays [12]. Trinucleotide repeat expansions in other genes are associated with neurological diseases such as Huntington's disease, the fragile X syndrome, and myotonic dystrophy, where repeat sequences occur in coding or noncoding regions depending on the gene [9]. The AR CAG trinucleotide repeat length has undergone an exponential increase through evolution of the primates back to the lemur [13]. The rat AR sequence has a short glutamine repeat at the position of the human AR CAG repeat, but remarkably, has a CAG repeat expansion at another region of the AR NH_2-terminal domain [14].

The AR has multiple phosphorylation sites that occur predominantly in the NH_2-terminal region and include Ser-Pro sequences [15]. Thus far, specific functions have not been attributed to AR phosphorylation or to phosphorylation of the steroid hormone receptors in general. While all of the steroid receptors contain multiple phosphorylation sites, most studies have failed to define their functional significance.

DNA-binding domain

The central region of the AR is the DNA-binding domain. It consists of amino acid residues 559–624 in human AR. The DNA-binding domain is comprised of two zinc-finger structures typical of steroid receptors [16] and mediates AR binding to DNA as a dimer [17]. The DNA-binding domain binds androgen response element DNA to mediate androgen-induced gene transcription [18,19]. Some of the androgen response elements associated with androgen-regulated genes are simple elements that conform to the consensus sequence 5′GGTACAnnnTGTTCT3′. Others have more complex sequences. A number of androgen-regulated genes have been identified that contain simple or complex androgen response elements [20–23], including the androgen responsive regions of prostate specific antigen [24], probasin [25,26], and the *Slp* gene [27].

The DNA-binding domain also functions in interactions with coactivators such as protein inhibitor of activated STAT-1, which increases AR-mediated gene activation [28,29]. Another DNA-binding domain interacting protein is a small nuclear RING finger protein which increased AR transactivation [30].

The DNA-binding domain is contiguous with a hinge region that spans amino acid residues 625–675. The function of the hinge region is not well understood. Spanning the DNA binding and hinge regions is a bipartite nuclear targeting signal composed of two clusters of basic amino acids [31]. In transiently transfected COS cells, androgen binding activates the nuclear targeting signal to direct subcellular localization of the AR from the cytoplasm to the nucleus [7,31]. The subcellular localization of the endogenous AR has been difficult to establish, however. In studies using tissue sections, endogenously expressed AR is nuclear in the presence of androgen, but is not easily detected in the absence of androgen [32], most likely due to its rapid degradation [6]. In normal skin fibroblasts and some cancer-derived cell lines, the AR is nuclear in the absence of dihydrotestosterone (DHT) [33].

Ligand-binding domain

The carboxyl-terminal region from amino acid residues 676–919 makes up the ligand-binding domain. Structural aspects of the ligand-binding domain of many steroid receptors have been elucidated through crystallization studies as recently reported for the AR [34,35]. Crystal structure analysis has revealed multiple alpha helices folded to create a hydrophobic pocket for hormone binding. Binding of the steroid completes the hydrophobic center of the ligand-binding domain and repositions helix 12 [36]. In the presence of agonist binding, a hydrophobic binding surface forms in the ligand-binding domain known as activation function 2 (AF2). AF2 binds the p160 family of coactivators which have histone acetyltransferase activity. Binding occurs through the L*XX*LL motifs of the p160 coactivators, where L is leucine and *X* is any amino acid. The AF2 binding surface for some receptors also serves as the binding site for the L*XX*LL-like sequences present in corepressors. The nature of the ligand bound in the ligand-binding pocket, whether agonist or antagonist, determines the conformation of AF2 to favor either coactivator or corepressor binding.

Steroid-binding specificity and kinetics

The AR binds the biologically active androgens, testosterone, and DHT, with the same apparent equilibrium binding affinity of Kd 0.1–0.3 nM. Yet, DHT is a more effective androgen than testosterone in part because it dissociates more slowly from the AR. The dissociation half-time of DHT is about 3 times slower than that of testosterone [37]. The similar binding affinity and different dissociation rates was supported by differences in androgen association rates [37]. Other steroids such as estradiol and progesterone bind the AR with lower affinity, causing AR nuclear transport, but are less effective in activating reporter vectors in transient transfection studies, requiring high concentrations [6]. Recent studies indicate that the slow dissociation rate of bound androgen results in part from an NH_2-terminal and carboxyl-terminal (N/C) interaction discussed below.

Androgen antagonists bind the AR and inhibit androgen-induced gene activation. The most notable AR antagonist is hydroxyflutamide, which is the active metabolite of flutamide, a pharmaceutical drug administered for the treatment of prostate cancer. Hydroxyflutamide binds AR with an apparent binding affinity of 175 nM [38], which is 3 orders of magnitude weaker than the binding affinity of testosterone and DHT. Other pharmaceutical antagonists in use in prostate cancer therapy include casodex (bicalutamide) and nilutamide. The mechanism of inhibition of these antagonists is competition for androgen binding and inhibition of AR DNA binding, as shown for hydroxyflutamide [39]. One antagonist, RU56187, binds the AR with an affinity (Kd 0.39 nM) similar to that for testosterone and DHT [38]. However, the dissociation half-time of [^{3}H]RU56187 is 5 min at 35 °C compared to 2.5 h for [^{3}H]R1881 [38]. Thus, one distinguishing feature of AR agonists and antagonists is a more rapid dissociation rate of antagonists. This occurs in part because antagonists fail to induce the N/C interaction [40]. Depending on the binding affinity, antagonists require sufficiently high concentrations to compete for binding of the active androgens to block agonist-induced gene transcription. Environmental antiandrogens bind the AR with relatively low affinity and would be expected to have rapid binding and dissociation kinetics. Environmental antiandrogens, therefore, require relatively high concentrations in accordance with their lower binding affinity to elicit an antagonistic effect.

Androgen receptor stabilization and the N/C interaction

Several years ago, we discovered an NH_2-terminal and carboxyl-terminal (N/C) interaction in the AR [17,40]. The N/C interaction is remarkably specific in that it is induced by ligands that display androgen agonist activity in vivo, such as testosterone and DHT [41]. Furthermore, the agonist-induced N/C interaction is inhibited by antagonist binding [40]. Thus, the ligand requirements for the N/C interaction reflect the biological activity of AR as observed in vivo suggesting it is important to AR function. Based on studies using mutant receptors that cause the androgen insensitivity syndrome, we proposed a model for an antiparallel dimer of the AR where the N/C interaction directly participates in AR dimerization in association with DNA binding [40,42]. Studies by other laboratories have confirmed the N/C interaction in the AR [43–49] and for other nuclear receptors [50–53].

Recent studies have localized the regions in the AR ligand-binding domain and NH_2-terminal region required for the N/C interaction. Through the use of selective AR mutants that cause the androgen insensitivity syndrome, we showed that the binding site in the ligand-binding domain for the NH_2-terminal region overlaps with the binding site for the p160 coactivators [54]. The coactivator binding site in the ligand-binding domain is known as activation function 2 (AF2). Interaction of the p160 coactivators with AF2 is mediated by multiple L*XX*LL motifs. We searched for similar motif sequences in the AR NH_2-terminal region. Two motifs were identified, F*XX*LF and W*XX*LF, with the sequences ^{23}FQNLF^{27} and ^{433}WHTLF^{437}, where F is phenylalanine and W is tryptophan. Each binding motif is predicted to form an amphipathic alpha helix and interact with AF2 in the presence of androgen to mediate the N/C interaction [55]. An interesting aspect of these studies is that the F*XX*LF motif is present in several previously reported AR coactivators [56]. The studies indicated that the F*XX*LF interacting motif is specific for the AR.

Because of the overlapping binding sites present in AF2 for the L*XX*LL motif of p160 coactivators and the F*XX*LF motif in the AR, we determined whether competitive inhibition occurs. We showed that the N/C interaction inhibits the recruitment of p160 coactivators [57]. Thus, one function of the agonist-induced N/C interaction may be to limit AR activation through AF2. On the other hand, there may be a temporal sequence of interactions between AF2 and the F*XX*LF and L*XX*LL motifs that does not preclude AR activation by p160 coactivators through the AF2 domain. Other coactivators and interacting proteins that contain the F*XX*LF motif may interact with the AF2 binding surface at different times during the process of AR-mediated gene activation.

Recent studies showed that the androgen-dependent N/C interaction mediated by the F*XX*LF and W*XX*LF motifs is required for the activation of several naturally occurring androgen-regulated genes including PSA and probasin [58]. This was in striking contrast to the mouse mammary tumor virus long terminal repeat and *Slp* androgen-responsive regions, which showed less dependence on the N/C interaction for activation. It is not known what distinguishing properties of these androgen-responsive regions require an AR N/C interaction for transactivation.

With few exceptions, compounds with AR agonist activity in vivo induce the N/C interaction. This includes the lower-affinity anabolic steroids such as oxandrolone and fluoxymesterone, which bind AR with an equilibrium binding affinity of 62 and 44 nM, respectively [38]. These are potent androgens in vivo, supporting an in vivo role for the N/C interaction in AR activity. This contrasts other steroids such as estradiol and progesterone, which bind AR with a similar moderate affinity, but which are ineffective in activating the AR in vivo. Specificity of the N/C interaction for agonists and antagonists makes it a useful assay to screen compounds for endocrine disruptor activity. The two hybrid mammalian cell assay used for the N/C interaction was recently described in detail [59]. It utilizes the GAL4 DNA-binding domain fusion protein with the AR ligand-binding domain and the VP16 transactivation domain fusion protein with the AR NH_2-terminal and DNA-binding domains. The reporter vector contains multiple binding sites for the GAL4 DNA-binding domain and the luciferase gene for detection.

Nuclear coactivators of the androgen receptor

Many proteins have been reported to interact with the AR and function as AR coactivators. Some of these include Ets [60], c-Jun [61,62], TFIIF [63], TFIIH [64], retinoblastoma protein [65], CREB-binding protein [66], BAG-1L [67], ARA70/ELE1α [68–70], ARIP3 [71], p160s [72,73], ARA160 [74], ARA54 [75], ARA55/Hic5 [76,77], β-catenin [78,79], FHL2 [80], HBO1 [81], cyclin E [82], cyclin D1 [83], PDEF [84], BRCA1 [85], P-TEFb [86], caveolin-1 [87], and ARA24 [88]. Many of the reports utilized yeast two hybrid screening methods to detect interacting proteins and made use of transient transfection studies to demonstrate increased transcriptional activity mediated by the AR. There are as yet few or no studies that directly link an AR interacting protein with AR functional activity in vivo. Knock-out studies of the p160 coactivators in mice have failed to significantly impact male reproductive development and function [89–91]. This type of study is complicated, however, by the compensating effects of other coactivators that may become expressed at increased levels. Another complication in determining the functional significance of an AR interacting protein results from limitations of conditions used for transient transfection assays. We [56] and others [92] have observed that the use of balancing DNA additions in transient cotransfection assays inhibit AR transcriptional activity in controls, but not necessarily in samples expressing the coactivator being tested. The mechanism of inhibition has not been defined, but may include inhibition of AR expression [92] or inhibition of AR functional activity. The result is an apparent increase in AR-mediated transactivation by the putative coactivator under study, which may not be observed without the inhibitory effect. Conclusions regarding many of the interacting proteins that show apparent increases in AR transcriptional activity, therefore, require further study to clearly establish their role in AR function.

ANDROGEN RECEPTOR GENE MUTATIONS

Androgen insensitivity syndrome

Male sexual development depends on androgen-induced gene transcription. This occurs in the human embryo between 8 and 12 weeks of gestation. At this critical period, the external genitalia of the male forms, and interference with androgen action at this time results in different degrees of incomplete masculinization. At birth, the androgen-insensitive genetic male external genital phenotype can vary from near complete masculinization to ambiguous genitalia, or an apparent normal female. The requirement for the AR in this process has been well documented in the androgen insensitivity syndrome. In this syndrome, 46XY genetic males have a mutation in the AR gene, which results in different degrees of loss of AR function depending on the specific mutation [93]. There have been approximately 230 different mutations identified in the AR gene that cause androgen insensitivity. These are summarized at <www.mcgill.ca/androgendb/>.

AR mutations that cause androgen insensitivity include partial or complete AR gene deletions [94,95] and nonsense or missense mutations [96–99]. Frameshift mutations that result in the premature stop codon sequence destabilize the messenger RNA and can result in a truncated AR protein [100]. In cases of partial androgen insensitivity, the function of the AR is not entirely disrupted, resulting in different degrees of incomplete virilization in the newborn [101]. Thus, the full function of the AR is required for male sexual development to occur. Exposure to sufficient levels of environmental antiandrogens could have a similar effect of inhibiting AR-mediated gene transcription with partial or complete interference of male sexual development in utero. The androgen insensitivity syndrome, therefore, unequivocally demonstrates the requirement for the AR in male sexual development. Similar phenotypic disruptions could result by interrupting AR-mediated gene activation by the presence of environmental androgen antagonists.

Androgen receptor in prostate cancer

Androgens are required for prostate growth and likely have role in excessive tissue growth associated with benign prostate hyperplasia and prostate cancer [33]. Accordingly, antiandrogens are frequently used in the treatment of prostate cancer. Antiandrogens at moderate concentrations block androgen-induced stimulation of the prostate [102]. However, the lower-affinity environmental antiandrogens or androgens likely have little impact on benign prostatic hyperplasia or prostate cancer. Some reports have suggested that inappropriate exposure in utero to environmental estrogenic toxicants can increase prostate growth and development later in life [103], although the significance of lose-dose effects to environmental contaminants has recently been challenged [104]. Adult males are exposed to relatively high circulating androgen levels that would not be influenced by environmental antiandrogens or androgens. It should nevertheless be considered that some stages of prostate cancer are sensitive to low concentrations of androgen through mechanisms not yet understood [33,105]. In patients treated with androgen withdrawal therapy, exposure to androgens could be detrimental. High p160 coactivator expression has been reported in recurrent prostate cancer in the androgen-withdrawn patient, which may contribute to an increased sensitivity to androgen [105]. It is unlikely that environmental androgens or antiandrogens would be in sufficiently high concentrations to impact these examples of increased prostate growth.

RISK ASSESSMENT OF ANTIANDROGENS AND ANDROGENS AS ENVIRONMENTAL ENDOCRINE DISRUPTORS

Environmental antiandrogens

A number of chemicals used in agricultural weed and pest control have been identified as precursors to environmental antiandrogens. The pesticides *p,p′*-DDT [1,1,1-trichloro-2,2-bix(*p*-chlorophenyl)ethane] [106] and methoxychlor [2,2-bis (*p*-methoxyphenol)-1,1,1-trichloroethane] [107] are metabolized to more potent compounds with androgen antagonist activity. The AR antagonist activity of *p,p′*-DDT derives from its metabolite *p,p′*-DDE [1,1-dichloro-2,2-bis(*p*-chlorophenyl)ethylene], which binds the AR with an apparent equilibrium binding affinity of Kd 3.5 μM [106]. The potency of *p,p′*-DDE as an endocrine disruptor is increased because of its long biological half-life, which was estimated to be only slightly less than the average human life span. The long half-life of *p,p′*-DDE derives from its solubility in fat stores from where it is released during mobilization of fat and lactation. The active metabolite of methoxychlor is HPTE [2,2-bis (*p*-hydroxyphenol)-1,1,1-trichloroethane], which binds AR with an equilibrium binding affinity of Kd 1.4 μM [107]. HPTE has a high turnover rate in the body and thus a lower potential as an endocrine disruptor in human development and health. While *p,p′*-DDT is no longer in use in the United States as a pesticide, methoxychlor is an actively used pesticide and thus could be a source of antiandrogen activity in the environment. The fungicide vinclozolin is currently in use in the United States. Vinclozolin [3-(3,5-dichlorophenyl)-5-methyl-5-vinyloxazolidine-2,4-dione] is metabolized to M2 (3′,5′-dichloro-2-hydroxy-2-methylbut-3-enanilide), an androgen antagonist that binds the AR with an affinity of Kd 9.7 μM [39,108]. Administration of vinclozolin to pregnant rats inhibits male reproductive tract development in male offspring, indicating its androgen antagonist activity [109,110].

Other potential environmental antiandrogen precursors include procymidone [*N*-(3,5-dichlorophenyl)-1,2-dimethylcyclopropane-1,2-dicarboximide], a fungicide with antiandrogen activity in experimental animals [111]. The active metabolite of procymidone remains uncharacterized. Other environmental contaminants disrupt male reproductive development through mechanisms that are thought to be independent of the AR. These include di(*n*-butyl) phthalate, a widely used commercial plasticizer present in fatty foods [112]. Dibutylphthalate is metabolized to the active mono (*n*-butyl) phthalate and inhibits testosterone biosynthesis [112,113].

The reported increased incidence of male reproductive tract abnormalities in the human population in some parts of the world has raised concerns about the role for environmental contaminants with antiandrogen activity. The question of declining sperm counts worldwide or regionally was recently addressed. Most evidence supports regional declines, but controversy remains over the reliability of historical data [114]. The role of environmental antiandrogens in inhibiting spermatogenesis has not been addressed, but could reflect global effects of environmental contaminants. Another male developmental abnormality of incomplete masculinization with reported increased frequency is hypospadias deformity, where the urethra opens on the shaft or base of the penis rather than at the tip [115]. Hypospadias is a relatively common developmental abnormality of the human male fetus, occurring at a rate as high as 40 in 10 000 newborn male infants [115,116]. In the United States, the incidence of hypospadias reached a high level in the 1980s, but has since declined [117]. It is not clear what role endocrine disruptors have in the etiology of hypospadias in the newborn.

Environmental androgens

Effluent from paper mills flowing into a Florida river was postulated to contain substances with androgen activity based on the masculinization of entire populations of *Gambusia affinis holbrooki* female mosquitofish [118]. One active substance was identified as androstenedione present in the river at concentrations of 0.14 nM [119]. Previous studies showed that short exposure to higher concentrations of androstenedione masculinized the anal fin or gonopodium used in male reproduction [118]. Thus, aquatic wildlife in areas of paper mill discharge are at increased risk of reproductive tract abnormalities due to the presence of androgen precursors in the river water [120]. It was postulated that the abundant phytosterols from pine tree oils, including β-sitosterol, campesterol and stigmastanol, act as substrates for bacteria like the common bacterium *Mycobacterium sp.* [119], which was shown to metabolize phytosterols to androstenedione [121,122]. Other studies confirmed the presence of androgen activity in river water downstream of a paper mill, but as yet no additional specific compounds have been identified [123]. Ongoing studies of river water extracts indicate the presence of additional compounds with androgen activity. Preliminary evidence indicates that their structure differs from the known steroids. There is also recent evidence that other substances with androgen activity are present in the control river [124].

Another potential source for environmental androgens is excreted anabolic steroids administered to cattle to improve beef production [125]. The anabolic steroid trenbolone acetate is rapidly converted in vivo to 17β-trenbolone, an active metabolite with androgen activity. In a number of in vitro assays, trenbolone has a potency similar to the most active androgens known such as DHT and the synthetic androgen R1881 (methyltrienolone) [125]. Detection of reproductive abnormalities in fish exposed to effluent from feed lots has raised questions whether the masculinizing effects result from excretion products that contaminate the water. This type of environmental contamination by pharmaceutical byproducts in water sources may be widespread resulting from human or livestock drug use. Such contamination allows for possible exposure to high-affinity compounds, which if not eliminated by water purification procedures, can expose wildlife and human populations.

An environmental threat, therefore, exists, especially to aquatic wildlife that live in contaminated waters that contain effluent from paper mills or feedlots, as evidenced by the increased incidence of reproductive abnormalities. It is yet unclear to what extent these androgen precursors, anabolic steroids, or metabolites impact the human population. The presence of high-affinity compounds or androgen precursors lowers the exposure requirements for a detrimental effect. Further study of these contaminants with high affinity for the AR will reveal the overall threat to wildlife and human populations.

Risk assessment

Several criteria must be considered in the risk assessment of environmental antiandrogens and androgens. These include (a) exposure levels, (b) in vivo metabolic activation, (c) biological half-life of the active metabolites, (d) binding affinity and activity for the AR, and (e) the developmental stage of exposure.

(a) The level of exposure of the human population to environmental androgens or antiandrogens in most cases is not known. The relatively low binding affinities of the environmental antiandrogens discovered thus far would require significant exposure levels for an antagonistic effect to occur. Exposure to endocrine disruptors with antiandrogen activity may be an occupational hazard and thus limited to select populations involved in chemical production or farming practices. In contrast, the presence of high-affinity compounds derived from pharmaceutical and personal care products reduces the exposure requirements for a detrimental impact.

(b) Metabolic activation was observed with all of the environmental antiandrogens derived from pesticides and herbicides. This indicates the importance of screening with assay techniques that allow for metabolic processes to occur. In vitro binding assays using inactive parent compounds and isolated receptor preparations could preclude the identification of androgen antagonists or agonists since metabolic conversion may not occur. Metabolic activation necessitates the use of biological systems such as mammalian cell lines that utilize the correct metabolic enzyme pathways. For example, in yeast-based assays, metabolic conversion may not occur or may occur through alternative pathways not applicable to mammalian cells. Ideally, in vivo whole animal experiments may be required to demonstrate the activity of a compound. However, such an initial screening approach would be expensive and laborious. In the case of vinclozolin, mammalian cell-based assays displayed the same metabolic conversions that occur in the whole animal [39]. For a clear documentation of the inhibitory effects, the active metabolites must be identified and confirmed.

(c) Long half-lives of active endocrine disruptors can result from high fat solubility, a property applicable to most steroidal compounds that interact with the ligand-binding domain of steroid receptors. When fat stores are mobilized, the potential impact to human health increases. In addition, steroids or steroid-like compounds can accumulate in wildlife species to be passed up the food chain or to increase the body load to levels sufficient for inhibitory or stimulatory effects to occur.

(d) The environmental antiandrogens discovered thus far have relatively low binding affinities for the AR. This necessitates relatively high exposure levels for an inhibitory effect, which may only occur during occupational exposure. On the other hand, the presence of pharmaceutical and personal care products such as cholesterol-lowering drugs and steroids like estrone in water sources, raises new threats of the presence of high-affinity steroids and ligands that would require lower exposure levels for a detrimental effect. It was shown that some pharmaceutical products detected in the environment are associated with populations of feminized male fish, thus broadening the involvement of environmental contaminants in male reproductive abnormalities [126]. The detection of pharmaceutical and personal care products in the environment, which characteristically have high binding affinities for the AR, reduces the exposure levels required for a detrimental effect.

(e) Risk assessment should consider the developmental period of exposure. The developing male fetus is sensitive to the presence of antiandrogens since the male embryo depends on androgen-induced gene transcription during the critical period of male sexual development [127]. Periods of increased fetal sensitivity to exposure to antiandrogen activity from the fungicide vinclozolin were identified in a study of rats. The most sensitive period to the antiandrogenic effects of vinclozolin was rat gestational day 16 and 17, with the sensitive period ranging from gestational day 14 to 19, the time of male sexual development [128]. The relatively weak binding affinities of the known environmental antiandrogens for the AR necessitates relatively high exposure levels for a biological impact. It remains to be established whether any of the environmental antiandrogens identified thus far reach levels that are sufficient to interfere with normal human reproductive development. It was recently suggested that the lack of feedback regulation at the hypothalamus and pituitary by fetal androgens during development may increase the inhibitory effects of an antiandrogen [129]. In this case, androgen production by the fetal testis is induced by placental human chorionic gonadotropin. This could prevent a compensatory increase in androgen production in the presence of an antiandrogen, which could result in a more pronounced inhibitory effect.

REFERENCES

1. D. B. Lubahn, D. R. Joseph, P. M. Sullivan, H. F. Willard, F. S. French, E. M. Wilson. *Science* **240**, 327–330 (1988).
2. C. J. Brown, S. J. Goss, D. B. Lubahn, D. R. Joseph, E. M. Wilson, F. S. French, H. F. Willard. *Am. J. Hum. Genet.* **44**, 264–269 (1989).
3. D. B. Lubahn, D. R. Joseph, M. Sar, J. A. Tan, H. N. Higgs, R. E. Larson, F. S. French, E. M. Wilson. *Mol. Endocrinol.* **2**, 1265–1275 (1988).
4. C. W. Gregory, B. He, E. M. Wilson. *J. Mol. Endocrinol.* **27**, 309–319 (2001).
5. E. M. Wilson and F. S. French. *J. Biol. Chem.* **254**, 6310–6319 (1979).
6. J. A. Kemppainen, M. V. Lane, M. Sar, E. M. Wilson. *J. Biol. Chem.* **267**, 968–974 (1992).
7. J. A. Simental, M. Sar, M. V. Lane, F. S. French, E. M. Wilson. *J. Biol. Chem.* **266**, 510–518 (1991).
8. Z. X. Zhou, C. I. Wong, M. Sar, E. M. Wilson. *Rec. Prog. Hormone Res.* **49**, 249–274 (1994).
9. C. S. Choong and E. M. Wilson. *J. Mol. Endocrinol.* **21**, 235–257 (1998).
10. P. M. Bingham, M. O. Scott, S. Wang, M. J. McPhaul, E. M. Wilson, J. Y. Garbern, D. E. Merry, K. H. Fischbeck. *Nat. Genet.* **9**, 191–196 (1995).
11. A. R. La Spada, E. M. Wilson, D. B. Lubahn, A. E. Harding, K. H. Fischbeck. *Nature* **352**, 77–79 (1991).
12. C. S. Choong, J. A. Kemppainen, Z. X. Zhou, E. M. Wilson. *Mol. Endocrinol.* **10**, 1527–1535 (1996).
13. C. S. Choong, J. A. Kemppainen, E. M. Wilson. *J. Mol. Evol.* **47**, 334–342 (1998).
14. J. A. Tan, D. R. Joseph, V. E. Quarmby, D. B. Lubahn, M. Sar, F. S. French, E. M. Wilson. *Mol. Endocrinol.* **2**, 1276-–1285 (1988).
15. Z. X. Zhou, J. A. Kemppainen, E. M. Wilson. *Mol. Endocrinol.* **9**, 605–615 (1995).
16. L. P. Freedman. *Endocrine Rev.* **13**, 129–145 (1992).
17. C. I. Wong, Z. X. Zhou, M. Sar, E. M. Wilson. *J. Biol. Chem.* **268**, 19004–19012 (1993).
18. K. C. Ho, K. B. Marschke, J. A. Tan, S. G. A. Power, E. M. Wilson, F. S. French. *J. Biol. Chem.* **268**, 27226–27235 (1993).
19. J. A. Tan, K. B. Marschke, K. C. Ho, S. T. Perry, E. M. Wilson, F. S. French. *J. Biol. Chem.* **267**, 4456–4466 (1992).
20. K. C. Ho, V. E. Quarmby, F. S. French, E. M. Wilson. *J. Biol. Chem.* **267**, 12660–12667 (1992).
21. K. C. Ho, R. Snoek, V. E. Quarmby, D. H. Viskochil, P. S. Rennie, E. M. Wilson, F. S. French, N. Bruchovsky. *Biochemistry* **28**, 6367–6373 (1989).
22. N. J. Charest, D. R. Joseph, E. M. Wilson, F. S. French. *Mol. Endocrinol.* **2**, 999–1004 (1988).
23. D. H. Viskochil, S. T. Perry, O. A. Lea, D. W. Stafford, E. M. Wilson, F. S. French. *J. Biol. Chem.* **258**, 8861–8866 (1983).
24. K. B. J. M. Cleutjens, H. A. G. M van der Korput, C. C. Ehren-van Eekelen, R. A. Sikes, C. Fasciana, L. W. Chung, J. Trapman. *Mol. Endocrinol.* **11**, 1256–1265 (1997).
25. N. M. Greenberg, F. J. DeMayo, P. C. Sheppard, R. Barrios, R. Lebovitz, M. Finegold, R. Angelopoulou, J. G. Dodd, M. L. Duckworth, J. M. Rosen, R. J. Matusik. *Mol. Endocrinol.* **8**, 230–239 (1994).

26. S. Kasper, P. S. Rennie, N. Bruchovsky, P. C. Sheppard, H. Cheng, L. Lin, R. P. C. Shiu, R. Snoek, R. J. Matusik. *J. Biol. Chem.* **269**, 31763–31769 (1994).
27. A. Scheller, R. I. Scheinman, E. Thompson, C. O. Scarlett, D. M. Robins. *Mol. Cell. Endocrinol.* **121**, 75–86 (1996).
28. J. A. Tan, S. H. Hall, K. G. Hamil, G. Grossman, P. Petrusz, J. Liao, K. Shuai, F. S. French. *Mol. Endocrinol.* **14**, 14–26 (2000).
29. J. A. Tan, S. H. Hall, K. G. Hamil, G. Grossman, P. Petrusz, F. S. French. *J. Biol. Chem.* **277**, 16993–17001 (2002).
30. A. M. Moilanen, H. Poukka, U. Karvonen, M. Häkli, O. A. Jänne, J. J. Palvimo. *Mol. Cell. Biol.* **18**, 5128–5139 (1998).
31. Z. X. Zhou, M. Sar, J. A. Simental, M. V. Lane, E. M. Wilson. *J. Biol. Chem.* **269**, 13115–13123 (1994).
32. M. Sar, D. B. Lubahn, F. S. French, E. M. Wilson. *Endocrinology* **127**, 3180–3186 (1990).
33. C. W. Gregory, R. T. Johnson, J. L. Mohler, F. S. French, E. M. Wilson. *Cancer Res.* **61**, 2892–2898 (2001).
34. P. M. Matias, P. Donner, R. Coelho, M. Thomaz, C. Peixoto, S. Macedo, N. Otto, S. Joschko, P. Scholz, A. Wegg, S. Bäsler, M. Schäfer, U. Egner, M. A. Carrondo. *J. Biol. Chem.* **275**, 26164–26171 (2000).
35. J. S. Sack, K. F. Kish, C. Wang, R. M. Attar, S. E. Kiefer, Y. An, G. Y. Wu, J. E. Scheffler, M. E. Salvati, S. R. Krystek, R. Weinmann, H. M. Einspahr. *Proc. Natl. Acad. Sci. USA* **98**, 4904–4909 (2001).
36. J. P. Renaud, N. Rochel, M. Ruff, V. Vivat, P. Chambon, H. Gronemeyer, D. Moras. *Nature* **378**, 681–689 (1995).
37. E. M. Wilson and F. S. French. *J. Biol. Chem.* **251**, 5620–5629 (1976).
38. J. A. Kemppainen, E. Langley, C. I. Wong, K. Bobseine, W. R. Kelce, E. M. Wilson. *Mol. Endocrinol.* **13**, 440–454 (1999).
39. C. I. Wong, W. R. Kelce, M. Sar, E. M. Wilson. *J. Biol. Chem.* **270**, 19998–20003 (1995).
40. E. Langley, Z. X. Zhou, E. M. Wilson. *J. Biol. Chem.* **270**, 29983–29990 (1995).
41. Z. X. Zhou, M. V. Lane, J. A. Kemppainen, F. S. French, E. M. Wilson. *Mol. Endocrinol.* **9**, 208–218 (1995).
42. E. Langley, J. A. Kemppainen, E. M. Wilson. *J. Biol. Chem.* **273**, 92–101 (1998).
43. P. Doesburg, C. W. Kuil, C. A. Berrevoets, K. Steketee, P. W. Faber, E. Mulder, A. O. Brinkmann, J. Trapman. *Biochemistry* **36**, 1052–1064 (1997).
44. T. Ikonen, J. J. Palvimo, O. A. Jänne. *J. Biol. Chem.* **272**, 29821-29828 (1997).
45. C. A. Berrevoets, P. Doesburg, K. Steketee, J. Trapman, A. O. Brinkmann. *Mol. Endocrinol.* **12**, 1172–1183 (1998).
46. C. L. Bevan, S. Hoare, F. Claessens, D. M. Heery, M. G. Parker. *Mol. Cell. Biol.* **19**, 8383–8392 (1999).
47. T. Slagsvold, I. Kraus, T. Bentzen, J. Palvimo, F. Saatcioglu. *Mol. Endocrinol.* **14**, 1603–1617 (2000).
48. J. Thompson, F. Saatcioglu, O. A. Jänne, J. J. Palvimo. *Mol. Endocrinol.* **15**, 923–935 (2001).
49. A. Bubulya, S. Y. Chen, C. J. Fisher, Z. Zheng, X. Q. Shen, L. Shemshedini. *J. Biol. Chem.* **48**, 44704–44711 (2001).
50. W. L. Kraus, E. M. McInerney, B. S. Katzenellenbogen. *Proc. Natl. Acad. Sci. USA* **92**, 12314–12318 (1995).
51. D. Shao, S. M. Rangwala, S. T. Bailey, S. L. Krakow, M. J. Reginato, M. A. Lazar. *Nature* **396**, 377–380 (1998).
52. M. J. Tetel, P. H. Giangrande, S. A. Leonhardt, D. P. McDonnell, D. P. Edwards. *Mol. Endocrinol.* **13**, 910–924 (1999).
53. R. Métivier, G. Penot, G. Flouriot, F. Pakdel. *Mol. Endocrinol.* **15**, 1953–1970 (2001).

54. B. He, J. A. Kemppainen, J. J. Voegel, H. Gronemeyer, E. M. Wilson. *J. Biol. Chem.* **274**, 37219–37225 (1999).
55. B. He, J. A. Kemppainen, E. M. Wilson. *J. Biol. Chem.* **275**, 22986–22994 (2000).
56. B. He, J. T. Minges, L. W. Lee, E. M. Wilson. *J. Biol. Chem.* **277**, 10226–10235 (2002).
57. B. He, N. T. Bowen, J. T. Minges, E. M. Wilson. *J. Biol. Chem.* **276**, 42293–42301 (2001).
58. B. He, L. W. Lee, J. T. Minges, E. M. Wilson. *J. Biol. Chem.* **277**, 25631–25639 (2002).
59. E. M. Wilson, B. He, E. Langley. *Meth. Enzym.* **364** (2003).
60. J. Schneikert, H. Peterziel, P. A. Defossez, H. Klocker, Y. de Launoit, A. C. B. Cato. *J. Biol. Chem.* **271**, 23907–23913 (1996).
61. A. Bubulya, S. C. Wise, X. Q. Shen, L. A. Burmeister, L. Shemshedini. *J. Biol. Chem.* **271**, 24583–24589 (1996).
62. S. C. Wise, L. A. Burmeister, X. F. Zhou, A. Bubulya, J. L. Oberfield, M. J. Birrer. L. Shemshedini. *Oncogene* **16**, 2001–2009 (1998).
63. I. J. McEwan, J. A. Gustafsson. *Proc. Natl. Acad. Sci. USA* **94**, 8485–8490 (1997).
64. D. K. Lee, H. O. Duan, C. Chang. *J. Biol. Chem.* **275**, 9308–9313 (2000).
65. J. Lu and M. Danielsen. *J. Biol. Chem.* **273**, 31528–31533 (1998).
66. K. Fronsdal, N. Engedal, T. Slagsvold, F. Saatcioglu. *J. Biol. Chem.* **273**, 31853–31859 (1998).
67. B. A. Froesch, S. Takayama, J. C. Reed. *J. Biol. Chem.* **273**, 11660–11666 (1998).
68. S. Yeh and C. Chang. *Proc. Natl. Acad. Sci. USA* **93**, 5517–5621 (1996).
69. P. Alen, F. Claessens, E. Schoenmakers, J. V. Swinnen, G. Verhoeven, W. Rombauts, B. Peeters. *Mol. Endocrinol.* **13**, 117–128 (1999).
70. Z. X. Zhou, B. He, S. H. Hall, E. M. Wilson, F. S. French. *Mol. Endocrinol.* **16**, 287–300 (2002).
71. A. M. Moilanen, U. Karvonen, H. Poukka, W. Yan, J. Toppari, O. A. Jänne, J. J. Palvimo. *J. Biol. Chem.* **274**, 3700–3704 (1999).
72. H. Ma, H. Hong, S. M. Huang, R. A. Irvine, P. Webb, P. J. Kushner, G. A. Coetzee, M. R. Stallcup. *Mol. Cell. Biol.* **19**, 6164–6173 (1999).
73. T. Slagsvold, I. Kraus, K. Fronsdal, F. Saatcioglu. *J. Biol. Chem.* **276**, 31030–31036 (2001).
74. P. W. Hsiao and C. Chang. *J. Biol. Chem.* **274**, 22373–22379 (1999).
75. H. Y. Kang, S. Yeh, N. Fujimoto, C. Chang. *J. Biol. Chem.* **274**, 8570–8576 (1999).
76. N. Fujimoto, S. Yeh, H. Y. Kang, S. Inui, H. C. Chang, A. Mizokami, C. Chang. *J. Biol. Chem.* **274**, 8316–8321 (1999).
77. L. Yang, J. Guerrero, H. Hong, D. B. DeFranco, M. R. Stallcup. *Mol. Biol. Cell* **11**, 2007–2018 (2000).
78. C. I. Truica, S. Byers, E. P. Gelmann. *Cancer Res.* **60**, 4709–4713 (2000).
79. J. E. Pawlowski, J. R. Ertel, M. P. Allen, M. Xu, C. Butler, E. M. Wilson, M. E. Wierman. *J. Biol. Chem.* **277**, 20702–20710 (2002).
80. J. M. Müller, U. Isele, E. Metzger, A. Rempel, M. Moser, A. Pscherer, T. Breyer, C. Holubarsch, R. Buettner, R. Schüle. *EMBO J.* **19**, 359–369 (2000).
81. M. Sharma, M. Zarnegar, X. Li, B. Lim, Z. Sun. *J. Biol. Chem.* **275**, 35200–35208 (2000).
82. A. Yamamoto, Y. Hashimoto, K. Kohri, E. Ogata, S. A. Kato, K. Ikeda, M. Nakanishi. *J. Cell Biol.* **150**, 873–879 (2000).
83. A. T. Reutens, M. Fu, C. Wang, C. Albanese, M. J. McPhaul, Z. Sun, S. P. Balk, O. A. Janne, J. J. Palvimo, R. G. Pestell. *Mol. Endocrinol.* **15**, 797–811 (2001).
84. P. Oettgen, E. Finger, Z. Sun, Y. Akbarali, U. Thamrongsak, J. Boltax, F. Grall, A. Dube, A. Weiss, L. Brown, G. Quinn, K. Kas, G. Endress, C. Kunsch, T. A. Libermann. *J. Biol. Chem.* **275**, 1216–1225 (2000).
85. S. Yeh, Y. C. Hu, M. Rahman, H. K. Lin, C. L. Hsu, H. J. Ting, H. Y. Kang, C. Chang. *Proc. Natl. Acad. Sci. USA* **97**, 11256–11261 (2000).
86. D. K. Lee, H. O. Duan, C. Chang. *J. Biol. Chem.* **276**, 9978–9984 (2001).

87. M. L. Lu, M. C. Schneider, Y. Zheng, X. Zhang, J. P. Richie. *J. Biol. Chem.* **276**, 13442–13451 (2001).
88. P. W. Hsiao, D. L. Lin, R. Nakao, C. Chang. *J. Biol. Chem.* **274**, 20229–20234 (1999).
89. J. Xu, Y. Qiu, F. J. DeMayo, S. Y. Tsai, M. J. Tsai, B. W. O'Malley. *Science* **279**, 1922–1925 (1998).
90. J. Xu, L. Liao, G. Ning, H. Yoshida-Komiya, C. Deng, B. W. O'Malley. *Proc. Natl. Acad. Sci. USA* **97**, 6379–6384 (2000).
91. Z. Wang, D. W. Rose, O. Hermanson, F. Liu, T. Herman, W. Wu, D. Szeto, A. Gleiberman, A. Krones, K. Pratt, R. Rosenfeld, C. K. Glass, M. G. Rosenfeld. *Proc. Natl. Acad. Sci. USA* **97**, 13549–13554 (2000).
92. K. Hofman, J. V. Swinnen, F. Claessens, G. Verhoeven, W. Heyns. *Mol. Cell. Endocrinol.* **168**, 21–29 (2000).
93. C. A. Quigley, A. De Bellis, K. B. Marschke, M. K. El-Awady, E. M. Wilson, F. S. French. *Endocrine Rev.* **16**, 271–321 (1995).
94. C. A. Quigley, B. A. J. Evans, J. A. Simental, K. B. Marschke, M. Sar, D. B. Lubahn, P. Davies, I. A. Hughes, E. M. Wilson, F. S. French. *Mol. Endocrinol.* **6**, 1103–1112 (1992).
95. C. A. Quigley, K. J. Friedman, A. Johnson, R. G. Lafreniere, L. M. Silverman, D. B. Lubahn, T. R. Brown, E. M. Wilson, H. F. Willard, F. S. French. *J. Clin. Endocrinol. Metab.* **74**, 927–933 (1992).
96. C. S. Choong, C. A. Quigley, F. S. French, E. M. Wilson. *J. Clin. Invest.* **98**, 1423–1431 (1996).
97. A. De Bellis, C. A. Quigley, N. F. Cariello, M. K. El-Awady, M. Sar, M. V. Lane, E. M. Wilson, F. S. French. *Mol. Endocrinol.* **6**, 1909–1920 (1992).
98. W. G. Yarbrough, V. E. Quarmby, J. A. Simental, D. R. Joseph, M. Sar, D. B. Lubahn, K. L. Olsen, F. S. French, E. M. Wilson. *J. Biol. Chem.* **265**, 8893–8900 (1990).
99. D. B. Lubahn, T. R. Brown, J. A. Simental, H. N. Higgs, C. J. Migeon, E. M. Wilson, F. S. French. *Proc. Natl. Acad. Sci. USA* **86**, 9534–9538 (1989).
100. N. J. Charest, Z. X. Zhou, D. B. Lubahn, K. L. Olsen, E. M. Wilson, F. S. French. *Mol. Endocrinol.* **5**, 573–581 (1991).
101. A. De Bellis, C. A. Quigley, K. B. Marschke, M. K. El-Awady, M. V. Lane, E. P. Smith, M. Sar, E. M. Wilson, F. S. French. *J. Clin. Endocrinol. Metab.* **78**, 513–522 (1994).
102. J. A. Kemppainen and E. M. Wilson. *Urology* **48**, 157–163 (1996).
103. F. S. vom Saal, B. G. Timms, M. M. Montano, K. A. Thayer, S. C. Nagel, M. G. Dhar, V. K. Ganjam, S. Parmigiani, W. V. Welshons. *Proc. Natl. Acad. Sci. USA* **94**, 2056–2061 (1997).
104. S. H. Safe. *Environ. Health Perspect.* **108**, 487–493 (2000).
105. C. W. Gregory, B. He, R. T. Johnson, O. H. Ford, J. L. Mohler, F. S. French, E. M. Wilson. *Cancer Res.* **61**, 4315–4319 (2001).
106. W. R. Kelce, C. R. Stone, S. C. Laws, L. E. Gray, J. A. Kemppainen, E. M. Wilson. *Nature* **375**, 581–585 (1995).
107. W. R. Kelce, C. R. Lambright, T. E. Wiese, L. E. Gray, C. I. Wong, E. M. Wilson. (2003). In preparation.
108. W. R. Kelce and E. M. Wilson. *J. Mol. Med.* **75**, 198–207 (1997).
109. L. E. Gray, J. S. Ostby, W. R. Kelce. *Toxicol. Appl. Pharmacol.* **129**, 46–52 (1994).
110. W. R. Kelce, E. Monosson, M. P. Gamcsik, S. C. Laws, L. E. Gray. *Toxicol. Appl. Pharmacol.* **126**, 276–285 (1994).
111. S. Hosokawa, M. Murakami, M. Ineyama, T. Yamada, Y. Koyama, Y. Okuno, A. Yoshitake, H. Yamada, J. Miyamoto. *J. Toxicol. Sci.* **18**, 111–124 (1993).
112. E. Mylchreest, D. G. Wallace, R. C. Cattley, P. M. D. Foster. *Toxicol. Sci.* **55**, 143–151 (2000).
113. E. Mylchreest, M. Sar, R. C. Cattley, P. M. D. Foster. *Toxicol. Appl. Pharmacol.* **156**, 81–95 (1999).
114. S. H. Swan, E. P. Elkin, L. Fenster. *Environ. Health Perspect.* **108**, 961–966 (2000).

115. M. Aho, A. M. Koivisto, T. L. J. Tammela, A. Auvinen. *Environ. Health Perspect.* **108**, 463–465 (2000).
116. L. J. Paulozzi, J. D. Erickson, R. J. Jackson. *Pediatrics* **100**, 831–834 (1997).
117. L. J. Paulozzi. *Environ. Health Perspect.* **107**, 297–302 (1999).
118. W. Howell, A. Black, S. Bortone. *Copeia* **4**, 676–681 (1980).
119. R. Jenkins, R. A. Angus, H. McNatt, W. M. Howell, J. A. Kemppainen, M. Kirk, E. M. Wilson. *Environ. Toxicol. Chem.* **20**, 1325–1331 (2001).
120. J. Raloff. *Science News* **159**, 8–10 (2001).
121. W. J. Marsheck, S. Kraychy, R. D. Muir. *Appl. Microbiol.* **23**, 72–77 (1972).
122. A. Conner, M. Nagaoka, J. W. Rowe, D. Perlman. *Appl. Environ. Microbiol.* **32**, 310–311 (1976).
123. L. G. Parks, C. S. Lambright, E. F. Orlando, L. J. Guillette, G. T. Ankley, L. E. Gray. *Toxicol. Sci.* **62**, 257–267 (2001).
124. A. L. Edwards, R. L. Jenkins, L. J. Davenport, J. A. Duke. *Economic Botany* **56**, 204–206 (2002).
125. V. S. Wilson, C. Lambright, J. Ostby, L. E. Gray. *Toxicol. Sci.* **70**, 202–211 (2002).
126. C. Potera. *Environ. Health Perspect.* **108**, A446 (2000).
127. S. Y. Euling and C. A. Kimmel. *Sci. Total Environ.* **274**, 103–113 (2001).
128. C. J. Wolf, G. A. LeBlanc, J. S. Ostby, L. E. Gray. *Toxicol. Sci.* **55**, 152–161 (2000).
129. S. Y. Euling, C. Gennings, E. M. Wilson, J. A. Kemppainen, W. R. Kelce, C. A. Kimmel. *Toxicol. Sci.* **69**, 332–343 (2002).

Pure Appl. Chem., Vol. 75, Nos. 11–12, pp. 1699–1707, 2003.

Topic 1.5

Genetic dissection of gluco- and mineralocorticoid receptor function in mice*

E. F. Greiner, S. Berger, and G. Schütz‡

Department of Molecular Biology of the Cell I, German Cancer Research Center, Im Neuenheimer Feld 280, 60120 Heidelberg, Germany

Abstract: Nuclear hormone receptors function to transduce hormonal signals into transcriptional responses by controlling the activity of specific target genes. These target genes comprise a genetic network whose coordinate activity defines the physiological responses to hormonal signals. Dissecting nuclear hormone receptor functions in vivo by gene inactivation and transgenic strategies represents an invaluable and powerful approach to increase our knowledge of these genetic networks and their physiological functions. Glucocorticoids and mineralocorticoids are involved in numerous physiological processes important to maintain metabolic, cardiovascular, central nervous, and immune system homeostasis. Germline and somatic gene targeting as well as an increased dosage of the glucocorticoid receptor (GR) allows the characterization of the various functions and molecular modes of action of this receptor. Most of the effects of the GR are mediated via activation and repression of gene expression. To separate activating from repressing functions of the GR, a point mutation was introduced which allowed us to characterize and distinguish functions dependent on GR binding to DNA from those mediated by protein/protein interaction. Cell/tissue-specific mutations of the gluco- and mineralocorticoid receptor is the basis for the evaluation of their cell-specific functions, including the characterization of target genes of the receptors in order to describe their specific effects on different targets.

INTRODUCTION

Glucocorticoids and mineralocorticoids regulate diverse functions important to maintain central nervous system, cardiovascular, metabolic, and immune homeostasis [1]. The corticosteroid hormones produced in the adrenal cortex are ideal candidates for integrating such a complex array of physiological functions. According to the lipophilic nature of their steroid structure, they diffuse from their source, permeate to their targets, and exert their functions by binding two closely related corticosteroid receptors: the mineralocorticoid receptor (MR, type-I receptor) and the glucocorticoid receptor (GR, type-II receptor) [2]. Under physiological conditions, glucocorticoids are able to activate both receptors, whereas mineralocorticoids are specific for the MR. Mineralocorticoid-responsive cells (e.g., the principal cells of the collecting duct in kidney) express the enzyme 11-β-hydroxysteroiddehydrogenase, type 2. Conversion of cortisol to cortisone by 11-β-hydroxysteroiddehydrogenase type 2 prevents glucocorticoids from binding to the MR, and thus this pre-receptor specificity mechanism ensures mineralocorticoid-specific effects.

---

*Report from a SCOPE/IUPAC project: Implication of Endocrine Active Substances for Human and Wildlife (J. Miyamoto and J. Burger, editors). Other reports are published in this issue, *Pure Appl. Chem.* **75**, 1617–2615 (2003).

‡Corresponding author

Mineralocorticoids control genes involved in transepithelial sodium transport and thereby regulate sodium and potassium homeostasis and water balance [3]. Aldosterone, the main circulating mineralocorticoid, has a major role in blood-pressure volume regulation in normal subjects, and is thought to be involved in the pathogenesis of hypertension and target-organ damage.

Glucocorticoids regulate genes involved in the production and mobilization of glucose. Glucocorticoids increase the breakdown of fat and proteins and induce the synthesis of gluconeogenic enzymes to generate blood glucose [3]. In addition, glucocorticoids promote maturation of the lung and are highly beneficial in the treatment of the acute respiratory distress syndrome in newborn infants [4]. Glucocorticoids and their receptor also play a role in maturation of erythroid progenitors [5]. Furthermore, glucocorticoids can induce apoptosis of thymocytes; they are important anti-inflammatory and immunosuppressive agents [1]. Their immunomodulatory activity has been exploited for the treatment of diverse inflammatory and allergic disorders including asthma [1].

Levels of mineralocorticoids and glucocorticoids are controlled by tight regulatory mechanisms [3]. The renin-angiotensin-aldosterone system (RAAS) controls the mineralocorticoid release, while the hypothalamic-pituitary-adrenal (HPA) axis regulates glucocorticoid secretion. The hypothalamus secretes corticotropin-releasing hormone (CRH), which acts on the anterior pituitary gland to secrete the adrenocorticotropin hormone (ACTH). ACTH stimulates the release of glucocorticoids from the cortex of the adrenal gland. Increased blood glucocorticoid levels in turn repress the transcription and inhibit the release of both, CRH and ACTH. Glucocorticoids are released in response to various stressors (physical, emotional, chemical, physiological) [6]. Stress is considered a protective mechanism that prepares the organism to react to threatening stimuli in an appropriate way. These stimuli activate the HPA axis, which in conjunction with other physiological adaptations coordinate the behavioral responses of the organism. In addition, the levels of glucocorticoids show a diurnal rhythm reaching peak levels at the onset of the active phase of the organism. Chronic changes in control of the HPA axis may have pathological consequences, since it is now well established that depression- and anxiety-related disorders are associated with dysregulation of the HPA axis.

Mouse molecular genetic approaches led the foundation for the enormous progress toward understanding the various functions of steroid hormone receptors. Conventional knock-out technology was supplemented by more sophisticated gene targeting strategies. Exploiting the Cre/loxP recombination system from bacteriophage λ in mice allowed the generation of conditional mutations, cell- and tissue-specific mutations, and function-selective mutations [7,8]. Genetic manipulations of the corticosteroid receptors in mice provided important information on both, the molecular mechanism of receptor function and their role in development and physiology. Often, these analyses have resulted in unexpected findings, that is, the liver-specific function of the glucocorticoid receptor (GR) in growth control (see below). Figure 1 summarizes the various types of mutations ranging from null and hypomorphic mutations to cell/tissue-specific and function-selective mutations, which have been generated with the help of the Cre/loxP recombination system.

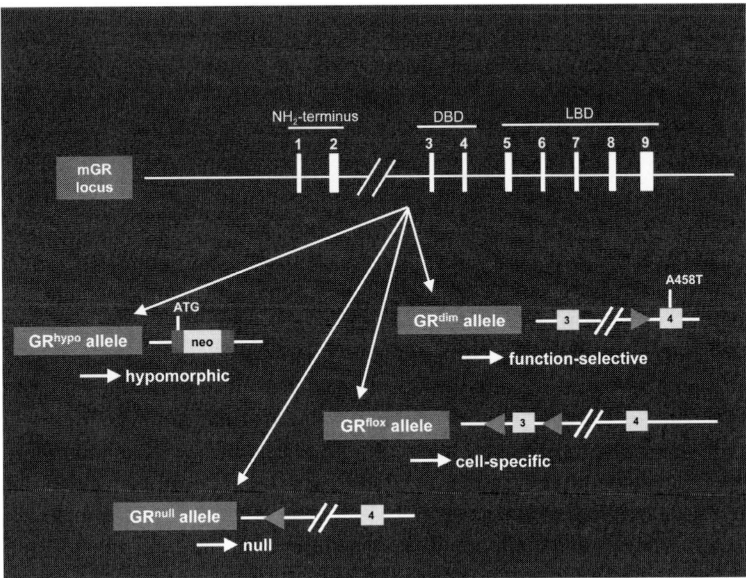

Fig. 1 Analysis of GR gene function by null, cell-specific, and function-selective mutations. The organization of the GR gene is shown at the top. The four types of mutations of the GR gene are shown in relation to the endogenous GR gene.

LOSS-OF-FUNCTION MOUSE MUTANTS FOR THE GLUCOCORTICOID AND THE MINERALOCORTICOID RECEPTOR

Two GR-inactivating mutations have been generated: a hypomorphic allele that resulted from insertion of a PGK-neomycin-resistance cassette right after the ATG of the receptor coding sequence [9] and a null allele that has been generated by Cre-loxP mediated excision of exon 3 (F. Tronche and G. Schütz, unpublished). In the hypomorphic allele, alternative splicing results in a truncated mRNA, which leads to the synthesis of a shortened protein lacking most of the NH_2-terminal part of the protein before the DNA-binding domain causing 20 % survival of the homozygous mutant animals. The allele with deletion of exon 3 represents a loss-of-function mutation with a complete penetrance of the phenotype. Mice without functional GR die from atelectasis of the lungs shortly after birth. GR function in lung development and perinatal survival is poorly understood, but dimeric DNA binding of the receptor is not required (see below). Primary GR target genes essential for lung development remain to be established. Analysis of these mice with regard to receptor function revealed, among other things, that the GR is required for rapid expansion of erythroid progenitors under hypoxic conditions in vivo [5]. Table 1 lists a summary of the phenotypic consequences of the GR null/hypomorphic mutation [9,10].

Mineralocorticoid receptor (MR)-deficient mice have a normal prenatal development. During the first week of life, MR-deficient mice develop symptoms of pseudohypoaldosteronism with hyponatremia, hyperkalemia, high renal sodium loss, and a strongly activated renin-angiotensin-aldosterone system (RAAS) [11]. MR knockout mice die in the second week after birth but daily subcutaneous injections of isotonic NaCl solution until weaning and continued oral NaCl supply lead to survival [11]. The NaCl-rescued MR knockout mice display a strongly enhanced fractional renal sodium excretion, hyperkalemia, and persistently high levels of renin and aldosterone [12]. The activity of the amiloride-sensitive epithelial sodium channel (ENaC) is strongly reduced in colon and kidney, but at day 8 after birth there is no down-regulation of the mRNA abundance of the three ENaC subunits. Therefore, mineralocorticoids do not regulate the sodium reabsorption by transcriptional control of ENaC but of other yet unidentified MR target genes [11].

Table 1 Alterations of physiological functions found in mice carrying a targeted disruption of the GR gene.

Lung	Perinatal death due to respiratory failure in newborn mice
Liver	Reduced expression of genes encoding gluconeogenic enzymes
Adrenals	Hypertrophy and hyperplasia of the cortex with increased expression of steroidogenic enzymes; impaired chromaffin cell differentiation in the medulla, and absence of epinephrine synthesis
HPA-axis	Elevated serum levels for ACTH and CORT; increased expression of POMC and CRH
Bone marrow	Impaired proliferation of erythoid precursor cells
Thymus	Loss of glucocorticoid-dependent apoptosis in thymocytes

FUNCTION-SELECTIVE MOUSE MUTANTS: THE DNA-BINDING DEFICIENT GR REVEALS THE IMPORTANCE OF GR CROSS-TALK WITH OTHER TRANSCRIPTION FACTORS

The GR mediates the effects of glucocorticoids by positive and negative regulation of gene transcription in two modes of action, DNA binding-dependent and DNA binding-independent. Binding of GR to a glucocorticoid response element (GRE) activates transcription [4], binding to so-called negative GREs (nGRE) represses transcription [13]. To study activities of the receptor that might be independent of DNA-binding, a function-selective mouse mutant was generated by a knock-in strategy. By exploiting the Cre/loxP system a point-mutation was introduced into the mouse genome resulting in the amino acid exchange A458T within the D-loop, one of the important dimerization interfaces of the receptor [14]. This function-selective mutation abolishes homodimerization of the receptor and thereby impairs binding of GR to DNA. However, regulation of gene activity by protein/protein interaction remains intact.

The dimerization-defective mutant referred to as GR^{dim} represents a powerful tool for studying the molecular mechanisms of glucocorticoid action in vivo. Surprisingly, homozygous GR^{dim} mutant mice are viable, indicating that dimeric DNA-binding of the receptor is not essential for survival. Activating and inhibitory functions of the GR could be demonstrated (Table 2).

Table 2 GR^{dim} mice allow to decipher processes which require DNA binding of GR and those which are mediated by protein-protein interactions of GR with other transcription factors.

DNA binding-dependent processes (impaired in GR^{dim} mice)	**DNA binding-independent processes (not impaired in GR^{dim} mice)**
Induction of GRE-dependent reporters in embryonic fibroblasts and gluconeogenic enzymes in liver	Repression of TPA-induced expression of collagenase-3 and gelatinase B in embryonic fibroblasts and skin
Glucocorticoid-dependent apoptosis of thymocytes	Lung maturation, survival at birth
Expression of pro-opiomelanocortin (POMC) in the anterior pituitary	Immunoreactivity of corticotropin releasing hormone (CRH) in the median eminence
Long-term proliferation of erythroid progenitor cells in vitro, stress erythropoiesis in vivo	Repression of proinflammatory cytokines in thymocytes and macrophages, repression of inflammatory responses

Animal studies supported by the analysis of cells derived from GR^{dim} mice demonstrated that important anti-inflammatory activities of the receptor are maintained in this GR dimerization-deficient mouse model [15,16]. Inflammatory responses including the induced ear edema following TPA treatment and phorbolester-induced local inflammation of the skin could be repressed by glucocorticoids in GR^{dim} mice. DNA binding-independent GR functions are demonstrated to be sufficient for antagonizing AP-1-dependent gene activation in skin [17]. Intraperitoneal injection with lipopolysaccharide from gramnegative bacteria (LPS from E. coli) stimulates cytokine synthesis and release from thymocytes and macrophages. Interestingly, glucocorticoids suppress the cytokine response in both wild-type and GR^{dim} mice. Glucocorticoids also repress the proinflammatory potential of NFκB activity even when the receptor is impaired for DNA-binding [15]. Findings based on GR^{dim} mice promise that selective glucocorticoid receptor mediators (SGRMs) with potent anti-inflammatory activity can be found. Glucocorticoids are among the most potent immunosuppressive and anti-inflammatory drugs, used both locally and systemically in the treatment of inflammation, leukemia and autoimmune diseases [18]. However, long-term therapies are usually accompanied by severe side effects such as osteoporosis, atrophy of the skin, myopathy, and psychosis. One approach to improve the treatment would be to develop new GR ligands that show highly potent immunosuppressive and anti-inflammatory activity but reduced side effects. Given that many immunoregulatory genes, such as cytokines, are controlled by the AP-1 and NFκB families of transcription factors, it can be assumed that part of the immunosuppressive and anti-inflammatory activity of GR is mediated via its DNA binding-independent cross-talk function. It is thus conceivable that at least some of the side effects require the DNA binding-dependent function of GR.

To provide the proof-of-concept for this model and at the same time to elucidate the molecular mode of action of GR in more detail, regulation of cytokine expression and inflammation was studied in GR^{dim} mice. Analysis of some important pro-inflammatory cytokines (IL-1, IL-6, TNFα) in thymocytes and macrophages showed that suppression is largely independent of the DNA binding function of GR. Taken together, this shows that by using GR^{dim} mice it is possible to discriminate the molecular mode of action of GR in the immune system and to exploit this information for the development of improved therapeutic approaches.

Synthesis and release of glucocorticoids from the adrenal cortex are regulated by negative feedback loops in the HPA axis. Increase in the secretion of glucocorticoids cause a decrease in synthesis and secretion of its regulating substances CRH and ACTH [1,3]. GR^{dim} mice enabled us to investigate the molecular mechanisms underlying this negative feedback loop. Immunohistochemical analysis revealed that CRH levels in the hypothalamus of GR^{dim} mice do not differ significantly from those of wild-type mice. Therefore, glucocorticoids control hypothalamic CRH synthesis by protein/protein interactions between the GR and other factors [14]. The transcriptional mechanism of this control is however not understood. Expression of the gene encoding POMC, the precursor of ACTH, is elevated by almost an order of magnitude demonstrating loss of negative control of POMC transcription. These findings illustrate the different modes of action of GR in HPA-axis control and strongly support the hypothesis of negative GREs to which multimers of the receptor have to bind in order to exert this feedback control [13].

CELL/TISSUE-SPECIFIC MUTATIONS OF THE GR AND MR GENES

Mice without a functional GR or MR die after birth. Therefore, in order to define the function of these two receptors in the adult organism we had to develop cell/tissue-specific mutations. We anticipated that the generation of somatic mutations would not only avoid the lethality of germline mutant mice, but that it would also allow us to define the function of this receptor in a particular cell or tissue. This analysis would then also allow to estimate to which extent this function would contribute to the physiology of the entire organism. To analyze the function of these two receptors in specific cells and tissues, exon 3 of the GR and exon 3 of the MR gene were flanked by loxP sites using homologous recombination in embryonic stem cells. To attain a specific mutation, it is required that the Cre recombinase is expressed in the desired cells or tissues. Since we have observed that the penetrance of the mutation depends to a large extent on the expression pattern of the recombinase we use either bacterial artificial chromosomes or yeast artificial chromosomes for the expression of the Cre recombinase. In this way, a high degree of specificity and selectivity and copy number-dependent expression could be achieved. Figure 2 gives an overview of the specific GR mutations that we have generated or that are in the process of their development.

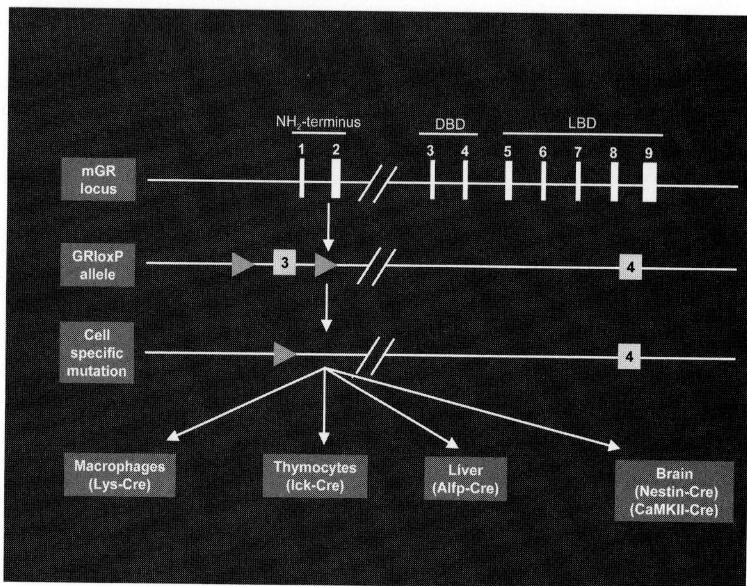

Fig. 2 Summary of cell- and tissue-specific mutations of the GR gene. Specific mutations in the GR locus were obtained by expressing the Cre recombinase in monocytes/macrophages under control of the regulatory elements of the lysozyme gene, in thymocytes under control of the lck gene, in hepatocytes under control of the regulatory elements of the albumin gene, and in brain with regulatory sequences of the nestin and CaMKIIα gene, respectively.

The GR and MR are both highly expressed in the central nervous system, in particular in the hippocampus. The pre-receptor specificity mechanism that prevents activation of the MR by corticosteroids in mineralocorticoid target tissues is not operative in the CNS. It is an intriguing question what the function of these two receptors might be. The GRs and MRs are thought to regulate an overlapping set of target genes since DNA-binding specificity of the two receptors is identical in vitro. We therefore think that specific inactivation of the MR and GR gene in the central nervous system might allow us to define their specific functions in greater detail.

In order to define GR function in the nervous system, we used the nestin gene promoter and enhancer to drive Cre recombinase expression [19]. The nestin gene is active in neuronal and glial cell precursors and leads to the deletion of the GR gene in the entire nervous system. In contrast with the germline mutation, lack of the GR in the central nervous system is not lethal. The animals have a Cushing-like syndrome with elevated glucocorticoids, altered fat distribution, and reduced bone density. Inactivation of the GR in the central nervous system reduces anxiety-related behavior. These mice are less anxious in tasks that exploit the behavioral conflict between exploring and avoiding an aversive compartment, the elevated zero maze, and the dark light box. These findings indicate an important role of GR signaling in emotional behavior [10,19]. To generate mutations lacking both receptors in the forebrain region, we used a bacterial artificial chromosome carrying the CaMKIIα gene to drive Cre recombinase expression. Mice with loss of the MR as far as analyzed show no alteration in HPA axis activity. Using the same Cre-expressing mice, we have generated a novel brain-specific GR mutation.

To study the role of GR in liver functions of the adult, we wished to inactivate the GR selectively in parenchymal cells of the liver. This was achieved by choosing gene regulatory sequences of the albumin gene for expression of the Cre recombinase [20]. Mice with a hepatocyte-specific alteration displayed after four weeks a severe growth deficit. Since no alterations in the serum levels of growth hormone and glucocorticoids could be found in these animals, we reasoned that the growth deficit resides in growth hormone signaling. Growth hormone is known to affect growth through stimulation of synthesis of insulin-like growth factor 1 (IGF-1) [21]. Interestingly, when we determined the mRNA levels of IGF-1 and growth hormone-regulated genes in the liver we found that the level was drastically reduced in the mutant. The level of Stat5α and β, which are thought to be crucial for mediation of the growth hormone signaling was not altered nor was their phosphorylation status. Mice with the dimerization-defective mutation are of normal size, and the expression of IGF-I and other growth hormone-regulated genes is unaltered. These results demonstrate that GR function in hepatocytes is crucial for body growth. The different response in GR^{dim} mice in comparison to mice with a hepatocyte-specific mutation provides strong evidence that the growth-promoting activity of the receptor does not require binding to a glucocorticoid-responsive element. It rather appears to function as a coactivating molecule thus synergizing with Stat5 activity, supporting previous in vivo experiments, which demonstrated a requirement of GR for Stat5 activation by prolactin [22].

To study the function of GR in the immune system, we inactivated the GR in thymocytes. Inactivation of GR function in thymocytes was achieved by using the lck gene to drive the Cre recombinase expression. The lck gene is activated during the early differentiation of thymocytes. Thymocytes of these mutant animals are entirely resistant to glucocorticoid-induced apoptosis. To generate mice lacking the receptor in monocytes/macrophages, we used the lysozyme gene promoter for expression of Cre recombinase. When these mice were challenged by treatment with a given dose of LPS, it was observed that all mutant mice died within 36 hours. Wild-type mice survived this challenge. Thymocytes and macrophages from these two mutants are now being analyzed with regard to their gene expression profiles. It will be of interest to define a gene or genes which are required for the glucocorticoid-induced apoptosis of thymocytes. Furthermore, the basis for the protective effect of glucocorticoids with regard to the LPS challenge will certainly be much better understood once LPS-induced genes which are affected by glucocorticoid treatment, have been defined.

To investigate the effects of an increased gene dosage of the GR, we have generated transgenic mice carrying two additional copies of the GR gene by using a yeast artificial chromosome (YAC) carrying the mouse GR gene [23]. Expression of plasmid transgenes in mice is often variable and does not necessarily reflect the expression pattern of the endogenous gene, disadvantages which can be circumvented by YAC transgenesis. Interestingly, overexpression of the GR from a 290 kb YAC alters the basal regulation of the HPA, resulting in reduced expression of C2H and ACTH and a fourfold reduction in the level of circulating glucocorticoids. Animals with an increased gene dosage of GR are more resistant to stress and endotoxic shock. In addition, primary thymocytes obtained from these transgenic mice show an enhanced sensitivity to glucocorticoid-induced apoptosis. These results underscore the importance of tight regulation of GR expression for the control of physiological and pathological processes. Furthermore, they may explain differences in the susceptibility of humans to inflammatory diseases and stress, depending on individual prenatal and postnatal experiences known to influence the expression of the GR.

CONCLUSIONS

Exploiting the Cre/loxP recombination system in the mouse a series of alleles of the GR has been generated. These mutants have been instrumental in the molecular genetic analysis of GR functions in vivo.

They allow us to define four different modes of action of the receptor as illustrated in Fig. 3. The receptor is able to activate gene transcription by binding as a dimer to a GR element in the control region of a regulated gene. The growth-promoting activity of the receptor does not require binding of the receptor to a GRE, but is mediated by interaction of the receptor with Stat5. The receptor functions rather as a coactivator in Stat5 signaling. The receptor inhibits expression of genes such as the POMC gene and the prolactin gene in the anterior pituitary by binding to so-called negative GREs in the control regions of these genes. This hypothesis is strongly supported by our observations of elevated levels of the mRNAs encoding these proteins in mice with the dimerization-defective and thus DNA binding-deficient GR. The receptor is also able to modulate target gene activity by a DNA binding-independent mechanism. It thus is able to enhance the transcriptional activity of the Stat5 proteins and inhibit the activity of AP-1- and NFκB-dependent transcription. It is a fascinating challenge to define the multiple activities of GR in molecular detail in the future.

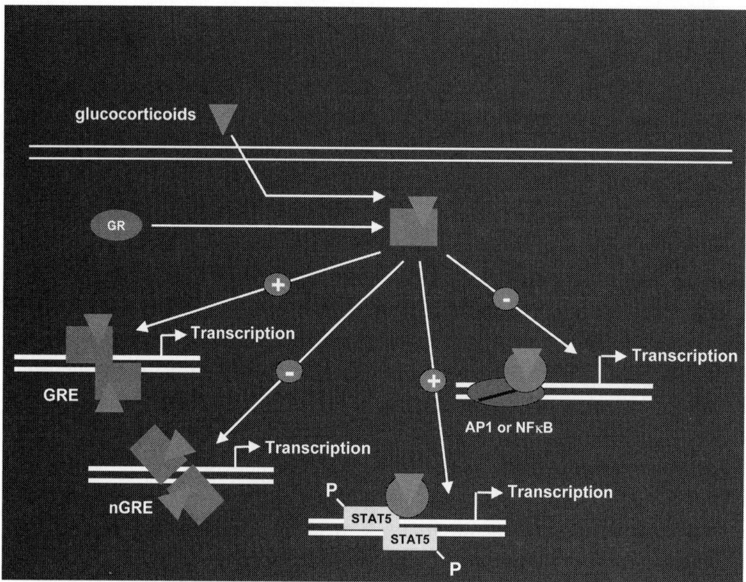

Fig. 3 The GR modulates transcription by different modes of action. The GR is able to influence transcription by binding to a GRE or nGRE. It is assumed that binding of GR to a nGRE leads to an altered GR conformation. The GR can also activate or inhibit transcription by protein/protein interaction as shown on the right.

REFERENCES

1. W. L. Miller and J. Blake Tyrrel. In *Endocrinology and Metabolism*, P. Felig, J. D. Baxter, I. A. Frohman (Eds.), pp. 555–711, McGraw-Hill, New York (1995).
2. J. W. Funder, P. T. Pearce, R. Smith, A. I. Smith. *Science* **242**, 583–585 (1988).
3. D. N. Orth and W. J. Kovacs. In *Williams Textbook of Endocrinology*, J. D. Wilson, D. W. Foster, H. M. Kronenberg, P. R. Larsen (Eds.), pp. 517–664, W. B. Saunders, New York (1998).
4. M. Beato, P. Herrlich, G. Schütz. *Cell* **83**, 851–857 (1995).
5. A. Bauer, F. Tronche, O. Wessely, C. Kellendonk, H. M. Reichardt, P. Steinlein, G. Schütz, H. Beug. *Genes Dev.* **13**, 2996–3002 (1999).
6. M. Joels and E. R. De Kloet. *Prog. Neurobiol.* **43**, 1–36 (1994).
7. H. Gu, J. D. Marth, P. C. Orban, H. Mossmann, K. Rajewsky. *Science* **265**, 103–106 (1994).
8. B. Sauer and N. Henderson. *Proc. Natl. Acad. Sci. USA* **85**, 5166–5170 (1988).

9. T. J. Cole, J. A. Blendy, A. P. Monaghan, K. Krieglstein, W. Schmid, A. Aguzzi, G. Fantuzzi, E. Hummler, K. Unsicker, G. Schutz. *Genes Dev.* **9**, 1608–1621 (1995).
10. F. Tronche, C. Kellendonk, H. M. Reichardt, G. Schütz. *Curr. Opin. Genet. Dev* **8**, 532–538 (1998).
11. S. Berger, M. Heich, T. J. Cole, J. Peters, H. Watanabe, W. Kriz, R. Warth, R. Greger, G. Schütz. *Proc. Natl. Acad. Sci. USA* **95**, 9424–9429 (1998).
12. M. Bleich, R. Warth, M. Schmidt-Hieber, A. Schulz-Baldes, P. Hasselblatt, D. Fisch, S. Berger, K. Kunzelmann, W. Kriz, G. Schütz, R. Greger. *Pflügers Arch. Eur. J. Physiol.* **438**, 245–254 (1999).
13. J. Drouin, Y. L. Sun, M. Chamberland, Y. Gauthier, A. De Lean, M. Nemer, T. J. Schmidt. *EMBO J.* **12**, 145–156 (1993).
14. H. M. Reichardt, K. H. Kaestner, O. Wessely, J. Tuckermann, P. Angel, O. Kretz, R. Bock, W. Schmid, P. Herrlich, G. Schütz. *Cell* **93**, 1–20 (1998).
15. H. M. Reichardt, J. P. Tuckermann, M. Göttlicher, M. Vujic, F. Weih, P. Angel, P. Herrlich, G. Schütz. *EMBO J.* **20**, 7168–7173 (2001).
16. H. M. Reichardt, F. Tronche, S. Berger, C. Kellendonk, G. Schütz. *Hormones & Signalling, Advances in Pharmacology 47* B. O'Malley (Ed.), pp. 1–21, Academic Press, San Diego (2000).
17. J. P. Tuckermann, H. M. Reichardt, R. Arribas, K. H. Richter, G. Schütz, P. Angel. *J. Cell Biol.* **147**, 1–6 (1999).
18. P. J. Barnes. *Clin. Sci.* **94**, 557–572 (1998).
19. F. Tronche, C. Kellendonk, O. Kretz, P. Gass, K. Anlag, P. C. Orban, R. Bock, R. Klein, G. Schütz. *Nat. Genet.* **23**, 99–103 (1999).
20. C. Kellendonk, C. Opherk, K. Anlag, G. Schütz, F. Tronche. *Genesis* **26**, 151–153 (2000).
21. M. H. MacGillivray. In *Endocrinology and Metabolism*, P. Felig, J. D. Baxter, L. A. Frohman (Eds.), pp. 1619–1673, McGraw-Hill, New York (1995).
22. E. Stocklin, M. Wissler, F. Gouilleux, B. Groner. *Nature* **383**, 726–728 (1996).
23. A. Schedl, L. Montoliu, G. Kelsey, G. Schutz. *Nature* **362**, 258–261 (1993).

Pure Appl. Chem., Vol. 75, Nos. 11–12, pp. 1709–1732, 2003.

Topic 1.6

Functions of RARs and RXRs in vivo: Genetic dissection of the retinoid signaling pathway*

Manuel Mark and Pierre Chambon‡

Institut de Génétique et de Biologie Moléculaire et Cellulaire (IGBMC) (CNRS/INSERM/ULP/Collège de France), Institut Clinique de la Souris (ICS) BP 10142, 67404 Illkirch Cedex, France

Abstract: Retinoids, the active metabolites of vitamin A, regulate complex gene networks involved in vertebrate morphogenesis, growth, cellular differentiation, and homeostasis. They are used for the treatment of skin disorders and as chemopreventive agents for certain cancers. Molecular biology and genetic studies performed during the last 15 years in vitro, using either acellular systems or transfected cells, have shown that retinoid actions are mediated through heterodimers between the 8 major RARα, β, and γ isoforms and the 6 major RXRα, β, and γ isoforms that belong to the nuclear receptor (NR) superfamily, and act as ligand-dependent transcriptional regulators. Furthermore, RXRs not only heterodimerize with RARs, but also with numerous other members of the NR superfamily. As in vitro studies are carried out under nonphysiological conditions, they only indicate what is possible, but not necessarily what is actually occurring in vivo. Therefore, mutations have been introduced by homologous recombination (HR) in F9 embryonal carcinoma (EC) cells, a cell-autonomous system that differentiates in the presence of RA, in order to disrupt RAR and RXR genes and establish their cellular and molecular functions in RA-induced differentiation. However, genetic approaches in the animal should be used to determine the function of retinoid receptors under truly physiological conditions. HR in embryonic stem (ES) cells, has therefore been used to generate null mutations of the various RARs and RXRs in the mouse germline. As reviewed here, the generation of such RAR and RXR germline mutations, combined with pharmacological approaches to block the RA signaling pathway, has provided many valuable insights on the developmental functions of RA receptors. However, due to (i) the complexity in "hormonal" signaling through transduction by the multiple RARs and RXRs, (ii) the functional redundancies (possibly artefactually generated by the mutations) within receptor isotypes belonging to a given gene family, and (iii) in utero or postnatal lethality of certain germline null mutations, these genetic studies through germline mutagenesis have failed to reveal many of the physiological functions of RARs and RXRs, notably in adults. We conclude that spatio-temporally controlled somatic mutations generated in animal models in given cell-types/tissues and at chosen times during pre- and postnatal life, are required to reveal the physiological and pathophysiological functions of the receptor genes involved in the retinoid signaling pathway throughout the life of the mouse.

---

*Report from a SCOPE/IUPAC project: Implication of Endocrine Active Substances for Human and Wildlife (J. Miyamoto and J. Burger, editors). Other reports are published in this issue, *Pure Appl. Chem.* **75**, 1617–2615 (2003).

‡Corresponding author: Tel.: 33 3 88653213, Fax: 33 3 88 653203, E-mail: chambon@igbmc.u-strasbg.fr.

INTRODUCTION: THE BASICS OF RETINOID SIGNALING

Both clinical and experimental approaches have revealed that vitamin A (retinol) and its biologically active derivatives (collectively referred as to retinoids), notably retinoic acids, exert a wide variety of profound effects on vertebrate embryonic morphogenesis and organogenesis, cell proliferation, differentiation and apoptosis, homeostasis, as well as in their disorders (reviewed in refs. [1–6]). In the 1930s, Hale initially showed that vitamin A deficiency (VAD) induces ocular malformations in newborn pigs [7]. Subsequently, the group headed by Warkany demonstrated that fetuses from VAD rats exhibit a large array of congenital malformations affecting the eyes, the myocardium and heart outflow tract, the diaphragm, as well as the respiratory and urogenital systems (the fetal VAD syndrome, see Tables 1 and 2; reviewed in ref. [8]). Dietary deficiency studies also showed that vitamin A is indispensable throughout postnatal development and adult life for growth, survival, reproduction, vision, and also for the homeostasis of numerous tissues. Indeed, widespread squamous metaplasia of glandular and transitional epithelia, degeneration of the seminiferous tubules and of the retina, are hallmarks of the postnatal VAD syndrome [9]. Retinoic acid (RA), the most biologically active naturally occurring retinoid, can both prevent and rescue the defects caused by a VAD diet in adult animals with the exceptions of night-blindness and retinal degeneration [3,10]. It was also shown recently that RA could replace vitamin A during embryogenesis, at least at certain stages and in certain organs [11,12].

How these structurally simple molecules can exert such pleiotropic effects was a long-standing question which found its solution with the discovery of two classes of nuclear ligand-dependent transcriptional regulators that belong to the superfamily of nuclear receptors (NRs), the three retinoic acid receptors isotypes (RARα, β, and γ, that bind all-trans and 9-cis RAs) and the three retinoid X receptor isotypes (RXRα, β, and γ, that selectively bind 9-cis RA), (for refs. see [13]). RARs and RXRs exhibit the conserved modular structure of NRs (reviewed in refs. [14,15], which can be divided into six variably conserved homology regions A to F. In vitro studies, performed with either cell-free systems or cultured cells co-transfected with vectors overexpressing the different RARs and RXRs together with recombinant reporter genes, demonstrated that RARs and RXRs bind as RAR/RXR heterodimers to DNA response elements of RA-responsive genes, and also allowed the determination of the functions of the A to F regions [13]. It was notably shown that the highly conserved centrally located region C corresponds to the core of the DNA-binding domain (DBD), while region E is functionally complex, as it contains the ligand-binding domain (LBD), a surface for RAR/RXR heterodimerization and, in the case of RARs, a transcriptional silencing domain that binds corepressors. The LBD also contains the ligand-dependent transactivation function AF-2 that requires the integrity of a highly conserved amphipathic helix, the AF-2AD core that corresponds to helix 12 located at the C-terminal end of the LBD [13]. Agonistic ligand binding induces a major structural change in the conformation of helix 12, which creates a new LBD surface for binding of coactivators, while corepressors are released [13,16–19 and refs. therein]. In addition, the amino-terminal A/B region contains the ligand-independent transcriptional activation function AF-1.

For each RAR isotype, there are several isoforms that differ from one another in their N-terminal region A. These isoforms arise from the differential usage of two promoters (of which the downstream one, P2, is RA-inducible) and alternative splicing. There are two major isoforms for RARα (α1 and α2) and for RARγ (γ1 and γ2), and four major isoforms for RARβ (β1 and β3 initiated at the P1 promoter, and β2 and β4 initiated at the P2 promoter) [13,14, and refs. therein]. Similarly, several major isoforms differing from another in their N-amino terminal region have been identified for RXRα (α1 and α2), RXRβ (β1 and β2). and RXRγ (γ1 and γ2) [for refs., see 13]. For a given receptor isoform, the AF-1 and AF-2 activities synergize and exhibit some specificity that is dependent on both the cell-type and the promoter-context of RA-responsive genes [20,21]. Synergistic transcriptional activation has also been observed between RAR and RXR partners, indicating that RXRs are not a priori transcriptionally silent partners in RAR/RXR heterodimers [13,22, and refs. therein]. However, at least in some instances the ligand-dependent activity of RXR is "subordinated" to ligand binding to its RAR partner [23]. Such an RXR subordination may also apply to other NRs that use RXRs promiscuously as heterodimerization partners, in order to avoid confusion between retinoid and other signaling pathways. Interestingly, this appear to be the case for the thyroid hormone and vitamin D3 signaling pathways, as ligand binding to TR and VDR is a prerequisite for the RXR partner to respond to its agonistic ligand. However, in the case of other NRs for which RXRs act as heterodimeric partners (e.g., FXR, LXRs, and PPARs), the RXR ligand-induced activity appears to be "permissive", i.e., RXR agonists can autonomously activate transcription through the corresponding heterodimers [see 13,15].

The high degree of conservation of the various RAR or RXR isoforms across vertebrate evolution, as well as their selective spatio-temporally expression patterns in developing embryos and adult tissues (for refs., see [14,24,25]), led to the initial suggestion that each RAR and RXR isoform may perform unique functions, thus accounting for the highly diverse effect of retinoic acid throughout vertebrate life [14]. Results of cell-free and cellular in vitro studies led to the further proposal that the highly pleiotropic effects of RA reflect a highly combinatorial mechanism in which the multiple actors (the heterodimers and their coregulators) differentially transduce retinoid signals to selectively control the expression of numerous sets of RA target genes [13,14].

A genetic dissection of the retinoid signaling pathway was obviously required to investigate the in vivo relevance of the above in vitro-characterized molecular mechanisms, and to determine the physiological functions of the multiple retinoid receptors. As genetic analyses in the mouse are tedious, and also because their interpretation at the molecular level could be equivocal due to difficulties in discriminating between cell-autonomous and non-cell-autonomous events in the intact animal, we first chose the RA-responsive F9 murine embryonal carcinoma (EC) cell line as a cell-autonomous model system for analyzing RA signaling under in vitro conditions that mimic, at least to some extent, physiological processes occurring during early embryogenesis (reviewed in ref. [26]). Combining a genetic strategy (targeted mutagenesis of RARs and RXRs through homologous recombination, followed by re-expression of wild-type or mutant receptors in rescued lines), and a pharmacological strategy using RAR isotype- and panRXR-selective synthetic retinoids, established that (i) RXR/RAR heterodimers are the functional units that selectively mediate RA-induced differentiation, growth arrest, and target gene expression in F9 cells; (ii) the AF-1 and AF-2 activation functions of RARs and RXRs act synergistically and selectively to transcriptionally control the physiological and molecular responses of F9 cells to RA; (iii) the AF-2 ligand-dependent transcriptional activity of RXRs is subordinated to ligand binding to their RAR heterodimeric partners; (iv) RAR and RXR gene knock-out may generate artefactual conditions unmasking potential functional redundancies between RAR or RXR isotypes, which do not exist under wild-type conditions. In other words, in these cases, suppression or mutation of a given RAR or RXR can be functionally compensated by another RAR(s) or RXR(s), respectively. In addition, it was found that phosphorylation within the AF-1 activation domain of RARγ is required for primitive endodermal differentiation of F9 EC cells, while phosphorylation of RARα in the LBD is required for parietal endodermal differentiation, leading to the conclusion that, through binding of cognate ligands and phosphorylation of their activation domains, retinoid receptors act as highly sophisticated signal transducers, integrating signals belonging to distinct signaling pathways involving both membrane and nuclear receptors.

In the present review, we summarize and discuss what has been learnt from the genetic dissection of the retinoid signaling pathway under truly physiological conditions, i.e., at the organismal level, in the mouse, during development, and postnatally. We focus on two main questions: (i) Where and when are RARs and RXRs involved in the transduction of the retinoid signals in the mouse? To that end, we review the phenotypes of mice in which the various RARs and RXRs have been ablated, and compare them to those of mice which have been under a VAD diet, in order to determine in which developmental or homeostatic events the various RARs and RXRs are instrumental. (ii) To what extent do the results of the genetic dissection of the retinoid signaling pathways in the mouse support the molecular mechanisms underlying the transduction of the RA signal by retinoid receptors, as they have been deduced from in vitro acellular and cellular studies?

GENETIC EVIDENCE THAT RARs TRANSDUCE RETINOID SIGNALS IN VIVO: RAR (α, β, AND γ) NULL MUTANT MICE DISPLAY SEVERAL ASPECTS OF THE POSTNATAL VAD SYNDROME AS WELL AS SOME CONGENITAL ABNORMALITIES

RARα inactivation results in a spermatogenetic impairment and reappearance of an atavistic trait

While RARα1 and RARα2 null mutants are apparently normal, RARα null mutants males are sterile due to a low production of spermatozoa [27,28, and our unpublished results] (Table 1). The Sertoli cell is the main target of the RARα null mutation, as the RARα protein is normally readily detectable in these cells, but not in germ cells [28,29]. Spermatogenesis is complete in adult mutants testes, which nevertheless display a degeneration characterized by the presence of large and empty vacuoles located between Sertoli cells. These vacuoles probably appear as a consequence of a chronic loss of round spermatids, which are prematurely released into the lumen of the mutant seminiferous tubules, before the completion of their maturation phase (spermiogenesis). The nature of the seminiferous epithelium lesions suggests that a signaling pathway mediated by RARα favors cell cohesiveness within the seminiferous epithelium. Interestingly, a RA-inducible gene, *Stra6*, encodes an integral membrane protein of Sertoli cells whose spermatogenetic cycle-dependent expression is lost in testes of RARα null mutants [30]. The onset of the testicular degeneration is variable: in some RARα null testes, numerous vacuoles are already present at P10 [postnatal day 10, i.e., before completion of puberty (around P35)], whereas other mutant testes still appear histologically normal at six weeks of age (our unpublished results). A second intriguing feature of adult RARα null testes is the patchy distribution of the lesions: seminiferous tubules containing a full complement of germ cells are often found adjacent to tubules composed almost exclusively of Sertoli cells [28]. This variability suggests that RARβ and RARγ, which are also detected in Sertoli cells [31,32], partially compensates for the lack of RARα, and that their activity varies stochastically among RARα null mutant males, and also between the seminiferous tubules of a given RARα null testis. Functional compensation by RARβ and RARγ probably also accounts for the observation that the RARα null testicular phenotype mimics a state of mild VAD, as opposed to a severe and prolonged VAD, which yields testes devoid of germ cells [33].

Aside from this completely penetrant testis degeneration, some RARα null mutants show some congenital defects (Table 1), which include interdigital webbing and a supernumerary skeletal element connecting the incus (a middle ear ossicle) with the alisphenoid bone (a braincase bone). This latter congenital abnormality corresponds to an atavistic trait, defined as the reappearance of a character that was lost during evolution, namely the upper jaw (or pterygoquadrate) cartilage present in reptilian ancestors of mammals. The occurrence of several other atavistic features in RAR single and double null mutants strongly support the view that modulation of RA signaling has been employed during vertebrate evolution as a mean to modify skull shapes and functions (see refs. [34–37]).

RARγ inactivation causes epithelial defects and some congenital malformations

RARγ males are sterile but, contrary to RARα null mice, their spermatogenesis is normal [38] (Table 1). This sterility results from the transformation of the glandular epithelia of the seminal vesicle and prostate (which normally produce essential fractions of the seminal fluid) into epithelia resembling epidermis. Aside from sterility, this aberrant cell differentiation process known as "squamous metaplasia" is responsible for severe genito-urinary tract infections, and extends, upon aging, to other epithelia, namely those of the epididymis, urinary bladder, urethra, and salivary glands [33]. As squamous metaplasia is one of the hallmarks of the postnatal VAD syndrome (see Introduction), RARγ probably plays a crucial role in the maintenance of all epithelia that require RA in the mouse. Along these lines, it is noteworthy that keratinization of the tracheal and conjunctival epithelia, which are hallmarks of rat VAD, are not seen in models of VAD mice, such as CRBPI (cellular retinol binding protein one) and RBP (retinol binding protein) null mutants raised on VAD diets, possibly reflecting species-specific differences in RA tissue requirements [33, and our unpublished results]. Likewise, RARγ null mice and VAD mice, surprisingly, do not develop the corneal ulcerations found in the human VAD [39].

Table 1 Postnatal manifestations of RAR and RXR knock-outs. CD: congenital defects; PnVAD: abnormalities present in postnatal VAD.

Genotypes	Abnormalities	Refs.
RARα1$^{-/-}$	None	[27,28]
RARα2$^{-/-}$	None	our unpublished results
RARα$^{-/-}$	Growth retardation (PnVAD); male sterility (testis degeneration[a]; PnVAD); impaired alveolar formation. CD: webbed digits; homeotic transformations and malformations of cervical vertebrae; pterygoquadrate cartilage; malformation of the squamosal bone; malformed laryngeal (i.e., cricoid) cartilage	[28,35,38,42,101]
RARβ1$^{-/-}$/RARβ3$^{-/-}$	None	[44]
RARβ2$^{-/-}$/RARβ4$^{-/-}$	CD: persistence and hyperplasia of the primary vitreous body (fetal VAD)	[43]
RARβ$^{-/-}$	Growth retardation (PnVAD); behavioral defects; altered alveolar formation. CD: homeotic transformations and malformations of cervical vertebrae; persistence and hyperplasia of the primary vitreous body (fetal VAD)	[44–46,81,98,99]
RARγ1$^{-/-}$	Growth deficiency (PnVAD). CD: malformations of cervical vertebrae[a]; malformed laryngeal (i.e., cricoid) cartilage; abnormal differentiation of granular keratinocytes[a]	[42], our unpublished results
RARγ2$^{-/-}$	None	[38]
RARγ$^{-/-}$	Growth deficiency (PnVAD); male sterility (squamous metaplasia of the seminal vesicle and prostate gland epithelia[a] (PnVAD); squamous metaplasia of other epithelia (PnVAD); impaired alveolar formation. CD: webbed digits; homeotic transformations and malformations of cervical vertebrae[a]; malformed laryngeal cartilages and tracheal rings[a]; agenesis of the Harderian glands; agenesis of the metoptic pillar of the skull; abnormal differentiation of granular keratinocytes[a]	[35,38,41,100], our unpublished results
RARβ/RARγ1$^{-/-}$	Hydronephrosis	[42]
RARβ2$^{-/-}$/RARγ2$^{-/-}$	Dysplasia and degeneration[a] of the retina. CD: persistence and hyperplasia of the primary vitreous body (fetal VAD); blepharophimosis[a]; partial agenesis of the sclera and choroid[a]	[43]
RARβ1$^{-/-}$/RARβ3$^{-/-}$ /RARγ$^{-/-}$	Degeneration of the retina. CD: persistence and hyperplasia of the primary vitreous body (fetal VAD); partial agenesis of the sclera and choroid[a]	[44]
RXRα$^{+/-}$	Growth retardation (PnVAD). CD: webbed digits	[24,47]
RXRβ$^{-/-}$	Male sterility[a] (defective spermatogenesis); abnormal lipid metabolism in Sertoli cells; behavioral defects	[45,83]
RXRγ$^{-/-}$	Metabolic and behavioral defects	[45,46,81,84]

[a]These abnormalities are completely penetrant. RARα1$^{-/-}$, RARα2$^{-/-}$, RARβ1$^{-/-}$, RARβ2$^{-/-}$, RARβ3$^{-/-}$, RARβ4$^{-/-}$, RARγ1$^{-/-}$, and RARγ2$^{-/-}$ refer to isoform-specific knock-outs.

RARγ null newborns are readily distinguishable from their littermates by the glossy appearance of their skin [40]. Their flat epidermal surface, accounting for this characteristic external aspect, likely reflects a deficit in corneodesmosomes which normally provide cell cohesion within the stratum corneum. Moreover, lamellar bodies in the mutant epidermal granular layer are morphologically abnormal and fail to be properly exocytosed, resulting in an accumulation of abnormal vesicles in extracellular spaces and uneven deposition of lipids in the stratum corneum (our unpublished results). These defects persist in adults, and are reminiscent of those observed in cases of human congenital ichthyoses and in elderly skin xerosis. RA is a widely used therapeutic agent for skin diseases. In this context, it is noteworthy that the subtle role of RARγ in the differentiation of granular keratinocytes, which is revealed only by careful electron microscopic analysis of the RARγ null mutants, probably represents the only physiological functions of the whole RAR family in the homeostasis of the mouse resting epidermis [41].

RARγ null newborns also display laryngeal and tracheal cartilage malformations, as well as interdigital webbing (Table 1). The Harderian glands, which provide for lubrification of the eyelids are often missing in RARγ null mutants. Moreover, all of these mutants (as well as some RARα and RARβ mutants) display homeotic transformations of cervical vertebrae, indicating that signaling through RARs is required for the patterning of the antero-posterior body axis during somite formation [35,38] (Table 1).

Interestingly, mice lacking specifically the RARγ1 isoform display the congenital skin defect, as well as vertebral and laryngeal malformations characteristic of the RARγ null mutation (all isoforms disrupted), but not the squamous metaplasia, interdigital webbing, and Harderian gland agenesis of RARγ null mutants [38,42]. In contrast, RARγ2 null mutants are apparently normal [38,42] (Table 1). Thus, RARγ1 is the main RARγ isoform involved in several developmental processes, postnatal growth and skin homeostasis, whereas any of the two isoforms can perform the RARγ functions required for the maintenance of genito-urinary tract and glandular epithelia, and for the involution of the interdigital mesenchyme .

RARβ null mutants display the most frequent congenital VAD abnormality and a locomotor deficiency

Almost all RARβ null mutants display a persistent hyperplastic primary vitreous body (PHPV or retrolenticular membrane), consisting in an abnormal mass of fibro-vascular tissue which is interposed between the lens and the retina, and representing the most frequent malformation of the fetal VAD syndrome [8,35] (Table 1). The RARβ null mutants PHPV can be ascribed to a lack of RARβ2/β4 isoform activity, as it is also observed in almost all RARβ2/β4 null mutants [43]. In contrast, specific inactivations of RARβ1 and RARβ1/β3 isoforms do not yield morphological defects [44].

Abundant RARβ protein is detected in the adult central nervous system, notably in the striatum which is involved in the control of voluntary movements. Despite the apparent normality of this structure at the morphological level, all RARβ mutants display severely impaired abilities in locomotion and motor coordination tests [45]. Performance deficits in spatial learning, possibly unrelated to the visual impairment caused by the PHPV, and alterations of hippocampal synaptic plasticity have also been reported in RARβ null mutants [46].

SIGNALING THROUGH RARs IS INDISPENSIBLE, AT MANY DISTINCT DEVELOPMENTAL STEPS, FOR EMBRYONIC PATTERNING AND SHAPING, AND ORGANOGENESIS

As reviewed above, RARα, RARβ, and RARγ null mutant mice altogether display some aspects of the postnatal and fetal VAD syndromes (Table 1). However, RAR (α, β, or γ) single null mutant mice abnormalities are confined to a small subset of the tissues normally expressing these receptors during embryogenesis and adulthood, probably reflecting the existence of functional redundancies between RARs [discussed in 3,22,47,48]. To test this hypothesis, mutants lacking a pair of RAR isotypes (RARαβ, RARαγ and RARβγ double null mutants) or two or more isoforms belonging to distinct isotypes were generated. For the sake of clarity, only abnormalities displayed by double null mutants lacking a couple of RAR isotypes (all isoforms deleted) are listed in Tables 2 and 3. Similar abnormalities, albeit often less penetrant, which are displayed by "isoform-specific" double null mutants are listed in [3] (RARα1/RARβ2/4, RARα1/RARγ, RARα1α2$^{+/-}$/RARγ and RARβ2/4/RARγ mutants), in [42] (RARα/RARγ1 and RARα/RARγ2 mutants), in [49] (RARα1/RARβ mutants), [44] (RARα/RARβ1/3 and RARβ1/3/RARγ mutants), and in [43] (RARβ2/RARγ2 mutants).

Table 2 Abnormalities of the fetal VAD syndrome present in RARβ null mutants (β), RXRα null mutants, and RAR(α, β, and γ) double null mutants (α/β, α/γ, and β/γ).

Abnormalities of the fetal VAD syndrome	Genotypes of RAR null mutants showing similar abnormalities	Abnormalities observed in RXRα null mutants
Respiratory system defects		
• Agenesis or hypoplasia of the left lung	α/β[a]	No
• Hypoplasia of the right lung	α/β[a]	No
• Agenesis of the oesophago-tracheal septum	α/β[a]	
• Diaphagmatic hernia	α/β	No
Hypoplasia of the ventricular myocardium	α/γ	Yes[a]
Heart outflow tract defects		
• Persistent truncus arteriosus	α/β[a], α/γ[a]	No
• High interventricular septal defect or double outlet right ventricle (E18.5)/conotruncal septum defect (E14.5)	α/β[a], α/γ[a], β/γ	Yes
• Abnormal great arteries derived from aortic arches	α/β[a], α/γ[a], β/γ	No
Kidney hypoplasia	α/β[a], α/γ[a]	No
Ureteral defects		
•Agenesis	α/β, α/γ[a]	No
•Ectopia	α/β, β/γ	No
Genital tract defects		
– **Female** • Agenesis of the oviduct and uterus (E18.5)/agenesis of the Müllerian duct (E14.5)		
– Complete	α/β[a]	No
– Partial	α/γ[a]	No
• Agenesis of the cranial vagina	α/β[a], α/γ[a]	NA
– **Male** • Agenesis or dysplasia of the vas deferens	α/γ[a]	NA
• Agenesis of the seminal vesicles	α/γ[a]	NA
Ocular defects		
• Coloboma of the retina	α/γ[a]	No
• Coloboma of the optic disc	β/γ, α/γ[a]	Yes
• Persistence et hyperplasia of the primary vitreous body (PHPV)	β, α/β[a], β/γ[a], α/γ[a]	Yes[a]
• Hypoplasia of the conjunctival sac	β/γ[a], α/γ[a]	Yes[a]
• Thickening of the corneal stroma	β/γ[a], α/γ	Yes[a]
• Ventral rotation of the lens	β/γ[a]	Yes[a]
• Shortening of the ventral retina	β/γ[a]	Yes[a]

[a]This abnormality is completely penetrant. NA not applicable as the corresponding structure is normally not found at E14.5, the time around which RXRα null mutants die. From refs. [24,34,35]. Note that most of the abnormalities seen in RARα/RARβ double null mutants occur at similar frequencies in RARα/RARβ2 mutants [50].

Table 3 Abnormalities absent from the fetal VAD syndrome are found in RAR(α, β, or γ) single null mutants (α, β, γ), and RAR(α, β, and γ) double null mutants (α/β, α/γ, and β/γ).

Congenital abnormalities not associated with the fetal VAD syndrome	Genotypes of RAR null fetuses showing these defects
Nervous system defects	
• Exencephaly	α/γ
• Agenesis of the corpus callosum	α/γ[a]
Skeletal defects	
• Agenesis or multiple malformations of cranial skeletal elements	α/γ[a]
• Homeotic transformation and malformations of cervical vertebrae	α, β, γ, α/γ[a], α/β[a], β/γ[a]
• Agenesis and malformations of limb bones	α/γ[a]
• Reappearence of atavistic skeletal elements	
– Antotic pillar	α/γ[a]
– Ptrerygoquadrate cartilage	α, α/γ[a], α/β
Eye defects	
• Corneal-lenticular stalk	α/γ
• Agenesis of the lens	α/γ
Glandular defects	
• Agenesis or dysplasia, of the sub-maxillary and sub-lingual glands and/or of their excretory ducts	α/γ[a], β/γ[a]
• Agenesis of the Harderian gland	γ, α/γ[a], β/γ[a]
• Agenesis or ectopia of the thyroid, thymus and parathyroid glands	α/β, α/γ
Other defects	
• Webbed digits	α, γ, $\beta^{+/-}$/γ[a], β/γ[a]
• Abnormal laryngeal cartilages and tracheal rings	α, γ[a], α/γ[a], α/β[a], β/γ[a]
• Kidney agenesis	α/γ
• Agenesis of the anal canal	α/β[a]

[a]This abnormality is completely penetrant. From refs. [34,35,38,42]. Note that most of the abnormalities seen in RARα/RARβ double null mutants occur at similar frequencies in RARα/RARβ2 mutants [50].

RARαβ, RARαγ and RARβγ double null mutants, as well as double mutants for RAR isoforms belonging to different isotypes (with the exception of RARβ2/RARγ2) die in utero or at birth because of severe developmental defects that altogether include the complete spectrum of malformations corresponding to the classical fetal VAD-induced syndrome reported by Warkany's group 50 years ago [8] (Table 2). As RAR single mutants (see Table 1 and above), RAR double null mutants (Table 3) also exhibit congenital abnormalities that were not described in the classical fetal VAD studies, ranging from ageneses of the Harderian glands to skeletal defects of the skull, face, vertebrae and limbs [34,35,38,42,49,50]. The occurrence of these "non-VAD" defects in RAR double null mutant mice is most probably accounted by the difficulty to achieve, by dietary deprivation, a state of profound VAD compatible with pregnancy (see above). In fact, almost all these "non-VAD" defects have been subsequently produced in rodent embryos (i) deficient in vitamin A, but supplemented with RA [11,12,51]; (ii) lacking the RA synthesising enzymes RALDH2 (retinaldehyde dehydrogenase 2) or RALDH3 [52–56; and our unpublished results] or (iii) treated with synthetic retinoids possessing panRAR antagonistic activities [57–59].

The comparison of the RAR double null mutants phenotypes with those of rodents carrying the aforementioned blocks in RA signal transduction, demonstrate that liganded RARs play crucial roles at many distinct stages of the development of numerous organs [3, and refs. therein]. For example, the severe malformations found in RARαγ double null embryos are similar to those of RALDH2 null embryos, and reflect early roles of RAR signaling in axial rotation, mesoderm segmentation and closure of the hindbrain, formation of otocysts, pharyngeal arches, and forelimb buds, as well as in closure of the primitive gut [58]. RARs are also indispensable for the ontogenesis of (almost) all the anatomical structures that are derived from mesectodermal cells, i.e., the cranial neural crest cells (NCCs) that give rise to mesenchymal derivatives (reviewed in refs. [3,36,37], and see below). RARs are involved in antero-posterior patterning of the somitic mesoderm and hindbrain neurectoderm [34,35,38,58,60] through controlling homeobox gene expression [60–63], as well as in the establishment of the antero-posterior axis of the limbs [22,34,64]. RARs are required for the development of a large number of eye structures (Tables 2 and 3) and for retinal histogenesis [34,35,43], cardiomyocyte differentiation [24,65], as well as for the control of apotosis in the retina [43], the frontonasal and interdigital mesenchymes [34,35,64], and conotruncal segment of the embryonic heart [44]. In the embryonic urogenital tract, RARs control epithelial-mesenchymal interactions in the kidney through expression of the receptor tyrosine kinase Ret [66–68], as well as the formation of the genital ducts and ureters [35,50,68]. RARs also regulate distinct steps of lung morphogenesis and are required for the partitioning of the primitive foregut into oesophagus and trachea [69, and see below].

Due to the apparent functional redundancy between RARs in RAR knocked-out mice (discussed in refs. [3,22,47,48]), the number of organs that require RA for their development might be underestimated from the panel of malformations displayed by RARαβ, RARαγ, and RARβγ isotype or isoform double mutants (Tables 2 and 3). In this respect, it is noteworthy that the developing tooth, one of the favorite models for the study of morphogenetic epithelial-mesenchymal interactions, expresses a specific RA-metabolizing enzyme (P. Dollé, personal communication) and critically requires RA in organ cultures [70,71], although it is not obviously altered in any of the RAR double null embryos.

RXRα IS ESSENTIAL FOR TRANSDUCING RA SIGNALS NECESSARY FOR THE DEVELOPMENT OF THE MYOCARDIUM AND EYE STRUCTURES

All RXRα null mutants display an hypoplasia of the compact layer of the ventricular myocardium manifested on histological sections by thin and spongy ventricular walls. This defect appears to be the main cause of mutant death, occurring by cardiac failure around E14.5 [24,65,72] (Table 2). A similar hypoplasia of the myocardium is observed in VAD fetuses, and in some RAR double null fetuses [8,50], suggesting that RXRα is involved in the transduction of an RA signal required for myocardial growth. Additionally, as early as E8.5, the most peripheral ventricular cardiomyocytes in the heart of RXRα null mutants show precocious features of differentiation, such as myofibril striation, indicating that RXRα normally plays an early role in the differentiation of these cells. However, the aforementioned cell differentiation defect, which is also exhibited by RARα null mutants, is not sufficient to account for the hypoplasia of the myocardium [65]. The requirement of RXRα for myocardial growth is unlikely to be cell-autonomous. Indeed, breeding of a transgenic mouse line specifically overexpressing an RXRα protein in cardiomyocytes onto an RXRα null genetic background does not prevent the hypoplasia of the ventricular myocardium and fetal lethality associated with the RXRα null genotype, even though the transgene is expressed in the ventricles as early as E10.5 [73]. Recent data suggest that RXRα located in the adjacent epicardium may act on myocardial growth through a paracrine mechanism [74]. In addition to the abnormal histogenesis of the ventricular myocardium, downregulation of genes involved in general metabolism in RXRα null embryos might participate in their death by cardiac failure [75].

About one-third of the RXRα null mutants lack the conotruncal septum, which normally divides the embryonic heart outflow tract (or conotruncus) into the intracardiac portions of the aorta and pulmonary trunk [24]. Interestingly, deficiencies of this septum represent both a classical VAD defect in rodents and a leading cause of human congenital heart defects, ranging from high interventricular septal defects to double outlet right ventricle (DORV). The agenesis of the conotruncal septum in RXRα null mutants appears secondary to an enhanced rate of cell death in both the mesenchymal cells of the conotruncal ridges and the parietal conotruncal cardiomyocytes, therefore indicating that RXRα is required for the transduction of the RA signal that controls apoptosis in the conotruncal segment of the embryonic heart [44].

In addition to heart defects, all fetuses lacking RXRα show a characteristic ocular syndrome associating a PHPV, closer eyelid folds, thickened ventral portion of the corneal stroma, ventral rotation of the lens, agenesis of the sclera, and a shorter ventral retina [24] (Table 2). As similar defects are present in VAD fetuses and in RARβγ double null mutants [8,35] (Table 2), RXRα appears essential to transduce the RA signals required for several ocular morphogenetic processes, notably the formation of the ventral retinal field. Interestingly, this later event critically requires the activity of a specific retinaldehyde dehydrogenase, RALDH3 [76, and our unpublished results].

That all the congenital defects exhibited by RXRα null fetuses are also observed in RAR single or double null mutants provided the first genetic evidence of a convergence between RAR and RXR signaling pathways, and also gave the first clue that RXRα/RAR heterodimers are the functional units that transduce RA signals in vivo (see below).

RXRs CAN BE TRANSCRIPTIONALLY ACTIVE, AND THE AF-1-CONTAINING A/B REGION AND THE AF-2 TRANSCRIPTIONAL ACTIVATION FUNCTION OF RXRα ARE DIFFERENTIALLY INVOLVED IN OCULAR MORPHOGENESIS AND IN MYOCARDIAL GROWTH

Whether RXRs are transcriptionally active within RAR/RXR heterodimers has been a controversial issue in in vitro studies (see Introduction). To determine the roles played by RXRα AF-1 and AF-2 activities in vivo, mouse mutants were engineered that express truncated RXRα proteins lacking either (i) most of the RXRα N-terminal A/B region that includes AF-1 (RXRα*af1*^{o} mutants [77]), or (ii) the C-terminal 18 amino acid-long sequence of the RXRα protein that includes the core of the activating domain of the activation function 2 (AF-2AD core) (RXRα*af2*^{o} mutants [22]), or (iii) both AF-1 and AF-2 activities (RXRα*af*^{o} mutants; Mascrez et al., unpublished results).

RXRα*af2*^{o} mutant mice occasionally display the hypoplasia of the myocardium and the ocular syndrome characteristic of the RXRα null syndrome, whereas RXRαaf1^{o} mutants never display RXRα-like developmental defects apart from a small and weakly penetrant PHPV [22,77]. The low frequency in RXRα*af2*^{o} mutants, and near-absence in RXRα*af1*^{o} mutants, of defects that are fully penetrant in RXRα null mutants could reflect a functional compensation by RXRβ and/or RXRγ (note that RXRβ null, RXRγ null, as well as RXRβγ double null mutants develop normally; see below). This seems, indeed, to be the case for RXRα*af2*^{o} mutants as (i) the frequency of their ocular syndrome increases from less than 15 to 100 % upon further inactivation of RXRβ (which has no effect on its own), and (ii) the frequency of their myocardial defect increases from 5 to about 50 % upon additional inactivation of either the RXRβ gene, or of both the RXRβ and the RXRγ genes, yielding RXRα*af2*^{o}/RXRβ null and RXRα*af2*^{o}/RXRβ null/RXRγ null fetuses, respectively [22]. On the other hand, PHPV is the only defect of the RXRα null syndrome whose frequency is increased in RXRα*af1*^{o}/RXRβ null/RXRγ null mutants, when compared to RXRα*af1*^{o} mutants [77].

Even though it can be achieved only in an impaired genetic background, the full penetrance of the RXRα null ocular phenotype that is obtained in RXRα*af2*^{o}/RXRβ null mutants (and also in RXRα*af*^{o} mutants; [22] and unpublished results), supports the view that the RXRα AF-2 activity (and thus possibly 9-cis RA) is indispensible for ocular morphogenesis. In contrast, involution of the primary vitreous body, the developmental process which is likely to require the highest concentration of RA-liganded receptor (as it is the event the most sensitive to VAD) [8], requires the additional integrity of the RXRα A/B domain, while in a wild-type this latter domain is dispensible for the other RA-dependent ocular morphogenesis event. These interpretations are supported by the observations that RXRα null-like ocular abnormalities: (i) affect 100 % of RXRα*af1*^{o}/RAR(β null or γ null) double mutants, whereas (with the exception of the PHPV) they are all absent in the corresponding single null mutants and (ii) are rare in RXRα*af2*^{o} mutants, but fully penetrant in RXRα*af2*^{o} mutants, which additionally lack the AF-1 containing RXRα A/B region. Altogether, these data strongly support the view that the activation functions of RXRα are required for normal eye development, and also that, due to functional redundancy, the role played by RARα AF-1 and AF-2 can be revealed only in certain impaired genetic backgrounds (or RA insufficiency conditions).

On the other hand, the rare occurrence of hypoplasia of the myocardium in RXRα*af2*^{o} and RXRα*af1*^{o}/RXRβ null/RXRγ null fetuses, and only in half of the RXRα*af2*^{o}/RXRβ null/RXRγ null fetuses, suggests that the transcriptional activity of RXRα is necessary for myocardial growth only in "unfavorable" genetic backgrounds. Along the same lines, we recently found that the vast majority (80 %) of RXRα*af*^{o} fetuses display a normal heart histology (our unpublished results), thereby demonstrating that a transcriptionally "silent" RXRα can efficiently promote myocardial growth and ruling out the requirement for an RXR ligand in this developmental process.

Interestingly, RXRα*af2*^{o} and RXRα*af*^{o} mutants die at birth even though only few of them display a hypoplasia of the myocardium. Thus, both their lethality and severe growth retardation may be at least in part secondary to metabolic problems arising from placental defects (see below).

SPECIFIC FUNCTION OF THE RXRα AF-1 DOMAIN-CONTAINING A/B REGION IN INVOLUTION OF INTERDIGITAL MESENCHYME

AF-2 appears to be more important than AF-1 for the function of RXR during embryonic development: RXRα*af2*^{o}/RXRβ null/RXRγ null fetuses all die in utero and display a large array of congenital defects, whereas RXRα*af1*^{o}/RXRβ null/RXRγ null mutants are often viable and display few congenital defects. Moreover, AF-2 but not AF-1 of RXRα is crucial for transcription of a RA responsive *lacZ* reporter transgene in the mouse [22,77]. Note, however, that all RXRα*af1*^{o}/RAR (α, β, or γ) compound mutants die in utero and exhibit a large array of malformations that nearly recapitulate the full spectrum of the fetal VAD syndrome, indicating that RXR AF-1 could nevertheless be instrumental to the transcriptional activity of RAR/RXR heterodimers, particularly under conditions of limiting concentrations of RA and/or RA receptors [77]. In this respect, it is noteworthy that RXRα*af2*^{o}/RXRβ double mutants die in utero at a late fetal stage (E14.5 to birth), whereas RXRα*af*^{o}/RXRβ mutants, that additionally lack AF-1, die at midgestation (E10.5) (our unpublished results). Interestingly, the RXRα A/B region, which supports the AF-1 activity, has also a unique role in the RA-dependent disappearance of the interdigital mesenchyme.

The first evidence implicating RA in the involution of the interdigital mesenchyme was provided by whole limb cultured in a RA-deprived medium [78]. Subsequently, it was shown that mice lacking both alleles of either the RARα or the RARγ genes, as well as mice heterozygous for the RXRα null mutation occasionally exhibit mild forms of interdigital webbing (soft tissue syndactyly) ([24,28,35,38]; Table 1). Surprisingly, this defect was absent in RARβ null mutants, even though RARβ is strongly and specifically expressed in interdigital necrotic zones (INZs) ([35]; and references therein). However, interdigital webbing is severe and completely penetrant upon disruption of one (or both) allele(s) of the RARβ gene in a RARγ null genetic background [35]. The persistence of the fetal interdigital mesenchyme is caused by marked decrease in programmed cell death, as well as by an increase of cell proliferation in the mutant INZs [64]. As RARβ and RARγ are not co-expressed in the INZs, involution of the interdigital mesenchyme must involve paracrine interactions between the interdigital mesenchyme (the site of RARβ expression), and either the cartilaginous blastema of the digits or the surface epidermis, which both express RARγ. RARβ/RARγ (RARβγ) compound mutants also display a specific downregulation of tissue transglutaminase (tTG) promoter activity and of stromelysin 3 expression in the interdigital mesenchyme [64]. The presence of putative retinoic acid response elements in the promoter regions of both tTG and stromelysin 3 genes suggests that RA might promote cell death in the INZs through a direct increase of tTG expression, and could also contributes to the tissue remodeling, which accompanies cell death through an increase of stromelysin 3 expression [64]. It has also been shown that the expression of the anti-death gene *BAG-1*, which is normally downregulated upon initiation of interdigital apoptosis, remains unaltered in the limbs of RARβγ double null mutants [79].

The RXRα A/B region is indispensable for the function of RXRα/RARβ and/or RXRα/RARγ heterodimers in the involution of the interdigital mesenchyme, as the majority of RXRα*af1*^{o} mutants and all RXRα*af1*^{o}/RXRβ null/RXRγ null mutants display soft tissue syndactyly [77]. Moreover, this function selectively requires the RXRα A/B region, as RXRα*af2*^{o} and RXRα*af2*^{o}/RXRβ null/RXRγ null mutants never display this defect [22]. Interestingly, phosphorylation of RXRα at a specific serine residue located in the A domain is necessary for the antiproliferative response of F9 teratocarcinoma cells to RA [26,80]. Therefore, within RXRα/RAR (β and γ) heterodimers, phosphorylation of the RXRα A domain may play an important function in the cascade of molecular events that, in vivo, leads to the normal disappearance of the interdigital mesenchyme.

RETINOIC ACID SIGNALS ARE TRANSDUCED DURING DEVELOPMENT BY SPECIFIC RXRα/RAR(α, β, OR γ) HETERODIMERS

Compound mutants in which a null mutation of a given RAR isotype (α, β, or γ) is associated either with (i) a RXRα null mutation, (ii) a RXRα*af1*^{o} mutation, or (iii) a RXRα*af2*^{o} mutation, altogether recapitulate the abnormalities exhibited by RAR double null mutants [22,24,47,77] (Table 4, and see examples below). This synergism between RAR and RXRα loss-of-function mutations support the conclusion that RXRα/RAR heterodimers are the functional units that transduce RA signals during embryonic development. Moreover, RXRα is the functionally most important RXR during development, as the development of RXRβ/RXRγ double null mutants appears to be normal [81]. This last conclusion is further supported by a lack of synergism during development between RAR (either α, β, or γ) and RXRβ or RXRγ inactivations [47].

The analysis of the various RXRα/RAR compound mutants led to the identification of the heterodimers, which, in a given developmental process, are preferentially involved in transducing RA signals. For instance, hypoplasia of the myocardium is never seen in RAR(α, β, and γ) single null mutants nor in RXRα*af1*^{o} single mutants, while it is found in less than 5 % of RXRα*af2*^{o} single mutants. On the other hand, 45 % of RXRα*af1*^{o}/RARα null mutants (but none of the RXRα*af1*^{o}/RARβ null or RARγ null mutants) and 80 % of RXRα*af2*^{o}/RARα null mutants (but only 20 % of RXRα*af2*^{o}/RARβ null or RARγ null mutants) display this defect [22,77]. Thus, these genetic analyses indicate that RXRα acts on myocardial growth preferentially in the form of heterodimers with RARα.

Similarly, several lines of evidence indicate that, although all three RARs are expressed in developing ocular structures [82], RXRα acts on eye morphogenesis in the form of heterodimers with either RARβ or RARγ, but not with RARα. Firstly, there is a strong synergism between RXRα and RARβ or RARγ inactivations, which is manifested by a marked increase in the severity of the RXRα null ocular defects in RXRα/RARβ and RXRα/RARγ double null fetuses (Table 4) [47]. Secondly, there is also a strong synergism for the generation of RXRα null-like ocular defects between RARβ or RARγ inactivations and ablations of either RXRα AF-1 or RXRα AF-2. Indeed, apart from the PHPV, the RXRα null ocular syndrome is never present in RAR(α, β, and γ) single null mutants nor in RXRα*af1*^{o} single mutants, and is found in only a minority (less than 15 %) of RXRα*af2*^{o} single mutants. On the other hand, this ocular syndrome is observed in 100 % of RXRα*af1*^{o}/RAR(β null or γ null) and RXRα*af2*^{o}/RAR(β null or γ null) double mutants, whereas it is absent in RXRα*af1*^{o}/RARα null mutants, as well as in RXRα*af2*^{o}/RARα null mutants [22,77]. These results indicate that RXRα/RARβ and RXRα/RARγ heterodimers are instrumental in ocular morphogenesis.

Table 4 Evidence that RXRα and RAR act synergistically on embryonic development: similar congenital defects absent (or very rare) in RXRα null, RARα*af2*^{o} RXRα*af1*^{o} and RAR (α, β, or γ) null single mutants are observed in RXR (α null, α*af2*^{o}, or α*af1*^{o})/RAR (α null, β null or γ null) and in RAR/RAR double null mutants.

Abnormalities	Genotypes of RXR/RAR and RAR/RAR compound mutants showing similar defects		
	RXRα/RAR (X/A) double null mutants	RAR/RAR (A/A) double null mutants	RXRα*af1*^{o}or RXRα*af2*^{o}/RAR(α, β, and γ) mutants
Ocular defects (VAD)			
Severe shortening or agenesis of the ventral retina	Xα/Aβ^{b}; Xα/Aγ $^{+/- b}$; Xα/Aγ^{b}	Aβ/Aγ^{b}	Xα*af2*^{o}/Aβ^{b}; Xα*af2*^{o}/Aγ; Xα*af1*^{o}/Aβ; XX*af1*^{o}/Aγ^{b}
Respiratory system defects (VAD)			
• Lung hypoplasia	Xα/Aα^{a}	Aα/Aβ^{b}	Xα*af2*^{o}/Aα^{b}; Xα*af2*^{o}/Aβ; Xα*af1*^{o}/Aα^{a}
• Agenesis of the esophagotracheal septum	Xα/Aα^{a}	Aα/Aβ^{b}	Xα*af2*^{o}/Aα; Xα*af2*^{o}/Aβ; Xα*af1*^{o}/Aα
Heart outflow tract defects (VAD)			
• Persistent truncus arteriosus	Xα/Aα $^{+/-}$; Xα/Aα^{b}, Xα/Aβ; Xα/Aγ	Aα/Aβ^{b}; Aα/Aγ^{b}	Xα*af2*^{o}/Aα; Xα*af1*^{o}/Aα
• Abnormal arteries derived from aortic arches	Xα/Aα$^{+/-}$; Xα/Aα^{a}, Xα/Aβ; Xα/Aγ	Aα/Aβ^{b}; Aα/Aγ^{b}	Xα*af2*^{o}/Aα; Xα*af2*^{o}/Aβ; Xα*af1*^{o}/Aα
Urogenital system defects (VAD)			
• Kidney hypoplasia	Xα/Aα^{b}	Aα/Aβ^{b}	Xα*af2*^{o}/Aα; Xα*af1*^{o}/Aα
• Complete agenesis of Müllerian ducts	Xα/Aα^{b}	Aα/Aβ^{b}	Xα*af1*^{o}/Aα
Hypoplasia of the sub-maxillary gland	Xα/Aγ^{b}	Aα/Aγ^{b}	Xα*af2*^{o}/Aγ^{b}
Skeletal defects			
• Multiple cranio-facial defects	Xα/Aγ	Aα/Aγ^{b}	Xα*af2*^{o}/Aγ^{b}
• Limb defects	Xα/Aγ	Aα/Aγ^{b}	Xα*af2*^{o}/Aγ

[a]This abnormality is present in a majority of the mutants.
[b]This abnormality is completely penetrant. VAD, these abnormalities belong to the fetal vitamin A deficiency syndrome. From refs. [22,24,47,77].

SELECTIVE FUNCTIONS OF RXRβ AND RXRγ IN THE CONTROL OF FERTILITY, METABOLIC PROCESSES, AND BEHAVIOR

RXRβ null mutant males are sterile due to abnormal spermatid maturation and release, leading to a severe number reduction, decreased mobility and high percentage of abnormalities of spermatozoa. As the RXRβ protein is only detectable in Sertoli cells, a dysfunction of these cells most probably accounts for the spermiogenetic defects in RXRβ null mutants. Moreover, large lipid droplets accumulate in the cytoplasm of RXRβ null Sertoli cells, suggesting functional interactions between RXRβ and nuclear receptor signaling pathways controlling lipid metabolism [83]. Interestingly, males lacking only the AF-2 activity of RXRβ (RXRβ*af2*^{o}) are fertile, but display a Sertoli cell-restricted lipid metabolic defect identical to that of the RXRβ null mutant males (our unpublished results). Altogether, these data demonstrate that within the Sertoli cell, RXRβ exert independent functions in spermiogenesis and lipid metabolism and, as lipid accumulations in tissues are never observed under VAD conditions, they also provide the first in vivo evidence that the AF-2 activation function of an RXR can be important for the transcriptional activity of heterodimers other than RXR/RAR.

RXRγ null mutants are fertile and morphologically indistinguishable from wild-type littermates. They exhibit a mild thyroid hormone resistance and an increased metabolic rate [84], and also show defects in cognitive functions [46]. Moreover, the recent observation that growth plate development is more severely impaired in RXRγ/VDR double null mutants than in VDR single null mutants, has suggested that RXRγ/VDR heterodimers may be involved in the differentiation of hypertrophic chondrocytes [85].

Similarly to RARβ, RXRβ and RXRγ are expressed at high levels in the striatum, the main dopaminergic signaling organ controlling coordination of movements. Abnormal locomotor behaviors in RXRβ/RARβ, RXRγ/RARβ, and RXRβ/RXRγ double null mutant mice, and to a lesser extent in single null mutants of these receptors, are correlated with dysfunction of the mesolimbic dopaminergic signaling pathway. Indeed, expression of dopamine receptors types 1 and 2 (DR1 and DR2) in the ventromedial regions of the striatum of these double mutants are significantly reduced, and their response to cocaine, a modulator of dopamine signaling, is blunted [45]. The expression of D2R in the striatum of the mutant animals may be altered at the transcriptional level, as a functional RA response element has been characterized in the promoter of the corresponding gene [86]. That RXRβ/RARβ and RXRγ/RARβ heterodimers mediate retinoid signals required for the function of the mesolimbic dopaminergic system suggests that RA signaling defects may contribute to pathologies such as Parkinson's disease and schizophrenia.

PLACENTATION FIRST REQUIRES A TRANSCRIPTIONALLY "SILENT" RXR, THEN AN ACTIVE RXRα

Embryos carrying null mutations of both RXRα and RXRβ (RXRαβ) double null mutants display a wide range of abnormalities resembling those of embryos carrying blocks in RA signaling, e.g., RALDH2 null and RARα/RARγ double null mutants (see above). However, they exhibit a unique lethality that occurs at midgestation and appears to be caused by a labyrinthine agenesis, i.e., the lack of formation of the labyrinthine zone of the chorioallantoic placenta [87]. In the normal placenta, this zone represents the main site of exchanges between mother and embryo. Labyrinthine agenesis is never associated with RAR single or double null mutant embryos, but a similar, although less severe abnormality (i.e., a labyrinthine hypoplasia) is seen in embryos lacking either PPARβ or PPARγ [88,89]. At later, fetal stages of gestation, the placenta of RXRα single null mutant displays a thickening of the labyrinthine trabeculae, which are interposed between maternal blood sinuses and fetal capillaries, and represent the placental barrier across which nutrient and gas exchanges between the maternal and fetal blood occur [90].

Altogether, these results indicate that RXRs are involved in placentation at two distinct steps. At E8.5, RXRα and/or RXRβ are required for the initial formation of the placental labyrinthine trabeculae from the chorionic plate. Between E14.5 and the term of pregnancy (E19.0), RXRα is required for the proper differentiation of the trophoblast cells forming the labyrinthine trabeculae. The early RXR-dependent step of placentation (formation of the labyrinth), does not involve RXR/RAR heterodimers, and as RXRα/RAR (α, β, or γ) double null fetuses do not die earlier than RXRα single null fetuses, this probably also applies to the later step [47]. Moreover, the early step is taking place normally in RXRα*af*^{o}/RXRβ null placentas, as well as in placentas lacking all RXR AF-1 activities (RXRα*af1*^{o}/RXRβ null/RXRγ null placentas) or all AF-2 activities (RXRα*af2*^{o}/RXRβ null/RXRγ null placentas) ([77], and unpublished results). Therefore, the heterodimers involved in the initial stages of labyrinthine formation, presumably RXR(α and/or β)/PPAR(β and/or γ) heterodimers (see above), do not require a transcriptionally active RXR. In contrast, as RXRα*af2*^{o} placentas reproduce with a complete penetrance the histological defects of RXRα null placentas, the later RXRα-dependent step of placentation (differentiation of the labyrinthine trophoblast), depends critically on the AF-2 ligand binding-dependent transactivation function of RXRα [77].

COMBINING GENETIC AND PHARMACOLOGICAL APPROACHES PROVIDE CLUES ON RAR-DEPENDENT CELLULAR MECHANISMS OPERATING DURING EMBRYOGENESIS

The endoderm of branchial arches is a major target of RA action mediated by RARα and/or RARβ

Mutants carrying targeted inactivations of both the RARα and RARβ genes (RARαβ mutants), analyzed at fetal stages of gestation, display the complete set of defects generated in the chick by surgical ablation of large portions of the post-otic neural crest, namely thymus and parathyroid gland ageneses or ectopias, aberrant pattern of the great cephalic arteries, absence of the pulmonary arteries, and aorticopulmonary septum [35,37,50, and refs. therein]. These defects are also present in the DiGeorge syndrome, which is an archetype of human neurocristopathy, i.e., "a condition arising from aberrations of the early migration, growth, and differentiation of neural crest cells (NCC)" [91]. These and other observations led to the proposal that cranial NCC fated to give rise to mesenchymal derivatives (i.e., the mesectodermal cells) are major targets of RA action [37]. Unexpectedly, RARαβ double null mutants analyzed at embryonic stages of gestation do not show NCC alterations, but their caudal branchial arches (BAs) are very small [60]. BAs are transient bulges of the embryonic head and neck, partially filled with NCC and separated from one another by evaginations of the endoderm, the pharyngeal pouches. Caudal BA and pouches give rise to the adult organs affected in the aforementioned NCC ablation experiments. As BA defects of RARαβ embryos are less severe than those of RALDH2 null embryos, which are devoid of RA [52], they do not reflect a complete block in RA signal transduction. To analyze NCC migration and formation of BA and pharyngeal pouches in a situation where the degree of the block in RA signal transduction could be modulated and its timing precisely controlled, a culture system was designed in which wild-type embryos are exposed to a panRAR antagonist, BMS493 [57].

Treatment with the panRAR antagonist induces a complete lack of caudal BA and pharyngeal pouches, and disturbs the paths of post-otic NCC migration, however, without affecting the amount of NCC. Moreover, and most interestingly, this treatment inhibits caudal BA development only during a narrow window of time which does not correspond to the period of post-otic NCC migration. Both the nature and time of appearance of the defects in panRAR antagonist-treated embryos indicate that, contrary to what was expected from the set of abnormalities displayed by RARαβ double null fetuses, migrating NCC destined to the caudal BA do not represent primary targets of RA action. On the other hand, the antagonist-induced alterations in endodermal expression of "patterning" genes (e.g., *Hoxa1, Hoxb1, Pax1*, *Pax9*) and of genes encoding signal peptides (*Fgf3* and *Fgf8*), indicate that RA signaling (i) is required to specify the pharyngeal endoderm, and (ii) may provide a permissive environment for NCC migration through endodermal secretion of specific paracrine factors [57]. These data also raise the possibility that genes deleted in the human DiGeorge syndrome are actually expressed in the endoderm under the control of RA as early as the fourth week of gestation.

RARs act on top of a genetic cascade controlling hindbrain segmentation

The hindbrain of vertebrate embryos is transiently divided into segments (rhombomeres), of which seven (R1 to R7) are visible in mammals. Although early and transient, hindbrain segmentation is instrumental in organizing adult structures, such as cranial nerves. The coordinated expression of several transcriptional regulators is required to pattern the embryonic hindbrain from an initially unsegmented "naïve" neural plate. These transcriptional regulators: (i) control the segmentation process, through which the cells located within the nascent rhombomeres acquire distinct adhesive properties preventing them to mix together, and/or (ii) impart segmental identity, i.e., the acquisition by these cells for specific molecular addresses that will determine their definitive fates. These transregulators also commonly serve as molecular markers of specific rhombomeres in in situ hybridization assays [58, and refs. therein]

The hindbrain of RARα/RARγ double null mutant (RARαγ) embryos shows a posterior expansion of R3 and R4 markers, but fails to express *kreisler*, a specific marker of R5 and R6. In contrast, the neurectodermal territory corresponding to R5 and R6 is markedly enlarged in RARα/RARβ double null mutant (RARαβ) embryos. Treating E7.0 wild-type embryos with the panRAR antagonist BMS493 produces a phenocopy of the RARαγ hindbrain abnormal phenotype, whereas this treatment started at E8.0 results in a RARαβ-like phenotype. Thus, distinct hindbrain phenotypes in RARαβ and RARαγ null embryos are related to different time windows of RA action: at E7.5 (the time at which embryonic RA synthesis begins), RARγ (and RARα) transduce a signal required to specify the R5/R6 territory; at E8.0, a RARβ- (and RARα)-mediated local increase in RA signaling in the posterior portion of the hindbrain controls the position of the R6 caudal boundary, thus allowing the next caudal rhombomere, R7, to be specified.

That the expression domains of several important hindbrain patterning genes are altered in RARαβ and RARαγ mutant embryos [58] provides evidence that RA acts on top of the genetic hierarchy controlling hindbrain patterning. Moreover, generation of a graded embryonic block in RA signal transduction through varying the concentrations of the panRAR antagonist in the culture, demonstrates that individual rhombomeres are specified by distinct thresholds of RA signaling, and support the view that RA acts as a posteriorizing signal for the patterning of the embryonic hindbrain [58,92, and refs. therein]. Thresholds levels of RA signaling could be set up through modulations of RAR levels, and/or of RA-synthesizing and metabolizing enzyme expression domains [56,93–96, and refs. therein].

RA signals mediated by RXRα/RARα and RXRα/RARβ heterodimers have opposite effects on lung-branching morphogenesis

The discovery that VAD rat fetuses often display bilateral lung hypoplasia as well as oesophagotracheal septum agenesis provided the first indication that RA signaling is important for the development of the respiratory system [8]. VAD-related lung and tracheal malformations are absent in RAR (α, β, and γ) single null mutants, but are seen in all RARαβ double null mutants [35,50]. Although RARβ can apparently efficiently compensate for the loss of RARα in RARα single null mutants, the functions of these two RARs in prenatal lung growth are not equivalent. Indeed, RARα/RXRα double null mutants constantly display severe lung hypoplasia and oesophagotracheal septum agenesis, whereas RXRα/RARβ double null mutants never show these defects. This observation confirms that functional compensation within the RAR family is much less efficient when RXRα, the main heterodimeric partner of RARs during embryonic development, is absent [see ref. 47]. As mentioned above, the genetic dissection of the retinoid signaling pathway strongly suggests that the functional units involved in the primary lung bud and trachea formation are RXRα/RARα heterodimers [47]. Thus, formation of the primary lung buds provides another example of developmental processes that seemingly involve specific heterodimers.

To investigate the role of RA signaling pathways during primary lung bud formation and subsequently during branching morphogenesis, wild-type embryos or lung explants were cultured in the presence of RA and of the specific panRAR antagonist, BMS493 [69]. PanRAR antagonist treatment of embryos at E8.0, prior to the first appearance of the primary lung buds, inhibits their outgrowth and causes a failure of oesophagotracheal fold formation, indicating that (i) a RA signaling is required for the formation of the primary lung buds from the primitive foregut and (ii) the severe lung and tracheal defects observed at fetal stages of gestation in RARαβ and RXRα/RARα double null mutant mice, as well as in VAD rats, are determined prior or at the onset of lung development [69, and refs. therein].

PanRAR antagonist treatments of explants collected at a later stage of lung-branching morphogenesis, increase the number of distal buds, the sites of lung branching, whereas RA administered during this period has an inverse effect [69,97]. This RA-induced inhibition of lung branching is apparently mediated by RARβ. RARβ transcripts are strictly confined to the morphogenetically stable proximal bronchi during in vivo lung development [44]. However, the branching inhibition induced by RA correlates with an ectopic expression of RARβ in distal buds, whereas stimulation of lung branching caused by the panRAR antagonist correlates with a decrease of RARβ expression in proximal bronchi. Additionally, RA treatments that decrease the number of distal buds in wild-type explants do not affect explants from RARβ null mutants. Collectively, these findings support the view that activation of RARβ by RA favors morphogenetic stabilization of the developing pulmonary tree [69]. Such a negative control of embryonic lung growth by a RARβ signaling pathway might be disturbed in the congenital cystic adenomatoid malformation, a human pathology characterized by an overgrowth of bronchial tissue at the expenses of the alveolar tissue.

Interestingly, RARβ-transduced RA signals, in addition to negatively regulate lung-branching morphogenesis in utero, also negatively regulate, after birth, the process of alveolar septation [98,99]. Suppressing perinatal RARβ signaling by selective antagonists may thus offer a novel mean of preventing, or curing, failed septation in prematurely born children suffering from bronchopulmonary dysplasia. In contrast, RXRα/RARγ heterodimers appear to be positively involved in alveolar morphogenesis during the perinatal period [100], while RARα may regulate alveolar formation after the perinatal period [99,101].

COMBINING GENETIC AND PHARMACOLOGICAL APPROACHES PROVIDE CLUES ON RAR AND RXRα-MEDIATED TERATOGENIC EFFECTS

RA is a potent teratogen which, at pharmacological concentrations, can induce congenital defects in all vertebrate species as well as in certain invertebrates [102–104]. RA treatments of animals overexpressing, or carrying null mutations of retinoid receptors, have provided evidence that teratogenic effects of retinoids are receptor-mediated. For instance, it was shown that RARγ null embryos are resistant to RA-induced caudal truncations, whereas RXRα null embryos are resistant to RA-induced limb defects and cleft palate [38,105–107].

In humans, oral intake of Accutane (13-cis RA) during gastrulation and early organogenesis (gestational weeks 2–5) results in a spectrum of congenital malformations collectively referred to as the retinoic acid embryopathy (RAE) [108]. It has been assumed that the branchial arch (BA) defects observed in RA-exposed embryos at the equivalent of E8.0 and E9.0 in the mouse, can account for alterations displayed in newborns, and that neural crest is the primary target tissue of RA-induced teratogenesis in the BA region of the embryo [109,110].

Fusion and hypoplasia of the first two BA, a hallmark of RAE, is generated in E8.0 cultured mouse embryos upon treatment with BMS453, a synthetic compound exhibiting RARβ agonistic properties in transfected cells [111]. In contrast, no BA defects are observed following treatment with synthetic retinoids exhibiting RARα or RARγ agonistic properties. These BMS453-induced BA defects are enhanced in the presence of a panRXR agonist, that is not teratogenic on its own, and they are accompanied by ectopic expression of RARβ and of several other direct RA target genes in the morphologically altered region. On the other hand, BA defects and ectopic expression of RA target genes cannot be induced in RARβ null embryos upon BMS453 treatment. Altogether, these data indicate that craniofacial abnormalities characteristic of RAE are mediated through ectopic activation of RXR/RARβ heterodimers, in which the ligand-dependent activity of RXR is subordinated to that of RARβ.

RAE apparently meets the criteria for a neurocristopathy [91; and see above]. However, NCC do not appear to be primary targets of RA-induced teratogenicity as: (i) retinoid-induced fusion of the first and second BA occurs without alterations of NCC migration or apoptosis [110] and (ii) contrary to other embryonic tissues, NCC do not express a RA-responsive transgene upon treatment with BMS453. In contrast, treatment with BMS453 triggers a RARβ-dependent RA signaling in the endoderm lining the first two BA, manifested by rostral shifts of the expression domains of RA-responsive patterning genes, such as *Hoxa1* and *Hoxb1*. *Hox* gene expression is thought to play an important role in antero-posterior regionalization of the pharyngeal endoderm [57; and refs. therein], while pharyngeal endoderm plays a seminal role in the formation of BA, through imparting patterning information to NCC [112]. Thus, the current data support the view that many RAE defects can result from an abnormal function of the pharyngeal endoderm [111,113].

CONCLUSIONS AND PERSPECTIVES

We have reviewed here the results of phenotypic analyses of single and compound germline mutants lacking RAR and RXR isotypes and/or isoforms, and of mutants lacking the RXRα transactivation functions. These analyses have provided the first compelling evidence that RA is actually the active metabolite of vitamin A during embryonic development, which was subsequently confirmed by the demonstration that RA synthesized by the retinal deydrogenases 2 and 3 (RALDH2 and RALDH3) acts as an indispensable developmental hormone [52–54,56,114, and our unpublished results]. These genetic studies have also led to equally important conclusions concerning the physiological functions played by the multiple RAR and RXR receptors in vivo, notably during early embryogenesis and organogenesis, but also postnatally. Moreover, they strongly supports the conclusion that the molecular mechanisms underlying the transduction of the RA signal by retinoid receptors, as they have been deduced from in vitro studies in acellular and cellular systems (see Introduction), are also instrumental in RA signaling under truly physiological conditions, i.e., at the organismal level.

Several lines of evidence lead to the conclusion that RXR/RAR, notably RXRα/RAR heterodimers are the main functional units that transduce RA signals during development, and that specific RXR/RAR pairs are involved in given developmental processes. This strongly supports the initial proposal [14] that the highly pleiotropic effects of RA reflect sophisticated combinatorial mechanisms through which multiple RXR/RAR heterodimers differentially transduce retinoid signals to selectively control the expression of numerous sets of RA target genes controlling the shaping and axial patterning of the early embryo, and subsequently multiple aspects of organogenesis.

The mouse genetic studies also demonstrate that within RXR/RAR heterodimers, the RXR partner can be either transcriptionally active (and thus synergistically acting with its RAR partners) or inactive, depending on the developmental event under consideration. However, as previously demonstrated in vitro (see Introduction), it appears that the transcriptional activity of the RXR partner is subordinated to ligand binding to the RAR partner [111,115]. Moreover, when RXRα is transcriptionally active, either one (AF-1 or AF-2) or both activation functions can be instrumental, and their activity also depends on the nature of the RA-controlled event. The frequent requirement of RXRα AF-2, whose transcriptional activity is known to be dependent on 9-cis RA from studies in vitro, strongly suggests that this retinoid or a similar ligand could be instrumental in transactivation by RXR heterodimers in vivo (for further discussions, see ref. [22]). Note that, as the role of RXRα AF-1 has been inferred from deletion of the whole A/B region [77], and as phosphorylation of the RXRα A region has been shown to be required for AF-1 activity in studies in vitro [80], it will be interesting to genetically investigate the function of this phosphorylation in the mouse.

The genetic study of the physiological roles of RAR and RXR has revealed an extensive functional redundancy within the members of each family (RARs or RXRs), although in all cases each of these members appears to individually exert at least one specific physiological function. Even though this functional redundancy is not surprising, as the members of each family share a common ancestor, it raises the question as to whether it is physiologically relevant or artefactually generated when a given RAR or RXR is knocked out, as it has been shown to be the case in cellular studies in vitro [26,116]. It is not unlikely that, in most instances, the functional redundancy does not exist under wild-type conditions. In fact, redundancy is frequent within the RAR or the RXR families, i.e., a given defect is very frequently or exclusively seen in either RAR/RAR or RXRα/RXR double null mutants, while it is weakly penetrant or absent in the corresponding single null mutants). In striking contrast, redundancy is less frequent in the case of the RXRα/RAR heterodimer knock-out which generates this defect in a fully penetrant manner. As discussed extensively elsewhere [22,47,77], assuming that RXRα/RAR heterodimers are indeed the functional transducing units, the easiest way to interpret these observations is to postulate that redundancy can occur when only one of the two partners of the physiological heterodimer is ablated. In other words, the activity of a given heterodimer selectively involved in the control of a given event may still be above a physiological threshold level when either one of the two partners

(RXR or RAR) is ablated, but not when both are missing. Therefore, the selective involvement of a given RAR or RXR could be revealed only under conditions where the threshold level is not reached, which would also account for the observations that single isoform knock-outs are often phenotypically normal, and that the role of the RXRα AF-1 or AF-2 functions cannot be fully revealed unless the activity of the heterodimer is altered by the additional mutation of the RAR partner, or by knock-out of one or both of the potentially redundant RXR isotypes. Thus, any conditions (e.g., a decrease availability of intracellular RA) that would lower the activity of RXR/RAR heterodimers and bring it close to physiological threshold levels, may reduce or abrogate functional redundancy. As the actual intracellular concentration of RA could be more limiting in the wild than in animal facilities, functional redundancy may be less prominent in natural environments. In this respect, it is worth mentioning that, even though phosphorylation of either RAR AF-1 or AF-2 domains can be required for RA-induced differentiation of F9 EC cells in vitro [26], mouse mutants bearing mutations in the phosphorylated amino acid residues are apparently normal (our unpublished results). Therefore, it will be worth investigating whether the function of these phosphorylations can be revealed under conditions of limited RA supply. Similar conditions could also reveal selective functions of the various RAR and RXR isoforms.

Genetic analysis of RAR and RXR functions have also demonstrated that the teratogenic effects resulting from administration of exogenous RA to embryos do not reflect a physiological role of endogenous RA in the corresponding developmental processes. Indeed, in two instances in which an involvement of a given RAR or RXR in the mediation of a teratogenic event was demonstrated, the same receptor was clearly not required for the development of the corresponding structure during embryogenesis. This is the case for the RA excess-induced lumbosacral truncation that is mediated by RARγ [38] and the RA excess-induced limb truncations that do not occur in RXRα mutants [106].

Clearly, the generation of RAR and RXR germline mutations, combined with pharmacological approaches to block the RA signaling pathway, have provided many valuable insights on the developmental functions of RA receptors. However, this strategy has intrinsic limitations that are mostly due to the introduction of the mutation in the germline. First, the effect of a germline mutation may be functionally compensated during development, thus precluding the appearance of a defect in the adult animal. On the other hand, the mutation can be lethal in utero or postnatally, thus preventing analysis of the functions of the gene at later developmental or postnatal stages. This is the case for RXRα knockout. Along the same lines, germline mutations can arrest the development of a given organ at an early stage, thus preventing further analysis of the gene functions at later stages of organogenesis. For instance, the function of RARβ in lung-branching morphogenesis is not revealed in RARβ single mutants, and therefore remains cryptic in RARαβ double null mutants in which a very early step of embryonic lung formation is impaired. Moreover, introducing mutations in the germline often makes it very difficult to distinguish cell-autonomous from non-cell-autonomous functions of a gene belonging to a family involved in highly pleiotropic signaling pathways. Thus, in many instances, these limitations of germline mutations prevent the determination of the function of a given gene product in a defined tissue at a given time of the animal life. This is obviously the case for RARs and RXRs.

To overcome these limitations, strategies for targeted spatio-temporally controlled somatic mutagenesis of RARs and RXRs in the mouse have been designed, which are based on cell-type-specific expression and inducible activity of the bacteriophage P1 Cre recombinase. To that end, conditional tamoxifen-inducible Cre recombinases (called Cre-ER^T and Cre-ER^{T2}) have been generated by fusing Cre with a mutated ligand-binding domain of the estrogen receptor ERα which binds tamoxifen, but not estrogens [117,118]. Selective ablation of the RXRα and RARγ genes in keratinocytes indicate that RXRα has key roles in hair cycling, most probably through RXR/VDR heterodimers, as well as in homeostasis of proliferation/differentiation of epidermal keratinocytes and of the skin immune system though non-RXRα/RAR(α, β, or γ) heterodimers [41,119,120]. The same approach applied to adipocytes and hepatocytes has demonstrated that RXRα is involved both in preadipocyte differentiation, adipogenesis, and lipolysis, probably in the form of RXRα/PPARγ heterodimers [121], and plays important cell-autonomous functions in mechanisms that control the lifespan of hepatocytes and are involved in liver regeneration [122]. These examples show that the combined use of transgenic mouse lines expressing tamoxifen-inducible chimeric Cre recombinases in specific cell-types and of mouse lines harboring "floxed" receptor genes will provide invaluable mouse models to further elucidate the in vivo functions of retinoid receptors.

ACKNOWLEDGMENTS

We thank all past and present members of the retinoid groups of the IGBMC for useful discussions. Our work was supported by funds from the Centre National de la Recherche Scientifique (CNRS), the Institut National de la Santé et de la Recherche Médicale (INSERM), the Hôpital Universitaire de Strasbourg, the Collège de France, the Institut Universitaire de France, the Association pour la Recherche sur le Cancer (ARC), the Fondation pour la Recherche Medicale (FRM), the Ligue Nationale Contre le Cancer, the Human Frontier Science Program, and Bristol Myers Squibb.

REFERENCES

1. R. Blomhoff. *Nutr. Rev.* **52**, S13–S23 (1994).
2. M. B. Sporn, A. B. Roberts, D. S. Goodman. *The Retinoids: Biology, Chemistry and Medicine*, Raven Press, New York (1994).
3. P. Kastner, M. Mark, P. Chambon. *Cell* **83**, 859–869 (1995).
4. A. Agadir et al. *Retinoids*: *The Biochemical and Molecular Basis of Vitamin A and Retinoid Action*, Springer-Verlag, Berlin (1999).
5. M. A. Livrea. *Vitamin A and Retinoids: An Update of Biological Aspects and Clinical Applications,* Birkhäuser Verlag, Basel (2000).
6. G. M. Morriss-Kay and S. J. Ward. *Int. Rev. Cytol.* **188**, 73–131 (1999).
7. F. Hale. *J. Hered.* **24**, 105–106 (1933).
8. J. G. Wilson, C. B. Roth, J. Warkany. *Am. J. Anat.* **92**, 189–217 (1953).
9. S. B. Wolbach and P. R. Howe. *J. Exp. Med.* **42**, 753–777 (1925).
10. J. N. Thompson, J. McC. Howell, G. A. J. Pitt. *Proc. Royal Soc.* **159**, 510–535 (1964).
11. E. D. Dickman, C. Thaller, S. M. Smith. *Development* **124**, 3111–3121 (1997).
12. J. C. White, V. N. Shankar, M. Highland, M. L. Epstein, H. F. DeLuca, M. Clagett-Dame. *Proc. Natl. Acad. Sci. USA* **95**, 13459–13464 (1998).
13. P. Chambon. *FASEB J.* **10**, 940–954 (1996).
14. M. Leid, P. Kastner, P. Chambon. *Trends Biochem. Sci.* **17**, 427–433 (1992).
15. P. Germain, J. Iyer, C. Zechel, H. Gronemeyer. *Nature* **15**, 187–192 (2002).
16. D. Moras and H. Gronemeyer. *Curr. Opin. Cell. Biol.* **10**, 384–391 (1998).
17. C. K. Glass and M. G. Rosenfeld. *Genes Dev.* **14**, 121–141 (2000).
18. P. F. Egea, B. P. Klaholz, D. Moras. *FEBS Lett.* **476**, 62–67 (2000).
19. H. Gronemeyer and V. Laudet. *Protein Profile* **2**, 1173–1308 (1995).
20. S. Nagpal, M. Saunders, P. Kastner, B. Durand, H. Nakshatri, P. Chambon. *Cell* **70**, 1007–1019 (1992).
21. S. Nagpal, S. Friant, H. Nakshatri, P. Chambon. *EMBO J.* **12**, 2349–2360 (1993).
22. B. Mascrez, M. Mark, A. Dierich, N. Ghyselinck, P. Kastner, P. Chambon. *Development* **125**, 4691–4707 (1998).
23. B. Durand, M. Saunders, C. Gaudon, B. Roy, R. Losson, P. Chambon. *EMBO J.* **13**, 5370–5382 (1994).
24. P. Kastner, J. M. Grondona, M. Mark, A. Gansmuller, M. LeMeur, D. Décimo, J. L. Vonesch, P. Dollé, P. Chambon. *Cell* **78**, 987–1003 (1994).
25. R. Mollard, S. Viville, S. J. Ward, D. Decimo, P. Chambon, P. Dollé. *Mech. Dev.* **94**, 223–232 (2000).

26. C. Rochette-Egly and P. Chambon. *Histol. Histopathol.* **16**, 909–922 (2001).
27. E. Li, H. M. Sucov, K. F. Lee, R. M. Evans, R. Jaenisch. *Proc. Natl. Acad. Sci. USA* **90**, 1590–1594 (1993).
28. T. Lufkin, D. Lohnes, M. Mark, A. Dierich, P. Gorry, M. P. Gaub, M. LeMeur, P. Chambon. *Proc. Natl. Acad. Sci. USA* **90**, 7225–7229 (1993).
29. B. Chapellier, M. Mark, J. M. Garnier, M. LeMeur, P. Chambon, N. B. Ghyselinck. *Genesis* **32**, 87–90 (2002).
30. P. Bouillet, V. Sapin, C. Chazaud, N. Messaddeq, D. Decimo, P. Dollé, P. Chambon. *Mech. Dev.* **63**, 173–186 (1997).
31. J. M. Dufour and K. H. Kim. *Biol. Reprod.* **61**, 1300–1308 (1999).
32. B. Boulogne, C. Levacher, P. Durand, R. Habert. *Biol. Reprod.* **61**, 1548–1557 (1999).
33. N. B. Ghyselinck, C. Bavik, V. Sapin, M. Mark, D. Bonnier, C. Hindelang, A. Dierich, C. B. Nilsson, H. Hakansson, P. Sauvant, V. Azais-Braesco, M. Frasson, S. Picaud, P. Chambon. *EMBO J.* **18**, 4903–4914 (1999).
34. D. Lohnes, M. Mark, C. Mendelsohn, P. Dollé, A. Dierich, P. Gorry, A. Gansmuller, P. Chambon. *Development* **120**, 2723–2748 (1994).
35. N. B. Ghyselinck, V. Dupé, A. Dierich, N. Messaddeq, J. M. Garnier, C. Rochette-Egly, P. Chambon, M. Mark. *Int. J. Dev. Biol.* **41**, 425–447 (1997).
36. M. Mark, D. Lohnes, C. Mendelsohn, V. Dupé, J. L. Vonesch, P. Kastner, F. Rijli, A. Bloch-Zupan, P. Chambon. *Int. J. Dev. Biol.* **39**, 111–121 (1995).
37. M. Mark, N. Ghyselinck, P. Kastner, V. Dupé, O. Wendling, W. Krezel, B. Mascrez, P. Chambon. *Eur. J. Oral Sci.* **106** (suppl. 1), 24–31 (1998).
38. D. Lohnes, P. Kastner, A. Dierich, M. Mark, M. LeMeur, P. Chambon. *Cell* **73**, 643–658 (1993).
39. B. A. Underwood and P. Arthur. *FASEB J.* **10**, 1040–1048. (1996).
40. N. B. Ghyselinck, B. Chapellier, C. Calleja, A. Kumar Indra, M. Li, N. Messaddeq, M. Mark, D. Metzger, P. Chambon. *Ann. Dermatol. Venereol.* **129**, 793–799 (2002).
41. B. Chapellier, M. Mark, N. Messaddeq, C. Calleja, X. Warot, J. Brocard, C. Gerard, M. Li, D. Metzger, N. B. Ghyselinck, P. Chambon. *EMBO J.* **21**, 3402–3413 (2002).
42. V. Subbarayan, P. Kastner, M. Mark, A. Dierich, P. Gorry, P. Chambon. *Mech. Dev.* **66**, 131–142 (1997).
43. J. M. Grondona, P. Kastner, A. Gansmuller, D. Décimo, P. Chambon, M. Mark. *Development* **122**, 2173–2188 (1996).
44. N. B. Ghyselinck, O. Wendling, N. Messaddeq, A. Dierich, C. Lampron, D., Décimo, S. Viville, P. Chambon, M. Mark. *Dev. Biol.* **198**, 303–318 (1998).
45. W. Krezel, N. Ghyselinck, T. A. Samad, V. Dupé, P. Kastner, E. Borrelli, P. Chambon. *Science* **279**, 863–867 (1998).
46. M. Y. Chiang, D. Misner, G. Kempermann, T. Schikorski, V. Giguere, H. M. Sucov, F. H. Gage, C. F. Stevens, R. M. Evans. *Neuron.* **21**, 1353–1361 (1998).
47. P. Kastner, M. Mark, N. B. Ghyselinck, W. Krezel, V. Dupé, J. M. Grondona, P. Chambon. *Development* **124**, 313–326 (1997).
48. M. Mark, P. Kastner, N. B. Ghyselinck, W. Krezel, V. Dupé, P. Chambon. *C.R. Soc. Biol.* **191**, 77–90 (1997).
49. J. Luo, H. M. Sucov, J. A. Bader, R. M. Evans, V. Giguere. *Mech. Dev.* **55**, 33–44 (1996).
50. C. Mendelsohn, D. Lohnes, D. Décimo, T. Lufkin, M. LeMeur, P. Chambon, M. Mark. *Development* **120**, 2749–2771 (1994).
51. J. C. White, M. Highland, M. Kaiser, M. Clagett-Dame. *Dev. Biol.* **220**, 263–284 (2000).
52. K. Niederreither, V. Subbarayan, P. Dollé, P. Chambon. *Nat. Genet.* **21**, 444–448 (1999).
53. K. Niederreither, J. Vermot, B. Schuhbaur, P. Chambon, P. Dollé. *Development* **127**, 75–85 (2000).

54. K. Niederreither, J. Vermot, N. Messaddeq, B. Schuhbaur, P. Chambon, P. Dollé. *Development* **128**, 1019–1031 (2001).
55. K. Niederreither, V. Fraulob, J. M. Garnier, P. Chambon, P. Dollé. *Mech. Dev.* **110**, 165–71 (2002).
56. K. Niederreither, J. Vermot, B. Schuhbaur, P. Chambon, P. Dollé. *Development* **129**, 3563–3574 (2002).
57. O. Wendling, C. Dennefeld, P. Chambon, M. Mark. *Development* **127**, 1553–1562 (2000).
58. O. Wendling, N. B. Ghyselinck, P. Chambon, M. Mark. *Development* **128**, 2031–2038 (2001).
59. D. M. Kochhar, H. Jiang, J. D. Penner, A. T. Johnson, R. A. Chandraratna. *Int. J. Dev. Biol.* **42**, 601–608 (1998).
60. V. Dupé, N. B. Ghyselinck, O. Wendling, P. Chambon, M. Mark. *Development* **126**, 5051–5059 (1999).
61. V. Dupé, M. Davenne, J. Brocard, P. Dollé, M. Mark, A. Dierich, P. Chambon, F. M. Rijli. *Development* **124**, 399–410 (1997).
62. M. Houle, P. Prinos, A. Lulianella, N. Bouchard, D. Lohnes. *Mol. Cell. Biol.* **20**, 6579–6586 (2000).
63. D. Allan, M. Houle, N. Bouchard, B. I. Meyer, P. Gruss, D. Lohnes. *Dev Biol.* **240**, 46–60 (2001).
64. V. Dupé, N. B. Ghyselinck, V. Thomazy, L. Nagy, P. J. A. Davies, P. Chambon, M. Mark. *Dev. Biol.* **208**, 30–43 (1999).
65. P. Kastner, N. Messaddeq, M. Mark, O. Wendling, J. M. Grondona, S. Ward, N. Ghyselinck, P. Chambon. *Development* **124**, 4749–4758 (1997).
66. C. Mendelsohn, E. Batourina, S. Fung, T. Gilbert, J. Dodd. *Development* **126**, 1139–1148 (1999).
67. E. Batourina, C. Choi, N. Paragas, N. Bello, T. Hensle, F. D. Costantini, A. Schuchardt, R. L. Bacallao, C. L. Mendelsohn. *Nat. Genet.* **32**, 109–115 (2002).
68. E. Batourina, S. Gim, N. Bello, M. Shy, M. Clagett-Dame, S. Srinivas, F. Costantini, C. Mendelsohn. *Nat. Genet.* **27**, 74–78 (2001).
69. R. Mollard, N. B. Ghyselinck, O. Wendling, P. Chambon, M. Mark. *Int. J. Dev. Biol.* **44**, 457–462 (2000).
70. M. P. Mark, A. Bloch-Zupan, J. V. Ruch. *Int. J. Dev. Biol.* **36**, 517–526 (1992).
71. A. Bloch-Zupan, M. P. Mark, B. Weber, J. V. Ruch. *Arch. Oral Biol.* **39**, 891–900 (1994).
72. H. M. Sucov, E. Dyson, C. L. Gumeringer, J. Price, K. R. Chien, R. M. Evans. *Genes Dev.* **8**, 1007–1018 (1994).
73. V. Subbarayan, M. Mark, N. Messadeq, P. Rustin, P. Chambon, P. Kastner. *J. Clin. Invest.* **105**, 387–394 (2000).
74. T. Chen, T. Chang, J. Kang, B. Choudhary, T. Makita, C. Tran, J. Burch, H. Eid, H. Sucov. *Dev. Biol.* **250**, 198 (2002).
75. P. Ruiz-Lozano, S. M. Smith, G. Perkins, S. W. Kubalak, G. R. Boss, H. M. Sucov R. M. Evans, K. R. Chien. *Development* **125**, 533–544 (1998).
76. F. A. Mic, A. Molotkov, X. Fan, A. E. Cuenca, G. Duester. *Mech. Dev.* **97**, 227–230 (2000).
77. B. Mascrez, M. Mark, W. Krezel, V. Dupe, M. LeMeur, N.B. Ghyselinck, P. Chambon. *Development* **128**, 2049–2062 (2001).
78. M. Lussier, C. Canoun, C. Ma, A. Sank, C. Shuler. *Int. J. Dev. Biol.* **37**, 555–564 (1993).
79. A. Crocoll, U. Herzer, N. B. Ghyselinck, P. Chambon, A. C. Cato. *Mech. Dev.* **111**, 149–152 (2002).
80. J. Bastien, S. Adam-Stitah, J. L. Plassat, P. Chambon, C. Rochette-Egly. *J. Biol. Chem.* **277**, 28683–28689 (2002).
81. W. Krezel, V. Dupé, M. Mark, A. Dierich, P. Kastner, P. Chambon. *Proc. Natl. Acad. Sci. USA* **93**, 9010–9014 (1996).
82. M. Mori, N. B. Ghyselinck, P. Chambon, M. Mark. *Invest. Ophthalmol. Vis. Sci.* **42**,1312–1318 (2001).

83. P. Kastner, M. Mark, M. Leid, A. Gansmuller, J. M. Grondona, D. Décimo, W. Krezel, A. Dierich, P. Chambon. *Genes Dev.* **10**, 80–96 (1996).
84. N. S. Brown, A. Smart, V. Sharma, M. L. Brinkmeier, L. Greenlee, S. A. Camper, D. R. Jensen, R. H. Eckel, W. Krezel, P. Chambon, B. R. Haugen. *J. Clin. Invest.* **106**, 73–79 (2000).
85. N. Yagishita, Y. Yamamoto, T. Yoshizawa, K. Sekine, Y. Uematsu, H. Murayama, Y. Nagai, W. Krezel, P. Chabon, T. Matsumoto, S. Kato. *Endocrinology* **142**, 5332–5341 (2001).
86. T.A. Samad, W. Krezel, P. Chambon, E. Borrelli. *Proc. Natl. Acad. Sci. USA* **94**, 14349–14354 (1997).
87. O. Wendling, P. Chambon, M. Mark. *Proc. Natl. Acad. Sci. USA* **96**, 547–551 (1999).
88. Y. Barak, M. C. Nelson, E. S. Ong, Y. Z. Jones, P. Ruiz-Lozano, K. R. Chien, A. Koder, R. M. Evans. *Mol. Cell* **4**, 585–595 (1999).
89. Y. Barak, D. Liao, W. He, E. S. Ong, M. C. Nelson, J. M. Olefsky, R. Boland, R. M. Evans. *Proc. Natl. Acad. Sci. USA* **99**, 303–308 (2002).
90. V. Sapin, P. Dollé, C. Hindelang, P. Kastner, P. Chambon. *Dev. Biol.* **191**, 29–41 (1997).
91. R. P. Bolande. *Pediatr. Pathol. Lab. Med.* **17**, 1–25 (1997).
92. V. Dupé and A. Lumsden. *Development* **128**, 2199–2208 (2001).
93. H. L. Ang and G. Duester. *Dev. Dyn.* **208**, 536–543 (1997).
94. K. Niederreither, S. Abu-Abed, B. Schuhbaur, M. Petkovich, P. Chambon, P. Dollé. *Nat. Genet.* **31**, 84–88 (2002).
95. G. MacLean, S. Abu-Abed, P. Dollé, A. Tahayato, P. Chambon, M. Petkovich. *Mech. Dev.* **107**, 195–201 (2001).
96. S. Abu-Abed, P. Dollé, D. Metzger, C. Wood, G. MacLean, P. Chambon, M. Petkovich. *Development* **130**, 1449–1459 (2003).
97. S. Malpel, C. Mendelsohn, W. V. Cardoso. *Development* **127**, 3057–3067 (2000).
98. G. D. Massaro, D. Massaro, W. Y. Chan, L. B. Clerch, N. Ghyselinck, P. Chambon, R. A. Chandraratna. *Physiol. Genomics* **4**, 51–57 (2000).
99. D. Massaro and G. D. Massaro. *Am. J. Respir. Cell. Mol. Biol.* **28**, 271–274 (2003).
100. S. Mc Gowan, S. K. Jackson, M. Jenkins-Moore, H. H. Dai, P. Chambon, J. M. M. Snyder. *Am. J. Respir. Cell. Mol. Biol.* **23**, 162–167 (2000).
101. G. D. Massaro, D. Massaro, P. Chambon. *Am. J. Physiol.* **284**, 431–433 (2003).
102. D. R. Soprano and K. J. Soprano. *Annu. Rev. Nutr.* **15**, 111–132 (1995). Retinoids as teratogens.
103. M. D. Collins and G. E. Mao. *Annu. Rev. Pharmacol. Toxicol.* **39**, 399–430 (1999).
104. H. Escriva, N. D. Holland, H. Gronemeyer, V. Laudet, L. Z. Holland. *Development* **129**, 2905–2916 (2002).
105. A. Lulianella and D. Lohnes. *Dev. Dyn.* **209**, 92–104 (1997).
106. H. M. Sucov, J. C. Izpisua-Belmonte, Y. Ganan, R. M. Evans. *Development* **21**, 3997–4003 (1995).
107. P. Nugent, H. M. Sucov, M. Pisano, R. M. Green. *Int. J. Dev. Biol.* **43**, 567–570 (1999).
108. E. J. Lammer, D. T. Chen, R. M. Hoar, N. D. Agnish, P. J. Benke, J. T. Braun, C. J. Curry, P. M. Fernhoff, A. W. Grix, I. T. Lott, J. M. Richard, S. C. Sun. *N. Engl. J. Med.* **313**, 837–841 (1985).
109. R. M. Pratt, E. H. Goulding, B. D. Abbott. *J. Craniofac. Genet. Dev. Biol.* **7**, 205–217 (1987).
110. Y. M. Lee, N. Osumi-Yamashita, Y. Ninomiya, C. K. Moon, U. Eriksson, K. Eto. *Development* **121**, 825–837 (1995).
111. N. Matt, N. B. Ghyselinck, O. Wendling, P. Chambon, M. Mark. *Development* **130**, 2083–2093 (2003).
112. G. Couly, S. Creuzet, S. Bennaceur, C. Vincent, N. M. Le Douarin. *Development* **129**, 1061–1073 (2002).
113. G. B. Mulder, N. Manley, L. Maggio-Price. *Teratology* **58**, 263–275 (1998).
114. J. Vermot, K. Niederreither, J. M. Garnier, P. Chambon, P. Dollé. *Proc. Natl. Acad. Sci. USA* **100**, 1763–1768 (2003).

115. M. M. Elmazar, R. Ruhl, H. Nau. *Toxicol. Appl. Pharmacol.* **170**, 2–9 (2001).
116. R. Taneja, B. Roy, J. L. Plassat, C. F. Zusi, J. Ostrowski, P. R. Reczek, P. Chambon. *Proc. Natl. Acad. Sci. USA* **93**, 6197–6202 (1996).
117. D. Metzger and P. Chambon. *Methods* **24**, 71–80 (2001).
118. D. Metzger, A. Indra Kumar, M. Li, B. Chapellier, C. Calleja, N. B. Ghyselinck, P. Chambon. *Methods Enzymol.* (2003). In press.
119. M. Li, A. K. Indra, X. Warot, J. Brocard, N. Messaddeq, S. Kato, D. Metzger, P. Chambon. *Nature* **407**, 633–636 (2000).
120. M. Li, H. Chiba, X. Warot, N. Messaddeq, C. Gérard, P. Chambon, D. Metzger. *Development* **128**, 675–688 (2001).
121. T. Imai, M. Jiang, P. Chambon, D. Metzger. *Proc. Natl. Acad. Sci. USA* **98**, 224–228 (2001).
122. T. Imai, M. Jiang, P. Kastner, P. Chambon, D. Metzger. *Proc. Natl. Acad. Sci. USA* **98**, 4581–4586 (2001).

Pure Appl. Chem., Vol. 75, Nos. 11–12, pp. 1733–1742, 2003.

Topic 1.7

Biological function and mode of action of nuclear xenobiotic receptors*

Junichiro Sonoda and Ronald M. Evans‡

Howard Hughes Medical Institute, Gene Expression Laboratory, The Salk Institute for Biological Studies, 10010 North Torrey Pines Road, La Jolla, CA 92037, USA

Abstract: Two related nuclear receptors, the pregnane X receptor (PXR) and the constitutive androstane receptor (CAR), act as xenobiotic sensors that protect the body from a multitude of foreign chemicals (xenobiotics) and play a central role in the metabolism and clearance of steroids and toxic endogenous lipids (endobiotics). A structurally diverse array of chemicals including pharmaceutical drugs, steroids, herbal extracts, and pesticides activate PXR or CAR. This activation results in induction of overlapping, but yet distinct drug clearance pathways consisting of cytochrome P450 enzymes, conjugating enzymes, drug transporters, and other related proteins. Similar pathways are also utilized to protect the body from toxic compounds of endogenous origin. Thus, the xenobiotic regulatory circuit contributes both to drug–drug and food–drug interactions as well as endocrine disruption. Consistent with the notion that xenobiotic receptors regulate drug clearance, single nucleotide polymorphisms (SNPs) in either the receptors themselves or receptor-binding sites in the regulatory region of genes encoding metabolic enzymes appear to contribute to the polymorphic expression of components of drug clearance pathways. Together, the xenobiotic receptors PXR and CAR confer metabolic immunity via the ability to control an integrated array of target genes.

XENOBIOTIC METABOLISM AND CLEARANCE: OVERVIEW

In the process of consuming foods, numerous foreign compounds enter our body that are neither used as dietary energy sources nor as building blocks for biological matrices. Uptake of such xenobiotics occurs mainly with food and water consumption, but also by inhalation or transdermally. In addition, normal metabolism and residing microbes, such as intestinal bacteria, can produce de novo toxins. Unless metabolized and eliminated, these substances may accumulate, giving rise to harmful effects by a variety of mechanisms including interaction with hormone receptors (endocrine active substances, i.e., EASs) or reaction with nucleic acids (genotoxic carcinogens) and membrane solubilization (cytoxic lipids). Thus, the entero-hepatic system has evolved to efficiently detoxify and eliminate unwanted chemicals. This system consists of microsomal cytochrome P450 enzymes (CYPs) and other oxidating and hydroxylating enzymes (phase I response), conjugation enzymes such as glucuronosyl- and sulfotransferases (phase II response) and membrane-bound drug pumps such as MDR1 (phase III), that function in a concerted fashion to inactivate and clear chemical compounds (reviewed in ref. [1]). The same system is also utilized to metabolize endogenous compounds such as steroids, bile acids, thyroid hormone, retinoids, cytokines, and fatty acids. To cover the potential diversity of molecular structure that is present within a given ecosystem, metabolic enzymes and drug transporters must have broad speci-

*Report from a SCOPE/IUPAC project: Implication of Endocrine Active Substances for Human and Wildlife (J. Miyamoto and J. Burger, editors). Other reports are published in this issue, *Pure Appl. Chem.* **75**, 1617–2615 (2003).

‡Corresponding author

ficities. Furthermore, each class and subclass of enzymes forms a large family with overlapping, yet distinct substrate specificity.

One characteristic of xenobiotic metabolizing enzymes and transporters is their inducibility by their substrates. This allows enhanced production of these proteins only as needed. For example, the antibiotic rifampicin induces CYP3A isozymes, whereas the anti-epileptic drug phenobarbital (PB) or the planar hydrocarbon 1,4-bis[2-(3,5-dichloropyridyloxy)]benzene (TCPOBOP) induces CYP2B isozymes. These compounds also induce a variety of other metabolic enzymes and transporters. While the aryl hydrocarbon receptor (AhR) mediates induction of CYP1A, CYP1B, and glutathione-*S*-transferase (GST) by dioxin and other structurally similar polycyclic aromatic hydrocarbons, the molecular basis through which the vast majority of drugs and toxins induce the xenobiotic response was only recently uncovered.

Orphan nuclear receptors are structurally related to known nuclear hormone receptors, but lack previously identified physiological ligands or activators. In the last decade, the biological role of several of these orphans, in particular those which act as heterodimers with the retinoid X receptor (RXR), have been revealed through the isolation of relevant endogenous ligands as well as by generation of knock-out mouse models that lack the functional receptors. This work has led to the realization that many orphans act as sensors for dietary lipids as opposed to high affinity endogenous hormones. For example, liver X receptor (LXR), peroxisome proliferator activated receptor (PPAR), and farnesoid X receptor (FXR) have been identified as sensors for cholesterol, fatty acids, and bile acids, respectively, to cooperatively regulate lipid homeostasis (reviewed in ref. [2]). In addition, two closely related receptors, the pregnane X receptor (PXR) and the constitutive androstane receptor (CAR), have emerged as xenobiotic sensors that mediate induction of drug clearance pathways to ensure rapid detoxification of potentially harmful substances. In this chapter, we will focus on PXR and CAR, their role in drug clearance, molecular mechanisms of their action, and the implication of xenobiotic regulation in pathophysiological conditions and endocrine disruption.

PXR, A MEDIATOR OF CYP3A INDUCTION

The rodent receptor PXR was originally identified as candidate xenobiotic receptor that mediates induction of CYP3A by known chemical inducers [3] (for review, see ref. [4]). The human homolog was first isolated as the steroid and xenobiotic receptor (SXR) [5] and the pregnane-activated receptor (PAR) [6]. Several lines of evidence indicate that mammalian PXRs mediate CYP3A regulation by xenobiotics. First, both PXR and CYP3A are expressed most abundantly in the liver and the intestine, the tissues most involved in drug metabolism. Second, PXR, together with its obligatory partner RXR, binds to the sites in the CYP3A promoter that are known to mediate drug-inducibility and transactivates gene expression. These sites form either DR3 (direct repeats of AGGTCA or closely related sequence with the spacing of three nucleotides) or ER6 (everted repeats with six nucleotides spacing) type elements depending on the species and isoforms. Third, PXR is activated by numerous structurally unrelated drugs including those known to induce CYP3A expression. The activation, in most cases, involves direct binding of inducing chemicals to the ligand-binding domain (LBD), resulting in dissociation of corepressor molecules such as the silencing mediator for retinoid and thyroid hormone receptor (SMRT) and the nuclear receptor corepressor (NcoR) and simultaneous recruitment of coactivator molecules, including members of the p160 family (SRC-1, GRIP, and ACTR), RIP140, and PBP (DRIP205 or TRAP220) [7]. Together, these observations suggest PXR may be the central factor contributing to CYP3A regulation. This idea is furthered by the species specificity of the CYP response. Notably, the PXR LBD from different species is considerably divergent. Within the LBD, amino acid identity of human PXR and mouse PXR is only 76 %, whereas the DNA-binding domain is highly conserved (96 % identity). Accordingly, PXRs from different species display different ligand-binding specificities. For example, the rodent orthologue can be activated by the glucocorticoid antagonist, pregnenolone-

16α-carbonitrile (PCN), while the human receptor cannot. In contrast, the human receptor is activated by the antibiotic rifampicin, a potent drug interactor, while the rodent ortholog is not. This is consistent with the long-known species differences in drug inducibility of CYP3A and suggests that the receptor the organism employs rather than the CYP3A gene promoter determines the response profile. Genetic evidence for this idea came from creation of loss of function and gain of function mice [8,9]. PXR null mice are both viable and fertile, indicating that in the absence of toxic insults the xenobiotic response is dispensable. However, PXR null mice completely lack inducibility of CYP3A by PCN or PCN-mediated induction of resistance against multiple toxins [8–11]. Furthermore, a transgenic mouse strain in which PXR was replaced by its human ortholog in the liver shows a "humanized" response to species-specific inducers [8]. These results together unequivocally establish PXR as the central mediator of CYP3A induction.

The creation of mouse models with humanized xenobiotic responses offers a standardized in vivo system for predicting potential human drug–drug interactions and may thus aid pharmaceutical development. Historically, drug-induction of CYP3A has been considered an unexplained adverse side effect associated with drug–drug interactions. For decades, rodent models have been standard components in the assessment of potential toxicities in the development of candidate human drugs. However, the reliability of rodents as predictors of the human xenobiotic response is compromised due to species-variation. Cultured human primary hepatocytes are valuable alternative tools, but are compromised by inter-individual variability, limited and unpredictable availability as well as high cost. Thus, the generation of the transgenic mice expressing human PXR and deficient in rodent PXR represents a major step toward generating a standardized humanized toxicological model.

CAR, A MEDIATOR OF CYP2B INDUCTION

The nuclear receptor CAR was identified as a candidate xenobiotic receptor that mediates induction of CYP2B by PB-type inducers [12] (for review, see ref. [13]). Expressed predominantly in the liver, CAR, together with its heterodimeric partner RXR, binds to the DR4 element in the CYP2B promoter that is known to mediate inducibility by PB. In most transiently and stably transfected cells, CAR shows constitutive activity and induces expression of endogenous CYP2B. Its constitutive activity is repressed by antagonistic steroids such as androstanol and androstenol [14], whereas treatment with PB and PB-type compounds reactivates CAR [15]. The role of CAR in mediating CYP2B induction by PB-type inducers was confirmed by generating a knock-out mouse strain which lacks the CAR gene locus [16]. Like PXR null mice, CAR null mice are viable and fertile, indicating the CAR function is also dispensable in the absence of toxic insult. As expected, CAR null mice completely lack inducibility of CYP2B by PB and a PB-type inducer TCPOBOP as well as PB-mediated resistance to zoxazolamine or sensitivity to cocaine or acetaminophen [16,17]. Together, these results unequivocally show that CAR is a mediator of CYP2B induction.

Like PXR, the LBD of CAR is also divergent among species (72 % identity between human and mouse). Not surprisingly, CAR also exhibits a strong species specificity for activators. For example, a potent mouse CAR agonist TCPOBOP and the reverse agonist androstanol are inactive for human CAR, whereas the potent human CAR reverse agonist clotrimazol is inactive for the mouse receptor [18]. However, since most of published studies are with mouse CAR or mouse liver cells, the extent of the species variation and the role of the receptor in the process is not clear.

A potential major distinction of CAR from other receptors is its constitutive activity in the absence of ligand in nonhepatic cells. Indeed, only TCPOBOP and androstanes are known to bind mouse CAR directly and modulate interaction with coactivator molecules such as SRC-1 and GRIP1 [14,18–20]. Rather, an alternative activation pathway appears to be employed that is indirect and does not involve ligand binding [21]. In vivo or in cultured hepatic cells, CAR is maintained in an inactive state by being localized to the cytoplasm. The inducing compound, while not binding, triggers a cytoplasm to nuclear translation resulting in activation of target genes. Although, the mechanism by which

CAR activators stimulate nuclear translocation is not yet clear, it is likely that some ligands including PB may act indirectly by triggering a juxtamembrane cascade, perhaps by a kinase, that modulates the LBD or an associated protein [22]. Defining the molecular basis of this process is important in understanding the signaling cascade of CAR xenobiotic activators.

PXR AND CAR CROSS-TALK

As discussed above, PXR and CAR were originally characterized as independent regulators of the CYP3A and 2B genes, respectively, presumably through distinct classes of drugs. The xenobiotic response elements in these two classes of CYP genes are also distinct, furthering the concept of independent regulation. Surprisingly, recent observations suggest that there is significant cross regulation of CY2Bs and 3As by these two receptors [8], which appears to be achieved by adaptive recognition of the opposing response element.

As the identity of PXR and CAR targets increases, it becomes progressively clear that they induce largely overlapping, if not identical, sets of genes. In addition to CYP3A and CYP2B, PXR appears to regulate the entire array of genes involved in metabolism and clearance of xenobiotics. These include: (1) members of cytochrome P450 enzymes, CYP2C, CYP1A, CYP1B, CYP2A, and CYP4F, which play central roles in inactivating endogenous hormones or exogenous toxins; (2) other phase I reductases and hydrolases such as carboxylesterase, monoamine oxidase, catalase, flavin-containing monooxygenases (FMOs); (3) conjugating enzymes such as UDP-glucuronosyltransferase (UGT), cytosolic sulfotransferase (SULT), and GST, which solubilize hydrophobic compounds to prepare for clearance; (4) finally, the membrane-bound transporters such as MDR1 and MRP2, which act as efflux pumps to clear drugs and drug conjugates (reviewed in [4], also see refs. [7,23–25]). In addition, PXR may regulate inducible nitric oxide synthase (iNOS) involved in the inflammatory response [26]. Like PXR, CAR has recently been shown to regulate a similar array of xenobiotic genes including several CYP enzymes, aldehyde dehydrogenase, esterase, FMO, methyl transferase, GST [27], SULT [28], UGT [29], and MRP2 [30] as well as iNOS [26]. Analysis of the promoter regions of these xenobiotic genes and identification of receptor binding sites reveals that both PXR and CAR can adaptively bind to common response elements [26,30,31], indicating that the DNA-binding specificities of these xenobiotic receptors are somewhat promiscuous. Regulation of a network of metabolic genes by a xenobiotic receptor suggests the existence of coordinated molecular cascade of drug clearance (Fig. 1). Furthermore, the ability of the receptors to respond to an overlapping set of drugs suggests a metabolic fail-safe system that confers a second layer of metabolic immunity from potentially toxic foreign compounds.

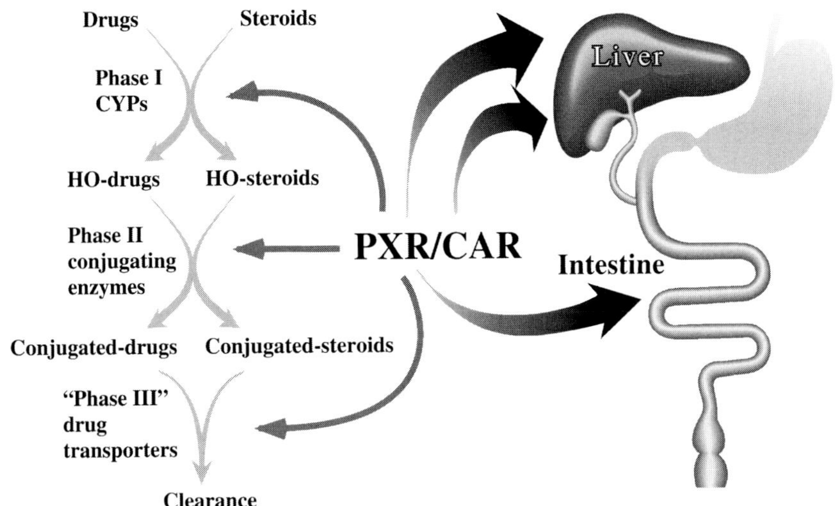

Fig. 1 Schematic representation of the mammalian xenobiotic response. PXR and CAR function as master regulators of xenobiotic response by activating both phase I and II xenobiotic enzymes, as well as the drug transporters. PXR is expressed throughout the enterohepatic axis, whereas CAR is predominantly expressed in the liver. Abbreviations: HO, hydroxylated.

MULTIPLICITY OF LIGANDS FOR XENOBIOTIC RECEPTORS

As a xenobiotic receptor, it is not surprising that PXR binds a diversity of structurally unrelated chemicals. In fact, X-ray crystal structures of the PXR LBD have revealed that its ligand-binding pocket is relatively large (1150 $Å^3$) compared to most other nuclear receptor LBDs, and can accommodate even a single hydrophobic ligand in multiple configurations [32]. The absence of a highly constrained pocket allows for molecular flexibility and plasticity of ligand recognition, in a fashion that is somewhat reminiscent of the low substrate specificity of xenobiotic enzymes. For example, the human receptor is activated by antibiotic macrolide rifampicin, certain cholesterol lowering drugs such as SR12813 and the statins, the antidepressant herb St. John's Wort, the antineoplastic drug paclitaxel, and the antimycotic clotrimazole (see Fig. 2, reviewed in ref. [4]). With relevance to endocrine disruption, human PXR is activated by numerous endogenous steroids including corticosterone and estradiol as well as other estrogenic chemicals including diethylstilbestrol, the phytoestrogen coumestrol [5], bisphenol A, a substance widely used in the food industry and in dentistry [33], and organochlorine pesticides such as chlordane, dieldrin, and endosulfan [34]. Other environmental contaminants including endocrine-disrupting chemicals such as nonylphenol and phthalic acid, the nonplanar polychlorinated biphenyls (PCBs), and thc organochloride pepticides such as *trans*-nonachlor and chlordane have been shown to activate mouse PXR [35]. In addition, an antagonistic ligand that can block PXR-mediated induction of CYP3A, ecteinascidin-734, has also been identified [7]. Thus, in principle, it should be possible to design specific drugs that could selectively inhibit or promote the xenobiotic response.

The list of compounds that elicit CAR activity is small compared to the multiplicity of ligands that activate PXR. As mentioned above, mouse CAR is indirectly activated by PB and directly activated by TCPOBOP. Other mouse CAR activators include the antipsychotic chlorpromazine, plant products picrotoxin and camphor, and pesticides including PCBs, dieldrin, DDT, and methoxycholar and its metabolites (reviewed in ref. [13]). In addition, mouse CAR is repressed strongly by progesterone and androstanes while it is activated by estrogen [36]. Human CAR does not respond to these steroids and the difference has been attributed to residue 350 in the LBD (threonine in mouse and methionine in human) [37]. As mentioned above, human CAR is activated by PB, but not by TCPOBOP. To date, there is no known agonistic ligand that directly binds to the human receptor.

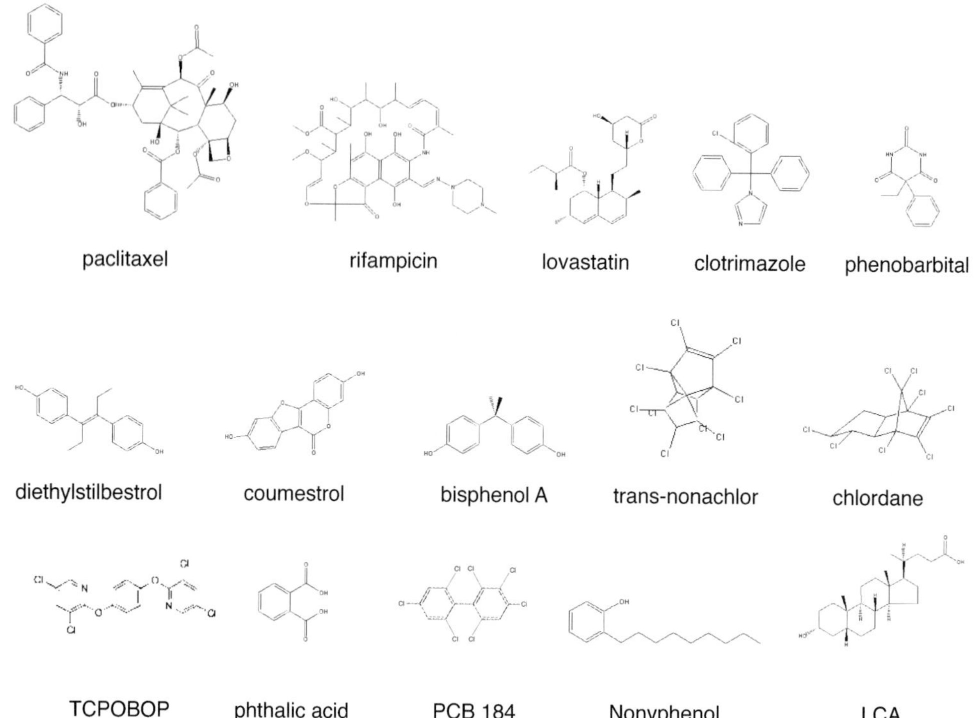

Fig. 2 Structures of representative compounds that activate PXR. Note that phenobarbital and TCPOBOP also activate CAR, whereas clotrimazole represses its constitutive activity.

The number of PXR and CAR ligands will continue to grow and the identification of endogenous as well as environmental chemicals that modulate the xenobiotic receptors will advance our understanding of the mechanisms by which environmental compounds affect our endocrine balance and may offer novel strategies for anticipating or preventing chemical toxicity.

BILE ACID REGULATION BY PXR/FXR/VDR

In addition to environmental toxins, our body is continuously exposed to a variety of toxic endogenous chemicals. For example, the secondary bile acid, lithocholic acid (LCA), is generated from non-toxic bile acids by intestinal bacteria and its elevation is implicated in pathogenesis of cholestatic liver disease and colon cancer. Because LCA is generated in an external cavity by a foreign organism, it is electively a xenobiotic compound.

The observation that LCA can induce CYP3A expression led to the suggestion that PXR may modulate this induction to reduce hepatotoxicity [9,10]. Three lines of evidence support this notion. First, LCA and its direct metabolite 3-keto LCA directly bind to and activate PXR. Second, in vivo activation of PXR by administration of an agonistic ligand (PCN) or by expressing a constitutively active form of PXR in the liver of transgenic mice results in marked resistance to LCA toxicity in rodents. Third, the potent CYP3A inducer and agonist for human PXR, rifampicin, has been reported to be effective in treating pruritus associated with chronic cholestasis. Detoxification of LCA by PXR appears to be mediated by the combined induction of CYP3A and the cytosolic sulfotransferase ST2A, both of which convert LCA to nontoxic metabolites [9,10,24]. Thus, the drug clearance pathway regulated by PXR can be utilized to detoxify endogenously produced toxins.

In addition to PXR, at least two other closely related nuclear receptors appear to contribute to bile acid regulation. Like PXR, the vitamin D receptor (VDR), was recently shown to bind secondary bile acids such as LCA and can induce CYP3A in the intestine [38]. This may help to explain why vitamin D3 is protective in certain animal models of LCA-induced colon cancer. Studies from a number of laboratories have shown that primary bile acids are under homeostatic control by the farnesoid X receptor, FXR, that activates a distinct set of target genes controlling bile acid synthesis and transport (reviewed in ref. [39]). Thus, the nuclear receptors, PXR, VDR and FXR, appear to collaborate to protect our body from accumulation of toxic bile acids via the control of their synthesis and degradation. It is currently unknown whether CAR plays any role in bile acid clearance.

GENETIC POLYMORPHISM IN XENOBIOTIC RECEPTORS AND RECEPTOR BINDING-SITES

Genetic polymorphisms or single nucleotide polymorphisms (SNPs) that alter expression level or activity of gene products are likely to have a large influence on the susceptibility of individuals to environmental toxicants such as carcinogens or EAS. Such polymorphisms could also predispose individuals to unexpected adverse drug reactions that create major clinical risks. In relation to the xenobiotic response, changes in metabolic enzymes or drug transporters can have a large impact on inactivation and elimination of toxins. In fact, polymorphisms have widely been detected in genes encoding drug clearance components including cytochrome P450 enzymes, phase II conjugating enzymes and drug transporters (reviewed in refs. [40,41]). Emergence of PXR as a master regulator of these pathways has led to the speculation as to whether polymorphisms in PXR or PXR-binding sites in target genes could be, in part, responsible for variations in drug response. Recent studies indicate that this may be the case.

In one study, PXR coding exons and flanking regions from over 200 patients were sequenced and a total of 28 SNP sites were identified including six (three in the N-terminal region and three in the LBD) that alter amino acid coding. Interestingly, one allele with an amino acid alteration in the LBD shows 3-fold higher activity compared to the normal allele when rifampicin was used as an agonist, but appears to be slightly less active when corticosterone was used. This observation is consistent with the idea that polymorphisms in the LBD could directly contribute to individual differences in the xenobiotic response profile [42]. In an independent study, the promoter, coding exon and flanking region sequences were analyzed in over 100 patients and 38 SNP sites including three non-synonymous changes in the coding region were identified. Interestingly, several of the noncoding SNPs located within the promoter or introns correlated with either enhanced or reduced inducibility of PXR target genes, suggesting that such SNPs could indeed alter drug response in human populations [43].

The molecular mechanisms of polymorphic expression of adult CYP3A7 represent an example of polymorphisms in xenobiotic receptor binding sites. While initially characterized as a fetal-specific isoform, CYP3A7 expression was also detected in adult livers but in a highly polymorphic manner. A survey of more than 300 patients has identified the CYP3A7*1C allele as a consistent marker of increased CYP3A7 expression both in the liver and the intestine [44,45]. By recombining a part of the CYP3A7 promoter with the corresponding region of CYP3A4, the CYP3A7*1C allele acquires the proximal ER6 response element of CYP3A4. Both PXR and CAR efficiently bind and activate the CYP3A4-ER6 element providing a molecular basis for increased response associated with the CYP3A7*1C allele [45].

REGULATION OF XENOBIOTIC RECEPTOR EXPRESSION

Several compounds are known to induce PXR or CAR mRNA indicating another level of control. For example, the glucocorticoid receptor (GR) can induce expression of PXR, CAR, and their heterodimeric partner RXR in cultured cells [46,47]. In addition, in the rodent liver PXR expression is autoinduced by PCN and induced by PPARα specific drugs such as perfluorodecanoic acid and clofibrate [48]. In theory, induction of the xenobiotic receptors could potentiate induction of downstream target genes. However, as both PXR and CAR are abundantly expressed in the liver, the effect of further induction is not clear. Perhaps, this regulation is more significant for those individuals who have alleles of xenobiotic receptors with a low constitutive level or weak activity. The levels of hepatic PXR and CAR mRNA have also been reported to be down-regulated in response to inflammatory signals [49,50]. Further studies are expected to reveal the relevance of xenobiotic receptor regulation, and its impact on drug metabolism.

PERSPECTIVE: THE POTENTIAL INFLUENCE OF XENOBIOTIC RECEPTOR ON THE ACTION OF EASs

From the available studies, it is likely that many industrial and natural EASs bind and activate either PXR, CAR, or both. How might xenobiotic receptors and the pathways regulated by these receptors influence the action of EASs? PXR and CAR most likely play a protective role against most EASs by promoting their detoxification and/or clearance. However, in rare, but relevant cases, PXR and CAR may inadvertently promote the deleterious effects exerted by some EASs. First, detoxification reactions, such as CYP3A induction, are also known to activate certain substrates to carcinogenic and/or cytotoxic products. For example, the first step in aflatoxin metabolism creates a reactive adduct for DNA, enabling its activity as a hepato-carcinogen. Thus, activation of PXR or CAR may enhance the toxicity of some environmental chemicals. Second, because steroids and thyroid hormones are metabolized and typically inactivated by enzymes such as CYPs, UGTs, and SULTs, constitutive activation of PXR or CAR by environmental chemicals could alter endocrine systems. In fact, the chronic activation of PXR in transgenic mouse livers results in increased corticosterone in serum and urine and by extension most likely stimulates the production of gonadal steroids [31]. Furthermore, prototypical chemical activators for PXR or CAR induce increased metabolism and decreased level of thyroid hormones, consequential increase in thyroid stimulating hormone and thyroid hypertrophy in rats [51]. Further studies are required to clarify the effect of chronic activation or inactivation of the xenobiotic receptors on global endocrine physiology and disease progression.

In summary, we believe that the xenobiotic regulation of drug clearance by nuclear receptors will be an emerging and exciting field of research in the coming years. The results of these studies will greatly advance our understanding of the complexity of xenobiotic regulation and their implication in human physiology, pathology, pharmaceutical development as well as enable a broad assessment of environmental risks.

NOTE ADDED IN PROOF

A specific human CAR agonist has recently been identified [52].

ACKNOWLEDGMENTS

J. S. is a Damon Runyon Alumni fellow of the Damon Runyon Cancer Research Foundation, DRG 1711-02. R. M. E. is an Investigator of the Howard Hughes Medical Institute at the Salk Institute for Biological Studies and March of Dimes Chair in Molecular and Developmental Biology.

REFERENCES

1. F. Oesch and M. Arand. "Xenobiotic metabolism", in *Toxicology*, H. Marquardt, S. G. Schafer, R. McClellan, F. Welsch (Eds.), pp. 83–109, Academic Press, New York (1999).
2. A. Chawla, J. J. Repa, R. M. Evans, D. J. Mangelsdorf. *Science* **294** (5548) 1866–1870 (2001).
3. S. A. Kliewer, J. T. Moore, L. Wade, J. L. Staudinger, M. A. Watson, S. A. Jones, D. D. McKee, B. B. Oliver, T. M. Willson, R. H. Zetterstrom, T. Perlmann, J. M. Lehmann. *Cell* **92** (1), 73–82 (1998).

4. B. Goodwin, M. R. Redinbo, S. A. Kliewer. *Annu. Rev. Pharmacol. Toxicol.* **42**, 1–23 (2002).
5. B. Blumberg, W. Sabbagh, Jr., H. Juguilon, J. Bolado, Jr., C. M. van Meter, E. S. Ong, R. M. Evans. *Genes Dev.* **12** (20), 3195–3205 (1998).
6. G. Bertilsson, J. Heidrich, K. Svensson, M. Asman, L. Jendeberg, M. Sydow-Backman, R. Ohlsson, H. Postlind, P. Blomquist, A. Berkenstam. *Proc. Natl. Acad. Sci. USA* **95** (21), 12208–12213 (1998).
7. T. W. Synold, I. Dussault, B. M. Forman. *Nat. Med.* **7** (5), 584–590 (2001).
8. W. Xie, J. L. Barwick, M. Downes, B. Blumberg, C. M. Simon, M. C. Nelson, B. A. Neuschwander-Tetri, E. M. Brunt, P. S. Guzelian, R. M. Evans. *Nature* **406** (6794), 435–439 (2000).
9. J. L. Staudinger, B. Goodwin, S. A. Jones, D. Hawkins-Brown, K. I. MacKenzie, A. LaTour, Y. Liu, C. D. Klaassen, K. K. Brown, J. Reinhard, T. M. Willson, B. H. Koller, S. A. Kliewer. *Proc. Natl. Acad. Sci. USA* **98** (6), 3369–3374 (2001).
10. W. Xie, A. Radominska-Pandya, Y. Shi, C. M. Simon, M. C. Nelson, E. S. Ong, D. J. Waxman, R. M. Evans. *Proc. Natl. Acad. Sci. USA* **98** (6), 3375–3380 (2001).
11. J. Staudinger, Y. Liu, A. Madan, S. Habeebu, C. D. Klaassen. *Drug Metab. Dispos.* **29** (11), 1467–1472 (2001).
12. P. Honkakoski, I. Zelko, T. Sueyoshi, M. Negishi. *Mol. Cell Biol.* **18** (10), 5652–5658 (1998).
13. T. Sueyoshi and M. Negishi. *Annu. Rev. Pharmacol. Toxicol.* **41**, 123–143 (2001).
14. B. M. Forman, I. Tzameli, H. S. Choi, J. Chen, D. Simha, W. Seol, R. M. Evans, D. D. Moore. *Nature* **395** (6702), 612–615 (1998).
15. T. Sueyoshi, T. Kawamoto, I. Zelko, P. Honkakoski, M. Negishi. *J. Biol. Chem.* **274** (10), 6043–6046 (1999).
16. P. Wei, J. Zhang, M. Egan-Hafley, S. Liang, D. D. Moore. *Nature* **407** (6806), 920–923 (2000).
17. J. Zhang, W. Huang, S. S. Chua, P. Wei, D. D. Moore. *Science* **298** (5592), 422–424 (2002).
18. L. B. Moore, D. J. Parks, S. A. Jones, R. K. Bledsoe, T. G. Consler, J. B. Stimmel, B. Goodwin, C. Liddle, S. G. Blanchard, T. M. Willson, J. L. Collins, S. A. Kliewer. *J. Biol. Chem.* **275** (20), 15122–15127 (2000).
19. I. Tzameli, P. Pissios, E. G. Schuetz, D. D. Moore. *Mol. Cell. Biol.* **20** (9), 2951–2958 (2000).
20. G. Min, J. K. Kemper, B. Kemper. *J. Biol. Chem.* **277** (29), 26356–26363 (2002).
21. T. Kawamoto, T. Sueyoshi, I. Zelko, R. Moore, K. Washburn, M. Negishi. *Mol. Cell Biol.* **19** (9), 6318–22 (1999).
22. I. Zelko, T. Sueyoshi, T. Kawamoto, R. Moore, M. Negishi. *Mol. Cell. Biol.* **21** (8), 2838–2846 (2001).
23. J. M. Rae, M. D. Johnson, M. E. Lippman, D. A. Flockhart. *J. Pharmacol. Exp. Ther.* **299** (3), 849–57 (2001).
24. J. Sonoda, W. Xie, J. M. Rosenfeld, J. L. Barwick, P. S. Guzelian, R. M. Evans. *Proc. Natl. Acad. Sci. USA* **99** (21), 13801–13806 (2002).
25. J. M. Maglich, C. M. Stoltz, B. Goodwin, D. Hawkins-Brown, J. T. Moore, S. A. Kliewer. *Mol. Pharmacol.* **62** (3), 638–646 (2002).
26. A. Toell, K. D. Kroncke, H. Kleinert, C. Carlberg. *J. Cell Biochem.* **85** (1), 72–82 (2002).
27. A. Ueda, H. K. Hamadeh, H. K. Webb, Y. Yamamoto, T. Sueyoshi, C. A. Afshari, J. M. Lehmann, M. Negishi. *Mol. Pharmacol.* **61** (1), 1–6 (2002).
28. C. Garcia-Allan, P. G. Lord, J. M. Loughlin, T. C. Orton, J. E. Sidaway. *J. Biochem. Mol. Toxicol.* **14** (2), 65–72 (2000).
29. J. Sugatani, H. Kojima, A. Ueda, S. Kakizaki, K. Yoshinari, Q. H. Gong, I. S. Owens, M. Negishi, T. Sueyoshi. *Hepatology* **33** (5), 1232–1238 (2001).
30. H. R. Kast, B. Goodwin, P. T. Tarr, S. A. Jones, A. M. Anisfeld, C. M. Stoltz, P. Tontonoz, S. Kliewer, T. M. Willson, P. A. Edwards. *J. Biol. Chem.* **277** (4), 2908–2915 (2002).

31. W. Xie, M. F. Yeuh, A. Radominska-Pandya, S. P. Saini, Y. Negishi, B. S. Bottroff, G. Y. Cabrera, R. H. Tukey, R. M. Evans. *Proc. Natl. Acad. Sci. USA* **100** (7), 4150–4155 (2003).
32. R. E. Watkins, G. B. Wisely, L. B. Moore, J. L. Collins, M. H. Lambert, S. P. Williams, T. M. Willson, S. A. Kliewer, M. R. Redinbo. *Science* **292** (5525), 2329–2333 (2001).
33. A. Takeshita, N. Koibuchi, J. Oka, M. Taguchi, Y. Shishiba,Y. Ozawa. *Eur. J. Endocrinol.* **145** (4), 513–517 (2001).
34. X. Coumoul, M. Diry, R. Barouki. *Biochem. Pharmacol.* **64** (10), 1513–1519 (2002).
35. E. G. Schuetz, C. Brimer, J. D. Schuetz. *Mol. Pharmacol.* **54** (6), 1113–1117 (1998).
36. T. Kawamoto, S. Kakizaki, K. Yoshinari, M. Negishi. *Mol. Endocrinol.* **14** (11), 1897–1905 (2000).
37. A. Ueda, S. Kakizaki, M. Negishi, T. Sueyoshi. *Mol. Pharmacol.* **61** (6), 1284–1288 (2002).
38. M. Makishima, T. T. Lu, W. Xie, G. K. Whitfield, H. Domoto, R. M. Evans, M. R. Haussler, D. J. Mangelsdorf. *Science* **296** (5571), 1313–1316 (2002).
39. A. Chawla, E. Saez, R. M. Evans. *Cell* **103** (1), 1–4 (2000).
40. M. V. Relling and T. Dervieux. *Nature Rev. Cancer* **1** (2), 99–108 (2001).
41. M. Pirmohamed and B. K. Park. *Trends Pharmacol. Sci.* **22** (6), 298–305 (2000).
42. E. Hustert, A. Zibat, E. Presecan-Siedel, R. Eiselt, R. Mueller, C. Fuss, I. Brehm, U. Brinkmann, M. Eichelbaum, L. Wojnowski, O. Burk. *Drug Metab. Dispos.* **29** (11), 1454–1459 (2001).
43. J. Zhang, P. Kuehl, E. D. Green, J. W. Touchman, P. B. Watkins, A. Daly, S. D. Hall, P. Maurel, M. Relling, C. Brimer, K. Yasuda, S. A. Wrighton, M. Hancock, R. B. Kim, S. Strom, K. Thummel, C. G. Russell, J. R. Hudson, Jr., E. G. Schuetz, M. S. Boguski. *Pharmacogenetics* **11** (7), 555–572 (2001).
44. P. Kuehl, J. Zhang, Y. Lin, J. Lamba, M. Assem, J. Schuetz, P. B. Watkins, A. Daly, S. A. Wrighton, S. D. Hall, P. Maurel, M. Relling, C. Brimer, K. Yasuda, R. Venkataramanan, S. Strom, K. Thummel, M. S. Boguski, E. Schuetz. *Nat. Genet.* **27** (4), 383–391 (2001).
45. O. Burk, H. Tegude, I. Koch, E. Hustert, R. Wolbold, H. Glaeser, K. Klein, M. F. Fromm, A. K. Nuessler, P. Neuhaus, U. M. Zanger, M. Eichelbaum, L. Wojnowski. *J. Biol. Chem.* **277** (27), 24280–24288 (2002).
46. J. M. Pascussi, S. Gerbal-Chaloin, J. M. Fabre, P. Maurel, M. J. Vilarem. *Mol. Pharmacol.* **58** (6), 1441–1450 (2000).
47. J. M. Pascussi, L. Drocourt, J. M. Fabre, P. Maurel, M. J. Vilarem. *Mol. Pharmacol.* **58** (2), 361–372 (2000).
48. H. Zhang, E. LeCulyse, L. Liu, M. Hu, L. Matoney, W. Zhu, B. Yan. *Arch. Biochem. Biophys.* **368** (1), 14–22 (1999).
49. J. M. Pascussi, S. Gerbal-Chaloin, L. Pichard-Garcia, M. Daujat, J. M. Fabre, P. Maurel, M. J. Vilarem. *Biochem. Biophys. Res. Commun.* **274** (3), 707–713 (2000).
50. A. P. Beigneux, A. H. Moser, J. K. Shigenaga, C. Grunfeld, K. R. Feingold. *Biochem. Biophys. Res. Commun.* **293** (1), 145–149 (2002).
51. C. D. Klaassen and A. M. Hood. *Toxicol. Pathol.* **29** (1), 34–40 (2001).
52. J. M. Maglich, D. J. Parks, L. B. Moore, J. L. Collins, B. Goodwin, A. N. Billin, C. A. Stoltz, S. A. Kliewer, M. H. Lambert, T. M. Willson, J. T. Moore. *J. Biol. Chem.* **278** (19), 17277–17283 (2003).

Pure Appl. Chem., Vol. 75, Nos. 11–12, pp. 1743–1756, 2003.

Topic 1.8

Molecular mechanisms of cross-talk between growth factors and nuclear receptor signaling*

Didier Picard‡

Département de Biologie Cellulaire, Université de Genève, Sciences, III 30 quai Ernest-Ansermet, CH - 1211 Genève 4, Switzerland

Abstract: Signaling pathways can be linear, but more complex patterns are common. Growth factors and many other extracellular signals cannot directly enter cells and transduce their information via membrane-bound receptors. In contrast, steroid receptors are members of the nuclear receptor superfamily and await their cognate hormones inside the cells. These two types of signaling pathways are extensively intertwined and cross-talk at many different levels. A wide range of extra- and intracellular signals, including a variety of growth factors, can activate the transcriptional activity of steroid receptors in the absence of their cognate hormones. Conversely, steroid receptors lead a double life. By coupling to signaling molecules that mediate signal transduction of extracellular factors, they can elicit very rapid nongenomic responses. The signaling pathways of steroid-independent activation of steroid receptors, on the one hand, and of nongenomic signaling by steroid receptors, on the other, display a remarkable reciprocal relationship suggesting that these two modes of signaling cross-talk may be two faces of the same coin.

DEFINITION AND OVERVIEW

Extracellular signals modify intracellular processes through cognate receptors that elicit a cascade of events. However, a linear view of signal transduction falls short of describing all effects. Instead, branching, feedback, integration, and networking are characteristics of most if not all signal transduction pathways. Signaling cross-talk refers to a situation where one signal affects the output of another, seemingly distinct, signal transduction pathway.

There is signaling cross-talk at all levels of signaling, from affecting availability of the signal to modifying the regulation of expression of target genes. For example, at the level of an organism, steroid hormone concentrations are influenced by signals that affect the levels of their serum binding proteins. Signals can stimulate the biosynthesis of other signals and their receptors. In breast cancer cells, estrogen induces the expression of the progesterone receptor (PR) [1], the EGF receptor [2], and several members of the EGF ligand family [3,4]. With respect to a target cell, signaling cross-talk may take place in the extracellular space or within the cell, inside or outside the nucleus.

In this review, I will focus on the molecular mechanisms by which a variety of extra- and intracellular signals can modulate the activities of steroid receptors, in particular the receptors for the sex steroids, that is, the estrogen receptors (ERs) α and β, PR, and the androgen receptor (AR). These are all members of the nuclear receptor superfamily. Moreover, I will discuss how the very same nuclear receptors appear to elicit very rapid nongenomic effects by feeding into other signaling pathways. There is mounting evidence that these two nuclear receptor activities may be two faces of the same coin

*Report from a SCOPE/IUPAC project: Implication of Endocrine Active Substances for Human and Wildlife (J. Miyamoto and J. Burger, editors). Other reports are published in this issue, *Pure Appl. Chem.* **75**, 1617–2615 (2003).

‡Tel.: +41 22 379 6813; Fax: +41 22 379 6928; E-mail: Picard@cellbio.unige.ch

(Fig. 1), and that they may be intricately linked in positive feedback loops in some cases. A better understanding of the molecular mechanisms of signaling cross-talk involving steroid receptors will contribute to a framework for assessing the potential dangers of endocrine active substances (EASs).

I will limit my review to posttranslational events, and refer the reader to the extensive literature on cross-talk at the level of the regulation of the biosynthesis of ligands and receptors. Likewise, I will not be able to discuss another platform for extensive signaling cross-talk, the transcriptional regulatory elements themselves. These are typically composite and recruit several transcription factors, which can influence the DNA binding and transcriptional regulatory functions of each other, and be themselves regulated by signaling pathways. The interested reader will easily find both primary and review articles on individual genes and transcription factors.

STEROID-INDEPENDENT ACTIVATION OF STEROID RECEPTORS

Phenomena

Originally, steroid receptors were thought to be exclusively activated as transcription factors by binding cognate ligands. Agonistic ligands induce a conformational change in the hormone binding domain that allows the recruitment of transcriptional coactivators [5]. Assuming that the switch from inactive aporeceptor to active receptor can be described as a chemical equilibrium, it is perhaps not a surprise that the equilibrium can be shifted by altering the concentrations or activities of factors that interact with the active or inactive forms of steroid receptors. Indeed, the overexpression of transcriptional coactivators has often been found to increase the basal activity of steroid receptors. And yet, it came as a surprise when it was discovered in the late 1980s and early 1990s that sex steroid receptors can be activated by a host of extracellular signals in the absence of their cognate ligands [for reviews, see refs. 6–8]. This clearly represents an extreme form of signaling cross-talk. Over the last 10 years, more phenomena have been reported, signaling pathways have begun to be elucidated, but the molecular mechanisms remain poorly understood. Deciphering them should also help to explain the more typical and probably more physiological form of signaling cross-talk: synergy between low levels of steroids and other signals.

Table 1 provides a list of signals that have been reported to activate ER, PR, and AR in the absence of cognate ligand. In addition to a number of growth factors and the neurotransmitter dopamine, it includes several intracellular proteins and drugs. The activation by intracellular factors is of interest for two reasons: (i) It may mimic the stimulation by certain extracellular factors by plugging into their signaling pathways, and (ii) it may be physiologically relevant since intracellular signaling components can be overexpressed or constitutively activated in certain tumors. For example, cyclin D1 overexpression is common in breast tumors [47], and the phosphatidylinositol-3-OH kinase (PI3K)/Akt pathway is activated in many cancers [28,48,49]. In some cases, it is activated indirectly because of the loss of the tumor suppressor PTEN, which is a negative regulator of this pathway [48]. These pathways would be particularly relevant for cancers and diseases that are dependent on ERα.

The phenomenon of steroid-independent activation of steroid receptors may also be relevant to the cell-specific partial agonism of many anti-hormones. Steroid receptors have two main transcriptional activation functions (AFs). AF-1 and AF-2 are associated with the N-terminal domain and the hormone-binding domain, respectively. Antagonistic ligands of steroid receptors bind in the hormone-binding pocket and block AF-2 in an inactive conformation. However, partial antagonists display cell- and promoter-specific agonistic effects that are thought to be due, at least in part, to AF-1 activity. Interestingly, some of the signals that are able to activate unliganded steroid receptors can further increase the agonistic effects of partial antagonists, in some cases by further stimulating AF-1 (see below). In transfection experiments, the ER antagonist tamoxifen is switched to an agonist in the presence of growth factors or their downstream signaling molecules [29,50–53]. Similarly, the PR antagonist RU486 becomes an agonist when the protein kinase A (PKA) pathway is stimulated [44–46].

Table 1 Steroid-independent activators of unliganded sex steroid receptors[a].

Factor[b]	ER[c]	PR[d]	AR
Dopamine	[9]	[9]	
Epidermal growth factor (EGF)	[10,11]	[12]	[13]
Heregulin	[14]		
TGFα	[11]		
Insulin and insulin-like growth factors (IGF)	[15–17]		[13]
Keratinocyte growth factor			[13]
Fibroblast growth factor 2 (FGF-2 = bFGF)	[18]		
Gonadotropin release hormone (GnRH)	[19]	[20]	
Sex hormone binding globulin			[21]
Interleukin-6	[22]		[23,24]
erbB2 = HER-2	[14]		[25]
PI3K	[26]		
Protein kinase B (= Akt)	[26–28]		
v-Src	[29]		
Ras (constitutive mutant)	[30]		
MEKK1 (constitutive mutant)			[31]
MAPKK (constitutive mutant)	[32]		
Cyclin D1	[33,34]		
Cyclin A-Cdk2	[35]		
Ets-1	[36]		
Activators of protein kinase A	[15]	[37]	[38]
Inhibitors of protein phosphatases 1 and 2A	[9]	[9,37]	
Inhibitor of phosphotyrosine phosphatases		[12]	
Activator of protein kinase C	[30,39]		[40]

[a]For a regularly updated table, see <http:/www.picard.ch/>.

[b]This list may be incomplete as the extent of steroid-independent activation varies widely; moreover, several reports have indicated that some of the effects may be cell- and/or promoter-specific [see, e.g., refs. 25,41].

[c]Almost all publications have examined ERα. ERβ has only been shown to be activated by EGF [42,43].

[d]The response of PR displays marked species differences: chicken and rodent PRs can be activated in the absence of cognate hormone, whereas human PR is affected only in the presence of a ligand such as the partial antagonist RU486 [44–46].

Signaling pathways

Efforts have been made to dissect signaling pathways but many gaps remain (Fig. 1). Peptide growth factors signal to ERα through their respective tyrosine kinase receptor and the Ras-MAPK pathway [30,32,54]. Constitutive activation of the MAPK pathway mimics activation of ERα by the extracellular factor, and its inhibition blocks it [30,32]. Activation of ERα by GnRH requires protein kinase C in addition to the MAPK pathway [19]. Interestingly, constitutive activation of the PI3K-Akt pathway is not only able to activate ERα [26–28,55], but this pathway is also required for activation of ERα by IGF-I [27,55]. This suggests that peptide growth factors may signal to steroid receptors through multiple or branched pathways. Further studies are needed to explain why MAPK activity can be sufficient in one case whereas in another activation by growth factors can be blocked by a PI3K inhibitor. Tissue-specific differences in the signaling circuitry may provide part of the explanation [see, e.g., refs. 30,56].

The direct phosphorylation of steroid receptors by some of these pathways is important. MAPK phosphorylates a serine residue in the AF-1 domain of the ERs. This modification is necessary for ac-

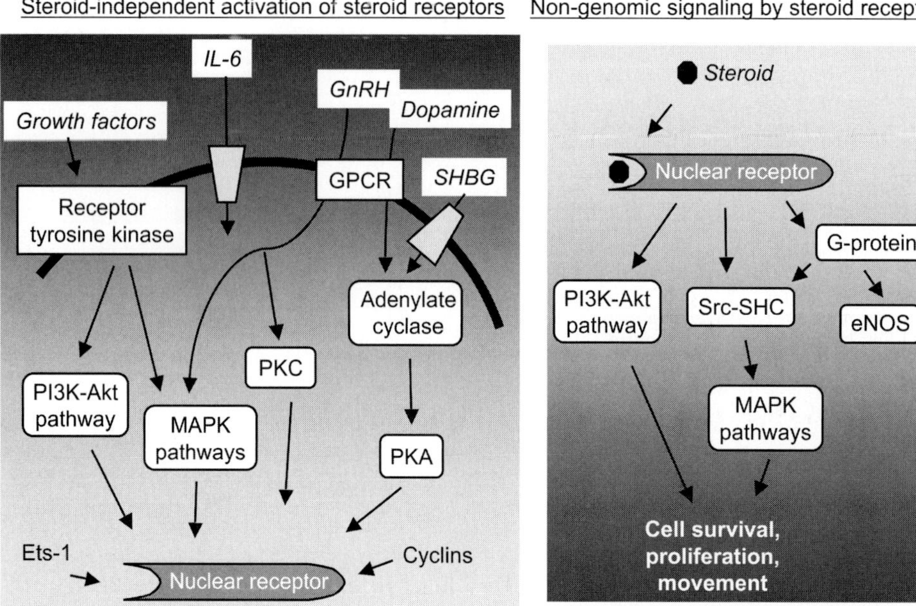

Fig. 1 Schematic comparison of steroid-independent activation of steroid receptors and nongenomic signaling by steroid receptors. Signaling pathways are simplified and do not necessarily apply as shown to all sex steroid receptors (ERα, ERβ, PR, AR). The shading of the two boxes illustrates the reciprocal relationship of the two signaling modes. GPCR, G-protein coupled receptor; SHBG, sex hormone binding globulin.

tivation by peptide growth factors [30,32,42,54]. Since simply putting a negative charge at the AF-1 associated serine 118 of ERα is not sufficient for ligand-independent activation [32], other sites of ERα or other factors may need to be modified as well. Other activating signals and pathways also lead to the hyperphosphorylation of ERα. Heregulin, potentially through the MAPK pathway, leads to ERα hyperphosphorylation [14]. The stimulation of ERα by the PI3K-Akt pathway appears to depend on the direct phosphorylation of yet other sites of the ERα N-terminus, notably S167 [26,27]. The enhancement of ERα activity by the cyclin A-Cdk2 complex, in particular in the presence of estrogen, is linked to the phosphorylation of S104 and S106, but not S118 [57].

There are alternative activation pathways that do not seem to involve the direct phosphorylation of ERα. One and the same signal can target different ERα domains in different cell types: whereas insulin targets S118 in most cell types, in neuronal cells it signals in a S118-independent fashion to the hormone binding domain [58]. Moreover, the same signal can target the same ERα domain through different pathways in different cell types: MAPK and S118 phosphorylation are not involved in the activation of ERα by EGF in vascular cells even though the same domain, AF-1, is the target [56]. Likewise, the stimulatory effect of v-Src on AF-1 activity is only in part due to the phosphorylation of S118 by MAPK; v-Src also stimulates AF-1 activity indirectly through the Rac-MEKK-JNKK-JNK pathway [29]. However, in all of these cases one cannot exclude a role for minor phosphorylation sites of ERα itself.

Dopamine functions through a G-protein coupled membrane receptor. Its activation results in the stimulation of adenylate cyclase, elevation of cAMP, and activation of PKA. Despite the fact that both cAMP and PKA mimic the ability of dopamine to activate steroid receptors in the absence of ligand, there is no evidence that PKA elicits the steroid-independent activity by directly phosphorylating the steroid receptors [59,60]. Rather, the phosphorylation of S236 in the DNA-binding domain of ERα by

PKA is inhibitory for dimerization and DNA binding [61]. The issue is complicated by the fact that cAMP signaling can either inhibit or activate the MAPK pathway depending on cell type [62]. However, cAMP-mediated activation of ERα also works in cells where PKA inhibits the MAPK pathway (our unpublished results). Moreover, EGF and cAMP appear to activate ERα through distinct pathways targeting different domains; cAMP may target AF-2 [63], or possibly both AF-1 and AF-2 (our unpublished results). Further progress towards dissecting steroid-independent activation of steroid receptors by cAMP has been made in whole animal experiments with rodents. Both dopamine and progesterone elicit the lordosis response of females in a PR-dependent fashion [64]. Elegant studies have placed the dopamine- and cAMP-regulated phosphoprotein-32 (DARPP-32) and protein phosphatase 1 (PP1) downstream of PKA. In the presence of either dopamine or progesterone, DARPP-32 is hyperphosphorylated and inhibits PP1 [65,66]. This finding correlates with the original observation that the phosphatase inhibitor okadaic acid activates apo-PR [9].

And yet, the role of direct phosphorylation of PR remains confusing. The mutation of the C-terminal S628 in chicken PR selectively abolishes the response to dopamine while leaving unaffected the response to hormone and okadaic acid [9]. A more recent survey of the PR sites that are hyperphosphorylated in response to either hormone or increased levels of cAMP revealed differences, but failed to provide evidence for a critical requirement of any of the major sites [60]. The role of phosphorylation will have to be reexamined in comparable biological systems, and it will have to be ascertained that the lordosis response to dopamine entails the steroid-independent activation of the transcriptional function of PR.

Molecular mechanisms

The challenge is to explain how the transcriptional activity of aporeceptors is stimulated in the absence of steroid at the molecular level. As alluded to above, the role of phosphorylation of the ERs for transcriptional coactivator recruitment has received particular attention. The EGF-induced phosphorylation of the ERβ AF-1 domain promotes its recruitment of the coactivator SRC1 [42], and transcriptional activity is further increased in the presence of the CREB binding protein CBP [43]. In contrast to the p68 RNA helicase [67], SRC1 recruitment to the ERα AF-1 domain is not stimulated by growth factor signaling and phosphorylation of S118 [42,68]. However, the stimulatory effect of the p68 RNA helicase is weak and not observed in HeLa cells [67] where EGF has been reported to work [32].

The steroid-independent activation of ERα by cyclin D1 overexpression represents an interesting paradigm. Cyclin D1 functions independently of its kinase partner to provide a bridge between p160 coactivators and the ERα, which is unable to bind these coactivators directly in the absence of hormone [69]. Remarkably, cAMP signaling promotes the interaction between ERα and cyclin D1, and this correlates with increased transcriptional activity of apo-ERα [70]. It remains to be determined how cAMP stimulates this interaction and whether the activation of apo-ERα by cAMP (see above) or by other signals requires cyclin D1.

In principle, almost any of the close to 100 ER interacting proteins (for a regularly updated list, see <http://www.picard.ch/>) or proteins interacting with the other nuclear receptors could serve as a platform or conduit for signaling cross-talk and many are [71]. MAPK phosphorylates the p160 coactivators AIB1 [72] and Grip1 [73], and stimulates their intrinsic transcriptional activity and coactivation function. Likewise, MAPK signaling leads to the phosphorylation of SRC1 [74], and thereby stimulates its ability to coactivate the chicken PR in the presence of either cAMP or progesterone [75]. Interestingly, activation of AR by IL-6 is MAPK-dependent, stimulated by MAPK-phosphorylated SRC1, but the IL-6 induced recruitment of SRC1 to AR depends on yet another unidentified signaling pathway [76].

The induced release of a repressor would constitute an alternative mechanism for steroid-independent activation. In the absence of ligand, steroid receptors are associated with molecular chaperones, which are thought to keep them transcriptionally inactive [77,78]. Other targets might be Brca1, which can function as a repressor of apo-ERα [79], and the transcriptional corepressors NCoR and Smrt. NCoR appears to interact with ERα in the presence of the antagonist tamoxifen, and EGF stimulates the release of NCoR and concomitant activation of ERα [52]. Similarly, cAMP decreases the association of NCoR and Smrt with antagonist-occupied human PR [80].

The association with other transcription factors might contribute to steroid-independent effects. In addition to coactivators (see above), the stimulation of AR by interleukin-6 (IL-6) involves the recruitment of IL-6-activated Stat3 to AR [24]. Even more remarkable is the observation that the transcription factor Ets-1, upon association with nuclear receptors, induces their transcriptional activity in the absence of ligand [36]. Ets-1 appears to induce a conformational change in the unliganded nuclear receptor that allows it to bind coactivators.

Physiological significance

The physiological significance of steroid-independent activation of the ERα is best documented for the growth regulation of the rodent uterus. Both estrogen and peptide growth factors stimulate uterine growth, and both types of signals depend on ERα function [10,81]. In ERα knock-out mice, IGF-I signaling normally activates Akt and MAPK but fails to stimulate uterine growth; moreover, an ERE-luciferase transgene is activated by either IGF-I or estrogen [82]. As discussed above, steroid-independent activation of PR by dopamine may be involved in the lordosis response in female rodents [64]. This behavioral response can also be elicited by EGF, but in this case it depends on ERα rather than PR [83]. But despite the elegance of these in vivo experiments, there is no formal proof yet in this system that ERα and PR are transcriptionally activated by stimulation with EGF and dopamine, respectively. It is still possible that these nuclear receptors must be there as competence factors at some point preceding the stimulation.

Essentially all of the other evidence comes from studies with tissue culture cell lines. For example, androgen- and estrogen-dependent prostate and breast cancer cell lines can lose their steroid- but not their steroid receptor-dependence upon overexpression of the tyrosine kinase receptor HER-2 [14,25].

Considerable circumstantial evidence has led to the speculation that steroid-independent activation of ERα contributes to breast cancer progression, notably to estrogen-independence and tamoxifen resistance of ERα-positive tumors [for discussion, see refs. 6,84]. A large number of growth factors are indeed expressed in breast tumors, and act in both autocrine and paracrine fashions [85]. In addition to extracellular factors, HER-2 and cyclin D1 play a role and are often overexpressed in breast tumors [47,86]. In transfection experiments, both are able to activate the ERα in the absence of estrogen (see Table 1 and above), and overexpression of HER-2 renders breast cancer cell lines less dependent on estrogen and reduces their tamoxifen sensitivity [14,87]. Thus, these pathways may not only stimulate the proliferation of ERα-dependent breast tumor cells by themselves, but they must also be considered for their potential to pervert the action of estrogen antagonists by switching them to agonists [29,50–53]. By stimulating the AF-1 function of ERα, growth factors reduce the requirement for AF-2, and render ERα "resistant" to antagonists such as tamoxifen, which primarily block AF-2. Even worse, tamoxifen may then even facilitate the activation of ERα. Remarkably, this phenomenology in tissue culture seems to correlate with the observation that tamoxifen, in a small percentage of breast cancer patients, ends up stimulating tumor progression. In any case, HER-2 overexpression seems to predict a poor response to endocrine therapy [88,89], but the weakness of the correlation, not surprisingly, emphasizes that signaling cross-talk is but one of several causes of tamoxifen resistance.

NONGENOMIC SIGNALING BY STEROID RECEPTORS

A new role for old receptors

Evidence has accumulated over several decades that steroid hormones also signal through nongenomic pathways with very rapid cellular effects [90–95]. Within seconds or minutes, steroids can induce an increase in several second messengers such as inositol triphosphate, cAMP, Ca^{2+}, and the activation of MAPK and PI3 kinase [e.g., see refs. 96–100]. Biological responses attributed to short-term effects range from vasorelaxation of endothelial cells, neuroprotection, and bone protection to the stimulation of proliferation of carcinoma cells.

The molecular mechanisms and notably the receptor(s) have only recently begun to be unraveled. It now appears that at least some of the reported effects are mediated by the same steroid receptors that are known as nuclear receptors. This alternative lifestyle of the known steroid receptors will be the focus of the following paragraphs. However, it is important to note that they clearly do not account for all nongenomic phenomena elicited by steroids. Several unrelated membrane receptors contribute to a large diversity of rapid responses [90,94,101–103].

The conclusion that the known steroid receptors can themselves be at the top of a signaling cascade for rapid responses is primarily based on the following findings: (i) Rapid nongenomic responses to estrogen, progesterone, and androgen can be reconstituted in unresponsive cell lines by transient expression of ER, PR, and AR, respectively, and these responses seem to be independent of the transcriptional competence of the steroid receptor [97,98,104–115]. (ii) A small fraction of the total cellular (and mostly nuclear) ER complement associates with the cell membrane [110,116–119], extranuclear localization is necessary [109], and cytoplasmic sequestration by a variant of the metastatic tumor antigen 1 promotes nongenomic signaling [120]; in the case of ER signaling to eNOS in endothelial cells, this nongenomic response can even be demonstrated with isolated caveolae, a specialized membrane compartment [110,118,121]. (iii) "Nuclear" steroid receptors can be immunoprecipitated with a whole series of signaling molecules that are typical of signaling pathways triggered by membrane receptors (see below). (iv) Antibodies to "nuclear" ERα added to the culture medium can modulate estradiol responses [119,122].

Signaling pathways

In line with their proposed membrane association, the ERs, AR, and PR have been shown to couple to a large number of signaling molecules (Fig. 1). Signaling of ERα to eNOS takes place in caveolae [110,118,123] and involves the recruitment of a $G_{\alpha i}$ protein to ERα [124]. ERs are also coupled to other types of G proteins [106]. PR may be coupled to G proteins in frog oocytes, in which the rapid response to progesterone leads to the inhibition of adenylate cyclase [125–127]. Many groups have reported that activation of steroid receptors can activate the MAPK pathway. ERα, AR, and PR can elicit this rapid response by binding and activation of c-Src [97,104,109,112,113]. The molecular mechanism of this activation remains unclear. For human PR, it is particularly controversial. One group has provided evidence that suggests PR binds and displaces the inhibitory SH3 domain of c-Src [113], whereas another group has argued that the activation by PR is indirect and depends on the association of PR with ERα [104]. Additional adaptor proteins may also be involved in MAPK activation by steroid hormones since several reports demonstrate a role for SHC and its adaptor proteins Grb2 and Sos [109,128], and the modulator protein MNAR promotes the formation and activity of ternary complexes between ERα and c-Src [129]. The formation of ternary ERα-Src-PI3K complexes [114] may explain why both PI3K and the Src-MAPK pathway have been placed downstream of ERα [28,98,130] and AR [115]. Surprisingly, the kinetics of activation of PI3K by estradiol has been shown to be slower than that of MAPK [98]. As a result of the activation of PI3K by ERα, Akt is activated [28,98,114,119,131]. In adipocytes, MAPK activation by estradiol requires a G protein, Src, PI3K as well as protein kinase C (PKC) [132].

Signaling pathways are likely to differ substantially between cell types and experimental systems. For example, the requirement for c-Src for estradiol signaling mentioned above has not always been observed [128,133]. Moreover, the existence of negative and positive feedback mechanisms may often render a linear view of signaling inappropriate. Indeed, the IGF-I receptor is both upstream and downstream of ERα in a pathway leading to the activation of MAPK [111]. Likewise, Akt is both upstream and downstream of ERα [28, see also ref. 129]. A prediction from this type of positive feedback is that the system should respond to lower levels of stimuli.

Nongenomic signaling: Open questions

Many mechanistic details remain poorly understood. The evidence is confusing regarding the membrane association of steroid receptors. How do they get there? Do they insert into the membrane, and if so on which side? Do they actually expose the ligand-binding pocket of the hormone-binding domain on the outside of the cell? Signaling can be triggered by membrane-impermeable forms of steroids such as BSA conjugates [see, e.g., refs. 108,132], but internalization of the conjugated or deconjugated steroid has not been rigorously excluded [see also ref. 134]. The minimal signaling domain of ERα seems to be the hormone-binding domain [see, e.g., refs. 109,112], but according to another report this domain is not even required [128]. Could different domains couple to different signaling pathways? The findings with PR [113] would seem to support such a view.

Pharmacology has been used and abused extensively to draw conclusions about the type of receptor involved in particular responses. However, there is no reason whatsoever to assume that the ligand specificity and response of a nuclear receptor should be the same once it associates with the membrane or other signaling molecules. Thus, antagonists of the transcriptional function may be either antagonists or agonists for nongenomic signaling by the very same receptor (and even more so of course for unrelated membrane receptors), and vice-versa for antagonists of nongenomic signaling. An intriguing pharmacological study has revealed a correlation between the ability of synthetic estrogens to elicit nongenomic as opposed to genomic signaling in tissue culture and bone protection in whole animals [109,135], but the target cells and molecular mechanisms in vivo remain to be elucidated. Establishing the phenomenology is important, but more caution should be used in interpreting it.

What are the biological consequences of nongenomic signaling? Apart from the stimulation of frog oocyte maturation, little is formally established. And even there, a very recent reassessment demonstrates that it is androgens that signal through the frog AR rather than progesterone through PR; while implicating nongenomic signaling by a nuclear receptor was correct, both the hormone and the receptor had been mistaken [127]. In MCF-7 breast cancer cells, estradiol induces ruffles and pseudopodia in cell peripheral areas where ERα and actin colocalize [128]. A consequence of the activation of MAPK is the activation of their target transcription factors including Elk-1 and the serum response factor [115,128,130,131] leading again to delayed genomic responses. In combination, these responses may contribute to stimulate the motility and proliferation of estrogen-dependent breast cancer cells. In pituitary cells, estradiol modulates the prolactin release from stores [122].

Despite the demonstration that nongenomic signaling by nuclear receptors occurs in a wide variety of different cell types in vitro, the physiological significance beyond the frog oocyte system remains very speculative [for discussion, see ref. 93]. Based on where nongenomic signaling of estrogen has been described, it has been argued that it is neuroprotective, beneficial in the cardiovascular system, antiapoptotic in the bone, and stimulatory for the progression of breast cancer. However, it remains a major challenge for the future to discriminate between genomic and nongenomic effects of steroid hormones in whole organisms and to ascribe them to the correct receptor.

TWO FACES OF THE SAME COIN?

I have mentioned evidence in support of the notion that nongenomic signaling by steroid receptors and steroid-independent activation of steroid receptors could be intricately linked. When ERα stimulates MAPK, MAPK can in turn phosphorylate ERα to influence its subcellular localization and to increase its AF-1 activity. The same can be said for the Akt pathway. Perhaps one of the main functions of nongenomic signaling by steroid receptors may be to provide a rapid boost to their classical function as transcription factors. In turn, by stimulating the expression of signaling molecules, steroid receptors might contribute to more sustained changes in signaling patterns. A comparative analysis of the cross-talk potential of other members of the nuclear receptor family might help to determine which role came first.

SIGNALING CROSS-TALK AND ENDOCRINE ACTIVE SUBSTANCES: RELEVANCE AND RECOMMENDATIONS

Signaling cross-talk modifies and enhances the response to extracellular signals. In this context, two types of interactions with EAS that are steroid receptor ligands must be considered. First, their agonistic or antagonistic genomic effects could be altered by cross-talk with other signals. Second, in addition to their well-documented genomic effects, EAS could elicit nongenomic responses mediated by membrane-associated steroid receptors, a possibility that has not yet received much attention [136].

By acting synergistically, extracellular factors might be expected to lower the EC_{50} and/or to increase the maximal transcriptional response of xenosteroids. Thus, signaling cross-talk must be considered as an additional risk factor because the disruptive potential of xenosteroids may otherwise be underestimated. Remarkably, this has recently been modeled with estrogen-dependent breast cancer cells. They have been shown to display an increased proliferative response to low levels of estrogens upon constitutive activation of the MAPK pathway [137]. Augmented MAPK activity may also contribute to the estrogen hypersensitivity that ERα-dependent breast cancer cells acquire upon long-term estrogen deprivation [138]. Moreover, as in the case of tamoxifen and RU486, signaling cross-talk could convert weakly antagonistic xenosteroids into weak agonists.

RECOMMENDATIONS

The following recommendations primarily call for additional basic research and modifications in the protocols of large-scale screening programs.

- The transcriptional effects of xenosteroids should be examined both in the absence and in the presence of additional stimuli such as growth factors.
- Dose–response curves of xenosteroids should be established both in the absence and in the presence of additional stimuli such as growth factors.
- Screens for xenosteroids should be done in the presence of factors that can synergize with weak agonists.
- Antagonistic xenosteroids should be tested for agonism under conditions of signaling cross-talk.
- Xenosteroids should be systematically assessed for their potential in eliciting nongenomic responses through the classical nuclear receptors as well as unrelated membrane receptors.

ACKNOWLEDGMENTS

I am grateful to members of my group for continuous discussions and critical reading of the manuscript. My laboratory is supported by the Swiss National Science Foundation, Krebsforschung Schweiz, the Fondation Medic, and the Canton de Genève.

REFERENCES

1. F. E. May, M. D. Johnson, L. R. Wiseman, A. E. Wakeling, P. Kastner, B. R. Westley. *J. Steroid Biochem.* **33**, 1035–1041 (1989).
2. R. I. Yarden, A. H. Lauber, D. El-Ashry, S. A. Chrysogelos. *Endocrinology* **137**, 2739–2747 (1996).
3. I. Martinez-Lacaci, M. Saceda, G. D. Plowman, G. R. Johnson, N. Normanno, D. S. Salomon, R. B. Dickson. *Endocrinology* **136**, 3983–3992 (1995).
4. D. El-Ashry, S. A. Chrysogelos, M. E. Lippman, F. G. Kern. *J. Steroid Biochem. Mol. Biol.* **59**, 261–269 (1996).
5. N. J. McKenna and B. W. O'Malley. *Cell* **108**, 465–474 (2002).
6. B. Cenni and D. Picard. *Trends Endocrinol. Metab.* **10**, 41–46 (1999).
7. S. Mani. *Horm. Behav.* **40**, 183–190 (2001).
8. N. L. Weigel and Y. Zhang. *J. Mol. Med.* **76**, 469–479 (1998).
9. R. F. Power, S. K. Mani, J. Codina, O. M. Conneely, B. W. O'Malley. *Science* **254**, 1636–1639 (1991).
10. D. M. Ignar-Trowbridge, K. G. Nelson, M. C. Bidwell, S. W. Curtis, T. F. Washburn, J. A. McLachlan, K. S. Korach. *Proc. Natl. Acad. Sci. USA* **89**, 4658–4662 (1992).
11. D. M. Ignar-Trowbridge, C. T. Teng, K. A. Ross, M. G. Parker, K. S. Korach, J. A. McLachlan. *Mol. Endocrinol.* **7**, 992–998 (1993).
12. Y. Zhang, W. Bai, V. E. Allgood, N. L. Weigel. *Mol. Endocrinol.* **8**, 577–584 (1994).
13. Z. Culig, A. Hobisch, M. V. Cronauer, C. Radmayr, J. Trapman, A. Hittmair, G. Bartsch, H. Klocker. *Cancer Res.* **54**, 5474–5478 (1994).
14. R. J. Pietras, J. Arboleda, D. M. Reese, N. Wongvipat, M. D. Pegram, L. Ramos, C. M. Gorman, M. G. Parker, M. X. Sliwkowski, D. J. Slamon. *Oncogene* **10**, 2435–2446 (1995).
15. S. M. Aronica and B. S. Katzenellenbogen. *Mol. Endocrinol.* **7**, 743–752 (1993).
16. Z. Q. Ma, S. Santagati, C. Patrone, G. Pollio, E. Vegeto, A. Maggi. *Mol. Endocrinol.* **8**, 910–918 (1994).
17. C. J. Newton, R. Buric, T. Trapp, S. Brockmeier, U. Pagotto, G. K. Stalla. *J. Steroid Biochem. Mol. Biol.* **48**, 481–486 (1994).
18. R. S. Piotrowicz, L. Ding, P. Maher, E. G. Levin. *J. Biol. Chem.* **16**, 16 (2000).
19. F. Demay, M. De Monti, C. Tiffoche, C. Vaillant, M. L. Thieulant. *Endocrinology* **142**, 3340–3347 (2001).
20. J. L. Turgeon and D. W. Waring. *Mol. Endocrinol.* **8**, 860–869 (1994).
21. A. M. Nakhla, N. A. Romas, W. Rosner. *J. Biol. Chem.* **272**, 6838–6841 (1997).
22. V. Speirs, M. J. Kerin, D. S. Walton, C. J. Newton, S. B. Desai, S. L. Atkin. *Br. J. Cancer* **82**, 1312–1316 (2000).
23. A. Hobisch, I. E. Eder, T. Putz, W. Horninger, G. Bartsch, H. Klocker, Z. Culig. *Cancer Res.* **58**, 4640–4645 (1998).
24. T. Chen, L. H. Wang, W. L. Farrar. *Cancer Res.* **60**, 2132–2135 (2000).
25. N. Craft, Y. Shostak, M. Carey, C. L. Sawyers. *Nat. Med.* **5**, 280–285 (1999).
26. R. A. Campbell, P. Bhat-Nakshatri, N. M. Patel, D. Constantinidou, S. Ali, H. Nakshatri. *J. Biol. Chem.* **276**, 9817–9824 (2001).
27. M. B. Martin, T. F. Franke, G. E. Stoica, P. Chambon, B. S. Katzenellenbogen, B. A. Stoica, M. S. McLemore, S. E. Olivo, A. Stoica. *Endocrinology* **141**, 4503–4511 (2000).
28. M. Sun, J. E. Paciga, R. I. Feldman, Z. Yuan, D. Coppola, Y. Y. Lu, S. A. Shelley, S. V. Nicosia, J. Q. Cheng. *Cancer Res.* **61**, 5985–5991 (2001).
29. W. Feng, P. Webb, P. Nguyen, X. Liu, J. Li, M. Karin, P. J. Kushner. *Mol. Endocrinol.* **15**, 32–45 (2001).

30. C. Patrone, Z. Q. Ma, G. Pollio, P. Agrati, M. G. Parker, A. Maggi. *Mol. Endocrinol.* **10**, 499–507 (1996).
31. M. T. Abreu-Martin, A. Chari, A. A. Palladino, N. A. Craft, C. L. Sawyers. *Mol. Cell. Biol.* **19**, 5143–5154 (1999).
32. G. Bunone, P.-A. Briand, R. J. Miksicek, D. Picard. *EMBO J.* **15**, 2174–2183 (1996).
33. R. M. Zwijsen, E. Wientjens, R. Klompmaker, J. van der Sman, R. Bernards, R. J. Michalides. *Cell* **88**, 405–415 (1997).
34. E. Neuman, M. H. Ladha, N. Lin, T. M. Upton, S. J. Miller, J. DiRenzo, R. G. Pestell, P. W. Hinds, S. F. Dowdy, M. Brown, M. E. Ewen. *Mol. Cell. Biol.* **17**, 5338–5347 (1997).
35. J. M. Trowbridge, I. Rogatsky, M. J. Garabedian. *Proc. Natl. Acad. Sci. USA* **94**, 10132–10137 (1997).
36. R. M. Tolón, A. I. Castillo, A. M. Jimenez-Lara, A. Aranda. *Mol. Cell. Biol.* **20**, 8793–8802 (2000).
37. L. A. Denner, N. L. Weigel, B. L. Maxwell, W. T. Schrader, B. W. O'Malley. *Science* **250**, 1740–1743 (1990).
38. L. V. Nazareth and N. L. Weigel. *J. Biol. Chem.* **271**, 19900–19907 (1996).
39. D. M. Ignar-Trowbridge, M. Pimentel, C. T. Teng, K. S. Korach, J. A. McLachlan. *Environ. Health Perspect.* **103**, 35–38 (1995).
40. C. Darne, G. Veyssiere, C. Jean. *Eur. J. Biochem.* **256**, 541–549 (1998).
41. B. D. Gehm, J. M. McAndrews, V. C. Jordan, J. L. Jameson. *Mol. Cell. Endocrinol.* **159**, 53–62 (2000).
42. A. Tremblay, G. B. Tremblay, F. Labrie, V. Giguère. *Mol. Cell* **3**, 513–519 (1999).
43. A. Tremblay and V. Giguère. *J. Steroid Biochem. Mol. Biol.* **77**, 19–27 (2001).
44. C. A. Beck, N. L. Weigel, M. L. Moyer, S. K. Nordeen, D. P. Edwards. *Proc. Natl. Acad. Sci. USA* **90**, 4441–4445 (1993).
45. C. A. Sartorius, L. Tung, G. S. Takimoto, K. B. Horwitz. *J. Biol. Chem.* **268**, 9262–9266 (1993).
46. C. A. Sartorius, S. D. Groshong, L. A. Miller, R. L. Powell, L. Tung, G. S. Takimoto, K. B. Horwitz. *Cancer Res.* **54**, 3868–3877 (1994).
47. Q. Zhou, T. Hopp, S. A. Fuqua, P. S. Steeg. *Cancer Lett.* **162**, 3–17 (2001).
48. L. Simpson and R. Parsons. *Exp. Cell Res.* **264**, 29–41 (2001).
49. R. Katso, K. Okkenhaug, K. Ahmadi, S. White, J. Timms, M. D. Waterfield. *Annu. Rev. Cell. Dev. Biol.* **17**, 615–675 (2001).
50. N. Fujimoto and B. S. Katzenellenbogen. *Mol. Endocrinol.* **8**, 296–304 (1994).
51. B. A. Ince, M. M. Montano, B. S. Katzenellenbogen. *Mol. Endocrinol.* **8**, 1397–1406 (1994).
52. R. M. Lavinsky, K. Jepsen, T. Heinzel, J. Torchia, T. M. Mullen, R. Schiff, A. L. Del Rio, M. Ricote, S. Ngo, J. Gemsch, S. G. Hilsenbeck, C. K. Osborne, C. K. Glass, M. G. Rosenfeld, D. W. Rose. *Proc. Natl. Acad. Sci. USA* **95**, 2920–2925 (1998).
53. H. Lee, F. Jiang, Q. Wang, S. V. Nicosia, J. Yang, B. Su, W. Bai. *Mol. Endocrinol.* **14**, 1882–1896 (2000).
54. S. Kato, H. Endoh, Y. Masuhiro, T. Kitamoto, S. Uchiyama, H. Sasaki, S. Masushige, Y. Gotoh, E. Nishida, H. Kawashima, D. Metzger, P. Chambon. *Science* **270**, 1491–1494 (1995).
55. A. Stoica, M. Saceda, A. Fakhro, M. Joyner, M. B. Martin. *J. Cell. Biochem.* **76**, 605–614 (2000).
56. R. H. Karas, E. A. Gauer, H. E. Bieber, W. E. Baur, M. E. Mendelsohn. *J. Clin. Invest.* **101**, 2851–2861 (1998).
57. I. Rogatsky, J. M. Trowbridge, M. J. Garabedian. *J. Biol. Chem.* **274**, 22296–22302 (1999).
58. C. Patrone, E. Gianazza, S. Santagati, P. Agrati, A. Maggi. *Mol. Endocrinol.* **12**, 835–841 (1998).
59. G. Lazennec, J. A. Thomas, B. S. Katzenellenbogen. *J. Steroid Biochem. Mol. Biol.* **77**, 193–203 (2001).
60. W. Bai, B. G. Rowan, V. E. Allgood, B. W. O'Malley, N. L. Weigel. *J. Biol. Chem.* **272**, 10457–10463 (1997).

61. D. Chen, P. E. Pace, R. C. Coombes, S. Ali. *Mol. Cell. Biol.* **19**, 1002–1015 (1999).
62. P. J. Stork and J. M. Schmitt. *Trends Cell Biol.* **12**, 258–266 (2002).
63. M. K. K. El-Tanani and C. D. Green. *Mol. Endocrinol.* **11**, 928–937 (1997).
64. S. K. Mani, J. M. Allen, J. P. Lydon, B. Mulac Jericevic, J. D. Blaustein, F. J. DeMayo, O. Conneely, B. W. O'Malley. *Mol. Endocrinol.* **10**, 1728–1737 (1996).
65. S. K. Mani, A. A. Fienberg, J. P. O'Callaghan, G. L. Snyder, P. B. Allen, P. K. Dash, A. N. Moore, A. J. Mitchell, J. Bibb, P. Greengard, B. W. O'Malley. *Science* **287**, 1053–1056 (2000).
66. P. Greengard, P. B. Allen, A. C. Nairn. *Neuron* **23**, 435–447 (1999).
67. H. Endoh, K. Maruyama, Y. Masuhiro, Y. Kobayashi, M. Goto, H. Tai, J. Yanagisawa, D. Metzger, S. Hashimoto, S. Kato. *Mol. Cell. Biol.* **19**, 5363–5372 (1999).
68. P. Webb, P. Nguyen, J. Shinsako, C. Anderson, W. Feng, M. P. Nguyen, D. Chen, S. M. Huang, S. Subramanian, E. McKinerney, B. S. Katzenellenbogen, M. R. Stallcup, P. J. Kushner. *Mol. Endocrinol.* **12**, 1605–1618 (1998).
69. R. M. Zwijsen, R. S. Buckle, E. M. Hijmans, C. J. Loomans, R. Bernards. *Genes Dev.* **12**, 3488–3498 (1998).
70. J. Lamb, M. H. Ladha, C. McMahon, R. L. Sutherland, M. E. Ewen. *Mol. Cell. Biol.* **20**, 8667–8675 (2000).
71. O. Hermanson, C. K. Glass, M. G. Rosenfeld. *Trends Endocrinol. Metab.* **13**, 55–60 (2002).
72. J. Font de Mora and M. Brown. *Mol. Cell. Biol.* **20**, 5041–5047 (2000).
73. G. N. Lopez, C. W. Turck, F. Schaufele, M. R. Stallcup, P. J. Kushner. *J. Biol. Chem.* **276**, 22177–22182 (2001).
74. B. G. Rowan, N. L. Weigel, B. W. O'Malley. *J. Biol. Chem.* **275**, 4475–4483 (2000).
75. B. G. Rowan, N. Garrison, N. L. Weigel, B. W. O'Malley. *Mol. Cell. Biol.* **20**, 8720–8730 (2000).
76. T. Ueda, N. R. Mawji, N. Bruchovsky, M. D. Sadar. *J. Biol. Chem.* **277**, 38087–38094 (2002).
77. W. B. Pratt and D. O. Toft. *Endocr. Rev.* **18**, 306–360 (1997).
78. D. Picard. *Cell. Mol. Life Sci.* **59**, 1640–1648 (2002).
79. L. Zheng, L. A. Annab, C. A. Afshari, W. H. Lee, T. G. Boyer. *Proc. Natl. Acad. Sci. USA* **98**, 9587–9592 (2001).
80. B. L. Wagner, J. D. Norris, T. A. Knotts, N. L. Weigel, D. P. McDonnell. *Mol. Cell. Biol.* **18**, 1369–1378 (1998).
81. S. W. Curtis, T. Washburn, C. Sewall, R. DiAugustine, J. Lindzey, J. F. Couse, K. S. Korach. *Proc. Natl. Acad. Sci. USA* **93**, 12626–12630 (1996).
82. D. M. Klotz, S. C. Hewitt, P. Ciana, M. Raviscioni, J. K. Lindzey, J. Foley, A. Maggi, R. P. DiAugustine, K. S. Korach. *J. Biol. Chem.* **277**, 8531–8537 (2002).
83. E. M. Apostolakis, J. Garai, J. E. Lohmann, J. H. Clark, B. W. O'Malley. *Mol. Endocrinol.* **14**, 1086–1098 (2000).
84. D. Picard, G. Bunone, J. W. Liu, O. Donzé. *Biochem. Soc. Trans.* **25**, 597–602 (1997).
85. S. P. Ethier. *J. Natl. Cancer Inst.* **87**, 964–973 (1995).
86. R. M. Neve, H. A. Lane, N. E. Hynes. *Ann. Oncol.* **12**, S9–13 (2001).
87. Y. Liu, D. El-Ashry, D. Chen, I. Y. Ding, F. G. Kern. *Breast Cancer Res. Treat.* **34**, 97–117 (1995).
88. M. Dowsett. *Endocr. Relat. Cancer* **8**, 191–195 (2001).
89. M. Piccart, C. Lohrisch, A. Di Leo, D. Larsimont. *Oncology* **61**, 73–82 (2001).
90. M. Wehling. *Annu. Rev. Physiol.* **59**, 365–393 (1997).
91. M. J. Kelly and E. R. Levin. *Trends Endocrinol. Metab.* **12**, 152–156 (2001).
92. M. A. Valverde, P. Rojas, J. Amigo, D. Cosmelli, P. Orio, M. I. Bahamonde, G. E. Mann, C. Vergara, R. Latorre. *Science* **285**, 1929–1931 (1999).
93. M. E. Mendelsohn. *Circ. Res.* **87**, 956–960 (2000).
94. D. Picard. *Nature* **392**, 437–438 (1998).
95. A. C. Cato, A. Nestl, S. Mink. *Sci. STKE* **2002**, RE9 (2002).

96. M. Singh, G. Setalo, Jr., X. Guan, M. Warren, C. D. Toran-Allerand. *J. Neurosci.* **19**, 1179–1188 (1999).
97. A. Migliaccio, M. Di Domenico, G. Castoria, A. de Falco, P. Bontempo, E. Nola, F. Auricchio. *EMBO J.* **15**, 1292–1300 (1996).
98. T. Simoncini, A. Hafezi-Moghadam, D. P. Brazil, K. Ley, W. W. Chin, J. K. Liao. *Nature* **407**, 538–541 (2000).
99. T. Improta-Brears, A. R. Whorton, F. Codazzi, J. D. York, T. Meyer, D. P. McDonnell. *Proc. Natl. Acad. Sci. USA* **96**, 4686–4691 (1999).
100. S. M. Aronica, W. L. Kraus, B. S. Katzenellenbogen. *Proc. Natl. Acad. Sci. USA* **91**, 8517–8521 (1994).
101. R. Rupprecht and F. Holsboer. *Trends Neurosci.* **22**, 410–416 (1999).
102. F. Wunderlich, W. P. Benten, M. Lieberherr, Z. Guo, O. Stamm, C. Wrehlke, C. E. Sekeris, H. Mossmann. *Steroids* **67**, 535–538 (2002).
103. E. J. Filardo. *J. Steroid Biochem. Mol. Biol.* **80**, 231–238 (2002).
104. A. Migliaccio, D. Piccolo, G. Castoria, M. Di Domenico, A. Bilancio, M. Lombardi, W. Gong, M. Beato, F. Auricchio. *EMBO J.* **17**, 2008–2018 (1998).
105. G. Castoria, M. V. Barone, M. Di Domenico, A. Bilancio, D. Ametrano, A. Migliaccio, F. Auricchio. *EMBO J.* **18**, 2500–2510 (1999).
106. M. Razandi, A. Pedram, G. L. Greene, E. R. Levin. *Mol. Endocrinol.* **13**, 307–319 (1999).
107. Z. Chen, I. S. Yuhanna, Z. Galcheva-Gargova, R. H. Karas, M. E. Mendelsohn, P. W. Shaul. *J. Clin. Invest.* **103**, 401–406 (1999).
108. C. B. Wade, S. Robinson, R. A. Shapiro, D. M. Dorsa. *Endocrinology* **142**, 2336–2342 (2001).
109. S. Kousteni, T. Bellido, L. I. Plotkin, C. A. O'Brien, D. L. Bodenner, L. Han, K. Han, G. B. DiGregorio, J. A. Katzenellenbogen, B. S. Katzenellenbogen, P. K. Roberson, R. S. Weinstein, R. L. Jilka, S. C. Manolagas. *Cell* **104**, 719–730 (2001).
110. K. L. Chambliss, I. S. Yuhanna, C. Mineo, P. Liu, Z. German, T. S. Sherman, M. E. Mendelsohn, R. G. Anderson, P. W. Shaul. *Circ. Res.* **87**, E44–52 (2000).
111. S. Kahlert, S. Nuedling, M. van Eickels, H. Vetter, R. Meyer, C. Grohe. *J. Biol. Chem.* **275**, 18447–18453 (2000).
112. A. Migliaccio, G. Castoria, M. Di Domenico, A. de Falco, A. Bilancio, M. Lombardi, M. V. Barone, D. Ametrano, M. S. Zannini, C. Abbondanza, F. Auricchio. *EMBO J.* **19**, 5406–5417 (2000).
113. V. Boonyaratanakornkit, M. P. Scott, V. Ribon, L. Sherman, S. M. Anderson, J. L. Maller, W. T. Miller, D. P. Edwards. *Mol. Cell* **8**, 269–280 (2001).
114. G. Castoria, A. Migliaccio, A. Bilancio, M. Di Domenico, A. de Falco, M. Lombardi, R. Fiorentino, L. Varricchio, M. V. Barone, F. Auricchio. *EMBO J.* **20**, 6050–6059 (2001).
115. H. Peterziel, S. Mink, A. Schonert, M. Becker, H. Klocker, A. C. Cato. *Oncogene* **18**, 6322–6329 (1999).
116. C. S. Watson, C. H. Campbell, B. Gametchu. *Steroids* **67**, 429–437 (2002).
117. Q. Lu, H. Ebling, J. Mittler, W. E. Baur, R. H. Karas. *FEBS Lett.* **516**, 1–8 (2002).
118. K. L. Chambliss, I. S. Yuhanna, R. G. Anderson, M. E. Mendelsohn, P. W. Shaul. *Mol. Endocrinol.* **16**, 938–946 (2002).
119. D. C. Márquez and R. J. Pietras. *Oncogene* **20**, 5420–5430 (2001).
120. R. Kumar, R. A. Wang, A. Mazumdar, A. H. Talukder, M. Mandal, Z. Yang, R. Bagheri-Yarmand, A. Sahin, G. Hortobagyi, L. Adam, C. J. Barnes, R. K. Vadlamudi. *Nature* **418**, 654–657 (2002).
121. P. W. Shaul. *Annu. Rev. Physiol.* **64**, 749–774 (2002).
122. A. M. Norfleet, C. H. Clarke, B. Gametchu, C. S. Watson. *FASEB J.* **14**, 157–165 (2000).
123. H. P. Kim, J. Y. Lee, J. K. Jeong, S. W. Bae, H. K. Lee, I. Jo. *Biochem. Biophys. Res. Commun.* **263**, 257–262 (1999).

124. M. H. Wyckoff, K. L. Chambliss, C. Mineo, I. S. Yuhanna, M. E. Mendelsohn, S. M. Mumby, P. W. Shaul. *J. Biol. Chem.* **276**, 27071–27076 (2001).
125. M. Bayaa, R. A. Booth, Y. Sheng, X. J. Liu. *Proc. Natl. Acad. Sci. USA* **97**, 12607–12612 (2000).
126. J. Tian, S. Kim, E. Heilig, J. V. Ruderman. *Proc. Natl. Acad. Sci. USA* **97**, 14358–14363 (2000).
127. L. B. Lutz, L. M. Cole, M. K. Gupta, K. W. Kwist, R. J. Auchus, S. R. Hammes. *Proc. Natl. Acad. Sci. USA* **98**, 13728–13733 (2001).
128. R. X. Song, R. A. McPherson, L. Adam, Y. Bao, M. Shupnik, R. Kumar, R. J. Santen. *Mol. Endocrinol.* **16**, 116–127 (2002).
129. C. W. Wong, C. McNally, E. Nickbarg, B. S. Komm, B. J. Cheskis. *Proc. Natl. Acad. Sci. USA* **99**, 14783–14788 (2002).
130. R. Duan, W. Xie, R. C. Burghardt, S. Safe. *J. Biol. Chem.* **276**, 11590–11598 (2001).
131. R. Duan, W. Xie, X. Li, A. McDougal, S. Safe. *Biochem. Biophys. Res. Commun.* **294**, 384–394 (2002).
132. E. G. Dos Santos, M. N. Dieudonne, R. Pecquery, V. Le Moal, Y. Giudicelli, D. Lacasa. *Endocrinology* **143**, 930–940 (2002).
133. J. J. Watters, T. Y. Chun, Y. N. Kim, P. J. Bertics, J. Gorski. *Mol. Endocrinol.* **14**, 1872–1881 (2000).
134. P. E. Stevis, D. C. Deecher, L. Suhadolnik, L. M. Mallis, D. E. Frail. *Endocrinology* **140**, 5455–5458 (1999).
135. S. Kousteni, J. R. Chen, T. Bellido, L. Han, A. A. Ali, C. A. O'Brien, L. Plotkin, Q. Fu, A. T. Mancino, Y. Wen, A. M. Vertino, C. C. Powers, S. A. Stewart, R. Ebert, A. M. Parfitt, R. S. Weinstein, R. L. Jilka, S. C. Manolagas. *Science* **298**, 843–846 (2002).
136. A. Nadal, A. B. Ropero, O. Laribi, M. Maillet, E. Fuentes, B. Soria. *Proc. Natl. Acad. Sci. USA* **97**, 11603–11608 (2000).
137. N. Atanaskova, V. G. Keshamouni, J. S. Krueger, J. A. Schwartz, F. Miller, K. B. Reddy. *Oncogene* **21**, 4000–4008 (2002).
138. W. S. Shim, M. Conaway, S. Masamura, W. Yue, J. P. Wang, R. Kmar, R. J. Santen. *Endocrinology* **141**, 396–405 (2000).

Pure Appl. Chem., Vol. 75, Nos. 11–12, pp. 1757–1769, 2003.

Topic 1.9

Estrogen receptor action through target genes with classical and alternative response elements*

Peter J. Kushner‡, Paul Webb, Rosalie M. Uht, Meng-Min Liu, and Richard H. Price, Jr.

Department of Medicine, University of California, San Francisco, CA 94115-1640, USA

Abstract: The estrogen receptors alpha and beta (ERα and ERβ) mediate the changes in gene expression from physiological and environmental estrogens. Early studies identified classical estrogen response elements (EREs) in the promoter region of target genes whose expression is regulated by estrogen and to which the ERs bind via their DNA-binding domain (DBD). EREs in the pituitary prolactin promoter, for example, mediate an activation by both ERα and ERβ albeit with different affinities for different ligands. Full activation in most cell types requires the integrity of the activation function 2 (AF-2) in the receptors ligand binding domain (LBD), which is engaged by estrogens and disengaged by tamoxifen, raloxifene, and other antiestrogens. However, in some cells and ERE contexts, the AF-1 in the ERα amino terminal domain (NTD) is sufficient.

We now know that ERs also regulate expression of target genes that do not have EREs, but instead have various kinds of alternative response elements that bind heterologous transcription factors whose activity is regulated by interactions with ERs. Thus, ERα activates genes, including collagenase and cyclin D1, an important mediator of cellular proliferation, by AP-1 and CRE sites, which bind Jun/Fos or Jun/ATF-2 transcription factors. ERα also activates gene expression through GC-rich elements that bind the SP1 transcription factor. Finally, we also know that ERs mediate inhibition of the expression of many genes. In one well-studied instance, ERs counterexpression of genes involved in the inflammatory response by inhibiting the action at tumor necrosis factor response elements (TNF-REs) that bind the NFkappaB transcription factor. ERβ is especially efficient at this inhibition.

ERα activation of AP-1/CRE target genes is of special interest because of the putative role of these target genes in mediating proliferation. The AF-1 and AF-2 functions of ERα are both needed for this activation in most cell types. However, in uterine cells, the AF-1 function is sufficient. Thus, the antiestrogen tamoxifen, which allows AF-1, mimics estrogen and drives activation of AP-1/CRE target genes and proliferation of uterine cells. This estrogen-like action, which can increase the risk of uterine cancer, complicates the use of tamoxifen to prevent breast cancer. Surprisingly, ERβ inhibits AP-1/CRE target genes in the presence of estrogen. When both receptors are present, ERβ efficiently opposes activation by ERα. Moreover, ERβ activates the AP-1/CRE target genes in the presence of antiestrogens especially so-called "complete" antiestrogens raloxifene, and ICI 182, 780.

---

*Report from a SCOPE/IUPAC project: Implication of Endocrine Active Substances for Human and Wildlife (J. Miyamoto and J. Burger, editors). Other reports are published in this issue, *Pure Appl. Chem.* **75**, 1617–2615 (2003).
‡Corresponding author: Tel.: (415) 476 6790; Fax: (415) 885 7724; E-mail: kushner@itsa.ucsf.edu

We here review the evidence for different kinds of promoter elements that mediate ER action, for the differential ligand preferences of ERα and ERβ at these different elements, and the potential mechanisms by which they are mediated. One attractive strategy for the investigation and comparison of potential environmental estrogens is to assay their activity in cell culture systems using reporter genes with simplified promoter elements. Thus, the findings of complexity in ERα and ERβ activation at different types of response elements needs to be taken into account in the development and interpretation of assays using simplified promoter elements systems.

TWO ESTROGEN RECEPTORS, ALPHA AND BETA, MEDIATE THE ACTION OF ENDOGENOUS AND ENVIRONMENTAL ESTROGENS, INCLUDING THE EFFECTS ON CELLULAR PROLIFERATION AND CANCER INDUCTION

Estrogen has profound effects on the development and function of reproductive organs. Among its other effects, it stimulates proliferation of epithelial cells in the reproductive tract and mammary gland of females and in the prostate of males. In the female tissues, and most likely in the prostate as well, it also plays a role in the development of cancer [for recent reviews, see 12,23,57,63]. This proliferative effect of estrogens is, of course, a primary parameter of concern in the study of environmental estrogens.

Two related proteins, the estrogen receptor alpha (ERα) and beta (ERβ), which function as transcription factors to regulate expression of target genes, carry out and modulate the effects of estrogen [23,43]. Studies of mice with disrupted estrogen receptors indicate that ERα mediates the major proliferative effects of estrogen [10]. Thus, female mice in which ERα has been knocked-out (αERKOs) lack estrogen-provoked proliferation of the uterus, cervix and vagina, and have rudimentary mammary glands [6,10,29]. Male αERKO mice are completely resistant to estrogen-provoked prostate proliferation, hyperplasia, and cancer [47,48]. In contrast, ERβ knock-out (βERKO) female mice have full estrogen responses of the reproductive tract, and males have a full response of the prostate.

There is suggestive evidence that ERβ may modulate the proliferative effects of ERα. Thus, the βERKOs are reported to have exaggerated estrogen responses in the uterus, and to have spontaneous hyperplasia of the prostate, although the latter observation is not without controversy and needs to be confirmed [14,74,75]. Consistent with this, there is a progressive loss of ERβ expression in prostate cancer [18,41]. In mammary gland, ERβ is present in high amounts in normal epithelium and progressively lost in multi-stage carcinogenesis, an observation that suggests a potential loss of an inhibitor of proliferation [49].

ERs REGULATE TRANSCRIPTION AT TWO TYPES OF TARGET GENES, THOSE WITH CLASSICAL ESTROGEN RESPONSE ELEMENTS AND THOSE WITH AP-1/CRE SITES, THE LATTER OF WHICH MAY BE IMPORTANT FOR PROLIFERATIVE EFFECTS

ERs are transcriptional activation proteins that bind estrogen and acquire the ability to stimulate transcription of estrogen-responsive target genes [for review, see 32,37,40]. ERs work through at least two different pathways to affect transcription. In one pathway, the ERs act at target genes that have classical estrogen response elements (EREs) within the promoter region that allow the ER to bind DNA and regulate transcription [33]. Examples of these classical target genes with EREs include the vitellogenin genes of birds and frogs and the mammalian prolactin, pS2, cathepsin D, and lactoferrin genes.

In a second pathway of action, ERs regulate transcription at promoter elements that directly bind heterologous transcription factors. These promoter elements include AP-1 sites that bind Jun/Fos [26], variant cyclic-AMP response elements (CREs) that bind c-Jun/ATF-2 proteins [50,69,72], and Sp1 sites [51,52]. ERs have also been reported to regulate through other sites whose binding proteins have not yet been identified [34–36,79]. Neither AP-1, CRE, or Sp1 sites bind ERs and regulation is presumed to occur through protein–protein interactions, in which the proteins that do bind these sites somehow allow for ER recruitment.

ER regulation of target genes with AP-1 and CRE sites is typically less dramatic than at EREs, yet such AP-1/CRE targets may be important for proliferative effects of estrogen. Two sorts of evidence suggest this. First many of the target genes that are regulated through AP-1 and CRE elements seem to be involved in proliferation or tissue remodeling. These include, for example, the collagenase gene, other genes for matrix metalloproteinases, and the IGF-1 gene, which are regulated by estrogen through AP-1 sites [65,69,72]. Above all, the cyclin D1 gene, an estrogen-induced gene which has been implicated as a central node for estrogen-mediated growth both of human breast cancer cells in culture [46], and of mammary epithelial cells in vivo [61], does not have an ERE. Instead, estrogen regulation of the cyclin D1 gene has been ascribed to a variant CRE element that binds Jun/ATF-2 and does not bind ERs [1,50]. Second, there is an excellent correlation between the ability of selective estrogen receptor modulators (SERMs) to mimic estrogen and stimulate proliferation and the ability of SERMs to mimic estrogen and activate target genes with AP-1/CRE elements. Thus tamoxifen, but not raloxifene, or ICI 182,780 (hereafter ICI) can mimic estrogen and stimulate uterine (but not mammary) cell growth. In parallel, tamoxifen, but not raloxifene or ICI, activates AP-1/CRE transcription in uterine cells, but not mammary cells [69]. This ability of tamoxifen, but not raloxifene, to activate IGF-1, and c-Myc (another estrogen induced gene without an ERE, [13].) and only in uterine cells has recently been confirmed with the endogenous genes in human endometrial cells [53]. Finally it should be mentioned that in some mammary cancer cells the introduction of ERα inhibits proliferation, and in these cells there is a corresponding inhibition of AP-1 regulated target genes [44].

In addition to AP-1/CRE elements, it appears that GC-rich elements that bind the Sp1 family of transcription factors can also mediate estrogen induction via ERα in the context of some promoters. Sp1 sites contribute to estrogen response in the promoters of a large number of genes both with EREs and with alternative response elements [51,52]. Indeed, GC-rich elements that bind Sp1 contribute to the estrogen induction of cyclin D1 [8]. There is also some evidence that Sp1 sites can by themselves with the appropriate core promoter confer an estrogen response.

ERs also downregulate some target genes with estrogen. One well-studied example are target genes with elements for the NF-κB transcription factors, which include those for TNFα and other cytokines including il-1 and il-6 [2]. This mechanism is believed to underlie some of the anti-inflammatory, and possibly some of the anti-osteoclastogenic activities of estrogen.

ERα AND ERβ HAVE PARALLEL ACTION AT ERE TARGET GENES

ERα and ERβ each activate ERE containing target genes in the presence of estrogen. In most ERE contexts ERβ tends to be a weaker activator than ERα, and the weaker activation is dominant in cells with both receptors, as illustrated in Fig. 1 [11,16,42]. In some contexts, ERβ also requires higher estrogen concentrations for activation than ERα, and thus, the dominant weaker activation by ERβ is exaggerated under conditions where estrogen is limiting [16]. ERβ has preferential binding to xeno-estrogens, most notably to genistein and other phytoestrogens, which bind and activate ERβ at lower concentrations than at ERα [5,25].

How ERs work at EREs is covered in other articles in this topic section (see article by Donald McDonnell et al.), and a simple outline is indicated in Fig. 1b. Estrogen binding to the ER releases it from complexes with chaperone proteins, allowing it to dimerize through the C-terminal ligand-binding domain (LBD), and bind to the ERE via the centrally located DNA-binding domain (DBD). The ERE is itself a palindromic dimer of two half-sites (sequence AGGTCA) separated by three base pairs. Once situated on the ERE, the ER works as a molecular tether for coactivators, which it recruits through two activation functions, AF-1 and AF-2. The AF-1 in the amino terminal domain of ERα is constitutive and relatively weak in most cell types. No AF-1 is found in the somewhat shorter amino terminal

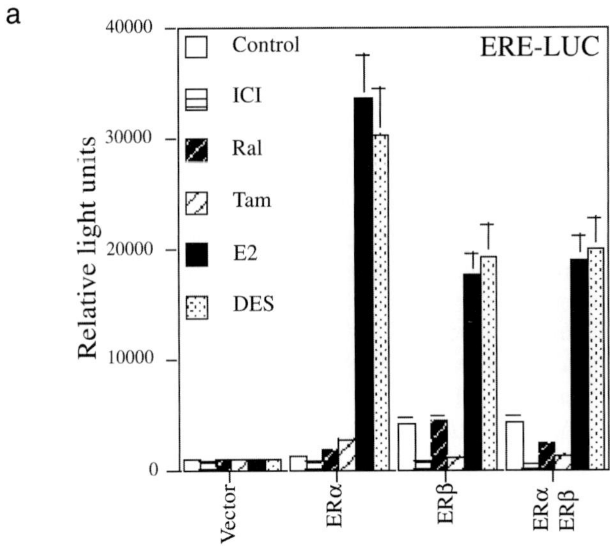

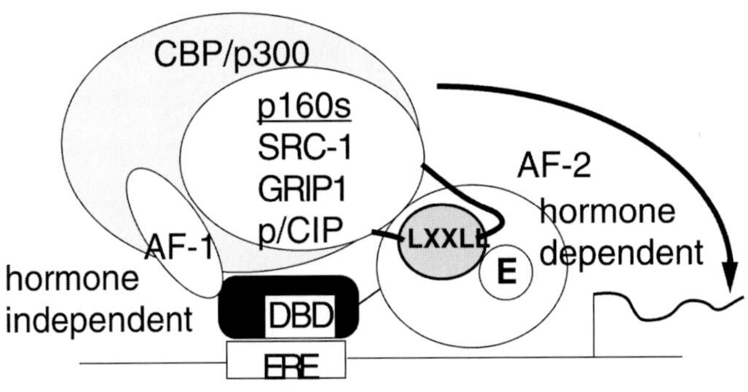

Fig. 1 (a) *ERβ modulates ERα activation at the ERE elements.* Activity of the ERE-II-LUC reporter in HeLa cells transfected with expression vectors for ERα, ERβ, both, or empty vector and treated with ligands as indicated. (b) *How ERs activate at an ERE.* ERs bind to the ERE with their DBD, ERα recruits coactivators with the hormone independent AF-1 and both ERs with their hormone dependent AF-2. The coactivators then stimulate transcription. AF-1 binds both components of the p160/CBP coactivator complex, and AF-2 forms a hydrophobic cleft (stippled) into which NR boxes (LXXLL) of the p160s bind.

domain of ERβ. The powerful and hormone-activated AF-2 function is found in the LBDs of both ERs. AF-2 is active when bound to estrogen, DES or other agonist and is inactive when bound to tamoxifen, raloxifene, ICI, or other antiestrogens.

A family of related proteins SRC-1, GRIP1, p/CIP, here referred to as p160s, mediate both AF-2 activity and a substantial amount of AF-1 activity [71]. The p160s bind to the ER-LBD only when it is liganded to estrogen or other agonist, but not when liganded to antiestrogens (for review, see ref. [56]). p160 Binding to AF-1 is hormone-independent. That p160 proteins mediate AF-1 or AF-2 function is shown in part because increasing their abundance potentiates AF-1 and AF-2, and interfering with them abolishes AF-1 and AF-2 [38,71]. Other proteins including the CBP/p300 family of coactivators complex with the p160s and also appear to play a role in AF-1 and AF-2 function [9,17,21,24,58]. Antiestrogens block AF-2 because they block recruitment to the ER of the p160-CBP coactivator complex that mediates AF-2 function. Interestingly, AF-1 and AF-2 are modular, and can each be inactivated without inactivating the other. Thus, tamoxifen blocks AF-2 but allows AF-1.

Once recruited to the ERE by ER, the coactivator complex is believed to mediate transcriptional activation by two means. First, the complex remodels chromatin in part through the strong histone acetyl transferase activity (HAT) in CBP that acetylates lysines in the N-terminal tails of histones H3 and H4 [77] (see also [54]). CBP may also make direct contact with components of the transcriptional apparatus. The p160s, in addition to their binding to CBP, bind CARM1/PRMT1 two histone arginine methyltransferases. CARM1/PRMT1 enhances p160 action, possibly by contributing to chromatin remodeling through methylation of arginines in H3 and H4 [60].

Exactly how estrogens allow the binding of p160s, and how antiestrogens block the binding has been revealed by 3D X-ray structures and structure-guided mutational analysis and is shown in Fig. 2 [7,55]. Estrogen is bound within the LBD and allows a hydrophobic cleft made up of helices 3, 5, and 12 to form on the surface of the LBD. The hydrophobic cleft is the docking surface for p160s [15,30]. GRIP1 and other p160 coactivators have multiple nuclear receptor boxes (NR boxes) with the sequence LXXLL. The leucines of the NRboxes project into the hydrophobic cleft to couple the two proteins [55]. Key residues in helix 3, 5, and especially 12, of the ER are needed for both this coupling, for AF-2 function, and for gene activation. Tamoxifen blocks AF-2 function and binding of p160s because it projects from the LBD displacing helix 12. In the tamoxifen structure, helix 12 rotates from its normal position into the hydrophobic cleft, thereby occluding the cleft from coactivators [55]. Notice that in the ta-

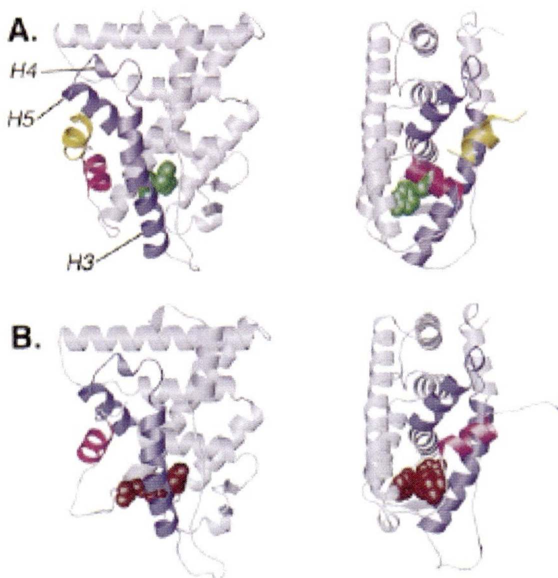

Fig. 2 *How tamoxifen blocks AF-2. Overall structures of the DES-ERα LBD-GRIP1 NR box II peptide complex and of the OHT-ERα LBD complex. (a) Two orthogonal views of the DES-ERα LBD-NR box II peptide.* The coactivator peptide and the LBD are shown as ribbon drawings. The peptide is colored gold, and helix 12 (residues 538–546) is colored magenta. Helices 3, 4, and 5 (labeled H3, H4, and H5, respectively) are colored blue. DES, colored green, is shown in space-filling representation. (b) *Two orthogonal views of the OHT-ERα LBD complex similar to those of the agonist complex in (a).* The LBD is depicted as a ribbon drawing. As in (A), helix 12 (residues 536–544) is colored in magenta, and helices 3, 4, and 5 are colored blue. OHT, in red, is shown in space-filling representation.

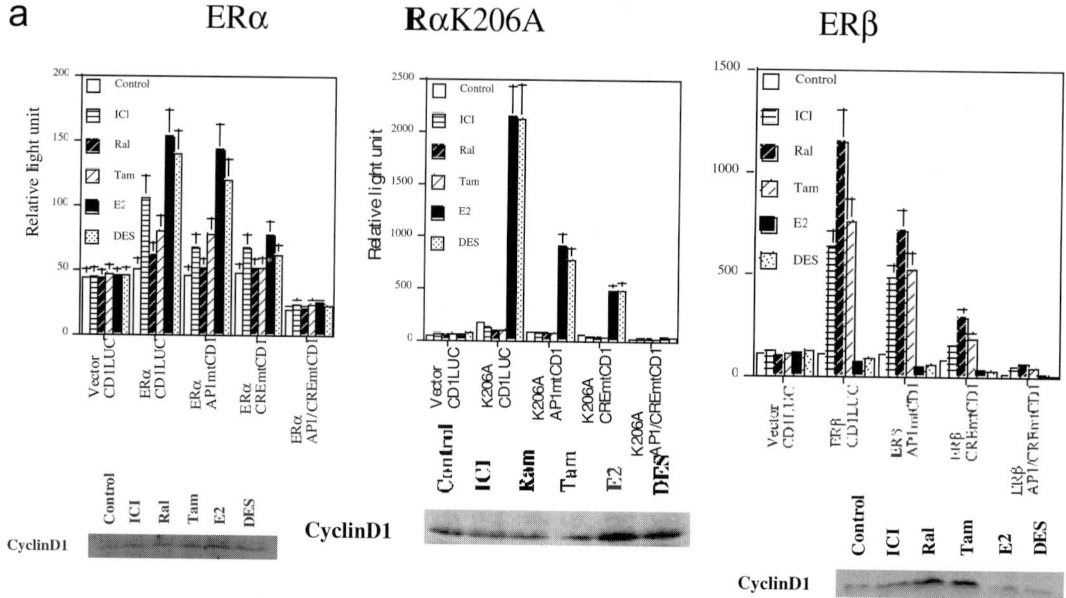

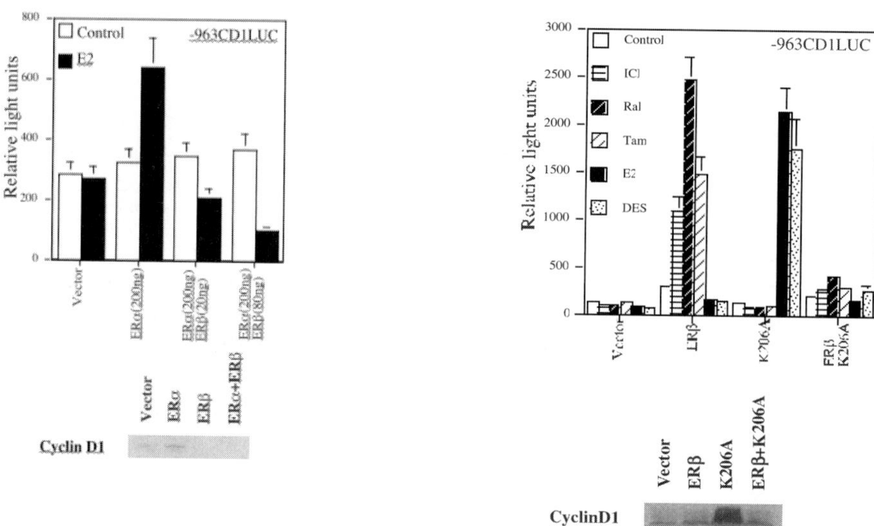

Fig. 3 *Opposing action of ERα and ERβ on the cyclin D1 promoter. (a) Opposite ligand preferences of ERα, the superactive ERαK206A versus ERβ.* The upper panels show the response to various ligands of the cyclin D1 promoter driving luciferase in HeLa cells with the indicated ER (left ERα, middle K206A, right ERβ). Mutants in the AP-1 element, CRE element, or both are also shown. The lower panel shows the response of the endogenous HeLa cell cyclin D1 gene by Westerns. *(b) ERβ inhibits ERα or even K206A activation of cyclin D1 with estrogen.* Upper panels show the response of the cyclin D1 promoter luciferase reporter gene to activated by ERα (on the left) or by K206A (on the right), and then inhibited by ERβ. The lower panels show endogenous HeLa cell cyclin D1 gene expression first activated by ERα or K206A and then inhibited by ERβ.

moxifen structure helix 12 of the receptor is in the space that is occupied by the NR boxes of the coactivator in the structure with the estrogen diethylstilbestrol (DES). In summary then, ERα and ERβ both activate at EREs, and each works by recruiting a CBP-p160 coactivator complex through AF-2.

ERα has a potent AF-1 in the first 109 amino acids that also works by contacting p160s, but through a different surface than AF-2 [70]. ERβ lacks this conventional AF-1 but seems to share with ERα an AF-1 that is inducible by MAP kinase cascades and subsequent phosphorylation in the NTD [22,64,66].

Corepressors also may modulate the actions of ERs especially in the presence of antagonist of AF-2 (antiestrogens). The corepressors N-CoR and SMRT bind directly to the ERα antiestrogen complexes and inhibit AF-1 action [19,27,78]. It is known that N-CoR and SMRT recognize the thyroid and other nuclear receptors through hydrophobic helices (called ID domains) which couple with the hydrophobic cleft in the LBD [68]. The binding is similar to that of coactivators except that the corepressor ID domains displace helix 12 and bind both to the newly exposed surface and the hydrophobic cleft [76]. A structure of the ERs with corepressors has not yet been solved, but genetic analysis suggests a similar mode of binding with key contacts underneath helix 12. Mutations of these contacts (L379) eliminates corepressor binding, whereas eliminating helix 12 increases N-CoR binding and allows all ligands to promote binding. Thus, while the molecular details are still unknown it appears that corepressors modulate ER action, especially AF-1.

ERα AND ERβ HAVE OPPOSING ACTION AT AP-1/CRE TARGET GENES, AND SEPARATE MODELS FOR THEIR ACTION ARE PROPOSED

The action of ERα and ERβ at AP-1 target genes could not be more different from their action at classical EREs. ERα activates AP-1/CRE target genes primarily with estrogen, just as it does at an ERE. In uterine cells ERα can also activate AP-1 targets to with tamoxifen, but not with raloxifene, or ICI [26,69,72]. Estrogen, however, is still the preferred ligand for ERα activation. ERβ, in contrast activates AP-1/CRE target genes with antiestrogens, especially with ICI and raloxifene, and not with estrogen [28,39]. The contrast between the ligand preferences for ERα and ERβ activation of the AP-1/CRE regulated cyclin D1 promoter is illustrated in Fig. 4. ERα also prefers estrogens and ERβ antiestrogens for activating retinoic acid receptor α-1 gene expression. Interestingly, the RARα promoter has no ERE, and the reported ERβ preference requires the integrity of Sp1 sites in the promoter [80].

These contrasting ligand preferences have suggested that ERα and ERβ might have opposing action on the expression of target genes that are regulated via alternative response elements such as AP-1/CRE sites and Sp1 sites [31,39]. This has been confirmed for cyclin D1 gene expression which is under the regulation of AP-1/CRE and Sp1 elements, and an example of whose regulation by different ligands is illustrated. Notice that whereas ERα activates cyclin D1 expression with estrogen, ERβ inhibits with estrogen (and activates only with antiestrogens) [28]. When both receptors are present ERβ efficiently blocks the activation by ERα. The observed opposing effects of ERα and ERβ on cyclin D1 gene expression are consistent with the reports noted above that suggest an inhibitory role for ERβ in estrogen-mediated cellular proliferation, a process in which cyclin D1 appears to play a central role.

Each of the ligands discussed here binds well to both ERα and ERβ. Thus, it is suggested that the dramatic difference in the ligand preference for activation at AP-1/CRE, and possibly Sp1, reflects different modes of action of the two liganded receptors [26]. ERα and ERβ do indeed have very different modes of action at AP-1/CRE [72]. The actions of ERα at AP-1 are primarily mediated by the activation functions AF-1 and AF-2 and their ability to dock to p160 coactivators. When the AF1 and AF-2 surfaces that contact p160 coactivators are disrupted by mutation, ERα cannot activate with estrogen. Increasing the levels of p160s increases the ability of ERα to activate with estrogen, whereas mutation of the p160 coactivator surfaces (the NR boxes and the AF-1 interaction domain) that contact ERα disrupts the ability to enhance ERα activation with estrogen. Moreover, tamoxifen response is completely

a

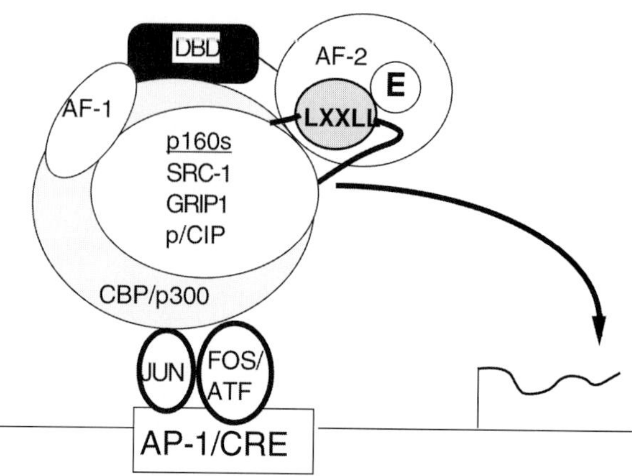

b

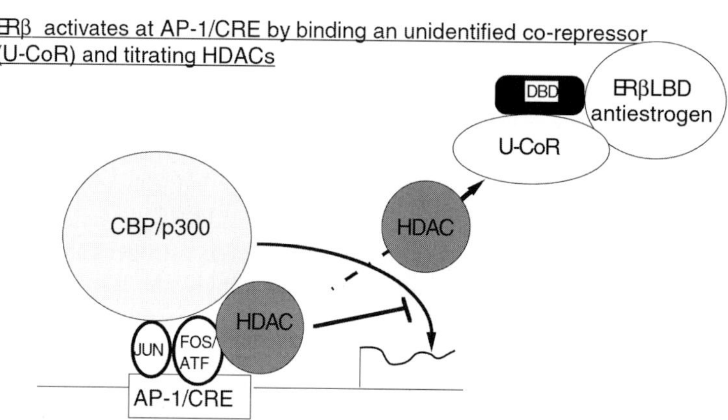

Fig. 4 *How ERs activate at AP-1/CRE. (a) ERα.* At AP-1/CRE sites, ERα uses its AF functions, to nucleate formation of a complex of CBP/p300 with p160s. The complex is tethered at the promoter by contacts with Jun/Fos (or in the case of the CRE with Jun/ATF-2). In the absence of ERα only CBP/p300 is recruited. *(b) ERβ.* ERβ activates at AP-1/CRE independently of AF functions by tethering to a sequestration site, binding N-CoR or some other unidentified co-repressor (U-CoR), and then titrating HDACs away from the promoter and thereby allowing full action of CBP/p300.

dependent on the integrity of the NTD, which harbors AF-1. The above studies suggest that ERα works at AP-1 via AF-1 and AF-2 contacts with p160s. Indeed, Myles Brown and colleagues have recently demonstrated with chromatin immunoprecipitation assays that both ERα and p160 coactivators are present at the IGF-1 (an AP-1-regulated gene ref), c-MYC (a non-ERE-regulated gene with Sp1 sites), and cyclin D1 promoters within a few minutes of estrogen addition [45,53]. Furthermore, in uterine but not in mammary cells, the p160 coactivator SRC3 is present at these non-ERE promoters after tamoxifen exposure [53].

ERβ activation with antiestrogens at AP-1/CRE and Sp1 targets, by contrast, is completely independent of AF function. ERβ does not have a conventional AF-1, and antiestrogens do not allow ERβ AF-2. Neither does mutation of the AF-2 surface of ERβ impede its robust ability to activate at AP-1 with antiestrogens [72]. The amino terminal domain of ERβ does contribute to activation with antiestrogens [67,80], but the nature of this contribution is unknown. Certainly, it is not a conventional AF-1 function.

As noted earlier, the primary structural difference between ERα and ERβ appears to be the presence of an AF-1 function in the amino terminal domain of ERα, and the absence of an AF-1 in ERβ. The presence or absence of an AF-1 function appears to be a key feature in controlling the ligand preference of ERs at AP-1. Thus, replacing the amino terminal domain of ERβ, with that of ERα completely prevents ERβ from activating with antiestrogens and allows it to activate with estrogens [67]. Similarly, the progressive loss of AF-1 function by deletion of the N-terminal domain of ERα progressively allows ERα to activate AP-1 with raloxifene and ICI [72]. Complete deletion of the N-terminal domain of ERα converts it into a pheno-copy of ERβ; ERα deltaAF-1 activates at AP-1/CRE sites only with raloxifene and ICI. These studies thus indicate that the presence of an efficient AF-1 prevents ERs from activating at AP-1 with antiestrogens such as raloxifene and ICI.

We propose that there are two different pathways whereby ERs act at AP-1/CRE sites [26,69,72]. The ERα pathway requires estrogens and AF functions and is illustrated in Fig. 2 below. We propose that the AP-1 element is bound by Jun/Fos and that ERα is not bound at the AP-1 DNA. Jun/Fos recruit their coactivator CBP/p300, which may also bring in a p160 such as GRIP1 (illustrated). ERα could join this complex through the AF-1 and AF-2 surfaces that contact GRIP1, or by direct contacts with Jun [44,62]. In doing so, ERα could either increase the amount of the CBP-p160 complex, or modulate its activity, thereby enhancing transcription. ERα, in brief, is a coactivator for Jun/Fos family proteins at AP-1/CRE sites.

Our model for ERβ action at AP-1 is completely different than that for ERα. We presume, although our evidence at the moment is indirect, that ERβ acts, not at the complex of proteins at the promoter, but at a sequestration site away from the promoter, for example, on bulk DNA. ERβ at the sequestration site and in the presence of antiestrogens such as raloxifene associates with an unidentified corepressor, possibly N-CoR or SMRT that we here call "U-CoR". U-CoR in turn binds histone deactylases (HDACs) which are titrated away from the AP-1 regulated promoter. Since HDACs in general counter the HAT activity of coactivators and repress transcription, their removal from the promoter leads to activation of transcription. Tests of this model are in progress.

ERα AND ERβ ACTION AT Sp1 SITES RESEMBLES, BUT IS SUBTLY DIFFERENT FROM ACTION AT AP-1/CRE ELEMENTS

Binding sites for the Sp1 transcription factor (GC-rich elements) are present in many estrogen response gene promoter regions especially those involved in regulation of proliferation: cyclin D1, E2F1, DNA polymerase, IGFBP4, telomerase, and possibly c-myc [51]. These elements cooperate with EREs, ERE half-sites, and binding sites for AP-1/CRE factors in promoting estrogen response. Moreover, simplified reporter genes with tandem Sp1 binding sites respond to estrogen. A particular interesting example of Sp1 contribution is in the cyclin D1 promoter where a tandem repeat of Sp1 elements cooperates with the CRE element to yield a full estrogen response [8]. ERα activates at Sp1 elements, but unlike activation at AP-1/CRE elements the activation is strongest in mammary cells and weak or nonexistent in HeLa and other uterine cells [52]. Furthermore, both estrogens and antiestrogens tend to activate. ERb, in contrast, inhibits expression at Sp1 elements with estrogen, and activates with antiestrogens [52,80].

The functions of ERα involved in activation of Sp1 elements are similar to those involved in activation at AP-1/CRE elements: the AF-1 domain plays a central role, whereas the ERE recognition function of the DNA-binding domain does not [20,52,69,73]. ERs and Sp1 bind to each other, and the presence of ER can promote recruitment of Sp1 to GC-rich elements [8]. This promotion of binding to the GC-rich elements may underlie ERα activation. However, the complexities of ligand spectrum have suggested that ERs may also behave as coactivators and corepressors at the complex of proteins on the GC-rich elements [52].

ERα AND ERβ DOWNREGULATION OF TARGET GENES THAT HAVE BEEN ACTIVATED BY CYTOKINES

ERs also downregulate many genes as have become very evident with RNA expression micro-array studies [59]. A well studied instance is ERα and ERβ inhibition of TNFa action at so-called TNF response elements that are found in the promoters of many cytokine genes including TNF itself [3,4]. The NFκ-B, AP-1, and Ets1 transcription factors bind to these elements, but ERs do not. Instead, ERs appear to inhibit the ability of these factors to stimulate transcription via protein–protein interactions. Interestingly, inhibition requires the ability of ERs to interact with the p160 coactivators, which appear to play a key role in mediating inhibition, by a mechanism that is presently unknown. ERβ is especially efficient at the inhibition of the TNFa response, which is also produced by phytoestrogens. This may reflect a potential function of ERβ in suppression of inflammation. SERMs reverse the estrogen-dependent inhibition and, in fact, further activate transcription from this type of response element, much as they do at AP-1 sites.

SPECULATIONS ON THE PHYSIOLOGICAL RELEVANCE OF ANTIESTROGEN (SERM) INDUCTION OF TARGET GENES

As described above, SERMs can activate AP-1, SP-1, and EphRE-responsive genes in conditions in which estrogens either elicit a weaker response, or actively repress transcription. Why would a gene respond to an artificial compound that is not found in nature, yet not respond to the cognate receptor ligand? We envision at least two possible explanations. The ERs can be activated both by ligands and, in the absence of ligands, by second messenger inputs (see Topic 1.8). Perhaps SERMs create an ER conformation that resembles that of unliganded ERs and somehow mimics these ligand independent modes of activation. Second, it is possible that there are natural ligands that resemble SERMs and that SERM responses mimic a natural response to these unspecified ligands. One possibility is that the SERMs mimic natural steroid metabolites [79]. Alternatively, there may be unspecified sources of SERM-like compounds in the environment and SERM-dependent gene induction might represent a response to exposure to these environmental SERMs. In any of these cases, one of the implications of the findings presented above is that environmental ER interacting compounds could show estrogen-like or SERM-like activities. It is therefore necessary to monitor each of these activities.

IMPLICATION FOR DEVELOPING SYSTEMS TO MONITOR AND EVALUATE ENVIRONMENTAL ESTROGENS

It would be highly desirable to have a simple and uniform system to monitor environmental estrogens in a laboratory setting, and that would be some indication, even if imperfect, as to how the environmental estrogens would affect human health. Systems using reporter gene response in cultures of human cells expressing ERs are very attractive. The above considerations put a limit on how simple these systems can be. First, both ERα and ERβ must be included, because they have both different binding affinities for different ligands, and more importantly, different preferences for activation of transcription at some response elements. Thus, and secondly, both reporter genes with EREs, and reporter genes with alternative response elements will have to be included. The AP-1/CRE elements are among the most important alternative response elements, but for some purposes Sp-1 and NFkappaB reporter genes may have to be included as well. Of course, the development of RNA expression arrays with every gene represented is ideal in one way. But even complete microarrays do not substitute for the use of idealized reporter genes with simplified response elements, which gives information of its own sort.

ACKNOWLEDGMENTS

Thanks to members of the Kushner lab for discussion and criticism. Supported by grants from the NIH to PJK.

REFERENCES

1. L. Altucci, R. Addeo, L. Cicatiello, S. Dauvois, M. G. Parker, M. Truss, M. Beato, V. Sica, F. Bresciani, A. Weisz. *Oncogene* **12**, 2315–2334 (1996).
2. J. An, R. C. Ribeiro, P. Webb, J. A. Gustafsson, P. J. Kushner, J. D. Baxter, D. C. Leitman. *Proc. Natl. Acad. Sci. USA* **96**, 15161–15166 (1999).
3. J. An, R. C. J. Ribeiro, P. Webb, J.-A. Gustafsson, P. J. Kushner, J. D. Baxter, D. C. Leitman. *Proc. Natl. Acad. Sci. USA* **96**, 15161–15166 (1999).
4. J. An, C. Tzagarakis-Foster, T. C. Scharschmidt, N. Lomri, D. C. Leitman. *J. Biol. Chem.* **276**, 17808–17814 (2001).
5. T. Barkhem, B. Carlsson, Y. Nilsson, E. Enmark, J.-A. Gustafsson, S. Nilsson. *Mol. Pharmacol.* **54**, 105–112 (1998).
6. W. P. Bocchinfuso, J. K. Lindzey, S. C. Hewitt, J. A. Clark, P. H. Myers, R. Cooper, K. S. Korach. *Endocrinology* **141**, 2982–2994 (2000).
7. A. M. Brzozowski, A. C. Pike, Z. Dauter, R. E. Hubbard, T. Bonn, O. Engstrom, L. Ohman, G. L. Greene, J. A. Gustafsson, M. Carlquist. *Nature* **389**, 753–758 (1997).
8. E. Castro-Rivera, I. Samudio, S. Safe. *J. Biol. Chem.* **276**, 30853–30861 (2001).
9. D. Chakravarti, V. LaMorte, M. Nelson, T. Nakajima, I. Schulman, H. Juguilon, M. Montminy, R. Evans. *Nature* **383**, 99–103 (1996).
10. J. F. Couse and K. S. Korach. [published erratum appears in *Endocr. Rev.* **20** (4), 459 (1999)]. *Endocr. Rev.* **20**, 358–417 (1999).
11. S. M. Cowley and M. G. Parker. *J. Steroid Biochem. Mol. Biol.* **69**, 165–175 (1999).
12. R. B. Dickson and G. M. Stancel. *J. Natl. Cancer Inst. Monographs* 135–145 (2000).
13. D. Dubik and R. P. Shiu. *Oncogene* **7**, 1587–1594 (1992).
14. S. Dupont, A. Krust, A. Gansmuller, A. Dierich, P. Chambon, M. Mark. *Development* **127**, 4277–4291 (2000).
15. W. Feng, R. C. Ribeiro, R. L. Wagner, H. Nguyen, J. W. Apriletti, R. J. Fletterick, J. D. Baxter, P. J. Kushner, B. L. West. *Science* **280**, 1747–1749 (1998).
16. J. M. Hall and D. P. McDonnell. *Endocrinology* **140**, 5566–5578 (1999).
17. B. Hanstein, R. Eckner, J. DiRenzo, S. Halachmi, H. Liu, B. Searcy, R. Kurokawa, M. Brown. *Proc. Natl. Acad. Sci. USA* **93**, 11540–11545 (1996).
18. L. G. Horvath, S. M. Henshall, C. S. Lee, D. R. Head, D. I. Quinn, S. Makela, W. Delprado, D. Golovsky, P. C. Brenner, G. O'Neill, R. Kooner, P. D. Stricker, J. J. Grygiel, J. A. Gustafsson, R. L. Sutherland. *Cancer Res.* **61**, 5331–5335 (2001).
19. T. A. Jackson, J. K. Richer, D. L. Bain, G. S. Takimoto, L. Tung, K. B. Horwitz. *Mol. Endocrinol.* **11**, 693–705 (1997).
20. M. Jakacka, M. Ito, J. Weiss, P.-Y. Chien, B. D. Gehm, J. L. Jameson. *J. Biol. Chem.* **276**, 13615–13621 (2001).
21. Y. Kamei, L. Xu, T. Heinzel, J. Torchia, R. Kurokawa, B. Gloss, S. C. Lin, R. A. Heyman, D. W. Rose, C. K. Glass, M. G. Rosenfeld. *Cell* **85**, 403–414 (1996).
22. S. Kato, H. Endoh, Y. Masuhiro, T. Kitamoto, S. Uchiyama, H. Sasaki, S. Masushige, Y. Gotoh, E. Nishida, H. Kawashima, et al. *Science* **270**, 1491–1494 (1995).

23. B. S. Katzenellenbogen and J. A. Katzenellenbogen. *Breast Cancer Res.* **2**, 335–344 (2000).
24. Y. Kobayashi, T. Kitamoto, Y. Masuhiro, M. Watanabe, T. Kase, D. Metzger, J. Yanagisawa, S. Kato. *J. Biol. Chem.* **275**, 15645–15651 (2000).
25. G. G. J. M. Kuiper, B. Carlsson, K. Grandian, E. Enmark, J. Haggblad, S. Nilsson, J.-A. Gustafsson. *Endocrinology* **138**, 863–870 (1997).
26. P. J. Kushner, D. A. Agard, G. L. Greene, T. S. Scanlan, A. K. Shiau, R. M. Uht, P. Webb. *J. Steroid Biochem. Mol. Biol.* **74**, 311–317 (2000).
27. R. M. Lavinsky, K. Jepsen, T. Heinzel, J. Torchia, T. M. Mullen, R. Schiff, R. A. Del, M. Ricote, S. Ngo, J. Gemsch, S. G. Hilsenbeck, C. K. Osborne, C. K. Glass, M. G. Rosenfeld, D. W. Rose. *Proc. Natl. Acad. Sci. USA* **95**, 2920–2925 (1998).
28. M. M. Liu, C. Albanese, C. M. Anderson, K. Hilty, P. Webb, R. M. Uht, R. H. Price, R. G. Pestell, P. J. Kushner. *J. Biol. Chem.* **277**, 24353–24360 (2002).
29. D. B. Lubahn, J. S. Moyer, T. S. Golding, J. F. Couse, K. S. Korach, O. Smithies. *Proc. Natl. Acad. Sci. USA* **90**, 11162–11166 (1993).
30. H. Y. Mak, S. Hoare, P. M. A. Henttu, M. G. Parker. *Mol. Cell. Biol.* **19**, 3895–3903 (1999).
31. S. Maruyama, N. Fujimoto, K. Asano, A. Ito. *J. Steroid Biochem. Mol. Biol.* **78**, 177–184 (2001).
32. D. P. McDonnell, S. L. Dana, P. A. Hoener, B. A. Lieberman, M. O. Imhof, R. B. Stein. 1995. *Ann. NY Acad. Sci.* **761**, 121–137 (1995).
33. N. J. McKenna, J. Xu, Z. Nawaz, S. Y. Tsai, M. J. Tsai, B. W. O'Malley. *J. Steroid Biochem. Mol. Biol.* **69**, 3–12 (1999).
34. M. M. Montano, K. Ekena, M. R. Delage, W. Chang, P. Martini, B. S. Katzenellenbogen. *Proc. Natl. Acad. Sci. USA* **96**, 6947–6952 (1999).
35. M. M. Montano and B. S. Katzenellenbogen. *Proc. Natl. Acad. Sci. USA* **94**, 2581–2586 (1997).
36. M. M. Montano, W. L. Kraus, B. S. Katzenellenbogen. *Mol. Endocrinol.* **11**, 330–341 (1997).
37. S. Nilsson, S. Makela, E. Treuter, M. Tujague, J. Thomsen, G. Andersson, E. Enmark, K. Pettersson, M. Warner, J.-A. Gustafsson. *Physiol. Rev.* **81**, 1535–1565 (2001).
38. S. A. Onate, S. Y. Tsai, M. J. Tsai, B. W. O'Malley. *Science* **270**, 1354–1357 (1995).
39. K. Paech, P. Webb, G. G. Kuiper, S. Nilsson, J. Gustafsson, P. J. Kushner, T. S. Scanlan. *Science* **277**, 1508–1510 (1997).
40. M. G. Parker, N. Arbuckle, S. Dauvois, P. Danielian, R. White. *Ann. NY Acad. Sci.* **684**, 119–126 (1993).
41. D. Pasquali, V. Rossi, D. Esposito, C. Abbondanza, G. A. Puca, A. Bellastella, A. A. Sinisi. *J. Clin. Endocrinol. Metab.* **86**, 2051–2055 (2001).
42. K. Pettersson, F. Delaunay, J. A. Gustafsson. *Oncogene* **19**, 4970–4978 (2000).
43. K. Pettersson and J. A. Gustafsson. *Annu. Rev. Physiol.* **63**, 165–192 (2001).
44. A. Philips, C. Teyssier, F. Galtier, C. Rivier-Covas, J. M. Rey, H. Rochefort, D. Chalbos. *Mol. Endocrinol.* **12**, 973–985 (1998).
45. M. D. Planas-Silva, Y. Shang, J. L. Donaher, M. Brown, R. A. Weinberg. *Cancer Res.* **61**, 3858–3862 (2001).
46. O. W. Prall, E. M. Rogan, R. L. Sutherland. *J. Steroid Biochem. Mol. Biol.* **65**, 169–74 (1998).
47. G. S. Prins, L. Birch, J. F. Couse, I. Choi, B. Katzenellenbogen, K. S. Korach. *Cancer Res.* **61**, 6089–6097 (2001).
48. G. Risbridger, H. Wang, P. Young, T. Kurita, Y. Z. Wang, D. Lubahn, J. A. Gustafsson, G. Cunha, Y. Z. Wong. {Erratum in: *Dev. Biol.* **231** (1), 289 (2001); Note: Wong YZ [corrected to Wang YZ]}. *Dev. Biol.* **229**, 432–442 (2001).
49. P. Roger, M. E. Sahla, S. Makela, J. A. Gustafsson, P. Baldet, H. Rochefort. *Cancer Res.* **61**, 2537–2541 (2001).
50. M. Sabbah, D. Courilleau, J. Mester, G. Redeuilh. *Proc. Natl. Acad. Sci. USA* **96**, 11217–11222 (1999).
51. S. Safe. *Vit. Horm.* **62**, 231–252 (2001).

52. B. Saville, M. Wormke, F. Wang, T. Nguyen, E. Enmark, G. Kuiper, J. A. Gustafsson, S. Safe. *J. Biol. Chem.* **275**, 5379–5387 (2000).
53. Y. Shang and M. Brown. *Science (Washington DC)* **295**, 2465–2468 (2002).
54. Y. Shang, X. Hu, J. DiRenzo, M. A. Lazar, M. Brown. *Cell* **103**, 843–852 (2000).
55. A. K. Shiau, D. Barstad, P. M. Loria, L. Cheng, P. J. Kushner, D. A. Agard, G. L. Greene. *Cell* **95**, 927–937 (1998).
56. H. Shibata, T. E. Spencer, S. A. Onate, G. Jenster, S. Y. Tsai, M. J. Tsai, B. W. O'Malley. *Recent Prog. Horm. Res.* **52**, 141–164; discussion 164–165 (1997).
57. S. Signoretti and M. Loda. [Comment on: *Am. J. Pathol.* **159** (1), 79–92 UI: 21331060 (2001)]. *Am. J. Pathol.* **159**, 13–16 (2001).
58. C. Smith, S. Onate, M.-J. Tsai, B. O'Malley. *Proc. Natl. Acad. Sci. USA* **93**, 8884–8888 (1996).
59. M. Soulez and M. G. Parker. *J. Mol. Endocrinol.* **27**, 259–274 (2001).
60. M. R. Stallcup. *Oncogene* **20**, 3014–3020 (2001).
61. P. S. Steeg and Q. Zhou. *Breast Cancer Res. Treatment* **52**, 17–28 (1998).
62. C. Teyssier, K. Belguise, F. Galtier, D. Chalbos. *J. Biol. Chem.* **276**, 36361–36369 (2001).
63. D. A. Tonetti and V. C. Jordan. *J. Mammary Gland Biol. Neoplasia* **4**, 401–413 (1999).
64. A. Tremblay, G. B. Tremblay, F. Labrie, V. Giguere. *Mol. Cell* **3**, 513–519 (1999).
65. Y. Umayahara, R. Kawamori, H. Watada, E. Imano, N. Iwama, T. Morishima, Y. Yamasaki, Y. Kajimoto, T. Kamada. *J. Biol. Chem.* **269**, 16433–16442 (1994).
66. M. Watanabe, J. Yanagisawa, H. Kitagawa, K. Takeyama, S. Ogawa, Y. Arao, M. Suzawa, Y. Kobayashi, T. Yano, H. Yoshikawa, Y. Masuhiro, S. Kato. *EMBO J.* **20**, 1341–1352 (2001).
67. R. V. Weatherman and T. S. Scanlan. *J. Biol. Chem.* **276**, 3827–3832 (2001).
68. P. Webb, C. M. Anderson, C. Valentine, P. Nguyen, A. Marimuthu, B. L. West, J. D. Baxter, P. J. Kushner. *Mol. Endocrinol.* **14**, 1976–1985 (2000).
69. P. Webb, G. N. Lopez, R. M. Uht, P. J. Kushner. *Mol. Endocrinol.* **9**, 443–456 (1995).
70. P. Webb, P. Nguyen, J. Shinsako, C. Anderson, W. Feng, M. P. Nguyen, D. Chen, S.-M. Huang, S. Subramanian, E. McKinerney, B. S. Katzenellenbogen, M. R. Stallcup, P. J. Kushner. *Mol. Endocrinol.* **12**, 1605–1618 (1998).
71. P. Webb, P. Nguyen, J. Shinsako, C. M. Anderson, M. P. Nguyen, E. McKinerney, B. S. Katzenellenbogen, M. Stallcup, P. J. Kushner. *Mol. Endocrinol.* **12**, 1605–1618 (1998).
72. P. Webb, P. Nguyen, C. Valentine, G. N. Lopez, G. R. Kwok, E. McInerney, B. S. Katzenellenbogen, E. Enmark, J. A. Gustafsson, S. Nilsson, P. J. Kushner. *Mol. Endocrinol.* **13**, 1672–1685 (1999).
73. P. Webb, P. Nguyen, C. Valentine, R. V. Weatherman, T. S. Scanlan, P. J. Kushner. *J. Biol. Chem.* **275**, 37552–37558 (200).
74. Z. Weihua, S. Makela, L. C. Andersson, S. Salmi, S. Saji, J. I. Webster, E. V. Jensen, S. Nilsson, M. Warner, J. A. Gustafsson. *Proc. Natl. Acad. Sci. USA* **98**, 6330–6335 (2001).
75. Z. Weihua, S. Saji, S. Makinen, G. Cheng, E. V. Jensen, M. Warner, J. A. Gustafsson. *Proc. Natl. Acad. Sci. USA* **97**, 5936–5941 (2000).
76. H. E. Xu, T. B. Stanley, V. G. Montana, M. H. Lambert, B. G. Shearer, J. E. Cobb, D. D. McKee, C. M. Galardi, K. D. Plunket, R. T. Nolte, D. J. Parks, J. T. Moore, S. A. Kliewer, T. M. Willson, J. B. Stimmel. *Nature (London)* **415**, 813–817 (2002).
77. L. Xu, C. K. Glass, M. G. Rosenfeld. *Curr. Opin. Genet. Dev.* **9**, 140–147 (1999).
78. Y. Yamamoto, O. Wada, M. Suzawa, Y. Yogiashi, T. Yano, S. Kato, J. Yanagisawa. *J. Biol. Chem.* **276**, 42684–42691 (2001).
79. N. Yang, M. Venugopalan, S. Hardikar, A. Glasebrook. *Science* **273**, 1222–1225 (1996).
80. A. Zou, K. B. Marschke, K. E. Arnold, E. M. Berger, P. Fitzgerald, D. E. Mais, E. A. Allegretto. *Mol. Endocrinol.* **13**, 418–430 (1999).

Pure Appl. Chem., Vol. 75, Nos. 11–12, pp. 1771–1784, 2003.

Topic 1.10

Nuclear receptor action involved with sex differentiation*

Ieuan A. Hughes‡, Howard Martin, Jarmo Jääskeläinen, and Carlo L. Acerini

Department of Paediatrics, University of Cambridge, Addenbrooke's Hospital, Box 116, Cambridge CB2 2QQ, UK

Abstract: Sex determination and differentiation in the male is an orderly sequence of events coordinated by genetic and hormonal factors operating in a time- and concentration-dependent manner. The constitutive sex in mammals is female. Disorders of fetal sex development have provided the means to identify testis-determining genes and the molecular mechanisms of hormone action. Thus, the androgen receptor, a nuclear hormone receptor critical for androgen-induced male sex differentiation, displays unique intra-receptor and protein–protein interactions which, when disturbed, can result in extreme forms of sex reversal. Polymorphic variants are associated with milder disorders of sex development. Against this genetic background, endocrine active substances may further contribute to the underlying causes of an increase in male reproductive tract disorders.

INTRODUCTION

The androgen receptor (AR) is a key mediator in the control of male sexual differentiation. It is expressed in fetal tissues, and activated in a ligand-dependent manner to coordinate expression of suitably responsive genes. The classic experiments of Jost [1], and observations in syndromes such as Turner, show that the ovary, and therefore estrogens, are not required for prenatal sex differentiation of the female. In contrast, male differentiation requires prior formation of the testis and subsequent testosterone production to masculinize the embryo [2]. This is mediated by the AR as a ligand-activated transcription factor, directing both internal and external male gonadal development.

Terminology must be clearly defined. Sex determination refers to formation of a testis or an ovary from the indifferent, bipotential gonad. Sex differentiation is the phenotypic expression of male or female development in the formation of internal and external genitalia. In the male, this requires the active intervention of androgens. The terms, determination and differentiation, refer to prenatal events. Sex development continues postnatally with the onset of secondary sexual characteristics appropriate for gender at puberty, and the acquisition of reproductive capacity.

SEX DETERMINATION

The process of sex determination in mammals represents a genetically regulated choice between divergent pathways of development. The result of this choice is the programming of the bi-potential gonad to either an ovarian or a testicular fate. A number of genes critical to these processes have been isolated,

---

*Report from a SCOPE/IUPAC project: Implication of Endocrine Active Substances for Human and Wildlife (J. Miyamoto and J. Burger, editors). Other reports are published in this issue, *Pure Appl. Chem.* **75**, 1617–2615 (2003).

‡Corresponding author: E-mail: iah1000@cam.ac.uk

but mutation analysis in syndromes of sex reversal indicate that a large number of additional genes in the pathways that regulate sex determination and facilitate gonad differentiation remain to be identified. It has been assumed that the default pathway of gonad determination was female, suggesting a male dominant genetic control. By implication, mutation in key testis regulatory genes and abnormal regulation would redirect the pathway toward a female endpoint.

This original view is now being modified according to an understanding that the regulation of sex determination must also involve anti-testis genes or ovarian determinants, in addition to the currently known male determinants.

SRY (SEX-RELATED GENE ON THE Y CHROMOSOME)

For male sex determination, probably the most significant regulator is *SRY*, located on the short arm of the Y chromosome. Despite the fact that it is a small single exon gene, transgenic studies in female mice have clearly demonstrated that it is the only gene required to initiate development of the testis [3]. Studies in humans have confirmed the testis-determining role of *SRY* by the appearance of a complete sex reversal phenotype in XY individuals with mutant versions of the gene [4].

The critical region of the SRY protein is believed to be the HMG (high mobility group) box motif. This is highly conserved across species and within which the majority of sex-reversing mutations are located [5]. It is this motif that allows SRY to bind DNA and induce a sharp structural bend [6]. The kinetic stability of the bended HMG box-DNA complex appears to regulate the transcription of downstream genes [7].

SRY has a small window of male specific expression in the developing genital ridge which is programmed to form the testis or ovary [8]. This takes place in the mouse between 10.5 dpc (days post coitum), with *Sry* expression having ceased by 12.5 dpc. What regulates *Sry* remains unknown. *Sry* expression spreads through the genital ridge in a defined pattern, resulting in the programming of supporting cells with a bi-potential ability to develop as Sertoli cells. Subsequent cellular events include an increase in Sertoli cell mass, the attraction of cells from the mesonephros and the outer surface of the genital ridge, and the formation of testicular cords by 13.5 dpc. The cords comprise Sertoli cells surrounding germ cells. The somatic cell migrations are male specific and occur as a response to *Sry* expression [9]. Germ cells arise during early embryogenesis at the base of the allantois. They become incorporated into the hindgut during their migration to colonize the developing gonads in the genital ridge [10]. The newly programmed Sertoli cells express another HMG box-containing protein, SOX9 (Sry like HMG box), one of the earliest markers of Sertoli cell differentiation. *Sox9* has a significant role in the control of male sex determination, and is capable of programming the bi-potential genital ridge to a male fate in genetic females expressing *Sox9* as a transgene [11].

SOX9

Targets of SOX9 have been identified. It has a role in bone morphogenesis, whereby the syndrome of skeletal campomelic dysplasia and associated sex reversal results from mutations in human *SOX9* [12]. The gene is expressed in developing chondrocytes, and regulates the control of anti-Müllerian hormone (AMH) expression in Sertoli cells [13]. This peptide causes regression of the female internal genital structures (Müllerian ducts) in a time-dependent fashion. *SOX9* is located on chromosome 17q24 and is initially expressed in both the male and female genital ridge. Recent studies have suggested how specific regulation ensures that inappropriate expression of *SOX9* does not occur in the female pathway. Long-range repressor elements have been detected by fortuitous deletion upstream of *Sox9*, located on mouse chromosome 11 [14]. This deletion removed repressors of *Sox9* expression so that in XX mice, *Sox9* remained active resulting in the development of testes and complete sex reversal. Regulation of this repressor function is believed to involve both *Sry* and *Dax1*, an orphan nuclear receptor with a known role in the antagonism of *Sry* [15].

DAX1 (DOSAGE-SENSITIVE SEX REVERSAL–AHC CRITICAL REGION ON THE X, GENE 1)

DAX1 is an unusual member of the nuclear receptor superfamily. The C-terminus is similar to the ligand-binding domain (LBD) of nuclear hormone receptors, but the N-terminal region bears no similarity to known proteins. There is no central DNA-binding domain. Powerful transcriptional repression activity is present in the putative LBD, although no specific ligand has been identified [16,17].

Human *DAX1* is located on chromosome Xp21. Mutations of this gene in affected males causes congenital X-linked adrenal hypoplasia congenita, which is also associated with hypogonadotropic hypogonadism. Most are nonsense mutations affecting the N-terminus, whereas missense mutations cluster within the putative LBD [18]. All severely impair the function of DAX1 as a transcriptional repressor. Affected males are not sex reversed. Duplication of *DAX1* in males, however, does lead to sex reversal, confirming *Dax1* as a dose-dependent sex reversal locus [19]. Further characterization suggests that *Dax1* acts to antagonize *Sry*, as a putative anti-testis gene [15]. In the mouse genital ridge, *Dax1* is expressed in both the male and female genital ridges at about 11.5 dpc; thereafter expression is downregulated in the male after the start of testis determination (12.5 dpc) but expression persists in the female genital ridge [20].

SF1 (STEROIDOGENIC FACTOR 1)

SF1 is also an orphan nuclear receptor that is expressed in the genital ridge, adrenals, and the hypothalamo-pituitary region. Targeted mutagenesis in mice and rare natural mutations in humans indicate the role this receptor plays in the control of a myriad of processes involving male sex differentiation, steroidogenesis, and regression of Müllerian structures via AMH [21]. Typically, loss of function *SF1* mutations in humans causes XY sex reversal and adrenal insufficiency.

Recent studies in mice identified a highly conserved *SF1* consensus binding site upstream of *Dax1* [22]. *Sf1*-deficient mice display much reduced expression of *Dax1* in the developing gonad, evidence of the role of *Sf1* in regulating *Dax1* gonadal expression. Two unique, novel orphan nuclear receptors appear to interact critically in the early events of gonadogenesis.

WNT4 (WINGLESS-TYPE MOUSE MAMMARY TUMOR VIRUS INTEGRATION SITE)

WNT4 is a member of a large family of signaling glycoprotein molecules which is expressed in the developing mesonephros. *Wnt4* is downregulated in the testis at 11.5 dpc but persists in the developing ovary [23]. Targeted deletion in mice results in the masculinization of XX pups, with stabilization of male internal genitalia (the Wolffian ducts) and degeneration of the female equivalent, Müllerian ducts [23]. Duplication of *WNT4* results in XY sex reversal and in vitro analysis suggests that *Dax1* is upregulated by *Wnt-4* expression [24]. These observations imply that *Dax1* and *Wnt-4* may represent anti-testis genes, contributing to ovarian development by preventing testis formation [25,24].

Several other genes with roles in the control of mammalian gonad development have been identified, primarily from targeted gene disruption experiments. None are members of the nuclear receptor family. They can be broadly grouped according to their function in the early stages of differentiation of the urogenital system. A prime example is *WT1* (Wilm's tumor suppressor gene). This is characterized by a proline and glutamine-rich transregulatory region in the N-terminus of the protein and four zinc fingers in a C-terminal DNA-binding domain [26]. There is thus similarity with the nuclear receptor family as far as the zinc finger moieties, but there is no LBD. Mutations in *WT1* in humans gives rise to syndromes such as the Denys-Drash and Frasier syndromes, characterized by gonadal dysgenesis, nephropathy, and Wilm's tumor of the kidney [27,28]. A number of useful reviews may be referred to for a description of other genes that may play key roles in the development of the genital ridge and testis determination [29,30].

SEX DIFFERENTIATION

Fetal sex differentiation proceeds in an orderly sequence in the male following testis determination (Fig. 1). The first event is regression of the Müllerian ducts mediated by anti Müllerian hormone (AMH) secreted from the Sertoli cells. AMH is a member of the transforming growth factor β family and acts through an AMH Type II receptor expressed in the mesenchyme of the Müllerian duct [31]. This results in regression of the Müllerian ducts by 10 weeks of fetal age. Mutations in either the *AMH* gene or the AMH Type II receptor gene lead to persistence of uterus and Fallopian tube development in otherwise normally differentiated males [32]. The most frequent mutation is a 27 bp deletion in the *AMH Type II* receptor gene [33]. During this period of regression, Leydig cell activity is initiated resulting in fetal testosterone synthesis. Androgen production is initially gonadotrophin-independent before becoming placental hCG-dependent and then later in gestation, under the control of fetal pituitary LH secretion. Fetal serum testosterone concentrations increase to 10 nmol/L or more, levels comparable to adult males. Testosterone stabilizes and differentiates the Wolffian ducts to form the male internal genitalia. The central role of androgens in stabilization of the Wolffian ducts to form the vas deferens, epididymis, and seminal vesicles was originally proposed based on the results of the classic fetal rabbit castration experiments of Jost [1]. However, the observation of Wolffian duct stabilization in males with testosterone biosynthetic effects (17β-hydroxysteroid dehydrogenase deficiency) and resistance to the action of androgens suggests non-androgen dependent effects may have a role in Wolffian duct stabilization [34]. The external male genitalia are also masculinized by androgens although this is primarily mediated by dihydrotestosterone (DHT), a metabolite of testosterone which is more potent. Development of the prostate is also DHT-dependent.

The final step in the sequence of male fetal development is the descent of the testes into the scrotum. The testes migrate from the lower pole of the kidney to the scrotum in a two-stage process of transabdominal migration and inguinoscrotal descent [35,36]. The testis is initially attached to the lower pole of the kidney by the cranial suspensory ligament. This persists in the female to maintain the ovary on the pelvic wall. In the male, the suspensory ligament degenerates by apoptosis in a part-androgen dependent manner. The gubernaculum is also critical for testicular descent. It develops from the caudal suspensory ligament; contraction and thickening produces a bulbar outgrowth to guide the migrating testis towards the internal inguinal ring. The gubernaculums persists as a thin cord in the female. Gubernacular development may be partly androgen dependent, but also appears to be influenced by Insl3 (insulin-like 3) and its receptor [37–39]. Failure of the testes to descend by birth in boys is a com-

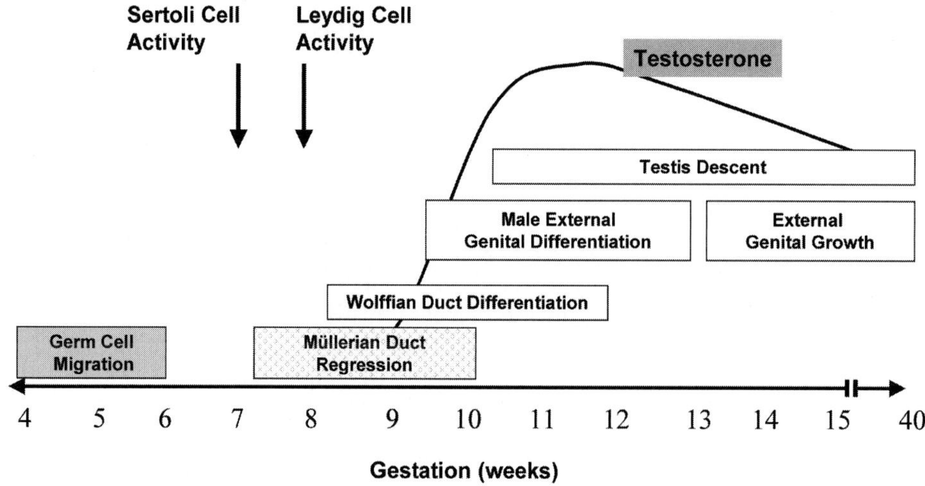

Fig. 1 A schematic of the embryology fetal male development.

mon problem. However, mutations of *INSL3* in boys with bilateral cryptorchidism are rare [40]. The final phase of inguinoscrotal descent of the testes is androgen-dependent, as shown by the gonadal positioning in disorders of androgen insensitivity and hypogonadotrophic hypogonadism [41].

The ontogeny of fetal male sex development involving differentiation of the internal and external genitalia is under the trophic control of androgens. This process is time- and concentration-dependent. Critical to stabilization of the Wolffian ducts and differentiation and growth of the external genitalia is ligand-activation of the nuclear AR. The receptor is ubiquitously expressed, but particularly in sex-dependent tissues. The role of the AR as a transcription factor and its importance in male development can be observed by the study of patients with the androgen insensitivity syndrome (AIS).

ANDROGEN INSENSITIVITY SYNDROME

The clinical manifestation of AIS varies in severity from completely female external genitalia (CAIS) to varying degrees of partial masculinization (PAIS). Testis determination and function is normal during fetal life in AIS. Thus, Sertoli cell production of AMH causes Müllerian duct regression and Leydig cells produce an abundant amount of testosterone with normal conversion to DHT. However, there is resistance to the action of androgens. In postnatal life, CAIS can present in infancy with bilateral inguinal swellings due to herniated testes. Otherwise, presentation is delayed until adolescence with primary amenorrhoea being the stimulus for investigation. This would include a karyotype that reveals the XY chromosomes. Breast development is normal, as androgens are readily aromatized to estrogens. Pubic and axillary hair is absent or scanty because of the androgen resistance. Internal female genitalia are absent, but there is a shortened, blind-ending vagina. In PAIS, there is some responsiveness to androgens which typically is manifest with the appearance of micropenis, bifid scrotum, and a perineoscrotal hypospadias. In some cases, the resistance to androgens is very minimal with a mild, isolated hypospadias or even oligospermia in an otherwise normally masculinized male [42]. Serum concentrations of LH are generally elevated in AIS, despite the increased testosterone levels. Histology of the testes shows absence of spermatogenesis, but often interstitial cell hyperplasia. There is an increased risk of gonadoblastoma formation.

ANDROGEN RECEPTOR

The AR is encoded within a 90 kb region on Xq11-12. It consists of 8 exons and is translated from a 10.5 kb mRNA comprising 919 amino acids [43]. The AR protein comprises three functional domains, in common with other nuclear receptors. The N-terminal domain is the least conserved amongst members of the nuclear receptor family. The AR contains an activation function domain AF-1 comprised of two subdomains, AF-1a and AF-1b. It has a transcriptional activation role, regulating ligand-dependent activation of responsive genes. This regulation is attenuated by polymorphic regions in the N-terminus, with a variable number of glutamine repeats in the 5′ terminus and glycine repeats in the 3′ end. A link between variation in transactivational capacity, and the number of glutamine repeats has been confirmed [44].

The central region of the AR protein comprises the DNA-binding domain (DBD) and represents the region of highest homology amongst the nuclear receptors. It contains two zinc fingers which each employ two pairs of cysteine residues to bind the zinc ions. The DBD encompasses 68 amino acids encoded by exons B and C. The C-terminal domain of the AR comprises the LBD containing a second activation function domain AF-2. A number of additional functions map to this domain and include heat shock protein interactions, dimerization, and nuclear localization signals. The role of the AF regions appears to entail interaction with coregulator proteins to regulate specific genes via the general transcriptional machinery [45]. This interaction is ligand-dependent and results in either induction or repression of transcription depending on whether coactivators or corepressors are complexed with the AR [46,47]. AF-1 functions in a ligand-independent manner, as shown by deletion mutants of the AR lacking the

LBD, with constitutive activity in vitro [48,49]. The AF-2 domain functions in a ligand-dependent manner and is located within helix 12 of the LBD. The AR, in common with several other nuclear receptors, has 12 alpha helices arranged in an anti-parallel fashion in three layers in the form of a sandwich fold. Helix 12 is the most C-terminal helix and forms a hydrophobic cleft in the presence of ligand, allowing coregulator binding via L*XX*LL (L, leucine; *X*, any amino acid) motifs. A number of coregulators are known to interact with the nuclear hormone receptors, a few being more AR specific. Examples include ARA70, which interacts with the AR in a ligand-dependent manner, even though it appears to act only as a weak activator in vitro [50]. Mutation analysis in XY patients with incomplete masculinization, however, has not identified mutations in ARA70 that may affect AR-dependent function [51].

N/C INTERACTION

N-terminal/C-terminal interaction has been demonstrated for the AR protein, as well as the estrogen and progesterone receptors [52]. The AF-2 region of the AR is required for N/C interaction, a function believed to be important for stabilization of the receptor by reducing ligand dissociation rate and degradation kinetics of the LBD/ligand complex [53]. The result is enhanced DNA-binding affinity and subsequent transactivation ability of the AR. A number of LBD mutants have been identified that disrupt N/C interaction potential but retain ligand-binding ability; these mutations can result in a phenotype consistent with AIS [54].

ANDROGEN-BINDING ASSAY

Androgen receptor binding ability is readily measured in cultured genital skin fibroblasts. A normal range for binding parameters can be derived by using circumcised foreskin explants, which give consistently reproducible results. There is considerable variation in receptor binding activity when using skin derived from other sites, such as suprapubic and scrotal skin (personal observations). A typical analysis of an androgen-binding assay is shown in Fig. 2. Primary cell cultures are established from 2–4 mm genital skin explants obtained at the time of surgery. Isolated cells are grown to confluency and starved of serum for 24 h prior to assay. For the binding assay, cells are seeded at a known density and incubated with increasing concentrations of ^{3}H-DHT, with or without a 200-fold excess of unlabeled ligand. An aliquot of cells is retained to quantify DNA content. Binding capacity (Bmax) and dissociation constant (Kd) are derived from Scatchard analysis using linear regression analysis to provide a

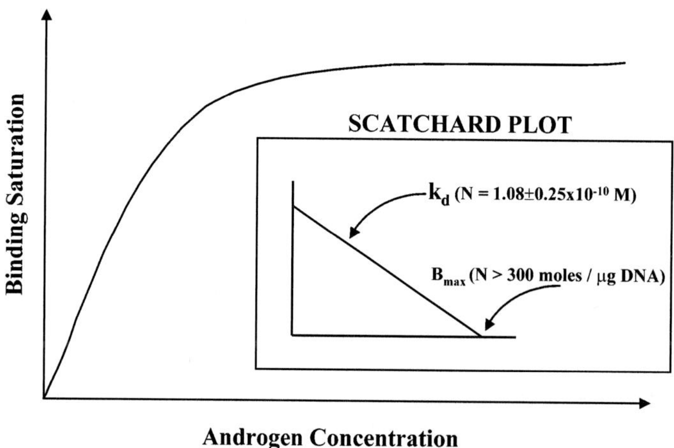

Fig. 2 Saturation curve of androgen binding in genital skin fibroblasts. The Scatchard analysis is based on bound/free hormone vs. bound.

best fit for the binding data [55]. AR expression in genital skin fibroblasts remains relatively constant, in contrast to 5 alpha reductase activity, which declines with age [56].

AR MUTATIONS

Numerous mutations have been identified throughout the AR and are detailed on an international database <http://www.mcgill.ca/androgendb>. Mutations associated with AIS cluster within the LBD, but are also located throughout the whole of the protein coding region, and less frequently within the intron/exon boundaries. A number of AR mutations associated with breast and prostate cancer are also detailed on this database.

The Cambridge Intersex Database contains a similar spread of mutations identified in a large number of patients with AIS (Fig. 3). The results are complemented by detailed clinical, biochemical, and histological information. Also included on the database is information on other sex reversal disorders whose phenotypes may be similar to CAIS and PAIS. Examples include complete and partial gonadal dysgenesis, androgen biosynthetic defects, sex chromosome abnormalities, hermaphroditism, and male factor infertility. This unique resource is now derived from information on more than 1000 cases.

Mutations of the AR in cases of CAIS are routinely identified in over 80 % of those analyzed, the majority of mutations located in the LBD. In contrast, only about 20 % of cases with findings consistent with PAIS have an identifiable *AR* gene mutation [57].

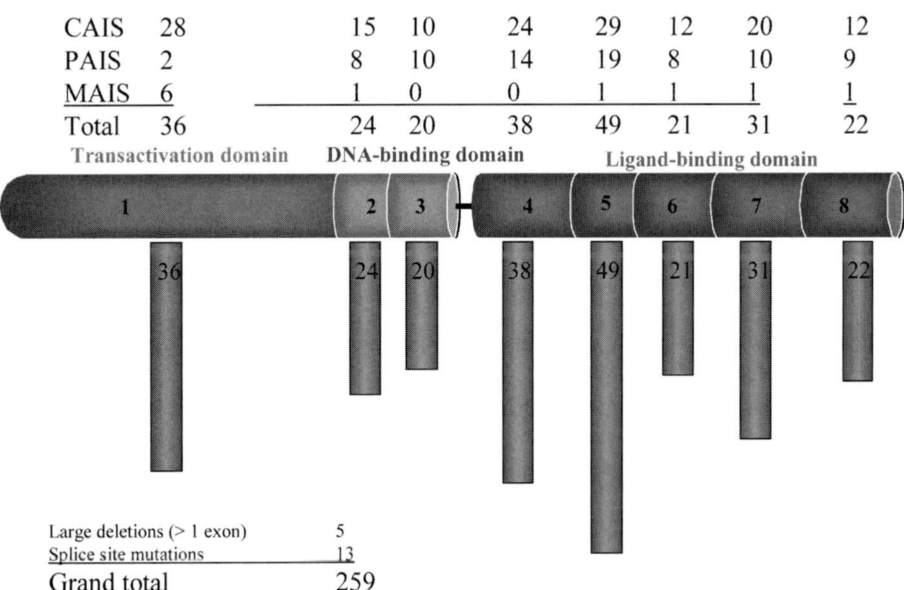

Fig. 3 The Cambridge AR mutation database. The number of mutations located in each exon (1–8) is shown. MAIS, mild androgen insensitivity syndrome.

Mutant AR proteins

Functional analysis of AR mutations can provide further understanding on which regions of the receptor are critical for activity. The crystal structure of the AR-LBD has now been elucidated [58]. This has enabled a number of residues with roles in ligand selectivity and stability of the AR protein to be identified. Mutations located within the most C-terminal alpha-helix 12 display ligand selectivity in functional studies. Thus, mibolerone, which is a potent and nonmetabolizable androgen, is transactivational in a reporter gene assay, whereas that is not the case with DHT (personal observations).

We have recently reported molecular studies on a pair of identical twins with CAIS in whom two novel *AR* gene mutations were identified [59]. In functional assays using HeLa cells, mutation Phe856Leu transactivated more than the wild-type, whereas mutation Ser865Pro completely ablated androgen-dependent transactivation (Fig. 4). Serine865 is located in helix10/11 within a region important for the positioning of helix 12. The latter is critical for coactivator binding. Thus, the functional consequences of mutation Ser865Pro are related to ligand binding, dimerization, and receptor activation; features manifest phenotypically as complete sex reversal in XY twins.

A number of mutations residing outside the LBD have been identified with relevance to N/C interaction and AIS. Specific mutations associated with the hinge region of the AR (between amino acids 621 and 671) have been shown to regulate N/C interaction ability and may have a repressive role on transactivation as shown by deletion mutants in this region [60].

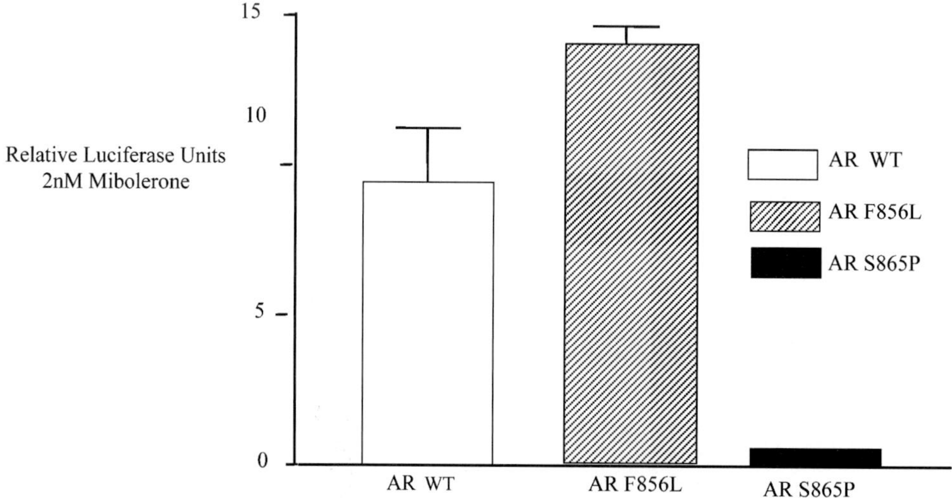

Fig. 4 Androgen-induced transactivation of wild-type AR and 2 mutant ARs transiently expressed in HeLa cells. A glucocorticoid response element-luciferase reporter system was used to measure transactivation capacity. Transfected cells were incubated for 48 h with 1 nM mibolerone.

AR polymorphisms

Located within the N-terminus of the AR is a CAG repeat trinucleotide producing a variable number of glutamine residues in the protein. This polyglutamine variation is an important modulator of receptor function, which may have a contributory role in disorders affecting the male reproductive tract. The neurodegenerative disorder, spinal and bulbar muscular atrophy (Kennedy's disease), is characterized by a hyper-expanded number of CAG repeats in the AR. Affected patients also show evidence of reduced AR function as mild androgen insensitivity and decreased spermatogenesis [61]. A longer (CAG)n tract, but within the normal population range (11–31 repeats in a Cambridge population) has been reported to be associated with reduced spermatogenesis in otherwise normal males in some Australian, North American, Japanese, and Singaporean populations [62,63]. However, this is not universally found, particularly in Scandinavian studies [64]. Shorter repeats are associated with an earlier age of onset of prostate cancer and the response to androgen ablation therapy [65,66]. A similar association was observed in men with androgenetic alopecia [67] and in older adults with indices of central obesity [68].

We have demonstrated that longer CAG repeats are associated with undermasculinization to varying degrees of unknown cause [69]. When the cause of a developmental disorder is multifactorial, the significance of a polymorphism such as the AR (CAG)n repeat may be more than just an association, but actually plays a causative role in the phenotype [70]. These observations in a heterogeneous group of pathophysiological situations (see Fig. 5) may be explained by the clear association found in vitro between the length of the CAG repeat and the activity of the AR as a transcription factor [71]. A number of control points operate in fetal male sex differentiation, several of the genes involved exhibiting polymorphisms. It is expected, therefore, that associations may be additive. This was the case when the cohort we studied with male undermasculinization of unknown cause was analyzed for the AR (CAG)n and a LH receptor polymorphism combined [72]. It would be anticipated that the effects of endocrine disruptors on fetal and early postnatal development may be more profound, depending on a particular genetic background.

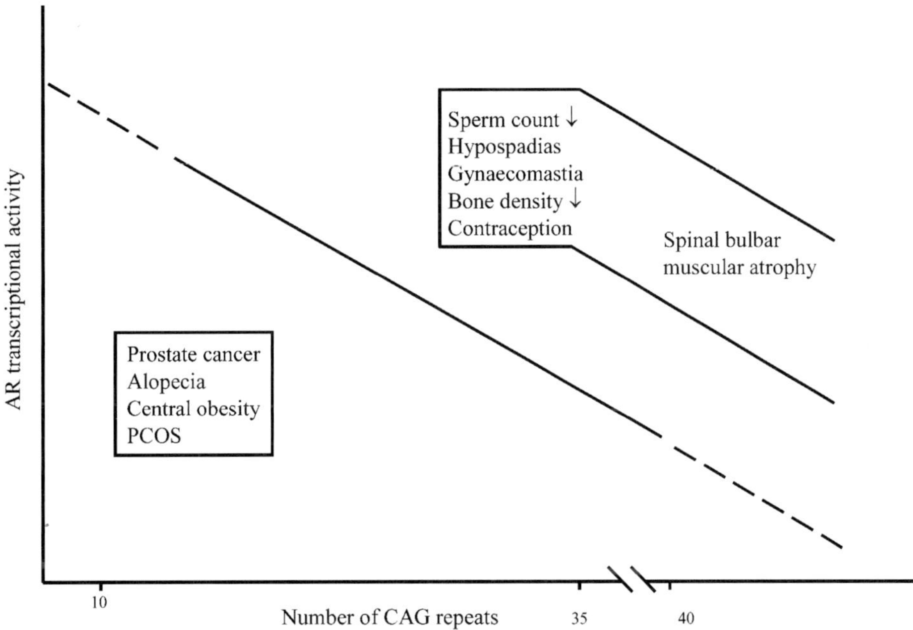

Fig. 5 Theoretical relationship between length of (CAG)n and AR transcriptional activity: relevance to a heterogeneous group of conditions. PCOS, polycystic ovarian syndrome.

AR cofactors

ARA70 has been identified as a coactivator specific to the AR in human prostate cells [73]. It is possible that this cofactor enhances transcription by weak androgens. Another cofactor, ARA24, is believed to interact with the AR polyglutamine repeat [74]. In our study of undermasculinized males, we found no evidence of mutations or polymorphisms in ARA24 associated with the patient phenotype [75]. In contrast, we did observe an association between a shorter number of glutamine repeats in SRC3, an AR coactivator [76,77]. It is quite plausible that these polymorphisms may account for the anti-androgen effects of some endocrine disruptors. AR transcription is regulated in an inhibitory fashion by a corepressor, SMRT (silencing mediator for retinoid and thyroid hormone receptors). Over-expression of SMRT inhibits androgen-dependent AR-mediated transactivation and enhances the effects of anti-androgens such as flutamide [78]. The mechanism of action probably involves inhibition of N/C terminal interaction.

Endocrine active substances

In the context of an apparent increase in reproductive disorders, AR function is relevant to the study of chemicals that may act as anti-androgens or directly as androgens [79,80]. To that end, we have developed an adenovirus-based assay using genital skin fibroblasts to assess AR function in response to test chemicals that may have androgenic or anti-androgenic properties (Fig. 6). The cell line derived from cultured foreskin explants, is telomerase immortalized (Clontech). It expresses the wild-type AR, binds androgens as a high affinity association, and displays a normally regulated AR signaling pathway (personal observations).

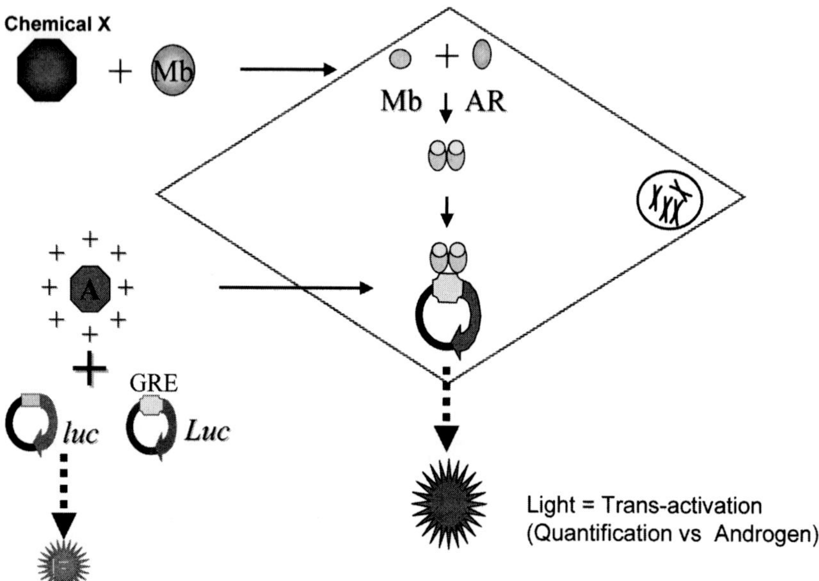

Fig. 6 Adenovirus based reporter gene assay to measure anti-androgenic activity of test chemicals. Genital skin fibroblast cells are transiently transfected with an androgen responsive—firefly luciferase reporter (GRE-*Luc*) and a control renilla luciferase plasmid (*luc*). Cells are exposed to test chemical either in isolation or in combination with 0.1 nM mibolerone. Antiandrogenic response is measured as a ratio of firefly to renilla activity, and compared to the AR transactivation ratio with 0.1 nM mibolerone. Androgenic activity is assessed by chemical-induced transactivation in the absence of mibolerone.

Figure 7 illustrates the effect of one chemical, bisphenol A, acting as an antiandrogen based inhibition of mibolerone-induced AR transactivation of luciferase. The methodology has now been validated for the study of a range of environmental chemicals, which may affect development of the reproductive system in fetal and early postnatal life.

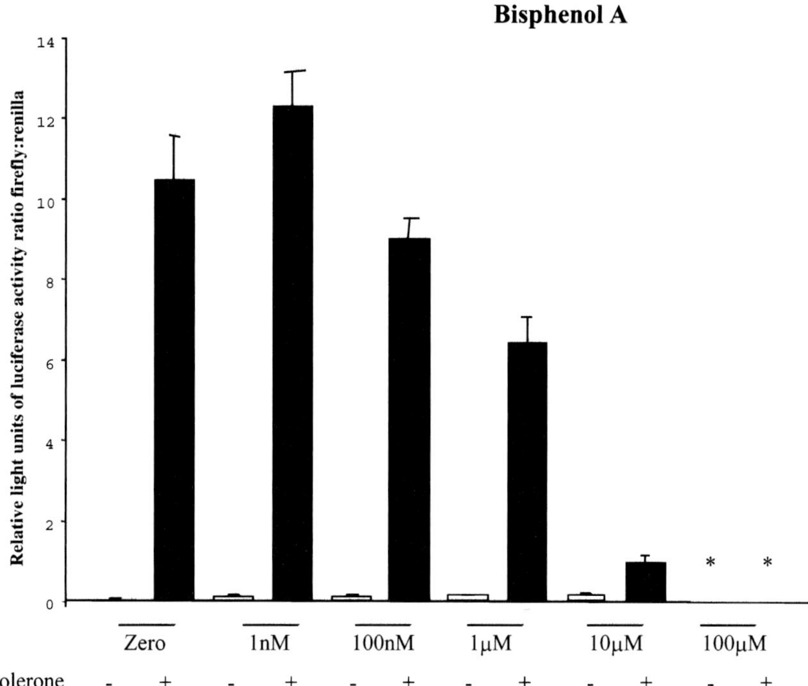

Fig. 7 Effect of bisphenol A as an antiandrogen. Increasing concentrations of bisphenol A demonstrate anti-androgenic activity by inhibition of mibolerone-induced AR transactivation. Lack of androgenic activity is confirmed by absence of bisphenol A-induced transactivation.

CONCLUSION

Two orphan nuclear receptors are inter-linked in contributing to the formation of the testis in the developing male. The nuclear AR, however, is the key molecular transducer of events in fetal male sex differentiation. The genetic and hormonal control of this process is time- and threshold-dependent. A number of single gene defects cause profound effects manifested sometimes as complete sex reversal. The exquisite sensitivity of the developing reproductive tract to hormones suggests that endocrine active substances may play a role in some disorders of human sex differentiation.

ACKNOWLEDGMENTS

Some of the studies described in this article were supported by grants from the Birth Defects Foundation and the European Union Framework V Project. Dr. J. Jääskeläinen was a European Society for Paediatric Endocrinology Research Fellow.

REFERENCES

1. A. Jost. *Rec. Prog. Horm. Res.* **8**, 379–418 (1953).
2. S. F. Ahmed and I. A. Hughes. *Clin. Endocrinol.* **56**, 1–18 (2002).
3. P. Koopman, J. Gubbay, N. Vivian, P. Goodfellow, R. Lovell-Badge. *Nature* **351**, 117–121 (1991).
4. P. Berta, J. R. Hawkins, A. H. Sinclair, A. Taylor, B. L. Griffiths, P. N. Goodfellow, M. Fellous. *Nature* **348**, 448–450 (1990).
5. J. R. Hawkins. *Hum. Mutat.* **2**, 347–50 (1993).

6. S. Ferrari, V. R. Harley, A. Pontiggia, P. N. Goodfellow, R. Lovell-Badge, M. E. Bianchi. *EMBO J.* **11**, 4497–4506 (1992).
7. E. Ukiyama, A. Jancso-Radek, B. Li, L. Milos, W. Zhang, N. B. Phillips, N. Morikaw, C.-Y. King, G. Chan, C. M. Haqq, J. T. Radek, F. Poulat, P. K. Donahoe, M. A. Weiss. *Mol. Endocrinol.* **15**, 363–377 (2001).
8. A. Hacker, B. Capel, P. Goodfellow, R. Lovell-Badge. *Development* **121**, 1603–1614 (1995).
9. B. Capel, K. H. Albrecht, L. L. Washburn, E. M. Eicher. *Mech. Dev.* **84**, 127–131 (1999).
10. R. Anderson, T. K. Copeland, H. Scholer, J. Heasman, C. Wylie. *Mech. Dev.* **91**, 61–68 (2000).
11. V. P. Vidal, M. C. Chaboissier, D. G. de Rooij, A. Schedl. *Nat. Genet.* **28**, 216–217 (2001).
12. J. W. Foster, M. A. Dominguez-Steglich, S. Guioli, G. Kowk, P. A. Weller, M. Stevanovic, J. Weissenbach, S. Mansour, I. D. Young, P. N. Goodfellow. *Nature* **372**, 525–530 (1994).
13. P. De Santa Barbara, N. Bonneaud, B. Boizet, M. Desclozeaux, B. Moniot, P. Sudbeck, G. Scherer, F. Poulat, P. Berta. *Mol. Cell. Biol.* **18**, 6653–6665 (1998).
14. C. E. Bishop, D. J. Whitworth, Y. Qin, A. I. Agoulnik, I. U. Agoulnik, W. R. Harrison, R. R. Behringer, P. A. Overbeek. *Nat. Genet.* **26**, 490–494 (2000).
15. A. Swain, V. Narvaez, P. Burgoyne, G. Camerino, R. Lovell-Badge. *Nature* **391**, 761–767 (1998).
16. M. Ito, R. Yu, J. L. Jameson. *Mol. Cell. Biol.* **17**, 1476–1483 (1997).
17. E. Zazopoulos, E. Lalli, D. M. Stocco, P. Sassone-Corsi. *Nature* **390**, 311–315 (1997).
18. J. C. Achermann, M. Ito, B. L. Silverman, R. L. Habiby, S. Pang, A. Rosler, J. L. Jameson. *J. Clin. Invest.* **105**, 321–328 (2000).
19. B. Bardoni, E. Zanaria, S. Guioli, G. Floridia, K. C. Worley, G. Tonini, E. Ferrante, G. Chiumello, E. R. McCabe, M. Fraccaro. *Nat. Genet.* **7**, 497–501 (1994).
20. A. Swain, e. Zanaria, A. Hacker, R. Lovell-Badge, G. Camerino. *Nat. Genet.* **12**, 404–409 (1996).
21. G. Ozisik, J. C. Achermann, J. J. Meeks, J. L. Jameson. *Horm. Res.* **59**, 94–98 (2003).
22. C. Hoyle, V. Narvaez, G. Alldus, R. Lovell-Badge, A. Swain. *Mol. Endocrinol.* **16**, 747–756 (2002).
23. S. Vainio, M. Heikkila, A. Kispert, N. Chin, A. P. McMahon. *Nature* **397**, 405–409 (1999).
24. B. K. Jordan, M. Mohammed, S. T. Ching, E. Delot, X. N. Chen, P. Dewing, A. Swain, P. N. Rao, B. R. Elejalde, E. Vilain. *Am. J. Hum. Genet.* **68**, 1102–1109 (2001).
25. P. N. Goodfellow and G. Camerino. *Cell. Mol. Life Sci.* **55**, 857–863 (1999).
26. N. D. Hastie. *Ann. Rev. Genet.* **28**, 523–528 (1994).
27. M. Little and C. Wells. *Hum. Mutat.* **9**, 209–225 (1997).
28. B. Klampt, A. Koziell, F. Poulat, P. Wiecracker, P. Scrambler, P. Berta, M. Gessle. *Hum. Molec. Genet.* **7**, 709–714 (1998).
29. P. Koopman. *Cell.* **105**, 843–847 (2001).
30. C. Tilmann and B. Capel. *Rec. Progr. Horm. Res.* **57**, 1–18 (2002).
31. N. Josso and N. di Clemente. *Trends Endocrinol. Metab.* **14**, 91–97 (2003).
32. N. Josso, J. Y. Picard, S. Imbeaud, N. di Clemente, R. Rey. *Clin. Endocrinol.* **47**, 137–144 (1997).
33. S. Imbeaud, C. Belville, L. Messika-Zeitoun, R. Rey, N. di Clemente, N. Josso, J. Y. Picard. *Hum. Mol. Genet.* **5**, 1269–1279 (1996).
34. J. Hodapp and I. A. Hughes. *Horm. Res.* **51**, 76 (1999).
35. J. M. Hutson and S. W. Beasley. *Aust. Paediatr. J.* **23**, 215–216 (1987).
36. J. M. Emmen, A. McLuskey, J. A. Grootegoed, A. O. Brinkmann. *Hum. Reprod.* **13**, 1271–1280 (1988).
37. S. Zimmermann, G. Steding, J. M. Emmen, A. O. Brinkmann, K. Nayernia, A. F. Holstein, W. Engel, I. M. Adham. *Mol. Endocrinol.* **13**, 681–691 (1999).
38. Y. Kubota, S. Nef, P. J. Farmer, C. Temelcos, L. Parada, J. M. Hutson. *J. Urol.* **165**, 1673–1675 (2001).
39. Y. Kubota, C. Temelcos, R. A. D. Bathgate, K. J. Smith, D. Scott, C. Zhao, J. M. Hutson. *Mol. Hum. Reprod.* **8**, 900–905 (2002).

40. H. N. Lim, E. Raiperts-de Meyts, N. E. Skakkebaek, J. R. Hawkins, I. A. Hughes. *Eur. J. Endocrinol.* **144**, 129–137 (2001).
41. H. N. Lim, I. A. Hughes, J. R. Hawkins. *Mol. Cell. Endocrinol.* **185**, 43–50 (2001).
42. C. A. Quigley, A. de Bellis, K. B. Marschke, M. K. el-Awady, E. M. Wilson, F. S. French. *Endocr. Rev.* **16**, 271–321 (1995).
43. W. D. Tilley, M. Marcelli, J. D. Wilson, M. J. McPhaul. *Proc. Natl. Acad. Sci USA* **86**, 327–331 (1989).
44. N. L. Chamberlain, E. D. Driver, R. L. Miesfeld. *Nucleic Acids Res.* **22**, 3181–3186 (1994).
45. R. G. Raeder. *Trends Biosci.* **21**, 327–335 (1996).
46. N. J. McKenna, R. B. Lanz, B. W. O'Malley. *Endocr. Rev.* **20**, 321–344 (1999).
47. D. Robyr, A. P. Wolffe, W. Wahli. *Mol. Endocrinol.* **14**, 329–347 (2000).
48. G. Jenster, H. A. van der Korput, C. van Vroonhoven, T. H. van der Kwast, J. Trapman, A. O. Brinkmann. *Mol. Endocrinol.* **5**, 1396–1404 (1991).
49. Z. X. Zhou, M. Sar, J. A. Simental, M. V. Lane, E. M. Wilson. *J. Biol. Chem.* **269**, 13115–13123 (1994).
50. T. Gao, K. Brantley, E. Bolu, M. J. McPhaul. *Mol. Endocrinol.* **13**, 1645–1656 (1999).
51. H. N. Lim, J. R. Hawkins, I. A. Hughes. *Clin. Genet.* **59**, 284–286 (2001b).
52. T. Ikonen, J. J. Palvimo, O. A. Jänne. *J. Biol. Chem.* **272**, 29821–29828 (1997).
53. E. Langley, J. A. Kemppainen, E. Wilson. *J. Biol. Chem.* **273**, 92–101 (1998).
54. J. Thomopson, F. Saatcioglu, O. A. Janne, J. J. Palvimo. *Mol. Endocrinol.* **15**, 923–935 (2001).
55. B. A. Evans, T. R. Jones, I. A. Hughes. *Clin. Endocrinol.* **20**, 93–105 (1984).
56. O. J. Hellwinkel, A. Muller, D. Struve, O. Hiort. *Eur. J. Endocrinol.* **143**, 217–225 (2000).
57. S. F. Ahmed, A. Cheng, L. A. Dovey, J. R. Hawkins, H. Martin, J. Rowland, N. Shimura, A. D. Tait, I. A. Hughes. *J. Clin. Endo. Metab.* **85**, 658–665 (2000).
58. P. M. Matias, P. Donner, R. Coelho, M. Thomaz, C. Peixoto, S. Macedo, N. Otto, S. Joschko, P. Scholz, A. Wegg, S. Basler, M. Schafer, U. Egner, M. A. Carrondo. *J. Biol. Chem.* **275**, 26164–26171 (2000).
59. N. P. Mongan, J. Jääskeläinen, K. Green, J. W. Schwabe, N. Shimura, M. Dattani, I. A. Hughes. *J. Clin. Endocrinol. Metab.* **87**, 1057–1061 (2002).
60. Q. Wang, J. Lu, E. L. Yong. *J. Biol. Chem.* **276**, 7493–7499 (2001).
61. A. R. La Spada, E. M. Wilson, D. B. Lubahn, A. E. Harding, K. H. Fischbeck. *Nature* **352**, 77–79 (1991).
62. E. L. Yong, C. J. Loy, K. S. Sim. *Hum. Reprod. Update* **9**, 1–7 (2003).
63. M. Zitzmann and E. Nieschlag. *Int. J. Androl.* **26**, 76–83 (2003).
64. E. M. Rajperts-De Meyts, H. Leffers, J. H. Petersen, A. G. Andersen, E. Carlsen, N. Jorgensen, N. E. Skakkebaek. *Lancet* **359**, 44–46 (2002).
65. J. L. Stanford, J. J. Just, M. Gibbs, K. G. Wicklund, C. L. Neal, B. A. Blumenstein, E. A Ostrander. *Cancer Res.* **57**, 1194–1198 (1997).
66. O. Bratt, A. Borg, U. Kristoffersson, R. Lundgren, Q. X. Zhang, H. Olsson. *Br. J. Cancer.* **81**, 672–676 (1999).
67. M. E. Sawaya and A. R. Shalita. *J. Cut. Med. Surg.* **3**, 9–15 (1998).
68. D. R. Gustafson, M. J. Wen, B. M. Koppanati. *Int. J. Obesity* **27**, 75–81 (2003).
69. H. M. Lim, H. Chen, S. McBride, A. M. Dunning, R. M. Nixon, I. A. Hughes, J. R. Hawkins. *Hum. Molec. Genet.* **9**, 829–834 (2000).
70. H. M. Lim, R. M. Nixon, H. Chen, I. A. Hughes, J. R. Hawkins. *J. Clin. Endocrinol. Metab.* **86**, 3207–3210 (2001).
71. T. G. Tut, F. J. Ghadessy, M. A. Trifiro, L. Pinsky, E. L. Yong. *J. Clin. Endocrinol. Metab.* **82**, 3777–3782 (1997).
72. N. P. Mongan, I. A. Hughes, H. M. Lim. *Eur. J. Endocrinol.* **147**, 103–107 (2002).
73. S. Yeh and C. Chang. *Proc. Natl. Acad. Sci. USA* **93**, 5517–5521 (1996).

74. P. W. Hsiao, D. L. Lin, R. Nakao, C. Chang. *J. Biol. Chem.* **274**, 20229–20234 (1999).
75. N. P. Mongan, H. N. Lim, I. A. Hughes. *Eur. J. Endocrinol.* **145**, 809–811 (2001).
76. N. P. Mongan, J. Jääskeläinen, S. Bhattacharyya, R. M. Leu, I. A. Hughes. *Eur. J. Endocrinol.* **148**, 277–279 (2003).
77. J. A. Tan, S. H. Hall, P. Petrusz, F. S. French. *Endocrinology* **141**, 3440–3445 (2000).
78. G. Liao, L-Y. Chen, A. Zhang, A. Godavarthy, F. Xia, J. C. Ghosh, H. Li, J. D. Chen. *J. Biol. Chem.* **278**, 5052–5061 (2003).
79. L. E. Gray, J. Ostby, J. Furr, C. J. Wolf, C. Lambright, L. Parks, D. N. Veeramachaneni, V. Wilson, M. Price, A. Hotchkiss, E. Orlando, L. Guillette. *Reprod. Update* **7**, 248–264 (2001).
80. C. J. Wolf, A. Hotchkiss, J. S. Ostby, G. A. LeBlanc, L. E. Gray, Jr. *Toxicol. Sci.* **65**, 71–86 (2002).

Pure Appl. Chem., Vol. 75, Nos. 11–12, pp. 1785–1796, 2003.

Topic 1.11

Human disorders caused by nuclear receptor gene mutations*

John C. Achermann[1] and J. Larry Jameson[2,‡]

[1]*Centre for Human Growth and Maturation, Institute of Child Health and Department of Medicine, University College London, London, UK;* [2]*Division of Endocrinology, Metabolism, and Molecular Medicine, Northwestern University Medical School, Chicago, IL 60611, USA*

Abstract: The identification of naturally occurring nuclear receptor mutations highlights the critical role that many of these transcription factors play in human endocrine development and function. Inactivating mutations in the ligand-dependent nuclear receptors (TRβ, VDR, ERα, GR, MR, AR) are well characterized in patients with conditions such as androgen insensitivity syndrome (AIS) and vitamin D resistance. On the other hand, mutations in TRβ act in a dominant negative manner to cause hormone resistance. Inactivating mutations in orphan nuclear receptors have also been identified (PPARγ2, HNF4α, PNR, NURR1, SF1, DAX1, SHP) and reveal important developmental and metabolic functions for this group of receptors with previously elusive physiologic roles. In addition to loss of function mutations, receptor activation can result from mutations that confer constitutive activity or altered ligand responsiveness to the receptor (MR, AR), or from genetic duplication (DAX1) or the expression of fusion proteins (RARA, PPARγ1). Together, these naturally occurring mutations provide fascinating insight into key structural and functional receptor domains to reveal the diverse role nuclear receptors play in human biology.

INTRODUCTION

Nuclear receptors are a family of transcription factors that play a crucial role in the development and function of many endocrine and non-endocrine tissues [1–3] (see also <http://bc.georgetown.edu/nrr/nrr.html>). The best-characterized of these are the ligand-dependent nuclear receptors (TRβ, VDR, ERα, GR, MR, AR). These receptors are activated by specific high-affinity hormone ligands such as thyroid hormone and estrogen. Inactivating mutations in the genes that encode these factors have been described in patients with a range of endocrine disorders (Table 1) [4].

In contrast, the majority of nuclear receptors identified to date are termed "orphan" nuclear receptors [5]. Although some "adopted" orphan receptors have been shown to respond to various low-affinity metabolic ligands, such as fatty acid derivatives and bile acids (PPARγ2, HNF4α) [6], most orphan nuclear receptors have no known natural ligands and may function in a ligand-independent manner. Orphan nuclear receptors are expressed in a wide range of endocrine and nonendocrine tissues, and inactivating mutations in several orphan nuclear receptors have now been identified in patients (Table 2).

*Report from a SCOPE/IUPAC project: Implication of Endocrine Active Substances for Human and Wildlife (J. Miyamoto and J. Burger, editors). Other reports are published in this issue, *Pure Appl. Chem.* **75**, 1617–2615 (2003).

‡Corresponding author: Address: Department of Medicine, The Feinberg School of Medicine, Northwestern Memorial Hospital Galter Pavilion, Suite 3-150, 251 East Huron Street, Chicago, IL 60611-2908, USA; Tel.: (312) 926-9436; Fax: (312) 926-7260; E-mail: ljameson@northwestern.edu

Table 1 Loss of function mutations in ligand-dependent nuclear receptors.

Receptor	Disorder	Features	Locus	Inherit.	Mutation	Number*
Thyroid β	Resistance to thyroid hormone	Small goiter Hyperactivity Tachycardia	3p24.3	AR, AD, S	P, D	>100
Vitamin D	Hereditary Vitamin D-resistant rickets	Hypocalcemia Rickets Short Statute Alopecia	12q12-14	AR	P	>20
Estrogen α	Estrogen resistance	Tall stature Delayed epiphyseal fusion Osteoporosis (m)	6p25.1	AR, S	P	1
Glucocorticoid	Glucocorticoid resistance	Hypertension Hyperandrogenism Infertility Fatigue	5q31	AR, AD, S	P	8
Mineralocorticoid	Pseudohypo-aldosteronism Type 1	Hypotension Salt loss (mild, remits with age)	4q31.1	AD	P	13
Androgen	Androgen insensitivity syndrome	Undermasculinization (complete, partial) Male infertility	Xcen-q13	XL, S	P, D	>300

AR = autosomal recessive, AD = autosomal dominant, S = somatic cell mutation, XL = X-linked, P = point mutation, D = deletion; *approximate numbers of different mutations in each receptor are shown.

Table 2 Loss of function mutations in "orphan" nuclear receptors.

Receptor	Features	Locus	Inherit.	Mutation	Number
PPARγ2	Obesity Insulin resistance	3p25	AD	P	4
HNF4α	MODY 1	20q12-13.1	AD	P	10
PNR	Enhanced S cone syndrome	15q23	AD, AR	P	12
NURR1	Schizophrenia/BPD	2q22-23	AD	P	4
SF-1	Primary adrenal failure XY sex-reversal Müllerian structures	9q33	AD, AR	P	3
DAX1	Adrenal hypoplasia congenita Hypogonadotropic hypogonadism Impaired spermatogenesis	Xp21.3-21.2	XL	P, D	>80
SHP	Mild obesity	1p36.1	AD	P	9

BPD = bipolar disorder

Inactivating mutations in nuclear receptors impair receptor function and gene transcription through mechanisms that include reduced ligand binding, abnormal cofactor interactions, loss of hetero- or homo-dimerization, abnormal nuclear localization or impaired DNA binding to target genes. In contrast, increased receptor activity can arise from alternative mechanisms such as nuclear receptor activation or altered ligand selectivity (e.g., MR, AR). Dominant negative mutations of a specific receptor isoform (e.g., TRβ), duplication of a gene that functions in a dosage-dependent manner (e.g., overexpression of *DAX1*), or the expression of gene fusion products (PPARγ1, RARA) can also lead to specific disease phenotypes (Table 3).

Table 3 Alternative mechanisms causing nuclear receptor disease.

Mechanism	Receptor	Features
Activating mutation	Mineralocorticoid	Hypertension, exacerbated by pregnancy
	Androgen	Prostate carcinoma
	PPARγ2	Obesity
Loss of function of a specific receptor isoform	Thyroid receptor β	Hyperactivity/ADHD Tachycardia
Overexpression due to gene duplication	DAX1	Dosage-sensitive sex reversal
Fusion proteins	FAS/ERα	Human cancer cell lines (prostate, breast, cervix, bladder)
	PAX8/PPARγ1	Thyroid follicular carcinoma
	EWS/TEC (NOR1)	Extraskeletal myxoid chondrosarcomas
	PML/RARA	Acute promyelocytic leukemia
CAG trinucleotide repeat	Androgen	↑ CAG: X-linked spinal and bulbar muscular atrophy ↓ CAG: Prostate cancer susceptibility

In this review, we provide an overview of hereditary diseases associated with nuclear receptor mutations in humans, and highlight several examples where naturally occurring mutations provide insight into receptor action. The effect of receptor polymorphisms on human health will be discussed only briefly, as the significance of many of these associations is still unclear.

MUTATIONS IN LIGAND-DEPENDENT NUCLEAR RECEPTORS

Thyroid receptor β (TRβ) (NR1A2): Resistance to thyroid hormone

The syndrome of resistance to thyroid hormone (RTH) is characterized by reduced target tissue responsiveness to circulating thyroid hormones, and results in elevated levels of serum T4 and T3 and an inappropriately nonsuppressed TSH [7,8]. Some of these patients are relatively asymptomatic, reflecting the fact that the elevation in circulating thyroid hormone levels is sufficient to partially compensate for the generalized resistance (GRTH). Other patients experience symptoms reminiscent of thyrotoxicosis, including weight loss, tremor, palpitations, insomnia, failure to thrive, growth abnormalities, and attention-deficit hyperactivity disorder [4]. These symptoms may reflect a predominant pituitary resistance (PRTH) due to TRβ mutations, allowing activation of normal TRα receptor isoforms in the heart and other tissues by high circulating thyroid hormone levels [9]. In practice, however, considerable overlap in clinical features is seen among individuals, whether they harbor the same or different mutations.

The vast majority of RTH cases have an autosomal dominant pattern of inheritance and heterozygous mutations within the ligand-binding domain (LBD) of the TRβ receptor [7,8]. Missense mutations in this region tend to cluster within three specific domains, and interfere with ligand binding and/or coactivator interaction to exert a "dominant negative" effect on wild-type receptor activity. Consequently, TRβ activation of target genes in the presence of ligand is impaired, but basal repression in the absence of ligand is unaffected. In some cases, TRβ mutations may enhance corepressor binding [10]. Further evidence of a dominant negative effect is evidenced by a severe phenotype (developmental delay and growth retardation) in a boy homozygous for a mutation in TRβ in comparison to individuals with a complete deletion of one or both TRβ alleles [11,12]. Somatic mutations and alternative splicing of TRβ has been described as the basis of resistance to feedback in a subset of TSH-secreting pituitary tumors.

Vitamin D receptor (VDR) (NR1I): Hereditary vitamin D-resistant rickets

The vitamin D receptor binds the active form of vitamin D (1,25-dihydroxyvitamin D) with high affinity to mediate its effects on calcium homeostasis, skeletal development and bone mineralization. Mutations in the VDR cause hereditary vitamin D-resistant rickets (HVDRR) [13]. Clinical features include bone pain, muscle weakness, hypotonia, convulsions, and clinical and radiological evidence of rickets within the first few months of life. Many children have sparse hair, while some have total alopecia. Laboratory investigations reveal hypocalcemia, elevated 1,25-dihydroxyvitamin D, and elevated PTH due to secondary hyperparathyroidism. More than 20 different homozygous mutations have been reported throughout the DNA- and ligand-binding domains of the VDR. Point mutations within the DNA-binding domain (DBD) tend to have a more severe effect and are usually associated with alopecia. Variations in the severity of LBD mutations may reflect differences in ligand-binding affinity and/or dimerization with RXR [14]. Rarely, partial deletions of the VDR locus or intronic mutations that cause exon skipping have been reported. As most patients with HVDRR do not respond to supraphysiological doses of vitamin D, adjunctive treatment with high dose intravenous calcium (to circumvent the impaired intestinal calcium absorption) has been used to treat this form of rickets. The relation between VDR polymorphisms and bone mineral density remains under investigation [15].

Estrogen receptor α (ERα) (NR3A1): Estrogen resistance

Only one human ER mutation has been described to date, in a male who presented at age 28 with tall stature (204 cm) and continued linear growth, and delayed epiphyseal fusion despite otherwise normal pubertal development [16]. He had increased serum estradiol, FSH and LH, and decreased sperm viability; impaired glucose tolerance and hyperinsulinemia with clinical evidence of acanthosis nigricans; abnormal serum lipids and evidence of early coronary atherosclerosis; a bone mineral density 3.1 SD below the mean; and resistance to estrogen treatment. Mutational analysis revealed a homozygous R157X mutation in ERα that caused premature truncation of the receptor, including the DNA- and ligand-binding domains. Other family members who are heterozygous for this mutation are phenotypically normal. Although this is the only report of an ERα mutation to date, these findings provide clear evidence for the profound influence of this receptor on many aspects of human reproduction, growth, bone mineralization, metabolism, and cardiovascular health. The effects of ERα mutations in women, or mutations in the ERβ isoform, are unknown. The influence of somatic point mutations, splice variants of ER isoforms and ER cofactors in tumors, such as breast cancer, is under active investigation [17,18], as are the effects of ER polymorphisms and variants on bone mineral density, and in cardiovascular and psychiatric disease [19].

Glucocorticoid receptor (GRα) (NR3C1): Familial glucocorticoid resistance

Mutations in GRα cause familial glucocorticoid resistance (FGR) [20]. This syndrome has a range of clinical features and can be inherited as an autosomal dominant or autosomal recessive condition, depending on the underlying molecular mechanism or severity of the mutation. Patients with FGR often feel fatigue, but other signs of glucocorticoid insufficiency are rare because the ACTH-driven elevation in cortisol compensates for receptor insensitivity. A consequence of this ACTH drive is the elevation of mineralocorticoids and androgens, resulting in (1) hypertension, hypokalemia, and metabolic alkalosis; and (2) hirsutism, acne, male-pattern baldness, oligomenorrhea, and infertility, respectively.

Most GRα mutations are located within the LBD of the receptor and affect ligand binding and transactivation. Patients or carriers with heterozygous mutations often have a milder clinical phenotype, although dominant negative LBD mutations appear to inhibit nuclear translocation of the wild-type receptor (I557N, I747N) or impair receptor/coactivator (p160) interactions (I749M) [21,22]. The clinical and biochemical effects of homozygous mutants in GRα tend to be more severe. For example, a homozygous V729I mutation has been reported in a boy with pseudoprecocious puberty, and a homozygous V571A mutation has been found in a girl with clitoral enlargement and labial fusion at birth (Note: this patient was also heterozygous for a CYP21/21-hydroxylase conversion) [23]. Polymorphisms in GRα have been reported in association with central obesity and somatic GRα mutations have been reported in a patient with glucocorticoid-resistant leukemia.

Mineralocorticoid receptor (MR) (NR3C2): Autosomal dominant or sporadic pseudohypoaldosteronism type I

To date, 13 heterozygous mutations have been described in the MR of patients with the autosomal dominant or sporadic forms of pseudohypoaldosteronism type 1 (PHA1) [24]. This condition is characterized by neonatal salt wasting (dehydration, hypotension, hyperkalemia), elevated aldosterone and plasma renin activity, and resistance to mineralocorticoid replacement therapy. Clinical and biochemical features tend to remit with age, in contrast to the autosomal recessive form of PHA1, due to mutations in the amiloride-sensitive epithelial sodium channel (EnaC). Most missense mutations in the MR reported to date are located within the LBD and abolish aldosterone-dependent transactivation in an in vitro assay system [25].

A fascinating report describes a heterozygous S810L mutation within the ligand-binding pocket of the MR in members of a kindred who developed early-onset hypertension (before age 20), and the affected females had an extreme elevation of blood pressure during pregnancy [26]. This missense mutation causes constitutive MR activity and altered receptor specificity, so that progesterone and other steroids lacking 21-hydroxyl groups become MR agonists (Fig. 1). This report illustrates how naturally occurring mutations or variations in receptor structure can alter ligand responsiveness, and has potential implications for the effect of environmental modulators on receptor function within a population.

Androgen receptor (AR) (NR3C4): Androgen insensitivity syndrome (AIS)

More than 200 different inactivating mutations in the AR have been reported in patients with various forms of the X-linked androgen insensitivity syndrome (see Topic 1.10, ref. 27; also <http://www.mcgill.ca/andogendb/>). These mutations can occur throughout the gene, although missense mutants usually affect critical amino acids in the DNA- or ligand-binding domains of the protein. Truncation mutants usually cause a severe phenotype of complete AIS (CAIS) in which genotypic males (46XY) show a total lack of virilization and have a female phenotype. These patients often present in adolescence with a failure of menstruation (no uterus is present) or inguinal masses (testes). Missense mutations in the AR are associated with both complete and partial forms of AIS (PAIS). Patients with PAIS have genital ambiguity or penoscrotal hypospadias at birth. The karyotype is 46XY, no Müllerian structures are detected, and biochemical tests (increased testosterone, increased LH) reflect androgen resistance. Finally, missense mutations with a milder loss of function have been identified in men with oligospermic infertility. These changes may have subtler effects, such as disruption of receptor/coactivator (e.g., TIF2) interactions [28].

An unusual feature of the AR is the variable polyglutamine and polyglycine tandem repeats in the amino-terminal region of the receptor. Expansion of the polyglutamine tract (from 20-23 to 43-65) triggers X-linked spinal and bulbar muscular atrophy (Kennedy disease) [29]. This condition is sometimes associated with decreased virilization, reduced sperm production, testicular atrophy, and infertility. Less severe repeat expansions (>28) have been associated with an increased risk of impaired spermatogenesis, and with moderate undermasculinization of males (46XY) compared to controls [30].

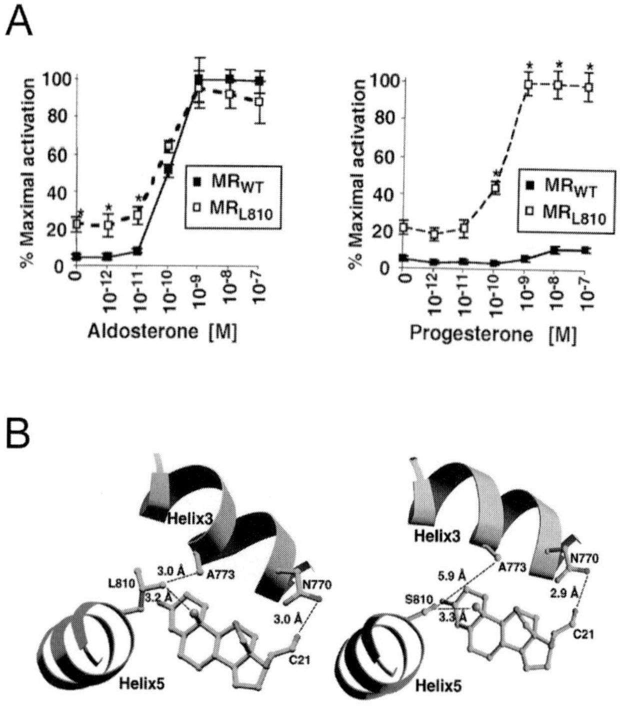

Fig. 1 An "activating" mutation in the mineralocortioid receptor has been described in a kindred with early-onset hypertension exacerbated during pregnancy. (A) The S810L mutant receptor has constitutive activity and inappropriate responsiveness to progesterone. (B) The wild-type serine (5.9 Å, right) to leucine (3.0 Å, left) change at position 810 results in a gain of van der Waals interaction between helix 5 and helix 3 that substitutes for interaction of the 21-hydroxyl group with helix 3 in the wild-type receptor (reproduced with permission from [26]).

Several studies have examined the role of the AR in prostate cancer. Shorter polyglutamine repeats can be associated with more aggressive forms of tumor, and several somatic AR mutations have been discovered in metastatic prostate tumors and human prostate cancer cell lines. The best-characterized of these is the T877A mutation, which alters the structure of the ligand-binding pocket and confers inappropriate responsiveness to progesterone, glucocorticoids, and other C17, C19, and C21 circulating steroids at concentrations found in vivo [31–33]. These studies have important implications for the treatment of prostate cancer, and once again highlight how receptor mutations can alter specificity to potential environmental modulators.

ORPHAN NUCLEAR RECEPTORS

Although the orphan nuclear receptors have no known natural ligands or respond with only low affinity to certain metabolic compounds, they are emerging as important targets for manipulation by environmental modulators and drugs. Mutations in several orphan nuclear receptors have now been identified.

Perioxisome proliferator-activated receptor-γ (PPARγ2) (NR1C3)

PPARγ2 is the target for the thiazolidinedione (TZD) group of drugs used in the treatment of type 2 diabetes. PPARγ2 is expressed in many tissues, including the liver, muscle and fat, and regulates adipocyte differentiation and atherogenesis as well as insulin sensitivity and lipid metabolism. PPARγ2 can be activated by eicosapentaenoic acid, 9-hydroxyoctadecadienoic acid and 15-deoxyprostaglandin J_2.

Given its role in the adipocyte, PPARγ2 was considered a candidate gene for human obesity, and the P115Q mutation was found in 4 out of 121 German patients with a body mass index greater than 29 [34]. This mutation disrupts phosphorylation of an adjacent serine residue, and results in a gain-of-function mutation that mediates enhanced adipocyte differentiation in vitro. In contrast, germline loss-of-function missense mutations (P467L, V290M) in PPARγ2 have been identified in patients from two kindred with severe insulin resistance, hirsutism, diabetes mellitus, and hypertension, but normal body mass index [35]. These heterozygous mutations destabilize helix 12 and impair transcriptional regulation in a dominant negative manner. Furthermore, a heterozygous R425C mutation in PPARγ2 has been identified in a woman who had early-onset diabetes mellitus, hypertriglyceridemia, and hirsutism, and developed lipodystrophy of the extremities and face at age 50 [36].

PPARγ may also have a role in tumor biology, as somatic inactivating mutations in PPARγ2 have been found in four out of 55 sporadic colon tumors [37]; a H449H polymorphism may be associated with a predisposition to glioblastoma; TZDs may modulate the growth of certain breast and prostate cancer cells; and a translocation resulting in a PAX8/PPARγ1 fusion protein has been detected in a subset of thyroid follicular carcinomas. The role of the P12A polymorphism in PPARγ2 in tumorigenesis, obesity, or metabolic dysfunction remains under investigation [for review, see ref. 38].

Hepatocyte nuclear factor-4 α (HNF4α) (NR2A1)

The orphan nuclear receptor HNF4α is expressed in the liver, pancreas, intestine, and kidney, where it is involved in the network of transcription factors that regulate hepatic gene expression and glucose, cholesterol, and fatty acid metabolism. Mutations in HNF4α have been found in patients with maturity-onset diabetes of the young, type 1 (MODY1), a relatively rare form of early-onset diabetes [39]. Most of these mutations are heterozygous changes that impair transcription through their effects on DNA binding, homodimerization, or nuclear localization, but do not appear to function in a dominant negative manner [40,41].

HNF4α can regulate transcription of target genes directly, but may exert many of its actions through the regulation of HNF1α (which is mutated in patients with MODY3). Indeed, the identification of a disrupted HNF4α binding site in the HNF1α promoter in an Italian family with MODY provides further evidence for the importance this pathway in hepatic gene regulation and glucose homeostasis [42]. This cascade of transcription factors likely involves intricate feedback mechanisms, as HNF1α, HNF1β, and IPF1 binding sites have been identified in the HNF4α promoter (P2), and a change in the IPF1 element cosegregates with diabetes in a large MODY kindred [43]. Furthermore, it is possible that HNF4α activity is modulated by metabolic intermediates, such as long-chain fatty acyl-CoA thioesters, which could represent an additional level of control in this complex regulatory network.

Photoreceptor-specific nuclear receptor (PNR) (NR2E3)

PNR is expressed in the outer nuclear layer of the neurosensory retina and plays a key role in human photoreceptor development. Several PNR mutations have been described in patients with enhanced S cone syndrome, a condition involving increased sensitivity to blue light, visual loss, night blindness and progressive retinal degeneration [44,45]. These patients have an increase in the number of S (short wavelength, blue) cones at the expense of other photoreceptor subtypes (L/M; red/green), suggesting that PNR has an important function in determining photoreceptor fate and maintaining retinal integrity [45]. Retinitis pigmentosa has been described in a cohort of Portuguese patients with PNR mutations [46], and features of retinal dysplasia and degeneration have also been identified in the rd7/rd7 mouse, which harbors a large deletion in the coding region of Nr2e3.

Nuclear receptor-related-1, NURR1 (NR4A2)

Homozygous NURR1 knockout mice fail to develop midbrain dopaminergic neurons and heterozygous animals have reduced dopamine levels. NURR1 was therefore considered a candidate gene in patients with neuropsychiatric disorders. Three partially inactivating missense mutations in NURR1 have been reported in a cohort of 324 patients with schizophrenia or manic-depression. An insertion in an untranslated exon was found in two of 177 schizophrenic patients, but not in the controls [47,48]. The true significance of NURR1 mutations in these patients, or in patients with Parkinson's disease, remains to be seen.

Steroidogenic factor-1 (SF1) (NR5A1)

SF1 regulates the transcription of an array of target genes involved in gonadal and adrenal development, steroidogenesis, and reproduction, by binding as a monomer to variations on an extended ER half-site (PyCA AGGTCA). Consistent with this pivotal role in gene transcription, homozygous deletion of SF1 in mice results in gonadal and adrenal agenesis, complete 46XY sex-reversal with persistent Müllerian structures in males, impaired gonadotropin release, abnormalities of the ventromedial hypothalamus and spleen, and obesity. Heterozygous animals have a milder adrenal phenotype.

SF1 mutations have been described in two patients (46XY) with complete sex-reversal, testicular dysgenesis, Müllerian structures, and primary adrenal failure. The first patient has a de novo heterozygous mutation (G35E) in the P-box of the first zinc finger of SF1 [49]. As this motif interacts with the major groove sequence (AGGTCA) to determine DNA-binding specificity, the P-box change affects SF1 binding and transactivation of many target genes [50]. Recently, a homozygous R92Q mutation in the A-box of SF1 has been reported in a patient with a similar phenotype [51]. Heterozygous carriers are normal. The A-box region stabilizes receptor monomer binding by forming a secondary interface with the minor groove of the DNA (PyCA). Thus, a *heterozygous* P-box mutation causes a severe clinical phenotype, whereas a *homozygous* A-box change is necessary for such a phenotype to occur (Fig. 2). Together, these cases confirm that SF1 plays a major role in the development and function of endocrine systems in humans, and reveal the importance of functional gene dosage effects when one factor regulates many different target genes.

The role of SF1 in ovarian function is less clear. The reported presence of ovaries in a girl with primary adrenal failure because of a heterozygous mutation in SF1 suggests that SF1 is not necessary

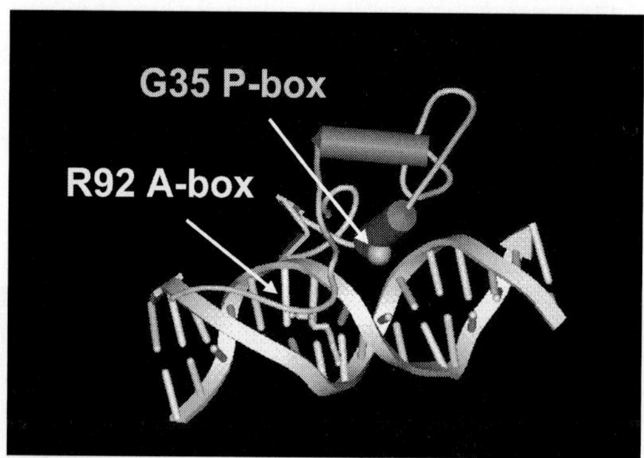

Fig. 2 Model of SF1 binding based on the crystal structure of nerve growth factor-induced-B (NGFI-B) bound to DNA as a monomer. The positions of amino acid 35, within the P-box and the amino acid 92, within the A-box, are indicated by arrows. The P-box amino acids bind to the half-site sequence (variations on AGGTCA) within the major groove of DNA, whereas the A-box is believed to bind to the 5′-flanking sequence (T/CCA) within the minor groove of DNA (reproduced with permission from [51]).

for ovarian differentiation [52]. Whether this SF1 mutation will impair estrogen biosynthesis at the time of puberty or folliculogenesis remains to be seen.

DAX-1 (NR0B1)

DAX1 is an atypical orphan nuclear receptor because of its conserved LBD, and it has an amino-terminal repeat motif structure instead of the classical DNA-binding domain. Functional studies show that DAX1 is a repressor of gene transcription and of SF1 mediated transcriptional regulation. Inactivating mutations or deletions in DAX1 cause X-linked adrenal hypoplasia congenita. More than 80 different mutations have been reported in more that 100 patients or families with this condition. Affected boys usually experience primary adrenal failure in early infancy or throughout childhood and hypogonadotropic hypogonadism (HH) emerges at puberty for boys treated with steroid replacement. DAX1 is expressed in the pituitary gonadotropes as well as the hypothalamus, and HH probably represents a combined defect at both these levels. Furthermore, studies of the *Ahch* (*Dax1*) knock-out mouse have revealed a crucial role for Dax1 in testis development and spermatogenesis. Limited data from patients with X-linked AHC suggest that DAX1 is involved in spermatogenesis in humans, too.

Most patients with X-linked AHC have nonsense or frameshift mutations that truncate the carboxyterminus of DAX1 and severely impair its function as a transcriptional repressor. Missense mutations in DAX1 are less common and cluster within the putative ligand binding domain [55]. Variant phenotypes associated with DAX1 mutations include extreme delayed puberty in female carriers of DAX1 mutations in one family [56], and HH in the absence of adrenal dysfunction in a woman homozygous for a truncation mutation in DAX1 through gene conversion [57]. An adult-onset form of X-linked AHC has also been described in two patients who experienced partial HH and mild adrenal failure [58,59]. The missense mutations found in these patients (Y380D, I439S) exhibit partial loss of function in transient gene expression assays, consistent with the mild clinical phenotype.

Small heterodimeric partner (SHP) (NR0B2)

SHP is an atypical orphan nuclear receptor that closely resembles DAX1 in its carboxy-terminal region, but lacks a conserved DNA-binding motif. SHP is expressed in the liver, intestine and pancreas, and is believed to play a role in lipid metabolism and bile acid synthesis through its interactions with related orphan nuclear receptors, such as FTF (NR5A2) and HNF4α (NR2A1). Given these associations, SHP was considered a candidate gene for patients with MODY. Five different missense mutations were found in six patients from a cohort of 173 Japanese patients with early-onset diabetes [60]. However, as all six individuals were obese, a cohort of 101 nondiabetic subjects with early-onset obesity were screened and SHP mutations were found in an additional six individuals. Thus, mutations in SHP were proposed as a cause of excess weight in a subset of Japanese subjects with mild obesity.

CONCLUDING REMARKS

Naturally occurring mutations in nuclear receptors cause a broad spectrum of endocrine, metabolic, psychiatric, and even ophthalmological conditions. These features highlight the important role this family of transcription factors play in human biology. Studying patients and families with nuclear receptor mutations provides insight into many key functional domains and mechanisms of action for these proteins, including variations in ligand responsiveness, cofactor interaction, nuclear trafficking, and DNA targeting. The combination of clinical investigation and an understanding of basic biology should help unravel whether mutations in other nuclear receptors are responsible for human disease phenotypes, as well as the possible implications of activating and inactivating mutations in the receptors described in this review. Finally, the recent reports of altered ligand-binding specificity due to mutations in the ligand-binding domains of the MR (S810L) and a somatic AR mutation found in prostate cancer (T877A) show that nuclear receptors are key targets for modulation not only by endogenous ligands, but may also be potential targets for modulation by a host of drugs and environmental agents.

ACKNOWLEDGMENTS

This work received funding from the National Cooperative Program for Infertility Research and was supported by NIH Grants U54-HD-29164, PO1 HD-21921 and GCRC grant MO1-RR-00048. JCA is the Fletcher Bequest Lecturer in Endocrinology at University College London and holds a Wellcome Trust Clinician Scientist Fellowship (068061).

REFERENCES

1. Nuclear Receptor Nomenclature Committee. *Cell* **97**, 161–163 (1999).
2. J. M. Olefsky. *J. Biol. Chem.* **276**, 36863–36864 (2001).
3. J. Duarte, G. Perrière, V. Laudet, M. Robinson-Rechavi. *Nucleic Acids Res.* **30**, 364–368 (2002).
4. J. L. Jameson. *Hormone Resistance Syndromes*, Humana Press, New Jersey (1999).
5. V. Giguère. *Endocr. Rev.* **20**, 689–725 (1999).
6. T. M. Willson and J. T. Moore. *Mol. Endocrinol.* **16**, 1135–1144 (2002).
7. K. Takeda, S. Balzano, A. Sakurai, L. J. DeGroot, S. Refetoff. *J. Clin. Invest.* **87**, 496–502 (1999).
8. A. Sakurai, K. Takeda, K. Ain, P. Ceccarelli, A. Nakai, S. Seino, G. I. Bell, S. Refetoff, L. J. DeGroot. *Proc. Natl. Acad. Sci. USA* **86**, 8977–8981 (1989).
9. J. D. Safer, M. G. O'Connor, S. D. Colan, S. Srinivasan, S. R. Tollin, F. E. Wondisford. *J. Clin. Endocrinol. Metab.* **84**, 3099–3109 (1999).
10. A. Marimuthu, W. Feng, T. Tagami, H. Nquyen, J. L. Jameson, R. J. Fletterick, J. D. Baxter, B. L. West. *Mol. Endocrinol.* **16**, 271–289 (2002).
11. S. J. Usala, J. B. Menke, T. L. Watson, F. E. Wondisford, D. B. Weintraub, J. Berard, W. E. C. Bradley, S. Ono, O. T. Mueller, B. B. Bercu. *Mol. Endocrinol.* **5**, 327–335 (1991).
12. K. Takeda, A. Sakurai, L. J. DeGroot, S. Refetoff. *J. Clin. Endocrinol. Metab.* **74**, 49–55 (1992).
13. M. R. Hughes, P. J. Malloy, D. G. Kieback, R. A. Keterson, J. W. Pike, D. Feldman, B. W. O'Malley. *Science* **242**, 1702–1705 (1988).
14. G. K. Whitfield, S. H. Selznick, C. A. Haussler, J.-C. Hsieh, M. A. Galligan, P. W. Jurutka, P. D. Thompson, S. M. Lee, J. E. Zerwekh, M. R. Haussler. *Mol. Endocrinol.* **10**, 1617–1631 (1996).
15. B. L. Riggs. *N. Engl. J. Med.* **337**, 125–126 (1997).
16. E. P. Smith, J. Boyd, G. R. Frank, H. Takahashi, R. M. Cohen, B. Speckler, T. C. Williams, D. B. Lubahn, K. S. Korach. *N. Engl. J. Med.* **331**, 1056–1061 (1994).
17. S. Sommer and S. A. Fugua. *Semin. Cancer. Biol.* **11**, 339–352 (2001).
18. Y. Shang and M. Brown. *Science* **295**, 2380–2381 (2002).
19. D. M. Herrington, T. D. Howard, G. A. Hawkins, D. M. Reboussin, J. Xu, S. L. Zheng, K. B. Brosnihan, D. A. Meyers, E. R. Bleecker. *N. Engl. J. Med.* **346**, 967–974 (2002).
20. D. M. Hurley, D. Accili, C. A. Stratakis, M. Karl, N. Vamvakopoulos, E. Rorer, K. Constantine, S. I. Taylor, G. P. Chrousos. *J. Clin. Invest.* **87**, 680–686 (1991).
21. T. Kino, R. H. Strauber, J. H. Resau, G. N. Pavlakis, G. P. Chrousos. *J. Clin. Endocrinol. Metab.* **86**, 5600–5608 (2001).
22. A. Vottero, T. Kino, H. Combe, P. Lecomte, G. P. Chrousos. *J. Clin. Endocrinol. Metab.* **87**, 2658–2667 (2002).
23. B. B. Mendoca, M. V. Leite, M. de Castro, T. Kino, L. L. K. Elias, T. A. S. Bachega, I. J. P. Arnhold, G. P. Chrousos, A. C. Latronico. *J. Clin. Endocrinol. Metab.* **87**, 1805–1809 (2002).
24. D. S. Geller, J. Rodriguez-Soriano, A. Vallo Boado, S. Schifter, M. Bayer, S. S. Chang, R. P. Lifton. *Nat. Genet.* **19**, 278–291 (1998).

25. T. Tajima, H. Kitagawa, S. Yokoya, K. Tachibana, M. Adachi, J. Nakae, S. Suwa, S. Katoh, K. Fujieda. *J. Clin. Endocrinol. Metab.* **85**, 4690–4694 (2000).
26. D. S. Geller, A. Farhi, N. Pinkerton, M. Fradley, M. Moritz, A. Spitzer, G. Meinke, F. T. F. Tsai, P. B. Sigler, R. P. Lifton. *Science* **289**, 119–123 (2000).
27. T. R. Brown, D. B. Lubahn, E. M. Wilson, D. R. Joseph, F. S. French, C. J. Migeon. *Proc. Natl. Acad. Sci. USA* **85**, 8151–8155 (1988).
28. F. J. Ghadessy, J. Lim, A. A. Abdullah, V. Panet-Raymond, C. K. Choo, R. Lumbroso, T. G. Tut, B. Gottlieb, L. Pinsky, M. A. Trifiro, E. L. Yong. *J. Clin. Invest.* **103**, 1517–1525 (1999).
29. A. R. La Spada, E. M. Wilson, D. B. Lubahn, A. E. Harding, K. H. Fischbeck. *Nature* **352**, 77–79 (1991).
30. L. Zhang, E. P. Leeflang, J. Yu, N. Arnheim. *Nat. Genet.* **7**, 531–535 (1994).
31. X.-Y. Zhao, P. J. Malloy, A. V. Krishnan, S. Swami, N. M. Navone, D. M. Peehl, D. Feldman. *Nat. Medicine* **6**, 703–706 (2000).
32. J. S. Sack, K. F. Kish, C. Wang, R. M. Attar, S. E. Kiefer, Y. An, G. Y. Wu, J. E. Sceffler, M. E. Salvati, S. R. Krystek, R. Weinmann, H. M. Einspahr. *Proc. Natl. Acad. Sci. USA* **98**, 4904–4909 (2001).
33. A. V. Krishnan, X.-Y. Zhao, S. Swami, L. Brive, D. M. Peehl, K. R. Ely, D. Feldman. *Endocrinology* **143**, 1889–1900 (2002).
34. M. Ristow, D. Muller-Wieland, A. Pfeiffer, W. Krone, C. R. Kahn. *N. Engl. J. Med.* **339**, 953–959 (1998).
35. I. Barroso, M. Gurnell, V. E. F. Crowley, M. Agostini, J. W. Schwabe, M. A. Soos, G. L. I. Masien, T. D. M. Williams, H. Lewis, A. J. Schafer, V. K. K. Chatterjee, S. O'Rahilly. *Nature* **402**, 880–883 (1999).
36. A. K. Agarwal and A. Garg. *J. Clin. Endocrinol. Metab.* **87**, 408–411 (2002).
37. P. Sarraf, E. Mueller, W. M. Smith, H. M. Wright, J. B. Kum, L. A. Aaltonen, A. de la Chapelle, B. M. Spiegelman, C. Eng. *Mol. Cell* **3**, 799–804 (1999).
38. D. Altschuler, J. N. Hirschhorn, M. Klannemark, C. M. Lindgren, M. C. Vohl, J. Nemesh, C. R. Lane, S. F. Schaffner, S. Bolk, C. Brewer, T. Tuomi, D. Gaudet, T. J. Hudson, M. Daly, L. Groop, E. S. Lander. *Nat. Genet.* **26**, 76–80 (2000).
39. K. Yamagata, H. Furuta, N. Oda, P. J. Kaisaki, S. Menzel, N. J. Cox, S. S. Fajans, S. Signorini, M. Stoffel, G. I. Bell. *Nature* **384**, 458–460 (1996).
40. M. Stoffel and S. A. Duncan. *Proc. Nat. Acad. Sci. USA* **94**, 13209–13214 (1997).
41. E. H. Hani, L. Suaud, P. Boutin, J.-C. Chevre, E. Durand, A. Philippi, F. Demenais, N. Vionnet, H. Furata, G. Velho, G. I. Bell, B. Laine, P. Frougel. *J. Clin. Invest.* **101**, 521–526 (1998).
42. C. Gragnoli, T. Lindner, B. N. Cockburn, P. J. Kaisaki, F. Gragnoli, F. Marozzi. G. I. Bell. *Diabetes* **46**, 1648–1651 (1997).
43. H. Thomas, K. Jaschkowitz, M. Bulman, T. M. Frayling, S. M. S. Mitchell, S. Roosen, A. Lingott-Frieg, C. J. Tack, S. Ellard, G. U. Ryffel, A. T. Hattersley. *Hum. Molec. Genet.* **10**, 2089–2097 (2001).
44. N. B. Haider, S. G. Jacobson, A. V. Cideciyan, R. Swiderski, L. M. Streb, C. Searby, G. Beck, R. Hockey, D. B. Hanna, S. Gorman, D. Duhl, R. Carmi, J. Bennett, R. G. Weleber, G. A. Fishman, A. F. Wright, E. M. Stone, V. C. Sheffield. *Nat. Genet.* **24**, 127–131 (2000).
45. A. H. Milam, L. Rose, A. V. Cideciyan, M. R. Barakat, W.-X. Tang, N. Gupta, T. S. Aleman, A. F. Wright, E. M. Stone, V. C. Sheffield, S. G. Jacobson. *Proc. Nat. Acad. Sci. USA* **99**, 473–478 (2002).
46. S. Gerber, J.-M. Rozet, S.-I. Takezawa, L. Coutinho dos Santos, L. Lopes, O. Gribouval, C. Penet, I. Perrault, D. Ducroq, E. Souied, M. Jeanpierre, S. Romana, J. Fezal, R. Yu-Umesono, A. Munnich, J. Kaplan. *Hum. Genet.* **107**, 276–284 (2000).

47. S. Buervenich, A. Carmine, M. Arvidsson, F. Xiang, Z. Zhang, O. Sydow, E. G. Jönsson, G. C. Sedvall, S. Leonard, R. G. Ross, R. Freedman, K. V. Chowdari, V. L. Nimgaonkar, T. Perlmann, M. Anvret, L Olson. *Am. J. Med. Genet.* **96**, 808–813 (2000).
48. Y.-H. Chen, M.-T. Tsai, C.-K. Shaw, C.-H. Chen. *Am. J. Med. Genet.* **105**, 753–757 (2001).
49. J. C. Achermann, M. Ito, M. Ito, P. C. Hindmarsh, J. L. Jameson. *Nat. Genet.* **22**, 125–126 (1999).
50. M. Ito, J. C. Achermann, J. L. Jameson. *J. Biol. Chem.* **275**, 31708–31714 (2000).
51. J. C. Achermann, G. Ozisik, M. Ito, U. A. Orun, K. Harmanci, B. Gurakan, J. L. Jameson. *J. Clin. Endocrinol. Metab.* **87**, 1829–1833 (2002).
52. A. Biason-Lauber and E. J. Schoenle. *Am. J. Hum. Genet.* **67**, 1563–1568 (2000).
53. F. Muscatelli, T. M. Strom, A. P. Walker, E. Zanaria, D. Recan, A. Meindl. B. Bardoni, S. Guioli, G. Zehetner, W. Rabl, H. P. Schwarz, J. C. Kaplan, G. Camerino, T. Meitinger, A. P. Monaco. *Nature* **372**, 672–676 (1994).
54. J. K. Phelan and E. R. McCabe. *Hum. Mutat.* **18**, 472–487 (2001).
55. J. C. Achermann, M. Ito, B. L. Silverman, R. L. Habiby, S. Pang, A. Rosler, J. L. Jameson. *J. Clin. Endocrinol. Metab.* **86**, 3171–3175 (2001).
56. S. B. Seminara, J. C. Achermann, M. Genel, J. L. Jameson, W. F. Crowley. *J. Clin. Endocrinol. Metab.* **84**, 4501–4509 (1999).
57. D. P. Merke, T. Tajima, J. Baron, G. B. Cutler. *N. Engl. J. Med.* **340**, 1248–1252 (1999).
58. A. Tabarin, J. C. Achermann, D. Recan, V. Bex, X. Bertagna, S. Christin-Maitre, M. Ito, J. L. Jameson, P. Bouchard. *J. Clin. Invest.* **105**, 321–328 (2000).
59. G. Mantovani, G. Ozisik, J. C. Achermann, R. Romoli, G. Borretta, L. Persani, A. Spada, J. L. Jameson, P. Beck-Peccoz. *J. Clin. Endocrinol. Metab.* **87**, 44–48 (2002).
60. H. Nishigori, H. Tomura, N. Tonooka, M. Kanamori, S. Yamada, K. Sho, I. Inoue, N. Kikuchi, K. Onigata, I. Kojima, T. Kohama, K. Yamagata, Q. Yang, Y. Matsuzawa, T. Miki, S. Seino, M.-Y. Kim, H.-S. Choi, Y.-K. Lee, D. D. Moore, J. Takeda. *Proc. Natl. Acad. Sci. USA* **98**, 575–580 (2001).

Pure Appl. Chem., Vol. 75, Nos. 11–12, pp. 1797–1817, 2003.

Topic 1.12

Interactions of exogenous endocrine active substances with nuclear receptors*

John A. Katzenellenbogen‡ and Rajeev Muthyala

Department of Chemistry, University of Illinois, Urbana, IL 61801, USA

Abstract: Nuclear receptors function as ligand-regulated transcription factors and modulate the expression of sets of genes in response to varying concentrations of ligands. The ligand modulators can be endogenous metabolites that function as hormones, or they can be exogenous substances, such as pharmaceutical agents or environmental substances of natural or man-made origin, which in some cases can cause endocrine disruption. Ligands modulate nuclear receptor activity by binding to their ligand-binding domains and stabilizing conformations that lead either to transcriptional activation or repression. The ligand-binding pocket is somewhat flexible, and binding affinities can be measured over a 10-million-fold range (i.e., with equilibrium dissociation constant values ranging from ca. 0.01 nM to 100 μM). Thus, it is not surprising that by binding a large variety of structures, some nuclear receptors can appear to be promiscuous; however, when affinity is considered, the binding patterns are more restricted. The spectrum of ligands that bind to the estrogen receptor has been most thoroughly investigated. Those from natural sources include natural products in food, such as soy isoflavones and whole grain lignans, as well as microbial products and components from wood. Aside from pharmaceuticals, man-made estrogen ligands can be found in industrial products, such as alkyl phenols from nonionic detergents, bisphenols from plastics, indicator dye impurities, polymer chemicals, and chlorinated aromatics and pesticides. Exogenous ligands are also known for the androgen and progesterone receptors. While it is possible that endocrine disruption can result from exogenous chemicals acting directly as ligands for the nuclear receptors, endocrine disruption needs to be considered in the broader context; thus, compounds also need to be assessed for their effects at other levels, such as on endogenous hormone production, transport, metabolism, and clearance, and at points in signal transduction cascades that are beyond the ligand–receptor interaction.

FUNDAMENTAL ASPECTS OF NUCLEAR RECEPTOR STRUCTURE AND DYNAMICS THAT DETERMINE LIGAND-BINDING AFFINITY, POTENCY, SPECIFICITY, AND EFFICACY

Nuclear receptors, sensors of endogenous small molecule ligands

Nuclear hormone receptors form a system of ligand-modulated transcription factors that control developmental programs and regulate physiological function (see ref. [1] and other articles in this topic section). The original members of this gene superfamily, the steroid, thyroid hormone, retinoid, and vita-

*Report from a SCOPE/IUPAC project: Implication of Endocrine Active Substances for Human and Wildlife (J. Miyamoto and J. Burger, editors). Other reports are published in this issue, *Pure Appl. Chem.* **75**, 1617–2615 (2003).

‡Corresponding author

min D receptors, were identified through their high-affinity binding of known endogenous hormones. The remaining members, identified by genetic sequence similarity of cloned receptors, were first considered orphan receptors, because their function and regulating ligands were unknown. Over the past several years, however, ligand regulators for many of these orphans have been discovered and their functions have been elucidated, so that most of them have now become adopted [2–5].

Most of the human nuclear hormone receptors are regulated by small molecule ligands [6,7], although some receptors can also be activated in a ligand-independent fashion that typically involves covalent modifications (i.e., phosphorylation, acetylation, glycosylation, etc.) regulated through other signal transduction pathways (i.e., growth factor receptors; see Topic 1.8). The focus of this review, however, will be strictly on nuclear receptor interaction and activation by the small molecule ligands.

The small molecule activators of most nuclear hormone receptors are generally endogenous hormonal substances, (although some receptors, the pregnane X receptor (PXR)/steroid and xenobiotic receptor (SXR) and the constitutive androstane receptor (CAR), are thought to function as xenobiotic sensors [5]). Variations in the concentration of these substances are sensed by their cognate nuclear receptors and used to regulate developmental or functional programs, or homeostatic metabolic feedback or feedforward cycles. Some of the receptors that normally respond to endogenous ligands, however, can also respond to exogenous substances, both of natural or man-made origin. It is this exogenously induced activation or inhibition that, in some cases, might negatively affect the normal pathways of development or physiological regulation, thereby resulting in endocrine disruption.

Ligand-binding affinity and the molecular basis of hormonal specificity

Some of the nuclear hormone receptors have evolved to be very discriminating, high-affinity sensors for their cognate ligands, binding them with nanomolar or subnanomolar equilibrium dissociation constants (K_d). This is the case with the steroid, thyroid hormone, retinoid and vitamin D receptors, and discrimination between these receptors for the heterologous or cross binding their ligands are typically very high. For example, the estrogen receptor discriminates against androgens with a factor of 100 000 [8]. These discrimination factors are less pronounced among receptors that share closer evolutionary relatedness, such as the members of the glucocorticoid family, which include the progesterone, androgen, glucocorticoid, and mineralocorticoid receptors. Still, discrimination factors between members of this family are typically in the range of 100–1000. Many of the adopted orphan nuclear receptors bind their endogenous ligands with much lower affinity, having K_d values typically in the micromolar range, yet they too show high specificity for interaction with their own ligands. It is of note that some orphan receptors that normally bind their endogenous ligands with micromolar affinities can, in fact, bind certain synthetic ligands with nanomolar affinities. Although the receptors have developed this high level of discrimination, many still bind a wide variety of exogenous ligands (see Topic 1.5).

Molecular basis of ligand efficacy: How ligand structure stabilizes particular receptor conformations and how this relates to agonism vs. antagonism

The manner in which ligand binding regulates the activity of nuclear receptors is intriguing: they induce distinct conformations that then engender a certain biocharacter of the receptor response. The ligand-binding domains of hormone receptors are large, consisting of around 250 amino acids [9–14]. The upper half of these domains (as they are typically oriented in representations of crystal structures) appears to be a stable, rather rigid structure, made up of three layers of criss-crossed α-helices. By contrast, the lower half of these domains, in which the ligand is accommodated, appears to be more flexible and dynamic [12,15–17].

In the absence of ligand, the lower half of the domain is shown, in different crystal structures, either to be collapsed (as in RXRγ) [12] or expanded (as in PPARγ) [16]. In either case, it is likely that apo ligand-binding domains have or can readily reach the character of a protein molten globule, that being a conformation that is condensed and partially folded, but much more conformationally dynamic a stable, fully folded protein [15]. Some receptors, in this apo state, have no effect on transcription, whereas others recruit corepressors proteins and actively repress transcription of target genes [18]. When ligands bind to this domain, they trigger the lower half of the ligand-binding domain to fold around them in various characteristic conformations that are most likely much more dynamically stable than are the apo-receptor forms [14].

The distinct conformation of the receptor that is induced by ligand binding reflects the size and shape of the ligand, and can—at least in large part—determine the agonist vs. antagonist nature of the response. Agonists typically stabilize a conformation in which the C-terminal helix, helix 12, folds back over the ligand-binding pocket and completes the formation of a deep but short hydrophobic coregulator groove into which coactivator proteins can bind via two alpha helical turns of short sequences called nuclear interaction domains [11,19]. Antagonists, by contrast, stabilize conformations in which helix 12 either is repositioned so as to block the coregulator binding groove or extends outward from the core of the domain, resulting in the expansion and widening of the coregulator binding groove in a manner that allows corepressors with three alpha helical turns of a more extended CoRNR box sequence to bind [20,21]. In these antagonist conformations, helix 12 may either be displaced actively, by a direct steric clash between large substituents on the ligand, or passively, by features of the ligand that indirectly interfere with a concatenation of surface residue interactions needed to hold helix 12 in the active conformation [22]. These agonist and antagonist conformations of helix 12, as viewed in various crystal structures, are most likely limiting conformations in a dynamic continuum that reflects the continuum of agonist vs. antagonist activities of various nuclear receptor ligands [22].

Modulation of ligand potency and efficacy by other cellular interactions

The binding affinity that a ligand has for the nuclear hormone receptor and the conformation that it induces in the receptor are probably the major determinants of its potency and biocharacter. Nevertheless, it is clear that other interactions can modulate both of these characteristics. For example, estradiol, which has a K_d for binding the estrogen receptor of ca. 0.2 nM, can stimulate the growth of MCF-7 cells at concentrations as low as 0.001 nM, whereas stimulation of the induction of progesterone receptor in these cells requires estradiol concentrations of 0.1–1 nM. Both responses require the binding of estradiol to the same estrogen receptor, but the potency of estradiol in effecting these responses, is very different. Thus, it is clear that the potency of an estrogen in inducing a specific biological response is not regulated simply by its binding affinity for the receptor, but can also be modulated by post-receptor interactions or other post-receptor rate-limiting events, which can collectively be considered part of the effector system for the estrogen receptor. The effector components are now appreciated to include a set of cellular coregulator proteins as well as promoter-specific factors [1,23]. Conceptually, the multicomponent nature of nuclear hormone receptors has been termed "tripartite receptor pharmacology" [23].

Tripartite receptor pharmacology is a complex issue, but one tenet is that the potency of a hormonal ligand is determined not only by its affinity for the receptor, but also by the manner in which the ligand-receptor complex is coupled with the effector system required to produce a given response. Coregulator binding has, in fact, been shown to modulate the kinetic stability of ligand-receptor complexes [24] and of receptor dimers (Tamarzi, et al., in press). Promoter context has an influence on receptor conformation [25] and coregulator selection by ligand-receptor complexes [26].

A second tenet of tripartite receptor pharmacology is that the biocharacter of a ligand, that is its activity as an agonist vs. antagonist, is likewise not simply a direct consequence of the nature of its interaction with its receptor, but can also be modulated by the cellular context of coregulator proteins and by the nature of the promoter system through which the response is mediated [23]. Again, this can be illustrated with ligands for the estrogen receptor, where the biocharacter of a compound such as tamoxifen varies from that of a nearly pure antagonist in the breast, to a partial agonist in the uterus, to a nearly pure agonist in liver, bone and the cardiovascular system [27]. Thus, the action of tamoxifen in these tissues is mediated through the estrogen receptor, yet the responses, in terms of agonism vs. antagonism, differ in a tissue/response-specific manner [26,28].

Thus, according to tripartite receptor pharmacology, both the potency and the biocharacter of a ligand are not fully determined just by the way in which the ligand interacts with the receptor, but they are also modulated by the manner in which the whole receptor-effector-response unit is set up to interpret the ligand.

Nature of the ligand-binding pocket and the molecular basis for hormone disruption

Whereas the effect that the ligand induces on the *exterior* topology of the ligand-binding domain is thought to be the principal determinant of ligand efficacy and its interaction with other cellular constituents, it is the details of ligand interaction with the *interior* of the ligand-binding pocket that underlies, fundamentally, ligand-binding affinity and hormonal potency. As was mentioned above, the ligand-binding pocket is ill formed in the absence of ligand [12,15,16], and the ligand itself forms a core unit—typically hydrophobic—around which the lower zone of the ligand-binding domain folds as the ligand is bound [14]. Ligand binding is obviously a complex and dynamic process that involves a lot of protein movement; this is reflected in the slow rates of ligand association with receptors. On the scale of protein–small molecule association rates, which are typically diffusion limited (k_{assoc} ca. 10^8 $M^{-1}s^{-1}$), the rate at which small molecule ligands associate with nuclear receptors is very slow (k_{assoc} ca. 10^4 $M^{-1}s^{-1}$) [29].

To some degree, the size and shape of the pocket that forms around the ligand is determined by the size and shape of the ligand. The interior of the ligand-binding pocket exhibits a considerable degree of plasticity, and reshapes itself around the contours of the ligand. The packing between protein and ligand, however, is not uniformly tight, and considerable voids can exist [9]. The sum of the volumes of these preformed or residual pockets can, in some cases, be comparable to that of the ligand. For example, with the estrogen receptor-α, X-ray structures show empty pockets above the 11β and below the 7α sites of the endogenous ligand estradiol that have a total volume of 200 $Å^3$, whereas the ligand volume is not much larger (250 $Å^3$) [9]. The existence of these preformed pockets around the ligand had been predicted by a global analysis of the binding affinity of substituted steroidal estrogens [30]. The ligand-binding pockets of nuclear hormone receptors are also typically very hydrophobic, with the few internal polar residues being positioned to interact with the few polar functions on the hormones.

It is not surprising, then, that the nuclear hormone receptors, with their large, flexible, and hydrophobic ligand-binding pockets, are able to interact with ligands that span a wide range of sizes and structures. It is this "specificity tolerance", which has sometimes been called "promiscuity" [31], that underlies the sensitivity of this receptor system to disruption by exogenous ligands [32]. The affinity with which exogenous ligands bind to these receptors depends, of course, on their structure and their functional nature, and the degree to which they deviate in size, shape, functional nature, and hydrophobicity from the natural high-affinity ligands. The relationship between ligand structure and binding affinity in nuclear hormone receptors, however, is not always a smooth one, where predictions can be made with confidence. In some cases, small changes in structure and stereochemistry can have large effects on ligand binding, yet molecules of different size and shape can have comparable affinities [30,33–36]. Thus, the structure-binding affinity relationships for these receptors are perhaps best termed "eclectic", rather than promiscuous.

IMPLICATIONS OF THE 10-MILLION-FOLD RANGE OF ENDOCRINE BINDING AND ACTIVITY ASSAYS

In considering the potential health effects of endocrine-disrupting substances on humans and wildlife, it is important to keep a focus on the issue of effective dose, which is, in essence, the product of potency and exposure. The exposure factor (see other contributions to this volume) will not be considered in this review, but one needs to appreciate that long and/or high exposure to a low-affinity ligand might be as effective as short or low exposure to a high-affinity ligand. Potency (as well as its biochemical cousin, affinity) is a critical factor to understand; it can be estimated by binding and bioactivity assays. These assays are not difficult to devise (for those receptors that bind ligands with nanomolar affinity). It is important to note that such assays are typically capable of detecting compounds that differ from one another in affinity and potency by a *factor of up to 10 000 000* (10^7) (Fig. 1).

For example, a simple competitive binding assay with the estrogen receptor might use [^{3}H]estradiol as tracer, a natural ligand that has a K_d of 0.2 nM or a relative affinity of 100 %. Such an assay can detect very high affinity estrogens, such as 11β-chloromethylestradiol, having a K_d of 6.7 pM and a relative affinity of 3000 %, or very low affinity estrogens, such as nonylphenol, with a K_d of 70 μM and a relative affinity of 0.0003 %. Similarly, a cell-based assay for estrogens based on the proliferation rate of breast cancer (MCF-7) cells shows a response to a potent compound such as estradiol at 1 pM, whereas the response to weak estrogens can often be measured by raising concentrations to as high as 10 μM without encountering nonspecific toxic effects. Certain physiological responses to estrogens that form the basis of whole animal assays, such as the uterine weight gain assay in immature female rats, are sensitive to potent estrogens such as 17α-ethinyl estradiol at doses as low as 0.01 μg, whereas doses as high as 10 mg of many compounds can be safely administered animals.

The point to reemphasize here is the following: The size of the universe of compounds that might be considered active "endocrine disruptors" or "exogenous endocrine substances" depends on the potency or affinity range that is considered. For example, the number of compounds and the range and diversity of structures of the compounds that are capable of activating the estrogen receptor are rather small if one considers only those that have a potency within a factor of 10 that of estradiol, whereas they are enormous if one considers all compounds whose potency is within a factor of 10^5 or 10^6 that of estradiol. Thus, comments that a receptor such as the estrogen receptor is "promiscuous" because it responds to such a wide range of compounds has little meaning if these statements are made without considering potency.

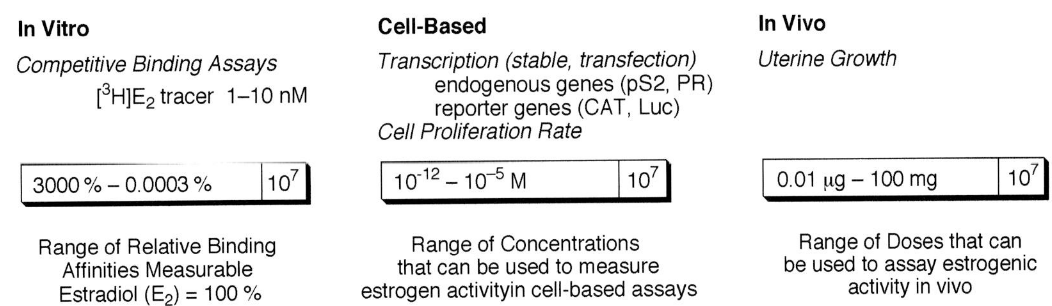

Fig. 1 Assays for estrogenic activity have high sensitivity and a wide working range.

SEX STEROID RECEPTOR INTERACTIONS WITH EXOGENOUS ENDOCRINE SUBSTANCES

In the sections that follow, we present a brief summary of compounds, from a variety of sources, which have been reported to be exogenous endocrine active substances through interaction with various nuclear hormone receptors. An attempt has been made to illustrate the range of structures of such substances and the diverse sources from which they arise. In this short review, however, it has not been possible to critically evaluate the validity of all of the potencies and each of the activities that are reported. Where information on potency or affinity is readily available (and appears to be reliable), some values are noted in the figures. For ease of comparison, binding affinity values have been converted to a relative binding affinity (RBA) scale, where the affinity of a compound is given as a percent of a reference ligand, which is considered to have a relative affinity of 100.

Estrogen receptor as a target for endocrine disruption

By far, the most intensively studied target for interaction with exogenous ligands and endocrine disruption is the estrogen receptor [32]. This receptor has a general tolerance for binding appropriately substituted phenols and other classes of lipophilic compounds [30,33–38]. There are two subtypes of the estrogen receptor (ERα and ERβ), but most studies on estrogens were done before the second subtype (ERβ) was known [39]. Binding affinity values of compound for ER are given on an RBA scale, and unless they are associated specifically with ERα or ERβ (i.e., numbers denoted with an α or β), they refer to studies done on extracts from estrogen target tissues (typically uterus), which is contains predominantly ERα.

Mammalian estrogens and estrogen pharmaceuticals

Estradiol, estrone, and estriol are the principal endogenous mammalian estrogens, though some more highly hydroxylated estrogens (estetrol, produced by the fetus) and some B-ring unsaturated estrogens (equine estrogens) are known. Many synthetic estrogens, both steroidal and nonsteroidal, have been prepared in the search for estrogen pharmaceuticals, and a variety of structural alterations have been made. The ER tolerates certain substitutions on the steroidal estrogen skeleton very well; in fact, some substitutions increase estrogenic potency by enhancing binding to the estrogen receptor (e.g., 11β-ethyl or 11β-chloromethyl), or by retarding metabolism (e.g., 17α-ethynyl or 11β-methoxy). The estrogen receptor will also tolerate very large substitutents at the 7α position, as is manifest in certain antiestrogens. Curiously, a phenolic ring is not an absolute requirement for an estrogen receptor ligand; certain androgen derivatives have significant binding affinity [40–42]. Examples of natural estrogens and estrogen pharmaceuticals are given in Fig. 2.

There are many classes of synthetic, nonsteroidal estrogens. From this variety of ligand structures, it may appear that the estrogen receptor system is unique among the steroid hormone receptors in its tolerance for nonsteroidal ligands, but this is more likely a quirk of history: It was shown as early as 1936 that the simple bisphenol diethystilbestrol was a potent estrogen [43], and this finding—nearly 60 years ago—prompted an active search for other nonsteroidal estrogens that has spanned many decades and produced many active agents, some estrogens and some antiestrogens [33–36]. In contast, the investigation of nonsteroidal ligands for other steroid hormone receptors (see below) has been a more recent endeavor, although it is clear that many potent nonsteroidal ligands exist for these other system, as well [44].

A number of nonsteroidal estrogens have been developed as tissue-selective estrogens, now better known as selective estrogen receptor modulators or SERMs [27]. These agents are used for breast cancer prevention and treatment [45], and they hold promise for menopausal hormone replacement with a better benefit/risk balance [27].

Because endogenous estrogens are hormones produced naturally and are essential for development, health, and reproduction, and because estrogen pharmaceuticals are administered for their hormonal effect in regulating fertility and for menopausal hormone replacement, these compounds are not typically thought of as potential endocrine disruptors in humans. Nevertheless, there is evidence that some of these compounds can find their way into rivers and streams and can affect aquatic wildlife [46].

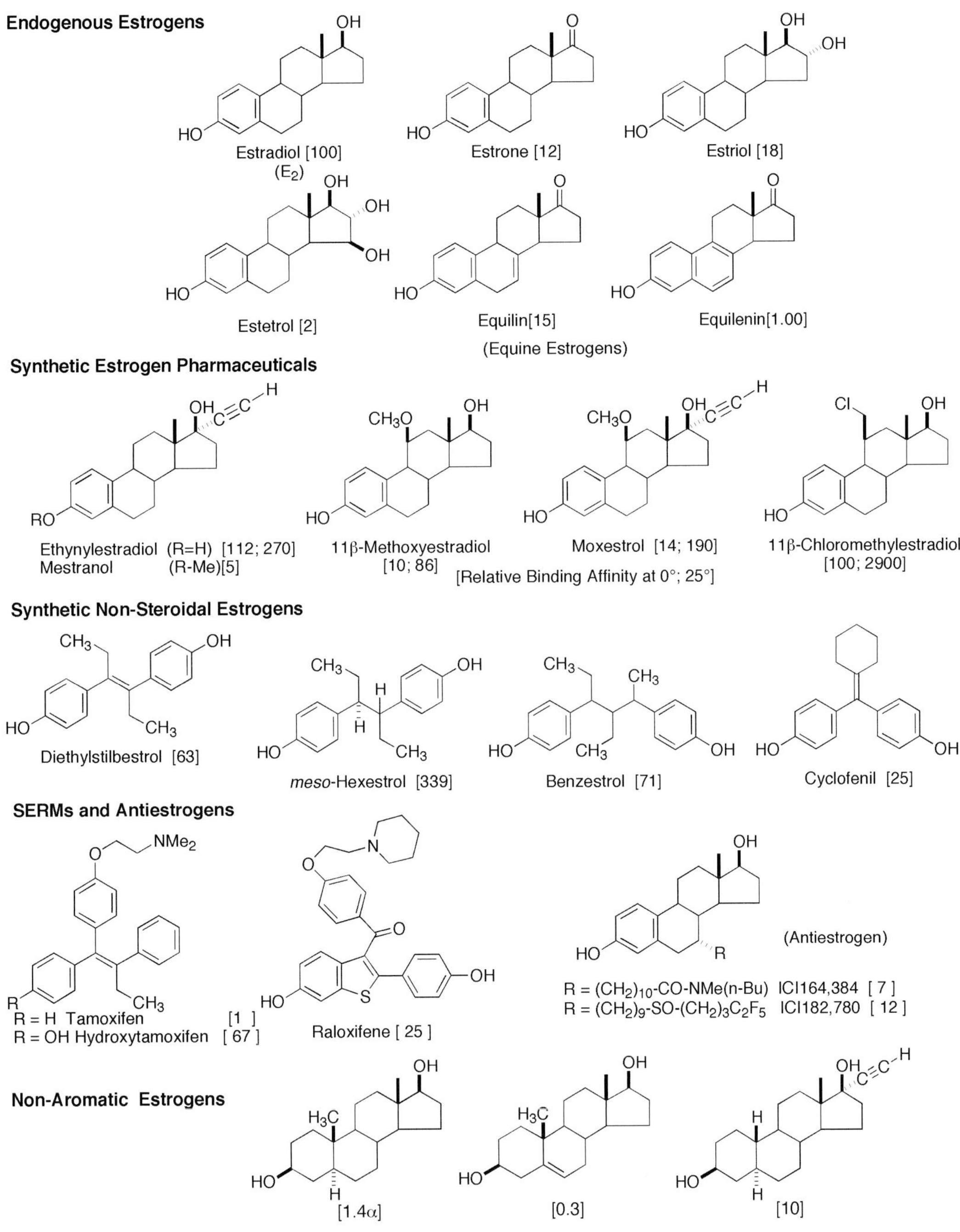

Fig. 2 Endogenous estrogens and estrogen pharmaceuticals.

Typical sewage treatment processes do not consume all estrogen pharmaceuticals (such as mestranol) and their metabolites that are present in human excrement; even endogenous human estrogens (such as estradiol and estriol) do not seem to be fully degraded by the standard methods of sewage treatment [46,47]. The use of certain other estrogen pharmaceuticals such as diethylstilbestrol and hexestrol and mycoestrogens (such as zeranol, see below) as anabolic agents in livestock [48,49] or as chemical castration agents in fowl can result in residual levels of hormone in meat or animal byproducts through which other animals and humans can be exposed [50,51].

Estrogen pharmacophore

A general estrogen pharmacophore might be considered to be a phenol substituted in the 3 and/or 4 positions with a variety of cyclic and acyclic carbon substituents and oxygen functionalities having a total carbon count in the range of 15–20 and generally lacking in highly polar or charged groups [30,33–38]. Within this paradigm, there are nearly endless varieties of polycyclic, acyclic, macrocyclic, and heterocyclic examples [33,52]. There are also estrogens that deviate from this general pharmacophore by lacking a phenol, but these typically have a phenol surrogate, such as an alcohol, hydrated carbonyl function, or sulfoxide group, and they are also usually very hydrophobic, often the result of multiple halogen substitution (see below).

Nonphenolic estrogens are typically of low potency, though some, especially the polychlorinated compounds, might be slow to be metabolized and cleared, and thus subject to bioconcentration, thereby amplifying the potential exposure of these agents to those at the end of the food chain [53]. In addition, because of their lipophilicity, these compounds can accumulate to high levels in lipids and membranes from which they can be released slowly, so as to provide a low, persistent level of compound in blood. Such continuous dosage might be particularly effective in stimulating certain estrogenic responses [54,55].

In the figures that follow are examples of compounds, reported to be estrogens, that come from a variety of natural and man-made sources and that are potential endocrine disruptors.

Exogenous estrogens from natural sources

Phytoestrogens, naturally occurring compounds in plants that mimic steroidal estrogens [56], are comprised of several classes, isoflavones, flavones, lignans, coumestans, and stilbenes (Fig. 3). Isoflavones are moderately potent estrogens found in legumes, the best known being genistein and daidzein from soy [57,58]. Genistein is one of the few ER ligands that shows rather selective binding for ERβ over ERα [59]. Considerable quantities of these agents are ingested by humans on "Asian diets", and they are believed to be primarily responsible for the beneficial effects of soy [60]. In fact, extracts that are very rich in soy isoflavones (e.g., NovaSoy®) are being promoted as nutritive supplements. Certain reduced isoflavones, such as equol, are estrogenic substances produced endogenously from isoflavones by the action of gut microflora [57].

Lignans are compounds that contain a 2,3-dibenzylbutyrolactone or butandiol structure and exist in minor quantities in plants that use them as building blocks for preparation of lignin, the major constituent in a plant cell wall [61]. The most abundant lignans, secoisolariciresinol and matairesinol, are found as glycosides in certain oilseeds, whole grains, berries, and some vegetables. In the colon, they are deglycosylated and further converted, by the action of intestinal microflora, to the weakly estrogenic enterodiol and enterolactone, which are reported to be very weak estrogens (no affinity values are available) [61]. Resveratrol, a stilbene triol found in the skin of grapes and in red wine, is a low potency estrogen [62]. The coumestans, coumestrol and mirestrol, found in certain clover, were first recognized as potent phytoestrogens as a result of endocrine disruption on sheep [63].

Human exposure to estrogens such as phytoestrogens that are natural components in healthy food is essentially unavoidable, and compounds such as genistein are generally considered to be a health benefit [60]. These natural estrogens are low-affinity, low-potency ligands for the estrogen receptor, and it is known that certain weak, short-acting estrogens, such as estriol, as well as other "impeded" estrogens, can reduce the effect of potent estrogens such as estradiol when they are given in an appropriate dose regimen. The net effect of such treatment can be the antagonism of an estrogen-responsive system that could be protective of health [54,55,64]. Nevertheless, the health consequences of exposure to

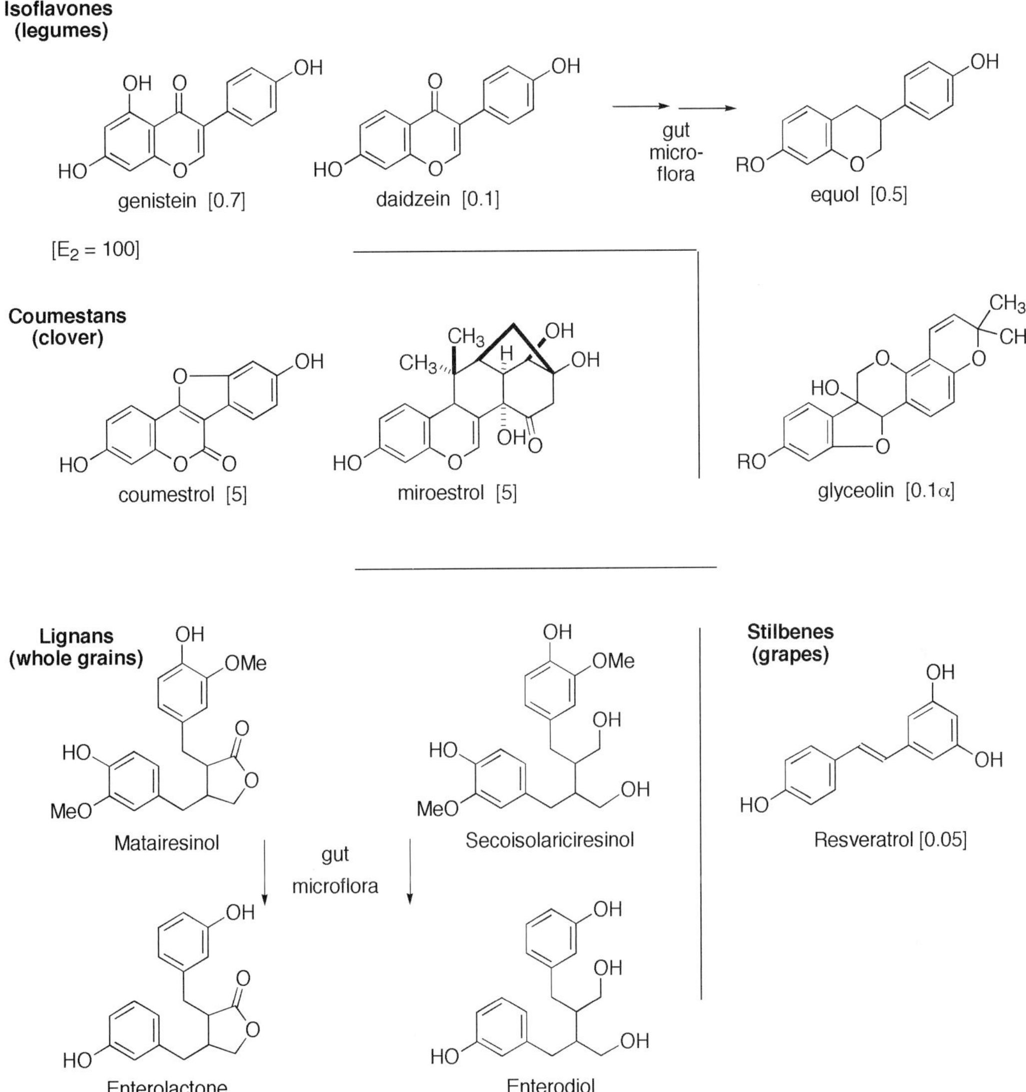

Fig. 3 Exogenous estrogens from natural foodstuff sources.

phytoestrogens is not likely to be a simple matter, and in certain situations, such as for women who have or are at high risk for breast cancer, phytoestrogen exposure could pose a risk, particularly if exposure is amplified by ingestion of large amounts of food extracts in which these substances are highly concentrated [65].

Some of these compounds also have other biological activities that might have health consequences. For example, genistein is a tyrosine kinase and topoisomoerase inhibitor and has some antiangiogenic activity so it that might be considered to have anticancer activity [60]. It should be noted, however, that the potency of genistein as an antiangiogenic agent is much less than its potency as an estrogen [66].

Certain microbial metabolites, termed "mycoestrogens", have estrogenic activity (Fig. 4) [67]. Most notable are resorcylic acid lactones produced by *Fusarium* [68], a mold that can infect corn and other forage crops, and has caused endocrine disruption in pigs. Zearalenone, the parent compound, is

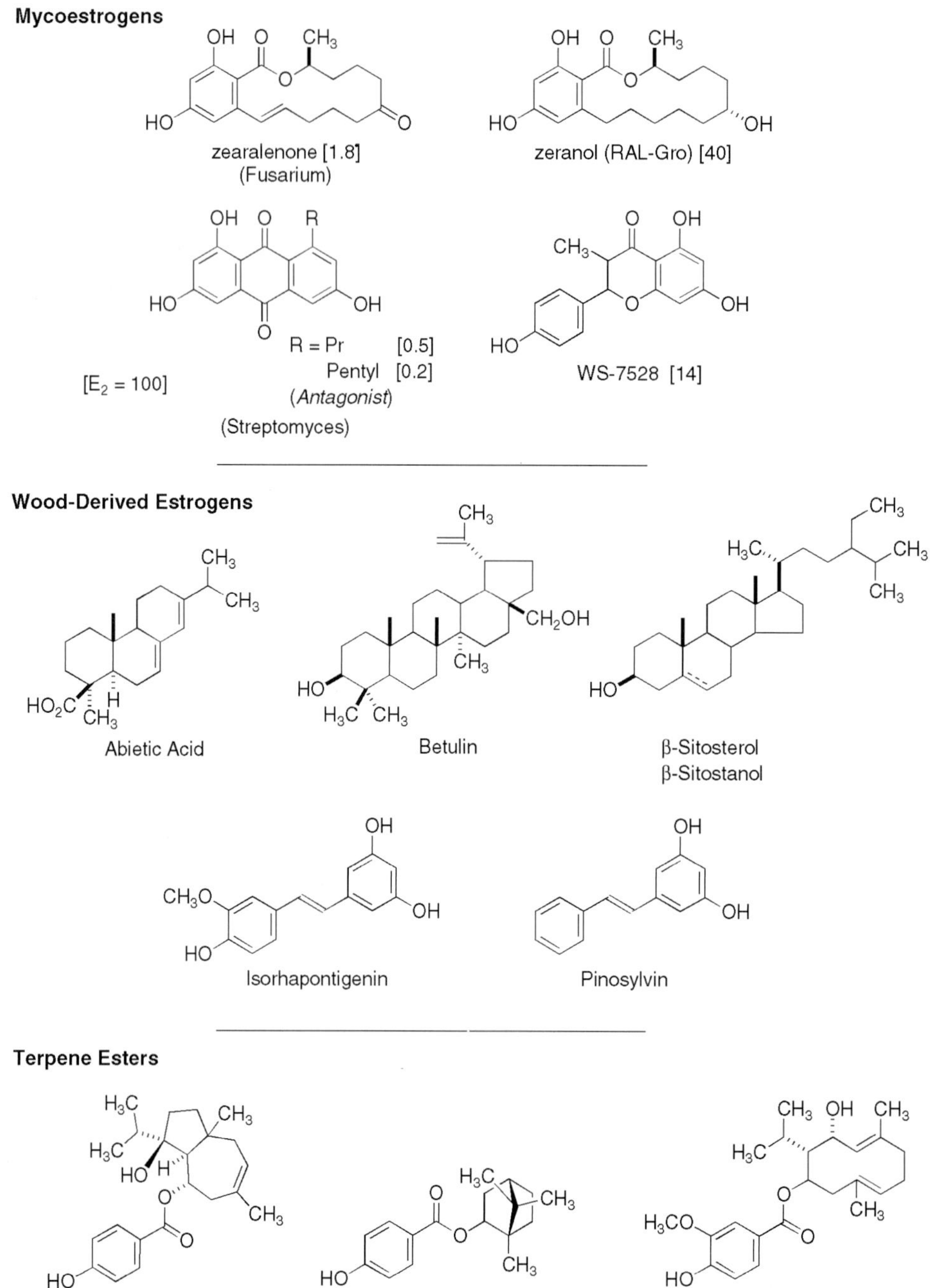

Fig. 4 Exogenous estrogens from other natural sources.

quite a potent estrogen [69], and some of its reduced analogs (Zeranol®, Ralgro®) have been used as anabolic agents in cattle [49]. Several anthraquinone alcohols from *Streptomyces* were found to bind to ER and to be antagonists (less potent than tamoxifen) [70].

Some phenolic and nonphenolic compounds derived from wood have been reported to be estrogenic in cell proliferation assays. Abietic acid and β-sitosterol are interesting nonphenolic examples, and the stilbenes isorhapontigenin and pinosylvin are close structural analogs of resveratrol (see above) [71]. The potency reported for some of these compounds as estrogens seems higher than expected, based on their structures (affinities are not reported) [71].

Terpenoids are less well known as estrogen receptor ligands, but recently, sesquiterpenoids derived from the umbelliferae family have been added to the growing list of phytoestrogens [72]. The terpene benzoate ester ferutinine had good affinity for ERα and ERβ, but others, tschimgine and tschmiganidine, have lower affinity. Interestingly, in transcription assays, ferutinine was a full agonist for ERα and an agonist/antagonist for ERβ, whereas tschmiganidine was an agonist only on ERα and an antagonist on ERβ.

Exogenous estrogens of man-made origin

An intriguing variety of man-made compounds, or their metabolites, have been found to have estrogenic activity. Though their potency is rarely high, some are very lipophilic and potentially environmentally persistent compounds; others come from surprising sources.

Estrogenic activity among certain commercial chemicals (Fig. 5) has been found, such as *p-tert*-alkyl phenols, most notably octylphenol (a single isomer) and nonylphenol (an isomer mixture). Such phenols form the hydrophobic core of the polyethoxylate nonionic detergents. Because the free phenols are released by the action of microbes during sewage treatment, they can be present into waste streams where they can be exposed to aquatic wildlife [73]. Estrogenicity is greatest when the substituent is branched at the benzylic position, para to the phenol, and of a certain size [73].

Certain components of plastics, dialkyl phthalates used as plasticizers and bisphenol A, a component of thermostable polycarbonate polymers, have been reported to be weak estrogens [74,75]. Bisphenol A was first prepared as an estrogen [43] and only later was used in the production of plastics. It is released into the contents of polycarbonate bottles when they are heated in an autoclave sterilizer, and at the concentration thus released, it can stimulate estrogen-responsive cells in culture [76]; its presence in polycarbonate baby bottles has raised concern [77]. Curiously, bisphenol A is found in recycled paper, but not in virgin paper [78]. Similarly, a bisphenolic impurity present at very low levels in commercial grade samples of phenol red, a widely used component of cell culture media, is a high-affinity estrogen and was found to be the cause of inadvertent stimulation of estrogen-responsive cells [79–81]. Hydroxylated azobenzenes bear structural resemblance to stilbene estrogens and have significant affinity for the estrogen receptor [82].

More curious are examples of estrogenic substances found in cosmetics. In vivo estrogenic activity was reported for a cyclic tetrasiloxane, which was a minor constituent of a polymer mixture prepared as a moisturizing cream [83]. A related linear siloxane, also containing a 1,3-diphenyl moiety, has been reported to be weakly estrogenic [84]. Certain ingredients in commercially available sunscreens, notably 4-methylbenzylidene camphor and octyl 4-methoxycinnmate, have been implicated as estrogens [85–88].

Quite a number of highly chlorinated aromatic substances, insecticides and industrial fluids, have been found to have estrogenic activity, and some have been implicated as endocrine disruptors (Fig. 6). *o,p*′-DDT, a minor isomer in commercial DDT, its metabolite *o,p*′-DDE, and the nonbioaccumulating analog methoxychlor and its phenolic metabolites are all weak estrogens [89,90]. The net estrogenic activity of *o,p*′-DDT and DDE and their endocrine disruptive effects (as well as that of the related *p,p*′ isomers) may, however, be the result of their antiandrogenic, rather than their estrogenic activity [91,92]. PCBs, which are typically mixtures of many isomers with varying degrees of chlorine substitution, are reported to be estrogenic [93–95]. The highest potency is found with PCB metabolites that have a *para*-hydroxy group [93,96,97], but RBA values are still quite low. Several hydroxylated PCBs, particularly those with chlorine substitutents at positions 2′, 4′ and 6′ on the nonphenolic ring, exhibit antiestrogenic effects in MCF-7 cells [98].

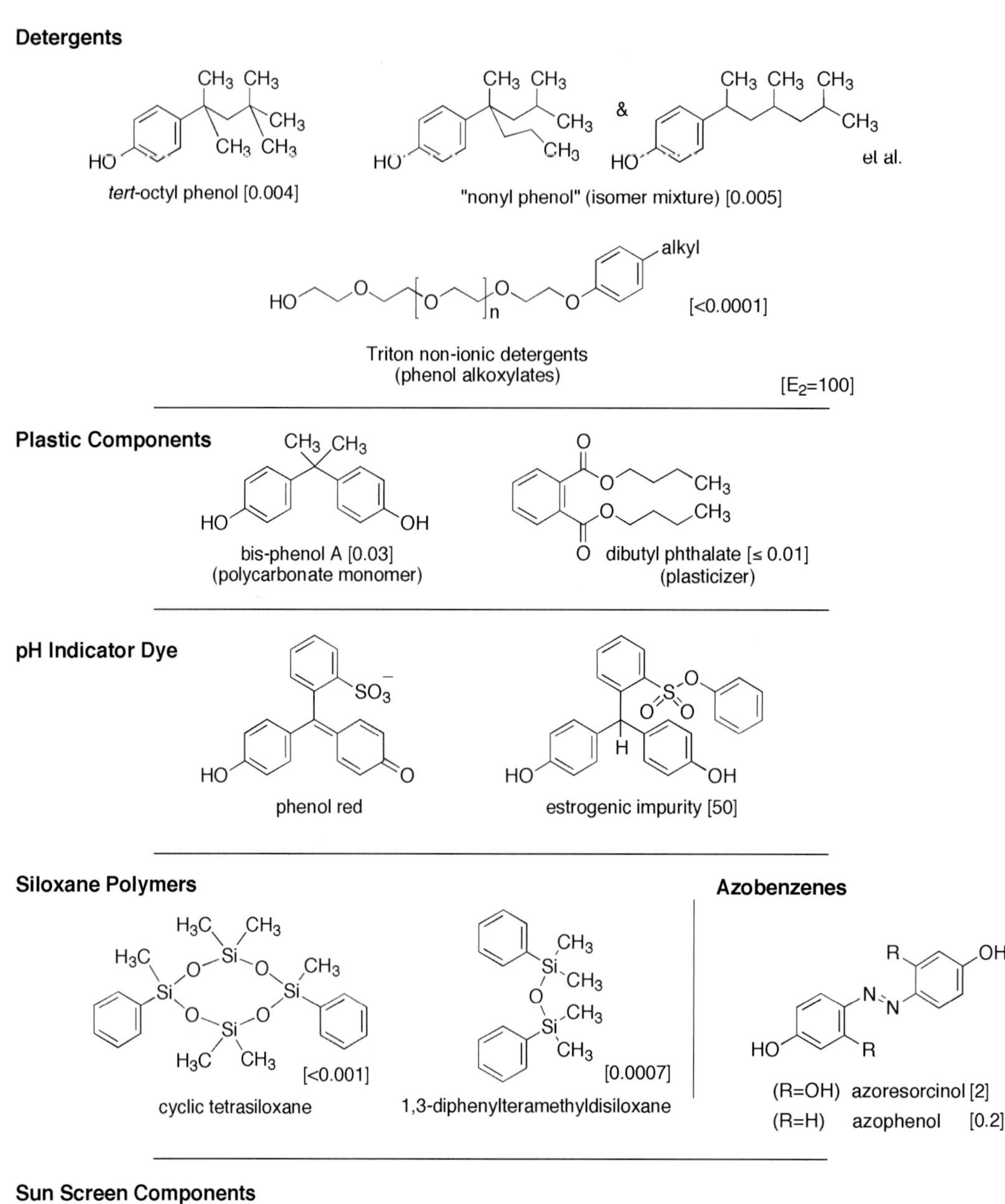

Fig. 5 Exogenous estrogens: Commercial chemicals and impurities.

A variety of nonaromatic chlorinated pesticides are reported to have estrogenic activity (Fig. 6). Most remarkable is that all of these lack not only the signature phenol of most estrogens, they are not even aromatic. Endosulfan, dieldrin, and kepone/chlordecane (a metabolite of the persistent fire ant insecticide mirex) are all polycyclic and heavily chlorinated [99,100], and though they are not phenolic,

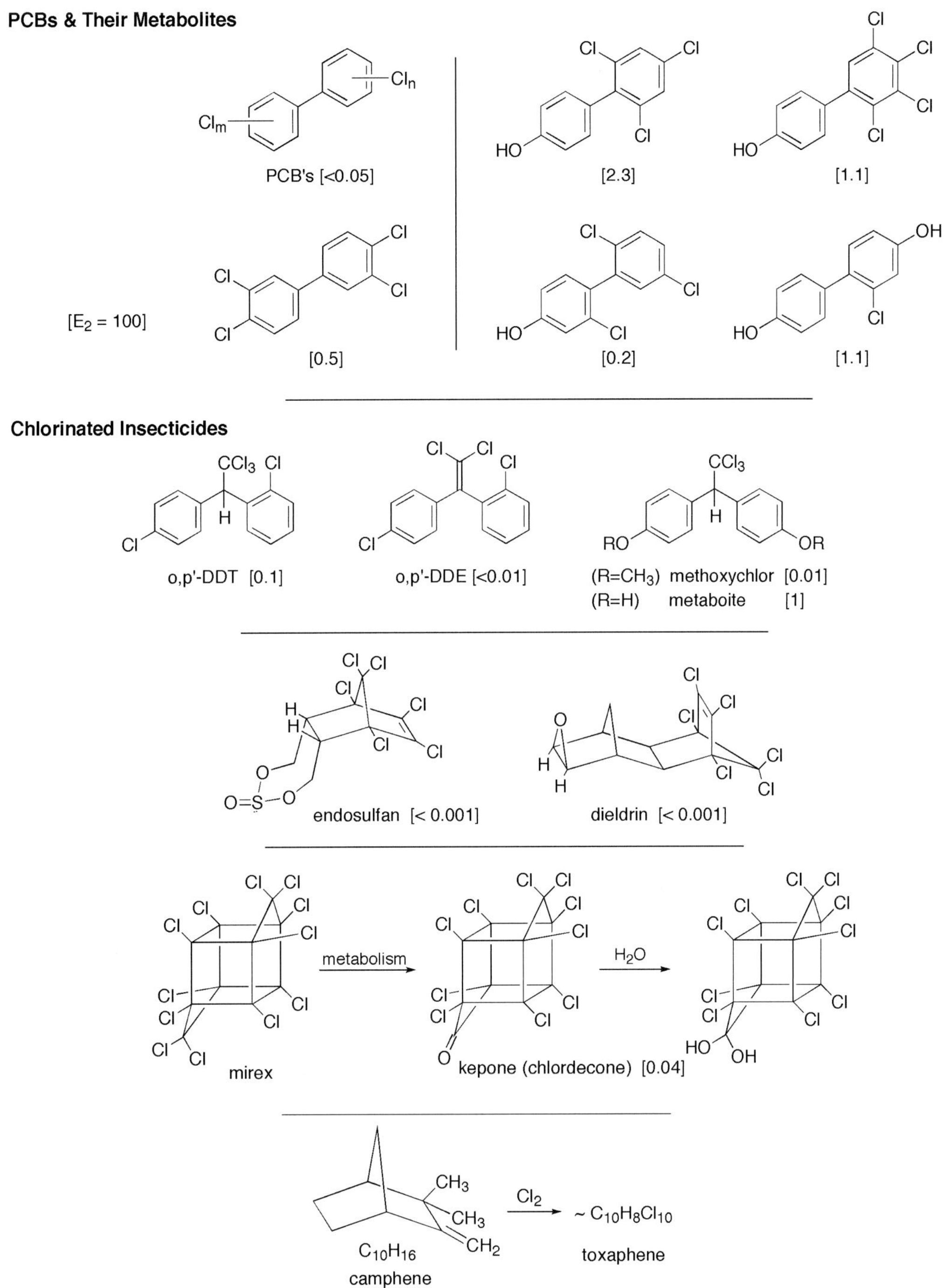

Fig. 6 Exogenous estrogens: Highly chlorinated.

they each have a polar function that might fulfill this function, e.g., a sulfite in endosulfan and an epoxide in dieldrin. The ketone function in kepone is most likely hydrated as a gem-diol, and this may be the functional equivalent of the phenol, although the overall structure of kepone certainly deviates markedly from that of typically estrogens. Lindane, a mixture of hexachlorocyclohexanes, and toxaphene, a complex mixture of chlorinated terpenes derived from camphene, have been reported to have estrogenic and antiestrogenic activity [101–103]; however, these materials bear no obvious structural or functional relationship to other estrogens, save that they are polychlorinated like some other compounds that are weak estrogens.

All of these compounds are very low potency estrogens, but they are exceedingly lipophilic agents (based on water–octanol partition coefficient estimates), and they are very persistent in the environment. It was proposed that these low-potency estrogens might engage in a potency-intensifying synergy [104]; however, this contention was not widely supported, and it has been withdrawn [105]

Androgen receptor as a target for endocrine disruption

The androgen receptor is also recognized as a target for endocrine disruption [91,92,106]. As with the estrogens, steroidal ligands for this receptor with improved potency and in vivo stability have been developed as pharmaceuticals agents, and a number of nonsteroidal ligands are known to be androgen antagonists, some of which are used clinically for the treatment of prostate cancer (Fig. 7) [44]. Certain phytochemicals have been shown to have weak androgenic and antiandrogenic activity [107]. Curiously, the fungicide vinclozolin is an androgen antagonist, and one of its metabolites whose struc-

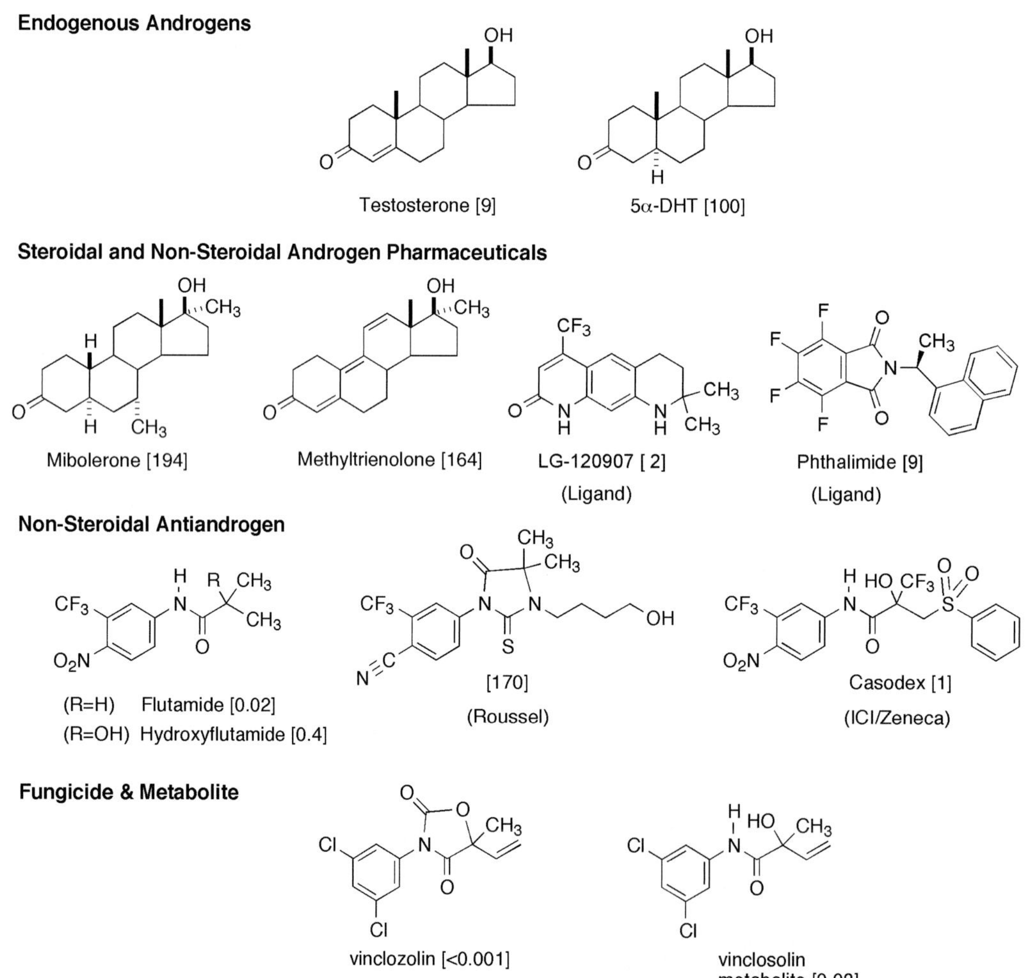

Fig. 7 Androgen receptor ligands.

ture resembles that of some nonsteroidal antiandrogens has significant affinity for the androgen receptor [108]. Also, as was mentioned above, the endocrine disruptive activity of the DDT and DDE isomers is thought to be the result of their antiandrogenic activity [92,106,108]. The waste water from wood pulping has been associated with androgenization of aquatic species [109], and the causative agents appears to be steroidal androgens produced by the action of bacteria on sterols from wood [110,111]. For illustrative purposes, various ligands for the progesterone receptor, both natural, synthetic, and nonsteroidal, are shown to illustrate the diversity of chemical structures that bind to this receptor with high affinity (Fig. 8). As far as we are aware, there are no examples of exogenous progestins causing endocrine disruption.

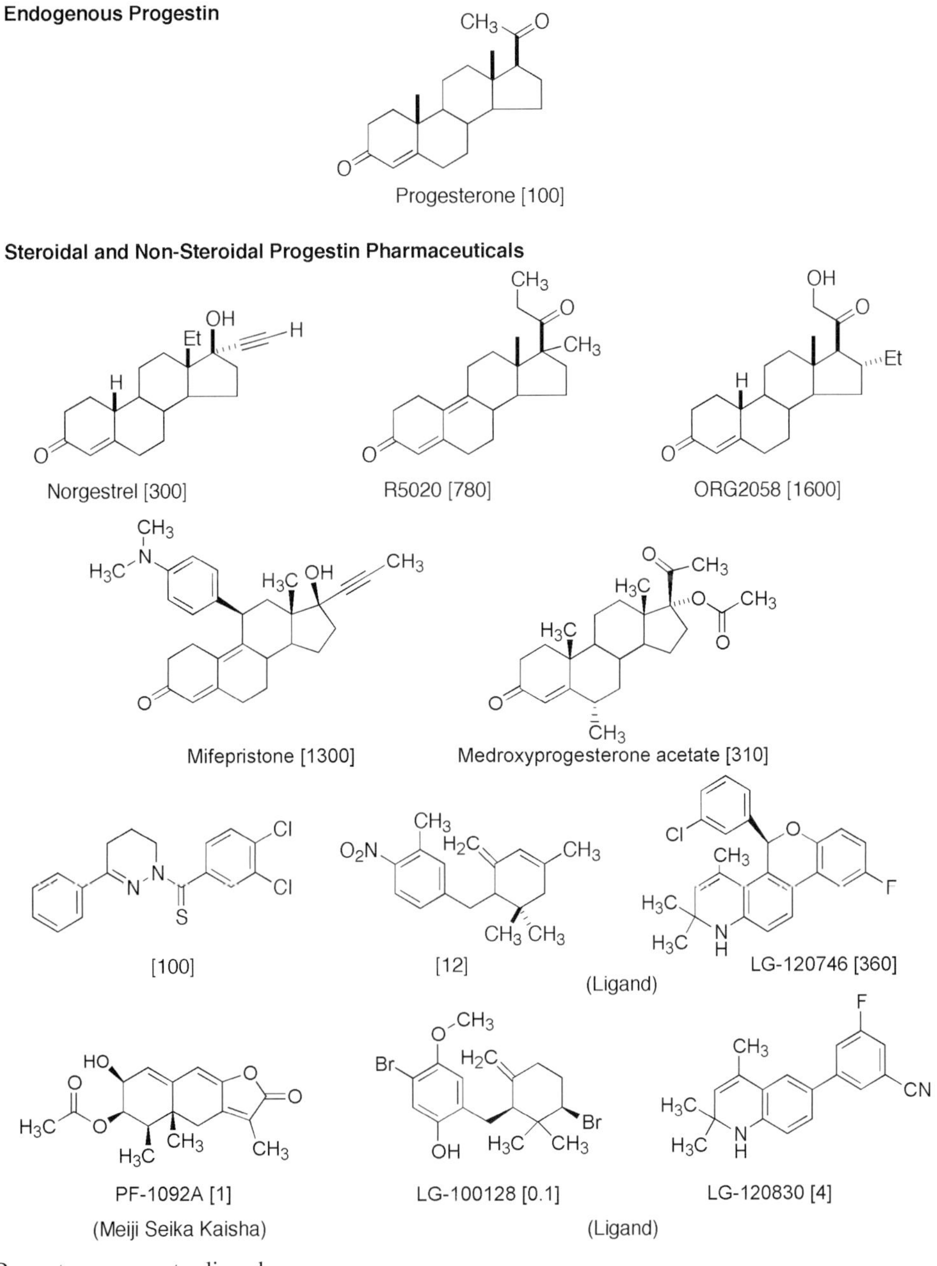

Endogenous Progestin

Progesterone [100]

Steroidal and Non-Steroidal Progestin Pharmaceuticals

Norgestrel [300]

R5020 [780]

ORG2058 [1600]

Mifepristone [1300]

Medroxyprogesterone acetate [310]

[100]

[12]

(Ligand)

LG-120746 [360]

PF-1092A [1]

(Meiji Seika Kaisha)

LG-100128 [0.1]

(Ligand)

LG-120830 [4]

Fig. 8 Progesterone receptor ligands.

Other nuclear receptors as targets for endocrine disruption

There is much less in the literature on endocrine disruption through other nuclear hormone receptors. PCBs and other halogenated aromatics affect the thyroid hormone system, but this appears to result from the displacement of endogenous thyroid hormones from plasma carrier proteins and activation of liver enzymes, both of which enhance the rate of hormone metabolism and clearance, rather than from direct binding of these agents to the thyroid hormone receptor [97,112–114]. Certain phytochemicals are reported to affect the progesterone receptor [107,115]. It has been proposed that developmental abnormalities in frogs arise from endocrine disruption acting through the retinoid receptors. Paper mill effluent is reported to inhibit retinoid signaling, but the causative agents have not been identified [116,117].

The Ah receptor is also a transcription regulator whose activity is regulated by small molecule ligands, though it is from a completely different structural class than the nuclear hormone receptors. It also appears to be a target for endocrine disruption, and there is evidence for negative cross talk between the Ah receptor and other nuclear hormone receptors, such as the estrogen receptor. (See Topic 1.13.)

NUCLEAR RECEPTORS IN CONTEXT: THE MULTIPLE LEVELS AT WHICH EXOGENOUS SUBSTANCES CAN EXERT AN EFFECT BY PERTURBING AN ENDOGENOUS ENDOCRINE SYSTEM

The preceding review displaying the wide variety of compounds—from different chemical classes and widely divergent sources—that have been reported to have activity on one or another nuclear hormone receptor should not cause one to become myopic and focus on these proteins as the only target for endocrine disruption. Hormone regulatory systems are complex, and though the receptors are at the center, the overall function of these interwoven webs also involves the processes of hormone production, transport in the blood, metabolism, and then clearance, as well as the cross talk that occurs between different cellular signal transduction pathways (Fig. 9).

In principle, hormone disruption could arise from the interaction of small molecules at any of these levels: by inhibiting the enzymes involved in hormone biosynthesis, displacing hormones from transport proteins, and activating or inhibiting the enzymes involved in hormone metabolism and clear-

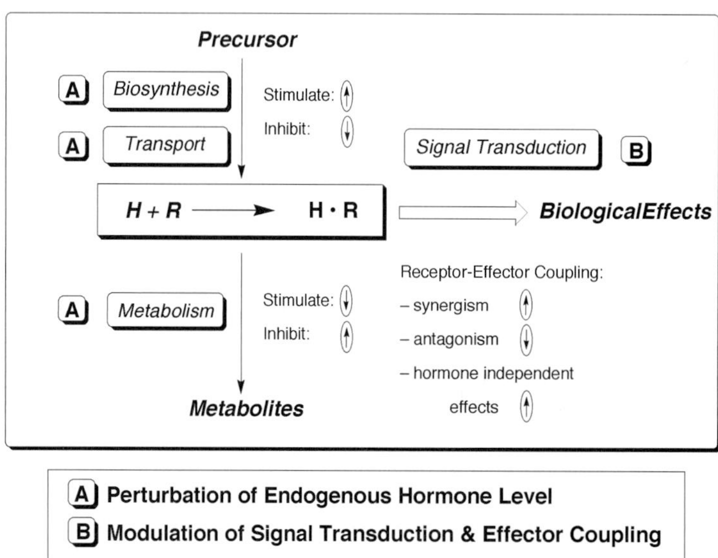

A **Perturbation of Endogenous Hormone Level**
B **Modulation of Signal Transduction & Effector Coupling**

Fig. 9 Modulation of endocrine activity not involving ligand binding to receptors.

ance. Certain pharmaceuticals target hormone production: aromatase inhibitors block the conversion of androgens to estrogens [118], 5α-reductase inhibitors prevent the bioactivation of testosterone to 5α-dihydrotestosterone [119], and aminoglutethamide blocks the early stage of the conversion of cholesterol to corticosteroids [120]. Other drugs, antibiotics, and natural products (such as hyperforin found in the widely consumed herb St. John's Wort) interact with the PXR/SXR system [121], activating a series of drug-metabolizing enzymes that can cause "drug interactions" [122]. In fact, the use of antibiotics in the treatment of human infections, which cause profound alterations in gut microflora, can affect the potency of endogenous and exogenous hormones, and this has been cited as a cause of the failure of estrogen oral contraceptives [123].

No doubt, there are other examples of endocrine disruption that occurs at "extra-receptor sites", such as was noted for the effect of halogenated aromatics on the thyroid system via disruption of hormone transport and acceleration of clearance [112–114] and as has been reported with some halogenated pesticides on estrogen biosynthesis [99]. The hormonal ("estrogenic") effects of certain hepatoxins, such as carbon tetrachloride, ethylene glycol, or dimethylformamide [124,125], are also most likely the indirect result of disrupted hormone clearance and elevated blood levels of endogenous hormones.

One must, therefore, keep a broad perspective in seeking the target or processes through which an endocrine disruptive event is mediated. The nuclear hormone receptors are just one of many possible targets. The regulation of homeostasis, development, and reproduction involves complex and autoregulated endocrine cycles whose sensitivity to exogenous hormonal substances may vary widely, depending on the phase of the cycle. Thus, one also needs to be most alert to assessing the effects of endocrine disruption at the most sensitive phases of these cycles [126].

Experiments can be carefully designed to evaluate whether a binding or biological effect of an endocrine disruptor is the result of interaction with a specific receptor: Specific receptor binding should be competitive with the binding of specific tracers, and agonism or antagonism through a specific receptor should be reversed by receptor-specific antagonist or agonists, respectively. Endocrine disruption that occurs indirectly, by alterations in hormone production, transport, or clearance, may be evident through altered levels of endogenous hormones, which can also be assayed by standard immunoassay and spectroscopic methods. Experiments in animals in which the organs of endogenous hormone production are removed can also be revealing of indirect endocrine disruption.

These varied considerations notwithstanding, one should keep in sight a number of simple facts about human exposure to hormone disruptors, and of estrogens in particular: Estrogens may be as good as they are bad, but human exposure to estrogenic substances (even if one considers only natural sources) is, by any reasonable standard, inescapable; this is true not only for adults but during embryonic development as well. Most human diets contain large amounts of weakly estrogenic substances (or their precursors) as natural components of food. Estrogens are produced endogenously by both males and females, and, aside from unusual cases, the major lifetime exposure that humans have to potent estrogens is to those of their own production.

The bottom line is simply this: On the one hand, one should remain aware of the widespread potential for compounds having varying structures, but sharing some common structural motifs and physical properties, as outlined above, to have nuclear hormone receptor activity. On the other hand, one must try—on a case-by-case basis—to come to a better understanding of whether the activity of these compounds and our exposure to them is a detriment or a benefit to animal and human health, or is, perhaps, inconsequential.

REFERENCES

1. N. J. McKenna and B. W. O'Malley. *Cell* **108**, 465–474 (2002).
2. D. J. Mangelsdorf, C. Thummel, M. Beato, P. Herrlich, G. Schutz, K. Umesono, B. Blumberg, P. Kastner, M. Mark, P. Chambon. *Cell* **83**, 835–839 (1995).
3. R. Sladek and V. Giguere. *Adv. Pharmacol.* **47**, 23–87 (2000).
4. S. A. Kliewer, J. M. Lehmann, T. M. Willson. *Science* **284**, 757–760 (1999).
5. W. Xie and R. M. Evans. *J. Biol. Chem.* **276**, 37739–37742 (2001).
6. T. M. Willson and J. T. Moore. *Mol. Endocrinol.* **16**, 1135–1144 (2002).
7. J. M. Maglich, A. Sluder, X. Guan, Y. Shi, D. D. McKee, K. Carrick, K. Kamdar, T. M. Willson, J. T. Moore. *Genome Biol.* **2**, RESEARCH0029 (2001).
8. K. Ekena, J. A. Katzenellenbogen, B. A. Katzenellenbogen. *J. Biol. Chem.* **273**, 693–699 (1998).
9. A. M. Brzozowski, A. C. Pike, Z. Dauter, R. E. Hubbard, T. Bonn, O. Engström, L. Öhman, G. L. Greene, J.-A. Gustafsson, M. Carlquist. *Nature* **389**, 753–758 (1997).
10. A. C. Pike, A. M. Brzozowski, R. E. Hubbard, T. Bonn, A. G. Thorsell, O. Engstrom, J. Ljunggren, J. Gustafsson, M. Carlquist. *EMBO J.* **18**, 4608–4618 (1999).
11. A. K. Shiau, D. Barstad, P. M. Loria, L. Cheng, P. J. Kushner, D. A. Agard, G. L. Greene. *Cell* **95**, 927–937 (1998).
12. W. Bourguet, M. Ruff, P. Chambon, H. Gronemeyer, D. Moras. *Nature* **375**, 377–382 (1995).
13. W. Bourguet, V. Vivat, J. M. Wurtz, P. Chambon, H. Gronemeyer, D. Moras. *Mol. Cell* **5**, 289–298 (2000).
14. R. L. Wagner, J. W. Apriletti, M. E. McGrath, B. L. West, J. D. Baxter, R. J. Fletterick. *Nature* **378**, 690–697 (1995).
15. A. C. Gee and J. A. Katzenellenbogen. *Mol. Endocrinol.* **15**, 421–428 (2001).
16. J. Uppenberg, C. Svensson, M. Jaki, G. Bertilsson, L. Jendeberg, A. Berkenstam. *J. Biol. Chem.* **273**, 31108–31112 (1998).
17. J. P. Renaud, N. Rochel, M. Ruff, V. Vivat, P. Chambon, H. Gronemeyer, D. Moras. *Nature* **378**, 681–689 (1995).
18. N. J. McKenna, R. B. Lanz, B. W. O'Malley. *Endocr. Rev.* **20**, 321–344 (1999).
19. R. T. Nolte, G. B. Wisely, S. Westin, J. E. Cobb, M. H. Lambert, R. Kurokawa, M. G. Rosenfeld, T. M. Willson, C. K. Glass, M. V. Milburn. *Nature* **395**, 137–143 (1998).
20. X. Hu and M. A. Lazar. *Nature* **402**, 93–96 (1999).
21. V. Perissi, L. M. Staszewski, E. M. McInerney, R. Kurokawa, A. Krones, D. W. Rose, M. H. Lambert, M. V. Milburn, C. K. Glass, M. G. Rosenfeld. *Genes Dev.* **13**, 3198–3208 (1999).
22. A. K. Shiau, D. Barstad, J. T. Radek, M. J. Meyers, K. W. Nettles, B. S. Katzenellenbogen, J. A. Katzenellenbogen, D. A. Agard, G. L. Greene. *Nat. Struct. Biol.* **9**, 359–364 (2002).
23. J. A. Katzenellenbogen, B. W. O'Malley, B. S. Katzenellenbogen. *Mol. Endocrinol.* **10**, 119–131 (1996).
24. A. C. Gee, K. E. Carlson, P. G. V. Martini, B. S. Katzenellenbogen, J. A. Katzenellenbogen. *Mol. Endocrinol.* **13**, 1912–1923 (1999).
25. J. R. Wood, V. S. Likhite, M. A. Loven, A. M. Nardulli. *Mol. Endocrinol.* **15**, 1114–1126 (2001).
26. Y. Shang and M. Brown. *Science* **295**, 2465–2468 (2002).
27. T. A. Grese and J. A. Dodge. *Curr. Pharm. Design* **4**, 71–92 (1998).
28. B. S. Katzenellenbogen and J. A. Katzenellenbogen. *Science* **295**, 2380–2381 (2002).
29. K. E. Carlson, I. Choi, A. Gee, B. S. Katzenellenbogen, J. A. Katzenellenbogen. *Biochemistry* **36**, 14897–14905 (1997).
30. G. M. Anstead, K. E. Carlson, J. A. Katzenellenbogen. *Steroids* **62**, 268–303 (1997).
31. R. Elsby, J. Ashby, J. P. Sumpter, A. N. Brooks, W. D. Pennie, J. L. Maggs, P. A. Lefevre, J. Odum, N. Beresford, D. Paton, B. K. Park. *Biochem. Pharmacol.* **60**, 1519–1530 (2000).
32. J. A. Katzenellenbogen. *Environ. Health Perspect.* **103**, 99–101 (1995).

33. H. Gao, J. A. Katzenellenbogen, R. Garg, C. Hansch. *Chem. Rev.* **99**, 723–744 (1999).
34. R. A. Magarian, L. B. Overacre, S. Singh, K. L. Meyer. *Curr. Med. Chem.* **1**, 61–104 (1994).
35. J. Grundy *Chem. Res.* **19**, 281–416 (1956).
36. U. V. Solmssen *Chem. Res.* **37**, 481–598 (1945).
37. L. Shi, W. Tong, H. Fang, Q. Xie, H. Hong, R. Perkins, J. Wu, M. Tu, R. M. Blair, W. S. Branham, C. Waller, J. Walker, D. M. Sheehan. *SAR QSAR Environ. Res.* **13**, 69–88 (2002).
38. H. Hong, W. Tong, H. Fang, L. Shi, Q. Xie, J. Wu, R. Perkins, J. D. Walker, W. Branham, D. M. Sheehan. *Environ. Health Perspect.* **110**, 29–36 (2002).
39. K. Pettersson and J. A. Gustafsson. *Annu. Rev. Physiol.* **63**, 165–192 (2001).
40. F. Larrea, R. Garcia-Becerra, A. E. Lemus, G. A. Garcia, G. Perez-Palacios, K. J. Jackson, K. M. Coleman, R. Dace, C. L. Smith, A. J. Cooney. *Endocrinology* **142**, 3791–3799 (2001).
41. H. Rochefort and M. Garcia. *Steroids* **28**, 549–560 (1976).
42. M. Garcia and H. Rochefort. *Steroids* **29**, 111–126 (1977).
43. E. C. Dodds and W. Lawson. *Nature* **137**, 996 (1936).
44. X. Lin and V. Huebner. *Curr. Opin. Drug Disc.* **3**, 383–398 (2000).
45. A. K. Salih and I. S. Fentiman. *Cancer Treat. Rev.* **27**, 261–273 (2001).
46. J. P. Sumpter. *Toxicol. Lett.* **102–103**, 337–342 (1998).
47. G. H. Panter, R. S. Thompson, N. Beresford, J. P. Sumpter. *Chemosphere* **38**, 3579–3596 (1999).
48. R. S. Baldwin, R. D. Williams, M. K. Terry. *Regul. Toxicol. Pharmacol.* **3**, 9–25 (1983).
49. S. F. Sundlof and C. Strickland. *Vet. Hum. Toxicol.* **28**, 242–250 (1986).
50. J. P. Fontenot and K. E. Webb, Jr. *J. Anim. Sci.* **40**, 1267–1277 (1975).
51. R. T. Duby and H. F. Travis. *Am. J. Vet. Res.* **32**, 1599–1602 (1971).
52. B. E. Fink, D. S. Mortensen, S. R. Stauffer, Z. D. Aron, J. A. Katzenellenbogen. *Chem. Biol.* **6**, 205–219 (1999).
53. E. E. Kenaga. *Residue Rev.* **44**, 73–113 (1972).
54. B. S. Katzenellenbogen, H. S. Iwamoto, D. F. Heiman, N. C. Lan, J. A. Katzenellenbogen. *Mol. Cell. Endocrinol.* **10**, 103–113 (1978).
55. B. S. Katzenellenbogen. *J. Steroid Biochem.* **20**, 1033–1037 (1984).
56. W. E. Ward and L. U. Thompson. *Handbook of Environmental Chemistry*, pp 101–128, Springer, Berlin, Germany (2001).
57. W. R. H. Shutt and D. A. Hogan. *Austr. J. Agric. Res.* **21**, 713–722 (1970).
58. M. E. Burow, S. M. Boue, B. M. Collins-Burow, L. I. Melnik, B. N. Duong, C. H. Carter-Wientjes, S. Li, T. E. Wiese, T. E. Cleveland, J. A. McLachlan. *J. Clin. Endocrinol. Metab.* **86**, 1750–1758 (2001).
59. G. G. Kuiper, B. Carlsson, K. Grandien, E. Enmark, J. Haggblad, S. Nilsson, J. A. Gustafsson. *Endocrinology* **138**, 863–870 (1997).
60. C. A. Lamartiniere. *Am. J. Clin. Nutr.* **71**, 1705S–1707S, discussion 1708S–1709S (2000).
61. W. Mazur and H. Adlercreutz. *Pure Appl. Chem.* **70**, 1759–1776 (1998).
62. B. D. Gehm, J. M. McAndrews, P. Y. Chien, J. L. Jameson. *Proc. Natl. Acad. Sci. USA* **94**, 14138–14143 (1997).
63. E. M. Bickoff, A. L. Linvingston, A. N. Booth. *Arch. Biochem. Biophys.* **88**, 262–266 (1960).
64. J. P. Raynaud, G. Azadian-Boulanger, M. M. Bouton, M. C. Colin, N. Faure, L. Fernand-Proulx, J. P. Gautray, J. M. Husson, A. Jolivet, P. Kelly, et al. *J. Steroid Biochem.* **20**, 981–993 (1984).
65. Y. H. Ju, D. R. Doerge, K. F. Allred, C. D. Allred, W. G. Helferich. *Cancer Res.* **62**, 2474–2477 (2002).
66. C. Y. Hsieh, R. C. Santell, S. Z. Haslam, W. G. Helferich. *Cancer Res.* **58**, 3833–3838 (1998).
67. *IARC Monogr. Eval. Carcinog. Risks Hum.* **56**, 397–444 (1993).
68. P. H. Hidy, R. S. Baldwin, R. L. Greasham, C. L. Keith, J. R. McMullen. *Adv. Appl. Microbiol.* **22**, 59–82 (1977).
69. B. S. Katzenellenbogen, J. A. Katzenellenbogen, D. Mordecai. *Endocrinology* **105**, 33–40 (1979).

70. Y. Hori, Y. Abe, M. Nishimura, T. Goto, M. Okuhara, M. Koshaka. *J. Antibiotics* **46**, 1069 (1992).
71. P. Mellanen, T. Petanen, J. Lehtimaki, S. Makela, G. Bylund, B. Holmbom, E. Mannila, A. Oikari, R. Santti. *Toxicol. Appl. Pharmacol.* **136**, 381–388 (1996).
72. K. Ikeda, Y. Arao, H. Otsuka, S. Nomoto, H. Horiguchi, S. Kato, F. Kayama. *Biochem. Biophys. Res. Commun.* **291**, 354–360 (2002).
73. R. White, S. Jobling, S. A. Hoare, J. P. Sumpter, M. G. Parker. *Endocrinology* **135**, 175–182 (1994).
74. C. A. Harris, P. Henttu, M. G. Parker, J. P. Sumpter. *Environ. Health Perspect.* **105**, 802–811 (1997).
75. J. A. McLachlan. *Endocr. Rev.* **22**, 319–341 (2001).
76. A. V. Krishnan, P. Stathis, S. F. Permuth, L. Tokes, D. Feldman. *Endocrinology* **132**, 2279–2286 (1993).
77. D. Feldman. *Endocrinology* **138**, 1777–1779 (1997).
78. A. M. Vinggaard, W. Korner, K. H. Lund, U. Bolz, J. H. Petersen. *Chem. Res. Toxicol.* **13**, 1214–1222 (2000).
79. Y. Berthois, J. A. Katzenellenbogen, B. S. Katzenellenbogen. *Proc. Natl. Acad. Sci. USA* **83**, 2496–2500 (1986).
80. R. D. Bindal and J. A. Katzenellenbogen. *J. Med. Chem.* **31**, 1978–1983 (1988).
81. R. D. Bindal, K. E. Carlson, B. S. Katzenellenbogen, J. A. Katzenellenbogen. *J. Steroid Biochem.* **31**, 287–293 (1988).
82. J. A. Katzenellenbogen. *Cancer Treat. Rep.* **62**, 1243–1249 (1978).
83. R. R. Levier and M. E. Jankowiak. *Biol. Reprod.* **7**, 260–266 (1972).
84. J. F. Hayden and S. A. Barlow. *Toxicol. Appl. Pharmacol.* **21**, 68–79 (1972).
85. J. R. Byford, L. E. Shaw, M. G. Drew, G. S. Pope, M. J. Sauer, P. D. Darbre. *J. Steroid Biochem. Mol. Biol.* **80**, 49–60 (2002).
86. M. Schlumpf, B. Cotton, M. Conscience, V. Haller, B. Steinmann, W. Lichtensteiger. *Environ. Health Perspect.* **109**, 239–244 (2001).
87. H. Tinwell, P. A. Lefevre, G. J. Moffat, A. Burns, J. Odum, T. D. Spurway, G. Orphanides, J. Ashby. *Environ. Health Perspect.* **110**, 533–536 (2002).
88. H. M. Bolt, C. Guhe, G. H. Degen. *Environ. Health Perspect.* **109**, A358–361 (2001).
89. W. H. Bulger and D. Kupfer. *Am. J. Ind. Med.* **4**, 163–173 (1983).
90. A. M. Cummings. *Crit. Rev. Toxicol.* **27**, 367–379 (1997).
91. L. E. Gray, J. Ostby, J. Furr, C. J. Wolf, C. Lambright, L. Parks, D. N. Veeramachaneni, V. Wilson, M. Price, A. Hotchkiss, E. Orlando, L. Guillette. *Hum. Reprod. Update* **7**, 248–264 (2001).
92. W. R. Kelce, C. R. Stone, S. C. Laws, L. E. Gray, J. A. Kemppainen, E. M. Wilson. *Nature* **375**, 581–585 (1995).
93. K. Connor, K. Ramamoorthy, M. Moore, M. Mustain, I. Chen, S. Safe, T. Zacharewski, B. Gillesby, A. Joyeux, P. Balaguer. *Toxicol. Appl. Pharmacol.* **145**, 111–123 (1997).
94. H. T. Jansen, P. S. Cooke, J. Porcelli, T. C. Liu, L. G. Hansen. *Reprod. Toxicol.* **7**, 237–248 (1993).
95. A. C. Layton, J. Sanseverino, B. W. Gregory, J. P. Easter, G. S. Sayler, T. W. Schultz. *Toxicol. Appl. Pharmacol.* **180**, 157–163 (2002).
96. K. S. Korach, P. Sarver, K. Chae, J. A. McLachlan, J. D. McKinney. *Mol. Pharmacol.* **33**, 120–126 (1988).
97. L. G. Hansen. *Environ. Health Perspect.* **106** (Suppl. 1), 171–189 (1998).
98. V. J. Kramer, W. G. Helferich, A. Bergman, E. Klasson-Wehler, J. P. Giesy. *Toxicol. Appl. Pharmacol.* **144**, 363–376 (1997).
99. H. R. Andersen, A. M. Vinggaard, T. H. Rasmussen, I. M. Gjermandsen, E. C. Bonefeld-Jorgensen *Toxicol. Appl. Pharmacol.* **179**, 1–12 (2002).
100. B. Hammond, B. S. Katzenellenbogen, N. Krauthammer, J. McConnell. *Proc. Natl. Acad. Sci. USA* **76**, 6641–6645 (1979).

101. A. Stelzer and H. M. Chan. *Toxicology* **138**, 69–80 (1999).
102. K. F. Arcaro, Y. Yang, D. D. Vakharia, J. F. Gierthy. *J. Toxicol. Environ. Health A* **59**, 197–210 (2000).
103. R. Steinmetz, P. C. Young, A. Caperell-Grant, E. A. Gize, B. V. Madhukar, N. Ben-Jonathan, R. M. Bigsby. *Cancer Res.* **56**, 5403–5409 (1996).
104. S. F. Arnold, D. M. Klotz, B. M. Collins, P. M. Vonier, L. J. Guillette, Jr., J. A. McLachlan. *Science* **272**, 1489–1492 (1996).
105. J. A. McLachlan. *Science* **277**, 462–463 (1997).
106. W. R. Kelce and E. M. Wilson. *J. Mol. Med.* **75**, 198–207 (1997).
107. R. S. Rosenberg, L. Grass, D. J. Jenkins, C. W. Kendall, E. P. Diamandis. *Biochem. Biophys. Res. Commun.* **248**, 935–939 (1998).
108. W. R. Kelce, E. Monosson, M. P. Gamcsik, S. C. Laws, L. E. Gray, Jr. *Toxicol. Appl. Pharmacol.* **126**, 276–285 (1994).
109. R. Jenkins, R. A. Angus, H. McNatt, W. M. Howell, J. A. Kemppainen, M. Kirk, E. M. Wilson. *Environ. Toxicol. Chem.* **20**, 1325–1331 (2001).
110. W. J. Marsheck, S. Kraychy, R. D. Muir. *Appl. Microbiol.* **23**, 72–77 (1972).
111. A. H. Conner, M. Nagaoka, J. W. Rowe, D. Perlman. *Appl. Environ. Microbiol.* **32**, 310–311 (1976).
112. D. Desaulniers, R. Poon, W. Phan, K. Leingartner, W. G. Foster, I. Chu. *Toxicol. Ind. Health* **13**, 627–638 (1997).
113. A. O. Cheek, K. Kow, J. Chen, J. A. McLachlan. *Environ. Health Perspect.* **107**, 273–278 (1999).
114. M. C. Lans, E. Klasson-Wehler, M. Willemsen, E. Meussen, S. Safe, A. Brouwer. *Chem. Biol. Interact.* **88**, 7–21 (1993).
115. D. T. Zava, C. M. Dollbaum, M. Blen. *Proc. Soc. Exp. Biol. Med.* **217**, 369–378 (1998).
116. S. J. Degitz, P. A. Kosian, E. A. Makynen, K. M. Jensen, G. T. Ankley. *Toxicol. Sci.* **57**, 264–274 (2000).
117. P. K. Schoff and G. T. Ankley. *Environ. Pollution* **119**, 1–4 (2002).
118. M. Recanatini, A. Cavalli, P. Valenti. *Med. Res. Rev.* **22**, 282–304 (2002).
119. O. W. Brawley, S. Barnes, H. Parnes. *Ann. NY Acad. Sci.* **952**, 145–152 (2001).
120. R. J. Santen. *J. Steroid Biochem. Mol. Biol.* **37**, 1029–1035 (1990).
121. L. B. Moore, B. Goodwin, S. A. Jones, G. B. Wisely, C. J. Serabjit-Singh, T. M. Willson, J. L. Collins, S. A. Kliewer. *Proc. Natl. Acad. Sci. USA* **97**, 7500–7502 (2000).
122. B. Goodwin, L. B. Moore, C. M. Stoltz, D. D. McKee, S. A. Kliewer. *Mol. Pharmacol.* **60**, 427–431 (2001).
123. B. D. Dickinson, R. D. Altman, N. H. Nielsen, M. L. Sterling. *Obstet. Gynecol.* **98**, 853–860 (2001).
124. V. Lopez del Pino and H. M. Bolt. *Arzneimittelforschung* **27**, 2117–2120 (1977).
125. L. Ren, A. Meldahl, J. J. Lech. *Chem. Biol. Interact.* **102**, 63–67 (1996).
126. B. C. Gladen and W. J. Rogan. *Am. J. Public Health* **85**, 504–508 (1995).

Pure Appl. Chem., Vol. 75, Nos. 11–12, pp. 1819–1826, 2003.

Topic 1.13

Transcriptional roles of AhR in expression of biological effects induced by endocrine disruptors*

Yoshiaki Fujii-Kuriyama[1,2,‡] and Junsei Mimura[1]

[1]*Center for Tsukuba Advanced Research Alliance, University of Tsukuba, Tennoudai, 1-1-1, Tsukuba, Ibaraki, 305-8577, Japan;* [2]*Core Research for Evolutional Science and Technology (CREST), Japan Science and Technology Corporation*

Abstract: 2,3,7,8-Tetrachlorodibenzo-*p*-dioxin (TCDD), one of most toxic man-made chemicals, binds arylhydrocarbon receptor (AhR or dioxin receptor), whose endogenous ligand remains unknown, with an extremely high affinity and expresses pleiotropic biological effects.

From analysis of the primary structures, AhR belongs to a distinct group of a supergene family from that of nuclear receptors. PCB, 3-methylcholanthrene, and benzo(a)pyrene are also ligands to the AhR, and these polycyclic aromatic chemicals are considered to display pleiotropic biological effects such as induction of a variety of drug-metabolizing enzymes, teratogenesis, tumor promotion, immunodeficiency due to thymic involution, and liver damage. Generation of the AhR-deficient mice by gene knock-out technology revealed that these biological effects described are mediated by AhR, because AhR($^{-}$/$_{-}$) mice lost susceptibility to these effects by TCDD and benzo(a)pyrene. It has recently been revealed that AhR is also involved in reproduction of female mice. Although the detailed mechanisms of involvement of AhR in exerting these effects are not always clarified, AhR is generally considered to function as a transcription factor, which activates the expression of genes by binding directly the XRE sequence in their promoter in a heterodimer form with Arnt. From analysis of DNA transfection and GST pull-down assays, it is revealed that AhR and Arnt interact with various coactivators such as RIP140, SRC-1/NcoA, and CBP/p300 to transmit their transactivation activity to general transcription factors (GTFs). AhR has also been shown to interact with various regulatory factors including Rb, NF-κB, and SP1, resulting in mutual inhibition or synergistic enhancement of their activities depending on the mode of localization of their cognate binding sequences in the target genes. Agonistic and antagonistic properties of various ligands to AhR also are discussed.

INTRODUCTION

When taken up into the animal bodies, 2,3,7,8-tetrachlorodibenzo-*p*-dioxin (TCDD) and its related compounds induce pleiotropic biochemical and toxicological effects, such as induction of drug-metabolizing enzymes, teratogenesis, immunosuppression due to thymic involution, tumor promotion, and

---

*Report from a SCOPE/IUPAC project: Implication of Endocrine Active Substances for Human and Wildlife (J. Miyamoto and J. Burger, editors). Other reports are published in this issue, *Pure Appl. Chem.* **75**, 1617–2615 (2003).

‡Corresponding author: Tel.: (81) 298-53-7323; Fax: (81) 298-53-7318; E-mail: ykfujii@tara.tsukuba.ac.jp.

liver damage. AhR (arylhydrocarbon receptor) is a receptor type transcription factor and has been believed to mediate these effects as a cellular factor. Under normal conditions, it exists in the cytoplasm in association with a Hsp90 (heat shock protein 90 kD) complex [1]. Upon binding TCDD with an extremely high affinity, the liganded AhR translocates to nuclei where it switches it partner molecule from Hsp90 complex to Arnt (AhR nuclear translocator). Thus, the formed AhR/Arnt heterodimer binds to the XRE (xenobiotic responsive element) sequence in the promoter region of the target genes to activate their expression [1–4]. Extensive studies of AhR have revealed that most, if not all, of the toxic effects of TCDD are mediated by AhR. Here, we briefly summarize recent studies on AhR. Because of limited space, we apologize that this review is not comprehensive.

MOLECULAR ANATOMY OF AhR AND MECHANISM OF TRANSACTIVATION

AhR and its partner molecule, Arnt, belong to a superfamily of structurally related transcription factors characterized by structural motifs designated bHLH and PAS [1,3] (Fig. 1). In the N-terminal region, these proteins contain a basic helix-loop-helix (bHLH) motif, which is shared by another family of transcription factors such as Myc and MyoD, and is involved in DNA binding and hetero- or homodimerization. Adjacent to the C-terminus of the bHLH domain, the sequence consisting of about 250 amino acids named as PAS continues. The PAS domain was originally identified as a conserved sequence between *Drosophila* PER, human ARNT, and *Drosophila* SIM, and includes two imperfect repeats of 50 amino acids, PAS-A and -B, the function of which is considered to be an interactive surface for hetero-

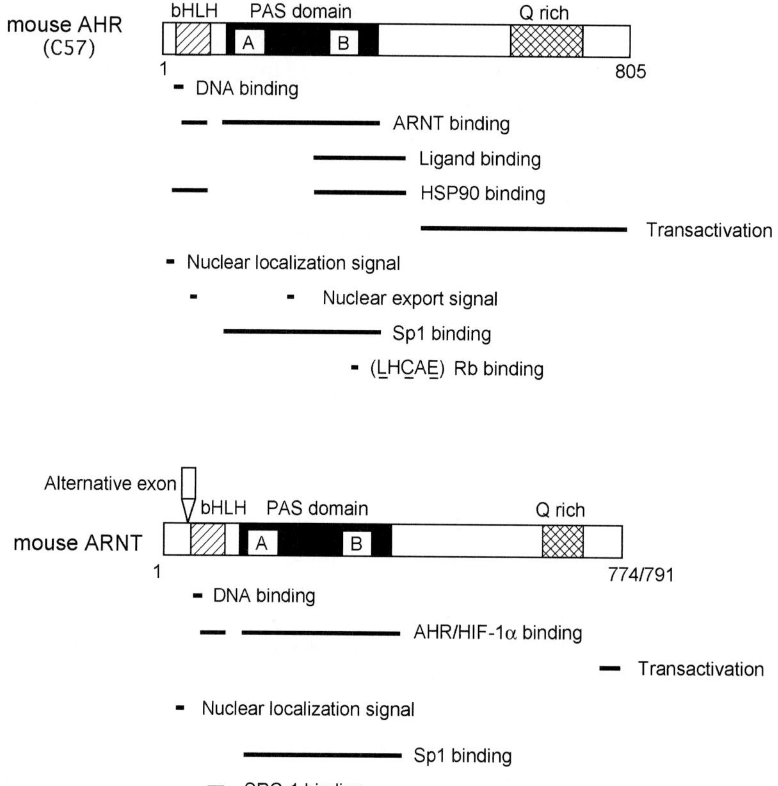

Fig. 1 Map of the functional domains of AhR bHLH: basic helix-loop-helix, PAS: Per/Arnt/Sim homology, Q rich: glutamine rich.

or homodimer formation, with Hsp90 complex and with ligands. The PAS domain is distributed in a wide variety of regulatory proteins involved in circadian rhythm (Per, Clock, BMAL1), hypoxic response (Hif-1α, Hif-2α/HLF), neurogenesis (Sim), and transcription coactivation (SRC-1, TIF2) in the animal kingdom and is also found in bacterial proteins that function as a light and oxygen sensor [4,5].

The ligand-binding domain of AhR is located in the sequence of 230-431 a. a. containing the PAS-B region, and this region overlaps the binding site for Hsp90, which makes it structurally competent to bind a ligand [6]. In addition, Hsp90 bound to the PAS-B domain is also known to interact with the bHLH region to mask the nuclear localization signal (NLS) therein, resulting in the cytoplasmic localization of AhR [7,8]. Under normal conditions, AhR exists in cytoplasm in a complex with Hsp90 containing XAP2 (hepatitis B virus X-associated protein) and P23. Upon binding with a ligand, AhR most probably changes its conformation to expose the NLS and translocates to the nucleus where it switches its partner molecule from Hsp90 to Arnt. As a result, the AhR/Arnt heterodimer recognizes and binds the XRE sequence in the promoter of the target genes to enhance their expression. A recent report has suggested that the Rb pocket sequence, LXCXE in the C-terminal region of PAS-B binds to Rb, thereby affecting the cell cycle in hepatoma cell lines [9]. The transactivation activity is distributed broadly in the C-terminal half of the AhR molecule, which has been reported to interact with multiple coactivators including Myb-binding 1a [10], BRG-1, RIP140 (receptor-interacting protein), and SRC-1/NcoA [11], while that of Arnt is localized at the very C-terminal 43-amino acid sequence [12] which interacts with CBP/p300 as coactivator [13,14]. Arnt contains a constitutive NLS in the N-terminal region of the bHLH domain [15]. As soon as it is synthesized in the cytoplasm, Arnt is transported into the nucleus. In addition, AhR contains a nuclear export signal (NES) in its second helix of the bHLH domain that mediates the nuclear export of the AhR protein, followed by proteasome degradation [16].

It has been reported that liganded AhR was rapidly degraded via the ubiquitin/proteasome pathway after transactivation of the target genes [17,18]. Nuclear localization and phosphorylation of AhR were necessary for the degradation. Interestingly, an inhibitor of protein synthesis, cycloheximide, inhibited the ligand-dependent degradation of AhR [18], although its precise mechanism remains unknown.

Phosphorylation of the AhR/Arnt heterodimer is reported to be also important for transactivation, because the binding activity of the AhR/Arnt heterodimer to the XRE sequence is abolished by the phosphatase treatment [19].

AhR HOMOLOGS

cDNAs for AhR were cloned from various species of animals ranging from mammals to flies and nematodes [3] (Fig. 2). cDNAs for two AhR homologs from *Fundulus* (FhAhR1 and FhAhR2) [20] have been cloned and sequenced, whereas only one AhR cDNA has been identified in mammalian species despite extensive efforts. These two homologs of fish AhR mRNAs are derived from their independent genes. Rainbow trout also expresses at least two isoforms of AhR (rtAhR2α and rtAhR2β). Interestingly, the expressions of rtAhR2α and rtAhR2β mRNAs were remarkably enhanced by TCDD, indicating that expression of AhR may be positively autoregulated in fish [21]. Concerning invertebrate species, structurally related AhR and Arnt homologs have been found in both *Drosophila* and *Caenorhabditis elegans* [22,23]. An AhR homolog in *Drosophila* is designated spineless. Deletion of the spineless gene caused morphological transformation in distal legs and antennae of adult flies. Spineless can form a heterodimer with Tango (an Arnt homolog of *Drosophila*), which recognizes the XRE sequence. AHR-1 is a *C. elegans* homolog of AhR and forms a heterodimer with an Arnt homolog of *C. elegans*, AHA-1, which binds the XRE. However, it remains unknown whether or not these invertebrate AhR homologs can mediate the toxicological effects of xenobiotics including dioxin, because nematode AHR-1 appears to exhibit no ability of binding to β-naphthoflavone like the mammalian AhR [23].

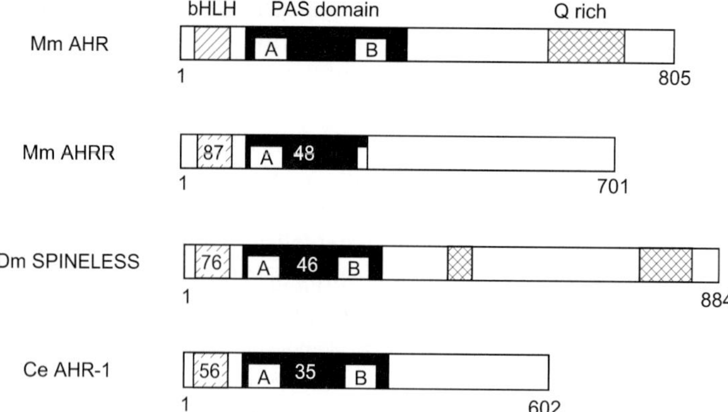

Fig. 2 Amino acid sequence comparison of mouse AhR (Mm AhR), mouse AhRR (Mm AhRR), fruit fly spineless (Dm spineless) and nematode AHR-1 (Ce AHR-1). Figures indicate the percent amino acid identity with mouse AhR.

TARGET GENES OF AhR

A number of phase I (CYP1A1, 1A2, 1B1, and 2A8) and phase II (NADPH-quinone xidoreductase, GST-Ya, UDP-glucronosyltransferase) drug-metabolizing enzymes (DMEs) are known to be induced by AhR ligands. In addition to these enzymes, AhR ligands upregulate expression of the genes involved in cell proliferation (TGF-β, IL-1β, and PAI-2), cell cycle regulation (p27 and jun-B), apoptosis (Bax), ADH3, and DNA polκ [1,24,25]. One or several copies of the XRE sequences are found in their regulatory region, indicating that ligand-activated AhR upregulates the expression of these genes by directly binding to the XRE sequences (i.e., primary effects). On the other hand, indirect mechanisms (i.e., secondary effects) for the induction of multidrug resistance gene mdr1 by 3-MC (3-methylchoranthrene) were also reported [26]. In HepG2 cells, the DNA microarray analysis revealed that 310 genes were either up- or down-regulated by the TCDD treatment [27]. Of the 310 genes whose expressions were affected by TCDD, 108 genes still were observed to be affected even by the simultaneous treatment with cycloheximide, while the altered expression of the remaining 202 genes by the TCDD treatment became unchanged by the treatment with the protein synthesis inhibitor. These results suggested that the expression of the former 108 genes was directly regulated by TCDD without protein synthesis, while regulation of the remaining 202 genes by TCDD required protein synthesis.

Safe et al. reported that the enhanced expression of cathepsin D and c-fos with estrogen treatment was inhibited in the presence of TCDD through the XRE sequence [28]. Although precise mechanisms of this inhibition still remain controversial, there is some possibility that AhR may play an inhibitory role in the transcription by binding to a specific sequence of XRE, designated as inhibitory XRE.

MODULATION OF AhR FUNCTION

As described, AhR exists in the cytoplasm in the complex of Hsp90, p23, and XAP2 in the absence of ligands. It is also reported that the pp60 src protein is also found in this AhR-chaperone complex [29]. As soon as AhR binds a ligand, src kinase is released from the AhR-Hsp90 complex and activates Ras-dependent signaling cascades [30]. At the initiation step of transcription, the AhR/Arnt heterodimer and Sp1 synergistically bind their respective cognate DNA sequences by interacting with each other [13]. In the promoter of the target genes of AhR, there usually exist one or several copies of the GC-box sequence together with the XRE. Synergistic transactivation by SP1 and AhR/Arnt heterodimer is considered to be a common mechanism for inducible expression of the drug-metabolizing enzymes by xenobiotics. Recent studies have demonstrated that Rb and NF-κB also interact with AhR. Liganded

AhR stabilizes the complex of Rb and E2F by binding to Rb, thereby enhancing the repression of the transcriptional activity of E2F and, hence, cell cycle is arrested [31]. It is also suggested that Rb functions as a coactivator of AhR by binding with AhR [32]. The ligand-activated AhR and the RelA subunit of NF-κB interacted with each other to interfere mutually in binding to their respective recognition sequence in the separate genes [33]. On the other hand, TCDD and TPA synergistically enhanced the expression of the AhRR gene driven by the XRE and NF-κB binding sequences in the promoter, probably through the interaction between AhR and NF-κB [34].

AhRR has a bHLH-PAS sequence with a high similarity to that of AhR and inhibits the transcription activity of AhR by competing with AhR for binding to Arnt and then to the XRE sequence [35] (Figs. 2 and 3). Since AhRR expression is induced by TCDD via AhR/Arnt heterodimer, AhR and AhRR form a regulatory feedback loop in the xenobiotic signal transduction.

A hypoxia signal also inhibits AhR-dependent transactivation [36]. In hypoxia, Hif-1α is stabilized and translocates to the nucleus where it heterodimerizes with Arnt in competition with AhR. Thus, xenobiotic and hypoxia signal transduction pathways interfere with each other by competing for a common partner molecule Arnt.

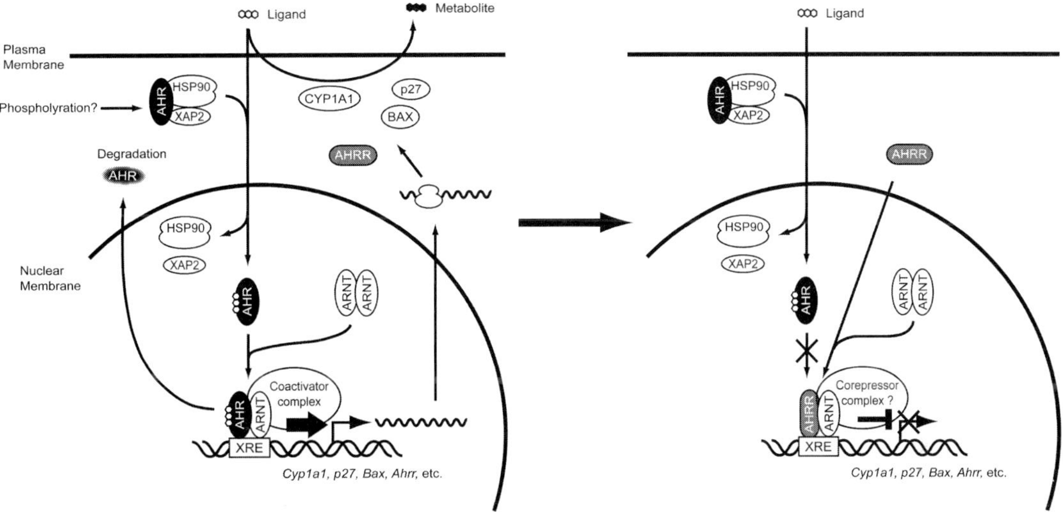

Fig. 3 AhR signaling pathway and negative feedback regulation of AhR function by AhRR.

POLYMORPHISM OF AhR

It is well known that there is a marked strain and species difference in sensitivity to TCDD [1,3]. LD_{50} values of TCDD varies over a 5000-fold range among different species. For example, LD_{50} values vary from 1 μg/kg for the guinea pig, the most sensitive animal to >5000 μg/kg for the hamster, the most resistant. In mice, difference in the responsiveness to TCDD among strains depends on the AhR alleles. cDNA cloning of a responder (C57BL) and nonresponder (DBA) mice revealed that Ala-to-Val substitution at codon 375 and the C-terminal extension due to a mutation in the termination codon in DBA mice reduced the binding affinity of AhR toward TCDD by about 10-fold [37]. Polymorphism in the AhR locus is also found in the rat. TCDD-sensitive Long–Evans (L-E) (LD_{50}, 10–20 μg/kg) and -insensitive Han/Wistar (H/W) (LD_{50}, >9600 μg/kg) rats show about 1000-fold difference in sensitivity. cDNA cloning indicated that L-E AhR cDNA sequence is the same as that of Sprague–Dawley rat previously determined [38]. In contrast, H/W AhR cDNA was revealed to carry a point mutation in the first base of intron 10, resulting in a frame shift with a different sequence of the C-terminal transactivation domain. In human, one genetic variation has been reported at codon 554, resulting in amino acid change from Arg to Lys [39,40]. However, this mutation does not seem to be related to any altered phenotype of TCDD toxicity in human.

AhR GENE-ENGINEERED MICE

In general, there are two ways to investigate the function of AhR—loss-of-function and gain-of-function studies. As loss-of-function study, three groups reported the generation of AhR-knock-out mice by using the gene targeting technology [41–43]. AhR-null mice were born in normal Mendelian genetics by cross-mating of the heterozygous AhR($^{+}$/$_{-}$) mice, but their growth was retarded for the first three weeks of life, but returned to normal thereafter. AhR-null mice were revealed to be defective in hepatic development, immune system development, and retinoic acid metabolism. It is reported that female AhR-null mice had difficulty in maintaining conception, lactation, and rearing pups to weaning [44]. Recently, AhR deficiency has been revealed to cause abnormal hepatic vascular structures [45]. Concerning xenobiotic metabolism, the lack of AhR abolished the inducible expression of CYP1A1, 1A2, and 1B1 in mice in response to polycyclic aromatic hydrocarbons, and lost the susceptibility to chemical carcinogenesis by benzo(a)pyrene and teratogenesis such as cleft palate and hydronephrosis caused by TCDD [43,46]. Although the genes responsible for TCDD-induced teratogenesis remain to be identified, the loss of the inducible expression of CYP1A1, 1A2, and 1B1 which metabolically activate various procarcinogens to the ultimate carcinogens, is considered to be a cause of resistance to chemical carcinogenesis. In addition, AhR together with Arnt has been shown to upregulate the expression of DNA polymerase κ, which duplicates DNA in an error-prone manner [47]. This is a good reason why AhR is also involved in the promotion of chemical carcinogens.

As gain-of-function study, analysis of transgenic mice with the expression of a constitutively active AhR has very recently been reported [48]. The constitutively active form of AhR was constructed by deleting a region containing the PAS-B domain and expressed under the control of a modified SV40 promoter and mouse IgH intron enhancer. In the transgenic mice, AhR was rather ubiquitously expressed and CYP1A1 was constitutively expressed in various tissues such as liver, lung, muscle, heart, and others. The constitutively active AhR expressed in the transgenic mice reduced the life span of the mice and induced tumors in glandular part of the stomach, indicating the oncogenic potential of the AhR and implicating the receptor in regulation of cell proliferation.

AhR LIGAND

TCDD is the most potent ligand of AhR with an extremely low K_d value. Other types of halogenated aromatic hydrocarbons such as PCB and polycyclic aromatic hydrocarbons such as benzo(a)pyrene and 3-MC also function as ligands to AhR. Some flavonoids function as an antagonist at low concentrations (~0.5 μM), while they are agonistic for AhR at higher concentrations (>10 μM) [49,50]. It has been reported that 7-ketocholesterol functions as AhR antagonist and actually antagonizes the TCDD effects in vivo [51]. This antagonistic effect of 7-ketocholesterol has been suggested to have physiological significance, when its physiological plasma concentration was taken into account. Resveratrol, a plant steroid, was also reported to be an antagonist for AhR [52]. It has been an interesting and long-lasting problem to identify a true and intrinsic ligand of AhR. Recently, a number of endogenous AhR ligands have been reported to have a high affinity toward AhR. Most of these chemicals are tryptophan derivatives. UV-irradiated or ozone-treated tryptophan products function as AhR agonist [53]. Some of the purified products were reported to function as potent ligand of AhR. Indirubin and indigo isolated from urine were reported to function as even more potent ligands of AhR than TCDD in yeast assay system [54]. It would be difficult to identify a true and intrinsic ligand of AhR, if any, until the intrinsic role of AhR other than induction of drug-metabolizing enzymes is clarified.

CONCLUSION

In summary, AhR is originally identified as a mediator of dioxin toxicities. However, recent studies have suggested that AhR plays a regulatory role in homeostasis and development of animals, and its functional roles in these processes remain to be clarified in future studies.

ACKNOWLEDGMENTS

This work was supported in part by a grant-in-aid for scientific research on priority area (A) from the Ministry of Education, Culture, Sports, Science and Technology, Japan, and grants from the Core Research for Evolutional Science and Technology (CREST), Japan Science and Technology Corporation, and Sankyo Co.

REFERENCES

1. O. Hankinson. *Annu. Rev. Pharmacol. Toxicol.* **35**, 307–340 (1995).
2. N. Matsushita, K. Sogawa, M. Ema, A. Yoshida, Y. Fujii-Kuriyama. *J. Biol. Chem.* **268**, 21002–21006 (1993).
3. M. E. Hahn. *Comp. Biochem. Physiol. C Pharmacol. Toxicol. Endocrinol.* **121**, 23–53 (1998).
4. Y. Z. Gu, J. B. Hogenesch, C. A. Bradfield. *Annu. Rev. Pharmacol. Toxicol.* **40**, 519–561 (2000).
5. B. L. Taylor and I. B. Zhulin. *Microbiol. Mol. Biol. Rev.* **63**, 479–506 (1999).
6. P. Coumailleau, L. Poellinger, J. A. Gustafsson, M. L. Whitelaw. *J. Biol. Chem.* **270**, 25291–25300 (1995).
7. T. Ikuta, H. Eguchi, T. Tachibana, Y. Yoneda, K. Kawajiri. *J. Biol. Chem.* **273**, 2895–2904 (1998).
8. M. J. Lees and M. L. Whitelaw. *Mol. Cell. Biol.* **19**, 5811–5822 (1999).
9. N. L. Ge and C. J. Elferink. *J. Biol. Chem.* **273**, 22708–22713 (1998).
10. L. C. Jones, S. T. Okino, T. J. Gonda, P. Whitlock, Jr. *J. Biol. Chem.* **277**, 22515–22519 (2002).
11. T. V. Beishlag, S. Wang, D. W. Rose, J. Torchia, S. Reisz-Porszasz, K. Muhammad, W. E. Nelson, M. R. Prost, M. G. Rosenfeld, O. Hankinson. *Mol. Cell. Biol.* **22**, 4319–4333 (2002).
12. K. Sogawa, K. Iwabuchi, H. Abe, Y. Fujii-Kuriyama. *J. Cancer Res. Clin. Oncol.* **121**, 612–620 (1995).
13. A. Kobayashi, K. Sogawa, Y. Fujii-Kuriyama. *J. Biol. Chem.* **271**, 12310–12316 (1996).
14. M. B. Kumar, R. W. Tarpey, G. H. Perdew. *J. Biol. Chem.* **274**, 22155–22164 (1999).
15. H. Eguchi, T. Ikuta, T. Tachibana, Y. Yoneda, K. Kawajiri. *J. Biol. Chem.* **272**, 17640–17647 (1997).
16. N. A. Davarinos and R. S. Pollenz. *J. Biol. Chem.* **274**, 28708–28715 (1999).
17. B. J. Roberts and M. L. Whitelaw. *J. Biol. Chem.* **274**, 36351–36356 (1999).
18. Q. Ma and K. T. Baldwin. *J. Biol. Chem.* **275**, 8432–8438 (2000).
19. S. Park, E. C. Henry, T. A. Gasiewicz. *Arch. Biochem. Biophys.* **381**, 302–312 (2000).
20. S. I. Karchner, W. H. Powell, M. E. Hahn. *J. Biol. Chem.* **274**, 33814–33824 (1999).
21. C. C. Abnet, R. L. Tanguay, M. E. Hahn, W. Heideman, R. E. Peterson. *J. Biol. Chem.* **274**, 15159–15166 (1999).
22. D. M. Duncan, E. A. Burgess, I. Duncan. *Genes Dev.* **12**, 1290–1303 (1998).
23. J. A. Powell-Coffman, C. A. Bradfield, W. B. Wood. *Proc. Natl. Acad. Sci. USA* **95**, 2844–2849 (1998).
24. S. K. Kolluri, C. Weiss, A. Koff, M. Gottlicher. *Genes Dev.* **13**, 1742–1753 (1999).
25. T. Matikainen, G. I. Perez, A. Jurisicova, J. K. Pru, J. J. Schlezinger, H. Y. Ryu, J. Laine, T. Sakai, S. J. Korsmeyer, R. F. Casper, D. H. Sherr, J. L. Tilly. *Nat. Genet.* **28**, 355–360 (2001).
26. M. C. Mathieu, I. Lapierre, K. Brault, M. Raymond. *J. Biol. Chem.* **276**, 4819–4827 (2001).
27. A. Puga, A. Maier, M. Medvedovic. *Biochem. Pharmacol.* **60**, 1129–1142 (2000).

28. S. Safe, F. Wang, W. Porter, R. Duan, A. McDougal. *Toxicol. Lett.* **102–103**, 343–347 (1998).
29. E. Enan and F. Matsumura. *Biochem. Pharmacol.* **52**, 1599–1612 (1996).
30. M. Schwarz, A. Buchmann, S. Stinchcombe, A. Kalkuhl, K. Bock. *Toxicol. Lett.* **112–113**, 69–77 (2000).
31. A. Puga, S. J. Barnes, T. P. Dalton, C. Chang, ES. Knudsen, M. A. Maicr. *J. Biol. Chem.* **275**, 2943–2950 (2000).
32. C. J. Elferink, N. L. Ge, A. Levine. *Mol. Pharmacol.* **59**, 664–673 (2001).
33. Y. Tian, S. Ke, M. S. Denison, A. B. Rabson, M. A. Gallo, *J. Biol. Chem.* **274**, 510–515 (1999).
34. T. Baba, J. Mimura, K. Gradin, A. Kursiwa, T. Watanabe, Y. Matsuda, J. Inazawa, K. Sogawa, Y. Fujii-Kuriyama. *J. Biol. Chem.* **276**, 33101–33110 (2001).
35. J. Mimura, M. Ema, K. Sogawa, Y. Fujii-Kuriyama. *Genes Dev.* **13**, 20–25 (1999).
36. K. Gradin, J. McGuire, R. H. Wenger, I. Kvietikova, M. L. Whitelaw, R. Toftgard, L. Tora, M. Gassmann, L. Poellinger. *Mol. Cell. Biol.* **16**, 5221–5231 (1996).
37. M. Ema, N. Ohe, M. Suzuki, J. Mimura, K. Sogawa, S. Ikawa, Y. Fujii-Kuriyama. *J. Biol. Chem.* **269**, 27337–27343 (1994).
38. R. Pohjanvirta, J. M. Wong, W. Li, P. A. Harper, J. Tuomisto, A. B. Okey. *Mol. Pharmacol.* **54**, 86–93 (1998).
39. K. Kawajiri, J. Watanabe, H. Eguchi, K. Nakachi, C. Kiyohara, S. Hayashi. *Pharmacogenetics* **5**, 151–158 (1995).
40. J. M. Wong, P. A. Harper, U. A. Meyer, K. W. Bock, K. Morike, J. Lagueux, P. Ayotte, R. F. Tyndale, E. M. Sellers, D. K. Manchester, A. B. Okey. *Pharmacogenetics* **11**, 85–94 (2001).
41. P. Fernandez-Salguero, T. Pineau, D. M. Hilbert, T. McPhail, S. S. Lee, S. Kimura, D. W. Nebert, S. Rudikoff, J. M. Ward, F. J. Gonzalez. *Science* **268**, 722–726 (1995).
42. J. V. Schmidt, G. H. Su, J. K. Reddy, M. C. Simon, C. A. Bradfield. **93**, 6731–6736 (1996).
43. J. Mimura, K. Yamashita, K. Nakamura, M. Morita, T. N. Takagi, K. Nakao, M. Ema, K. Sogawa, M. Yasuda, M. Katsuki, Y. Fujii-Kuriyama. *Genes Cells* **2**, 645–654 (1997).
44. B. D. Abbott, J. E. Schmid, J. A. Pitt, A. R. Buckalew, C. R. Wood, G. A. Held, J. J. Diliberto. *Toxicol. Appl. Pharmacol.* **155**, 62–70 (1999).
45. G. P. Lahvis, S. L. Lindell, R. S. Thomas, R. S. McCuskey, C. Murphy, E. Glover, M. Bentz, J. Southard, C. A. Bradfield. *Proc. Natl. Acad. Sci. USA* **97**, 10442–10447 (2000).
46. Y. Shimizu, Y. Nakatsuru, M. Ichinose, Y. Takahashi, H. Kume, J. Mimura, Y. Fujii-Kuriyama, T. Ishikawa. *Proc. Natl. Acad. Sci. USA* **97**, 779–782 (2000).
47. T. Ogi, J. Mimura, M. Hikida, H. Fujimoto, Y. Fujii-Kuriyama, H. Ohmori. *Genes Cells* **6**, 943–953 (2001).
48. P. Andersson, J. McGuire, C. Rubio, K. Gradim, M. L. Whitelaw, S. Pettersson, A. Hanberg, L. Poellinger. *Proc. Natl. Acad. Sci. USA.* In press.
49. J. J. Reiners, Jr., R. Clift, P. Mathieu. *Carcinogenesis* **20**, 1561–1566 (1999).
50. H. Ashida, I. Fukuda, T. Yamashita, K. Kanazawa. *FEBS Lett.* **476**, 213–217 (2000).
51. J. F. Savouret, M. Antenos, M. Quesne, J. Xu, E. Milgrom, R. F. Casper. *J. Biol. Chem.* **276**, 3054–3059 (2001).
52. R. F. Casper, M. Quesne, I. M. Rogers, T. Shirota, A. Jolivet, E. Milgrom, J. F. Savouret. *Mol. Pharmacol.* **56**, 784–790 (1999).
53. R. K. Sindhu, M. Mitshhashi, Y. Kikkawa. *J. Pharmacol. Exp. Ther.* **292**, 1008–1014 (2000).
54. J. Adachi, Y. Mori, S. Matsui, H. Takigami, J. Fujino, H. Kitagawa, C. A. Miller, 3rd, T. Kato, K. Saeki, T. Matsuda. *J. Biol. Chem.* **276**, 31475–31478 (2001).

Pure Appl. Chem., Vol. 75, Nos. 11–12, pp. 1827–1839, 2003.

Topic 1.14

Nonmammalian nuclear receptors: Evolution and endocrine disruption*

Joseph W. Thornton‡

Center for Ecology and Evolutionary Biology, University of Oregon, Eugene, OR 97403-5289, USA

Abstract. Most research to identify endocrine-disrupting chemicals and their impacts has relied on mammalian models or in vitro systems derived from them. But nuclear receptors (NRs), the proteins that transduce hydrophobic hormonal signals and are major mediators of endocrine disruption, emerged early in animal evolution and now play biologically essential roles throughout the Metazoa. Nonmammalian vertebrates and invertebrates, many of which are of considerable ecological, economic, and cultural importance, are therefore potentially subject to endocrine disruption by synthetic environmental pollutants.

Are methods that rely solely on mammalian models adequate to predict or detect all chemicals that may disrupt NR signaling? Regulation of NRs by small hydrophobic molecules is ancient and evolutionarily labile. Within and across genomes, the NR superfamily is very diverse, due to many lineage-specific gene and genome duplications followed by independent divergence. Receptors in nonmammalian species have in many cases evolved unique molecular and organismal functions that cannot be predicted from those of their mammalian orthologs. Endocrine disruption is therefore likely to occur throughout the metazoan kingdom, and a significant number of the thousands of synthetic chemicals now in production may disrupt NR signaling in one or more nonmammalian taxa. Many of these endocrine disruptors will not be detected by current regulatory/scientific protocols, which should be reformulated to take account of the diversity and complexity of the NR gene family.

ENDOCRINE DISRUPTION AND NUCLEAR RECEPTOR EVOLUTION

Nuclear receptors (NRs) are evolutionarily related multidomain proteins that bind to specific DNA motifs and up- or down-regulate transcription of specific genes. Most NRs bind tightly to hydrophobic hormones and other ligands, which effect a conformational change in the receptor, causing it to switch into transcriptionally active or repressive modes [2]. Many plant and synthetic chemicals, including pesticides, plastics components, and industrial by-products, have been found to bind to and thereby mimic, block, or otherwise disrupt the natural activity of vertebrate NRs [2,3].

The premise of this paper is that current biological systems carry the traces of the evolutionary history by which they were created. A historical perspective on the nuclear receptor superfamily allows us to organize and interpret the diversity of NRs and their properties in a biologically meaningful way, to discover patterns in the functional and evolutionary dynamics of NRs, and to predict the distribution of receptors in taxa that have not been well studied. Evolutionary analysis can, therefore, yield useful

---

*Report from a SCOPE/IUPAC project: Implication of Endocrine Active Substances for Human and Wildlife (J. Miyamoto and J. Burger, editors). Other reports are published in this issue, *Pure Appl. Chem.* **75**, 1617–2615 (2003).

‡E-mail: joet@darkwing.uoregon.edu

knowledge for understanding and predicting the effects of environmental endocrine disruptors (EDs). In this paper, I review knowledge of nuclear receptor evolution and survey the NR complement of major nonmammalian taxa, with a focus on implications for efforts to identify and manage endocrine-disrupting synthetic chemicals.

ANCIENT AND CONTINUING RECEPTOR DIVERSIFICATION

The NR superfamily, like other gene families, is the product of an ongoing process of genomic diversification within and among species. There are over 1000 nuclear receptor sequences in the SMART database of protein family domains [4]. The genomes of humans, *Drosophila melanogaster* and *Caenorhabditis elegans* contain 49, 21, and ~270 NRs, respectively [5]. NRs have a conserved domain structure and identifiable sequence homology to each other in their DNA-binding domains (DBDs) and ligand-binding domains (LBDs). BLAST searches indicate no statistically significant similarity to any other gene family. This pattern indicates that all NRs evolved by descent from a single ancestral NR—the origin of which remains unknown—through a series of gene duplications.

Because gene sequences diverge independently after duplication, the evolutionary relationships among NRs form a tree-like nested hierarchy, and the techniques of phylogenetic systematics can be used to reconstruct the history of NR diversification. As shown in Fig. 1, analysis of a large set of nuclear receptor sequences indicates that there are five evolutionary classes of nuclear receptors: (1) the vitamin D receptor (VDR), ecdysone receptor (EcR), retinoic acid receptors (RARs), thyroid hormone receptors (ThRs), and a number of receptors involved in cholesterol and xenobiotic metabolism; (2) the peroxisome proliferators activated receptors (PPARs) and several related orphans, which are involved in cell differentiation, development, and metabolism; (3) NGFI-B and related "orphan" receptors (receptors with no known ligand), which function in development and immunity; (4) the reproductive receptors, including steroidogenic factor 1 (SF-1), the gonadal and adrenal steroid and related receptors of vertebrates, and germ-cell nuclear factor (GCNF); and (5) the retinoid x receptors (RXRs) and related orphans, most of which are involved in development. Most of the groups are well supported, although certain relationships (such as placement of the RARs and ThRs) are sensitive to alignment and analysis parameters. This classification, like an earlier parsimony analysis [6], is different from the results of an earlier NR phylogeny/classification, which used a smaller number of sequences and the less reliable neighbor-joining method of phylogenetic inference [7]. The most important differences are the existence of distinct classes 1 and 2 and the close relationship of the gonadal/adrenal steroid receptors to SF-1 and GCNF.

NR diversity is ancient and therefore widespread. Five groups of receptors, all in the RXR and SF-1 classes, have been identified in both diploblastic Cnidaria (jellyfish, corals, and anemones) and triploblastic animals (Fig. 1). This finding indicates that the NR superfamily had already achieved considerable diversity before the advent of the bilaterally symmetric metazoa. By the time the lineages of protostomes (arthropods, nematodes, mollusks, annelids, etc.), and deuterostomes (chordates and echinoderms) split from each other at least 670 million years ago [8], the fundamental diversity of the nuclear receptor superfamily had been established: 15 groups of receptors, representing all five classes, are found in the genomes of both *D. melanogaster* and *Homo sapiens*. Only DAX-1, SHP, ThRs, and RARs of vertebrates have no known protostome orthologs; based on current evidence, these are "recent" chordate-specific novelties. These findings indicate that most NRs are ancient and distributed throughout the Eumetazoa, making the vast majority of animal species potential targets for endocrine disruption.

There have also been many lineage-specific gene duplication events that created new receptors in certain taxa. For example, there was considerable additional diversification in the vertebrate lineage, indicated by the presence of two to four retained duplicates in vertebrates for each arthropod receptor (Fig. 1). This pattern is consistent with the serial genome duplication hypothesis, which states that the vertebrate genome is the result of two serial rounds of genome duplication [9]. Duplications that generated taxon-specific receptors also occurred in other lineages, including teleosts, nematodes, and arthropods, as detailed later in this paper.

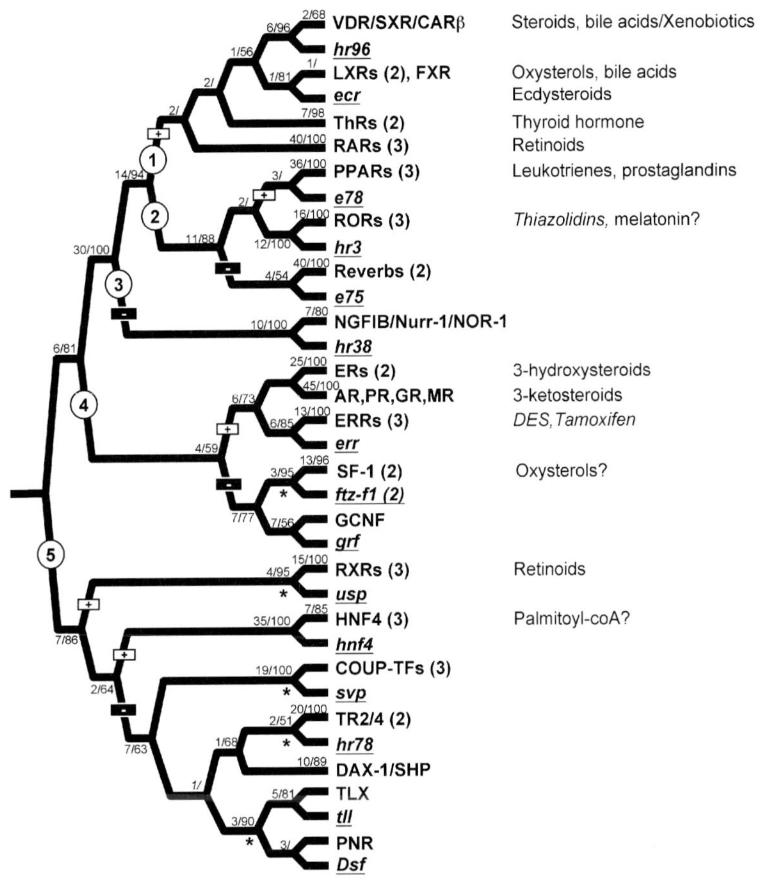

Fig. 1 Phylogeny of the nuclear receptors. A reduced consensus is shown of the 12 most parsimonious trees from analysis of 163 NR sequences of *H. sapiens, D. melanogaster, Bombyx morii, Gallus gallus, Brachydanio rerio, Xenopus laevis,* and *Acropora millepora.* Roman type indicates vertebrate sequences; insect NRs in underlined italic. Asterisks indicate the branch to which Cnidarian sequences connect. Numbers in parenthesis show the number of human paralogs in each group. Ligands, when known, are to the right of the tree; synthetic ligands are in italics, and question marks indicate unconfirmed results. On this phylogeny, an ancestral ligand-activated receptor is the most parsimonious reconstruction, as shown by four dark boxes (–) that indicate loss of ligand binding if the ancestral receptor was ligand-dependent, and five light boxes (+) that indicate gain of ligand binding if the ancestral receptor was ligand-independent. Node labels show unweighted branch support values [53]/bootstrap values >50 %. Sequences were downloaded from the EBI SMART database [4]; the DBDs and LBDs were extracted and then aligned using ClustalX 1.7 [54]. Parsimony analysis was conducted using PAUP* [55] with a stepmatrix that downweights conservative substitutions and a heuristic search strategy of 100 random taxon additions followed by TBR branch swapping. Reconstruction of change in ligand-binding status assumes parallelism is more likely than reversal; different assumptions would change the placement but not the number of changes.

THE ANCESTRAL NUCLEAR RECEPTOR: ORIGIN AND FUNCTION

When did the first nuclear receptor appear, and what was its function? No nuclear receptors have been found in any non-metazoan species, including the completely sequenced genomes of eubacteria, archaea, yeast, and plants. Because NRs have been identified in the three major divisions of eumetazoans—cnidarians, protostomes, and deuterostomes—we can conclude that the first NRs emerged some time after the divergence of metazoans from fungi but before the Cnidarian-Bilateria divergence (Fig. 2).

It has been argued that the ancestral NR was a ligand-independent receptor and that ligand binding was acquired later during NR evolution [10]. This scenario is possible, but there is no evidence to favor it over the alternative explanation of an ancestral ligand-activated receptor. Both ligand-binding and ligand-independent receptors have been identified in protostomes, deuterostomes, and cnidarians [11]; the inferred age of ligand-binding is therefore the same as the inferred age of ligand-independent receptors. Mechanistically, either scenario is plausible, although neither a complete loss of ligand activation nor a gain of ligand binding from a more ancient orphan has been unambiguously established in the evolution of NRs.

The most reliable means for inferring the characteristics of ancient proteins is by reconstructing ancestral functional states using phylogenetic methods. Based on the parsimony criterion, the best-supported hypothesis of ancestral function is the one that requires the fewest parallel gains or reversals on a gene family tree. As Fig. 1 shows, the NR phylogeny weakly supports an ancestral ligand-binding receptor if the tree is rooted between groups 4 and 5; the same would be true if it were rooted with an ancient ligand-binding NR like RXR in a basal position. If the tree were rooted on an ancient orphan receptor like COUP, an ancestral orphan receptor would be weakly favored. There is presently no way to determine which of these receptors should be used as the root, so current evidence does not support either scenario over the other.

Whatever the ligand-binding characteristics of the first NR were, there is no doubt that ligand binding is very ancient. Ligand binding must have evolved very early in the radiation of the animal phyla from an ancient eumetazoan ancestor, as shown by the presence of conserved ligand binding in Cnidaria and Bilateria. It also must have very early in the diversification of the NR superfamily into its five classes, as Fig. 1 shows. Virtually all receptors inherit the capacity to be ligand-activated and there-

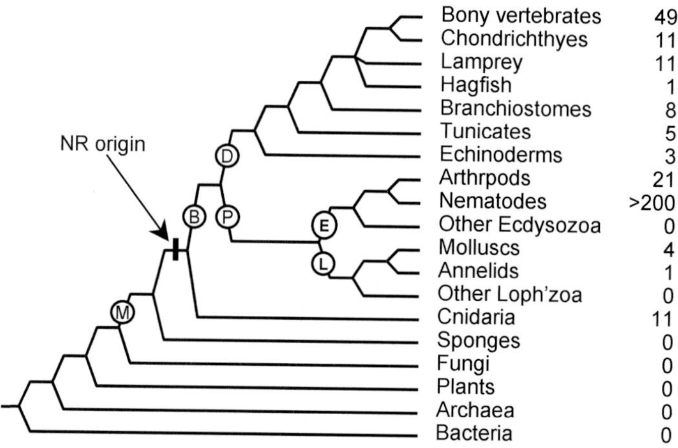

Fig. 2 Taxonomic distribution of nuclear receptors. A reduced phylogeny of living kingdoms and selected metazoan taxa is shown, with the number of NR paralogs that have been sequenced from each group. Number of NRs in jawed vertebrates, arthropods, and nematodes are derived from the complete genomes of *H. sapiens*, *D. melanogaster*, and *C. elegans*, respectively. Absence of NRs in plants, fungi, and prokaryotes is also supported by whole genome sequences. Other NR numbers in metazoa are inferred from screening efforts and are likely to underestimate the total NR complement of each group. M, Metazoa; B, Bilateria; P, Protostomia; D, Deuterostomia; E, Ecdysozoa; L, Lophotrochozoa, within the last century. These substances serve as NR ligands because their stereochemistry allows them to fit by chance into NR binding pockets, which have not been selected over the long term to exclude binding of these substances.

fore are potential targets for endocrine disruption. Further, the ancient nature of ligand binding makes it likely that endogenous ligands and endocrine disruptors will be discovered for many orphan receptors, as has already occurred for many [12,13].

PLASTICITY IN THE EVOLUTION OF LIGAND BINDING

Since the first ligand-binding receptor, there has been remarkable diversification of receptors' ligand-specifity. All NR ligands are hydrophobic and of relatively small size, but beyond these commonalities they are very diverse. High-affinity endogenous ligands for NRs include a variety of cholesterol-derivatives (steroid hormones, vitamin D, bile acids and other cholesterol metabolites), retinoids, modified amino acids (thyroid hormone), prostaglandins, leukotrienes, and several kinds of fatty acids and benzoates.

Many of the gene duplications by which NRs diversified have been followed by the evolution of novel ligand affinity. This is true even for the relatively recent chordate-specific duplications. For example, the vertebrate gonadal/andrenal steroid receptors and the members of the VDR and LXR/FXR groups are each activated by a unique hormone or group of hormones that do not bind to the others with high affinity. (The estrogen receptors α and β are an exception, although they do vary in their affinity for endocrine disruptors [14]).

Receptors have also evolved affinity for new ligands after speciation events. The ecdysone receptor of arthropods is orthologous to the FXR, LXRα, and LXRβ of vertebrates, which bind oxysterols and bile acids [15], but vertebrates do not respond to ecdysteroids. A similar pattern is found with more recent divergences: the principal ligand for the teleost androgen receptor, for example, is not testosterone (T) or 5α-dihydro-T as it is in tetrapods, but 11-ketoT [16]. It is thus impossible to predict the ligands or EDs that will bind to a receptor by extrapolation from other—even closely related —receptors or model species from other classes or phyla.

PARALLEL EVOLUTION OF LIGAND SPECIFICITY AND PROMISCUITY

There is a surprising lack of consistency in the evolution of certain types of ligands on the NR phylogeny. Steroids are bound by the adrenal/gonadal steroid receptors as well as by NRs in the distant VDR group, but the receptors that intervene in the tree between these groups are orphans or bind nonsteroidal ligands. Similarly, RXRs and the distantly related RARs are both activated by retinoic acids. The ability to bind steroids or retinoids must, therefore, be either an ancestral character lost and then regained in the more recent receptor group or a parallel acquisition in the NR classes in which it is found.

The use of steroids and similar molecules as high-affinity endocrine ligands has evolved in parallel in numerous other gene families. In plants, brassinosteroids bind to membrane receptor kinases and function as critical regulators of plant development [17]. Steroids regulate reproduction in the fungus *Achyla ambisexualis* [18], but fungi lack NRs. And in nitrogen-fixing symbiotic bacteria, the unrelated protein NodD is an intracellular transcriptional regulator that is activated by flavonoids excreted by the partner plant. NodD has a mode of action very similar to that of the NRs; NodD is antagonized by several pesticides and industrial chemicals that also disrupt vertebrate steroid receptors [19]. All these receptors are potential—and ecologically important—targets for environmental endocrine disruptors (EDs).

That several unrelated families of proteins independently evolved the ability to act as high-affinity receptors for steroids and chemically similar compounds suggests that these substances have intrinsic properties that pre-adapt them to long-distance signaling functions. Steroids are small and lipid soluble, which allows them to cross cell membranes. Their chair-like polycyclic structure makes them rigid enough to establish stereospecific binding relationships with receptors, and their structural complexity allows for the synthesis of hundreds of unique compounds by substitution of small moieties at various points around the steroid nucleus. Finally, steroids are metabolically cheap to make, since cholesterol and other steroid-precursors are essential to membrane biochemistry and other fundamental processes in eukaryotes.

NRs, in turn, have evolved their affinity for one or a few specific ligands in the context of the endogenous and exogenous chemical background of the organisms of which they are a part. Virtually all NRs are potentially promiscuous, as shown by the phenomenon of endocrine disruption and the existence of synthetic pharmaceuticals that bind receptors with very high affinity. This pattern of endogenous specificity and xenobiotic promiscuity is likely to have an evolutionary explanation. The function of most receptors is to transduce an endocrine signal, which indicates the internal state or environmental context of the organism, into a specific cellular response. (There are several broad-specificity receptors, but these regulate metabolism of the ligands that bind to them rather than more fundamental developmental or physiological responses). Receptors that are activated or antagonized by other compounds present in the organisms' endogenous chemistry or diet would add noise into the information processing capacity of the cell and organism, impairing its ability to regulate its development or respond to environmental conditions. Natural selection is, therefore, expected to drive receptors towards greater specificity for their primary ligands. In contrast, xenobiotic drugs and industrial chemicals have been created within the last century. These substances serve as NR ligands because their stereochemistry allows them to fit by chance into NR binding pockets, which have not been selected over the long term to exclude binding of these substances.

This promiscuity for evolutionarily novel ligands predicts that a nontrivial portion of recently developed hydrophobic synthetic chemicals of small to moderate size will bind to and disrupt nuclear receptors. There are an estimated 87 000 synthetic substances in commerce, plus thousands more formed and released as by-products of chemical production, use, and combustion [20]. Many of these—including polymers and hydrophilic molecules—are unlikely to be NR ligands. But a substantial number have shapes that may attract them to the hydrophobic cores of NRs. The affinity of novel ligands for receptors is often low, but pharmacokinetic differences and simultaneous exposure to many EDs may make environmental exposures significant [21].

DIVERSIFICATION OF ORGANISMAL FUNCTIONS

The regulatory roles played by NRs at the organismal level also reveals considerable evolutionary plasticity. The EcR of arthropods, the master regulator of insect metamorphosis, is orthologous to a group of three hepatic receptors (FXR, LXRα, and LXRβ), which regulate cholesterol homeostasis in vertebrates. It is not closely related to the thyroid hormone receptor, which regulates metamorphosis in amphibians. Similarly, the SEX-1 receptor of nematodes is involved in sex determination, but its ortholog E78a functions in insect metamorphosis [22].

It is, therefore, not possible to predict from distantly related mammalian models the phenotypic consequences that will ensue when a receptor is activated or disrupted by endocrine-disrupting chemicals. This makes the design of in vivo assays for environmental disruption of nuclear receptors difficult in taxa for which we do not yet know the natural functions of nuclear receptors.

BRIEF SURVEY OF NUCLEAR RECEPTOR DIVERSITY

In the following sections, I review the distribution of NRs in major nonmammalian taxa (Fig. 2). (Birds, reptiles, and teleosts are treated only briefly, because they are discussed in detail by Crews and colleagues in this volume.) For each group, I describe the identified NRs, based on taxonomically restricted protein BLAST searches of the National Center for Biological Information's nonredundant (NR) sequence database [23]. With the exception of species whose genomes have been completely sequenced, these descriptions are almost certainly incomplete: the methods used to isolate NRs—PCR and nucleic acid hybridization—are subject to false negatives, particularly when sequences have diverged greatly since an ancient common ancestor. Identified receptors therefore give a lower bound on the actual NR set in any taxon. I also describe what is known about the function of these receptors, emphasizing evolutionary implications, and briefly discuss knowledge of and potential for endocrine disruption in each taxon.

CNIDARIA

Cnidarian genomes have been found to contain 11 nuclear receptors, including phylogenetic orthologs of the vertebrate receptor groups RXR, COUP, tailless (TLL), the testicular receptors TR2 and TR4, and SF-1 [10,24]; several of the other Cnidarian genes are so divergent that they cannot be clearly assigned to ortholog groups. Only one of the Cnidarian receptors, the RXR of the jellyfish *Tripedalia cystophora*, has been functionally analyzed: like its mammalian ortholog, it activates gene express in the presence of very low concentrations of 9-cis retinoic acid [11], although its organismal role remains unknown. This receptor is therefore a candidate for endocrine disruption; if the others are also ligand-regulated, they may also be ED targets.

There may be other ligand-binding receptors in Cnidaria yet to be discovered. Vertebrate hormone-type steroids have been identified in mass coral spawns in concentrations that correlate with reproductive/developmental function [25], suggesting the possibility that these hormones are playing signaling functions in Cnidaria and that receptors that transduce their signals may yet be discovered. No studies of endocrine disruption in Cnidaria have been conducted, despite the fact that these animals play important roles in some marine food webs.

ECDYSOZOANS: ARTHROPODS, NEMATODES, AND OTHER PHYLA

Bilaterally symmetric animals are divided phylogenetically and taxonomically into two major groups: the protostomes (in which gut development occurs mouth first) and deuterostomes (in which the anus develops first). The prostostomes are in turn divided into two great clades: the Ecdysozoa (organisms that molt, including arthropods, nematodes, priapulids, and numerous others) and the Lophotrochozoa (organisms with specific kinds of larvae or mouth structures, including annelids, mollusks, kinorhyncs, sipunculans, and numerous other marine phyla) (Fig. 2).

The ecdysozoans include the model organisms *D. melanogaster* and *C. elegans*, the genomes of which have been completely sequenced. The fruit fly contains 21 nuclear receptors, most of which are involved in molting and metamorphosis, embryonic development, or behavior [26]. Several of these genes appear to encode arthropod or dipteran-specific receptors—including three developmental NRs that have no ligand-binding domains (Knirps, Kruppel, and Egon), as well as duplicates of the SF-1 gene and of the photoreceptor nuclear receptor gene (PNR).

The nematode *C. elegans* presents perhaps the most radical example of lineage-specific NR diversity. *C. elegans* possesses orthologs of most of the NRs identified in arthropods, with the notable exception of EcR and RXR [27], although an RXR ortholog has been detected in the nematode *Dirofilaria immitis* [28]. In addition to the previously recognized NRs, however, the *C. elegans* genome contains an additional >200 nuclear receptors, and a large number are also present in *Caenorhabditis briggsae* [27]. Phylogenetic analysis and gene mapping data show that the vast majority of these genes cluster together, indicating that the excess NRs are largely due to tandem gene duplications that occurred within the nematode lineage. Few of these receptors have been functionally assayed, but most have intact ligand-binding domains with hydrophobic pockets, suggesting that they may be regulated by ligands.

No information on the nuclear receptor content of other ecdysozoan phyla are available, and little research has been carried out on endocrine disruption in any ecdysozoans. It is likely, however, that signaling via some ecdysozoan receptors can be disrupted by synthetic chemicals: polycyclic aromatic hydrocarbons potentiate ecdysone signaling through the EcR, and numerous phytochemicals are EcR antagonists [29,30]. In addition, metabolites of the insect control chemical methoprene disrupt the action of juvenile hormone—another master regulator of moulting and metamorphosis, the receptor for which has not been identified—and are also potent agonists of the mammalian RXR [31]. Insects, crustaceans, spiders, nematodes, and other ecdysozoans play major roles in terrestrial and marine food webs, and some species in these groups—from butterflies to lobsters—are of considerable economic and cultural importance.

LOPHOTROCHOZOANS: MOLLUSKS, ANNELIDS, ETC.

The nuclear receptors of lophotrochozoans have not been well studied. Published complete or partial sequences are available only for a COUP of an annelid and a COUP-TF, an RXR, and an NGFI-B ortholog from a mollusk. No studies whatsoever are available on other lophotrochozoans, such as sipunculans, bryozoa, or echiura.

Recently, however, it has been discovered that one major group of receptors previously thought to be unique to vertebrates is in fact ancient and widespread. Thornton, Need, and Crews [56] have found that mollusk neural and reproductive tissues express a nuclear receptor gene that is unambiguously orthologous to the vertebrate estrogen receptor. This finding indicates that the gonadal and adrenal steroid receptor family—the estrogen, androgen, progestin, and corticoid receptors—are far more ancient than previously thought, with their origin predating the protostome-deuterostome divergence. The ER must have been lost in the ecdysozoan lineage leading to flies and nematodes, presumably due to evolutionary changes that made the receptor's functions dispensable. We can therefore predict that all lophotrochozoans and all deuterostomes—and possibly some ecdysozoans, as well—will possess estrogen receptor orthologs. Their endocrine systems may thus be subject to endocrine disruption by ER agonists and antagonists.

The discovery of an ER in mollusks provides a potential mechanism for the previously unexplained findings of vertebrate-type steroids, steroid-binding activity, and steroid-induced reproductive changes in mollusks (reviewed in ref. [32]). Further, endocrine disruption and reproductive impairment has been documented in mollusks after exposure to pollutants that are vertebrate estrogens [33]. If steroids do play a role in mollusk reproduction via the mollusk ER, then some of the xenobiotic estrogens and antiestrogens now circulating in the environment have the potential to cause significant damage to populations of mollusks and other ER-containing lophotrochozoan phyla, which are important in aquatic and terrestrial ecosystems.

INVERTEBRATE DEUTEROSTOMES

Deuterostomes are composed of two major phyla: the echinoderms (sea stars and urchins) and the chordates (tunicates, arrow worms, branchiostomes, and vertebrates). Very little work has been done to characterize the diversity, function, or disruption of NRs in echinoderms. Only three nuclear receptors—the orphans SF-1, COUP, and TR2/4—have been identified in echinoderms. Steroids, including testosterone and estradiol, however, have been found to be produced by sea stars and to vary with the animal's reproductive cycle, suggesting a signaling role in these organisms (reviewed in ref. [31]). The finding of an ancient origin for the estrogen receptor indicates that echinoderms are likely to possess ER orthologs. If they regulate the response to these steroids, these receptors may be targets for endocrine disruption.

A bit more information is available on NRs in urochordates, the most basal chordates. A preliminary study identified fragments that are apparently orthologs of RXR, SF-1, and estrogen-related receptors (ERR1) in urochordates [10]. In addition, full-length RXR and RAR sequences have been identified in the tunicate *Polyandrocarpa misakiensis*, and the expressed proteins, like their vertebrate orthologs, have been shown to activate transcription in the presence of retinoic acid [34]. In addition, a thyroid hormone receptor has been identified in the tunicate *Ciona intesitnalis*. Surprisingly, it did not activate transcription in the presence of triiodothyronine (T3), the vertebrate ThR ligand, despite the fact that thyroxine, the precursor of T3, plays a well-established role in *Ciona* metamorphosis [35]. If this result is not an artifact, it suggests a possible shift during urochordate or vertebrate evolution in the receptor's affinity for different metabolites of thyroxine.

Tunicates are the most basal taxa in which RAR and ThR have been identified, indicating that these receptor groups first emerged before tunicates split from the rest of the chordate lineage. Only a single member of each group has been recovered from species in this subphylum, a pattern consistent with the serial genome duplication hypothesis. RARs and ThRs are, therefore, expected to be present in all chordates—including hemichordates, urochordates, and cephalochordates—and these receptors may be targets for disruption by synthetic chemicals.

The cephalochordate genus *Branchiostoma*—including the species known as amphioxus—are the invertebrates most closely related to vertebrates. Single members of the NR groups SF-1, RXR, COUP, RAR, ERR, PPAR, FXR, and TR2/4 have been identified and sequenced in branchiostomes. Like its vertebrate ortholog, the amphioxus RAR activates transcription in the presence of retinoic acid [36]. Each receptor appears to be present in amphioxus in a single copy—again consistent with serial genome duplications, with both duplications occurring after the amphioxus-vertebrate divergence.

JAWLESS AND CARTILAGINOUS FISHES

Hagfish, lamprey, and elasmobranches are the most basally branching vertebrates. In hagfish, a PCR screen found one ortholog each of COUP-TFs, PPARs, ERRs, and an unidentified steroid receptor [10]; other nuclear receptors are likely present but were not identified.

Lamprey have been studied in more detail. One RXR, three PPARs, two ERRs, one NGFI-B and a COUP-TF ortholog were found in a preliminary screen [10]. Subsequently, an estrogen receptor, a progesterone-like receptor, and a corticoid receptor were identified in the sea lamprey *Petromyzon marinus*. Phylogenetic analysis indicated that lampreys are not expected to posses orthologs of the androgen, mineralocorticoid, and estrogen receptor beta, because these NRs were produced by duplications in the jawed vertebrate lineage after its divergence from jawless fishes about 435 million years ago [37]. This picture is consistent with the vertebrate serial genome duplication hypothesis, with one duplication occurring before the divergence of lamprey from jawed vertebrates and the other occurring after. The ancestral steroid receptor was inferred by phylogenetic methods and structure–function prediction to have functioned as an estrogen receptor, despite the fact that estrogens are the terminal hormones in the steroid biosynthetic pathway. The PR and AR are therefore duplicates of an ancient ER gene, which subsequently evolved high affinity for progestins and androgens, thereby turning intermediates in the estrogen synthesis pathway into bona fide hormones.

Very limited research has been conducted on the NRs of cartilagionous fishes. Sharks have an estrogen receptor, and also apparent orthologs of the androgen and glucocorticoid receptors, although the sequences of the latter are so short that orthology to other steroid receptors cannot be ruled out. In addition, two PPARs, two Reverbs, two ERRs, and two COUPs have been identified in a preliminary screen of shark cDNA [10]. The likely presence of orthologs to a variety of ligand-binding receptors indicates that cartilaginous fishes may be targets for a wide range of endocrine disruptors. It is also clear that some receptors in sharks have evolved lineage-specific ligand preferences. For example, the principle mineralocorticoid is aldosterone in mammals, cortisol in telesots, but 1α-hydroxycorticosterone in cartilaginous fishes [38].

TELEOST FISHES

Teleosts, the major group of aquatic bony fishes, appear to possess all the receptor types found in mammals, although some NRs have not been specifically studied. For example, teleosts have orthologs of all six steroid receptors found in humans and mice, and they have a full complement of RARs, RXRs, COUPs, and thyroid hormone receptors. In fact, due to an apparent whole-genome duplication that caused the expansion of a large number of gene families deep in the teleost lineage, followed by the retention of some duplicates and the apparent loss of others [39], bony fish have more NRs than most mammals, birds, and reptiles do. For example, several teleost species have been found to have two androgen receptors [40], two progesterone receptors [40], and three estrogen receptors [41] in contrast to the one, one, and two members of each group, respectively, that are found in tetrapods. As sequencing of teleost genomes continues, it is likely that most teleosts will have a set of NRs larger than tetrapods' by a factor of up to two.

There has been functional diversification of teleost NRs at both organismal and molecular levels. Estrogens and estrogenic pesticides, for example, can reverse the primary sexual differentiation of male fish, a phenomenon unknown in mammals [42]. Like the teleost androgen receptor, which has unique ligand affinity (see above), teleost progesterone receptors are activated primarily by 17α,20β-progesterone and 17α,20β,21-trihydroxypregnenone, not progesterone as they are in tetrapods [43,44]. Teleost receptors also have quantitatively different relationships with endocrine disruptors than their mammalian orthologs: endosulfan, dieldrin, and methoxychlor bind to the ERα of trout but not to those of human, mouse, chicken, or lizard [45].

BIRDS, REPTILES, AND AMPHIBIANS

The proliferation of nuclear receptors largely stopped before the divergence of tetrapods into its major classes, so most birds, reptiles, and amphibians have the same complement of NRs as mammals do. There are exceptions, however: the frog *Xenopus laevis*, an apparent tetraploid species, has retained duplicates of some mammalian NRs, such as the RARs, SF-1, and ERα [46,57].

Although the number of NRs has remained more or less stable, there has been considerable functional diversification of receptors in these vertebrate lineages. At the organismal level, thyroid hormone regulates metamorphosis in amphibians, but it regulates basal metabolism, behavior, and neurological development in nonmetamorphic vertebrates. At the molecular level, the benzoate X receptors (BXRs) of *X. laevis* are activated by benzoic acids, but their mammalian ortholog PXR/SXR binds steroids with no affinity for benzoates [47]. The potency of synthetic endocrine disruptors also varies considerably from that in mammals: chicken estrogen receptors, for example, have affinity for the pesticide chlordecone 100-fold higher than in the rat [3]. Endocrine disruption has already been documented in populations of nonmammalian vertebrates at environmental concentrations (i.e., refs [48,49]).

IMPLICATIONS FOR ENDOCRINE DISRUPTION

The evolutionary insights discussed in this paper have two major implications for endocrine disruption assessment and policy, which currently rely largely on testing in model vertebrates to identify individual EDs that interact with a few NRs. For example, U.S. EPA's Endocrine Disruptor Screening Program is using assays on mammals, amphibians, and teleosts to identify disruptors of the estrogen, androgen, and thyroid disruptors only [20].

The first insight is this: NRs have diversified so extensively in the many metazoan lineages that scientific and policy efforts focused on a few receptors in a few vertebrate classes cannot be expected to protect all species from all kinds of endocrine disruption. Since their origin, NRs have been diversifying independently in all the major metazoan lineages over many billions of years of parallel evolutionary time. Each receptor in each taxonomic lineage has the potential, often realized, to be functionally unique. It is therefore not possible to predict with confidence the existence, ligands, and functions of NRs in taxa that have not been specifically and empirically evaluated. The more distantly related a species is from mammalian models, the longer the evolutionary time over which its receptors have been diversifying, and the less reliable extrapolations from mammals will be. Programs to identify EDs using vertebrate models are therefore likely to falsely judge as non-EDs chemicals that are in fact ligands for nonmammalian NRs. The broad diversity of ligands for which NRs have evolved affinity implies that a nontrivial number of the many thousands of synthetic chemicals now in production are likely to have the potential to disrupt NR-mediated signaling.

Endocrine disruptor screening and testing programs should be expanded to evaluate a broader array of nonmammalian taxa. Such programs might be expanded to include, for example, testing of several classes each of arthropods, mollusks, annelids, cnidarians, echinoderms, and nonmammalian vertebrates. Such a program would be very demanding of time and resources, and it would leave numerous phyla and taxonomic classes empirically unevaluated. Still, it is the only way to gain even preliminary insight into the effects of synthetic chemicals on the endocrine systems of nonmammalian taxa and avoid false negative judgments of chemical safety.

The second insight relates to the impact of limited knowledge on ED science and policy. Current understanding of NRs and their functions throughout the Metazoa barely scratches the surface of their actual diversity. The full nuclear receptor complement is known for only two model invertebrates, and these are unfortunately relatively closely related to each other. In poorly studied taxa like echinoderms, mollusks, and annelids, our understanding of the NRs present and their functions ranges from very little to absolutely nothing. Even in such well-studied model organisms as humans, mice, and flies, we remain uncertain of the ligands and physiological functions of many receptors. To some extent, knowledge gaps can be repaired with more research. But the gaps are so great that we cannot even design an adequate program to identify EDs and their impacts in most nonmammalian taxa. The physiological/developmental roles of most NRs are unknown in the majority of invertebrate taxa; these roles have diversified so thoroughly that the endpoints used as indicators of endocrine disruption based on their function in mammals are seldom appropriate in invertebrates. At present, then, we lack the background knowledge to develop in vivo assays for endocrine disruption in most nonmammalian taxa, and extrapolation across phyla, classes, and receptors is often unreliable.

Efforts to identify as many endocrine disruptors as possible through broader testing are undoubtedly valuable for scientific and regulatory purposes. But we cannot be confident that any feasible testing program will be able to identify all—or even the majority—of EDs in animals. EDs that interact with receptors in bacteria, fungi and plants are even further beyond the ken of such a program. The likely outcome of current testing programs' focus on vertebrates and a small number of nuclear receptors is that a nontrivial number of EDs will be judged safe, though they may be capable of causing significant ecological damage. This situation will not be remedied soon, and exposure to endocrine-disrupting chemicals will continue as long as policies require each and every endocrine disruptor to be individually and positively identified before action can be taken to reduce its production or use.

To cope with our lack of knowledge on metazoan endocrinology and endocrine disruption, environmental policies can be based on the precautionary principle: action should be taken to prevent plausible environmental hazards even in the absence of specific and definitive evidence of damage [50]. Rather than requiring evidence of endocrine disruption for each and every chemical, as current policies do, a precautionary framework would prioritize classes of substances that are particularly likely to contain EDs, based on their structure, size, hydrophobicity, and persistence. Policies would then focus on avoiding the production and use of members of these classes by developing and implementing safer materials and processes [51], based on the principles of sustainability and green chemistry [52].

CONCLUSIONS

Over the course of billions of years, receptors and the organisms in which they function have been evolving endocrine systems that are astonishing in their complexity, diversity, and biological importance. Protecting the life forms and ecosystems that have emerged from this evolutionary process will require that our policies take account of these characteristics. Expanding testing programs is useful, but true prevention of endocrine disruption will require a new approach to chemicals management that begins with a recognition of the limits as well as the strengths of current scientific tools and knowledge in the face of nature's diversity and complexity.

ACKNOWLEDGMENTS

I am grateful to David Crews for sharing ideas, reviewing the manuscript, and providing the opportunity to contribute to this volume. I thank the Columbia Earth Institute and Darcy Kelley for hosting this research.

REFERENCES

1. A. Aranda and A. Pascual. *Physiol. Rev.* **81**, 1269–304 (2001).
2. J. A. McLachlan. *Endocr. Rev.* **22**, 319–341 (2001).
3. A. A. Rooney and L. J. Guillette. In *Environmental Endocrine Disrupters: An Evolutionary Perspective,* L. J. Guillette and A. D. Crain (Eds.), pp. 82–125, Taylor & Francis, New York (2000).
4. European Biotechnology Institute. SMART database, <http://smart.embl-heidelberg.de>.
5. J. M. Maglich, A. Sluder, X. Guan, Y. Shi. D. D. McKee, K. Carrick. K. Kamdar, T. M. Willson, J. T. Moore. *Genome Biol.* **2**, 0029.1–7 (2001).
6. V. Laudet. *J. Mol. Endocrinol.* **19**, 207–226 (1997).
7. J. W. Thornton and R. DeSalle. *Syst. Biol.* **49**, 183–201 (2000).
8. F. J. Ayala, A. Rzhetsky, F. J. Ayala. *Proc. Natl. Acad. Sci. USA* **95**, 606–611 (1998).
9. J. Spring. *Nat. Genet.* **31**, 128–129 (2002).
10. H. Escriva, R. Safi, C. Hanni, M-C Langlois, P. Saumitou-Laprade, D. Stehelin, A. Capron, R. Pierce, V. Laudet. *Proc. Natl. Acad. Sci. USA* **94**, 6803–6808 (1997).
11. Z. Kostrouch, M. Kostrouchova, W. Love, E. Janini, J. Piatigorsky, J. E. Rall. *Proc. Natl. Acad. Sci. USA* **95**, 13442–13447 (1998).
12. S. A. Kliewer, J. M. Lehmann, T. M. Willson. *Science* **284**, 757–760 (1999).
13. G. B. Tremblay, T. Kunath, D. Bergeron, L. Lapointe, C. Champigny, J. A. Bader, J. Rossant, V. Giguere. *Genes Dev.* **157**, 833–838 (2001).
14. T. Barkhem, B. Carlsson, Y. Nilsson, E. Enmark, J. A. Gustafsson, S. Nilsson. *Mol. Pharmacol.* **54**, 105–112 (1998).
15. M. L. Fitzgerald, K. J. Moore, M. W. Freeman. *J. Mol. Med.* **80**, 271–281 (2002).
16. T. Todo, T. Ikeuchi, T. Kobayashi, Y. Nagahama. *Biochem. Biophys. Res. Comm.* **254**, 378–383 (1999).
17. C. Mussig and T. Altmann. *Trends Endocrinol. Metab.* **12**, 398–401 (2001).
18. M. K. Agarwal. *FEBS Lett.* **322**, 207–210 (1993).
19. J. E. Fox, M. Starcevic, K. Y. Kow, M. E. Burow, J. A. McLachlan. *Nature* **413**, 128–129 (2001).
20. U.S. Environmental Protection Agency. *Fed. Reg.* **63** (248), 71542–71568 (1998).
21. E. Silva, N. Rajapakse, A. Kortenkamp. *Environ. Sci. Technol.* **36**, 1751–1756 (2002).
22. I. Carmi, B. Kopczynski, B. J. Meyer. *Nature* **396**, 168–173 (1998).
23. S. F. Altschul, T. L. Madden, A. A. Schäffer, J. Zhang, Z. Zhang, W. Miller, D. J. Lipman. *Nucl. Acids Res.* **25**, 3389–3402 (1997).
24. L. C. Grasso, D. C. Hayward, J. W. H. Trueman, K. M. Hardie, P. A. Janssens, E. E. Ball. *Mol. Phylogen. Evol.* **21**, 93–102 (2001).
25. S. Atkinson and M. J. Atkinson. *Coral Reefs* **11**, 33–35 (1992).
26. T. Kozlova and C. S. Thummel. *Trends Endocrinol. Metab.* **11**, 276–80 (2000).
27. A. E. Sluder and C. V. Maina. *Trends Genet.* **17**, 206–213 (2001).
28. U.S. National Center for Biotechnology Information. Genpept database sequence AAM08268; <www.ncbi.nlm.nih.gov/entrez>.
29. E. Oberdorster, D. M. Cottam, F. A. Wilmot, M. J. Milner, J. A. McLachlan. *Toxicol. Appl. Pharmacol.* **160**, 101–108 (1999).

30. E. Oberdorstor, M. A. Clay, D. M. Cottam, F. A. Wilmot, J. A. McLachlan, M. J. Milner. *J. Steroid Biochem. Mol. Biol.* **77**, 229–238 (2001).
31. M. A. Harmon, M. F. Boehm, R. A. Heyman, D. J. Manglesdorf. *Proc. Natl. Acad. Sci. USA* **92**, 6157–6160 (1995).
32. G. Leblanc. In *Endocrine Disruption: An Evolutionary Perspective,* L. J. Guillette and A. D. Crain (Eds.), pp. 126–154, Taylor & Francis, New York (2000).
33. J. Oehlemann, U. Schulte-Oehlmann, M. Tillmann, B. Markert. *Ecotoxicol.* **9**, 383-397 (2000).
34. M. Kamimura, S. Fujiwara, K. Kawamura, T. Yubisui. *Dev. Growth Differ.* **42**, 1–8 (2000).
35. E. Carosa, A. Fanelli, S. Ulisse, R. Di Lauro, J. E. Rall, E. A. Jannini. *Proc. Natl. Acad. Sci. USA* **95**, 11152–11157 (1998).
36. H. Escriva, N. D. Holland, H. Gronemeyer, V. Laudet, L. Z. Holland. *Development* **12**, 2905–2916 (2002).
37. J. W. Thornton. *Proc. Natl. Acad. Sci. USA*. **98**, 5671–5676 (2001).
38. H. A. Bern. *Science* **158**, 455–62 (1967).
39. J. S. Taylor, Y. Van de Peer, I. Braasch, A. Meyer. *Philos. Trans. R. Soc. London B Biol. Sci.* **356**, 1661–1679 (2001).
40. T. Ikeuchi, T. Todo, T. Kobayashi, Y. Nagahama. *Comp. Biochem. Physiol. B Biochem. Mol. Biol.* **129**, 449–455 (2001).
41. M. B. Hawkins, J. W. Thornton, D. Crews, J. K. Skipper, A. Dotte, P. Thomas. *Proc. Natl. Acad. Sci. USA*. **97**, 10751–10756 (2000).
42. J. S. Edmunds, R. A. McCarthy, J. S. Ramsdell. *Environ. Health Perspect.* **108**, 219–224 (2000).
43. J. Pinter and P. Thomas. *J. Steroid Biochem. Mol. Biol.* **60**, 113–119 (1997).
44. T. Todo, T. Ikeuchi, T. Kobayashi, H. Kajiura-Kopbayashi, K. Suzuki, M. Yoshikuni, K. Yamauchi, Y. Nagahama. *FEBS Lett.* **465**, 12–17 (2000).
45. J. Matthews, T. Celius, R. Halgren, T. Zacharewski. *J. Steroid Biochem. Mol. Biol.* **74**, 223–234 (2000).
46. M. K. Hughes and A. L. Hughes. *Mol. Biol. Evol.* **10**, 1360–1369 (1993).
47. L. B. Moore, J. M. Maglich, D. D. McKee, B. Wisely, T. M. Willson, S. A. Kliewer, M. H. Lambert, J. T. Moore. *Mol. Endocrinol.* **16**, 977–986 (2002).
48. T. B. Hayes, A. Collins, M. Lee, M. Mendozam, N. Noriega, A. A. Stuart, A. Vonk. *Proc. Natl. Acad. Sci. USA* **99**, 5476–5480 (2002).
49. G. A. Fox. In *Chemically-Induced Alterations in Sexual and Functional Development,* T. Colborn and C. Clement (Eds.), pp. 147–158, Princeton Scientific (1992).
50. J. Tickner and C. Raffensperger (Eds.). *Interpreting the Precautionary Principle*, Island Press, Washington, DC (2000).
51. J. Thornton. In *Science and the Precautionary Principle,* J. Tickner (Ed.), Island Press, Washington, DC. In press.
52. P. Anastas and J. Warner. *Green Chemistry,* Oxford University Press (1998).
53. K. Bremer. *Cladistics* **10**, 295–304 (1994).
54. D. Swofford. PAUP* (version 4.0b10). Sinauer, Sunderland, MA (2002).
55. J. D. Thompson, T. J. Gibson, F. Plewniak, F. Jeanmougin, D. G. Higgins. *Nucl. Acids Res.* **25**, 4876–4882 (1997).
56. J. W. Thornton, E. Need, D. Crews. *Science* **301**, 1714–1717 (2003).
57. K. H. Wu, M. L. Tobias, J. W. Thornton, D. B. Kelley. *Gen. Comp. Endocrinol.* **133**, 38–49 (2003).

TOPIC 2

ENVIRONMENTAL FATE AND METABOLISM OF ENDOCRINE ACTIVE SUBSTANCES

Pure Appl. Chem., Vol. 75, Nos. 11–12, pp. 1843–1857, 2003.

Topic 2.1

Analysis of endocrine active substances in food and the environment*

Patrick T. Holland

Cawthron Institute, 98 Halifax Street E, Nelson, New Zealand

Abstract: A critical review is made of techniques for analysis of residues of endocrine active substances (EASs) including sampling, extraction cleanup and determination based on GC/MS, LC/MS, ELISA, and bioassays. The growing importance of receptor-based in vitro bioassays is highlighted for integrated monitoring of environmental levels of certain classes of EASs and for establishing exposures. Some recent advances in methods of analysis for each of the key classes of EASs are summarized including for organochlorines, PCBs, dioxins and dioxin-like substances, polybrominated diphenyl ethers, phenolic xenoestrogens, phthalates, organotin compounds, steroidal hormones, and phytoestrogens. The issues raised in interpreting complementary chemical and bioassay data at an effects level are briefly discussed.

INTRODUCTION

This chapter provides a critical review of analytical methods that can detect, chemically characterize, and quantify levels of endocrine active substances (EASs) in foods and environmental matrices including water, effluents, sediments/soil, and wildlife. Biological monitoring is particularly important for the field of EASs. The concept is defined by IUPAC as: "The continuous or repeated measurement of potentially toxic substances or their metabolites or biochemical/physical effects in tissues, secreta, excreta and expired air in order to evaluate occupational or environmental exposure and health risk by comparison with appropriate reference values based on knowledge of the probable relationship between ambient exposure and resultant adverse health effects." Recent reviews have discussed aspects of this topic [1].

Risk analysis for endocrine-disrupting chemicals (EASs—here taken as proven or potential) in the environment or humans requires quantitative exposure estimates. These should be based on analytical data covering all compounds that may be expected to give rise to significant additive, antagonistic, or synergistic biological effects. EASs are represented in many chemical structure classes covering a range of polarities and other physical properties and including both synthetic organic chemicals and natural products. Continuing discoveries of subtle endocrine activities for many persistent organic pollutants (POPs) emphasize their particular significance as bioaccumulative EASs that require careful monitoring and control [2]. The biological activities of EASs with closely related structures may vary by several orders of magnitude with ppt-ppb levels of some chemicals being significant for endocrine effects while related compounds may only begin to exert effects at ppm levels. Therefore, general chemical screening for EASs must be based on methods that have detection capability for many specific com-

*Report from a SCOPE/IUPAC project: Implication of Endocrine Active Substances for Human and Wildlife (J. Miyamoto and J. Burger, editors). Other reports are published in this issue, *Pure Appl. Chem.* **75**, 1617–2615 (2003).

ponents, high sensitivity, and a wide dynamic range. This is a demanding task, made more difficult for food and many environmental matrices by their complex natures. In many cases, recourse must be had to several chemical assays each covering sub-sets of the required analytes. These studies may also extend to precursors of EASs such as the alkylphenol-ethoxylate surfactants or to metabolites such as hydroxylated PCBs, which may be more active than the parent.

Modern analytical chemistry has available a range of powerful technologies which are capable of accurately determining EASs to low levels in various food and environmental matrices. More rapid and less solvent-dependent techniques have been developed for extraction of analytes including accelerated solvent extraction (ASE) for solids and solid phase extraction (SPE) for aqueous samples. Preparative HPLC (silica, reversed phase, or size exclusion) provide universal clean-up systems that can be automated. Combined mass spectrometry with gas chromatography (GC/MS) or liquid chromatography (LC/MS) provide multi-residue capability, automation, and a high degree of confidence in the identification and quantification of EASs in complex matrices. LC combined with atmospheric pressure ionization has revolutionized the trace analysis of polar compounds that may be unstable in GC/MS [3].

Biochemical techniques such as receptor-linked expression assays for endocrine receptor agonists or immunoassays for biomarker proteins provide an alternative to chemical screening that allow biological monitoring of the integrated effects of diverse EASs with common modes of action [4]. They are increasingly also being used to provide estimates for environmental or food-monitoring purposes of the effective concentrations of particular classes of EASs (e.g., dioxins [5]) and to guide analyses in research to determine the specific chemicals causing adverse effects in an environment [6–8]. Assays in wildlife and man for biomarkers that are directly linked to EAS exposure levels (e.g., ELISA for the estrogenicity biomarker protein vitellogenin [9]) can also be used to establish the lower concentration ranges for particular EASs that are significant for biological effects. These bioassay results can assist setting of appropriate scope and limits of detection for chemical screening methods. Biosensors, where activation of a selective bioassay element is coupled by transducers to provide a concentration dependent electrical signal, are a significant research area. They provide analytical systems for continuous monitoring of temporal concentrations. Biosensors for EASs have been reported based on response elements for estrogens [10] and cytochrome P-450 [11].

The following sections summarize some of the key techniques for analysis of EASs and then review recent advances in methods of analysis for each of the most important classes of EASs.

OVERVIEW OF ANALYTICAL TECHNIQUES

Environmental sampling and sample preparation

Protocols for sampling of *water* emphasize the need for cleanliness of sample containers and adequate flushing of bores and sample lines where water is being pumped from wells or other remote locations [12]. Relatively large samples (0.5–2 l) are generally taken where possible to minimize walls effects in the sample container and provide laboratory flexibility. Sample preservation has mainly depended on cooling for multi-residue methods although specific compounds such as estrogens may tolerate low pH (2–3). Field extraction using SPE columns was shown to be a practical approach to sampling and preservation of labile surfactant and estrogen residues [13]. Field sampling of *soils* or *sediments* must ensure that that each sample is representative of the area of interest (plot, field, lake, estuary, etc.) and that cross-contamination is avoided. Where transport of EASs through soil or distribution in sediments is being studied, extreme care must be taken to avoid contamination of sub-surface core sections with upper sections where the majority of residues are typically found. Field sampling and laboratory preparation of organisms and wildlife are specialized activities, but biologists involved need to be aware of analytical requirements for representative and uncontaminated samples.

Extraction

Sample homogeneity is the major limitation to miniaturization of analytical methods for soil or sediment samples. Where analytes are stable to drying and fine-grinding, 2–5 g sub-samples may be adequate. However, overall cost-effectiveness of sample preparation and extraction will often dictate use of 10–50 g sub-samples of coarse-sieved (1–2 mm) moist field samples. ASE using elevated pressures and temperatures is an efficient and automated alternative to soxhlet extraction for intractable weathered residues of nonpolar contaminants such as organochlorine insecticides [14,15]. ASE is the basis for U.S. EPA SW-846 Method 3545 for extraction of contaminants from solid samples. The related technique of microwave-assisted solvent extraction also gave high recoveries of these contaminants, although degradation of some organophosphate compounds was observed [16,17]. Supercritical fluid extraction (SFE) using carbon dioxide requires expensive equipment but greatly reduced use of solvents [18]. Subcritical water (150 °C) has been shown to be an efficient extractant for POPs and some nonhydrolyzible pesticides [19]. Simple blending/ultrasonication/shaking methods using mixtures of water miscible (methanol, acetone, acetonitrile) and nonpolar (pentane, hexane) solvents either on field moist samples or with added water are effective for extraction of compounds of medium to low polarity from solids or biological specimens [20]. Water aids extraction by displacing analytes from polar binding sites and swelling clays. Buffers may be added to assist extraction of acidic compounds [21].

SPE techniques using C18 or divinylbenzene-styrene polymeric adsorbents have largely displaced liquid–liquid partitioning for extraction of water samples, even for samples with high bio-solids such as raw sewage waste [22]. Proprietary graphitized carbon columns have proved effective for extraction of polar compounds, including estrogens and surfactant degradates [23].

Clean-up

Humic substances, lipids, and pigments are among the many coextractives that can cause problems in analysis of EASs including direct interferences, matrix effects, and contamination of GC injectors. SPE columns in normal, reversed-phase, or ion-exchange formats are widely used in clean-up systems developed for specific classes of environmental contaminants [12b]. Careful optimization is required for multiresidue applications, particularly where HPLC-UVD is used for determination. Size-exclusion chromatography (gel permeation, GPC) clean-up systems have many advantages for multi-methods based on HPLC or HRGC determination. Automated GPC systems recover a wide range of low- to medium-MW compounds with high reproducibility and reliably remove higher molecular weight coextractives such as humic acids and pigments. Laboratory-packed Biobeads SX-3 columns (450 × 10 mm) are robust and of low cost [24], while rigid-bead commercial columns have some advantages in speed and resolution [25]. GPC followed by automated HPLC on silica was used to clean up low-level chlorobornane residues and remove PCBs prior to HRGC/MS (NCI) determination [26]. Ultra-trace analysis of dioxins and dioxin-like compounds requires specialized clean-ups with carbon-based column fractionation of the coplanar compounds [27]. Immuno- or receptor-based affinity columns have been used for highly selective cleanups for particular analyte classes [28].

GC/MS determination

GC with high-resolution capillary columns (HRGC) coupled to MS detection has developed into a primary technique for identification and quantitative determination of many EASs using small bench-top instruments with sophisticated data systems. Electron impact is the most common ionization method, although chemical ionization (positive or negative mode) has benefits for unstable or electron-capturing compounds, respectively. Single-quadrupole instruments rely on selected-ion monitoring (SIM) to achieve picogram sensitivities [29] while triple-quadrupole instruments can use MS/MS with selected reaction monitoring to enhance specificity. Ion-trap instruments can achieve high sensitivities for full-scan spectra and have MS/MS modes [30]. HRGC/MS with optimized capillary column separation could determine residues of 567 low- to medium-polarity compounds, including many EASs, within a 42-min run [31]. The very high effective scan rate of time-of-flight (TOF) instruments has enabled rapid GC/MS analysis of PCBs and other environmental contaminants [32]. The high selectivity of MS detection can reduce the need for sample clean-up, although coextractives may affect GC injector performance. Programmed temperature vaporizer (PTV) injection gave superior performance to splitless injection for trace analysis by HRGC/MS [33]. The direct sample introduction system of Amirav [34] allowed rapid screening with low detection limits when coupled to HRGC/MS/MS (ion-trap) for 22 pesticides in crude plant extracts [30] and alkylphenol polyethoxy-carboxylates (n = 1–9) in water [35]. The incorporation of isotope-labeled internal standards and surrogates increases the precision and accuracy of MS determination, although few labeled EASs are available. High-resolution MS techniques have become indispensable to achieving the lowest detection limits (low ppt) with high accuracy and precision for POPs such as dioxins [36] that occur as complex mixtures of congeners in association with other polychlorinated residues.

LC/MS determination

Advances in LC/MS interfacing (electrospray-ESI and atmospheric pressure chemical ionization-APCI) have enabled levels of sensitivity and reliability that are suitable for routine determination of EASs, particularly more polar compounds that would require derivatization for GC/MS. LC/MS can reduce clean-up requirements over HPLC-UVD based methods, although care must be taken with matrix effects on ESI responses that may affect quantitation. Spectra are often dominated by molecular cations or anions. The MS/MS (tandem-MS) modes available on ion-trap and triple analyzer instruments are valuable for confirmation of identity and to reduce typically high background signals. They utilize mass analysis of the collisional activation products of the predominant molecular anions or cations. The increased structural information and higher specificity of LC/MS/MS has been demonstrated for herbicide residues [37] and surfactants [38]. The techniques of GC/MS/MS, LC/MS (SIM), and LC/MS/MS were compared for determination of estrogenic hormones in environmental matrices and gave complementary information, but LC/MS/MS gave low detection limits without the need for derivatization [39].

Immunoassays

Radio-immunoassys have been the front-line technique for hormone determination in endocrinology for many years, and these assays have been adapted for food and environmental use. Enzyme-linked immunoassay (ELISA) has become an accepted technology for analysis of steroidal hormones and pesticide residues. Although it is beyond the capacity of most laboratories to develop their own antibodies for immunoassays, commercial kits are available for many compounds. ELISA can achieve remarkable sensitivities as well as high selectivity and adequate precision for screening of specific compounds [40]. A relatively simple extraction/clean-up is often adequate. IUPAC guidelines have been published for validation and operation of immunoassays for pesticide residue analysis [41]. The development of ELISAs for detection of EASs has been reviewed [42]. A high-sensitivity ELISA was shown to be a suitable basis for monitoring studies of specific estrogenic hormones at LODs of 0.05–0.1 ng/l for waste water and surface water respectively [43]. Results from an interlaboratory study comparing HPLC-UVD/FL with ELISA for determination of surfactants in wastewaters concluded that ELISA was an effective monitoring technique [44]. Immuno-affinity columns can provide very efficient extraction and clean-up of specific compounds from complex samples for standard instrumental analysis. The development of sol-gels as carriers for monoclonal antibodies in affinity columns provided greatly improved storage stability, high flows, and high specific binding capacities for s-triazine herbicides and dinitrophenol compounds [45]. Affinity media based on molecularly imprinted polymers (MIPS) [46] are also a development toward more stable, specific, and regenerable molecular recognition substrates for extraction/clean-up or biosensor applications.

Receptor-based bioassays

The wide range of potential EASs has led to the use of in vitro assays to initially characterize the type and levels of exposure on a more generic effects basis. This approach has been very successful for compounds with dioxin-like or estrogenic activities. Bioanalytical screening methods for dioxins and dioxin-like compounds have recently been critically reviewed [5,47] including those utilizing in vitro activation of Ah receptors such as CALUX assays (AhR coupled to luciferase gene expression) and the EROD assay (7-ethoxyresorufin-*O*-deethylase activity through cytochrome P-450 induction). The E-screen system for estrogens is based on proliferation of MCF-7 human breast cancer cell lines [48,49] and has been used to monitor estrogenic activities in STW [50]. The ER-YES-screen uses stable expression in recombinant yeast cells of the human estrogen receptor and plasmids for estrogen response elements and LacZ reporter leading to galactocidase production [51,52]. The estrogen receptor has also been coupled to the CALUX reporter gene [53]. The quantitative performance of the E-screen, YES screen, and in vitro estrogen receptor competitive binding assays were compared for detecting and identifying estrogenic EASs [54] with the conclusion that these assays provided complementary information [55]. In a comparison of in vitro bioassays for detection of estrogens in waste- and surface-waters, the ER-CALUX assay was more sensitive then the ER-YES screen [56]. Transformed yeast systems have also been produced to detect androgenic chemicals through expression of an androgen receptor with a response element coupled to a reporter plasmid [57]. Chemicals with anti-aromatase activity were also detectable by incorporating plasmids to create both aromatase and androgen receptor [58].

Bioactivity-guided fractionation and analysis

The identification of specific EASs in the environment has been guided by bioassays and receptor-assays. A common strategy was HPLC fractionation of extracts using preparative C18 reversed-phase chromatography. Collected fractions were bioassayed, and those from regions of significant activity were pooled and analyzed by GC/MS (following derivatization) or LC/MS. The ER-YES screen assay guided GC/MS identification of 17β-estradiol, estrone and 17α-ethynyl-estradiol in streams and sewage treatment waste (STW) at 1–50 ng/l [7]. These excreted estrogens were confirmed as significant EASs to fish in the streams using the vitellogenin biomarker [8]. The ER-YES screen was also used to guide the identification of nonylphenol and related compounds from wool-scouring effluent as significant estrogens in STW [59]. An androgen receptor-mediated transcription assay was used to guide identification of androstenedione in paper mill effluent and related to masculinization of fish in a river downstream of the mill [60]. Similarly, the ER-YES screen was used to guide the identification of estrogenic chemicals in STW and paper products by ELISA [61] or chromatographic analyses [62]. Changes in estrogenic and androgenic activities during wastewater treatment were monitored by YES screens [63]. The cystostatic and antiestrogenic metabolites of dietary indole-3-carbinol were isolated with the guidance of the MCF-7 cell proliferation assay [64].

Quality assurance for analysis of EASs

As with any analytical work, there is a need for QA/QC of the methods and results for EASs in food and the environment. This becomes even more important as the testing moves from an area of research to becoming a regulatory activity. QA/QC systems are well developed in areas such as pesticide residues and priority pollutants. However, the basic tenets of method validation, internal quality control, and interlaboratory sample exchange have not yet been rigorously applied in many other areas of EAS analysis, including several of those that are discussed in more detail below. In many cases, research reports lack basic information to support their quantitative findings such as the accuracy and precision of the methods used for the analytes and matrices under study. There are many related issues such as the handling, storage, and extraction of samples that require more investigation before there can be full confidence in quantitative data that will form the basis for risk assessments. These issues apply equally to the in vitro bioassay systems that are increasingly being used to provide quantitative estimates of EAS activities in complex matrices.

METHODS OF ANALYSIS FOR ENDOCRINE ACTIVE SUBSTANCES

Organochlorine insecticides (OCs)

Multiresidue analyses are well established for these POPs and extensive data is available on their environmental distribution, fate and bioaccumulation in wildlife, e.g., in bald eagles [65]. A recent study on bioaccumulation of pollutants in marine zooplankton illustrates some current analytical methodologies for wide-ranging screening of low-level OCs [66]. Recent trends in methods for toxaphene/chlorobornane analysis provide some paradigms for trace analysis. GC/ECD screens are of high sensitivity [67] although GC/MS (HRMS/SIM) [68] or GC/MS (NCI) [26] may be required to give greater certainty of determination, particularly at low levels where complex mixtures of PCBs and other OCs may dominate [69]. GC/MS was used to detect OCs and PCBs in human amniotic fluid with LODs of 0.01 ng/ml and 0.1 ng/ml respectively [70]. A simplified protocol for OCs in serum used SPE (disk) [71].

Polychlorinated- and polybrominated-biphenyls (PCBs, PBBs)

These POPs comprise complex mixtures of congeners with widely varying biological activities and environmental persistence. Recent reviews have emphasized that adequate risk characterization for PCBs requires exposure data for a wide range of congeners [72]. Although HRGC/ECD is suitable for congener specific analyses covering the majority of significant congeners in sediments or fatty tissues, mass spectrometric detection has become more common, especially for the dioxin-like non- and mono-ortho congeners. The hydroxylated metabolites of PCBs are potent thyroxin receptor agonists and were determined in the blood and fat of mammals using GC/ECD of the methyl derivatives [73]. Ion-trap GC/MS of the trifluoroacetyl derivatives has been optimized for their determination at trace levels [74]. Methylsulfonyl metabolites of PCBs also have significant endocrine activities and were determined in milk and human and marine mammal tissues by GC/MS after novel multiresidue clean-ups [75].

Dioxin and dioxin-like compounds

A variety of polyhalogenated environmental and food contaminants have been found to be potent AhR agonists and thus require careful monitoring and control. The most significant compounds are the polychlorinated dibenzodioxins and dibenzofurans (PCDDs, PCDFs), and coplanar congeners of PCBs. Accurate risk assessment for the complex mixtures requires congener specific analyses so that toxic equivalency factors (TEFs) relative to 2,3,7,8-TCCD can be applied for summing of activities to provide a toxic equivalency quotient (TEQ) [76]. There are a large number of other halogenated compounds that might require consideration in some situations [5]. For example, polychlorinated napthalenes were the main contributor to TEQs in wide-ranging analyses of sediments from the Detroit Basin [77] and Tokyo Bay [78]. The U.S. EPA ultra-trace (ppt LODs) analytical protocols for PCDDs and PCDFs (e.g., SW386 Method 8290 for solid waste) are based on highly selective clean-up followed by HRGC/HRMS (SIM). There is a trend to multiresidue protocols for extraction, and clean-up to streamline determination of PCBs and other chlorinated POPs as well as PCCDs and PCDFs, e.g., in food [79]. The use of isotope-labeled internal standards and well-developed quality control criteria [80] provide a paradigm for other instrumental protocols in environmental trace analysis. Ion-trap MS/MS detection provides an alternative to HRMS for screening of PCCDs and PCDFs with equivalent accuracy but higher limits of detection [81]. The most interesting development is bioassay techniques to estimate TEQs directly with adequate sensitivity for most screening purposes. ELISAs, AhR receptor-based assays, and cytochrome P-450 induction assays have been validated for dioxins and dioxin-like compounds [82,45]. The strategies have been explored for screening food and environmental samples using such bioanalytical tools and relying on the congener specific instrumental analyses only for confirmation. The AhR-CALUX assay was used to screen 1380 food and feed samples from recent contamination incidents in Brazil and Belgium. About 10 % of samples were positive (>5 pg TEQ/g) and half of these were tested by GC/MS with 50 % (24 samples) being confirmed as due to dioxins [83]. It should be noted that EAS endpoints such as relative estrogenic effects of dioxins may not be adequately represented by TEFs derived from AhR activity.

Polybrominated diphenyl ethers (PBDEs)

PBDEs are high-volume chemicals still in use as fire retardants. Current commercial products are highly brominated and contain less than 10 of the possible 209 congeners. They have some of the characteristics of other POPs, including EAS activity, and there is concern about the rising levels in fatty tissues of wildlife and humans [84]. Extraction and clean-up procedures similar to those for residues of persistent organochlorine compounds have been found to be effective [20c]. Methods have been reported with determination by HRGC with ECD [19a], MS (NCI) [20c], and HRMS [85]. Analysis of the highly brominated congeners is complicated by their high MWs and tendency to decomposition by light or heat. Many earlier methods only determined lower congeners. These difficulties were confirmed by an interlaboratory study where results for PBDEs in sediment and biota samples were very variable, especially for the deca-BDE (#209) [86].

Phenolic xenoestrogens and phthalates

Degradation of alkylphenol-ethoxylate surfactants and related high-volume industrial chemicals result in environmental contamination by phenolic degradates, the most prevalent being 4-nonylphenol (4NP, a mixture of 22 isomers). These compounds are also ubiquitous in food [87]. Analytical methods have been reviewed [88]. Bisphenol A, derived from plasticizers, also has significant estrogenicity and environmental persistence. Phenolic xenoestrogens were determined in aqueous samples by GC/MS following SPE (divinylbenzene-styrene polymer packing) and methylation [22]. ASE plus GC/MS [89] and microwave-assisted extraction plus LC/MS [90] have been used to determine alkylphenols in sediments. Steam distillation/solvent was used for extraction of 4NP from foods [87]. Isomer-specific analysis of 4NP was accomplished by capillary GC/MS and was enhanced by using derivatives (acetate, PFB, HFB, TMS) [87,91]. Alkylphenols have also been determined in human plasma samples by HPLC with electrochemical detection [92]. GC/MS formed the basis for detection of bisphenol A in hazardous landfill leachates [93].

Recently, there has been extensive research into the pathways and rates for breakdown of parent surfactants in the environment or during sewage treatment. Degradates of 4NP-ethoxylates (NPEs) with ecotoxic characteristics include 4NP, chain-shortened nonylphenol ethoxylates (mono- NP1EO and di-NP2EO), and carboxylate oxidation products. The high selectivity of tandem MS using precursor ion scanning and multiple reaction monitoring enabled a simple flow injection analysis of sewage treatment plant wastes for parent surfactants with LODs to 50 ng/l [94]. Generic extraction protocols for alkylphenols and intermediate polar degradation products used SPE for aqueous systems [95,96] and sonication for sediments [95]. LC/MS with ESI or APCI formed the basis for methods to determine NPEs and degradates in aqueous samples [38a,95–98], sediments [95], sewage sludge [99], and fish tissues [100]. A quantitative LC/MS method for metabolites of alkylphenol-ethoxylates in water used surrogate and isotope-labeled internal standards to correct for recovery and matrix effects during ESI [96]. HPLC with fluorescence detection formed the basis for simple and sensitive methods for 4NP, NP1EO, and NP2EO in aqueous samples [101], sediment [102] and aquatic biota [103]. Dicarboxylated metabolites of NPEs were analyzed in sewage waste using SPE (Carbopak) extraction and LC/MS and were found to be the predominant NPE degradates leaving sewage treatment plants [23,104]. Carboxylated metabolites have also been determined by GC/MS following methylation [105]. Urinary metabolites of 4NP in trout were determined by GC/MS following radio-HPLC isolation and derivatization and were found to be different from 4NP biotransformation products in sewage sludge or trout bile [106]. The matrix solid-phase dispersion technique was very efficient in extraction and clean-up of surfactants from biological tissues [107].

Well-established analytical protocols for phthalate ester plasticizers (PEPs) in environmental and food matrices are generally based on GC/MS, e.g., methods for PEPs in sewage sludge [62,108]. Their ubiquitous nature tends to result in high and variable laboratory blanks. Solid-phase microextraction (SPME) for analysis of aqueous samples has the advantage in this respect of minimizing sample handling and can achieve sub-ppb detection limits using GC/MS detection [109]. The monoester metabolites are the PEP compounds of greatest concern as EASs [110], and their excretion in urine is also a useful biomarker of overall human exposure to PEPs [111]. Seven of these compounds were determined in urine by an isotope dilution assay based on enzymatic deconjugation, SPE, and LC/MS/MS [112]. Monoesters were determined in river waters by GC/MS following SPE and methylation [113].

Organotin compounds

The high toxicity to mollusks of alkylated tin compounds, particularly tributyl-tin (TBT), necessitates sensitive and selective analyses in water and sediments. Total organic tin measurements can be reliably provided by graphite furnace atomic absorption spectroscopy (GF/AAS) or atomic emission spectroscopy (ICP/AES). However, speciation is required to understand the fate, exposures, and effects of tin compounds in marine environments. GC separation of mono-, di- and tri- alkyl tins can be achieved following peralkylation using $NaBH_4$ [114] or a Grignard reagent (commonly pentyl-magnesium bromide) [115]. Selective GC determination of the derivatives has been achieved using flame-photometric [116], atomic emission [104,117], or MS detection [115b,118]. ASE effectively extracted organotin compounds from sediments [119]. High laboratory blanks from ubiquitous tin compounds in plastics and other sources must be overcome to achieve high precision at the low concentrations in waters or sediments that are significant for ecotoxicity. Sulfur interferences must also be eliminated [120].

Steroidal hormones

Naturally occurring and synthetic steroid hormones are significant estrogenic components of animal and human waste that can act as EASs in the environment at very low concentrations. In-depth reviews of estrogen analysis, including sampling and extraction, has recently been published [22b,121]. Sensitive immunoassays in RIA [122] or ELISA [43] format are a convenient technique for screening of large numbers of samples for specific steroids. Chemical methods have been validated for steroid speciation in aqueous samples involving SPE, cleanup and GC/MS or GC/MS/MS determination of the steroids after derivatization [39,123–125]. Detection limits for 17β-estradiol, estrone, estriol and 17α-ethynylestradiol were 0.1–5 ng/l. The mono *tert*-butyl-dimethysilyl derivatives offer advantages of speed of reaction, stability, and good MS characteristics [112]. A comprehensive method using SPE and GC/MS (TMS derivatives) has been used to determine steroidal estrogens, phytoestrogens, and xenoestrogens in STW and the chemical data was correlated with E-Screen estrogenicity [126]. LC/MS (ESI or APCI; +ve or –ve ion) was found to be of superior specificity to HPLC/DAD for determination of estrogenic hormones in aqueous samples [127]. Others studies using LC/MS/MS (ESI-) gave sub-ng/l detection limits [13b,23b], comparable to the best GC/MS methods. Isotope labeled internal standards enhanced the accuracy and precision of MS-based methods by correcting for losses of the readily sorbed estrogens during extraction and workup of samples. Androstenedione was identified as a significant androgenic component in paper mill effluent using LC/MS [60].

Phytoestrogens

Many plants, particularly legumes, produce a range of sterols, isoflavonoids, and lignans with moderate to weak or antagonistic estrogenic activities. The *Fusarium* mycotoxin zearalenone and its synthetic relative zeranol are also estrogenic. A key focus of current research is to establish the protective or adverse effects of specific phytoestrogens on human health. Chemical assays provide complementary data to effects-based studies, particularly to establish exposure levels and metabolism. The main concern has been mammalian dietary intake in foods or feeds, but environmental issues arise with some compounds such as β-sitosterol, a significant by-product of pulp and paper production. Methods of analysis for food have been reviewed [128]. Most attention has been given to the isoflavone class, but lignans have also been studied [129]. Many of these compounds have native fluorescence, and HPLC with fluorescence detection provides the basis for specific and sensitive assays. HPLC with UV detection forms the basis for standard methods for soy-based foods [130], with either acid hydrolysis to fully release the aglycones or mild base hydrolysis to release ester forms of glycosides. Coulombic-array detection based on electrochemical oxidation of the phenolic groups provided very sensitive and selective detection of phytoestrogens [131]. Comprehensive protocols based on enzymatic and acidic hydrolysis to release the aglycones followed by GC/MS of the TMS derivatives have been used to elucidate the full suite of isoflavones and lignans present in diverse foods/feeds and where more detailed knowledge of metabolism and excretion was required [128,132]. LC/MS is also increasingly being applied to flavonoid determination. Derivatization is not required, and tandem MS techniques provide excellent selectivity with structural information, including on glycosidic linkages [133]. A study in mice showed substantial metabolism of estrogenic isoflavonoids to hydroxylated or dehydrogenated forms, some of which were more estrogenic than the parent [134]. A sensitive (5 nM) and rapid immunoassay has been developed for determination of enterolactone, genestein, and daidzein in urine using time-resolved fluorimetry to detect europium-labeled phytoestrogen tracers [135]. An estrogen receptor assay in microtitration plate format has been developed for rapid screening of soy-based foods for phytoestrogen content (genestein equivalents) and gave good correlation to HPLC data [136]. Affinity column separation and LC/MS/MS were used to establish the low estrogenic potential of some triterpene glycosides [137].

CONCLUSIONS

An array of chemical analytical methods (including ELISA) have been developed that provide adequate detection limits and precision for analysis residues of the most important classes of EASs in food and the environment at levels of biological significance. Many challenges remain to make many of these methods less complex and more robust.

The high degree of complexity in interpreting potential biological effects from chemical monitoring alone has been illustrated using organohalogen compounds as an example [4c]. The development of stable and highly sensitive recombinant receptor-based assays (EROD, CALUX, YES) has been very significant for EAS research. As well as providing in vitro test methods to assess endocrine activities of different chemicals, the assays have high utility for food and environment monitoring through their ability to integrate exposures at an effects level in a rapid and relatively inexpensive format. Obviously continued development of such assays into other endocrine activities beyond estrogen, ArH and androgen receptors, or into more specialized formats (e.g., ER-b) will continue to provide major benefits to studies of EAS fate and exposures.

Despite the benefits of integrating bioassays, there will always be a need for chemical speciation both for fundamental research to identify and understand the modes of action of particular EASs and for management and mitigation of problem chemicals. The receptor screens have proved powerful in some situations for directing isolation of specific chemicals causing adverse effects. However, the comparison of results between biological and chemical monitoring is providing interesting fundamental questions relating to a valid concept for joint toxicity. Receptor activities and other relatively short-term in vitro effects of chemical mixtures must be related to various longer-term toxicological endpoints. On the other hand, the integration of chemical concentration data through use of TEFs must use relative biological potencies that are relevant to particular endocrine effects (e.g., estrogenicity of dioxins). For example, CALUX-guided risk assessment of dioxins must incorporate the significant AhR activities of some PAHs. These issues have been carefully reviewed recently in relation to AhR responses [5,138] and are of very practical consequence. Discrepancies between AhR assay data and dioxin TEQs for biological or food samples have generally not been large [5]. However, measurements in contaminated sediments of alkylphenols, PAHs, OCs, and PCBs and associated estrogen receptor and AhR activities have shown that much of the in vitro bioassay responses could not be accounted for and were attributed to undetected compounds [139]. Similar conclusions have been drawn from studies on estrogenicity of STW, although the proportions of unaccounted for activity were generally lower [56,126,59] or attributed to antiestrogenic compounds [52]. Thus, chemical and biological analyses are both required to understand and manage the risks posed by particular chemicals in complex mixtures of contaminants (natural and man-made).

Some recommendations for future research arising from this review include:

- A high priority is required for study of contaminated sites, waste treatment processes, bioaccumulated mixtures of POPs in mammals, and EASs in human diets.
- Better integration of test methods for diverse classes of EASs.
- More automated and selective clean-up procedures, e.g., immuno-affinity columns.
- Methods should include more metabolites of xenobiotics as potential EASs,
- Increased use of LC/MS to supplant GC/MS methods that require derivatizations. This will require better confirmatory techniques such as tandem-MS and greater use of isotope-labeled internal standards to correct for matrix effects.
- Increased availability of ELISA and reporter gene assays in kit formats with well-defined specificity for selected AESs or classes of AESs.
- Development of bioanalytical instrumentation where separation, spectroscopic, and biological detector elements are linked to provide information on effects and chemical properties.
- Quantitative correlations of bioresponse and chemical data are powerful tools, but require careful consideration of all confounding factors.
- Greater conformity to the principles of laboratory quality assurance in terms of method validation, in-use performance verification, confirmatory tests, and interlaboratory studies.

REFERENCES

1. (a) M. Castillo and D. Barceló. *Trends Anal. Chem.* **16**, 574–583 (1997); (b) *Analysis of Environmental Endocrine Disruptors,* L. H. Keith, T. L. Jones, L. L. Needham (Eds.), ACS Symp. Ser. No. 747, American Chemical Society, Washington, DC (1999); (c) T. Jones-Lepp, C. L. Gerlach, E. J. Cooter. *Trends Anal. Chem.* **19**, 286–291 (2000); (d) M. Petrović, E. Eljarrat, M. López de Alda, D. Barceló. *Trends Anal. Chem.* **20**, 637–649 (2001).
2. (a) K. C. Jones and P. de Voogt. *Environ. Pollution* **100**, 209–221 (1999); (b) H. V. Vallack, D. J. Bakker, I. Brandt, E. Broström-Lunden, A. Brouwer, K. R. Bull, C. Gough, R. Guardans, I. Holoubek, B. Jansson, R. Koch, J. Kuylenstierna, A. Lecloux, D. Mackay, P. McCutcheon, P. Mocarelli, R. D. F. Taalman. *Environ. Toxicol. Pharmacol.* **6**, 143–175 (1998).

3. P. G. M. Kienhuis and R. G. Geerdink. *Trends Anal. Chem.* **19**, 460–474 (2000).
4. (a) M. C. Fossi. *Toxicol. Ind. Health* **14**, 291–309 (1998); (b) M. C. Fossi, S. Cassini, L. Marsili. *Chemosphere* **39**, 1273–1285 (1999); (c) A. Menditto and L. Turrio-Baldassarri. *Chemosphere* **39**, 1301–1307 (1999).
5. P. A. Behnisch, K. Hosoe, S. Sakai. *Environ. Int.* **27**, 495–519 (2001).
6. C. Desbrow, E. J. Routledge, G. C. Brighty, J. P. Sumpter, M. Waldock. *Environ. Sci. Tech.* **32**, 1549–1558 (1998).
7. E. J. Routledge, D. Sheahan, C. Desbrow, G. C. Brighty, M. Waldock, J. P. Sumpter. *Environ. Sci. Tech.* **32**, 1559–1565 (1998).
8. (a) P. Matthiessen, Y. T. Allen, C. R. Allchin, S. W. Feist, M. F. Kirby, R. J. Law, A. P. Scott, J. E. Thain, K. V. Thomas. *CEFAS Technical Report* No. 107, Centre for Environment, Fisheries & Aquaculture Science, MAFF, Lowestoft, UK (1998); (b) K. V. Thomas, M. R. Hurst, P. Matthiessen, D. Sheahan, R. J. Williams. *Water Res.* **35**, 2411–2416 (2001).
9. A. Marx, J. Sherry, P.-D. Hansen, B. Hock. *Chemosphere* **44**, 393–399 (2001).
10. B. Erickson. *Anal. Chem.* **70**, 444A (1998); (b) Z.-H. Mo, X.-H. Long, W.-L. Fu. *Anal. Commun.* **36**, 281-283 (1999); (c) M. Murata, M. Nakayama, H. Iric, K. Yakabe, K. Fukuma, Y. Katayama, M. Maeda. *Anal. Sci.* **17**, 387–390 (2001).
11. M. Hara. *Mat. Sci. Eng. C -Bio. Sci.* **12**, 103–109 (2000).
12. (a) R. L. Jones. *Chemistry of Plant Protection Vol. 11: Analysis of Pesticides in Ground and Surface Water I*, H. J. Stan (Ed.), pp. 3–18, Springer Verlag, Berlin (1995); (b) M. Ackerblom, ibid, pp. 19–66.
13. (a) M. Petrovic and D. Barceló. *Fresenius' J. Anal. Chem.* **368**, 676–683 (2000); (b) C. Baronti, C. Curini, G. D'Ascenzo, A. Di Corcia, A. Centili, R. Samperi. *Environ. Sci. Technol.* **34**, 5059–5066 (2000).
14. J. L. Ezzell, B. E. Richter, W. D. Felix, S. R. Black, J. E. Meikle. *LC-GC* **13**, 390–398 (1995).
15. Dionex Corporation Application Note 320: *Extraction of Chlorinated Pesticides Using ASE.*
16. V. Lopez -Avila, R. Young, N. Teplitsky. *J. AOAC Int.* **79**, 142–156 (1996).
17. V. Lopez -Avila, J. Benedicto, K. M. Bauer. *J. AOAC Int.* **81**, 1224–1228 (1998).
18. M. E. P. McNally. *Anal. Chem.* **67**, 308A–315A (1995).
19. (a) K. Kuosmanen, T. Hytötyläinen, K. Hartonen, M.-J. Riekkola. *J. Chromatogr. A.* **943**, 113–122 (2002); (b) M. S. Krieger, W. L. Cook, L. M. Kennard. *J. Agric. Food Chem.* **48**, 2178–2183 (2000); (c) C. Crescenzi, A. Di Corcia, M. Nazzari, R. Samperi. *Anal. Chem.* **72**, 3050–3055 (2000).
20. (a) P. T. Holland, C. W. Hickey, D. S. Roper, T. M. Trower. *Archives Environ. Contam. Toxicol.* **25**, 1251–1257 (1993); (b) C. W. Hickey, D. S. Roper, P. T. Holland, T. M. Trower. *Archives Environ. Contam. Toxicol.* **27**, 221–231 (1995); (c) J. de Boer, C. C. Allchin, R. Law, B. Zegers, J. P. Boon. *Trends Anal. Chem.* **20**, 591–599 (2001).
21. (a) C. R. Powley and P. A. de Bernard. *J. Agric. Food Chem.* **46**, 514–519 (1998); (b) A. C. Johnson, A. Belfroid, A. Di Corcia. *Sci. Total Environ.* **256**, 163–173 (2000).
22. (a) U. Bolz, W. Korner, H. Hagenmaier. *Chemosphere* **40**, 929–935 (2000); (b) M. J. López de Alda and D. Barceló. *J. Chromatogr. A* **938**, 145–153 (2002).
23. (a) J. C. Henion. *J. Chromatogr. A* **885**, 73–95 (2000); (b) A. Di Corcia, R. Cavallo, C. Crescenzi, M. Nazzari. *Environ. Sci. Tech.* **34**, 3914–3919 (2000).
24. P. Klaffenbach and P. T. Holland. *Biol. Mass Spectrom.* **22**, 565–578 (1993).
25. G. H. Dodo and M. M. Knight. *J. Chromatogr. A* **859**, 235–240 (1999).
26. C. Rappe, P. Hagland, R. Andersson, H. R. Buser. *Chemosphere* **20**, 1197–1204 (2000).
27. (a) E. Eljarrat, J. Saulo, A. Monjonell, J. Caixach, J. Rivera. *Fresenius' J. Anal. Chem.* **371**, 983–988 (2001); (b) N. Egashira, T. Shimamoto, K. Inoue, J. Piao, T. Uda. *Anal. Sci.* **17**, 783 (2001).
28. T. Tsikas. *J. Biochem. Biophys. Methods* **49**, 705–731 (2001).

29. H.-J. Stan and T Heberer. In *Chemistry of Plant Protection, Vol. 11: Analysis of Pesticides in Ground and Surface Water I*, H. J. Stan (Ed.), pp. 141–183, Springer Verlag, Berlin (1995).
30. S. J. Lehotay. *J. AOAC Int.* **83**, 680–697 (2000).
31. H. Prest, P. L. Wylie, K. Wiener, D. Agnew. *Application Note* 5968-4884E, Agilent Technol., Palo Alto, CA, USA (1999).
32. (a) J. M. Dimanda, J. Grainger, D. G. Patterson, Jr., W. E. Turner, L. L. Needham. *J. Expo. Anal. Environ. Epidemiol.* **10**, 761–768 (2000); (b) J. W. Cochran. *J. Chromatogr. Sci.* **40**, 254–268 (2002).
33. M. Hada, M. Takino, T. Yamagami, S. Daishima, K. Yamaguchi. *J. Chromatogr. A* **864**, 81–90 (1999).
34. A. Amirav and H. Jing. *J. Chromatogr. A* **814**, 133–150 (1999).
35. W.-H. Ding and C.-T. Chen. *J. Chromatogr. A* **862**, 113–120 (1999).
36. A. H. Grange and G. W. Sovocol. In *Analysis of Environmental Endocrine Disruptors,* L. H. Keith, T. L. Jones, L. L. Needham (Eds.), ACS Symp. Ser. No. 747, American Chemical Society, Washington, DC (1999).
37. S. J. Stout, A. R. daCunha, G. L. Picard, M. Safarpour. *J. AOAC Int.* **81**, 685–690 (1998).
38. (a) N. Jonkers, T. P. Knepper, P. de Voogt. *Environ. Sci. Technol.* **35**, 335–340 (2001); (b) F. Houde, C. DeBlois, D. Berryman. *J. Chromatogr. A* **961**, 245–256 (2002).
39. T. Croley, R. J. Hughes, B. G. Koenig, C. D. Metcalfe, R. E. March. *Rapid Commun. Mass Spectrom.* **14**, 1087–1093 (2000).
40. J. M. Van Emon, C. L. Geriach, J. C. Johnson (Eds.). *Environmental Immunochemical Methods: Perspectives and Applications*, American Chemical Society, Washington, DC (1996).
41. A. Krotzky and B. Zeeh. *Pure Appl. Chem.* **67**, 2065–2088 (1995).
42. Y. Goda, A. Yobayashi, K. Fukuda, S. Fujimoto, M. Ike, M. Fujita. *Water Sci. Tech.* **42**, 81–88 (2000).
43. C. H. Huang and D. L. Sedlak. *Environ. Toxicol. Chem.* **20**, 133–139 (2001).
44. M. Castillo, D. Barceló, et al. *J. Chromatogr. A* **889**, 195–209 (2000).
45. M. Altstein, N. Aharonson, G. Segewv, O. Ben-Aziz, D. Avnir, A. Turniansky, A. Bronshtein. *Italian J. Food Sci.* **12**, 191–206 (2000).
46. (a) T. Takeuchi and J. Haginaka. *J. Chromatogr. B* **728**, 1–20 (1999); (b) L. I. Andersson. *J. Chromatogr. B* **745**, 3–13 (2000).
47. P. A. Behnisch, K. Hosoe, S. Sakai. *Environ. Int.* **27**, 413–439 (2001).
48. A. M. Soto, K. L. Chung, C. Sonneschein. *Environ. Health Perspect.* **103**, 113–122 (1994).
49. W. Korner, U. Bolz, W. Submith, G. Hiller, W. Schuller, V. Hanf, H. Hagenmaier. *Sci. Total Environ.* **225**, 33–48 (1999).
50. W. Korner, U. Boltz, W. Submuth, C. Kempter, J. W. Metzger, H. Hagenmaier. *Chemosphere* **40**, 1131–1142 (2000).
51. S. Arnold, M. K. Robinson, L. K. Guillette, Jr., J. A. McLachlan. *Environ. Health Perspect.* **104**, 544–548 (1996).
52. N. Garcia-Reyoro, E. Grau, M. Castillo, M. J. López de Alda, D. Barceló, B. Piňa. *Environ. Toxicol. Chem.* **20**, 1152–1158 (2001).
53. J. Legler, C. E. van den Brink, A. Brouwer, A. J. Murk, P. T. van der Saag, A. D. Vethaak, B. van der Burg. *Toxicol. Sci.* **48**, 55–66 (1999).
54. H. Fang, W. D. Tong, R. Perkins, A. M. Soto, N. V. Prechtl, D. M. Sheehan. *Environ. Health Perspect.* **108**, 723–729 (2000).
55. T. Zacharewski. *Environ. Sci. Tech.* **31**, 613–625 (1997).
56. A. J. Murk, J. Legler, M. M. H. van Lipzig, J. H. N. Meerman, A. C. Belfroid, A. Spenkelink, B. van der Burg, G. B. J. Rijs, D. Vethaak. *Environ. Toxicol. Chem.* **21**, 16–23 (2002).
57. P. Sohoni and J. P. Sumpter. *J. Endocrinol.* **158**, 327–339 (1998).
58. P. Mak, F. Dela Cruz, S. Chen. *Environ. Health Perspect.* **107**, 855–860 (1999).

59. D. A. Sheahan, G. H. Brighty, M. Daniel, S. J. Kirby, M. R. Hurst, J. Kennedy, S. Morris, E. D. Routledge, J. P. Sumpter, M. J. Waldock. *Environ. Toxicol. Chem.* **21**, 507–514 (2002).
60. R. Jenkins, R. A. Angus, H. McNatt, W. M. Howell, J. A. Kemppainen, M. Kirk, E. M. Wilson. *Environ. Toxicol. Chem.* **20**, 1325–1331 (2001).
61. S. Matsui, H. Takigami, T. Matsuda, N. Taniguchi, J. Adachi, H. Kawami, Y. Shimizu. *Water Sci. Tech.* **42**, 173–179 (2000).
62. A. M. Vinggaard, W. Korner, K. H. Lund, U. Bolz, J. H. Petersen. *Chem. Res. Toxicol.* **13**, 1214–1222 (2000).
63. L. A. Kirk, C. R. Taylor, C. M. Lye, J. P. Sumpter. *Environ. Toxicol. Chem.* **21**, 972–979 (2002).
64. Y. C. Chang, J. Rigby, G. H. Chang, B. C. Peng, G. Firestone, L. K. Bjeldanes. *Biochem. Pharmacol.* **58**, 825–834 (1999).
65. W. W. Bowerman, D. A. Best, T. G. Grubb, G. M. Zimmerman, J. P. Giesy. *Environ. Monitor. Assess.* **53**, 197–212 (1998).
66. P. F. Hoekstra, T. M. O'Hara, C. Teixeira, S. Backus, A. T. Fisk, D. C. G. Muir. *Environ. Toxicol. Chem.* **21**, 575–583 (2002).
67. D. M. Whittle, R. M. Kirkluk, A. A. Carswell, M. J. Keir, D. A. MacEachen. *Chemosphere* **40**, 1221–1226 (2000).
68. A. Fromberg, T. Cederberg, G. Hilbert, A. Büchert. *Chemosphere* **40**, 1227–1232 (2000).
69. A. G. Frenich, J. L. M. Vidal, M. M. Frias, F. Olea-Serrano, N. Olea. *J. Mass Spectrom.* **35**, 967–975 (2000).
70. W. Foster, S. Chan, L. Platt, C. Hughes. *J. Clin. Endocrin. Metab.* **85**, 2954–2957 (2000).
71. A. Covaci and P. Schepens. *Chemosphere* **43**, 439–447 (2001).
72. L. G. Hanson. *Environ. Health Perspect.* **106**, 171–189 (1998); (b) J. P. Giesy and K. Kannan. *Crit. Rev. Toxicol.* **28**, 511–569 (1998).
73. (a) A. Bergman, E. Klasson-Wehler, H. Kurokai. *Environ. Health Perspect.* **102**, 464–469 (1994); (b) C. D. Sandau, P. Ayotte, E. Dewailly, J. Duffe, R. J. Norstrum. *Environ. Health Perspect.* **108**, 611–616 (2000).
74. V. M. Abraham and B. C. Lynn. *J. Chromatogr. A* **790**, 131–141 (1999).
75. (a) K. Noren, A. Lunden, E. Pettersson, A. Bergman. *Environ. Health Perspect.* **104**, 766–772 (1996); (b) C. Weistrand, K. Noren. *Environ. Health Perspect.* **105**, 644–649 (1997); (c) D. P. Herman, J. I. Effler, D. T. Boyd, M. Krahn. *Mar. Environ. Res.* **52**, 127–150 (2001).
76. WHO (Geneva). *Organohalogen Compd.* **34**, 237–240 (1997); (b) S. H. Safe. *J. Animal Sci.* **76**, 134–141 (1998); (c) S. H. Safe. *Environ. Health Perspect.* **106** (suppl. 4), 1051–1058 (1998).
77. K. Kannan, J. L. Kober, Y.-S. Kang, S. Masunaga, K. Nakanishi, A. Ostaszewski, J. P. Giesy. *Environ. Toxicol. Chem.* **20**, 1878–1889 (2001).
78. Y. Yao, S. Masunaga, H. Takada, J. Nakanishi. *Environ. Toxicol. Chem.* **21**, 991–998 (2002).
79. F. Krokos, C. S. Creaser, C. Wright, J. Startin. *Fresenius' J. Anal. Chem.* **357**, 732–742 (1997).
80. B. K. Gullett, J. V. Ryan, J. Tabor. *Chemosphere* **43**, 403–406 (2001); (b) B. van Bavel, C. Rappe, M. Tysklind, N. Takeda. *Organohalogen Compd.* **40**, 297–300 (1999).
81. D. G. Hayward, K. Hopper, D. Andrejweski. *Anal. Chem.* **71**, 212–220 (1999); (b) D. G. Hayward, J. Holcomb, R. Glidden, P. Wilson, M. Harris, V. Spencer. *Chemosphere* **43**, 407–415 (2001); (c) J.-F. Focant, G. Eppe, E. de Pauw. *Chemosphere* **43**, 417–424 (2001); (d) Y. Kemmochi and K. Tsutsumi. *Chemosphere* **43**, 433–437 (2001).
82. R. O. Harrison and R. E. Carlson. *Organohalogen Compd.* **40**, 31–34 (1999); (b) R. O. Harrison and G. H. Eduljee. *Sci. Total Environ.* **239**, 1–18 (1999); (c) G. Shan, W. R. Leeman, S. J. Gee, J. R. Sanborn, B. D. Hammock. *Organohalogen Compd.* **45**, 228–231 (2000).
83. R. Hoogenboom, L. Portier, C. Onstenk, T. Polman, A. Hamers, W. Traag. *Organohalogen Compd.* **45**, 180–183 (2000).
84. K. Hooper and T. A. McDonald. *Environ. Health Perspect.* **108**, 387–392 (2000).
85. M. Alaee, D. B. Sergeant, M. G. Ikonomou, J. M. Luross. *Chemosphere* **44**, 1489–1495 (2001).

86. J. de Boer and W. P. Cofino. *Chemosphere* **46**, 625–633 (2002).
87. K. Guenther, V. Heinke, B. Thiele, E. Kleist, T. Prast, T. Raecker. *Environ. Sci. Technol.* **36**, 1676–1680 (2002).
88. H. B. Lee. *Water Qual. Res. J. Canada* **34**, 3–35 (1999); (b) M. Petrović and D. Barceló. *J. AOAC Int.* **84**, 1074–1085 (2001).
89. K. R. Kim and H. Kim. *J. Chromatogr. A* **866**, 87–96 (2000).
90. S. N. Pederson and C. Lindholst. *J. Chromatogr. A* **864**, 17–24 (1999).
91. (a) T. D. Wheeler, J.R. Heim, M. R. LaTorre, A. B. Janes. *J. Chromatogr. Sci.* **35**, 19–30 (1997); (b) T. R. Croley and B. C. Lynn. *Rapid Commun. Mass Spectr.* **12**, 171–175 (1998).
92. K. Inoue, Y. Yoshimura, T. Makino, H. Nakazawa. *Analyst* **125**, 1959–1961 (2000).
93. T. Yamamoto, A. Yasuhara, H. Shiraishi, O. Nakasugi. *Chemosphere* **42**, 415–418 (2001).
94. J. B. Plomley, P. W. Crozier, V. Y. Taguchi. *J. Chromatogr. A* **854**, 245–257 (1999).
95. (a) M. Castillo, E. Martinez, A. Ginebreda, L. Tirapu, D. Barceló. *Analyst* **125**, 1733–1739 (2000); (b) M. Petrović, A. R. Fernadez-Alba, F. Borrull, R. M. Marce, E. Gonzalez-Mazo, D. Barceló. *Environ. Toxicol. Chem.* **21**, 37–46 (2002).
96. P. L. Ferguson, C. R. Iden, B. J. Brownawell. *Anal. Chem.* **72**, 4322–4330 (2000).
97. M. Takino, S. Daishima, K. Yamaguchi. *J. Chromatogr. A* **904**, 65–72 (2000).
98. M. Sole, M. J. L. de Alda, M. Castillo, C. Porte, K. Ladegaard-Petersen, D. Barceló. *Environ. Sci. Tech.* **34**, 5076–5083 (2000).
99. M. Petrović and D. Barceló. *Anal. Chem.* **72**, 4560–4567 (2000).
100. (a) S. A. Snyder, T. L. Keith, C. G. Naylor, C. A. Staples, J. P. Giesy. *Environ. Toxicol. Chem.* **20**, 1870–1873 (2001); (b) T. L. Keith, S. A. Snyder, C. G. Naylor, C. A. Staples, K. Kannan, J. P. Giesy. *Environ. Sci. Tech.* **35**, 10–13 (2001).
101. S. A. Snyder, T. L. Keith, D. A. Verbrugge, E. Snyder, T. S. Gross, K. Kannan, J. P. Giesy. *Environ. Sci. Tech.* **33**, 2814–2820 (1999).
102. P. de Voogt, K. de Beer, F. W. M. van der Wielen. *Trends Anal. Chem.* **16**, 584–595 (1997).
103. T. Tsuda, K. Suga, E. Kaneda, M. Ohsuga. *J. Chromatogr. B* **746**, 305–309 (2000).
104. C. Y. Hao, T. R. Croley, R. E. March, B. G. Koenig, C. D. Metcalfe. *J. Mass Spectrom.* **35**, 818–830 (2000).
105. W. H. Ding and C. T. Chen. *J. Chromatogr. A* **862**, 113–120 (1999).
106. R. Thibaut, L. Debrauwer, D. Rao, J. P. Cravedi. *Sci. Total Environ.* **233**, 193–200 (1999).
107. M. Zhao, F. W. M. van der Wielen, P. de Voogt. *J. Chromatogr. A* **837**, 129–138 (1999).
108. J. D. Berset and R. Etter-Holzer. *J. AOAC Int.* **84**, 383–391 (2001).
109. (a) Penlaver, E. Pocurull, F. Borrull, R. M. Marce. *J. Chromatogr. A* **872**, 191–201 (2000); (b) K. Luks-Beklej, P. Popp, B. Janoszka, H. Paschke. *J. Chromatogr. A* **938**, 93–101 (2001).
110. N. P. Moore. *Reprod. Toxicol.* **14**, 183–192 (2000).
111. J. W. Koo, F. Parkham, M. C. Kohn, S. A. Masten, J. W. Brock, J. L. Needham, C. J. Portier. *Environ. Health Perspect.* **110**, 405–410 (2002).
112. B. C. Blount, K. E. Milgram, M. J. Silva, N. A. Malek, J. A. Reidy, L. L. Needham, J. W. Brock. *Anal. Chem.* **72**, 4127–4134 (2000).
113. T. Suzuki, K. Yaguchi, S. Suzuki, T. Suga. *Environ. Sci. Tech.* **35**, 3757–3763 (2001).
114. (a) Rodriguez-Pereiro, V. O. Schmitt, J. Szpunar, O. F. X. Donard. *Anal. Biochem.* **68**, 4135–4140 (1996); (b) J.-P. Girard, J. Szpunar, M.-L. Pedrotti, D. Pesando. *Environ. Toxicol. Chem.* **19**, 1272–1277 (2000).
115. J. A. Staab, W. P. Cofino, B. van Hattum, U. A. Th. Brinkman. *Anal. Chim. Acta* **286**, 335–341 (1994); (b) S. Tsunoi, T. Matoba, H. Shioji, L. T. H. Giang, H. Harino, M. Tanaka. *J. Chromatogr. A* **962**, 197–206 (2002).
116. G. G. Salata, T. J. McDonalad, B. B. Bernard. *Application Note* AN9151, Thermoquest Corp., San Jose, CA (1999).
117. N. Folsik and E. M. Brevik. *J. High Resolut. Chromatogr.* **22**, 177–180 (1999).

118. J. A. Staab, B. Horst, J. van Kesteren, G. Stroomberg, I. Freriks, W. Cofino. *GC/MS Application Note* 31, Varian Chromatogr. Systems, Walnut Creek, CA, USA (1998).
119. US-EPA Method 3660 (1986).
120. P. Schubert. *J. Chromatogr. A* **810**, 245–251 (1999).
121. M. J. López de Alda and D. Barceló. *Fresenius' J. Anal. Chem.* **371**, 437–447 (2000).
122. (a) L. S. Shore, M. Gurevitch, M. Shemesh. *Bull. Environ. Contam. Toxicol.* **51**, 361–366 (1993); (b) S. A. Synder, T. L. Keith, D. A. Verbrugge, E. M. Snyder, T. S. Gross, K. Kainan, J. P. Geisy. *Environ. Sci. Technol.* **33**, 2814–2820 (1999).
123. H.-B. Lee and T. E. Peart. *J. AOAC Int.* **81**, 1209–1216 (1998).
124. C. Kelly. *J. Chromatogr. A* **872**, 309–314 (2000).
125. H. Kuch and K. Ballschmitter. *Fresenius' J. Anal. Chem.* **366**, 392–395 (2000).
126. (a) P. Spengler, W. Korner, J. W. Metzger. *Environ. Toxicol. Chem.* **20**, 2133–2141 (2001); (b) W. Korner, P. Spengler, U. Bolz, W. Schuller, V. Hanf, J. W. Metzger. *Environ. Toxicol. Chem.* **20**, 2142–2151 (2001).
127. M. J. López de Alda and D. Barceló. *J. Chromatogr. A* **892**, 391–406 (2000).
128. (a) W. Mazur and H. Adlercreutz. *Pure Appl. Chem.* **70**, 1759–1776 (1998); (b) H. Adlercreutz. *Environ. Toxicol. Pharm.* **7**, 201–207 (1999).
129. J. Liggins, S. A. Bingham, R. Grimwood. *Anal. Biochem.* **287**, 102–109 (2000).
130. (a) Y. Nakamura, S. Tsuji, Y. Tonogai. *J. AOAC Int.* **83**, 635–650 (2000); (b) S. P. Klump, M. C. Allred, J. L. MacDonald, J. M. Ballam. *J. AOAC Int.* **84**, 1865–1883 (2001).
131. T. Nurmi and H. Adlercreutz. *Anal. Biochem.* **274**, 110–117 (1999).
132. S. Heinonen, K. Wahala, H. Adlercreutz. *Anal. Biochem.* **274**, 211–219 (1999).
133. (a) N. Fang, S. Yu, T. M. Badger. *J. Agric. Food Chem.* **50**, 2700–2707 (2002); (b) N. G. Coldham and M. J. Sauer. *Food Chem. Toxicol.* **39**, 1211–1224 (2001); (c) L. Valentin-Blasii, B. C. Blount, H. S. Rogers, L. L. Needham. *J. Expo. Anal. Envir. Epid.* **10**, 799–807 (2000).
134. V. Breinholt, A. Hossaini, G. W. Svendsen, C. Broumer, S. E. Nielsen. *Food Chem. Toxicol.* **38**, 555–564 (2000).
135. M. Uehara, O. Lapcik, R. Hampl, N. Al-Maharik, T. Mäkelä, K. Wähälä, H. Mikola, H. Adlercreutz. *J. Steroid Biochem. Mol. Biol.* **72**, 273–282 (2000).
136. S. D. Garrett, H. A. Lee, P. M. K. Friar, M. A. Morgan. *J. Agric. Food Chem.* **47**, 4106–4111 (1999).
137. J. Onorato and J. D. Henion. *Anal. Chem.* **73**, 4704–4710 (2001).
138. W. Brack, H. Segner, M. Möder, G. Schürmann. *Environ. Toxicol. Chem.* **19**, 2493–2501 (2000).
139. J. S. Khim, D. L. Villeneuve, K. Kannan, K. T. Lee, S. A. Snyder, C. H. Koh, J. P. Giesy. *Environ. Toxicol. Chem.* **18**, 2424–2432 (1999).

Pure Appl. Chem., Vol. 75, Nos. 11–12, pp. 1859–1871, 2003.

Topic 2.2

Naturally produced steroid hormones and their release into the environment*

Laurence S. Shore‡ and Mordechai Shemesh

Kimron Veterinary Institute, P.O. Box 12, Bet Dagan, Israel

Abstract: Steroidal hormones produced by humans and animals are constantly excreted into the environment in their active forms. The primary steroid hormones are progesterone, estrone, estradiol, testosterone, and cortisol, all of which are lipophilic and poorly soluble in water. The steroids of major concern are estrone and estradiol-17β, since they exert their physiological effects at a lower concentration than other steroids and can be found in the environment in concentrations above their lowest observable effect level for fish and plants (10 ng/l). The steroid hormones can be readily measured in run-off, soil, and groundwater, but each steroid has its distinct pathway of transport. Since the major source of steroids in the environment appears to be cattle and chickens, the hormonal steroid input into the environment could be drastically reduced by well-established techniques such as buffer strips and composting.

INTRODUCTION

Hormones produced by humans and animals are constantly excreted into the environment. Many of these hormones are peptides and are rapidly destroyed. However, the steroid hormones are chemically very stable and are excreted in the free form or as conjugates, which very readily biotransform to the free form [1,2]. The primary steroid hormones are estrone, estradiol, progesterone, testosterone, and cortisol, all of which are lipophilic and poorly soluble in water (log P_{ow} between 3 to 4 [3]). The steroids of major concern are estrone and estradiol-17β, since they exert their physiological effects at lower concentrations than other steroids and can be found in the environment in concentrations above their lowest observable effect level (LOEL) for fish and plants (10 ng/l) [4–8]. Estradiol-17β is considered 10 times more potent than estrone as measured by uterotropic activity in the standard mouse assay [9]. However, estrone has about 1/5 the activity of estradiol-17β in inducing trout vitellogenin synthesis [10], and both compounds had the same activity in a mutagenesis test [11] and in increasing body weight in cattle [12]. Estrone is more potent than estradiol-17β in increasing alfalfa growth [8] and uterine imbibition test [13]. The other free estrogens excreted by animals and humans, estriol (significant quantities only in pregnant women), estradiol-17α (produced by cattle) and equilin (produced by pregnant horses) are considered weak estrogens, but estradiol-17α can be a potent estrogen in the neonatal mouse [14]. Therefore, unless otherwise stated, estrogens in this presentation is used to mean free estrone and estradiol-17β. Progesterone and testosterone are also excreted in the free active form but the concentrations measured in the environment (ppt) are at least a magnitude below their LOEL. Although comparable amounts of corticosterone and cortisol as other steroids are also produced by the various species, nearly all corticoids are excreted as inactive metabolites [15].

---

*Report from a SCOPE/IUPAC project: Implication of Endocrine Active Substances for Human and Wildlife (J. Miyamoto and J. Burger, editors). Other reports are published in this issue, *Pure Appl. Chem.* **75**, 1617–2615 (2003).

‡Corresponding author

PRODUCTION FROM HUMANS

Human females excrete about 5 µg/day each of estrone and estradiol and males about 10 mg/day of androgens (primarily testosterone and androstenedione) (Table 1) [16]. (A recent review has indicated the amounts shown for daily production in pre-pubertal children are substantial overestimations [17].) The amount of excreted estrogens of pregnant women can be 1000 times higher (late gestation production: 26 mg estradiol/24 h; 37 mg estrone/24 h [16]), depending on the stage of pregnancy. In addition, substantial amounts of natural estrogens are consumed as pharmaceuticals. This consumption has been calculated to be about 100 kg/yr/5 million inhabitants [18]. Using a formula to include the amount of estrogen produced by each segment of population (e.g., males, pregnant women), it has been calculated that estrone and estradiol excreted in human urine is in the order of 4.4 kg/yr/one million inhabitants. This estimation, which ignores the contribution of fecal estrogens or pharmaceutical consumption, could account for 50 % of the observed estrogen in the influents to sewage water plants [19].

Table 1 Human production and excretion of estrogens (from ref. [16]).

Sex steroid	Amount excreted in urine (µg/day)	Amount produced (µg/day)	Sex
17β-Estradiol	0.3–5	82–695	Female (cycling)
17β-Estradiol	–	13	Female (pre-pubertal)
17β-Estradiol	1.5	48	Male
17β-Estradiol	–	6.5	Male (pre-pubertal)
Estriol	3–65	–	Female (pregnant)
Estrone	2–20	110–497	Female (cycling)
Estrone	–	41	Female (pre-pubertal)
Estrone	3	88	Male
Estrone	–	35	Male (pre-pubertal)
Androgens	2100–23 100	6500 (testosterone)	Male
Androgens	800–10 500	240 (testosterone)	Female

PRODUCTION FROM LIVESTOCK

Animal manure is a major source of the natural steroids, estrogen and testosterone, reaching the environment. Manure is a mixture of feces, urine, and bedding. If the percentage of water is above 20 %, this mixture is referred to as slurry. When chicken manure has a high percentage of bedding, it is referred to as poultry litter. The use of animal manures for fertilization of fields and the production by concentrated animal feeding operations (CAFOs) has increased the impact of the manures on watersheds [8,20]. The animal manure most used for fertilization of fields in the United States is chicken manure. About 12 000 000 Mg/year of poultry litter was produced in the United States, most of which is applied to grasslands as fertilizer [21]. With the increased emphasis on manure management programs by various governments, it is expected that the use of poultry litter and cattle manure for fertilizing fields will increase substantially in the next decade. Some manure is used for feeding cattle in the form of silage (mixing manure with wheat or corn stalks under anaerobic conditions). Silaging increases the available protein as opposed to composting, which is a destructive process.

Production of estrogen by livestock

Most of the estrogen excreted in cow feces is in the last trimester of pregnancy, and nearly all of the estradiol-17β and estrone are in the free form [22] (Table 2). In addition to estradiol-17β and estrone, comparable amounts of estradiol-17α are produced. The contribution of urine where estrogen is present mostly in the form of conjugates to the total estrogen excreted is usually less than 20 %. However, these conjugates rapidly convert to the free active form after excretion [23]. Studies with injection of labeled steroid indicate that domestic animals differ widely in their routes of excretion [24] (Table 2).

Table 2 Percentage of infused labeled steroid excreted in feces (from ref. [24]).

Steroid	Sheep	Ponies	Pigs
Progesterone	77	75	34
Testosterone	44	28	14
Cortisol	28	59	7
Estrone	89	2	4

Similar to sheep in Table 2, in the cow actual measurement of daily sampling of urine and feces (Table 3) indicated the amount of steroidal estrogen excreted in the urine was less than 10 %. It has been calculated that a pregnant cow excretes 0.76 g of estrogen per pregnancy (mostly in the form of estradiol-17α) and that for Austria, which has about 700 000 milk cows, the yearly excrction of estrogen is in the order of 540 kg [25].

Table 3 Hormone content in cattle and swine manure.

Source	Estrone (μg/kg)	Estradiol 17β (μg/kg)	Comments	Ref.
Milk cows (slurry)	255–640	170–1230	Total solids	[2]
Bulls (slurry)	<2	<2	Total solids	[2]
Milk cows (feces, late gestation)	840 (estrone + estradiol)		Dry wt	[25]
Milk cows (manure pile)	700–1000 (estrone + estradiol)		Dry wt	[25]
Milk cows (feces)				
–100 days before parturition	0.9	9.0	Fresh wt	[22]
–60 days before parturition	0.1	13.9	Fresh wt	[22]
–30 days before parturition	4.1	19.1	Fresh wt	[22]
–10 days before parturition	9.4	42.2	Fresh wt	[22]
–5 days before parturition	11.4	60.0	Fresh wt	[22]
Swine (slurry)	<2–84	<2–64	Dry wt	[2]
Sow (feces, late gestation)	15–28		Dry wt	[32]
Mare (feces, late gestation)	50–200 mg		Dry wt	[33]
Milk cows (urine)	44 mg/24 h or 1.4 mg/kg	41 mg/24 h or 1.3 mg/kg	About 30 l urine/day	[31]
Pony mares (pregnant, urine)	200–800 mg/24 h (estrone+equilin) 400 mg/kg		About 2 l urine/day	[34]

The principle estrogens excreted by chickens are estrone and estradiol-17β. The excretion in urine of estrogen in laying and nonlaying hens was about 3 and 0.5 μg/d for estrone and 3 and 2 μg/d for estradiol-17β, respectively [27]. The clearance rate for estradiol in the laying hen has been calculated to be 6.3 ng/min [27] or 9 μg/d. This can be compared with a calculated excreted value of estrogen measured in manure (from Table 4) of 30 μg/d.

Hormones in manure

In slurries from dairy farms values of about 600–1600 μg estrogen/kg total solids have been reported [2]. This would agree with the value of 800–1300 μg of estrogen/kg dry matter (dm) manure found for pregnant cows [25,28]. Chicken manure contains up to 533 μg estrogen/kg dm and 670 μg testosterone/kg (Table 4) [29]. The amount of testosterone in chicken manure that was silaged or left in open pits was found to be constant over several months, and thermal processing had little effect on the hormone concentrations [29]. In contrast, in a study of a manure pile containing estrogen from cows in the peri-parturient period, it was found that after two months, the interior of the pile had little estrogen and estrogen was present in significant quantities only in the outer crust [25]. No published information for manure content is available for other livestock included in manure management programs (turkeys, ducks, sheep, horse), but turkey and ducks apparently produce much less estrogen/g manure than chickens (Shore, unpublished observations).

Table 4 Hormones content in chicken manure.

Source	T (µg/kg d.w.)	E (µg/kg d.w.)	Ref.
Immature broilers			
Females	133	65	[30]
Males	133	14	[30]
Laying Hens	254	533	[30]
Roosters	670	93	[30]
Chicken litter		133	[35]

Based on similar observations, Lange et al. [36] have calculated the total output of domestic animals in tons/yr (Table 5).

Table 5 Estimated yearly steroid hormone excretion by farm animals in the European Union and the United States—Year 2000 (from ref. [36]).

Species	European Union				USA			
	Million heads	Estrogens (tons)	Androgens (tons)	Gestagens (tons)	Million heads	Estrogens (tons)	Androgens (tons)	Gestagens (tons)
Cattle	82	26	4.6	185	98	45	1.9	253
Pigs	122	3.0	1.0	79	59	0.83	0.35	22
Sheep	112	1.3		58	7.7	0.092		3.9
Chickens	1002	2.8	1.6		1816	2.7	2.1	
Total	1318	33	7.1	322	1981	49	4.4	279

PRODUCTION FROM WILDLIFE

There is no information on the contribution of nondomestic animals to the environmental estrogen load. However, fields in which wild turkeys were present for several months showed high levels of estradiol and testosterone in the soil [21]. Water in fish aquariums reached equilibrium values ranging from 3.5 to 15 ng estradiol-17β/l [4], and fish ponds studied over a four-year period had equilibrium concentrations of 5 to 7 ng/l estrogens and comparable amounts of testosterone (Shore, unpublished observations). In a pond with 150 wild birds, it was found that the concentration in the pond of estrogen and testosterone remained at between 2–5 ng/l over a six-month period in spite of an input of 1250 µg estrogen/day (as measured in the feces and calculated from ref. [37]). The half-life of estrogen and testosterone in the water was in the order of an hour (Shore, unpublished observations).

Due to its importance as a noninvasive technique to monitoring wildlife, there is considerable information on the fecal concentrations of progesterone and estrogen in a variety of wildlife [38]. Concentrations of about 5 µg progestagens (20α progesterone, progesterone)/g dry feces is a common finding for the largest mammals during the luteal phase or late gestation, while estrogen concentrations vary widely between species from 0.1 µg/g in the elephant [39] to 17 µg/g in the musk ox [28].

HORMONES IN FOOD AND FEED

Hormones in food

Average adult consumption of hormones in food has been calculated to be 10 µg progesterone/day, 0.1 µg estrogen/day, and 0.05 µg testosterone/day [40], which is quite small compared to human endogenous production. The principle sources are meat and milk products. Meat can contains small amounts of steroids, e.g., 0.5 µg testosterone/kg for bulls [41], 7 µg progesterone/kg for heifers [42], but the levels of estrogen are barely detectable even in pregnant heifers (3–5 ng/kg, [16]). The highest levels of testosterone in edible tissues observed in the bulls were 3 µg/kg in kidney and 11 µg/kg in fat. Bovine milk is rich in a variety of hormones reflective of the plasma values, but steroid hormones, being lipophilic, can concentrate in the milk and milk products, depending on fat content [43,44] (Table 6). Although estrogenic hormones in milk from nonpregnant cows are in the pg/ml range, milk from pregnant cows can contain 500 ng estradiol/l, 1 µg estrone/l (mostly as conjugated sulfate), and 10 µg progesterone/l [43] (and approximately half of dairy herd is in late pregnancy during milking). Human breast milk for infants contains little estrogen or progesterone since nursing humans are generally not pregnant. Since young children consume about 300–700 ml/day of bovine milk, they may ingest 40 to 100 ng/day of estrogen (estradiol, estrone, and estrone sulfate), and whether this can be considered a safe level is a matter of debate [45]. However, there is little data on the incidence of the cardinal sign of hyperestrogenism, premature thelarche, except for Puerto Rico [46] where there is a high incidence of premature thelarche due to estrogenic substances in the environment.

Table 6 Estrogens (pg/ml or pg/g) in milk and milk products (from ref. [44]).

Source	Estrone	Estradiol	Estrone + estrone sulfate
Milk from cows			
Estrus	58	84	
Luteal	45	29	
Late pregnant	45	49	200–1000
Store milk	55	10	500
Butter	539	82	1470
Cheese	35	10	170
Cream		<30	260

Hormones in feed

Although plants contain a variety of compounds that are defined as steroids [47], finding of significant amounts of vetebrate steroids, such as estrone in apple seeds (130 µg estrone/kg [48]), is rare. However, plants produce phytoestrogens, which are well documented to cause reproductive problems in domestic animals [49]. The endogenous level of the phytoestrogens is raised when legumes are irrigated with sewage effluent as well as other forms of stress such as a fungal infection. An increase in endogenous phytoestrogen content can be induced with the levels of estrone and estradiol found in sewage effluent [50]. Chicken manure does have substantial estrogen and testosterone and when fed as silage can cause hyperestrogenism and delayed puberty in cattle [49,51].

INITIAL CONCENTRATIONS IN THE ENVIRONMENT

Initial concentrations in soil

Estradiol and testosterone both bind to the soil (at a depth of 5 cm) with a reported concentration of about 650 ng/kg on manured plots as opposed to 150 ng/kg on control fields [21]. However, only testosterone reaches the groundwater, while estrogen remains bound to the upper crust of the soil, probably because of its phenolic group binding to soil particles [8,52].

Run-off from manured fields and chicken manure stacks

Following rain events, both estrogen and testosterone are found in the run-off [20,21,35,53]. Run-off following a rain event from the fields contained substantial amounts of estrogen and testosterone (1–3 μg/l) [53]. Exposure of chum salmon to such levels (2 μg estradiol-17β/l for a month) has been reported to be lethal [54]. The amount of estradiol in the run-off increased linearly with the increasing application rate of the litter (1.76 to 7.05 Mg/hectacre), increase in pH, and TOC (total organic carbon) [35]. The concentration and mass losses of estradiol-17β in the run-off can be remarkably reduced using buffer strips or addition of alum to the litter. In a pond, which received run-off from a manured field, estrogen levels were above 5 ng/l (maximum 25 ng/l) for several months with a half-life of about 2.5 months [20]. This can be compared with the very short half-life of an hour in a wild bird pond (Shore, unpublished observations) and 6 days in English rivers [55]. Apparently, as described below for sludge, microbial adaptation plays a major role in destruction of the hormones.

Substantial quantities of estrogen can also elute from stack manure piles. It was found that eluants from chicken manure piles were found to contain 630 ng/l of testosterone and 730 ng/l of estrogen (L. Shore and C. Oshins, unpublished observations). Composting was found to virtually destroy the steroid hormone content in the eluant and in the pile itself.

Sewage

The levels of hormones in raw sewage are by nature highly variable depending on source and amount of rainfall.

Values of 40–130 ng estrogen/l over a six-month period were observed in sewage water consisting primary of septic tank effluents from a population of about 7000 inhabitants [20]. In an extensive survey of influent produced by populations of 3500 to 1.2 million, it was found the range of estradiol-17β was <0.5 to 48 ng/l, estrone 17–102 ng/l and estriol from <0.5 to 10 ng/l [19]. Sewage influent at five sewage treatment plants (STPs) was 1.5 ng/l for estradiol and 5.5 ng estrone/l. After hydrolysis, the levels were 3 ng/l estradiol and 13 ng/l estrone [56]. Conjugates can therefore contribute up to 50 % of the total concentration in the influent.

Treated sewage waste

Treatment of sewage water can be classified as primary, secondary, and tertiary. Primarily and secondary treatment (sedimentation and oxygenation ponds) are used for agricultural irrigation. Tertiary treatment is usually in an activated sludge (biosolids) plant that utilizes both anaerobic and aerobic digestion at high temperatures. Alternatively, constructed wetlands (which are considered to be more environmentally friendly) consisting primarily of bulrushes, are used. If the water is used to recharge the aquifer, ground filtration through sand, chalk, or other soils (geofiltration) results in further purification. The concentration of hormones found in the effluent depends on the concentration of hormones in the initial sewage, the type of treatment and physicochemical parameters such as flow rate and time of incubation. The solid materials left after the processing of the sewage water is termed sludge.

Irrigation water from sedimentation ponds

Water after treatment in sedimentation ponds was analyzed for the presence of testosterone and estrogen (Table 7). The source and season were factors in the hormone content. The irrigation water contained from 40 to 340 ng/l of estradiol-17β and estrone as measured by radioimmunoassay. This was well within the range shown to increase the phytoestrogen content and growth of alfalfa plants.

Table 7 Estrogen concentrations in primary treated sewage water used for irrigation (from ref. [7]).

Source	Season	Estradiol and estrone (ng/l)
Agricultural	Dry season	341
Agricultural	Wet season	152
Municipal	Dry season	116
Municipal	Wet season	39

Effluent from sewage treatment plants—artificial wetlands and activated sludge

Activated sludge

The concentrations of estrogen (estradiol+estrone) were studied in an activated sewage treatment plant over a one-year period [7]. During a high flow time the initial amount of estrogen and testosterone in the sewage water after primary and secondary treatment were 54 and 19 ng/l, respectively. In the supernatant of the digestion tank the values for both compounds were between 10 and 20 ng/l and the concentration in the interstitial water of the sludge was similar. In the effluent the values for estrogen and testosterone were 6.4 and 7.2 ng/l respectively. This corresponded to 90 % digestion for estrogen and 60 % digestion for testosterone. However, during low flow periods, where the initial concentration was between 50 and 140 ng estrogen/l and 200–300 ng testosterone/l, the resultant effluent contained 38–50 ng/l of estrogen and 46–121 ng testosterone/l testosterone indicating a lower (60 %) reduction in estrogen, but a comparable amount of digestion as observed at high flow for testosterone. Laboratory studies using active biosolids have shown that although 80–90 % of estrogens are destroyed in the first week, it requires three to four weeks to reduce the estrone, estradiol, and estriol levels to nondetectable levels [57]. The nature of the adapted microbiological populations is an important factor in the removal of testosterone and estrogen. Biosolids from municipal plants (84 %/24 h) are much more effective in reducing estrogen levels than those from industrial plants (4 %) [23]. The nature of the influent and degree of processing also have a effect. Comparison of Brazilian influent comparable influents (15–20 ng estradiol and 30–40 ng estrone/l) showed that elimination in Brazil was higher (99 and 83 % for estradiol and estrone, respectively) than in Germany (64 and 68 %) [58]. In 17 out of 38 sewage sludges natural estrogens could be detected, 17β-estradiol was found in 10 sludges in a concentration range between 4.2 and 111 μg/kg dm (median 12.7 μg/kg dm), estrone concentrations were detected between 3.3 and 328 μg/kg dm in 7 samples, and estriol could be analyzed in 3 samples at 18.1–31.4 μg/kg dm (mean 26 μg/kg dm) [2].

Artificial wetlands

An artificial wetland site was sampled five times over a 10-month period (Shore, unpublished observations). The water was first filtrated through gravel and sand (trickling filtration), which substantially reduced testosterone from 166 to 7 ng/l and estrogen (estradiol+estrone) from 73 to 2 ng/l. Passing through the artificial wetlands had little effect, and the effluent contained 5 ng testosterone/l and 2 ng estrogen/l. Passing through peat did reduce the testosterone further to 2 ng/l, but had no effect on the estrogen level.

Geofiltration

Percolation of tertiary treated sewage water through sandy soil to recharge an aquifer reduced the hormone concentration to undetectable levels (<0.1 ng/l) [7]. Water in springs from a mantled karst aquifer recharged after rain events contained levels of 6 to 66 ng/l, and the estradiol concentration correlated with the *E. coli* and fecal coliform counts [59].

Springs and wells

Although extensive surveys indicate that testosterone and estrogen can be detected in springs and wells used for drinking water, the levels are usually well under 1 ng/l. In an old survey of 64 wells in southern Germany in 1977, levels from <0.1 to 0.9 ng estradiol/l with an average of 0.2 were found [60]. Five wells from under a farm with extensive animal husbandry and manured fields had no detectable levels of estrogen (<0.1), but did have some testosterone (about 1 ng/l) [52]. In a survey of wells receiving contaminated water, it was found that six wells receiving water from rocky strata had detectable estrogen in the range of 1.7–5 ng/l. However, estrogen was undetectable (<0.5 ng/l) in six wells that received water filtered through sandy strata (Shore, unpublished observations). Municipal water supplies in Arkansas, a state with extensive use of poultry manure for land reclamation, had a concentration of 30 ng estradiol-17β/l [36].

Estrogen concentrations in surface waters

In an attempt to identify possible sources of steroid hormones, 17 streams in the Conestoga River Valley of the mid-Atlantic region of the United States were surveyed [20]. Results can be summarized as follows: For stream sampling, four of ten sites had testosterone concentrations of above 1 ng/l. Three of these sites were in areas with heavy use of chicken manure as fertilizer and one site received effluent from an STP. Comparison of a stream dominated by forest with a stream dominated by cropland indicated that there was a gradient of estrogen discharge downstream along the stream dominated by cropland (0.54–1.83 ng/l). Therefore, two sources of pollution were identified—run-off from fields fertilized with manure and discharge into streams from STPs. The levels in freely flowing streams apparently do not exceed 5 ng/l estradiol+estrone, but this level in the same magnitude of the LOEL and harbors a potential for environmental effects.

Several studies of estrogen content in surface waters have recently been reported [2,54]. In general, concentrations of estrogens in surface water were generally low (below 1 ng/l). In one study [2], estradiol-17β was detected in 6 out of 117 samples in a concentration range between 0.8 and 29 ng/l with a median concentration of the positive samples of 1.7 ng/l. In 14 samples, at least one of the metabolites estrone and estriol were determined. Estrone was found in 8 surface waters with a median concentration of the positive samples of 2.3 ng/l, and estriol was detectable in 7 samples with a median concentration of the positive samples of 3.0 ng/l. In marked contrast, an extensive reconnaissance study by the USGS (U.S. Geological Survey) [61] reported very high levels of hormones (<5 ng/l) in 10 to 20 % of the 139 stream studies. Specifically, in the positive samples, the following maximum and median values in ng/l were obtained: estrone 117, 27; 17 β estradiol 93, 9; estriol 51, 19; 17 α estradiol 74, 30; testosterone 214, 116 and progesterone 119, 111. Equilenin, a natural estrogen produced by horses and widely used as a replacement for estrogen, had a maximum of 278 and a median of 140 ng/l.

BIOABSORTION AND MINERALIZATION

Steroidal hormones leave the aqueous phase though absorption to particulate matter (sediment or sludge) or by mineralization (conversion of organic compounds to inorganic compounds). Mineralization (reduction of organic compounds to inorganic compounds) is usually the result of microbiologically activity, but photodegradation or other physicochemical reactions may take place. These compounds are heat stable (mp 175° for estradiol-17β, 252° for estrone).

Bioabsorption

Sediments

In experiments with river sediments, it was found that the initial rapid absorption of estrogen (4 μg/g/h) reached maximal absorption within an hour after which the rate of sorption remained the same or decreased. This absorption increased with the TOC of the sediments and the salinity of the water [62]. In a survey of 12 lake sediments, estradiol-17β was found in 3 sediments with a mean concentration of 8.5 μg/kg dm. Estrone was only found in one sediment at 13.7 μg/kg dm, and estriol was not detectable [2].

Soil

Estradiol rapidly dissipates into all the soils that have been tested (silt loam, sandy loam, and loam) [63]. In sterile soil, estradiol is converted abiotically to estrone with a half-life of about 50 h. The transformation of estradiol to estrone is apparently strongly favored as similar conversion takes place in sewage [6,58] and rivers [62].

Mineralization

Biosolids from sewage treatment plants

The importance of adapted microbiological populations was shown using biosolids from sewage treatment plants [23]. The mineralization of added estradiol-17β was 84 %/24 h by biosolids from a municipal plant, but only 4 % by biosolids from an industrial plant. Biosolids from municipal plants mineralized 70–80 % C^{14}-labeled estradiol to carbon dioxide within 24 h. Removal of estadiol from the aqueous phase by biogradation and/or biosorption to cell matter was greater than 90 %. Testosterone was mineralized in amounts ranging from 55–65 %/24 h, and its removal from the aqueous phase was also greater than 90 %. The rates of mineralization were first-order k's of 0.0042 min for estradiol and 0.0152 min for testosterone which could be calculated as an half-lives of 2.75 h for estradiol-17β and 46 min for testosterone. A 17β-estradiol degrading bacterium, which may be a new Novosphingobium species, has been isolated from activated sludge [64].

Soil

Although estradiol is converted to estrone in autoclaved soil, estrone remains stable. In nonsterile soils (loam, silt loam, and sand loam), by 72 h, both estrone and estradiol form nonextractable residues (57 to 90 %) which are only slowing mineralized. Using labeled steroids, it was found that after 61 d, only 10 to 15 % of the estrogen was mineralized as indicated by labeled CO_2 [63].

Rivers

Water samples from English rivers were studied to measure the biodegradation potential of the key steroid estrogen, estradiol-17β [55,62]. Microorganisms in the river water samples were capable of transforming estradiol-17β to estrone with half-lives of 0.2–9 days when incubated at 20 °C. Estrone was then further degraded at similar rates. Estradiol-17β degradation rates were similar for spiking concentrations throughout the range of 20 ng/l to 500 μg/l. Microbial cleavage of the steroid ring system was demonstrated by release of radiolabeled CO_2 from the aromatic ring of estradiol-17β (position 4). When estradiol-17β was degraded the loss of estrogenicity, measured by the yeast estrogen screen assay (YES), closely followed the loss of the parent molecule. Thus, apart from the transient formation of estrone, the degradation of estradiol-17β does not form other significantly estrogenic intermediates. Estradiol-17β could also be degraded when incubated with anaerobic bed-sediments. Estradiol-17β is susceptible to photodegradation, with half-lives in the order of 10 days under ideal conditions [55]. The half-life of estradiol in river sediments under aerobic conditions was 0.11 days (2.7 h), which is very close to that observed for digestion with biosolids, and under anaerobic conditions the half-life was 0.37 days.

IMPACT

Substantial quantities of estrogen are constantly excreted in the environment. The major source of measured environmental estrogen appears to be domestic animal manures as human sewage is generally degraded in sewage treatment plants. The contribution of wildlife and domestic pets is not known. However, equilibrium concentrations of 5 ng estrogen/l are found in duck and fish ponds and levels above this are not usually found in freely flowing streams. This level is not much below the NOEL 10–50 ng estrogens/l on plants and fish. Little is known of the environmental fate of these hormones, but they apparently do not accumulate in the environment. Any estimation of xenoestrogen impact on the environment must take into account the background level of natural estrogen, which can be significant in areas of concentrated animal husbandry and areas receiving sewage plant effluent from densely populated areas. The effects of the hormones are not necessarily direct. Estrogen in the irrigation water can cause legumes to produce high amounts of phytoestrogen, which in turn cause reproductive problems when ingested by cattle.

STATE OF THE ART

As opposed to many of the compounds discussed in this book, a great deal is known about the possible effects of estradiol and estrone on many animal species including multigenerational studies and NOEL levels. The metabolism of these compounds has also been extensively described in several species, particularly in humans. The amount released into the environment is also quantifiable as manure management programs extensively monitor the amount of manure produced by a variety of farm animals in many countries, especially the United States (e.g., ref. [37]) and Germany [65], and the amounts of sewage passing through sewage treatment plant are documented (e.g., for the United States, ref. [66]).

RECOMMENDATIONS FOR MANAGEMENT OF THE PROBLEM

The relevant environmental agencies should formulate predictable no-effect levels for steroid compounds similar to the proposed predicted-no-effect-concentrations (PNECs) for natural and synthetic steroid estrogens in surface waters by the UK Environmental Agency [67].

Simple low-technology processes such as buffer strips and composting can drastically reduce the amount of steroid hormones. Although not necessary economically justified for the sole purpose of removal of estrogens, buffer strips and composting have many other beneficial effects in protecting the environment. Therefore, composting and buffer strips should be an integral part of manure management programs.

RECOMMENDATIONS FOR FUTURE RESEARCH

The environmental fate and rate of mineralization of the steroid hormones is poorly characterized. Virtually no information is available on what bacteria digest the compound or how their physicochemical degradation are accomplished. Such information should be the subject of immediate research as (1) the technology to do it is readily available; (2) the effects of all estrogenmimetic compounds found in the environment need to be measured against the steroid estrogen background; and (3) since estrogen and testosterone are ubiquitous in human and animal excreta, it could be a standard monitor for sewage pollution and help identify the sources of various pollutants.

REFERENCES

1. G. H. Panter, R. S. Thompson, N. Beresford, J. P. Sumpter. *Chemosphere* **38**, 3576–3596 (1999).
2. A. Wenzel, Th. Kuechler, J. Mueller. *Konzentrationen Oestrogen Wirksamer Substanzen in Umweltmedien. Report.* Project sponsored by the German Environmental Protection Agency, Project No 216 02 011/11 (1998).
3. H. H. Tabak, R. N. Bloomhuff, R. L. Bunch. *Develop. Ind. Microbiol.* **22**, 497–519 (1981).
4. S. R. Miles-Richardson, V. J. Kramer, S. D. Fitzgerald, J. A. Render, B. Yamini, S. J. Barbee, J. P. Giesy. *Aquatic Toxicol.* **47**, 129–145 (1999).
5. G. H. Panter, R. S. Thompson, J. P. Sumpter. *Aquatic Toxicol.* **42**, 243–253 (1998).
6. E. J. Routledge, D. Sheahan, C. Desbrow, G. C. Brighty, M. Waldock, J. P. Sumpter. *Environ. Sci. Technol.* **32**, 1559–1565 (1998).
7. L. S. Shore, M. Gurevich, M. Shemesh. *Bull. Environ. Contam. Toxicol.* **51**, 361–366 (1993).
8. L. S. Shore, Y. Kapulnik, B. Ben-Dov, Y. Fridman, S. Wininger, M. Shemesh. *Physiol. Plant.* **84**, 217–222 (1992).
9. W. H. Perlman. In *The Hormones*, Vol. I, G. Pincus and K. V Thimann (Eds.), Chap. 10, pp. 351–405, Academic Press, New York (1948).
10. E. J. Routledge, D. Sheahan, C. Desbrow, G. C. Brighty, M. Waldock, J. P. Sumpter. *Environ. Sci. Technol.* **32**, 1559–1565, (1998).
11. R. M. Prudy, N. J. Maclusky, F. Naftolin. In *Estrogens in the Environment*, J. A. McLachlan (Ed.), pp. 145–165, Elsevier, Amsterdam (1985).
12. W. A. Green, L. G. Mogil, D. H. Lein, A. D. McCauley, F. H. Foote. *Cornell Vet.* **69**, 248–261 (1979).
13. J. S. Evans, R. F. Varney, F. C. Koch. *Endocrinology* **28**, 747–752 (1941).
14. R. A. Hajek, A. D. Rober, D. A. Johnston, N. T. Van, R. K. Tcholakian, L. A. Wagner, C. J. Conti, M. L. Meistrich, N. Contreras, C. L. Edwards, L. A. Jones. *Environ. Health Prosp.* **105**, Suppl. 3 (1997).
15. S. K. Wasser, K. E. Hunt, J. L. Brown, K. Cooper, C. M. Crockett, U. Bechert, J. J. Millspah, S. Larson, S. L. Monfort. *Gen. Comp. Endo.* **120**, 260–275 (2000).
16. B. Hoffmann and P. Evers. In *Drug Residues in Animals*, A. G. Rico (Ed.), pp. 111–146, Academic Press, New York (1986).
17. A.-N. Andsersson and N. E. Skakkebaek. *Eur. J. Endo.* **140**, 477–485 (1999).
18. F. Stuer-Lauridsen, M. Birkved, L. P. Hansen, H. C. Holten Lutzhoft, B. Halling-Sorensen. *Chemosphere* **40**, 783–793 (2000).
19. A. C. Johnson, U. A. Belfroid, A. Di Corcia. *Sci. Total Environ.* **256**, 163–173 (2000).
20. L. S. Shore, D. Correll, P. K. Chakroborty. In *Animal Waste and the Land-Water Interface*, K. Steele (Ed.), pp. 49–56, Lewis Publishers, Boca Raton, FL (1995).
21. O. Finlay-Moore, P. G. Hartel, M. L. Cabrera. *J. Environ. Qual.* **29**, 1604–1611 (2000).
22. B. Hoffmann, T. Gopes de Pinho, G. Shuler. *Exp. Clin. Endocrinol. Diabetes* **105**, 296–303 (1999).
23. A. C. Layton, B. W. Gregory, J. R. Seward, T. W. Schultz, G. S. Sayler. *Environ. Sci. Tech.* **34**, 3925–3931 (2000).
24. R. Palme, P. Fischer, H. Schildorfer, M. N. Ismail. *Anim. Reprod. Sci.* **43**, 43–63 (1996).
25. E. Möstl, A. Dorbretsberger, R. Palme. *Wien. Tierärztl. Mschr.* **84**, 140–143 (1997).
26. R. S. Mathur and R.H. Common. *Poultry Sci.* **48**, 100–104 (1969).
27. A. Johnson and A. Van Tienhoven. *Poultry Sci.* **60**, 2720–2723 (1981).
28. D. M. Desaulniers, A. K. Goff, K. J. Betteridge, J. E. Rowell, P. F. Flood. *Can. J. Zool.* **67**, 1148–1154 (1989).
29. L. S. Shore, E. Harel-Markowitz, M. Gurevich, M. Shemesh. *J. Environ. Sci. Health* **A28**, 1737–1749 (1993).

30. L. Shore and M. Shemesh. *Isr. J. Vet. Med.* **48**, 35–37 (1993).
31. T. N. Mellin, R. E. Erb, V. L. Estergreeen. *J. Dairy Sci.* **48**, 895–902 (1966).
32. B. Hoffmann. Personal communication.
33. F. Swarzenberger, E. Mostl, E. Bamberg, J. Pammer, O. Schmmehik. *J. Reprod. Fertil.* **44** (suppl.), 489–499 (1991).
34. J. Raeside and R. M. Liptrap. *J. Reprod. Fertil.* **23** (suppl.), 469–475 (1975).
35. D. J. Nichols, T. C. Daniel, A. Moore, D. R. Edwards, D. H. Pote. *J. Environ. Qual.* **26**, 1002–1006 (1997).
36. I. G. Lange, A. Daxenberger, B. Schiffer, H. Witters, D. Ibarreta, H. H. D. Meyer. *Anal. Chim. Acta* **471**, 27–37 (2002).
37. Ohio Livestock Manure and Wastwater Management Guide Bulletin 604 (1992). Available at <http://www.ag.ohio-state.edu/~ohioline/b604/index.html>.
38. F. Schwarzenberger, E. Möstl, R. Palme, E. Bamberg. *Anim. Reprod. Sci.* **42**, 515–526 (1996).
39 M. Fiess, M. Heistermann, J. K. Hodges. *Gen. Comp. Endo.* **115**, 76–89 (1999).
40. S. Fritsche and H. Steinhart. *Eur. Food Res. Tech.* **209**, 153–179 (1999).
41. B. Hoffmann and E. Rattenberger. *J. Anim. Sci.* **45**, 635–641 (1997).
42. L. Shore and M. Shemesh. *Isr. J. Vet. Med.* **47**, 22–23 (1992).
43. O. Koldovsky and W. Thornburg. *J. Pediatric. Gastroent. Nutr.* **6**, 172–196 (1987).
44. S. Zduncyk, J. Malecki-Tepicht, T. Jonwski. *Ubersichtsreferat Dtsch tierärztl. Wschr.* **108**, 174–178 (2001).
45. D. Ganmaa, P. Y. Wang, L. Q. Qin, K. Hoshi, A. Sato. *Med. Hypotheses* **57**, 510–514 (2001).
46. M. C. Larriuz-Serrano, C. M. Pérez, G. Ramos-Valencia, C. J. Bourdony. *Puerto Rico Health Sci. J.* **20**, 13–18 (2001).
47. L. Dinan, J. Harmatha, R. Lefont. *J. Chromatog. A* **935**, 105–123 (2001).
48. A. M. Gawienowski and C. C. Gibbs. *Phytochemistry* **8**, 685–686 (1969).
49. M. Shemesh and L. S. Shore. In *Factors Affecting Net Calf Crops*, M. J. Fields and R. S. Sand (Eds.), pp. 287–298, CRC Press, Boca Raton, FL (1994).
50. L. S. Shore, Y. Kapulnik, M. Gurevich, S. Wininger, H. Badamy, M. Shemesh. *Environ. Exper. Bot.* **35**, 363–369 (1995).
51. L. S. Shore, M. Shemesh, R. Cohen. *Aust. Vet. J.* **65**, 67 (1998).
52. L. S. Shore, D. W. Hall, M. Shemesh. *Dahlia Greidinger Inter. Symp. on Fertilization and the Environment*, pp. 250–255, Technion, Haifa, Israel (1997).
53. D. J. Nichols, T. C Daniel, P. A. Edwards, A. Moore, D. R. Pote. *J. Soil Water Cons.* **53**, 74–77 (1998).
54. M. Nakamuara. *Aquaculture* **48**, 83–90 (1984).
55. M. D. Jürgens, K. I. E. Holthaus, A. C. Johnson, J. J. L. Smith, M. Hetheridge, R. J. Williams. *Environ. Toxicol. Chem.* **21**, 480–488 (2002).
56. P. Adler, T. Steger-Hartmann, W. Kalbfus. *Acta Hydrochim. Hydrobiol.* **29**, 227–241 (2001).
57. H. Tabak and R. L. Bunch. In *Development in Industrial Microbiology*, pp. 367–376, Washington, DC (1970).
58. T. A. Ternes, M. Stumpf, J. Mueller, K. Haberer, R.-D. Wilken, M. Servos. *Sci. Total Envrion.* **225**, 81–90 (1999).
59. E. W. Peterson, R. K. Davis, H. A. Orndorff. *J. Environ. Qual.* **29**, 826–834 (2000).
60. R. D. Rurainski, H. D. Theiss, W. Zimmermann. *gwf-wasser/abwasser* **118**, 288–291 (1977).
61. D. W. Kolpin, E. T. Furlong, M. T. Meyer, E. M. Thurman, S. D. Zaugg, L. B. Barber, H. T. Buxton. *Environ. Sci. Technol.* **36**, 1202–1211 (2002).
62. K. M. Lai, K. L. Johnson, M. D. Scrimshaw, J. N. Lester. *Environ. Sci. Technol.* **34**, 3890–3894 (2000).
63. M. S. Colucci, H. Bork, E. Topp. *J. Environ. Qual.* **30**, 2070–2076 (2001).

64. F. Katsuhiko, S. Kikuchi, M. Satomi, N. Ushio-Sata, N. Morita. *Appl. Environ. Microbiol.* **68**, 2057–2060 (2002).
65. Ch. Brenk and W. Werner. *VDLUFA-Schiftenreihe* **40**, 333–336 (1995).
66. U.S. EPA. *1996 Needs Survey, Report to Congress*. Available at <www.epa.gov/owm> (1996).
67. UK Environmental Agency. W2-014/TR. March 2002.

Pure Appl. Chem., Vol. 75, Nos. 11–12, pp. 1873–1880, 2003.

Topic 2.3

Concentration of phytohormones in food and feed and their impact on the human exposure*

Ph. Verger‡ and J. C. Leblanc

INRA/DSNHSA – 147, rue de l'Université – 75338 Paris Cédex 07, France

Abstract: Phytohormones, which naturally occur in plants, must be taken into consideration for their contribution to the total human exposure to potential endocrine active substances (EASs). Phytohormones are usually divided into two families: phytoestrogens that mainly occur from soybeans and soy derivatives and phytosterols that occur from vegetables and vegetable oils. The present paper compiles different sources of information about the concentration of phytohormones in foodstuffs in order to assess the current human exposure to those substances via food. Particular attention is given to most exposed groups of consumers, on the one hand, infants and young children fed with soy-based infant formulas for their exposure to phytoestrogens and on the other hand, consumers of fortified foods for their exposure to phytosterols.

Available literature shows that the total dietary intake of isoflavones could reach 20 to 25 mg/day/person for the Japanese adult population. For infants and young children, the quantity of phytoestrogen ingested is likely to be 35–50 mg/day/person corresponding on a body weight basis to an exposure 7 to 11 times higher. Regarding phytosterols, an assessment of the exposure via food was done, considering both their natural occurrence and their potential concentration in fortified foods. Results shows that the "natural" exposure is estimated at 340 ± 440 mg/day/person at the mean and at 1040 mg/day/person at the 95th percentile. Considering the potential exposure via fortified foods, it is estimated at 2700 ± 1200 mg/day/person at the mean and at 4700 mg/day/person at the 95th percentile.

After their ingestion, isoflavones are absorbed from the intestinal tract before being excreted in the urine and feces. The increasing use of phytohormones in human foodstuffs could increase locally their release into the environment. Nevertheless, considering the weak estrogenic potential of phytohormones in relation to synthetic or endogenous steroids, any introduction of these substances into aquatic ecosystems would probably have comparatively minor effects on aquatic organisms.

INTRODUCTION

In men, hypospadias, cryptorchidism, cancer of prostate, testicular cancer, and semen quality and in women, breast cancer, cystic ovaries, and endometriosis have all been suggested as indicators of adverse trends in reproductive health [1]. The idea that these trends are real and are connected with environmental pollution is gaining credence internationally. The effect on human health of chemicals that are mediated through the endocrine system has generated huge interest and investment and even if it was

*Report from a SCOPE/IUPAC project: Implication of Endocrine Active Substances for Human and Wildlife (J. Miyamoto and J. Burger, editors). Other reports are published in this issue, *Pure Appl. Chem.* **75**, 1617–2615 (2003).

‡Corresponding author

postulated that synthetic chemicals in the environment are the prime source of the excessive oestrogenic simulation, it was also recognized that endocrine active substances are potentially present in food as natural phytohormones.

Even if they are a quantitatively less important source than the naturally or industrially produced hormones, phytohormones which naturally occur in plants must be taken into consideration for their contribution to the total human exposure to chemicals having potential hormonal effects. Phytohormones are usually divided in two families: at first, phytoestrogens (daidzein, genistein, coumesterol, formononetin and biochanin A), which are mainly occurring from soybeans and soy derivatives. Secondly, phytosterols (β-sitosterol, campestanol, stigmasterol and dihydrobrassicasterol) which occur from the fat-soluble membrane extracts of plants, including algae. Common known sources of phytosterols are vegetables and vegetable oils.

One of the aims of the present paper is to compile the different sources of information about the concentration of phytohormones in foodstuffs and to assess the current human exposure to those substances via food. It must be noted that when numerous scientists have concerns about endocrine active substances and their possible negative impact on human health, a parallel development of health claims about positive health effects of food fortified by phytohormones can contribute to a major exposure of certain groups of consumers. Therefore, a model of potential human exposure to phytosterols is described in the paper as an example.

DESCRIPTION OF PHYTOHORMONES

Phytoestrogens

A list of 94 plants reported to exhibit estrogenic activity was establish [2] and at the same time, 29 specific compounds were reported to have estrogenic activity. The main known phytoestrogens are the isoflavones daidzein, genistein, formononetin and biochanin A and the coumestan coumesterol. Daidzein and genistein exist in 4 related chemical structures, namely the aglycones, the 7-*O*-glucosides, the 6-*O*-acetylglucosides, and the 6-*O*-malonylglucosides. Formononetin is further metabolized to daidzein and biochanin A is a precursor of genistein.

Phytosterols

Phytosterols are C28 or C29 sterols, structurally very similar to cholesterol (C27), but differing in their nucleus and/or side chain configuration or polar groups. β-Sitosterol contains an ethyl group at C24, campesterol has a methyl group at C24 and stigmasterol is considered to be an unsaturated phytosterol because of the double bond at C22. Chemical saturation of the Δ-5 double bond leads to the 5-α position of the hydrogen atom (i.e., cholestanol, campestanol, sitostanol). Enzyme transformation by gut bacteria leads to epimerization of the H atom at position 5 to the β configuration as represented by coprostanol and its derivatives.

EFFECTS OF PHYTOHORMONES

Effects of phytoestrogens in humans

Phytoestrogens interest in isoflavones has exploded in the past five years after a wealth of scientific data showing that these phytoestrogens possess potent and wide-ranging activities. Most of these results were related to replacement of estrogens in women, which is an established therapy of climacteric symptoms like hot flashes or depressive moods. Soy consumption is suggested to contribute to the prevention of chronic diseases including cardiovascular disorders, osteoporosis, and cancer. Recent animal experiments show that only 3 doses of genistein given to newborn, prepubertal, or perinatal rats can reduce the incidence and number of breast tumors [3]. Beside these beneficial effects, the potential for isoflavones to create steroid hormone imbalances or to compete for the normal steroid, drug, and xenobiotics metabolizing enzymes is presently unknown. Moreover, because these compounds are metabolized differently by different animals (and humans) before binding to the estrogen receptor, the intrinsic estrogenic activity should be viewed with caution.

Effects of phytosterols in humans

Phytosterols in foods were identified as natural cholesterol-lowering agents in the 1950s. They are poorly absorbed in human (5 % for sitosterol, 15 % for campesterol). Emerging evidence associated with use of phytosterols in foods has been focus of both food and pharmaceutical industries seeking to provide the consumer with foods with a health-promoting component. Diets containing phytosterols at levels of 2–3 g/day can reduce total and LDL cholesterol levels to about 10 and 20 %, respectively.

Several animal studies indicated that, when used at high levels, plant sterols, especially sitosterol, might have estrogenic activity. Those studies show a significant decrease in testicular weight and sperm concentrations after long-term treatment in albino male rats [4] and clear estrogenic effects in fish [5,6].

Sources of phytoestrogens

The most important dietary source of isoflavones is soybeans and soy foods [7]. Those plants can contain relatively high amounts of phytoestrogens, up to several mg/g of dry weight [8]. Some other legumes contain very small amounts of phytoestrogens, like snow peas, Brussels sprouts, and spinach leaf [9]. In addition, it was suggested that onion, garlic, potato, cucumber, cabbage and coffee could exhibit estrogenic activity [10] even if the chemical nature of this active compounds remains unclear. There are also reports on alcoholic beverages showing that biochanin A and β-sitosterol were founded at levels between 7 to 21 μg/100 ml in bourbon. Daidzein and genistein were also identified in beer. A recent review of the literature [11] compiles quantitative data available for phytoestrogens naturally occurring in food. Nineteen scientific papers were analyzed; the median results are expressed on a wet weight basis (Table 1).

Table 1 Concentration of phytoestrogens in foodstuffs [11].

Food category	Number of samples	Daidzein μg/g	Genistein μg/g	Coumestrol μg/g	Formononetin μg/g	Biochanin A μg/g
Tofu	15	76	166	nd	nd	nd
Soy sauce	3	8	5	nd	nd	nd
Soy milk	10	18	26	nd	nd	nd
Soy based formula	3	<1	3	nd	nd	nd
Alfafa sprouts	1	nd	nd	47	3	nd
Mung bean sprouts	1	nd	nd	nd	trace	nd
Soybean sprouts	3	138	230	7	nd	nd
Soybean green	1	546	729	na	na	na
Tempeh	3	190	320	na	na	na
Soybean paste	6	159	171	nd	nd	nd
Miso paste	2	266	376	na	na	na
Miso paste (rice or barley)	3	79	260	na	na	na
Soy hot dog, tempeh burger	2	49	139	na	na	na

A special mention must be done about the concentration of phytoestrogens in infant formulas. Those products that have been used for more than 30 years are manufactured from soy protein isolates and contain significant amounts of phytoestrogens of the isoflavone class. The results of recent analytical results are expressed in Table 2.

Table 2 Concentration of phytoestrogens in infant formulas [12].

Food category	Number of samples –	Daidzein µg/g	Genistein µg/g	Total isoflavones (concentrate) mg/g	Total isoflavones (final product) mg/l
Powdered 1	1	7.4 ± 1	6.2 ± 2.1	307 ± 28	46
Powdered 2	1	10.9 ± 2.1	8.7 ± 1.2	317 ± 13	47
Concentrate 1	1	2.2 ± 0.6	1 ± 1	91 ± 18	45
Concentrate 2	1	0.9 ± 0.3	1.2 ± 0.3	64 ± 9	32
Ready to eat	1	1 ± 0.1	1 ± 0.1	43.5 ± 0.7	44

Sources of phytosterols

Phytosterols encompass the entire class of sterols found in the fat-soluble membrane extracts of plants, including algae. These include pure sterols, stanols, their phenolic acids and conjugated glucosides. Phytosterols are not synthesized endogenously in humans. Commonly known sources are vegetables, wood, and vegetable oils. Owing to the lack of substrate specificity of enzymes, plants can operate multiple sterol pathways that give rise to specific end-products. Over 250 naturally occurring sterols have been characterized from plant isolation studies. Despite this diversity, the most frequently occurring phytosterols in nature and thus in human diets include β-sitosterol, campestanol, stigmasterol, and dihydrobrassicasterol.

The concentration of phytosterols in food and feed was reviewed from the scientific literature. It must be stressed that if a considerable number of publications were produced during the last 20 years on the biological properties of phytosterols and on their interest for human health, very few of them provide quantitative estimation of these substances in products of vegetal origin. The most important and reliable source of data is the series of publications by Oka [13–16]. In addition, a compilation of data on the sterol content of foods of plant origin was done by U.S. Department of Agriculture in 1978 [17]. Five food categories were considered relatively to the concentration in phytosterols, namely edible oils, vegetables, fruits, cereals, and pulses. Besides those naturally occurring phytosterols and because of the beneficial effects of these substances in humans, food fortification must be considered as a potential way of exposure. Table 3 shows the mean and maximum concentration of phytosterols both in natural and fortified foods. The phytosterol concentrations of fortified foods are those proposed by food industry, but are not currently accepted (80 mg/g for margarine, 23 mg/g for biscuits, 16 mg/g for cheese and breakfast cereals, 13 mg/g for bread, 5.7 mg/g for yogurt, and 3.2 mg/g for milk).

Table 3 Concentration of phytosterols in foods.

Food category	Number of samples	Mean concentration mg/100 g	SD	Maximum concentration mg/100 g
Edible oils	56	500	982	5400
Vegetables	51	20	40	200
Fruits	22	15	10	60
Cereals	9	250	405	1325
Pulses	10	100	59	220
Spreads & margarines		8000		8000
Biscuits		2300		2300
Milk		320		320
Yogurt		570		570
Cheese		1600		1600
Bread		1300		1300
Breakfast cereals		1600		1600

Dietary exposure to phytohormones

Very few studies are actually available to assess the human exposure to phytoestrogens. In their recent report, WHO and IPCS [18] quoted preliminary results obtained in U.S. adults. Those results show that urinary levels of phytoestrogens did not differ significantly of those reported in the literature from western populations known to consume phytoestrogen supplements. In a recent estimate, Nagata [19] suggests a the total dietary intake of isoflavones of 20 to 25 mg/day/person for the Japanese adult population. Considering the higher amount of soya consumed in Japan, this exposure from naturally occurring phytoestrogen is likely to be higher than those of other countries and particularly of the United States and the European Union. For infants and young children, the most important source of exposure is soya-based infant formula. In case of use of these products, the quantity of phytoestrogen ingested is similar to the one used during postmenopausal substitution, e.g., 35–50 mg/day/person. Related to the respective body weights, the exposure of infants is consequently 7 to 11 times higher [12,20].

Regarding phytosterols, an assessment of the exposure via food was done, considering the available data on the natural occurrence of those substances in five relevant food categories (edible oils, vegetables, fruits, cereals, and pulses). Because only a relatively small number of analytical results were available, the data were included in the model assuming the log normality of the distribution of concentrations in each considered food category. In order to calculate the geometrical standard deviation (GSD), the data were log-transformed and a distribution curve was simulated using the mean and the GSD of the initial data set. Those distribution curves were combined with the intake distribution of these foods in France [21]. Results are shown in Fig. 1. In a second step, the possibility for a consumer to get foods fortified at the levels requested by industry was simulated using the following assumptions:

- The food intake data are from the French household survey (SECODIP panels) recording the purchases of 2759 subjects, consumers only, during one year.
- The phytosterol concentrations are from the industry request (80 mg/g for margarine, 23 mg/g for biscuits, 16 mg/g for cheese and breakfast cereals, 13 mg/g for bread, 5.7 mg/g for yogurt, and 3.2 mg/g for milk).
- All foods were considered fortified at the maximum authorized level.
- The baseline exposure from natural occurring phytosterols was not taken into consideration in the simulation.

Results shows that the "natural" exposure is estimated at 340 ± 440 mg/day/person at the mean and at 1040 mg/day/person at the 95th percentile (Fig. 1). Considering the potential exposure via forti-

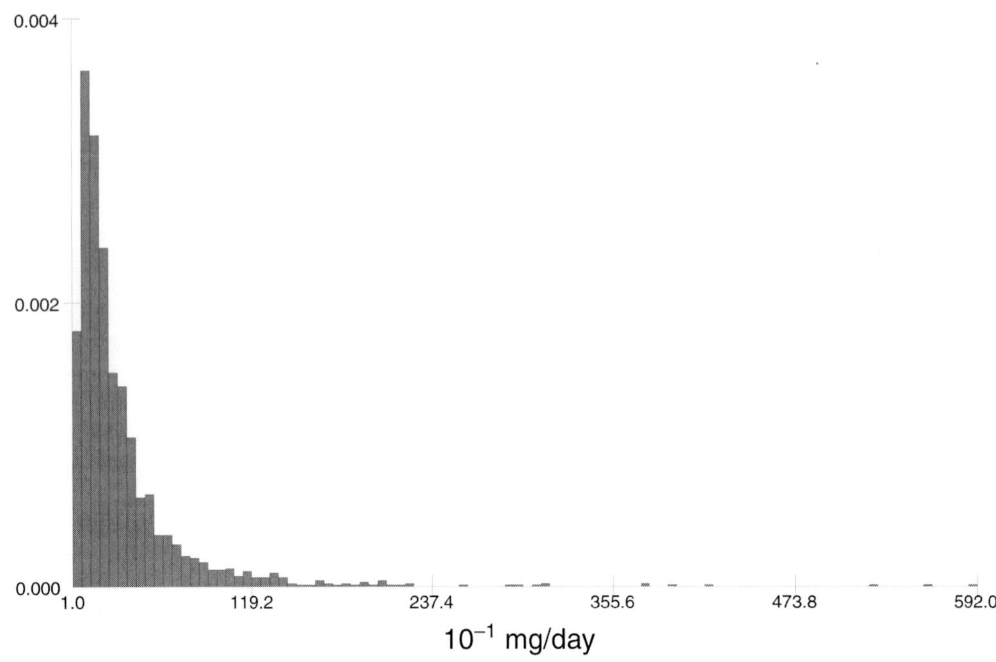

Fig. 1 Simulated exposure to naturally occurring phytosterols in food. (Display distribution as density.)

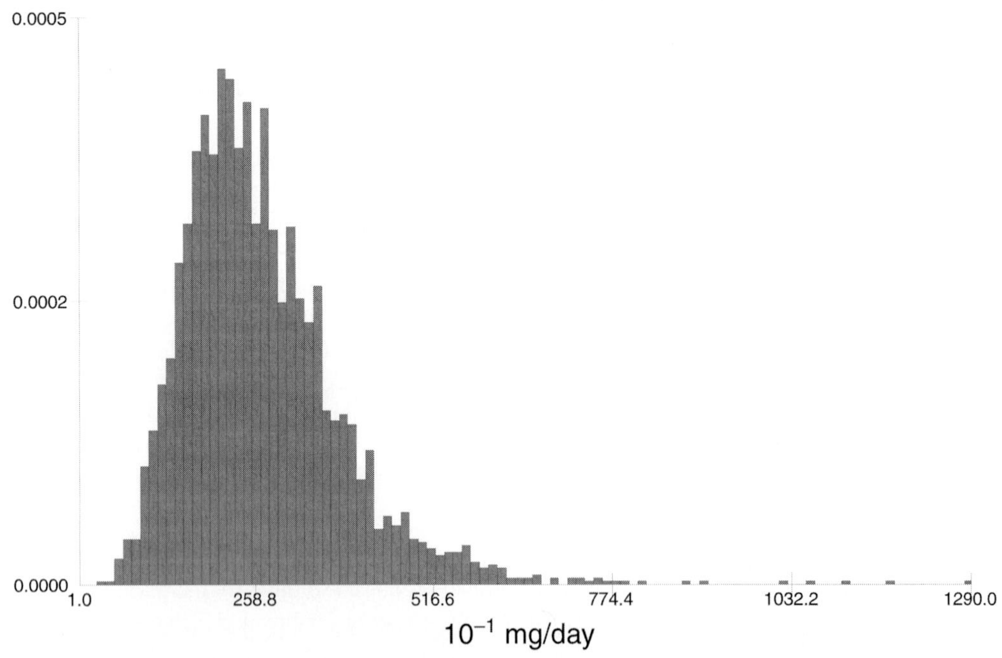

Fig. 2 Simulated distribution of the exposure via fortified foodstuffs. (Display distribution as density.)

fied foods, it is estimated at 2700 ± 1200 mg/day/person at the mean and at 4700 mg/day/person at the 95th percentile (Fig. 2). These results are consistent with a previous study using Australian food intake data [22] and estimating the mean exposure between 1600 and 3900 mg/person/day and the exposure at the 97.5th percentile between 3900 and 8700 mg/person/day.

Fate of phytohormones

Because phytohormones can be metabolized, at least in mammals, it presumably does not accumulate and an exposure of animals is probably limited to particular areas with single-crop farming or animals that are restricted in their choices. Available data indicate that phytohormones are biodegradable. Sufficient information is not available on the occurrence of a relevant exposure of aquatic organisms. In soy processing, phytoestrogens are entering the rinsing water [23,24]. Phytoestrogens are also excreted in the urine and feces of farm animals, and local introductions into surface waters cannot be excluded. Nevertheless, considering the weak estrogenic potential of phytohormones in relation to synthetic or endogenous steroids, any introduction of these substances into aquatic ecosystems would probably have comparatively minor effects on aquatic organisms, if at all [25,26,24].

In humans, after their ingestion, isoflavones are absorbed from the intestinal tract and a peak for plasma concentration is obtained after 5 to 10 h [27] before being excreted in the urine and feces. The increasing use of phytohormones in human foodstuffs because of their beneficial health effects and the elimination of those substances in urine could increase locally their release into the environment. Even if this phenomenon seems quantitatively relatively minor, it could be interesting to get an idea of the order of magnitude of the potential increase of human exposure via fortified foods in order to compare the resulting release into the environment with those from synthetic hormones.

RECOMMENDATIONS

- More specific and sensitive biomarkers should be developed to detect endocrine-mediated effects in individuals and populations.
- Further investigations are needed to assess the risk for aquatic organisms of increased phytohormone exposure.
- Long-term monitoring of wildlife species should be improved.
- More data should be obtained related to the in utero exposure of the fetus to phytohormones from the maternal diet.
- Exposure assessment and epidemiological studies based on specifically exposed groups consuming fortified foods in large amounts should be conducted.
- Investigations on the consequences of high exposure of the human population, including infants consuming soy-based milk substitutes, should be done.
- A full risk assessment vs. benefit of food fortification should be done in order to improve the management of food safety regulation.
- International collaboration and collaborative research should be encouraged.

REFERENCES

1. P. T. C. Harrisson, P. Holmes, C. D. N. Humphrey. *Sci. Total Environ.* **205**, 97–106 (1997).
2. N. R. Frasworth, A. S. Bingel, G. A. Cordell, F. A. Crane, H. H. S. Fong. *J. Pharm. Sci.* **64**, 717–745 (1975).
3. C. A. Lamartiniere, J. Moore, M. Holland, S. Barnes. *Proc. Soc. Exp. Biol. Med.* **208**, 120–123 (1995).
4. T. Malini and G. Vanithakumari. *J. Ethnopharmacol.* **35**, 149–153 (1991).
5. P. Mellanen, T. Petanen, J. Lehtimaki, S. Makela, G. Bylund, B. Holmbom, E. Mannila, A. Oikari, R. Santti. *Toxicol. Appl. Pharmacol.* **136**, 381–388 (1996).
6. D. L. MacLatchy and G. J. Van Der Kraak. *Toxicol. Appl. Pharmacol.* **134**, 305–312 (1995).
7. USDA-Iowa State University Database on the Isoflavone Content of Foods, Release 1.3. <http://www.nal.usda.gov/fnic/foodcomp/Data/isoflav/isoflav.html > (2002).
8. W. Mazur. *Baillère's Clin. Endocrinol. Metab.* **12**, 729–742 (1998).
9. B. E. Knuckles, D. de Fremy, G. O. Kohler. *J. Agric. Food Chem.* **24**, 1177–1180 (1976).

10. A. E. Jones, K. R. Price, G. R. Fenwick. *J. Sci. Food Agric.* **46**, 357–364 (1989).
11. K. Reinli and G. Block. *Nutr. Cancer* **26**, 123–148 (1996).
12. K. D. R. Setchell, L. Zimmer-Nechemias, J. Cai, J. E. Heubi. *J. Clin. Nutr.* **68** (suppl.), 1453–1461 (1998).
13. Y. Oka, S. Kiriyama, A. Yoshida. *J. Jpn. Soc. Food Nutr.* **25** (7), 543–549 (1972).
14. Y. Oka, S. Kiriyama, A. Yoshida. *J. Jpn. Soc. Food Nutr.* **26** (5), 317–327 (1973).
15. Y. Oka, S. Kiriyama, A. Yoshida. *J. Jpn. Soc. Food Nutr.* **26** (2), 121–128 (1973).
16. Y. Oka, S. Kiriyama, A. Yoshida. *J. Jpn. Soc. Food Nutr.* **27** (7), 347–355 (1974).
17. J. L. Weihrauch and J. M. Gardner. *J. Am. Dietetic Assn.* **73**, 39–47 (1978).
18. WHO/IPCS Global assessment of the state-of-the-science of endocrine disruptors </http://ehp.niehs.nih.gov/who/> (2001).
19. C. Nagata, N. Takatsuka, Y. Kurisu, H. Shimizu. *J. Nutr.* **128**, 209–213 (1998).
20. C Bennetau-Pelissero. *Cah. Nutr. Diét.* **36** (1), 25–38 (2001).
21. P. Combris, D. Delobel, C. Boizot. La consommation alimentaire en 1997: distribution des quantités consommées à domicile. Rapport du laboratoire de recherches sur la consommation, INRA (2000).
22. J. Baines. Phytosterol esters derived from vegetables oils: Dietary exposure report. Unpublished report (2001).
23. C. Wang, Q. Ma, S. Pagadala, M. S. Sherrard, P. G. Krishan. *J. Am. Oil Chem. Soc.* **75**, 337–341 (1998).
24. Genistein: Modellstoff zur Beschreibung endokriner Wirkungen von Phytoöstrogenen. BUA-Stoffbericht 222. Gesellschaft Deutscher Chemiker (2000).
25. C. E. Purdom, P. A. Hardiman, V. J. Bye, N. C. Eno, C. R. Tyler, J. P. Sumpter. *Chem. Ecol.* **8**, 275–285 (1994).
26. J. P. Sumpter, S. Jobling, C. R. Tyler. *Soc. Exp. Biol. Sem. Ser.* **57**, 205–224 (1996).
27. K. D. R. Setchell, N. M. Brown, P. Desai, L. Zimmer-Nechemias, B. E. Wolfe, W. T. Brashear, A. S. Kirschner, A. Cassidy, J. E. Heubi. *J. Nutr.* **131**, 1362S–1375S (2001).

Pure Appl. Chem., Vol. 75, Nos. 11–12, pp. 1881–1893, 2003.

Topic 2.4

Pharmaceuticals and personal care products: A source of endocrine disruption in the environment?*

Flemming Ingerslev[‡], Elvira Vaclavik, and Bent Halling-Sørensen

The Royal Danish School of Pharmacy, Universitetsparken 2, Copenhagen 2100, Denmark

Abstract: A wide variety of chemicals are used in pharmaceuticals. Most of these are already under thorough control for endocrine activity. The main causal agents recognized for endocrine disruption from sewage are substances used in medicine (sex hormones, glucocorticoids, and others), natural substances (estrone and 17β-estradiol), and synthetic estrogens (e.g., 17α-ethinylestradiol). Similar substances are used in anabolic agents (growth hormones) in livestock production in some countries. Although the estimated use of anabolic agents in livestock production is approximately one order of magnitude below the natural release of estrogens from farm animals, their possible significance remains unanswered.

At present, no other medical substances are recognized as endocrine disruptors in the environment. However, candidates may be identified on the basis of simple assumptions regarding their use and activity: (1) Nonestrogenic steroids may react with environmental endocrine receptors or metabolize on their way to the environment and thus form endocrine disruptors. (2) Many high-volume drugs released to the environment have not yet been tested for their endocrine properties, and some of these are known to interact with the human endocrine system. (3) Compared to medicinal substances, personal care products and additives in drugs are used in high amounts; from this group, parabens, siloxanes, and other substances are suspected of causing endocrine disruption in the environment.

INTRODUCTION

Since the discovery that pharmaceutical usage leads to the occurrence of medical substances in natural waters, numerous research activities have focused on revealing the extent and consequences of these increasingly emergent findings [1,2]. The occurrence of these substances in the environment is linked to usage in either a veterinary or human context. While the reasons for the occurrence of medical substances in the environment are well understood, the possible effect they may have is subject to discussion. A number of papers regarding general ecotoxicology show that these substances generally occur at nontoxic concentration levels [3,4]. But of even more interest may be research addressing questions related to specific pharmacological properties of these substances such as antibacterial resistance (antibiotics) [5] and endocrine disruption (endocrine active pharmaceuticals). At present, the knowledge regarding the possible endocrine effects in the environment that is related to pharmaceuticals is sparse.

*Report from a SCOPE/IUPAC project: Implication of Endocrine Active Substances for Human and Wildlife (J. Miyamoto and J. Burger, editors). Other reports are published in this issue, *Pure Appl. Chem.* **75**, 1617–2615 (2003).

‡Corresponding author

It is obvious that the use of pharmaceuticals is only acceptable if no side effects occur. Additionally, medical products are accepted if their side effects are tolerable when compared to the benefits associated with their use. In order to avoid safety problems, drugs must undergo a thorough approval procedure that aims to document their safety [6,7]. This approval practically covers all known aspects of health effects and drug behavior in the treated organism over both the short and long term. Approval also takes unwanted hormonal effects into account, and in this light, the risk that approved substances are unsafe to the organism under treatment is limited.

The hormonally related health problems associated with drug administration are under stringent control, however, an opposing situation exists in relation to the control and monitoring of environmental endocrine disruption due to the presence of medical substances. One may argue that it is unlikely that environmental effects can occur considering that health issues are under control. However, there are several reasons to believe that pharmaceuticals could be environmental endocrine-disrupting chemicals (EEDCs). Firstly, some drugs are in fact used with the purpose of controlling hormonal mechanisms (e.g., contraceptives). Secondly, most drugs that are only active in the organism under treatment for a few hours but may after excretion persist in the environment for a much longer period. Thirdly, hormonal systems of environmental target organisms may be different to those found within the organism under treatment and consequently react differently to the presence of a substance. Ultimately, hormonal systems in the environment may be more sensitive than human hormonal systems, thereby reacting to much lower concentrations of a substance.

In the current chapter, these and other aspects related to the possible endocrine effects of pharmaceuticals in the environment will be discussed in more detail. As endocrine effects in the environment have primarily been linked to the estrogen receptor, the term endocrine activity will relate to this receptor if nothing else is mentioned. Initially, a concept for assessing their relevance as EEDC is presented. The medical substances that are relevant in the current context will then be identified, and finally the present information about these substances will be used to identify future research needs.

LIFECYCLE OF ENVIRONMENTAL ENDOCRINE PHARMACEUTICALS

If the use of a medical compound should lead to adverse endocrine effects in the environment, the substance must gain or retain endocrine activity during numerous consecutive processes. These processes are conceptualized in Fig. 1, which shows that the properties of the medical substance determines its fate in three main compartments: (1) the organism under treatment, (2) the environment, and (3) absorption in the target organisms. It must be stressed that no environmental effects will occur if just one of these processes leads to degradation, deactivation, or immobilization of the chemical. Consequently, the assessment of whether or not a medical compound may have impact on natural endocrine systems should ideally include the whole lifecycle of the substance from its initial use to its fate within the target organism. Obviously, this is impossible for all substances, and a number of criteria for identifying relevant substances with potential risk of being EEDC should be considered in succession:

- The predicted environmental concentrations (PECs) or the measured environmental concentrations should be high enough to induce endocrine effects.
- The use of a medical compound should lead either to the excretion of compounds that are hormonally active or to excretion of nonactive substances that can be converted in the environment to such active substances.
- Excreted hormonally active substances should after excretion, be persistent and mobile in the environment.

A number of studies have published methods for estimating worst-case concentrations for human [4] and veterinary medical substances [8,9]. In the simplest form, these assume that no immobilization, degradation, or metabolism of the drugs occurs on their route to the environment. Rough estimates of the dilution in wastewater or manure and in the receiving environments (soil, waters receiving sewage

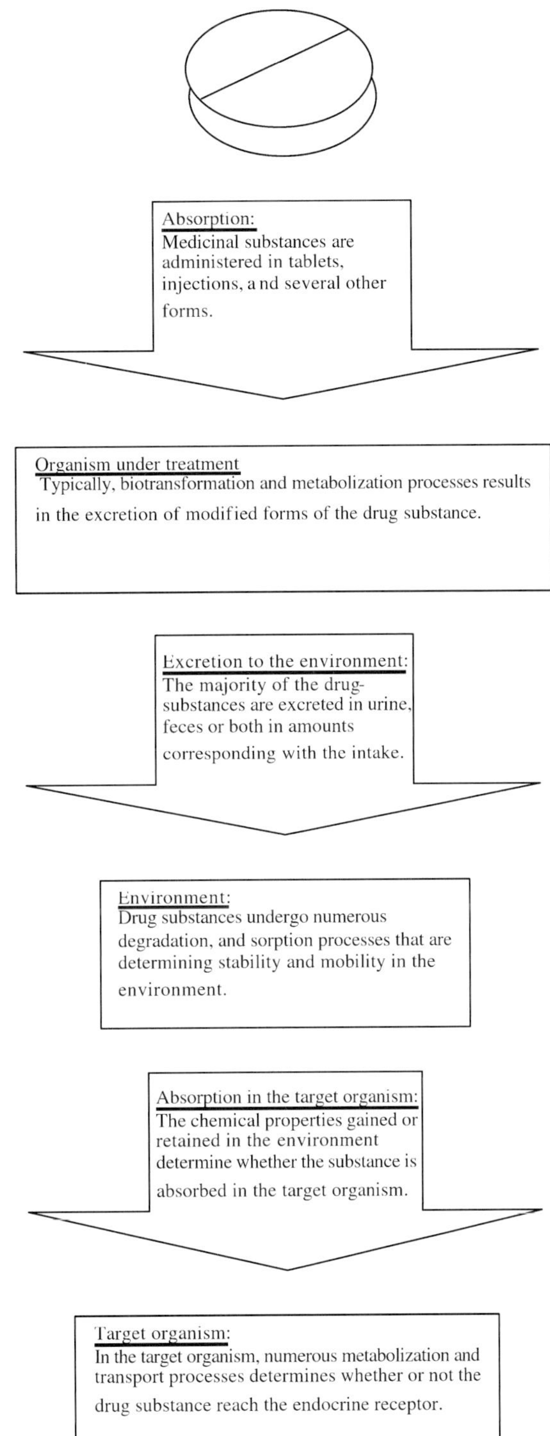

Fig. 1 Description of the route and processes of medicinal substances on their way to the target where they possibly result in an environmental endocrine effect.

effluent) is then used to calculate worst-case environmental concentrations. By implementing these estimation methods on commonly used human medical compounds; concentrations in sewage effluents ranging from a few ng/l (hormones) [10,11] to hundreds of μg/l (high-dose drugs such as painkillers) have been obtained [4,12].

As previously stated, it is assumed in the calculations above that the drugs in question can be fully recovered from excreta. The "bulk" intake of many compounds is equivalent to their environmental release, however these substances are rarely released in an unaltered state. In most cases, pharmaceuticals are excreted as one or several metabolites [1]. These metabolites are typically the result of either phase I or phase II metabolic reactions. Phase I reactions usually consist of oxidation, reduction or hydrolysis, and products are often more reactive and sometimes more toxic than the parent drug. Phase II reactions involve conjugation, which normally results in inactive compounds. Both phase I and phase II reactions change the physicochemical behavior of the substance and always render metabolites more water-soluble than parent compounds.

The behavior of a medical substance and its metabolic products in the environment should be assessed using the general procedures used for other environmental pollutants. However, some factors complicate this assessment. Firstly, the substances are present at extremely low concentrations, making them difficult to monitor. Secondly, investigation into the behavior of metabolites is complex. Despite this complexity, studies have showed that the phase II metabolites generally can be reactivated into the parent compounds [13]. Furthermore, many unconjugated metabolites have environmental properties similar to the parent compound. These results may justify that complex studies including all metabolites are limited to studies of the parent compound alone.

POTENTIAL ENDOCRINE-DISRUPTING, HIGH-VOLUME PHARMACEUTICALS

In theory, endocrine active pharmaceuticals are drug components that have impact on the endocrine system. This broad definition encompasses any compound possessing activity on endocrine receptors and originating from use in human or veterinary medicine. Substances with such properties are numerous and could be any type of endocrine agonist or antagonist (estrogens, androgens, thyroids, adrenocorticosteroids, etc.). It is, therefore, obvious that only those substances used in high volumes should be concentrated on here. The following will briefly introduce the most important classes of pharmaceuticals, and this will be followed by a discussion of their potential as environmental endocrine disruptors.

From sales statistics for human and veterinary drugs available in many countries, high-volume drugs have been identified by several authors [4,12,14,15]. With regard to endocrine activity, only certain groups are of interest. An overview is shown in Table 1. Generally, steroidal structures are the most relevant; as these are active in the endocrine system in humans, but other substances will also be discussed. Steroids can be characterized as lipophilic, nonvolatile substances. In the human organism, these steroids undergo various biotransformations usually leaving the steroid-structure intact [16]. The substances are then either excreted in urine as water-soluble conjugates or in feces in an unchanged state.

Estrogens and progestogens used for humans

Synthetic analogs of 17β-estradiol or the natural substance itself are primarily used for treatment of postmenopausal syndrome, breast cancer, or in contraceptives [17]. Contraceptives and postmenopausal drugs are extensively used in western countries [10–12]. Estrogen antagonists (e.g., tamoxifen) are also used in small quantities in the treatment of, for example, breast cancer.

The most important use of progesterone is in contraceptives where typical daily doses range from 0.25 to 2.5 mg. Progestogens are also used in the treatment of infertility, various cancer forms, and in combination with estrogens in the treatment of menstrual disorders and in contraceptives.

Table 1 Presentation of medicinal substances that could be relevant as endocrine disruptors.

Pharmaceutical group	Example	Main use in therapy
Estrogens	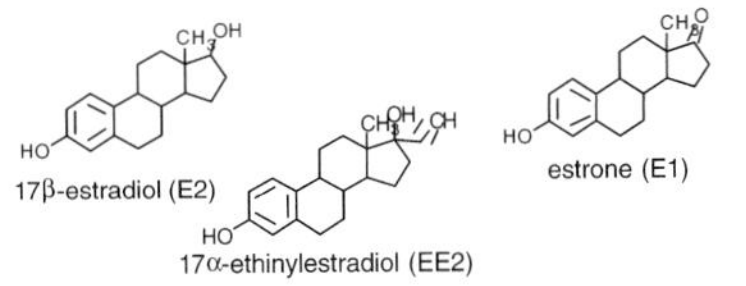 17β-estradiol (E2); 17α-ethinylestradiol (EE2); estrone (E1)	Contraception and hormone replacement therapy (treatment of postmenopausal syndrome).
Progestogens	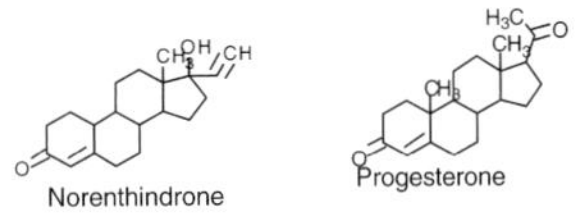 Norenthindrone; Progesterone	The main clinical use of progesterons is in oral contraceptives.
Androgens and glucocorticoids	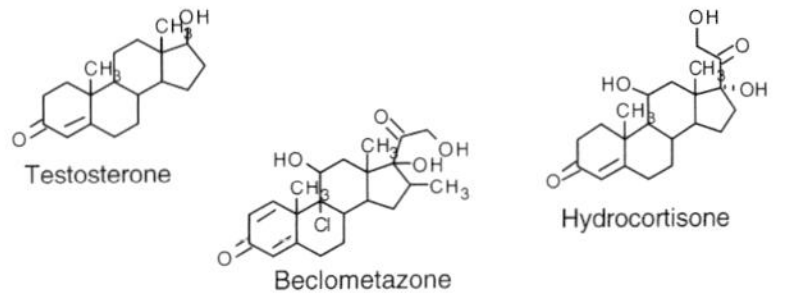 Testosterone; Beclometazone; Hydrocortisone	Androgens (testosterone) are used in small amounts in hormone replacement therapy. Glucocorticoids are used in the treatment of numerous diseases. Among these are treatments of inflammations, infections, asthma, allergy, skin diseases, and rheumatism.
Veterinary growth hormones	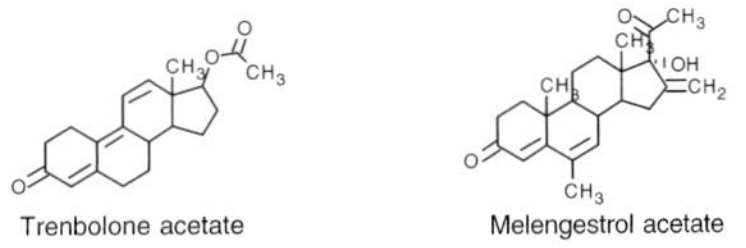 Trenbolone acetate; Melengestrol acetate	Used as growth promoters for meat-producing animals (anabolic steroids).
Nonsteroidal pharmaceutical substances	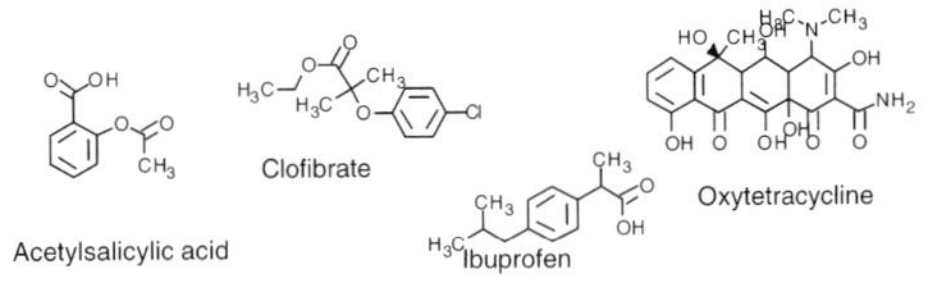 Acetylsalicylic acid; Clofibrate; Ibuprofen; Oxytetracycline	A multitude of nonsteroidal medicinal substances is used in high volumes. The examples given here are an agent against heart diseases (acetylsalicylic acid), a lipid-lowering agent (clofibrate), an antiinflammatoric agent (ibuprofen), and antibiotic (oxytetracycline).

Other steroids with impact on the endocrine system used in human medicine

Drugs with androgen activity consist of various testosterone formulations. These formulations differ in their ability to pass membranes, in their resistance to hydrolysis processes and in other factors related to their distribution in the organism. Modifications of androgenic substances (anabolic steroids) are used in large amounts in many countries as anabolic growth promoters for meat-producing animals. However, in comparison to estrogens, their use in human therapy is limited.

The adrenal cortex naturally secretes a number of steroids. The glucocorticoids are the most important of these due to the ability to affect carbon and protein metabolism. In addition, the glucocorticoids have anti-inflammatory and immunosuppressive activity. Therefore, both synthetic modifications and natural substances are widely used in drugs to control inflammatory diseases (hydrocortisone), asthma (e.g., beclometazone), infections, rheumatism, and other illnesses. The steroidal asthma products are extensively implemented and are typically among the 10 most used prescription drugs in industrialized countries [12]. The actions of the different steroids are overlapping indicating a broader spectrum than that observed for, e.g., sex hormones, which react very specifically with their respective hormone receptors.

Veterinary anabolic agents

In veterinary medicine, the two major classes of pharmaceuticals are antibiotics and anabolic growth hormones. As antibiotics are considered irrelevant in the context of this text, only anabolic agents appear to be relevant as environmental endocrine disruptors. At present the six hormones, 17β-estradiol, testosterone, zeranol, progesterone, trenbolone acetate (TBA), and melenogestrol acetate (MGA) are approved for use in the United States for enhancement of meat production in cattle and sheep [18]. Typically, these substances are administered by the use of growth implants as mixtures of steroidal androgens (progesterone, testosterone, TBA, and MGA) and estrogens (17β-estradiol and zeranol).

Nonsteroidal medical substances, personal care products, and drug additives

The number of substances from this group of substances is also numerous. Moreover, usage patterns differ highly between different countries. Therefore, a description of all the environmentally important substances is not possible here, however, several excellent reviews exist [1,2]. In these works, the most important medicinal substances are presented. These are from the following classes: agents used on the blood and blood-forming organs (e.g., acetylsalicylic acid), agents for the treatment of heart and circulatory diseases (e.g., clofibric acid), dermatological drugs (e.g., hydrocortisone), antibiotics (e.g., penicillin, amoxicyllin, tetracyclines), analgesics (e.g., paracetamol), anti-inflamatorics (e.g., ibuprofen), and agents used in treatment of allergy and asthma (e.g., budenoside).

Pharmaceuticals are administered in many different formulations and often in combination with various other chemicals that are included for purposes such as preservation, or as vehicles to aid administration/uptake in tablets, or for adjusting the ion-strength and pH in solutions. Many of these agents are also used in personal care products, and this contributes to the major release of these substances to the environment. In an environmental context, the major compounds are disinfectants, conservation agents, sun screens, musks, and other substances [2].

DISCUSSION OF PHARMACEUTICALS AS ENDOCRINE DISRUPTORS

Estrogens and progestogens used for humans

The most widely used pharmaceutical products for controlling the endocrine system in humans are the oral contraceptives and preparations used in human estrogen-replacement therapy (primarily for relieving postmenopausal syndrome and osteoporosis). Oral contraceptives contain typically 20–40 μg ethinylestradiol (EE_2) and 0.25–2.5 mg of various gestagens. Using these doses, the total consumption of contraceptives has been estimated for various countries (see Table 2). The active estrogenic components in hormone replacement drugs are conjugated estrogens, which are primarily the sulfate esters of estrone, equilin, and other naturally occurring compounds (in doses ranging from 0.3 to 2.5 mg). Depending on the individual circumstances, the substances are often used in combination with 2.5–5 mg progesterone [17]. In the United States, it is estimated that 12.5 to 33 % of the 40 million postmenopausal women are prescribed hormone replacement drugs [19]. In Table 2, estimated total annual consumption of estrogens in these products are reported for the United States and Denmark. As the population in the United States is only 55 times higher than in Denmark, it is clearly seen that the use of hormone replacement drugs per inhabitant varies considerably between different countries.

The endogenous excretion of hormones by healthy premenopausal women ranges from 10 to 100 μg estrogens per day. After menopause, women only excrete between 5 to 10 μg daily. The values for normal men average from 2 to 25 μg [17]. Pregnant women can excrete up to 30 mg per day, but average values are around 250 μg/day [20]. Using these data, total amounts of excreted estradiol in western countries were estimated using a method where the populations sex distribution was assumed as 50:50 male:female. Furthermore, it was assumed that 60 % of the female population was menstruating and that one out of every 75 females was pregnant. From these numbers the median excretion data and the amount of endogenous estrogens could be estimated. Table 2 lists such data for the world and a number of industrialized countries. From this table, it is generally seen that estrogenic pharmaceutical products may significantly contribute to the total estrogenic load from humans in the environment. Based on such consumption data, a number of authors have estimated worst-case concentrations of 17ß-estradiol up to approximately 100 ng/l in sewage effluents [19,21] and at levels from 0.01 to 0.1 ng/l in rivers [22]. In comparison, 17ß-estradiol has been shown to induce vitellogenin and inhibit testicular growth in male trout at concentrations of 2 ng/l [23] and other authors have obtained lowest observed effect concentrations of EE_2 at concentrations of 0.03 ng/l [24].

Table 2 Comparison of estimates of human consumption of pharmaceutical estrogens and the excretion of endogenous 17ß-estradiol. Data with no references are estimated as described in the text.

		World	US	UK	Denmark
Consumption of pharmaceuticals	E_2		1700[a]		0.7[b]
	EE_2	528[c]	88[a]	29[d]	1.9[b]
Endogenous excretion of E_2		54 504 kg	2575 kg	531 kg	45 kg

References: [a][10], [b][4], [c][17], [d][12].

These theoretical estimates have been validated in many research projects reporting monitoring data on estrogens in sewage water and effluents in many western countries [e.g., 19,25]. Concentrations of natural hormones (E_1 and E_2) range from 0 to 82 ng/l (E_1) and from 0 to 64 ng/l (E_2). For EE_2, typical concentrations are 5 to 10 times lower. E_1, E_2, and EE_2 have all been identified as the main causal agents explaining feminization observed on male fish in waters receiving sewage effluents [26]. Although, environmental concentrations of EE_2 suggest that it is less important than the natural hormones (E_1 and E_2), the relevance of this substance as an environmental endocrine disruptor is significant, because it has been shown that EE_2 is more persistent than E_2 [27,28] and often more potent [24]. Additionally it is well known that the ethinylation of the estradiol molecule in EE_2 decreases its affinity to the steroid binding proteins and as discussed by [29] this may lead to increased male sexual neural imprinting in female fetuses/larvae of vertebrates. In conclusion, it can stated there is strong evidence indicating that adverse endocrine effects on environmental organisms are a result of the presence of estrogens and additionally that the use of these substances in pharmaceuticals significantly adds to this pollution.

Although exogenous and endogenous estrogens are continually causing adverse effects in the environment, many details about their fate remain to be investigated. Early studies on the biodegradation of 17ß-estradiol at very high concentrations in sewage simulations systems showed that activated sludge microorganisms were able to mineralize estrogens at rates much slower than typical retention times in sewage treatment [28,30]. Although the primary degradation step for biodegradation of E_2 and EE_2 have been shown to occur within a few days and that mineralization occur within a week for E_2 [31], detailed studies on the biodegradation of estrogens at realistic concentrations are needed. In sewage, the source of these agents is excreta from humans containing sulfuric and glucuronic acid conjugates of estrogens. These conjugates do not possess a direct biological activity but there is strong evidence suggesting that they can be converted back to free estrogens by bacteria in the environment [32,33]. The kinetic details of these processes are practically unknown, and, furthermore, little is known about the other metabolic products formed as a part of the excretion of estrogens.

Considering that the most significant use of progestogens is in combination with EE_2 in contraceptives where the typical doses of progestogens are 5 to 15 times higher than EE_2, the detection of these substances in sewage effluent may be expected. However, these substances have only been detected in sewage in a few cases [34,35]. It has been proposed that because these substances are extremely lipophile, large portions are removed via sorptive processes in sewage treatment [20]. Another explanation may be that the significance of these substances as endocrine disruptors is still unclear and consequently they have been given little attention. [36].

Other steroids

It has been speculated that glucocorticoids may be relevant as environmental endocrine disruptors due to their chemical and biochemical links to the hormone system. The eventual transformation of cholesterol to estrogenic steroids due to activities of sewage bacteria has been investigated. However, these studies proved that this substance underwent rapid mineralization and no transformation to estrogens [37]. Considering that the glucocorticoids do not have the hydroxy group of the estrogens, it may be argued that the transformations may more likely result in the formation of substances with androgenic or other endocrine activities. This has, however, not yet been investigated.

The environmental relevance of steroids used in the treatment of asthma, skin diseases, and other diseases could be assessed by simple estimates of consumption. Based on the number of doses and estimates of the size of these doses the total consumption in Denmark was calculated as being 361 kg/year of glucocorticoids in pharmaceuticals. The significance of these substances as environmental endocrine disruptors seems important when compared to E_2 (45 kg/year) and EE_2 (0.7 kg/year) that have both been detected in ng/l-levels. However, as the natural release of steroids is at a level much higher (2–10 mg/l of cholesterol in sewage, [38]) it appears that the glucocorticoids used in medicine are relatively unimportant. However, their relative potency on environmental endocrine receptors is unknown and may differ strongly as it has been shown for receptors in the human organism [39].

Anabolic agents used in livestock production

The use of anabolic agents in meat production is banned in the European Union [40], however, in the United States and many other countries they are still used intensively. Similar to the excretion patterns observed in humans, studies on the metabolization and excretion of estrogens and androgens from agricultural animals have shown that the parent compound or its active metabolites are excreted mostly in urine as glucuronide conjugates [41]. After excretion, these substances may be cleaved, and, thus, the hormones may have impact on the environment when manure is used as fertilizer. The importance of hormones released from livestock farming is not yet clarified [42,43]. Although results are conflicting, it cannot be ruled out that this source of pollution with environmental endocrine disruptors is insignificant. Regarding the anabolic agents, the question of whether they contribute significantly to the total release of hormones from livestock can be answered by using simple mass balances. In cattle, implants with anabolic agents are typically used once or twice in a lifetime in doses of 100 to 200 mg androgens and 20 to 40 mg estrogens [44]. Using these numbers and assuming that all cattle slaughtered in the United States (33.5 million cattle were slaughtered in 2000 [45]) are treated once with growth implant, the total use of hormones can be roughly estimated to 671 to 1342 kg of estrogens and 3356 to 6712 kg of androgens. In comparison, concentrations of endogenously released E_1 and E_2 have been reported in cattle manure from animals used in both milk and beef production. In the six different farms studied, manure concentrations of E_1 and E_2 ranged from 52 to 640 and from 167 to 1229 μg/kg dry matter, respectively [46]. Using these data, and a typical manure production of 17.7 tons/year/animal with a dry matter content of 11.9 % [47], approximate releases of E_1 and E_2 from the 33.5 million beef cattle in the United States can be estimated to 4–45 and 12–87 tons, respectively. Although these numbers are significantly higher than the amount of anabolic agents used the present knowledge about the environmental fate of anabolic agents is too sparse to be used to conclude that these substances are environmentally unimportant. For example, it is possible that the synthetic anabolic agents, behave in similar ways as it shown for EE_2, and that they are more persistent than their respective endogenous hormones possibly possessing greater environmental impact than expected according to the concentrations in manure. Another example is that data exist showing that cattle subjected to certain anabolic agents generate urine with E_2 concentrations five- to six-fold over cattle not given these substances [48].

In view of the lack of scientific data on the fate of anabolic agents, the relevance of a recent study where the fate of the androgens trenbolone acetate and melengestrol acetate was studied after administration in cattle is obvious [49]. After eight weeks, approximately 10 ng/g of trenbolone acetate was detected in the manure collected during the experiment. During storage, the half-life of trenbolone was 260 days, and after spreading on a field the hormones were degraded more quickly. The question regarding the environmental importance of these compounds addressed by this study has been followed up by preliminary reports of endocrine effects on fish and Daphnia Magna due to trenbolone acetate [50].

Nonhormone pharmaceuticals

Apparently, no reports on nonhormonal drugs as environmental endocrine disruptors exist. This is not surprising, considering the low environmental concentrations and the thorough approval procedures that exist. On the other hand, drugs are biological active on many levels and are often used in the treatment of diseases related to the endocrine systems (e.g., metabolic disorders). Therefore, it is no surprise that they may have an impact on the endocrine system in the human body. However, as will be shown in the following the relevance of such data for environmental risk assessment is highly questionable and a number of examples will be given to illustrate some of these aspects.

Ibuprofen is an anti-inflammatory agent, which, due to its high use, has been the subject of studies on ecotoxicity [51], degradation [52], and environmental occurrence [53,54]. Based on such studies, the general conclusion is that ibuprofen is environmentally safe [55]. This conclusion may need to be reassessed considering that ibuprofen has been shown to block the effects of the estrogen agonist tamoxifen and partially to block the effects of 17ß-estradiol in a study of bone metabolism in rats [56] and thus may possibly be an environmental endocrine disruptor.

Clofibric acid, the active metabolite of the lipid-lowering agent, clofibrate is relevant because it is bound to a nuclear protein, the peroxisome proliferator activated receptor (PPAR). PPAR is thought to belong to the steroid receptor superfamily and therefore possibly influences the proliferation of cell organelles and lipid metabolism in organisms [29]. The endocrine effects of clofibric acid have been observed in different assays, however, results are often dubious. One example is a rat uterine assay where endocrine effects of clofibric acid were obtained; however, later studies could not confirm these data [57]. This compound is used in high volumes in many countries and is also a major pharmaceutical contaminant in the environment [58], therefore, it seems reasonable to investigate whether this substance may be an environmental endocrine disruptor [29].

In all the previous examples, the substances occurring in the environment are questioned regarding their safety due to results from the medical/pharmaceutical literature. Similarly, data revealing endocrine activity of nonsteroidal drugs can be found for other commonly used drugs. Examples are acetyl salicylic acid, which has been shown to affect the binding capacity of the estrogen receptor from MCF-7 cells [59] and oxytetracycline, which can interact with hormone metabolism [60]. In all cases, the data regarding the endocrine activity of the substance in question is based on assays relevant to human toxicology using concentrations much higher than those found in the environment. In other words, the comparability of these results is highly questionable, and results can only be used if they are confirmed in studies using more environmentally relevant assays. On the other hand, this information could be used as a starting point in the search of environmental endocrine disruptors and further, that future research should focus on extrapolating data from assays for human toxicology to environmental problems.

The final example of a pharmaceutical product being linked to adverse endocrine effects is a dental sealant showing estrogenic activity in the E-screen cell assay [61]. This effect was shown to be due to release in saliva of the well-known endocrine disruptor bisphenol A [62]. In this case, the substance in question is a well-known endocrine disruptor and its effects in nature is out of discussion. On the other hand, the question of whether it may reach the environment in concentrations high enough to cause damage in the receiving environments remains unanswered.

Personal care products and additives in pharmaceuticals

Among the personal care products (e.g., cosmetics or sunscreen agents), several substances are suspected as endocrine disruptors. A Swiss investigation [63] used the in vitro E-screen assay to prove the estrogenic activity of five UV screens [benzophenone-3 (Bp-3), homosalate (HMS), 4-methyl-benzylidene camphor (4-MBC), octyl-methoxycinnamate (OMC) and octyl-dimethyl-PABA]. These substances are lipophilic nonsteroidal substances that have been detected in fish from the Swiss Meerfelder Maar Lake at total concentrations of 2 mg/kg. Furthermore, there is evidence that Bp-3 is excreted in urine after dermal use [64] and in a rat uterotrophic estrogen assay, estrogenicity was observed in vivo after exposure to 4-MBC, OMC, Bp-3 [63]. In vitro estrogenicity of Bp-3 has been observed using the MCF-7 assay [65]. The interpretation of the data is unclear, and, therefore, the results have been subject of intense discussions in several environmental protection agencies.

Approximately 12 parabens are used commercially as preservatives in cosmetics, food, and pharmaceutical products [66]. The major source explaining the occurrence of parabens in sewage effluent at ng/l concentration levels is the use of these substances in shampoo and cream. Estrogenic effects of parabens have been demonstrated in vivo in mice [67], rats [68], and fish [69] and in various in vitro assays [68]. Using the E-screen assay, the estrogenic potency of the parabens was 5500 to 230 000 times lower than 17β-estradiol [66] indicating that this substance is unimportant as an environmental endocrine disruptor.

Certain phenyl-methyl substituted siloxanes used in cosmetics have been shown to exhibit high estrogenic potencies. The environmental significance of these substances has been questioned as these substances are considered as high-volume chemicals [70]. However, very little is presently known about these compounds as their analysis represents a major problem.

CONCLUSIONS

A few uses of human pharmaceuticals are directly linked to environmental endocrine disruption. Estrogens used in hormone replacement therapy of postmenopausal women (17β-estradiol and estrone) and synthetic hormones used in contraceptives (particularly 17α-ethinylestradiol) are recognized as the main causal agents for sexual disruption in fish in waters receiving effluents from sewage. It remains unclear whether or not the use of natural estrogens in human therapy adds significantly to the background level of natural hormones released from humans. However, due to their higher stability, the impact of synthetic estrogens compared to natural estrogens is greater than would be expected from the environmental concentrations. Technologies for removing these substances from wastewater should be developed further and implemented in waste management strategies.

The environmental importance associated with the use of veterinary anabolic agents for meat-producing animals in some countries is an emerging research subject. The endocrine effect of these synthetic steroids in the environment remains unanswered. Their relevance in comparison to naturally released hormones should be investigated.

In general, pharmaceutical compounds used in normal human therapy are not likely to be environmental endocrine disruptors. Firstly, most of the substances are used in very small amounts. Secondly, a range of processes determining the fate of the substance on its way from the use in medicinal products to the target should coincide. Thirdly, as a part of the approval procedure, medicinal substances undergo thorough testing for unwanted side effects including endocrine disruption on numerous levels.

On the other hand, there are many arguments suggesting that pharmaceutical compounds used in normal therapy may be environmental endocrine disruptors. Firstly, the medicinal substance may be transformed to endocrine active metabolites on its way to the target in the environment. Secondly, the endocrine system in the target may react differently than that of the patient where drug was used, and finally many drugs are used with the purpose of treating diseases related to the endocrine system (e.g., steroidal hormones).

In the search of whether other medicinal substances may be environmental endocrine disruptors, the following preliminary conclusions can be drawn:

- Even though the chemical structures of several widely used steroids are similar to estrogens (used in the treatment of asthma, skin diseases, etc.), no evidence indicating that these substances are endocrine disruptors in the environment exists. A hypothesis that such substances may metabolize in the environment and form estrogenic substances has been tested with negative result. This should however be investigated further especially with regard to nonestrogenic hormonal activity.
- In pharmaceutical and medical literature, numerous examples of substances with agonistic or antagonistic impacts on endocrine processes due to nonsteroidal drugs exist. These studies may be used as preliminary indicators of whether these substances could be environmental endocrine disruptors.
- Although many medical substances occur in the environment, they have generally not been tested for their endocrine activity. Such studies should be made using environmentally relevant assays.
- A number of substances used as the active substance in personal care products or as additives in medicinal products are of relevance in the current context. Of these substances, parabens and a number of sunscreens have been subject to concern due to their estrogenic activity. In addition, siloxanes (used in cosmetics) have been questioned as potential endocrine disruptors.

ACKNOWLEDGMENTS

We thank Anne Hougaard Buur for contributing estimates of consumption of steroidal drugs in Denmark.

REFERENCES

1. B. Halling-Sørensen, S. Nors Nielsen, P. F. Lanzky, F. Ingerslev, H. C. Holten Lützhøft, S. E. Jørgensen. *Chemosphere* **36**, 357–393 (1998).
2. C. G. Daughton and T. A. Ternes. *Environ. Health Perspect.* **107**, 907–937 (1999).
3. S. F. Webb. In *Pharmaceuticals in the Environment: Sources, Fate, Effects and Risks*, K. Kümmerer (Ed.), pp. 203–219, Springer Verlag, Berlin (2001).
4. F. Stuer-Lauridsen, M. Birkved, L. P. Hansen, H. C. Holten Lützhøft, B. Halling-Sørensen. *Chemosphere* **40**, 783–793 (2000).
5. S. B. Levy. *BMJ* **317**, 612–613 (1998).
6. R. A. Guarino. *New Drug Approval Process*, Marcel Dekker, New York (2000).
7. P. Evers. *Pharmaceutical Regulation in Europe*, Pearson Professional, London (1997).
8. M. H. M. M. Montforts, D. F. Kalf, P. L. A. van Vlaardingen, J. B. H. J. Linders. *Sci. Total Environ.* **225**, 119–133 (1999).
9. B. Halling-Sørensen, J. Jensen, J. Tjørnelund, M. H. M. M. Montforts. In *Pharmaceuticals in the Environment: Sources, Fate, Effects and Risks*, K. Kümmerer (Ed.), pp. 143–157, Springer Verlag, Berlin (2001).
10. L. D. Arcand-Hoy, A. C. Nimrod, W. H. Benson. *Int. J. Toxicol.* **17**, 139–158 (1998).

11. F. M. Christensen. *Regul. Toxicol. Pharmacol.* **28**, 212–221 (1998).
12. N. J. Ayscough, J. Fawell, G. Franklin, W. Young. R&D Technical Report P390. UK Environment Agency, Bristol, UK (2000).
13. K. V. Berger, B. Petersen, H. Bünung-Pfaue. *Arch. Lebensmittelhygiene* **37**, 85–108 (1986).
14. S. F. Webb. Technological Institute, Brussels, Belgium, *Proceedings of International Seminar on Pharmaceuticals in the Environment*, 9 March 2000 (2000).
15. B. Halling-Sørensen, S. Nors Nielsen, J. Jensen. Miljøprojekt 659. Miljøstyrelsen (2002).
16. S. B. Mahato and S. Garai. *Steroids* **62**, 332–345 (1997).
17. C. L. Williams, G. M. Stancel, J. G. Hardman, A. Goodman Gilman, L. E. Limbird (Eds.), *Goodman & Gillman's The Pharmacological Basis of Therapeutics*, Pergamon Press, New York (1996).
18. Drug Information Laboratory. Drug Information Laboratory, Virginia/Maryland Regional College of Veterinary Medicine, Blacksburg, VA (2001).
19. A. C. Belfroid, A. V. d. Horst, A. D. Vethaak, A. J. Schäfer, G. B. J. Rijs, J. Wegener, W. P. Cofino. *Sci. Total Environ.* **225**, 101–108 (1999).
20. A. C. Johnson, R. J.Williams, T. Ulahannan. *Environ. Sci. Technol.* **33**, 369–370 (1999).
21. A. C. Johnson, A. C. Belfroid, A. D. Corcia. *Sci. Total Environ.* **256**, 163–173 (2000).
22. R. J. Williams, M. D. Jürgens, A. C. Johnson. *Water Res.* **33**, 1663–1671 (1999).
23. S. Jobling, D. Sheahan, J. A. Osborne, P. Matthiessen, J. P. Sumpter. *Environ. Toxicol. Chem.* **15**, 194–202 (1996).
24. C. D. Metcalfe, T. L. Metcalfe, Y. Kiparissis, B. G. Koenig, C. Khan, R. J. Hughes, T. R. Croley, R. E. March, T. Potter. *Environ. Toxicol. Chem.* **20**, 297–308 (2001).
25. T. A. Ternes, M. Stumpf, J. Mueller, K. Haberer, R. D. Wilken, M. Servos. *Sci. Total Environ.* **225**, 81–90 (1999).
26. C. Desbrow, E. J. Routledge, G. C. Brighty, J. P. Sumpter, M. Waldock. *Environ. Sci. Technol.* **32**, 1549–1557 (1998).
27. D. G. J. Larsson, M. Adolfsson-Erici, J. Parkkonen, M. Pettersson, A. H. Berg, P.-E. Olsson, L. Förlin. *Aquat. Toxicol.* **45**, 91–97 (1999).
28. H. R. Tabak, R. N. Bloomhuff, R. L. Bunch. *Dev. Ind. Microbiol.* **22**, 497–519 (1981).
29. P. Stahlschmidt-Allner, B. Allner, J. Roembke, T. Knacker. *Environ. Sci. Pollut. Res. Int.* **4**, 155–162 (1997).
30. E. Stumm-Zollinger and G. M. Fair. *Journal WPCF* **37**, 1506–1510 (1965).
31. M. D. Jürgens, K. I. E. Holthaus, A. C. Johnson, J. J. L. Smith, M. Hetheridge, R. J. Williams. *Environ. Toxicol. Chem.* **21**, 480–488 (2002).
32. G. H. Panter, R. S. Thompson, N. Beresford, J. P. Sumpter. *Chemosphere* **38**, 3579–3596 (1999).
33. T. A.Ternes, P. Kreckel, J. Mueller. *Sci. Total Environ.* **225**, 91–99 (1999).
34. M. J. López de Alda and D. Barceló. *J. Chromatogr. A* **892**, 391-406 (2000).
35. H. M. Kuch and K. Ballschmiter. *Fresenius' J. Anal. Chem.* **366**, 392–395 (2000).
36. N. Garcia-Reyero, E. Grau, M. Castillo, M. J. Lopez de Alda, D. Barceló, B. Pina. *Environ. Toxicol. Chem.* **20**, 1152–1158 (2001).
37. S. J. Niven, J. Snape, M. Hetheridge, M. Evans, J. McEvoy, S. J. Rowland. *Sci. Total Environ.* **279**, 75–86 (2001).
38. S. J. Niven, J. Snape, M. Hetheridge, M. Evans, J. McEvoy, P. G. Sutton, S. J. Rowland. *Analyst* **126**, 285–287 (2001).
39. B. P. Schimmer, K. L. Parker, J. G. Hardman, A. Goodman Gilman, L. E. Limbird (Eds.), *Goodman & Gillman's The Pharmacological Basis of Therapeutics*, Pergamon Press, New York (1996).
40. EU. *Official Journal* **L125**, 10–32 (2001).
41. W. M. Knight. In *Estrogens in the Environment*, J. A. McLachlan (Ed.), p. 391, Elsevier, North Holland, New York (1980).

42. J. E. Drewes and L. S. Shore. In *Pharmaceutical and Personal Care Products in the Environment*, C. G. Daughton and T. L. Jones-Lepp (Eds.), American Chemical Society, Washington DC (2001).
43. L. S. Shore and M. Shemesh. *Pure Appl. Chem.* **75**, 1859–1871 (2003).
44. R. L. Preston. *Adv. Drug Delivery Rev.* **38**, 123–138 (1999).
45. National Agricultural Statistics Service (NASS). Mt An 1-2-1 (01)a. U.S. Department of Agriculture, Washington, DC (2001).
46. A. Wenzel, T. Küchler, K.-P. Henschel, W. Schnaak, M. Diedrich, J. Müller. Umweltforschungsplan des Bundesministers für umwelt, Naturschutz under Reaktorsicherheit Forschungsbericht 216 02 011/11 (1998).
47. H. Damgaard Poulsen, V. Friis Kristensen. 736. Ministeriet for Fødevarer, Landbrug og Fiskeri, Danmarks Jordbrugsforskning, Foulum, Tjele, Denmark (1997).
48. E. W. Peterson, R. K. Davis, H. A. Orndorff. *J. Environ. Qual.* **29**, 826–834 (2000).
49. B. Schiffer, A. Daxenberger, K. Meyer, H. H. D. Meyer. *Environ. Health Perspect.* **109**, 1145–1151 (2001).
50. R. Renner. *Environ. Sci. Technol.* **36**, 194A–197A (2002).
51. Knoll. Issue 6/4/1995. Knoll Pharmaceuticals, Nottingham, NG2 3AA, England (1995).
52. M. Winkler, J. R. Lawrence, T. R. Neu. *Water Res.* **35**, 3197–3205 (2001).
53. H.-R. Buser, T. Poiger, M. D. Müller. *Environ. Sci. Technol.* **33**, 2529–2535 (1999).
54. H.-J. Stan and T. Heberer. *Analusis* **25**, 20–23 (1997).
55. A. P. van Wezel and T. Jager. *Chemosphere* **47**, 11113–1128 (2002).
56. J. D. Sibonga, N. H. Bell, R. T. Turner. *J. Bone Mineral Res.* **13**, 863–870 (1998).
57. J. Ashby, P. A. Lefevre, J. Odum, H. Tinwell, S. J. Kennedy, N. Beresford, J. P. Sumpter. *Regul. Toxicol. Pharmacol.* **26**, 96–101 (1997).
58. H.-R. Buser, M. D. Müller, N. Theobald. *Environ. Sci. Technol.* **32**, 188–192 (1998).
59. C. van Aswegen, J. C. Dirksen van Schalkwyk, L. J. Roux, P. J. Becker, D. J. D. Plessis. *Clin. Physiol. Biochem.* **9**, 145–149 (1992).
60. E. Hamalainen, J. T. Korpela, H. Adlerkreutz. *Gut.* **28**, 439–445 (1987).
61. N. Olea, R. Pulgar, P. Pérez, F. Olea Serrano, A. Rivas, A. M. Soto, V. Pedraza, A. M. Soto, C. Sonnenschein. *Environ. Health Perspect.* **104**, 298–305 (1996).
62. C. A. Staples, P. B. Dorn, G. M. Klecka, S. T. O'Block, L. R. Harris. *Chemosphere* **36**, 2149–2173 (1998).
63. M. Schlumpf, B. Cotton, M. Conscience, V. Haller, B. Steinmann, W. Lichtensteiger. *Environ. Health Perspect.* **109**, 239–244 (2001).
64. T. Felix, B. J. Hall, J. S .Brodbelt. *Anal. Chim. Acta* **371**, 195–203 (1998).
65. Y. Nakagawa, T. Suzuki, S. Tayama. *Toxicol.* **156**, 27–36 (2000).
66. H. R. Andersen, M. M. Pedersen, M. Holst-Jørgensen, S. Nors Nielsen, T. H. Rasmussen, E. Sonnich Thomsen, B. Halling-Sørensen. Submitted for publication.
67. A. Hossaini, J.-J. Larsen, C. L. Larsen. *Fodd Chem. Toxicol.* **38**, 319–323 (2000).
68. E. J. Routledge, J. Parker, J. Odum, J. Ashby, J. P. Sumpter. *Toxicol. Appl. Pharmacol.* **153**, 12–19 (1998).
69. K. L. Pedersen, S. N. Pedersen, L. B. Christiansen, B. Korsgaard, P. Bjerregaard. *Pharmacol. Toxicol.* **86**, 110–113 (2000).
70. R. Nilsson. *Toxicol. Pathol.* **28**, 420–431 (2000).

Pure Appl. Chem., Vol. 75, Nos. 11–12, pp. 1895–1904, 2003.

Topic 2.5

Endocrine active industrial chemicals: Release and occurrence in the environment*

Andrew Johnson[‡] and Monika Jürgens

Centre for Ecology and Hydrology, Wallingford, OX10 8BB, UK

Abstract: Of the xenobiotic endocrine active substances (EASs), tributyltin (TBT) has had the clearest link to an impact on aquatic ecology. Its release from marine antifouling paints had a drastic impact on dogwhelk populations in polluted harbors due to a masculization effect. 4-*tert*-Nonylphenol is seen as the most significant of the industrial xenobiotic estrogen mimics, being implicated as the dominant endocrine disruptor in certain industrialized river reaches. Apart from hot spots associated with particular industries, the estrogenic alkylphenols, phthalates, and bisphenol A are present in effluent and receiving water at concentrations below that which would give cause for concern. Other more bioaccumulative compounds such as polybrominated flame retardants, dioxins, and furans may possess some endocrine active properties. The possibility of additivity effects may yet mean that low concentrations of xenobiotic EASs will need careful consideration. It is noted that considerable quantities of many of these compounds are often found in sewage sludge and sediments.

INTRODUCTION

Assessing whether any of the xenobiotic endocrine active substances (EASs) pose a threat to the natural environment requires balancing information on its potency against observed environmental concentrations. At least for fish, the overwhelming form of endocrine disruption (ED) observed in the aquatic environment has been estrogenic. Much effort has gone into screening xenobiotic compounds for estrogenic potential using in vitro tests. Thus, while we can report on the xenobiotic EASs of current concern, it is important to be aware of the following complicating factors that prevent an accurate assessment from being made with confidence:

- Exposure of fish or other animals to xenobiotic EASs may have a greater impact than predicted from in vitro screening tests, because the liver system may be able to clear estradiol (the reference compound in many in vitro tests) much quicker than a xenobiotic estrogen, which although it has less affinity for the estrogen receptor, may remain in the body for longer [1]. Possibly, they are responsible for the pathogenicities such as malformed sperm ducts which are of greatest consequence for fish populations [2].
- Additivity may mean that mixtures of low concentrations of xenobiotic EASs could yield overall significant ED to animals. Such summation effects have been shown in vitro [3] and in vivo [4]
- The impact may be greatest on species of aquatic fauna other than fish which we have yet to study. TBT, for example, had little effect on fish, but had a disastrous impact on the female mollusk population due to a masculization effect [5].

---

*Report from a SCOPE/IUPAC project: Implication of Endocrine Active Substances for Human and Wildlife (J. Miyamoto and J. Burger, editors). Other reports are published in this issue, *Pure Appl. Chem.* **75**, 1617–2615 (2003).

[‡]Corresponding author

- Perhaps we have yet to discover the most important xenobiotic EAS! This may be a breakdown product of an innocuous parent molecule formed in sediment, or transit down the river.

Similarly, care must be taken in the interpretation of water or soil concentrations of EASs. The concentration will in the first case largely depend on the local dilution available in the receiving water, and in the second case on the tons per hectare of sewage sludge spread to land, and its soil incorporation [6,7]. Sediment concentrations will be influenced initially by the octanol–water partition coefficient (K_{ow}) and recalcitrance of the compound, but sample location, such as distance from a discharge, will have an enormous impact on the concentrations reported.

CHEMICALS OR GROUPS OF INTEREST

The list of known or suspected EASs gets longer almost every day. For example, a working list produced for the European Union contains 564 chemicals or groups of which 66 were chosen as priority substances [8]. This chapter attempts to give an overview of some important EASs of industrial origin—for endocrine active natural compounds, pharmaceuticals, or pesticides please refer to Topics 2.2–2.4 and 2.6.

Tributyltin (TBT)

Sources and potencies

Tributyltin acts as a biocide and is mainly used as a slow-release antifouling agent incorporated in paints for ships. TBT has been linked with the breakdown of commercial oyster breeding in some areas and masculinization (imposex or intersex) of various marine invertebrates often to the point where they were unable to reproduce. This has led to the local extinction of the dogwhelk Nucella in many areas [5]. Because of these effects, which can be observed at concentrations as low as 1 ng/l [9], its use on pleasure boats is now banned in most countries and the International Maritime Organization (IMO) has adopted a convention, which prohibits the application of TBT-containing paints on all ships by 2003 and requires old paints to be removed or covered by 2008. Even though the use of TBT is now restricted, large amounts can still be released when old paint is stripped off prior to repainting a ship [10]. Polluted sediments may also release organotin compounds into the water column, especially during harbor dredging or storms. The use of organotins in stabilizers for plastics [5] and in wood preservatives [10] could lead to potentially more diffuse contamination of the environment including soil.

Environmental concentrations

Concentrations of several hundred to three thousand ng/l have been found in harbors and marinas within Germany, Switzerland, and Canada [5]. These concentrations exceed the threshold level for induction of imposex in marine neogastropods by over a hundred times. The log K_{ow} of TBT is 3–4, and therefore it should have a low to medium attraction to the organic fraction of sediments, but with its positive charge, TBT can also bind as a cation to sediments and biota, therefore partitioning to solids and bioaccumulation is higher than would be expected from the K_{ow} alone [10]. TBT levels in sediments can be up to µg/g levels especially in confined harbors and marinas [5,11].

Summary

The impact of TBT on marine and limnic organisms is very considerable. The measures restricting its use are showing some success, such as in some recently sampled North Sea sites [12], but TBT levels remain unacceptably high in many places. In contrast, a comparison of imposex parameters measured in 1995 and 2000 along the Portuguese coast found no improvements [13], and levels up to 200 ng/l were still found around Corsica in 1999 [14].

Alkylphenol polyethoxylates and their degradation products

Sources and potencies

The alkylphenol polyethoxylates (APEOs) parent compounds are commonly used nonionic surfactants which are partially degraded in sewage treatment works (STWs) to yield a wide variety of by-products [15]. These include the formation of APEOs with shorter ethoxylate (EO) chain lengths (such as the alkylphenol mono and diethoxylates). These undergo further degradation to form the alkylphenols, nonylphenol (NP) and octylphenol (OP). Log K_{ow} reported for OP and NP are 4.2–4.5 [16]. 4-*tert*-Octylphenol is the most potent of these compounds, being 1.5×10^3 less potent than 17β-oestradiol (E2), followed by 4-*tert*-nonylphenol at around 10^4 times less potent than E2 [17,18]. The NP1EO and 2EO metabolites are ten to a hundred times less potent than nonylphenol and so represent much less of a threat [19,20].

Environmental concentrations

Sewage treatment works effluent concentrations of NP have been reported at between 0.1 and 3.7 µg/l in Scotland (UK), Germany, Switzerland, and Italy [21–24]. In a recent monitoring exercise [25] of the UK rivers only in the Mersey and Aire were NP concentrations above 0.2 µg/l. High concentrations, up to 180 µg/l (total extractable NP), had been reported in the R. Aire (UK) [26] due to effluent from the local textile industries and in two rivers in Catalonia, Spain (0.15–644 µg/l) [27]. Concentrations of up to 0.4 mg/kg NP have been detected in river sediment in the UK [28] and up to 0.7 mg/kg in Canadian and 6.7 mg/kg in Italian marine sediments associated with cities [29,30]. The highest octylphenol value (13 µg/l) has been reported for the Tees estuary [26], which is heavily impacted by industry, however other values measured in the United Kingdom and Canada were <0.5 µg/l [26,31]. The hydrophobic NP and OP are readily sorbed to sludge and appear to be resistant to anaerobic digestion [32]. A study of sludge in Spain and Germany found NP concentrations ranging from 25–600 mg/kg [33]. There are some data indicating that NP does not persist in sludge-amended soil [34,35].

Summary

Potency studies would suggest that where concentrations greater than 1 µg/l of OP or NP in receiving waters occur then a danger of ED may exist. Usually, dilution in the receiving water would bring concentrations down to below 1 µg/l, but there are a few industrially impacted sites where NP concentrations of 10–100 µg/l have been measured, and ED dominated by these compounds. The other short chain EO metabolites are at least an order of magnitude less potent, but may be important if additivity is taken into account. It is not yet known whether alkylphenol binding to sediment or sludge has any ED significance.

Bisphenol A

Sources and potency

Bisphenol A (BPA, 4,4′-isopropylidenediphenol) is a widely used intermediate in the production of polycarbonate and epoxy resins. In vitro studies indicate BPA to be 10 000–30 000 times less potent than E2 [3,20,22]. A study using Japanese Medaka gave 10 µg/l as the lowest concentration where ED as induction of testis-ova could be observed [20]. One study on *Xenopus* frogs suggested an ED effect could be occurring at 23 µg/l [36], however, this was contradicted by a similar study [37]. Concentrations as low as 5 µg/l were reported as causing potentially fatal superfeminization in a species of prosobranch snails [38]. Staples [39] reported log K_{ow} values for BPA of 2.2–3.82.

Environmental concentrations

Median effluent concentrations of 0.14, 0.15, and 0.03 µg/l were observed in Canada [40], Germany [41], and Japan [42], respectively. Industrial effluent around Toronto had a median value of 11 µg/l [40]. BPA concentrations in the effluent of paper production (mean 41 µg/l), metal/wood production (17 µg/l), and the chemical industry (18 µg/l) have been reported [43]. In a survey of river systems in

Japan [44] and Germany [45], the majority of samples were below 0.1 µg/l with only one sample above 1 µg/l. An analysis of sewage sludge extracts taken from 18 treatment works in Canada gave a median concentration of 1.1 mg/kg [40]. High concentrations of BPA of 25–146 µg/l have been found in waste dump water and compost water [45].

Summary

Reported concentrations in domestic sewage effluent are typically below 1.5 µg/l. Based on the reported potency of BPA and receiving water dilution, this should not give rise to concern. Significantly higher concentrations (often 10s of µg/l) were observed in the effluent emanating from some specific industries which could lead to locally elevated sediment concentrations. In these cases, particularly where dilution is negligible, some ED of sensitive invertebrates may occur.

Phthalates

Sources and potencies

Phthalates are widely used in the manufacture of plastics. The phthalate esters with the most evidence for estrogenic activity in vitro are butyl benzyl phthalate (BBP), dibutyl phthalate (DBP), and di(2-ethylhexyl) phthalate (DEHP) with in vitro potencies from 10^{-5}–10^{-8} compared to E2 [46]. These in vitro results however could not be reproduced in vivo. Concentrations as high as 5 mg/l DEHP had no observable effect on Japanese Medaka fish [20]. The phthalates have reported log K_{ow} values in the range of 4–5 for DBP and BBP and >7 for DEHP [47].

Environmental concentrations

In UK and German sewage effluent studies DBP was detected at <1–14 µg/l; BBP <1–2.8 µg/l; and DEHP <2.4–182 µg/l [21,45,48]. Looking across Canadian, Michigan (USA), German, and Swiss municipal sludges, phthalates were found in the following ranges; DEHP 21–230 mg/kg; BBP 0.3–10.1 mg/kg; DBP 0.2–17 mg/kg [45,49,50]. Concentrations in sludges from industrial catchments in Canada had a median 10 mg/kg for DBP, and 80 mg/kg for DEHP [51]. DEHP was also found in German waste dump waters with concentrations between 17–169 µg/l [52]. In a recent review of thousands of surface water measurements [53], most values were below 1 µg/l for DBP and BBP, but a later German survey found higher concentrations of 0.3–98 µg/l (median 2.3 µg/l) DEHP and 0.12–8.8 µg/l DBP (median 0.5 µg/l) in surface water. [45]. Although some of these values exceed predicted no effect levels for toxicity [45], there is no evidence of endocrine effects at these levels.

Summary

The in vitro estrogenic potency of the phthalate esters is similar or weaker than BPA, with little evidence of in vivo effects except at exceptionally high concentrations. Concentrations in treated sewage effluents are at levels which would be considered unlikely to pose a significant risk to aquatic life. However, data indicates several phthalate esters can have a significant presence in sewage sludge as would be predicted from their high log K_{ow} values. It is unclear whether high sludge concentrations pose any endocrine disruptor risk.

Brominated flame retardants

Sources and potencies

Polybrominated biphenyls (PBBs) and polybrominated diphenylethers (PBDEs) are widely used as flame retardants in a wide variety of products [54]. Due to their persistence and hydrophobicity, brominated flame retardants can be detected in animal tissue from all over the world [54,55]. These compounds have a high hydrophobicity with log K_{ow} 6–10 [47]. Effects on thyroid function have been found at high concentrations in rodents [56,57] and possibly estrogenic or antiestrogenic effects as suggested by in vitro studies [58]. Because the compounds accumulate, exposure via the food chain is likely. This group of chemicals is discussed in detail in Topic 3.7.

Environmental concentrations

As may be expected with their hydrophobic nature, all data available for water or sewage concentrations is focused on the amount sorbed to suspended particles. In municipal sewage effluent from the Netherlands, 0.3–0.9 mg/kg BDE 209 has been found on the suspended particles [47,59]. High concentrations of BDE 209 (up to 4.6 mg/kg) were found in suspended matter in a Dutch estuary [47], but lower levels of 10 µg/kg or less for the individual PBDE in sediments of the river Elbe in Germany [60]. The highest levels found are associated with sediments downstream of plastics manufacturing sites or sewage works, with a clear trend over time for increasing concentrations found in sediments from the late 1970s onwards [55]

Summary

The potential widespread distribution in sediments, and its persistence in sediment is a matter for some concern. Their hydrophobic character and persistence will lead to some degree of bioaccumulation. However, it is still too early to determine whether these compounds could pose a realistic ED threat.

Polychlorinated biphenyls (PCBs), dibenzodioxins (PCDDs), dibenzofurans (PCDFs), and other industrial chemicals

Sources and potencies

PCBs dioxins and furans

PCBs were widely used in lubrication and in isolating, cooling, and hydraulic fluids. However, when their toxicity and potential to accumulate in the environment became known their use was phased out in many countries [61,62]. PCBs have been implicated in impaired reproduction and immune function in seals (reviewed in [63]). Various dioxins and furans, which can be generated for example during waste incineration, especially the Seveso poison 2,3,7,8-tetrachlorodibenzodioxin (2,3,7,8-TCDD) are arylhydrocarbon receptor (AhR) agonists and have been shown in vitro and in vivo to act as antiestrogens. However, the doses needed were very close to the lethal dose, so the acute and chronic toxic effects for this group of chemicals would probably be of greater concern than any suspected endocrine effects. Some coplanar and mono-orthocoplanar PCBs also have a weak dioxin-like effect but always less than 2,3,7,8-TCDD. Depending on the exact structure, PCBs can also have estrogenic effects (reviewed in [61]).

Styrene

Styrene is a monomer used for the production of polystyrene, synthetic rubber, etc. Monomers or small polymers such as dimers and trimers can leak from plastic containers. The highest estrogenic potential in vitro was found for some styrene trimers. The relative potency of these chemicals was 1.5×10^{-5}–2×10^{-6} [64]. Some endocrine related impacts on in female styrene-exposed workers have been reported [8], but so far it is not considered to possess a significant environmental threat.

Other polycyclic aromatic hydrocarbons

Many polycyclic hydrocarbons have been tested in vitro for estrogenic or dioxin-like activity. Generally small molecules with an unhindered phenolic moiety similar to the A-ring of estradiol and moderate hydrophobicity are likely to bind to the estrogen receptor and can act as weak estrogens [65], other hydrocarbons such as benzo(a)pyrene, benzo(a)anthrazene, and related compounds show dioxin-like activity with potencies generally at least a factor of a 1000 less than 2,3,7,8-TCDD [66].

Environmental concentrations

Due to their persistence and hydrophobicity (log P_{ow} 4.5–10 [61]), PCBs are widely distributed in sediments, soils, and animal tissue, but the measures replacing these chemicals are showing effects: for example, dated sediments from the Gulf of Finland show highest concentrations of PCBs (up to 57 μg/kg), polychlorinated dioxins and furans (PCDD/F up to 101 μg/kg) in sediments deposited in the 1960s and 1970s. This contamination was caused by a chemical plant producing a chlorophenol-based wood preservative until 1984 [67]. Surface sediments in the same area have lower concentrations (0.4–52 μg/kg PCDD/F and 1.85–39 μg/kg PCB). PCB concentrations in surface sediments from river and coastal regions of the United States were also in the low μg/kg range. [68]. However, recent studies have shown elevated PCB concentrations in marine top predators in the Mediterranean, illustrating the continuing hazard that these chemicals can pose due to bioaccumulation [69]. Low, to very low water concentrations, in the sub μg/l range are usually reported for compounds in the PAH family such as benzo(a)pyrene and naphthalene [28,70], but as observed previously, much higher concentrations, up to mg/kg levels can be found in river bed sediments [28].

Summary

While water concentrations of these compounds are rarely a matter of concern due to their poor solubility, the potential for accumulation along the food chain may be of importance especially for relatively long-lived animals. Whereas many of these compounds have toxic properties, they may also represent an ED hazard for top predators in water bounded by industries with, for example, historic PCB use.

CONCLUSIONS

In terms of the presence in freshwater receiving STW effluents, only NP occasionally reaches concentrations that might give rise to serious ED concern. The problem of TBT in certain marine and freshwater locations affecting dogwhelk populations seems set to continue due to the lag in implementing the new restrictions and its persistence in sediments. Concentrations of BPA in water and sediment may be an issue associated with a few industrial discharges. Industrial-derived endocrine disruptors can be found in high concentrations in sewage sludge. For example, a German survey found that alkylphenols and BPA contributed maximal 10 % to the estrogenicity of sewage effluents or surface water compared to 65 % for sewage sludge extracts [71], but there is no evidence yet that this will have an ED impact. The polybrominated flame retardants may become an increasingly undesirable constituent of some sediments and animal tissue, but its significance is still difficult to evaluate. The ED impact of such chemicals as dioxins, furans, and PCBs relative to their already known toxicity should not be ignored. There is still insufficient information on whether xenobiotic EASs have the potential to harm other forms of aquatic fauna, particularly invertebrates at concentrations below the no effect level in fish. Overall, the possibility of additivity effects of mixtures, which is discussed in Topic 3.11, may mean the concern of ED from these xenobiotic compounds may ratchet up in future years.

Future research needs

An exhaustive list could be drawn up of various aquatic fauna and EASs that should be tested. It is perhaps wiser to assess current ecosystem health in a range of representative river reaches to examine whether any problems actually exist in the first place. A toxicity identification and evaluation (TIE) technique with an in vitro estrogenicity test highlighted the importance of steroid estrogens compared to xenobiotic estrogen mimics in sewage effluent [72]. Could such a TIE approach be repeated using an in vivo test, with a suitable, easy-to-handle fish, such as Japanese medaka, or zebrafish (*Danio rerio*)? Understanding the role of mixtures of low concentrations of xenobiotic EASs on fish would help regulators in assessing how seriously these compounds should be taken.

Table 1 Examples of reported concentrations (after 1985).

Compound	In vitro potency comp. to E2	In vivo potency LOEL µg/l	Domestic (or not known) effluent conc. µg/l[a]	Industrial impacted effluent conc. µg/l[a]	Sewage sludge conc. µg/g dry wt	Surface water conc. µg/l[a]	Sediment conc. µg/g dry wt	Concern?
TBT	Not applicable	0.001 (imposex in marine snail) [61]	<0.03–35 [5,10]	<0.03–62 [5]	0.02–1.5 [5,73]	<0.005–3 [5]	0.001–53 [5,11,73,74]	Yes
Nonylphenol	10^{-3}–10^{-6} [19,46,47]	0.6, 6 (vtg fish) [4,75]	<0.02–69 [31,47]	142–330 [26,27]	8–4000 [31]	<0.01–644 [26,27,31,47]	<0.0015–72 [31,47]	Yes, in certain reaches with industrial im pact
Octylphenol	10^{-3}–10^{-6} [19,22,46,61]	5 (vtg, fish) [47]	0.005–1.7 [31,47]	0.26–9 [31]	<0.01–20 [31]	<0.005–13 [31,47]	<0.002–1.8 [31,47]	No, except for rare industrial impact
Bisphenol A	10^{-4}–10^{-5} [3,20,22]	1, 10 (snail; intersex fish) [20,76]	<0.0001–4.5 [22,47,52,77–79]	<0.01–8 [47,52,80]	0.004–1.36 [45]	<0.0001–1.4 [47,52,81]	<0.001–0.19 [47,52]	Possible in some industrial impact sites
Phthalates DEHP[b], DBP[c], BBP[d]	10^{-5}–10^{-8} [46,82]	>100 [20,75]	<0.1–30.5 [52]	<0.1–80.5 [52]	0.19–154 [52]	<0.1–98 [52]	<0.01–8.4 [52]	No, except for rare industrial impact sites
Brominated flame retardants BDE-47[e], BDE-99[f], BDE-100[g]	10^{-5}–10^{-7} [58]				0.015–0.12 [55]		<0.0002–0.9 [59,83,84]	Not until in vivo effect shown
PCBs	10^{-4}–<10^{-6} [61], But some are antiestrogenic				1.3 (sum of 52 peaks) [34]	<0.0001–0.03 (PCB153[h]) [61]	0.000002–9 (sum) [61,67]	Probably not
Dioxins and furans	Antiestrogenic				0.001–0.024 (sum) [85]	0.00009–0.0003 (sum) [61]	0.0002–0.10 (sum) [61,67]	Less than for toxicity

[a]May include chemical sorbed to suspended particles.
[b]di(2-Ethylhexyl)phthalate.
[c]di-*n*-Butylphthalate.
[d]Benzylbutylphthalate.
[e]2,2′,4,4′-Tetrabromo-diphenyl-ether (2,2′,4,4′-TeBDE).
[f]2,2′,4,4′,5-Pentabromo-diphenyl-ether (2,2′,4,4′,5-PeBDE).
[g]2,2′,4,4′,6-Pentabromo-diphenyl-ether (2,2′,4,4′,6-PeBDE).
[h]2,2′,4,4′,5,5′-Hexachlorobiphenyl (estrogenic in rat uterus test [61]).

REFERENCES

1. S. F. Arnold, M. K. Robinson, A. C. Notides, L. J. Guillette, J. A. McLachlan. *Environ. Health Perspect.* **104**, 544–548 (1996).
2. S. Jobling, M. Nolan, C. R. Tyler, G. C. Brighty, J. P. Sumpter. *Environ. Sci. Technol.* **32**, 2498–2506 (1998).
3. E. Silva, N. Rajapakse, A. Kortenkamp. *Environ. Sci. Technol.* **36**, 1751–1756 (2002).
4. K. Thorpe, T. Hutchinson, M. Hetheridge, M. Scholze, J. Sumpter, C. R. Tyler. *Environ. Sci. Technol.* **35**, 2476–2481 (2001).
5. U. Schulte-Oehlmann, J. Oehlmann, B. Bauer, P. Fiorini, M. Oetken, M. Heim, B. Markert. In *Ökotoxikologie: Ökosystemare Ansätze und Methoden,* B. Markert and J. Oehlmann (Eds.), pp. 350–363, Ecomed Verlagsgesellschaft mbh, München-Landsberg (1999).
6. R. C. Hale and M. J. La Guardia. *Sci. World* 10–13 (2002).
7. DEFRA. Agriculture and the Environment: R&D Newsletter **9** (April 2002) Department for Environment, Food and Rural Affairs (DEFRA). Anon. (2002).
8. BKH Consulting Engineers. *Towards the establishment of a priority list of substances for further evaluation of their role in endocrine disruption - preparation of a candidate list of substances as a basis for priority setting.* M0355008/1786Q/10/11/00. Final report, Delft, The Netherlands, in association with TNO Nutrition and Food Research, Zeist, The Netherlands for European Commission DG ENV (2000).
9. J. Oehlmann, E. Stroben, U. Schulte-Oehlmann, B. Bauer, P. Fiorini, B. Markert. *Fresenius' J. Anal. Chem.* **354**, 540–545 (1996).
10. World Wildlife Fund. *Tributyl Tin: The Case for Virtual Elimination in Canada.* Canada (1999).
11. A. O. Barakat, M. Kim, Y. Qian, T. L. Wade. *Environ. Toxicol. Chem.* **20**, 2744–2748 (2001).
12. A. C. Birchenough, N. Barnes, S. M. Evans, H. Hinz, C. Mothersill, C. Moss. *Mar. Pollut. Bull.* **44**, 534–543 (2002).
13. C. M. Barroso and M. H. Moreira. *Mar. Pollut. Bull.* **44**, 480–486 (2002).
14. P. Michel, B. Averty, B. Andral, J.-F. Chiffoleau, F. Galgani. *Mar. Pollut. Bull.* **42**, 1128–1132 (2001).
15. M. Ahel, D. Hrsak, W. Giger. *Arch. Environ. Contam. Toxicol.* **26**, 540–548 (1994).
16. M. Ahel and W. Giger. *Chemosphere* **26**, 1471–1478 (1993).
17. E. J. Routledge and J. P. Sumpter. *J. Biol. Chem.* **272**, 3280–3288 (1997).
18. S. Jobling, D. A. Sheahan, J. A. Osborne, P. Matthiessen, J. P. Sumpter. *Environ. Toxicol. Chem.* **15**, 194–202 (1996).
19. S. Jobling and J. P. Sumpter. *Aquat. Toxicol.* **27**, 361–372 (1993).
20. C. D. Metcalfe, T. L. Metcalfe, Y. Kiparissis, B. G. Koenig, C. Khan, R. J. Hughes, T. R. Croley, R. E. March, T. Potter. *Environ. Toxicol. Chem.* **20**, 297–308 (2001).
21. D. Pirie, L. Steven, S. McGrory, G. Best. *Survey of Hormone Disrupting Chemicals,* SEPA (Scottish Environment Protection Agency) (1996).
22. W. Körner, U. Bolz, W. Süssmuth, G. Hiller, W. Schuller, V. Hanf, H. Hagenmaier. *Chemosphere* **40**, 1131–1142 (2000).
23. M. Ahel, E. Molnar, S. Ibric, W. Giger. *Water Sci. Technol.* **42**, 15–22 (2000).
24. A. Di Corcia and R. Samperi. *Environ. Sci. Technol.* **28**, 850–858 (1994).
25. M. A. Blackburn, S. J. Kirby, M. J. Waldock. *Mar. Pollut. Bull.* **38**, 109–118 (1999).
26. M. A. Blackburn and M. Waldock. *Water Res.* **29**, 1623–1629 (1995).
27. M. Solé, M. J. L. de Alda, M. Castillo, C. Porte, K. Ladegaard-Pedersen, D. Barcelo. *Environ. Sci. Technol.* **34**, 5076–5083 (2000).
28. J. L. A. Long, W. A. House, A. Parker, J. E. Rae. *Sci. Total Environ.* **210**, 229–253 (1998).
29. D. Y. Shang, R. W. MacDonald, M. G. Ikonomou. *Environ. Sci. Technol.* **33**, 1366–1372 (1999).
30. A. Marcomini, B. Pavoni, A. Sfriso, A. A. Orio. *Mar. Chem.* **29**, 307–323 (1990).

31. D. T. Bennie. *Water Qual. Res. J. Canada* **34**, 79–122 (1999).
32. M. Ahel, W. Giger, M. Koch. *Water Res.* **28**, 1131–1142 (1994).
33. M. Petrovic and D. Barcelo. *Anal. Chem.* **72**, 4560–4567 (2000).
34. A. Marcomini, P. D. Capel, Th. Lichtensteiger, P. H. Brunner, W. Giger. *J. Environ. Qual.* **18**, 523–528 (1989).
35. H. Kirchmann, H. Astrom, G. Jonsall. *Swedish J. Agricultural Res.* **21**, 107–113 (1991).
36. W. Kloas, I. Lutz, R. Einspanier. *Sci. Total Environ.* **225**, 59–68 (1999).
37. D. Pickford, J. Caunter, M. Hetheridge, T. Hutchinson. Effects of bisphenol A on larval growth, development and sexual differentiation of *xenopus laevis*. Proceedings, 11th Annual Meeting of SETAC Europe, Madrid, Spain, 6–10 May 2001.
38. J. Oehlmann, U. Schulte-Oehlmann, M. Tillmann, B. Markert. *Ecotoxicology* **9**, 383–397 (2000).
39. C. A. Staples, P. B. Dorn, G. M. Klecka, S. T. O'Block, L. R. Harris. *Chemosphere* **36**, 2149–2173 (1998).
40. H. B. Lee and T. E. Peart. *Water Qual. Res. J. Canada* **35**, 283–298 (2000).
41. W. Körner, U. Bolz, W. Süssmuth, G. Hiller, V. Hanf, H. Hagemaier. *Organohalogen Compd.* 269–272 (1998).
42. M. Nasu, M. Goto, H. Kato, Y. Oshima, H. Tanaka. *Water Sci. Technol.* **43**, 101–108 (2001).
43. M. Fürhacker, S. Scharf, H. Weber. *Chemosphere* **41**, 751–756 (2000).
44. T. Tanaka, K. Yamada, T. Tonosaki, T. Konishi, H. Goto, M. Taniguchi. *Water Sci. Technol.* **42**, 89–95 (2000).
45. H. Fromme, T. Kuchler, T. Otto, K. Pilz, J. Muller, A. Wenzel. *Water Res.* **36**, 1429–1438 (2002).
46. A. J. Murk, J. Legler, M. M. H. van Lipzig, J. H. N. Meerman, A. Belfroid, J. Spenkelink, B. van der Burg, G. B. J. Rijs, D. Vethaak. *Environ. Toxicol. Chem.* **21**, 16–23 (2002).
47. A. D. Vethaak, G. B. J. Rijs, S. M. Schrap, S. M. Ruiter, A. Gerritsen, J. Lahr. *Estrogens and xeno-estrogens in the aquatic environment of the Netherlands.* RIZA/RIKZ report 2002.001 (2002).
48. O. S. Fatoki and F. Vernon. *Sci. Total Environ.* **95**, 227–232 (1990).
49. M. D. Webber, H. R. Rogers, C. D. Watts, A. B. A. Boxall, R. D. Davis, R. Scoffin. *Sci. Total Environ.* **185**, 27–44 (1996).
50. J. D. Berset, R. Etter-Holzer. *J. AOAC. Int.* **84**, 383–391 (2001).
51. M. D. Webber and S. Lesage. *Waste Manage. Res.* **7**, 63–82 (1989).
52. A. Wenzel, T. Kuechler, K.-P. Henschel, W. Schnaak, M. Diedrich, J. Mueller. *Konzentrationen oestrogen wirkender Substanzen in Umweltmedien (Concentration of estrogenic active substances in the environment).* Forschungsbericht 216 02 011/11. Umweltbundesamt Berlin, Germany, Berlin, Germany (2001).
53. C. A. Staples, T. F. Parkerton, D. R. Peterson. *Chemosphere* **40**, 885–891 (2000).
54. M. Alaee and R. J. Wenning. *Chemosphere* **46**, 579–582 (2002).
55. C. A. de Wit. *Chemosphere* **46**, 583–624 (2002).
56. T. Zhou, D. G. Ross, M. J. Devito, K. M. Crofton. *Toxicol. Sci.* **61**, 76–82 (2001).
57. T. Zhou, M. M. Taylor, M. J. Devito, K. A. Crofton. *Toxicol. Sci.* **66**, 105-116 (2002).
58. I. A. T. M. Meerts, R. J. Letcher, S. Hoving, G. Marsh, A. Bergman, J. G. Lemmen, B. van der Burg, A. Brouwer. *Environ. Health Perspect.* **109**, 399–407 (2001).
59. J. de Boer, A. Van den Horst, P. G. Wester. *Organohalogen Compd.* **47**, 85–88 (2000).
60. O. P. Heemken, J. Kuballa, B. Stachel. In *Endokrin wirksame Substanzen im Abwasser und Klärschlamm - Neueste Ergebnisse aus Wissenschaft und Technik,* Vol Band 23. B. Bilitewski, D. Weltin, P. Werner (Eds.), pp. 200–200, Forum für Abfallwirtschaft und Altlasten e.V. an der TU Dresden Aussenstelle Pirna Copitz, Dresden, Germany (2002).
61. M. Gülden, A. Turan, H. Seibert. *Endocrinically active chemicals and their occurance in surface waters.* UBA Texte 66/98. Umweltbundesamt, Berlin, Germany (1998).
62. *Endocrine-disrupting substances in the environment. What should be done?* Environment Agency, UK (1998).

63. *CSTEE Opinion on Human and Wildlife Health Effects of Endocrine Disrupting Chemicals, with Emphasis on Wildlife and on a Ecotoxicology Test Methods,* expressed at the 8th CSTEE plenary meeting, Brussels, 4 March 1999 (1999).
64. K. Ohyama, F. Nagai, Y. Tsuchiya. *Environ. Health Perspect.* **109**, 699–703 (2001).
65. H. Fang, W. Tong, L. M. Shi, R. Blair, R. Perkins, W. Branham, B. S. Hass, Q. Xie, S. L. Dial, C. L. Moland, D. M. Sheehan. *Chem. Res. Toxicol.* **14**, 280–294 (2001).
66. D. L. Villeneuve, J. S. Khim, K. Kannan, J. P. Giesy. *Environ. Toxicol.* **17**, 128-137 (2002).
67. P. Isosaari, H. Kankaanpää, J. Mattila, H. Kiviranta, M. Verta, S. Salo, T. Vartiainen. *Environ. Sci. Technol.* **36**, 2560–2565 (2002).
68. B. G. Loganathan, K. S. Sajwan, J. P. Richardson, C. S. Chetty, D. A. Owen. *Mar. Pollut. Bull.* **42**, 246–250 (2001).
69. L. Marsili and S. Focardi. *Environ. Monit. Assess.* **45**, 129–180 (1997).
70. D. W. Kolpin, E. T. Furlong, M. T. Meyer, E. M. Thurman, S. D. Zaugg, L. B. Barber, H. T. Buxton. *Environ. Sci. Technol.* **36**, 1202–1211 (2002).
71. U. Bolz, W. Körner, B. Kuch, J. W. Metzger. In *Endokrin wirksame Substanzen im Abwasser und Klärschlamm - Neueste Ergebnisse aus Wissenschaft und Technik,* Vol Band 23. B. Bilitewski, D. Weltin and P. Werner (Eds.), pp. 34–42, Forum für Abfallwirtschaft und Altlasten e.V. an der TU Dresden Aussenstelle Pirna Copitz, Dresden, Germany (2002).
72. C. Desbrow, E. J. Routledge, G. C. Brighty, J. P. Sumpter, M. Waldock. *Environ. Sci. Technol.* **32**, 1549–1558 (1998).
73. *Produktion und Verwendung zinnorganischer Verbindungen in Deutschland.* Umweltbundesamt, Berlin, Germany (2000).
74. M. M. C. Ko, G. C. Bradley, A. H. Neller, M. J. Broom. *Mar. Pollut. Bull.* **31**, 249–253 (1995).
75. J. E. Harries, T. Runnalls, E. Hill, C. A. Harris, S. Maddix, J. P. Sumpter, C. R. Tyler. *Environ. Sci. Technol.* **34**, 3003–3011 (2000).
76. J. Oehlmann, U. Schulte-Oehlmann, M. Duft, M. Tillmann. Effects of endocrine disruptors on prosobranch molluscs. Proceedings, Workshop: International Framework ARCEM - Bioindication, Federal Ministry of Agriculture, Forestry, Environment and Water Management, Hall, Stubenbastei 5, 1010 Vienna, Austria, 0 AD/10/12.
77. P. Spengler, C. Orth, C. Kempter, J. W. Metzger. Quantification of substances with estrogenic activity in effluents of municipal and industrial sewage plants. Proceedings, 7 March 1999.
78. P. Spengler, W. Körner, C. Orth, J. W. Metzger. Monitoring of substances with estrogenic activity in effluents of sewage treatment plants: Part 1: Chemical analysis. Proceedings, Third SETAC World Congress-Global Environmental Issues in the 21st Century: Problems, Causes and Solutions, Brighton UK, 21 May 2000 to 25/5/2000.
79. K. McTavish, H. Stech, F. Stay. *Environ. Toxicol. Chem.* **17**, 58–67 (1998).
80. C. A. Staples, P. B. Dorn, G. M. Klecka, S. T. O'Block, D. R. Branson, L. R. Harris. *Chemosphere* **40**, 521–525 (2000).
81. MOC Conducted First Endocrine Disruptors Study. *PWRI Newsletter* **80**, 5–6 (2000).
82. K. Thomas, M. Hurst, P. Matthiessen, M. Waldock. Characterisation of estrogenic compounds in UK estuaries. Proceedings, Third SETAC World Congress-Global Environmental Issues in the 21st Century: Problems, Causes and Solutions, Brighton, UK, 21 May 2000 to 25/5/2000.
83. C. R. Allchin, R. J. Law, S. Morris. *Environ. Pollut.* **105**, 197–207 (1999).
84. H. van Zeijl. *Report of the results of the on-off survey DIFFCHEM.* Report SIME 97/6/1-E. Oslo and Paris Commissions (1997).
85. E. Eljarrat, J. Caixach, J. Rivera. *Environ. Sci. Technol.* **33**, 2493–2498 (1999).

Pure Appl. Chem., Vol. 75, Nos. 11–12, pp. 1905–1916, 2003.

Topic 2.6

Release of pesticides into the environment and initial concentrations in soil, water, and plants*

K. D. Racke

Dow AgroSciences, 9330 Zionsville Road, Bldg. 308, Indianapolis, IN 46268, USA

Abstract: Considerable information exists as to the initial concentrations of pesticide residues to be expected in soils, plants, and water. Empirical or theoretical models have been developed for incorporating this data into exposure assessments for humans as well as terrestrial and aquatic wildlife. In addition, monitoring data exists for many older products, especially with respect to typical concentrations observed in food commodities for human consumption and in surface and ground waters. Estimated and observed concentrations of pesticides in these matrices have been routinely employed for more than 30 years in assessing the potential impacts of pesticides on a variety of biologically relevant endpoints. The same data will also prove useful for exposure assessments of endocrine active substances. There are some additional research needs, however. First, further research and development is needed to ensure that estimation and monitoring methods for pesticide concentrations in soil, water, and food are applicable and utilized for all important and relevant cultural, agronomic, and environmental conditions. This is especially true with respect to developing countries and tropical climates, which are often disproportionately ignored in favor of developing countries and temperate climates. Second, methodologies for collection of monitoring data and generation of modeled estimates for pesticide residues in soil, water, and food need to be carefully designed with the requirements of higher-tier, probabilistic exposure assessments in view. Although worst-case, point estimates or analyses may be useful for screening-level assessments, advanced assessments targeted at addressing the likelihood of biologically relevant exposures are urgently required by scientists and regulatory authorities for reaching sound risk assessment and management decisions.

INTRODUCTION

Pesticides are chemicals used to manage pest organisms in both agricultural and nonagricultural situations. By definition, a pesticide is a "substance or mixture of substances intended for preventing, destroying or controlling any pest, including vectors of human or animal disease, unwanted species of plants or animals causing harm or otherwise interfering with the production, processing, storage, transport, or marketing of food, agricultural commodities, wood, wood products or animal feedstuffs, or which may be administered to animals for the control of insects, mites/spider mites or other pests in or on their bodies" [1]. Although most attention for pesticides is generally directed at the active ingredients they contain (i.e., pesticidal), formulated pesticide products also contain a variety of so-called "inert ingredients" (i.e., nonpesticidal) intended to assist with such properties as handling, stability, de-

*Report from a SCOPE/IUPAC project: Implication of Endocrine Active Substances for Human and Wildlife (J. Miyamoto and J. Burger, editors). Other reports are published in this issue, *Pure Appl. Chem.* **75**, 1617–2615 (2003).

livery, application, and crop safety. Although abundant information is widely available on the use, properties, and behavior of active ingredients, the content and identity of inert ingredients is generally considered proprietary, trade secret information. Therefore, the primary focus of this paper will be on pesticidal active ingredients.

Government authorities involved in regulating the human safety aspects of chemical exposures are in the early stages of developing testing and evaluation criteria for endocrine active substances (EASs). Major programs are currently being developed in a number of regions including Europe [2], Japan [3], and the United States [4,5]. Pesticide products are some of the most widely tested chemicals with respect to toxicological hazard evaluation (e.g., multigeneration, developmental, and reproductive toxicology testing) and human health risk assessment (e.g., dietary, occupational). The exceptions to this rule are natural products and biochemicals (e.g., azadirachtin, rotenone, pyrethrum), for which some regulatory authorities have granted authorizations in the absence of availability of the same level of extensive testing. Although the popular literature and the Internet are replete with listings of pesticides and other chemicals "known" or "suspected" of causing endocrine disruption (many lists are merely compilations or modifications of earlier lists), uniform criteria for classifying products with respect to endocrine activity and the human or wildlife health risks they may pose have yet to be developed by regulatory authorities. The class of pesticides which contains several of the more widely recognized EASs is the chlorinated hydrocarbons (e.g., DDT), of which the most highly persisent members are in the process of being phased out by most national authorities. Regulatory authorities are in the early stages of developing testing and evaluation processes for endocrine-disrupting pesticides, and in some cases, prioritized listings of pesticides are being developed for purposes of screening and review [2,5].

Rather than narrow discussion to any one or two particular groups of pesticides, the focus of this paper will be on a review of general principles and considerations related to the environmental release of pesticides. Particular emphasis will be placed on crop protection products, for which both human and environmental risk assessment considerations are highly relevant. A general introduction of the major types and classes of pesticide products will be followed by discussion of sources of environmental entry. Summary information on initial concentrations of pesticides in environmental matrices will be presented, including residues that may be present in soil, water, target crops, and harvested food commodities. Finally, research needs related to availability of reliable estimates of pesticide environmental exposure levels and their appropriate use via the risk assessment paradigm will be presented.

PESTICIDES AND ROUTES OF ENVIRONMENTAL ENTRY

Pesticide classes and trends

Major types of pesticides include herbicides, insecticides, fungicides, and biocides (e.g., bactericides, algicides, slimicides). The number of existing active ingredients currently employed as pesticides is quite large (approx. 1000), and the great majority have been subject to extensive toxicological and environmental testing as part of government registration processes. Examples of major classes of existing products include organophosphate, pyrethroid, and neonicotinoid insecticides, phenoxy, dinitroaniline, and sulfonylurea herbicides, and dithiocarbamate and strobulirin fungicides.

There has been a general trend during the past 30–40 years toward introduction of products with lower application rates, decreased environmental persistence, and reduced nontarget organism toxicity. For example, whereas the average use rate for chlorinated hydrocarbon insecticides was estimated at 3 kg/ha, more recently introduced pyrethroid or neonicotinoid insecticides may be applied at 0.01 to 0.1 kg/ha [6]. Similar trends exist also for herbicides and fungicides, which for many crop uses had shifted from kg/ha in the 1950s to g/ha by the 1990s. As an example of this trend, Table 1 lists major classes of insecticide products along with their period of significant introductions and typical field use rates. Some of the trend toward lower-use rate and reduced impact products has been driven by more stringent testing criteria and increasing regulatory hurdles for new products. The U.S. EPA reduced risk pesticide program, which since the early 1990s has provided fast-track evaluation and approval for products demonstrating lower potential human and environmental impacts than currently available alternatives, is a different type of initiative nurturing this trend [7].

Table 1 Periods of major insecticide product introductions and typical use rates (adapted from [6]).

Insecticide class	Major introductions	Typical use rate (kg/ha)
Arsenical	1890–1940	4.00–60.00
Chlorinated hydrocarbon	1939–1956	1.00–4.00
Organophosphorus	1946–1986	0.50–2.00
Carbamate	1957–1984	0.50–2.00
Pyrethroid	1973–1992	0.01–0.20
Benzoylurea	1972–2002	0.01–0.05
Neonicotinoid	1990–2001+	0.01–0.10
Phenylpyrazole	1992–2001+	0.10–0.15

In contrast to active ingredients, the pool of inert formulation components is significantly larger (>2500), and there have been relatively fewer and less detailed toxicological or environmental studies generated. Examples of classes of inert formulation components include carrier solvents, emulsifiers, antifoamers, and preservatives [8]. Some inert ingredients, such as alkyl phenol ethoxylates, have been implicated as potential EASs [4]. As opposed to active ingredients, which may comprise from <0.1 to 50 % or more of the pesticide product on a wt/wt basis, inert ingredients generally form the bulk of formulated product. As with active ingredients, industrial development efforts and regulatory trends have been aimed toward introduction of less hazardous (e.g., low flammability) and more environmentally benign inert ingredients [9].

Sources of environmental entry

Pesticides are employed for pest management programs in both agricultural and nonagricultural settings. Agricultural pest management relies most heavily upon herbicides, with significant use of insecticides and fungicides in certain cropping situations. Row crop situations in particular (cereals, cotton, maize, soybean, rice) account for the bulk of pesticide applications on a yearly basis, with fruit and vegetable crops also important avenues for pesticide use. Summaries of world agricultural pesticide use by type of pesticide and crop group are listed in Table 2.

Table 2 World pesticide usage by class and crop (adapted from [10,11]).

World pesticide use by type		World pesticide use by crop	
Herbicides	10.2×10^5 MT	Fruits and vegetables	25.0 %
Insecticides	6.7×10^5 MT	Cereals	15.6 %
Fungicides	2.5×10^5 MT	Maize	14.2 %
Other	6.5×10^5 MT	Soybeans	10.9 %
		Rice	8.8 %
		Cotton	7.9 %

Intentional placement of pesticides into agricultural environments occurs by a variety of application methods including air and ground boom spraying. The highest initial concentrations of pesticides are generally present in plant foliage, soil, and water to which direct applications are made. Relatively few pesticide applications are made directly and exclusively to the target pest, and most application methods rely on application of enough pesticide to the environment so that exposure to the pest species reaches efficacious levels. Estimates for some scenarios indicate that less than 0.1 % of applied materials ever do so [12]. A recent and more precisely targeted method of pesticide application involves the use of transgenic crop plants, which have been engineered so as to synthesize insecticidal active ingredients or proteins within their tissues [13].

Unintentional entry of pesticides into the agricultural environment is associated with transport and offtarget deposition during or following application. For example, during spray application drift of airborne particles and/or volatilized pesticide to adjacent areas may occur. A literature survey revealed median offsite deposition of aerially applied pesticide at 30, 60, and 120 meters of 5, 2, and 0.8 % of the material, respectively [14]. Recent testing by the industry Spray Drift Task Force has confirmed levels of offsite drift from aerial and ground application equipment may be significant without appropriate management practices [15]. In addition, moving water may transport pesticides from treated areas to ground water via leaching or surface water via run-off [16]. The magnitude of transport to water may be fairly limited, but detection of trace quantities is made fairly routinely. It has been estimated that less than 0.5 % of applied pesticide may be lost from treated areas via surface run-off; losses of persistent chlorinated hydrocarbon pesticides, however, may be nearly twice this level [16]. Finally, inadvertent entry of pesticides into the environment may occur via accidental spills onto soil or into water bodies or via contamination associated with improper waste handling and disposal.

Nonagricultural use of pesticides may also represent significant sources of entry to the environment [17]. In the urban environment, significant pesticide uses include applications for turf and ornamental pests and for termite control. Aquatic weed management in lakes and recreational waters may involve direct application of herbicides to surface waters. Vector control programs rely in part on pesticide application to both surface waters and terrestrial environments to control insect larvae and adults, respectively, which may transmit diseases. Nonagricultural use of pesticides in the United States, for example, has been estimated to account for approximately 25 % of pesticide use by volume [10].

PESTICIDE CONCENTRATIONS IN THE ENVIRONMENT

Uses and sources of information

Major uses of data related to pesticide concentrations in environmental matrices include conduct of risk assessment, regulatory evaluation for product approval or reapproval, and regulatory compliance evaluation. The risk assessment process underlies most uses of this data, and it involves an evaluation of the likelihood of an adverse effect in an individual or population in light of known properties of hazard (effects) and estimated or actual exposure (magnitude, frequency, duration). Generally, the risk assessment process involves comparing some biological endpoint (e.g., NOEL, NOEC, NOAEL) with individual values or distributions of data related to pesticide concentration or intake. These evaluations can be focused on acute, short-term, or chronic biological endpoints, and thus different types of exposure value may need to be available.

For new pesticide products, environmental concentrations and potential exposures are most commonly estimated from laboratory and highly controlled field studies by use of empirical or mechanistic models. These data are generated by pesticide manufacturers and affiliated laboratories, and submitted to regulatory authorities for purposes of new product evaluation and approval. For existing pesticide products, in addition to modeled estimates there may be actual monitoring data available for various environmental matrices. Such monitoring data may be generated by manufacturers as well as by universities and government institutes. Routine environmental monitoring data is most widely available for human foods. Data on pesticide concentrations in raw surface and groundwater are also becoming more widely available, although monitoring of finished drinking water is still relatively difficult to obtain. Comprehensive data on pesticide residues in soils, sediments, and biota are not generally available. On a geographic basis, plentiful and reliable monitoring data is available only in some of the more highly developed countries of North America (e.g., U.S.), Europe (e.g., UK, Germany), and the Pacific (e.g., Australia, Japan). The bulk of the world's human population lives in countries in which monitoring data is sparse, unreliable, and/or nonexistent.

It should be pointed out that estimates or measurements of pesticide concentrations in environmental matrices are useful for a wide variety of human and ecological risk assessment processes and are not restricted with respect to endpoints of concern. Thus, the duration of exposure associated with a potential effect of interest (e.g., acute, short-term, chronic) is far more important than the type of effect (e.g., neurological damage, cancer, reproductive impairment, endocrine disruption) with respect to discrimination of environmental concentration data.

Pesticide concentrations in soil

Significant agricultural soil concentrations may result from either direct application to soil or drift and wash-off from foliar application. Depending on the rate employed, direct soil application may result in initial residues of less than 0.05 mg/kg to more than 10 mg/kg [18]. A common assumption for initial pesticide concentration in surficial (0–15 cm) soil is that it is approximately one-half that of the application rate (note: assumes 0.4 ha of 0–15 cm soil with a bulk density of 1.5 g/ml has a weight of 908 000 kg). For example, application of 1 kg (active ingredient) per hectare would result in approximately 0.5 mg/kg of initial pesticide concentration. A summary evaluation of initial soil pesticide concentrations from 184 field studies reported that measured concentrations averaged 74.9 % of the theoretical and were highly variable (SD = 43.2 %) [18]. Selected results are also summarized in Fig. 1. It should also be noted that a high level of variability is often observed in pesticide soil concentration even within a single agricultural field.

Some nonagricultural pesticide use scenarios may result in significantly higher initial pesticide concentrations. For example, soil drench barriers for termite protection around houses and other wooden structures may require initial deposit of 500 mg/kg or more [19,20]. Soil pesticide concentrations resulting from accidental spills or waste contamination are often very much higher than those concentrations resulting from intentional use, with deposits of 1000 mg/kg or higher not uncommon [21].

Pesticide concentrations in soil for risk assessment purposes are most often drawn from empirical estimates or limited numbers of field studies. Since most contemporary pesticides dissipate from the soil relatively quickly through various degradative or transport processes, initial soil concentrations may

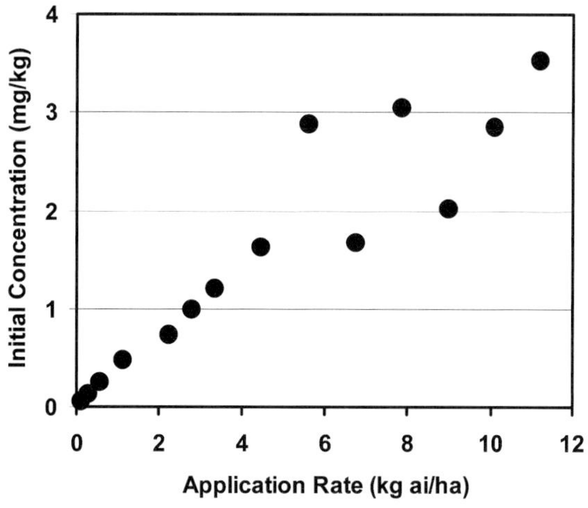

Fig. 1 Initial pesticide concentrations in 0–15 cm soil (adapted from [18]).

not be extremely useful for exposure evaluations. The most direct use may be for evaluation of effects to soil-dwelling arthropods or annelids. Indirectly, soil concentrations are often employed in a number of models that estimate crop plant uptake and run-off or leaching mobility with water. These models are most useful when the dissipation rate of pesticide is considered. Problems with respect to information on pesticide concentrations in soil include a high degree of spatial and temporal variability, accurate prediction of dissipation rate in light of variable environmental conditions, lack of routine monitoring data for most soils, and near total lack of pesticide residue data from tropical soils [22].

Pesticide concentrations in water

Surface water concentrations of pesticides are highest for those use patterns associated with direct application to water bodies such as for aquatic weed control or insect larvae management. For example, initial herbicide concentrations of 0.2 to 3 mg/l are generally targeted for aquatic weed control [23]. Indirect or unintentional entry of pesticides into water generally results in much lower concentrations, in the range of 0.001–0.01 mg/l or lower [24]. Such entry may occur due to accidental overspray or offtarget spray drift, or to surface run-off or leaching transport from a treated agricultural field.

Regarding offtarget drift of pesticide spray and deposition on nearby surface waters, initial concentrations are most commonly estimated from empirical or theoretical models. For example, spray drift tables of deposition with increasing distance from the edge of a treated field have been assembled based on experimental observations [25,26]. Example results of using such an approach to predict initial pesticide concentrations in water are presented in Table 3. Various spray drift models are also employed to predict surface water deposition and initial concentration under various application scenarios (e.g., ground boom, airblast sprayer, aerial application) and under different environmental conditions [14]. A key challenge for this type of estimation is the extreme variability in observed deposition of pesticide spray based on factors related to geography (e.g., size of field in relation to water body, distance of water from treated field), environmental conditions (e.g., wind direction, windspeed, humidity), and application parameters (e.g., sprayer type, nozzle type, formulation viscosity, spray boom height).

Table 3 Predicted offtarget spray drift as percent of applied and initial concentrations in water at the 95th percentile (assumes 1 kg ai/ha application rate and 1 m water depth) (adapted from [25,26]).

Distance from edge of field (m)	Vineyard		Orchard		Vegetable	
5	1.6–5.0 %	1.6–5.0 µg/l	10–20 %	10–20 µg/l	0.6–5 %	0.6–5 µg/l
10	0.4–1.5 %	0.4–1.5 µg/l	4.5–11 %	4.5–11 µg/l	0.4–1.5 %	0.4–1.5 µg/l
20	0.1–0.4 %	0.1–0.4 µg/l	1.5–4 %	1.5–4 µg/l	0.1–0.4 %	0.1–0.4 µg/l
30	0.1–0.2 %	0.1–0.2 µg/l	0.6–2 %	0.6–2 µg/l	0.1–0.2 %	0.1–0.2 µg/l

Surface or ground water pesticide concentrations may be measured experimentally in lysimeter or field studies. For groundwater concentration estimates, lysimeter studies in which small soil monoliths (0.5–2 m diameter) are maintained under actual or simulated environmental conditions have been employed. Prospective field monitoring studies, in which groundwater is sampled from within the soil profile or from well stations placed at various depths have also been commonly employed. For surface water evaluations, small plot (0.1–1 ha) or watershed-scale, site-specific studies have been used to measure pesticide concentrations in run-off water and sediments. These experimental approaches are useful for generating realistic data under a small set of experimental conditions, but are highly costly and cannot easily cover the range of environmental, geographic, and climatic variables present on a landscape-scale over time. For these reasons, theoretical surface run-off or groundwater leaching models are commonly employed to provide estimates of initial pesticide concentrations [27,28], and probabilistic approaches can accommodate the variability inherent for key input parameters.

In addition to site-specific field studies or modeling, large-scale surface or groundwater monitoring has been employed to a limited extent for assessing the concentrations and frequency of pesticide detections. For example, the National Water Quality Assessment Program (NAWQA) has been ongoing in the United States for the past 10 years and has involved analysis of thousands of samples across both agricultural and urban streams and groundwaters for a large number of pesticides and metabolites [29,30]. Some summary data is presented in Table 4 with respect to frequency of detection and 95th percentile water concentrations of some representative herbicides and insecticides. In general, these results indicate that the greatest frequency and highest concentrations are observed for several widely used, highly water-soluble, and loosely sorbed herbicides.

Table 4 USGS National stream and groundwater monitoring results for several common pesticides (1992–1998) (adapted from [29,30]).

Pesticide	Detection limit (mg/l)	Stream residue frequency	95th Percentile stream conc. (mg/l)	Groundwater residue frequency	95th Percentile groundwater conc. (mg/l)
HERB-1	0.002	39.0	0.21	2.8	0.50
HERB-2	0.001	85.0	3.51	44.0	3.00
HERB-3	0.005	59.0	0.22	12.0	1.30
HERB-4	0.15	15.0	0.37	0.6	4.54
CHL-1	0.006	5.5	0.00	3.6	0.00
CHL-2	0.001	5.5	0.00	1.1	0.05
CHL-3	0.004	0.6	<0.004	0.0	<0.004
OP-1	0.004	17.0	0.02	0.9	0.02
OP-1	0.002	15.0	0.02	0.5	0.07
OP-2	0.005	6.0	0.01	0.3	0.00
OP-3	0.001	1.9	<0.001	0.4	0.18
CB-1	0.003	9.2	0.01	0.4	0.02
CB-2	0.003	12.0	0.06	1.6	1.30
CB-3	0.017	1.2	<0.017	0.1	0.00
PYR	0.005	0.3	<0.005	0.0	0.00

*CHL = chlorinated hydrocarbon; OP = organophosphorus; CB = carbamate; PYR = pyrethroid; OTH = other

Problems with respect to usefulness of monitoring information on pesticide concentrations in water include a high degree of spatial and temporal variability, lack of availability of results for newly introduced pesticides, extremely high cost, and limited availability of data on a geographic basis.

Pesticide concentrations in target plants

Following many types of agricultural and urban (e.g., turfgrass) pesticide applications, plant foliage (crop or weed) often contains the highest initial concentration of pesticide residues among various environmental matrices. Depending on such factors as plant type and growth stage, application method, spray volume, and formulation characteristics, agricultural crops may intercept from 10 % to more than 80 % of the applied active ingredient [31,32]. For purposes of environmental modeling assessments, a set of harmonized foliar interception factors by crop and growth stage has been suggested, and some examples are listed in Table 5.

With respect to initial concentrations of pesticides in various plant components, in addition to percent foliar interception the application rate of the pesticide (kg ai/ha) is extremely important. Measurements indicate that, regardless of active ingredient, initial concentrations in plant components scale directly with application rate [33]. Initial concentrations of pesticides, assuming a 1 kg ai/ha application rate, may range from 1.5–7 mg/kg in fruit to 35–125 mg/kg in leaves. Based on a large series of field trials conducted with various insecticides, herbicides, and fungicides, an empirical distribution for maximum and typical initial concentrations in various plant components was developed during the early 1970s. The so-called Hoerger–Kenaga nomogram, with some slight modification, has been routinely employed by some regulatory authorities for wildlife exposure and risk assessments [33,34]. A summary of estimated mean and maximum pesticide concentrations in various types of plant components upon initial application is found in Table 6.

Table 5 Examples of harmonized foliar interception factors for various crops and stages (adapted from [32]).

Crop	Growth phase %	Foliar interception
Beans	Flowering	70 %
Cabbage	Flowering	90 %
Cereals	Tillering	50 %
Onions	Flowering	40 %
Pasture grass	All stages	40 %
Pome fruit	Bloom/leaf development	40 %
Rice	Flowering, fruit development	70 %
Stone fruit	Dormant	20 %
Stone fruit	Bloom/leaf development	40 %
Stone fruit	Full foliage	80 %
Vines	Leaf development	30 %
Vines	Inflorescence emergence	50 %
Vines	Flowering/fruiting	80 %
Vines	Senescence	60 %

Table 6 Estimated mean and maximum pesticide concentration on crop groups immediately following application of 1 kg ai/ha (adapted from [33,34]).

Plant Category	Estimated mean concentration (mg/kg)	Field data mean ± SD (mg/kg)	Estimated maximum concentration (mg/kg)	Estimated maximum concentration (mg/kg)
Short-range grass	112	76 ± 54	214	214
Long grass	82	32 ± 36	98	98
Leaves, leafy crops	31	31 ± 40	112	112
Forage legumes	30	40 ± 51	52	121
Pods and seeds	3	4 ± 5	11	11
Fruits	1	5 ± 9	6	13

Of critical importance from an environmental exposure standpoint are considerations related to dissipation of pesticide residues from plant parts due to such factors as degradation, volatilization, or wash-off. Many semivolatile or photolabile pesticides display dissipation half-lives of several hours to several days from exposed, foliar surfaces. For example, although Hoerger and Kenaga estimated typical initial concentrations of pesticide of 1.5–30 mg/kg, by six weeks after application these typical concentrations had decreased to <0.2 to 5 mg/kg [33].

Pesticide concentrations in harvested food commodities

Although generic estimates of pesticide residues in crop plants and weeds are often sufficient for assessment of wildlife exposure and risk, human dietary risk assessments have often relied upon more robust, product-specific residue data. Results of controlled field trials, in which maximum application rate/frequency and minimum preharvest intervals are employed, are relied upon by most regulatory authorities as well as the Codex Alimentarius Commission to establish maximum residue limits (MRLs) for pesticides in various agricultural commodities. For example, Table 7 lists established or proposed MRLs for a variety of insecticide products on selected food commodities [35].

Table 7 Examples of established or proposed Codex MRLs (mg/kg) for members of various insecticide classes (adapted from [35]).

Compound class*	First JMPR**	Apple	Cabbage	Grape	Orange	Potato	Tomato	Wheat
CHL-1	1965	1	1	1	0.5	0.2	0.5	0.2
CHL-2	1968			5	5			
OP-1	1965	2				0.05	1	
OP-2	1965	0.3	2			0.01	0.5	
OP-3	1965	1	2	1	2	0.05	2	0.2
OP-4	1965	2	8	8	4		3	0.5
OP-5	1972	1	1	1	1	0.05	0.5	0.5
OP-6	1974	2	2		2	0.05	1	5
CB-1	1965	5	5	5	7	0.2	5	2
CB-2	1975	2	5	5	1	0.1	1	0.5
CB-3	1979			0.2	0.2			0.02
CB-4	1980	2			5	0.1	2	
PYR-1	1979	2	3				1	2
PYR-2	1979	2	5	2		0.05	1	2
PYR-3	1980	0.1		0.05	0.05	0.01	0.2	1
PYR-4	1992				0.05	0.05		0.5
OTH-1	1981	1	1	1	1		1	
OTH-2	1992	0.02			0.01	0.01	0.02	
OTH-3	1996	1	5	2	2		1	
OTH-4	2000		0.02			0.02		0.002
OTH-5	2001	0.1	2		0.3	0.01	0.3	

*CHL = chlorinated hydrocarbon; OP = organophosphorus; CB = carbamate; PYR = pyrethroid; OTH = other

**1st evaluation by the FAO/WHO Joint Meeting on Pesticide Residues (JMPR)

Considerable monitoring of pesticide residues in foodstuffs also occurs through farmgate and market basket surveys [36–39]. Such monitoring data provide more realistic estimates for risk assessment purposes of the likely residues in the human diet. Examination of such data indicates that in many cases observed pesticide residue levels in consumed food are one to several orders of magnitude below maximum residue limits. As an example, α-endosulfan was detected in 43.7 % of cucumbers (322 out of 737 samples) in a U.S. survey, but the range of concentrations was 0.007–0.15 mg/kg as compared with an MRL of 2 mg/kg [37]. Even these values may be somewhat overpredictive of human consumption since research indicates that processing and cooking of foods may result in further decreases in residues due to dissipation or degradation [40]. Human dietary assessment methodologies for both chronic (lifetime) and acute (single-day) exposures have been well developed at the national and international level to make use of available residue trial and monitoring data [36,40].

DISCUSSION

Conclusions

Considerable information exists as to the initial concentrations of pesticide residues to be expected in soils, plants, and water. Empirical or theoretical models have been developed for incorporating this data into exposure assessments for humans as well as terrestrial and aquatic wildlife. In addition, monitoring data exists for many older products, especially with respect to typical concentrations observed in food commodities for human consumption and in surface and ground waters. Estimated and observed concentrations of pesticides in these matrices have been routinely employed for more then 30 years in assessing the potential impacts of pesticides on a variety of biologically relevant endpoints. The same data will also prove useful for exposure assessments of endocrine-disrupting substances.

There are some noteworthy problems, however, with the use of existing information on pesticide residues for exposure assessment purposes. One problem involves a lack of sufficient data for evaluation of all important and relevant cultural, agronomic, and environmental conditions. For example, whereas several of the countries with highly developed economies and agricultural systems (e.g., Australia, Japan, UK, USA) have plentiful data and calibrated models with respect to pesticide concentrations in key environmental matrices, many developing countries lack even the most rudimentary information or applicable models. Another problem involves the large degree of spatial variability that exists with respect to environmental concentrations of pesticides. Not only are concentrations of a given pesticide often orders of magnitude different depending on which soils or plants or waters are assayed, but even within a single field or a single fruit tree residue levels can vary dramatically. Finally, the temporal distribution of pesticide residues presents unique challenges for both monitoring and modeling evaluations. The concentration of a given pesticide should be viewed as a moving target, with factors related to transport and degradation ensuring that concentrations are in a constant state of flux. Although point estimates in time (e.g., initial concentrations) may provide suitable information for a tier I, screening level exposure evaluation, they are inadequate for the higher-tier (e.g., probabilistic) assessments that are necessary to answer questions related to the likelihood of an organism being exposed to a concentration of concern.

Research needs

- Recommendation 1: Further research and development is needed to ensure that estimation and monitoring methods for pesticide concentrations in soil, water, and food are applicable and utilized for all important and relevant cultural, agronomic, and environmental conditions. This is especially true with respect to developing countries and tropical climates, which are often disproportionately ignored in favor of developing countries and temperate climates.
- Recommendation 2: Methodologies for collection of monitoring data and generation of modeled estimates for pesticide residues in soil, water, and food need to be carefully designed with the requirements of higher-tier, probabilistic exposure assessments in view. Although worst-case, point estimates or analyses may be useful for screening-level assessments, advanced assessments targeted at addressing the *likelihood* of biologically relevant exposures are urgently required by scientists and regulatory authorities for reaching sound risk assessment and risk management decisions.

REFERENCES

1. P. T. Holland. *Pure Appl. Chem.* **68**, 1167–1193 (1996).
2. European Community. *Communication from the Commission to the Council and the European Parliament on the Implementation of the Community Strategy for Endocrine Disrupters.* p. 262, Brussels, Belgium (2001).
3. Japan Ministry of Health, Labor, and Welfare. *The Interim Risk Assessment Report by Review Committee Meeting of MHLW Regarding Hazards to Human Health of Endocrine Disrupting Chemicals,* Office of Chemical Safety, Pharamaceutical and Food Safety Bureau, Tokyo, Japan (2002).
4. U.S. National Research Council. *Hormonally Active Agents in the Environment,* National Academy of Sciences, National Academy Press, Washington, DC (1999).
5. U.S. Environmental Protection Agency. *U.S. Federal Register*, **67**, 79611–79629 (2002).
6. J. E. Casida and G. B. Quistad. *Ann. Rev. Entomol.* **43**, 1–16 (1998).
7. K. D. Racke. In *Chemistry of Crop Protection*, G. Voss and G. Ramos (Eds.), pp. 322–333, Wiley-VCH, Weinheim (2003).
8. U. S. Environmental Protection Agency. *List of Pesticide Product Inert Ingredients.* Washington, DC (1995).
9. T. S. Woods. In *Pesticide Chemistry and Bioscience: The Food-Environment Challenge,* G. T. Brooks and T. R. Roberts (Eds.), pp. 120–133, Royal Society of Chemistry, London (1999).
10. A. L. Aspelin and A. H. Grube. *Pesticides Industry Sales and Usage: 1996 and 1997 Market Estimates,* U.S. Environmental Protection Agency, 733-R-99-001, Washington, DC (1999).
11. Wood-Mackenzie. *Crop-Pesticide Sector Report* (1999).
12. D. Pimental and L. Levitan. *BioScience* **36**, 86–91 (1986).
13. A. D. Hammock, A. B. Inceoglu, W. Rajendra, J. R. Fuxa, N. Chejanovsky, D. Jarvis, T. N. Hanzlink. In *Pesticide Chemistry and Bioscience: The Food-Environment Challenge,* G. T. Brooks and T. R. Roberts (Eds.), pp. 73–99, Royal Society of Chemistry, London (1999).
14. S. L. Bird. In: *Agrochemical Environmental Fate: State of the Art,* M. L. Leng, E. M. K. Leovey, P. L. Zubkoff (Eds.). pp. 195–207, CRC/Lewis Press, Boca Raton, FL (1995).
15. Spray Drift Task Force. *A Summary of Aerial Application Studies,* Macon, Missouri (1997).
16. R. D. Wauchope. *J. Environ. Qual.* **7**, 459–472 (1978).
17. K. D. Racke. *Pesticides in Urban Environments: Fate and Significance.* American Chemical Society, Symposium Series No. 421, Washington, DC (1993).
18. R. S. McAllister. *Field Dissipation Studies: Zero-Time Recovery,* National Agricultural Chemicals Association, CL94-315.RSM, Washington, DC (1994).
19. K. D. Racke, D. D. Fontaine, R. N. Lubinski, J. R. Miller. *Pestic. Sci.* **42**, 43–51 (1994).
20. S. Baskaran, R. S. Kookana, R. Naidu. *Pestic. Sci.* **55**, 1222–1228 (1999).
21. A. S. Felsot, K. D. Racke, D. J. Hamilton. *Rev. Environ. Contam. Toxicol.* **177**, 123–200 (2003).
22. K. D. Racke, M. W. Skidmore, D. J. Hamilton, J. B. Unsworth, J. Miyamoto, S. Z. Cohen. *Pure Appl. Chem.* **69**, 1349–1371 (1997).
23. M. A. Ross and C. A. Lembi. *Applied Weed Science*, Macmillan, New York (1985).
24. R. D. Wauchope, D. B. Baker, K. Balu, H. Nelson. *Pesticides in Surface and Ground Water,* Council for Agricultural Science and Technology, Issue Paper No. 2, Ames, IA (1994).
25. H. J. Holterman, H. A. J. Porskamp, J. F. M Huijsmans. *Modelling Spray Drift from Boom Sprayers,* Institute of Agricultural and Environmental Engineering (IMAG-DLO). Wageningen, The Netherlands (1994).
26. H. Ganzelmeier et al. *Studies on the spray drift of plant protection products.* Mitteilungen aus der Biologischen Bundestanstalt fur Land- und Forstwirtschaft, Berlin, p. 305 (1995).
27. S. Z. Cohen, R. D. Wauchope, A. W. Klein, C. V. Eadsforth, R. Graney. *Pure Appl. Chem.* **67**, 2109–2148 (1995).

28. P. Adriaanse, R. Allen, V. Gouy, J. Hollis, J. Hosang, N. Jarvis, T. Jarvis, M. Klein, R. Layton, J. Linders, H. Schafer, L. Smeets, D. Yon. *Surface Water Models and EU Registration of Plant Protection Products: Final Report of the Work of the Regulatory Modelling Working Group on Surface Water Models of FOCUS*. 6476-VI-96 (1996).
29. A. W. Kolpin. *Pesticides in Ground Water: Summary Statistics, Results of the National Water Quality Assessment Program (NAWQA) 1992–1998,* U.S. Geological Survey, Washington, DC (2001).
30. J. Larson. *Pesticides in Streams: Summary Statistics, Results of the National Water Quality Assessment Program (NAWQA) 1992–1998,* U.S. Geological Survey, Washington, DC (2001).
31. J. A. R. Bates. *Pure Appl. Chem.* **62**, 337–350 (1990).
32. J. Linders, H. Mensink, G. Stephenson, D. Wauchope, K. Racke. *Pure Appl. Chem.* **72**, 2199–2218 (2000).
33. F. Hoerger, E. E. Kenaga. In *Environmental Quality and Safety: Chemistry, Toxicology, and Technology*, F. Coulston and F. Kort (Eds.), pp. 9–28, Georg Theime, Stuttgart (1972).
34. S. Fletcher, J. E. Nellessen, T. G. Pfleeger. *Environ. Toxicol. Chem.* **13**, 1383–1391 (1994).
35. Codex Alimentarius Commission, Joint FAO/WHO Food Standards Programme, Codex Committee on Pesticide Residues, Document CX/PR 02/6, Rome, Italy (2002).
36. D. J. Hamilton, P. T. Holland, B. Ohlin, W. J. Murray, A. Ambrus, G. C. De Baptista, J. Kovacicová. *Pure Appl. Chem.* **69**, 1373–1410 (1997).
37. U.S. Department of Agriculture. *Pesticide Data Program: Annual Summary Calendar Year 2000*, USDA Agricultural Marketing Service, Washington DC (2002).
38. European Commission. *Monitoring of Pesticide Residues in Products of Plant Origin in the European Union, Norway, Iceland, and Liechtenstein: 2000 Report*, Health & Consumer Protection Directorate-General, SANCO/687/02 (2002).
39. Australia New Zealand Food Authority. *The Australian Market Basket Survey 1996*, Melbourne (1998).
40. D. Hamilton, A. Ambrus, R. Dieterle, A. Felsot, C. Harris, B. Petersen, K. Racke, S. Wong, R. Gonzalez, K. Tanaka, M. Earl, G. Roberts, R. Bhula. *Pest Manag. Sci.* (2003). In press.

Pure Appl. Chem., Vol. 75, Nos. 11–12, pp. 1917–1932, 2003.

Topic 2.7

Role of metabolism in the endocrine-disrupting effects of chemicals in aquatic and terrestrial systems*

Martin van den Berg[1,‡], Thomas Sanderson[1], Norio Kurihara[2], and Arata Katayama

[1]*Institute for Risk Assessment Sciences, Utrecht University, P.O. Box 80176, 3508 TD Utrecht, The Netherlands;* [2]*Koka Laboratory, Japan Radioisotope Association, 121-19 Toriino, Koka, Shiga 520-3403, Japan;* [3]*Nagoya University, Research Center for Advanced Waste and Emission Management, Chikusa Nagoya 464-8603, Japan*

Abstract: This review describes the role of metabolism with endocrine active substances. Many modern synthetic compounds are readily metabolized to more polar forms that often contain hydroxy groups. This presence of polar groups and aromatic moieties in the parent compound or metabolite can play an important role in the mechanism of endocrine disruption. In addition, phase II metabolism (e.g., glucuronidation) can also lead to deactivation of the endocrine properties. In the case of bisphenol A and alkylphenols, metabolism can be considered as a detoxification mechanism as glucuronides decrease of inhibit binding to the estrogen receptors. In the case of phthalate esters, the primary metabolites, the monoesters, and further degraded metabolites do not interact with the estrogen receptor either. In contrast, the demethylation of methoxychlor in fish and other vertebrate species leads to metabolites with an increased affinity for the estrogen receptor. Certain PCB metabolites with hydroxy groups on the para position without vicinal chlorines have estrogenic activity, but these metabolites are not relevant for the environment. PCB metabolites with methylsulfonyl groups are commonly found in environmental biota and have been associated with several endocrine, developmental, and reproductive effects. Some DDT metabolites bind weakly to the estrogen receptor, but the major biotransformation product *p,p*-DDE is an androgen receptor (AR) antagonist. Vinclozolin is an anti-androgen and this effect appears to caused by two of its more water-soluble metabolites. The chloro-*s*-triazines exhibit an in vitro induction of aromatase, but their dealkylated metabolites show a decrease or lack of this effect.

It is recognized that common metabolic processes can differ strongly among species that complicates ecotoxicological risk assessment of endocrine active substances. In conclusion, the testing of metabolites for endocrine-disrupting properties should be encouraged in the future to establish a better risk assessment process.

An appendix containing levels and half-lives of various endocrine-disrupting chemicals in the environment and in wildlife is included at the end of this article.

*Report from a SCOPE/IUPAC project: Implication of Endocrine Active Substances for Human and Wildlife (J. Miyamoto and J. Burger, editors). Other reports are published in this issue, *Pure Appl. Chem.* **75**, 1617–2615 (2003).

‡Corresponding author

INTRODUCTION

Since the middle of the last century, many industrial chemicals and pesticides have been introduced into the environment. Many of these compounds are chlorinated aromatic structures, which are not easily metabolized, resulting in bioaccumulation or biomagnification. This bioaccumulation process has caused unwanted side effects in nontarget species, particularly those at higher trophic levels in the food chain. With growing awareness of these problems, various persistent halogenated compounds have gradually been banned from use. Since then, environmental levels in biota have decreased slowly during the last decades. As a replacement, new products have been developed that are less hydrophobic than organochlorines, and consequently do not accumulate in humans and wildlife.

The capacity of an organism to metabolize a compound to more polar products is often considered to be a detoxification mechanism. Furthermore, the presence of polar hydroxy groups in parent compounds often prevents bioaccumulation. In addition, introduction of even more polar groups such as glucuronides or sulfates further increases the ability of an organism to eliminate the compound [1,2].

A disadvantage of this approach is that the presence of hydroxy groups and complicated structures with aromatic moieties may bear resemblance to steroid structures. As a result, less bioaccumulative compounds such as alkylphenols, phthalate esters, and methoxychlor have shown biological activities similar to estrogens or androgens. These compounds can act as either (partial) agonists or antagonists for steroid receptors such as the estrogen receptor. Such interactions may have consequences for (sexual) development, reproduction, and the formation of hormone-dependent tumors. Thus, subtle differences among molecules, such as the presence or absence of an OH group, can lead to significant changes in their ability to bind to steroid receptors or inhibit steroidogenic enzymes. The presence of an OH and/or aromatic group plays a significant role in both steroid metabolism and receptor binding [3,4]. Thus, biotransformation plays a significant part in the endocrine-disrupting properties of a compound. On the one hand, the introduction of an OH group may bioactivate the parent compound by forming a metabolite that can interact with a steroid receptor or steroidogenic enzyme. On the other hand, a rapid phase II metabolism producing glucuronides or sulfates helps the organism to eliminate the parent compound from the body, reducing the opportunity for adverse effects.

This article summarizes the present state of knowledge of the role of metabolism in endocrine-related effects in aquatic and other terrestrial organisms. Although these effects are not discussed in view of results obtained in experimental laboratory systems with, for example, rodents, a comparison with experimental models is made if appropriate. In this topic, a selection of known endocrine active compounds (EASs) has been made that are presently considered to be most relevant for the environment. This overview focuses on the interactions of metabolites of EASs with the estrogen and androgen receptor or with steroidogenesis, because this has been the main emphasis of research over the last decade. It should however be recognized that these interactions are from a mechanistic point of view most firmly linked to reproduction, development, and hormone-dependent tumors. With increasing research efforts, undoubtedly other endocrine-related effects of metabolites of EASs will be uncovered. In addition, the authors have limited themselves to those EASs that are environmentally relevant. Pharmaceuticals with endocrine-disrupting properties have been excluded from this review. Nevertheless, these pharmaceutical compounds should not be ignored in the future as significant emissions from human and agricultural sources are to be expected.

BISPHENOL A

Most studies regarding the role of metabolism and metabolites in estrogenicity have been performed with rodents. In vitro studies with rat hepatocytes and in vivo studies with rats have shown that the major metabolite of bisphenol A (BPA) is a glucuronide. This phase II metabolite is predominantly formed in the liver and excreted in the bile [5–7]. In addition, at least four metabolites, among others a monosulfate and 3-OH BPA, were also formed in the rat, although quantitatively less important [6].

Experiments with different isoforms of UGT showed that UGT2B1 is probably most important glucuronidation enzyme in the rat [8].

In fish, the glucuronidation of BPA has also been reported to occur easily. This was illustrated by the presence of BPA glucuronide in the bile of caged fish that were exposed to sewage effluent [10]. Toxicokinetic studies with rainbow trout showed that the formation of BPA glucuronide can reach plasma concentrations that are about twice that of the parent compound (see Fig. 1) [11]. Thus, it can be concluded that in both mammalian and piscine systems the formation of glucuronides is the preferred metabolic pathway. This is to be expected as BPA and several of its analogs contain (several) hydroxy groups, which are highly susceptible to phase II metabolism.

Glucuronidation of BPA should be considered a detoxification process, as several investigators showed that these metabolites lacked or had decreased binding affinity for the ERα or ERβ in mammalian systems [7,9,12]. At present, there is no indication that these glucuronides would behave differently in piscine or avian systems.

However, in the rat certain minor metabolites of BPA do possess estrogenic activity, in some cases exceeding that of the parent compound. The biological relevance of these estrogenic metabolites is unknown. Incubations of BPA with rat liver S9 fraction showed that estrogenicity can increase several fold, which is possibly caused by activities of P450 enzymes [13]. Furthermore, hydroxylation of either the 3 or 5 position of BPA produces estrogenic metabolites that are 5 to 10 times less potent than the parent compound [6,9]. As several fish species exhibit significant cytochrome P450 activities the formation of hydroxylated BPA in fish is also likely. In view of the estrogenic activities of BPA observed in several fish species [11,14], it can not be excluded that certain hydroxylated metabolites of BPA may contribute.

Limited information is available about the biodegradation of BPA in the environment. Bacteria from sewage sludge were able to degrade BPA to an intermediary metabolite 4,4′-dihydroxy-α-methylstilbene, that has a structural resemblance to diethylstilbestrol (DES). However, this compound is easily further degraded to 4-hydroxybenzaldehyde and 4-hydroxyacetophenone [13,15]. The ecotoxicological significance of this formation process is presently unknown, but in general, BPA is considered as a readily biodegradable compound according to OECD standards [16].

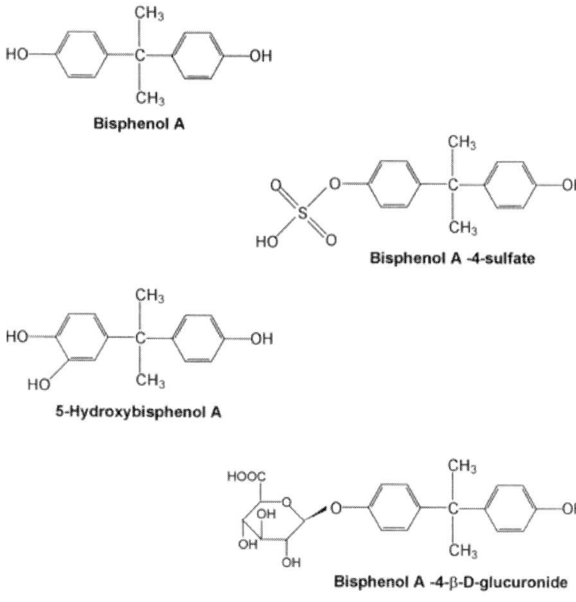

Fig. 1 Structures of BPA and metabolites from [9].

ALKYLPHENOLS

Alkylphenols (APs) are rapidly metabolized to either phase I or II metabolites in mammalian species such as rat and human, which has been reported from in vivo and in vitro studies [17–19]. Hydroxylation of the aromatic ring and the alkyl side chain, followed by glucuronidation are the major metabolic reactions in rat hepatocytes [19] one in vivo study with rats showed that glucuronidation of *p-tert*-octylphenol is the major metabolic pathway [17].

In fish, the metabolic pathways are basically similar to those in the rat. Glucuronidation of either the parent AP or the ω-hydroxylated metabolites is the major route of metabolism [20–22]. In rainbow trout, the major metabolite of 4-*n*-nonylphenol was the glucuronide of the parent compound, but hydroxylation of the ω-1 and 2 position of the alkyl chain also occurred. In addition, the rainbow trout was capable of oxidizing the ω-hydroxylated metabolites into carboxylic acid metabolites [21]. However, in the rainbow trout metabolism of 4-*n*-nonylphenol is also tissue-specific as distinct differences between the metabolic structures were found between the bile and the urine [23].

A study in rainbow trout of the metabolism of another AP congener, 4-*tert*-octylphenol, showed similar biotransformation pathways. The major metabolite was 4-*tert*-octylphenol-β-glucuronide, but hydroxylations of the ω (C2) or ω-3 (C4) positions of the alkyl chain were also found, beside ortho hydroxylation of the aromatic ring [24]. This rapid metabolism to more polar phase I and II metabolites leads to efficient elimination of this compound in the rainbow trout [22,25]. However, some tissue-specific retention of the parent compounds has been observed that may have some toxicological significance at higher exposure levels.

The possible biotransformation of nonylphenol diethoxylate to the more estrogenic nonylphenol was also studied in the rainbow trout, but no evidence could be found for such a metabolic pathway [26].

In Atlantic salmon, 4-*n*-nonylphenol was mainly metabolized to its glucuronide conjugate [27,28]. To a lesser extent, hydroxylated and oxidated metabolites were formed [27]. In vitro experiments with salmon hepatocytes did not show any significance difference in metabolite pattern when compared with in vivo results [28]. The major mono-hydroxylated metabolites were compounds with OH groups on the alkyl chain at the ω, ω-1, and ω-2 positions [29].

Studies with cytochrome P450-selective inhibitors suggest that the CYP2K and CYP2M enzymes play a significant role in the metabolism of 4-*n*-nonylphenol. These fish P450 enzymes show similarities with CYP3A and CYP2B enzymes in mammals [29]. In rats, some of these P450 enzymes are male-dominant and perform specific hydroxylations of testosterone. Several rodent studies have shown that APs can decrease the activity of CYP3A, CYP2B, but also the male-specific CYP2C11 [30–32]. In analogy with the results from rat studies, it may be possible for 4-*n*-nonylphenol to compete with steroids for CYP2K and CYP2M enzymes in fish. Such an interaction could influence the metabolism of steroids such as testosterone and progesterone [29]. In juvenile Atlantic salmon, it has recently been shown that 4-*n*-nonylphenol decreases the rate of the 6β, 16α, and 17α hydroxylation of progesterone [33].

The role of the major (glucuronide) metabolites of the APs was also studied in relation to in vitro binding with the estrogen or androgen receptor, including those in trout hepatocytes. None of the major metabolites of octyl- or nonylphenol showed significant interaction with these steroid receptors. This indicates that the role of metabolism is minor or absent with respect to the estrogenic activities of these compounds [34,35]. This is in agreement with a structure–activity relationships determined between the estrogen receptor and APs, which showed that binding of the para alkylphenols depends on covalent binding of the phenol and alkyl groups resembling the A ring and hydrophobic moiety of the steroids [36]. Thus, glucuronidation and hydroxylation would decrease the ER binding significantly. In summary, it can be concluded that metabolism of alkylphenols in fish, like in mammals, appears to be a detoxification process with respect to the endocrine-disrupting properties of the parent compounds.

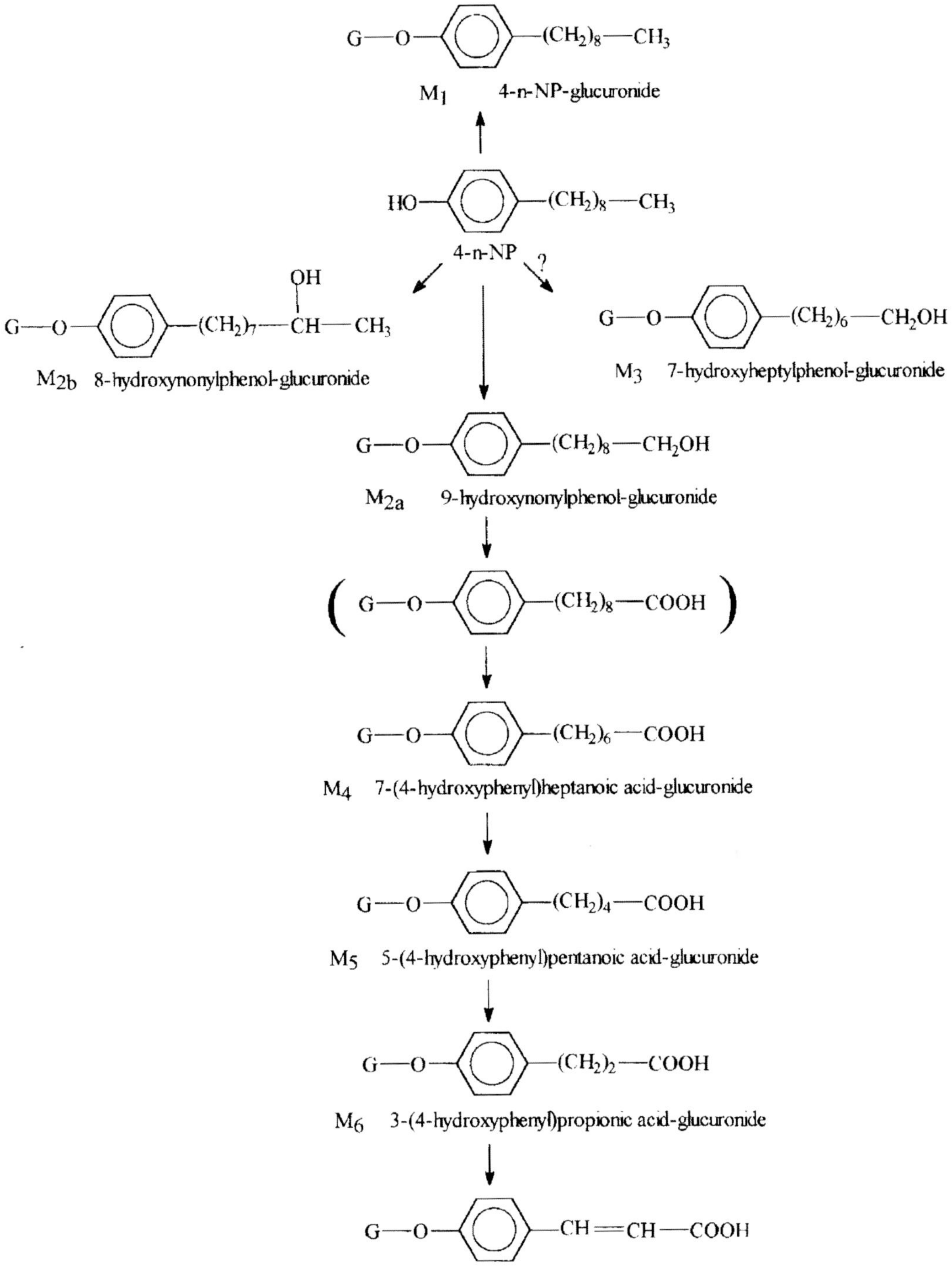

Fig. 2 Proposed biotransformation pathways for the in vivo formation of biliary metabolites from [21].

PHTHALATE ACID ESTERS

The esters of phthalic acid, either with dialkyl, aryl, or saturated cyclic hydrocarbon chains are important commercial compounds that are widely spread in the environment. Almost all phthalate acid esters

(PAEs) are not strongly bioaccumulative and metabolism occurs rapidly in all vertebrates [37]. In rats and fish, phthalate diesters are rapidly metabolized in the gastro-intestinal tract and gills, but also in liver and plasma [38–41]. The primary step in the metabolism of phthalate esters is conversion into monoesters, and occurs in mammals and fish (see Fig. 3) [37,42].

In mammals and fish, metabolic breakdown continues readily with phthalic acid as a major biotransformation product [43,44]. In the rainbow trout, extensive metabolism of di-2-ethylhexyl phthalate occurs in the gills where this compound is easily converted by esterases to the corresponding monoester [41]. Experiments with sheephead minnow and di-2-ethylhexyl phthalate combined with selective enzyme inhibitors showed that esterases and not P450 enzymes are responsible for the first step in metabolic breakdown [45]. In addition, the biological breakdown of phthalate esters in aquatic systems, including some invertebrates, is considered to be rapid, and no significant bioaccumulation has been reported [38,46].

As a few phthalate esters show some estrogenic activity [47] in certain in vitro systems, the role of metabolism was further examined. Most studies indicate that monoesters of phthalates and further degraded metabolites, such as phthalic acid, did not interact with the estrogen receptor [48,49]. The exceptions are the 4-hydroxylated metabolites, which have increased binding affinity for the estrogen receptor [50]. However, this metabolic pathway does not appear to be important because cytochrome P450 enzymes play only a minor role in the biotransformation of phthalates [45]. Thus, metabolic pathways in fish of phthalates should be considered as a detoxification process.

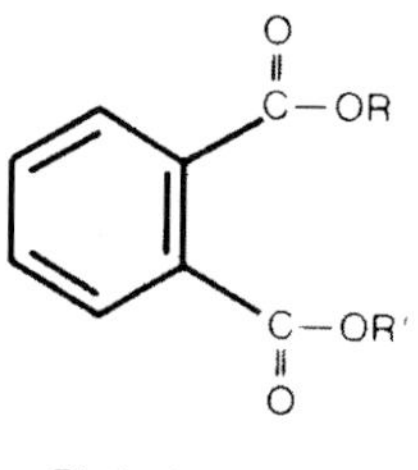

Phthalate Esters

Fig. 3 General structure of phthalate esters.

METHOXYCHLOR

In rodent methoxychlor (MXCL) is primarily metabolized to mono and bisphenol demethylated derivatives. In vitro studies with rat hepatic microsomal or S9 fractions in combination with specific P450 inhibitors or inducers implicated the involvement of P450 enzymes in this process (see Fig. 4) [13,51,52].

Information about the metabolism of MXCL in fish is scarce. Liver microsomes of catfish metabolized MXCL to mono- and bisdemethylated metabolites, but hydroxylation on the aromatic can also occur [53]. Thus, similar to mammals, the major metabolic pathway of MXCL in fish appears to be demethylation.

In rats, phenobarbital-inducible P450 enzymes play a role in the demethylation process, but enzymes from the CYP2B family are also responsible for aromatic hydroxylation [52,54]. Studies with

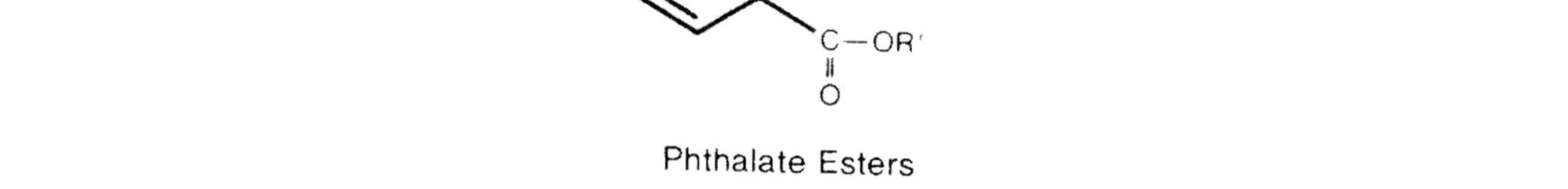

METHOXYCHLOR → MONO-OH-METHOXYCHLOR → BIS-OH-METHOXYCHLOR → TRIS-OH-METHOXYCHLOR

Fig. 4 Metabolism of methoxychlor in mammals (from Kupfer et al., 1990).

hepatic microsomes of the catfish and hepatocytes ruled out the involvement of CYP2K or CYP1A enzymes in the demethylation or aromatic hydroxylation of MXCL [14,53].

Structure–activity studies of the major metabolites of MXCL in rodents have shown that the mono- and bisdemethylated metabolites have agonist activities on the ERα, and antagonist activities on the ERβ and androgen receptor [55–57]. The question is to what extent results obtained with metabolites of MXCL and mammalian ERs are also applicable to other vertebrate groups such as fish, birds, and amphibians. One study in particular showed that different vertebrate species exhibit differential ligand preferences and binding affinities for estrogenic compounds [58].

In summary, it can be concluded that in analogy with the mammalian situation, MXCL most likely has to be bioactivated in fish and other vertebrate species to exert its endocrine-disrupting effects.

CHLORO-*S*-TRIAZINE HERBICIDES (TRZs)

Although the triazines, especially atrazine (ATR), have been shown to influence reproduction and development in multiple species and to stimulate the development of mammary tumors in one rat strain, the possible role of metabolism has not been fully elucidated [59–62]. Metabolism in fish, birds, mammals as well as microbes most commonly involves *N*-monodealkylation and hydroxylation processes [63–68]. For common ATR metabolites, see Fig. 5.

With respect to general toxicity, it is generally assumed that the metabolites of TRZs are substantially less active than their parent compounds [68]. Significant metabolic differences among mammalian species have been reported, but *N*-monodealkylation and isopropyl or ter-butyl hydroxylations followed by sulfoxidation remain the major pathways [63–65]. In rodent species, a distinct involvement of CYP1A and CYP2B enzymes in metabolism has been shown, but a role of CYP 2D1 and 2E1 has

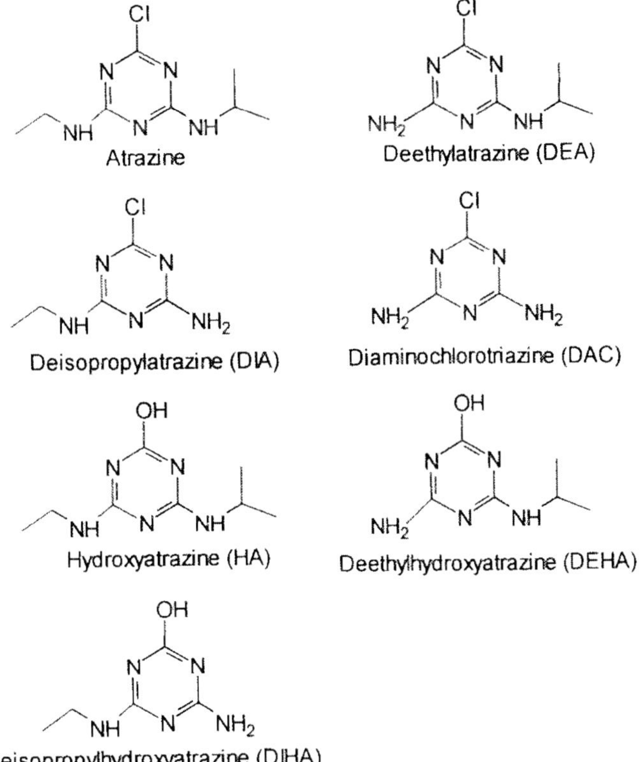

Fig. 5 The chemical structure of atrazine and the most common biodegradation products from [68].

also been found [64,69]. In humans, CYP1A2 specifically has been implicated to play a major role in the hepatic metabolism of atrazine [66]. Pre-induction with 3-methylcholantrene (3-MC) markedly induced the metabolism of TRZs. Therefore, in fish the major hepatic enzyme CYP1A1 that is 3-MC inducible may also be expected to be involved in the metabolism of atrazine. In addition, a study with zebra fish embryos indicated a possible role of GSH conjugation [70].

In general, triazine herbicides or their major metabolites do not show significant estrogen receptor binding in vitro or in vivo, but experimental results depend on the metabolite studied. The 2-chloro-4-amino-6-isopropylaminotriazine metabolites had a higher binding affinity for the ERα than the parent compounds, although many orders of magnitude lower than 17β-estradiol [63]. However, an earlier study with the fully dealkylated metabolite, diaminochlorotriazine, could not find any in vivo or in vitro estrogenic activity [71,72].

The possible (anti) estrogenic properties of the major metabolites of triazine herbicides were also studied in carp hepatocytes. None of the metabolites had any effect on the estrogen receptor mediated induction of vitellogenesis indicating a lack of interaction with the carp estrogen receptor, either as agonists or antagonists [73].

In vitro studies with the H295R human adrenocortical carcinoma cell line indicated that TRZs could induce the catalytic activity and mRNA expression of aromatase. Although mono *N*-dealkylated metabolites were still capable of inducing aromatase, this effect disappeared with full dealkylation or hydroxylation of the heterocyclic ring [73,74a]. Thus, regarding effects on the enzyme aromatase the metabolism of TRZs results in detoxification.

VINCLOZOLIN

The anti-androgenic activity of vinclozolin in mammals is well known [74b], and the activity depends on the binding of two vinclozolin metabolites, a butenanilide derivative and hydroxybutenoic acid derivative (Fig. 6), to the androgen receptor [74c]. A recent report [74d] on the demasculinization effects of these compounds on the adult male guppy seems to be caused by the similar mechanism. Thus, any endocrine activities displayed by chemicals in mammals via the target site that is ubiquitous among various organism species are likely to be also exhibited in various other species.

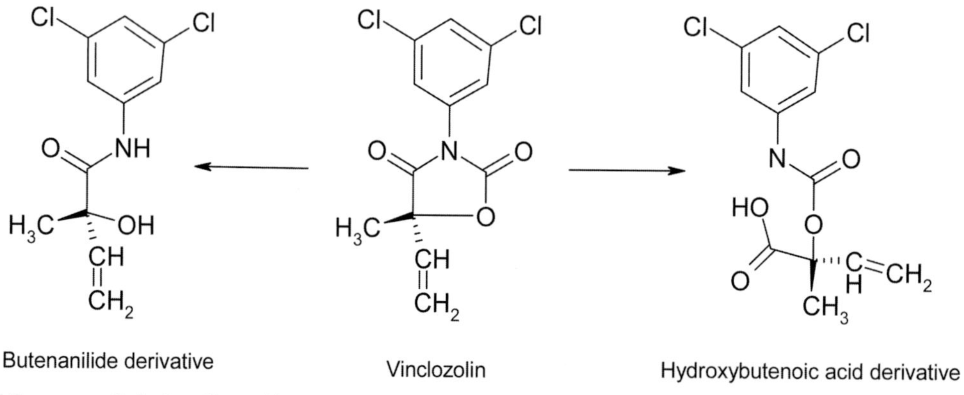

Butenanilide derivative Vinclozolin Hydroxybutenoic acid derivative

Fig. 6 Structure of vinclozolin and its two anti-androgenic metabolites.

DDT

This is one of the most persistent organohalogen pesticides known and occurs everywhere in the global environment. The major metabolite of DDT in vertebrates, DDE has an extremely long half-life in biota. DDE is formed by enzymatic dechlorination that involves a glutathione-dependent reaction. In addition, other degradation products such as DDD and DDA are formed by series of reductive dechlorination and oxidative processes in various vertebrate species. The general metabolism pathway of DDT in mammals and avian species is shown in Fig. 7.

The stability of DDT metabolites, including DDE and DDD, is notorious, and the question rises to what extent these metabolites are responsible for observed endocrine-disrupting properties in various vertebrate species. A large number of in vitro and in vivo studies have been performed to identify the endocrine-disrupting properties DDT and its metabolites, but mainly using mammalian systems. Nevertheless, a limited number of these studies used piscine or avian systems to elucidate the role of DDT and metabolites in this process. With respect to binding affinity of these compounds to the ER notable differences have been observed among species for *o,p*-DDT [75]. When evaluating results of these types of studies with DDT metabolites, the interaction appears to be rather species- and compound-specific.

Based on either the interaction with fish ER or the estrogen-mediated vitellogenin synthesis, it can be concluded that DDE is at best a weak ER agonist or antagonist differing many orders of magnitude with E2 in the tested mammalian, fish, amphibian, or reptile species [14,57,77,78]

Although some DDT metabolites bind weakly to the ER, the major biotransformation product *p,p*-DDE has been identified as an androgen receptor (AR) antagonist in mammalian systems [57,79,80]. Experiments with guppies showed that *p,p*-DDE had demasculinizing properties in fish consistent with AR antagonism [81]. At present, it is unclear whether the developmental and reproductive effects of DDT in fish or other vertebrates are caused by the interaction of DDT itself with the ER or that of its metabolite *p,p*-DDE with the AR.

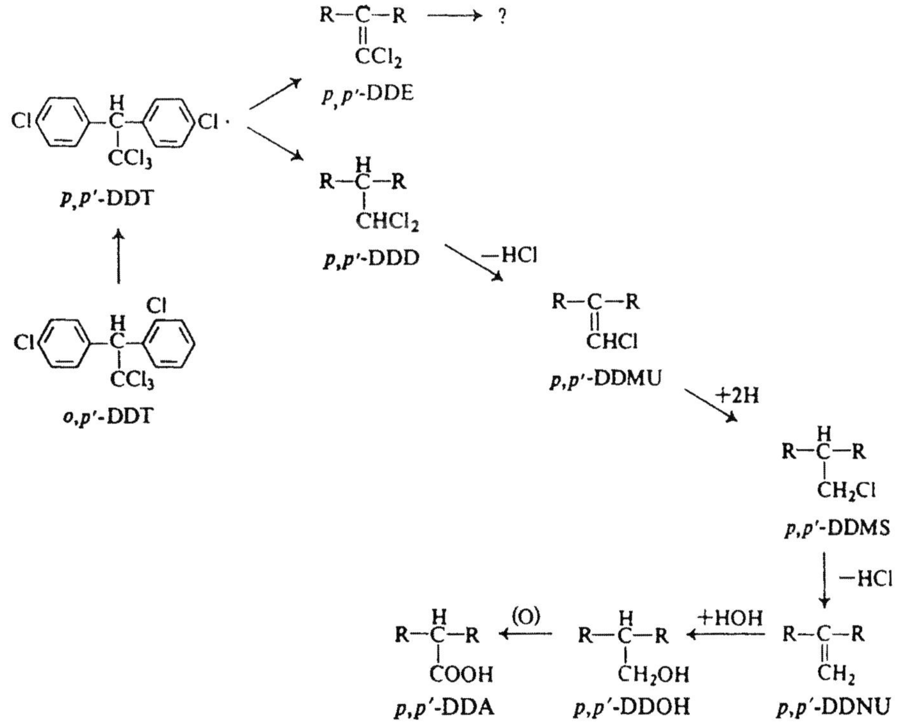

Fig. 7 Degradation of DDT by mammalian and avian tissues from [76].

POLYCHLORINATED BIPHENYLS

Although many polychlorinated biphenyls (PCBs) are retained as parent compounds in lipid-rich tissues of most vertebrate species, some of their effects have been attributed to the action of hydroxylated metabolites. In particular, the competitive binding of hydroxylated PCBs (OH-PCBs) to the thyroid hormone transporting protein, transthyretin (TTR), has been shown to have effect on the thyroid hormone levels in rodents. This competitive binding to TTR is most pronounced for those PCB metabolites that have a hydroxy group on the para position with two adjacent chlorine atoms [82]. In addition, some hydroxylated metabolites have been found to interact with the ERα. Certain PCB metabolites with hydroxy groups on the para position without vicinal chlorines have estrogenic activity. However, these OH-PCBs do not occur in the environment, where a preferred retention of metabolites with a para hydroxy group with vicinal chlorine atoms and five to seven chlorine atoms occurs [77,83,84]. The subsequent hydroxylation of intermediate epoxides of PCBs to, for example, catechols can also produce weak estrogenic metabolites that have an activity in the range of nonylphenol and *o,p*-DDT (see Fig. 8) [85].

All these experiments have been done with *either* in vivo or in vitro systems of mammalian origin, and the implications of these results for other groups of vertebrates such as fish or amphibian is unclear. Nevertheless, it should be recognized that fish species, just as mammals, are capable of metabolizing PCBs to hydroxylated forms, with a preference for the para position [86]. Thus interactions between (*di*) OH-PCBs and TTR or estrogen receptor may be expected in other than mammalian vertebrates [75].

Another group of relevant PCB metabolites are those with methylsulfonyl groups. These metabolites are commonly found in several species of fish and mammalian wildlife [87–90] in addition to human milk and plasma [91,92]. Structurally, the 3 and 4-methylsulfonyl metabolites appear to be most commonly found in humans, rodents, and wildlife [88,91–95]. These methylsulfonyl PCBs have been associated with developmental and reproductive effects in the mink [96]. Among the more subtle effects are hepatic P450 induction, modulation of thyroid hormone levels and inhibition of the glucocorticoid synthesis [94,95,97–100].

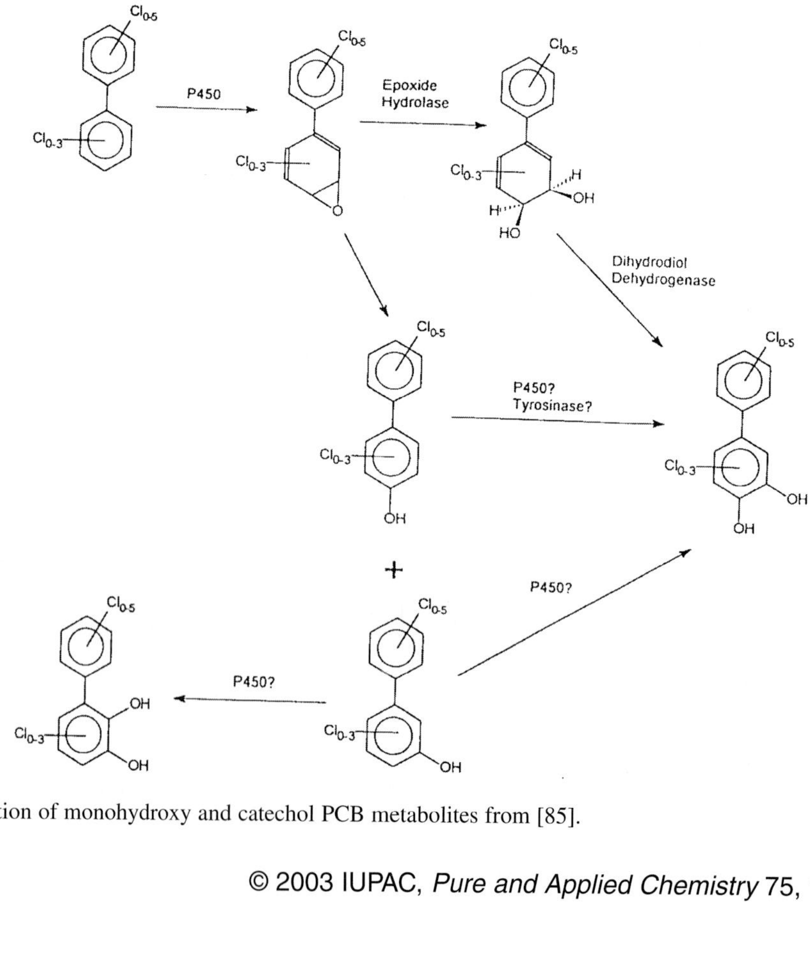

Fig. 8 Formation of monohydroxy and catechol PCB metabolites from [85].

CONCLUSIONS AND RECOMMENDATIONS

The information presented in this chapter illustrates that metabolism plays a significant role in bioactivation and/or deactivation of EASs. The information that is available for most well-known EASs have been derived from either in vivo or in vitro studies using material from humans or rodents. These studies have shown that particularly the presence of hydroxy or keton groups and aromatic moieties play an important role in the mechanism of action of EASs. This is not unexpected as in the steroid biosynthesis pathway these same functional groups are involved in the conversion of cholesterol to progestins, androgens, and estrogens. To prevent persistence and bioaccumulation/magnification, many modern synthetic compounds are readily metabolized to more polar forms often containing one or more hydroxy groups. This enhanced ability to undergo biotransformation also has increased the potential for the formation of more EASs. Consequently, the additional and consistent testing of metabolites for endocrine-disrupting properties should be encouraged in the future in order to establish a better risk assessment process for these types of compounds. Furthermore, it should be recognized that common metabolic processes such as hydroxylation, oxidation, and glucuronidation could differ strongly among species, complicating ecotoxicological risk assessment of EASs.

REFERENCES

1. M. Van den Berg, D. Van de Meent, W. J. G. M. Peijnenburg, D. T. H. M. Sijm, J. W. Tas. In *Risk Assessment of Chemicals: An Introduction*, C. J. Van Leeuwen and J. Hermens (Eds.), pp. 37–102, Kluwer, Dordrecht (1995).
2. A. Parkinson. In *Casarett & Doull's Toxicology: The Basic Science of Poisons*, C. D. Klaassen (Ed.), pp. 113–186, McGraw-Hill, New York (1995).
3. C. L. Williams and G. M. Sancel. In *Goodman & Gilman's The Pharmacological Basis of Therapeutics*, J. G. Hardman, A. Goodman Gilman, L. E. Limbird (Eds.), pp. 1411–1440, McGraw-Hill, New York. (1995).
4. J. D. Wilson. In *Goodman & Gilman's The Pharmacological Basis of Therapeutics*, J. G. Hardman, A. Goodman Gilman, L. E. Limbird (Eds.), pp. 1441 *Goodman & Gilman's The Pharmacological Basis of Therapeutics*, J. G. Hardman, A. Goodman Gilman, L. E. Limbird (Eds.), pp. 1411–1440, McGraw-Hill, New York (1995).
5. H. Inoue, H. Yokota, T. Makino, A. Yuasa, S. Kato. *Drug Metab. Dispos.* **29**, 1084–1087 (2001).
6. Y. Nakagawa and T. Suzuki. *Xenobiotica* **31**, 113–123 (2001).
7. R. W. Snyder, S. C. Maness, K. W. Gaido, F. Welsch, S. C. Sumner, T. R. Fennell. *Toxicol. Appl. Pharmacol.* **168**, 225–234 (2000).
8. H. Yokota, H. Iwano, M. Endo, T. Kobayashi, H. Inoue, S. Ikushiro, A. Yuasa. *Biochem. J.* **340**, 405–409 (1999).
9. R. Elsby, J. L. Maggs, J. Ashby, B. K. Park. *J. Pharmacol. Exp. Ther.* **297**, 103–113 (2001).
10. D. G. J. Larsson, M. Adolfsson-Erici, J. Parkkonen, M. Pettersson, A. H. Berg, P.-E. Olsson, L. Forlin. *Aquat. Toxicol.* **45**, 91–97 (1999).
11. C. Lindholst, S. N. Pedersen, P. Bjerregaard. *Aquat. Toxicol.* **55**, 75–84 (2001).
12. J. B. Matthews, K. Twomey, T. R. Zacharewski. *Chem. Res. Toxicol.* **14**, 149–157 (2001).
13. S. Yoshihara, M. Makishima, N. Suzuki, S. Ohta. *Toxicol. Sci.* **62**, 221–227 (2001).
14. J. M. Smeets, I. van Holsteijn, J. P. Giesy, W. Seinen, M. van den Berg. *Toxicol. Sci.* **50**, 206–213 (1999).
15. J. Spivack, T. K. Leib, J. H. Lobos. *J. Biol. Chem.* **269**, 7323–7329 (1994).
16. R. J. West, P. A. Goodwin, G. M. Klecka. *Bull. Environ. Contam. Toxicol.* **67**, 106–112 (2001).
17. H. Certa, N. Fedtke, H. J. Wiegand, A. M. Muller, H. M. Bolt. *Arch. Toxicol.* **71**, 112–22 (1996).
18. P. C. Lee. *Endocrine* **9**, 105–111 (1998).
19. R. T. Pedersen and E. M. Hill. *Chem. Biol. Interact.* **128**, 189–209 (2000).

20. A. C. Meldahl, K. Nithipatikom, J. J. Lech. *Xenobiotica* **26**, 1167–1180 (1996).
21. R. Thibaut, L. Debrauwer, D. Rao, J. P. Cravedi. *Xenobiotica* **28**, 745–757 (1998).
22. N. G. Coldham, S. Sivapathasundaram, M. Dave, L. A. Ashfield, T. G. Pottinger, C. Goodall, M. J. Sauer. *Drug Metab. Dispos.* **26**, 347–354 (1998).
23. R. Thibaut, L. Debrauwer, D. Rao, J. P. Cravedi. *Sci. Total Environ.* **233**, 193–200 (1999).
24. R. T. Pedersen and E. M. Hill. *Xenobiotica* **30**, 867–879 (2000).
25. A. M. Ferreira-Leach and E. M. Hill. *Mar. Environ. Res.* **51**, 75–89 (2001).
26. J. P. Cravedi, G. Boudry, M. Baradat, D. Rao, L. Debrauwer. *Aquat. Toxicol.* **53**, 159–172 (2001).
27. A. Arukwe, T. Celius, B. T. Walther, A. Goksoyr. *Aquat. Toxicol.* **49**, 159–170 (2000).
28. A. Arukwe, A. Goksoyr, R. Thibaut, J. P. Cravedi. *Mar. Environ. Res.* **50**, 141–145 (2000).
29. R. Thibaut, L. Debrauwer, E. Perdu, A. Goksoyr, J. P. Cravedi, A. Arukwe. *Aquat. Toxicol.* **56**, 177–190 (2002).
30. N. Hanioka, H. Jinno, Y. S. Chung, T. Nishimura, T. Tanaka-Kagawa, M. Ando. *Arch. Toxicol.* **73**, 625–631 (2000).
31. N. Hanioka, T. Tanaka-Kagawa, Y. S. Chung, T. Nishimura, H. Jinno, M. Ando. *Bull. Environ. Contam. Toxicol.* **64**, 804–810 (2000).
32. N. Hanioka, H. Jinno, Y. S. Chung, T. Tanaka-Kagawa, T. Nishimura, M. Ando. *Xenobiotica* **29**, 873–883 (1999).
33. A. Arukwe, F. R. Knudsen, A. Goksoyr. *Environ. Health Perspect.* **105**, 418–422 (1997).
34. G. J. Moffat, A. Burns, J. Van Miller, R. Joiner, J. Ashby. *Regul. Toxicol. Pharmacol.* **34**, 182–187 (2001).
35. T. Madigou, P. Le Goff, G. Salbert, J. P. Cravedi, H. Segner, F. Pakdel, Y. Valotaire. *Aquat. Toxicol.* **53**, 173–186 (2001).
36. Y. Tabira, M. Nakai, D. Asai, Y. Yakabe, Y. Tahara, T. Shinmyozu, M. Noguchi, M. Takatsuki, Y. Shimohigashi. *Eur. J. Biochem.* **262**, 240–245 (1999).
37. N. P. Moore. *Reprod. Toxicol.* **14**, 183–192 (2000).
38. J. A. Thomas and M. J. Thomas. *Crit. Rev. Toxicol.* **13**, 283–317 (1984).
39. I. R. Rowland, R. C. Cottrell, J. C. Phillips. *Food Cosmet. Toxicol.* **15**, 17–21 (1977).
40. R. D. White, D. E. Carter, D. Earnest, J. Mueller. *Food Cosmet. Toxicol.* **18**, 383–386 (1980).
41. M. G. Barron, I. R. Schultz, W. L. Hayton. *Toxicol. Appl. Pharmacol.* **98**, 49–57 (1989).
42. M. J. Melancon and J. J. Lech. *Drug Metab. Dispos.* **5**, 29–36 (1977).
43. T. Mettang, D. M. Alscher, C. Pauli-Magnus, R. Dunst, U. Kuhlmann, A. W. Rettenmeier. *Adv. Perit. Dial.* **15**, 229–233 (1999).
44. C. Nativelle, K. Picard, I. Valentin, J. C. Lhuguenot, M. C. Chagnon. *Food Chem. Toxicol.* **37**, 905–917 (1999).
45. A. H. Karara and W. L. Hayton. *Drug Metab. Dispos.* **16**, 146–150 (1988).
46. R. L. Metcalf, G. M. Booth, C. K. Schuth, D. J. Hansen, P. Y. Lu. *Environ. Health Perspect.* **4**, 27–34 (1973).
47. S. Jobling, T. Reynolds, R. White, M. G. Parker, J. P. Sumpter. *Environ. Health Perspect.* **103**, 582–587 (1995).
48. C. A. Harris, P. Henttu, M. G. Parker, J. P. Sumpter. *Environ. Health Perspect.* **105**, 802–811 (1997).
49. K. Picard, J. C. Lhuguenot, M. C. Lavier-Canivenc, M. C. Chagnon. *Toxicol. Appl. Pharmacol.* **172**, 108–118 (2001).
50. D. Asai, Y. Tahara, M. Nakai, Y. Yakabe, M. Takatsuki, T. Nose, T. Shinmyozu, Y. Shimohigashi. *Toxicol. Lett.* **118**, 1–8 (2000).
51. G. D. Charles, M. J. Bartels, C. Gennings, T. R. Zacharewski, N. L. Freshour, B. Bhaskar Gollapudi, E. W. Carney. *Reprod. Toxicol.* **14**, 207–216 (2000).
52. D. Kupfer, W. H. Bulger, A. D. Theoharides. *Chem. Res. Toxicol.* **3**, 8–16 (1990).

53. D. Schlenk, D. M. Stresser, J. C. McCants, A. C. Nimrod, W. H. Benson. *Toxicol. Appl. Pharmacol.* **145**, 349–356 (1997).
54. S. S. Dehal and D. Kupfer. *Drug Metab. Dispos.* **22**, 937–946 (1994).
55. K. Sumida, N. Ooe, H. Nagahori, K. Saito, N. Isobe, H. Kaneko, I. Nakatsuka. *Biochem. Biophys. Res. Commun.* **280**, 85–91 (2001).
56. K. M. Waters, S. Safe, K. W. Gaido. *Toxicol. Sci.* **63**, 47–56 (2001).
57. K. W. Gaido, S. C. Maness, D. P. McDonnell, S. S. Dehal, D. Kupfer, S. Safe. *Mol. Pharmacol.* **58**, 852–858 (2000).
58. J. Matthews, T. Celius, R. Halgren, T. Zacharewski. *J. Steroid Biochem. Mol. Biol.* **74**, 223–34 (2000).
59. J. C. Eldridge, M. K. Tennant, L. T. Wetzel, C. B. Breckenridge, J. T. Stevens. *Environ. Health Perspect.* **102** (Suppl 11), 29–36 (1994).
60. C. Wiegand, E. Krause, C. Steinberg, S. Pflugmacher. *Ecotoxicol. Environ. Saf.* **49**, 199–205 (2001).
61. J. T. Stevens, C. B. Breckenridge, L. Wetzel. *J. Toxicol. Environ. Health A* **56**, 69–109 (1999).
62. R. L. Cooper, T. E. Stoker, L. Tyrey, J. M. Goldman, W. K. McElroy. *Toxicol. Sci.* **53**, 297–307 (2000).
63. N. Hanioka, H. Jinno, T. Tanaka-Kagawa, T. Nishimura, M. Ando. *Xenobiotica* **29**, 1213–1226 (1999).
64. N. Hanioka, H. Jinno, T. Tanaka-Kagawa, T. Nishimura, M. Ando. *Toxicol. Appl. Pharmacol.* **156**, 195–205 (1999).
65. D. Lang, D. Criegee, A. Grothusen, R. W. Saalfrank, R. H. Bocker. *Drug Metab. Dispos.* **24**, 859–865 (1996).
66. D. H. Lang, A. E. Rettie, R. H. Bocker. *Chem. Res. Toxicol.* **10**, 1037–1044 (1997).
67. S. U. Khan and T. S. Foster. *J. Agric. Food Chem.* **24**, 768–771 (1976).
68. K. R. Solomon, D. B. Baker, R. P. Richards, K. R. Dixon, S. J. Klaine, T. W. La Point, R. J. Kendall, C. P. Weiskopf, J. M. Giddings, J. P. Giesy, W. J. J. Hall, W. M. Williams. *Environ. Toxicol. Chem.* **15**, 31–76 (1996).
69. N. Hanioka, H. Jinno, K. Kitazawa, T. Tanaka-Kagawa, T. Nishimura, M. Ando, K. Ogawa. *Chem. Biol. Interact.* **116**, 181–198 (1998).
70. C. Wiegand, S. Pflugmacher, M. Giese, H. Frank, C. Steinberg. *Ecotoxicol. Environ. Saf.* **45**, 122–131 (2000).
71. M. K. Tennant, D. S. Hill, J. C. Eldridge, L. T. Wetzel, C. B. Breckenridge, J. T. Stevens. *J. Toxicol. Environ. Health.* **43**, 197–211 (1994).
72. M. K. Tennant, D. S. Hill, J. C. Eldridge, L. T. Wetzel, C. B. Breckenridge, J. T. Stevens. *J. Toxicol. Environ. Health.* **43**, 183–196 (1994).
73. J. T. Sanderson, R. J. Letcher, M. Heneweer, J. P. Giesy, M. van den Berg. *Environ. Health Perspect.* **109**, 1027–1031 (2001).
74. (a) J. T. Sanderson, W. Seinen, J. P. Giesy, M. van den Berg. *Toxicol. Sci.* **54**, 121–127 (2000); (b) W. H. Bulger, V. J. Feil, D. Kupfer. *Mol. Pharmacol.* **27**, 115–124 (1985); (c) C. Wong, W. R. Kelce, M. Sar, E. M. Wilson. *J. Biol. Chem.* **270**, 19998–20003 (1995); (d) E. Baatrup and M. Junge. *Environ. Health Perspect.* **109**, 1063–1070 (2001).
75. K. Fent. *Toxicol. in Vitro* **15**, 477–488 (2001).
76. D. J. Ecobichon. In *Casarett & Doull's Toxicology: The Basic Science of Poisons*, C. D. Klaassen (Ed.), pp. 643–689, McGraw-Hill, New York (1995).
77. V. J. Kramer and J. P. Giesy. *Sci. Total Environ.* **233**, 141–161 (1999).
78. E. J. Clark, D. O. Norris, R. E. Jones. *Gen. Comp. Endocrinol.* **109**, 94–105 (1998).
79. W. R. Kelce, C. R. Stone, S. C. Laws, L. E. Gray, J. A. Kemppainen, E. M. Wilson. *Nature* **375**, 581–585 (1995).
80. P. Sohoni and J. P. Sumpter. *J. Endocrinol.* **158**, 327–339 (1998).

81. E. Baatrup and M. Junge. *Environ. Health Perspect.* **109**, 1063–1070 (2001).
82. A. Brouwer, U. G. Ahlborg, M. Van Den Berg, L. S. Birnbaum, E. R. Boersma, B. Bosveld, M. S. Denison, L. E. Gray, L. Hagmar, E. Holene, M. Huisman, S. W. Jacobson, J. L. Jacobson, C. Koopman-Esseboom, J. G. Koppe, B. M. Kulig, D. C. Morse, G. Muckle, R. E. Peterson. *Eur. J. Pharmacol., Environ. Toxicol. Pharmacol. Sect.* **293**, 1–40 (1995).
83. V. J. Kramer, W. G. Helferich, A. Bergman, E. Klasson-Wehler, J. P. Giesy. *Toxicol. Appl. Pharmacol.* **144**, 363–376 (1997).
84. D. D. Vakharia and J. F. Gierthy. *Toxicol. Lett.* **114**, 55–65 (2000).
85. C. E. Garner, W. N. Jefferson, L. T. Burka, H. B. Matthews, R. R. Newbold. *Toxicol. Appl. Pharmacol.* **154**, 188–197 (1999).
86. G. Mehrtens and F. Laturnus. *Chemosphere* **38**, 2995–3002 (1999).
87. R. Lawung, V. Prachayasittikul, L. Bulow. *Protein Expr. Purif.* **23**, 151–158 (2001).
88. R. J. Letcher, R. J. Norstrom, A. Bergman. *Sci. Total Environ.* **160–161**, 409–420 (1995).
89. H. M. Stapleton, R. J. Letcher, J. E. Baker. *Environ. Sci. Technol.* **35**, 4747–4752 (2001).
90. K. Wiberg, R. Letcher, C. Sandau, J. Duffe, R. Norstrom, P. Haglund, T. Bidleman. *Anal. Chem.* **70**, 3845–3852 (1998).
91. K. Noren, A. Lunden, E. Pettersson, A. Bergman. *Environ. Health Perspect.* **104**, 766–772 (1996).
92. K. Noren, C. Weistrand, F. Karpe. *Arch. Environ. Contam. Toxicol.* **37**, 408–414 (1999).
93. Y. Kato, K. Haraguchi, K. Tomiyasu, Hiroyuki Saito, M. Isogai, Y. Masuda, R. Kimura. *Environ. Toxicol. Pharmacol.* **3**, 137–144 (1997).
94. Y. Kato, K. Haraguchi, T. Shibahara, Y. Masuda, R. Kimura. *Arch. Toxicol.* **72**, 541–544 (1998).
95. Y. Kato, K. Haraguchi, T. Shibahara, S. Yumoto, Y. Masuda, R. Kimura. *Toxicol. Sci.* **48**, 51–54 (1999).
96. B.-O. Lund, J. Orberg, A. Bergman, C. Larsson, A. Bergman, M. Backlin B.-H. Hakansson, A. Madej, A. Brouwer, B. Brunstrom. *Environ. Toxicol. Chem.* **18**, 292–298 (1999).
97. Y. Kato, K. Haraguchi, T. Shibahara, Y. Shinmura, Y. Masuda, R. Kimura. *Chem.-Biol. Interact.* **125**, 107–115 (2000).
98. M. Johansson, C. Larsson, A. Bergman, B.-O. Lund. *Pharmacol. Toxicol.* **83**, 225–230 (1998).
99. M. Johansson, S. Nilsson, B. O. Lund. *Environ. Health Perspect.* **106**, 769–772 (1998).
100. B.-O. Lund. *Environ. Toxicol. Chem.* **13**, 911–917 (1994).

APPENDIX

Table 1 Half-lives of some endocrine active substances in air, water, and soil[a].

Chemicals	Environments		
	Air	Water	Soil/sediments
2,3,7,8-TCDD	1.2–9.6 h		1.5–10 years (O)[b]
2,3,7,8-TCDF	2.1–11.5 h		–20 years (O)
2,3,3′,4,4′-PCB	8–80 days	>56 days	0.91–7.25 years (O)
p,p′-DDT	<3 days (P)[b]		45–420 days (O), 7–45 days (R)[b]
Gamma-HCH			1–10 months (O), 10–20 days (R)
Chlodane	12 h–5 days	239 days	0.4–8 years (O)
Benzo(b)fluoranthene	3.4 h–1.4 days	>100 days	14.2 years–87 days (O)
Benzo(a)pyrene	2.4 h	5.4–17.3 years	14.6 years–151 days (O)
Benzo(g,h,I)perylene	0.31–10.0 h	>100 days	1.8 years–173 days (O)
Tributyltin		3–20 days	1–6 months (O), >2 years (R)
Triphenyltin		180 days	Longer than tributyltins
Bisphenol A	0.74–7.4 h (P)	2.5–5 days	
Nonylphenols	10–15 h (P)	2.5–20 days	10–104 days (O), >60 years (R)
4-*tert*-Octylphenol		7–50 days	>60 years (R)
Ethinylestradiol	10 days (P)	20–40 days	Much longer than estradiol
17β-Estradiol	10 days (P)	0.2–9 days	0.11 days (O), 0.37–0.66 days (R)
Estrone		0.2–9 days	0.4 days (O), 11–14 days (R)

[a]It should be noted that half-lives shown here can be varied widely due to the environmental conditions, such as light intensity, temperature, initial concentration of compounds, soil, and water properties, etc., although representative data was collected from the references.

[b]Symbols, (P) denotes photolysis (not always in air), (O) aerobic degradation, (R) anaerobic degradation.

TABLE 1 REFERENCES

- M. Ahel, F. E. Scully Jr. et al. *Chemosphere* **28**, 1361–1368 (1994).
- I. Angelidaki, A. S. Mogensen et al. *Biodegradation* **11**, 377–383 (2000).
- R. Ekelund et al. *Environ. Pollut.* **79**, 59–61 (1993).
- C. A. Staples, P. B. Dorn et al. *Chemosphere* **36**, 2149–2173 (1998).
- S. Kuwatsuka. In *Pesticides: Design and Development Guide* (in Japanese), I. Yamamoto and J. Fukami (Eds.), p. 1222, Softscience, Tokyo (1979).
- J. Kanazawa. In *Environmental Science of Pesticides* (in Japanese), Godo Shuppan, Tokyo (1992).
- S. Prakash, G. S. Tandon et al. *Biochem. Biophys. Res. Commun.* **199**, 1284–1288 (1994).
- D. Y. Shang, R. W. Macdonald et al. *Environ. Sci. Technol.* **33**, 1366–1372 (1999).
- C. A. Staples, J. B. Williams et al. *Chemosphere* **38**, 2029–2039 (1999).
- U.S. Environmental Protection Agency. Persistent bioaccumulative toxic (PBT) chemicals; Final rule, Federal Register, 29 Oct. 1999, pp. 58665–58753 (1999).
- M. Hesselsoe, D. Jensen et al. *Environ. Sci. Technol.* **35**, 3695–3700 (2001).
- C. A. Staples, C. G. Naylor et al. *Environ. Sci. Technol.* **20**, 2450–2455 (2001).
- M. D. Jurgens, K. I. E. Holthaus et al. *Environ. Sci. Technol.* **21**, 480–488 (2002).

Table 2 Selected results of Japanese 1998–1999 survey of EDs in the environment.

EDs	Environmental samples	Sample numbers	Detected numbers	Concentration range
PCDDs + PCDFs + coplanar PCB	Air	100	100	0.0017–0.70 pg TEQ/m^3
	Surface water	204	204	0.0014–13 pg TEQ/l
	Ground water	188	–	ND-5.4 pg TEQ/l
	Sediments	205	–	ND-260 pg TEQ/kg-dry
	Soil	286	286	0.0015–61 pg TEQ/kg
	Fish & aquatic invertebrates	368	368	0.0022–30 pg TEQ/kg
	Carp	48	48	0.20–5.9 pg TEQ/kg
	Frogs	80	80	0.20–7.5 pg TEQ/kg
	Pigeons	8	8	0.99–10 pgTEQ/kg
	Kites	20	20	22–220 pg TEQ/kg
	Predators (mainly owls)	9	9	14–530 pg TEQ/kg
	Whales	22	22	1.3–200 pg TEQ/kg
	Seals	13	13	8.6–27 pg TEQ/kg
	Apodemus speciosus	37	37	0.52–120 pg TEQ/kg
	Japanese monkey	6	6	0.82–9.4 pg TEQ/kg
	Bears	6	6	0.18–1.1 pg TEQ/kg
	Raccoon dogs	5	5	13–100 pg TEQ/kg
Tributyltin	Surface water	170	23	ND(<0.002)–0.008 ug/l
	Sediments	48	44	ND(<0.2)–170 ug/kg-dry
	Soil	7	0	ND(<20) ug/kg
	Fish	141	113	ND(<1)–120 ug/kg
	Carp	145	92	ND(<0.3)–75 ug/kg
	Whales	26	18	ND(<20–50)–330 ug/kg
	Seals	19	1	ND(<20–50)–110 ug/kg
	Pigeon	31	0	ND(<200) ug/kg
	Kites	26	2	ND(2–200)–8 ug/kg
	Owls	5	0	ND(<2) ug/kg
	Predators	30	0	ND(<200) ug/kg
	Wild mice	30	0	ND(<200) ug/kg
	Rhesus monkey	41	0	ND(<200) ug/kg
	Bears	17	0	ND(<50–200) ug/kg
	Raccoon dogs	15	0	ND(<50–200) ug/kg
Nonylphenol	Surface water	431	80	ND(0.1)–4.6 ug/l
	Sediments	51	23	ND(<0.1)–2700 ug/kg-dry
	Soil	94	0	ND(<50) ug/kg
	Fish	141	42	ND(<15)–780 ug/kg
	Carp	145	0	ND(<50) ug/kg
	Pigeons	31	16	ND(<15)–113 ug/kg
	Apodemus speciosus	30	22	ND(<15)–190 ug/kg
	Raccoon dogs	15	14	ND(<15)–2000 ug/kg
17β-estradiol	Water	130	79	ND(<0.001)–0.035 ug/l
	Sediments	20	19	ND(<0.001)–0.3 ug/kg-dry
Ethinylestradiol	Water	197	2	ND(<0.0001)–0.0002 ug/l
	Sediments	20	4	ND(<0.01)–0.34 ug/kg-dry

Data was obtained from the following Japanese Ministry of Environment publications (in Japanese):

- *Advanced Report of Survey on Environmental Active Substances in Water* (1999)
- *Results of Survey on Accumulation of Dioxins in Wildlife* (1999)
- *Results of Survey on Effects of Endocrine Active Substances on Wildlife* (1999)
- *Results of 1999 Survey of Chemical Emission into the Environment* (2000)
- *Results of Survey on Dioxin-Related Compounds in Arable Soils and Crops* (2000)

Pure Appl. Chem., Vol. 75, Nos. 11–12, pp. 1933–1948, 2003.

Topic 2.8

Critical factors in exposure modeling of endocrine active substances*

Pim de Voogt[1,‡] and Bert van Hattum[2]

[1]Environmental and Toxicological Chemistry, Institute for Biodiversity and Ecosystem Dynamics, University of Amsterdam, Nieuwe Achtergracht 166, 1018 WV Amsterdam, The Netherlands; [2]Institute for Environmental Studies, Vrije Universiteit, De Boelelaan 1087, 1081 HV, Amsterdam, The Netherlands

Abstract: Multimedia transport, partitioning, and degradation pathways are key processes in the probability of a substance to interact with target organisms. Biotic factors such as toxicokinetics, biotransformation capacity, and behavioral and life-cycle aspects of the organisms are determinants for final concentrations at target organs. The role of metabolites in endocrine disruption can be quite different from those of the parent compounds, and often this requires separate toxicokinetic evaluation.

The exposure assessment of endocrine active substances (EASs) suffers from a huge lack of reliable data, of both values that are used as input parameters in exposure models, and field data that are needed for validation purposes. In general, for the more classic EASs, such as PCBs, *p,p′*-DDE, chlorinated dioxins, some pesticides, and organotins, reliable data are sufficiently available, but careful evaluation of the quality of databases is necessary. Several data quality evaluation systems have been proposed. For the "newer" compounds, only few data have been gathered so far. The latter compounds include alkylphenols, bisphenol A, brominated flame retardants, phytoestrogens, and in particular natural and synthetic hormones, which in view of their high estrogenic potency could be the most important compounds in terms of risk.

The suitability of current exposure assessment models for EASs at this moment seems to be restricted to the persistent compounds such as PCBs, PCDDs, and PCDFs. Especially for the compounds subject to biodegradation and biotransformation, the lack of experimental data to derive model-input parameters and perform validation studies at this moment is one of the main obstacles for the further application of generic exposure models to other EASs. Most of the current models do not allow life stage-specific predictions.

Although the mechanisms of endocrine disruption involve different types of action, the principle of additivity, based on the equivalent toxicity approach (using estrogen equivalent potencies relative to 17ß-estradiol) seems promising for the design of integrated exposure and effect models for EASs.

Research programs aimed at the endocrine disruption issue must focus on promoting experimental studies for generation of reliable, high-quality parameter data on the one hand, and surveys or monitoring campaigns for collection of representative field data on the other. The non-specificity of possible effects caused through endocrine mechanisms implies that in

*Report from a SCOPE/IUPAC project: Implication of Endocrine Active Substances for Human and Wildlife (J. Miyamoto and J. Burger, editors). Other reports are published in this issue, *Pure Appl. Chem.* **75**, 1617–2615 (2003).

‡Corresponding author

order to reveal dose–response relationships all potentially active agents, or at least as many as feasible, need be included in the risk assessments. Current regulatory monitoring programs should further be evaluated and harmonized with validation requirements of models used in exposure assessment.

INTRODUCTION

Exposure is a result of the emission of a chemical into the environment and its subsequent fate. Both emission and fate depend on many factors, which have been discussed extensively by many authors (see, e.g., [1,2]).

Abiotic factors include the amount of substance involved, its persistence, mobility and availability, and the duration of exposure. Multimedia transport, partitioning, and degradation pathways are key processes in the probability of a substance to interact with target organisms. Biotic factors include toxicokinetics, physiological, behavioral and life-cycle aspects of the organisms, predator–prey relationships, and biotransformation capacity. Dietary exposure is an important aspect, in particular for the exposure assessment of birds, mammals, and humans. The role of metabolites in endocrine disruption can be quite different from those of the parent compounds, and often this requires separate toxicokinetic evaluation. While these abiotic and biotic factors by no means are specific to the biological action of EASs, organisms in their normal development travel through various distinct stages of sensitivity accompanied or even directed by hormonal control. Examples are early life stages, and more specifically development of gonads, or egg maturation; many aspects of reproduction, such as (temperature-dependent) gender determination (in amphibians); and migration. Exposure to endocrine disrupters during such stages can result in highly specific effects. Therefore, exposure characterization needs to take into account such stages, e.g., through measuring or predicting the dose in target tissues. Measuring actual levels and tissue residues has improved our understanding of the relationships between external and internal doses. Although the mechanisms of endocrine disruption involve different types of action, the principle of additivity has been shown to hold in estrogenic activity.

The purpose of the present chapter is to highlight some critical aspects in environmental exposure assessment of endocrine disruptors and to indicate which type of research is needed to fill existing gaps in this field. The focus of this contribution is on the aquatic environment.

ABIOTIC FATE PROCESSES

The principle environmental processes that govern the abiotic fate of chemicals once brought into the environment include transport, distribution, and transformation processes.

Key parameters in environmental fate modeling therefore relate to these processes and include two types of parameters. System-dependent parameters include temperature, advection rates (e.g., wind speed, air trajectories, water currents, particle deposition rates in air or water), acidity, organic carbon contents, and particle size. Substance-related parameters include partition coefficients (e.g., sorption K_d, Henry's Law coefficients), mass transfer and diffusion rates, persistence, transformation rates (photolysis, hydrolysis, etc.), and speciation.

In general the variability in system related parameter values appears to be much lower than that in substance-related parameters. This is due in part because the former are more easy to measure (e.g., acidity, temperature, wind speed, compared to K_d or H) and also because the latter may depend on system parameters themselves. In other words, the conditions under which the substance parameters may have been determined may vary from experiment to experiment, and therefore such data have a larger intrinsic variability.

BIOTIC FACTORS

Residue levels of contaminants in organisms are a result of a variety of processes, the most important of which are the partitioning between and within biotic and abiotic compartments, and simultaneous transformation reactions such as biodegradation and biotransformation. Order of magnitude differences are present between individual compounds in both physicochemical characteristics (water solubility, volatility, hydrophobicity, photodegradation) and biological parameters and activity, e.g., microbial degradability, toxicokinetics, biotransformation, and endocrine-disrupting potency.

Uptake of compounds may take place from aqueous systems, via gills or the skin, and from dietary sources via the gastrointestinal tract. Epibenthic and sediment-inhabiting invertebrates may have additional uptake from ingested sediments or from porewater.

In aquatic environments, direct aqueous uptake of compounds (bioconcentration) seems to be dominant in most invertebrates and fish for compounds with a log $K_{ow} < 4$ [3].

The fraction of a contaminant concentration that is available for uptake by aquatic organisms (i.e., the bioavailable fraction) varies between species and depends on the relative significance of different uptake pathways [4]. Following uptake, compounds may be subject to biotransformation, to internal distribution among tissues and organs, and to elimination of parent compounds or biotransformation products. Within organisms, probably the role of metabolism is the most essential in homeostasis [2]. Although generally metabolism leads to degradation, detoxification, and elimination, products may be generated with enhanced endocrine potential. Especially in vertebrates the biotransformation route is more predominant in comparison to invertebrates and phytoplankton [5]. Elimination may be the result of redistribution at respiratory surfaces or in the gastrointestinal tract or via excretion products. Especially the ability of biotransformation of EASs seems to be of prime importance in explaining variability in tissue residues among different species in a specific habitat. For natural and synthetic hormones, alkylphenols and phthalates, only limited information is available on biotransformation in aquatic food chains [6].

DATA AVAILABILITY FOR ENVIRONMENTAL MODELING

When considering the availability of reliable data on physicochemical and environmental properties for EASs, there is a large discrepancy between the compounds previously included in priority lists, such as, e.g., PCBs, PCDDs, PCDFs, phthalates and organotin compounds, and compounds that were identified after the problem of endocrine disruption was acknowledged as an environmental issue. For many of the "old" compounds experimentally determined values are available, while for the "new" environmental agents, such as alkylphenol ethoxylates (APEOs), natural and synthetic hormones, phytoestrogens, and brominated compounds, only limited data are available.

For a preliminary exposure assessment of these compounds, quantitative structure activity relationship (QSAR)-based predictions (see, e.g., [7–9]) of the main environmental properties usually are applied, in order to allow a first screening of chemical fate and hazards with risk-assessment models, such as, e.g., EUSES [10] in the European regulatory context or the EPIWIN suite of models applied and distributed by the U.S. EPA [11].

Examples of important databases of environmental properties of compounds are the well-known publications of Mackay et al. [12], the Environmental Fate database of Syracuse [13], or the EINECS database of the European Commission [14]. These databases contain evaluated data from experimental studies and predicted values from QSAR studies. When possible, preference should be given to validated data from experimental studies.

As an example to illustrate to which extent experimental data are available, a summary is given in Tables 1 and 2 of available experimental data for 31 EASs in the Syracuse EFDB database and predicted values with the software in the EPIWIN program [11]. As can be seen for the physicochemical data, experimental data are available for 28 % (BP) to 59 % (solubility) of the compounds. Measured

Table 1 Summary of measured (bold) physicochemical data and environmental properties of EASs in the EFDB database and QSAR-based predictions with EPIWIN.*

Substance	CAS no.	Mol wt	BP °C	MP °C	VP mm Hg	Log K_{ow}	S mg/l	HLC P.m^3/mol	Log K_{oc}	Log BCF experimental	Log BCF predicted
4-Nonylphenol (branched)	84852-15-3	220.36	**295**	89.94	**9.42E-05**	5.92	**5000**	**4.22E-01**	4.708	**1.1**	3.85
Nonylphenol (linear)	25154-52-3	220.36	**293**	**42**	**9.42E-05**	**5.76**	**6.35**	**3.33E+00**	4.785		2.74
4-Nonylphenol (linear)	104-40-5	220.36	**293**	**42**	**9.42E-05**	**5.76**	**6.35**	**3.33E+00**	4.785		2.74
Bisphenol A	80-05-7	228.29	363.54	**153**	2.27E-07	**3.32**	**120**	8.98E-07	4.876	**1.0–2.0**	1.86
Tributyltin	56573-85-4	325.51	273.72	**–19**	1.30E-02	**4.76**	0.7478		4.176		3.37
Tributylstannane	688-73-3	291.07	250	28.89	3.99E-02	7.35	0.007302	1.49E+05	3.998		4.23
Tributyltin hydroxide	1067-97-6	307.07	318.06	70.99	1.55E-05	**4.09**	3.589	8.74E+01	4.176		2.85
Triphenyltin	668-34-8	351.04	358.52	94.01	1.78E-05	6.58	0.01464	7.69E+00	5.648		5.77
Estradiol (beta)	50-28-2	272.39	395.47	**221.5**	1.99E-09	**4.01**	**3.9**	3.57E-06	4.205		2.39
Estradiol (alpha)	57-91-0	272.39	395.47	**221.5**	1.99E-09	**4.01**	**3.9**	3.57E-06	4.205		2.39
Estrone	53-16-7	270.37	**154**	**260.2**	5.09E-03	**3.13**	**30**	3.72E-05	4.477		1.71
Ethynyl estradiol	57-63-6	296.41	411.21	**183**	1.95E-09	**3.67**	**11.3**	7.78E-07	4.678		2.13
DBP Dibutyl phthalate	84-74-2	278.35	**340**	**–35**	**2.01E-05**	**4.5**	**11.2**	**1.78E-01**	3.164	**1.1–3..8**	2.76
DEHP di-*sec*-octyl phthalate	117-81-7	390.57	**384**	**–55**	**1.42E-07**	**7.6**	**0.27**	**2.65E-02**	5.219	**2.1–4.1**	2.49
Di-*n*-propyl phthalate	131-16-8	250.3	**317.5**	**18.19**	**1.32E-04**	**3.27**	**108**	**3.95E-02**	2.633		1.82
Diethyl phthalate	84-66-2	222.24	**295**	**–40.5**	**2.10E-03**	**2.42**	**1080**	**5.98E-02**	2.101	**1.1–2.1**	1.16
Genistein	446-72-0	270.24	464.36	**301.5**	5.18E-12	2.84	257.7	5.02E-12	3.815		0.65
Formononetin	485-72-3	268.27	418.34	**256.5**	1.71E-09	3.11	157.3	2.18E-08	3.24		0.85
Dihydroxyisoflavone (daidzein)	486-66-8	254.24	429.52	**323**	2.77E-11	2.55	568.4	3.83E-11	3.597		0.42
Diosgenin	512-04-9	414.63	469.45	**205.5**	1.94E-11	6.34	**0.02**	1.56E-04	4.162		4.18
ß-Sitosterol	83-46-5	414.72	448.98	**140**	4.51E-10	9.65	4.63E-05	2.89E+01	6.656		1.18
Hexabromocyclododecane	25637-99-4	641.7	462.03	180.03	1.68E-08	7.74	2.09E-05	1.68E-01	5.07		3.79
Tetrabromo diphenylether	40088-47-9	485.8	405.51	161.73	2.41E-07	6.77	0.001461	2.91E-01	4.274		4.51
Decabromodiphenyl oxide	1163-19-5	959.17	**425**	**295**	2.01E-09	12.11	**0.025**	1.16E-03	5.611		*7.2*
Hexabromodiphenyl ether	36483-60-0	643.59	466.91	197.14	2.87E-09	8.55	4.15E-06	4.62E-02	4.702		2.69
Pentabromodiphenyl ether	32534-81-9	564.69	436.21	182.8	2.44E-08	7.66	7.86E-05	1.16E-01	4.484		3.91
PCBs	1336-36-3	291.99	359.51	122.32	**8.63E-05**	**6.29**	**0.277**	**3.36E+01**	4.651		4.76
3,3′,4,4′-tetrachloro-1,1′-Biphenyl	32598-13-3	291.99	359.51	122.32	**1.64E-05**	**6.63**	**0.000569**	**9.22E-01**	4.651		5.03
3,3′,4,4′,5-pentachlorobiphenyl	57465-28-8	326.44	378.21	134.6	2.22E-06	6.98	0.009394	9.06E+00	4.87		5.29
3,3′,4,4′,5,5′-Hexachlorobiphenyl	32774-16-6	360.88	396.9	146.34	5.81E-07	**7.41**	**0.00051**	6.71E+00	5.088		4.87
2,3,7,8-TCDD	1746-01-6	321.98	379.17	**305**	**1.50E-09**	**6.8**	**0.0002**	**4.90E+00**	5.165		4.54
2,3,7,8-TCDF	51207-31-9	305.98	382.92	138.99	1.53E-06	**6.53**	**0.000692**	1.52E+00	4.908		4.33
Experimental values			28 %	56 %	31 %	56 %	59 %	31 %		15 %	

[a]Measured values as derived from the EFDB database [13]; predicted values derived with the EPIWIN suite [11].

Table 2 Biodegradation evaluation in the EFDB database and rating of reliability of data.

Substance	Screen test	Biol. treatment simulation	Grab sample soil	Grab sample water	Field test	Aerobic summary	Anaerobic summary	No. of references	EPIWIN predicted[a]
Nonylphenol (linear?)	BST-2	BST-3				BST-1		3	w
Bisphenol A	BS-2	BFA-3		BFA-1		BST-1		4	w-m
DBP Dibutyl phthalate	BF-1	BF-1	BFA-1	BF-1		BF-1	BF-1	21	d-w
DEHP di-*sec*-octyl phthalate	BST-1	BFA-1	BFA-1	BST-1		BFA-1	BSA-1	28	w
Diethyl phthalate	BF-1	BF-3		BFA-1		BFA-1	BST-1	9	w
2,3,7,8-TCDD	BSA-3		BSA-1			BSA-1	BSA-1	3	r

Source: EFDB database [13]; codes: BF = biodegradation at a fast rate, BFA = fast rate with acclimation, BS = slow rate, BSA = slow rate with acclimation, BST = biodegrades sometimes; reliability ratings: 1 = tested in 3 or more tests with consistent results, 2 = in two tests, or >2 interpretable tests with some confilicting data, 3 = only 1 test or uninterpretable conflicting data.
[a]Predicted with BIOWIN—ultimate survey model: w = weeks, m = months, r = recalcitrant.

BCF data derived form experimental studies are included in the EFDB database for only 15 % of the compounds. Very few data are available for the biodegradability of EDC compounds. For only 6 out of the 31 EASs, data from experimental studies are available, mainly for phthalates. Some of the studies were conflicting or uninterpretable. For none of the compounds information from field tests on biodegradation under natural conditions was available in the EFDB database.

As an illustration of the large variation in physicochemical characteristics among EASs, a frequency distribution was made of the wide range of hydrophobicity (Fig. 1). The frequency distribution shows two maximums: one at log K_{ow} = 3–5, including the natural and synthetic hormones and phytoestrogens, and a second at log K_{ow} = 7–8 including the polyhalogenated aromatic compounds (PHAHs).

The wide range in hydrophobicity (K_{ow}) values and those of several other properties, such as the Henry's law constant (H), implies large differences in environmental behavior and chemical fate. This is reflected in the results of the outcome of preliminary calculations with a screening model (level III, based on the approach of Mackay et al. [15]), included in the EPIWIN programme. In the level III approach, used for this preliminary screening, the distribution over the media water, soil, air, and sediment

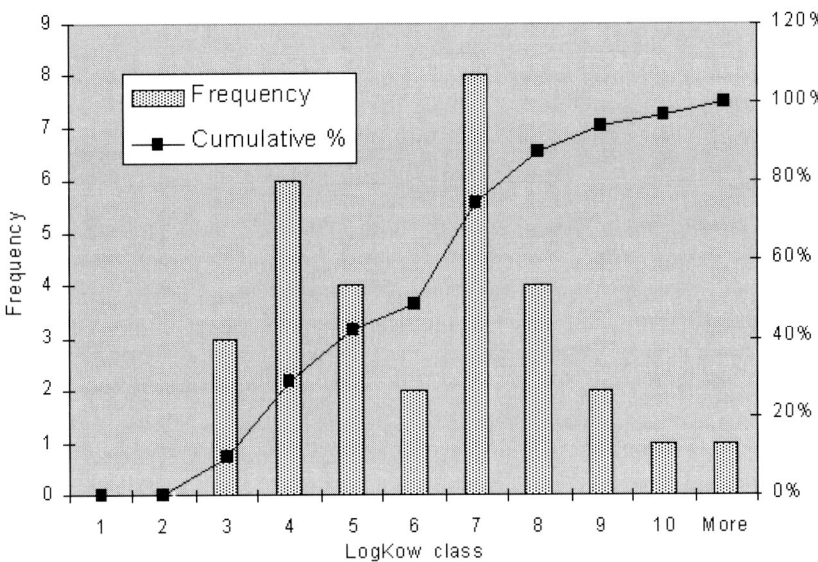

Fig. 1 Distribution of predicted log K_{ow} value classes for endocrine-disrupting compounds.

is calculated, based on equal emissions to air, water, and soil. The results are indicated in Fig. 2. Sediments and soil are identified as major sinks (total 50–95 %) in the level III model, especially for the PHAH compounds (>90 %). The water phase is an important compartment for the phenols (10–18 %), the organotins (5–30 %), natural and synthetic hormones (2–20 %), and the phthalates (10–40 %). The air compartment is only significant for some of the phthalates and TBT.

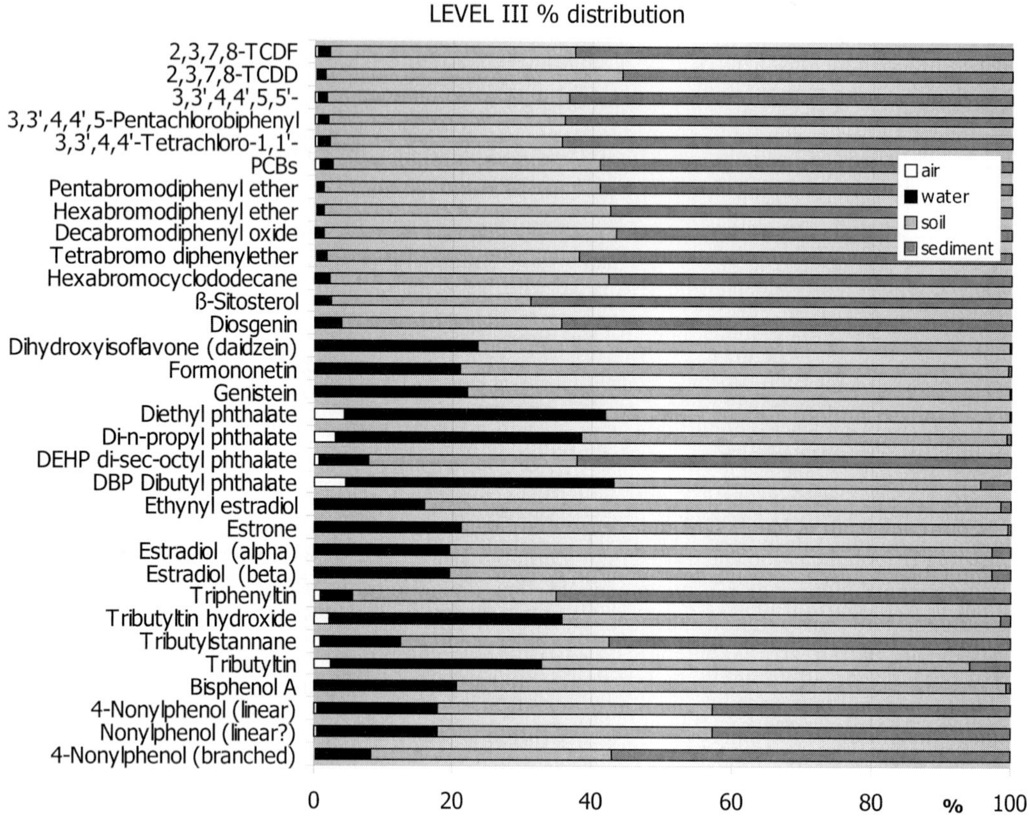

Fig. 2 Predicted environmental fate of EASs in the level III model included in EPIWIN.

DATA QUALITY

The quality and reliability of physicochemical properties data is a critical aspect in chemical fate modeling and hazard and risk assessment. Whenever a compilation of such data is made from literature reports, large discrepancies in experimental values for properties become evident. While it is obvious that different test methods used to obtain a certain physicochemical compound property will result in different outcomes, well-conducted experiments should result in values within ranges of statistical method variability. As indicated above, apart from experimental data-generating studies, (mathematical) estimation methods (e.g., QSARs) for properties can be used for obtaining data.

Kollig [16,17] and Klimisch et al. [18] have developed indicators for data quality evaluation of experimental studies. Kollig distinguished four categories of criteria: (1) analytical information; (2) experimental information; (3) statistical information; and (4) corroborative information. Each category contained subcriteria that were developed for various properties to make it possible to estimate the reliability of the measurement within one category. The data reliability indicator (DRI) consists of the relative reliability for all four categories. Klimisch et al. [18] used four reliability scores for experimental data-generating studies: (1) reliable without restrictions; (2) reliable with restrictions; (3) not reliable; and (4) not assignable.

In a Dutch database of physicochemical properties, a modification of the Kollig approach was used to evaluate experimental as well as estimated (QSAR) data. Values were selected on the basis of three sets of general criteria [19]. For experimental studies, analytical, methodological, and statistical aspects, respectively, of the study were scored. Analogously, for QSAR studies descriptors, methodology and statistics were evaluated.

The EFDB database (which is included in the EPIWIN suite of models) contains three categories of reliability for evaluating tests: these are based on the number of times independent tests were carried out for a certain property, combined with agreement of results within that number [13]. Especially the evaluation of biodegradability data is complex, as is illustrated by the example in Table 2. The evaluation criteria used for the quality assessment of data in the European EINECS database are described in the technical guidance documents [8].

Whereas the Kollig method does not yield a final judgement of reliability, the Dutch database study has modified the Kollig procedure to provide a reliability score. The Klimisch method does provide a final judgement, but it does not give a detailed set of criteria. Finally the EFDB and EINECS categories are qualitative rather than quantitative.

MODELS

A large number of mathematical models (rate constant-, clearance-, fugacity-, physiological-, and pharmacokinetics-based) has been applied to laboratory bioconcentration and bioaccumulation studies [20,21] and food chain transfer in the field situation [22,23]. In Table 3, we have listed a selection of models and their mean features, as described in recent literature. The focus is on generic screening and aquatic models. The models range from simple screening models (EUSES, EQC level I III), which include only abiotic compartments, and are mainly used to identify the environmental compartment of concern (see example calculated with EPIWIN software shown in Fig. 2), to more elaborate multicompartment chemical fate models and biotic models including food chain transfer.

Table 4 provides an overview (noncomprehensive) of the main compounds tested or validated with the different models. With respect to compounds with ED potency, most of the information is available for di- and mono-*ortho*-substituted PCBs and dioxins. In most of the validation studies the non-*ortho*-substituted PCBs with ED potency were not addressed. In two studies, TBT was considered. Natural or synthetic estrogens were addressed only in the study of Lai and coworkers [6]. Predicted concentrations were 2 to 3 orders of magnitude below reported concentrations from experimental studies. The authors attributed this to the lack of reliable data on biodegradation rates in sediments and biotransformation rates in biota for a proper estimation of model input-parameters.

No studies could be identified, in which other than screening level models were applied to the compound categories of phthalates and brominated compounds.

With respect to the modeling of EASs, it is important to recognize that organisms in their normal development travel through various distinct stages of sensitivity accompanied or even directed by hormonal control. Examples are early life stages, and more specifically development of gonads, or egg maturation; many aspects of reproduction, such as (temperature-dependent) gender determination (in amphibians); and migration. Exposure to endocrine disruptors during such stages can result in highly specific effects. Therefore, exposure models for aquatic organisms should take into account such stages, e.g., through measuring or predicting the dose in target tissues.

Table 3 Summary and main features of recent environmental exposure models.

	C/f	Nr. of abitoic media	Dimensions	Steady state/ dynamics	Hydro dynamics	No. of trophic levels	Generic	Validated	Skills required	Availability of model	Refs.
Screening models											
EUSES	c	4	1D	s	–		g		low	+	[10]
EQC-based models levels I-III	f	>4	1D	s	–		g		low	++	[15,24]
Simplebox 2.0	c	4	1D	s	–		g		low	+	[25]
Chemical fate models											
QWASI	f	>4	2D	s/d	+		l	+	high	+	[26]
ECOS	c	>4	3D	s/d	+		g	+	high	+	[27]
DELWAQ	c	>3	3D	s/d	+		g	+	high	+	[28]
EXAMS	c	>4	3D	s/d	+		g	+	high	+–	[29]
Models including food web											
Thomann model	c	3	1D	s	–	4	g	+	high	–	[30]
TOXFATE	f	>4	2D	s/d	+	3	l	+	high	–	[31]
EMC food web model	f	3	1D	s	–	4	g	+	high	++	[6,32]
ECOFATE	f	3	1D	s/d	–	4	g	+	high	++	[33,34]
TBT freshwater	c	3	1D	s	–	4	l	+	high	–	[35]

Explanation: C/f: concentration or fugacity-based model; No. of media: number of abiotic main and subcompartments included (water column, sediment, air, soil, particulate matter); Hydrodynamics: ability to cope with more complex marine hydrodynamic features in estuarine enviornments; Generic: generic model (g) or location specific (l); Availability: ++ freeware downloadable, + commercially available, – not easily available

Although some examples exist of age specific models, e.g., for PCB biomagnification in the food web in the Seine [36], life-stage specific predictions are not possible in most of the current models.

Table 4 Overview of compounds tested or validated in chemical fate or food web models.

Model	Compounds tested or validated	Refs.
Screening models		
EQC-based models levels I–III	Chlorobenzenes, LAS	[15]
Simplebox 2.0	Triazines, BTEX, vinyl chloride, TCA, TCE	[25]
Chemical fate models		
QWASI	(v) PCBs, PAHs, antifoulants	[26,37]
ECOS	(t) Cd, PAHs, APEO, pesticides	[27,38,39]
DELWAQ	(t) Trace metals, PCBs	[28]
EXAMS	(t) PCBs, TBT, Seanine	[40]
Models including food web		
Thomann model	(v) PCBs, chlorinated pesticides, PAHs	[3,4,30]
TOXFATE	(v) PCBs, Mirex	[31]
EMC food web model	(v) PCBs, dioxins, (t) natural and synthetic estrogens	[6,32]
ECOFATE	(v) PCBs, dioxins, chlorinated pesticides	[33,34]
TBT Freshwater	(v) TBT, TPT	[35,41]

(t) Model tested; (v) model validated with reasonable match between predicted and measured values.

RISK ASSESSMENT

Environmental risk assessment of substances is based on an evaluation of exposure pathways and concentrations on the one hand and identification and selection of sensitive endpoints on the other. The concept is operationalized by comparing real or estimated (predicted) exposure concentrations (PECs) with calculated no-effect concentrations (NECs). The basic approach is similar in European and North American legislation, and has been adopted by industry [42,43]. The comparison can be implemented by calculating the quotient of exposure and NEC. If the quotient is less than, e.g., one, then the substance poses no significant risk to the environment. If the quotient is greater than, e.g. one, the substance may pose a risk, and further action is required, e.g., a more thorough analysis of probability and magnitude of effects will be carried out. Other thresholds than unity may be defined. The difference between the actual ratio and a chosen threshold is often referred to as the margin of safety. The principle outlined above assumes that threshold doses exist for endocrine-disrupting compounds, an assumption that has been questioned recently [44].

Miyamoto and Klein [2] have reviewed the risk assessment procedure and its pitfalls in particular for endocrine disruptors. Critical aspects are the assessment factors used, the poor understanding of endocrine-disrupting mechanisms, the huge differences in potencies, and the conflict of high potency even at levels below analytical detection limits (as may occur for, e.g., the synthetic hormones). Despite these pitfalls and the limited database available for EDC, current risk assessment methods are believed to be valid for estrogenic agents [43].

For the derivation of the NEC, several approaches have been proposed. Generally, these can be categorized into three distinct assessments: a conservative, a distributional, and a mixture toxicity approach. In conservative approaches, usually the most (realistic) sensitive endpoint (e.g., LC_{50}, NOEC) known is taken and divided by an uncertainty factor (e.g., 10 or 100 or 1000). The uncertainty factor value selected depends on the type of endpoint and the number of available endpoint data, and is applied to account for laboratory to field extrapolations, species differences in sensitivities, and similar uncertainties.

In distributional approaches, a series of, or all available literature data are taken and a selected cut-off value is applied to the distribution of these data. The cut-off value may be, e.g., the concentration value that will protect 95 % of the species (tested) [45]. In general, again an uncertainty factor (usually of 10) is then applied to take into account species differences.

In the mixture toxicity approach, a similar mode of action is assumed for the assessment of the combined (additive) effect of different compounds present in a medium. All relevant mixture components are scaled relative to the most potent one. This results in relative potencies for each component. The total effect of the mixture is then calculated by summing the products of concentration and relative potency for each component.

Mixture toxicities

It is obvious that endocrine disruptors can occur in the environment as complex mixtures. In risk characterization studies the toxicity of individual constituents of such mixtures, whether assayed in acute and chronic toxicity, or in estrogenicity tests, is being considered to occur—for each separate endpoint—through the same separate mode of action, and consequently to be additive.

Thus, relative potencies can be established for each individual constituent of a mixture, referring to, e.g., the toxicity of a single compound. For example, for estrogens in general, this would be 17ß-estradiol. The relative potencies are often referred to as toxic equivalent factors (TEFs), analogous to the concept that has first been applied to the toxicity of chlorinated aromatic hydrocarbons (e.g., chlorinated dioxins) that exhibit Ah-receptor mediated toxicity. A criticism of such additive approaches is that antagonistic activities are not accounted for [43].

Calculation of the estrogen equivalent concentration (EEQ) of a chemically determined mixture is based on all measured xenoestrogens with a known estradiol equivalency factor (EEF; Table 6) according to

$$\mathrm{EEQ}_i = \mathrm{C}_i \times \mathrm{EEF}_i, \text{ and } \mathrm{EEQ}_t = \Sigma\, \mathrm{EEQ}_i$$

where *i* refers to compound *i* in the mixture with concentration C, and $\mathrm{EEQ_t}$ is the total EEQ. The EEFs are usually expressed on a molar basis because this is toxicologically more relevant than expressing concentrations on a weight basis.

Several studies have shown that additivity is indeed observed for estrogenic compounds (see, e.g. [46]). As an example, the assessment of nonionic surfactants of the alkylphenol ethoxylate type is discussed below. The reference compound in this case is usually nonylphenol, NP, which is one of the degradation products of NPEOs.

NPEOs are surfactants used in industrial and—formerly—household cleaning products. Commercially, they are manufactured and supplied as mixtures of oligomers and isomers.

Estrogenic activity has been observed for NP, t-octylphenol (OP), NPEO_{1-2} (nonylphenol mono- and diethoxylate) and some of the carboxylated degradation products (NPECs) [47].

Until now, there is no direct evidence in the literature whether or not the mode of toxic action for NP, NPEO, and NPEC is the same. In fact, higher NPEO oligomers may well have a different mode of action than NP, because the mechanism is likely a physical surfactant effect [47]. Despite this, for, e.g., acute or chronic toxicity endpoints, NP equivalent factors have been proposed for NPEO and NPEC.

In one approach [48], NPEO and NP toxicities were collected and combined into a quantitative structure-activity relationship (QSAR). To that end, 26 available acute LC_{50} (48 or 96 h) and EC_{50} data for aquatic organisms (including fish, insects, algae) were regressed versus the number of EO units, $\mathrm{N_{EO}}$.

The QSAR thus derived is shown in eq. 1:

$$\mathrm{Ln}\ \{\mathrm{LC}_{50} \text{ or } \mathrm{EC}_{50}\} = 0.27\ \mathrm{N_{EO}} + 0.17 \qquad n = 26 \qquad \mathrm{r}^2 = 0.9331 \qquad (1)$$

The relative potency can now be derived from this equation. The risk characterization then proceeds by calculating the total concentration of the sample (expressed in equivalents of NP) as follows (eq. 2):

$$\text{Total concentration} = \text{PEC} = \Sigma\{[\text{OP}] + [\text{NP}] + [\text{NPEO}_1]/1.3 + [\text{NPEO}_2]/1.7 + [\text{NPEO}_3]/2.3 + \ldots[\text{NPEO}_{15}]/57.6\} \quad (2)$$

As eq. 2 shows, the relative potency of $NPEO_1$, as derived from eq. 1, is (1/1.3) = 0.77 times that of NP. The toxicity of octylphenol is considered equivalent to that of NP in this approach, and hence its relative potency is equal to 1. NPEC were not included in this assessment.

In the Environment Canada distributional approach to characterize risks of NP, NPEO, and NPEC [47], relative toxicities were proposed based on categorizing acute and chronic toxicities. A similar approach was used in that study to provide relative estrogenicities. Both are listed in Table 5.

Table 5 Relative potencies of alkylphenol ethoxylates and their carboxylated degradation products proposed by Environment Agency UK and Environment Canada.

Substance	Relative toxicity to NP[a]	Relative toxicity to NP[b]	Relative estrogenicity to NP[b]
NP	1	1	1
OP	1	1	4.1
$NPEO_1$	0.77	0.5	0.67
$NPEO_2$	0.59	0.5	0.67
$NPEO_{3-17}$	0.017–0.43 ($NPEO_{15}$–$NPEO_3$)	0.005	0 (0.02)
NPEC	–	0.005	0.63
$NPEC_2$	–	0.005	0.63
OPEC	–	0.005	0.63
$OPEC_2$	–	0.005	0.63

[a]From Environment Agency UK [48].
[b]From Environment Canada [47].

It can be concluded that, concerning aquatic toxicities, in the Environment Agency (UK) assessment higher relative potencies were attributed to the longer chain ethoxylates ($N_{EO} = 4$ to $N_{EO} = 15$) than in the assessment made by Environment Canada. This may be due to several reasons: the UK assessment considered only acute toxicities, whereas the Canadian considered both acute and chronic ones. Moreover, in the Canadian assessment many more data (>200) were considered than in the UK assessment, and weighing factors were applied related to the confidence in studies.

In the preceding derivation of relative potencies, aquatic toxicity was considered as an endpoint. The finding that environmental contaminants including AP and APEO, can bind to the estrogen receptor (ER) in various species and thus regulate the activity of estrogen-responsive genes has raised concern. Similarly to the relative potencies for toxicity, relative potencies for receptor binding, vitellogenin induction, and several other ER-mediated responses, such as those from the in vitro tests presented in Table 6, have been derived. In the Canadian assessment the relative estrogenic potencies shown in Table 5 have been used, which are a weighted mean of several ER-mediated responses [47]. Such relative potencies must be used with some care, as they depend of course on the endpoints considered. Moreover, it has been shown that in vitro potencies can differ substantially from in vivo data, also for APEO [49]. In Table 6, examples are given of the several relative estrogenic potencies (estradiol equivalent factors, EEFs) of different estrogens and xenoestrogens, observed in three in vitro assays.

Table 6 The mol-based estradiol equivalency factors (EEF, estrogenic potency relative to estradiol) used for calculation of the estradiol equivalents (EEQs) in mixtures of chemically analyzed xenoestrogens.

Code	Compound	ER-CALUX		YES assay	ER binding
		EEF[a]	REF	EEF[b]	EEF[b]
E2	17β-Estradiol	1	[50]	1	1
E2-17α	17α-Estradiol	0.016	[50]	0.01	0.11
E1	Estrone	0.056	[50]	0.1	0.07
EE2	17α-Ethynylestradiol	1.2	[50]	1.2	0.8
BPA	Bisphenol A	7.8E-06	[50]	1.0E-05	1.0E-03
DMP	Dimethylphthalate	1.1E-05	[50]	1.0E-06	0
DEP	Diethylphthalate	3.2E-08	[50]	5.0E-07	5.0E-07
DBP	Di-*n*-butylphthalate	1.8E-08	[50]	1.0E-07	
BBP	Butylbenzylphthalate	1.4E-06	[50]	1.0E-06	
DEHP	Di(2-ethylhexyl)phthalate	<6.0E-07	[49]		
DOP	Dioctylphthalate	<6.0E-07	[49]		
NPEO	Nonylphenol ethoxylates	3.8E-06	[50]	4.0E-06	1.0E-05
OPEO	Octylphenol ethoxylates	<6.0E-07	[49]	4.0E-06	4.0E-06
NP	4-Nonylphenol	2.3E-05	[50]	5.7E-04	5.0E-04
OP	4-t-Octylphenol	1.4E-06	[50]	1.0E-05	5.0E-05
BDE47	2,4,2′,4′-Tetrabromodiphenylether	2.0E-07	[51]		
BDE85	2,3,4,2′4′-Pentabromodiphenylether	2.0E-07	[51]		
BDE99	2,4,5,2′4′-Pentabromodiphenylether	2.0E-07	[51]		
BDE100	2,4,6,2′4′-Pentabromodiphenylether	2.0E-05	[51]		
T3-like OH-BDE	1,3,5,3′-Tetrabromo-4′-hydroxy-diphenylether	1.0E-04	[51]		
o,p′-DDT	2-(*o*-Chlorophenyl),2-(*p*-chlorophenyl)-1,1,1-trichloroethane	9.1E-06	[52]		
o,p′-DDE	2-(*o*-Chlorophenyl),2-(*p*-chlorophenyl)-1,1-dichloroethene	2.3E-06	[49]		
	Methoxychlor	1.0E-06	[52]		
	Dieldrin	2.4E-07	[52]		
	Endosulfan	1.0E-06	[52]		
	Chlordane	9.6E-07	[52]		
	Genistein	6.0E-05	[52]		

[a]Ratio of EC_{50} (17β-estradiol)/EC_{50} (compound) in the ER-CALUX in vitro test.
[b](From ref. [50]); blank values correspond to nontested chemicals.

VALIDATION

Undoubtedly, one of the weakest links in environmental fate modeling and risk assessment is validation. There are several reasons for this. First, monitoring of environmental exposure concentrations can be very expensive, certainly in the case of identification and quantification of transformation products. Such measurements require expensive analytical equipment with trained personnel. In remote areas, such as open oceans, the Arctic and Antarctic, or high-altitude locations, sampling equipment (e.g., shipping time, airplanes) becomes a limiting factor. Second, in the case of endocrine disruptors, a multitude of compound mixtures appears to be active either as estrogens, androgens, or their anti-active counterparts. Therefore, complex mixtures have to be identified and included in the assessments including congeneric and isomeric compounds (as in, e.g., halogenated aromatics, surfactants) and their transformation products.

Seemingly small differences in chemical structure may have large consequences for endocrine-disrupting potency. Fingerprint patterns of such mixtures in real field samples may be quite different from the ones for which test results are available.

Third, the sampling design and parameter choices in regulatory monitoring programs in general are not in line with requirements of model validation studies. Various organizations such as the European Chemical Industry (CEFIC) have addressed this issue [53]. With the increased use of models in risk assessment, harmonization of regulatory monitoring programs and requirements for model validation studies is of great importance. Most aquatic models use freely dissolved water concentrations as one of the major input variables. In monitoring programs, usually only total water concentrations are measured, which also include the fractions bound to suspended matter, phytoplankton, and dissolved organic carbon. For hydrophobic compounds, this overestimates the freely dissolved fraction.

In a recent study, the bioaccumulation of steroid estrogens in aquatic organisms was estimated with a food web model [6,54]. The model calculations resulted in much lower predicted bioconcentration factors than those that are actually observed in the field. The authors concluded that the models required further development and additional data, including accurate transformation rates and understanding of factors controlling uptake. Moreover, they stated that the current absence of field data on estrogens in sediments might lead to inaccurate model outputs as experimental partitioning data are being used instead [6,54].

In reviews of evidence of endocrine disruption in wildlife (e.g., [55,56]), it was concluded that reported studies on wildlife are limited to very few animal species, and that effects at the population level are scarcely observed. Recent monitoring surveys in Europe have confirmed that evidence for endocrine disruption in wildlife is probably confined to hot spots [57,58].

CONCLUSION

The exposure assessment of endocrine disruptors suffers from a huge lack of reliable data, of both values that are used as input parameters in exposure models, and field data that are needed for validation purposes. In general, for the more classic EASs, such as PCBs, *p,p′*-DDE, chlorinated dioxins, some pesticides, and organotins, reliable data are sufficiently available, but careful evaluation of the quality of databases is necessary. Several data quality evaluation systems have been proposed. For the "newer" compounds, only few data have been gathered so far. The latter compounds include alkylphenols, bisphenol A, brominated flame retardants, phytoestrogens, and, in particular, natural and synthetic hormones, which in view of their high estrogenic potency could be the most important compounds in terms of risk. Research programs aimed at the endocrine disruption issue must focus on filling these gaps by promoting experimental studies for generation of reliable, high-quality parameter data on the one hand, and surveys or monitoring campaigns for collection of representative field data on the other. The nonspecificity of possible effects caused through endocrine mechanisms implies that in order to reveal dose–response relationships all, or at least as many as feasible, potentially active agents need be included in the risk assessments. Hence, it is obvious that comparable quality (i.e., number of data as well as reliability) of parameter and validation data for each of these agents is indispensable. EASs show a large variety in properties critical for environmental fate and therefore may elicit totally different environmental behavior. This may lead in turn to large differences not only in environmental exposure concentrations, but also in which target organisms can be potentially at risk.

The suitability of current exposure assessment models for EASs at this moment seems to be restricted to the persistent compounds such as PCBs, PCDDs, and PCDFs. Especially for the compounds subject to biodegradation and biotransformation the lack of experimental data to derive model-input parameters and perform validation studies at this moment is one of the main obstacles for the application of generic exposure models to other EASs. Current regulatory monitoring programs should be evaluated and harmonized with validation requirements of models used in exposure assessment.

Finally, most of the current models do not allow life stage-specific predictions. As EASs may affect early life stages (development of gonads, or egg maturation) there is a need to include life stage-specific predictions in exposure models.

REFERENCES

1. R. P. Schwarzenbach, P. M. Gschwend, D. M. Imboden. *Environmental Organic Chemistry*, Wiley, New York (1996).
2. J. Miyamoto and W. Klein. *Pure Appl. Chem.* **70**, 1829–1845 (1998).
3. R. V. Thomann and J. Komlos. *Environ. Toxicol. Chem.* **18**, 1060–1068 (1999).
4. R. V. Thomann, J. P. Conolly, T. F. Parkerton. *Environ. Toxicol. Chem.* **11**, 615–629 (1992).
5. T. C. Van Brummelen, A. G. M. van Hattum, T. Crommentuijn, D. F. Kalf. *Bioavailability and Toxicity of PAHs. Handbook of Environmental Chemistry*, Vol. 3, Part J, pp. 203–263, Springer Verlag, Berlin (1998).
6. K. M. Lai, M. D. Scrimshaw, J. N. Lester. *Sci. Total Environ.* **289**, 159–168 (2002).
7. R. S. Boethling and D. Mackay (Eds.). *Handbook of Property Estimation Methods for Chemicals – Environmental Health Sciences*. Lewis Publishers, New York (2000).
8. TGD. Technical guidance document in support of commission directive 93/67/EEC on risk assessment for new notified substances and commission regulation (EC) No. 1488/94 on risk assessment for existing substances. Office for Official Publications of the European Communities, Luxembourg (1996).
9. OECD. *Application of Structure–Activity Relationships to the Estimation of Properties Important in Exposure Assessment,* Organisation for Economic Cooperation and Development, OECD Environment Monograph No. 67, Paris (1993).
10. ECB. *EUSES - The European Union System for the Evaluation of Substances.* Joint Research Centre European Commission Environment Institute, European Chemicals Bureau, Ispra (1997).
11. W. Meylan and P. Howard. *User's guide for EPIWIN*^{C} *- EPI Suite - Estimation Programs Interface for Microsoft Windows.* Syracuse Research Corporation, North Syracuse, NY (1999). Available at <www.epa.gov/opptintr/exposure/docs/episuite.htm>.
12. D. Mackay, W. Y. Shiu, K. C. Ma. *Illustrated Handbook of Physical-chemical Properties of Environmental Fate for Organic Chemicals*, Springer Verlag, Berlin/Heidelberg (1997).
13. Syracuse Research Corporation. Environmental Fate Data Base (EFDB), North Syracuse, NY (1998). Online version accessible at: <www.esc.syrres.com/EFDBInfo.htm>.
14. F. Geiss, G. del Bino, G. Blech, O. Norager, E. Orthmann, G. Mosselmans, G. Powell, R. Roy, T. Smyrniotis, W. G. Town. *Toxicol. Environ. Chem.* **37**, 21–33 (1992).
15. D. Mackay et al. *Environ. Toxicol. Chem.* **15**, 1638–1648 (1996).
16. H. P. Kollig. *Toxicol. Environ. Chem.* **17**, 287–311 (1988).
17. H. P. Kollig and B. F. Kitchens. *Toxicol. Environ. Chem.* **28**, 95–103 (1990).
18. H.-J. Klimisch, M. Andreae, U. Tillmann. *Regulat. Toxicol. Pharmacol.* **25**, 1–5 (1997).
19. H. B. Krop, M. van Velzen, P. de Voogt, H. A. J. Govers, E. H. G. Evers, L. M. van der Heijdt, J. J. G. Zwolsman. *Data Quality Indicators in the Database Aquapol,* RIKZ report OS-0012, The Hague (1996).
20. A. Spacie and J. L. Hamelink. In *Fundamentals of Aquatic Toxicology,* G. M. Rand and S. R. Petrocelli (Eds.), pp. 495–525, Hemisphere Publishing, Washington, DC (1985).
21. M. G. Barron, G. R. Stehly, W. L. Hayton. *Aquat. Toxicol.* **18**, 61–68 (1990).
22. D. Mackay and A. Fraser. *Environ. Pollut.* **110**, 375–391 (2000).
23. F. A. P. C. Gobas and H. Morrison. In *Handbook of Property Estimation Methods for Chemicals – Environmental Health Sciences*, R. S. Boethling and D. Mackay (Eds.), Lewis Publishers, New York (2000).

24. D. Mackay. *Multimedia Environmental Models. The Fugacity Approach*, Lewis Publishers, Chelsea, MI (1991).
25. B. M. Pedersen, L. J. Thibodeaux, K. T. Valsaraj, D. Reible. *Environ. Toxicol. Chem.* **20**, 2114–2121 (2001).
26. H. Ling, M. Diamond, D. Mackay. *J. Great Lakes Res.* **19**, 582–602 (1993).
27. J. R. W. Harris, R. N. Gorley, C. A. Bartlett. *ECOS Version 2 User Manual - An Estuarine Simulation Shell,* Plymouth Marine Laboratory, Plymouth (1993).
28. P. M. A. Boderie. *DELWAQ 4.0: Technical Reference Manual.* WL/Delft Hydraulics, Delft (1994).
29. Anonymous *Manual of EXAMS II - 2.96*, CEAM, US-EPA, Athens GA (1997).
30. R. V. Thomann. *Environ. Sci. Technol.* **23**, 699–707 (1989).
31. E. Halfon and R. J. Allan. *Environ. Internat.* **21**, 557–569 (1995)
32. J. Camphens and D. Mackay. *Environ. Sci. Technol.* **31**, 577–583 (1997).
33. F. A. P. C. Gobas. *Ecol. Modelling* **69**, 1–17 (1993).
34. H. A. Morrison, F. A. P. C. Gobas, R. Lazar, D. M. Whittle, G. D. Haffner. *Environ. Sci. Technol.* **30**, 3377–3384 (1996).
35. T. P. Traas, R. Luttik, O. Klepper, J. E. M. Beurskens, M. D. Smit, P. E. G. Leonards, A. G. M. van Hattum, T. Aldenberg. *Environ. Toxicol. Chem.* **20**, 205–212 (2001).
36. V. Loizeau, A. Abarnou, P. Cugier, A. Jaouen-Madoulet, A. M. Le Guellec, A. Menesguen. *Mar. Poll. Bull.* **43**, 242–255 (2001).
37. HSE *REMA – Regulatory Environmental Modelling of Antifoulants,* Biocides and Pesticides Assessment Unit, Health and Safety Executive, London (1999).
38. N. Jonkers, R. W. P. M. Laane, P. de Voogt. *Proceedings 12th Annual Meeting SETAC Europe*, Vienna, 46-25, p. 185 (2002).
39. R. J. C. A. Steen, E. H. G. Evers, B. van Hattum, W. P. Cofino, U. A. Th. Brinkman, *Environ. Pollut.* **116**, 75–84 (2002).
41. G. L. Willingham and A. H. Jacobson. In *Designing Safer Chemicals: Green Chemistry for Pollutant Prevention,* S. C. De Vito and R. L. Garrett (Eds.), ACS Symposium Series 640, pp. 225–233, American Chemical Society, Washington, DC (1996).
41. J. A. Stäb, T. P. Traas, G. Stroomberg, J. van Kesteren, P. Leonards, A. G. M. van Hattum, U. A. Th. Brinkman, W. P. Cofino. *Arch. Environ. Contam. Toxicol.* **31**, 319–328 (1996).
42. T. H. Hutchinson, R. Brown, K. E. Brugger, P. M. Campbell, M. Holt, R. Laenge, P. M. McCahon, L. J. Tattersfield, R. van Egmond. *Environ. Health Perspect.* **108**, 1007–1015 (2000).
43. G. P. Daston, J. W. Gooch, W. J. Breslin, D. L. Shuey, A. I. Nikiforov, T. A. Fico, J. W. Gorsuch. *Reprod. Toxicol.* **11**, 465–481 (1997).
44. D. M. Sheehan, E. Willingham, D. Gaylor, J. M. Bergeron, D. Crews. *Environ. Health Perspect.* **107**, 155–159 (1999).
45. N. M. Van Straalen and C. A. J. Denneman. *Ecotoxicol. Environ. Saf.* **18**, 241–251 (1989).
46. J. Payne, N. Rajapakse, M. Wilkins, A. Kortenkamp. *Environ. Health Perspect.* **108**, 983–987 (2000).
47. M. R. Servos, R. J. Maguire, D. T. Bennie, H.-B. Lee, P. M. Cureton, N. Davidson, R. Sutcliffe, D. F. K. Dawn. *Supporting Document for Nonylphenol and its Ethoxylates*, National Water Research Institute 00-029, pp. 1–217, Environment Canada, Burlington (2000).
48. P. Whitehouse and G. Brighty. *Proceedings 3rd SETAC World Congress,* Brighton, p. 115 (2000).
49. J. Legler. Ph.D. thesis, Wageningen Universiteit, Ch. 3, p. 50 (2001).
50. M. S. Holt, K. Fox, E. Griessbach, S. Johnsen, J. Kinnunen, A. Lecloux, R. Murray-Smith, D. R. Peterson, R. Schroder, M. Silvani, W. F. J. Ten Berge, R. J. Toy, T. C. M. Feijtel. *Chemosphere* **41**, 1799–1808 (2000).
51. I. A. T. M. Meerts, R. J. Letcher, S. Hoving, G. Marsh, A. Bergman, J. G. Lemmen, B. van den Burg, A. Brouwer. *Environ. Health Perspect.* **109**, 399–407 (2001).

52. J. Legler, C. E .van den Brink, A. Brouwer, A. J. Murk, P. T. van der Saag, A. D. Vethaak, B. van den Burg. *Toxicol. Sci.* **48**, 55–66 (1999).
53. M. S. Holt, K. Fox, E. Griessbach, S. Johnsen, J. Kinnunen, A. Lecloux, R. Murray-Smith, D. R. Peterson, R. Schroder, M. Silvani, W. F. J. Ten Berge, R. J. Toy, T. C. M. Feijtel. *Chemosphere* **41**, 1799–1808 (2000).
54. K. M. Lai, M. D. Scrinshaw, J. N. Lester. *Crit. Rev. Toxicol.* **32**, 113–132 (2002).
55. C. R. Tyler, S. Jobling, J. P. Sumpter. *Crit. Rev. Toxicol.* **28**, 319–361 (1998).
56. P. Matthiessen. *Ecotoxicology* **9**, 21–24 (2000).
57. J. G. Vos, E. Dybing, H. A. Greim, O. Ladefoged, C. Lambre, J. V. Tarazona, I. Brandt, A. D. Vethaak. *Crit. Rev. Toxicol.* **30**, 71–133 (2000).
58. A. D. Vethaak et al. (Eds.). *Endocrine Disrupting Compounds in Aquatic Ecosystems in The Netherlands,* RIZA/RIKZ report 2002.001, The Hague (2002).

Pure Appl. Chem., Vol. 75, Nos. 11–12, pp. 1949–1953, 2003.

Topic 2.9

Environmental fate and metabolism: Issues and recommendations*

Pim de Voogt[1], Bent Halling-Sørensen[2], Bert van Hattum[3], Patrick T. Holland[4], Flemming Ingerslev[2], Andrew Johnson[5], Monika Jürgens[5], Arata Katayama[6], Werner Klein[7,‡], Norio Kurihara[8], J. C. Leblanc[9], K. D. Racke[10], Thomas Sanderson[11], Mordechai Shemesh[12], Laurence S. Shore[12], Elvira Vaclavik[2], Martin van den Berg[11], and Ph. Verger[9]

[1]Environmental and Toxicological Chemistry, Institute for Biodiversity and Ecosystem Dynamics, University of Amsterdam, Nieuwe Achtergracht 166, 1018 WV Amsterdam, The Netherlands; [2]The Royal Danish School of Pharmacy, Universitetsparken 2, Copenhagen 2100, Denmark; [3]Institute for Environmental Studies, Vrije Universiteit, De Boelelaan 1087, 1081 HV, Amsterdam, The Netherlands; [4]Cawthron Institute, 98 Halifax Street E, Nelson, New Zealand; [5]Centre for Ecology and Hydrology, Wallingford, OX10 8BB, UK; [6]Nagoya University, Research Center for Advanced Waste and Emission Management, Chikusa Nagoya 464-8603, Japan; [7]Fraunhofer Institute for Molecular Biology and Applied Ecology, 57392 Schmallenberg, Germany; [8]Koka Laboratory, Japan Radioisotope Association, 121-19 Toriino, Koka, Shiga 520-3403, Japan; [9]INRA/DSNHSA – 147, rue de l'Université – 75338 Paris Cédex 07; [10]Dow AgroSciences, 9330 Zionsville Rd, Bldg. 308, Indianapolis, IN 46268, USA; [11]Institute for Risk Assessment Sciences, Utrecht University, P.O. Box 80176, 3508 TD Utrecht, The Netherlands; 12 Kimron Veterinary Institute, Israel

Abstract: This main topic of the project and symposium includes the issue of releases of endocrine active substances (EASs) and their monitoring in the environment, food, and feed in order to provide the full set of criteria relevant for exposure assessment. Much less research has been devoted to these areas as compared to investigations on the effects. Issues of special importance regarding exposure to EASs, both from a research and risk management point of view, predominantly result from the fact that high-potency natural products are released as well as anthropogenic substances.

In order to provide reliable information for risk assessment and management, substantial research, methodological improvements, and improvements in data interpretation are needed regarding the following: releases and technologies for their mitigation; monitoring; establishment of background levels; transport, partitioning, persistence, degradation, and metabolism of EASs; dealing with "joint toxicity"; and providing reliable analytical methodology conforming to the principles of quality assurance.

Phytohormones play a special role in this whole area since they may be used as food amendments.

*Report from a SCOPE/IUPAC project: Implication of Endocrine Active Substances for Human and Wildlife (J. Miyamoto and J. Burger, editors). Other reports are published in this issue, *Pure Appl. Chem.* **75**, 1617–2615 (2003).

‡Corresponding author

BACKGROUND AND OBJECTIVES

Within risk assessment, exposure assessment is of equal importance as effects assessment. Within exposure assessment—the broader field of this article—a multitude of factors and criteria including the release of EASs into the environment, their presence in the aquatic and terrestrial environment, in food, or other consumer products, as well as issues on environmental fate and metabolism, are of crucial importance.

The challenge posed to science by the "appearance" of chemicals causing endocrine disruption in the environment has primarily led to basic research on the biochemical and biological processes explaining the effects observed. These areas are dealt with in the other topics of the current project. The studies of the fate and metabolism of endocrine substances have, however, not been as intensive. One reason is that the issue does not differ specifically for EASs as compared to other chemicals. In other words, the existing body of theory regarding the fate and metabolism of environmental chemicals is applicable for EASs. However, a number of issues are of particular importance with regard to EASs, and these will be emphasized in the current section.

Fate and metabolism depend on intrinsic substance properties, including physicochemical characteristics, but also on interacting environmental and technological (food processing) variables. In general, the major criteria can be classified into four areas:

1. Releases: industry sectors including food; agriculture; food supplements and drugs; humans and livestock; and wildlife
2. Transport and partitioning phenomena, which include mobility; bio- and geoaccumulation; bioavailability; and long-range transport in atmosphere and surface waters
3. Persistence in soil, water, and air, and either resistance to biotic or abiotic degradation
4. Metabolic pathways and metabolic patterns and their species variation. These may result in detoxification (inactivation), but also activations or reactivations. Information on metabolic pathways and metabolic patterns is the basis for the analysis of the hazard of metabolites, which may have the same or a different mode of action and potency as compared to the parent compound. Hazards of metabolites are relevant for the organisms in which they are formed, but also for other organisms, which are exposed to them.

For the purposes of risk assessment, the approaches to exposure assessment comprise a tiered system, which is governed by relevance. For generic assessments, compound properties and representative environmental variables play the major role, whereas for real situations (e.g., local or regional), exposure analyses, site-specific or regionally representative environmental variables have to be included, and especially monitoring of concentrations of the EASs in addition to the fate information plays an important role. Attempts to link these approaches have recently led to significant advances in this area resulting in stochastic and specific assessments (e.g., sensitive groups of populations) which, however, are not yet fully implemented. In all areas, especially for EASs, there is usually greater uncertainty in exposure assessment, compared with hazard characterization.

ISSUES DEALT WITH

Following the criteria and objectives on releases and the initial environmental behavior, the state of knowledge with identification of data gaps and further information needs has been discussed for the relevant groups of EASs—natural steroid hormones, phytohormones, pharmaceuticals and personal care products, industrial chemicals, and pesticides. Degradation and accumulation of EASs in the environment, metabolism in mammals, in aquatic and terrestrial systems have been addressed specifically in addition. Since integration of data and their wider use proceeds upon exposure modeling usually, special attention was given to the critical factors involved. It was not the objective of this topic to give an overview on available and missing information on the hundreds of substances, which upon hazard characterization show endocrine effects, but to focus on those with high effects potential and/or having high exposure potential.

RECOMMENDATIONS RELATED TO RISK MANAGEMENT

Monitoring programs

Whereas reliable and valid information on the fate and metabolism of substances in the environment and humans is a prerequisite to perform the generic evaluation, it is also essential for the design of monitoring programs (e.g., selection of substances, sites/locations of sampling, frequency). Especially regarding the endocrine disruptor issue, the usual evaluation of monitoring programs by averaging is not optimal, but should be targeted on population groups and for the differentiation of regions. Therefore, it is essential to extend monitoring to developing countries. In this context, it should be mentioned that a further instrument, which allows for retrospective exposure assessment, is to use appropriately stored environmental and human samples (specimen cryobanking). This instrument is a valuable tool also for EASs, since it allows for the assessment of time trends, whenever a newly identified potentially high-risk substance requires assessment.

High-priority EASs, for which the likelihood of a risk for humans or the environment has been sufficiently substantiated on a scientific basis, should be included in monitoring programs. This also includes metabolites of released chemicals for which this risk has been established.

Existing monitoring programs should be adopted to the requirements of models used in exposure assessment in order to better combine the two approaches in exposure assessment and to achieve comprehensive risk assessment.

Release mitigation

Technologies are available to remove EASs from effluents and solid waste for further use. These include buffer strips, biological treatment, composting, etc. These should be used as far as possible.

Assessment concept

It seems appropriate to assess anthropogenic EASs vs. natural hormones and phytohormones. Therefore, it would be advisable to establish background levels for environment and food regarding these as reference. For the human risk assessment, these reference substances could be 17ß-estradiol and other steroid hormones. The comparison should preferably be done using "toxicity" equivalents.

Responsible use of data

The relation of exposure to effects data regarding EASs is an issue requiring special scientific responsibility, although there is a vast amount of information from in vitro studies and receptor-related investigations. These effects data cannot be directly related to environmental occurrence, external exposure, or food concentration data. Distribution and metabolism in the organisms of concern qualitatively and quantitatively change the exposure of the target. Consequently, external exposure data should be assessed reliably, only vs. corresponding in vivo effect information. This applies also for in vivo testing where still frequently nominal concentrations are used and which give rise to substantial errors

Phytohormones

In the light of the whole EAS issue, a new risk/benefit assessment of fortified food should consider other exposure routes to these EASs. This is of high priority as the uses are increasing.

Sources of endocrine potency in food and environment: TIE concept and joint toxicity

Within the issue of endocrine active substances, TIE studies with chemical analyses and bioprobe analyses have been done on a number of samples. Frequently, there is a substantial fraction of the endocrine potential, not covered by the results of chemical analyses. In sewage effluents, however, frequently 90 % of the estrogenic activity can be explained by natural hormones and 17α-ethinylestradiol from contraceptives. Nevertheless, a completion of the chemical analyses (including natural hormones) in an in vivo TIE approach will help to set priorities for management.

Considering that, apart from the different endpoints, which need to be assessed, there remains the issue of interactions (additivity, antagonism, etc.) between the EASs to which organisms are exposed simultaneously. Although it is well understood, that substantial research is still needed regarding "mixture toxicity" also with respect to exposure, a valid concept for *joint toxicity* should now be developed and become the basis for management decisions soon.

RESEARCH PRIORITIES TO IMPROVE EXPOSURE ASSESSMENT

Identification of high-potency EASs

Realizing that EASs may be formed by metabolic hydroxylation of nonendocrine active substances (e.g., from some isoflavones), and that these might become an exposure issue for a wider range of organisms through environmental occurrence, substantial research is requested to systematically investigate this type of metabolic activation. This issue includes also endocrine activity, which is not directly hormone receptor-related.

Groups of chemicals that need systematic screening for endocrine activity and risks are the anabolic agents and nonhormonal steroids used in medical treatment. For these substances, the metabolites excreted in urine and/or feces should be identified and assessed for their environmental presence along with the parent compound. An important question in this context also is the reactivation of conjugated metabolites in the environment.

Modeling exposure

In order to expand the potential of exposure modeling, high-quality parameter data and representative field data are required. Due to the specificity of possible effects within the range of EASs, a large variety of properties needs to be investigated. Especially for the compounds subject to result in endocrine active metabolites, lack of respective data is at present one of the main obstacles for the application of generic exposure models to these compounds.

As EASs may affect different life stages, there is a need to develop the information required for life stage-specific evaluations by exposure modeling.

Phytohormones

Apart from research required to investigate potential environmental risks, especially to aquatic organisms, by increased releases of phytohormones (e.g., from processing fortified food), data to develop environmental and food PNECs should be elaborated.

There are additional specific questions to be addressed in research and monitoring programs, one important issue is the in utero exposure of the fetus to phytoestrogens depending on maternal diet.

Mineralization of EASs and metabolic pathways

Considering that ED active metabolites may present an increasing problem, it would be helpful to also investigate the mineralization (total degradation) of precursors to identify EAS persistence.

Furthermore, it is suggested to establish full metabolic pathways of those chemicals that from structure–activity predictions might result in endocrine active metabolites.

To improve human safety, it has been suggested to perform metabolism studies on EASs in human volunteers with very low doses and using AMS/HPLC analytical methodology. Ethical issues in those cases need careful risk benefit evaluation. This would provide an explanation also on the occasionally reported inverted U-shape dose–response curve, which might be due to a minor metabolite, only formed in a definite concentration range.

Analytical methods

Chemical analytical methods have been developed providing adequate detection limits and precision for the analysis of the most important groups of EASs in food and the environment at the requested levels—that is, the levels of biological significance. Research to make these methods less complex and more robust includes:

- provision of more specific and sensitive biomarkers,
- better integration for diverse classes of EASs,
- more automated and selective clean-up procedures,
- increased use of LC/MS and LC/MS/MS,
- increased availability of sensitive and stable ELISA tests, and
- greater conformity to the principles of quality assurance.

SUMMARY OF RECOMMENDATIONS

Risk management

- Environmental monitoring programs should, on a global scale,
    - be focused on high priority EASs,
    - include relevant metabolites, and
    - be designed to support exposure assessment.
- Quantitative correlations for chemical analyses and bioassays (TIE) should be used to reevaluate the biological relevance of target EASs for monitoring programs.
- In addition to source control, available technologies for reducing environmental entry should also be considered.

Research priorities

- Increased reliability of detection methods
    - development of robust and economical endocrine receptor bioassays
    - greater conformity to quality-assurance principles
- Elucidation of metabolic pathways, including potential activation vs. detoxification
- Key environmental fate parameters should be generated for highly active EASs (e.g., steroid hormones, industrial chemicals, and drugs).
- Improved models for exposure assessment
    - accounting for variability over time and space
    - including less well-characterized exposure scenarios and all potential exposure pathways (e.g., wastewater treatment plant, farm animal effluent)
- Development and validation of more efficient processes for reducing environmental loading.

TOPIC 3

EFFECTS OF ENDOCRINE ACTIVE CHEMICALS IN RODENTS AND HUMANS, AND RISK ASSESSMENTS FOR HUMANS

Pure Appl. Chem., Vol. 75, Nos. 11–12, pp. 1957–1971, 2003.

Topic 3.1

Interactions of xenobiotics with the steroid hormone biosynthesis pathway*

Thomas Sanderson‡ and Martin van den Berg

Institute for Risk Assessment Science, Utrecht University, P.O. Box 8017, 3508 TD Utrecht, The Netherlands

Abstract: Environmental contaminants can potentially disrupt endocrine processes by interfering with the function of enzymes involved in steroid synthesis and metabolism. Such interferences may result in reproductive problems, cancers, and toxicities related to (sexual) differentiation, growth, and development. Various known or suspected endocrine disruptors interfere with steroidogenic enzymes. Particular attention has been given to aromatase, the enzyme responsible for the conversion of androgens to estrogens. Studies of the potential for xenobiotics to interfere with steroidogenic enzymes have often involved microsomal fractions of steroidogenic tissues from animals exposed in vivo, or in vitro exposures of steroidogenic cells in primary culture. Increasingly, immortalized cell lines, such as the H295R human adrenocortical carcinoma cell line are used in the screening of effects of chemicals on steroid synthesis and metabolism. Such bioassay systems are expected to play an increasingly important role in the screening of complex environmental mixtures and individual contaminants for potential interference with steroidogenic enzymes. However, given the complexities in the steroid synthesis pathways and the biological activities of the hormones, together with the unknown biokinetic properties of these complex mixtures, extrapolation of in vitro effects to in vivo toxicities will not be straightforward and will require further, often in vivo, investigations.

INTRODUCTION

There is increasing evidence that certain environmental contaminants have the potential to disrupt endocrine processes, which may result in reproductive problems, certain cancers, and other toxicities related to (sexual) differentiation, growth, and development. Research has focused mainly on interactions with sex hormone receptors, particularly the estrogen receptor. Numerous chemicals have been shown to be agonists (or antagonists) for the estrogen receptor, although usually with very low affinities relative to 17β-estradiol. Many of these chemicals, except for certain halogenated compounds, do not bioaccumulate in the environment. The resultant biological potencies of these chemicals are invariably low, and, in most situations, it appears unlikely that environmental concentrations are sufficiently high to compete effectively with 17β-estradiol and other endogenous estrogens for the receptor. However, other mechanisms of interference with endocrine functions are increasingly being considered, including effects on enzymes involved in steroid hormone synthesis and metabolism. Particularly, the cytochrome P450 (CYP) enzymes responsible for the highly specific reactions in the steroid biosynthetic pathway

*Report from a SCOPE/IUPAC project: Implication of Endocrine Active Substances for Human and Wildlife (J. Miyamoto and J. Burger, editors). Other reports are published in this issue, *Pure Appl. Chem.* **75**, 1617–2615 (2003).

‡Corresponding author

[1] are of interest as potential targets, given their key role in the formation of various highly potent endogenous steroid hormones. It is possible for certain chemicals to cause or contribute to hormonal disruption and subsequent reproductive and developmental toxicities by interfering with the function of key enzymes involved in steroid synthesis and breakdown. This proceeding discusses the potential targets in the steroid biosynthetic pathway; the development of sensitive techniques and bioassays for the rapid screening of chemicals; the types of chemicals that may be expected to interfere with steroidogenesis, with the emphasis on the enzyme aromatase (CYP19); and the implications of such interferences. A final section will briefly discuss potential effects of chemicals on steroid metabolism. The emphasis throughout will lie on the synthesis and metabolism of estrogens from androgens.

ENZYMES INVOLVED IN STEROID SYNTHESIS

Steroidogenic enzymes are responsible for the biosynthesis of various steroid hormones, including glucocorticoids, mineralocorticoids, progestins, and sex hormones (Fig. 1), and consist of various specific CYPs, and several hydroxysteroid dehydrogenases (HSDs) and reductases [1]. De novo synthesis of estrogens starts with the conversion of cholesterol to pregnenolone by CYP11A (cholesterol side-chain cleavage). In the subsequent steps, 3β-HSD, CYP17 (17α-hydroxylase and 17,20 lyase activity), 17β-HSD and CYP19 (aromatase) are involved. Several key enzymes in steroidogenesis are described in more detail below.

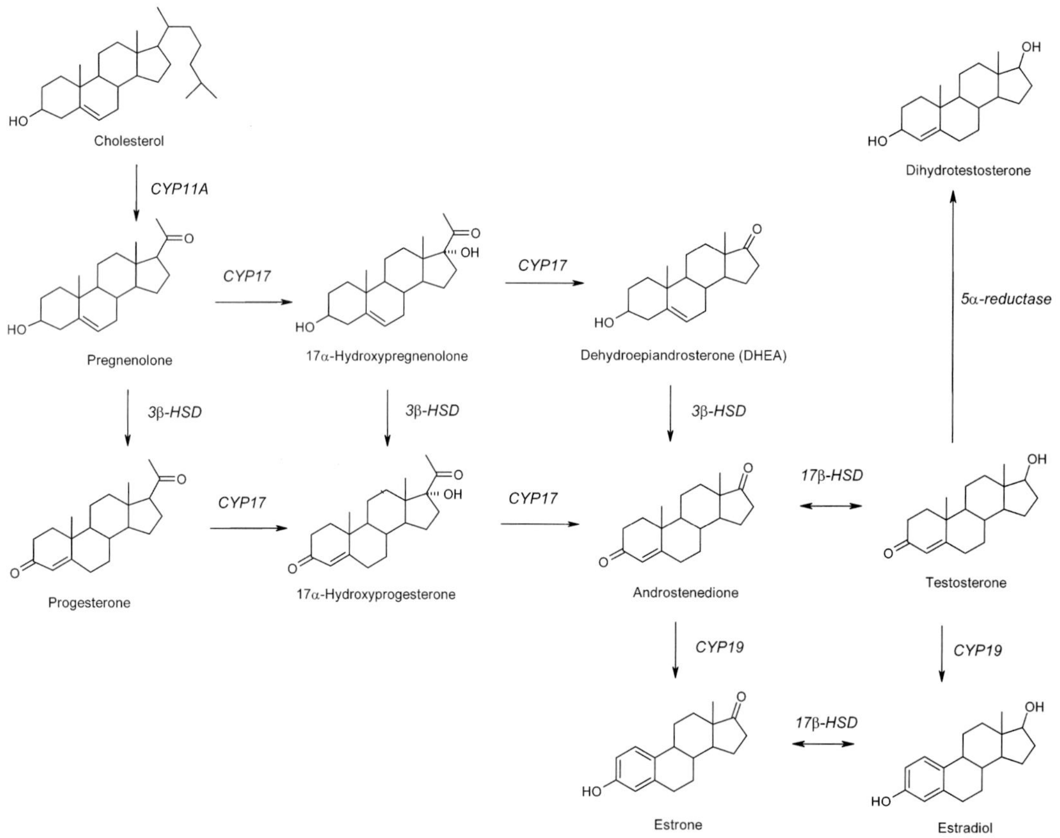

Cholesterol
CYP11A
Pregnenolone
CYP17
17α-Hydroxypregnenolone
CYP17
Dehydroepiandrosterone (DHEA)
Dihydrotestosterone
5α-reductase
3β-HSD
3β-HSD
3β-HSD
Progesterone
CYP17
17α-Hydroxyprogesterone
CYP17
Androstenedione
17β-HSD
Testosterone
CYP19
CYP19
Estrone
17β-HSD
Estradiol

Fig. 1 Enzymes involved in de novo synthesis of androgens and estrogens.

Cholesterol side-chain cleavage (CYP11A)

CYP11A converts cholesterol to pregnenolone via a three-step reaction: 20α-hydroxylation, 22R-hydroxylation, and C20-C22 lyase, which occurs at a single catalytically active site on the CYP450 molecule [2]. CYP11A is bound to the inner membrane of the mitochondrion and is found in all steroidogenic tissues, but is not expressed in nonsteroidogenic tissues. The CYP11A product pregnenolone is converted to progesterone by 3β-hydroxysteroid dehydrogenase (3β-HSD), one of the few non-CYP450 enzymes involved in steroidogenesis and which is found in both mitochondria and microsomes (smooth endoplasmic reticulum). 3β-HSD is widely distributed in steroidogenic and nonsteroidogenic tissues.

Steroid 17α-hydroxylase/17,20-lyase

CYP17 is responsible for the 17α-hydroxylation and the C17-20 lysis of steroid structures, which takes place in the endoplasmic reticulum [1]. Its C17-20 lyase activity is the key activity involved in directing the biosynthesis of steroids toward sex steroids and is highly expressed in testicular Leydig and ovarian follicle cells. In the adrenal cortex, CYP17 is responsible for the 17α-hydroxylation of pregnenolone and progesterone, which are found specifically in the zonae reticularis and fasciculata, but not in the zona glomerulosa. The C17-20 lyase activity is low in the adrenal cortex. Both activities appear to be due to distinct catalytic sites on the enzyme. CYP17 hydroxylates the 17α position of pregnenolone or progesterone to form the respective 17α-hydroxysteroids. The two 17α-hydroxylated steroids can be converted by CYP17 to the weak androgens dehydroepiandrosterone and androstenedione, respectively.

Aromatase

The 17α-hydroxysteroids are further converted to androstenedione and testosterone in the testis and ovary. The balance between these androgens depends on the activity of 17β-HSD; high 17β-HSD (type 1) activity favors the formation of testosterone. In both these organs, CYP19 or aromatase is expressed in the endoplasmic reticulum and is capable of converting androstenedione and testosterone into estrone and 17β-estradiol, respectively.

Implications of interferences with aromatase

CYP19 is of particular interest as it is the rate-limiting catalyst in the formation of estrogens, not only in cells engaged in de novo synthesis of estrogens, such as ovarian granulosa cells and the human adrenal cortex, but also in tissues such as the brain, adipose, and placenta, which utilize circulating levels of androstenedione or testosterone as precursors. Aromatase plays an important role in sexual differentiation, development, reproduction, and behavior, particularly in the gonads and the brain [3], but is also involved in diseases such as estrogen-dependent tumors [4]. Thus, interferences with the catalytic activity or expression of aromatase activity may be expected to result in disruptions of endocrine-regulated processes, such as estrous cycle, sperm production and maturation, development of puberty, masculinization/feminization of (sexual) behavior, and the inhibition or stimulation of the development and growth of hormone-dependent tumors of the breast, ovary, and prostate.

However, the prediction of biological effects of interferences with steroidogenic enzymes in intact organisms is highly complex. Induction of steroidogenic enzymes is highly tissue- and cell-type-specific and is controlled by different promoters and second messenger pathways, which, in turn, provide various targets for interaction with xenobiotics. Inhibition of steroidogenic enzymes may occur by mechanisms such as substrate competition, or mechanism-based inactivation and other forms of noncompetitive inhibition. Although inhibition by xenobiotics is likely to be less cell-type- and organism-dependent, it may be differentially influenced by cell-type- and organism-specific biokinetics.

Further complicating matters, the resultant hormone products have various physiological functions dependent on the tissue of formation and the stage of development of the organism. For example, estrogens are involved in determining sex-dependent behavior in the brain, whereas, peripherally, they control growth of bone, lipid metabolism and distribution, and the reproductive cycle in tissues such as the ovaries and uterus. In some tissues, aromatase plays a crucial role by forming the required estrogens locally using circulating levels of androgens; in others, circulating estrogens are required that originate mainly from the ovaries. Thus, effects of chemicals on aromatase activity may result in altered estrogen synthesis, but the ultimate tissue and organism responses will be harder to predict.

MECHANISMS OF INTERFERENCE WITH STEROID SYNTHESIS AND TECHNIQUES OF DETECTION

In contrast to a weak interaction with a sex hormone receptor, an interaction with a rate-limiting enzyme involved in sex hormone synthesis has the potential to profoundly affect the function and homeostasis of the highly potent endogenous steroid hormones. Among interactions with steroidogenic enzymes, various mechanisms can play a role. These include direct reversible or irreversible catalytic inhibition, and up- or down-regulation of enzyme expression (e.g., induction or inhibition of gene expression). Other, less direct effects on steroidogenic enzyme activities, such as modulation by the hypothalamic-pituitary-gonadal axis, may also occur.

To investigate effects on steroidogenic enzymes, various methods are available, starting with the choice of a biological system. This can be the simple isolated enzyme, if available, or a microsomal fraction of tissues that express the enzyme of interest. On an increasingly more complex level, cell lines, primary cells in culture, tissues slices, or whole animals may be used. To answer the straightforward question whether a compound can inhibit a specific enzymatic reaction, simple systems, such as purified enzyme, microsomes, or cell lines, may suffice. For questions regarding the effects of chemicals on the expression of steroidogenic enzymes, more complex systems are required. Cell lines and primary cultures may provide information on intracellular regulation, co-cultures may shed light on intra and intercellular regulation, and in vivo studies will be necessary to investigate possible effects of chemicals on steroidogenesis by affecting the hypothalamic-pituitary-gonadal/adrenocortical axes.

Once the choice of a suitable biological system has been made, various aspects of enzyme function can be investigated. Catalytic activity is one of the most functional endpoints that can be measured using selective substrates for the enzyme. To obtain reliable estimates of catalytic activity, disappearance of substrate or formation of product may be determined. However, disappearance of substrate does not necessarily implicate the involvement of a specific enzyme, and estimates of product formation can be hampered by further metabolism of the product, particularly in the steroidogenesis pathway. Ideally, catalytic measurements should involve the measurement of product formation in the presence of selective inhibitors that block further metabolism. In the case of aromatase, its catalytic activity is determined by measuring the release of tritiated water during the aromatization of 1β-^{3}H-androstenedione [5,6], in combination with the measurement of estrone formation, in the presence and absence of the selective aromatase inhibitor 4-hydroxyandrostenedione [6,7].

Up- and/or down-regulation of enzyme expression can be determined using techniques such as northern blotting or RT-PCR to determine levels of mRNA expression. RT-PCR and real-time RT-PCR are particularly powerful methods as they are highly selective and sensitive once optimized appropriately. Another approach is to use immunoblotting techniques, such as western blotting, that require selective antibodies developed to detect levels of enzyme protein quantitatively.

An indirect way to measure effects on steroidogenic enzyme function is to measure alterations in the ability of cell lines to excrete certain steroid products as an indicator of potential effects of xenobiotics on steroidogenesis. Such experiments would involve measuring a large number of steroid products in the presence of the xenobiotic of interest, possibly in combination with a stimulating concentration of an early precursor in the steroidogenic pathway, such as pregnenolone. An advantage of this approach is that alterations in the profile of the steroid hormones secreted provide an indication of the identity of the enzymes affected by the xenobiotic treatment, without the need to examine each enzyme activity individually.

IN VITRO BIOASSAYS TO SCREEN FOR INTERFERENCES WITH ENZYMES INVOLVED IN STEROID SYNTHESIS AND METABOLISM

Our laboratory has evaluated several human cell lines for the investigation of effects of xenobiotics on steroidogenic enzymes, such as MCF-7 breast tumor, JEG-3 and JAR placental choriocarcinoma, and H295R adrenocortical carcinoma cells. We also have experience with rat cell lines, such as R2C and LC540 Leydig tumor cells. Each cell line has its advantages and disadvantages. MCF-7 cells are not capable of de novo synthesis of estrogens and generally do not express aromatase, although there have been conflicting reports [8–11]. They are useful for the study of estrogen hydroxylations (discussed later in this chapter) as they express relatively high levels of CYP1A1 and 1B1, which are inducible by 2378-tetrachlorodibenzo-*p*-dioxin (TCDD). The JEG-3 and JAR cells express high levels of aromatase, but appear relatively sensitive to cytotoxic effects of chemicals and seem more prone to apoptosis, rendering these cell systems difficult to use for screening purposes [12,13]. H295R cells are somewhat less sensitive to cytotoxicity and have the major advantage that they express a wide range of steroidogenic enzymes, including all of the enzymes required to produce mineralocorticoids, glucocorticoids, androgens, and estrogens [14,15]. Experiments in our laboratory have recently shown that they also express (TCDD-inducible) CYP1A1 and 1B1, albeit at lower levels than in MCF-7 cells [7].

H295R human adrenocortical carcinoma cell line

The H295R human adrenocortical carcinoma cell line has demonstrated to be a useful bioassay to screen for interferences with steroidogenesis. The H295 and H295R (a subpopulation of H295 that forms a monolayer in culture) human adrenocortical carcinoma cell lines have been characterized in detail and shown to express all the key enzymes necessary for steroidogenesis [14–17]. These include CYP11A (cholesterol side-chain cleavage), CYP11B1 (steroid 11β-hydroxylase), CYP11B2 (aldosterone synthetase), CYP17 (steroid 17α-hydroxylase and/or 17,20 lyase), CYP19 (aromatase), CYP21B2 (steroid 21-hydroxylase), and 3β-hydroxysteroid dehydrogenase. The cells have the physiological characteristics of zonally undifferentiated human fetal adrenal cells, with the ability to produce the steroid hormones of each of the three phenotypically distinct zones found in the adult adrenal cortex [14,16]. It expresses numerous steroidogenic enzymes, including aromatase [6,14,15].

EFFECTS OF VARIOUS CLASSES OF COMPOUNDS ON AROMATASE

Pesticides

Several classes of (relatively) persistent pesticides, such as organotin compounds, DDT, and several metabolites, a number of azole fungicides, and several 2-chloro-*s*-triazine herbicides, are suspected or have been shown to interfere with steroidogenesis. Particular attention has been given to the enzyme aromatase (CYP19), which catalyzes the final, rate-limiting step in the conversion of androgens to estrogens.

Organotin compounds

Organotin compounds are highly toxic chemicals and ubiquitous environmental contaminants due to their persistence and wide use in industry, agriculture and antifouling paints. It has been postulated that organotin compounds may cause endocrine disruptive effects such as "imposex" (penis development in females) in mollusks by inhibiting aromatase (CYP19) activity [18]. Organotins have further been reported to inhibit cytochrome P450 (CYP) activities, such as CYP1A1 and aromatase (CYP19), in fish. However, little evidence supports aromatase inhibition as a mechanism of organotin-mediated imposex. Inhibitory effects on CYPs in vitro [19,20] and in vivo [21] were generally reported at environmentally unrealistic concentrations (upper micromolar range). Recently, we have published detailed concentration-response experiments in H295R cells demonstrating that, although the organotin compounds dibutyl-, tributyl-, and triphenyltin chloride decreased the activities of both CYP1A and CYP19 in the upper nanomolar range, the decrease occurred concomitantly with quantitatively similar decreases in various measures of cell viability [22] (Fig. 2). Thus, it could not be concluded that the organotin compounds could selectively inhibit aromatase activity. Several recent publications also do not support the aromatase inhibition hypothesis of imposex. A field study in gastropods (*Bolinus brandaris*) in Spain [23] showed that a population highly polluted with organotin compounds (100 % incidence of imposex in females) and having strongly decreased estradiol levels compared to a relatively uncontaminated population (37 % imposex) did not have altered aromatase activities compared with the less polluted population. Furthermore, a recent report points out that the reductions in steroid levels occur in the later stages of imposex development and appear to be a consequence rather than a cause of imposex [24]. Instead, it is suggested that certain peptide hormones are more likely to play an important role in masculinization of mollusks [24]. The above studies indicate that the development of imposex and the action of organotin compounds occur via mechanism(s) other than inhibition of aromatase activity.

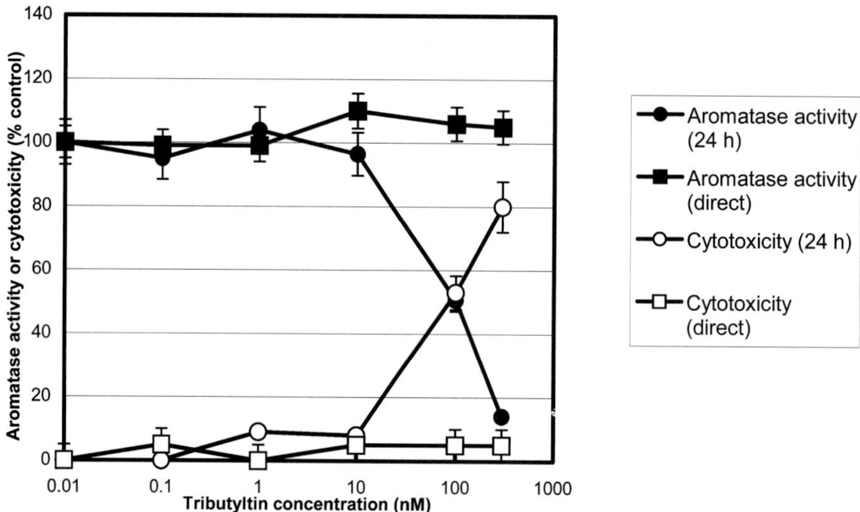

Fig. 2 Effects of tributyltin (TBT) on aromatase activity and cytotoxicity in H295R human adrenocortical carcinoma cells. Cells were exposed to TBT either for 24 h or only for the duration of the aromatase assay (1.5 h; direct catalytic inhibition).

Azole fungicides and vinclozolin

The class of azole fungicides targets the ergosterol biosynthesis pathway in yeasts and fungi by inhibiting the CYP enzyme 14α-lanesterol demethylase. The selectivity of these fungicides is variable, and some are known to inhibit several human CYP enzymes, including the steroidogenic CYP enzymes CYP17 (steroid 17β-hydroxylase activity) [25] and aromatase [25–27]. They have also been shown to inhibit aromatase activity in rainbow trout ovarian microsomes [28], indicating their potential to block natural estrogen-mediated responses, such as vitellogenin synthesis in female oviparous species during reproduction.

Several studies have shown that the commonly used fungicides imazalil and prochloraz (for various structures, see Fig. 3) are potent inhibitors of human, rodent, and fish aromatase activity [25–27]. A recent study using H295R cells [22] demonstrated that these two imidazole fungicides are mixed-type inhibitors ($K_i/K_I' = 0.04/0.3$ and 0.02/0.3 μM, respectively), whereas the triazole fungicides propiconazole, difenoconazole, and penconazole were less potent competitive inhibitors (K_i = 1.9, 4.5, and 4.7 μM, respectively). In the same study, the fungicide vinclozolin, which is structurally unrelated to the azoles, was identified as an inducer of aromatase activity and mRNA expression. Although vinclozolin has been shown to weakly antagonize the androgen receptor [29], it is possible that vinclozolin may exert additional anti-androgenicity via aromatase induction if this mechanism were to occur in vivo.

A study examining the effects of aromatase inhibition on embryonic development found that exposure in ovo of chickens to a nonsteroidal azole aromatase inhibitor led to "masculinization" of females [30]. In other words, aromatase inhibition during the critical time of embryonic development causes genotypical females to develop as fenotypical males. Also, coadministration of exogenous estrogen prevented the observed masculinization of females but led to "feminization" of males, which are the "default" sex in avian species [30]. Estrogens are key hormones involved in feminization of the central nervous system in birds, while, in contrast, they lead to defeminization and masculinization of the mammalian central nervous system [31]. Thus, it can be suggested that, during critical (irreversible) developmental periods, such as embryonic, perinatal, and pubertal development, aromatase induction may result in inappropriate (de)feminizing responses, depending on the type of species, gender, and the tissues in which local estrogen concentrations have been increased.

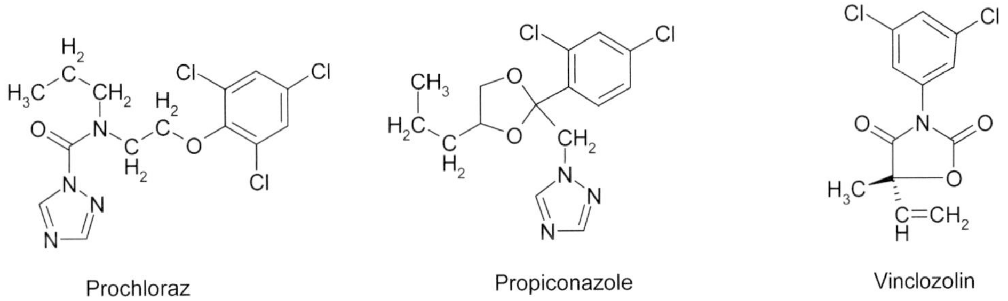

Prochloraz Propiconazole Vinclozolin

Fig. 3 The structures of the fungicides prochloraz (imidazole), propiconazole (triazole), which are aromatase inhibitors, and vinclozolin, which is an aromatase inducer.

Triazines

The 2-chloro-*s*-triazine family of herbicides is used in large amounts to control weeds, particularly on maize crops, in North America and Europe. The estimated use of atrazine in the United States was almost 35 000 tons in 1993 [32]. As a result, it is found in relatively high concentrations in surface waters in large parts of the North American continent [33]. It is relatively persistent to abiotic and biotic breakdown [33,34]. Epidemiological studies have associated long-term exposures to triazine herbicides with increased risk of ovarian cancer in female farm workers in Italy [35] and increased risk of breast cancer in the general population of Kentucky in the United States [36]. In experiments with female F344 rats, atrazine has been shown to induce tumors of the mammary gland and reproductive organs [37]. In female Sprague–Dawley rats, atrazine caused lengthening of estrous cycle and a dose-dependent increase in plasma levels of estradiol [38]. Atrazine also resulted in an earlier onset of the incidence of mammary and pituitary tumors [38], responses typical of exposure to exogenously administered estrogens [39,40]. Recently, atrazine exposure during lactation has been shown to suppress suckling-induced prolactin release in female Wistar rats [41]. Also, the lactationally exposed male offspring of the atrazine-exposed dams had an increased incidence of prostatitis [41], an effect known to be induced by exposure to exogenous 17β-estradiol [42]. A subsequent study in Long–Evans and Sprague–Dawley rats has attributed the effects of atrazine on serum prolactin levels to alterations in the hypothalamic control of the release of this hormone by the pituitary [43].

It has further been observed that atrazine causes various endocrine-disrupting effects in Florida alligators in areas contaminated with numerous pesticides including atrazine, DDT, dicofol, and vinclozolin. Male and female alligators from the contaminated Lake Apopka had elevated estradiol to testosterone plasma concentration ratios relative to a control site (Lake Woodruff), indicating a disturbance of the balance of androgens and estrogens, which is partly regulated by the activity of aromatase. In addition, females from Apopka had an abnormal ovarian morphology, with increased numbers of polyovular follicles and polynuclear oocytes. Apopka males had poorly organized testes and abnormally small penises.

Initially, investigations into the mechanism of these apparent estrogenic effects were directed toward the estrogen receptor. However, consistent interactions of triazine herbicides with the estrogen receptor or effects on receptor-mediated responses were never demonstrated [44–46]. Effects on enzymes involved in steroid metabolism have been limited to a study of the inhibition of testosterone metabolism in the anterior pituitary of rats exposed in vivo or of whole anterior pituitaries exposed in vitro to atrazine [47]. Weak inhibitory effects were observed on testosterone 5α-reductase (20–37 %) at an atrazine concentration of 0.5 (mM); a similar observation was made for the deethylated metabolite atrazine-desethyl. Studies in our laboratory of effects on enzymes involved in steroid synthesis demonstrated that several 2-chloro-*s*-triazine herbicides (atrazine, simazine, and propazine) and a number of their common metabolites (atrazine-desethyl and atrazine-desisopropyl) (Fig. 4) induced human aromatase activity and gene expression in vitro in H295R adrenocortical carcinoma cells [6,48]. It was fur-

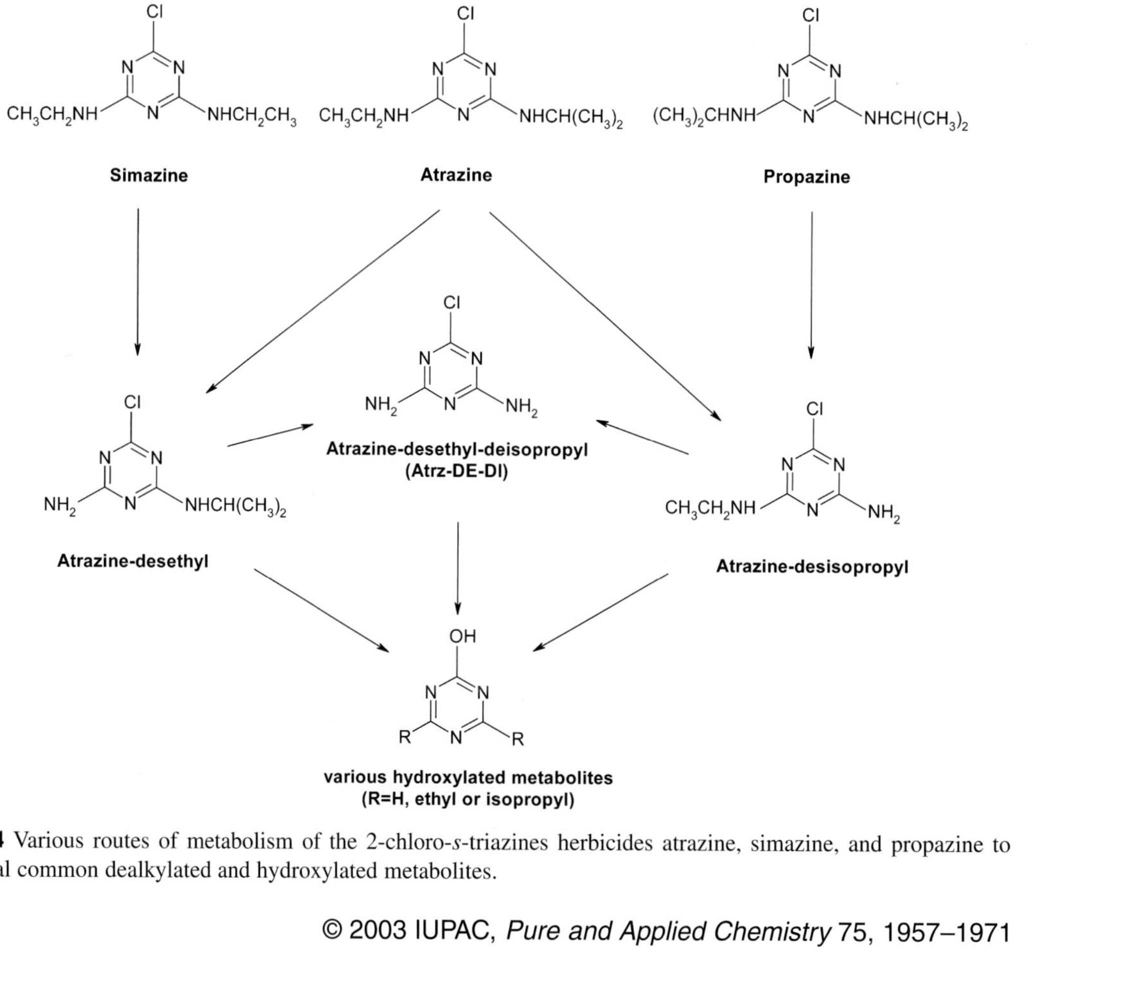

Simazine

Atrazine

Propazine

Atrazine-desethyl-deisopropyl (Atrz-DE-DI)

Atrazine-desethyl

Atrazine-desisopropyl

various hydroxylated metabolites (R=H, ethyl or isopropyl)

Fig. 4 Various routes of metabolism of the 2-chloro-*s*-triazines herbicides atrazine, simazine, and propazine to several common dealkylated and hydroxylated metabolites.

ther shown that none of the triazine herbicides nor their metabolites induced estrogen-dependent vitellogenin production in male carp hepatocytes. Nor did they antagonize the induction of vitellogenin by 17β-estradiol. Increased synthesis of vitellogenin, a yolk-precursor protein in fish and birds, is a response highly sensitive to estrogens and occurs after exposure to xenobiotics that are agonists for the estrogen receptor. Together, these experimental findings indicate that the estrogenic effects associated with triazine herbicides or their major metabolites in vivo are unlikely to be estrogen receptor-mediated, but may be partly explained by their observed ability to induce aromatase in vitro.

More experimental evidence is necessary to support the hypothesis that aromatase induction may play a role in vivo to explain the estrogenic effects of various chemicals. It is not clear whether aromatase induction occurs in vivo, nor in which target tissues it would occur. Given the recent evidence that plasma estradiol and estrone levels are increased about two-fold in atrazine-treated male Wistar rats [49], it is apparent that the presence of ovarian aromatase is not essential for the effects of atrazine. The further observation that estrone levels appear to be preferentially increased in vivo [49] may be an indication of a tissue-specific effect on aromatase. If aromatase induction is shown to play a role in vivo, it may be hypothesized that the induction would occur in tissues, such as adrenal cortex and adipose, that contain relatively greater levels of androstenedione than testosterone as precursor. Preliminary results from a collaboration between the laboratory of Dr. Susan Laws (U.S. Environmental Protection Agency) and our own laboratory suggest that, although slight increases in aromatase activity are measured, they are not statistically significant due to high interindividual variability. Nevertheless, good correlations ($p < 0.05$; $n = 24$) existed between aromatase activity and either estradiol or estrone plasma levels when using the entire data-set of 12 control and 12 atrazine-exposed animals, thus indicating that aromatase activity is a good predictor of the variation in plasma estrogen concentration even after two-fold induction by atrazine.

ORGANOCHLORINES AND METABOLITES

Polyhalogenated aromatic hydrocarbons

Highly persistent contaminants, such as polychlorinated dibenzo-*p*-dioxins (PCDDs), dibenzofurans (PCDFs), and biphenyls (PCBs) and the pesticides DDT and its metabolites, have undergone extensive toxicological studies over the last few decades. Certain members of theses classes of contaminants are potent endocrine disruptors. PCDDs, such as TCDD and several TCDD-like PCDFs and PCBs, are known to cause reproductive toxicities and various forms of endocrine disruption, including disturbances in sex steroid homeostasis and action, most notable anti-estrogenicity. Most of these effects are mediated by binding of TCDD-like compounds to the aryl hydrocarbon (Ah) receptor [50,51]. Ah receptor activation results in the induction of various enzymes, including CYP1A1, 1A2, and 1B1, which are involved in estrogen metabolism, and in "negative cross-talk" with the estrogen receptor-mediated pathway [52]. Effects of TCDD-like chemicals on steroidogenic enzymes, however, have either shown very weak or inconsistent effects.

TCDD has been shown to inhibit adrenocorticotropin hormone- (ACTH-) stimulated adrenocortical steroid synthesis in bovine primary cultures, although not by altering the activity of steroidogenic enzymes. TCDD appeared to interfere with the availability of cholesterol to CYP11A [53]. In another report, TCDD was shown to reduce the conversion of precursor steroids to progesterone in porcine luteinizing granulosa cells [54]. TCDD decreased estradiol synthesis in human luteinizing granulosa cells without affecting the activity of aromatase [55]. A decrease in aromatase activity was observed in vitro in TCDD- and PCB-126-exposed JEG-3 cells, although possible aselective cytotoxic effects could not be ruled out [13]. In H295R cells, TCDD had no effect on basal aromatase or cholesterol side-chain cleavage activity, but did reduce the inducibility of both activities by 8-bromo-cyclic AMP [7,56].

Exposures of JEG-3 or JAR cells to various organochlorines and some of their hydroxylated or methylsulfonated metabolites, such as MeSO2-PCBs, have demonstrated no specific effects on the aromatase enzyme [12]. However, 4-hydroxy-2,4,6-trichlorobiphenyl, tris-(4-chlorophenyl)-methanol, and several MeSO2-PCBs, such as the 3- and 4-MeSO2-PCB-52, -70, -87 and -101, were highly cytotoxic to the placental cells, possibly via an apoptotic mechanism [12]. Several methylsulfonated metabolites of PCBs were found to inhibit CYP11B1, the enzyme responsible for glucocorticoid synthesis, in Y 1 mouse adrenocortical tumor cells [57]. It was demonstrated that the methylsulfone group on the 3- or 4-position of the PCB molecule (Fig. 3) was essential for this interaction.

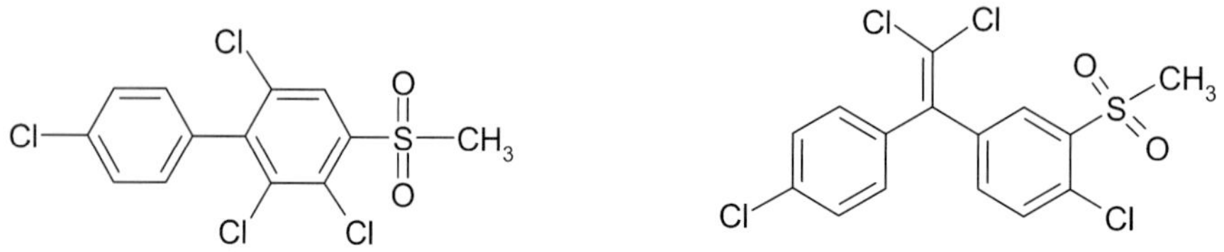

2,3,4',6-tetrachloro-4-methylsulfonyl-biphenyl   3-Methylsulfonyl-DDE

Fig. 5 The structures of two methylsulfonated metabolites of persistent organochlorine chemicals that have inhibitory effects on the glucocorticoid synthesizing enzyme CYP11B1.

DDT and metabolites

Another methylsulfonated compound, the DDT metabolite 3-$MeSO_2$-2,2-bis(4-chlorophenyl)-1,1-dichloroethene (MeSO2-DDE), is a relatively potent inhibitor of CYP11B1 (11β-steroid hydroxylase) in Y-1 mouse adrenocortical tumor cells at concentrations above 3 μM [57,58]. Here, it acts as a substrate for the enzyme, inhibiting glucocorticoid synthesis, and is bioactivated to a reactive intermediate that binds to proteins, ultimately resulting in adrenocortical cytotoxicity.

In addition, *p,p′*-DDE, which has anti-androgenic properties by acting as an antagonist for the androgen receptor [59], has been reported to increase aromatase protein in rat liver [60]. However, this induction was not confirmed in rat hepatocytes in primary culture [60]. This study did not rule out the possibility that the increase in immunoreactive protein observed in vivo may have been due to the known ability of *p,p*-DDE to induce hepatic CYP2B and 3A in rat liver but not in rat hepatocytes. Also, the tritiated water release assay used to measure aromatase activity can be confounded by hydroxylation of 1β-^{3}H-androstenedione at the 1β-position. This reaction is catalyzed partly by CYP3A1 and possibly CYP2B1 [61], enzymes that are highly induced by *p,p′*-DDE in rat liver [62]. It was recently also shown that *p,p*-DDE, unlike compounds such as atrazine and vinclozolin, was not capable of inducing aromatase activity in H295R cells [22].

Industrial chemicals

Various mass-produced industrial chemicals are known to cause endocrine disruption and reproductive toxicities in laboratory studies. Phthalates are used on a large scale as plasticizers and are found in various polyvinylchloride packages used in the medical and food industry. Phthalates leach out of the packages, contaminating the surrounding area. Particular attention has been given to the endocrine-disrupting properties of di-(2-ethylhexyl) phthalate (DEHP) and its metabolite, mono-(2-ethylhexyl) phthalate (MEHP). DEHP and MEHP are not [63] or are very weakly [64] estrogenic in vitro, but demonstrate ovarian [65] and testicular [66–68] toxicities in rats at very high concentration. Several studies have indicated that DEHP and, more effectively, MEHP interfere with estradiol synthesis, disrupting the estrous cycle in rats. MEHP, but not DEHP, was shown to decrease the activity of aromatase in rat granulosa cells in vitro. The mechanism of this decrease did not appear to be catalytic inhibition, but down-regulation of CYP19 mRNA expression [69].

Several high-volume production chemicals are suspected or known endocrine disruptors, including the alkylphenols and bisphenol A. 4-*tert*-octylphenol, nonylphenol, and bisphenol A are weakly estrogenic and appear to have inhibitory properties toward the 17α-steroid hydroxylase activity of CYP17 [70]; the same appeared to be true for octylphenol and bisphenol A [71,72]. However, a drawback of these studies is that the potential for aselective cytotoxic effects by the chemicals was not accounted for. Octyl- and nonylphenol, and bisphenol A exhibit signs of decreased cell function at concentrations between 1 and 30 mM in vitro, dependent on the cell system and exposure time used [73].

Effects of chemicals on enzymes involved in steroid metabolism

In addition to steroidogenic enzymes, steroid hormone homeostasis is regulated by enzymes that metabolize steroid hormones [74]. Although not the focus of this review chapter, chemically induced alterations in steroid metabolism represent an important mechanism of endocrine disruption and warrant a brief introduction, using estrogen metabolism as example.

In human females, the main sites of synthesis of 17β-estradiol are the ovaries, whereas the liver is one of the major sites of metabolism of estrogens [75]. However, estrogens are synthesized in various other organs and are metabolized extensively throughout the body. Circulating estrogens, such as 17β-estradiol and estrone, are metabolized in the liver either by direct conjugation to sulfate to form estrone- or 17β-estradiol-2-sulfate, or by sulfate conjugation subsequent to hydroxylation by cytochrome P450 enzymes (CYP) [74]. Although the majority of estrogens are conjugated to sulfate or glucuronide, rendering them hormonally inactive, a relatively small amount of estrogens are converted by CYPs to catechol estrogens. CYP enzymes can hydroxylate estrogens at various positions, most commonly the 2-, 4-, and 16α-position of the 17β-estradiol or estrone molecule, to form 2,3- and 3,4-catechol estrogens, and estriol (16α-OH-estradiol) or 16α-OH-estrone, respectively. Some of these metabolites, such as 4-OH-estradiol and 16α-OH-estrone, are considered genotoxic and mutagenic and have been implicated in estrogen-mediated carcinogenesis [76]. 16α-OH-estrone can undergo covalent binding to proteins such as the estrogen receptor, albumin, and various proteins containing free amino-groups, and to DNA [74]. 4-OH-estradiol appears to be just as carcinogenic as 17β-estradiol in Syrian hamster kidney,

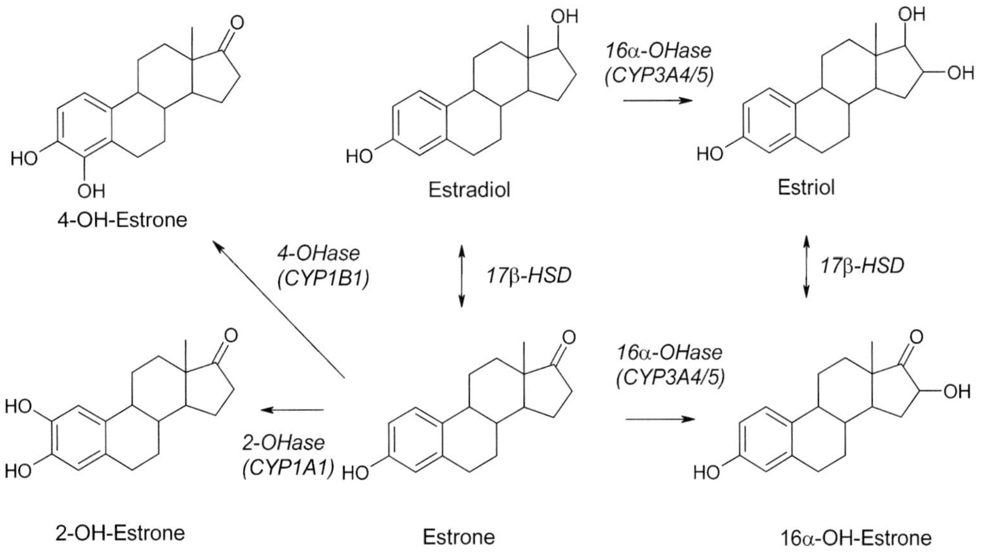

4-OH-Estrone

Estradiol

16α-OHase (CYP3A4/5)

Estriol

4-OHase (CYP1B1)

17β-HSD

17β-HSD

2-OH-Estrone

2-OHase (CYP1A1)

Estrone

16α-OHase (CYP3A4/5)

16α-OH-Estrone

Fig. 6 Enzymes involved in estrogen hydroxylations. Catechol estrogens can undergo methylation by catechol *O*-methyltransferase (COMT), and all hydroxyl groups can be conjugated to glucuronide or sulfate.

whereas 2-OH-estradiol is not carcinogenic [77,78]. The same may be true for mammary carcinogenesis in humans [79]. 4-OH-estrogens can be converted to quinone structures that may undergo redox cycling to produce reactive free radicals and oxygen species, resulting in genotoxic damage [80]. The catechol estrogens undergo further metabolism by catechol *O*-methyltransferase (COMT) to methoxylated metabolites or conjugation by sulfotransferase and UDP-glucuronidyltransferase, all resulting in inactivated (nonestrogenic and nongenotoxic) products. The isoforms of CYP enzymes involved in estrogen hydroxylation are species- and tissue-dependent. In the human liver, which is the main site of estrogen metabolism, 2- and 16α-hydroxylations are catalyzed by CYP3A4 and/or 3A5, with a smaller contribution from CYP1A2 [74,81]. In extrahepatic tissues, where CYP3A expression is much lower [82], but expression of CYP1A1 and 1B1 is relatively higher, CYP1A1-mediated 2- and CYP1B1-mediated 4-hydroxylation of estrogens occurs. Thus, certain routes of estrogen metabolism, such as 4-hydroxylation, that are minor in the liver, may play a considerably greater role in tissues such as the breast or uterus.

Various environmental chemicals can affect the catalytic activity of the cytochrome P450 enzymes involved in estrogen metabolism. The most prominent groups of compounds are polyhalogenated and polycyclic aromatic hydrocarbons. Many of these compounds are more or less potent inducers of the enzymes CYP1A1, 1A2, and 1B1 in humans, via activation of the Ah receptor. Most potent Ah receptor agonists, such as TCDD, 23478-PCDF, and PCB-126, generally induce CYP1A1 to a greater extent than CYP1A2 or 1B1, thus favoring the induction of estrogen 2-hydroxylation above 4- and/or 16α-hydroxylation in extrahepatic cell systems. It was further found that, in MCF-7 breast cancer cells, PCB 169, although an Ah receptor agonists capable of inducing CYP1A1 and 1B1 gene expression [83], is a selective inhibitor of the 4-hydroxylation of estradiol [84].

In human liver, where the majority of estrogen hydroxylations are performed by CYP3A4 and 3A5 [74,85], it would be expected that inducers of these enzymes would enhance oxidative estrogen metabolism. Typically, human CYP3A4 and 3A5 inducers belong to the class of phenobarbital-type inducers, which includes environmental contaminants such as di-*ortho*-PCBs (non-TCDD-like), and organochlorine pesticides, such as aldrin, dieldrin, DDT and certain metabolites, and toxaphene.

COMT, a crucial enzyme in the detoxification of catechol estrogens and catecholamines [86], has been reported to be inhibited by hydroxylated metabolites of PCBs [87]. The PCB metabolites 2,4,6-trichloro-3′,4′-dihydroxybiphenyl, 2,5-dichloro-3′,4′-dihydroxybiphenyl, and 2,4-dichloro-3′,4′-dihydroxybiphenyl were substrates for methylation by COMT with K_m and V_{max} values similar to those for 2- and 4-hydroxyestradiol, and adrenaline (epinephrine). The study further showed that these three catechol PCBs inhibited COMT-catalyzed methylation of 2- and 4-hydroxyestradiol with mixed-type kinetics. K_i values were estimated to be around 0.3 to 0.5 micromolars (μM).

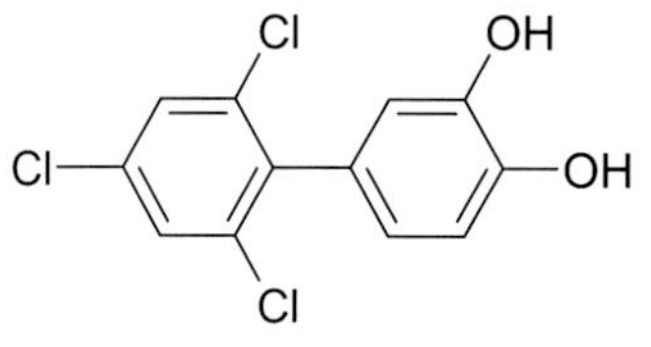

2,4,5,-trichloro-3',4'-dihydroxybiphenyl

Fig. 7 The structure of the catechol PCB metabolite 2,4,6-trichloro-3′,4′-dihydroxybiphenyl, which is an inhibitor of catechol *O*-methyl transferase and competes with catechol estrogens for methylation.

CONCLUSIONS AND RECOMMENDATIONS

Steroidogenic enzymes are important targets for chemicals that can interfere with the endocrine system. Until recently, this mechanism of endocrine disruption has received relatively little attention in comparison with steroid hormone receptors. Nevertheless, several classes of environmental contaminants have been shown to inhibit the function of certain enzymes involved in either steroid hormone synthesis or metabolism, based on the results of in vivo and in vitro studies. Induction of certain steroidogenic enzymes has also been demonstrated, but only in a limited number of in vitro studies.

So far, relatively few chemicals have been screened for their potential to interfere with steroid hormone synthesis and/or metabolism, and relatively few steroidogenic enzymes have been investigated. There are relatively few bioassays available to perform such initial screenings, but the H295R cell line and possibly other in vitro systems may prove to be a useful tools for future investigations in this area.

Very little is also is known about the consequences in vivo of interferences with steroid biosynthesis. Future animal studies are required to interpret and extrapolate in vitro studies of xenobiotic interferences with steroidogenic enzymes to the in vivo situation. Given the complexities in the steroid synthesis pathways and the numerous biological activities of the resultant steroid hormones, together with the often unknown biokinetic properties of the xenobiotics, this will be a formidable but important task.

REFERENCES

1. W. L. Miller. *Endocrine Rev.* **9**, 295–318 (1988).
2. K. L. Parker and B. P. Schimmer. *Vitamins Hormones* **51**, 339–370 (1995).
3. E. R. Simpson, M. S. Mahendroo, G .D. Means, M. W. Kilgore, M. M. Hinshelwood, S. Graham-Lawrence, B. Amarneh, Y. Ito, C. R. Fisher. *Endocrine Rev.* **15**, 342–355 (1994).
4. W. R. Miller and J. O'Neill. *Steroids* **50**, 537–548 (1987).
5. E. D. Lephart and E. R. Simpson. *Methods Enzymol.* **206**, 477–483 (1991).
6. J. T. Sanderson, W. Seinen, J. P. Giesy, M. Van den Berg. *Toxicol. Sci.* **54**, 121–127 (2000).
7. J. T. Sanderson, L. Slobbe, G. D. Lansbergen, S. Safe, M. Van den Berg. *Toxicol. Sci.* **61**, 40–48 (2001).
8. L. Jorgensen, N. Brunner, M. Spang-Thomsen, M. R. James, R. Clarke, P. Dombernowsky, B. Svenstrup. *J. Steroid Biochem. Mol. Biol.* **63**, 275–281 (1997).
9. L. A. Castagnetta, O. M. Granata, V. Bellavia, R. Amodio, E. Scaccianoce, M. Notarbartolo, M. R. Follari, M. D. Miceli, G. Carruba. *J. Steroid Biochem. Mol. Biol.* **61**, 287–292 (1997).
10. H. P. Ciolino, T. T. Wang, N. Sathyamoorthy. *Br. J. Cancer.* **83**, 333–337 (2000).
11. M. Kudoh, Y. Susaki, Y. Ideyama, T. Nanya, M. Mori, H. Shikama, T. Fujikura. *J. Steroid Biochem. Mol. Biol.* **58**, 189–194 (1996).
12. R. J. Letcher, I. van Holsteijn, H.-J. Drenth, R. J. Norstron, A. Bergman, S. Safe, R. Pieters, M. van den Berg. *Toxicol. Appl. Pharmacol.* **160**, 10–20 (1999).
13. H.-J. Drenth, C. A. Bouwman, W. Seinen, M. Van den Berg. *Toxicol. Appl. Pharmacol.* **148**, 50–55 (1998).
14. B. Staels, D. W. Hum, W. L. Miller. *Mol. Endocrinol.* **7**, 423–433 (1993).
15. W. E. Rainey, I. M. Bird, J. I. Mason. *Mol. Cell. Endocrinol.* **100**, 45–50 (1994).
16. A. F. Gazdar, H. K. Oie, C. H. Shackleton, T. R. Chen, T. J. Triche, C. E. Myers, G. P. Chrousos. *Cancer Res.* **50**, 5488–5496 (1990).
17. W. E. Rainey, I. M. Bird, C. Sawetawan, N. A. Hanley, J. L. McCarthy, E. A. McGee, R. Wester, J. I. Mason. *J. Clin. Endocrinol. Metab.* **77**, 731–737 (1993).
18. K. Fent. *Crit. Rev. Toxicol.* **26**, 1–117 (1996).
19. K. Fent, B. R. Woodin, J. J. Stegeman. *Comp. Biochem. Physiol.* **C121**, 277–288 (1998).
20. K. Fent and J. J. Stegeman. *Aquat. Toxicol.* **20**, 159–168 (1991).
21. K. Fent and J. J. Stegeman. *Aquat. Toxicol.* **24**, 219–240 (1993).
22. J. Sanderson, J. Boerma, G. W. A. Lansbergen, M. van den Berg. *Toxicol. Appl. Pharmacol.* **182**, 44–54 (2002).

23. Y. Morcillo and C. Porte. *Environ. Res.* **81**, 349–354 (1999).
24. E. Oberdorster. *Ann. NY Acad. Sci.* **948**, 75–79 (2001).
25. J. I. Mason, B. R. Carr, B. A. Murry. *Steroids* **50**, 179–189 (1987).
26. A. M. Vinggaard, C. Hnida, V. Breinholt, J. C. Larsen. *Toxicol. in Vitro* **14**, 227–234 (2000).
27. M. Ayub and M. J. Levell. *J. Steroid Biochem.* **31**, 65–72 (1988).
28. G. Monod, A. De Mones, A. Fostier. *Marine Environ. Res.* **35**, 153–157 (1993).
29. W. R. Kelce, C. R. Lambright, L. E. Gray, Jr., K. P. Roberts. *Toxicol. Appl. Pharmacol.* **142**, 192–200 (1997).
30. A. Elbrecht and R. G. Smith. *Science* **255**, 467–470 (1992).
31. A. Jost. *Psychoneuroendocrinology* **8**, 183–193 (1983).
32. USEPA. *Pesticides Industrial Sales and Usage*, US Environmental Protection Agency: Washington, DC (1994).
33. K. R. Solomon, D. B. Baker, R. P. Richards, K. R. Dixon, S. J. Klaine, T. W. La Point, R. J. Kendall. *Environ. Toxicol. Chem.* **15**, 31–76 (1996).
34. S. U. Khan and T. S. Foster. *J. Agric. Food Chem.* **24**, 768–771 (1976).
35. A. Donna, P. Crosignani, F. Robutti, P. G. Betta, R. Bocca, N. Mariani, F. Ferrario. *Scand. J. Work Environ. Health* **15**, 47–53 (1989).
36. M. A. Kettles, S. R. Browning, T. S. Prince, S. W. Horstman. *Environ. Health Perspect.* **105**, 1222–1227 (1997).
37. A. Pinter, G. Torok, M. Borzsonyi, A. Surjan, M. Calk, Z. Kelecsenvi, Z. Kocsis. *Neoplasma.* **37**, 533–544 (1990).
38. L. T. Wetzel, L. G. Luempert III, C. B. Breckenridge, M. O. Tisdel, J. T. Stevens. *J. Toxicol. Environ. Health* **43**, 169–182 (1994).
39. C. F. Geschickter and E. W. Byrnes. *Arch. Pathol.* **33**, 334–356 (1942).
40. J. R. Brawer and C. Sonnenschein. *Am. J. Anat.* **114**, 57 (1975).
41. T. E. Stoker, C. L. Robinette, R. L. Cooper. *Toxicol. Sci.* **52**, 68–79 (1999).
42. L. Tangbanluekal and C. L. Robinette. *Endocrinology* **132**, 2407–2416 (1993).
43. R. L. Cooper, T. E. Stoker, L. Tyrey, J. M. Goldman, W. K. McElroy. *Toxicol. Sci.* **53**, 297–307 (2000).
44. K. Connor, J. Howell, I. Chen, H. Liu, K. Berhane, C. Sciarretta, S. Safe, T. Zacherewski. *Fundam. Appl. Toxicol.* **30**, 93–101 (1996).
45. M. K. Tennant, D. C. Hill, J. C. Eldridge, L. T. Wetzel, C. B. Breckenridge, J. T. Stevens. *J. Toxicol. Environ. Health* **43**, 197–211 (1994).
46. M. K. Tennant, D. C. Hill, J. C. Eldridge, L. T. Wetzel, C. B. Breckenridge, J. T. Stevens. *J. Toxicol. Environ. Health* **43**, 183–196 (1994).
47. T. Babic-Gojmerac, Z. Kniewald, J. Kniewald. *J. Steroid Biochem.* **33**, 141–146 (1989).
48. J. T. Sanderson, R. L. Letcher, M. Heneweer, J. P. Giesy, M. Van den Berg. *Environ. Health Perspect.* **109**, 1027–1031 (2001).
49. T. E. Stoker, S. C. Laws, D. L. Guidici, R. L. Cooper. *Toxicol. Sci.* **58**, 50–59 (2000).
50. A. B. Okey, D. S. Riddick, P. A. Harper. *Toxicol. Lett.* **70**, 1–22 (1994).
51. S. H. Safe. *CRC Crit. Rev. Toxicol.* **21**, 51–71 (1990).
52. S. Safe, F. Wang, W. Porter, R. Duan, A. McDougal. *Toxicol. Lett.* **102–103**, 343–347 (1998).
53. R. W. Moore, C. R. Jefcoate, R. E. Peterson. *Toxicol. Appl. Pharmacol.* **109**, 85–97 (1991).
54. E. L. Gregoraszczuk, E. Zabielny, D. Ochwat. *J. Physiol. Pharmacol.* **52**, 303–311 (2001).
55. F. M. Moran, A. J. Conley, C. J. Corbin, E. Enan, C. VandeVoort, J. W. Overstreet, B. L. Lasley. *Biol. Reprod.* **62**, 1102–1108 (2000).
56. J. T. Sanderson and M. Van den Berg. *Organohalogen Compd.* **37**, 77–80 (1998).
57. M. Johansson, C. Larsson, A. Bergman, B. O. Lund. *Pharmacol. Toxicol.* **83**, 225–230 (1998).
58. B.-O. Lund and J. Lund. *J. Biol. Chem.* **270**, 20895–20897 (1995).
59. W. R. Kelce and E.M. Wilson. *J. Mol. Med.* **75**, 198–207 (1997).

60. L. You, M. Sar, E. Bartolucci, S. Ploch, M. Whitt. *Mol. Cell. Endocrinol.* **178**, 207–214 (2001).
61. D. J. Waxman, G. A. Dannan, F. P. Guengerich. *Biochemistry* **24**, 4409–4417 (1985).
62. L. You, S. K. Chan, J. M. Bruce, S. Archibeque-Engle, M. Casanova, J. C. Corton, H. Heck. *Toxicol. Appl. Pharmacol.* **158**, 197–205 (1999).
63. C. A. Harris, P. Henttu, M. G. Parker, J. P. Sumpter. *Environ. Health Perspect.* **105**, 802–811 (1997).
64. A. Blom, E. Ekman, A. Johannisson, L. Norrgren, M. Pesonen. *Arch. Environ. Contam. Toxicol.* **34**, 306–310 (1998).
65. D. K. Agarwal, W. H. Lawrence, J. E. Turner, J. Autian. *J. Toxicol. Environ. Health* **26**, 39–59 (1989).
66. T. J. Gray and S. D. Gangolli. *Environ. Health Perspect.* **65**, 229–235 (1986).
67. P. Sjoberg, U. Bondesson, T. J. Gray, L. Ploen. *Acta Pharmacol. Toxicol. (Copenh).* **58**, 225–233 (1986).
68. L. E. Gray, J. Ostby, J. Furr, M. Rice, D. N. R. Veeramachaneni, L. Parks. *Toxicol. Sci.* **58**, 350–365 (2000).
69. T. N. Lovekamp and B. J. Davis. *Toxicol. Appl. Pharmacol.* **172**, 217–224 (2001).
70. T. Niwa, Y. Maekawa, M. Fujimoto, K. Kishimoto, Y. Yabusaki, F. Ishibashi, M. Katagiri. *Biol. Pharm. Bull.* **25**, 235–238 (2002).
71. E. P. Murono, R. C. Derk, J. H. de Leon. *Reprod. Toxicol.* **14**, 275–288 (2000).
72. E. P. Murono, R. C. Derk, J. H. de Leon. *Reprod. Toxicol.* **15**, 551–560 (2001).
73. S. S. Raychoudhury. *Toxicol. Appl. Pharmacol.* **157**, 192–202 (1999).
74. C. P. Martucci and J. Fishman. *Pharmac. Ther.* **57**, 237–257 (1993).
75. C. L. Williams and G. M. Stancel. In *Goodman & Gilman's The Pharmacological Basis of Therapeutics*, J. G. Hardman and L. E. Limbird (Eds.), pp. 1411–1440, McGraw-Hill, New York (1996).
76. J. G. Liehr. *Mutat. Res.* **238**, 269–276 (1990).
77. J. G. Liehr, W. F. Fang, D. A. Sirbasku, A. Ari-Ulubelen. *J. Steroid Biochem.* **24**, 353–356 (1986).
78. J. Weisz, Q. D. Bui, D. Roy, J. G. Liehr. *Endocrinology* **131**, 655–661 (1992).
79. J. G. Liehr and M. Ricci. *Proc. Natl. Acad. Sci. USA* **93**, 3294–3296 (1996).
80. J. G. Liehr, A. A. Ulubelen, H. W. Strobel. *J. Biol. Chem.* **261**, 16865–16870 (1986).
81. V. Kerlan, Y. Dreano, J. P. Bercovici, P. H. Beaune, H. H. Floch, F. Berthou. *Biochem. Pharmacol.* **44**, 1745–1756 (1992).
82. I. De Waziers, P. H. Cugnenc, C. S. Yang, J.-P. Leroux, P. H. Beaune. *J. Pharmacol. Exp. Ther.* **253**, 387–394 (1990).
83. M. Van Duursen, J. T. Sanderson, M. Van den Berg. *Toxicologist* **66**, 169 (2002).
84. S. Pang, J. Q. Cao, B. H. Katz, C. L. Hayes, T. R. Sutter, D. C. Spink. *Biochem. Pharmacol.* **58**, 29–38 (1999).
85. A. J. Lee, L. H. Mills, J. W. Kosh, A. H. Conney, B. T. Zhu. *J. Pharmacol. Exp. Ther.* **300**, 838–849 (2002).
86. P. T. Mannisto and S. Kaakkola. *Pharmacol. Rev.* **51**, 593–628 (1999).
87. C. E. Garner, L. T. Burka, A. E. Etheridge, H. B. Matthews. *Toxicol. Appl. Pharmacol.* **162**, 115–123 (2000).

Pure Appl. Chem., Vol. 75, Nos. 11–12, pp. 1973–2012, 2003.

Topic 3.2

Organochlorine compounds and breast cancer risk*

Michelle A. Mendez and Lenore Arab

School of Public Health, University of North Carolina, Chapel Hill, NC 27599-7435, USA

Abstract: A number of epidemiologic studies on organochlorines (OCs) and breast cancer risk have been published. The majority (n = 18) measured OCs in adipose tissue, primarily from mammary biopsies in cancer cases and controls with benign breast disease, and studied incident disease. Seven of these studies each included fewer than 50 cases and controls and had limited capacity for covariate adjustment. Eleven studies used serum samples collected from 6 months to 25 years prior to diagnosis. An additional 13 studies (2 with some overlap) used serum collected at or after diagnosis. Regardless of the medium used to measure OC levels, studies conducted to date do not provide consistent evidence that any of the OCs examined thus far play a role in the initial breast cancer risk. This paper provides a compound specific review with discussion of how the lack of evidence for adverse effects might be explained by factors related to study design, or by variation in risk across subgroups.

The sum of the evidence does not implicate any OC compound as significantly related to risk of occurrence of breast cancer. The evidence base is greatest for dichlorodiphenyltrichloroethane (DDT) and dichlorodiphenyldichloroethylene (DDE). Limited research has been done on individual polychlorinated biphenyls (PCBs) and their isomers. The studies of OC exposure reflect current exposure levels of chemicals banned as long ago as 20 years. Although the information is extremely limited and not without major design flaws, the association between OCs and disease severity and progression is interesting and worthy of further examination.

More studies are needed on OCs other than DDTs in developing countries where use is more recent or continuing, especially given that most estrogenic OCs are not persistent. It is possible that other pathways may be involved, including activity related to cytochrome P450 (CYP) and glutathione-*S*-transferase (GST); there is limited research to date on this hypothesis. In developed countries, a body mass index (BMI)/weight loss model may warrant further analysis, perhaps using existing data.

OC exposures cannot reliably be related to trends in breast cancer incidence, as other known risk factors for breast cancer, such as childbearing and lactation, have changed along with OC residues during this timeframe. Changes in screening and treatment over time also complicate making such links. Ecologic data relating high-exposure countries to high breast cancer mortality rates do not suggest a strong link. Breast cancer rates are not generally higher in parts of the world with high DDE levels. Countries with relatively similar levels of PCBs, such as Great Britain and Japan, have very different breast cancer rates.

---

*Report from a SCOPE/IUPAC project: Implication of Endocrine Active Substances for Human and Wildlife (J. Miyamoto and J. Burger, editors). Other reports are published in this issue, *Pure Appl. Chem.* **75**, 1617–2615 (2003).

‡Corresponding author

As OCs are present in the environment as mixtures of correlated isomers and metabolites, it may be difficult to distinguish possible causal links from associations in which measured compounds are merely markers of other underlying exposures. For highly correlated compounds, traditional adjustment strategies may not be feasible. More complex analytical strategies may help to isolate potentially relevant isomers.

INTRODUCTION

Although a variety of other compounds, including pharmaceutical agents (e.g., oral contraceptives, tamoxifen) and phytoestrogen-containing foods (e.g., soybeans, cabbage) may mimic or modulate estrogen activity, chemicals produced for industrial or agricultural use have been the primary focus of research on endocrine disruptors and human health [1]. These chemicals, known as organochlorine compounds (OCs), comprise a diverse group of compounds, including: (i) pesticides (e.g., DDT, aldrin, dieldrin, chlordane, heptachlor epoxide, hexachlorocyclohexane, hexachlorobenzene, methoxychlor); (ii) industrial chemicals (e.g., PCBs used in lubricants, coolants, sealants, and pesticides; mirex used as a fire retardant coating and a pesticide); and (iii) breakdown products of these chemicals. OCs are lipophilic and persist in the environment. Thus, despite the bans, exposure to many OC isomers and metabolites endures decades later as a result of residues in soil, water, dust, and foods, most notably in fat-containing animal products such as meats, fish, and dairy products [2–4]. Environmental exposure to these organochlorines has been postulated to increase the risk of breast cancer through potential estrogenic activity [5–7]. Not all OCs possess estrogenic activity, and some, depending upon the internal milieu, may have antiestrogenic or anticarcinogenic effects [8–14]. Nonetheless, OCs may also increase the risk of breast cancer via other mechanisms, including promotion of tumor growth [15–17] and modulation of enzyme activity [8,18,19].

Peak use of most OC chemicals occurred in the 1950s and 1960s, some three decades after their introduction [20,21]. Most were banned in the United States and other industrialized countries in the early 1970s (DDT) or late 1970s (PCBs, mirex, benzene hexachloride [BHC], and hexachlorobenzene [HCB]) [4,22]. Use of a number of these chemicals continues in many developing countries. OC levels in human tissues and foods in these countries are several times higher than in the United States and Europe [2,23].

Data from North America and Europe indicate that residues of most OCs in human adipose tissue, breast milk, and serum have declined substantially over the past 20 years, but that more persistent isomers and metabolites are still present in detectable amounts [4,24–28]. DDT-related isomers appear to have decreased more rapidly than PCBs and some other OCs such as HCB [24,28]. Given their ubiquitous presence in the environment and the continued use of some OCs in developing countries, it is of interest to determine whether these compounds are indeed related to the risk of breast cancer. Early studies yielded conflicting results. They also generated hypotheses that might help to explain inconsistencies, such as the possibility of variation related to subgroups defined by tumor estrogen receptor (ER) status [29], lactation history [22], or ethnicity, perhaps as a result of higher exposure [30]. This paper will assess whether various aspects of study design may be related to inconsistent results, and will review whether studies published to date support the hypotheses related to at-risk subgroups. We emphasize compounds that have been the focus of most research to date, namely: (i) DDT-related compounds, including DDE, a common persistent metabolite of found in the food chain and the environment; and (ii) PCBs. We also review the more limited evidence available for several other OCs.

OVERVIEW OF RESULTS

Epidemiologic studies of OCs and breast cancer risk published to date are summarized in Table 1. The majority (n = 20; seven of which incorporate populations at least partially included in other reports)

Text continues on p. 1984.

Table 1 Studies on organochlorines and breast cancer: Design and summary results.

1st author, yr (place) [ref.]	Design	Years: diagnosis	# Cases/ controls	Response rate [other attrition]	Key covariates excluded?	Positive & significant associations	NS or inverse associations
1A. ADIPOSE TISSUE, RETROSPECTIVE (Studies with >50 cases/controls)							
Aronson et al., 2000 (Canada) [50; see also 43]	Hospital-based *Cases*: Incident in situ or invasive BC *Controls*: BBD, biopsy. Frequency matched-age.	1995–97	217/213	80.5 % overall [sufficient tissue for 81.3 cases, 100 % controls]	Excluded: wt Δ. (ER status included in [14])	– PCBs: 2a (105, 118, 156); 2b (170); 3 (180) OTHER: Mirex in parous/never-lactated	DDT, DDE PCBs: 1b (187); 2b (138); 3 (99, 153, 183) OTHER: CN, TN, OCD, HCB, b-HCH
Bagga et al., 2000 (US) [34]	HMO-based *Cases*: Any BC *Controls*: Breast reduction surgery	1995–96	73/73	Not reported.	Excluded: lactation hx, wt Δ, dietary fat, OC/HRT use, ER status. (ORs age-adj.)		DDT, DDD, DDE
Holford et al., 2000 (US) [83; data = 57]	See [85]	1994–97	304/186	See [57]	See [57]	PCBs: 3 (180, 183)	PCBs: 1b (187); 2a (74, 118, 156); 2b (138, 170); 3 (153, 180, 183)
Lucena et al., 2001 (Spain) [84]	Hospital-based *Cases*: Any BC *Controls*: BBD	1997	65/69	n/a	Results not shown except PCB 28.	PCBs: U (28)	PCBs: 1b (187); 2 (118, 170); 3 (153, 180, 183); U (52, 101, 138, 188).
Stellman et al., 2000 (US) [32]	Hospital-based *Cases*: Incid. invasive (n = 199) or in situ (33) *Controls*: BBD [<4 % atypia] (n = 250) + non-breast surgery (n = 73) w/ adipose tissue from other sites.	1994–96	232/323	95 % overall [adipose tissue from 86 %]	Excluded: wt Δ, dietary fat. Age- & BMI-adjusted means presented, & detailed analysis of DDE, PCBs and total OCs.	*p,p*-DDT PCBs: 2a (74); 2b (138); 3 (183) OTHER: OCD	*o,p*-DDT, *p,p*-DDE PCBs: 1b (187); 2a (118, 156, 167); 2b (170); 3 (99, 146, 153, 172, 178, 180) OTHER: TN, b-HCH, HCB

(*continues on next page*)

Table 1 (*Continued*).

1st author, yr (place) [ref.]	Design	Years: diagnosis	# Cases/ controls	Response rate [other attrition]	Key covariates excluded?	Positive & significant associations	NS or inverse associations
			1A. ADIPOSE TISSUE, RETROSPECTIVE (Studies with >50 cases/controls)				
van't Veer et al., 1997 (5 countries, Europe) [56]	Pop'n.-based (2 ctries.) Hospital (3 ctries.) *Cases*: Incident, postmenopausal BC *Controls*: Matched-age, ctr, menopausal status. All adipose = buttocks.	1991–92	265/341	75–97 % cases 22–91 % ctrls. (varied by ctr.) [adipose tissue from 91 %cases, 98 % ctrls.]	Excluded: lactation hx, dietary fat, ER status; stable wt = eligibility requirement	–	DDE
Woolcott et al., 2001 (Canada) [43; data = 50]	Hospital-based *Cases*: Incident BC, sufficient tissue *Controls*: BBD. Matched-age, study site.	See [50]	See [50]	See [50]	Excluded: wt Δ	DDE (in ER+ tumors only)	*p,p*-DDT PCBs: Σ estimated as (138+153] x 5.2 ≈ arochlor; 1 (187); 2 (118, 156, 170); 3 (99, 153, 180, 183); U (138). OTHER: CN, TN, HCB, Mirex, b-HCH
Zheng 1999a (US) [57; see also 52,58–60,83]	Hospital-based *Cases*: Incident primary BC w/ surgery *Controls*: Incident proliferative BBD excl. atypical hyperplasia	1994–97	304/186	79 % cases, 74 % controls	Excluded: wt Δ, ER status; perhaps OC/HRT use.	–	DDT, DDE
Zheng 1999b (US) [58; data = 57]	See [57]	See [57]	See [57]	See [57]	See [57]	–	OTHER: b-benzene hexachloride (b-BHC) (ns OR in nulliparous)
Zheng 1999c (US) [59; data = 57]	See [57]	See [57]	See [57]	See [57]	See [57]	–	HCB
Zheng 2000a (US) [52; data = 57]	See [57]	See [57]	See [57]	See [57]	See [57]	–	PCBs: 1b (187); 2a (74, 118, 156); 2b (138, 170); 3 (153, 180, 183)

Table 1 (*Continued*).

1st author, yr (place) [ref.]	Design	Years: diagnosis	# Cases/ controls	Response rate [other attrition]	Key covariates excluded?	Positive & significant associations	NS or inverse associations
1A. ADIPOSE TISSUE, RETROSPECTIVE (Studies with >50 cases/controls)							
Zheng 2000b (US) [52; data = 57]	See [57]	See [57]	See [57]	See [57]	See [57]	–	OTHER: TN, OCD
1B. ADIPOSE TISSUE, RETROSPECTIVE (Studies with <50 cases/controls)							
Charles et al., 2001 (US) [44]	Hospital-based *Cases*: Any *Controls*: BBD	1987–89	46/21	n/a	Unadjusted means	*o,p′*-DDE	*o,p′*-DDT, *p,p′*-DDT, *p,p′*-DDE PCBs: 1a (49, 52, 70); 1b (101, 177, 187); 2a (105, 118, 156, 169); 2b (138, 170); 3 (153, 180, 183, 187); U (137, 110, 189)
Dewailly et al., 1994 (Canada) [29]	Hospital-based *Cases*: Incident BC *Controls*: BBD	1991–92	20/17 *[Cases: 9 = ER+ 9 = ER–]*	n/a	Age-adjusted ORs (DDE) & unadjusted means only. [Similar wt Δ, parity; cases > age, < lactation reported]	DDE (in ER+ only) PCBs: 3 (99) in ER+ OTHER: TN in ER-, Mirex in ER+	– PCBs: Σ; 1b (187); 2a (105, 118, 156); 2b (138, 170); 3 (153, 180, 183) OTHER: HCB, b-HCH, OCD
Falck et al., 1992 (US) [39]	Hospital-based; stored specimens *Cases*: Incident BC *Controls*: BBD	1987	20/20	n/a	Age-adjusted ORs (DDE, PCBs) & unadjusted means only [Similar BMI; cases > age]	DDE (ns after adjust for smoking) PCBs: Σ	DDT – OTHER: HCB, TN, Σ(HCE+OCD)
Guttes et al., 1998 (Germany) [85]	Hospital-based *Cases*: Any BC (levels measured both in & distant from tumor) *Controls*: BBD	1993–94	45/20	n/a	Age-adjusted means only. *[Cases > age; other potential confounder data n/a]*	DDE PCBs: 2 (118); 3 (153)	DDT PCBs: 2a (156); 2b (138, 170); 3 (180) OTHER: b-HCH, HCB

(*continues on next page*)

Table 1 (*Continued*).

1st author, yr (place) [ref.]	Design	Years: diagnosis	# Cases/ controls	Response rate [other attrition]	Key covariates excluded?	Positive & significant associations	NS or inverse associations
1B. ADIPOSE TISSUE, RETROSPECTIVE (Studies with <50 cases/controls)							
Hardell et al., 1996 (Sweden) [86; see also 49]	Hospital-based *Cases*: Incident BC *Controls*: BBD	1993–95	22/19	n/a	Excluded: wt Δ, dietary fat.	OTHER: Octachloro-dibenzo-*p*-dioxin (OCDD)	OTHER: Various PCDD & PCDF isomers
Liljegren et al., 1998 (Sweden) [49; data partially used in 86]	Hospital-based *Cases*: Incident invasive BC *Controls*: BBD	1993–95	43/35	n/a	Excluded: wt Δ, dietary fat.	– PCBs: 2a (77) among ER+ postmenopausal women OTHER: HCB (ER+)	DDE PCBs: Σ; 2a (126, 169) (see below: add'l unadjusted means compared)
Mussalo-Rauhamaa et al., 1990 (Finland) [41]	Hospital-based *Cases*: Any BC *Controls*: Breast tissue from accident fatalities	1985–86	44/33	n/a	Excluded: wt Δ, dietary fat, menop. status, OC use, HRT use, ER status.	OTHER: b-BHC	DDT, DDD, DDE PCBs: Σ OTHER: HCB, HCE
Unger et al., 1984 (Denmark) [42]	Hospital-based *Cases*: (i) BC deaths; (ii) Any BC *Controls*: (i) Non-cancer deaths, (ii) BBD, matched-age.	<1984	(i) 18/35 (ii) 14/21	Not reported	Excluded: lactation hx, BMI, wt Δ, dietary fat, menop. status, OC use, HRT use, ER status.	–	DDE PCBs: Σ
2. SERUM, PROSPECTIVE (nested case-control)							
Dorgan et al., 1999 (US) [33]	Volunteer serum bank; ≥4 ml samples *Cases*: Incident BC *Controls*: Matched-age, blood yr, BBD hx.	1977–89 {1977–87} [max 9.5 yr lag, medn 2.7 yr]	105/208	n/a [70 % followed only thru 1983]	Excluded: lactation hx, wt Δ, dietary fat, ER status.	– OTHER: HCB	DDT, DDE PCBs: Σ; 2a (118); 2b (138) OTHER: b-HCH, dieldrin

Table 1 (*Continued*).

1st author, yr (place) [ref.]	Design	Years: diagnosis	# Cases/ controls	Response rate [other attrition]	Key covariates excluded?	Positive & significant associations	NS or inverse associations
			2. SERUM, PROSPECTIVE (nested case-control)				
Helzsouer et al., 1999 [35] (US)	Volunteer serum bank (1 = 1974, 2 = 1989) *Cases*: BC incidence or death (registry, hospital, death cert.) *Controls*: Matched - age, race, menop status, blood yr.	1974/ 1989–94 {1974, 1989 or both} [max 20 yr lag]	1974: 235/235 1989: 105/105	89 % cases, 76 % controls (for fu questionnaire)	Excluded: dietary fat.	–	DDE PCBs: Σ (see below)
Hoyer et al., 1998 (Denmark) [36]	Population-based (1 = 1976–78, 2 = 1981–83) *Cases*: Invasive incident (registry) *Controls*: Matched-age, vital status	1976–92 {1976–78 & 1981–83} [max 17 yr lag]	240/477	75.7 % exam I [78.3 % fu, exam II] [sufficient serum from 89 %]	Excluded: lactation hx, wt Δ, dietary fat, OC use, HRT use, ER status.	– OTHER: Dieldrin	*p,p′*-DDT, *o,p′*-DDT, DDE PCBs: Σ (see below) OTHER: b-HCH
Hoyer et al., 2000a (Denmark) [31; subset used in 36]	Population-based *Cases*: Invasive incident w/ repeated serum available. *Controls*: Matched-age, vital status	See [33] [mean lag 8.1 yr exam 1, 4.8 yr exam 2]	155/274	See [36]	See [36]; adjustment for wt Δ included.	*p,p′*-DDT PCBs: 2b (138)	DDT Σ, *p,p*-DDE PCBs: Σ (see below); 2a (118); 3 (153, 180) OTHER: b-HCH, dieldrin
Hunter et al., 1997 (US) [74]	Nurses Health Study *Cases*: Incident BC (83 % invasive) *Controls*: Matched-age, menopause, HRT, time/ fasting status of blood	1989–92 {1989–90}	240/240	72 % [serum from 27 %; 95 % overall follow-up rate]	Excluded: wt Δ, dietary fat, OC use.	–	DDE PCBs: Σ (defined as sum of higher congeners)

(*continues on next page*)

Table 1 (*Continued*).

1st author, yr (place) [ref.]	Design	Years: diagnosis	# Cases/ controls	Response rate [other attrition]	Key covariates excluded?	Positive & significant associations	NS or inverse associations
			2. SERUM, PROSPECTIVE (nested case-control)				
Krieger et al., 1994 (US) [30]	Nested HMO cohort *Cases*: Incident BC diagnosed >6 mo after serum sample. Random 50/ethnic gp. *Controls*: Matched-age, ethnicity, date joined HMO, exam yr.	1964–90 {1964–90} [min 6 mo lag; max >25 yr]	150/150 (50 black, 50 white, 50 Asian)	n/a n/a	Excluded: lactation hx, wt Δ, dietary fat, OC use, HRT use, ER status.	DDE in blacks only	DDE overall, Asians, whites PCBs: Σ (ns ↑ in blacks)
Laden et al., 2001 (US) [46; subset used in 74]	Nurses Health Study *Cases*: Incident any <1992; incident post-menop. invasive >1992. *Controls*: See [23]	1989–94 {1989–90}	381/381	See [74]	See [74]	– PCBs: Σ; 2a (118); 2b (138); 3 (153) (in nulliparous only)	DDE PCBs: 3 (180) (ns ↑)
Ward et al., 2000 (Norway) [24]	Serum bank (donors, routine health exams) *Cases*: Incident BC (registry) ≥2 yr after serum collection. *Controls*: Matched-age, sample date	1975–93 {1973–91} *[min 2-yr lag]*	150/150	n/a	Excluded: lactation history, BMI, wt Δ, dietary fat, menopausal status, OC/HRT use.	–	DDT, DDE PCBs: Σ; Gps 1, 2, 3 (see below) OTHER: OCD, TN, HCE, b-HCCH. (Others had limited detection; see below)
Wolff et al., 1993 (US) [40]	NYU Wom Hlth Study *Cases*: Incident BC <6 months after enrollment. *Controls*: Matched-age, menopause, day of menstrual cycle, dates of blood samples.	1985–91 {1985–91} [max 6 mo lag]	58/171	n/a	Excluded: wt Δ, dietary fat; no OC or HRT use in past 6 mo	DDE PCBs: Σ (higher congeners)	–
Wolff et al., 2000a (US) [25; see also 40]	See [27] *Cases*: Incident BC ≥6 mo after enrollment. *Controls*: Matched-age, menopause, dates of blood samples.	1987–94 {1985–94} [min 6 mo lag; up to 9 yr]	148/295	n/a [lipid-adjusted estimates for 74 % cases, 72 % ctrls.]	Excluded: wt Δ, dietary fat; no OC or HRT use in past 6 mo.	–	DDE PCBs: Σ

Table 1 (*Continued*).

1st author, yr (place) [ref.]	Design	Years: diagnosis	# Cases/ controls	Response rate [other attrition]	Key covariates excluded?	Positive & significant associations	NS or inverse associations
				3. SERUM, RETROSPECTIVE			
Dello Iacova et al., 1999 (Italy) [87]	Community *Cases*: Any BC, 1st tx, single hospital *Controls*: Ongoing cohort study, healthy	1997–98	170/195	n/a	Unadjusted means except DDE; wt Δ, OC use, HRT use, ER status excluded.	OTHER: Heptachlor (unadj. means only)	DDT, DDE OTHER: b-BHC, endrin aldeide
Demers et al., 2000 (Canada) [51]	Hosp./pop'n. controls. *Cases*: Incident 1ary invasive BC *Controls*: (i) hosp. pts; (ii) insurance registry; matched-age, region.	1994–97	315/ 219-hosp 307-pop'n.	91 % cases 89 % hosp. ctrls. 47 % pop'n. ctrls.	Excluded: wt Δ, dietary fat, menop status, ER status.	–	DDT, DDE PCBs: 3 (153) OTHER: b-HCH, TN, OCD
Gammon et al., 2002 [54]	Population-based *Cases*: Incident in situ or invasive *Controls*: Frequency matched: age	1996–97	646/429	83 % cases 68 % ctrls.	Excluded: Dietary fat. (BMI current and at age 20 included as estimate of wt Δ.)	–	DDT, DDE PCBs: 2a (118), 2b (138), 3 (153, 180) OTHER: OC+TN, dieldrin
Lopez-Carillo et al., 1997 [61] (Mexico)	Hospital-based *Cases*: Any BC, ≥20 yr residents, Mexico City *Controls*: Patients; matched-age.	1994–96	141/141	81 % cases 72 % ctrls.	Excluded: wt Δ, dietary fat, OC use, HRT use, ER status.	–	*p,p*-DDT, DDE
Mendonca et al., 1999 [62] (Brazil)	Hospital-based *Cases*: Any BC *Controls*: Hospital visitors, matched-age.	1995–96	177/350	99 % cases 79 % ctrls. [serum from 92 % cases, 95 % ctrls.]	Excluded: wt Δ, dietary fat, OC use, HRT use, ER status	–	DDE
Millikan et al., 2000 (US) [45]	Population-based *Cases*: Incident invasive BC *Controls*: Population	1993–96	889/841	74 % cases 53 % ctrls. [serum from 98 %]	Excluded: wt Δ, dietary fat, ER status. (Fruit, vegetable, fish consumption included.)	– PCBs: Σ (defined as 118, 138, 153, 180) in African Americans; ns ↑ in nulliparous & low BMI gps.	DDE (ns, low BMI) PCBs: Σ in whites, Asians

(*continues on next page*)

Table 1 (*Continued*).

1st author, yr (place) [ref.]	Design	Years: diagnosis	# Cases/ controls	Response rate [other attrition]	Key covariates excluded?	Positive & significant associations	NS or inverse associations
				3. SERUM, RETROSPECTIVE			
Moysich et al., 1998 (US) [22; see also 76]	Population–based *Cases*: Postmenopausal incident (hospital); blood before chemo or radiation + w/in 3 mo of surgery. *Controls*: Matched-age, county of residence, date of blood draw.	1986–1991	154/191	57 % cases 47 % ctrls. [serum from 63 %]	Excluded: wt Δ, dietary fat, ER status. Fruit & vegetable intake included.	– PCBs: Σ (73 congeners, see below) in parous never lactators	DDE OTHER: Mirex, HCB
Moysich et al., 1999b (US) [76; data used in 22]	See [34] By cytochrome P450 polymorphism (drug-metabolism)	See [22]	See [22]	See [22]	See [22]	PCBs: Σ in variants of CYP1A1 with valine substitutions.	PCBs: Σ in common variant
Olaya-Contreras et al., 1998 [88] (Columbia)	Hospital-based *Cases*: *Controls*: Matched, age	1995–96	153/153	93 % overall	Excluded: wt Δ, dietary fat, OC/HRT use, ER status.	DDE	
Romieu et al., 2000 [23] (Mexico)	Population-based *Cases*: Incident *Controls*: Age-stratified random population sample	1990–95	126/120	94 % cases, 89 % ctrls.	Excluded: wt Δ, dietary fat, OC use, ER status.	DDE (if adjusted for *p,p′*-DDT)	*p,p′*-DDT
Schecter et al., 1997 [37] (Vietnam)	Hospital-based *Cases*: Incident invasive; continuous residence N. Vietnam *Controls*: BBD pts; matched-age, area of residence	1994	21/21	Not reported	Excluded: wt Δ dietary fat, menopausal status, OC use & HRT use, ER status.	–	*p,p*-DDT, total DDT, *p,p*-DDE (*p,p*-DDT ns OR)

Table 1 (*Continued*).

1st author, yr (place) [ref.]	Design	Years: diagnosis	# Cases/ controls	Response rate [other attrition]	Key covariates excluded?	Positive & significant associations	NS or inverse associations
				3. SERUM, RETROSPECTIVE			
Wolff et al., 2000b (US) [53]	Hospital-based *Cases*: Incident BC *Controls*: Gp1 = BBD (n = 181), gp 2 = routine screenings & minor surgeries (n = 175) no hx of BBD; matched-age, race.	1994–96	175/356 [181 ctrls. = BBD]	65 % overall	Excluded: wt Δ, dietary fat	–	*p,p*-DDT, DDE PCBs: Σ; High chlorination (Σ 118, 153, 141, 138, 183, 187, 167, 174, 177, 156, 180, 170, 201, 203); low chlorination (Σ 28, 66, 74, 99, 101) OTHER: TN
Zheng 2000c (US) [89; sample overlap with 57]	Hospital & population *Cases*: Incident BC from hospital or 1 county (tx initiated 1st) *Controls*: (1) Hospital = BBD excluding atypical hyperplasia, (2) County residents. Frequency matched-age.	1995–97	326/347- hosp. 149/155- pop'n.	Hospital: 77 % cases, 71 % cntrls. County: 74 % cases, 61 % cntrls.	Excluded: wt Δ	–	*p,p*-DDE PCBs: Σ (74, 118, 138, 153, 156, 170, 180, 183, 187)

Abbreviations: *Chemicals*: CN = *cis*-nonachlor; TN = *trans*-nonachlor; AC = a-chlordane; YC = y-chlordane; OCD = oxychlordane; HCB = hexachlorobenzene; HCE = heptachlorepoxide; b-HCH = beta-hexachlorocyclohexane; OCD = oxychlordane; PCBs = polychlorinated biphenyls; PCDDs = poly-chlorinated dibenzo-*p*-dioxins; PCDFs = poly-chlorinated di-benzofurans. *Other:* BBD = benign breast disease; BC = breast cancer; OC = oral contraceptives; HRT = hormone replacement therapy; wt Δ = weight change. ER = estrogen receptors. Conversions/estimates: mcg/kg = ng/g; mg/kg × 1000 = ng/g; mcg/g × 1000 = ng/g. Wet wt: ppb = ng/ml; ppm × 1000 = ng/ml. Midpoints used to estimate medians when not given. Other notes: *Additional PCBs examined*: Helzsouer [35] reported NS differences (not shown) in: 1b (177, 187, 201); 2a (74, 118, 156); 2b (138,170); 3 (153,180, 183, 203); U (28, 146, 172, 178, 189, 193, 194, 195, 206). Ward [24] also examined (ns): 1b (177, 187, 201) 2a (74, 105, 118, 126, 156, 157, 167, 169, 189); 2b (170); 3 (99, 146, 153, 172, 178, 180, 183, 194, 195, 206, 209); several U. Liljegren [49] showed ns mean differences for: 1a (52), 1b (101, 177, 187, 201), 2a (66, 74, 105, 118, 156), 2b (138, 170), 3 (99, 153, 183, 203/196), U (28, 47, 110, 114, 157, 128/167, 171, 172, 180, 193, 189, 194, 195, 202, 206, 207, 208, 209). *Other compounds examined*: Ward [24] found limited detection of dieldrin, aldrin, endrin, *o,p′*-DDT, Mirex, OCDD, OCDF and other dioxins; further analysis was not possible.

measured OCs in adipose tissue, primarily from mammary biopsies in cancer cases and controls with benign breast disease (BBD). Eight of these studies each included fewer than 50 cases and controls and had limited capacity for covariate adjustment. Ten studies (three with overlap in populations) used serum samples collected from 6 months to 25 years prior to diagnosis. An additional 13 studies (two with some overlap) used serum collected at or after diagnosis.

As summarized in Table 1 and detailed below, regardless of the medium used to measure OC levels, studies conducted to date do not provide consistent evidence that any of the OCs examined thus far play a role in the initial breast cancer risk. A compound-specific review follows with discussion of how the lack of evidence for adverse effects might be explained by factors related to study design, or by variation in risk across subgroups.

DDT/DDE

p,p′-DDT is only weakly estrogenic, but persists in the food chain [2,11,13]. Of 10 studies reporting multivariate-adjusted ORs, only one [31] found elevated risk associated with adipose or serum levels of *p,p′*-DDT (Table 2). Another eight studies compared mean levels of *p,p′*-DDT but incorporated limited covariate adjustment. One of these [32] reported significantly higher levels in cases, after adjusting for age and BMI. *o,p′*-DDT—a more estrogenic but less persistent isomer—has not been studied epidemiologically because even studies using serum collected in the 1970s, when levels were higher, had too few participants with detectable levels to analyze associations separately [33]. A number of studies [31,33–37] examined associations with total DDT, estimated as the sum of *p,p′*-DDT and several metabolites. None reported evidence of adverse effects in relation to high levels of the sum of these DDT metabolites.

Most studies of DDE have examined *p,p′*-DDE, the main persistent metabolite of DDT. Since this compound, which is not estrogenic, is found in foods, elevated levels may not be the result of direct exposure to *p,p′*-DDT [11,37,38]. Although a few studies conducted from 1992 to 1994 [29,39], in ER positive [ER+] tumors; [40] suggested possible increased risk associated with *p,p′*-DDE, other early exploratory studies did not find increased risk [41,42], and the majority of studies since that time have reported inverse or null associations (Table 3). Two of the three initial studies reporting elevated *p,p′*-DDE in subjects with breast cancer were very small ($\leq$20 cases) and unable to conduct adequate multivariate adjustment [29,39]. Positive associations reported by [40] were based on cases diagnosed within 6 months of serum collection and were not confirmed in a follow-up using cases with similar OC levels, diagnosed >6 months after enrollment [25]. Adjustment for serum lipids, omitted in the first analysis, did not influence results of the second study.

To date, 27 moderate to large (i.e., at least 50 cases) studies with multivariate adjustment have explored associations between *p,p′*-DDE and breast cancer. Other than the early studies referenced above, only one other study ([23] in a Mexico City study of parous women) has reported significant positive associations overall. Two additional studies reported significant positive associations in different subgroups ([43] in ER negative [ER–] tumors; [30] in blacks but not whites or Hispanics). One small study [44], using adipose tissue samples collected from 1987 to 1989 and analyzed without adjustments did report significantly higher means of the more estrogenic but short-lived metabolite *o,p′*-DDE in cases than controls.

In summary, the majority of studies, regardless of medium or study size, showed no negative impact of DDT or DDE on risk of breast cancer. Risk was not more likely to be apparent in populations with higher mean concentrations of DDT or DDE. In fact, many studies show risk ratios below 1, as above 1 (as seen on Figs. 1a and 1b).

Text continues on p. 1990.

Table 2 DDT and breast cancer risk.

1st author, year published [ref.]	+ve	ns	–ve	Years	Place	# Cases/ controls	Mean DDT		Adjusted?	ORs				
							Cases	Controls		1	2	3	4	5
1A. ADIPOSE TISSUE, RETROSPECTIVE (>50 cases/controls)														
Aronson [50][a]			–	1995–97	Canada	217/213	22.0	19.3 ng/g	y	1.0	0.8	0.9	1.2	
Bagga [34][†]			–	1995–96	US	73/73	261.6	267.3 ng/g	age	–				
Stellman [32][†,1]	↑			1994–96	US	232/323	12.3	12.1* ng/g	age, bmi	–				
Woolcott [43]		ns		1995–97	Canada	217/213	23.5 = ER– 21.3 = ER+	19.3 ng/g	not shown	– –				
Zheng [57][†]			–	1994–97	US	304/186	51.8	55.6 ng/ml	y	1.0	0.8	0.6	0.8	
1B. ADIPOSE TISSUE, RETROSPECTIVE (<50 cases/controls)														
Charles [44]														
p,p′-DDT		ns		1987–89	US	43/21	102.0	77.8 ng/g	n	–				
o,p′-DDT		ns		″	″	″	15.9	10.4 ng/g	n	–				
Falck [39]		ns		1987	US	20/20	216	148 ng/g	n	–				
Guttes [85][†]			–	1993–94	Germany	45/20	30	28 ng/g	age	–				
Mussalo-Rauhamaa [41]		ns		1985–86	Finland	44/33	70	60 ng/g	n	–				
2. SERUM, PROSPECTIVE														
Dorgan [33]														
p,p′-DDT			–	1977–87	US	105/208	–	–	y	1.0	1.0	1.1	0.4	
Total DDT			–	″	″	″	–	–	y	1.0	1.1	0.3	0.8	
Hoyer [36][1]														
p,p′-DDT			–	1976–83	Denmark	240/477	pooled	141 ng/g	y	1.0	1.1	0.9	1.2	
Total DDT			–	″	″		pooled	1326 ng/g	y	1.0	0.8	0.9	0.8	
Hoyer [31]														
p,p′-DDT[1]	↑			1976–78 1981–83	Denmark	155/274	pooled	144 ng/g (bs) 46 ng/g (fu)	y	1.0	1.3	2.1	3.6*	
Total DDT[1]		ns					pooled	1350 ng/g (bs) 1191 ng/g (fu)	y	1.0	1.1	1.4	2.4	
Ward [24]			–	1973–91	Norway	150/150	119.5	137.7 ng/g	y	1.0	0.2	0.5	0.3	

(*continues on next page*)

Table 2 (*Continued*).

1st author, year published [ref.]	+ve	ns	–ve	Years	Place	# Cases/ controls	Mean DDT		Adjusted?	ORs				
							Cases	Controls		1	2	3	4	5
						3. SERUM, RETROSPECTIVE								
Dello Iacova [87][1]		ns		1997–98	Italy	170/195	2.47	1.77 ng/ml	n	–				
Demers [51]														
Pop'n. controls			–	1994–97	Canada	315/307	12.7	11.0 ng/g	y	1.0	0.6	0.5	0.7	0.8
Hosp. controls			–	″	″	315/219	12.7	12.5 ng/g	y	1.0	0.9	1.1	1.1	1.4
Gammon [54]			–	1996–97	US	633/418	69.0	69.3 ng/g	y	1.0	0.7	1.0	1.2	1.2
Lopez-Carillo [61]			–	1994–96	Mexico	141/141	62	85 ng/g	n	–				
Romieu [23]			–	1990–95	Mexico	126/120	150	230 ng/g	n	–				
Premenopausal			–	″	″	56/64	220	130 ng/g	n	–				
Postmenopausal		ns		″	″	64/62	250	180 ng/g	n	–				
Schecter [37][2]														
p,p′-DDT		ns		1994	Vietnam	21/21	2.33	2.37 ng/ml	y	1.0	2.2	1.2		
Total DDT			–	″	″		15.90	20.95 ng/ml	y	1.0	0.4	1.1		
Wolff [53][†]			–				30	28 n/g	y	1.0	1.2	1.3		

DDT = *p,p*-DDT unless otherwise noted; total DDTs are sum of several isomers. Statistical significance defined as $p < 0.05$/CI excludes null or $p < 0.10$. bs = baseline; fu = follow-up.
[1]Medians vs. means.
[2]Means are wet wt rather than lipid basis.
[a–e]Supplementary analysis gave similar results when stratified by: a: pre/post-menopausal; b-ER status; c: lactation/parity group; d: BMI group.
[†]Other notes: Zheng [57] are age- and lipid-adjusted means; Wolff [53]: Results similar for blacks, hispanics, and whites.

Table 3 DDE and breast cancer risk.

1st author, year published [ref.]	+ve	ns	–ve	Years	Place	# Cases/ controls	Mean DDE Cases	Mean DDE Controls	Adjusted?	ORs 1	ORs 2	ORs 3	ORs 4	ORs 5
1A. ADIPOSE, RETROSPECTIVE: STUDIES WITH >50 CASES/CONTROLS														
Aronson [50][a,c]		ns		1995–97	Canada	217/213	693	596 ng/g	y	1.0	1.0	0.9	1.6	
Bagga [34][†]		ns		1995–96	US	73/73	800*	709 ng/g	age					
Stellman [32][†1,b]			–	1994–96	US	232/323	419	374 ng/g	y	1.0	1.1	0.7		
van't Veer [56]			–	1991–92	Europe	265/341	1350	1510 ng/g	y	1.0	1.1	0.7	0.5*	
Woolcott [43; = 50]														
ER+			–	″	″	″	638	596 ng/g	y	1.0	0.9	1.1		
ER–	↑			1995–97	Canada	217/213	906	596 ng/g	y	1.0	0.8	2.4*		
Zheng [57][†,1]			–	1994–97	US	304/186	736.5	784.1 ng/ml	y	1.0	1.3	0.9	0.9	
1B. ADIPOSE, RETROSPECTIVE: STUDIES WITH <50 CASES/CONTROLS														
Charles [44]														
p,p′-DDE		ns		1987–89	US	43/21	1472.3	1387.7 ng/g	n	–				
o,p′-DDE	↑			″	″	″	1.4	0.5 ng/g	n	–				
Dewailly [29]														
ER+	↑			1991–92	Canada	9/17	2132*	765 ng/g	age					
ER–			–	″	″	″	609	765 ng/g	age					
Falck [39]	↑			1987	US	20/20	2200*	1487 ng/g	age					
Guttes [85][†]	↑			1993–94	Germany	45/20	805*	496 ng/g	age					
Liljegren [49][a,b]			–	1993–95	Sweden	43/35	767	1026 ng/g	y	1.0	0.4			
Mussalo-Rauhamaa [41]			–	1985–86	Finland	44/33	96	98 ng/g	n					
Unger (i) [42]			–	<1984	Denmark	18/35	< detection	197 ng/g	n					
Unger (ii)			–	″	″	14/21	123	125 ng/g	n					
2. SERUM, PROSPECTIVE														
Dorgan [33]			–	1977–87	US	105/208	–	–	y	1.0	0.9	0.4	0.8	
Helzsouer '74 [35][a,b]			–	1974	US	235/235	1698.9	1920.3 ng/g	y	1.0	1.2	1.0	0.9	0.7
ER+			–	″	″	″	–	–	y	1.0	0.9	0.8		
ER–		ns		″	″	″	–	–	y	1.0	1.1	1.7		

(*continues on next page*)

Table 3 (*Continued*).

1st author, year published [ref.]	+ve	ns	–ve	Years	Place	# Cases/ controls	Mean DDE		Adjusted?	ORs				
							Cases	Controls		1	2	3	4	5
2. SERUM, PROSPECTIVE														
Helzsouer '89 [35]-All			–	1989	US	105/105	1311.9	1586.3 ng/g	y	1.0	1.2	0.6		
Premenopausal		ns		″	″	″	–	–	y	1.0	4.3	1.4		
Postmenopausal		ns		″	″	″	–	–	y	1.0	1.6	0.5		
ER+		ns		″	″	″	–	–	y	1.0	2.3	0.6		
ER–			–	″	″	″	–	–	y	1.0	0.7	0.2		
Hoyer [36][1]			–	1976–83	Denmark	240/477	pooled	1183 ng/g	y	1.0	0.8	0.8	0.9	
Hoyer [31][1]			–	″	″	155/274	pooled	1197 ng/g (bs) 1169 ng/g (fu)	y	1.0	1.0	0.8	1.4	
Hunter [74][1,2]			–	1989–90	US	240/240	6.01	6.97 ng/ml	y	1.0	0.8	0.5	0.7	0.7
Laden [46][1,c,d]			–	″	″	381/381	768	817 ng/g	y	1.0	1.0	0.5	0.9	0.8
Krieger [30][1]-All			–	1964–90	US	150/150	43.3	43.1 ng/ml	y	1.0	1.3	1.3		
White		ns		″	″		35.7	35.0 ng/ml	y	1.0	1.9	2.4		
Black	↑			″	″		49.2	43.4 ng/ml	y	1.0	2.3	3.9*		
Asian			–	″	″		45.1	50.8 ng/ml	y	1.0	0.9	0.7		
Ward [24][b]			–				1230	1260 ng/g	y	1.0	0.7	1.0	1.2	
Wolff [40][1]	↑						11.0	7.7* ng/ml	y	1.0	1.7	4.4*	2.3	3.7*
Wolff [25][b]			–				977	1097 ng/g	y	1.0	0.8	0.6	1.3	
3. SERUM, RETROSPECTIVE														
Dello Iacova [87]			–	1997–98	Italy	170/195	9.55	8.98 ng/ml	y	1.0	0.8	1.2		
Demers [51]														
Pop'n. controls			–	1994–97	Canada	315/307	509	480 ng/g	y	1.0	0.8	1.1	0.9	1.0
Hosp. controls		ns		″	″	315/219	509	463 ng/g	y	1.0	0.9	0.7	1.5	1.4
Gammon [54][a,b,c,d]			–	1996–97	US	643/427	672	646 ng/g	y	1.0	0.9	0.9	0.9	1.2
Lopez-Carillo [61][a]			–	1994–96	Mexico	141/141	563	506 ng/g	y	1.0	0.6	0.8		
Mendonca [62][2]			–	1995–96	Brazil	177/350	5.1	4.8 ng/ml	y	1.0	1.0	1.3	1.1	0.8
Millikan [45][a,b,c,e]			–	1993–96	US	889/841	1960 660	1690 (Black) 760 (White) ng/g	y	1.0	1.1	1.1		
Black, BMI <25.0		ns							y	1.0	2.7	3.8		
White, BMI <25.0			–						y	1.0	0.7	0.9		

Table 3 (*Continued*).

1st author, year published [ref.]	+ve	ns	–ve	Years	Place	# Cases/ controls	Mean DDE Cases	Mean DDE Controls	Adjusted?	ORs 1	ORs 2	ORs 3	ORs 4	ORs 5
3. SERUM, RETROSPECTIVE														
Moysich [76] – All			–				11.5	10.8 ng/g	y	1.0	1.0	1.3		
Never lactated, parous		ns					13.2	10.8 ng/g	y	1.0	2.0	1.8		
Ever lactated			–				10.4	10.4 ng/g	y	1.0	0.8	1.3		
Olaya-Contreras [88]	↑			1995–96	Columbia	153/153	3.30	2.50 ng/ml	y	1.0	1.2	2.0*		
Premenopausal		ns		″	″	60/60	3.02	2.1 ng/ml	y	1.0	1.4	2.5		
Postmenopausal		ns		″	″	93-93	3.45	3.0 ng/ml	y	1.0	1.1	1.9		
Romieu [23] – All	↑			1990–95	Mexico	126/120	3840	2510 ng/g	y	1.0	1.2	2.3	3.8*	
Premenopausal		ns		″	″	56/64	2400	1930 ng/g	y	1.0	1.4	2.5	2.4	
Postmenopausal		ns		″	″	64/62	5100	3120 ng/g	y	1.0	1.1	2.4	5.3	
Schecter [37][2]			–	1994	Vietnam	21/21	12.17	16.67 ng/ml	y	1.0	0.5	1.1		
Wolff [53][b,e]			–	1994–96	US	175/356	610	660 ng/g	y	1.0	0.8	0.9		
Zheng [89][†,b,c]			–	1995–97	US	475/502	460.1	456.2 ng/g	y	1.0	1.1	1.0		

DDE = p,p-DDE unless otherwise noted. Statistical significance defined as $p < 0.05$/CI excludes null or $p < 0.10$. bs = baseline fu = follow-up

[1]Medians rather than means.

[2]Means are wet weight rather than lipid basis.

[a–e]Supplementary analysis found no meaningful differences when stratified by: a: pre/post-menopausal; b: ER status; c: lactation/parity group; d: BMI group; e: ethnic groups.

[†]Other notes: Stellman [32] are age/BMI-adjusted means. Zheng [57] are age- and lipid-adjusted means.

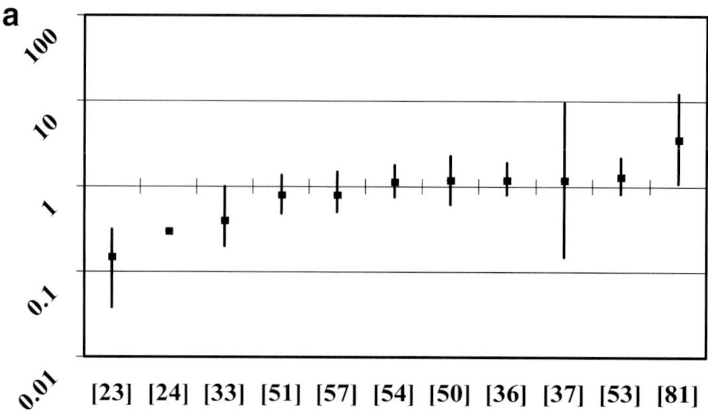

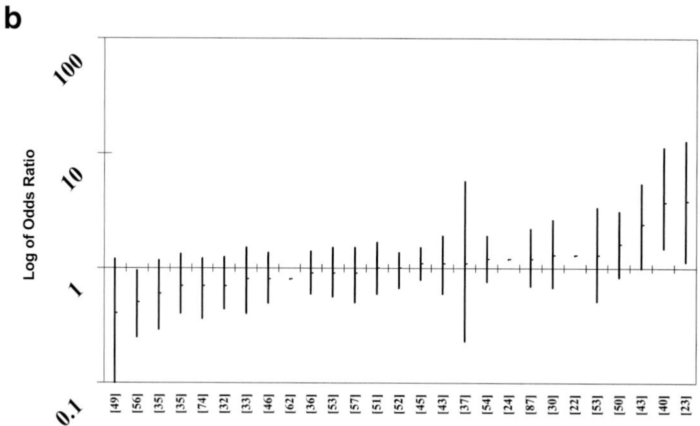

Fig. 1 (a) DDT and breast cancer risk. (b) DDE and breast cancer risk.

PCBs

Four of 17 published studies on total PCBs and breast cancer risk reported significant positive associations (Table 4). The report of Wolff et al. [40] showed a significant positive association that was not replicated in the follow-up study [25]. In the three other studies, elevated risk was limited to various diverse population subgroups (Moysich et al. [22] in parous never-lactators; Millikan et al. [45] in blacks, particularly if obese; Laden et al. [46] in nulliparous women).

Examining total PCBs may mask variation in congener-specific effects, as some of the 209 congeners are thought to be estrogenic (type 1a), while others appear to be antiestrogenic (types 2a and b), or to have no estrogenic activity (type 1b, based on groupings proposed by Wolff and Toniolo [8]; Wolff et al. [47]). Net effects of PCB mixtures may vary depending on relative concentrations of these different congeners [47,48]. The low-chlorination, potentially estrogenic congeners of group 1a are not persistent, as they are readily metabolized and eliminated [48]. However, the two studies [44,49] reporting means for congeners in this group did not find significant differences between cases and controls (Table 5). Given their rapid elimination, retrospective studies may not be reliable for assessing any long-term effects of these congeners. The prospective studies have had few subjects with detectable levels of 1a congeners [33].

Text continues on p. 2000.

Table 4 Total PCBs and breast cancer risk.

1st author [ref.]	+ve	ns	–ve	Years	Place	# Cases/ controls	Mean PCB Σ Cases	Mean PCB Σ Controls	Adjusted?	ORs 1	2	3	4	5
IA. ADIPOSE TISSUE, RETROSPECTIVE (>50 CASES)														
Stellman [32][†,1,b]			–	1994–96	US	232/323	294.7	257.1 ng/g	y	1.0	1.1	1.0		
Woolcott [43; data = 50]														
ER+			–	″	″	″	920		y	1.0	1.3	1.3		
ER–		ns		1995–97	Canada	217/213	1,020	870 ng/g	y	1.0	1.1	1.7		
Zheng [60][†,a,b,c]			–	1994–97	US	304/186	494.1	478.6 ng/g	y	1.0	0.6	0.7		
IB. ADIPOSE TISSUE, RETROSPECTIVE (< 50 CASES)														
Dewailly [29]														
ER+			–	″	″	″	404.7	397.0 ng/g	age	–				
ER–			–	1991–92	Canada	9/17	331.5	397.0 ng/g	age	–				
Falck [39]	↑			1987	US	20/20	1965*	1395 ng/g	age	–				
Liljegren [49][a,b]														
All			–	1993–95	Sweden	43/35	1205	1149 ng/g	y	1.0	0.7			
Postmenop, ER+		ns		″	″	″	–	–	y	1.0	1.8			
Mussalo-Rauhamaa [41]			–	1985–86	Finland	44/33	1050	1300 ng/g	n	–				
Unger [42] (i)		ns		<1984	Denmark	18/35	6470	5120 ng/ml	n					
Unger [42] (ii)			–	14/21	123	125 ng/g	3890	3930 ng/ml	n					
II. SERUM, PROSPECTIVE														
Dorgan [33]		ns		1977–87	US	105/208	–	–	y	1.0	0.7	1.1	0.7	
Helzsouer 1974 [35][b]														
All			–	1974	US	235/235	735.3	663.6 ng/g	y	1.0	1.4	0.9	1.1	1.1
Premenopausal		ns		″	″	″	–	–	y	1.0	0.7	2.2		
Postmenopausal			–	″	″	″	–	–	y	1.0	0.7	0.6		
Helzsouer 1989 [35][a,b]														
All			–	1989	″	105/105	327.7	332.9 ng/g	y	1.0	0.8	0.8		
Premenopausal		ns		″	″	″	–	–	y	1.0	1.5	2.1		
Postmenopausal			–	″	″	″	–	–	y	1.0	0.8	0.7		
Hoyer [36][1]			–	1976–84	Denmark	240/477	pooled	1100 ng/g	y	1.0	0.9	0.8	1.1	
Hoyer [31][1]			–	″	″	155/274	pooled	1102 ng/g (bs) 979 ng/g (fu)	y	1.0	0.8	0.8	1.6	

(*continues on next page*)

Table 4 (*Continued*).

1st author [ref.]	+ve	ns	–ve	Years	Place	# Cases/ controls	Mean PCB Σ Cases	Mean PCB Σ Controls	Adjusted?	ORs 1	ORs 2	ORs 3	ORs 4	ORs 5
II. SERUM, PROSPECTIVE														
Hunter [74][†,1,2,b]			–	1989–90	US	240/240	5.1	5.2 ng/ml	y	1.0	0.6	0.5	0.5	0.7
Laden [46][b,d]														
All			–	1989–94	US	381/381	544	543 ng/g	y	1.0	0.7	0.8	0.9	0.8
Nulliparous	↑						–		y	1.0	0.8	5.3*	–	–
Krieger [30][2]														
All			–	1964–90	US	150/150	4.4	4.8 ng/ml	y	1.0	1.2	0.9		
White			–	″	″	50/50	3.6	4.2 ng/ml	y	1.0	1.0	0.5		
Black		ns		″	″	50/50	4.8	4.5 ng/ml	y	1.0	1.7	2.2		
Asian			–	″	″	50/50	4.9	5.6 ng/ml	y	1.0	1.2	0.8		
Ward [24][b]			–	1973–91	Norway	150/150	776.1	806.6 ng/g	y	1.0	0.6	0.8	0.5	
Wolff [40][1]	↑			1985–91	US	58/171	8.0	6.7 ng/ml	y	1.0	5.2*	7.0*	4.1	4.4
Wolff [25][b]		ns		1985–94	US	148/295	683	663 ng/g	y	1.0	1.6	1.2	2.0	
III. SERUM, RETROSPECTIVE														
Millikan [45][†,a,b]														
All			–	1993–96	US	889/841	560 (black) 380 (white)	510 ng/g 380 ng/g	y	1.0	1.3	1.1		
Black	↑			″	″		–	–	y	1.0	1.4	1.7*		
Black, BMI >30	↑			″	″		–	–	y	1.0	1.7	4.3*		
White			–	″	″		–	–	y	1.0	1.3	1.0		
White, BMI >30		ns		″	″		–	–	y	1.0	2.0	1.5		
Moysich [22][2,†]														
All			–	1986–91	US	154/191	4.3	4.1 ng/g serum	y	1.0	0.7	1.1		
Never lact, parous	↑			″	″		4.6	4.0 ng/g serum	y	1.0	1.7	2.9*		
Ever lactated			–	″	″		4.3	4.3 ng/g serum	y	1.0	0.4	0.7		
Lo chlor, never lact.	↑			″	″		–	–	y	1.0	0.7	3.6		
Hi chlor, never lact.			–	″	″		–	–	y	1.0	0.5	1.5		
Wolff [53][b,e]														
HPCB			–	1994–96	US	175/356	600	620 ng/g	y	1.0	0.9	0.8		
LPCBs		ns		″	″		110	110 ng/g	y	1.0	1.5	1.0		

Table 4 (*Continued*).

1st author [ref.]	+ve	ns	–ve	Years	Place	# Cases/ controls	Mean PCB Σ		Adjusted?	ORs				
							Cases	Controls		1	2	3	4	5
III. SERUM, RETROSPECTIVE														
Zheng [89][†,2,b,c]														
All			–	1994–97	US	304/186	733.1	747.6 ng/ml	y	1.0	1.0	1.0		
Nulliparous		ns		″	″				y	1.0	1.5	0.8		

HPCBs = high-chlorination PCBs; LPCBs = low-chlorination PCBs. Statistical significance defined as $p < 0.05$/CI excludes null or $p < 0.10$.

[1]Medians presented rather than means.

[2]Means are wet weight rather than lipid basis.

[a–e]Supplementary analysis found no meaningful differences when stratified by: a: pre/post-menopausal; b: ER status; c: lactation/parity group; d: BMI group; e: ethnic groups.

[†]Other notes: Millikan [45] found lower ORs in low-moderate chlorination congeners than in high chlorination congeners; as well as NS > risk in nulliparous. Moysich [22] reported higher ORs in low-moderate chlorination congeners, particularly among parous women who never lactated (cite OR). Stellman [32] are age/BMI-adjusted means. Zheng [89] present age- and lipid-adjusted means. Helzsouer [35] examined the following PCBs individually and by group; results were NS and not shown: 1b (177, 187, 201); 2a (74, 118, 156); 2b (138,170); 3 (153,180,183, 203); U (28, 146, 172,178, 189,193, 194, 195, 206). Hoyer [36] – PCBs are sum of: 1 (177, 187, 201); 2 (74, 105, 118, 156, 170, 189); 3 (99, 146, 153, 172, 178, 180,183, 194, 195, 206); U (28, 52, 56, 66, 101, 110, 138, 193, 203).

Table 5 Individual PCB congeners and breast cancer risk.

PCB # (GRP)	+ve, Significant 1st author, yr	Mean: cases	Controls	+ve, NS 1st author, yr	Mean: cases	Controls	Inverse/none 1st author, yr	Mean: cases	Controls
Σ, gp1	*Serum, prospective*								
	None	–	–	None	–	–	Ward [24]	62.6	65.0 ng/g
49 (1a)	*Adipose tissue*								
	None	–	–	None	–	–	Charles [44]	0.5	1.0 ng/g
52 (1a)	*Adipose tissue*								
	None	–	–	Charles [44]	2.9	2.0 ng/g	Liljegren [49]	0.8	0.8 ng/g
70 (1a)	*Adipose tissue*								
	None	–	–	Charles [44]	3.7	2.0 ng/g	None	–	–
101 (1b)	*Adipose tissue*								
	None	–	–	Charles [44]	4.6	3.8 ng/g	None	–	–
177 (1b)	*Adipose tissue*								
	None	–	–	None	–	–	Liljegren [49]	14.5	14.0 ng/g
							Charles [44]	7.5	6.7 ng/g
	Serum, prospective								
	None	–	–	None	–	–	Ward [24]	12.1	12.4 ng/g
187 (1b)	*Adipose tissue*								
	None	–	–	Charles [44]	29.4	24.7 ng/g	Aronson [50]	25.7	24.2 ng/g
				Stellman [32]	16.2	12.8 ng/g	Dewailly [29] [ER+/–]	19.4/18.7	20.4 ng/g
				Woolcott [43] ER+	27.1	24.2 ng/g	Liljegren [49]	54.9	55.9 ng/g
				Woolcott [43] ER–	25.3				
							Zheng [60]	28.8	ng/g pooled
	Serum, prospective								
	None	–	–	None	–	–	Ward [24]	37.2	38.2 ng/g
							Zheng [89]	61.5	ng/g pooled
201 (1b)	*Adipose tissue*								
	None	–	–	None	–	–	Liljegren [49]	21.2	21.6 ng/g
	Serum, prospective								
	None	–	–	None	–	–	Ward [24]	14.0	14.7 ng/g

Table 5 (*Continued*).

PCB # (GRP)	+ve, Significant 1st author, yr	Mean: cases	Controls	+ve, NS 1st author, yr	Mean: cases	Controls	Inverse/none 1st author, yr	Mean: cases	Controls
Σ, gp2	*Adipose tissue*								
	None	–	–	None	–	–	Zheng [60]	269.9	ng/g pooled
	Serum, prospective								
	None	–	–	None	–	–	Ward [24]	165.2	169.8 ng/g
	Serum, retro.								
	None	–	–	None	–	–	Zheng [89]	247.0	ng/g pooled
74 (2a)	*Adipose tissue*								
	Stellman [32]	29.6	26.7 ng/g	None	–	–	Liljegren [49] Zheng [60]	17 307	18 ng/g ng/g pooled
	Serum, prospective								
	None	–	–	None	–	–	Ward [24]	27.0	28.7 ng/g
77 (2a)	*Adipose tissue*								
	None	–	–	None	–	–	Liljegren [49]	5.2	4.9 pg/g
105 (2a)	*Adipose tissue*								
	Aronson [50]	7.1	6.3 ng/g				Charles [44] Dewailly [29] Liljegren [49]	17.7 7.3/3.9 [ER+/–] 8	19.6 ng/g 6.0 ng/g 7 ng/g
	Serum, prospective								
	None	–	–	None	–	–	Ward [24]	11.1	11.2 ng/g
118 (2a)	*Adipose tissue*								
	Aronson [50] [in premenopausal]	30.3	24.7 ng/g, (all)	Charles [44]	55.1	46.5 ng/g	Dewailly [29]	37.7/19.1 [ER+/–]	34.9 ng/g
	Guttes [85]	81	65 ng/g	Stellman [32]	30.4	24.0 ng/g	Liljegren [49] Zheng [60]	41 44.1	35 ng/g ng/g pooled
	Serum, prospective								
	Laden [46] [in nulliparous]	67	68 ng/g (all)	Hoyer [31]	63.9/42.5 [bs/fu]	ng/g pooled	Dorgan [33] Ward [24]	75 50.7	ng/g pooled 52.8 ng/g
	Serum, retrospective								
	None	–	–	None	–	–	Gammon [54]	55.1	56.4 ng/g

(*continues on next page*)

Table 5 (*Continued*).

PCB # (GRP)	+ve, Significant 1st author, yr	Mean: cases	Controls	+ve, NS 1st author, yr	Mean: cases	Controls	Inverse/none 1st author, yr	Mean: cases	Controls
126 (2a)	*Adipose tissue*								
	Liljegren [49] [>postmen, ER+]	0.237	0.164 ng/g (postmen)				Ward [24]	0.164	0.166 ng/g
138 (2b)	*Adipose tissue*								
	Stellman [32]	28.7	21.7 ng/g	Aronson [50]	73.8	66.8 ng/g	Zheng [60]	83.9	ng/g pooled
				Guttes [85]	228	194 ng/g	Dewailly [29]	78.1/59.6 [ER+/–]	70.1 ng/g
				Woolcott [43] ER+	71.7	66.8 ng/g	Liljegren [49]	235	234 ng/g
				Woolcott [43] ER–	81.8				
	Serum, prospective								
	Laden [46] [in nulliparous]	95	97 ng/g (all)	None	–	–	Dorgan [33]	94	ng/g pooled
	Hoyer [31] [bs only]	176.1	ng/g	Charles [44]	94.8	84.7 ng/g			
	Serum, retrospective								
	None	–	–	None	–	–	Gammon [54]	82.1	84.9 ng/g
156 (2a)	*Adipose tissue*								
	Aronson [50]	18.6	17.2 ng/g	Stellman [32]†	11.2	9.1 ng/g	Zheng [60]	21	ng/g pooled
				Woolcott [43] ER+	18.3	17.2 ng/g	Charles [44]	2.2	17.6 ng/g
				Woolcott [43] ER–	20.1				
							Dewailly [29]	15.7/16.8 [ER +/–]	17.9 ng/g
							Guttes [85]	61	76 ng/g
							Liljegren [49]	26	24 ng/g
	Serum, prospective								
	None	–	–	None	–	–	Ward [24]	15.5	16.6 ng/g
157 (U)	*Serum, prospective*								
	None	–	–	None	–	–	Ward [24]	3.8	3.7 ng/g
67 (2a)	*Adipose tissue*								
	None	–	–	None	–	–	Liljegren [49]	10	9 ng/g
	Serum, prospective								
	None	–	–	None	–	–	Ward [24]	6.4	6.7 ng/g

Table 5 (*Continued*).

PCB # (GRP)	+ve, Significant 1st author, yr	Mean: cases	Controls	+ve, NS 1st author, yr	Mean: cases	Controls	Inverse/none 1st author, yr	Mean: cases	Controls
69 (2a)	*Adipose tissue*								
	None	–	–	Liljegren [49]	0.118	0.105 ng/g	Charles [44]	0.6	0.7 ng/g
	Serum, prospective								
	None	–	–	None	–	–	Ward [24]	0.080	0.084 ng/g
170 (2b)	*Adipose tissue*								
170 (2b)	Aronson [50] [postmenopausal]	34.3	32.0 ng/g (all)	Charles [44]	27.3	22.5 ng/g	Dewailly [29]	30.6/28.2 [ER+/–]	36.7 ng/g
				Woolcott [43] ER+	33.8	32.0 ng/g	Guttes [85]	245	267 ng/g
				Woolcott [43] ER–	36.1				
							Liljegren [49]	106	105 ng/g
							Stellman [32]	13.5	11.2 ng/g
							Zheng [60]	33.2	ng/g pooled
	Serum, prospective								
	None	–	–	None	–	–	Ward [24]	49.2	52.0 ng/g
189 (U)	*Adipose tissue*								
	None	–	–	None	–	–	Charles [44]	1.9	1.8 ng/g
	Serum, prospective								
	None	–	–	None	–	–	Ward [24]	3.02	2.79 ng/g
Σ, gp3	*Serum, retro.*								
	None	–	–	None	–	–	Zheng [89] (Σ 153, 180)	256.g	ng/g pooled
99 (3)	*Adipose tissue*								
	Aronson [50]	19.5	17.7 ng/g	Stellman [32]	19.3	13.9 ng/g	Liljegren [49]	19.1	19.4 ng/g
	Dewailly [29]	31/15 [ER+/–]	20.5 ng/g	Woolcott [43] ER+	18.9	17.7 ng/g			
				Woolcott [43] ER–	21.9				
	Serum, prospective								
	None	–	–	None	–	–	Ward [24]	25.7	28.8 ng/g
146 (3)	*Adipose tissue*								
	None	–	–	Stellman [32]	9.2	6.9 ng/g	Ward [24]	21.8	23.4 ng/g

(*continues on next page*)

Table 5 (*Continued*).

PCB # (GRP)	+ve, Significant 1st author, yr	Mean: cases	Controls	+ve, NS 1st author, yr	Mean: cases	Controls	Inverse/none 1st author, yr	Mean: cases	Controls
153 (3)	*Adipose tissue*								
	Guttes [85]	624*	505 ng/g	Charles [44]	129.2	110.7 ng/g	Aronson [50]	105.2	98.3 ng/g
				Stellman [32]	76.1	63.1 ng/g	Dewailly [29]	100/82.2 [ER+/–]	95.6 ng/g
				Woolcott [43] ER+	102.8	98.3 ng/g	Liljegren [49]	289	291 ng/g
				Woolcott [43] ER–	114.6				
	Serum, prospective								
	Laden [46] [nulliparous]	107	106 ng/g	None	–	–	Hoyer [31]	223/204 [bs/fu]	ng/g (pooled)
							Ward [24]	183.4	195.6 ng/g
	Serum, retro.								
	None	–	–	None	–	–	Demers [51] [pop'n/hosp ctrls.]	58.7	55.6/53.3 ng/g
							Gammon [54]	151.7	153.2 ng/g
172 (3)	*Adipose tissue*								
	None	–	–	Stellman [32]	2.4	1.6 ng/g	Liljegren [49]	11	11 ng/g
	Serum, prospective								
	None	–	–				Ward [24]	6.41	6.93 ng/g
178 (3)	*Adipose tissue*								
	None	–	–	Stellman [32]	3.9	3.0 ng/g	Ward [24]	8.4	8.5 ng/g
180 (3)	*Adipose tissue*								
	Aronson [50] [postmenopausal]	71.9	65.7 ng/g (all)	Charles [44]	77.6	65.1 ng/g	Dewailly [29]	74.6/80.0 [ER+/–]	86.2
				Guttes [85]	375	301 ng/g	Liljegren [49]	204	198 ng/g
				Stellman [32]	42.4	33.7 ng/g			
				Woolcott [43] ER+	71.4	65.7 ng/g			
				Woolcott [43] ER–	75.0				
	Serum, prospective								
	None	–	–	Laden [46] [nulliparous]	74	75 ng/g (all)	Gammon [54]	79.7	78.6 ng/g
							Hoyer [31]	85.8/82.2 [bs/fu]	ng/g (pooled)
							Ward [24]	107.7	115.8 ng/g

Table 5 (*Continued*).

PCB # (GRP)	+ve, Significant 1st author, yr	Mean: cases	Controls	+ve, NS 1st author, yr	Mean: cases	Controls	Inverse/none 1st author, yr	Mean: cases	Controls
183 (3)	*Adipose tissue*								
	Stellman [32]	5.8	4.0 ng/g	Charles [44]	13.8	10.7 ng/g	Aronson [50]	10.3	9.5 ng/g
				Dewailly [29]	10/6.8 [ER+/–]	7.5 ng/g	Liljegren [49]	25	25 ng/g
				Woolcott [43] ER+	9.9	9.5 ng/g			
				Woolcott [43] ER–	11.8				
	Serum, prospective								
	None	–	–	None	–	–	Ward [24]	16.2	16.4 ng/g
194 (3)	*Adipose tissue*								
							Liljegren [49]	28	30 ng/g
	Serum, prospective								
	None	–	–	None	–	–	Ward [24]	17.2	18.3 ng/g
195 (3)	*Adipose tissue*								
	None	–	–	None	–	–	Liljegren [49]	6	7 ng/g
	Serum, prospective								
	None	–	–	None	–	–	Ward [24]	4.84	5.43 ng/g
206 (3)	*Adipose tissue*								
	None	–	–	None	–	–	Liljegren [49]	8	7 ng/g
	Serum, prospective								
	None	–	–	None	–	–	Ward [24]	5.20	5.61 ng/g
209 (3)	*Adipose tissue*								
	None	–	–	None	–	–	Liljegren [49]	8	9 ng/g
	Serum, prospective								
	None	–	–	None	–	–	Ward [24]	3.97	4.52* ng/g

Group 1b PCBs are weak phenobarbital inducers and are more persistent. Published studies have not reported positive associations of these congeners as a group [24] or individually (multivariate results—[24,50–52]; means—[29,32,43,49]).

Congeners in group 2a are thought to be antiestrogenic and immunotoxic; those in group 2b appear to have limited dioxin activity. The highly persistent congeners in group 3 are thought to induce cytochrome P450 and phenobarbital activity. No studies have reported elevated risk associated with the sum of the congeners from these groups. A few studies have reported elevated risk associated with individual congeners in each group (e.g., PCBs 118, 138, 99), but results have been inconsistent (Table 5).

Two studies compared PCB congeners classified as high- vs. low-chlorination in relation to breast cancer and again found ambiguous results. Wolff et al. [53] reported evidence of elevated risk only for low-chlorination congeners. In contrast, Millikan et al. [45] reported greater risk associated with high-chlorination congeners.

Other OCs

The few studies examining other OCs in relation to breast cancer risk have, for the most part, found no association (Table 6). No studies have reported significant increased risk associated with chlordane metabolites, including *cis*-nonachlor (CN), *trans*-nonachlor (TN), oxychlordane (OXY), and heptachlor epoxide (HCE) [24,43,50–54]. Mixed results have been reported for dieldrin, mirex, HCB, and HCH (or b-BHC; see Table 6). No studies on endosulfan were found.

STUDY DESIGN FACTORS

Adipose tissue vs. blood

As OCs are lipophilic, adipose tissue sample stores are thought to provide a better index of long-term exposure than blood-based measures. Strong correlations between adipose tissue and lipid-adjusted serum levels of DDE have been reported in some (r = 0.8, [55]) but not all (r = 0.4, [64]) studies. Similar, low correlations between adipose and lipid-adjusted serum levels of PCBs have been seen (e.g., r = 0.4 for PCB 153; [55]).

Differences in these measures do not, however, appear to explain study inconsistencies, as results using each medium have been heterogeneous. For example, significant positive associations with DDE and breast cancer were reported for: (i) one of six reports from large studies using adipose tissue samples; (ii) two of 10 studies using prospective serum samples; and (iii) none of 10 studies based on retrospective serum (Table 2).

Study populations—using benign breast disease (BBD) controls

Most studies based on adipose tissue used BBD patients as controls because it is more difficult to obtain adipose tissue samples from disease-free subjects (Table 1). If OCs contribute to the risk of BBD, this strategy may bias results toward the null. However, studies using alternative control groups also failed to observe positive associations. Using patients presenting for breast reduction surgery, Bagga et al. [34] found null associations for various DDTs. Inverse associations with DDE were reported by van't Veer et al. [56] in a study based on gluteal adipose tissue samples and population-based controls. Studies that excluded [57–60] or limited [32] the proportion of BBD controls with evidence of atypia found no strong evidence of adverse effects. Wolff et al. [53] also found no difference in results between BBD and non-BBD non-cancer controls, suggesting that the predominance of BBD controls is not the reason for lack of a measurable detrimental effect.

Table 6 Other organochlorines and breast cancer.

1st author [ref.]	+ve	ns	–ve null	Years	Place	# Cases/ controls	Mean		Adjusted?	ORs				
							Cases	Controls		1	2	3	4	5
I. CHLORDANE METABOLITES														
Cis-nonachlor: *Adipose*														
Aronson [50][a]			–	1995–97	Canada	217/213	6.0	6.0 ng/g	y	1.0	0.8	0.5	0.8	
Woolcott [43; data = 50][†]														
ER+			–	″	″	″	5.9	6.0 ng/g	y	1.0	0.9	0.7		
ER–			–	″	″	″	6.2	″	y	1.0	0.5	0.5		
Trans-nonachlor: *Adipose*														
Aronson [50][a]			–	″	″	″	40.4	41.1 ng/g	y	1.0	0.9	0.7	0.8	
Woolcott[†] [43; data = 50]														
ER+			–	″	″	″	39.1	41.1 ng/g	y	1.0	0.8	0.5		
ER–			–	″	″	″	43.3		y	1.0	1.0	1.0		
Zheng [52][a,b,c]			–	1994–97	US	304/186	55.5	58.1 ng/g	y	1.0	1.2	0.7	1.1	
Serum, prospective														
Ward [24]				1973–91	Norway	150/150	117	104 ng/g	y	1.0	1.0	0.9	1.0	
Serum, retrospective														
Demers [51][†]														
Pop'n. controls		ns		1994–97	Canada	315/307	16.6	16.0 ng/g	y	1.0	0.8	1.5	0.7	1.2
Hosp. controls		ns		″	″	315/219	″	16.7 ng/g	y	1.0	1.3	1.5	0.6	0.7
Wolff [53][e]				??	US	175/356	0.035	0.036 ng/g	y	1.0	1.0	0.7		
Oxychlordane: *Adipose*														
Aronson [50][a]			–	1995–97	Canada	217/213	30.4	30.5 ng/g	y	1.0	0.7	0.6	0.6	
Zheng [52][a,b,c]			–	1994–97	US	304/186	36.4	38.0 ng/g	y	1.0	0.7	0.7	0.7	
Serum, prospective														
Ward [24]			–	1973–91	Norway	150/150	10.0	10.9 ng/g	y	1.0	1.0	1.0	0.9	
Serum, retrospective														
Demers [51]														
Pop'n. controls		ns		1994–97	Canada	315/307	12.9	12.2 ng/g	y	1.0	1.1	1.0	1.3	1.5
Hosp. controls			–	″	″	315/219	″	13.0 ng/g	y	1.0	1.1	1.0	0.8	0.6
Heptachlor epoxide: *Serum, prospective*														
Ward [24]		ns		1973–91	Norway	150/150	7.1	8.5 ng/g	y	1.0	1.5	1.8	1.0	

(*continues on next page*)

Table 6 (*Continued*).

1st author [ref.]	+ve	ns	–ve null	Years	Place	# Cases/ controls	Mean		Adjusted?	ORs				
							Cases	Controls		1	2	3	4	5
I. CHLORDANE METABOLITES														
Oxychlordane + *trans*-nonachlor: *Serum, retrosp.*														
Gammon [54]: All[a,b,d]			–	1996–97	US	597/397	97.6	95.9 ng/g	y	1.0	0.9	1.0	1.2	1.0
Nulliparous		ns		″	″	″	–	–	y	1.0	2.4	2.8		
II. DIELDRIN:														
Serum, prospective														
Dorgan [33]			–	1977–87	US	105/208			y	1.0	0.7	0.8	0.7	
Hoyer [36][1]	↑			1976–83	Denmark	240/477	pooled	24 ng/g	y	1.0	1.6	2.0*	2.1*	
Ward [24]														
Serum, retrospective														
Gammon [54]			–	1996–97	US	597/397	97.6	95.9 ng/g	y	1.0	0.9	1.0	1.2	1.0
III. MIREX														
Adipose														
Aronson [50][a] – All			–	1995–97	Canada	217/213	9.0	9.9 ng/g	y	1.0	1.2	1.4	1.2	
Never lactators, parous	↑			″	″	″	–	–	y	1.0	2.6	1.9	4.2*	
Ever lactators, parous			–	″	″	″	–	–	y	1.0	1.5	0.9	0.3	
Serum, retrospective														
Moysich [22][†,2] – All			–	1986–91	US	154/191	0.043	0.037 ng/g serum	y	1.0	1.4			
Parous, never lactated		ns			″	″	0.083	0.046 ng/g serum	y	1.0	2.4			
IV. HEXACHLOROBENZENE:														
Adipose														
Aronson [50][a]			–	1995–97	Canada	217/213	32.0	30.1 ng/g	y	1.0	1.0	0.8	1.2	
Liljegren [49] – All			–	1993–95	Sweden	43/35	73	48 ng/g		1.0	1.3			
Premenopausal			–	″	″	″	33	35 ng/g		–				
Postmenopausal		ns		″	″	″	87	56 ng/g	y	1.0	1.9			
ER–		ns		″	″	″	–	–	y	1.0	2.0			
ER+	↑			″	″	″	–	–	y	1.0	7.1*			
Zheng [59][a,b] – All			–	1994–97	US	304/186	21.0	19.1	y	1.0	0.7	0.7	0.9	
Nulliparous		ns		″	″	″			y	1.0	0.5	2.1		

Table 6 (*Continued*).

1st author [ref.]	+ve	ns	–ve null	Years	Place	# cases/ controls	Mean Cases	Mean Controls	Adjusted?	ORs 1	2	3	4	5
IV. HEXACHLOROBENZENE:														
Serum, retrospective														
Dorgan [33]	↑			1977–87	US	105/208	73 (est)	ng/g pooled	y	1.0	2.5*	1.9	2.3*	
Moysich-postmen [22][2,†]			–	1986–91	US	154/192	0.41	0.42 ng/g serum	y	1.0	0.6	0.8		
Parous, never lactated							0.45	0.39 ng/g serum	y	1.0	1.3	1.8		
V. HEXACHLOROCYCLOHEXANE (b-HCH/b-BHC):														
Adipose tissue														
Aronson [50][a]			–	1995–97	Canada	217/213	43.1	41.5 ng/g	y	1.0	0.7	1.0	0.7	
Zheng [58][a,b] – All			–	1994–97	US	304/186	27.1	26.3 ng/g	y	1.0	0.7	0.8	0.6	
Nulliparous		ns		″	″	″			y	1.0	1.4	3.1		
Woolcott[†] [43; data = 1]														
ER–			–	″	″	″	56.2	41.5 ng/g	y	1.0	0.8	1.4		
ER+			–	″	″	″	39.3	″	y	1.0	0.8	0.7		
Serum, prospective														
Dorgan [33]			–	1977–87	US	105/208	–	–	y	1.0	0.5*	0.5*	0.6	
Hoyer [36][1]			–	1976–83	Denmark	240/477	pooled	119 ng/g	y	1.0	1.1	1.4	1.4	
Hoyer [31][1]			–	1976–83	Denmark	155/274	pooled	119 (bs) ng/g 60 (fu)	y	1.0	1.3	1.2	1.2	
Ward [24]			–	1973–91	Norway	150/150	60.0	63.4 ng/g	y	1.0	1.0	0.7	0.7	
Serum, retrospective														
Demers [51]														
Pop'n. controls			–	1994–97	Canada	315/307	21.1	19.4 ng/g	y	1.0	0.6	0.6	0.9	0.8
Hosp. controls			–	″	″	315/219	″	17.5 ng/g	y	1.0	0.7	0.9	0.7	0.8

Statistical significance defined as $p < 0.05$/CI excludes null or $p < 0.10$.

[1]Medians presented rather than means.

[2]Means are wet weight rather than lipid basis.

[a–e]Supplementary analysis found no meaningful differences when stratified by: a: pre/post-menopausal; b: ER status; c: lactation/parity group; d: BMI group; e: ethnic groups.

[†]Other notes: Moysich [22] found elevated ns ORs for HCB only in parous women who never lactated. Zheng [52,58,59] are age- and lipid-adjusted means.

Hoyer [36] –Results not shown for mirex, aldrin, endrin, AC, YC, heptachlor, heptachlor oxide, OCD, TN, y-HCH, HCB—presumed NS.

Timing of exposure measurement vs. use of OC chemicals

The majority of studies in industrialized countries have measured OC residues in biological samples collected from adults in the 1980s and 1990s, well after most source chemicals were banned. Given half-life estimates on the order of 7 to 11 years for DDE and PCBs [53], residues measured in these studies may largely reflect exposure to persistent but less estrogenic OCs (such as *p,p′*-DDE and high-chlorination PCB congeners), rather than long-term exposure to estrogenic but less persistent compounds such as DDT, dieldrin, and low-chlorination PCBs [11,13,48]. Therefore, recent measures of OCs may not be good estimates of past exposure to estrogenic compounds.

Direct estimates using samples obtained closer to the period when chemical sources remained in use should be more reliable for identifying any adverse effects of less persistent OCs, as long as samples were stable and analytic measures were sensitive enough. In industrialized countries, four of five prospective studies used blood collected as early as the 1960s and 1970s. These did not find overall adverse effects of DDTs, PCBs, or several other OCs [24,30,33,35]. Concentrations of DDE and DDT in these samples were relatively high (Tables 2 and 3). The one study showing effects that examined *p,p′*-DDT in a population-based Danish study [31] used repeated serum measures and adjusting for weight change. No other studies have incorporated repeated measures of OC exposure, and few have included measures of weight change (see Table 1).

Of the studies examining associations between DDT, DDE, and breast cancer risk in countries where use of DDT continues (Table 2), some based in Mexico City [23,61] and in Vietnam [37] found lower mean levels of *p,p′*-DDT in breast cancer cases vs. controls. No association between DDE and breast cancer in hospital-based case-control studies were noted in Vietnam [37], Mexico City [61], or in a population-based study in Brazil [62]. An anomalous result was seen in a population-based study of parous women in Mexico City [23] that reported strong and significant positive associations with DDT-adjusted levels of serum DDE. The authors hypothesized that adjusting for DDT provided better estimates of long- vs. short-term exposure to DDE. It is not clear whether they limited their adjustment model to low-persistence DDT isomers. Without this adjustment, multivariate associations were not significant (OR for highest vs. lowest quartile 2.16, CI 0.85–5.50 vs. 3.81, CI 1.14–2.8).

It is surprising that, although timeframes for the Mexico City studies overlapped (1990 to 1995 for Romieu et al. [23]; 1994 to 1996 for Lopez-Carillo et al. [61]), serum DDE was five times higher in the study reporting positive associations (lipid-adjusted means in cases/controls were 3840/2510 vs. 563/506 nanograms per gram [ng/g]). It is not clear to what extent discrepancies between the two Mexico City studies may reflect true differences in DDT and DDE exposure, metabolism, temporal changes, or differences in laboratory techniques.

Mobilization and elimination of OCs—weight change, adiposity, and lactation

A number of factors, including body fat burden, weight changes, and lactation history, may influence the dilution, mobilization, and elimination of fat schedule substances stored in the body [25,31,63,64].

Weight loss over a 5-year period was associated with significant declines in DDE and several PCBs, while weight gain was associated with larger increases in PCBs as compared to weight-stable subjects [31]. Self-reported losses or gains of >5 pounds in the past year was associated with significantly lower levels of DDE as compared to weight-stable women when examined cross-sectionally [63].

Greater adiposity may result in more dilute residues, particularly in adipose tissue [25,53,65]. Wolff and Anderson [65] proposed a pharmacokinetic model in which one or two half-lives after significant uptake elimination may become more complete in lean subjects than in those with a greater adipose tissue reservoir. Under this model, BMI should be treated as a modifier rather than as a confounder.

Lactation may also influence OC body burden, as breast milk is thought to be an important route of excretion of OCs. High concentrations of OC residues are detected in breast milk, and lactation has been associated with lower OCs [40,64,66]. Indeed, Wolff et al. [40] reported that adjusting for lactation status increased coefficients for DDE exposure by 57 % (from 0.051 ± 0.022 to 0.080 ± 0.030). Limited data also suggest that higher DDE is associated with reduced duration of breastfeeding [67], and several studies found that OCs were associated with higher risk of breast cancer among nulliparous women, or parous women who never breastfed than lactators ([22]–HCB; [50]–Mirex; [45,46]–total PCBs; [54]–chlordane). These findings are consistent with the hypothesis that breastfeeding may help to reduce the OC burden during a period of increased susceptibility to breast cancer.

Dietary confounders

Few studies have included dietary covariates (see Table 1). Dietary fat may be a confounder, given that fat-containing foods are important sources of exposure to persistent OC metabolites [3]. However, results of studies that adjusted for fat intakes did not differ from those that did not [50,52,57–59]. Furthermore, the role of dietary fat in breast cancer is unclear [68]. Dietary phytoestrogens may modify or confound associations, perhaps by inhibiting estrogenic action of OCs [69,70] A number of dietary phytoestrogens (e.g., genistein) appear to be more potent estrogens than typical OCs [71]. No studies to date have reported accounting for dietary phytoestrogens, whose role in breast cancer etiology is also uncertain [72,73].

Consideration of adjustment for these variables is essential for approximating cumulative or past exposure based on single measures of OC residues. While most studies have adjusted for BMI, few have accounted for lactation history or weight change (see Table 1 for details) [29,31,35,56]. No study has simultaneously considered all three factors.

POSSIBLE HIGH-RISK SUBGROUPS

Menopausal status: It has been suggested that postmenopausal women because of low circulating estrogen levels may be more susceptible to risk factors that influence the hormonal milieu [72]. A few studies have observed stronger associations in post-menopausal than in pre-menopausal women for DDT ([23] for DDT) or total PCBs ([49], especially if ER+). The majority of studies have not found strong associations among postmenopausal women for DDT/DDE [22,25,30,32,33,45,49,50,53, 54,56,57,60,61,74], total PCBs [22,25,32,45,53,60] or other OCs such as HCB, mirex, and chlordane metabolites [22,40,50,52,54,58,59]. In fact, several studies found stronger associations among premenopausal than postmenopausal women for DDE [35], total PCBs [35], and selected PCB congeners [50].

Assessing the significance of any differences in risk by menopausal status is complicated by differences in exposure patterns of younger vs. older women. A cohort effect in older women having experienced greater childhood and lifetime exposures may be responsible for what might be considered an age-related or lifetime accumulation effect when compared with younger women. Among younger women, whose exposure during peak use of these chemicals was limited to childhood and adolescence, adult OC residues may largely reflect ongoing food-borne exposure to persistent metabolites. Direct exposure to industrial and agricultural chemicals in the past likely explains why OC levels are higher in older than in younger women (e.g., [27]). However, exposure in postmenopausal women may have been qualitatively different from that in younger women: during peak chemical use in the 1950s and 1960s, these women were adults.

ER status

Consistent with the hypothesis that the hormones may be more involved in the etiology of ER+ than ER– tumors [75], Dewailly et al. [29] found that DDE, PCB 99, and *trans*-nonachlor were associated with ER+ but not ER– tumors (see Tables 3, 4, and 7). Liljegren et al. [49] also found significantly higher risk associated with HCB and several PCBs (77, 126, 169) among ER+ tumors, especially in postmenopausal women. Other studies have not confirmed these findings. Woolcott et al. [43] found somewhat stronger associations between DDE, total PCBs, *trans*-nonachlor, b-HCH, and several PCB congeners in ER– tumors (see Tables 3–5). Although Helzsouer et al. [35] found higher associations with DDE in ER+ tumors using serum collected in 1989, associations were higher in ER– tumors using serum from 1974. Numerous studies reported similar associations with DDE [24,25,45,49,52–54,74], PCBs [24,25,35,45,52–54,60,74], and other OCs [43,52,54,58,59] regardless of ER status.

Ethnicity and threshold effects: Krieger et al. [30] and Millikan et al. [45] reported somewhat stronger associations between serum DDE, total PCBs, and breast cancer risk among African Americans than other ethnic groups (whites and Asians). They hypothesized that the higher associations might be attributable at least in part to higher exposure levels. Yet there were no differences in associations among African Americans, whites, and Hispanics in a report by Wolff et al. [53], in which OC levels were similarly elevated among blacks. There was also no association with breast cancer risk in other studies in whites with similar serum OC levels [25,35]. However, associations between breast cancer and DDE were significant in the studies with the highest serum levels [23], after adjustment for DDT) and adipose levels [29,39]. Research in multiethnic settings with high levels of DDE is needed to confirm this hypothesis.

Genetic polymorphisms

In addition to mimicking estrogen activity, OCs are inducers of detoxification and drug-metabolizing enzymes [76,77]. Variation in genotypes for drug-metabolism genotypes such as cytochrome P4501A1 (CYP1A1) and detoxification enzymes such as glutathione-*S*-transferase μ (GSTM1) may be associated with susceptibility to OCs. Moysich et al. [76] found elevated breast cancer risk among subjects with high total PCBs and either heterozygous (isoleucine: valine) or homozygous valine subtypes for CYP1A1. Compared to the isoleucine homozygous group with low PCBs, this group was 2.9 times more likely to have breast cancer after multivariate adjustment (95 % CI 1.2–7.5). For the isoleucine homozygous group with high PCB exposure, the OR was 1.08 (CI 0.6–1.9). However, Helzsouer et al. [35] did not find strong evidence of modified susceptibility to the effects of DDE or total PCBs associated with polymorphisms in GSTM1, GSTT1, GSTP1, COMT, or CYP17.

OCCUPATIONAL STUDIES

Few studies on occupational exposure and breast cancer among women have been published. Both the exposure route and the estrogenicity of isomers to which women are exposed occupationally are likely to differ substantially from food-borne exposure to persistent OCs. Studies of pesticide workers have not reported an increased breast cancer incidence or mortality cancer associated with this industry ([78,79]; earlier studies reviewed by Adami et al. [80]). While farmworking per se was not associated with increased risk, Duell et al. [21] reported increased risk of breast cancer associated with women who were present in fields during or shortly after pesticide application (OR = 1.8, 95 % CI = 1.1–2.8). Risk was also elevated among women who reported not using protective clothing while applying pesticides (OR = 2.0; 95 % CI = 1.0–4.3) but not among those who reported using protective clothing (OR = 0.8; 95 % CI = 0.4–1.8). As no consistent exposure-related effects are seen in other studies, this may reflect other farm-based exposures or characteristics of the lifestyle, rather than OC exposure.

OCS AND BREAST CANCER SEVERITY AND PROGRESSION

Although laboratory data suggest that OCs may act as tumor promoters [16], the hypothesis that OCs may be related to the progression rather than the incidence of breast cancer has been explored in very few studies, so less is known on this topic [51]. Associations with various measures of stage, aggressiveness, and tumor markers have been reported in six studies to date (see Table 7). These studies—all

Table 7 Organochlorines and breast cancer survival, severity, and progression.

	DDT	DDE	PCBs	Other compounds
I. Survival	*ns* ↑	*null*	*mixed*	*mixed*
Hoyer et al., 2000b [81] – *Serum, prospective*				
	OR = 1.6 ns (*q4 vs. 1*) OR = 2.1* (#138, gp2b) (*q4 vs. 1*)	OR = 1.1 ns (*q4 vs. 1*)	OR = 0.9 ns (ΣPCBs)	Dieldrin: OR = 2.6* HCB: OR = 0.7 ns (*q4 vs. 1*)
II. Severity, among cases	*ns* ↑	↑	↑	↑
Demers [51] – *Serum, retrospective*				
Tumor ≥ v <2cm	OR = 1.6 ns (*q3 vs. 1*)	OR = 1.6 ns (*q3 vs. 1*)	OR = 1.5 ns (# 153, gp3) (*q3 vs. 1*)	TN: OR = 2.3* b-HCH: OR = 2.3* OCD: OR = 1.7 ns (*q3 vs. 1*)
Lymph yes/ no	OR = 1.5 ns (*q3 vs. 1*)	OR = 2.9* (*q3 vs. 1*)	OR = 2.1* (#153, gp3) (*q3 vs. 1*)	TN: OR = 2.0 ns b-HCH: OR = 2.0* OCD: OR = 2.3* (*q3 vs. 1*)
III. Severity at diagnosis	–	ns ↑/-ve	ns ↑/-ve	
Woolcott [43] – *Adipose, retrospective*				
Tumor < 2cm vs. none	–	OR = 1.6 ns (*q3 vs. 1*)	OR = 1.5 ns (Σ PCBs) OR = 0.9 ns (#187, gp1b) (*q3 vs. 1*)	
Tumor ≥ 2cm vs. none	–	OR = 1.1 ns (*q3 vs. 1*)	OR = 1.5 ns (Σ PCBs) OR = 1.6 ns (#187, gp1b) (*q3 vs. 1*)	
Tumor grade I/II vs. none	–	OR + 1.1 ns	OR = 1.2 ns (Σ PCBs)	
Tumor grade III vs. none		OR = 1.4 ns	OR = 1.5 ns (Σ PCBs)	
Zheng [89] – *Serum, retrospective. Means (adj. age & lipids)*				
Tumor grade I/II	–	Mean = 455.9 ng/ml	719.8 ng/ml (Σ PCBs)	
Tumor grade III/IV	–	Mean = 402.1 ng/ml	638.9 ng/ml (Σ PCBs)	
IV. Tumor markers (cases)	*ns* ↑	*mixed*		
Wolff et al. [53] – *Serum, retrospective.Means (adjusted for age, race, menopausal status)*				
erbB-2–	33 ng/g	740 ng/g	590 ng/g (HPCBs)	– –
erbB-2+	35 ng/g	800 ng/g	600 ng/g (HPCBs)	– –
p53–	35 ng/g	730 ng/g	600 ng/g (HPCBs)	
p53+	30 ng/g	820 ng/g	580 ng/g (HPCBs)	

†Notes: Demers [51]: Stronger associations also found for DDE using combined tumor size and lymph node involvement as marker of severity. Millikan [45] also examined associations stratified by stage at diagnosis and reported no differences in associations (results not shown). Zheng [52] also reported no association between TN and OCD and breast cancer histology or stage at diagnosis (data not shown).

conducted in industrialized countries using recent OC samples—have not reported associations between DDT/DDE and breast cancer survival, stage, aggressiveness, or tumor marker (p53 or erbB-2) [45,51,53,81]. Positive associations with PCBs and various measures of progression, severity, or survival have been reported in some studies [51,81], but weak or null associations have been reported in other [43,45,52,53] studies. Gammon et al. [54] did not find positive associations between stage at diagnosis (in situ vs. invasive disease) and concentrations of DDE, PCBs, or chlordane. Limited data suggest that other OCs may be associated with severity [51] and survival [81]. More data are needed before conclusions can be drawn, but this is an area deserving further study.

SUMMARY AND RECOMMENDATIONS

The sum of the evidence does not implicate any OC compound as significantly related to risk of occurrence of breast cancer. The evidence base is greatest for DDT and DDE. More limited research has been done on individual PCBs and their isomers. The studies of OC exposure reflect current exposure levels of chemicals that may have been banned for 20 years. Despite long half-lives (of 10 years) in adipose tissue, samples were generally collected 30 years after the peak exposure times. However, given ongoing exposure to persistent, nonestrogenic metabolites in foods, it is not certain that ranking of exposures using these samples is an accurate measure of past exposure to those compounds thought to increase risk.

More studies are needed on OCs other than DDTs in developing countries where use is more recent or continuing, especially given that most estrogenic OCs are not persistent. The possibility that OC exposure in excess of certain thresholds may be related to risk cannot be ruled out based on existing data. It is possible that other pathways that influence susceptibility may be involved, including activity related to CYP and GST; limited research is available to date on this hypothesis. Additionally, the BMI/weight loss model may warrant further analysis, perhaps using existing data.

OC exposures cannot reliably be related to trends in breast cancer incidence, as other known risk factors for breast cancer, such as childbearing and lactation. Changes in screening and treatment over time also complicate making such links. Ecologic data relating high-exposure countries to breast cancer mortality rates do not suggest a strong link. As noted by Rogan [82], "in general, breast cancer rates are not higher in parts of the world with high DDE levels (the People's Republic of China, India, and Guatemala), and countries with relatively similar levels of PCBs (Great Britain and Japan) have very different breast cancer rates."

Although the information is extremely limited and not without major design flaws, the association between disease severity and progression is interesting and worthy of further examination.

As OCs are present in the environment as mixtures of correlated isomers and metabolites, it may be difficult to distinguish possible causal links from associations in which measured compounds are merely markers of other underlying exposures. Few studies [23,36,40,52,83] have reported adjusting for other OCs. For highly correlated compounds, traditional adjustment strategies may not be feasible. More complex analytical strategies, such as those used by Holford et al. [83], may help to isolate potentially relevant isomers.

REFERENCES

1. D. R. Juberg. *Ecotoxicol. Environ. Saf.* **45** (2), 93–105 (2000).
2. K. Kannan, S. Tanabe, J. P. Giesy, R. Tatsukawa. *Rev. Environ. Contam. Toxicol.* **152**, 1–55 (1997).
3. E. Devoto, L. Kohlmeier, W. Heeschen. *Arch. Environ. Health* **53** (2), 147–155 (1998).
4. F. W. Kutz, P. H. Wood, D. P. Bottimore. *Rev. Environ. Contam. Toxicol.* **120**, 1–82 (1991).
5. D. L. Davis, H. L. Bradlow, M. Wolff, T. Woodruff, D. G. Hoel, H. Anton-Culver. *Environ. Health Perspect.* **101** (5), 372–377 (1993).

6. D. M. Klotz, B. S. Deckman, S. M. Hill, J. A. McLachlan, M. R. Walters, S. F. Arnold. *Environ. Health Perspect.* **104** (10), 1084–1089 (1996).
7. J. D. McKinney and C. L. Waller. *Environ. Health Perspect.* **102** (3), 290–297 (1994).
8. M. S. Wolff and P. G. Toniolo. *Environ. Health Perspect.* **103** (Suppl. 7), 141–145 (1995).
9. C. Sonnenschein and A. M. Soto. *J. Steroid Biochem. Mol. Biol.* **65** (1–6), 143–150 (1998).
10. K. Ramamoorthy, M. S. Gupta, G. Sun, A. McDougal, S. H. Safe. *Carcinogenesis* **20** (1), 115–123 (1999).
11. K. W. Gaido, L. S. Leonard, S. Lovell, J. C. Gould, D. Babai, C. J. Portier, D. P. McDonnell. *Toxicol. Appl. Pharmacol.* **143** (1), 205–212 (1997).
12. H. T. Jansen, P. S. Cooke, J. Porcelli, T. C. Liu, L. G. Hansen. *Reprod. Toxicol.* **7** (3), 237–248 (1993).
13. P. V. Shekhar, J. Werdell, V. S. Basrur. *J. Nat. Cancer Inst.* **89** (23), 1774–1782 (1997).
14. K. C. Silinskas and A. B. Okey. *J. Nat. Cancer Inst.* **55** (3), 653–657 (1975).
15. D. Desaulniers, K. Leingartner, J. Russo, G. Perkins, B. G. Chittim, M. C. Archer, M. Wade, J. Yang. *Environ. Health Perspect.* **109** (7), 739–747 (2001).
16. J. D. Scribner and N. K. Mottet. *Carcinogenesis* **2** (12), 1235–1239 (1981).
17. A. K. Robison, D. A. Sirbasku, G. M. Stancel. *Toxicol. Lett.* **27** (1–3), 109–113 (1985).
18. D. L. Eaton. *Neurotoxicology* **21** (1–2), 101–111 (2000).
19. M. H. Kester, S. Bulduk, D. Tibboel, W. Meinl, H. Glatt, C. N. Falany, M. W. Coughtrie, A. Bergman, S. H. Safe, G. G. Kuiper, A. G. Schuur, A. Brouwer, T. J. Visser. *Endocrinology* **141** (5), 1897–1900 (2000).
20. U.S. Environmental Protection Agency, Office of Pollution Prevention and Toxics. (2002). <http://www.epa.gov/opptintr/pbt/cheminfo.htm>
21. E. J. Duell, R. C. Millikan, D. A. Savitz, B. Newman, J. C. Smith, M. J. Schell, D. P. Sandler. *Epidemiology* **11** (5), 523–531 (2000).
22. K. B. Moysich, C. B. Ambrosone, J. E. Vena, P. G. Shields, P. Mendola, P. Kostyniak, H. Greizerstein, S. Graham, J. R. Marshall, E. F. Schisterman, J. L. Freudenheim. *Cancer Epidemiol., Biomarkers Prevent.* **7** (3), 181–188 (1998).
23. I. Romieu, M. Hernandez-Avila, E. Lazcano-Ponce, J. P. Weber, E. Dewailly. *Am. J. Epidemiol.* **152** (4), 363–370 (2000).
24. E. M. Ward, P. Schulte, B. Grajewski, A. Andersen, D. G. Patterson Jr., W. Turner, E. Jellum, J. A. Deddens, J. Friedland, N. Roeleveld, M. Waters, M. A. Butler, E. DiPietro, L. L. Needham. *Cancer Epidemiol., Biomarkers Prevent.* **9** (12), 1357–1367 (2000).
25. M. S. Wolff, A. Zeleniuch-Jacquotte, N. Dubin, P. Toniolo. *Cancer Epidemiol., Biomarkers Prevent.* **9** (3), 271–277 (2000a).
26. G. Schade and B. Heinzow. *Sci. Total Environ.* **215** (1–2), 31–39 (1998).
27. A. M. Sweeney, E. Symanski, K. D. Burau, Y. J. Kim, H. E. Humphrey, M. A. Smith. *Environ. Res.* **86** (2), 128–139 (2001).
28. P. E. Robinson and G. Mack. *Environ. Res.* **53**, 175–192 (1990).
29. E. Dewailly, S. Dodin, R. Verreault, P. Ayotte, L. Sauve, J. Morin, J. Brisson. *J. Nat. Cancer Inst.* **86** (3), 232–234 (1994a).
30. N. Krieger, M. S. Wolff, R. A. Hiatt, M. Rivera, J. Vogelman, N. Orentreich. *J. Nat. Cancer Inst.* **86** (8), 589–599 (1994).
31. A. P. Hoyer, T. Jorgensen, P. Grandjean, H. B. Hartvig. *Cancer Causes Control* **11** (2), 177–184 (2000a).
32. S. D. Stellman, M. V. Djordjevic, J. A. Britton, J. E. Muscat, M. L. Citron, M. Kemeny, E. Busch, L. Gong. *Cancer Epidemiol., Biomarkers Prevent.* **9** (11), 1241–1249 (2000).
33. J. F. Dorgan, J. W. Brock, N. Rothman, L. L. Needham, R. Miller, H. E. Stephenson Jr., N. Schussler, P. R. Taylor. *Cancer Causes Control* **10** (1), 1–11 (1999).

34. D. Bagga, K. H. Anders, H. J. Wang, E. Roberts, J. A. Glaspy. *J. Nat. Cancer Inst.* **92** (9), 750–753 (2000).
35. K. J. Helzsouer, A. J. Alberg, H. Y. Huang, S. C. Hoffman, P. T. Strickland, J. W. Brock, V. W. Burse, L. L. Needham, D. A. Bell, J. A. Lavigne, J. D. Yager, G. W. Comstock. *Cancer Epidemiol., Biomarkers Prevent.* **8** (6), 525–532 (1999).
36. A. P. Hoyer, P. Grandjean, T. Jorgensen, J. W. Brock, H. B. Hartvig. *Lancet* **352** (9143), 1816–1820 (1998).
37. A. Schecter, P. Toniolo, L. C. Dai, L. T. Thuy, M. S. Wolff. *Arch. Environ. Contam. Toxicol.* **33** (4), 453–456 (1997a).
38. A. Schecter and L. Li. *Chemosphere* **34** (5–7), 1449–1457 (1997b).
39. F. Falck, Jr., A. Ricci Jr., M. S. Wolff, J. Godbold, P. Deckers. *Arch. Environ. Health* **47** (2), 143–146 (1992).
40. M. S. Wolff, P. G. Toniolo, E. W. Lee, M. Rivera, N. Dubin. *J. Nat. Cancer Inst.* **85** (8), 648–652 (1993).
41. H. Mussalo-Rauhamaa, E. Hasanen, H. Pyysalo, K. Antervo, R. Kauppila, P. Pantzar P. *Cancer* **66** (10), 2124–2128 (1990).
42. M. Unger, H. Kiaer, M. Blichert-Toft, J. Olsen, J. Clausen. *Environ. Res.* **34** (1), 24.2–24.8 (1984).
43. C. G. Woolcott, K. J. Aronson, W. M. Hanna, S. K. SenGupta, D. R. McCready, E. E. Sterns, A. B. Miller. *Cancer Causes Control* **12** (5), 395–404 (2001).
44. M. J. Charles, M. J. Schell, E. Willman, H. B. Gross, Y. Lin, S. Sonnenberg, M. L. Graham. *Arch. Environ. Contam. Toxicol.* **41** (3), 386–395 (2001).
45. R. Millikan, E. DeVoto, E. J. Duell, C. K. Tse, D. A. Savitz, J. Beach, S. Edmiston, S. Jackson, B. Newman. *Cancer Epidemiol., Biomarkers Prevent.* **9** (11), 1233–1240 (2000).
46. F. Laden, S. E. Hankinson, M. S. Wolff, G. A. Colditz, W. C. Willett, F. E. Speizer, D. J. Hunter. *Int. J. Cancer* **91** (4), 568–574 (2001).
47. M. S. Wolff, D. Camann, M. Gammon, S. D. Stellman. *Environ. Health Perspect.* **105** (1), 13–14 (1997).
48. K. B. Moysich, P. Mendola, E. F. Schisterman, J. L. Freudenheim, C. B. Ambrosone, J. E. Vena, P. G. Shields, P. Kostyniak, H. Greizerstein, S. Graham, J. R. Marshall. *Am. J. Ind. Med.* **35** (3), 223–231 (1999a).
49. G. Liljegren, L. Hardell, G. Lindstrom, P. Dahl, A. Magnuson. *Eur. J. Cancer Prevent.* **7** (2), 135–140 (1998).
50. K. J. Aronson, A. B. Miller, C. G. Woolcott, E. E. Sterns, D. R. McCready, L. A. Lickley, E. B. Fish, G. Y. Hiraki, C. Holloway, T. Ross, W. M. Hanna, S. K. SenGupta, J. P. Weber. *Cancer Epidemiol., Biomarkers Prevent.* **9** (1), 55–63 (2000).
51. A. Demers, P. Ayotte, J. Brisson, S. Dodin, J. Robert, E. Dewailly. *Cancer Epidemiol., Biomarkers Prevent.* **9** (2), 161–166 (2000).
52. T. Zheng, T. R. Holford, S. T. Mayne, J. Tessari, B. Ward, D. Carter, P. H. Owens, P. Boyle, R. Dubrow, S. Archibeque-Engle, O. Dawood, S. H. Zahm. *Cancer Epidemiol., Biomarkers Prevent.* **9** (2), 167–174 (2000b).
53. M. S. Wolff, G. S. Berkowitz, S. Brower, R. Senie, I. J. Bleiweiss, P. Tartter, B. Pace, N. Roy, S. Wallenstein, A. Weston A. *Environ. Res.* **84** (2), 151–161 (2000b).
54. M. D. Gammon, M. S. Wolff, A. I. Neugut, S. M. Eng, W. L. Teitelbaum, J. W. Britton, M. C. Terry, B. Levin, S. D. Stellman, G. C. Kabat, M. Hatch, R. Senie, G. Berkowitz, H. L. Bradlow, G. Garbowski, C. Maffeo, P. Montalvan, M. Kemeny, M. Citron, F. Schnabel, A. Schuss, S. Hajdu, V. Vinceguerra, N. Niguidula, K. Ireland, R. M. Santella. *Cancer Epidemiol., Biomarkers Prevent.* **11**, 686–697 (2002).
55. S. L. Archibeque-Engle, J. D. Tessari, D. T. Winn, T. J. Keefe, T. M. Nett, T. Zheng. *J. Toxicol. Environ. Health* **52** (4), 285–293 (1997).

56. P. van't Veer, I. E. Lobbezoo, J. M. Martin-Moreno, E. Guallar, J. Gomez-Aracena, A. F. Kardinaal, L. Kohlmeier, B. C. Martin, J. J. Strain, M. Thamm, P. van Zoonen, B. A. Baumann, J. K. Huttunen, F. J. Kok. *BMJ* **315** (7100), 81– (1997).
57. T. Zheng, T. R. Holford, S. T. Mayne, B. Ward, D. Carter, P. H. Owens, R. Dubrow, S. H. Zahm, P. Boyle, S. Archibeque-Engle, J. Tessari. *Am. J. Epidemiol.* **150** (5), 453–458 (1999a).
58. T. Zheng, T. R. Holford, S. T. Mayne, P. H. Owens, B. Ward, D. Carter, R. Dubrow, S. H. Zahm, P. Boyle, J. Tessari. *Cancer* **85** (10), 2212–2218 (1999b).
59. T. Zheng, T. R. Holford, S. T. Mayne, J. Tessari, P. H. Owens, S. H. Zahm, B. Zhang, R. Dubrow, B. Ward, D. Carter, P. Boyle. *Cancer Epidemiol., Biomarkers Prevent.* **8** (5), 407–411 (1999c).
60. T. Zheng, T. R. Holford, J. Tessari, S. T. Mayne, P. H. Owens, B. Ward, D. Carter, P. Boyle, R. Dubrow, S. Archibeque-Engle, S. H. Zahm. *Am. J. Epidemiol.* **152** (1), 50–58 (2000a).
61. L. Lopez-Carrillo, A. Blair, M. Lopez-Cervantes, M. Cebrian, C. Rueda, R. Reyes, A. Mohar, J. Bravo. *J. Cancer Res.* **57** (17), 3728–3732 (1997).
62. G. A. Mendonca, J. Eluf-Neto, M. J. Andrada-Serpa, P. A. Carmo, H. H. Barreto, O. N. Inomata, T. A. Kussumi. *Int. J. Cancer* **83** (5), 596–600 (1999).
63. J. M. Schildkraut, W. Demark-Wahnefried, E. DeVoto, C. Hughes, J. L. Laseter, B. Newman. *Cancer Epidemiol., Biomarkers Prevent.* **8** (2), 179–183 (1999).
64. L. Lopez-Carrillo, L. Torres-Sanchez, J. Moline, K. Ireland, M. S. Wolff. *Environ. Res.* **87** (3), 131–135 (2001).
65. M. S. Wolff and H. A. Anderson. *Cancer Epidemiol., Biomarkers Prevent.* **8** (10), 951–952 (1999).
66. E. Dewailly, P. Ayotte, J. Brisson. *J. Nat. Cancer Inst.* **86** (10), 803 (1994b).
67. B. C. Gladen and W. J. Rogan. *Am. J. Public Health* **85** (4), 504–508 (1995).
68. M. M. Lee and S. S. Lin. *Annu. Rev. Nutrition* **20**, 221–248 (2000).
69. S. P. Verma and B. R. Goldin. *N. Engl. J. Med.* **338** (14), 990 (1998).
70. S. P. Verma, E. Salamone, B. Goldin. *Biochem. Biophys. Res. Commun.* **233** (3), 692–696 (1997).
71. G. G. Kuiper, J. G. Lemmen, B. Carlsson, J. C. Corton, S. H. Safe, P. T. van der Saag, B. van der Burg, J. A. Gustafsson. *Endocrinology* **139** (10), 4252–4263 (1998).
72. A. Cassidy and S. Milligan. *Climacteric* **1** (3), 229–242 (1998).
73. S. Barnes. *Baillieres Clin. Endocrinol. Metabol.* **12** (4), 559–579 (1998).
74. D. J. Hunter, S. E. Hankinson, F. Laden, G. A. Colditz, J. E. Manson, W. C. Willett, F. E. Speizer, M. S. Wolff. *N. Engl. J. Med.* **337** (18), 1253–1258 (1997).
75. J. D. Potter, J. R. Cerhan, T. A. Sellers, P. G. McGovern, C. Drinkard, L. R. Kushi, A. R. Folsom. *Cancer Epidemiol., Biomarkers Prevent.* **4** (4), 319–326 (1995).
76. K. B. Moysich, P. G. Shields, J. L. Freudenheim, E. F. Schisterman, J. E. Vena, P. Kostyniak, H. Greizerstein, J. R. Marshall, S. Graham, C. B. Ambrosone. *Cancer Epidemiol., Biomarkers Prevent.* **8**, 41–44 (1999b).
77. E. G. Schuetz. *Curr. Drug Metab.* **2** (2), 139–147 (2001).
78. L. E. Fleming, J. A. Bean, M. Rudolph, K. Hamilton. *J. Occup. Environ. Med.* **41** (4), 279–288 (1999a).
79. L. E. Fleming, J. A. Bean, M. Rudolph, K. Hamilton. *Occup. Environ. Med.* **56** (1), 14–21 (1999b).
80. H. O. Adami, L. Lipworth, L. Titus-Ernstoff, C. C. Hsieh, A. Hanberg, U. Ahlborg, J. Baron, D. Trichopoulos. *Cancer Causes Control* **6** (6), 551–666 (1995).
81. A. P. Hoyer, T. Jorgensen, J. W. Brock, P. Grandjean. *J. Clin. Epidemiol.* **53** (3), 323–330 (2000b).
82. W. J. Rogan. *Arch. Pediatric Adolescent Med.* **150** (9), 981–990 (1996).
83. T. R. Holford, T. Zheng, S. T. Mayne, S. H. Zahm, J. D. Tessari, P. Boyle. *Int. J. Epidemiol.* **29** (6), 975–982 (2000).
84. A. R. Lucena, M. F. Allam, I. H. Costabeber, M. L. Villarejo, R. F. Navajas. *Eur. J. Cancer Prevent.* **10** (1), 117–119 (2001).

85. S. Guttes, K. Failing, K. Neumann, J. Kleinstein, S. Georgii, H. Brunn. *Arch. Environ. Contam. Toxicol.* **35** (1), 140–147 (1998).
86. L. Hardell, G. Lindstrom, G. Liljegren, P. Dahl, A. Magnuson. *Eur. J. Cancer Prevent.* **5** (5), 351–357 (1996).
87. R. Dello Iacovo, E. Celentano, A. M. Strollo, G. Iazzetta, I. Capasso, G. Randazzo. *Adv. Exper. Med. Biol.* **472**, 57–66 (1999).
88. P. Olaya-Contreras, J. Rodriguez-Villamil, H. J. Posso-Valencia, J. E. Cortez. *Caldernos de Saude Publ.* **14** (Suppl. 3), 125–132 (1998).
89. T. Zheng, T. R. Holford, J. Tessari, S. T. Mayne. S. H. Zahm, P. H. Owens, B. Zhang, B. Ward, D. Carter, Y. Zhang, W. Zhang, R. Dubrow, P. Boyle. *J. Epidemiol. Biostat.* **5** (3), 153–160 (2000c).

Pure Appl. Chem., Vol. 75, Nos. 11–12, pp. 2013–2022, 2003.

Topic 3.3

Prevention of ambiguous genitalia by prenatal treatment with dexamethasone in pregnancies at risk for congenital adrenal hyperplasia*

Maria I. New

Pediatric Endocrinology, New York Presbyterian Hospital–Weill Cornell Medical Center, New York, NY 10021, USA

Abstract: Congenital adrenal hyperplasia (CAH) refers to a family of monogenic inherited disorders of adrenal steroidogenesis most often caused by a deficiency of the 21-hydroxylase enzyme. In the classic forms of CAH (simple virilizing and salt-wasting), androgen excess causes external genital ambiguity in newborn females and progressive postnatal virilization in males and females. Prenatal treatment of CAH with dexamethasone has been successfully utilized for over a decade. This article reports on 595 pregnancies prenatally diagnosed using amniocentesis or chorionic villus sampling between 1978 and 2002 at the New York Presbyterian Hospital–Weill Medical College of Cornell University. No significant or enduring side effects were noted in the fetuses, indicating that dexamethasone treatment is safe. Prenatally treated newborns did not differ in weight from untreated, unaffected newborns. Based on our experience, prenatal diagnosis and treatment of 21-hydroxylase deficiency is effective in significantly reducing or eliminating virilization in the newborn female. Prevention of genital virilization in female newborns with classic CAH avoids the risk of sex misassignment and diminishes the need for corrective surgery and the resulting psychological impact that may extend into adulthood.

INTRODUCTION

Congenital adrenal hyperplasia is a family of inherited disorders of adrenal steroidogenesis [1]. Each disorder results from a deficiency in one of the five enzymatic steps necessary for normal cortisol synthesis (Fig. 1). Deficiency of the 21-hydroxylase enzyme (21-OHD) accounts for over 90 % of CAH cases. There is a wide range of clinical presentations in classic and nonclassic CAH, from virilization with labial fusion to precocious adrenarche, pubertal or postpubertal virilization, and reduced fertility. A five-stage classification by Prader [2] is used to represent different degrees of virilization (Fig. 2), where on a scale of 1 to 5 (I–V) the genitalia can be scored from slightly virilized (e.g., mildly enlarged clitoris) to indistinguishable from a male. Most classical cases of 21-hydroxylase deficiency are born with Prader IV genitalia.

*Report from a SCOPE/IUPAC project: Implication of Endocrine Active Substances for Human and Wildlife (J. Miyamoto and J. Burger, editors). Other reports are published in this issue, *Pure Appl. Chem.* **75**, 1617–2615 (2003).

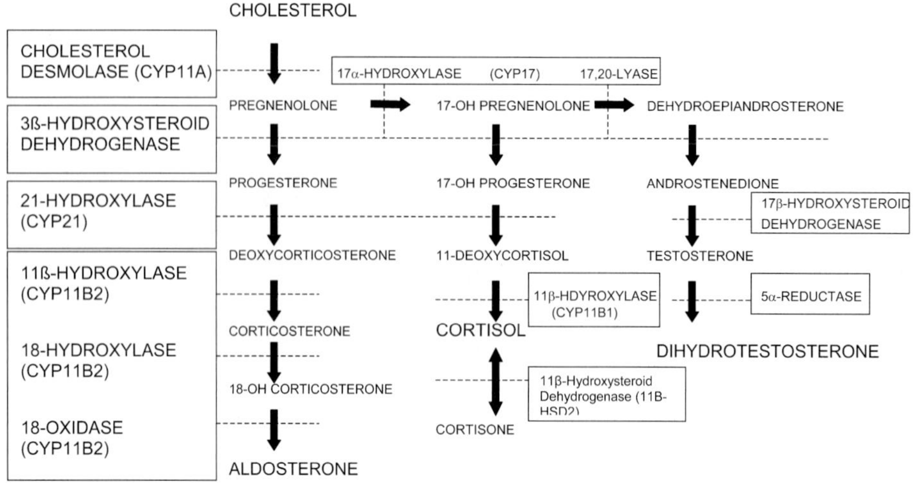

Fig. 1 Simplified scheme of adrenal steroidogenesis.

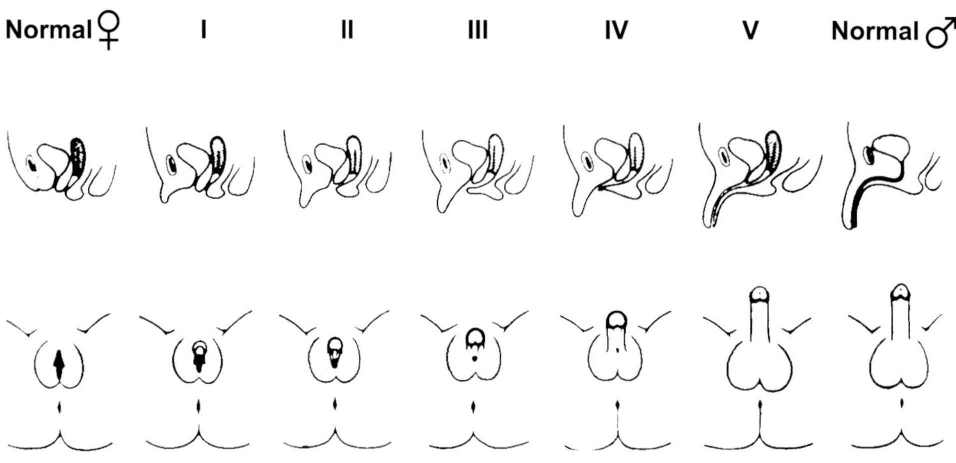

(Courtesy of Andrea Prader. Reprinted from *Helvetica Paediatrica Acta.)*

Fig. 2 Diagrams of different degrees of virilization using the Prader Scale [3].

In the classic form of 21-OHD, prenatal androgen excess causes external genital ambiguity in newborn females (female pseudohermaphroditism) (Fig. 3). Postnatally, males and females exhibit progressive virilization, which may include progressive penile or clitoral enlargement, precocious pubic hair, hirsutism, acne, advanced somatic and epiphyseal development, and central precocious puberty. Reduced fertility and menstrual abnormalities have been observed in women, and testicular adrenal rests in untreated men [3–6]. There are two types of classical steroid 21-OHD, simple virilizing and salt wasting. Three-fourths of classical cases are salt-wasting [1]. To some extent, the symptoms can be arrested or reversed by treatment with glucocorticoids, which suppresses ACTH stimulation of the adrenal cortex. Those patients with aldosterone deficiency require treatment with salt-retaining steroids as well. Analysis of CAH incidence data from almost 6.5 million newborns screened in the general population worldwide has demonstrated an overall incidence of between 1:13 000 and 1:15 000 live births for the classic form of CAH [7–9].

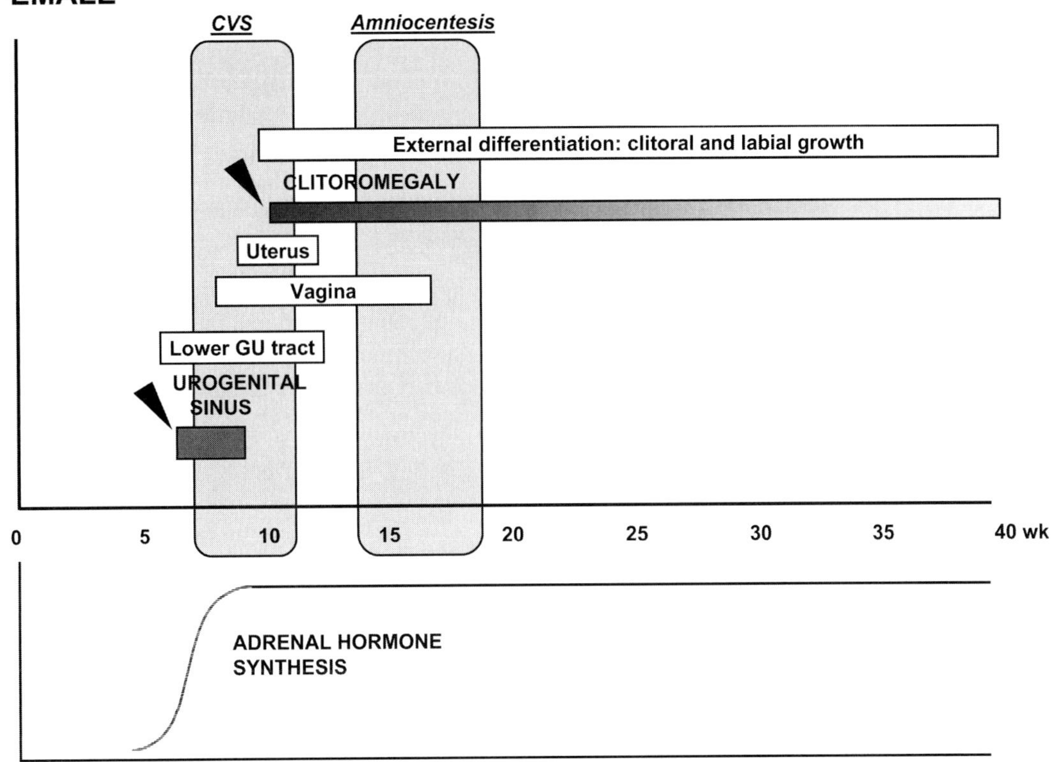

Fig. 3 Timetable of female sexual differentiation.

The most reliable hormonal diagnostic test for 21-OHD has proven to be the ACTH (Cortrosyn, 0.25 mg) stimulation test measuring the serum concentration of 17-hydroxyprogesterone (17-OHP). After intravenous bolus ACTH administration (preferably in the morning due to the diurnal variation of 17-OHP), 17-OHP is measured at 0 and 60 min. A logarithmic nomogram provides hormonal standards for assignment of the 21-OHD type by relating baseline to ACTH-stimulated serum concentrations of 17-OHP [10] (Fig. 4).

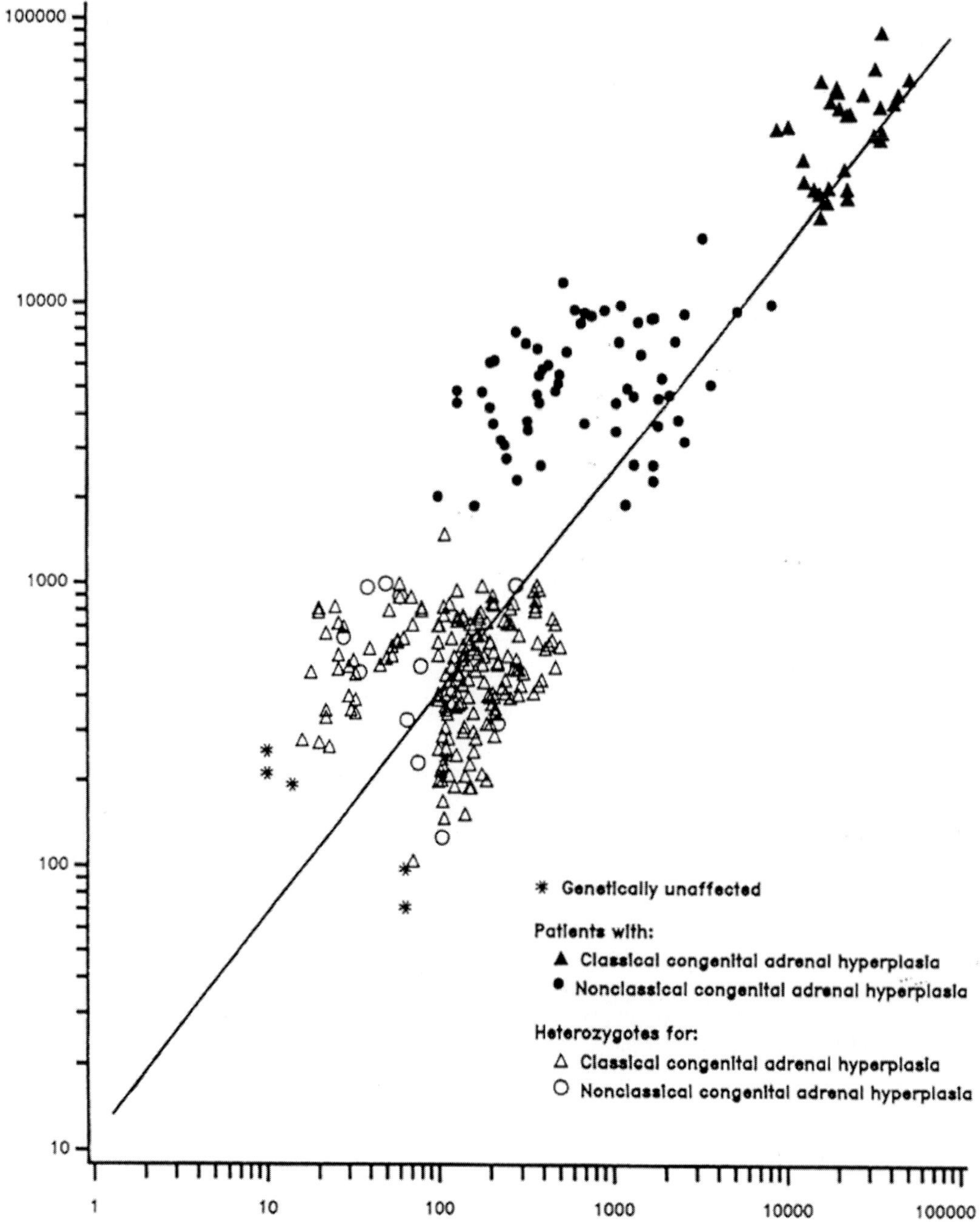

Fig. 4 Nomogram relating baseline to 60′ ACTH-stimulated serum concentrations of 17-hydroxyprogesterone (17-OHP). The scales are logarithmic.

MOLECULAR GENETICS

Congenital adrenal hyperplasia due to 21-hydroxylase deficiency is a monogenic autosomal recessive disorder. The gene for adrenal 21-hydroxylase, CYP21, is located about 30 kb from a pseudogene, CYP21P, on chromosome 6p, adjacent to the HLA genes. The high degree of sequence similarity (96–98 %) between CYP21 and CYP21P permits two types of recombination events: (1) unequal crossing-over during meiosis, which results in complete deletions and duplications of CYP21 and the possible transmission of a null allele, and (2) and gene conversion events that transfer deleterious mutations present in the pseudogene to CYP21 [11–13].

Specific mutations may be correlated with a given degree of enzymatic compromise and a clinical form of 21-OHD [14–18]. The genotype for the classical form of CAH is predicted to be a severe mutation on both alleles at the 21-OH locus, with markedly decreased enzymatic activity generally associated with salt wasting. The point mutation A (or C) to G near the end of Intron 2, which is the single most frequent mutation in classic 21-OHD, causes premature splicing of the intron and a shift in the translational reading frame [11,15]. Most patients who are homozygous for this mutation have the salt-wasting form of the disorder [19,20]. Recent studies, however, have demonstrated that there is occasionally a divergence in phenotypes within mutation-identical groups, the reason for which requires further investigation [20,21].

PRENATAL DIAGNOSIS AND TREATMENT

While it was known that CAH-affected fetuses exhibit elevated 17-OHP and Δ^4-androstenedione concentrations in their amniotic fluid, the differentiation of external genitalia and the urogenital sinus begins at approximately the 9th week of gestation. Thus, if prenatal treatment with dexamethasone must begin no later than the 9th week of gestation to effectively prevent virilization of a female fetus, an amniocentesis performed in the second trimester would show suppressed 17-OHP levels and be unreliable for diagnosis. The later use of human leukocyte antigen (HLA) genetic linkage marker analysis was also unsuccessful because recombination or haplotype sharing lead to many diagnostic errors. The method currently used is direct DNA analysis of the 21-OH gene (CYP21) with molecular genetic techniques [22–24]. Chorionic villus sampling (CVS) can also be used to obtain fetal tissue for prenatal diagnosis by molecular genetic analysis at 9–11 weeks gestation.

An algorithm was first published in 1990 for the prenatal diagnosis of 21-OHD congenital adrenal hyperplasia using direct molecular analysis of the 21-OH locus and dexamethasone treatment [25] (Fig. 5). Unlike hydrocortisone, dexamethasone crosses the placenta. Because maternally administered dexamethasone is not bound by corticosteroid-binding globulin, (also termed transcortin), and dexamethasone is not converted by 11B-hydroxysteroid dehydrogenase to cortisone it undergoes efficient placental transfer to the fetus.

Since the first report of prenatal therapy by David and Forest in 1984 [26], there have been several large studies of prenatal treatment [24,27,28]. When the proper protocol is followed, these studies have proved extremely successful in preventing virilization in females affected with classic CAH. Of the 595 pregnancies prenatally diagnosed using amniocentesis or chorionic villus sampling between 1978 and 2002 at the New York Presbyterian Hospital–Weill Medical College of Cornell University, 281 were prenatally treated for CAH, owing to a risk for 21-hydroxylase deficiency.

The prevention of genital virilization in female newborns with classic CAH has significant implications. For the prenatally treated child, there is the benefit of unambiguous genitalia, including the diminished need for future vaginoplasty and the resulting psychological impact of genital surgery. Prenatal treatment also avoids a potential male sex assignment to virilized female newborns and has been suggested to prevent gender ambiguity sometimes seen in CAH females, attributed to the high levels of androgens exposed to the brain during differentiation [29].

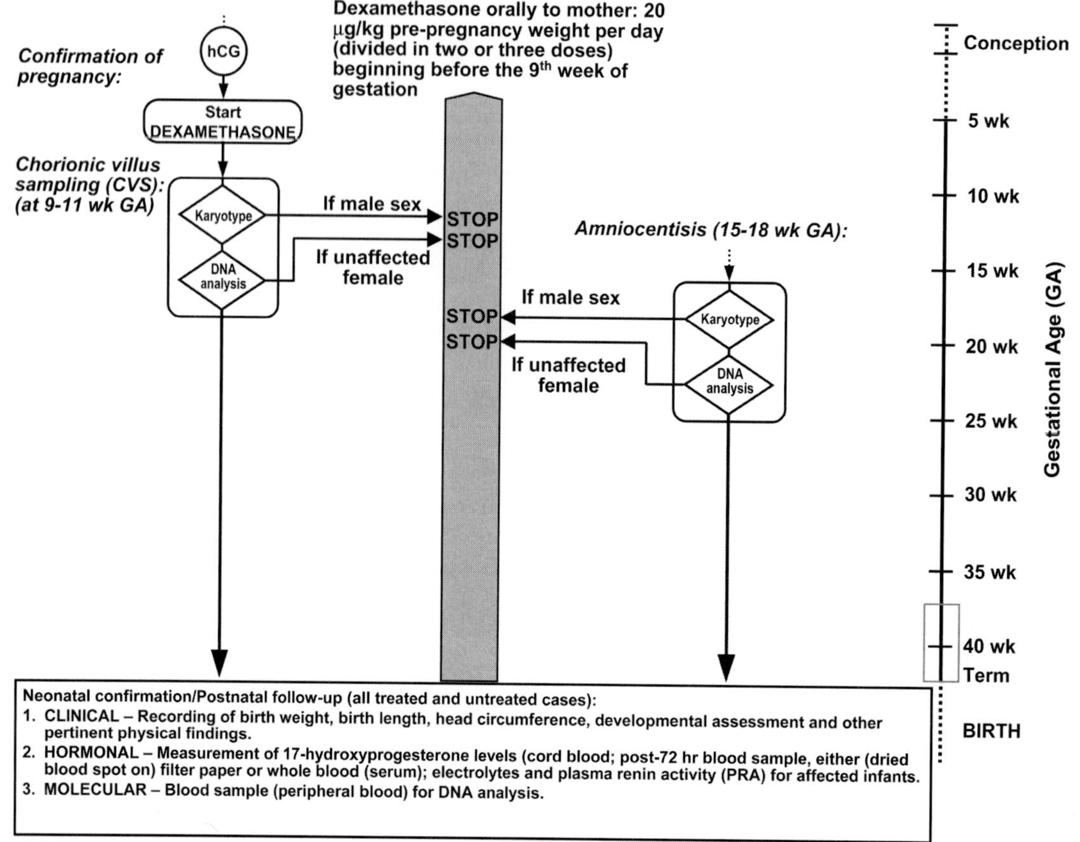

Fig. 5 Algorithm depicting prenatal management of pregnancy in families at risk for a fetus affected with 21-OHD [25].

THE PROTOCOL

Dexamethasone (20 μg/kg/day in 3 divided doses) was administered to the pregnant mother before the 9th week of gestation, blind to the affected status of the fetus, to suppress excess adrenal androgen secretion and prevent virilization should the fetus be an affected female (Fig. 6). Diagnosis by DNA analysis requires chorionic villus sampling in the 9th to 11th week gestation or sampling of amniotic fluid cells obtained by amniocentesis in the second trimester. The fetal DNA is used for specific amplification of the CYP21 gene utilizing polymerase chain reaction (PCR) [30]. If the fetus is determined to be an unaffected female upon DNA analysis or a male upon karyotype analysis, treatment is discontinued. Otherwise, treatment is continued to term.

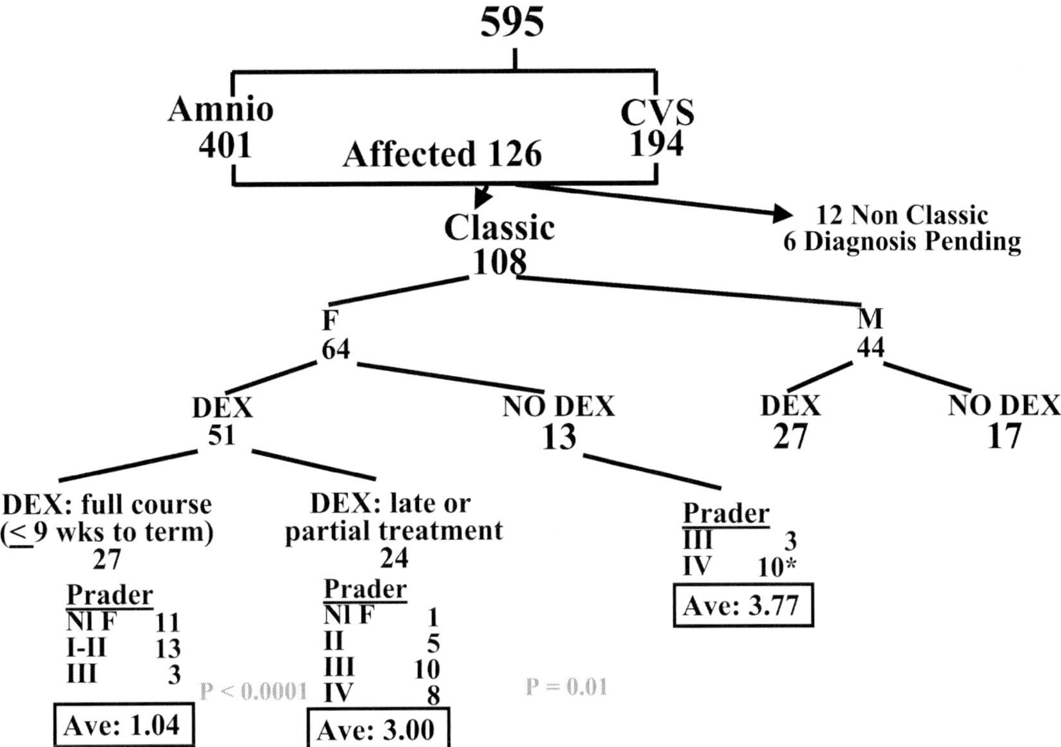

Fig. 6 Diagram depicting prenatal dexamethasone treatment outcome by Prader scores in fully and partially treated affected newborns. Partial treatment includes initiation of dex >9 weeks gestation, noncompliance, inadequate dosage for body weight. Dex = dexamethasone; Amnio = amniocentesis; CVS = chorionic villus sampling; F = female; M = male; Nl = normal.

RESULTS

Of 595 pregnancies evaluated, 126 fetuses were found to be affected with 21-OHD, of whom 108 were classical cases. Of the classical cases, 64 were female, 51 of whom were treated prenatally with dexamethasone. Dexamethasone administered at or before 9 weeks of gestation (27 affected female fetuses) was effective in reducing virilization. Of these 27, 11 fetuses were born with entirely normal female genitalia, while 13 were significantly less virilized (Prader stages 1–2) than those untreated (Fig. 7). Sixteen affected females treated with dexamethasone full-term had untreated affected female sibs. In all 16 cases, the external genitalia of the treated females were less virilized than the genitalia of the untreated sibs. Most of the newborn females whose genitalia were rated Prader stages 3–4 who had been treated prenatally with dexamethasone, started treatment late, were under-treated by referring physician, or were noncompliant.

Overall for affected females, the average Prader score for those treated prenatally at or before 9 weeks gestation was 1.04 while those with no prenatal treatment had an average Prader score of 3.77 ($P \ll 0.003$). The external genitalia of affected females prenatally treated at or before 9 weeks were less virilized than the partially treated affected females (mean Prader score of 3.00; $P < 0.0001$). The Prader scores of partial dexamethasone treatment and no treatment in affected females (mean Prader score 3.77) are also statistically significant ($P = 0.01$).

No significant or enduring side effects were noted in newborns and children who were prenatally treated. Fetal wastage did not differ statistically between dexamethasone treated (partial- and full-term)

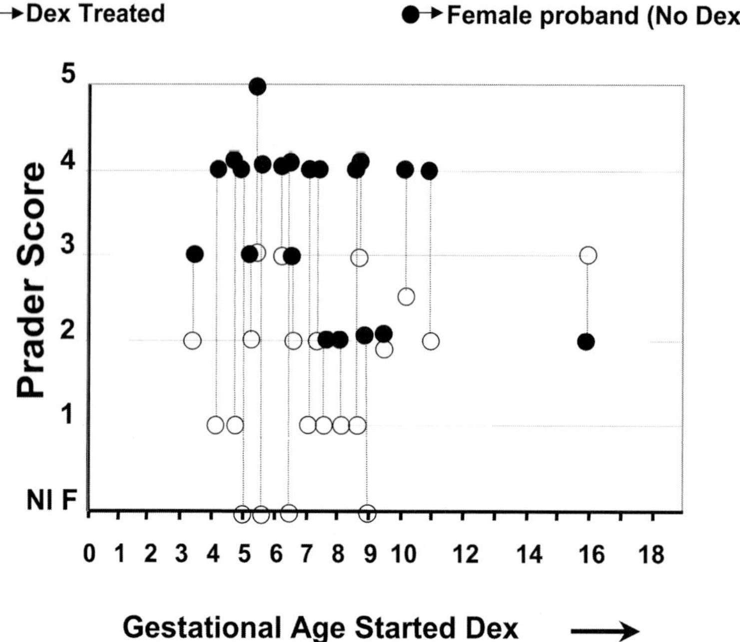

Fig. 7 Diagram depicting Prader stages of affected female infants in monitored, dexamethasone prenatally treated pregnancies, in relation to gestational age when dexamethasone was started. Affected untreated sibs are shown attached by a dotted line.

pregnancies from untreated. As previously reported [23], prenatally treated (partial- and full-term) newborns do not differ significantly in birth weight from untreated newborns. Mean birth weight for dexamethasone prenatally treated fetuses was 3.39 kg, while for untreated it was 3.46 kg ($P = 0.17$; Table 1). The birth weight of 59 affected infants treated with dexamethasone is significantly lower than affected infants not prenatally treated with dexamethasone ($P = 0.02$). Despite statistical significance, it is unlikely that there is any clinical significance between the birth weights of the treated (mean wt. 3.28 kg) and untreated affected (mean wt. 3.60 kg) groups. Infants not affected with CAH whether prenatally treated with dexamethasone (mean wt. 3.42 kg) or not (mean wt. 3.44 kg), had similar birth weights ($P = 0.74$). The birth length and head circumference (data not shown) were normal in offspring of dexamethasone-treated pregnancies compared to those not treated, which is consistent with other studies where patients and physicians adhered to the recommended therapeutic protocol [22–24,27,28]. A large quantitative follow-up study is currently in progress regarding cognition, gender, temperament, and

Table 1 Birth weights (kg) in dexamethasone treated vs. untreated fetuses at risk for classic CAH. Excludes premature infants, twins with suspected feto-fetal transfusion.

	Dexamethasone	No dexamethasone	T-test
Affected	3.28 ($n = 59$)	3.60 ($n = 20$)	$P = 0.02$
Not Affected	3.42 ($n = 171$)	3.44 ($n = 143$)	$P = 0.74$

handedness (an indicator of prenatal androgen effect) in children and adults who were prenatally treated with dexamethasone.

The authors did not find significant differences in side effects between the mothers who were treated with dexamethasone and the mothers who were not treated, except in weight gain, edema, and striae. By report, mothers who were not treated with dexamethasone gained an average of 29.7 lbs., while treated mothers gained an average of 36.8 lbs., which was statistically significant ($P < 0.005$). There was a statistically significant difference found for the presence of striae ($P = 0.01$) and edema ($P = 0.02$). There was not a statistically significant difference found for hypertension ($P = 0.5$), or gestational diabetes ($P = 0.34$) in the treated or untreated pregnancy groups. All mothers who took prenatal dexamethasone (partial- and full-term) treatment stated they would take dexamethasone again in the event of a future pregnancy.

DISCUSSION

Controversy of prenatal treatment

One report in 1997 questioned the safety of long-term prenatal glucocorticoid treatment of fetuses potentially affected with congenital adrenal hyperplasia [31]. The authors claimed that prenatal treatment with dexamethasone contributes to low birth weight, fetal demise, serious maternal complications, and cognitive and developmental deficiencies. However, the cited references are predominately based on animal studies, in which excess glucocorticoid dosages are used. We find these claims to be unfounded based on our experience with the largest number of treated human pregnancies in the world, in addition to the results of other large studies [27,32,33].

The risk-to-benefit ratio in view of no enduring side effects in mother or child favors prenatal treatment. Additionally, males and unaffected females treated with short-term dexamethasone show no side affects [28]. Treatment of affected females alleviates potential sex mis-assignment, repeated genital surgeries that cannot easily recreate natural genital structures, and psychological effects. We agree that long-term studies are needed to conclusively determine outcome of treatment, which are currently in progress.

Based on the experience at New York Presbyterian Hospital–Weill Cornell Medical College, proper prenatal diagnosis and treatment of 21-OHD is safe and is effective in significantly reducing or eliminating virilization in the affected female. Of the monogenic disorders, steroid 21-OHD is one of the few in which prenatal treatment is effective and influences postnatal life.

ACKNOWLEDGMENTS

I wish to express my appreciation to Brian Betensky for his editorial assistance in the preparation of this manuscript.

REFERENCES

1. H. F. L. Meyer-Bahlburg, R. Gruen, M. I. New, J. J. Bell, A. Morishima, M. Shimshi, Y. Bueno, I. Vargas, S. W. Baker. *Hormon. Behav.* **30**, 319 (1996).
2. M. I. New and P. C. White. In *Genetic Disorders of Steroid Metabolism*, R. V. Thakker (Ed.), pp. 525–554, Bailliere-Tindall, London (1995).
3. A. Prader. *Helv. Paediatr. Acta* **9**, 231 (1954).
4. M. Urban, P. Lee, C. Migeon. *New Engl. J. Med.* **299**, 1392 (1978).
5. R. Clark, B. Albertson, A. Munabi, F. Cassorla, G. Aguilera, D. Warren, R. Sherins, D. Loriaux. *J. Clin. Endocrinol. Metab.* **70**, 1408 (1990).

6. M. Srikanth, B. West, M. Ishitani, H. J. Isaacs, H. Applebaum, G. Costin. *J. Pediatr. Surg.* **27**, 639 (1992).
7. L. Premawardhana, I. Hughes, G. Read, M. Scanlon. *Clin. Endocrinol. (Oxf.)* **46**, 327 (1997).
8. S. Y. Pang, M. A. Wallace, L. Hofman, H. C. Thuline, C. Dorche, I. C. Lyon, R. H. Dobbins, S. Kling, K. Fujieda, S. Suwa. *Pediatrics* **81**, 866 (1988).
9. S. Pang and A. Clark. *Trends Endocrinol. Metab.* **1**, 300 (1990).
10. S. Pang and A. Clark. *Screening* **2**, 105 (1993).
11. M. I. New and R. C. W. Wilson. *Proc. Natl. Acad. Sci. USA* **96**, 12790 (1999).
12. Y. Higashi, A. Tanae, H. Inoue, T. Hiromasa, Y. Fujii-Kuriyama. *Proc. Natl. Acad. Sci. USA* **85**, 7486 (1988).
13. M. Tusie-Luna and P. White. *Proc. Natl. Acad. Sci. USA* **92**, 10796 (1995).
14. P. C. White, M. I. New, B. Dupont. *Proc. Natl. Acad. Sci. USA* **83**, 5111 (1986).
15. J. W. Werkmeister, M. I. New, B. Dupont, P. C. White. *Am. J. Hum. Genet.* **39**, 461 (1986).
16. Y. Higashi, T. Hiromasa, A. Tanae, T. Miki, J. Nakura, T. Kondo, T. Ohura, E. Ogawa, K. Nakayama, K. Y. Fujii. *J. Biochem.* **109**, 638 (1991).
17. E. Mornet, P. Crete, F. Kuttenn, M. C. Raux-Demay, J. Boue, P. C. White, A. Boue. *Am. J. Hum. Genet.* **48**, 79 (1991).
18. P. W. Speiser, J. Dupont, D. Zhu, J. Serrat, M. Buegeleisen, L. M. Tusie, M. Lesser, M. I. New, P. C. White. *J. Clin. Invest.* **90**, 584 (1992).
19. A. Wedell, E. M. Ritzen, S. B. Haglund, H. Luthman. *Proc. Natl. Acad. Sci. USA* **89**, 7232 (1992).
20. P. W. Speiser, M. I. New, G. M. Tannin, D. Pickering, S. Y. Yang, P. C. White. *Hum. Genet.* **88**, 647 (1992).
21. R. C. Wilson, A. B. Mercado, K. C. Cheng, M. I. New. *J. Clin. Endocrinol. Metab.* **80**, 2322 (1995).
22. N. Krone, A. Braun, A. Roscher, D. Knorr, H. Schwarz. *J. Clin. Endocrinol. Metab.* **85**, 1059 (2000).
23. M. Forest. *Curr. Opin. Endocrin. Diabet.* **4**, 209 (1998).
24. A. D. Carlson, J. S. Obeid, N. Kanellopoulou, R. C. Wilson, M. I. New. X^{th} International Congress on Hormonal Steroids, Quebec, Canada, p. 19 (1999).
25. S. Lajic, A. Wedell, T. Bui, E. Ritzen, M. Holst. *J. Clin. Endocrinol. Metab.* **83**, 3872 (1998).
26. P. W. Speiser, N. Laforgia, K. Kato, J. Pareira, R. Khan, S. Y. Yang, C. Whorwood, P. C. White, S. Elias, E. Schriock, E. Schriock, J. L. Simpson, M. Taslimi, J. Najjar, S. May, G. Mills, C. Crawford, M. I. New. *J. Clin. Endocrinol. Metab.* **70**, 838 (1990).
27. M. David and M. G. Forest. *J. Pediatr.* **105**, 799 (1984).
28. M. G. Forest, H. Betuel, M. David. *Endocr. Res.* **15**, 277 (1989).
29. A. B. Mercado, R. C. Wilson, K. C. Cheng, J. Q. Wei, M. I. New. *J. Clin. Endocrinol. Metab.* **80**, 2014 (1995).
30. R. C. Wilson, J. Q. Wei, K. C. Cheng, A. B. Mercado, M. I. New. *J. Clin. Endocrinol. Metab.* **80**, 1635 (1995).
31. J. Seckl and W. Miller. *JAMA* **277**, 1077 (1997).
32. A. D. Carlson, J. S. Obeid, N. Kanellopoulou, R. C. Wilson, M. I. New. In *Prenatal Treatment and Diagnosis of Congenital Adrenal Hyperplasia Owing to Steroid 21-Hydroxylase Deficiency*, M. I. New and F. L. Reddick (Eds.) (1999).
33. M. G. Forest, M. David, Y. Morel. *J. Steroid Biochem. Mol. Biol.* **45**, 75 (1993).

Pure Appl. Chem., Vol. 75, Nos. 11–12, pp. 2023–2038, 2003.

Topic 3.4

Male reproductive disorders and the role of endocrine disruption: Advances in understanding and identification of areas for future research*

Richard M. Sharpe[1] and Niels E. Skakkebaek[2,‡]

[1]MRC Human Reproductive Sciences Unit, Centre for Reproductive Biology, The University of Edinburgh Academic Centre, 49 Little France Crescent, Old Dalkeith Road, Edinburgh EH16 4SB, UK; [2]University Department of Growth & Reproduction, Rigshospitalet, Blegdamsvej 9, DK-2100 Copenhagen OE, Denmark

Abstract: This review addresses whether there is a secular increasing trend in male reproductive developmental disorders (cryptorchidism, hypospadias, testis cancer, low sperm counts), and highlights the limitations of available data and how these issues are being addressed. These disorders are considered to represent a syndrome of disorders [testicular dysgenesis syndrome (TDS)] with a common origin in fetal life, and in which "endocrine disruption" plays a central role. The potential involvement of environmental estrogens in the etiology of these disorders is reviewed in light of new understanding about the pathways and dose–effect relationships of estrogen action on male reproductive development. Several new pathways of estrogen action have been identified, including suppression of the production of testosterone and insulin-like factor-3 by fetal/neonatal Leydig cells and suppression of androgen receptor expression in androgen target tissues. It is tentatively concluded that identified environmental chemicals are unlikely to activate these pathways because of their intrinsically weak estrogenicity. However, chemicals that may alter endogenous estrogen production, bioavailability, or inactivation represent a new focus of concern. Additionally, environmental chemicals that alter endogenous levels of androgens in the rat fetus (certain phthalates) induce a similar collection of disorders to TDS. Whether human exposure to such compounds might contribute to TDS remains to be shown, but studies in animals should help to define susceptible pathways for induction of TDS.

INTRODUCTION

World-wide interest in male reproductive disorders probably stems from the publication in 1992 of the paper by Carlsen et al. [1] suggesting that sperm counts in human males might have declined by nearly half during the previous 50 to 60 years. This interest became more intense when we published our hypothesis paper in the *Lancet* in the following year [2], in which we argued that the fall in sperm counts, increase in incidence of testis cancer, and in other reproductive abnormalities (cryptorchidism, hypospadias) in the human male might be related to increased estrogen exposure in utero. In our paper, we

*Report from a SCOPE/IUPAC project: Implication of Endocrine Active Substances for Human and Wildlife (J. Miyamoto and J. Burger, editors). Other reports are published in this issue, *Pure Appl. Chem.* **75**, 1617–2615 (2003).

‡Corresponding author: Tel: +44-131-229-257; Fax: +44-131-228-5571; E-mail: r.sharpe@hrsu.mrc.ac.uk

went on to identify several mechanisms by which this increased exposure could have occurred—one of which was exposure to environmental chemicals, in particular to environmental estrogens. The coincidence of this publication with an explosion of interest in environmental estrogens meant that a link between human exposure to environmental chemicals and the occurrence of male reproductive disorders was widely discussed [3]. There is still nothing other than circumstantial evidence to support this possibility. Though this may mean that environmental chemicals are not important in this context, it is becoming increasingly obvious that provision of definitive proof, one way or the other, is a demanding task. As this review will show, there have been many problems to overcome that relate to both the quality of the data and to our lack of understanding of the underlying biology. However, it is reassuring that these deficits have been recognized (though not by all!) and appropriate responses and changes made to overcome these shortcomings. These are helping to accurately define the reproductive health problem, identify how the search for "causes" may be addressed, and to redefine the causal role (if any) that environmental chemicals might play.

An important development in recent years, and one that forms the core around which this review is built, is the realization that the human male reproductive disorders of concern—namely cryptorchidism, hypospadias, testis germ cell cancer, and low sperm counts—may form a syndrome of interconnected disorders with a common origin in fetal life. This has been termed "testicular dysgenesis syndrome" (TDS) [4]. It is argued that there may be more than one cause of this syndrome and that its manifestation may include one or more of the aforementioned disorders. Most importantly, from the perspective of this review, is the recognition that "endocrine disruption" (altered androgen and/or estrogen levels/action) is a central feature of this syndrome. Below, we discuss the current views on secular trends in male reproductive disorders, the pathways via which endocrine disruption may occur, and the possibility that environmental chemicals may impact these pathways.

MALE REPRODUCTIVE DISORDERS—SECULAR TRENDS AND THE ROLE OF ENDOCRINE DISRUPTION

There are two major issues concerning human male reproductive disorders (testicular germ cell cancer, cryptorchidism, hypospadias, low sperm counts), namely, the evidence that the disorders are increasing in incidence with time (secular trends) and, second, the role of endocrine disruptors in this (increase in) incidence. These two issues are clearly linked in many people's minds, but it is more balanced to consider them as separate problems *that may be linked.* As this short review will emphasize, the evidence that "endocrine disruption" plays a central role in the origin of these disorders is becoming ever clearer (Fig. 1). However, what factors can cause or trigger endocrine disruption remains unclear, but is by no means confined to exposure to environmental endocrine disruptors. Endocrine disruption (i.e., hormonal imbalance) underlies many of the most common human disorders, but this disruption can have many causes such as infection, diet/bodyweight, lifestyle, inherited genetics, or environmental exposures, though there is a paucity of data for the latter when compared with the other listed triggers [5]. Below, we consider what new pieces of evidence have emerged in recent years to help redefine both the concerns and the potential causes.

Sperm counts—secular trends

As already indicated, the evidence that sperm counts in Western countries might have fallen by approximately 50 % since the 1930s to 1940s [1] was a driving impetus for concerns about environmental effects on the male. The initial study by Carlsen and colleagues was subjected to much criticism and reanalysis, many of which attempts appeared contrived [reviewed in: 3,6–8]. Independent reanalysis [8] reached exactly the same conclusions as had the original study, and a recent updated analysis [9] that included semen analysis data up to 1996 (101 studies in all) again confirmed the trends and conclusions of the original study. However, these studies are all based on meta-analyses of retrospective data, and can only raise the *possibility* that sperm counts have fallen. Numerous studies were prompted by the original findings, nearly all retrospective, and their results divided into two camps, those showing evidence of a secular trend related to year of birth and those showing no such trend [see 3,6–8]. From these studies, it also appeared that significant differences in sperm counts, both within and between countries, might exist. If these differences are real, global analysis of sperm counts to discern secular trends must consider these geographical differences, and this cannot be done retrospectively with any confidence. This reality prompted a major effort at the European level to obtain prospective data for sperm counts in various European countries using standardized methods of subject recruitment and semen analysis [10,11]. These studies have had three main aims:

1. establish robust methodology that could be used in all semen analysis laboratories;
2. establish a solid data platform from which future trends in sperm counts can be discerned and followed with confidence; and
3. enable a cross-sectional comparison of sperm counts in similar groups of men in European (and other) countries at the present time.

The initial group chosen in which to analyze sperm counts were "recently fertile" men whose partners were currently pregnant, recruitment taking place via parentcraft classes. Some 250 to 350 men were recruited from four centers in Denmark, Finland, France, and Scotland, and their semen analyzed using methods standardized throughout the four centers. Sperm counts were then standardized for a 30-year-old man with an abstinence period of four days in each country by allowing for known confounders such as period of abstinence. Median sperm counts were found to be 34 % higher in Finland than in Denmark with Scotland and France intermediate between these two extremes [11]. A seasonal difference in sperm counts was also evident in all four countries, sperm counts being approximately 30 % lower in summer than in winter, confirming other studies [reviewed in ref. 12]. Although not all relevant studies have reported such a seasonal difference in sperm counts, season, as well as age, abstinence, and geographical location should be considered when undertaking studies related to secular trends in sperm counts.

For the reasons outlined above, retrospective studies of sperm count data, which have been extensively reviewed [3,6,7], cannot be considered definitive. We will therefore not consider this data further here. There is, however, one new approach that is relevant to the secular trends in sperm counts issue, and that is the analysis of sperm counts in young men aged 18 to 20 years. These were prompted by the demonstration in several of the retrospective studies in older men that sperm counts appeared to decline in relation to when a man was born rather than according to when the semen sample was actually provided, consistent with fetal origins of the problem [3,6]. If later year of birth is associated with declining sperm counts, then analysis of sperm counts in the latest birth cohort, namely young men, should provide evidence to support or refute this thinking. If the young men have normal sperm counts (i.e., typical of those reported in the literature) then a secular decrease appears unlikely, whereas if they have "low" sperm counts a secular decrease in sperm counts would appear more likely. For solidity and comparability of these studies, they have been undertaken in conjunction with the European multicenter study of recently fertile men, outlined above, and have used the same standardized methods. Semen analysis studies are always plagued by "biased" recruitment, so the principal approach has been to undertake the studies in countries in which there is still compulsory military service for young men. To date, only data for Danish young men are available, and these indicate a median sperm count of only 41 million/ml) [13], a value that is substantially lower than in older men from the same country [11]. With such low sperm counts, effects on time to achieve a pregnancy become more likely [14–16]. This finding, which has been confirmed by other as yet unpublished studies in Denmark, supports the hypothesis of a secular decrease in sperm counts related to later year of birth. It remains to be shown if this is a change specific to Denmark, and as similar studies in other European countries on 18- to 20-year-old men are nearing completion, it is prudent to wait for these results before conclusions are drawn. In this regard, a recent study of 408 young (18 years of age) men from the Czech Republic reported median sperm counts of 44 million/ml [17], though it is emphasized that this study was not part of the European initiative outlined above, but was a study of the impact of seasonal air pollution.

Sperm counts—role of endocrine disruption

Though sperm are not made until later in puberty, capacity to make sperm is determined by the numbers within the testes of Sertoli cells, and as these cells proliferate in fetal, neonatal (in particular), and peripubertal life [18], inhibition of Sertoli cell proliferation during any of these phases in life could affect sperm counts in adulthood. This was the thinking in the original "estrogen hypothesis" in which it was proposed that perinatal estrogen exposure could reduce Sertoli cell proliferation, probably by suppressing follicle-stimulating hormone (FSH) secretion from the pituitary gland [2]. Studies in rats have since shown that neonatal estrogen exposure does indeed reduce Sertoli cell number, testis size, and sperm production in adulthood dose-dependently [19,20]. Though a direct effect of estrogens on the Sertoli cells is possible [20], it is likely that suppression of FSH secretion is the most important mechanism behind the reduction in sperm counts (unpublished data). However, whether estrogen suppression of FSH levels perinatally in humans would result in a *permanent* reduction in Sertoli cell numbers is questionable. This is because studies in the marmoset have shown that although suppression of FSH levels neonatally does reduce Sertoli cell proliferation/number by approximately 30 % at the end of the neonatal period, by the time such treated animals have reached adulthood Sertoli cell number has recovered to normal [21]. As the marmoset appears similar to man in terms of when Sertoli cell proliferation occurs [18,20], this might account for the rather modest changes in semen quality reported in men exposed to DES in utero [22] and the associated absence of any change in fertility [23]. Note that this explanation does not rule out the possibility that DES exposure in utero did reduce Sertoli cell numbers (which is highly likely based on data from the rat), only that if such a reduction did occur it was probably compensated for at some time postnatally.

Testis cancer—secular trends

Testis cancer has continued to increase in incidence in Caucasian men in many countries [22,24–27]. Numerous studies have demonstrated that later year of birth is associated with a progressively increasing risk of developing testis cancer [22,28,29], implicating environmental, as opposed to genetic, factors in the etiology of this increase. New data has also confirmed that men with testis cancer exhibit reduced fertility prior to the occurrence of their tumor [30,31]. This reinforces the hypothesis that there may be a causal link between occurrence of testis cancer and low sperm counts due to abnormal germ cell and/or Sertoli cell development in fetal life (see Fig. 1 and below).

Testis cancer—role of endocrine disruption

It has long been established that disorders of sexual differentiation, a hormonally mediated process in fetal life, are associated with an exceptionally high risk of developing testis cancer in young adulthood [32]. Other studies had suggested an increased risk of testis cancer in men exposed in utero to DES or other estrogens, though not all such studies concurred [33]. A meta-analysis of published studies concluded that estrogen exposure in the first trimester of pregnancy was associated with a small but significant increase in risk of developing testis cancer [22]. Three new studies have readdressed this issue and reached somewhat opposing conclusions [34–36], consistent with earlier data. Overall, it seems fair to conclude that exposure to *exogenous* estrogens in early pregnancy results in only a modest increase in risk of developing testis cancer. However, two new studies also raise the possibility that an increased risk of testis cancer might stem from increased exposure to *endogenous* (maternal) estrogens. These showed that twins of brothers who had developed testis cancer had a 12- to 37-fold increased risk of developing testis cancer themselves [37,38]. Twin pregnancies are associated with higher estrogen levels

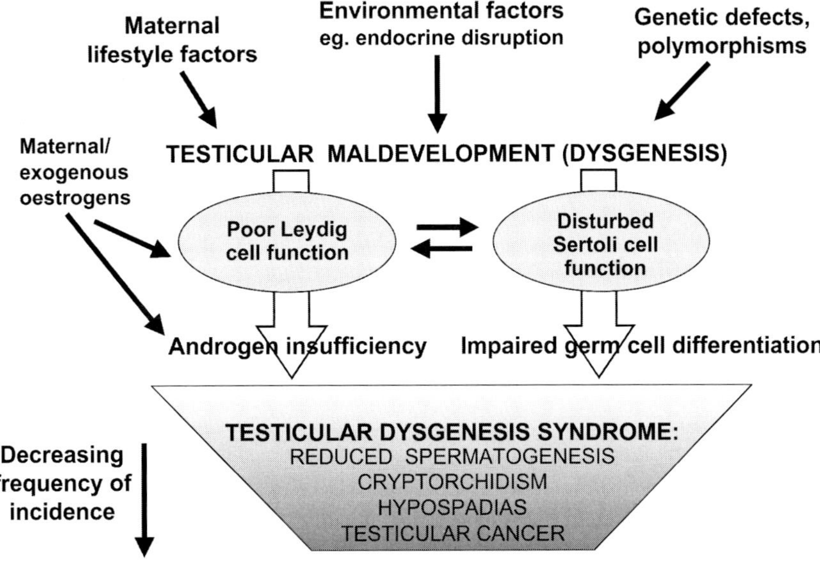

Fig. 1 TDS in the human. Schematic diagram to illustrate the central features and general pathways via which the disorders that may comprise this syndrome are likely to arise. Note that although abnormal testicular cell differentiation appears to be at the heart of this syndrome of disorders, numerous pathways might lead to this occurrence, including genetic, environmental, and lifestyle factors. It is tentatively suggested that any effect of maternal or exogenous estrogens on the induction of TDS is most likely to result from impairment of testosterone production or action, though impairment at an earlier step (e.g., cellular differentiation) cannot be ruled out. Note also that the disorders comprising TDS occur with differing frequency, varying from quite common (reduced sperm production, cryptorchidism) to rare (testis germ cell cancer). Note also that some of the disorders may occur for reasons other than TDS (e.g., low sperm counts).

than in single pregnancy and this is especially so for dizygotic twin pregnancies—accordingly, testis cancer risk is 50 to 100 % higher in dyzgotic than in monozygotic twins [38,39]. Of course, increased risk of testis cancer in twins might also indicate evidence of "genetic predisposition" to testis cancer, though there is a dearth of evidence to support such a view. Indeed, as the main identified risk factors for testis cancer all relate to fetal/pregnancy factors (see Table 1), it is most logical to explain the "twin data" on the basis that twins and siblings share a common intrauterine environment and that maternal/placental estrogens are one component of this, though not necessarily the most important component (Fig. 1).

It has become ever clearer that the premalignant germ cells from which testis cancer arises, namely carcinoma in situ (CIS) cells [40–42], themselves have their origins in fetal life [42–44]. It is presumed that the CIS cells arise because of failure of normal differentiation of fetal germ cells, but how and why this should occur remains unclear (Fig. 1). However, based on the identified risk factors, it is clear that subnormal androgen exposure [42–44] and/or increased estrogen exposure (see above) are potentially important factors (see Table 1). As these hormonal changes are also key risk factors for cryptorchidism (see below), it is noteworthy that cryptorchidism has been reconfirmed as the numerically most important risk factor for testis cancer [35], as well as being an important risk factor for low sperm counts, infertility, and hypospadias (Table 1).

Table 1 Common risk factors for testis germ cell cancer, cryptorchidism, hypospadias, and low sperm counts/infertility, based on numerous epidemiological studies. These support the suggestion that these four disorders may form a syndrome (testicular dysgenesis syndrome) with a common origin during the period of testicular/sexual differentiation in fetal life.

Testis germ cell cancer	Cryptorchidism	Hypospadias	Low sperm counts/ infertility
Cryptorchidism	Cryptorchidism in a sibling	Cryptorchidism	Cryptorchidism
Hypospadias	Hypospadias	Hypospadias in a sibling	
Low sperm count/ infertility	Low sperm count/ infertility		Low sperm count/ infertility in father
Testis cancer in a sibling			Testis cancer
Low birthweight (IUGR)	Low birthweight (IUGR)	Low birthweight (IUGR)	Low birthweight (IUGR)
First pregnancy	First pregnancy		
Impaired androgen production/action in fetal life[a]	Impaired androgen production/action in fetal life[a]	Impaired androgen production/action in fetal life[a]	Impaired androgen production/action in fetal life[a]
Increased DES exposure during mother's pregnancy	Increased DES exposure during mother's pregnancy	Increased DES exposure during mother's pregnancy	Increased DES exposure during mother's pregnancy

IUGR = intra-uterine growth restriction; DES = diethylstilboestrol
[a]Includes genetic disorders of androgen insensitivity.

Cryptorchidism—secular trends

The evidence for a secular increase in the incidence of cryptorchidism has been limited by inherent problems related to diagnosis and reporting [22]. Indeed, it is now accepted that registry data is unreliable and if any secular trends are to be discerned, then a structured and standardized diagnostic approach is essential and the studies should be prospective [45]. Such studies were initiated several years ago in Europe and involve a structured comparison of incidence of cryptorchidism at birth and at 3 months in consecutive series of babies born in various European countries. This data is likely to confirm that differences between countries in incidence of cryptorchidism are as marked as differences in incidence of testis cancer [22,24] and sperm counts [11], and will provide a sound baseline from which to monitor future trends in incidence. Despite the reservations about the accuracy of registry data for cryptorchidism, it is clear that this remains by far the most common congenital abnormality (2 to 4 % incidence at birth) in babies of either sex and surgical correction is required in 30 to 50 % of affected boys.

Cryptorchidism—role of endocrine disruption

Boys born to women treated with DES in early pregnancy have an increased incidence of cryptorchidism [see 22 for references], and new pathways via which cryptorchidism can be induced by estrogens in rodents have now been identified (see below). Nevertheless, it appears that exposure to hormones other than DES in early pregnancy, including those in the oral contraceptive pill, is not associated with increased risk of either cryptorchidism or hypospadias, according to meta-analysis [46].

Hypospadias—secular trends

Data from the United States Birth Defects Monitoring Program has shown a rate of increase of 2 to 3 % per year in incidence of hypospadias in all regions [47]. As with earlier studies that suggested similar trends [22], this is based on registry data, raising questions as to whether this increase is real or might be explained by reporting differences [48]. However, the study showed that the most severe forms of hypospadias (those requiring surgery), and which are less likely to go unreported, showed the same or even a larger temporal trend than did minor cases. A recent prospective study from the Netherlands that used a structured diagnostic procedure, reported a 4- to 6-fold *higher* incidence (0.7 % of boys at birth) than did the official registry data for the same region [49]. Remarkably, the ratio of severe to minor cases of hypospadias was 3:1 in the prospective study and the mirror image (1:3) in the registry data! This study therefore confirms that registry data for hypospadias is highly unreliable [45], and more importantly suggests that it underestimates the true incidence, especially for severe cases. Further evidence for an increasing incidence of hypospadias comes from neonatal intensive care units in the United States, which reported a 10-fold higher incidence (4 vs. 0.4 %) of hypospadias in 2000 compared with 1987 [50]. Though these new studies fuel concerns about the accuracy of registry data for hypospadias, and thus cast doubt on the accuracy of secular trends based on such data, they also suggest strongly that hypospadias is far more common than is generally perceived. If the new data from the Netherlands is representative, it suggests that hypospadias is the second most common congenital abnormality in children of either sex, after cryptorchidism [49].

Hypospadias—role of endocrine disruption

It was recently pointed out [51] that the widely disseminated belief that DES exposure during pregnancy led to an increased incidence of hypospadias in humans, was incorrect as the original study [52] did not specify hypospadias but "urethral abnormalities". In fact, these arguments have in turn been refuted, as a recent cohort study from the Netherlands has shown that boys born to mothers treated with DES during pregnancy show a massive 20-fold increased incidence of hypospadias [53]. Additionally, new studies in pregnant rats have uncovered the mechanism via which this might occur as DES treatment results in gross suppression of fetal testosterone levels (see below), which would be expected to increase risk of hypospadias (Table 1). In contrast, a meta-analysis of studies in which pregnant women were exposed to estrogens (other than DES) or other hormones in early pregnancy found no increased risk of hypospadias in the offspring [46], raising the possibility that induction of hypospadias is specific to DES, rather than estrogen, exposure per se (i.e., DES is an atypical estrogen). This seems unlikely based on two new pieces of data from animal studies. First, it has been shown in mice that a functional ERα receptor is essential for DES to induce reproductive tract abnormalities in either males or females [54,55]. Second, several of the abnormalities of development that are induced in animals treated in utero with DES, such as cryptorchidism and other abnormalities of the reproductive tract, are also induced in a transgenic mouse that overexpresses aromatase [56].

The apparent conflict between the increased risk of reproductive tract abnormalities in boys exposed in utero to DES and the lack of risk in those exposed to other hormones, including other synthetic estrogens, requires explaining, as it has important implications with regard to assessment of risk from endocrine disruptors. One obvious possibility is that the conflict relates to the dose and/or duration of estrogen exposure. When DES was administered to pregnant mothers in early gestation, extraordinarily high doses were used (80–>2000 µg per kilogram per day [µg/kg/day]; see ref. [22]). In contrast, in studies in which pregnant women were exposed to "hormones" other than DES during pregnancy, this usually involved either brief or single exposure to high levels during hormone administration as a pregnancy test or inadvertent exposure to oral contraceptives before the pregnancy was diagnosed [46], most of which involved exposure to lower doses of "estrogen" or for relatively shorter periods than occurred with DES exposure. In studies in rats and mice in which adverse effects on male reproductive tract development have been induced by DES, ethinyl oestradiol, or oestradiol, very high doses of estrogen (>50 μg/kg/day) have been used [20]. Indeed, studies in neonatal rats suggest that only doses of estrogen that induce associated reductions in testosterone levels and in androgen receptor expression will cause reproductive developmental abnormalities, and for this to happen doses in excess of 100 μg/kg/day are required [20,57]. This suggests that only exposure to very high estrogen levels poses a major risk to male reproductive tract development, though this interpretation requires formal testing.

MECHANISMS VIA WHICH ALTERED HORMONE EXPOSURE (ENDOCRINE DISRUPTION) COULD INDUCE MALE REPRODUCTIVE DISORDERS

Adequate androgen production and action are essential prerequisites for normal male reproductive tract development. Gene mutations or other factors that interfere with such processes result inevitably in abnormalities of reproductive development (Fig. 1). At its most extreme, in "complete androgen-insensitivity syndrome", complete failure of phenotypic masculinization occurs, despite the fact that the affected individuals have testes. In such cases, the testes are abdominal and are at high risk of testis cancer [32]. Though this has long been established, it is worth restating as it emphasizes the central role that androgens play in normal male development and perhaps questions the logic behind the focus on environmental estrogens, or indeed even on potent estrogens such as DES. However, perusal of Table 1 clearly shows that *reductions* in androgen production or action or *increases* in estrogen exposure during male development are both associated with increased risk of all of the disorders listed, and the same is true for experimental studies in animals [20]. This similarity suggests a link, and recent data from animal studies appears to have uncovered four mechanistic links that may explain this phenomenon.

Suppression of androgen production

Administration of potent estrogens such as DES or ethinyl estradiol to the rat *in high doses* (>100 μg/kg/day), drastically suppresses Leydig cell function and, as a result, testosterone levels in both the testis and/or blood are dose-dependently suppressed. Treatment of pregnant females in this way results in gross suppression of testosterone levels in the male fetus [58], and direct administration to male rat pups neonatally has similar effects [20] that may persist through to adult life [19]. The pathways via which estrogens reduce testosterone production may involve suppression of expression of steroidogenic factor-1 [59] and 17α-hydroxylase-$C_{17\text{-}20}$-lyase [60].

Suppression of androgen receptor expression

Coincident with suppression of testosterone production, neonatal treatment of rats with DES or ethinyl estradiol also grossly suppresses expression of the androgen receptor (AR) in the testis and throughout the developing reproductive tract [20,57,61]. This suppression only occurs at very high doses (>100 μg/kg), similar to the effects on suppression of testosterone levels reported above. However, the suppression of testosterone production and loss of AR expression are separate effects as experimental suppression of testosterone levels by other mechanisms (switching off pituitary gonadotrophin secretion by treatment with a GnRH antagonist) is unable to induce loss of AR expression [20,57,61]. Dose–response studies indicate that estrogen-induction of testicular and reproductive tract abnormalities is coincident with suppression of AR expression, so this change is presumably involved in the aetiology of these disorders [57].

Distortion of the androgen–estrogen balance

Though the findings in 1 and 2 described above imply that estrogens may simply function as effective anti-androgens, at least at very high doses, other findings suggest that the relationship between androgens and estrogens is more complex than this. For example, it has been shown that most, and perhaps all, of the adverse effects induced by high doses of DES on male reproductive development when administered neonatally, can be prevented by coadministration of testosterone, even though the particular adverse efects induced by DES cannot be induced by simply suppressing testosterone production or action [57]. More recently, it has been shown that administration of a 100-fold lower dose of DES, which is largely ineffective on its own, can induce a similar spectrum of reproductive tract abnormalities if testosterone production or action is suppressed at the same time [62]. These findings have been interpreted as evidence that it is the balance between androgens and estrogens that is important for normal reproductive tract development rather than the absolute levels of either hormone [20,62]. If this is true, it has considerable implications for the assessment of mixtures of endocrine disruptors, especially mixtures of compounds with anti-androgenic and estrogenic activity. However, in the study by Rivas et al. [62], administration of a high dose of bisphenol A was still unable to induce any "estrogenic" reproductive tract abnormalities when testosterone production was inhibited at the same time.

Suppression of secretion of insulin-like factor-3 by fetal Leydig cells

In the last few years, a completely new pathway involved in testicular descent has been discovered and has been shown to be suppressible by estrogens, namely production of insulin-like factor-3 (InsL3), also termed relaxin-like factor, by fetal Leydig cells. InsL3 acts on the gubernaculum of the testis, which plays an important role in guiding the testis during its phase of transabdominal descent. Administration of estrogens to pregnant rats during the time of sexual differentiation of the male fetus results in suppression of InsL3 expression and failure of testis descent [63,64]. The importance of InsL3 was confirmed by transgenesis, as mice homozygous for a knock-out of the InsL3 gene exhibit bilateral cryptorchidism [65,66]. Indeed, in transgenic mice that overexpress InsL3, ovarian "descent" is induced in the female offspring [67]. These findings demonstrate that production of InsL3 by fetal Leydig cells plays a key role in the transabdominal phase of testis descent, and that InsL3 secretion can be suppressed by estrogens. It remains unknown what relationship, if any, there is between estrogen-induced suppression of testosterone and InsL3 production, or what physiological role estrogens might play in regulating InsL3 secretion.

These four new pieces of data from animal studies go a long way toward explaining why suppression of androgen production/action or overexposure to estrogens may have very similar effects on the developing male reproductive system [20], and probably explain the coincidence of these two factors as risk factors for reproductive developmental disorders in humans (Table 1, Fig. 1). They also have other important implications. First, estrogen exposure *on its own*, either during fetal or neonatal life, only appears to cause adverse effects that impact health/function when administered at very high doses (doses at which effects on androgen production/action occur). At face value, this implies that weakly estrogenic compounds will be incapable of inducing such effects, at least because of their intrinsic estrogenicity. In this regard, it is emphasized that studies showing effects of extremely low doses of estrogens or estrogenic compounds (so-called "low-dose effects") on male reproductive development have used endpoints that are fundamentally different from those being discussed here (see below).

RELEVANCE OF ENVIRONMENTAL COMPOUNDS WITH ESTROGENIC OR ANTIANDROGENIC ACTIVITY TO HUMAN MALE REPRODUCTIVE DEVELOPMENTAL DISORDERS

In view of the improvement in understanding the mechanisms via which estrogens can induce male reproductive abnormalities, and the demonstration that only very high doses of (potent) estrogens are apparently able to activate these pathways, it is possible to make a fairly radical reassessment of the hazard posed by estrogenic and anti-androgenic chemicals. For example, it is clear that all of the identified "environmental estrogens" possess weak or very weak intrinsic estrogenic activity when measured by conventional in vitro and in vivo assays for estrogenicity [68–70], with the possible exception of zeranol (which is used as a growth promoter in livestock in the United States [71]). Based on intrinsic estrogenic potency of these chemicals, it seems unlikely that any of the identified environmental compounds could induce either cryptorchidism, hypospadias, or testis germ cell cancer, and only a tiny possibility that such compounds could affect sperm counts/sperm production.

We reason this conclusion as follows. For induction of male reproductive tract abnormalities in *male* animals, exposure to doses of DES higher than or equal to 50 μg/kg/day are required, and these are the doses that also induce major suppression of androgen production/androgen receptor expression [20,58]. Similarly high amounts were administered to pregnant women and were associated with increased incidence of reproductive developmental abnormalities (Table 1), whereas exposure to lower estrogen doses or to high doses for only short periods did not induce such effects (at least not cryptorchidism or hypospadias) [46]. Based on estrogenic potency, human exposure to the most potent environmental estrogens would need to be >1,000-fold higher than 50 μg/kg/day for adverse effects relevant to the human male to be induced, and such levels of exposure are remote or impossible [69]. Zeranol, in contrast to other identified environmental estrogens, is considerably more potent based on in vitro tests [71], but even in this case it is inconceivable that exposure in this range due to residues in meat could occur, as the available evidence points to levels of exposure that are in the pg/kg/day range, or at the very most the low ng/kg/day range [72,73]. Consistent with this interpretation, we have been unable to induce "DES-like" reproductive tract abnormalities in rats treated neonatally with zeranol at a high dose of 300 μg/kg/day (unpublished data) or with either of two environmental estrogens, bisphenol A or octylphenol, even when these were administered at levels approximating 37 to 150 mg/kg/day [20,61,74].

In contrast, other studies have reported adverse effects of various environmental estrogens such as bisphenol A and octylphenol on male rodents. These effects have included impaired development/proliferation of germ cells and altered steroidogenesis [75–78]. In most studies, relatively high doses that are probably not relevant to human exposure were used to induce such effects. However, extremely low levels were used in the studies by vom Saal et al. [79] in which sperm production and prostate weight were both altered significantly (sperm production downwards) in adulthood in mice after in utero exposure to levels of bisphenol A in the range 2 to 20 μg/kg. Such "low-dose" effects can also be induced by 0.02 to 2 μg/kg/day ethinyl estradiol in this model system [80], though it should be noted that comparable and more detailed studies have been unable to repeat these findings using doses of bisphenol A in the range 0.2 to 200 μg/kg/day [81,82; reviewed in ref. 70]. In the studies by vom Saal and colleagues [79,80], it is impossible to account for the effects of bisphenol A in terms of its intrinsic estrogenic potency, so an estrogen-independent mechanism of action needs to be invoked, and this remains to be identified [70]. However, it must be kept firmly in mind that estrogenic or anti-androgenic environmental chemicals may exhibit other biological activities that might still be relevant to human health, and some examples of this are discussed below.

Though the disparate findings just described are difficult to reconcile, it is clear that none of the disorders that comprise the "testicular dysgenesis syndrome" in humans, with the possible exceptions of reduced sperm counts, have been reported in any study in which an environmental estrogen has been administered to rats or mice. As has been suggested above, it may be that in humans, exposure to very high levels of potent estrogens during a relatively long period of time may induce congenital abnormalities such as cryptorchidism and hypospadias, and the animal data are largely consistent with this hypothesis. Therefore, based on present understanding, it seems unlikely that altered human exposure to weak estrogenic compounds can account for the possible increasing incidence of male reproductive-tract disorders, though this must be considered a tentative conclusion. Nevertheless, this does not mean that exposure to environmental chemicals can be ruled out as being involved etiologically in "testicular dysgenesis syndrome", as in utero exposure of rats to certain phthalates has been shown to induce a remarkably similar constellation of disorders [83–85]. These effects are attributable to impaired testosterone production by fetal Leydig cells in phthalate-exposed rats [86]. Such findings, together with the understanding that lowered androgen production/action (or altered androgen-estrogen balance) are involved in DES-/estrogen-induction of similar disorders, suggests strongly that in the search for environmental causes of human male reproductive developmental disorders, the focus should be very much on factors that can lower endogenous androgen production/ action or on factors that can elevate endogenous (potent) estrogens in the fetus. This assumes that exposure to exogenous estrogens will be insufficient, for reasons argued above. Any factor(s) that can both lower androgen production/action and elevate endogenous estrogens, thereby drastically altering the androgen–estrogen balance, would be of greatest concern.

Alteration of fetal androgen production/action during human pregnancy

As has already been discussed, and as illustrated in Table 1, gene mutations or chromosomal disorders that result in lowered androgen production or action increase the risk of male offspring having one or more of the disorders that comprise testicular dysgenesis syndrome (Fig. 1). Other than rare genetic changes, no other causes of lowered androgen production or action in the human fetus have been identified. However, one important observation is that African-American males have a substantially lower risk of testis cancer than do Caucasian Americans [27], and there is also a reported three-fold lower incidence of cryptorchidism in African-American babies compared with Caucasian babies [87]. The demonstration that pregnant African-American women in early gestation have 55 % higher blood levels of testosterone than do pregnant Caucasian women in the United States, when matched for gestational stage and body weight [88], has therefore been mooted as potentially playing a role in ensuring "high androgen status in the male fetus" [88]. One study in the United Kingdom has also suggested a link between low maternal testosterone levels during early gestation and the risk of cryptorchidism in the male offspring [89]. However, it is not clear what relationship, if any, there is between maternal and fetal androgen levels during early gestation, though it is clear that overproduction of androgens by the mother can partially masculinize the female fetus. The factors that might affect maternal androgen levels, other than SHBG (see below) remain to be defined.

A new possibility is that exposure of the pregnant mother to certain phthalate esters might lower fetal production of testosterone, based on their ability in studies of rats to do this and to cause a syndrome of disorders that are similar to those of concern in the human [83–85]. Human exposure to phthalates is extensive, and recent data (based on measurement of urinary phthalate metabolites), has identified that a subset of humans may have very much higher levels of exposure than the majority [90]. These are mainly women of reproductive age, and the source of their high exposure remains to be defined. In this subset, exposure ranges up to 160 μg/kg/day [91] whereas the effects in rats have been demonstrated in the range 100 to 750 mg/kg/day. This suggests that the risk to humans from phthalates is perhaps more theoretical than real, unless the human male fetus is more susceptible to the testosterone-lowering effects of phthalates than is the fetal rat; or differences in the ability of individuals to absorb, metabolize, or excrete phthalates makes them unusually susceptible to effects of phthalates. Studies that assess the relative sensitivity in vitro of the rat and human fetal testis to phthalate monoesters should help to address this possibility, and such studies are in progress.

Alteration of fetal exposure to endogenous estrogens during human pregnancy

An obvious way in which exposure of the human fetus to potent estrogens could be increased is via increased bioavailability and, thus, increased transplacental transfer of the enormously high estradiol levels that occur during normal human pregnancy. This is more than a theoretical concern, as it is established that first pregnancies or twin pregnancies, when estradiol levels are accepted to be unusually high, are risk factors for testis cancer and cryptorchidism (Table 1). The bioavailability of estradiol is determined by three factors:

1. level of production;
2. levels of sex hormone-binding globulin (SHBG) to which most estradiol and testosterone are bound; and
3. rate of metabolism/excretion of estradiol.

All three factors are potentially modulable, but the available evidence points to the second and third possibilities as likely to be the most susceptible to environmental/lifestyle influence. For example, insulin is a powerful, physiological suppressor of SHBG production, which will increase the bioavailability of estradiol and alter the androgen–estrogen balance [92]. The Western trend toward obesity and associated insulin resistance (= raised insulin levels) in women could therefore be an increasingly influential factor [93] to consider.

Reduced ability to inactivate or excrete estrogens could be another mechanism via which "pregnancy estrogens" might get transferred to the fetus in abnormally high amounts. One possible mechanism is altered inactivation of estradiol via estrogen sulphotransferase (SULT1E1), which is the primary mechanism via which estradiol is inactivated and excreted [94]. Polychlorinated biphenyls (PCBs) have been shown to suppress activity of SULT1E1 [95], while another study has shown even more potent suppressive effects of a range of environmental polyhalogenated hydrocarbons [96]. Human exposure to PCBs has been widespread, as is that to polyhalogenated hydrocarbons, so these new findings identify at least one route via which environmental chemicals could affect sex steroid levels or balance in the fetus, independent of there being any intrinsic hormonal activity in the compounds in question. It will prove of interest to see whether human exposure to such compounds during early pregnancy can be accurately determined and related to the occurrence of cryptorchidism or hypospadias in the offspring.

CONCLUSIONS AND FUTURE PERSPECTIVES

Though some of the data regarding the incidence and time trends in human male reproductive developmental disorders are less than robust, the widest perspective indicates that such disorders are extremely common and are becoming more common, at least in certain countries. Male reproductive health receives far less attention and healthcare investment than does female reproductive health, yet cryptorchidism and hypospadias are the two most common congenital malformations (affecting 2 to 5 % of boys, many of whom will require surgery), testis cancer is the most common cancer of young men (lifetime risk 0.3 to 0.8 % in most countries), and poor semen quality is the biggest defined cause of couple infertility and affects 6 to 8 % of men. The recognition that all of these disorders may form a hitherto unrecognized syndrome of disorders, "testicular dysgenesis syndrome", with a common origin in fetal life [4], has also emerged recently and provides an important focus for research to identify the pathways involved (Fig. 1). The fact that each of the component disorders of this "syndrome" are important risk factors for each other (Table 1), and that a similar "syndrome" can be induced in rats by phthalate exposure, strongly supports this contention. It is becoming increasingly clear that lowered androgen production/action is of central importance in "testicular dysgenesis syndrome", but it should also be kept in mind that lowered androgen production/action may itself be only a downstream manifestation of an earlier change, for example incomplete differentiation of Sertoli, Leydig, peritubular, and/or fetal germ cells with consequent impairment of function. More than one path (cause) may lead in this direction (Fig. 1).

Research in experimental animals during the last decade has helped to redefine the risk to male reproductive health from estrogens, and the current perspective is that "environmental estrogens" at least probably pose little, if any, risk to human male reproductive development, based on their intrinsic estrogenicity. However, the very fact that an "estrogen hypothesis" was first proposed in the late 1970s to early 1980s as potentially explaining some cases of testicular cancer and cryptorchidism [33], and that it can still be restated today (though in somewhat different terms), emphasizes that it must contain a kernel of truth. The emergence of new data and new mechanisms via which estrogens can induce cryptorchidism (suppression of InsL3; suppression of androgen production/action) or hypospadias (suppression of androgen production/action) in experimental animals is exciting, and provides a path forward to even better understanding. This data, coupled with the findings from studies with phthalates, also emphasizes the important role that experimental studies in animals are likely to play in dissecting apart the pathways that lead to male reproductive developmental abnormalities. Identifying these pathways and their experimental manipulation will provide an important means of pinpointing potential environmental or lifestyle factors that might impact them, which can then be factored into human epidemiological and prospective clinical studies.

If the incidence of testicular cancer is a beacon, then it must be accepted that environmental and/or lifestyle changes must play an important role in the increasing prevalence of male reproductive developmental disorders. Although we argue that exposure to environmental estrogens are probably unimportant in this context, an important new focus should be on chemicals that alter endogenous hormone production, action, or inactivation. Perusal of any clinical endocrinology textbook shows the catastrophic effects when normal endogenous hormonal status of an individual is altered, and male reproductive development provides some prime examples. The discovery that the most ubiquitous and/or persistent environmental chemicals, DDT/DDE, PCBs, and phthalates, can all perturb endogenous hormones in this way, serves as a timely reminder that the possibility that environmental chemicals play a role in the aetiology of human male reproductive disorders cannot be dismissed, only redefined.

ACKNOWLEDGMENT

This work is supported by the Svend Andersen Foundation and EU Grant Envir. Reprod. Health, Contract no.: QLK4-CT-1999-01422.

REFERENCES

1. E. Carlsen, A. Giwercman, N. Keiding, N. E. Skakkebaek. *Br. Med. J.* **305**, 609–613 (1992).
2. R. M. Sharpe and N. E. Skakkebaek. *Lancet* **341**, 1392–1395 (1993).
3. R. M. Sharpe. *Pure Appl. Chem.* **70**, 1685–1701 (1998).
4. N. E. Skakkebaek, E. Rajpert-de Meyts, K. M. Main. *Hum. Reprod.* **16**, 972–978 (2001).
5. R. M. Sharpe and S. Franks. *Nature Cell Biol.* (2002). In press.
6. D. S. Irvine. *Bailliere's Clinics Obstet. Gynaecol.* **11**, 655–671 (1997).
7. B. Jégou, J. Auger, L. Multigner, C. Pineau, P. Thonneau, A. Spira, P. Jouannet. In *The Male Gamete: From Basic Science to Clinical Applications*, C. Gagnon (Ed.), pp. 446–454, Cache River Press, Clearwater, Florida (1999).
8. S. H. Swan, E. P. Elkin, L. Fenster. *Environ. Health Perspect.* **105**, 1228–1232 (1997).
9. S. H. Swan, E. P. Elkin, L. Fenster. *Environ. Health Perspect.* **108**, 961–966 (2000).
10. N. Jorgensen, J. Auger, A. Giwercman, D. S. Irvine, T. K. Jensen, P. Jouannet, N. Keiding, C. Le Bon, E. MacDonald, A. M. Pekuri, T. Scheike, M. Simonsen, J. Suominen, N. E. Skakkebaek. *Int. J. Androl.* **20**, 201–208 (1997).

11. N. Jorgensen, A.-G. Andersen, F. Eustache, D. S. Irvine, J. Suominen, J. H. Petersen, A. N. Andersen, J. Auger, E. H. Cawood, A. Horte, T. K. Jensen, P. Jouannet, N. Keiding, M. Vierula, J. Toppari, N. E. Skakkebaek. *Hum. Reprod.* **16**, 1012–1019 (2001).
12. R. M. Sharpe. *Baillieres Clinics Endocrinol. Metab.* **14**, 489–503 (2000).
13. A. G. Andersen, T. K. Jensen, E. Carlsen, N. Jorgensen, A.-M. Andersson, T. Krarup, N. Keiding, N. E. Skakkebaek. *Hum. Reprod.* **15**, 366–372 (2000).
14. J. P. Bonde, E. Ernst, T. K. Jensen, N. H. Hjollund, H. Kolstad, T. B. Henriksen, T. Scheike, A. Giwercman, J. Olsen, N. E. Skakkebaek. *Lancet* **352**, 1172–1177 (1998).
15. M. J. Zinaman, C. C. Brown, S. G. Selevan, E. D. Clegg. *J. Androl.* **21**, 145–153 (2000).
16. R. Slama, F. Eustache, B. Ducot, T. K. Jensen, N. Jorgensen, A. Horte, D. S. Irvine, J. Suominen, A. G. Andersen, J. Auger, M. Vierula, J. Toppari, A. N. Andersen, N. Keiding, N. E. Skakkebaek, A. Spira, P. Jouannet. *Hum. Reprod.* **17**, 503–515 (2002).
17. S. G. Selevan, L. Borkovec, V. L. Slott, Z. Zudova, J. Rubes, D. P. Evenson, S. D. Perrault. *Environ. Health Perspect.* **108**, 887–894 (2000).
18. R. M. Sharpe. In *Fetal Programming: Influences on Development and Disease in Later Life,* P. M. S. O'Brien, T. Wheeler, D. J. P. Barker (Eds.), pp. 187–194, Royal College of Obstetricians & Gynaecologists Press, London (1999).
19. N. Atanassova, C. McKinnell, M. Walker, K. J. Turner, J. S. Fisher, M. Morley, M. R. Millar, N. P. Groome, R. M. Sharpe. *Endocrinology* **140**, 5364–5373 (1999).
20. K. Williams, C. McKinnell, P. T. K. Saunders, M. Walker, J. S. Fisher, K. J. Turner, N. Atanassova, R. M. Sharpe. *Hum. Reprod. Update* **7**, 236–247 (2001).
21. R. M. Sharpe, M. Walker, M. R. Millar, K. Morris, C. McKinnell, P. T. K. Saunders, H. M. Fraser. *Biol. Reprod.* **62**, 1685–1693 (2000).
22. J. Toppari, J. C. Larsen, P. Christiansen, A. Giwercman, P. Grandjean, L. J. Guillette, Jr., B. Jégou, T. K. Jensen, P. Jouannet, N. Keiding, H. Leffers, J. A. McLachlan, O. Meyer, J. Müller, E. Rajpert-De Meyts, T. Scheike, R. M. Sharpe, J. S. Sumpter, N. E. Skakkebaek. *Environ. Health Perspect.* **104** (Suppl. 4), 741–803 (1996).
23. A. J. Wilcox, D. D. Baird, C. R. Weinberg, P. P. Hornsby, A. L. Herbst. *New Engl. J. Med.* **332**, 1411–1416 (1995).
24. H. Adami, R. Bergstrom, M. Mohner, W. Zatonski et al. *Int. J. Cancer* **59**, 33–38 (1994).
25. I. Dos Santos Silva, A. J. Swerdlow, C. A. Stiller, A. Reid. *Int. J. Cancer* **83**, 630–634 (1999).
26. H. Moller. *Hum. Reprod.* **16**, 1007–1011 (2001).
27. SEER (Surveillance, Epidemiology and End Results; National Cancer Institute, USA). Incidence and age-adjusted rates of testis cancer 1973–1999. Web site address: <http://seer.cancer.gov> (2002).
28. R. Bergstrom, H. Adami, M. Mohner, W. Zatonski et al. *J. Natl. Cancer Inst.* **88**, 727–733 (1996).
29. T. Zheng. T. R. Holford, Z. Ma, B. A. Ward, J. Flannery, P. Boyle. *Int. J. Cancer* **65**, 723–729 (1996).
30. H. Moller and N. E. Skakkebaek. *Br. Med. J.* **318**, 559–562 (1999).
31. R. Jacobsen, E. Bostofte, G. Engholm, J. Hansen, J. H. Olsen, N. E. Skakkebaek, H. Moller. *Br. Med. J.* **321**, 789–792 (2000).
32. M. O. Savage and D. G. Lowe. *Clin. Endocrinol.* **32**, 519–533 (1990).
33. B. E. Henderson, R. Ross, L. Bernstein. *Cancer Res.* **48**, 246–253 (1988).
34. K. P. Dieckmann, T. Endsin, U. Pichlmeier. *Eur. Urol.* **40**, 677–683 (2001).
35. H. K. Weir, L. D. Marrett, N. Kreiger, G. A. Darlington, L. Sugar. *Int. J. Cancer* **87**, 438–443 (2000).
36. W. C. Strohsnitter, K. L. Noller, R. N. Hoover, S. J. Robboy, J. R. Palmer, L. Titus-Ernstoff, R. H. Kaufman, E. Adam, A. L. Herbst, E. E. Hatch. *J. Natl. Cancer Inst.* **93**, 545–551 (2001).
37. T. Westergaard, J. H. Olsen, M. Frisch, N. Kroman, J. W. Nielsen, M. Melba. *Int. J. Cancer* **66**, 627–635 (1996).

38. A. J. Swerdlow, B. L. De Stavola, M. A. Swanwick, N. E. S. Maconochie. *Lancet* **350**, 1723–1728 (1997).
39. M. M. Braun, A. Ahlbom, B. Floderus, L. A. Brinton, R. N. Hoover. *Cancer Causes Control* **6**, 519–524 (1995).
40. N. E. Skakkebaek, J. G. Berthelsen, A. Giwercman, J. Muller. *Int. J. Androl.* **10**, 19–28 (1987).
41. K. P. Dieckmann and N. E. Skakkebaek. *Int. J. Cancer* **83**, 815–822 (1999).
42. M. Rorth, E. Rajpert-de Meyts, L. Andersson, K. P. Dieckmann, S. D. Fossa, K. M. Grigor, W. F. Hendry, H. W. Herr, L. H. Looijenga, J. W. Osterhuis, N. E. Skakkebaek. *Scand. J. Urol. Nephrol.* **205**, 166–186 (2000).
43. E. Rajpert-de Meyts, N. Jorgensen, K. B. Nielsen, J. Muller, N. E. Skakkebaek. *APMIS* **106**, 198–206 (1998).
44. J. Slowikowska-Hilczer, R. Walczak-Jedrzejowska, K. Kula. *Folia Histochem. Cytobiol.* **39**, 67–72 (2001).
45. J. Toppari, M. Kaleva, H. E. Virtainen. *Hum. Reprod. Update* **7**, 282–286 (2001).
46. L. Raman-Wilms, A. L. Tseng, S. Wighardt, T. R. Einarson, G. Koren. *Obstet. Gynecol.* **85**, 141–149 (1995).
47. L. J. Paulozzi, J. D. Erickson, R. J. Jackson. *Pediatrics* **100**, 831–834 (1997).
48. L. J. Paulozzi. *Environ. Health Perspect.* **107**, 297–302 (1999).
49. F. H. Pierik, A. Burdoff, J. M. Rien Nijman, S. M. de Muinck Keizer-Schrama, R. E. Juttmann, R. F. A. Weber. *Hum. Reprod.* **17**, 1112–1115 (2002).
50. N. Hussain, A. Chaghtai, C. D. Herndon, V. C. Herson, T. S. Rosenkrantz, P. H. McKenna. *Pediatrics* **109**, 473–478 (2002).
51. M. Joffe. *Hum. Reprod.* **17**, 520–523 (2002).
52. B. E. Henderson, B. Benton, J. Jing, M. C. Yu, M. C. Pike. *Int. J. Cancer* **23**, 598–602 (1979).
53. H. Klip, J. Verloop, J. D. van Gool, M. E. Koster, C. W. Burger, F. E. van Leeuwen. *Lancet* **359**, 1102–1107 (2002).
54. J. F. Couse, D. Dixon, M. Yates, A. B. Moore, L. Ma, R. Maas, K. S. Korach. *Devel. Biol.* **15**, 224–238 (2001).
55. G. S. Prins, L. Birch, J. F. Couse, I. Choi, B. Katzenellenbogen, K. S. Korach. *Cancer Res.* **61**, 6089–6097 (2001).
56. X. Li, E. Nokkala, W. Yan, T. Streng, N. Saarinen, A. Warri, I. P. Huhtaniemi, R. Santti, S. Makela, M. Poutanen. *Endocrinology* **142**, 2435–2442 (2001).
57. C. McKinnell, N. Atanassova, K. Williams, J. S. Fisher, M. Walker, K. J. Turner, P. T. K. Saunders, R. M. Sharpe. *J. Androl.* **22**, 323–338 (2001).
58. T. Haavisto, K. Numela, R. Pohjanvirta, H. Huuskonen, F. El-Gehani, J. Paranko. *Mol. Cell. Endocrinol.* **178**, 169–179 (2001).
59. G. Majdic, M. Sharpe, M. R. Millar, P. T. K. Saunders. *Mol. Cell. Endocrinol.* **127**, 91–98 (1997).
60. G. Majdic, R. M. Sharpe, P. J. O'Shaughnessy, P. T. K. Saunders. *Endocrinology* **137**, 1063–1070 (1996).
61. K. Williams, J. S. Fisher, K. J. Turner, C. McKinnell, P. T. K. Saunders, R. M. Sharpe. *Environ. Health Perspect.* **109**, 1227–1235 (2001).
62. A. Rivas, J. S. Fisher, C. McKinnell, N. Atanassova, R. M. Sharpe. *Endocrinology* (2002). In press.
63. J. M. Emmen, A. McLuskey, I. M. Adham, W. Engel, J. M. Verhoef-Post, A. P. Themmen, J. A. Grootegoed, A. O. Brinkmann. *Endocrinology* **141**, 846–849 (2000).
64. S. Nef, T. Shipman, L. F. Parada. *Devel. Biol.* **224**, 354–361 (2000).
65. S. Nef and L. F. Parada. *Nat. Genet.* **22**, 295–299 (1999).
66. S. Zimmermann, G. Steding, J. M. Emmen, A. O. Brinkmann, K. Nayernia, A. F. Holstein, W. Engel, I. M. Adham. *Mol. Endocrinol.* **13**, 681–691 (1999).

67. I. M. Adham, G. Steding, T. Thamm, E. E. Bullesbach, C. Schwabe, I. Paprotta, W. Engel. *Mol. Endocrinol.* **16**, 244–252 (2002).
68. T. M. Crisp, E. D. Clegg, R. L. Cooper, W. P. Wood, D. G. Anderson, K. P. Baetcke, J. L. Hoffmann, M. S. Morrow, D. J. Rodier, J. E. Schaeffer, L. W. Touart, M .G. Zeeman, Y. M. Patel. *Environ. Health Perspect.* **106** (Suppl. 1), 11–56 (1998).
69. H. M. Bolt, P. Janning, H. Michna, G. H. Degen. *Arch. Toxicol.* **74**, 649–662 (2001).
70. R. Melnick, G. Lucier, M. Wolfe, R. Hall, G. Stancel, G. Prins, M. Gallo, K. Reuhl, S. M. Ho, T. Brown, J. Moore, J. Leakey, J. Haseman, M. Kohn. *Environ. Health Perspect.* **110**, 427–431 (2002).
71. H. Leffers, M. Naesaby, B. Vendelbo, N. E. Skakkebaek, M. Jorgensen. *Hum. Reprod.* **16**, 1037–1045 (2001).
72. D. M. Henricks, S. L. Gray, J. J. Owenby, B. R. Lackey. *APMIS* **109**, 273–283 (2001).
73. I. G. Lange, A. Daxenberger, H. H. D. Meyer. *APMIS* **109**, 53–65 (2001).
74. J. S. Fisher, K. J. Turner, D. Brown, R. M. Sharpe. *Environ. Health Perspect.* **107**, 397–405 (1999).
75. E. P. Murono, R. C. Derk, J. H. de Leon. *Reprod. Toxicol.* **13**, 451–462 (1999).
76. H. Nikula, T. Talonpoika, M. Kaleva, J. Toppari. *Toxicol. Appl. Pharmacol.* **157**, 166–173 (1999).
77. S. S. Raychoudhury, C. A. Blake, C. F. Millette. *Toxicol. Appl. Pharmacol.* **157**, 192–202 (1999).
78. E. Bendsen, S. Laursen, C. Olesen, L. Westergaard, C. Andersen, A. Byskov. *Hum. Reprod.* **16**, 236–243 (2001).
79. F. S. Vom Saal, P. S. Cooke, D. L. Buchanan, P. Palanza, K. A. Thayer, S. C. Nagel, S. Parmigiani, W. V. Welschons. *Toxicol. Ind. Health* **14**, 239–260 (1998).
80. K. A. Thayer, R. L. Ruhlen, K. L. Howdeshell, D. L. Buchanan, P. S. Cooke, D. Preziosi, W. V. Welshons, J. Haseman, F. S. vom Saal. *Hum. Reprod.* **16**, 988–996 (2001).
81. M. Ema, S. Fujii, M. Furukawa, M. Kiguchi, T. Ikka, A. Harazano. *Reprod. Toxicol.* **15**, 505–523 (2001).
82. T. Nagao, Y. Saito, K. Usumi, S. Yoshimura, H. Ohno. *Reprod. Toxicol.* **16**, 123–130 (2002).
83. E. Mylchreest, M. Sar, R. C. Cattley, P. M. D. Foster. *Toxicol. Appl. Pharmacol.* **156**, 81–95 (1999).
84. E. Mylchreest, D. G. Wallace, R. C. Cattley, P. M. D. Foster. *Toxicol. Sci.* **55**, 143–151 (2000).
85. R. M. Sharpe. *Toxicol. Lett.* **120**, 221–232 (2001).
86. L. G. Parks, J. S. Ostby, C. R. Lambright, B. D. Abbott, G. D. Klinefelter, N. J. Barlow, L. E. Gray, Jr. *Toxicol. Sci.* **58**, 339–349 (2000).
87. O. P. Heinonen, D. Slone, S. Shapiro. *Birth Defects and Drugs in Pregnancy,* p. 176, John Wright, Littleton, MA, USA (1977).
88. B. E. Henderson, L. Bernstein, R. K. Ross, R. H. Depue, H. L. Judd. *Br. J. Cancer* **57**, 216–218 (1988).
89. T. J. Key, D. Bull, P. Ansell, A. R. Brett, G. M. Clark, J. W. Moore, C. E. Chilvers, M. C. Pike. *Br. J. Cancer* **73**, 698–701 (1996).
90. B. C. Blount, M. J. Silva, S. P. Caudhill, L. L. Needham, J. L. Pirkle, E. J. Sampson, G. W. Lucier, J. Jackson, J. W. Brock. *Environ. Health Perspect.* **108**, 979–982 (2000).
91. M. C. Kohn, F. Parham, S. A. Masten, C. J. Portier, M. D. Shelby, J. W. Brock, L. L. Needham. *Environ. Health Perspect.* **108**, A440–A442 (2000).
92. S. M. Haffner. *Hormone Res.* **45**, 233–237 (1996).
93. P. Preziosi, E. Barrett-Connor, L. Papoz, M. Roger, M. Saint-Paul, K. Nahoul, D. Simon. *J. Clin. Endocrinol. Metabol.* **76**, 283–287 (1993).
94. W. C. Song. *Ann. NY Acad. Sci.* **948**, 43–50 (2001).
95. M. H. A. Kester, S. Bulduck, D. Tibboel, W. Meinl, H. Glatt, C. N. Falany et al. *Endocrinology* **141**, 1897–1900 (2000).
96. M. H. A. Kester, S. Bulduck, H. van Toor, D. Tibboel, W. Meinl et al. *J. Clin. Endocrinol. Metabol.* **87**, 1142–1150 (2002).

Pure Appl. Chem., Vol. 75, Nos. 11–12, pp. 2039–2046, 2003.

Topic 3.5

Brominated flame retardants and endocrine disruption*

Joseph G. Vos[1,‡], Georg Becher[2], Martin van den Berg[3], Jacob de Boer[4], and Pim E. G. Leonards[4]

[1]*National Institute for Public Health and the Environment, 3720 BA Bilthoven, The Netherlands;* [2]*Norwegian Institute of Public Health, Oslo, Norway;* [3] *Institute for Risk Assessment Sciences, Utrecht University, The Netherlands;* [4]*Netherlands Institute for Fisheries Research, IJmuiden, The Netherlands*

Abstract: From an environmental point of view, an increasing important group of organohalogen compounds are the brominated flame retardants (BFRs), which are widely used in polymers and textiles and applied in construction materials, furniture, and electronic equipment. BFRs with the highest production volume are the polybrominated diphenyl ethers (PBDEs), tetrabromobisphenol A (TBBP-A), and hexabromocyclododecane (HBCD). Because of their persistence and low biodegradation profile, several of the PBDE congeners accumulate in biota and are widely found in the aquatic food chain. Their levels in the environment and in humans have increased during the last decades, in contrast to compounds such as polychlorinated biphenyls (PCBs) and dichlorodiphenyltrichloroethane (DDT), for example. Humans may be exposed to PBDEs mainly through consumption of fatty food of animal origin (e.g., fish), but exposure through skin contact with textiles protected with flame retardants or through inhalation of BFRs volatilized from electronic and electric equipment may also occur. The levels of PBDEs in Swedish human milk showed a doubling in concentration every five years over the period 1972 to 1997 (2,2′,4,4′-tetraBDE being the predominant congener). The levels of penta- and hexa-BDEs increased at the same rate in ringed seals collected in the Canadian Arctic from 1981 to 2000. PBDEs exhibit a great variety of biological effects, depending on the bromine substitution pattern. PBDEs are potential endocrine disrupters, based on shared toxicity with the structurally related PCBs, polychlorinated dibenzofurans (PCDFs), and polychlorinated dibenzo-*p*-dioxins (PCDDs) (partial aryl hydrocarbon- [Ah-] receptor agonist and antagonist activity in vitro, thyroid toxicity, and immune effects), including developmental toxicity. The potency of TBBP-A to interact with thyroid hormone homeostasis is indicated from in vitro studies in which the compound competes with thyroxin (T4) for binding to transthyretin (TTR). So far, the toxicological profile of many BFRs is too incomplete and insufficient to perform an adequate risk assessment, and further information is required regarding the potential for endocrine disruption of these compounds that are of increasing environmental concern.

*Report from a SCOPE/IUPAC project: Implication of Endocrine Active Substances for Human and Wildlife (J. Miyamoto and J. Burger, editors). Other reports are published in this issue, *Pure Appl. Chem.* **75**, 1617–2615 (2003).

‡Corresponding author

INTRODUCTION

There is growing concern about the possible harmful consequences of exposure to xenobiotic compounds that are capable of modulating or disrupting the endocrine system. This concern for endocrine-disrupting chemicals (EDCs) is directed at both humans and wildlife [4,38]. Several expert working groups [1,21,44,47,48] have concluded that there is increasing evidence of adverse effects in human and wildlife reproductive health, and have discussed the hypothesis that chemicals in the environment have caused these endocrine-mediated adverse effects. Endocrine disruption is a complex area to address and it is difficult to establish causal links between exposure to suspected EDCs and any measured effects. The most prominent and persistent organic pollutants that are associated or even causally linked with endocrine disruption in wildlife and in human individuals are the organohalogen compounds (OHCs), including dichlorodiphenyltrichloroethane (DDT) and metabolites, polychlorinated biphenyls (PCBs), polychlorinated dibenzo-*p*-dioxins (PCDDs), and polychlorinated dibenzofurans (PCDFs). From an environmental point of view, an increasing important group of OHCs are the brominated flame retardants (BFRs)

Flame retardants constitute a diverse group of compounds used to prevent fires or minimize the extent of a fire. The worldwide annual production of flame retardants was estimated as 600 000 tons in 1992 [23]. There are three main categories of chemical flame retardants: halogenated hydrocarbons, organophosphorous compounds, and inorganic products often based on metallic hydroxides; these represent 45, 24, and 27 %, respectively, of the total market of flame retardants in 1999 (see ref. [41]). The BFRs are made up of structurally very different chemicals with a wide variety in physicochemical and reactivity characteristics [2]. Important BFRs are the polybrominated diphenyl ethers (PBDEs), tetrabromobisphenol A (TBBP-A), and hexabromocyclododecane (HBCD) (Fig. 1), high-production-volume chemicals that are widely used in polymers and textiles and applied in construction materials, furniture, and electric and electronic equipment. The annual market demand in 1999 has been estimated as 67 000 tons for PBDEs and 121 000 tons for TBBP-A [41]. The properties of PBDE congeners are variable as well [7,20]. There are three groups of industrial products of PBDEs with an average of five (pentaBDE), eight (octaBDE), or ten bromine atoms (decaBDE) in the molecule, while the theoretical number of possible congeners is 209. Further, photolytic degradation of decaBDE forms by debromination a large number of PBDE congeners not found in technical products of PBDEs [37]. PBDEs are highly lipophilic compounds and generally have a low biodegradation profile. In contrast to PCBs and DDT, the levels of some BFRs, such as PBDEs, show an increasing trend in wildlife and humans dur-

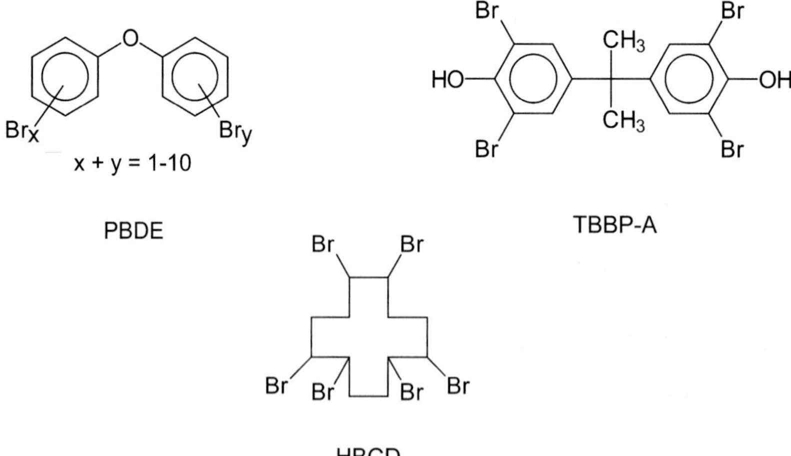

Fig. 1 Chemical structures of polybrominated diphenyl ethers (PBDE), tetrabromobisphenol A (TBBP-A), and hexabromocyclododecane (HBCD).

ing the last decades. Based on shared toxicity with the structurally related PCBs, PCDFs, and PCDDs, PBDEs are potential endocrine disruptors. Moreover, PBDEs may be contaminated with polybrominated dibenzo-*p*-dioxins and dibenzofurans (PBDDs/PBDFs), compounds that also can be formed during combustion of PBDEs. These substances are regarded as equally toxic as their chlorinated counterparts [26].

EXPOSURE DATA

PBDEs accumulate in environmental biota and are widely found in the aquatic food chain [24,36,49], including the Canadian Arctic [22]. Notably, they were found in top predators, such as the harbor seal and in the sperm whale; the latter finding indicates that PBDEs also occur in deep-sea food chains [8]. In the abiotic environment, a wide variety of PBDEs have been found, ranging from congeners with three to ten bromine atoms. From a quantitative point of view, decaBDE is the most dominant PBDE in sediments, up to milligram per kilogram (mg/kg) levels. Debromination of decaBDE may yield PBDE congeners with higher bioavailability and toxicity. Other important lower brominated PBDE congeners in sediment are 2,2′,4,4′-tetraBDE (BDE-47), 2,2′,4,4′,5-pentaBDE (BDE-99), 2,2′,4,4′,6-pentaBDE (BDE-100), and 2,2′,4,4′,5,5′-hexaBDE (BDE-153). These congeners can be found, for example, in sewage sludge and sediment up to several hundreds of micrograms per kilogram (μg/kg) dry weight [9,17], while, in addition, decabrominated BDE can be present into the mg/kg dry weight range [9].

Tissue analysis of wildlife and fish has shown that PBDEs can accumulate, with BDE-47, BDE-99, BDE-100, and BDE-153 being the dominant congeners present [7,11]. The first three are the predominant congeners present in ringed seal from the Canadian Arctic. In humans, these PBDEs have also been found in blood, adipose tissue, and milk with mean levels ranging between 4 and 16 ng/g lipid [10,32,39,42], and levels of approximately 200 ng/g lipid were reported recently in a pooled sample of human milk from the United States (levels of 132, 27, and 15 ng/g lipid of BDE-47, BDE-99, and BDE-153, respectively) [33]. Humans may be exposed to PBDEs mainly through the consumption of fatty food from animal origin (e.g., fish), but exposure through skin contact with textiles protected with flame retardants or through inhalation of BFRs volatilized from hot electrical equipment may also occur. For example, subjects working at an electronics dismantling plant had elevated plasma levels of higher brominated BDEs and TBBP-A [42].

An important observation is that, in contrast to PCBs and DDT, for example, the levels of PBDEs are increasing in human milk: a study in Sweden showed a doubling in concentration every five years over the period 1972 to 1997, with BDE-47 being the predominant congener [31] (Table 1). From 1998 to 2000, a decrease in PBDE levels was noticed that can be a consequence of the phase-out of commercial pentaBDE in Sweden [15]. The temporal trends and influence of age and gender on six BDE congeners was investigated on archived serum samples from Norway [43]. The sum of the BDEs increased from 0.44 ng/g lipids in 1977 to 3.3 ng/g in 1999, with BDE-47 being the most abundant congener. BFR levels in the different age groups were relatively similar, except for the age group of 0 to 4 years, which had 1.6 to 3.5 times higher serum concentrations, with breast milk considered the main source.

Table 1 Temporal trend of PBDEs in Swedish human milk[a].

	1972	1976	1980	84/85	1990	1994	1996	1997	1998	1999	2000
BDE-47	0.06	0.18	0.28	0.49	0.81	1.48	2.08	2.28	2.29	1.97	1.70
Sum of PBDE congeners	0.07	0.35	0.48	0.73	1.21	2.17	3.11	4.02	3.90	3.47	2.80

[a]Levels in ng/g lipid of pooled milk samples (after [15]).

PBDE concentrations showed an exponential increase in ringed seals collected from subsistence hunts in the Canadian Arctic in 1981, 1991, 1996, and 2000, with doubling of the penta- and hexa-BDE in less than five years; the current PBDE concentrations are 50 times lower than those of mono-ortho and non-ortho PCBs [22]. In fish and birds from the Baltic region, increasing levels of BDE-47, -99, and -100 have been reported since the 1970s but have begun to decline or level off in the 1990s [11], which can be a consequence of the phase-out of commercial pentaBDE in Sweden. Increasing levels of PBDEs in fish and birds are also reported for the U.S. Great Lakes, while PCBs levels have decreased during the same time frame. In different fish species collected in 2000 from the North Sea and the Celtic Sea, levels of BDE-47 were similar to levels of PCB 153 and *p,p*-DDE, whereas levels of hexachlorobenzene (HCB) and toxaphene (CHB-50) were lower (Fig. 2). The occurrence and fate of decaBDE is less well known, partly due to analytical difficulties and the fact that the compound has not been prioritized as an analyte, but also due to the different environmental fate of decaBDE as compared to other PBDEs. Further, little is known about other BFRs concerning environmental persistence, bioaccumulation, and toxicological effects, in spite of the fact that these compounds are also high-production-volume chemicals. TBBP-A has been detected in occupationally exposed persons, while HBCD has been found in wildlife and in environmental samples; no HBCD data on humans have so far been presented [2,3,42]. TBBP-A is rapidly excreted by mammals [16] while nothing is known about TBBP-A derivatives in this context. Thus, PBDEs have been steadily increasing over the last decades in biota (including humans), while much less has been published on other BFRs, such as TBBP-A and HBCD. Consequently, the question arises as to what extent these BFRs pose a risk to species higher in the food chain, in particular to top predators and humans. Human exposure probably occurs mainly via food in analogy to PCBs and related compounds, but occupational exposure (e.g., through handling electronic equipment) may also play a significant role. However, it should be noted that detailed information regarding the routes of human exposure to BFRs is presently lacking.

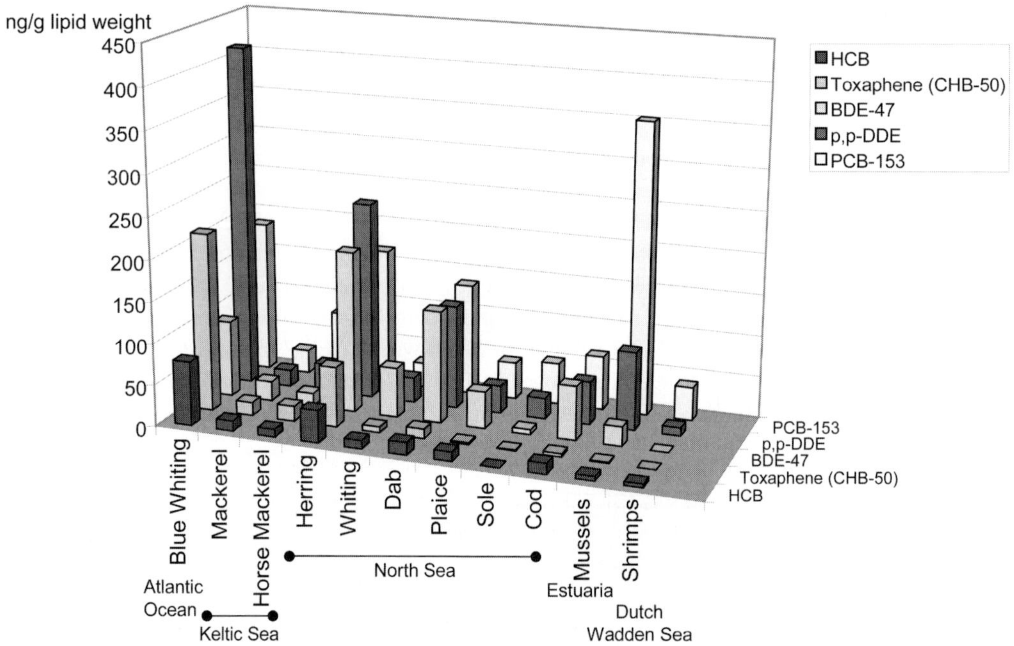

Fig. 2 Contaminant profile of fish, mussels and shrimp caught in 2000 in different waters (pooled samples of 25 animals each).

EFFECT DATA

For a review of the toxicity of BFRs, see refs. [7,19,24,25,27]. From these studies, it can be concluded that there is a lack of information regarding the potential for BFRs to cause endocrine disruption. Alteration of endocrine function by BFRs is realistic as there are striking resemblances in toxicological effects between the chlorinated aromatic hydrocarbons PCBs, PCDFs, and PCDDs and the PBDEs (i.e., partial aryl hydrocarbon- (Ah-) receptor agonist and antagonist activity in vitro, thyroid toxicity, and immune effects). Thus, PBDEs may produce toxic effects in a number of ways. The structural resemblance to thyroxin may explain interaction with the thyroid hormone system, and effects on thyroid function seem to be a sensitive endpoint. Decreased serum T4 levels were shown in mice exposed to a penta-BDE mixture [14] as well as to the 2,2′,4,4′-BDE congener (BDE-47) [18]. In addition, it was found that PBDEs after metabolic activation compete with the thyroid hormone (T4) for binding to transthyretin, the T4 transporting protein [29]. Following perinatal maternal exposure of rats to DE-71 (a commercial tetra- and penta-BDE mixture), reduced serum T4 was measured in the offspring showing its developmental toxicity [51]. Neonatal exposure to BDE-47 and BDE-99 has been found to induce neurotoxic effects in the adult animal [12]. In mice, the penta-BDE mixture also produced a decrease in the thymus/body weight ratio and in the antibody response to sheep red blood cells, and increased the activity of the hepatic cytochrome P450 mixed function oxidase system [14].

Using the in vitro CALUX-assay, pure PBDE-congeners appeared to act via the Ah-receptor signal transduction pathway as agonists, but mainly as antagonists [28]. A recent in vivo study with rats and commercial PBDE mixtures also indicated that hepatic induction of CYP1A1 and CYP2B activities could occur [50]. The former activity is of special relevance as this is an Ah-receptor-mediated process, which is common for dioxin-like compounds and PCBs, and most of the toxic responses (including the immunotoxicity) of dioxin-like compounds are mediated through binding to this receptor [34,35,46]. Persistent compounds with this type of mechanism are included in the toxic equivalency concept (TEF) and are now generally used in risk assessment procedures [45]. The CYP2B induction found in in vivo studies is also of toxicological interest, as this indicates the activation of multiple genes (e.g., glucocorticoid) associated with the phenobarbital responsive unit [50]. Finally, limited in vitro data indicate that some PBDEs and their hydroxylated metabolites can act as estrogenic compounds—data that suggest that in vivo metabolism of PBDEs might produce more potent pseudoestrogens [30]. The information about the toxicological effects of TBBP-A and HBCD is even more limited [25]. In vivo studies with TBBP-A and rats indicated that this compound can cause hepatotoxicity and disturbance of the haem synthesis [40]. In addition, it was shown that TBBP-A and lower brominated analogs can be potent competitors with T4 for binding to transthyretin (TTR) in vitro [30].

DISCUSSION

The toxicological profile of many BFRs is too incomplete and insufficient to perform an adequate human and ecological risk assessment. For a selected number of BFRs (including PBDEs and TBBPA), interactions with thyroid hormone homeostasis, estrogen, and Ah (dioxin) receptor have been reported. However, these studies are based only on in vitro or short-term experiments; therefore, significant gaps in knowledge exist for the situation of chronic and low-level exposure of humans and wildlife. Most notably, the PBDEs exhibit a high variety of biological effects, depending on the bromine substitution pattern. As these PBDEs occur in the environment in complex mixtures, the identification of biological and toxicological structure–activity relationships (SARs) is of great importance for understanding the mixture toxicity of this group of compounds. In addition, there is very little information available regarding useful biological or toxicological markers of low-level exposure to these compounds. Because of this lack of information and the apparently similar toxicity of structurally related PCBs and dioxins (partial Ah-receptor agonist and antagonist activity in vitro, thyroid toxicity, and immune effects) the European Union (EU) Scientific Committee for Toxicity, Ecotoxicity and the Environment (CSTEE) concluded that further information is required regarding the potential of (PeBDE) for endocrine disruption and/or dioxin-like effects. This was concluded for both human and environmental risk assessment [5,6]. In response to the need for research on endocrine disruptors, the European Union allocated a budget of ¤20 million, and proposed four projects for funding, including a project on BFRs [13].

ACKNOWLEDGMENT

In part, this article was supported by the European Union project FIRE, contract no. QRLT-2001-00596.

REFERENCES

1. G. Ankley, E. Mihaich, R. Stahl, D. Tillitt, T. Colborn, S. McMaster, R. Miller, J. Bantle, P. Campbell, N. Denslow, R. Dickerson, L. Folmar, M. Fry, J. Giesy, L. E. Gray, P. Guiney, T. Hutchinson, S. Kennedy, V. Kramer, G. LeBlanc, M. Mayes, A. Nimrod, R. Patino, R. Peterson, R. Purdy, R. Ringer, P. Thomas, L. Touart, G. Van Der Kraak, T. Zacharewski. *Environ. Toxicol. Chem.* **17**, 68–8 (1998).
2. Å. Bergman. *Organohalogen Compd.* **47**, 36–40 (2000).
3. S. Bouma, D. Vethaak, P. Meininger, A. Holland. *De visdiefkolonie (Sterna hirundo) bij Terneuzen: blijven er problemen?* Rijksinstituut voor Kust en Zee/RIKZ, Middelburg, Rapport 2000.045 (2001).
4. T. Colborn, F. Vom Saal, A. Soto. *Environ. Health Perspect.* **101**, 378–384 (1993).
5. CSTEE. *Opinion of the EU Scientific Committee for Toxicity, Ecotoxicity and the Environment on the Results of the Human Risk Assessment of Pentabromodiphenyl Ether* [CAS N° 32534-81-9] carried out in the framework of Council Regulation (EEC) 793/93 on the evaluation and control of the risks of existing substances1 - Opinion expressed at the 16th CSTEE plenary meeting, Brussels, 19 June 2000.
6. CSTEE. *Opinion of the EU Scientific Committee for Toxicity, Ecotoxicity and the Environment on the Results of the Environmental Risk Assessment of Pentabromodiphenyl Ether* [CAS N° 32534-81-9], carried out in the framework of Council Regulation (EEC) 793/93 on the evaluation and control of the risks of existing substances - Opinion expressed at the 13th CSTEE plenary meeting, Brussels, 4 February 2000.
7. P. O. Darnerud, G. S. Eriksen, T. Johannesson, P. B. Larsen, M. Viluksela. *Environ. Health Perspect.* **109** (Suppl. 1), 49–68 (2001).
8. J. de Boer, P. G. Wester, H. J. Klamer, W. E. Lewis, J. P. Boon. *Nature* **394**, 28–29 (1998).
9. J. de Boer, K. de Boer, J. P. Boon. In *Handbook of Environmental Chemistry,* Vol. 3, Part K, *New Types of Persistent Halogenated Compounds*, J. Paasivirta (Ed.), pp. 61–95, Springer Verlag, Berlin (2000).
10. J. de Boer, A. van der Horst, P. G. Wester. *Organohalogen Compd.* **47**, 85–88 (2000).
11. C. A. de Wit. *Chemosphere* **46**, 583–624 (2002).
12. P. Eriksson, E. Jakobsson, A. Frederiksson. *Environ. Health Perspect.* **109**, 903–908 (2001).
13. European Union, 2002. Health and the Environment: European research on endocrine disrupters receives major boost (<http://europa.eu.int/comm/research/press/2002/pr1505en.html>).
14. J. R. Fowles, A. Fairbrother, L. Baecher-Steppan, N. I. Kerkvliet. *Toxicol.* **86**, 49–61 (1994).
15. D. Guvenius Meironyté. *Organohalogen contaminants in humans with emphasis on polybrominated diphenyl ethers*, Doctoral dissertation, University of Stockholm, Sweden (2002).
16. H. Hakk, G. Larsen, Å. Bergman. *Xenobiotica* **30**, 881–890 (2000).
17. R. C. Hale, M. J. La Guardia, E. P. Harvey, M. O. Gaylor, T. M. Mainor, W. H. Duff. *Nature* **412**, 140–141 (2001).
18. S. Hallgren and P. O. Darnerud. *Organohalogen Compd.* **35**, 391–394 (1998).
19. M. L. Hardy. *Chemosphere* **46**, 757–777 (2002).

20. M. Harju and M. Tysklind. *Organohalogen Compd.* **41**, 529–532 (1999).
21. P. T. C. Harrison, C. D. N. Humfrey, M. Litchfield, D. Peakall, L. K. Shuker. *IEH Assessment on Environmental Oestrogens: Consequences to Human Health and Wildlife*. Medical Research Council, Institute for Environment and Health, Page Bros., Norwich (1995).
22. M. G. Ikonomou, S. Rayne, R. F. Addison. *Environ. Sci. Technol.* **36**, 1886–1892 (2002).
23. IPCS, Environmental Health Criteria 192. *Flame Retardants, A General Introduction*, World Health Organization, International Programme on Chemical Safety, Geneva (1997).
24. IPCS, Environmental Health Criteria 162. *Brominated Diphenylethers*, World Health Organization, International Programme on Chemical Safety, Geneva (1994).
25. IPCS, Environmental Health Criteria 172. *Tetrabromobisphenol A and Derivatives*, World Health Organization, International Programme on Chemical Safety, Geneva (1995).
26. IPCS, Environmental Health Criteria 205. *Polybrominated Dibenzo-p-dioxins and Dibenzofurans*, World Health Organization, International Programme on Chemical Safety, Geneva (1998).
27. T. A. McDonald. *Chemosphere* **46**, 745–755 (2002).
28. I. A. T. M. Meerts, E. A. C. Luijks, G. Marsh, E. Jakobsson, Å. Bergman, A. Brouwer. *Organohalogen Compd.* **37**, 147–150 (1998).
29. I. A. Meerts, J. J. van Zanden, E. A. Luijks, I. van Leeuwen-Bol, G. Marsh, E. Jakobsson, Å. Bergman, A. Brouwer. *Toxicol. Sci.* **56**, 95–104 (2000).
30. I. A. T. Meerts, R. J. Letcher, S. Hoving, G. Marsh, Å. Bergman, H. G. Lemmen, B. van der Burg, A. Brouwer. *Environ. Health Perspect.* **109**, 399–407 (2001).
31. D. Meironyté, K. Norén, A. J. Bergman *Toxicol. Environ. Health Part A* **58**, 101–113 (1999).
32. M. Meneses, H. Wingfors, M. Schuhmacher, J. L. Domingo, G. Lindström, B. V. Bavel. *Chemosphere* **39**, 2271–2278 (1999).
33. O. Päpke, L. Bathe, Å. Bergman, P. Fürts, D. M. Guvenius, T. Herrmann, K. Norén. *Organohalogen Compd.* **52**, 197–200 (2001).
34. S. Safe. *CRC Crit. Rev. Toxicol.* **21**, 51–88 (1990).
35. S. Safe. *CRC Crit. Rev. Toxicol.* **24**, 87–149 (1994).
36. U Sellström. *Determination of some polybrominated flame retardants in biota, sediment and sewage sludge.* Doctoral dissertation, Stockholm University (1999).
37. U. Sellström, G. Soderstrom, C. de Wit, M. Tysklind. *Organohalogen Compd.* **35**, 447–450 (1998).
38. R. M. Sharpe and N. E. Skakkebaek. *Lancet* **341**, 1392–1395 (1993).
39. A. Sjödin, L. Hagmar, E. Klasson-Wehler, K. Kronholm-Diab, E. Jakobsson, A. Bergman. *Environ. Health Perspect.* **107**, 643–648 (1999).
40. J. A. Szymanska, J. K. Piotrowski, B. Frydych. *Toxicol.* **142**, 87–95 (2000).
41. C. Thomsen. *Brominated flame retardants-development and application of analytical methods for characterization of human exposure*, Doctoral dissertation, University of Oslo (2002).
42. C. Thomsen, E. Lundanes, G. Becher. *J. Environ. Monitor.* **3**, 366–370 (2001).
43. C. Thomsen, E. Lundanes, G. Becher. *Environ. Sci. Technol.* **36**, 1414–1418 (2002).
44. J. Toppari, J. C. Larsen, P. Christiansen, A. Giwercman, P. Grandjean, L. J. Guillette, B. Jegou, T. K. Jensen, P. Jouannet, N. Keiding, H. Leffers, J. A. McLachlan, O. Meyer, E. Müller, E. Rajpert-De Meyts, T. Scheike, R. Sharpe, J. Sumpter, N. Skakkebaek. *Male Reproductive Health and Environmental Chemicals with Estrogenic Effects,* Danish Environmental Protection Agency, Copenhagen (1995).
45. M. Van den Berg, L. S. Birnbaum, A. T. C. Bosveld, B. Brunström, Ph. Cook, M. Feeley, J. P. Giesy, H. Hanberg, R. Hasegawa, S. W. Kennedy, T. Kubiak, J. Ch. Larsen, F. X. R. van Leeuwen, A. K. D. Liem, C. Nolt, R. E. Peterson, L. Poellinger, S. Safe, D. Schrenk, D. Tillitt, M. Tysklind, M. Younes, F. Wærn, T. Zacharewski. *Environ. Health Perspect.* **106**, 775–792 (1998).
46. J. G. Vos, C. de Heer, H. Van Loveren. *Teratogenesis, Carcinogenesis, Mutagenesis* **17**, 275–284 (1997/1998).

47. J. G. Vos, E. Dybing, H. A. Greim, O. Ladefoged, C. Lambré, J. V. Tarazona, I. Brandt, A. D. Vethaak. *CRC Crit. Rev. Toxicol.* **30**, 71–133 (2000).
48. Weybridge, *Proceedings European Workshop on the Impact of Endocrine Disrupters on Human Health and Wildlife*, 2–4 December 1996, Weybridge, UK (1996).
49. B. N. Zegers, W. E. Lewis, M. R. Tjoen-A-Choy, C. Smeenk, U. Siebert, J. P. Boon. Levels of some polybrominated diphenyl ether (PBDE) flame retardants in animals of different trophic levels of the North Sea food web. In: Å. Bergman (Ed.), Stockholm, 14–16 May 2001, *The Second International Workshop on Brominated Flame Retardants*, pp. 143–147 (2001).
50. T. Zhou, D. G. Ross, M. J. DeVito, K. M. Crofton. *Toxicol. Sci.* **61**, 76–82 (2001).
51. T. Zhou, M. M. Taylor, M. J. DeVito, K. M. Crofton. *Toxicol. Sci.* **66**, 105–116 (2002).

Pure Appl. Chem., Vol. 75, Nos. 11–12, pp. 2047–2053, 2003.

Topic 3.6

Toxicity vs. beneficial effects of phytoestrogens*

Hideki Wanibuchi‡, Jin Seok Kang, Elsayed I. Salim, Keiichirou Morimura, and Shoji Fukushima

Department of Pathology, Osaka City University Medical School, Osaka 545-8585, Japan

Abstract: Phytoestrogens are nonsteroidal plant-derived compounds possessing estrogenic activity. These include two major classes: isoflavonoids and lignans. Phytoestrogens have have received recently great attention because of their beneficial effects, which include the prevention of cancer, atherosclerosis, and bone density loss. However, they have estrogenic activity and may affect as endocrine disruptors. In this review, we pay attention to both the toxic and beneficial effects of phytoestrogens.

Epidemiological data support that isoflavonoids help prevent cancer of breast, prostate, stomach, and lung. However, there have been some reports about a positive association between some phytoestrogens (e.g., campesterol and stigmasterol) and prostate cancer risk. Animal experiments and in vitro experiments have shown that the biological effects of phytoestrogens may be organ-specific, inhibiting cancer development in some sites, yet showing no effect or an enhancing effect on tumorigenesis at other sites. Also, their effects may be dependent on the timing and duration of exposure. For example, several studies in rodents have established that the favorable effect of an isoflavone-rich diet on breast cancer risk may be significant only if consumption occurs before puberty or during adolescence. On the other hand, prenatal exposure of genistein was reported to have a carcinogenic effect on the uterus of rodents. Phytoestrogen, especially soy products, has been used as hormone-replacement therapy, reducing such symptoms as hot flashes, vaginal dryness, and mood changes while protecting women from osteoporosis and heart disease. However, the animal data suggest that the timing of exposure to such compounds is crucial, with neonatal exposure having the most pronounced effects. Given the exposure of neonates to phytoestrogens, this should be a cause for concern.

INTRODUCTION

Phytoestrogens are nonsteroidal plant-derived compounds that may mimic or interact with estrogen hormones in mammals. Phytoestrogens include two main groups: isoflavones and lignans. For isoflavones, the main dietary sources for humans are soybeans and other legumes. Mammalian lignans are products of intestinal microbial breakdown of plant lignans found in whole grains, fibers, and flax seeds, and many fruits and oil.

---

*Report from a SCOPE/IUPAC project: Implication of Endocrine Active Substances for Human and Wildlife (J. Miyamoto and J. Burger, editors). Other reports are published in this issue, *Pure Appl. Chem.* **75**, 1617–2615 (2003).

‡Corresponding author: Department of Pathology, Osaka City University Medical School, 1-4-3, Asahi-machi, Abeno-ku, Osaka 545-8585, Japan; Tel.: +81-6-6645-3736; Fax: +81-6-6646-3093; E-mail: wani@med.osaka-cu.ac.jp

Epidemiological studies suggest that the consumption of a phytoestrogen-rich diet may have beneficial effects on health. As phytoestrogens can have both agonist and antagonist action on estrogenic activity, it has also been proposed that dietary phytoestrogens could play a role in the prevention of certain cancers (breast, uterus, and prostate) in humans. Phytoestrogens were also found to modulate other estrogen-related disorders when consumed in ordinary diet, namely cardiovascular diseases, menopausal symptoms, and post-menopausal osteoporosis. On the other hand, consumption of relatively high levels of some phytoestrogens may pose some health risk. Reproductive problems have been documented in laboratory animals, farm animals, and wild animals that consumed very high amounts of phytoestrogen-rich plants. In this review, we pay attention to the beneficial effects and toxicity of phytoestrogens.

BENEFICIAL EFFECTS

Phytoestrogens and cancer

Diets rich in phytoestrogens (plant estrogens, particularly soy and unrefined grain products) may be associated with low risk of some hormone-dependent cancers.

Breast cancer

Reduced risk of breast cancer in premenopausal women was found to correlate significantly with high soy intake in Singapore [1,2]. Another experiment showed that soy intake during adolescence conferred a strong significant protective effect on breast cancer [3]. In addition, most recent reports revealed that high soy intake in childhood in Asian-Americans is associated with reduced breast cancer risk, which may be further reduced by intake as an adult. These benefits do not extend to Westerners, however, who don't traditionally eat these products [4].

Numerous in vitro studies on breast cancer cells in culture have shown that phytoestrogens stimulate tumor growth when administered at low concentrations but inhibit tumor growth at high concentrations. Genistein, an isoflavone found in soy products, was found to exert antiproliferative effects when administered at concentrations above 10 micromolars (μM) to both positive estrogen receptor (ER+) and negative estrogen receptor (ER–) human breast cancer cells in vitro [5,6]. But it exerted marked hyperproliferative effects when administered at concentrations below 100 nM due to its estrogen agonistic actions [6]. At higher concentrations, the chemoprotective effects of genistein are mediated by multiple mechanisms independent of ERs, such as inhibition of tyrosine kinase activity of growth factor receptors and oncogene products [5].

Genistien markedly inhibited dimethylbenazanthracene- (DMBA-) and methyl-nitrosourea- (MNU-) induced rat mammary carcinogenesis during the neonatal or prepubertal and perinatal periods [7–11]. Several studies in rodents established that the favorable effects of isoflavone-rich diet on breast cancer may be significant only if administration occurred before puberty or during adolescence [8–10]. These results demonstrate that the timing of exposure to the isoflavones is important for breast cancer chemoprevention [12]. These important experimental studies have now gained support from a recent human study in which high soy intake during adolescence was found to reduce the risk of breast cancer later in life [3,4].

Soy isoflavones may reduce breast cancer risk by affecting endogenous sex hormone concentrations and the menstrual cycle [13]. There is evidence from early studies in vitro and in vivo that suggests that phytoestrogens stimulate the production of sex hormone binding globulin (SHBG) in the liver cells [13,14]. An increase in SHBG, leading to lower free-sex-hormone concentrations and a longer menstrual cycle, which would lower breast cancer risk, was seen in some, but not all, studies [13,15,16].

Endometrial cancer

A case-control study in Hawaii [17] showed that high consumption of soy products and other legumes was associated with a decreased risk of endometrial cancer (p for trend = 0.01; odds ratio = 0.46).

However, animal experiments do not support the epidemiological data (see toxicity of phytoestrogen, endometrial cancer).

Prostate cancer

Epidemiological data of phytoestrogens and prostate cancer strongly support the cancer protective effects of isoflavones found in soy products [18].

In experimental animal studies, many of the studies with dietary administration of soy, pure isoflavones, or rye bran reduced the incidence of prostatic carcinomas [19–22].

Prostate tissues have both estrogen receptor ERα and ERβ [23]. ERβ binds phytoestrogens and may inhibit ERα-mediated estrogen action through heterodimerization [13]. It could be postulated that the mechanisms of genistein could help prevent prostate cancer by regulating steroid-receptor pathways [24]. It inhibits the expression of epidermal growth factor and the erbB2 receptors in rat dorsolateral prostate, and could, therefore, block the growth factor-mediated stimulation of cell proliferation [25].

Colon cancer

A number of epidemiological studies, primarily of Asian origin, have examined the relationship between soy intake and the risk of colorectal cancer; however, these studies provide little support for a protective effect of soy [26]. In animal experiments, rye bran, flax seed, and purified lignans seemed to protect against colon carcinogenesis [27,28]. However, some studies provided little support for a protective effect of soy, rich in phytoestrogens [29,30].

Other cancers

Recently, a prospective study has reported that soy intake may reduce the mortality rate of stomach cancer [31]. Another case control study of the relationship between soy products and cancer, conducted in the Far East, reported that increased soy consumption is also protective in lung and stomach cancers [32].

Phytoestrogens as hormone replacement therapy

Selective estrogen receptor modulators (SERMs) have been developed to restore the benefits of traditional hormone therapy while avoiding undesired side effects. Phytoestrogens may be added to the list of SERMs, given their agonist/antagonist properties in estrogen-sensitive tissue [33]. The ideal SERMs would treat postmenopausal symptoms, such as hot flashes, vaginal dryness, and mood changes, while protecting women from osteoporosis and heart disease. More women have looked to phytoestrogen, such as the isoflavones found in the soy plant, to tailor their menopausal therapy in a "natural" way.

The most widely studied phytoestrogens are genistein and coumestrol, which both exerted typical estrogen-like actions in female and male experimental animals [34–36]. Animal studies suggest that soy protein and/or isoflavones may prevent cardiovascular disease by multiple mechanisms [37,38] and inhibit ovariectomy-induced bone loss in rats [39]. The protective effect could be related to reduction in androgen production, either via central nervous system-gonadal axis, or through inhibition of 5-reductase and, consequently, a decrease in the synthesis of dihydrotestosterone, which is the most active androgen in the prostate [40,41].

Reports concerning broad tissue distribution of ERβ [42,43] that show that phytoestrogens usually bind with ERβ [44] indicate that it may be possible to develop new natural substances from phytoestrogens for treatment and prevention of menopausal symptoms, including osteroporosis, and cardiovascular disease.

Toxicity of phytoestrogens

Endometrial cancer

Newbold et al. (2001) showed that uterine adenocarcinoma was induced in mice that were treated neonatally with diethylstilbestrol (DES) or genistein [45]. In this experiment, newborn female CD-1 mice were treated on days 1 through 5 with DES at the dose of 0.001 mg/kg/day subcutaneously (s.c.) or with genistein at the dose of 50 mg/kg/day, s.c. At 18 month of age, incidence of uterine adenocarcinoma was 31 % for DES-treated mice, 35 % for genistein-treated mice, and 0 % for control mice. Such a close correlation of activities indicates that the critical events in the carcinogenic effect of these two chemicals on the mouse uterus resulted from ER-mediated interactions in the neonates [46].

Prostate cancer

Although most of the epidemiological studies of phytoestrogens and prostate cancer indicated the protective effects of isoflavones on prostate cancer, an epidemiological study indicates an increase of prostate cancer risk due to some phytoestrogens (campesterol and stigmasterol) [47].

Colon cancer

In a report promoting the potential of genistein in colon tumor model [48], Rao et al. investigated the effects of genistein on azoxymethane- (AOM-) induced colon carcinogenesis in male F344 rats, and revealed that administration of genistein (250 ppm in diet) significantly increased noninvasive and total adenocarcinoma multiplicity in the colon, as compared to the control diet [48]. The results of this investigation emphasize that the biological effects of genistein may be organ-specific, inhibiting cancer development in some sites yet showing no effect or an enhancing effect on tumorigenesis at other sites, such as the colon [48].

Other cancers

Some recent reports revealed strong in vitro genotoxicity, as well as in vivo carcinogenic potential of soy isoflavones in p53 knock-out mice [49]. In a recent experiment in our lab, genistein enhanced lung carcinogenesis in rat medium-term multi-organ carcinogenesis bioassay at dietary doses of 25 and 250 ppm [50].

PROBLEMS OF PHYTOESTROGENS AS A SOURCE OF HORMONE REPLACEMENT THERAPY

Although phytoestrogens are thought to contribute to the prevention of hormone-dependent cancers, compounds with estrogenic activity also have the capability of causing endocrine disruption, particularly when exposure occurs prior to puberty [51]. Phytoestrogens are also capable of altering the toxicological behaviors of other endocrine active compounds, and the interactions of these compounds may involve complexities that are difficult to predict based on their in vitro steroid receptor reactivities. The primary concern of phytoestrogens as dietary supplements stems from the reports that estrogens are associated with perinatal toxicity to the reproductive tract. During the fetal and early period, when reproductive organs are developing, changes in the hormonal milieu can induce dramatic structural and functional alterations in the reproductive tract of both male and females [52]. You et al. investigated the potential of the phytoestrogen genistein to influence the reproductive developmental toxicity of the endocrine active pesticide, methoxychlor [53]. The estrogenic responses to genistein and methoxychlor administered together were apparently accumulative of the effects associated with each compound alone. Soy-based formula for human infant nutrition is widely used, with approximately 25 % of formula-fed infants in the United States consuming soy-based formula [54]. Infants consuming soy formula are exposed to high levels of genistein and daidzein. On average, infants fed soy-based formula consume isoflavones an order of magnitude greater than adults eating high-soy diets. Total plasma levels of isoflavones and genistein in soy-fed infants are 10-fold greater than levels in Japanese adults whose diets have traditionally induced soy, and 200-fold greater than plasma levels in infants fed cow's milk formula or human breast milk. Total plasma isoflavone levels in soy-fed infants are up to 22 000 times greater than 17β-estradiol levels [55]. However, estrogenicity of genistein is only one 1000th to one 10 000th that of 17β-estradiol [56]. Yellayi et al. [57] recently have examined thymic and immune effects of genistein in mice, and reported the possibility that serum genistein concentrations found in soy-fed infants may be capable of producing thymic and immune abnormalities, as suggested by previous reports of immune impairments in soy-fed human infants [58,59]. For the phytoestrogen genistein, it has been shown that injections of neonatal female rats with high doses of genistein (500 or 1000 μg/day on days 1 through 10 after birth) decreased basal luteinizing hormone (LH) levels and pituitary responsiveness to gonadotrophin-releasing hormone (GnRH) and increased the volume of the sexually dimorphic nucleus of the hypothalamus [60]. On the other hands, lower doses of genistein (10 to 100 μg) had opposite effects on LH secretion and increased pituitary response to GnRH. More recent developmental toxicology studies have demonstrated that dietary exposure to physiological concentrations of genistein yields little or no toxicity [61]. Most recent studies revealed that supraphysiological concentrations of daizein administered via the diet did not cause significant toxicity to the female reproductive tract or provide a protective effect against chemically induced breast cancer [62]. Recently, Burton and Well reported on the effect of phytoestrogens on the female genital tract [63]. This review discussed the evidence from both animal studies and humans of an effect of these ubiquitous compounds on the development of the human female genital tract, in addition to prolonging the menstrual cycle, alleviating symptoms of menopause, and protecting against the development of endometrial carcinoma.

CONCLUSION

In conclusion, the animal data suggest that the timing of exposure to such compounds is crucial, with neonatal exposure having the most pronounced effects. Given the exposure of neonates to phytoestrogens, this should be a cause for concern. It has been assumed that exposure to these compounds is always good, but inappropriate or excessive exposure may be detrimental [64]. Many natural compounds, especially natural hormones, can be potent and can have both good and bad health affects, because, like any other hormone, too much or too little can alter hormone-dependent cellular and tissue functions. These substances should be used in moderation to avoid any unintentional health consequences. At present, because there is too little information to make any firm conclusions on the impact of phytoestrogens on human health, we need more information on well-designed animal studies.

REFERENCES

1. H. P. Lee, L. Gourley, S. W. Duffy, J. Esteve, J. Lee, N. E. Day. *Lancet* **331**, 1197–1200 (1991).
2. H. P. Lee, L. Gourley, S. W. Duffy, J. Esteve, J. Lee, N. E. Day. *Cancer Causes Control* **3**, 313–322 (1992).
3. Q. Dai, X-O. Shu, F. Jin, J. D. Potter, L. H. Kushi, J. Teas, Y.-T. Gao, W. Zheng. *Br. J. Cancer* **85**, 372–378 (2001).
4. A. H. Wu, P. Wan, J. Hankin, C. C. Tseng, M. C. Yu, M. C. Pike. *Carcinogenesis* **23**, 1491–1496 (2002).
5. M. C. Pagliacci, M. Smacchia, G. Migliorati, F. Grignani, C. Riccardi, I. Nicoletti. *Eur. J. Cancer* **30**, 1675–1616 (1994).
6. C. Y. Hsieh, R. C. Santell, S. Z. Haslam, W. G. Helferich. *Cancer Res.* **58**, 3833–3838 (1998).
7. A. I. Constantinou, R. G. Mehta, A. Vaughan. *Anticancer Res.* **16**, 3293–3298 (1996).
8. W. B. Murrill, N. M. Brown, J. X. Zhang, P. A. Manzolillo, S. Barnes, C. A. Lamartiniere. *Carcinogenesis* **17**, 1451–1457 (1996).
9. C. A. Lamartiniere, J. B. Moore, N. M. Brown, R. Thompson, M. J. Hardin, S. Barnes. *Carcinogenesis* **16**, 2833–2840 (1995).

10. W. A. Fritz, L. Coward, J. Wang, C. A. Lamartiniere. *Carcinogenesis* **19**, 2151–2458 (1998).
11. S. Barnes, G. Peterson, C. Grubbs, K. Setchell. *Adv. Exp. Med. Biol.* **354**, 135–147 (1994).
12. C. A. Lamartiniere, M. S. Cotroneo, W. A. Fritz, J. Wang, R. Mentor-Marcel, A. Elgavish. *J Nutr.* **132**, 552S–558S (2002).
13. H. Adlercreutz. *Lancet Oncol.* **3**, 364–373 (2002).
14. H. Adlercreutz. In *Reproductive and Developmental Toxicology*, K. S. Korach (Ed.), pp. 299–371, Marcel Dekker, New York (1998).
15. L.-E.W. Lu, K. E. Anderson, J. J. Grady, M. Nagamani. *Cancer Epidem. Biom. Prev.* **5**, 63–70 (1996).
16. A. H. Wu, F. Z. Stanczyk, S. Hendrich et al. *Br. J. Cancer* **82**, 1879–1886 (2000).
17. M. T. Goodman, L. R. Wilkens, J. H. Hankin, L. C. Lyu, A. H. Wu, L. N. Kolonel. *Am. J. Epidemiol.* **146**, 294–306 (1997).
18. E. P. Castle and J. B. Thrasher. *Urol. Clin. North Am.* **29**, 71–81 (2002).
19. M. Onozawa, T. Kawamori, M. Baba, K. Fukuda, T. Toda, H. Sato, M. Ohtani, H. Akaza, T. Sugimura, K. Wakabayashi. *Jpn. J. Cancer Res.* **90**, 393–398 (1999).
20. K. Kato, S. Takahashi, L. Cui, T. Toda, S. Suzuki, M. Futakuchi, S. Sugiura, T. Shirai. *Jpn. J. Cancer Res.* **91**, 786–791 (2000).
21. R. Mentor-Marcel, C. A. Lamartiniere, I. E. Eltoum, N. M. Greenberg, A. Elgavish. *Cancer Res.* **61**, 6777–6782 (2001).
22. A. Bylund, J. X. Zhang, A. Bergh, J. E. Damber, A. Widmark, A. Johansson, H. Adlercreutz, P. Aman, M. J. Shepherd, G. Hallmans. *Prostate* **42**, 304–314 (2000).
23. K. Pettersson and J. A. Gustafsson. *Annu. Rev. Physiol.* **63**, 165–192 (2001).
24. W. A. Fritz, J. Wang, I. E. Eltoum, C. A. Lamartiniere. *Mol. Cell. Endocrinol.* **186**, 89–99 (2002).
25. A. Dalu, J. F. Haskell, L. Coward, C. A. Lamartiniere. *Prostate* **37**, 36–43 (1998).
26. M. Messina and M. Bennink. *Baillieres Clin. Endocrinol. Metab.* **12**, 707–728 (1998).
27. M. J. Davies, E. A. Bowey, H. Adlercreutz, I. R. Rowland, P. C. Rumsby. *Carcinogenesis* **20**, 927–931 (1999).
28. M. Mutanen, A. M. Pajari, S. I. Oikarinen. *Carcinogenesis* **21**, 1167–1173 (2000).
29. I. K. Sorensen, E. Kristiansen, A. Mortensen, G. M. Nicolaisen, J. A. Wijnands, H. J. van Kranen, C. F. van Kreijl. *Cancer Lett.* **130**, 217–225 (1998).
30. M. J. Davies, E. A. Bowey, H. Adlercreutz, I. R. Rowland, P. C. Rumsby. *Carcinogenesis* **20**, 927–931 (1999).
31. C. Nagata, N. Takatsuka, N. Kawakami, H. Shimizu. *Br. J. Cancer* **87**, 31–36 (2002).
32. M. Messina. *Am. J. Clin. Nutr.* **62**, 645 (1995).
33. D. Carusi. *Prim. Care Update Ob. Gyns.* **7**, 253–259 (2000).
34. R. C. Santell, Y. C. Chang, M. G. Nair, W. G. Helferich. *J. Nutr.* **127**, 263–269 (1997).
35. S. R. Milligan, A. V. Balasubramanian, J. C. Kalita. *Environ. Health Perspect.* **106**, 23–6 (1998).
36. L. Strauss, R. Santti, N. Saarinen, T. Streng, S. Joshi, S. Makela. *Toxicol. Lett.* **102–103**, 349–54 (1998).
37. M. S. Anthony, T. B. Clarkson, C. L. Hughes Jr., T. M. Morgan, G. L. Burke. *J. Nutr.* **126**, 43–50 (1996).
38. M. S. Anthony, T. B. Clarkson, B. C. Bullock, J. D. Wagner. *Arterioscler. Thromb. Vasc. Biol.* **17**, 2524–2531 (1997).
39. J. J. Anderson, W. W. Ambrose, S. C. Garner. *Proc. Soc. Exp. Biol. Med.* **217**, 345–50 (1998).
40. B. A. Evans, K. Griffiths, M. S. Morton. *J. Endocrinol.* **147**, 295–302 (1995).
41. Y. C. Kao, C. Zhou, M. Sherman, C. A. Laughton, S. Chen. *Environ. Health Perspect.* **106**, 85–92 (1998).
42. P. T. Saunders, S. M. Maguire, J. Gaughan, M. R. Millar. *J. Endocrinol.* **154**, R13–6 (1997).
43. A. H. Taylor and F. Al-Azzawi. *J. Mol. Endocrinol.* **24**, 145–55 (2000).

44. G. G. Kuiper, B. Carlsson, K. Grandien, E. Enmark, J. Haggblad, S. Nilsson, J. A. Gustafsson. *Endocrinology* **138**, 863–70 (1997).
45. R. R. Newbold, E. P. Banks, B. Bullock, W. N. Jefferson. *Cancer Res.* **61**, 4325–4328 (2001).
46. J. Ashby. *Mutat. Res.* **483**, 107–8 (2001).
47. S. S. Strom, Y. Yamamura, C. M. Duphorne, M. R. Spitz, R. J. Babaian, P. C. Pillow, S. D. Hursting. *Nutr. Cancer* **33**, 20–25 (1999).
48. C. V. Rao, C. X. Wang, B. Simi, R. Lubet, G. Kelloff, V. Steele, B. S. Reddy. *Cancer Res.* **57**, 3717–3722 (1997).
49. R. R. Misra, S. D. Hursting, S. N. Perkins, N. Sathyamoorthy, J. C. Mirsalis, E. S. Riccio, J. A. Crowell. *Int. J. Toxicol.* **21**, 277–285 (2002).
50. N. Seike, H. Wanibuchi, K. Morimura, T. Nishikawa, H. Kishida, D. Nakae, K. Hirata, S. Fukushima. *Cancer Lett.* **175**, 113–119 (2002).
51. M. S. Cotroneo, J. Wang, I. A. Eltoum, C. A. Lamartiniere. *Mol. Cell. Endocrinol.* **173**, 135–145 (2001).
52. R. R. Newbold, B. C. Bullock, J. A. McLachlan. *Cancer Res.* **45**, 5145–50 (1985).
53. L. You, M. Casanova, E. J. Bartolucci, M. W. Fryczynski, D. C. Dorman, J. Everitt, K. W. Gaido, S. M. Ross, H. Heck. *Toxicol. Sci.* **66**, 91–104 (2002).
54. American Academy of Pediatrics Committee on Nutrition. *Pediatrics* **101**, 148–153 (1998).
55. K. D. Setchell, L. Zimmer-Nechemias, J. Cai, J. E. Heubi. *Lancet* **350**, 23–27 (1997).
56. P. L. Whitten and H. B. Patisaul. *Environ Health. Perspect.* **109**, Suppl 5–20 (2001).
57. S. Yellayi, A. Naaz, M. A. Szewczykowski, T. Sato, J. A. Woods, J. Chang, M. Segre, C. D. Allred, W. G. Helferich, P. S. Cooke. *Proc. Natl. Acad. Sci. USA* **99**, 7616–7621 (2002).
58. G. Zoppi, F. Gerosa, A. Pezzini, N. Bassani, P. Rizzotti, P. Bellini, G. Todeschini, G. Zamboni, G. Vazzoler, G. Tridente. *J. Pediatr. Gastroenterol. Nutr.* **1**, 175–182 (1982).
59. D. J. Jenkins, C. W. Kendall, P. W. Connelly, C. J. Jackson, T. Parker, D. Faulkner, E. Vidgen. *Metabolism* **51**, 919–924 (2002).
60. K. A. Faber and C. L. Hughes Jr. *Biol. Reprod.* **45**, 649–653 (1991).
61. K. M. Flynn, S. A. Ferguson, K. B. Delclos, R. R. Newbold. *Toxicol. Sci.* **55**, 311–319 (2000).
62. C. A. Lamartiniere, J. Wang, M. Smith-Johnson, I. E. Eltoum. *Toxicol. Sci.* **65**, 228–238 (2002).
63. J. L. Burton and M. Wells. *J. Clin. Pathol.* **55**, 401–407 (2002).
64. S. R. Davis, F. S. Dalais, E. R. Simpson, A. L. Murkies. *Recent Prog. Horm. Res.* **54**, 185–210 (1999).

Pure Appl. Chem., Vol. 75, Nos. 11–12, pp. 2055–2068, 2003.

Topic 3.7

Evaluation of thyroid function in neonatal and adult rats: The neglected endocrine mode of action*

Mildred S. Christian‡ and Nancy A. Trenton

Argus Research, Division of Charles River Laboratories, 905 Sheehy Drive, Bldg. A, Horsham, PA 19044, USA

Abstract: Although known to regulate growth and development, cellular metabolism, the use of oxygen, and basal metabolic rate, thyroid hormones have been only minimally evaluated in neonatal rodents at critical times of development. Despite some modulation of metabolic rate by other hormones, such as testosterone, growth hormone, and norepinephrine, 3,5,3′-triiodothyronine (T3) and 3,5,3′,5′-tetraiodothyronine (T4) are the most important metabolic rate modulators. Endpoints used for thyroid function assessment in neonatal and adult rats include thyroid-stimulating hormone (TSH), T3, and T4 levels and histopathology. In rodents, decreased serum levels of T3 and T4 and increased serum TSH levels, with sustained release of TSH and resultant follicular cell hypertrophy/hyperplasia, are typical hormonal and histopathological findings attributable to compounds altering thyroid function. Hypothyroidism early in the neonatal period can affect reproductive endpoints in both male and female rats, with the critical period of exposure being the first two weeks postnatal. Hypothyroidism has been shown to reduce gonadotrophin levels and delay pubertal spermatogenesis in male rats and to block gonadotropin-induced first ovulation in immature female rats by decreasing FSH and luteinizing hormone (LH) serum concentrations. Inclusion of evaluations of TSH, T3, and T4 assays in multigeneration and developmental neurotoxicity protocols may assist in risk assessments.

INTRODUCTION

Since the 1920s, hypothyroidism in infants has been noted as a cause of mental retardation. This endemic hypothyroidism resulted from absence of sufficient dietary iodine in the diet. Since that time, multiple causes of infant hypothyroidism have been identified. In order of frequency, these include endemic hypothyroidism (1:7), thyroid dysgenesis (1:4000), thyroid dyshormonogenesis (1:30 000), transient hypothyroidism (1:40 000), and anomalies of the hypothalamic-pituitary axis (1:100 000) [1].

In humans, hypothyroidism can affect brain function and eventually produce histopathological findings [2]. Permanent damage may be prevented by treatment with thyroid hormones immediately after birth, and, due to recent concern for loss of potential due to severe hypothyroidism at birth, treatment is often initiated in utero in suspected cases derived from ultrasound studies showing goiters and a maternal history of Graves' disease [2,3].

*Report from a SCOPE/IUPAC project: Implication of Endocrine Active Substances for Human and Wildlife (J. Miyamoto and J. Burger, editors). Other reports are published in this issue, *Pure Appl. Chem.* **75**, 1617–2615 (2003).

‡Corresponding author

THYROID GLAND AND ITS FUNCTIONS

Histologically, the thyroid gland primarily consists of spherical follicles that surround colloid. In addition to these components, parafollicular cells are present between the thyroid follicles and secrete calcitonin [4].

As noted by Farwell and Braveman [5], the main function of the thyroid is to take iodine, generally from food, convert it to iodide, and uptake the iodide into the thyroid gland via the sodium-iodide symporter (NIS) system, which is stimulated by thyroid-stimulating hormone (TSH). Iodide is oxidized by thyroid peroxidase and yields the amino acid residues monoiodotyrosyl and diiodotyrosyl. These amino acid residues of thyroglobulin, a glycoprotein made by the cuboidal follicular cells [4], make up and are stored in the colloid. Two diiodotyrosyls, or one monoiodotyrosyl and one diiodotyrosyl, are combined via an oxidative reaction to form thyroxine [3,5,3′,5′-tetraiodothyronine (T4)] and 3,5,3′-triiodothyronine (T3). Approximately 80 % of the hormones secreted by the thyroid are T4; 99 % of secreted T4 is bound to plasma proteins [6]. After release into the bloodstream, T3 and T4 bind to albumin and prealbumin (transporting proteins) via noncovalent bonds and are distributed to peripheral tissues [7].

Thyroid hormones are transported throughout the body in the bloodstream by an acidic glycoprotein, thyroxine-binding globulin (TBG). This bound transport prevents the thyroid hormones from being metabolized and excreted. As shown on Fig. 1, once these hormones reach the peripheral tissues, T4 is converted to the active form of T3 and binds to the intracellular nuclear receptor, which, in turn, stimulates gene transcription [5,8]. The remaining T4, or free T4, is biologically active, exerts negative feedback (inhibition) on pituitary TSH secretion, and is capable of entering cells [9]. Once free T4 enters the cells, it is deiodinated to form T3 or rT3. T3 is generally being produced during normal circumstances, while reverse T3 (rT3) is generally produced during times of illness, starvation, or excessive endogenous catabolism [9].

Metabolism of T4 and T3 primarily occurs in the liver. However, local metabolism by target tissues, such as the brain, can also occur. T4 is metabolized in the liver by removal of the 5′ iodide, which yields T3, or by removal of the 5 iodide, which yields inactive T3 (rT3) [5]. Once in the liver, T3 and T4 are combined with glucuronic and sulfuric acids via the phenolic hydroxyl group and then are ex-

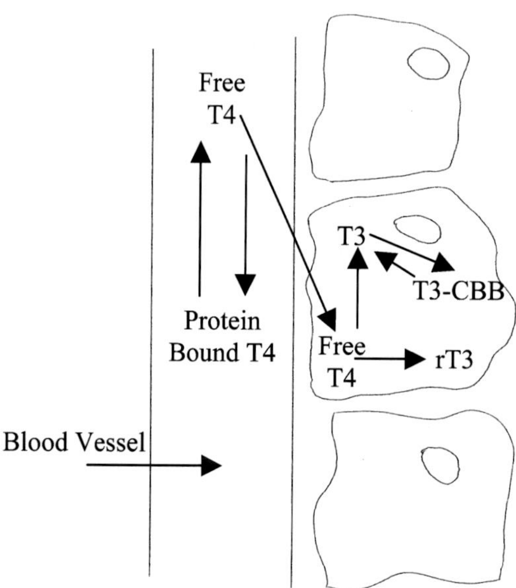

Fig. 1 Cellular metabolism of T3 and T4 in the peripheral tissues (adapted from ref. [8]).

creted in the bile [5]. Thyroid hormones can also be metabolized locally by target tissues such as the brain [5].

The thyroid gland operates via a negative feedback mechanism and is regulated by the pituitary gland, which, in turn, is regulated by the hypothalamus gland [10]. As shown on Fig. 2, the hypothalamus gland releases thyroid-stimulating hormone releasing factor (TSHRF), which stimulates the pituitary gland to release TSH and subsequently stimulates the thyroid gland to release T3 and T4 into the bloodstream. When T3 and T4 levels are too low or too high, the hypothalamus and pituitary glands function to regulate production [10].

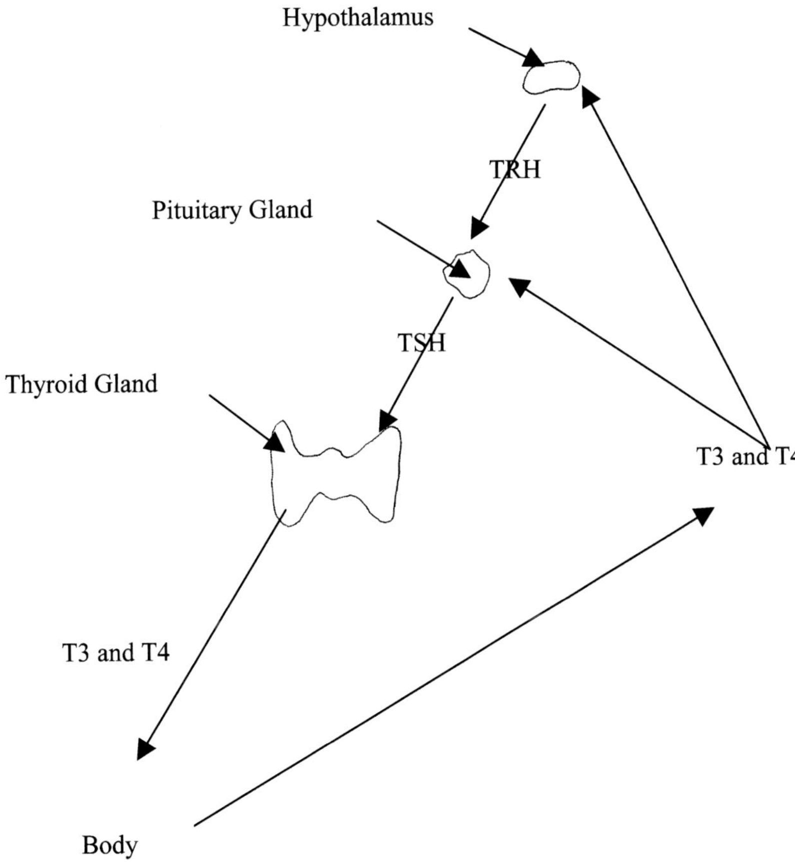

Fig. 2 Negative feedback mechanism for the thyroid gland (adapted from ref. [10]).

ROLE OF THE THYROID IN FETAL DEVELOPMENT

Porterfield and Hendrich [11] describe brain development relative to thyroid function in rats as divided into three phases. Phase I occurs on days 0 to 17 of gestation; during this phase, the fetus is solely dependent on maternal thyroid function. Phase II occurs from day 17 of gestation until birth, with the fetus dependent on maternal and fetal thyroid function (radioiodine levels can be detected in the thyroid follicle at day 17 of gestation; by day 20 of gestation, fetal rats can produce thyroid hormones). Phase III occurs postnatally, with the neonatal rat totally dependent on its own thyroid function.

Neuro-endocrine development interactions occur in utero in humans but during the first 2 to 3 weeks postnatal in rats [12,13]. At 10 days postpartum, the rat brain is equivalent in development to the human brain at birth [12]. Conversely, the human brain at 5 to 6 months of gestation is equivalent to the rat brain at birth [13]. Neurogenesis occurs in rats from day 12 of gestation through birth, at which time cerebral neurogenesis is generally complete [14–16]. Gliogenesis is initiated at birth in rats and continues until adulthood, with the highest amount of glyosis occurring between days 10 to 45 postpartum [15]. Myelinogenesis and development of neuronal processes in rats occurs from birth to 30 days postpartum [14]. In contrast, neurogenesis in humans is generally completed by the seventh month of pregnancy, with gliogenesis initiated in utero [17,18].

Thyroid hormones are critical for normal neural development in both rats and humans. As demonstrated by Eayrs [19], thyroid deficiency requires early remediation for normal development in the rat, as in humans. The degree of cerebral impairment produced in rats was inversely related to the age at thyroidectomy. Thyroidectomy of weanling rats on or after day 25 postpartum did not adversely affect neural development. Withholding T3 and T4 hormone treatment until after day 24 postpartum resulted in irreversible effects on neural development, including a decrease in the number of cerebral vascular connections [20] and alterations of the negative feedback system [21].

In addition to adverse effects on brain development resulting from hypothyroidism, brain development can be affected by excess thyroid hormones. As noted by Best, Duncan, and Best [22], administration of excessive levels of thyroid hormones to neonatal rats over the first 2 weeks postnatal permanently impaired growth and resulted in a resetting of the activity level of the thyroid axis, with permanent reduction of pituitary and plasma TSH concentrations [21,23].

Most experimental data regarding the thyroid's effect on fetal development has been obtained in rodent and bovine models (rats and sheep, respectively) [24]. When interpreting this experimental data and correlating it to human data, it is important to compare the developmental stages of each species because thyroid hormones play a critical role in fetal growth, tissue differentiation, and metabolism. This role of thyroid hormones can be demonstrated by removing the thyroid gland from the fetus before birth. Such studies can be complicated to conduct and interpret because removal of the thyroid gland from the fetus as early as possible is not only difficult but confounded by the fetus' ability to utilize maternal T3 and T4 for growth.

The circulating half-life of maternal thyroid hormones is 12 to 24 h; the half-life of tissue-stored thyroid hormones is greater then 24 h, and a pool of protein-bound thyroid hormones is present that can last for several days [2]. Although iodine freely crosses the placenta and occurs at similar concentrations in the mother and fetus, not all chemicals and hormones are transported equally across the placenta. T4 and T3 cross the placenta but occur at higher concentration in the mother. TSH does not cross the placenta. TSHRF occurs at higher concentrations in the fetus than the mother [11].

Growth

Hopkins and Thorburn [24] clearly demonstrated the in utero effects of thyroid hormones on growth in a thyroidectomized lamb model. Removal of the fetal thyroid between days 81 to 96 of gestation resulted in a malformed lamb with short limbs and a birth weight only 33 % of the normal body weight of a newborn lamb. In Long–Evans rats, postnatal daily subcutaneous administration of propylthiouracil (PTU), an agent used to decrease thyroid function, from birth until day 24 or 30 postpartum caused growth impairment, including reduced body weight and retarded brain growth [25].

Bone development

Thyroid hormones have been demonstrated to affect bone development in fetal rats, including delays in ossification centers, reduced longitudinal bone growth, and retarded development of the bones of the skull [2].

Sexual maturation

Puberty is the stage between the juvenile and adult stages when secondary sex characteristics and fertility develop. Puberty begins in the hypothalamic-pituitary-gonadal (HPG) axis [26].

Males

Sexual maturation in male rats is considered to occur in four phases: (1) neonatal, days 1 to 7 postpartum; (2) infantile, days 8 to 21 postpartum; (3) juvenile, days 22 to 35 postpartum; and (4) peripubertal, days 36 to 55/60 postpartum [27,28]. Additional landmarks of sexual maturation in male rats include testicular descent, at approximately day 15 postpartum [28], and preputial separation, at approximately day 45 postpartum [28,29].

The thyroid plays an essential role in male sexual maturation. Induced hypothyroidism, both chemically and via thyroidectomy, of male rats results in atrophy of the testes and secondary sex glands; reduced gonadotrophin, prolactin, and testosterone levels; and inhibition of gametogenesis and interstitial cell development [30,31]. In addition, T3 receptors have been recognized in Sertoli cells, a finding suggestive of T3's having a role in Sertoli cell function [32].

Females

Puberty in female rats is considered to occur at vaginal patency, which is present between days 30 to 37 postpartum [29,33] and is strain-dependent. Vaginal patency occurs approximately 8 days after late follicular growth resulting from the first ovulation [34].

Adverse affects of hypothyroidism on puberty in female rats include changes in folliculogenesis, formation of corpora lutea, irregular estrous cycles, and ovarian atrophy [6].

RELEVANCE OF EVALUATING POTENTIAL EFFECTS OF ENDOCRINE DISRUPTORS ON THYROID FUNCTION

Xenobiotics may affect thyroid function by altering the biosynthesis, secretion, absorption, or metabolism of the thyroid hormones. Such agents are generally categorized as thioamides, aminoheterocyclic compounds, or substituted phenols [35]. Many pharmaceutical agents alter peripheral thyroid function, including calcium channel blockers, steroids, retinoids, chlorinated hydrocarbons, and polyhalogenated biphenyls [36]. Table 1 lists some of the agents that have been identified as altering thyroid function in rats and/or humans.

Because the thyroid has regulatory activity on essentially all tissues, thyroid function should be considered when determining the risk potential of an agent. Thus, the U.S. Environmental Protection Agency's (USEPA's) Endocrine Disruptor Screening and Testing Advisory Committee (EDSTAC) included evaluation of thyroid function in its two-tiered screening and testing program for estrogenic, androgenic, and thyroid effects. Tier 1 tests, designed to be extremely sensitive and to minimize false negatives. These tests include a diverse range of organisms used to evaluate all types of metabolism and to detect all modes of action of the endocrine endpoints within a range of taxonomic groups that incorporate diverse endpoints. Proposed Tier 1 in vivo screens, including evaluation of thyroid function, are the rodent 20-day pubertal female assay, the 14-day intact adult male with thyroid assay, and the 20-day thyroid/pubertal assay. The Tier 2 screen, including thyroid evaluation, is the two-generation mammalian reproduction and development study. These EDSTAC testing guidelines are currently undergoing an extensive review and validation process by the Endocrine Disruptor Methods Validation Subcommittee (EDVMS).

Table 1 Agents producing thyroid alterations in rats [7,48,49].

Environmental chemicals	
Propylthiouracil (PTU)	Aroclor 1242
Extracts of subsurface soil	Perchlorates
with 2.5 % PCBs	Brominated flame retardants
	Nitrofen
Pharmaceuticals decreasing TSH function	
Dopamine	Octreotide
Glucocorticoids	
Pharmaceuticals altering thyroid hormone secretion	
Reduced thyroid hormone secretion	
Lithium	Amiodarone
Iodide	Aminoglutethimide
Increased thyroid hormone secretion	
Iodide Amiodarone	
Pharmaceuticals reducing T4 absorption	
Colestipol	
Cholestyramine	Ferrous sulfate
Aluminum hydroxide	Sucralfate
Pharmaceuticals altering T4 and T3 transport in serum	
Increased serum TBG concentration	
Estrogens	Methadone
Tamoxifen	Mitotane
Heroin	Fluorouracil
Reduced serum thyroxine binding globulin concentration	
Androgens	Slow-release nicotinic acid
Anabolic seroides	Glucocorticoids
Displacement from protein binding sites	
Furosemide	Mefenamic acid
Fenclofanac	Salicylates
Pharmaceuticals altering T4 and T3 metabolism	
Increased hepatic metabolism	
Phenobarbital	Phenotoin
Rifampin	Carbamazepine
Decreased T4 5′-deiodinase activity	
Propylthiouracil	Beta-adrenergic-antagonist drugs
Amiodarone	Glucocorticoids
Cytokines	
Interferon alpha	Interleukin-2

THYROID FUNCTION TESTS

Weights and histological evaluations of the thyroid are classically required for evaluation of thyroid function. Additional evaluations include biomarkers demonstrating effects on metabolism, thyroid function, and blood levels of these hormones.

In addition, interspecies differences in protein binding capabilities, metabolism, production, and sex-related differences should be considered. For example, TBG is present in humans but essentially absent in rats. The half-life of both T4 and T3 is remarkably longer in humans than in rats. Production of both T4 and TSH is much higher in rats than in humans, and rats also show a sex-related difference, with TSH levels in female rats approximately twice that observed in male rats. In addition to the sex-related differences, multiple other factors are also known to affect thyroid hormone levels in rats. These include common factors such as strain, sex, and age, and, in females, estrous cycling, circadian rhythms, room temperature, stress associated blood collection techniques or animal handling, and the relative activity of the animal. Serum thyroid hormone radioimmunoassay (RIA) tests are currently available for measurement of T4, free T4, T3, free T3, rT3, and TSH concentrations.

Distinct differences exist between male and female rats with regard to TSH, T3, and T4 levels. TSH and T3 occur at higher concentrations in female rats than in male rats, while T4 occurs at higher concentrations in male rats than in female rats. Cycling female rats have a cycle-associated biorhythm for TSH, which peaks during proestrus. In rats, T4 has predominate binding to albumin, with lower amounts bound to prealbumin and postalbumin, and T3 bound only to albumin. This is because rats do not have TBG, which is the principal transport protein for T4 in man [37].

The first test developed to measure thyroid hormone levels was serum iodine concentration [9]. Thyroid function is now usually measured by comparing baseline serum thyroid hormone concentrations with levels attained after exposure to a xenobiotic.

Because most T4 and free T4 are present in the bloodstream, the easiest and least expensive way to assess thyroid function is to determine serum T4 concentration, the sum of T4 and free T4. This is now generally accomplished in the laboratory by use of RIA kits. RIA tests provide an evaluation of the competitive binding of free T4 with an isotope-labeled T4 to IgG molecules in the antiserum [9].

An alternate method for determining thyroid hormone levels is equilibrium dialysis, which is considered to be the "gold standard". However, the complexities of this method and its expense have eliminated it from common use since the development of the more sensitive RIA kits. Serum measurements of T3 and free T3 are not recommended because most T3 is stored in cells.

RIA kits developed for human use can be used across species to measure T4 and T3 levels. However, RIA kits for measuring TSH levels must be specific to the species tested, because human, rat, mouse, and rabbit TSH are not equivalent.

Manson et al. [38] published control thyroid hormone levels for Long–Evans sometimes hyphenated hooded pregnant female rats and their fetuses in a study in which Nitrofen, an herbicide similar in structure to thyroid hormones, significantly altered thyroid function. Amerlex RIA kits (Amersham Corporation, Arlington Heights, Illinois) were used for evaluation of plasma T3 and T4 levels, and a rat TSH assay kit, provided by the National Institutes of Health's Rat Pituitary Hormone Distribution Program (National Institute of Arthritis, Diabetes, Digestive and Kidney Diseases [NIADDK]) was used for evaluation of plasma TSH. Further description of the methodology used by Manson et al. [38] follows.

Thyroidectomized nonpregnant female rats were assigned to an untreated (control) group or administered T4 or nitrofen orally (gavage) for 2 weeks, after which, blood samples were collected by cardiac puncture. As shown in Table 2, T4 levels obtained from control nonpregnant female rats (5.10 μg/100 ml ± 0.61) [38] were similar to those obtained by Lu and Staples [39] in this strain of rats (6.2 μg/100 ml). Treatment with 15 or 30 mg/kg/day of nitrofen suppressed T4 plasma levels and increased TSH levels, as compared to the sham-operated rats which were the expected findings for a negative feedback system.

Table 2 Control data for T3, T4, and TSH levels in Long–Evans hooded nonpregnant female rats and DG 22 fetuses as maternal peak level [42].

	Nonpregnant female rats	DG 22 fetuses
N	6	5 (litters)
T3 levels (ug/100 ml)	0.47 ± 0.06	NA
T4 levels (ug/100 ml)	5.10 ± 0.61	4.45 ± 0.55
TSH levels (ng/ml)	125 ± 26	365 ± 60

NA = Not available

TSH levels in pregnant rats were studied by obtaining a baseline blood sample by cardiac puncture on day of gestation (DG) 11 (sperm observed = DG 1) and subsequent blood samples 6 h after oral (gavage) treatment of the rats with corn oil (control), or nitrofen and on DGs 12, 14, 17, and 22. Blood samples for T3 and T4 evaluations were obtained via the tail vein from nonanesthetized pregnant rats before treatment with corn oil or nitrofen, at 2, 4, 6, and 8 h post-treatment on DGs 11, 12, 13, 14, 18, 20, and 22. DG 22 fetal blood samples were obtained via a neck incision, using heparin-coated capillary tubes. Fetal blood was pooled by litter for analysis for T3 and T4 levels.

The results of these studies by Manson et al. [38] demonstrated that, in Long–Evans hooded female rats, maternal TSH levels increased from DGs 11 to 17 and peaked at DG 22, while T4 levels were reduced from DGs 1 to 21 (data not shown), replicating results described by Fukuda et al., in which nitrofen treatment of Sprague–Dawley rats reduced plasma T4 levels in the maternal rats and in their fetuses [40]. Values for the nonpregnant female rats and DG 22 fetuses at the maternal peak level are presented in Table 2 [38].

Studies by Gray et al. [41,42] demonstrated that nitrofen also affects thyroid function in CD-1 mice. In a study in which nitrofen was administered to pregnant CD-1 mice on days 7 to 17 of gestation, oral (gavage) dosages of 6.25 to 100 mg/kg/day reduced the weight of the Harderian gland, and the Harderian glands were absent or visibly reduced in size in groups given 25, 50, and 100 mg/kg/day. Dosages off 12.5 to 100 mg/kg/day retarded pup growth rates, and groups given 150 and 200 mg/kg/day dosages had no surviving pups by day 3 postpartum. In groups with surviving pups, lung and liver weights were reduced (≤6.25 mg/kg/day), seminal vesicle weights were reduced (≤12.5 mg/kg/day) and testes weights were reduced (100 mg/kg/day). Puberty was delayed in the offspring in the 50 and 100 mg/kg/day dosage groups [41]. In the study by Gray in which mice were administered 500 and 1000 mg/kg/day of nitrofen for 3 consecutive days, statistically significant reductions in serum T4 levels in both treatment groups were observed. Control values for serum T3 and T4 levels for these mice were 43.2 ng/dl and 2.15 ng/dl, respectively.

THYROID HORMONES AND THEIR DEVELOPMENT IN RATS

Little information regarding age-related differences in thyroid hormone levels in rats is provided in the literature.

In a study of Sprague–Dawley male rats, serum TSH levels generally increased on days 30 to 50 postpartum, with a subsequent decrease by day 60 postpartum [43], as shown on Fig. 3. Trends for T3 and T4 levels in neonatal male rats were not noted. In Sprague–Dawley female rats, TSH levels were reported to have an initial peak during the first 2 weeks postnatal [44] with a second elevation on days 40 and 50 postnatal, followed by a decline on days 60 to 80 postnatal [43]. T3 and T4 serum levels are reported to be low at birth in female rats, with the T4 level increasing on days 4 through 16 postpartum, and the T3 level increasing on days 10 to 30 postpartum [44,45], as shown on Fig. 4.

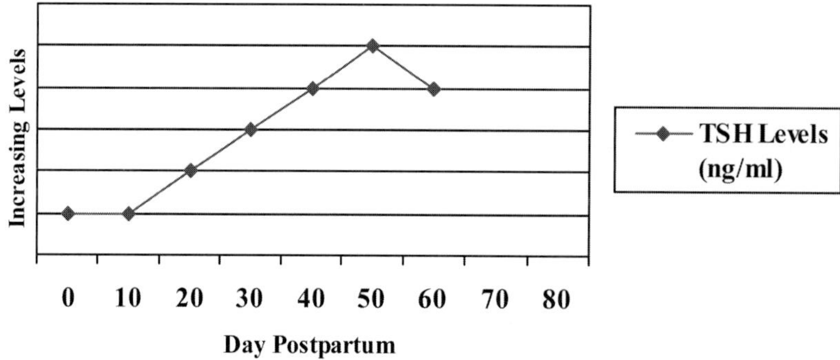

Fig. 3 Serum TSH levels obtained from Sprague–Dawley male rats following birth (adapted from ref. [43]).

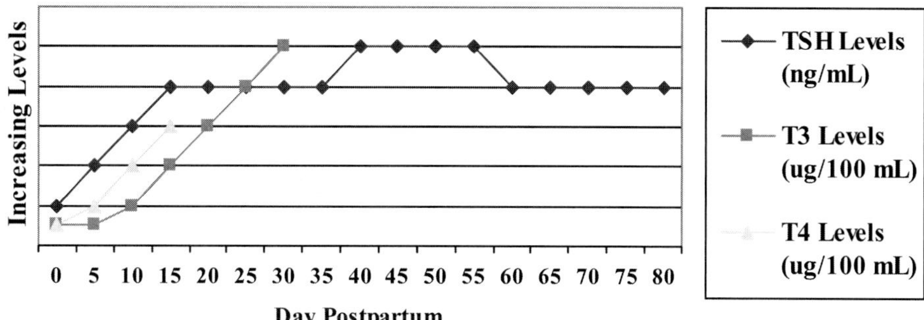

Fig. 4 Serum TSH, T3, and T4 levels obtained from female Sprague–Dawley rats following birth (adapted from refs. [44,45]).

METHODS

The thyroid hormone values that are the subjcct of this publication are control values for Charles River Crl:CD®BR VAF/Plus® (Sprague–Dawley) rats used in several different EPA define- (GLP-) compliant studies [46,47]. These include a two-generation study (Study A), a hormone evaluation study (Study B), and a neurobehavioral developmental toxicity study (Study C). The vehicle (control article) used in each of these studies was reverse osmosis processed deionized water (R.O. water); in each study, the test substance was provided to the rats in the drinking water. All blood samples were obtained from the inferior vena cava; obtained serum was allocated into three vials, immediately frozen on dry ice, and subsequently analyzed for serum T3, T4, and TSH levels, using either RIA kits from Diagnostic Product Corporation (T3 and T4) or a double antibody RIA procedure developed by Analytics Inc., Gaithersburg, Maryland. The thyroid/parathyroid glands were weighed after fixation in neutral buffered 10 % formalin for at least 48 h. Additional pertinent information regarding these study designs is presented in Table 2.

Study A

At weaning on lactation day (DL) 21, 30 rats per sex per dosage group were randomly selected for continued analysis of potential behavioral changes. This cohort was not evaluated for thyroid function. Of the remaining pups, at least three pups/sex/litter were necropsied and examined for gross lesions; blood samples were pooled by sex and litter. After fixation in neutral buffered 10 % formalin, the thyroids/parathyroids were weighed, trimmed, and retained for histopathological evaluation.

Study B

This study had three subsets: (1) DG 21 fetuses; (2) DL 10 pups; and (3) DL 22 pups. Fetuses or pups were weighed and examined for sex and gross external alterations. Serum was pooled by litter (DG 21 fetuses and DL 10 pups) and by sex and litter (DL 22 pups). The thyroid and brain were retained from each subset for morphometric and histopathological evaluations. Table 3 describes additional methodology pertaining to the fetal evaluations.

Table 3 Summary of study designs in which thyroid function was evaluated in adult Sprague–Dawley rats.

Parameter	Study A[a]	Study B[a]	Study C[a]
N	30/sex/group	15/16 per group	25/group
Treatment period	70 days prior to cohabitation until sacrifice	14 days prior to cohabitation until sacrifice	DG 0 until DL 10
Method of sacrifice	Exsanguination following CO_2 anesthetization	Exsanguination following CO_2 anesthetization	CO_2 asphyxiation
Day of sacrifice	DL 21	DG 21, DL 10 or DL 22	DL 10

[a]See following text for fetal analyses.
DG—Day of gestation.
DL—Day of lactation.

Study C

All pups not selected for continued observation were sacrificed via carbon dioxide asphyxiation and necropsied on day 5 postpartum (DP 5). Blood samples were pooled by litter. The thyroids were fixed in situ in neutral buffered 10 % formalin before histological examination. On DPs 90 to 92, blood was obtained from the inferior vena cava of one F1 generation rat per sex per litter in each dosage group after sacrifice, and serum samples were evaluated for thyroid hormone levels.

RESULTS

Data obtained from control animals are presented in Tables 4 through 7. In each study, statistically significant, dosage-dependent effects on thyroid function were observed; only hormone levels will be described. These effects were detected by serum analyses of T3, T4, and TSH levels made using RIA kits. Study A (York et al.) [48] demonstrated that exposure of rats to ammonium perchlorate in the drinking water at doses of 0.3, 3.0, and 30.0 mg/kg/day significantly increased relative thyroid weights in the female rats in each ammonium perchlorate-exposed group and in the male rats in the 3.0 and 30.0 mg/kg/day dosage groups. Dosage-dependent increases in thyroid hypertrophy and hyperplasia were also evident, and statistically significant differences in TSH, T3, and T4 levels were present in the 30 mg/kg/day dosage group. Further information regarding the results obtained for studies B and C are proprietary and not currently available.

Table 4 shows that control T3, T4, and TSH levels obtained from adult male rats in two studies were comparable, although the younger male rats (21–23 weeks, 147–161 days of age) had slightly lower T4 levels and slightly higher T3 and TSH levels than the older male rats (26 weeks, 182 days, of age).

Table 4 Summary of subsets in Study B in which thyroid function was evaluated in Sprague–Dawley fetal and neonatal rats.

Subset	1	2		3	
Day sacrificed	DG 21	DL 5	DL 10	DL 5	DL 22
N	16 (T3 = 8)	12–16	16	15 (T3 = 11)	15/sex
Method of sacrifice	Beuthanasia®-D special[a] (histo.) Decapitation (thyroid hormone)	Exsanguination following CO_2 anesthetization		CO_2 asphyxiation	
Blood sampling	Decapitation	Cardiac puncture		Cardiac puncture	
Sample pooling	By litter	By litter	By litter	By litter	By litter and sex

[a]Manufactured by Schering-Plough Animal Health, Union, New Jersey.

Table 5 shows control data for adult female rats on DG 21, DL 10, or DLs 21/22. Generally, T3 levels were increased on DG 21 and DL 10, as compared with DLs 21/22 values. TSH values were generally decreased on DLs 21/22, as compared with DG 21 or DL 10 values. T4 values were similar on DG 21 and DLs 21/22 and increased on DL 10.

Table 5 Control data for T3, T4 and TSH levels in Sprague–Dawley adult male rats.

Study	A	A
Age	21–23 wks	26 wks
N	30	29
T3 Levels (ng/dl)[a]	82.5 ± 8.7	72.5 ± 11.2
T4 Levels (ug/dl)[a]	3.8 ± 0.5	4.6 ± 0.5
TSH Levels (ng/ml)[a]	2.5 ± 1.0	1.5 ± 1.0

[a]Data provided as means and standard deviations.

Table 6 shows control data for fetuses on DG 21 and for pups on DLs 5 and 10. Table 7 shows control data for male and female pups on DLs 21/22. T3 and T4 levels tended to increase from DG 21 to DL 10, while TSH levels were highest on DG 21 and lowest on DL 5. T3 and T4 levels on DLs 21/22 were generally increased, as compared to those obtained from fetuses on DGs 21 or pups on DLs 5 and 10. TSH levels were generally decreased for male and female pups on DLs 21/22, as compared to fetal values on DGs 21 or values for pups on DLs 5 and 10.

Table 6 Control data for T3, T4, and TSH levels in Sprague–Dawley adult female rats.

Study	B	B	C	A	A	B
DG/DL	DG 21	DL 10	DL 10	DL 21	DL 21	DL 22
Age	16 wks	18 wks	16 wks	23–24 wks	20–22 wks	19 wks
N	16 (T3 = 15)	16	3	30	30	15
T3 Levels (ng/dl)[a]	99.7 ± 6.0	99.9 ± 10.9	79.0 ± 2.2	57.8 ± 28.2	61.5 ± 25.2	120.3 ± 12.4
T4 Levels (ug/dl)[a]	2.3 ± 0.1	4.0 ± 0.4	3.0 ± 0.1	2.1 ± 0.7	2.2 ± 1.0	3.5 ± 0.4
TSH Levels (ng/ml)[a]	6.0 ± 0.6	5.1 ± 0.5	5.0 ± 0.3	2.0 ± 0.9	1.6 ± 1.0	6.9 ± 0.7

[a]Data provided as means and standard deviations.

Table 7 Control data for T3, T4, and TSH levels in Sprague–Dawley fetuses.

Study	B	B	B
DG/DL	DG 21	DL 5	DL 10
M/F	M/F	M/F	M/F
N	16 (T3 = 8)	a	16
T3 Levels (ng/dl)[b]	23.4 ± 3.3	62.1 ± 6.6 (15)	79.5 ± 6.6
T4 Levels (ug/dl)[b]	1.6 ± 0.2	2.1 ± 0.2 (12)	3.7 ± 0.4
TSH Levels (ng/ml)[b]	7.2 ± 0.7	4.9 ± 0.5 (16)	6.6 ± 0.7

[a]N value listed in parentheses following the value.
[b]Data provided as means and standard deviations.

Table 8 Control data for T3, T4, and TSH levels in day 21/22 postpartum Sprague–Dawley pups.

Study	A	A	B	A	A	B
Male/female	Male	Male	Male	Female	Female	Female
N	27	20	15	28	20	15
T3 Levels (ng/dl)[a]	105.9 ± 10.0	106.3 ± 18.3	191.0 ± 20.0	106.0 ± 13.1	108.4 ± 21.1	169.3 ± 15.6
T4 Levels (ug/dl)[a]	4.4 ± 1.0	3.2 ± 0.8	4.0 ± 0.5	4.3 ± 1.0	3.4 ± 0.7	3.6 ± 0.4
TSH Levels (ng/ml)[a]	1.2 ± 0.4	0.8 ± 0.2	3.6 ± 0.4	1.1 ± 0.5	0.9 ± 0.3	5.3 ± 0.5

[a]Data provided as means and standard deviations.

Follicular cell hypertrophy/hyperplasia are typical hormonal and histopathological findings attributable to agents altering thyroid function. These findings occurred in the treated groups in these studies, as described previously. All control instances of these findings were minimal, mild, or moderate (data not shown). No control rats had marked hypertrophy and hyperplasia of the follicular epithelium.

Control data obtained for the two generations of adult Sprague–Dawley male rats were similar and within the ranges previously identified for Sprague–Dawley male rats.

DISCUSSION

In adult female rats, T3 levels during gestation and the early period of lactation were generally above the range observed for the adult male rats, as would be expected, based on the sex-related difference. T3 levels decreased as the lactation period continued. T4 values were generally lower than the range observed for the male rats. TSH levels were relatively variable in the female rats but were generally higher than the values for the male rats.

Male and female fetuses and pups showed the expected increase in T3 values with age, reaching maximum levels that were higher than adult levels by day 21 postnatal. T4 levels showed a similar pattern of gradual increase. As would be expected, TSH levels decreased as the rats aged, and were lower than adult levels by day 21 postnatal.

CONCLUSION

Minimal historical experience exists in evaluation of thyroid hormone levels in pregnant rats and their offspring in regulatory compliant studies. Although relatively small in number, the results obtained for CRL Sprague–Dawley rats in USEPA-design multigeneration and developmental neurotoxicity studies conducted at our laboratory indicate that the use of RIA kits for analyses for serum TSH, T4, and T3 levels in fetal, neonatal, juvenile, and adult rats provides a relatively consistent, sensitive, and appropriate biomarker for detecting functional changes in the thyroid.

REFERENCES

1. D. A. Fisher. In M. A. Sperling. *Pediatric Endocrinology*, pp. 51–70, W. B. Saunders, Philadelphia (1996).
2. R. W. Beard and P. W. Nathanielsz. *Fetal Physiology and Medicine* **12**, pp. 216–231, W. B. Saunders, Philadelphia (1976).
3. J. B. Henry. *Clinical Diagnosis and Management by Laboratory Methods*, 19th ed., pp. 333–342, W. B. Saunders, Philadelphia (1996).
4. A. Stevens and J. Lowe. *Human Histology*, pp. 258–261, Times Mirror, Barcelona, Spain (1997).
5. J. G. Hardman and L. L. Limbird. *Goodman and Gilman's The Pharmacological Basis of Therapeutics* **57**, pp. 1563–1593, McGraw-Hill, New York (2001).

6. K. Tamura, M. Hatsuta, G. Watanabe, K. Taya, H. Kogo. *Biochem. Biophys. Res. Comm.* **242**, 102–108 (1998).
7. M. Larsson, T. Petterson, A. Carlstrom. *Crit. Rev. Toxicol.* **30** (2), 135–196 (2000).
8. J. W. Apriletti, R. C. Ribeiro, R. L. Wagner, W. Feng, P. Webb, P. J. Kushner, B. L. West, S. Nilsson, T. S. Scanlan, R. J. Fletterick, J. D. Baxter. *Crit. Rev. Toxicol.* **30** (2), 135–196 (2000).
9. R. W. Nelson and C. G. Couto. *Small Animal Internal Medicine* **51**, pp. 703–733, Mosby, St. Louis, MO (1998).
10. M. E. Hadley. *Endocrinology*, pp. 290–313, Prentice Hall, New Jersey (1996).
11. S. P. Porterfield and C. E. Hendrich. *Endocrine Rev.* **1**, 94–106 (1993).
12. N. H. Bass, E. W. Pelton, E. Young. In *Thyroid Hormones and Brain Development*, G. D. Grave (Ed.), pp. 199–214, Raven Press, New York (1977).
13. J. T. Eayrs. In *Endocrinology and Human Behavior*, R. P. Michael (Ed.), pp. 239–255, Oxford University Press, London (1968).
14. R. Balazs. In *Biochemistry of the Developing Brain,* W. Himivich (Ed.), **1**, 39–63, Marcel Dekker, New York (1973).
15. M. Berry. In *Studies on the Development of Behavior and the Nervous System: Aspects of Neurogenesis,* G. Gottlieb (Ed.), **2**, 7–67, Academic Press, New York (1974).
16. S. A. Stein, D. R. Shanklin, P. M. Adams, G. M. Mihailoff, M. B. Palnitkar, B. Anderson. In *Iodine and the Brain,* G. R. Delong, J. Robbins, P. G. Condliffe (Eds.), pp. 59–78, Plenum, New York (1989).
17. S. Zamenhof and E. Van Marthens. In *Cellular Aspects of Neuronal Growth and Differentiation*, D. C. Pease (Ed.), pp. 329–355, University of California Press, Berkeley (1971).
18. J. Dobbing and J. Sands. *Arch. Dis. Child* **48**, 757–767 (1973).
19. J. T. Eayrs. *J. Endocrinol.* **22**, 409–419 (1961).
20. J. T. Eayrs. *J. Anat.* **88**, 164–173 (1954).
21. J. L. Bakke, R. J. Gellert, N. Lawrence. *J. Lab. Clin. Med.* **76**, 25–33 (1976).
22. M. M. Best and C. H. Duncan. *J. Lab. Clin. Med.* **73**, 135–143 (1969).
23. F. Azzi, A. G. Vagenakis, J. Bollinger, S. Reichlin, L. E. Braverman, S. M. Ingbar. *Endocrinology* **94**, 1681–1688 (1974).
24. P. S. Hopkins and G. D. Thorburn. *J. Endocrinol.* **54**, 55–56 (1972).
25. J. L Nicholson and J. Altman. *Brain Res.* **44**, 13–23 (1972).
26. E. Clegg. *J. Reproduct. Fertil.* **1**, (1960).
27. S. R. Ojeda, W. W. Andrews, J. P. Advis, S. S. White. *Endocr. Rev.* **1**, 228–257 (1980).
28. C. C. Korenbrot, I. T. Huhtaniemi, R. I. Weiner. *Biol. Reprod.* **17**, 298–303 (1977).
29. E. M. Lewis, J. F. Barnett Jr., L. Freshwater, A. M. Hoberman, M. S. Christian. *Drug Chem. Toxicol.* **25** (4), 437–458 (2002).
30. L. B. Valle, R. M. Oliveira-Filho, J. H. Romaldini, P. F. Lara. *J. Steroid Biochem.* **23**, 253–257 (1985).
31. A. Chowdury, A. K. Gautam, B. B. Chatterjee. *Arch. Androl.* **13**, 233–239 (1984).
32. S. Palmero, M. Prati, P. De Marco, P. Trucchi, E. Fugassa. *J. Endocrinol.* **136** (2), 277–282 (1993).
33. R. W. Rivest. Cited in J. M. Goldman, S. C. Laws, S. K. Balchak, R. L. Cooper, R. J. Kavlock. *Crit. Rev. Toxicol.* **30** (2), 135–196 (2000).
34. H. M. A. Meijs-Roelofs, P. Osman, P. Kramer. *J. Endocrinol.* **92**, 341–349 (1982).
35. T. M. Crisp, E. D. Clegg, R. L. Cooper, W. P. Wood, D. G. Anderson, K. P. Baetcke, J. L. Hoffmann, M. S. Morrow, D. J. Rodier, J. E. Schaeffer, L. W. Touart, M. G. Zeeman, Y. M. Patel. *Environ. Health Perspect.* **106** (Suppl. 1), 11–36 (1998).
36. C. C. Capen. *Toxicol. Pathol.* **25**, 39–48 (1997).
37. W. F. Loeb and F. W. Quimby (Eds.). *The Clinical Chemistry of Laboratory Animals,* 2nd ed., pp. 33–48, Taylor & Francis, Philadelphia (1999).

38. J. M. Manson, T. Brown, D. M. Baldwin. *Toxicol. Appl. Pharmacol.* **73**, 323–335 (1984).
39. M. H. Lu and R. E. Staples. *Teratology* **17**, 171–178 (1978).
40. H. Fukuda, K. Oshima, M. Mori, I. Kovayashi, M. A. Greer. *Endocrinology* **107**, 1711–1716 (1980).
41. L. E. Gray Jr, R. J. Kavlock, N. Chernoff, J. Ostby, J. Ferrell. *Toxicol. Appl. Pharmacol.* **67** (1), 1–14 (1983).
42. L. E. Gray and R. J. Kavlock. *Toxicol. Lett.* **15**, 231–235 (1983).
43. J. W. Simpkins, J. F. Bruni, R. J. Mioduszewski, J. Meites. *Endocrinology* **98** (6), 1365–1369 (1998).
44. H. Fukuda and M. A. Greer. *J. Endocrinol. Invest.* **1**, 311–314 (1978).
45. J. H. Dussault and F. Labrie. *Endocrinology* **97**, 1321–1324 (1975).
46. U.S. Environmental Protection Agency. Federal Insecticide, Fungicide and Rodenticide Act (FIFRA); Good Laboratory Practice Standards; Final Rule. 40 CFR Part 160.
47. U.S. Environmental Protection Agency. Toxic Substances Control Act (TSCA); Good Laboratory Practice Standards; Final Rule. 40 CFR Part 792.
48. R. G. York, W. R. Brown, M. F. Girard, J. S. Dollarhide. *Int. J. Toxicol.* **20** (4), 183–197 (2001).
49. M. I. Surks and R. Sievert. *N. Engl. J. Med.* **333** (25), 1691 (1995).

Pure Appl. Chem., Vol. 75, Nos. 11–12, pp. 2069–2079, 2003.

Topic 3.8

Modification of endocrine active potential by mixtures*

Kevin Gaido[1,‡], Li You[1], and Steve Safe[2]

[1]*CIIT Centers for Health Research, 6 Davis Drive, Research Triangle Park, NC 27709, USA;* [2]*Department of Veterinary Physiology and Pharmacology, Texas A&M University, College Station, TX 77843, USA*

Abstract: Wildlife and humans are exposed to a complex mixture of endocrine active chemicals. The activity of a specific chemical in any mixture can be modified through interactions with other components of the mixture. The toxic equivalency factor (TEF) approach for risk assessment was developed for chemicals such as halogenated aromatics that induce their effects through ligand-activated receptors. For persistent halogenated aromatic AhR agonists, this approach has some utility. However, the use of the TEF approach for endocrine active compounds is confounded by the unique tissue- and response-specific activities of these structurally diverse compounds. The term "selective receptor modulator" describes the ability of a natural or synthetic receptor ligand to manifest agonist activity in one tissue or for one response and antagonist activity in other tissues or for another response in the same tissue. Thus, it is possible for chemicals in a mixture to behave in an additive manner for one response and an antagonist manner for another response. A mechanisms-based hazard risk assessment of endocrine active chemical mixtures must account for these multiple variables.

INTRODUCTION

Risk assessment of chemicals has primarily focused on individual compounds where overall risks are estimated from human exposure data and toxicity data obtained in laboratory animal studies. In reality, wildlife and humans are exposed to complex mixtures of endocrine active chemicals that include endogenous steroid hormones, environmental contaminants such as pesticides and plasticizers, and natural substances such as phytoestrogens and fungal metabolites. These endocrine active chemicals act through multiple pathways, and the activity of a specific chemical in any mixture can be modified through interactions with other components of the mixture [1]. Figure 1 illustrates examples of different environmental contaminants that bind directly to receptors and activate intracellular signaling pathways; these include 2,3,7,8-tetrachlorodibenzo-*p*-dioxin (TCDD), an aryl hydrocarbon receptor (AhR) agonist [2]; 2,2-bis(*p*-chlorophenyl)(*o*-chlorophenyl)-1,1,1-trichloroethane (*o,p*′-DDT), an estrogen receptor (ER) agonist [3]; 1,1-bis(*p*-chlorophenyl)-1,1-dichloroethylene (DDE), an androgen receptor antagonist [4]; 2,3,3′,4′,5′,6-hexachlorobiphenyl (HCB), an agonist for constitutive androstane receptor (CAR) [5]; diethylhexylphthalate (DEHP), an agonist for the pregnane X receptor (PXR) [6]; and chlordane, an estrogen receptor-related-α (ERRα) antagonist [7].

*Report from a SCOPE/IUPAC project: Implication of Endocrine Active Substances for Human and Wildlife (J. Miyamoto and J. Burger, editors). Other reports are published in this issue, *Pure Appl. Chem.* **75**, 1617–2615 (2003).

‡Corresponding author: E-mail: Gaido@ciit.org

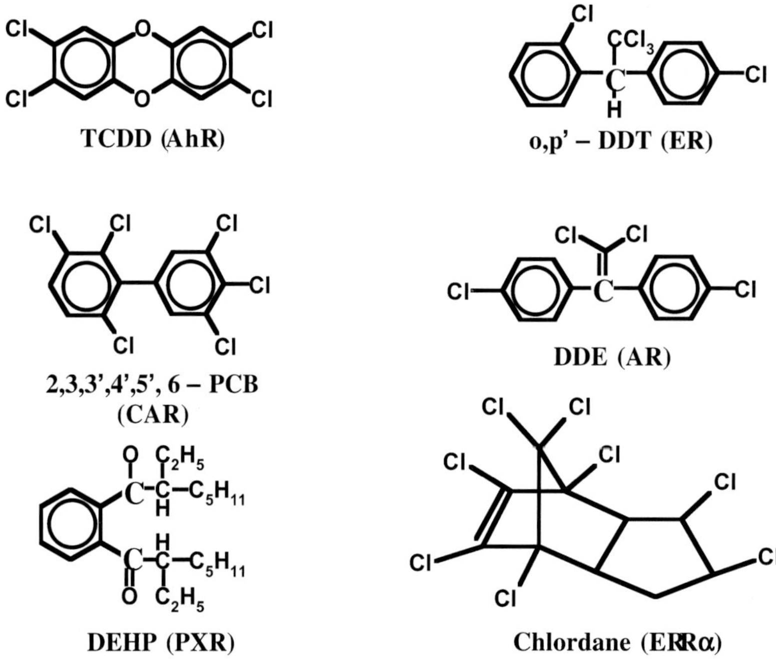

TCDD (AhR)

o,p' – DDT (ER)

2,3,3',4',5', 6 – PCB (CAR)

DDE (AR)

DEHP (PXR)

Chlordane (ERRα)

Fig. 1 Environmental contaminants that activate multiple receptor-mediated signaling pathways.

TOXIC EQUIVALENCY FACTOR APPROACH FOR RISK ASSESSMENT OF HALOGENATED AROMATICS

The toxic equivalency factor (TEF) approach for hazard and risk assessment can be used for chemical mixtures that act through a common pathway, where the toxic equivalents (TEQs) of a mixture are equal to the sum of the concentrations of the individual compounds (C_i) times their fractional potency (TEF_i) relative to a common standard.

$$TEQ = \Sigma [C_i \cdot TEF_i]$$

The TEF approach has been used for TCDD and related 2,3,7,8-substituted polychlorinated dibenzo-*p*-dioxins (PCDDs) and dibenzofurans (PCDFs) where TCDD is assigned a potency of 1 [8–11]. All of these compounds bind and activate the AhR and induce a broad spectrum of toxic responses, including a wasting syndrome, developmental and reproductive toxicity, thymic atrophy, impaired immune responses, chloracne, hepatic responses and porphyria, carcinogenic and anticarcinogenic responses, and induction of CYP1A1 and other drug-metabolizing enzymes. Development of TEFs for individual PCDD and PCDF congeners used data from multiple quantitative structure–activity studies that reported response-specific potencies of congeners relative to TCDD [11]. For each compound, a range of TEFs was obtained, and scientific judgment was used to select individual TEF values for PCDD and PCDF congeners (Table 1).

However, environmental samples containing PCDDs and PCDFs also contain higher levels (>100-fold) of polychlorinated biphenyl (PCB) mixtures [12–14]. Two structural classes of PCB congeners also bind the AhR and induce prototypical AhR-mediated biochemical and toxic responses [15,16]. Coplanar or non-ortho substituted PCBs that include 3,3′,4,4′-tetrachlorobiphenyl (tetraCB), 3,4,4′,5-tetraCB, 3,3′,4,4′,5-pentaCB (PCB 126), and 3,3′,4,4′,5,5′-hexaCB are the most active AhR agonists, and their mono-ortho derivatives containing one ortho-chloro substituent also exhibit weak AhR agonist activities. TEFs have been assigned to the most environmentally relevant non-ortho and mono-ortho substituted PCB congeners (Table 1) [17]. These values have been extensively used by scientists to calculate TEQs in environmental samples, and, in many samples, the TEQs-PCBs are comparable or higher than the TEQs-PCDDs/PCDFs [14,18].

Table 1 Toxic equivalency factors for the 2,3,7,8-substituted PCDDs and PCDFs and selected polychlorinated biphenyl congeners [37,50].

Congener		TEF
PCDD	2,3,7,8-TCDD	1.0
	1,2,3,7,8-PentaCDD	0.5
	1,2,3,4,7,8-HexaCDD	0.1
	1,2,3,6,7,8-HexaCDD	0.1
	1,2,3,7,8,9-HexaCDD	0.1
	1,2,3,4,6,7,8-HeptaCDD	0.01
	OCDD	0.001
PCDF	2,3,7,8-TCDF	0.1
	2,3,4,7,8-PentaCDF	0.5
	1,2,3,,7,8-PentaCDF	0.1/0.05
	1,2,3,4,7,8-HexaCDF	0.1
	2,3,4,6,7,8-HexaCDF	0.1
	1,2,3,6,7,8-HexaCDF	0.1
	1,2,3,7,8,9-HexaCDF	0.1
	1,2,3,4,6,7,8-HeptaCDF	0.01
	1,2,3,4,7,8,9-HeptaCDF	0.01
	OCDF	0.001
PCB	3,3′,4,4′,5-PentaCB	0.1
	3,3′,4,4′,5,5′-HexaCB	0.01
	3,3′,4,4′-TetraCB	0.0005
	2,3,3′,4,4′-PentaCB	0.0001
	2,3,3′,4,4′,5-HexaCB	0.0005
	2,3,4,4′,5-PentaCB	0.0001
	2,3,3′,4,4′,5′-HexaCB	0.0005
	2′,3,4,4′,5-PentaCB	0.0001
	2,3,4,4′,5-PentaCB	0.0005

Several studies report non-additive antagonistic interactions between AhR-active PCBs, PCDDs, PCDFs, and AhR-independent PCB congeners such as 2,2′,4,4′,5,5′-hexachlorobiphenyl (PCB 153) [19–24]. For example, in cells or animals treated with an effective dose of TCDD or 3,3′,4,4′,5-pentaCB, cotreatment with 2,2′,4,4′,5,5′-hexaCB inhibits induction of CYP1A1 and related activities, porphyria, teratogenicity (murine fetal cleft palate and hydronephrosis), chick embryotoxicity, malformations, liver lesions, and edema. Moreover, the ratios of 2,2′,4,4′,5,5′-hexaCB/TCDD (or TEQ) required for non-additive antagonist interactions are in the range of those observed in environmental samples.

Another serious problem associated with the TEF approach for risk assessment of halogenated aromatic compounds is the effects of concurrent exposures to AhR-active phytochemicals. The TEF approach was initially developed at a time when only a limited number of synthetic industrial compounds or combustion by-products were characterized as AhR agonists. However, more recent studies have demonstrated that structurally diverse phytochemicals and endogenous biochemicals also interact with the AhR; these include flavonoids, indole-3-carbinol and related heteroaromatics, carotenoids, 7-keto-cholesterol, compounds in herbal extracts, bilirubin, biliverdin, and resveratrol [25–34]. It is estimated that serum levels of "natural" AhR agonists/antagonists (e.g., flavonoids) are in the nM to low µM range [26]. In contrast, serum levels of TEQs (for TCDD and related compounds) are in the subpicomolar range suggesting that ratios of phytochemical AhR agonists/TEQs are 10^4 to 10^6. In most studies with phytochemicals that exhibit AhR antagonist activity, their inhibitory effects would be observed at antagonist/agonist ratios of 10^4 to 10^6, whereas their levels in serum are below concentrations for agonist-induced responses. Thus, the TEF approach for hazard and risk assessment of halogenated aromatics has significant deficiencies due to interactions of "TEQ-compounds" (Table 1) with PCBs and phytochemicals that inhibit AhR-mediated responses.

USING THE TEF APPROACH FOR ENDOCRINE ACTIVE CHEMICALS

The U.S. Environmental Protection Agency's (USEPA's) Endocrine Disruptor Screening Program (EDSP) has outlined several in vitro and in vivo bioassays for identifying endocrine active chemicals (EACs). These assays can provide data on relative activity for mixtures [35–38]. For example, we have used multiple bioassays to show that the estrogen equivalents (EQs) in 200 milliliters (ml) of red wine are at least 1000 times higher than the EQs for the average daily intake of a mixture of known estrogenic pesticides in the diet (Table 2) [38]. The use of individual bioassays and EQs is comparable to the TEF/TEQ method for hazard and risk assessment of TCDD and related halogenated aromatics and is based on their common mechanism of action through binding to common steroid hormone receptors.

Table 2 Estimated daily intake of estrogen equivalents.

	Estrogen equivalents (μg)
Wine (200 ml)	1.87
Pesticide mix (2.44 μg*)	0.00021

*Estimated daily intake of organochlorine pesticides determined in a 1995–1996 Food and Drug Administration market survey for contaminants in food.
From [38]

However, results of ongoing studies indicate that risk assessment of AhR agonists and other EACs using an additive approach may be too simplistic [1,39–42]. Studies in our laboratories have demonstrated that structurally diverse synthetic and naturally occurring ER agonists differentially activate various estrogen-responsive constructs in cancer cell lines cotransfected with wild-type or variant forms of ERα [43–47].

LIGAND-DEPENDENT REGULATION OF STEROID HORMONE RECEPTOR STRUCTURE AND FUNCTION

Selective receptor modulator (SRM) describes the ability of a ligand to manifest receptor agonist activity in some tissues but block receptor activity in other tissues [48]. The properties of selective receptor modulators are due in part to unique ligand-induced conformational changes in the steroid hormone receptor that affect the subsequent tissue-specific recruitment of other nuclear factors required for ligand-induced gene expression [49,50]. The antiestrogenic drug tamoxifen is an example of a selective estrogen receptor modulator (SERM). Tamoxifen functions as an estrogen receptor antagonist in breast cancer cells, but behaves as an agonist in the uterus and bone [51–56]. This tissue-specific estrogen receptor activity of tamoxifen is likely related to its differential interactions, relative to estradiol, with domains of the estrogen receptor [57].

Figure 2 illustrates structurally diverse synthetic and natural estrogenic compounds used in our studies to investigate the effects of ligand structure on estrogen receptor function [43–46]. We hypothesized that the diversity in structure of these compounds would likely result in induction of unique conformations of the estrogen receptor that would ultimately affect estrogen receptor function in a unique gene- and tissue-specific manner similar to the pharmaceutical SRMs. Results of studies in HepG2

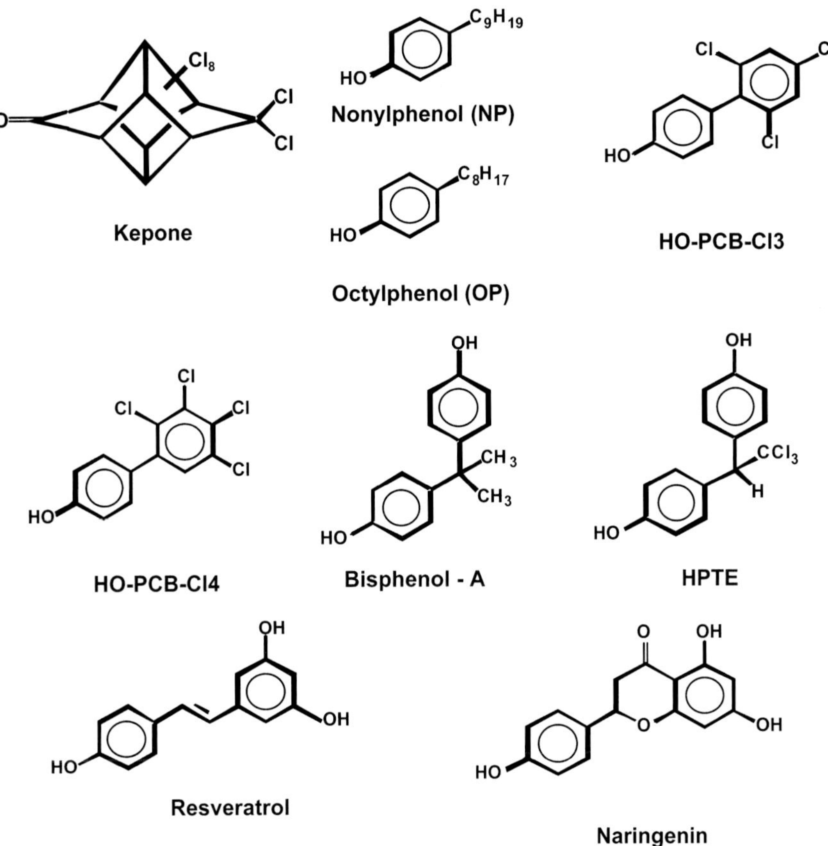

Fig. 2 Structures of naturally occurring and synthetic compounds.

human hepatoma cells confirmed our hypotheses and showed that the phenolic compounds (mono- and dihydroxy) gave similar but not identical patterns of induced gene expression that were clearly different from those observed for the phytoestrogens naringenin and resveratrol and the chlorinated hydrocarbon kepone [43,45].

Cell context is also an important determinant. For example, resveratrol induced reporter gene activity in U2 human osteogenic sarcoma cells transfected with hERα-AF1 while naringenin was inactive, whereas these activities were reversed in HepG2 cells [45]. In HepG2 cells cotreated with estradiol plus synthetic and natural estrogens, bisphenol A (BPA) and naringenin exhibited partial antiestrogenic activity with one or more forms of wild type or variant hERα [47,58]. BPA also exhibited SERM-like activity in an estrogen-responsive human endometrial carcinoma cell line [59]. In this cell line, estradiol induced both progesterone receptor expression and cell proliferation, whereas BPA induced only progesterone receptor expression. In combination studies, BPA inhibited the induction of cell proliferation by estradiol. These results indicate that BPA can act as both an estrogen and an antiestrogen in the same cell, depending on the response being examined. The inhibitory effects of BPA have also been observed in vivo [47]. Ongoing studies using wild-type and variant forms of hERα in HepG2, U2, and MDA-MB-231 breast cancer cells show that activation of receptor activity by natural and synthetic estrogens depends on ligand structure, cell context, and form of hERα [43–45,47]. Moreover, most of the test compounds exhibit antiestrogenic activity in one or more of these assays.

The pattern of ERα activation by BPA and 2,2′-bis(*p*-hydroxyphenyl)-1,1,1-trichloroethane (HPTE) in HepG2 cells was similar; however, results obtained using U2 and MDA-MBA-231 cells clearly distinguish between the two 4,4′-dihydroxydiphenylmethane analogs that differ only in their methylene bridge substituents (Fig. 2). Recent studies in our laboratories also show that BPA, not HPTE, activates ERß (Table 3) [44,46]. In contrast, HPTE acts as an ERß antagonist and an androgen receptor (AR) antagonist. BPA does not interact with the AR.

Table 3 Differential interaction of structural analogs with ERα, ERß, and AR.

	ERα	ERß	AR
HPTE	+++	– – –	– – –
Bisphenol A	++	++	
p,p′-DDE			– – –
Di-hydroxy DDE	+++	– – –	– – –

From [44]

Together, these results indicate that structurally diverse natural and synthetic compounds can have selective receptor modulating activity capable of having additive or antagonistic properties when combined, making it difficult, if not impossible, to predict the activity of a mixture of these chemicals in target tissues in vivo.

MODIFICATION OF ENDOCRINE ACTIVE POTENTIAL IN VIVO BY CHEMICAL MIXTURES

We investigated the combined effect of estradiol and HPTE on gene expression in the reproductive tissues of male and female mice [60]. Alone, estradiol and HPTE acted similarly on expression of most genes in the ovaries, uterus, testes, and prostate (Table 4). However, in each tissue, there were subsets of genes differentially regulated by these two compounds. In the uterus, progesterone receptor, ERα, AR, insulin-like growth factor 1, insulin-like growth factor binding protein 5, and clusterin mRNAs were significantly reduced with both E2 or HPTE treatments, whereas cathepsin B was induced. Conversely, induction of cathepsin B by E2 in the ovary was reversed after cotreatment with HPTE, and ERβ expression was induced by HPTE but not E2. In addition, E2 uniquely upregulated glutathione peroxidase 3, glutathione *S*-transferase, and cytochrome P450 17α-hydroxylase, with no effect of HPTE. In male mice, mast cell growth factor, clusterin, cyclin A2, and glutathione peroxidase 3 (GPX3) mRNAs were significantly induced with either E2 or HPTE treatments in the testes, whereas insulin-like growth factor 1A (IGF-IA) and UDP-glucuronosyltransferase mRNAs were decreased. In the prostate, IGF-IA, clusterin, and GPX3 mRNAs were induced by E2 and HPTE. IGF binding protein 3 was induced by E2 but not by HPTE in the testes, and this E2-specific induction was blocked by cotreatment with HPTE. Cytochrome P450 17α-hydroxylase mRNA was downregulated by E2, and AR was uniquely upregulated by HPTE in the testes. These results demonstrate that E2 and HPTE induce both common and also unique patterns of tissue-specific and receptor-dependent gene expression.

We also investigated the combined effect of genistein and methoxychlor, the parent compound of HPTE, on reproductive development in Sprague–Dawley rats [61]. Sprague–Dawley rats were exposed to the compounds either alone or in combinations through dietary administration during pregnancy and lactation; the offspring were exposed after lactation. Both compounds singly and in combination accelerated vaginal opening and altered estrous cyclicity in female offspring, and these estrogenic responses to genistein plus methoxychlor were additive. Methoxychlor but not genistein delayed preputial separation in male rats, an indication of antiandrogenic action. When administered in combination with methoxychlor, genistein enhanced the effects of methoxychlor (Fig. 3). While the estrogenic responses are supported by in vitro estrogen-receptor-based transcriptional activational assays, the potentiation of

Table 4 Effect of estradiol and HPTE on gene expression.

		Uterus[2]	Ovary[2]	Testis	Prostate
Estrogen receptor α	Estradiol	↓	NC	NC	
	HPTE	↓	NC	NC	
	Estradiol + HPTE	+	NC	NC	
Progesterone receptor	Estradiol	↓	↑		
	HPTE	↓	NC		
	Estradiol + HPTE	+	+		
Androgen receptor	Estradiol	↓	NC	NC	NC
	HPTE	↓	NC	↑	NC
	Estradiol + HPTE	+	NC	*	NC
Insulin-like growth factor 1A	Estradiol	↓	NC	↓	↓
	HPTE	↓	NC	↓	↓
	Estradiol + HPTE	+	NC	+	+
IGF binding protein	Estradiol	↓	↓	↑	↑
	HPTE	↓	NC	↑	NC
	Estradiol + HPTE	+	–	–	–
Clusterin	Estradiol	↓	NC		↑
	HPTE	↓	NC		↑
	Estradiol + HPTE	+	NC		+[1]
Glutathione peroxidase 3	Estradiol	↑	↑	↑	↑
	HPTE	NC	NC	↑	↑
	Estradiol + HPTE	–	–	+	+
Glutathione *S*-transferase	Estradiol	↓	↑	NC	NC
	HPTE	NC	NC	NC	NC
	Estradiol + HPTE	–	–	NC	NC
Cathepsin B	Estradiol	↑	↑		NC
	HPTE	↑	NC		NC
	Estradiol + HPTE	+	–[1]		NC
Cytochrome P45017	Estradiol		↓	↓	
	HPTE		↓	NC	
	Estradiol + HPTE		+[1]	+[1]	
Mast cell growth factor	Estradiol			↑	
	HPTE			↑	
	Estradiol + HPTE			+	
Cyclin A2	Estradiol			↑	NC
	HPTE			↑	NC
	Estradiol + HPTE			+[1]	NC

↑ Indicates an increase in gene expression.
↓ Indicates a decrease in gene expression.
+ The actions of estradiol plus HPTE were additive.
– HPTE antagonized the actions of estradiol.
* Indicates activity unique to HPTE.
[1]Indicates cases where HPTE also displayed antiandrogenic activity.
[2]From [60]

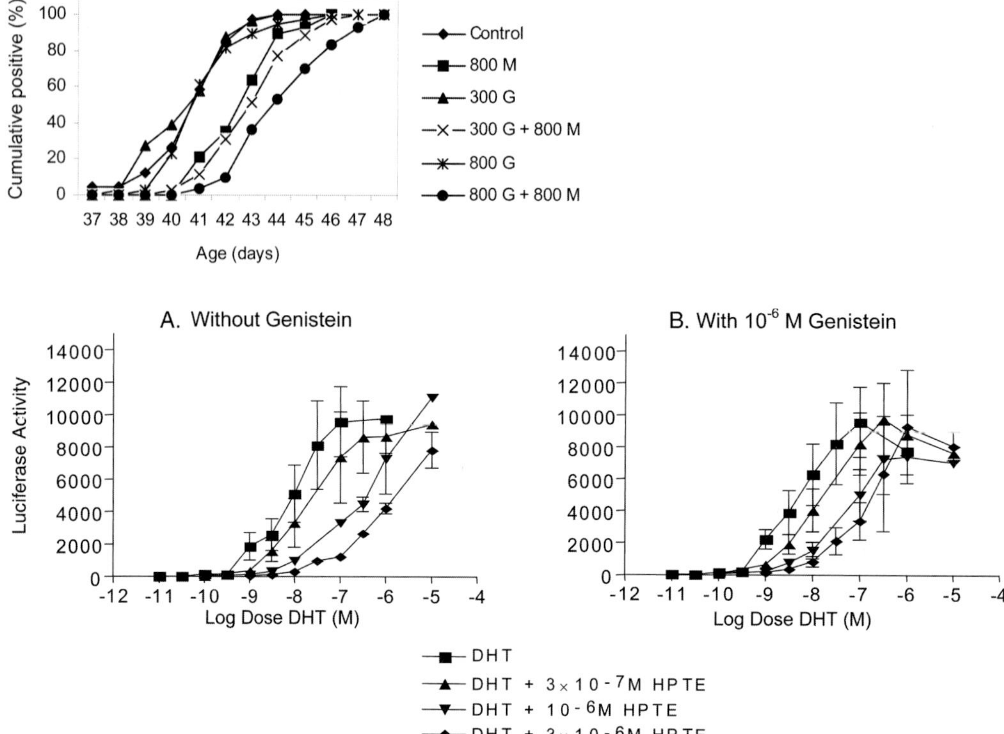

Fig. 3 *Antiandrogenic interaction of the binary mixture of genistein and methoxychlor.* Upper panel: Time course of cumulative percentage positive for preputial separation (PPS), a developmental landmark of androgenic action, in male rats treated with the xenochemicals genistein and methoxychlor. The age at which PPS occurred was delayed by methoxychlor. Genistein by itself did not have an effect, but the presence of 800-ppm genistein enhanced the effect of methoxychlor. Lower panel: Androgen receptor- (AR-) dependent transcriptional activation assay using expression plasmid-transfected HepG2 cells. The methoxychlor metabolite 2,2-bis(*p*-hydroxyphenyl)-1,1,1-trichloroethane (HPTE) antagonized the AR-activating effect of dihydrotestosterone (DHT) by shifting the dose–response curves to the right (A). The presence of genistein had no effect on the AR antagonism of HPTE (B). The *p*Kb value for HPTE was 6.42 ± 0.17 in the absence of genistein and 6.64 ± 0.17 in the presence of genistein [from 61].

methoxychlor antiandrogenicity by genistein is not predicted based on in vitro androgen receptor transactivation assays (Fig. 3).

In a parallel set of experiments, we examined interactions of genistein and methoxychlor on mammary gland development in juvenile rats. Inguinal mammary glands were obtained from both female and male pups. In male rats, methoxychlor caused elongation of the glandular ducts, while genistein enhanced ductile branching. The two compounds together promoted development of alveolar-lobular structure, an effect not observed with either compound alone.

Together, these studies highlight the complexity of steroid hormone receptor-mediated responses and the difficulties for predicting endocrine activities of chemical mixtures based on the actions of these chemicals individually in short-term in vitro and in vivo bioassays.

SUMMARY

The TEF/TEQ approach for risk assessment was developed for chemicals such as halogenated aromatics that induce their effects through ligand-activated receptors. For persistent halogenated aromatic AhR agonists, this approach has some limited utility. However, the structure-dependent interactions of SRMs have been extensively investigated, and the results suggest that a TEF/TEQ approach for these compounds is not appropriate due to their unique tissue-specific agonist and antagonist activities. Ligands that bind AhR and other nuclear receptors also induce tissue-, species-, and age-dependent responses. Therefore development of mechanisms-based hazard risk assessment of receptor agonists/antagonists must account for these multiple variables.

ACKNOWLEDGMENTS

S. Safe acknowledges financial support from the National Institutes of Health (ES09106 and ES04917).

REFERENCES

1. S. H. Safe. *Environ. Health Perspect.* **106**, 1051–1058 (1998).
2. A. Poland and J. C. Knutson. *Annu. Rev. Pharmacol. Toxicol.* **22**, 517–554 (1982).
3. J. Bitman and H. C. Cecil. *J. Agric. Food Chem.* **18**, 1108–1112 (1970).
4. W. R. Kelce, C. R. Stone, S. C. Laws, L. E. Gray, J. A. Kemppainen, E. M. Wilson. *Nature* **375**, 581–585 (1995).
5. T. Sueyoshi, T. Kawamoto, I. Zelko, P. Honkakoski, M. Negishi. *J. Biol. Chem.* **274**, 6043–6046 (1999).
6. H. Masuyama, Y. Hiramatsu, M. Kunitomi, T. Kudo, P. N. MacDonald. *Mol. Endocrinol.* **14**, 421–428 (2000).
7. C. Yang and S. Chen. *Cancer Res.* **59**, 4519–4524 (1999).
8. L. S. Birnbaum and M. J. Devito. *Toxicology* **105**, 391–401 (1995).
9. J. Bellin and D. Barnes. "Interim procedures for estimating risks associated with exposures to mixtures of chlorinated dibenz-p-dioxins and dibenzofurans (CDDs and CDFs)", U.S. EPA (1989).
10. Nato/CCMS. "Method of risk assessment for complex mixtures of dioxins and related compounds", Tech. Report No. 176 (1988).
11. U. G. Ahlborg, A. Brouwer, M. A. Fingerhut, J. L. Jacobson, S. W. Jacobson, S. W. Kennedy, A. A. Kettrup, J. H. Koeman, H. Poiger, C. Rappe, et al. *Eur. J. Pharmacol.* **228**, 179–199 (1992).
12. V. A. McFarland and J. U. Clarke. *Environ. Health Perspect.* **81**, 225–239 (1989).
13. S. Tanabe, H. Iwata, R. Tatsukawa. *Sci. Total Environ.* **154**, 163–177 (1994).
14. K. Noren and D. Meironyte. *Chemosphere* **40**, 1111–1123 (2000).
15. S. Safe. *Crit. Rev. Toxicol.* **21**, 51–88 (1990).
16. S. H. Safe. *Crit. Rev. Toxicol.* **24**, 87–149 (1994).
17. U. G. Ahlborg, G. C. Becking, L. S. Birnbaum, A. Brouwer, H. J. G. M. Derks, M. Feeley, G. Golor, A. Hanberg, J. C. Larsen, A. K. D. Liem, S. Safe, C. Schlatter, F. Wærn, M. Younes, E. Yrjänheikki. *Chemosphere* **28**, 1049–1067 (1994).
18. N. Kannan, S. Tanabe, R. Tatsukawa. *Arch. Environ. Health* **43**, 11–14 (1988).
19. R. Bannister and S. Safe. *Toxicology* **44**, 159–169 (1987).
20. L. Biegel and S. Safe. *J. Steroid Biochem. Mol. Biol.* **37**, 725–732 (1990).
21. L. Biegel, M. Harris, D. Davis, R. Rosengren, L. Safe, S. Safe. *Toxicol. Appl. Pharmacol.* **97**, 561–571 (1989).
22. J. M. Haake, S. Safe, K. Mayura, T. D. Phillips. *Toxicol. Lett.* **38**, 299–306 (1987).
23. D. Davis and S. Safe. *Toxicology* **63**, 97–111 (1990).
24. D. Davis and S. Safe. *Toxicol. Lett.* **48**, 35–43 (1989).
25. H. P. Ciolino, T. T. Wang, G. C. Yeh. *Cancer Res.* **58**, 2754–2760 (1998).
26. H. Ashida, I. Fukuda, T. Yamashita, K. Kanazawa. *FEBS Lett.* **476**, 213–7 (2000).

27. L. F. Bjeldanes, J. Y. Kim, K. R. Grose, J. C. Bartholomew, C. A. Bradfield. *Proc. Natl. Acad. Sci. USA* **88**, 9543–9547 (1991).
28. R. F. Casper, M. Quesne, I. M. Rogers, T. Shirota, A. Jolivet, E. Milgrom, J. F. Savouret. *Mol. Pharmacol.* **56**, 784–790 (1999).
29. I. Chen, S. Safe, L. Bjeldanes. *Biochem. Pharmacol.* **51**, 1069–1076 (1996).
30. H. P. Ciolino and G. C. Yeh. *Mol. Pharmacol.* **56**, 760–767 (1999).
31. T. A. Gasiewicz, A. S. Kende, G. Rucci, B. Whitney, J. J. Willey. *Biochem. Pharmacol.* **52**, 1787–803 (1996).
32. Y. J. Chun, S. Y. Ryu, T. C. Jeong, M. Y. Kim. *Drug Metab. Disposition* **29**, 389–393 (2001).
33. S. A. Quadri, A. N. Qadri, M. E. Hahn, K. K. Mann, D. H. Sherr. *Mol. Pharmacol.* **58**, 515–25 (2000).
34. D. Phelan, G. M. Winter, W. J. Rogers, J. C. Lam, M. S. Denison. *Arch. Biochem. Biophys.* **357**, 155–163 (1998).
35. K. Ramamoorthy, F. Wang, I. C. Chen, S. Safe, J. D. Norris, D. P. McDonnell, K. W. Gaido, W. P. Bocchinfuso, K. S. Korach. *Science* **275**, 405 (1997).
36. K. Ramamoorthy, F. Wang, I. C. Chen, J. D. Norris, D. P. McDonnell, L. S. Leonard, K. W. Gaido, W. P. Bocchinfuso, K. S. Korach, S. Safe. *Endocrinology* **138**, 1520–1527 (1997).
37. K. Ramamoorthy, C. Vyhlidal, F. Wang, I. C. Chen, S. Safe, D. P. McDonnell, L. S. Leonard, K. W. Gaido. *Toxicol. Appl. Pharmacol.* **147**, 93–100 (1997).
38. K. Gaido, L. Dohme, F. Wang, I. Chen, B. Blankvoort, K. Ramamoorthy, S. Safe. *Environ. Health Perspect.* **106** (Suppl 6), 1347–51 (1998).
39. S. H. Safe. *J. Anim. Sci.* **76**, 134–41 (Jan, 1998).
40. S. Safe, K. Connor, K. Ramamoorthy, K. Gaido, S. Maness. *Regul. Toxicol. Pharmacol.* **26**, 52–58 (1997).
41. S. Safe. *Teratogenesis, Carcinogenesis, Mutagenesis* **17**, 285–304 (1997).
42. S. H. Safe, L. Pallaroni, K. Yoon, K. Gaido, S. Ross, B. Saville, D. McDonnell. *Reprod. Fertil. Develop.* **13**, 307–315 (2001).
43. K. Yoon, L. Pallaroni, M. Stoner, K. Gaido, S. Safe. *J. Steroid Biochem. Mol. Biol.* **78**, 25–32 (2001).
44. K. W. Gaido, S. C. Maness, D. P. McDonnell, S. S. Dehal, D. Kupfer, S. Safe. *Mol. Pharmacol.* **58**, 852–858 (2000).
45. K. Yoon, L. Pellaroni, K. Ramamoorthy, K. Gaido, S. Safe. *Mol. Cell. Endo.* **162**, 211–220 (2000).
46. K. W. Gaido, L. S. Leonard, S. C. Maness, J. M. Hall, D. P. McDonnell, B. Saville, S. Safe. *Endocrinology* **140**, 5746–5753 (1999).
47. J. C. Gould, L. S. Leonard, S. C. Maness, B. L. Wagner, K. Conner, T. Zacharewski, S. Safe, D. P. McDonnell, K. W. Gaido. *Mol. Cell. Endo.* **142**, 203–214 (1998).
48. B. S. Katzenellenbogen and J. A. Katzenellenbogen. *Science* **295**, 2380–2381 (2002).
49. L. A. Paige, D. J. Christensen, H. Gron, J. D. Norris, E. B. Gottlin, K. M. Padilla, C. Chang, L. M. Ballas, P. T. Hamilton, D. P. McDonnell, D. M. Fowlkes. *Proc. Natl. Acad. Sci. USA* **96**, 3999–4004 (1999).
50. J. M. Hall, D. P. McDonnell, K. S. Korach. *Mol. Endocrinol.* **16**, 469–486 (2002).
51. A. Goulding, E. Gold, W. Feng. *Bone Miner.* **18**, 143–152 (1992).
52. R. R. Love, R. B. Mazess, H. S. Barden, S. Epstein, P. A. Newcomb, V. C. Jordan, P. P. Carbone, D. L. DeMets. *N. Engl. J. Med.* **326**, 852–856 (1992).
53. I. S. Fentiman, Z. Saad, M. Caleffi, M. A. Chaudary, I. Fogelman. *Eur. J. Cancer* **28**, 684–685 (1992).
54. V. C. Jordan. *Cancer* **70**, 977–982 (1992).
55. S. E. Reis, J. P. Costantino, D. L. Wickerham, E. Tan-Chiu, J. Wang, M. Kavanah. *J. Natl. Cancer Inst.* **93**, 16–21 (2001).

56. R. P. Kedar, T. H. Bourne, T. J. Powles, W. P. Collins, S. E. Ashley, D. O. Cosgrove, S. Campbell. *Lancet* **343**, 1318–21 (1994).
57. M. T. Tzukerman, A. Esty, D. Santiso-Mere, P. Danielian, M. G. Parker, R. B. Stein, J. W. Pike, D. P. McDonnell. *Mol. Endocrinol.* **8**, 21–30 (1994).
58. M. F. Ruh, T. Zacharewski, K. Connor, J. Howell, I. Chen, S. Safe. *Biochem. Pharmacol.* **50**, 1485–1493 (1995).
59. R. M. Bergeron, T. B. Thompson, L. S. Leonard, L. Pluta, K. W. Gaido. *Mol. Cell. Endo.* **150**, 179–187 (1999).
60. K. M. Waters, S. Safe, K. W. Gaido. *Toxicol. Sci.* **63**, 47–56 (2001).
61. L. You, M. Casanova, E. J. Bartolucci, M. W. Fryczynski, D. C. Dorman, J. I. Everitt, K. W. Gaido, S. M. Ross, H. D. Heck. *Toxicol. Sci.* **66**, 91–104 (2002).

Pure Appl. Chem., Vol. 75, Nos. 11–12, pp. 2081–2098, 2003.

Topic 3.9

Experience with new testing guidelines with endocrine-sensitive endpoints*

Rochelle W. Tyl‡

Center for Life Sciences and Toxicology, RTI International, 3040 Cornwallis Road, P.O. Box 12194, Research Triangle Park, NC 27709-2194, USA

Abstract: Concerns about the effects of endocrine disruptors on humans and wildlife have resulted in revised governmental testing guidelines (e.g., U.S. Environmental Protection Agency, Organization for Economic Cooperation and Development, U.S. Food and Drug Administration), adding endpoints to enhance their capability to detect endocrine active compounds. Based on experience with these testing guidelines, I present my opinions and data on study design, performance, results, endpoints, interpretation, and recommendations for improvement. New regulatory endpoints must be reproducible, robust, sensitive, relevant, and consistent. These new endpoints are appropriate: anogenital distance, examination of culled pups on postnatal day 4, examination for retained nipples/areolae in preweanling males, weanling necropsy, acquisition of puberty in offspring, prebreed estrous cyclicity, reproductive organ weights and histopathology, and andrology. Endpoints considered not appropriate are ovarian primordial follicle counts, stage of estrus for parental females at demise, and single blood sample at necropsy to measure circulating levels of relevant hormones. In conclusion: (1) regulatory guidelines represent only the minimum requirements and should/will change as science improves and scientific and societal concerns arise; (2) scientists must interact to identify and refine new methods, endpoints, and mechanisms; (3) the objective of these studies is to provide good science and useful information for risk assessment. Regulatory agencies should convene workshops with stakeholders to discuss issues such as new fundamental/applied information, mechanisms, and current and new endpoints. Interactions will lead to enhanced science, periodic revisions of testing guidelines, and improved risk assessment.

BACKGROUND

During the late 1980s, circumstantial evidence was accumulating that humans and wildlife were exhibiting effects consistent with alterations in endocrine function. These included decreasing sperm counts worldwide, increasing incidences of prostate and breast cancer, accelerated puberty, and increased incidences of male reproductive malformations in humans. For wildlife, these effects included reductions in reproductive success, increases in malformed offspring, alterations in sexual and reproductive behaviors in fish-eating birds of the Great Lakes, alterations in external reproductive organs in alligators in Lake Apopka (Florida) and in Florida panthers, accelerated reductions (or total losses) of localized populations of amphibians, imposexes in harbor snails, inappropriate secondary and tertiary

---

*Report from a SCOPE/IUPAC project: Implication of Endocrine Active Substances for Human and Wildlife (J. Miyamoto and J. Burger, editors). Other reports are published in this issue, *Pure Appl. Chem.* **75**, 1617–2615 (2003).

‡Tel.: 919-541-5972; Fax: 919-541-5956; E-mail: rwt@rti.org

sex characteristics (i.e., male structures in females, female structures in males) in various fish populations, and other effects. Dr. Theo Colborn of the World Wildlife Fund recognized these events and coined the term "endocrine disruption" to describe effects from endogenous (e.g., phytoestrogens) and anthropogenic (i.e., manmade; examples include phthalate plasticizers, Kraft pulp paper mill effluents, synthetic hormones in waste water from use of contraceptive pills, nonylphenol, octylphenol, and bisphenol A-based plastics) materials in the environment. Almost single-handedly, she organized a series of conferences/workshops (Wingspread conferences) on various aspects of the effects and mechanisms of endocrine classes (e.g., steroids, thyroid hormone) on humans and wildlife, with participants from various areas of expertise, including academic endocrinologists, researchers using animal models, wildlife biologists, epidemiologists, neurologists, and andrologists. Dr. Colborn also testified before the U.S. Congress.

The U.S. Environmental Protection Agency (USEPA), in response to the growing evidence for effects of endocrine active compounds, revised its original testing guideline for Reproductive Toxicity Effects (Fig. 1; 40 Code of Federal Regulations [CFR] 798.4700 [1]), with a draft of a new guideline with limited circulation in 1994 and a public draft in 1996 for comments, and established the Endocrine Disruptors Screening and Testing Advisory Committee (EDSTAC) in 1996. The USEPA Toxic Substance Control Act (TSCA) regulators finalized the new testing guidelines in 1997 (799.9380; USEPA TSCA [2]), and the USEPA Office of Prevention, Pesticides and Toxic Substances (OPPTS) regulators, including both TSCA and Federal Insecticide, Fungicide and Rodenticide Act (FIFRA), finalized the new testing guidelines in 1998 (870.3800; USEPA OPPTS; Fig. 2 [3]). The OECD promulgated its new reproductive toxicity testing guideline in 2001 (OECD, No. 416; Fig. 3 [4]), and the U.S. Food and Drug Administration (FDA) presented its new reproductive toxicity testing guideline in the Redbook (Fig. 4) [5].

All of the new guidelines retained the original study design in terms of exposures and matings (Fig. 1), but added endocrine-sensitive endpoints to enhance the efficacy of the testing guideline to detect endocrine-mediated effects (Fig. 2). My laboratory has completed ten studies under the new OPPTS

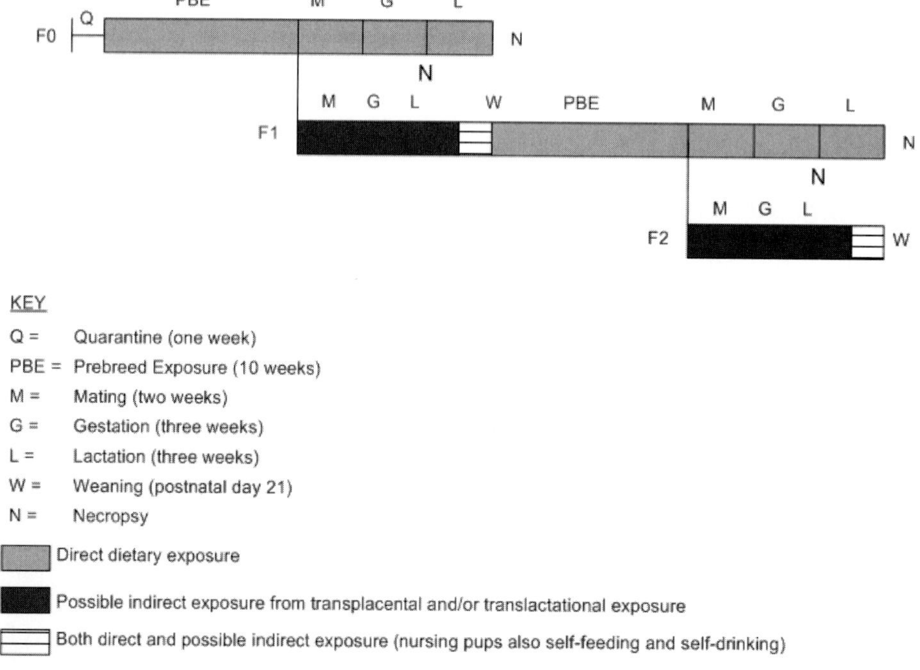

Fig. 1 Previous USEPA (TSCA and FIFRA) Testing Guidelines since 1978.

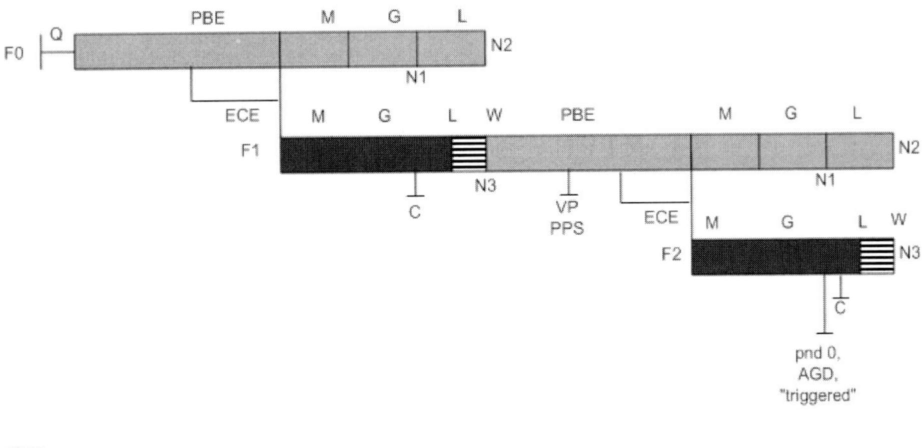

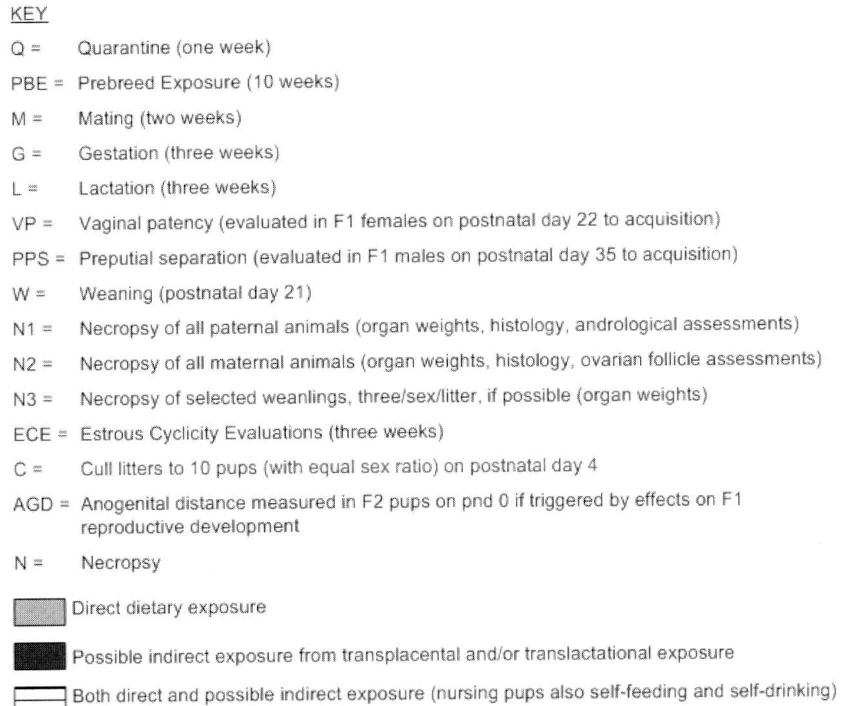

KEY
Q = Quarantine (one week)
PBE = Prebreed Exposure (10 weeks)
M = Mating (two weeks)
G = Gestation (three weeks)
L = Lactation (three weeks)
VP = Vaginal patency (evaluated in F1 females on postnatal day 22 to acquisition)
PPS = Preputial separation (evaluated in F1 males on postnatal day 35 to acquisition)
W = Weaning (postnatal day 21)
N1 = Necropsy of all paternal animals (organ weights, histology, andrological assessments)
N2 = Necropsy of all maternal animals (organ weights, histology, ovarian follicle assessments)
N3 = Necropsy of selected weanlings, three/sex/litter, if possible (organ weights)
ECE = Estrous Cyclicity Evaluations (three weeks)
C = Cull litters to 10 pups (with equal sex ratio) on postnatal day 4
AGD = Anogenital distance measured in F2 pups on pnd 0 if triggered by effects on F1 reproductive development
N = Necropsy
Direct dietary exposure
Possible indirect exposure from transplacental and/or translactational exposure
Both direct and possible indirect exposure (nursing pups also self-feeding and self-drinking)

Fig. 2 Most recent USEPA TSCA Final Test Guidelines (USEPA 799.9380, 1997) and OPPTS (FIFRA/TSCA) Final Testing Guidelines (USEPA 870.3800, 1998), "Reproductive and Fertility Effects".

testing guidelines, as well as FDA Segment I and III studies, OECD 415 and 416 studies, and OECD-modified 422 and 421 studies, with inclusion of the new endocrine-sensitive endpoints.

Based on our experience, I offer you my thoughts on the study design, performance, endpoints, interpretation, and reporting of study results, including my recommendations to correct perceived deficiencies.

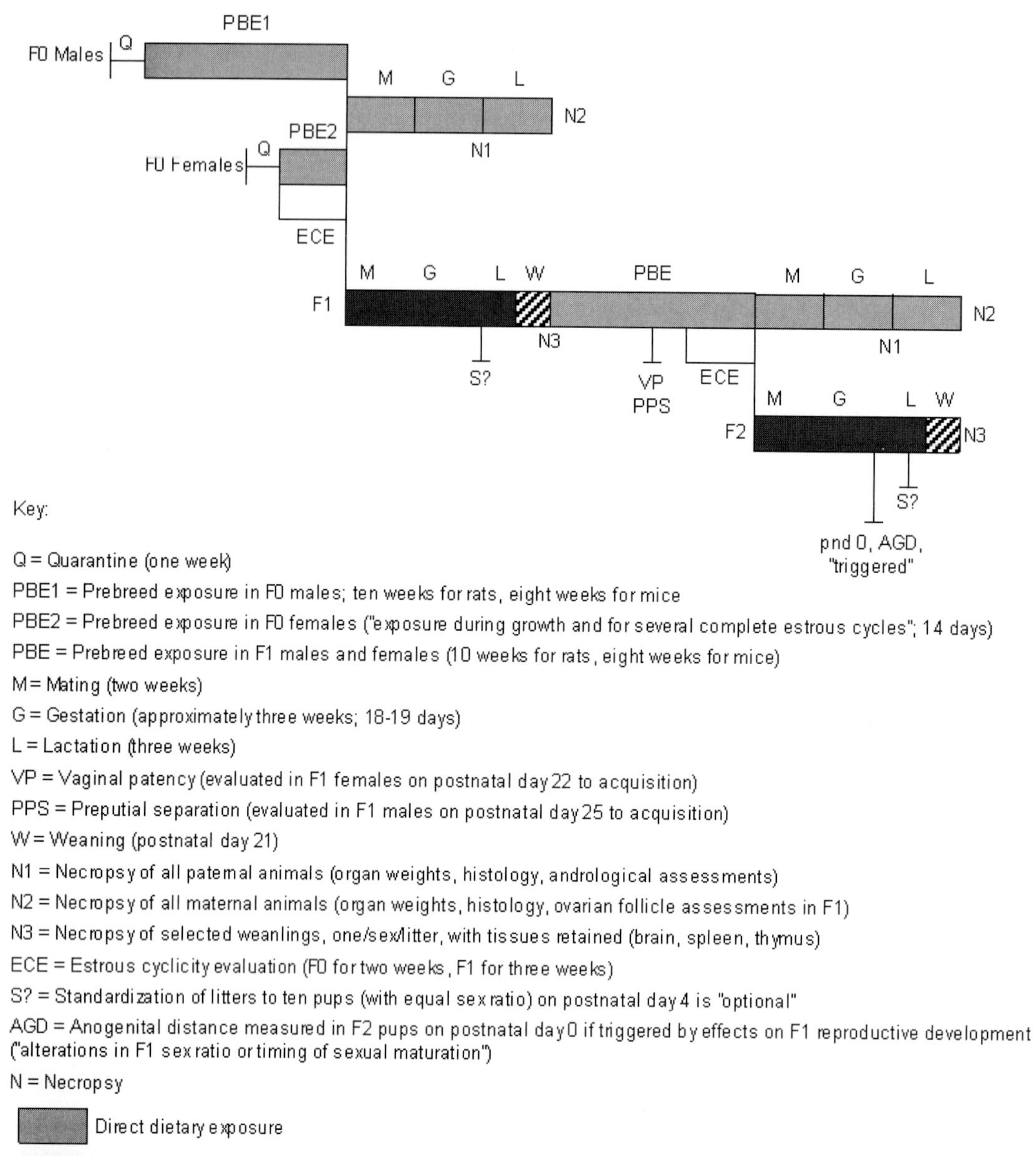
PBE1
F0 Males
Q
M G L
N2
PBE2
N1
F0 Females
Q
ECE
M G L W
PBE
M G L
F1
N2
N3
N1
S?
VP
PPS
ECE
M G L W
F2
N3
S?
pnd 0, AGD, "triggered"
Key:
Q = Quarantine (one week)
PBE1 = Prebreed exposure in F0 males; ten weeks for rats, eight weeks for mice
PBE2 = Prebreed exposure in F0 females ("exposure during growth and for several complete estrous cycles"; 14 days)
PBE = Prebreed exposure in F1 males and females (10 weeks for rats, eight weeks for mice)
M = Mating (two weeks)
G = Gestation (approximately three weeks; 18-19 days)
L = Lactation (three weeks)
VP = Vaginal patency (evaluated in F1 females on postnatal day 22 to acquisition)
PPS = Preputial separation (evaluated in F1 males on postnatal day 25 to acquisition)
W = Weaning (postnatal day 21)
N1 = Necropsy of all paternal animals (organ weights, histology, andrological assessments)
N2 = Necropsy of all maternal animals (organ weights, histology, ovarian follicle assessments in F1)
N3 = Necropsy of selected weanlings, one/sex/litter, with tissues retained (brain, spleen, thymus)
ECE = Estrous cyclicity evaluation (F0 for two weeks, F1 for three weeks)
S? = Standardization of litters to ten pups (with equal sex ratio) on postnatal day 4 is "optional"
AGD = Anogenital distance measured in F2 pups on postnatal day 0 if triggered by effects on F1 reproductive development ("alterations in F1 sex ratio or timing of sexual maturation")
N = Necropsy
Direct dietary exposure
Possible indirect exposure from transplacental and/or translactational exposure
Both direct and possible indirect exposure (nursing pups also self-feeding and self-drinking)

Fig. 3 Most recent OECD Guideline for the Testing of Chemicals (No. 416), Two-Generation Reproductive Toxicity Study (adopted 22 January 2001).

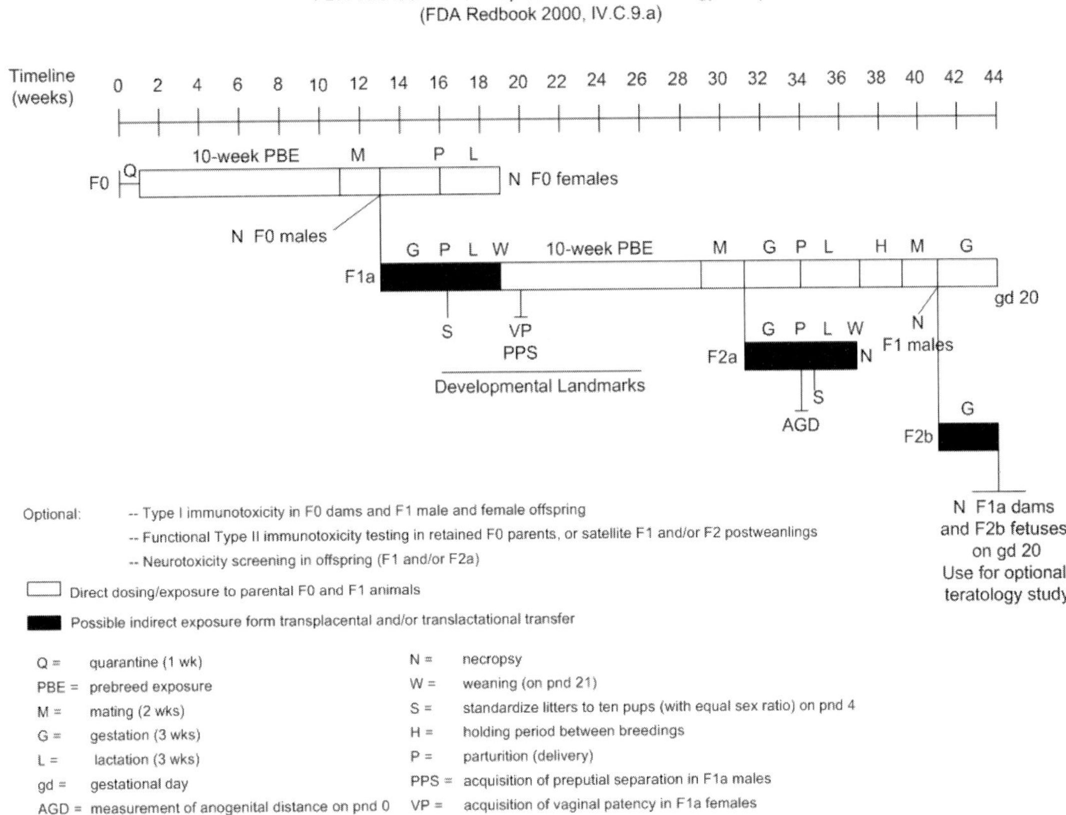

Fig. 4 Most recent FDA Two-Generation Reproductive and Teratology Study (FDA Redbook 2000, IV.C.9.a).

STUDY DESIGN

The multigeneration study design is "apical". Intact animals are exposed beginning as postpubescent animals through adulthood and reproduction (F0 generation), beginning as gametes through adulthood and reproduction (F1 generation—the most important generation), and beginning as gametes to weaning (F2 generation). Therefore, a number of critical, sensitive life stages are exposed, including prenatal, perinatal, lactational, prepubescent, peripubescent, adult, and reproductive. Route of administration mimics the route known, anticipated, or modeled for the target species. Doses (or dietary or drinking-water concentrations) are usually selected to include a clear adult effect level (usually exceeding environmental exposure by order[s] of magnitude) to very low environmentally relevant doses. There are many endpoints to detect endocrine-mediated outcomes, from centrally mediated effects (e.g., brain, hypothalamus) to locally mediated effects (e.g., on gonads, sex accessory organs, other glands and organs). As an apical test (and not a "specific" one), it will not necessarily identify the specific mechanism(s), but its design evaluates multiple possible mechanisms. The endpoints chosen become very important (endpoints are discussed in more detail below).

PERFORMANCE

Staff must be well trained and competence reaffirmed, with intra- and inter-technician variability ascertained and shown to be low. Historical control data from the performing laboratory are also very important to provide context and interpretation for observed effects (e.g., the concurrent control value is very low, but the dose group value is well within the historical control range).

The choice of methodologies used to evaluate endpoints is also critical. For example, anogenital distance is measured in some laboratories with a ruler and a hand-held animal, resulting in large variability. In my laboratory, we use an ocular diopter with a stage micrometer attached to a dissecting microscope for newborn anogenital distances. The selection of which instruments will determine the accuracy, with some models accurate to 0.2 mm (when newborn control rat values are typically 2.0 mm for males and 1.0 mm for females) and some accurate to 0.01 mm. This means that differences of 0.2 to 0.3 mm may be statistically significantly different but not necessarily biologically significant. For weanling and adult anogenital distances, my laboratory uses digital vernier calipers (these work best on animals after euthanasia prior to necropsy). Other simple things that may affect the study performance include checking balance calibrations before and after weighing animals and organs, keeping procedures consistent across generations and studies, and documenting clinical observations carefully, completely, and consistently. These studies, typically used for risk assessment by cognizant governmental agencies, are ideally performed in compliance with stringent good laboratory practice (GLP) principles, regulations, and standards. This adherence to GLPs will ensure appropriate procedures and training for the study performance.

ENDPOINTS

It is my thesis that endpoints in this type of study (for risk assessment purposes) must first be shown to be robust, reproducible, appropriately sensitive, biologically plausible, and relevant to the adverse outcomes of concern. Definitions of the attributes of such endpoints are as follows:

1. *Reproducible*: These endpoints must be reliable: the same findings occur under the same conditions within the initial reporting laboratory (intra-laboratory) and among other laboratories (inter-laboratory). If the results from endpoints are not reproducible, they cannot form the basis for future research and are most likely not useful for risk assessment.
2. *Robust*: These endpoints must be present after comparable routes of exposure (e.g., dosed feed or dosed water). The use of oral gavage, a bolus dose once per day, may result in exacerbation of the effect on an endpoint if the parent material is the proximate toxicant and is metabolized to a non-toxic metabolite, and, if bolus dosing overwhelms the metabolic capacity of the organism or preparation, it may result in diminution or loss of the effect on an endpoint if the parent compound must be metabolized to the active form. Different effects may be observed by non-oral routes, such as inhalation, topical application, and injection, since these routes bypass "first-pass" metabolism by the liver. The findings from routes unrelated to human or environmental exposures may not be useful for risk assessment. These findings must also be present at the same routes and doses over time.
3. *Sensitive*: These endpoints should not be dependent on unique conditions (e.g., intrauterine position [IUP]), especially those that are not relevant to the species at risk. These endpoints should not exhibit high variability (insensitive) or be greatly affected by confounders (too sensitive).
4. *Relevant*: These endpoints must be biologically plausible and related to "adverse" effects of interest/concern. If there are no "adverse" effects at the dose/duration/route evaluated, these endpoints should be predictive of other "adverse" effects at higher doses, after longer exposure duration, and/or by different routes, etc.
5. *Consistent*: These endpoints must occur in the presence of effects in other related, relevant endpoints, if possible, at the same dose, timing, duration, routes of exposure, etc.

Examples of such appropriate and inappropriate endpoints for risk assessment (in my opinion) are as follows (data from my laboratory and elsewhere are also presented):

1. *Anogenital distance (both sexes at birth, at weaning, at adulthood; Table 1)*: This is considered a very appropriate endpoint. It is dihydrotestosterone- (DHT-) mediated, and endocrine-mediated effects persist into adulthood. However, it is confounded by body weight. Therefore, the current practice is to present the data as mm, mm/gram (mm/g) body weight, mm/cube root of the body weight, and/or to analyze the data by analysis of covariance (ANCOVA), with the body weight at measurement (e.g., birth, weaning) as the covariate, to account for differences in body weight (especially in groups where there is systemic toxicity, such as reduced parental and offspring body weights). Very small changes in this parameter (e.g., ≤0.05 mm on postnatal day [pnd] 0) may indicate only body weight-related delays in development, vs. large changes (e.g., ≥0.15 mm on pnd 0) more likely indicative of effects from endocrine disruption. Male pups with demasculinized (feminized) anogenital distance are more likely to exhibit reproductive system malformations, but the correlation is not perfect (i.e., some males with shortened anogenital distance exhibit no malformations, and some males with normal anogenital distance do exhibit malformations). It is also a reasonable predictor at lower doses of increased incidence of male reproductive malformations from perinatal exposure at higher doses. Current OECD and USEPA guidelines trigger anogenital distance in newborn F2 offspring only if effects on reproductive development are detected in F1 offspring (e.g., delays in acquisition of puberty, alterations in sex ratio). Since F2 offspring are terminated at weaning, effects on anogenital distance in F2 newborn pups cannot be related to any postwean effects, such as puberty, estrous cycling, adult reproductive system structures, or func-

Table 1 Anogenital distance in rats.

			Female anogenital distance (mm)				**Male anogenital distance (mm)**			
Study	PND	Gen.	Control		High dose		Control		High dose	
Code			BW (g)	AGD	BW (g)	AGD	BW (g)	AGD	BW (g)	AGD
A	0	F2	6.08	0.73	6.02*	0.63**	6.47	2.25	5.64*	2.11*
B	0	F2	5.93	0.76	5.85	0.79*	6.27	2.01	6.22	2.00
C	0	F2	5.97	0.95	5.99	0.96	6.28	1.98	6.38	2.00
	0	F3	5.98	0.92	6.14	0.94	6.36	1.97	6.49	1.96
D	0	F2	6.09	0.96	5.81	0.96	6.43	2.11	6.20	2.00*
E	0	F2	6.00	0.97	6.12	0.98	6.34	2.05	6.46	2.06
F[a]	0	F1	6.35	0.96	5.91*	0.92	6.76	2.06	6.15**	1.71***
	0	F2	6.21	0.98	5.93	0.99	6.63	2.05	6.29	1.77***
G (E2)	0	F1	6.17	0.95	6.56*	0.95	6.47	2.00	6.74	2.03
H (F)[b]	1	F1	NOT DONE					2.9		1.4*
I (L)[c]	1	F1	NOT DONE					2.9		2.7*
J (TP)[d]	2	F1		1.72		1.64	[3.07–3.33][e]	3.80		2.92*
K (TB)[f]	2	F1		1.5		2.2**		UNAFFECTED		

* = $p < 0.05$; ** = $p < 0.01$; *** = $p < 0.001$ (study codes A through G are from the author's laboratory)

[a]An antiandrogen (Tyl, unpublished observations).

[b]Flutamide [6].

[c]Linuron [7].

[d]Testosterone propionate [5].

[e]At the top dose, external sex distinction could not be made for most offspring; this is the range of AGD values for these pups [5].

[f]17ß-trembolone [8].

tions. We recommend measuring anogenital distance in newborn F1 animals as well, a generation that is thoroughly evaluated through sexual maturity and reproduction. This will enable interpretation of the consequences or predictability, if any, of altered anogenital distance.

2. *Necropsy of culled pups on pnd 4*: Neither the OECD nor OPPTS guidelines specify what to do with the culled pups (standardizing litters is optional in OECD 416). My strong recommendation is to necropsy the culled pups. Testes, epididymides, ovaries, and uteri can be weighed to detect changes in size better than by visual examination alone. Ectopic/undescended testes can be detected, as well as major malformations (e.g., missing parts of epididymides, missing testes and/or epididymides, necrosis, swelling, flaccidity of testes) in these organs.
3. *Retained/reduced nipples and areolae in preweanlings (Table 2, A and C)*: This is not included in the current guidelines. Involution of fetal nipples in males is DHT-mediated, and retained nipples persist into adulthood. In the author's laboratory, retained nipples have never been observed in control preweanling CD® (Sprague–Dawley [SD]) male rats, although areolae are present in our laboratory in 0 to 2.7 % of control males on pnd 11–13 (based on examination of over 3000 males in toto). Reduced nipples and/or areolae in female preweanlings are observed when they are exposed in utero to a strong androgen (Table 2, B and D). This is a sensitive indicator of altered testosterone (T) and/or DHT levels (e.g., effects on synthesis, degradation, receptor binding, transcriptional activation). Male pups with retained nipples are more likely to exhibit reproductive system malformations, but the correlation is not perfect (i.e., some males with nipples exhibit no malformations, and some males with no nipples do exhibit malformations). It is also a reasonable predictor at lower doses of male reproductive malformations from perinatal exposures at higher doses. Reduction in areolae/nipples in preweanling females was associated with reproductive system abnormalities [6]. In my laboratory, we examine preweanling rat offspring on pnd 11–13 and preweanling mouse offspring on pnd 9–11.

Table 2 Nipple and areolae retention/reduction in preweanling rats (pnd 11–13).

A. Nipple retention in male rats

		Control		High dose	
Study code	Gen.	% Males	No./male	% Males	No./male
A[a]	F1	0.00	0.00	19.23***	0.72***
	F2	0.00	0.00	16.46	0.51*
B[b]	F1	0.00	0.00	0.00	0.00
	F2	0.00	0.00	0.00	0.00
	F3	0.00	0.00	0.00	0.00
C (L)[c]	F1	–	0.8	–	3.3*[c]
D (F)[d]	F1	16	0.5[c]	100.0	12.0*

B. Nipple reduction in female rats

		Control		High dose	
Study code	Gen.	% Females	No./female	% Females	No./female
A (TP)[e]	F1	100.0	12.0	6.67****	0.27****
B (TB)[f]					

(*continues on next page*)

Table 2 (*Continued*).

C. Areolae retention in male rats

Study code	Gen.	Control % Males	Control No./male	High dose % Males	High dose No./male
A[a]	F1	2.63	0.07	32.5***	1.29**
	F2	2.13	0.05	72.15***	3.14***
B[b]	F1	0.00	0.00	0.83	0.01
	F2	2.26	0.05	2.26	0.05
	F3	0.00	0.00	0.00	0.00

D. Areolae reduction in female rats

Study code	Gen.	Control % Females	Control No./female	High dose % Females	High dose No./female
A (TP)[e]	F1	100.0	12.03	30.0***	0.53****
B (TB)[f]	F1 (total areolae)		12.0		1.0**
	F1 (normal areolae)		12.0		0.0**

Values represent incidence of nipples and/or areolae on pnd 13: * = $p < 0.05$, ** = $p < 0.01$, *** = $p < 0.001$, **** = $p < 0.0001$.
[a]An antiandrogen (Tyl, unpublished observations).
[b]Bisphenol A [9].
[c]L = linuron [7].
[d]Flutamide [6]
[e]Testosterone proprionate [5].
[f]17ß-Trenbolone [8].

4. *Weanling necropsy*: As currently specified in both OECD and OPPTS guidelines, limited organ weights (brain, spleen, thymus) are collected from only 1 pup/sex/litter. In the OPPTS guideline, up to 3/sex/litter are necropsied, but organ weights are taken for only 1 pup/sex/litter. In my laboratory, we weigh organs from all 3/sex/litter; it does not increase the power for statistical analysis since the number of litters is the same, but it better characterizes the parameters per litter. We also recommend weighing ovaries, uterus with cervical vagina, testes, epididymides, seminal vesicles, and liver (and possibly kidneys and adrenals), especially if these are possible target organs.
5. *Acquisition of puberty (in both sexes; Table 3)*: In females, it is indicated by vaginal opening or patency (VP); in males, it is indicated by preputial separation (PPS; balanopreputial separation). Within one or a few days post VP, the female exhibits her first estrus, so age at first estrus (absolute age and/or interval from VP to first estrus) is also useful. VP is dependent on 17ß-estradiol, and PPS is dependent on testosterone (T). Less useful in males is testes descent (into the scrotal sacs from the abdominal cavity through the inguinal canal and ring), which occurs during lactation (pnd 15–20) and may be mediated by T and/or DHT. In control CD® (SD) rats in the author's laboratory, the grand mean age at VP is 31.1 days and 41.9 days at PPS (based on 20 studies from 1996 to 2002). Acquisition of puberty in both sexes is affected by body weight, so the current approach is to covary the age at acquisition by the body weight at acquisition (so that pups are at equivalent physiological states, regardless of age), or by an arbitrary calendar date or age, preferably an age that is during the time of acquisition of puberty (so that pups are at equivalent ages, regardless of physiological state). Other suggested body weights as covariates include those at weaning, birth (the last is least useful in the author's opinion; it also requires that the pups are individually identified at birth), or by some measure of weight gain during the postlactational,

Table 3 Acquisition of puberty in rats.

Study code	Gen.	Acquisition of vaginal patency (VP)				Acquisition of preputial separation (PPS)			
		Control		High dose		Control		High dose	
		BW (g)	VP age (days)	BW (g)	VP age (days)	BW (g)	PPS age (days)	BW (g)	PPS age (days)
A	F1	–	32.1	–	31.2[NS,a]	–	42.2	–	41.7[NS]
B	F1	–	31.6	–	31.1[NS]	–	42.4	–	44.0[NS]
C	F1	103.12	30.0	96.87	30.05[NS]	223.65	41.9	219.59	42.70**[,b]
D	F1	99.52	32.5	91.97††[,c]	36.0**	210.72	43.6	184.83††	47.8**
	F2	104.24	31.7	88.43††	33.8**	201.98	41.2	191.67	44.9**
E	F1	97.83	30.5	91.91	31.8**	220.07	43.1	207.01	44.7**
	F2	97.03	30.6	92.51	31.3**	202.75	42.2	197.74	43.6**
Fd	F1	102.52	30.5	92.32††	33.0***	215.70	41.9	194.02†††	45.8***
	F2	105.04	31.0	102.50	34.5***	219.74	42.1	200.13††	47.9***
	F3	105.59	31.3	99.04	33.8***	209.33	42.1	186.76†††	45.2***
G	F1	110.03	29.9	112.02	30.5[NS]	235.5	41.5	236.8	41.8[NS]
H	F1	116.19	32.3	102.26	33.5**	214.49	41.1	209.51	44.8**
I	F1	104.60	30.6	93.30††	32.03***	213.56	41.6	198.76††	43.7***
J	F1	–	31.15	–	35.04***	–	41.31	–	44.61***
K	F1	110.88	31.3	96.91††	32.0***	206.86	40.9	205.35	43.6***
L (AA)	F1	108.25	31.4	106.30	34.1***	208.17	40.9	207.63	45.2***
M (E2)	F1	104.16	31.2	53.94***	24.8***	212.04	41.7	241.04***	48.7***
N (F)[e]	F1			NOT DONE				NOT DONE[f]	
O (TP)[g]	F1		34.74		34.05		43.62		44.28

[a]NS = Not statistically significant (by ANOVA and pairwise tests); Study codes A through M from the author's laboratory (AA = antiandrogen, E2 = 17ß-estradiol).
[b]** = $p < 0.01$; *** = p < 0.001; by ANOVA or ANCOVA (body weight at acquisition is covariate).
[c]†† = p < 0.01; ††† = $p < 0.001$; by Dunnett's test.
[d]Bisphenol A [9].
[e]Flutamide [6].
[f]100 % of the F1 male offspring at the high dose exhibited hypospadias, so PPS could not be determined [6].
[g] Testosterone propionate [5].

prepubertal period (the selection of the end date for weight gain is problematic). Small delays in acquisition (≤3 days) may only indicate body weight-related delays in development from systemic toxicity. Large changes in delays (≥5 days) and any acceleration more likely indicate effects from endocrine disruption.

6. *Reproductive organ weights (in both sexes at adulthood; Table 4):* These should be presented as absolute and relative to terminal body weight and should include: (a) ovaries with oviducts and uterus with cervix and vagina for females; (b) testes, epididymides, prostate (whole, and dorsolateral and ventral lobes separately; dissection should be postfixation, if possible), seminal vesicles, coagulating glands, preputial glands, bulbourethral (Cowper's) glands, and levator ani/bulbocavernosus (LABC) complex for males; and (c) systemic organs in both sexes (e.g., liver, thyroid, adrenal glands, pituitary, brain [regions]). Relative weights will correct for effects on body weights (i.e., systemic toxicity). Reproductive organ weights (testes and epididymides and ovaries and uterus) can also be collected in the weanling animals selected for necropsy.

Table 4 Reproductive organ weights in male rats.

Study code	Gen.	Organ		Control	High dose
MALES					
A (AA)	F0	Paired testes:	A	3.447 ± 0.058	3.570 ± 0.064
			R	0.578 ± 0.010	0.611 ± 0.013
		Paired epididymides:	A	1.424 ± 0.019	1.424 ± 0.008
			R	0.238 ± 0.004	0.244 ± 0.005
		Prostate:	A	0.846 ± 0.034	0.821 ± 0.039
			R	0.142 ± 0.006	0.042 ± 0.005
		SV/CG:	A	2.362 ± 0.059	2.303 ± 0.063
			R	0.396 ± 0.010	0.396 ± 0.003
	F1	Paired testes:	A	3.598 ± 0.050	2.858 ± 0.179***
			R	0.600 ± 0.010	0.521 ± 0.030
		Paired epididymides:	A	1.351 ± 0.028	1.208 ± 0.054*
			R	0.226 ± 0.006	0.217 ± 0.009
		Prostate:	A	0.756 ± 0.037	0.563 ± 0.029***
			R	0.126 ± 0.006	0.103 ± 0.005**
		SV/CG:	A	2.145 ± 0.052	1.752 ± 0.094***
			R	0.358 ± 0.009	0.316 ± 0.016
B (TP)	F1	LABC (mg):	A	162.9 ± 9.1	512.8 ± 67.2*
		V. prostate (mg)	A	16.6 ± 3.0	206.7 ± 21.3*
		SV/CG (mg)	A	46.5 ± 3.0	776.5 ± 116.2*
C(TB)	F1	LABC (mg):	A	173.7 ± 9.3	458.7 ± 13.9*
		V. prostate (mg)	A	19.3 ± 0.6	38.7 ± 4.2*
		SV/CG (mg)	A	39.3 ± 4.9	90.9 ± 16.4*
D	F0	Paired testes:	A	3.48 ± 0.06	3.43 ± 0.04
			R	0.63 ± 0.01	0.80 ± 0.01***
		Paired epididymides:	A	1.46 ± 0.03	1.36 ± 0.02
			R	0.27 ± 0.01	0.32 ± 0.01***
		Prostate:	A	1.05 ± 0.06	0.73 ± 0.04***
			R	0.19 ± 0.01	0.17 ± 0.01
		SV/CG:	A	2.24 ± 0.08	1.79 ± 0.07***
			R	0.41 ± 0.01	0.41 ± 0.02
E	F1	Paired testes:	A	3.72 ± 0.05	3.24 ± 0.005***
			R	0.66 ± 0.01	0.78 ± 0.02***
		Paired epididymides:	A	1.45 ± 0.02	1.30 ± 0.02***
			R	0.26 ± 0.01	0.31 ± 0.01***
		Prostate:	A	0.73 ± 0.04	0.58 ± 0.02*
			R	0.13 ± 0.01	0.14 ± 0.01
		SV/CG:	A	2.19 ± 0.07	1.82 ± 0.07***
			R	0.39 ± 0.01	0.44 ± 0.02*
F	F2	Paired testes:	A	3.71 ± 0.05	3.27 ± 0.05***
			R	0.63 ± 0.01	0.78 ± 0.01***
		Paired epididymides:	A	1.45 ± 0.03	1.31 ± 0.02***
			R	0.25 ± 0.01	0.31 ± 0.00***
		Prostate:	A	0.65 ± 0.03	0.49 ± 0.02***
			R	0.11 ± 0.01	0.12 ± 0.01
		SV/CG:	A	2.29 ± 0.07	1.76 ± 0.07***
			R	0.39 ± 0.01	0.42 ± 0.02

(*continues on next page*)

Table 4 (*Continued*).

Study code	Gen.	Organ		Control	High dose
MALES					
G	F3	Paired testes:	A	3.65 ± 0.04	3.19 ± 0.07***
			R	0.23 ± 0.01	0.87 ± 0.02***
		Paired epididymides:	A	1.34 ± 0.02	1.20 ± 0.02***
			R	0.27 ± 0.01	0.33 ± 0.01***
		Prostate:	A	0.56 ± 0.02	0.42 ± 0.02***
			R	0.11 ± 0.00	0.11 ± 0.00
		SV/CG:	A	1.79 ± 0.01	1.43 ± 0.05***
			R	0.36 ± 0.01	0.39 ± 0.01
H (CA)	Adult exposure	Testes		3.2 ± 0.1	3.0 ± 0.1
		Epididymides		1.16 ± 0.02	0.87 ± 0.02*
		ASG		0.587 ± 0.022	0.227 ± 0.017*
		SV		0.412 ± 0.017	0.139 ± 0.009**
		Prostate		0.172 ± 0.010	0.087 ± 0.009
I (DBP)	Adult exposure	Testes		3.3 ± 0.1	3.2 ± 0.1
		Epididymides		1.02 ± 0.03	1.09 ± 0.02
		ASG		0.532 ± 0.016	0.558 ± 0.018
		SV		0.392 ± 0.011	0.417 ± 0.016
		Prostate		0.143 ± 0.006	0.144 ± 0.005
J (F)	Adult exposure	Testes		3.3 ± 0.1	3.3 ± 0.1
		Epididymides		1.26 ± 0.03	0.79 ± 0.03*
		ASG		0.588 ± 0.021	0.200 ± 0.01*
		SV		0.390 ± 0.025	0.119 ± 0.009*
		Prostate		0.190 ± 0.013	0.082 ± 0.007*
K (L)	Adult exposure	Testes		3.3 ± 0.1	3.3 ± 0.1
		Epididymides		1.16 ± 0.03	1.05 ± 0.03*
		ASG		0.597 ± 0.016	0.532 ± 0.026*
		SV		0.443 ± 0.014	0.407 ± 0.020
		Prostate		0.153 ± 0.008	0.124 ± 0.007*
L (V)	Adult exposure	Testes		3.1 ± 0.1	3.3 ± 0.0
		Epididymides		1.12 ± 0.03	1.02 ± 0.02**
		ASG		0.562 ± 0.015	0.474 ± 0.024*
		SV		0.387 ± 0.013	0.334 ± 0.019
		Prostate		0.172 ± 0.007	0.138 ± 0.008*
M (F)	Adult exposure	Testes		3.3 ± 0.1	3.4 ± 0.1
		Epididymides		1.16 ± 0.02	1.17 ± 0.02
		ASG		0.596 ± 0.016	0.526 ± 0.018*
		SV		0.437 ± 0.012	0.372 ± 0.014*
		Prostate		0.157 ± 0.008	0.152 ± 0.013

(*continues on next page*)

Table 4 (*Continued*).

Study code	Gen.	Organ	Control	High dose
MALES				
N (K)	Adult exposure	Testes	3.4 ± 0.1	3.3 ± 0.1
		Epididymides	1.22 ± 0.02	1.11 ± 0.05*
		ASG	0.578 ± 0.020	0.475 ± 0.026*
		SV	0.410 ± 0.015	0.357 ± 0.020*
		Prostate	0.166 ± 0.009	0.119 ± 0.010*

AA = antiandrogen (Tyl, unpublished observations).
TP = testosterone propionate [5].
TB = 17ß-trenbolone [8].
BPA = bisphenol A [9].
CA = cyproterone acetate.
DBP = di-*n*-butyl phthalate [10].
Fl = flutamide [10].
L = linuron [10].
V = vinclozolin [10].
Fa = fadrazole; aromatase inhibitor [11].
K = ketaconazole; testosterone biosynthesis inhibitor [11].
*, **, *** = $p < 0.05$, 0.01, 0.001 vs. control group value.
SV/CG = seminal vesicles/coagulating glands.
ASG = accessory sex glands.
LABC = levator ani bulbocavernosus complex.
A, D–G = multigeneration studies with exposure in the feed; exposures to F0 animals began at puberty, exposure to F1, F2, and F3 animals began as gametes (F3 animals not bred).
B and C = immature castrated male rats were dosed for 8–13 days by subcutaneous injection.
H–N = 15-day exposure by oral gavage to adult intact male rats.
A = absolute organ weight in grams (g).
R = relative organ weight as percentage of terminal body weight.

7. *Gross Necropsy Observations (Fig. 5)*: Careful dissections under a dissecting microscope by experienced technical staff can identify alterations (e.g., size, shape, components) in male and female reproductive systems on gestation day (gd) 20 and pnd 4 (ovaries, oviducts, uterus, cervix and vagina, cranial suspensory ligament in females; testes, epididymides [caput, corpus, cauda], afferent and efferent ducts [vas deferens], seminal vesicles, gubernaculum in males), on pnd 21 (same plus male coagulating glands, prostate, bulbourethral glands), and as adults (same plus male preputial glands, LABC complex, penis, prepuce [foreskin]), plus functional assessments (see item 9).
8. *Histopathology*: Organs with gross lesions, changes in absolute and relative weight, or identified as target organs from the list in item 7 above. Systemic organs should not be ignored, as effects on reproduction or development may be mediated by systemic toxicity.
9. *Andrology*: Includes epididymal (cauda) sperm number, motility (and progressive motility), morphology, and testicular homogenization-resistant spermatid head counts (SHC) to calculate daily sperm production (DSP) and efficiency of DSP. Epididymal sperm motility must be done immediately after demise (within 2 min in the author's laboratory); the other endpoints can be evaluated at a later date. The age of the male at demise is critical. With an average duration of spermatogenesis in rats of approximately 70 days and of epididymal transit time in rats of 10 to 14 days, there are essentially no sperm in the epididymides in a 60-day old rat, very little in the cauda of a 70-day old rat, and close to adult values in rats at and above 85 days of age.

A.

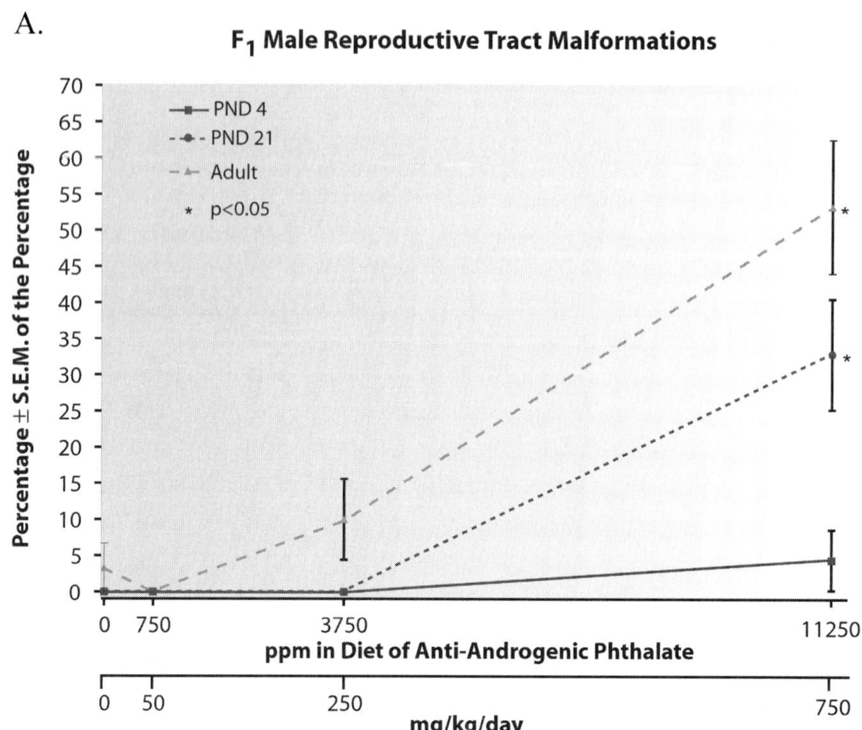

B.

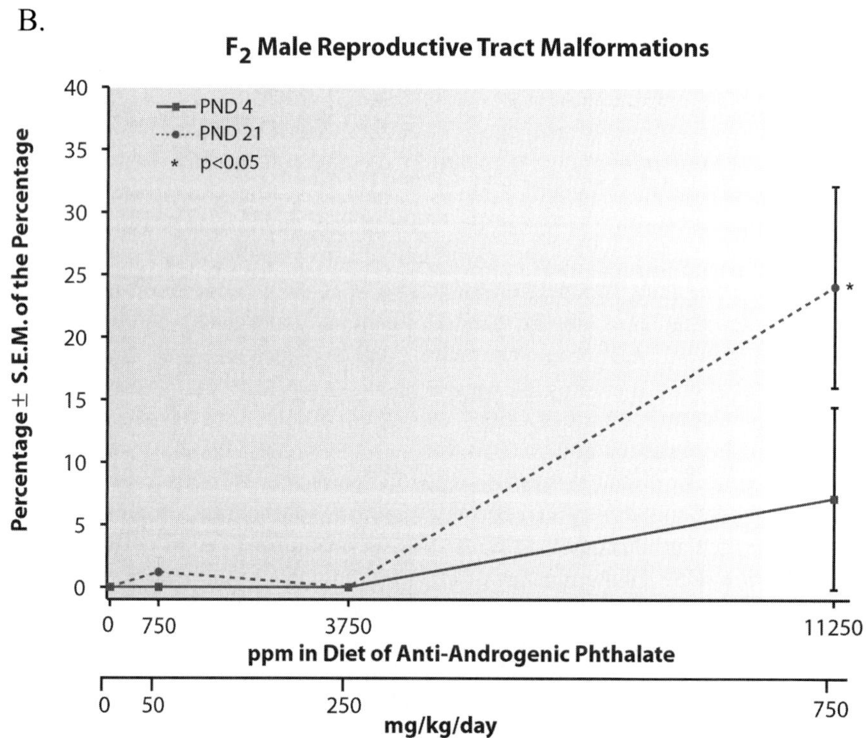

Fig. 5 Male Reproductive Tract Malformations by Gross Necropsy From an Antiandrogen (Tyl, unpublished). A. F1 males at pnd 4 and 21, and as adults; B. F2 males at pnd 4 and 21.

10. *Other*: Such as estrous cyclicity (evaluated daily for at least 2 weeks; 3 weeks is better), precoital interval, gestational length, and circulating hormone concentrations (basal and after challenge); the last is best done by longitudinal evaluations (in satellite groups of cannulated rats), etc.
11. Current regulatory guidelines specify selection of one F1 pup/sex/litter at weaning to be retained to adulthood for breeding to generate F2 offspring. Will we miss subtle lesions or effects at low incidence? Should more pups per litter be retained postwean? This will not improve statistical power since that is based on the number of litters/group, but it will improve the characterization and sensitivity of parameters for each litter. This concern is being evaluated under the USEPA EDSP contract.

Examples of inappropriate endpoints (the author's current list) are:

1. *Ovarian primordial follicle counts (Table 5)*: As currently done, this endpoint is too "noisy" (insensitive), with large variance terms. Interpretation is also a problem. A better approach would be to quantify all stages of oocyte development: primordial, small antral, large antral, and mature (Graäfian) follicles. This is very labor intensive (and, therefore, expensive), but it provides a complete assessment of oogenesis. Therefore, it should be done (in my opinion) only if other endpoints indicate possible ovarian toxicity (e.g., changes in absolute and relative ovarian weight, total implants per litter).

Table 5 Ovarian primordial follicle counts.

Study code	Gen.	Control	High dose
A	F0	140.7 ± 11.6[a]	146.1 ± 10.8
	F1	134.0 ± 8.8	135.3 ± 9.4
B	F0	132.3	114.4
	F1	150.0	148.75
C (BPA)	F0	315.9 ± 41.6	453.2 ± 26.3*
	F1	353.0 ± 35.4	409.7 ± 46.8
	F2	409.2 ± 32.7	378.0 ± 25.5
	F3	384.6 ± 55.7	355.4 ± 38.3
D	F0	242.1 ± 11.6	216.2 ± 14.1
	F1	182.2 ± 14.9	164.5 ± 9.1
E	F0	351.6 ± 38.6	405.3 ± 45.7
	F1	391.2 ± 41.4	385.5 ± 57.1
F	F0	281.3 ± 24.9	321.1 ± 51.2
	F1	357.9 ± 38.7	368.8 ± 30.0
G (AA)	F0	281.1 ± 34.1	272.9 ± 35.1
	F1	368.4 ± 26.3	414.9 ± 56.5

[a]Data presented as mean ± SEM.
* = $p < 0.05$ (in the presence of reduced absolute and relative ovarian weights).
C (BPA) = bisphenol A [9].
G (AA) = antiandrogen (Tyl, unpublished observations).
A – G is from the author's laboratory.

2. *Stage of estrus at scheduled necropsy (Table 6)*: This is a compromise endpoint; earlier draft versions of the current USEPA OPPTS reproductive toxicity testing guidelines specified that the parental females were all to be necropsied at the same stage of estrus. This endpoint, as currently specified in the finalized guidelines (1998), is considered insensitive. This is also very difficult to analyze statistically. In my laboratory, we analyze the percentage of females in each estrous stage for each group. Some laboratories only identify three stages (proestrus [P], estrus [E], and diestrous [D]); other laboratories also identify metestrus (M) for a total of four stages. Females iden-

tified as in transition (e.g., P/E, E/M, M/D) in my laboratory are included in the earlier stage (e.g., P, E, M); otherwise, the statistical analyses are close to impossible. The specified 3-week premating evaluation of estrous cyclicity is much more informative.

Table 6 Stage of estrus at scheduled necropsy and in-life estrus cycle length[a,b].

	Bisphenol A (ppm in the feed)						
	0	0.015	0.3	4.5	75	750	7500
A. F0 Females							
No. females evaluated	30	29	30	30	29	29	29
No. in proestrus	$11^{\pounds\pounds}$	4^{Φ}	10	3^{Φ}	5	3^{Φ}	$1^{\Phi\Phi}$
% in proestrus	$40.74^{\Psi\Psi\Psi}$	13.79	35.71	10.34	18.52	10.34	3.70
No. in estrus	4	4	5	9	5	6	0
% in estrus	14.81	13.79	17.86	31.03	18.52	20.69	0.00
No. in metestrus	3	4	3	1	3	0	1
% in metestrus	11.11	13.79	10.71	3.45	11.11	0.00	3.70
No. in diestrus	$9^{\pounds\pounds}$	17	10	16	14	20^{Φ}	$25^{\Phi\Phi\Phi}$
% in diestrus	$33.33^{\Psi\Psi\Psi}$	58.62	35.71	55.17	51.85	68.97	92.59
Estrous cycle length (days)	4.58 ± 0.25	4.41 ± 0.09	4.48 ± 0.20	4.50 ± 0.11	4.57 ± 0.14	4.45 ± 0.18	4.26 ± 0.09
B. F1 Females							
No. females evaluated	27	30	29	30	28	30	28
No. in proestrus	5	8	3	3	5	4	4
% in proestrus	20.00	26.67	10.34	10.34	18.52	14.29	15.38
No. in estrus	0	5	4	4	2	2	0
% in estrus	0.00	16.67	13.79	13.79	7.41	7.14	0.00
No. in metestrus	2	0	2	0	0	0	0
% in metestrus	8.00	0.00	6.90	0.00	0.00	0.00	0.00
No. in diestrus	18	17	20	22	20	22	22
% in diestrus	72.00^{Ψ}	56.67	68.97	75.86	74.07	78.57	84.62
Estrous cycle length (days)	4.41 ± 0.10	4.47 ± 0.13	4.19 ± 0.09	4.70 ± 0.23	4.94 ± 0.21	4.40 ± 0.18	4.54 ± 0.11
C. F2 Females							
No. females evaluated	30	30	29	29	30	30	28
No. in proestrus	2	6	1	3	4	3	0
% in proestrus	7.14	20.00	3.45	10.34	13.33	10.71	0.00
No. in estrus	$8^{\pounds\pounds}$	6	9	5	13	11	0^{Φ}
% in estrus	28.57	20.00	31.03	17.24	43.33	39.29	0.00
No. in metestrus	3	1	4	3	2	0	1
% in metestrus	10.71	3.33	13.79	10.34	6.67	0.00	3.70
No. in diestrus	$15^{\pounds\pounds}$	17	15	18	11	14	$26^{\Phi\Phi}$
% in diestrus	53.57	56.67	51.72	62.07	36.67	50.00	96.30
Estrous cycle length (days)	4.54 ± 0.21	4.61 ± 0.24	4.39 ± 0.14	4.47 ± 0.23	4.17 ± 0.07	4.56 ± 0.24	4.56 ± 0.11
D. F3 Females[c]							
No. females evaluated	30	30	30	29	30	30	30
No. in proestrus	3	2	4	3	5	9	5
% in proestrus	10.34^{Ψ}	6.67	13.33	10.34	16.67	31.03	16.67
No. in estrus	9	12	12	13	15	4	10
% in estrus	31.03	40.00	40.00	44.83	50.00	13.79	33.33
No. in metestrus	2	1	4	3	2	3	3
% in metestrus	6.90	3.33	13.33	10.34	6.67	10.34	10.00
No. in diestrus	15	15	10	10	8	13	12
% in diestrus	51.72	50.00	33.33	34.48	26.67	44.83	40.00
Estrous cycle length (days)	4.32 ± 0.18	4.34 ± 0.12	4.32 ± 0.09	4.39 ± 0.18	4.66 ± 0.21	4.59 ± 0.21	4.31 ± 0.09

[a]Stage of estrus (Tyl, unpublished observations); in-life estrus cycle length [9].

(*continues on next page*)

Table 6 (*Continued*).

[b]For presentation and statistical analysis purposes those females in two stages were pooled in the following manner: proestrus/estrus was considered proestrus; estrus/metestrus and estrus/diestrus were considered estrus; metestrus/diestrus was considered metestrus; and diestrus/proestrus was considered diestrus. The females for which the stage could not be determined or no cells were present were not included in the statistical analysis.
[c]F3 females were not mated prior to scheduled necropsy as adults.
[££]$p < 0.01$; Chi-square test.
[Φ]$p < 0.05$; Fisher Exact test.
[ΦΦ]$p < 0.01$; Fisher Exact test.
[ΦΦΦ]$p < 0.001$ Fisher Exact test.
[Ψ]$p < 0.05$; Cochran–Armitage test.
[ΨΨ]$p < 0.01$; Cochran–Armitage test.

3. *Single sample of circulating hormone concentrations*: Due to the pulsatile nature of circulating hormone levels and complex positive and negative feedback loops, a single blood sample from adult animals at necropsy (i.e., a "single snapshot in time") is considered insensitive. It would only detect profound changes in the animals that would also be detectable by results in other parameters (see above for suggestions for longitudinal evaluation).

CONCLUSIONS

1. Regulatory guidelines represent the minimum requirements. There is no reason not to enhance them if and as appropriate.
2. Regulatory guidelines will change over time as the science improves and new societal, regulatory, and scientific concerns arise. The sooner we begin discussions on how to improve them, the sooner the regulatory agencies will consider amending them.
3. Basic scientists will continue to provide new endpoints, mechanisms, and approaches for consideration of inclusion into the guidelines. Applied scientists will continue to identify new needs and to refine parameters from the basic researcher. This interaction between basic and applied science (between specific and apical investigations) must be encouraged, supported, and valued.
4. The objectives in performing these studies are to perform sound science and to provide useful information to the regulatory agencies for risk assessment. The better the studies, the better the risk assessment. My last recommendation is for the regulatory agencies to periodically convene workshops with basic (academic, governmental) scientists, industrial scientists, contract performing laboratories, and regulators to discuss the rapid increase in information (e.g., target tissues, critical life phases of exposure, endpoints, interpretation), and the assessment of current and proposed endpoints. My hope is that this will lead to periodic revision and improvement of the regulatory testing guidelines.

ACKNOWLEDGMENTS

The author wishes to thank Ms. Christina Myers (supervisor for reproductive toxicity studies and data analyst), Ms. Melissa Marr (laboratory supervisor), and her reproductive and developmental toxicology staff at RTI International who did the work and made this presentation possible. Thanks to Ms. Cathee Winkie (Dr. Tyl's administrative coordinator) for her patient typing (and retyping…) of the presentation, poster, and manuscript. Thanks to Ms. Christina Myers and Ms. Cathee Winkie for their assistance in generating the tables and figures.

REFERENCES

1. U.S. Environmental Protection Agency (EPA). Toxic Substances Control Act Test Guidelines; Final Rule, Section 798.4700, reproduction and fertility effects, *Federal Register* **50** (188), 43834–43838 (Friday, 27 September 1985).
2. U.S. Environmental Protection Agency (EPA). 40CFR Part 799, Toxic Substances Control Act Test Guidelines; Final Rule, Section 799.9380, TSCA reproduction and fertility effects, *Federal Register* **62** (158), 43834–43838 (Friday, 15 August 1997).
3. U.S. Environmental Protection Agency (EPA). Office of Prevention, Pesticides and Toxic Substances (OPPTS), Health Effects Testing Guidelines, OPPTS 870.3800, reproduction and fertility effects (Final Guideline, August 1998).
4. Organization for Economic Cooperation and Development, *OECD Guideline for the Testing of Chemicals; Proposal for Updating Guideline 414: Prenatal Developmental Toxicity Study*, pp. 1–11 (adopted 22 January 2001).
5. U.S. Food and Drug Administration (FDA). "Redbook 2000", Toxicological Principles for the Safety of Food Ingredients, Center for Food Safety and Applied Nutrition, Office of Premarket Approval (CFSAN, OPA) 20 July 2000.
6. C. J. Wolf, A. Hotchkiss, J. S. Ostby, G. A. LeBlanc, L. E. Gray, Jr. *Toxicol. Sci.* **65** (1), 71–86 (2002).
7. B. S. McIntyre, N. J. Barlow, P. M. D. Foster. *Toxicol. Sci.* **62**, 236–249 (2001).
8. B. S. McIntyre, N. J. Barlow, P. M. D. Foster. *Toxicol. Sci.* **65** (1), 62–70 (2002).
9. V. S. Wilson, C. Lambright, J. Ostby, L. E. Gray, Jr. *Toxicol. Sci.* **70** (2), 202–211 (2002).
10. R. W. Tyl, C. B. Myers, M. C. Marr, B. F. Thomas, A. R. Keimowitz, D. R. Brine, M. M. Veselica, P. A. Fail, T. Y. Chang, J. C. Seely, R. L. Joiner, J. H. Butala, S. S. Dimond, S. Z. Cagen, R. N. Shiotsuka, G. D. Stropp, J. M. Waechter. *Toxicol. Sci.* **68** (1), 121–146 (2002).
11. J. C. O'Connor, S. R. Frame, G. S. Ladics. *Toxicol. Sci.* **69** (1), 92–108 (2002a).
12. J. C. O'Connor, S. R. Frame, G. S. Ladics. *Toxicol. Sci.* **69** (1), 79–91 (2002b).

Pure Appl. Chem., Vol. 75, Nos. 11–12, pp. 2099–2123, 2003.

Topic 3.10

Critical evaluation of observed adverse effects of endocrine active substances on reproduction and development, the immune system, and the nervous system*

John C. O'Connor[‡,1] and Robert E. Chapin[2]

[1]DuPont Haskell Laboratory for Health and Environmental Sciences, P.O. Box 50, Elkton Road, Newark, DE 19714, USA; [2]Pfizer, Inc., Drug Safety Evaluation, MS 8274-1336, Eastern Point Road, Groton, CT 06340-8014, USA

Abstract: The last 40 years have seen many reports that man-made chemicals and environmental pollutants cause adverse effects in humans and wildlife; however, actually linking an exposure with a mechanism and an effect has yet to be done for endocrine disruption. Certainly, studies in experimental animals have shown that sufficient doses of select compounds can disrupt the endocrine system and produce the attendant adverse outcomes. The purpose of this contribution is to evaluate some of the recent reports of the adverse effects on reproduction and development, the immune system, and the nervous system that have been observed in experimental animals after treatment with man-made chemicals and environmental pollutants. Space limitations prevent us from presenting a comprehensive review of all reported endocrine active chemicals and their effects. Instead, we have focused on drawing conclusions as to the scope and etiology of the adverse effects in experimental animals using examples from the scientific literature, and on suggesting a path forward for further work.

INTRODUCTION

The purpose of this contribution is to evaluate the adverse effects on reproduction and development, the immune system, and the nervous system that have been observed in experimental animals after treatment with man-made chemicals and environmental pollutants. Space limitations prevent a comprehensive review of all reported endocrine active chemicals (EACs) and their effects; we have focused on drawing conclusions as to the scope and etiology of the adverse effects in experimental animals, and on suggesting a path forward for further work.

REGULATION OF REPRODUCTION AND DEVELOPMENT IN MAMMALS

The endocrine and nervous systems are the major mechanism(s) by which the body communicates information between cells and/or organ systems, and both are critical for the regulation of growth and development, reproduction, and maintaining metabolic processes. The endocrine system is a highly com-

*Report from a SCOPE/IUPAC project: Implication of Endocrine Active Substances for Human and Wildlife (J. Miyamoto and J. Burger, editors). Other reports are published in this issue, *Pure Appl. Chem.* **75**, 1617–2615 (2003).

‡Corresponding author: Tel: (302) 366-6169; Fax: (302) 366-5003; E-mail: john.c.oconnor@usa.dupont.com

plex and integrated system of glands that secrete hormones into the circulatory system, ultimately regulating the function of specific target tissues/organs. The main components of the endocrine system are the hypothalamus, pituitary, and a variety of endocrine glands (e.g., testis, ovary, thyroid, adrenals, pancreas) that each participates in regulating numerous physiological processes. The inherent design of the endocrine system allows the body to react to acute changes in homeostasis through the positive and negative feedback loops that control hormone production and release. However, the complexity of the endocrine system also provides many potential sites for endocrine disruption (ED) to occur. Figure 1 summarizes the basic regulation of the reproductive axes in males and females, and illustrates several potential sites of ED. Examples of potential mechanisms of ED include [1,2]:

- alterations in receptor-mediated signaling (e.g., agonism and antagonism);
- alterations in hormone synthesis;
- alterations in hormone storage and/or release;
- alterations in hormone transport;
- alterations in hormone metabolism; and
- alterations in post-receptor activation.

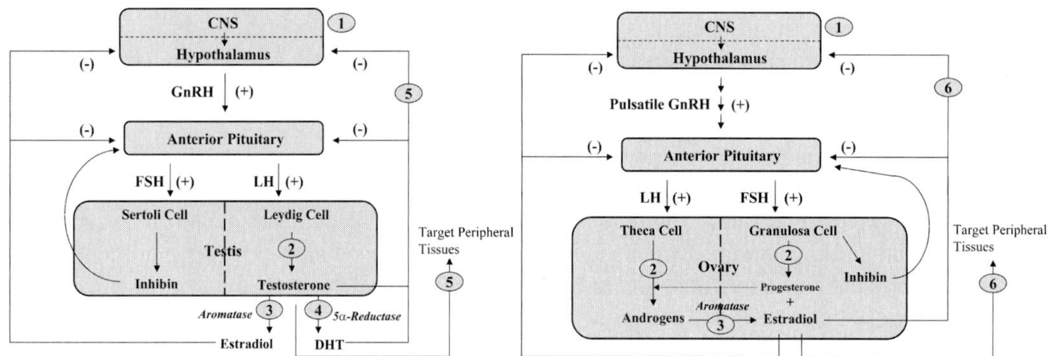

Fig. 1 *Regulation of the hypothalamic-pituitary-gonadal axis in male (a) and female (b) mammals.* Abbreviations: CNS: central nervous system; GnRH: gonadotropin releasing hormone; FSH: follicle-stimulating hormone; LH: luteinizing hormone; DHT: dihydrotestosterone. Potential sites of ED include: (1)dopamine agonists would act on the CNS to affect GnRH release; (2)steroid biosynthesis inhibitors would inhibit testosterone production, reducing the amount of testosterone and perhaps DHT or estradiol; (3) aromatase inhibitors would inhibit the conversion of testosterone to estradiol; (4) 5α-reductase inhibitors would inhibit the conversion of testosterone to DHT; (5) androgen receptor blockers would interfere with the normal androgen feedback to the pituitary and brain, as well as decreasing androgen action peripherally; (6) estrogen receptor blockers would interfere with the normal estrogen feedback to the pituitary and brain, as well as decreasing estrogen action peripherally. Inhibin has not yet been rigorously evaluated in environmental toxiciology.

Unfortunately, for most compounds, the specific mechanism of action is unknown, or is confounded by the ability of the substance to affect multiple sites of endocrine control (e.g., binding to more than one receptor). Although not all-inclusive, Table 1 summarizes some of the EACs that have been identified in experimental animals.

Although adult animals are susceptible to ED (as described below), the developing fetus is uniquely sensitive to alterations in endocrine status, and the spectrum of effects that are observed after in utero exposure to EACs is a reflection of the complexities of reproductive development and differentiation. For this reason, selected aspects of mammalian differentiation will be summarized here; many thorough reviews are available on the subject of mammalian differentiation, to which the reader is referred [3,4].

Table 1 Examples of known or suspected endocrine active substances.

ER agonists
DDT (*o,p'*-DDE)
Methoxychlor
Chlordecone
Bisphenol A
PCBs
Endosulfan
Dieldrin
Dicofol
Chlordane
Toxaphene
Lindane
Butyl benzyl phthalate (BBP)
Alkylphenols (nonylphenol, octylphenol)
Endogenous estrogens (estradiol, estrone)
Pharmaceuticals (DES, ethiynl estradiol)
Phytoestrogens (coumestrol, genistein)
Mycoestrogens (zearalenone)
ER antagonists
ICI-182,780
ICI-182,164
AR antagonists
Vinclozolin
(DDT) *p,p'*-DDE
PCBs
Linuron
Cyproterone acetate
Procymidone
Steroid biosynthesis inhibitors
Finasteride
Fenarimol
Exemestane
Ketoconazole
Di-*n*-butyl phthalate (antiandrogen-like)
Other
Dioxin (TCDD, Ah receptor agonist)

In rats, development of the endocrine system starts at approximately gestation day 8 with the first steps in the differentiation of the bipotential gonad into the testis or the ovary (Fig. 2). The activation of the *Sry* gene on the Y chromosome triggers a cascade of events that result in the development of the male phenotypic traits, whereas in the absence of the *Sry* gene, the embryo develops into a female. In this respect, the female phenotype is considered the "default" pathway for reproductive development in mammals. Activation of *Sry* induces the differentiation of the bipotential embryonic gonads into the testes. As the testes develop, two main cell types produce hormones that ultimately drive reproductive development and differentiation. The Sertoli cells, the cells that will ultimately support the maturation of the germ cells, produce Müllerian inhibiting substance (MIS), also known as anti-Müllerian hormone (AMH), which causes regression of the Müllerian ducts. In the absence of MIS, the Müllerian ducts give rise to the female genitalia, uterus, and vagina. Concurrently, the Leydig cells of the embryonic testes secrete testosterone (T), which supports the differentiation of the Wolffian ducts, which give rise to the male epididymis, seminal vesicles, and vas deferens. Dihydrotestosterone (DHT), the major metabolite of

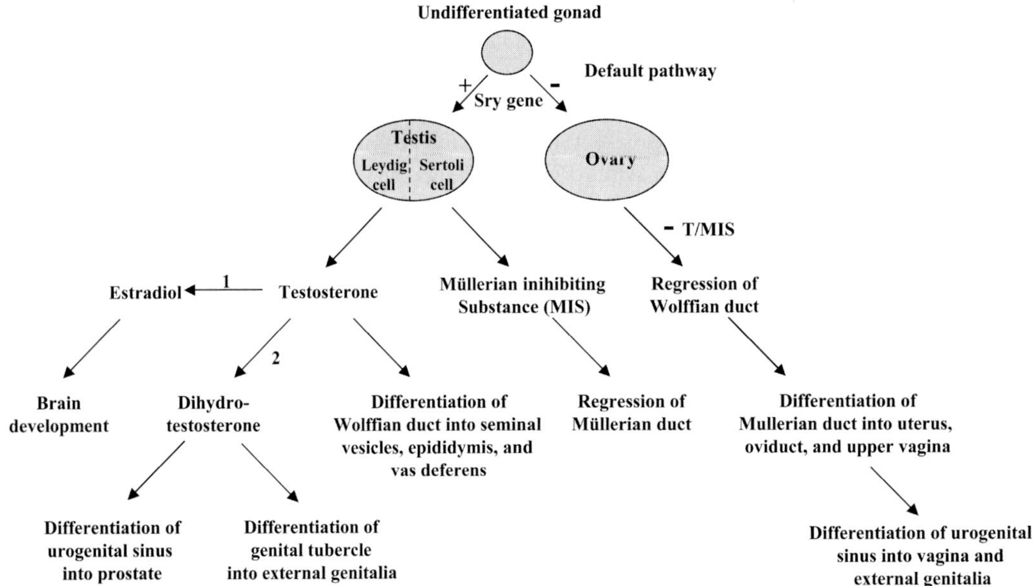

Fig. 2 *Regulation of mammalian reproductive differentiation and development.* Abbreviations: T: testosterone; MIS: Müllerian inhibiting substance; 1 (aromatase enzyme); 2 (5α-reductase enzyme).

T, induces the formation of the external genitalia and also participates in the descent of the testes into the scrotum (along with MIS). Estradiol, the aromatization product of T, is required for normal brain sexual/behavioral development (see discussion below). Each of these complex steps is under the control of androgens, and requires proper androgen signaling for normal development to occur.

In both males and females, differentiation of the reproductive organs continues throughout gestation, while reproductive maturation (e.g., masculinization of external genitalia, anogenital distance, behavioral development) continues throughout the first few weeks of postnatal life [3,4] until puberty, which is under hormonal control (i.e., androgen-dependent in males and estrogen-dependent in females). After puberty, processes such as spermatogenesis [5,6] or ovarian function remain under hormonal control [3,4]. This is a greatly simplified description of the very complex processes of mammalian sexual development, a process rife with many potential sites for ED. As described below, perturbations in the endocrine-signaling pathways during development lead to very distinct developmental abnormalities.

ADVERSE EFFECTS OF ENDOCRINE ACTIVE SUBSTANCES ON REPRODUCTION AND DEVELOPMENT

In *adult* animals, adverse effects to EACs are typically transient; that is, the effects subside if chemical treatment is withdrawn. This is a function of an endocrine system that developed normally, and that can maintain hormonal homeostasis via the built-in feedback loops in the presence of external challenges. Transient effects can include changes in weight and morphology of target organs, and alterations in reproductive capacity. For example, exposure of adult rats to high levels of endogenous estrogens such as 17β-estradiol or estrone [7–12]; synthetic estrogens such as diethylstilbestrol (DES) [13–16]; or environmental estrogens such as methoxychlor, chlordecone, and octylphenol [17–23] leads to decreased reproductive capacity in both males and females. In males, these effects are characterized by decreased reproductive organ weights and abnormal reproductive tract morphology, often accompanied by impaired spermatogenesis resulting in decreased sperm count, decreased sperm motility, and/or altered sperm morphology. In female rats, alterations in estrous cyclicity and evidence of ovarian malfunction (e.g., decreased corporea lutea, decreased ova count) are commonly reported. In both cases, reproductive capacity is compromised due to disruption of the hormone feedback loops resulting in decreased gonadotropin release from the pituitary, and ultimately altered function of the male and female gonads. Activation of estrogen receptor (ER)-mediated events likely contributes to the effects. While these examples represent compounds that act via binding to the ER, there are similar reports in adult animals exposed to EACs with other mechanisms of action such as antiandrogens [1,2], aromatase inhibitors [24,25], and testosterone biosynthesis inhibitors [26,27]; and the list of suspected EACs continues to grow. In cases where prolonged exposures occur, EACs often induce neoplasia of hormone-responsive tissues. The particular type of neoplasia is dependent upon the mechanism of action of the EAC [28].

While sustained alterations in hormonal homeostasis at any point during life can result in adverse effects as discussed above, even small transient alterations in hormonal homeostasis *during development* can be detrimental since the developing organism is uniquely sensitive to hormonal perturbations. The inherent sensitivity of the fetus is due to the reproductive and behavioral "programming" that occurs during development of the endocrine system in the fetus and neonate [3,4,29,30]. Even small perturbations in the endocrine axes during this period of development may result in permanent alterations in the way the affected cells respond to hormones at any time in the future. Thus, there may be lasting impacts on the reproductive and/or behavioral capacity of the animal. Table 2 summarizes some examples of chemicals that are known EACs in experimental animals when administered in utero, and the adverse effects that are associated with each.

Overall, male progeny of pregnant xenobiotic-treated dams seem to be more susceptible than females to perturbations in endocrine signaling during reproductive development, whereas both males and females are equally susceptible to alterations in behavioral development (see below). This is true in both the scope and severity of the effects, as well as the number of EACs that have been shown to adversely impact experimental animals. This is not surprising when one considers the series of events that is required for reproductive development of males versus females (Fig. 2). In general, any EAC with the ability to alter androgen signaling has the potential to cause adverse effects in males. Three classes of EACs appear particularly important in male reproductive development: ER agonists, androgen receptor (AR) antagonists, and arylhydrocarbon (Ah) receptor agonists [1]. Any EAC with the ability to disrupt steroidogenesis also has the potential to induce adverse effects in males. Insufficient androgen signaling during reproductive development manifest in a pseudohermaphrodite condition of the male offspring, the scope and severity dependent upon a variety of factors including the period (i.e., gestation days) and duration of exposure, the mechanism of action of the EAC, and the level of exposure (i.e., dose). Surprisingly, given the different mechanisms of action of the EACs that have been evaluated in experimental animals, the scope of effects induced by different EACs on male reproductive development are remarkably similar.

One of the most well-studied disruptors of androgen signaling in developing males is vinclozolin, which is metabolized to two chemicals that have been shown to act as AR antagonists [31,32]. The sequelae of effects induced by vinclozolin exposure include decreased (i.e., female-like) anogenital distance (AGD), delayed puberty (delayed preputial separation), presence of female reproductive tissues (e.g., vaginal pouch), decreased sperm production, and a variety of malformations of the reproductive tract, from small/atrophied to completely absent male reproductive organs [31,33–36]. These effects impair the reproductive performance and success of the affected animal. All of the effects that are observed in the male progeny are the result of insufficient androgen exposure during development as a result of blockage of the AR by vinclozolin. T- and DHT-dependent events are both adversely effected (Fig. 2), resulting in incomplete differentiation of the male reproductive tissues and/or incomplete regression of the female reproductive tissues. The results observed with vinclozolin are representative of those observed for a variety of other AR antagonists including flutamide, procymidone, linuron, and *p,p′*-DDE [31,33,36–40]. The effects are likely the result, at least in part, of the attenuated transcription of AR-mediated genes during development and/or perturbations in other signaling pathways as a result of permanent alterations in the hormone-feedback loops during reproductive imprinting [4].

Table 2 Examples of endocrine active substances and their effects on reproduction and development in experimental animals exposed in utero.

Compound (mode of action)	Exposure	Effect	Ref.
Testosterone (AR agonist)	Rats; subcutaneous injection; 0 (corn oil), 0.1, 0.5, 1, 2, 5, 10 mg/0.1 ml on GD14–19.	Males: ↓ anogenital distance; ↓ glans penis weight (persisted at adulthood). Females: ↑ anogenital distance (persisted at adulthood); delayed puberty (vaginal opening); ↓ number and/or absent areolas/nipples; presence of male reproductive tissues and/or no vaginal orifice; altered estrous cyclicity.	[84]
Vinclozolin (AR antagonist)	Rats; gastric gavage; 0 (corn oil), 100, 200 mg/kg/day on GD14–PND3.	Males: ↓ anogenital distance; retained areolas/nipples (persisted at adulthood); cleft phallus with hypospadias; ectopic/undescended testes; blind vaginal pouch; small or absent accessory sex glands; ↓ sperm concentration. Females: no effects.	[33,34]
	Rats; gastric gavage; 0 (corn oil), 10, 30, 100 mg/kg/day from age 22–56 days.	Males: delayed puberty (age of preputial separation); ↓ organ weights (accessory sex gland unit and epididymis weight); hormonal alterations (increased testosterone and luteinizing hormone).	[35]
Flutamide (AR antagonist)	Rats; gastric gavage; 0 (corn oil), 6.25, 12.5, 25, 50 mg/kg/day on GD12–21.	Males: ↓ anogenital distance (persisted at adulthood); retained nipples/areolas (persisted at adulthood); hypospadias; ectopic/undescended testes; prostate agenesis; ↓ organ weights (seminal vesicles, testes, epididymides, levator ani/bulbocavernosus); epididymal malformations.	[36]
	Rats; subcutaneous injection; 0 (ethanol/corn oil), 18, 24, 50, 75, 100, 200, 250, 300 mg/kg/day on GD12–21.	Males: severe feminization of external genitalia; absent vas deferens and prostate; malformation of epididymidis.	[37]
Procymidone (AR antagonist)	Rats; gastric gavage; 0 (corn oil), 25, 50, 100, 200 mg/kg/day on GD14–PND3.	Males: ↓ anogenital distance; hypospadias; retained areolas/nipples (persisted at adulthood); ectopic/undescended testes; vaginal pouch; ↓ organ weights (ventral prostate, levator ani/bulbocavernosus, Cowper's glands, seminal vesicles, glans penis); histological lesions in accessory sex glands.	[38]
	Rats; gastric gavage; 0 (corn oil) or 100 mg/. kg/day on GD14–PND3	Males: ↓ anogenital distance; hypospadias; retained areolas/nipples (persisted at adulthood); vaginal pouch; ↓ weights (ventral prostate and seminal vesicles); prostate agenesis.	[33]
Linuron (AR antagonist)	Rats; gastric gavage; 0 (corn oil) or 100 mg/kg/day on GD14–18.	Males: delayed puberty; hypospadias; ↓ anogenital distance; retained areolas/nipple; malformations of epididymis and testes; ↓ organ weights (testes, epididymis, seminal vesicles); agenesis of testes and epididymis.	[33]

Table 2 (*Continued*).

Compound (mode of action)	Exposure	Effect	Ref.
p,p′-DDE (AR antagonist)	Rats; gastric gavage; 0 (corn oil), 100 mg/kg/day on GD14–18 or PND25–57, or 200 mg/kg/day for 4 days (adult).	Males (GD14-18): retained areolas/nipples; ↓ anogenital distance. Males (PND25-57): delayed puberty. Males (adult): ↓ organ weights (seminal vesicle, ventral prostate).	[39]
	Rats; gastric gavage; 0 (corn oil), 10, 100 mg/kg/day on GD14–18.	Males: retained areolas/nipples, ↓ anogenital distance. Females - no effects.	[40]
	Rats; gastric gavage; 0 (corn oil) or 100 mg/kg/day on GD14–PND3.	Males: ↓ anogenital distance; hypospadias; retained areolas/nipples (persisted at adulthood); ↓ organ weights (ventral prostate, levator ani/bulbocavernosus, epididymis).	[33]
Diethylstilbestrol (ER agonist)	Mice; gastric gavage; 0 (corn oil), 0.1, 1, 10, 100 mg/kg/day on GD9-16.	Males: ↓ fertility; abnormal sperm motility and morphology; ↓ sperm number; ectopic/undescended testes; retained Mullerian duct remnants; small phallus; hypospadias; histological lesions of the reproductive tract; neoplasia of reproductive tissues at adulthood.	[15]
	Mice; subcutaneous injection; 0 (corn oil), 0.01, 1, 2.5, 5, 10, 100 μg/kg/day on GD9-16.	Females: ↓ fertility; ovarian malfunction (decreased ova count); structural abnormalities of reproductive tract (oviduct, uterus, cervix, vagina); altered estrous cyclicity; hypospadias; neoplasia of reproductive organs at adulthood.	[16]
	Mice; subcutaneous injection; 0 (corn oil), 0.01, 1, 2.5, 5, 10, 100 μg/kg/day on GD9–16.	Females: abnormalities of reproductive tract (vagina, cervix, uterus, ovaries, oviduct); hypospadias; neoplasia of reproductive organs at adulthood.	[14]
	Mice; subcutaneous injection; 0 (corn oil) or 100 μg/kg/day on GD9–16.	Males: sterility; ectopic/undescended testes; histological lesions of testes, epididymis, seminal vesicles.	[13]
17β-Estradiol (ER agonist)	Rats; dietary; 0, 0.05, 2.5, 10, 50 ppm; one-generation reproduction study.	Males: ↓ sperm counts; delayed puberty; feminization of male reproductive organs (mammary gland); atrophy of reproductive organs; hormonal alterations. Females: accelerated puberty; ↓ fertility; altered estrous cyclicity; ovarian malfunction (↓ corporea lutea); hormonal alterations.	[10–12]
Nonylphenol (ER agonist)	Rats; gastric gavage; 0 (peanut oil), 3, 15, 75 mg/kg/day on GD11–18	Males: ↓ epididymal weight.	[53]
	Rats; dietary; 0, 200, 650, 2000 ppm; three-generation reproduction study.	Females: accelerated puberty; altered estrous cyclicity; ↓ fertility; ↓ ovary weight.	[54]
		Males: ↓ sperm number in F1 adults (questionable effect).	[55]

(*continues on next page*)

Table 2 (*Continued*).

Compound (mode of action)	Exposure	Effect	Ref.
	Rats; gastric gavage; 0, 2, 10, 50 mg/kg/day; two-generation reproduction study.	Males: ↑ pituitary weight. Females: accelerated puberty; ↓ ovary weight; ↓ number of implantation sites.	[55]
Butylbenzyl phthalate (ER agonist)	Rats; gastric gavage; 0, 20, 500 mg/kg/day; two-generation reproduction study.	Males: ↓ anogenital distance; delayed puberty; testicular atrophy; ↓ sperm counts; hormonal alterations. Females: ↑ anogenital distance, ↓ ovary weight.	[55]
Methoxychlor (ER agonist/ AR antagonist)	Rats; gastric gavage; 0 (corn oil), 25, 50, 100, 200 mg/kg/day; various study designs.	Males: delayed puberty; ↓ organ weights (seminal vesicles, testes, epididymis); ↓ sperm count; histological lesions in testes. Females: accelerated puberty (age at vaginal opening); accelerated time to first estrus; ↓ fertility; altered estrous cyclicity; ↓ ovary weight.	[82]
	Rats; review paper.	Males: ↓ organ weights (seminal vesicles, prostate, epididymis); delayed puberty; ↓ sperm count; testicular atrophy. Females: accelerated puberty; accelerated time to first estrous; ↓ fertility; altered estrous cyclicity; ↓ implantation sites; ↓ corporea lutea; ↓ serum progesterone.	[159]
	Rats; 0, (corn oil), 5, 50, 150; maternal gavage GD14–PND7 and direct dose PND7–42.	Males: delayed puberty; ↓ organ weights (seminal vesicles, testes, epididymis, prostate); ↓ sperm count and sperm motility Females: accelerated puberty; altered estrous cyclicity; ↓ corporea lutea; ↓ fertility	[23]
Ketoconazole (steroidogenesis inhibitor)	Rats; 0 (corn oil), 12.5, 25, 50 mg/kg/day on GD14–PND3.	Males: ↓ organ weights (testes, seminal vesicles, epididymis); no effect on areolas/nipples, reproductive tract malformations, or anogenital distance.	[33]
	Rats; review paper.	Females: ↓ implantation sites; ↑ ovary weight; ↓ serum progesterone	[88]
Di-*n*-butyl phthalate (steroidogenesis inhibitor)	Rats; gastric gavage, corn oil vehicle; various exposures from GD3–PND20; doses from 0.5–750 mg/kg/day.	Males: retained areolas/nipples; delayed puberty; malformations of epididymis; malformations or missing prostate and seminal vesicles; histological lesions in testes; ↓ anogenital distance; ectopic/undescended testes; hypospadias. Females: no effects.	[50,160, 161]
	Rats; 0 (corn oil) or 500 mg/kg/day on GD14–PND3.	Males: retained areolas/nipples; hypospadias; ↓ anogenital distance; ↓ organ weights (ventral prostate, epididymis, testes, levator ani/bulbocavernosus); delayed puberty; ↓ sperm count.	[33]

Table 2 (*Continued*).

Compound (mode of action)	Exposure	Effect	Ref.
	Rats; dietary; 0, 0.1, 0.5, 1 %; continuous breeding protocol.	Males: ↓ fertility; ↓ organ weights (testes, seminal vesicles); ↓ sperm count; ectopic/undescended testes; reproductive tract malformations (epididymis, external genitalia). Females: no effects.	[162]
Finasteride (5α-reductase inhibitor)	Rats; gastric gavage; 0.5 % methylcellulose vehicle; various exposures from GD6–20; doses from 0.03–300 mg/kg/day.	Males: ↓ anogenital distance (reversed at adulthood); hypospadias; retained areolas/nipples (reversed at adulthood). Females: no effects.	[41,42]
	Rats; subcutaneous injection; 0 (ethanol/corn oil), 25, 50, 120, 160, 320 mg/kg/day on GD 12–21.	Males: feminization of external genitalia; ↓ organ weights (seminal vesicles and prostate); ↓ anogenital distance; hypospadias	[37]
Fenarimol (Aromatase inhibitor)	Rats; dietary and gastric gavage; various exposures–reproduction studies; doses from 5–350 ppm (diet) or 35 mg/kg/day (gavage).	Males: ↓ fertility due to feminized behavior.	[24,25]
Exemestane (Aromatase inhibitor)	Rats; gastric gavage; various exposures–reproduction & developmental studies; doses from 52–1000 mg/kg/day.	Females: ↑ anogenital distance; no effects on fertility; delayed parturition.	[89,163]
TCDD (Ah receptor agonist)	Rats; review paper.	Males: ↓ anogenital distance; ectopic/undescended testes; impaired spermatogenesis (persisted at adulthood); delayed puberty; ↓ organ weights (prostate and seminal vesicle); feminization of male sexual behavior. Females: cleft phallus; hypospadias; incomplete vaginal opening; ↓ ovary weight.	[164]
	Rats; 0 (DMSO/corn oil); various exposures and doses.	Males: ↓ organ weights (ventral prostate, testis, epididymis); ↓ anogenital distance.	[60]
	Rats; gastric gavage; 0 (corn oil), 0.05, 0.2, 0.8 μg/kg/day on GD15.	Males: ↓ fertility; delayed puberty; ↓ sperm count; ↓ organ weights (ventral prostate and seminal vesicles). Females: cleft phallus; hypospadias; incomplete vaginal opening; ↓ ovary weight; delayed puberty; malformations of the reproductive tract at adulthood.	[59]

Abbreviations: GD: gestational day; PND: postnatal day.

As stated previously, the severity of the effects is dependent on several factors. In some instances, many of the effects are irreversible (e.g., AGD, retained areolas/nipples, reproductive tract malformations, hormonal alterations) [31,33,34,36]. For example, the AR antagonists flutamide, vinclozolin, *p,p′*-DDE, and procymidone all induce permanent retention of areolas/ nipples at adulthood in male offspring [31,33,34,36,38], and flutamide permanently decreases AGD in male offspring [36]. In contrast, the vinclozolin-induced effects on AGD were reversible in a dose- and time-dependent manner if treatment was stopped on postnatal day 3 [31]. It has been hypothesized that the reversibility of the AGD effects after vinclozolin exposure may be due to a decrease in cell number in the AGD region of male rats during development that is not reversible, but the remaining cells may grow in size enough to regrow the region, and allow room for scrotal development [31]. Regardless of the molecular events that control reversibility/irreversibility of the secondary sex characteristics described above, in the most severe cases, the affected males permanently resemble phenotypic females, although they still possess testes (undescended) and do not have a full compliment of female reproductive organs.

The spectrum of the effects that is observed also reflects specific events that are interrupted during mammalian development. For example, AR antagonists (e.g., flutamide or vinclozolin), which decrease both T- and DHT-dependent signaling pathways and reproductive processes, induce adverse effects (i.e., malformations) of the reproductive structures arising from the Wolffian ducts, urogenital sinus, and genital tubercle (Fig. 2) [31,33–37], while also resulting in incomplete Müllerian duct regression [31,33,34]. In contrast, the effects induced by 5α-reductase inhibitors primarily reflect attenuation of the DHT-dependent pathways (i.e., the urogenital sinus and genital tubercle) [37,41–46]. Compounds that have more broad inhibitory effects on steroid hormone synthesis can induce a similar profile of effects on both T- and DHT-dependent pathways [33,47–52]. Interestingly, ER agonists [10,11,13,15,23,53–56] and Ah receptor agonists (e.g., TCDD) [57–60] also induce a very similar pattern of effects to compounds that interfere with androgen signaling, although the exact mechanism of action of TCDD on the reproductive tract remains elusive. This underscores the complexity of mammalian reproductive development, and illustrates the importance that both estrogens and androgens play in the developmental process, a hypothesis that is supported by the presence of both receptor types throughout the reproductive tract in males and females [61–63].

Thyroid hormones control multiple physiological processes, and compounds that alter thyroid hormone homeostasis have the ability to affect both reproductive and behavioral development. A wide range of compounds (e.g., PCBs, dioxins, thionamides, phenobarbital) have been shown to alter thyroid homeostasis in experimental animals [64,65]. In fact, thyroid disorders are among the most common of endocrine-related disorders. Disruption of thyroid homeostasis during mammalian development exerts its most striking effects on behavioral development [66,67], however, effects on reproductive development have also been observed. For example, altered thyroid hormone homeostasis (either hypo- or hyperthyroidism) during male sexual development can result in altered testicular development, and as a result, quantitatively altered spermatogenesis at adulthood [5,68–72]. Surprising given the role of thyroid involvement in many physiological processes, few reports of altered reproductive development are found in the scientific literature.

Other EACs can affect reproduction and/or development without inducing any noticeable lesions of the reproductive tract. For example, aromatase inhibitors such as fenarimol cause no noticeable effects on the reproductive tract, but due to alterations in brain development, the sexual behavior of the males is affected and fertility is compromised (see additional discussion below) [24,25,73,74]. In addition, other compounds (e.g., atrazine) may induce adverse effects, while the exact mechanism(s) of action remains elusive [75]. Compounds that depress the central nervous system (CNS) will also have the potential to impact reproduction and development, as well as behavioral development. For example, phenobarbital, a pharmaceutical agent that is used as a sedative, induces a spectrum of effects in male rats that is consistent with a general depression of the CNS: decreased AGD, decreased seminal vesicle weights, delayed testicular descent, and altered reproductive hormone levels [76–78]. These examples illustrate that although receptor agonists and antagonists have received the most attention regarding their potential to disrupt reproduction and development, there are clearly EACs that will effect reproduction and development through nontraditional mechanisms of toxicity.

In females, a more limited number of EACs have been shown to induce adverse effects on reproduction and development. This is in part a reflection of the fact that differentiation of the female phenotype occurs in the absence of androgens; hence, EACs that impact androgen signaling do not *usually* manifest in noticeable effects in females, although androgens are involved in regulating ovarian function [79–81]. With the exception of alterations in the age of puberty, ED in females often results in long-term alterations (e.g., altered estrous cyclicity and ovarian function) that are not easily observed in the neonate, and therefore are typically not detected in short-term studies.

The effects of EACs on female reproduction and development have primarily been evaluated after exposure to ER agonists, of which DES is probably the most well-studied example [14,16]. Adult effects of perinatal exposure to DES, and other estrogens, appear to be primarily on the CNS systems controlling gonadotropin secretion. These permanent alterations in gonadotropin secretion and control produce numerous downstream effects, first in the ovary (i.e., decreased ovary weight, ova count, and corporea lutea), and subsequently in the various estrogen or progesterone-responsive tissues [11,12,14,16,54–56,82]. These impacts are most easily identified as decreased reproductive success in mating studies with the gestationally exposed females. In addition, some instances of reproductive tract malformation (structural abnormalities of oviduct, uterus, cervix, vagina; hypospadias) have also been observed with DES [14,16], indicating that estrogens can also derange reproductive tract development in both female and male offspring.

In females, AR agonists produce masculinizing effects, both morphologically and behaviorally [83–85]. Adverse effects associated with androgenic exposure in females include increased AGD (i.e., male-like), delayed puberty, altered estrous cyclicity, and masculinization (e.g., decreased number of areolas/nipples, presence of male reproductive tissues) [83–85]. Numerous documented cases of alterations in female sexual behavior are also common [83,86,87]. In contrast, prenatally administered anti-androgens do not typically cause adverse effects on reproduction or development in females (Table 2).

Depending on specificity, steroid biosynthesis inhibitors can also affect females, and in cases where female sex steroid production is attenuated (e.g., ketoconazole treatment), alterations in ovarian function, estrous cyclicity, and/or delayed puberty occur [88,89]. The possibility exists for several other mechanisms to alter reproductive capacity and/or development in females, although documented examples are limited. For example, the herbicide atrazine has been shown to alter estrous cyclicity and gonadotropin release in females via a neuroendocrine mechanism [90–92]. Similar to the effects observed in males, phenobarbital induces a spectrum of effects in female rats that is consistent with a general depression of the CNS: lowered gonadotropin release leading to reduced steroid levels and delayed puberty, altered estrous cyclicity, and infertility [76–78]. Prolactin, an endogenous hormone, or compounds that alter dopamine signaling and therefore prolactin levels, can also affect female reproductive development (e.g., enhanced puberty) [93]. The number of compounds with nontraditional mechanisms of action (i.e., not receptor-mediated) continues to grow [1].

The following paragraphs summarize the state of the science regarding the effects of EACs on reproduction and development:

- There is clear evidence that man-made chemicals and environmental pollutants induce adverse effects in experimental animals. While most of the focus has centered on receptor-mediated mechanisms of ED, and more specifically on ER agonist and AR antagonists, a large number of potential mechanisms of ED exist. As research on ED continues, the number of EACs and the variety of mechanisms of action are certain to increase.
- The data collected for most of the compounds evaluated to date suggests that males are more susceptible to ED than females due to the events involved in mammalian sexual differentiation and development. In males, adverse effects generally include decreased reproductive organ weight and function, altered morphology, altered age of puberty, and compromised reproductive capac-

ity (i.e., spermatogenesis); with alterations in androgen signaling, incomplete masculinization (e.g. retention of the female Müllerian ducts) also occurs. Interestingly, compounds with a variety of endocrine mechanisms (i.e., AR antagonists, ER agonists, Ah receptor agonists) induce a common profile of effects in males. In females, adverse effects generally include altered age of puberty, ovarian function (e.g., altered estrous cyclicity), and in some cases, morphological alterations of the reproductive tract. Prenatal exposures can result in permanent alterations in reproduction (e.g., morphology and behavior) and development. In both male and female animals, early changes that produce long-term increases in gonadotropin levels often produce neoplasia.

- Current guideline studies, for example, the current U.S. Environmental Protection Agency (USEPA) multigeneration reproduction study design, have endpoints recently added for evaluating potential ED, and many proposed screening studies will also evaluate potential endocrine activity of man-made chemicals. These data will prove critical in evaluating the potential effects of a wide variety of compounds on reproduction and development in experimental animals.
- While numerous cases of the adverse effects of EACs have been documented in experimental animals and wildlife species, data showing altered human reproductive system structure or function after environmental exposures are still lacking. The pharmaceutical DES is, of course, the best example of a human endocrine disruptor, although not an "environmental" exposure. In addition, most of the documented cases of ED in experimental animals involve doses greater than those encountered in the environment, or are high-level exposure for long durations.
- In addition, physiological differences between species can confound human risk assessment. The nascent National Children's Study, currently being planned in the United States, could be the first study to really address the issue of early exposures and reproductive system function in humans. Until these data come in, epidemiological studies on existing populations and exposures should be performed to determine whether ED is a true health problem for humans.

ADVERSE EFFECTS OF ENDOCRINE ACTIVE SUBSTANCES ON THE IMMUNE SYSTEM

The reproductive system is not the only body system affected by developmental exposure to EACs. As is clear from Table 3, the immune system is also a target, and these changes can last for extended durations (more than half of the total lifespan). This interaction between the reproductive and immune systems has been well known for several decades (Fig. 3). There appear to be several ways these interactions can occur. There are several recent reviews of this topic, to which the reader is referred for more in-depth analysis and examples [94,95].

To begin with, the immune system carries significant sexual dimorphism: numerous characteristics of the immune system vary significantly between adult males and adult females. Broadly speaking, compared to males, females (humans and rodents) generate a stronger immune response, are more resistant to immune tolerance, have greater levels of immunoglobulins, and have a higher incidence of certain forms of autoimmune diseases, including induced experimental forms [96]. While these statements seem to imply that female hormones are supportive of immune function, it is interesting to note that ovariectomy allows thymic hypertrophy to occur, while administration of estradiol causes thymic involution [97,98]. This involution appears to be more long-lasting than that produced by hydrocortisone. Thus, estradiol has both inhibitory and stimulatory effects upon different parts of the immune system. Meanwhile, male rats have greater thymic weight and thymocyte cellularity than females [99], T administration generally causes less thymic involution than estrogen [94], and T also seems to limit the immune responsiveness of males [96]. Thus, even at the basic descriptive level, there are differences between the genders in terms of basal functioning and set-points in the adult immune system.

Aside from sex differences in immune system measures, we should also note the apparent dichotomy in sex hormone effects: estradiol both involutes the thymus and allows for increased immune

Table 3 Examples of endocrine active substances and their effects on the immune system in experimental animals exposed in utero.

Compound	Exposure	Effect	Ref.
Lead	Rats; lead acetate in drinking water; 0, 25, 50 ppm; dams exposed 7 weeks prior to mating, during gestation, and lactation; offspring exposed to same levels from weaning to evaluation on PND35–45.	↓ thymus weight; ↓ mitogen-stimulated splenic lymphocyte proliferation.	[165]
	Sprague–Dawley rats; lead acetate in drinking water; 500 ppm; GD3–9 or 15–21; evaluated 12 weeks postpartum (adult).	Males, late gestation exposure: ↑ IL-12 production, ↓ IL-10 production. Females, late gestation exposure: reduced DTH, ↑ IL-10 production, ↑ monocyte count.	[166]
Chlordane	Mice; analytical chlordane free-fed in peanut butter during gestation only; 0.16 or 8.0 mg/kg/day; pups randomized within treatment groups for rearing; evaluated on PND101.	Males and females: ↓ contact hypersensitivity (high dose); no change in primary antibody response or secondary antibody response.	[167]
	BALB/c mice; technical chlordane free-fed in peanut butter during gestation; 4 or 16 mg/kg/day; no postpartum randomization; evaluated on PND30 and 100.	Males: DTH measures NS; splenic lymphocyte proliferation in response to ConA NS. Females: ↓ DTH at PND100, at other times NS; splenic lymphocyte proliferation in response to ConA increased at PND30.	[168]
	Mice; technical chlordane free-fed in peanut butter during gestation; 4 or 8 mg/kg/day; evaluated on PND100 and 200.	Both doses had ↓ colony-forming units in bone marrow and spleen at both time points; same 18-day treatment in adults gave no effect; indicates altered myeloid lineage rather than T-cell function.	[169]
TCDD	Rats; TCDD (>99 % pure) dissolved in acetone/diluted in corn oil; gastric gavage; 0, 1, 5 μg/kg on GD11, 18, PND4, 11, 18; evaluated on PND25.	Complete mortality at 5 μg/kg; ↓ organ weights (thymus and adrenal) at PND25 at 1 μg/kg; ↓ thymic PHA stimulation at PND25.	[170]
	Rats; TCDD (>99 % pure) dissolved in acetone/diluted in corn oil; gastric gavage; 0, 5 μg/kg on PND0, 7, 14; evaluated on PND25.	↓ organ weights (thymus and spleen) PND25; ↓ thymic PHA stimulation; ↓ host-resistance at PND25.	[170]
	Mice; 0, 2, 5 μg/kg on GD14, 17, PND1, 8, 15; evaluated on PND23.	↓ host rejection on PND23.	[170]
	F344 rats; gastric gavage; 5 μg/kg in corn oil on GD18 and/or PND 0, 7, 14; evaluated on PND25, 39, 59, 145.	↓ thymus weight at all time points; ↑ spleen weight at PND25 and 39; ↓ response to mitogens in vitro, ↓ DTH responses at all time points.	[171]

(continues on next page)

Table 3 (*Continued*).

Compound	Exposure	Effect	Ref.
	BALB/c mice; gastric gavage; 10 µg/kg on GD14; evaluated on PND4, 11, 18.	↓ TdT synthesis in liver and bone marrow.	[172]
	C57Bl/6N mice; gastric gavage; 0, 1.5, 3 µg/kg on GD6–14; cross-fostered after birth; evaluated on GD18, PND6, 14, 21	↓ thymus weight on GD18 at both doses; ↓ thymic cellularity on GD18 and PND6 at both doses; altered T-cell differentiation.	[113]
	C57Bl/6 mice; gastric gavage; 0, 1.5, 3 µg/kg on GD6–14; evaluated on GD18.	↓ thymus weight and cellularity on GD18; inhibited prenatal thymocyte differentiation.	[173]
	F344 rats; 1 or 3 µg on GD14; evaluated on GD19 or PND1.	↓ thymus weight on GD19 at 3 µg; changes in surface markers on t hymocytes on PND1; ↑ liver weight; surface marker changes both dose groups.	[174]
	F344 rats; (a) 3 µg/kg on GD14; evaluated DTH in males and females at 4, 8, 12, and 19 months old. (b) 0.1, 0.3, or 1 µg/kg on GD14; evaluated at 4 or 14 months old.	(a) Males: ↓ DTH at all time points. Females: *↓ DTH at 4 months only; ↓ thymic cellularities and phenotypic changes. (b) Males: ↓ DTH at all doses. Females: ↓ DTH at 0.3 and 1 µg, greater response at 14 months than at 4; response to both KLH and BSA was reduced.	[114]
Heptachlor	Sprague–Dawley rats; 0, 30, 300, or 3000 µg/kg/day; maternal gavage GD12–PND7 and direct dose PND8–42; evaluated PND46 and 156.	Males: ↓ primary antibody response to sheep red blood cells at all dose levels at 8 weeks and at the middle dose at 21 weeks.	[175]
Methoxychlor	Sprague–Dawley rats; 5, 50, 150 mg/kg/day; maternal gavage GD12–PND7 and direct dose PND8–42; evaluated PND46 and [a]156.	Males: ↓ thymus weight at PND46. Females: ↓ thymus weight at PND46 and 156; reduced plaque forming cells/ spleen at PND156.	[23]
DMSA (meso-2,3-dimercapt-osuccinic acid)	Fisher 344 rats; 30 or 60 mg/kg/day; gastric gavage on GD6–21; evaluated females only at 13 weeks of age.	Males: ↑ IL-2 after ConA stimulation in vitro. Females, high dose: ↓ DTH response to KLH and ↓ serum levels of monocyte chemoattractant protein-1; ↑ IL-2 after ConA stimulation in vitro.	[176]

Abbreviations: GD: gestational day; PND: postnatal day; NS: not significant; PHA: phytohemagglutinin; TdT: terminal deoxynucleotidyl transferase; DTH: delayed type hypersensitivity; KLH: keyhole lympet hemocyanin; ConA: concanavalin A.

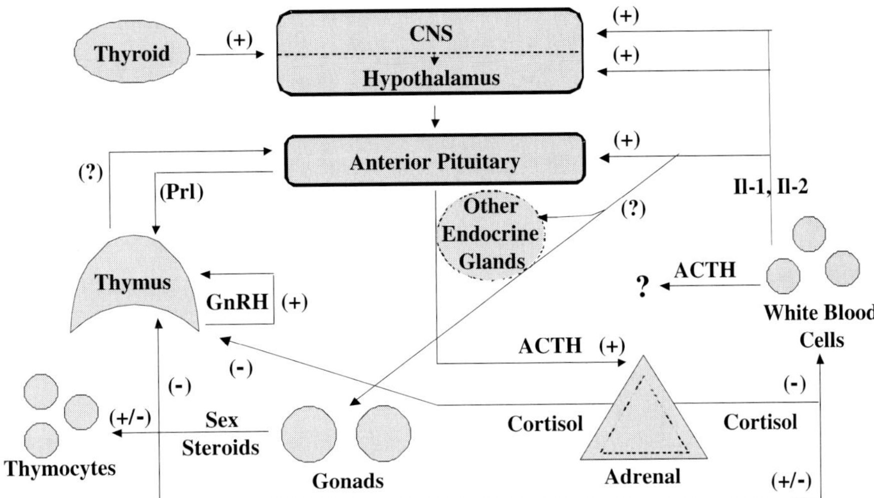

Fig. 3 *Interaction of the endocrine, immune, and nervous systems in mammals.* Overview of interrelation of endocrine, immune, and nervous signaling pathways.

responses, and T can be immunosuppressive but also confers more immunocompetence [100]. This illustrates the complex responses that could be seen after developmental exposure to EACs. That is to say, two different EACs, both with apparent estrogen-like activity, may produce different effects on the immune system, which, upon further investigation may be entirely consistent with their estrogenic activities.

Sex steroid effects are often mediated by the nuclear steroid receptors, which are specific for each steroid [101]. Steroid receptors have been found in various parts of the immune system including the individual cells as well as the generative epithelium [95,97,102–106]. It is important to note that not all of the effects seen in the immune system correlate with the distribution of steroid receptors [107]. Steroids also have numerous effects that are independent of nuclear receptors [108,109], and are believed to be mediated by membrane-bound receptors. Nuclear receptor-mediated or not, sex steroids are likely to have direct effects on the immune system. If a xenobiotic binds to either nuclear or membrane receptors (i.e., ER, AR, or progesterone receptor) as part of its effects on the endocrine system, then effects on the immune system should also be expected.

Additionally, other nonsteroid reproductive-related hormones exert profound effects on immune functions. Prolactin has been shown to promote lymphocyte growth and differentiation [110] by acting on the thymus as well as directly on the lymphocytes themselves [111,112]. One current view is that prolactin operates as an immunomodulator: while not playing a key role in immune function, it is one of several hormones (including growth hormone, thyroid hormone, and insulin-like growth factor [1]) that appear to "fine tune" cellular responses, primarily when the organism is under stress [112].

An important point to remember is that the immune system, largely unlike the reproductive system, can be modulated up or down. A hyperactive immune system can be just as adverse for the organism as a repressed immune system. The development of autoimmune disease in women is one example of this, and any exposure-related increase in this condition would be cause for real concern. Currently, too little is known about the range of postnatal immunologic consequences of perinatal exposure to EACs to confidently predict the outcome, but the *potential* impact, given the literature reviewed above and the data in the accompanying table certainly suggests that adverse effects can occur.

Since females are more prone to autoimmune disease, since this tends to happen more in aged individuals, and since many EACs appear to carry some component of estrogenic activity, one could plausibly hypothesize that an increase in autoimmune dysfunction would be observed if studies were performed on animals of sufficient age. This has been noted for TCDD [113], which increases the level of concern for estrogenic compounds. Because of the length of time required for such studies, they have been infrequently performed in the past. Given the potential health impacts in humans, more attention should be paid in the future to age-related autoimmunity.

Finally, delayed-type hypersensitivity (DTH) reactions help protect against infectious and neoplastic challenges, and appear most important in the very old and very young [114]. The ability of modest amounts of TCDD (1–3 µg/kg, once on GD14) to reduce DTH reactions even in aged rats [114] shows again that long-term health consequences are possible from pre/perinatal exposures. A careful analysis of this endpoint after specifically altering sex steroid levels would help put some perspective around the ability of EACs to cause similar changes, and would help set our concern at an appropriate level.

The following paragraphs summarize the state of the science regarding the effects of EACs on the immune system:

- A review of the available papers indicates significant variability in the hypotheses being tested, the periods of exposure, the ages of the animals at evaluation, and the methods of evaluation. The issue of EAC-induced changes in immune structure and function would benefit greatly from a consensus about the important questions to be addressed, followed by a focus on generating a common database for several EACs. This would be an excellent project for a multinational collaboration, one that could take advantage of existing networks of professional collaborations and organizations.
- The field will be hampered by the current uneven definition of the roles of sex steroids in setting baseline values for the immune system cells and tissues. A thorough and systematic evaluation of the effects of specific ER and AR agonists and antagonists will set the stage for a cleaner interpretation of toxicant exposures and greatly facilitate the identification of possible mechanisms. This is not an open-ended call for more research, but for work tightly focused on the short- and long-term effects of prenatal steroid administration on every aspect of immune system function.
- Alterations that either reduce immunocompetence or cause hyper-reactivity are of concern. Such effects are biologically plausibly after prenatal exposure, yet no studies to date were found that have evaluated both possible changes in the same study.
- There are multiple known mechanisms that EACs could use to impinge on the immune system. The implications are that: (1) more mechanisms will be found with further work in this area, (2) more effects will be found, and (3) a logical approach to generating a comprehensive data set, instituted very soon, would save years of environmental damage, regulatory uncertainty, and industrial and public anxiety.

ADVERSE EFFECTS OF ENDOCRINE ACTIVE SUBSTANCES ON THE NERVOUS SYSTEM

The CNS is not only a key part of reproduction, but it also has much in common with the immune system. Cytokines play a central role in the function of both the CNS and immune systems, and both are targets for steroid action (Fig. 3) [115–117], which makes them readily susceptible to xenobiotics that modulate steroid activity.

The determination of effects on the function of the CNS after prenatal exposure can be considered functional, or behavioral, teratology. Indeed, the field of behavioral teratology is a recognized specialty, with its own methods [118,119], society, and journal (*Neurotoxicology and Teratology*). The evaluation of endocrine-mediated effects on CNS organization and/or function after developmental exposure is really a subspecialty of this complicated area.

Like the immune system, many parts of the brain have steroid hormone receptors, which make the brain a target for xenobiotics that interact with receptors or alter hormone concentrations [120–122].

The forebrain (association), hippocampus (memory), most areas of the hypothalamus (endocrine physiology and control), as well as the midbrain (integrative function) and cortex (cognition and processing) possess steroid receptors at some point during development. Both glucocorticoids and sex steroids (as exemplified by estradiol) appear to act on multiple cell types, and through both nuclear receptors and non-receptor-mediated mechanisms [120,121]. Thyroid hormones also act on most parts of the adult and developing brain, and appear to act as each part of the brain passes through a specific developmental window [123,124]. All these hormones modulate the differentiation of neurons by affecting cellular migration, death, and synapse formation and pruning [123–125]. These fundamental changes in cellular structure impact neuronal function, and thus overall CNS output. This structure–function relationship is slightly more challenging to prove in the CNS, where output is an amalgamated result of the output of millions of neurons, modulated by an additional overwhelming number of cells, but there is considerable support for this structure–function link. For example, the increased neuronal spine density in neurons in the male hippocampus, and the fewer spines in females, has been shown to correlate positively with spatial learning ability (i.e., more synapses on spines confers greater spatial ability). This is corroborated by the data from males castrated neonatally (who have decreased spine density and reduced spatial learning ability), and the increases in spine density in T-treated females (who have increased spatial learning capabilities) [126]. These correlations are one example that supports the relationship between microscopic structure and overall CNS function. Thus, hormonally induced changes in neuronal structure will have some effect on function, at the cellular and/or the organismal level.

While receptors are dispersed differentially throughout the brain, several areas appear particularly sexually dimorphic (different in males and females). These include the sexually dimorphic nucleus of the preoptic area (SDN-POA), the ventromedial nucleus of the hypothalamus, and the hippocampus. The SDN and hippocampal dentate gyrus are larger in males. The hypothalamic nucleus appears important in sexual behavior, while the hippocampus subserves spatial learning and memory [126]. The hippocampus is also vulnerable to changes in thyroid hormone levels and glucocorticoids during development [126,127]. Levels of these hormones correlate positively with the number of cells and the number of synapses per cell in these areas.

Not only can steroids modify neuronal structure and function, but neurotransmitters, the means by which neurons communicate and process depolarizations, biochemically converge with steroid receptors to control neuron function and signaling [128,129]. Thus, compounds that alter neurotransmitter activity can produce effects on gene transcription and neuronal activity that are seen in other situations to be caused by hormones, and endocrine alterations can perturb neurotransmitter-mediated functions. This demonstrates a nonhormonal means by which an exposure can produce a change in the brain that may appear hormonally mediated.

One counter-intuitive fundamental that underlies gender differences in brain development is the nature of the developmental steroid dependence. Prenatally, estradiol is primarily bound to α-fetoprotein; very little is free to diffuse into cellular compartments. T, which is unbound, is taken up by the developing CNS cells and converted by the enzyme aromatase to estradiol, which subsequently *masculinizes* the brain. Increasing the circulating levels of T or estradiol will masculinize that animal, through the result of increased local levels of estradiol, while decreasing levels of T will feminize the brain and behaviors of that animal [126]. This paradox of increasing estradiol resulting in a more masculinized brain exemplifies again the complexities in this area.

Effects on the CNS can manifest at multiple levels, from cellular fine structure and light-microscope-level morphology, to electrophysiology, control of secretion, sensory control and perception, biochemical processes, or that most integrative of measures, behavior. The best evaluation of exposure-related effects will include all of these levels. Some themes emerge from the recent literature that is discussed below.

EACs affect sex-specific CNS endpoints. The SDN-POA is larger in males than in females, because it contains more cells due to the rescue effect of fetal T, whose unbound concentrations in plasma are greater than estradiol because of its inability to bind to α-fetoprotein [130]. Fetal treatment with T, high levels of estradiol, or DES will increase the size of the SDN-POA [131,132]. Females have higher levels of dopamine (DA) in their striatum, and estradiol increases the amount of DA released by striatal neurons [133]. Although exposure to 500 parts per million (ppm) genestein during gestation and into adulthood did not change baseline release of striatal DA, amphetamine-stimulated DA release was significantly increased in males but not females. TCDD, administered to pregnant rat dams on gestation day (GD) 15, selectively affected female brains, reducing the amount of gamma-aminobutyric acid decarboxylase gene expression in female POA to the male level, with no detectable change in the male POA, or in other brain areas examined [134]. Methoxychlor, whose metabolites are estrogenic and antiandrogenic, reduces follicle-stimulating hormone (FSH) release in only perinatally treated adult female rats [23], while it decreases mating behavior in male rats and increases those behaviors in perinatally exposed females [135,136]. Bisphenol A, when administered to the dams prenatally and lactationally at 1.5 mg/kg/day, eliminated gender differences normally seen in rats in open field and passive avoidance tests [137]. Bisphenol A also eliminated the gender differences in the size of the locus coeruleus. All the CNS effects described above were seen in the absence of detectable changes in male or female reproductive organ weights or serum concentrations of luteinizing hormone (LH), FSH, T, or estradiol.

Gender-specific behaviors can be affected by EACs. Weiss [138] provides a useful review of the benefits of evaluating gender-specific behaviors that are not related to reproduction, using the polychlorinated aromatics TCDD and PCBs using spatial learning and operant behaviors as measures of effect. Play behavior is organized by postnatal T action mediated via AR in the amygdala [83]. Administering T to female rats increases the amount of male-specific rough-and-tumble play [83]. Exposure to bisphenol A (an estrogenic mimic at high doses) during either development or adulthood (but curiously, not both) reduced maternal care and nursing behavior in treated mice [139], though this was without detectable effect on the offspring using the measures employed by the authors. A large dose of bisphenol A given to mice on GD11–19, masculinized some peripubertal behaviors in females, as did estradiol [140]. Sociosexual play and exploration was reduced in male rats and increased in females after perinatal bisphenol A exposure [141]. This can also be demonstrated by the work of Roegge and coworkers [142] who showed that radial arm maze performance was impaired by Aroclor 1254 in male rats, but not in females.

However, gender-specific behaviors can also be untouched by otherwise active EACs. For example, prenatal and developmental genestein exposure caused only very modest effects on the consumption of salt water (a sexually dimorphic behavior), and there was no treatment-by-gender effect [143]. Many other gender-specific behaviors remained unaffected: open field activity, play behavior, running wheel activity, and saccharin fluid consumption were untouched by genestein. Using a similar paradigm, these same investigators noted that *p*-nonylphenol exposure during gestation, maturation, and testing increased the intake of a sodium solution in females but not males, while other sexually dimorphic behaviors were unchanged [144]. These and other related studies were recently reviewed by these authors [145].

EACs also affect non-gender-specific endpoints in the nervous system. Neurite formation and MAPK-pathway activation were reduced by hexachloro-biphenyl and endosulfan in neuronal stem cells in vitro [146]. Aroclor 1254 can induce a low-frequency hearing loss in adult rats after perinatal exposure [147]. There were no effects in other sensory organ systems, as measured by evoked potentials. The effect on hearing loss has been recently shown to be due to fewer outer hair cells in the organ of Corti [148]. This is believed to be due to the hypothyroidism experienced by these animals during treatment with the Aroclor 1254 [147].

Gender-specific behaviors can also be affected by endocrine-**in**active substances. Locomotor activity is one example: treatment with a variety of endocrine inactive substances (e.g., nitrofen, cytosize, arabinoside) can increase locomotor activity in both genders in the absence of any discernible endocrine change [149].

There can be occasional disconnects between biochemistry and output that resolve with further investigation. For example, levels of thyroid hormone during perinatal brain development are known to stimulate cerebellar granule cell proliferation, cell migration, synapse formation and densities, myelination, and glial cell proliferation and maturation [123]. From this, one would expect that anything that reduced circulating thyroid hormone levels in the dam or fetus would adversely impact brain function. Interestingly, developing rats exposed to Aroclor 1254 (a PCB), which reduced fetal thyroid hormone levels, had no effect on spatial learning or baseline electrophysiologic measures in the hippocampal dentate gyrus [150]. They were able to identify changes in long-term potentiation, but did not observe the expected effect on spatial activity in a water maze. Similarly, Taylor and coworkers [151] exposed developing rats to a polybrominated diphenyl ether, which also reduces neonatal thyroid hormone levels, and found no change in motor activity or auditory startle. Recently, Zoeller and coworkers [124] have proposed a possible mechanism by which PCBs might reduce thyroid hormone yet not impact many thyroid-dependent processes: they review evidence that is consistent with the ability of certain PCBs to bind to the thyroid receptor. This would explain the data of Crofton's group [150,151], and could comprehensively explain many of the seemingly contradictory effects seen after PCB exposure, where low thyroid hormone levels are not automatically followed by a complete hypothyroid-like picture.

Behavior is complex, and must be approached thoughtfully. Behavioral scientists assert that behavior is uniquely sensitive to ED because it integrates the output from many cells and organizational levels in the brain and periphery, and so provides more targets for toxicant action. It is also affected by a wide variety of external factors [152]. In addition, this integration implies that these multiple inputs bring along many ways to change gender-specific behaviors that have no detectable link to endocrine mechanisms.

Ethologists would have us believe that it is more appropriate to assess the behavioral effects of a compound using a behavior that is relevant to the species in question [153]. This leaves an investigator or regulator trying to interpret whether a change in grooming behavior, though statistically significant, is meaningful to any other species, including humans. Reduced ability to solve problems is a clear adverse effect; a delay in the achievement of a righting reflex that becomes equivalent to controls a few days later is less demonstrably adverse. Although the ethological approach appeals to one's sense of the appropriate, and while clearly the right approach in identifying ecologic hazards, it also leaves more ambiguity in the risk assessment application of such data.

In characterizing the toxicity of any suspected EAC, the evaluation of behavioral changes is likely to occur relatively late in the process. The authors are partial to an approach where one first queries steroid- or thyroid-dependent CNS functions, and only later moves on to hormone-independent functions. This maximizes the chances of finding the most likely effects first. Conversely, it also means that a compound that affects cognitive development but not, say, mating behavior, would be less likely to be identified. Still, such an approach has been productive in the past [73,154,155]. The prior determination of actual changes in a hormone-related endpoint means that one can then go on to ask what those changes would mean to other functions of the animal, such as behavior. However, because of the complexity involved, changes in behavior should not be expected to initially identify a compound as an endocrine disruptor.

SUMMARY

The reproductive, immune, and central nervous systems all depend on endogenous steroids for correct prenatal programming and postnatal function. Therefore, exogenous treatments that interfere with such signaling systems also disrupt the structure and function of all these systems. We are just now learning the scope and dose-response nature of these effects. Numerous pressing questions remain: Which of these organ systems is more sensitive? Are humans being affected by environmental levels of EACs?

What are the full impacts of such exposures on the lifespan of both test species and wildlife? Have we found all the effects of EACs?

The small amount of data that have been generated to specifically address the relative organ system sensitivity issue (the National Institute of Environmental Health Sciences [NIEHS]/USEPA collaborative Juvenile Pesticide Studies) indicates that the most sensitive body system varies by compound when tested in rats. It is also likely that the most sensitive system will vary in different species. This greatly complicates any testing strategy, and suggests that the initial evaluations be apical tests of integrated function performed for several organ systems within one study. Further studies are warranted only if a change is seen in these more integrative initial tests.

While there are a large number of studies in experimental animals that link exposure to EACs to reproductive, developmental, and immune abnormalities and increased cancer incidence, the link to human health is still unclear. There is still no conclusive evidence that EACs that produce adverse effects in experimental animals will also do so in humans and/or wildlife. However, the conserved function of endocrine signaling pathways across numerous species, including humans, raises a legitimate concern that effects in laboratory species or wildlife, and humans, will be comparable [156,157].

Only by looking at function throughout the lifespan of an animal will we really be able to determine the "reach" of neonatal alterations. This affects all the organ systems described here: the reproductive system could cease functioning earlier, leading to premature menopause and shortened reproductive lifespan; there could be an increase in late-onset autoimmune diseases like lupus; and the changes that occur in aging brain function [158] and structure (i.e., plaque aggregation in Alzheimer's) could be exacerbated by neonatal exposure to EACs. These lifecycle studies will be done in rodents first, and should be planned as holistic investigations and commenced without delay.

The surprises of the past suggest that we have not found all the effects of neonatal exposure to EACs. Not only are effects in other organ systems likely, but other hormonal activities are probable. For example, while the presence of environmental androgens has not been well studied or substantiated, recent cases in wildlife suggest that environmental androgens do exist [83–85]. A logical, prospective design that systematically evaluates those tissues that we know are steroid responsive would be necessary and not overly difficult. This, too, should be commenced without delay.

In conclusion, the last 40 years have seen many reports that man-made chemicals and environmental pollutants cause adverse effects in humans and wildlife; however, few instances of anthropogenic ED have been scientifically substantiated. Certainly, studies in experimental animals have shown the potential for adverse effects in humans and wildlife. The task before us is to understand how big the issue of ED is to human health and the environment. Science and ignorance created this issue; ultimately, science and wisdom will solve it.

REFERENCES

1. T. M. Crisp, E. D. Clegg, R. L. Cooper, W. P. Wood, D. G. Anderson, K. P. Baetcke, J. L. Hoffmann, M. S. Morrow, D. J. Rodier, J. E. Schaeffer, L. W. Touart, M. G. Zeeman, Y. M. Patel. *Environ. Health Perspect.* **106**, 11–56 (1998).
2. R. J. Kavlock, G. P. Daston, C. DeRosa, P. Fenner-Crisp, L. E. Gray, S. Kaattari, G. Lucier, M. Luster, M. J. Mac, C. Maczka, R. Miller, J. Moore, R. Rolland, G. Scott, D. M. Sheehan, T. Sinks, H. A. Tilson. *Environ. Health Perspect.* **104**, 715–740 (1996).
3. E. Knobil and J. D. Neill. *The Physiology of Reproduction*, Raven Press, New York (1994).
4. L. J. DeGroot. *Endocrinology*, W. B. Saunders, Philadelphia (1995).
5. R. M. Sharpe. In *The Physiology of Reproduction,* E. Knobil and J. D. Neill (Eds.), pp. 1363–1434, Raven Press, New York (1994).
6. B. R. Zirkin. *Cell Mol. Biol. Testis* 166–188 (1993).
7. J. Yochim and V. J. D. Feo. *Endocrinology* **72**, 317–326 (1963).
8. J. Yochim and V. J. D. Feo. *Endocrinology* **71**, 134–142 (1962).

9. V. Haddad and M. M. Ketchel. *Int. J. Fert.* **14**, 56–63 (1969).
10. J. C. Cook, L. Johnson, J. C. O'Connor, L. B. Biegel, C. H. Krams, S. R. Frame, M. E. Hurtt. *Toxicol. Sci.* **44**, 155–168 (1998).
11. L. B. Biegel, J. A. Flaws, A. N. Hirshfield, J. C. O'Connor, G. S. Elliott, G. S. Ladics, E. K. Silbergeld, C. S. VanPelt, M. E. Hurtt, J. C. Cook, S. R. Frame. *Toxicol. Sci.* **44**, 116–142 (1998).
12. L. B. Biegel, J. C. Cook, M. E. Hurtt, J. C. O'Connor. *Toxicol. Sci.* **44**, 143–154 (1998).
13. J. A. McLachlan, R. R. Newbold, B. Bullock. *Science* **190**, 991–992 (1975).
14. J. A. McLachlan, R. R. Newbold, B. C. Bullock. *Cancer Res.* **40**, 3988–3999 (1980).
15. R. R. Newbold and J. A. McLachlan. In *Estrogens in the Environment II,* J. A. McLachlan (Ed.), pp. 288–318, Elsevier, New York (1985).
16. J. A. McLachlan, R. R. Newbold, H. C. Shah, M. D. Hogan, R. L. Dixon. *Fertil. Steril.* **38**, 364–371 (1982).
17. P. S. Guzelian. *Annu. Rev. Pharmacol. Toxicol.* **22**, 89–113 (1982).
18. H. C. Hodge, E. A. Maynard, J. F. Thomas, H. J. Blanchet, Jr., W. E. Wilt, Jr., K. E. Mason. *J. Pharmacol. Exptl. Therap.* **99**, 140–148 (1950).
19. S. J. Harris, H. C. Cecil, J. Bitman. *J. Agric. Food Chem.* **22**, 969–973 (1974).
20. H. S. Bal. *Proc. Soc. Exp. Biol. Med.* **176**, 187–196 (1984).
21. F. R. Boockfor and C. A. Blake. *Biol. Reprod.* **57**, 267–277 (1997).
22. C. A. Blake and F. R. Boockfor. *Biol. Reprod.* **57**, 255–266 (1997).
23. R. E. Chapin, M. W. Harris, B. J. Davis, S. M. Ward, R. E. Wilson, M. A. Mauney, A. C. Lockhart, R. J. Smialowicz, V. C. Moser, L. T. Burka, B. J. Collins. *Fundam. Appl. Toxicol.* **40**, 138–157 (1997).
24. K. S. Hirsch, E. R. Adams, D. G. Hoffman, J. K. Markham, N. V. Owen. *Toxicol. Appl. Pharmacol.* **86**, 391–399 (1986).
25. K. S. Hirsch, D. E. Weaver, L. J. Black, J. F. Falcone, N. J. MacLusky. *Toxicol. Appl. Pharmacol.* **91**, 235–245 (1987).
26. S. Bhasin, S. Sikka, T. Fielder, U. Sod-Moriah, H. B. Levine, R. S. Swerdloff, J. Rajfer. *Endocrinology* **118**, 1229–1232 (1986).
27. W. R. Heckman, B. R. Kane, R. E. Pakyz, M. J. Cosentino. *J. Androl.* **13**, 191–198 (1992).
28. NRC. In *Hormonally Active Agents in the Environment*, pp. 210–242, National Academy Press, Washington DC (2000).
29. K. D. Dohler. *Int. Rev. Cyt.* **131**, 1–57 (1991).
30. R. A. Gorski. In *Pharmacology, Biology and Clinical Applications of Androgens,* A. Bhasin (Ed.), pp. 159–168, Wiley-Liss, New York (1996).
31. L. E. Gray, Jr., J. Ostby, E. Monosson, W. R. Kelce. *Toxicol. Ind. Health* **15**, 48–64 (1999).
32. W. R. Kelce, E. Monosson, M. P. Gamcsik, S. C. Laws, L. E. Gray, Jr. *Toxicol. Appl. Pharmacol.* **126**, 276–285 (1994).
33. L. E. Gray, Jr., C. Wolf, C. Lambright, P. Mann, M. Price, R. L. Cooper, J. Ostby. *Toxicol. Ind. Health* **15**, 94–118 (1999).
34. L. E. Gray, J. Ostby, W. R. Kelce. *Toxicol. Appl. Pharmacol.* **129**, 46–52 (1994).
35. E. Monosson, W. R. Kelce, C. Lambright, J. Ostby, L. E. Gray, Jr. *Toxicol. Ind. Health* **15**, 65–79 (1999).
36. B. S. McIntyre, N. J. Barlow, P. M. D. Foster. *Toxicol. Sci.* **62**, 236–249 (2001).
37. J. Imperato-McGinley, R. S. Sanchez, J. R. Spencer, B. Yee, E. D. Vaughan. *Endocrinology (Baltimore)* **131**, 1149–1156 (1992).
38. J. Ostby, W. R. Kelce, C. Lambright, C. J. Wolf, P. Mann, L. E. Gray, Jr. *Toxicol. Ind. Health* **15**, 80–93 (1999).
39. W. R. Kelce, C. R. Stone, S. C. Laws, L. E. Gray, J. A. Kemppainen, E. M. Wilson. *Nature* **375**, 581–585 (1995).

40. L. You, M. Casanova, S. Archibeque-Engle, M. Sar, L.-Q. Fan, H. d. A. Heck. *Toxicol. Sci.* **45**, 162–173 (1998).
41. R. L. Clark, J. M. Antonello, S. J. Grossman, L. D. Wise, C. Anderson, W. J. Bagdon, S. Prahalada, J. S. MacDonald, R. T. Robertson. *Teratology* **42**, 91–100 (1990).
42. C. A. Anderson and R. L. Clark. *Teratology* **42**, 483–496 (1990).
43. B. S. McIntyre, N. J. Barlow, P. M. D. Foster. *Toxicol. Sci.* **65**, 62–70 (2002).
44. J. Imperato-McGinley, Z. Binienda, A. Arthur, D. T. Mininberg, E. D. Vaughan, Jr., F. W. Quimby. *Endocrinology (Baltimore)* **116**, 807–812 (1985).
45. J. Imperato-McGinley, Z. Binienda, J. Gedney, E. D. Vaughan, Jr. *Endocrinology (Baltimore)* **118**, 132–137 (1986).
46. F. W. George and K. G. Peterson. *Endocrinology* **122**, 1159–1164 (1988).
47. E. Bloch, M. Lew, M. Klein. *Endocrinology* **89**, 16–31 (1971).
48. A. S. Goldman, R. D. Eavey, M. K. Baker. *J. Endocrinol.* **71**, 289–297 (1976).
49. J. M. Goldman, L. E. Gray, R. L. Cooper. *Principles and Processes for Evaluating ED in Wildlife, Proceedings from a Conference on ED in Wildlife, Kiawah Island, S. C., Mar. 1996*, 311–333 (1998).
50. E. Mylchreest, R. C. Cattley, P. M. Foster. *Toxicol. Sci.* **43**, 47–60 (1998).
51. E. Mylchreest, M. Sar, D. G. Wallace, P. M. D. Foster. *Reprod. Toxicol.* **16**, 19–28 (2002).
52. L. G. Parks, J. S. Ostby, C. R. Lambright, B. D. Abbott, G. R. Klinefelter, N. J. Barlow, L. E. Gray, Jr. *Toxicol. Sci.* **58**, 339–349 (2000).
53. A. Hossaini, M. Dalgaard, A. M. Vinggaard, H. Frandsen, J.-J. Larsen. *Reprod. Toxicol.* **15**, 537–543 (2001).
54. R. E. Chapin, J. Delaney, Y. Wang, L. Lanning, B. Davis, B. Collins, N. Mintz, G. Wolfe. *Toxicol. Sci.* **52**, 80–91 (1999).
55. T. Nagao, R. Ohta, H. Marumo, T. Shindo, S. Yoshimura, H. Ono. *Reprod. Toxicol.* **14**, 513–532 (2000).
56. T. Nagao, K. Wada, H. Marumo, S. Yoshimura, H. Ono. *Reprod. Toxicol.* **15**, 293–315 (2001).
57. R. L. Cooper and R. J. Kavlock. *J. Endocrinol.* **152**, 159–166 (1997).
58. L. S. Birnbaum. *Environ. Health Perspect.* **102**, 676–679 (1994).
59. L. E. Gray, Jr., J. S. Ostby, W. R. Kelce. *Toxicol. Appl. Pharmacol.* **146**, 11–20 (1997).
60. S. Ohsako, Y. Miyabara, M. Sakaue, R. Ishimura, M. Kakeyama, H. Izumi, J. Yonemoto, C. Tohyama. *Toxicol. Sci.* **66**, 283–292 (2002).
61. R. M. Sharpe. *Pure Appl. Chem.* **70**, 1685–1701 (1998).
62. R. M. Sharpe. *Trends in Endocrinology and Metabolism* **9**, 371–377 (1998).
63. K. Williams, C. McKinnell, P. T. K. Saunders, M. Walker, J. S. Fisher, K. J. Turner, N. Atanassova, R. M. Sharpe. *Hum. Reprod. Update* **7**, 236–247 (2001).
64. C. C. Capen. In *Casarett and Doull's Toxicology: The Basic Science of Poisons,* C. D. Klaassen (Ed.), pp. 617–640, McGraw-Hill, New York (1996).
65. C. C. Capen. *Toxicol. Pathol.* **25**, 39–48 (1997).
66. S. Darbra, F. Balada, A. Garau, P. Gatell, J. Sala, M. A. Marti-Carbonell. *Behav. Brain Res.* **68**, 159–164 (1995).
67. E. A. Attree, A. K. Sinha, M. J. Davey, M. R. Pickard, F. D. Rose, R. P. Ekins. *Med. Sci. Res.* **20**, 197–199 (1992).
68. E. A. Jannini, S. Ulisse, M. D'Armiento. *Endocr. Rev.* **16**, 443–459 (1995).
69. J. D. Kirby, A. E. Jetton, P. S. Cooke, R. A. Hess, D. Bunick, J. F. Ackland, F. W. Turek, N. B. Schwartz. *Endocrinology* **131**, 559–565 (1992).
70. P. S. Cooke, Y.-D. Zhao, L. G. Hansen. *Toxicol. Appl. Pharmacol.* **136**, 112–117 (1996).
71. D. Canale, M. Agostini, G. Giorgilli, C. Caglieresi, G. Scartabelli, V. Nardini, E. A. Jannini, E. Martino, A. Pinchera, E. Macchia. *J. Androl.* **22**, 284–288 (2001).

72. S. Francavilla, G. Cordeschi, G. Properzi, L. Di Cicco, E. A. Jannini, S. Palmero, E. Fugassa, B. Loras, M. D'Armiento. *J. Endocrinol.* **129**, 35–42 (1991).
73. L. E. Gray and J. Ostby. *Toxicol. Ind. Health* **14**, 159–184 (1998).
74. Y. Takahashi, H. Yamanaka, I. Akiyama, Y. Sakuma. *Endocrinol. Jpn.* **36**, 29–36 (1989).
75. T. E. Stoker, S. C. Laws, D. L. Guidici, R. L. Cooper. *Toxicol. Sci.* **58**, 50–59 (2000).
76. C. Gupta, B. H. Shapiro, B. R. Sonawane, S. J. Yaffe. *Colloq. - Inst. Natl. Sante Rech. Med.* **89**, 89–93 (1980).
77. C. Gupta, B. H. Shapiro, S. J. Yaffe. *Pediatr. Pharmacol.* **1**, 55–62 (1980).
78. C. Gupta and S. J. Yaffe. *Pediatr. Res.* **15**, 1488–1491 (1981).
79. C. Beyer, M. L. Cruz, V. L. Gay, R. B. Jaffe. *Endocrinology* **95**, 722–727 (1974).
80. G. L. Kumari, J. K. Datta, R. P. Das, S. Roy. *Horm. Res.* **9**, 112–120 (1978).
81. J. P. Louvet, S. M. Harman, J. R. Schreiber, G. T. Ross. *Endocrinology* **97**, 366–372 (1975).
82. L. E. Gray, J. Ostby, J. Ferrell, G. Rehnberg, R. Linder, R. Cooper, J. Goldman, V. Slott, J. Laskey. *Fundam. Appl. Toxicol.* **12**, 92–108 (1989).
83. A. K. Hotchkiss, Jr. *Environ. Health Perspect.* **110**, 435–439 (2002).
84. C. J. Wolf, A. Hotchkiss, J. S. Ostby, G. A. LeBlanc, L. E. Gray, Jr. *Toxicol. Sci.* **65**, 71–86 (2002).
85. S. J. McCoy and B. A. Shirley. *Life Sci.* **50**, 621–628 (1992).
86. R. W. Rhees, B. A. Kirk, S. Sephton, E. D. Lephart. *Dev. Neurosci. (Basel)* **19**, 430–437 (1997).
87. L. Huffman and S. E. Hendricks. *Physiol. Behav.* **26**, 773–778 (1981).
88. A. M. Cummings, J. L. Hedge, J. Laskey. *Fundam. Appl. Toxicol.* **40**, 238–246 (1997).
89. D. Beltrame, E. DiSalle, E. Giavini, K. Gunnarsson, M. Brughera. *Reprod. Toxicol.* **15**, 601–602 (2001).
90. J. C. Eldridge, D. G. Fleenor-Heyser, P. C. Extrom, L. T. Wetzel, C. B. Breckenridge, J. H. Gillis, L. G. Luempert, III, J. T. Stevens. *J. Toxicol. Environ. Health* **43**, 155–167 (1994).
91. R. L. Cooper, T. E. Stoker, J. M. Goldman, M. B. Parrish, L. Tyrey. *Reprod. Toxicol.* **10**, 257–264 (1996).
92. R. L. Cooper, T. E. Stoker, W. K. McElroy, J. Hein. *Toxicologist* **42**, 160 (1998).
93. J. P. Advis and S. R. Ojeda. *Endocrinology* **103**, 924–935 (1978).
94. H. O. Besedovsky and A. DelRey. *Endocr. Rev.* **17**, 64–102 (1996).
95. D. Tomaszewska and F. Przekop. *J. Physiol. Pharmacol.* **48**, 139–158 (1997).
96. S. C. Wilcoxen, E. Kirkman, K. C. Dowdell, S. A. Stohlman. *J. Immunol.* **164**, 6237–6243 (2000).
97. A. Martin, L. M. Alonso, M. Gomez del Moral, A. G. Zapata. *Tissue Cell* **26**, 169–179 (1994).
98. A. E. Silverstone, D. E. Frazier, Jr., N. C. Fiore, J. A. Soults, T. A. Gasiewicz. *Toxicol. Appl. Pharmacol.* **126**, 248–259 (1994).
99. G. Leposavic, B. Karapetrovic, S. Obradovic, B. Vidic Dankovic, D. Kosec. *Pharmacol. Biochem. Behav.* **54**, 269–276 (1996).
100. A. Peters. *Proc. Roy. Soc. London.* **267**, 883–889 (2000).
101. M. A. Carson-Jurica, W. T. Schrader, B. W. O'Malley. *Endocr. Rev.* **11**, 201–220 (1990).
102. I. Screpanti, A. Gulino, J. R. Pasqualini. *Endocrinology* **111**, 1552–1561 (1982).
103. P. T. Pearce, B. A. K. Khalid, J. W. Funder. *Endocrinology* **113**, 1287–1291 (1983).
104. W. J. Kovacs and N. J. Olsen. *J. Immunol.* **139**, 490–493 (1987).
105. A. Gulino, I. Screpanti, M. R. Torrisi, L. Frati. *Endocrinology* **117**, 47–54 (1985).
106. T. C. Chao, P. J. Van Alten, R. J. Walter. *Am. J. Reprod. Immunol.* **32**, 43–52 (1994).
107. C. J. Grossman. *Endocr. Rev.* **5**, 435–455 (1984).
108. M. Wehling. *Trends Endocrinol. Metab.* **5**, 347–353 (1994).
109. A. Revelli, M. Massobrio, J. Tesarik. *Endocr. Rev.* **19**, 3–17 (1998).
110. A. R. Buckley. *Lupus* **10**, 684–690 (2001).
111. V. De Mello-Coelho, W. Savino, M.-C. Postel-Vinay, M. Dardenne. *Dev. Immunol.* **6**, 317–323 (1998).
112. K. Dorshkind and N. D. Horseman. *Endocr. Rev.* **21**, 292–312 (2000).

113. S. D. Holladay, P. Lindstrom, B. L. Blaylock, C. E. Comment, D. R. Germolec, J. J. Heindell, M. I. Luster. *Teratology* **44**, 385–393 (1991).
114. B. C. Gehrs and R. J. Smialowicz. *Toxicology* **134**, 79–88 (1999).
115. D. A. Weigent and J. E. Blalock. *Immunol. Rev.* **100**, 79–108 (1987).
116. M. Dardenne. *Ann. Med. (Helsinki)* **31**, 34–39 (1999).
117. R. Mentlein and M. D. Kendall. *Immunol. Today* **21**, 133–140 (2000).
118. E. P. Riley and C. V. Vorhees. *Behavioral Toxicol.* 522 (1986).
119. W. P. Weisenburger. In *Toxicology Testing Handbook,* D. Jacobson-Kram, K. A. Keller (Eds.), pp. 255–290, Marcel Decker, New York (2001).
120. E. R. De Kloet, S. A. B. E. Van Acker, R. M. Sibug, M. S. Oitzl, O. C. Meijer, K. Rahmouni, W. De Jong. *Kidney Intl.* **57**, 1329–1336 (2000).
121. B. McEwen, K. Akama, S. Alves, W. G. Brake, K. Bulloch, S. Lee, C. Li, G. Yuen, T. A. Milner. *Proc. Natl. Acad. Sci. USA* **98**, 7093–7100 (2001).
122. P. J. Shughrue and I. Merchenthaler. *Frontiers Neuroendocrinol.* **21**, 95–101 (2000).
123. K. L. Howdeshell. *Environ. Health Perspect.* **110**, 337–348 (2002).
124. R. T. Zoeller, A. L. S. Dowling, C. T. A. Herzig, E. A. Iannacone, K. J. Gauger, R. Bansal. *Environ. Health Perspect.* **110**, 355–361 (2002).
125. N. J. MacLusky, D. A. Bowlby, T. J. Brown, R. E. Peterson, R. B. Hochberg. *Neurochem. Res.* **22**, 1395–1414 (1997).
126. S. L. Schantz and J. J. Widholm. *Environ. Health Perspect.* **109**, 1197–1206 (2001).
127. E. S. Sher, X. M. Xu, P. M. Adams, C. M. Craft, S. A. Stein. *Toxicol. Ind. Health* **14**, 121–158 (1998).
128. S. K. Mani, J. M. C. Allen, J. H. Clark, J. D. Blaustein, B. W. O'Malley. *Science* **265**, 1246–1249 (1994).
129. K. D. Dohler. *Arch. Toxicol.* **20**, 131–141 (1998).
130. A. C. Scallet and J. M. Meredith. *Neurotoxicol. Teratol.* **24**, 81–85 (2002).
131. K. A. Faber and C. L. Hughes, Jr. *Reprod. Toxicol.* **7**, 35–39 (1993).
132. K. A. Faber, L. Ayyash, S. Dixon, C. L. Hughes, Jr. *Biol. Reprod.* **48**, 947–951 (1993).
133. J. B. Becker. *Pharmacol. Biochem. Behav.* **64**, 803–812 (1999).
134. L. E. Hays. *Environ. Health Perspect.* **110**, 369–376 (2002).
135. L. E. Gray, Jr., J. Ostby, R. L. Cooper, W. R. Kelce. *Toxicol. Ind. Health* **15**, 37–47 (1999).
136. L. E. Gray, Jr., J. S. Ostby, J. M. Ferrell, E. R. Sigmon, J. M. Goldman. *Toxicol. Appl. Pharmacol.* **96**, 525–540 (1988).
137. K. Kubo, O. Arai, R. Ogata, M. Omura, T. Hori, S. Aou. *Neurosci. Lett.* **304**, 73–76 (2001).
138. B. Weiss. *Environ. Health Perspect.* **110**, 387–391 (2002).
139. P. L. Palanza. *Environ. Health Perspect.* **110**, 415–422 (2002).
140. M. Sato, M. Shimada, Y. Sato. *Congenital Anomalies* **41**, 187–193 (2001).
141. F. Dessi-Fulgheri, S. Porrini, F. Farabollini. *Environ. Health Perspect.* **110**, 403–407 (2002).
142. C. S. Roegge, B.-W. Seo, K. M. Crofton, S. L. Schantz. *Toxicol. Sci.* **57**, 121–130 (2000).
143. K. M. Flynn, S. A. Ferguson, K. B. Delclos, R. R. Newbold. *Toxicol. Sci.* **55**, 311–319 (2000).
144. S. A. Ferguson, K. M. Flynn, K. B. Delclos, R. R. Newbold. *Neurotoxicol. Teratol.* **22**, 583–591 (2000).
145. S. A. Ferguson, A. C. Scallet, K. M. Flynn, J. M. Meredith, B. A. Schwetz. *Neurotoxicology* **21**, 947–956 (2000).
146. K.-S. Kang, J.-E. Park, D.-Y. Ryu, Y.-S. Lee. *J. Vet. Med. Sci.* **63**, 1183–1190 (2001).
147. E. S. Goldey and K. M. Crofton. *Toxicol. Sci.* **45**, 94–105 (1998).
148. K. M. Crofton, D. L. Ding, R. Padich, M. Taylor, D. Henderson. *Hearing Res.* **144**, 196–204 (2000).
149. L. E. Gray, Jr., R. J. Kavlok, J. Ostby, J. Ferrell, J. Rogers, K. Gray. *Neurotoxicology* **7**, 449–462 (1986).

150. M. E. Gilbert, W. R. Mundy, K. M. Crofton. *Toxicol. Sci.* **57**, 102–111 (2000).
151. M. Taylor, J. Hedge, M. DeVito, K. Crofton. *Toxicologist* **66**, 133 (2002).
152. P. Clausing, H. K. Mothes, B. Opitz. *Neurotoxicol. Teratol.* **22**, 113–123 (2000).
153. P. Palanza, F. Morellini, S. Parmigiani, F. S. vom Saal. *Neurosci. Behav. Rev.* **23**, 1011–1027 (1999).
154. F. D. Farabollini. *Environ. Health Perspect.* **110**, 409–414 (2002).
155. E. D. Lephart, T. W. West, K. S. Weber, R. W. Rhees, K. D. R. Setchell, H. Adlercreutz, T. D. Lund. *Neurotoxicol. Teratol.* **24**, 5–16 (2002).
156. D. O. Norris, *Vertabrate Endocrinology*. Academic Press, San Diego (1996).
157. G. Ankley, E. Mihaich, R. Stahl, D. Tillitt, T. Colborn, S. McMaster, R. Miller, J. Bantle, P. Campbell, N. Denslow, R. Dickerson, L. Folmar, M. Fry, J. Giesy, L. E. Gray, P. Guiney, T. Hutchinson, S. Kennedy, V. Kramer, G. Leblanc, M. Mayes, A. Nimrod, R. Patino, R. Peterson, R. Purdy, R. Ringer, P. Thomas, L. Touart, G. Van Der Kraak, T. Zacharewski. *Environ. Toxicol. Chem.* **17**, 68–87 (1998).
158. L. Helmuth. *Science* **296**, 2131–2133 (2002).
159. A. M. Cummings. *Crit. Rev. Toxicol.* **27**, 367–379 (1997).
160. E. Mylchreest, M. Sar, R. C. Cattley, P. M. Foster. *Toxicol. Appl. Pharmacol.* **156**, 81–95 (1999).
161. E. Mylchreest, D. G. Wallace, R. C. Cattley, P. M. Foster. *Toxicol. Sci.* **55**, 143–151 (2000).
162. R. N. Wine, L.-H. Li, L. H. Barnes, D. K. Gulati, R. E. Chapin. *Environ. Health Perspect.* **105**, 102–107 (1997).
163. D. Beltrame, E. DiSalle, E. Giavini, K. Gunnarsson, M. Brughera. *Reprod. Toxicol.* **15**, 195–213 (2001).
164. L. S. Birnbaum. *Environ. Health Perspect.* **103**, 89–94 (1995).
165. R. E. Faith, M. I. Luster, C. A. Kimmel. *Clin. Exp. Immunol.* **35**, 413–420 (1979).
166. T. L. Bunn, P. J. Parsons, E. Kao, R. R. Dietert. *Toxicol. Sci.* **64**, 57–66 (2001).
167. J. M. Spyker-Cranmer, J. B. Barnett, D. L. Avery, M. F. Cranmer. *Toxicol. Appl. Pharmacol.* **62**, 402–408 (1982).
168. J. B. Barnett, L. S. F. Soderberg, J. H. Menna. *Toxicol. Lett.* **25**, 173–183 (1985).
169. J. B. Barnett, B. L. Blaylock, J. Gandy, J. H. Menna, R. Denton, L. S. F. Soderberg. *Fundam. Appl. Toxicol.* **14**, 688–695 (1990).
170. J. G. Vos and J. A. Moore. *Int. Arch. Allergy Appl. Immunol.* **47**, 777–794 (1974).
171. R. E. Faith and J. A. Moore. *J. Toxicol. Environ. Health* **3**, 451–464 (1977).
172. J. S. Fine, T. A. Gasiewicz, A. E. Silverstone. *Mol. Pharmacol.* **35**, 18–25 (1989).
173. B. L. Blaylock, S. D. Holladay, C. E. Comment, J. J. Heindel, M. I. Luster. *Toxicol. Appl. Pharmacol.* **112**, 207–213 (1992).
174. B. C. Gehrs, M. M. Riddle, W. C. Williams, R. J. Smialowicz. *Toxicology* **130**, 219–228 (1998).
175. R. J. Smialowicz, W. C. Williams, C. B. Copeland, M. W. Harris, D. Overstreet, B. J. Davis, R. E. Chapin. *Toxicol. Sci.* **61**, 164–175 (2001).
176. S. Chen, K. A. Golemboski, F. S. Sanders, R. R. Dietert. *Toxicology* **132**, 67–79 (1999).

Pure Appl. Chem., Vol. 75, Nos. 11–12, pp. 2125–2141, 2003.

Topic 3.11

Significance of experimental studies for assessing adverse effects of endocrine-disrupting chemicals*

L. Earl Gray, Jr.[1,‡] and Paul M. D. Foster[2]

[1]2525 Highway 54, MD 72, NHEERL, Endocrinology Branch, Reproductive Toxicology Division, ORD, USEPA, Research Triangle Park, NC 27711, USA; [2]Environmental Toxicology Program, NIEHS, MD E1-06, 111 TW Alexander Drive, Research Triangle Park, NC 27709, USA

Abstract: The U.S. Environmental Protection Agency (USEPA) is developing an endocrine disruptor screening and testing program to detect chemicals that alter hypothalamic-pituitary-gonadal (HPG) function, estrogen, androgen, and thyroid (EAT) hormone synthesis or metabolism and induce androgen (AR) and estrogen (ER) receptor-mediated effects in mammals and other animals. The utility of this approach is based upon the knowledge that mechanisms of endocrine-disrupting chemical (EDC) action are highly conserved at the cellular and molecular levels among vertebrates. Some EDC mechanisms also are shared with invertebrates. High-priority chemicals would be evaluated in a Tier 1 screening (T1S) battery, and chemicals that are positive in T1S would then be tested in Tier 2 (T2). T1S includes in vitro ER and AR receptor binding and/or gene expression, an assessment of steroidogenesis and mammalian (rat) and nonmammalian (fish) in vivo assays. In vivo, the uterotropic assay detects estrogens and antiestrogens, while steroidogenesis, antithyroid activity, antiestrogenicity, and HPG function are assessed in a pubertal female assay. Antiandrogens are detected in the Hershberger assay (weight of androgen-dependent tissues in castrate-immature-male rats). Fish and amphibian assays are also being developed to identify EDCs. Several alternative mammalian in vivo assays have been proposed. Of these, a short-term pubertal male rat assay appears most promising. T1S is designed to be sensitive to EAT activities, but many of the effects detected at the screening level would not be considered adverse, the dosage levels may be high, and the route of administration used may not be the most relevant. However, issues of adversity, dose response, and route(s) of exposure would be resolved in the testing phase. In addition to using an enhanced multigenerational test for Tier 2, an in utero-lactational screening protocol is also being evaluated by the USEPA for use in T2 or T1S. For T2, the numbers of endocrine-sensitive endpoints and offspring (F1) examined in multigenerational tests need to be expanded for EDCs in a thoughtful manner, based in part upon the results of T1S. In addition, for some chemicals histological examination of 10 adult F1 per sex in only the control and high-dose groups provides inadequate

---

*Report from a SCOPE/IUPAC project: Implication of Endocrine Active Substances for Human and Wildlife (J. Miyamoto and J. Burger, editors). Other reports are published in this issue, *Pure Appl. Chem.* **75**, 1617–2615 (2003).

‡Corresponding author: Tel.: 919-541-7750; E-mail: gray.earl@epa.gov

DISCLAIMER. The research described in this article has been reviewed by the National Health Environmental Effects Research Laboratory, U.S. Environmental Protection Agency, and approved for publication. Approval does not signify that the contents necessarily reflect the views and policies of the Agency nor does mention of trade names or commercial products constitute endorsement or recommendation for use.

statistical power to detect low-dose lesions induced during development. In these cases, we propose that all the offspring be examined after puberty for gross and histological reproductive abnormalities. Since EDCs, like the phthalates and AR-antagonists, produce characteristic profiles, or syndromes, of adverse effects, data need to be reported in a manner that clearly identifies the proportion of animals displaying one or more of the abnormalities in a syndrome. Consideration should be given to tailoring T2, based on the results of T1S, to assure that all of the effects in such chemically induced developmental syndromes are included in the study.

INTRODUCTION

The potential effects of endocrine-disrupting chemicals (EDCs) on human health and the proven effects of EDCs on wildlife are a major concern among the public and the scientific communities. Due to gaps in the current testing of chemicals, the U.S. Environmental Protection Agency (USEPA) was given a mandate in 1996 under the Food Quality Protection and Safe Drinking Water Acts to develop a screening and testing program for endocrine effects. Some of the impetus for these actions arose from a Work Session in 1991 on "Chemically Induced Alterations in Sexual Development: The Wildlife/Human Connection" [1], which stated that "many compounds introduced into the environment by human activity are capable of disrupting the endocrine system of animals, including fish, wildlife, and humans. Endocrine disruption can be profound because of the crucial role hormones play in controlling development." [2]. Scientists at this meeting "estimated with confidence" that impairments in humans have resulted from exposure to endocrine disruptors. Laboratory animal studies corroborate the abnormalities of reproductive development observed in the field and, in some cases, define toxic mechanisms causing adverse effects. Among chemicals of concern, are pesticides, "inert" ingredients, industrial chemicals, pharmaceuticals, phytochemicals, food supplements, personal care products, and "natural" products or nutraceuticals.

In response to the 1996 legislative mandate for an endocrine screening and testing program, the USEPA formed the Endocrine Disruptors Screening and Testing Advisory Committee (EDSTAC), which proposed a tiered screening and testing strategy for EDCs in its final report (1998) (<http://www.epa.gov/scipoly/oscpendo/history/finalrpt.htm>). The EDSTAC proposal includes (i) a process to prioritize chemicals for evaluation and (ii) tiered screening (Tier 1) and (iii) testing (Tier 2) batteries. The chemical "universe" to be considered includes over 80 000 chemicals, of which only a subset of high priority chemicals would initially enter the screening program. Prioritization would include an estimation of the chemicals' ability to interact with steroid hormone receptors using either quantitative structure–activity relationships (QSARs) for chemicals that bind steroid receptors or high-throughput prescreening (HTPS) using hormone-dependent gene expression assays, among several other factors. However, to date validated QSAR models or HTPS have not yet been developed for either estrogens or androgens by the USEPA.

The screening battery recommended by EDSTAC was designed to detect chemicals that alter hypothalamic-pituitary-gonadal (HPG) function, estrogen, androgen, and thyroid hormone synthesis or metabolism, or induce androgen (AR) and estrogen (ER) receptor-mediated effects in mammals and other taxa. Chemicals positive in Tier 1, based upon a "weight-of-evidence" analysis, would be considered as potential EDCs and subjected to testing (Tier 2). Tier 1 should include assays sensitive enough to detect EDCs while issues of "dose-response, relevance of the route of exposure, sensitive life stages and adversity" would be resolved in the testing phase. Although the USEPA has not yet endorsed this approach, we believe that equivocal effects in Tier 1 assays should be replicated or evaluated further in additional short-term assays before more extensive Tier 2 testing (T2T) is initiated (Fig. 1). This could prevent unnecessary testing in some cases, saving time and money and reducing animal use, and would provide useful information for tailoring Tier 2 tests.

Based in large part upon the EDSTAC recommendations, the Office of Prevention, Pesticides and Toxic Substances (OPPTS) of the USEPA designed the EPA Endocrine Disruptors Screening Program (EDSP), and the agency has begun implementing elements of this program (report to Congress, available at <http://www.epa.gov/scipoly/oscpendo/index.htm#currentstatus>), and their efforts are being coordinated internationally with the Office of Economic Cooperation and Development (OECD) (described at the following Web site: <http://www.oecd.org//ehs/ENDOCRIN.HTM>).

The first portion of this review will briefly summarize the status of some of the screening assays, while the second part will discuss approaches and limitations of different methods for testing EDCs identified as potential endocrine disruptors in T1S.

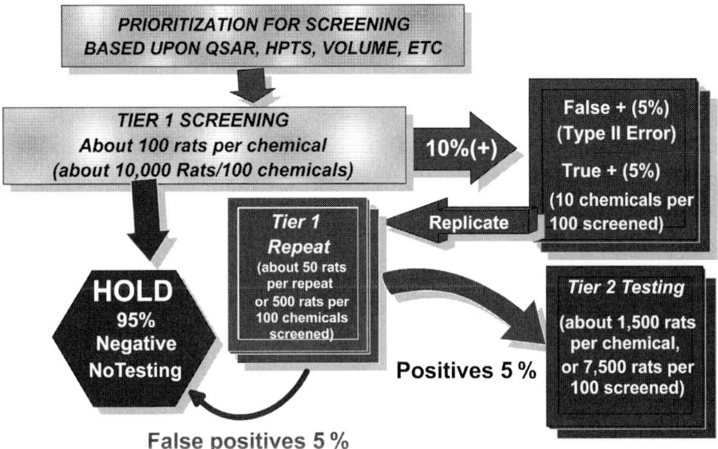

Fig. 1 In this figure, we provide an estimate the animal use in the Screening and Testing Program based upon the numbers of animals used per hundred chemicals screened. This estimate is based upon several assumptions. T1S would use about 100 animals per chemical screened. T1S includes three in vivo assays [uterotropic (n = 6/group), Hershberger (n = 6/group) and the pubertal female rat assay (n = 15/group) or a total of 27 rats per group]. We assumed that each chemical would be evaluated using three groups (control, medium and high dosage levels = 81 rats; the high dosage level should not exceed an MTD, less than a 10 % body weight reduction vs. control) with a few animals being used for limited dose-range finding for T1S. In this figure, we assumed that about 5 % of the chemicals screened would be positive because they act in an antiestrogenic, antiandrogenic, antithyroid manner or inhibit steroidogenesis or HPG maturation. We also assumed that another 5 % would be statistical false positives. We propose that false positives could be virtually eliminated from entering T2T and sent to the hold box by repeating or expanding positive T1S results before proceeding to the testing phase of the battery. T2T is the most intensive phase of the battery in terms of animal use, using about 1500 animals in a multigenerational test per chemical. Note that for each 5 % increase in the rate of false positives there is an increase of 250 rats used per 100 chemicals in the "repeat" phase we have proposed as a transition from T1S to either T2T or the hold box. Hence, if the false positive rate was 50 % rather than 5 %, animal use would increase from 18 000 only by 2250 rats to 20 250 per 100 chemicals. We suspect that initially only a few hundred high-priority chemicals will be evaluated in the EDSP Tier 1 screening battery.

TIER 1 SCREENING: IN VITRO

In vitro, cell-free, or whole-cell mammalian AR or ER binding assays were proposed to detect compounds that act as hormone agonists or antagonists [3]. The use of whole cells or recombinant human receptors for binding assays [as described by refs. 4,5,6] eliminates the need to isolate receptors from animal tissues and may be amenable to high-throughput screening. Gene expression assays also were proposed because they can distinguish receptor agonists from antagonists. The use of cell lines containing AR or ER stably or transiently transfected with a reporter gene, or cell lines cotransfected with the reporter gene and the steroid receptor are now available to rapidly discriminate receptor agonists from antagonists [7–10]. Stable cell lines are likely to be less variable than transient transfections. and they do not require repeatedly transfecting cells, while new transfection methods have improved the utility of the transient transfection assays [11]. Recently, CV-1 and MDA-453 cells were transduced with hAR and reporter genes using adenovirus to detect antiandrogens [12]. These transduction assays provided low background and 80–100-fold induction with 1–10 nM dihydrotestosterone. Adenovirus transduced cell lines and stably transfected cells are amenable to high-throughput screening. In addition, transduction assays using adenovirus could also be used to deliver other genes to different cell types to screen for additional EDC mechanisms beyond EAT. Optimized and standardized protocols along with performance criteria (intra- and inter-assay and interlaboratory CVs) for binding and transcriptional activation assays need to be developed. While the ER and AR binding and transcriptional activation assays appear to be quite sensitive, except for some chemicals that require metabolic activation, the specificity of responses at the high concentrations that are used for weak ligands appears poor. The fact that a chemical induces a response in vitro that reaches an IC_{50} in the high micromolar to millimolar range may be due to disruption of the assay conditions for receptor stability or cell viability, rather than competitive binding to the receptor. Improving the specificity of these assays will require further in vitro experimentation to separate receptor ligands from false positives. Cytotoxicity (for whole-cell assays), and Ki [13] or Kb [14] determinations are required to enhance the specificity of these assays to acceptable levels.

Some EDCs inhibit reproduction and development by altering steroid hormone synthesis. While the presence of ER and AR in invertebrates has yet to be established, chemicals that alter steroid hormone synthesis in vertebrates can also inhibit ecdysteroid synthesis in invertebrates. Chemicals that inhibit steroidogenesis can be detected in vitro or ex vivo using a variety of methods [briefly reviewed by refs. 15,16]. Potential protocols for these assays are being evaluated under within the USEPA and in contract laboratories. A T1S in vitro assessment of steroidogenesis would complement the in vivo effects of altered steroidogenesis in the pubertal female rat assay.

Scientists are attempting to use as few animals as possible in the most precise and sensitive assays, by incorporating sensitive in vitro assays in T1S and by using QSAR models or HTPS in the prioritization of chemicals for screening. Unwarranted animal use also can be avoided as chemicals negative in T1S will not be subject to T2T (Fig. 1). In addition, testing T1S statistical "false positives" in T2 can be almost entirely eliminated by assuring that T1S assay results are replicated or further investigated in supplemental studies, before moving to T2T. Attempts to enhance T2T by adding more sensitive endpoints and a more thorough evaluation of the animals already on study should lead to improved hazard characterization and result in reductions in animal use.

TIER 1 SCREENING: IN VIVO

Three short-term in vivo mammalian assays were included in the EDSTAC Tier 1 screening battery [EDSTAC, 1998; 15,17,18]. In vivo assays are required in T1S because in vitro assays cannot account for absorption, distribution, metabolic activation, and excretion of xenobiotics, potentially resulting in "false negative" responses (positive in vivo, but negative in vitro). For example, purified methoxychlor is inactive or weakly active in vitro whereas in vivo it is almost as potent as is 17β estradiol when both are given orally. In addition, the integrated nature of the endocrine system in the developing organism and the relationship of the endocrine toxicity to other systemic effects cannot be simulated in vitro.

Fish and amphibian assays also are being developed for T1S [19–21]. The fathead minnow assay, developed by USEPA scientists, can identify EDCs displaying several mechanisms of concern, including androgen and estrogen receptor agonists and antagonists and inhibitors of steroid hormone synthesis. An amphibian metamorphosis assay is being developed to detect thyroid-active substances.

UTEROTROPIC ASSAY

In T1S, estrogen agonists and antagonists are detected in a 3-day uterotropic assay using subcutaneous (sc) administration of the test compound. Several variations of the uterotropic assay protocol are being examined by OECD, and several phases of the interlaboratory study have been completed and published [22–24]. The described uterotropic assays for estrogens and antiestrogens uses either intact juvenile or ovariectomized adult/juvenile female rats and was found by all laboratories to be responsive to estrogenic toxicants over a wide range of potencies.

The second in vivo assay in the EDSTAC T1S is designed to detect antiandrogenic activity simply by weighing androgen-dependent tissues following 10 days of oral treatment in the immature castrated male rat [15,17,25] (Fig. 2). Both the uterotropic and Hershberger assays were standardized and validated for screening chemicals by an expert committee of endocrinologists in 1962 [26]. In this assay, weights of the ventral prostate, Cowper's glands, seminal vesicle (with coagulating glands and fluids), glans penis and levator ani/bulbocavernosus muscles are measured in castrated, sc testosterone-treated (or untreated) male rats after 10 days of oral treatment with the test compound. This is a very sensitive assay for detection of androgens and antiandrogens. Other useful endpoints that help reveal the mechanism of action include dorsolateral prostate, adrenal, liver, and kidney weights and serum (collected by cardiac puncture) levels of testosterone and LH. The Hershberger assay has been shown to be much more sensitive and specific to AR-mediated alterations than assessment of endocrine activity in the intact adult male rat, which fails to consistently detect the antiandrogenic activity of several weakly antiandrogenic pesticides like *p,p*-DDE and linuron that are easily detected in the Hershberger assay [6,27]. Chemicals that induce testosterone metabolism also can reduce androgen-dependent tissue weights while those that inhibit 5α reductase activity, can dramatically reduce male accessory sex gland weight with lessor effect on testosterone-mediated muscle. It appears that steroidogenesis in the

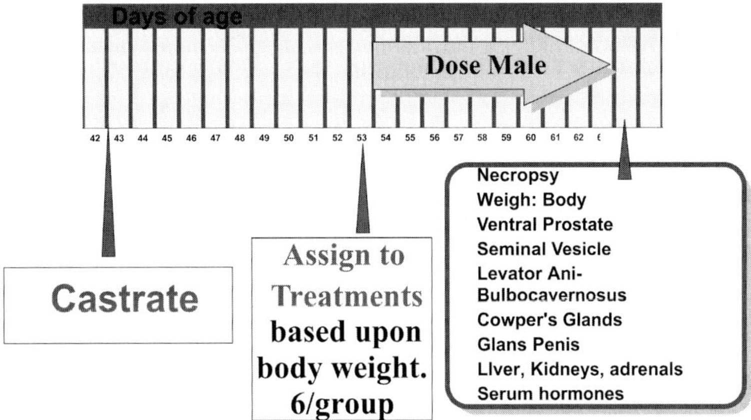

Fig. 2 OECD Hershberger assay protocol for antiandrogens, testosterone propionate (TP sc at 0.4 mg/kg/d for detection of an antiandrogen) is coadministered to male rats with the oral treatment for 10 days. To detect androgens, omit concurrent TP treatment. Castrate-immature male rats are dosed sc with TP and orally with the toxicant for 10 days, after which the animals are necropsied and androgen-dependent tissues (ventral prostate, seminal vesicles, Cowper's glands, glans penis and levator ani-bulbocavernosus muscles) and other organs (liver, kidney, adrenals) are weighed. Serum (collected by cardiac puncture) can also be taken for analysis of testosterone levels. Detects androgen receptor (AR) agonists and antagonists, and chemicals that alter AR levels, inhibitor of 5α reductase or stimulate testosterone metabolism.

adrenals together with thyroid function could also be evaluated in the Hershberger assay, but these endpoints are not routinely included by most investigators using this assay. Although changes in androgen-dependent organ weights in this assay are not necessarily considered as being adverse, we have found that chemicals that are positive in this assay often produce adverse effects during puberty and after in utero exposure. Members of the Endocrinology Branch of NHEERL, ORD, USEPA are working as the lead laboratory supporting the OECD-led effort to standardize and validate this assay in an interlaboratory (17 laboratories) study of androgens and antiandrogens of varying potencies. To date, phase 1 [an interlaboratory study using testosterone propionate sc (17 laboratories, 6 dosage levels), and flutamide (6 laboratories, 6 dosage levels)] and phase 2 (17 laboratories, 7 chemicals at several dosage levels, with 3–8 laboratories per chemical) have been completed, reported to the OECD, and the results are being prepared for publication.

PUBERTAL FEMALE RAT ASSAY

The "pubertal female rat assay" in T1S (Fig. 3) [28] was originally developed in the Endocrinology Branch in our laboratory along with the "pubertal male assay" nearly two decades ago [29,30]. In the pubertal female assay, weanling rats are dosed daily by gavage for about 20 days and the age at vaginal opening (puberty) and estrous cycles are monitored and the females are necropsied at about 42 days of age. The age at vaginal opening appears to be exquisitely sensitive to xenoestrogens like methoxychlor [30]. Additional measurements include serum thyroid hormones, and uterine and ovarian weight and histology. This assay detects alterations in thyroid hormone status, HPG function, inhibition of steroidogenesis (including aromatase), estrogens, and antiestrogens [31,32].

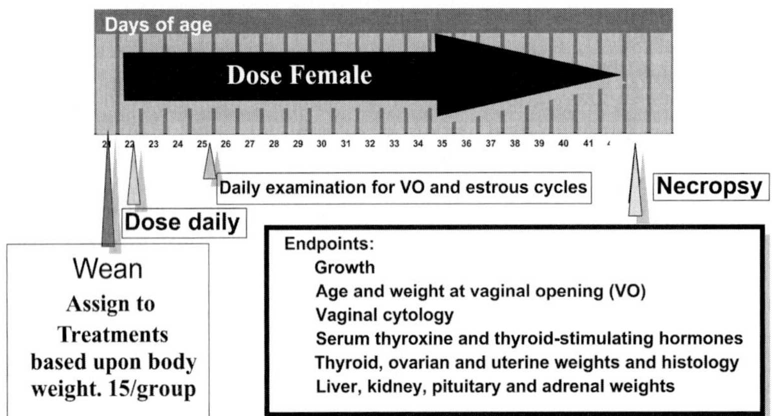

Fig. 3 Pubertal female rat protocol. Originally developed by the USEPA, ORD, NHEERL, Endocrinology Branch [28–31,33,35,37,40]. Intact weanling females are dosed by gavage from 22–43 days of age). The age and weight at vaginal opening (puberty) is determined, estrous cycles are monitored after puberty and the females are necropsied at 43 days of age. Necropsy includes an assessment of reproductive (uterus and ovaries) and nonreproductive (liver, kidney, adrenals, thyroid) organ weights and histology (uterus, ovaries, and thyroid). Serum is taken for T4 and TSH analyses. This protocol is designed to detect inhibition of steroidogenesis including aromatase, antithyroid, and antiestrogenic activities and altered HPG maturation.

PUBERTAL MALE ASSAY: AN ALTERNATIVE SCREENING ASSAY

Alternative T1 and T2 in vivo assays were discussed by EDSTAC, and some of these are being evaluated by the USEPA in the EDSP. If these assays prove to be of sufficient sensitivity and equivalent utility, they could replace or augment current T1S assays. One promising alternative assay is the "pubertal male rat assay" (Fig. 4) [15,29,30,34–37], which detects alterations in thyroid function, HPG maturation, steroidogenesis (but not aromatase), and steroid hormone function (androgen). Intact weanling males are exposed orally to the test substance for about 30 days, the age at puberty is determined, and reproductive tissues are evaluated and serum taken for hormonal analyses. Although this is a screening assay, the USEPA has previously used altered puberty [a two-day delay in preputial separation (vinclozolin) or a similar delay in vaginal opening in the female] as adverse effects in their risk assessments. In addition, for vinclozolin the extra 10 X factor mandated by 1996 Food Quality Protection Act for chemicals of special concern for the child was retained in the risk assessment. In this case, a delay in preputial separation was accompanied by other endocrine and reproductive alterations, in the absence of an effect on body weight or growth. Furthermore, all of the in vivo reproductive effects of vinclozolin are consistent with the AR antagonism seen in vitro and altered androgen-dependent gene expression in vivo [13,38] and other in vivo effects [39,40].

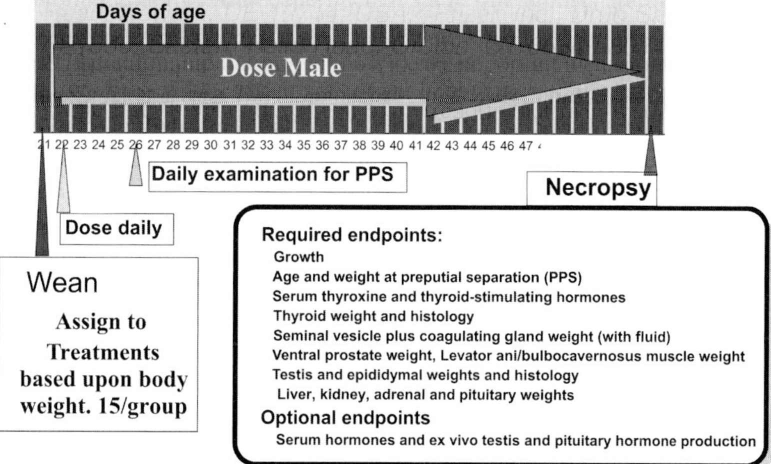

Fig. 4 Pubertal male rat protocol. Originally developed by the USEPA, ORD, NHEERL, Endocrinology Branch [28–31,33,35,37,40]. Weanling males are dosed by gavage from 22–53 days of age. The age and weight at preputial separation (puberty) is determined and the males are necropsied at 52–55 days of age. Necropsy includes an assessment of reproductive (testes, epididymides, ventral prostate, seminal vesicles, Cowper's glands, levator ani plus bulbocavernosus muscles) and nonreproductive (liver, kidney, adrenals), pi organ weights and histology. Serum is taken for T4 and TSH analyses. Thyroid weight and histology also are evaluated. This protocol is designed to detect inhibition of steroidogenesis, antithyroid, and antiandrogenic activities and altered HPG maturation.

TIER 2 TESTING

In a tiered screening and testing approach, only chemicals displaying positive, reproducible responses in T1S would be evaluated further in full life cycle, or multigenerational tests (Fig. 1). It is in T2 testing, not T1S, that issues of dose response, relevance of the route of exposure, sensitive life stages, and adversity are resolved [15,41,42]. For some endocrine activities, the numbers of sensitive endpoints and (F1) offspring examined in these assays should be expanded on a case-by-case basis. For example, endpoints like anogenital distance at birth and nipple/areola retention in infant female and male rats should be included in testing of androgens and antiandrogens, respectively, because they are sensitive, potentially permanent effects that are highly correlated with malformations and reproductive organ weight changes later in life. These early alterations constitute part of antiandrogen-induced developmental syndromes. As described in detail earlier [40–47], a careful evaluation of the male rat offspring allows one to distinguish the "phthalate syndrome" where effects on reproductive development involve a decrease in fetal testicular testosterone and insulin like-3 peptide hormone biosynthesis [46,48,49] from the "AR antagonist syndrome" typified by vinclozolin or flutamide [50]. The main distinction being that the phthalate syndrome includes testicular, epididymal, and gubernacular cord agenesis [40,47,51]. These lesions are rarely seen in the AR antagonist syndrome even when all the males display hypospadias. In the AR-antagonist syndrome, DHT-dependent tissues are more severely than T-dependent tissues and the dose–response curves display an orderly cascade of effects in the male offspring (listed in order of low to high dosage levels of 3–200 mg vinclozolin kg/d) including reduced anogenital distance, retained areolas, reduced androgen-dependent organ weights and retained nipples, hypospadias, vaginal pouch, agenesis of the sex accessory glands, and undescended testes) [15,39,44,50,52]. In the risk assessment on vinclozolin, the acute dietary risk was based upon a permanent reduction in ventral prostate weight (retaining an extra 10 X factor under FQPA). Of the all the chemicals that interfere with androgen signaling pathway in the fetal male rat, only the phthalates affect Leydig cell insl-3 peptide hormone synthesis and cause undescended testes with gubernacular agenesis [53].

When conducting a transgenerational study, one must evaluate all the components of a syndrome so that affected animals are not misidentified and classified as "normal" and the data should be summarized in a manner that clearly delineates the proportion of animals that are affected, indicating that they display any lesion (histological or gross pathology) consistent with the syndrome. Teratology/developmental toxicity studies typically present and analyze data in this manner, indicating the number malformed/number observed on an individual and litter basis, while multigenerational studies frequently do not, even when clear teratogenic and other developmental responses are noted postnatally. In the multigeneration study, one cannot easily determine how many animals are adversely affected because the histopathological data (usually on limited numbers of F1 animals from some, but not all of the treatments) are not analyzed with the gross pathology data. It is important to note that the endpoints altered by exposure to classes of antiandrogens are not necessarily as sensitive to disruption by other important endocrine mechanisms. For EDCs that display androgenic or antithyroid activity, alter steroidogenesis, estrogenic or display antiestrogenic activity, the syndromes (diagnostic profiles of developmental effects) will differ markedly from those described for the AR-antagonist and phthalate syndromes.

As discussed above, if a chemical produced positive results in T1S and T1-Repeat (as described in Fig. 1) it would proceed to T2T in "enhanced" multigenerational tests. Multigenerational protocols are used in T2 because these are the only protocols that expose the animals during all critical stages of development and examine reproductive function of offspring after they mature (Fig. 5). In contrast, the developmental toxicity/teratology test is too insensitive to be used to detect the reproductive effects of EDCs (Fig. 6) because it only examines fetal animals a period when most reproductive tract malformations cannot be detected by a gross examination. For example, trenbolone-induced vaginal agenesis and linuron-induced epididymal agenesis were not detected in developmental toxicity tests on these chemicals. This can be problematic because this assay is often used by the USEPA to establish NOAELs, RfDs, and aPADs for acute dietary exposures.

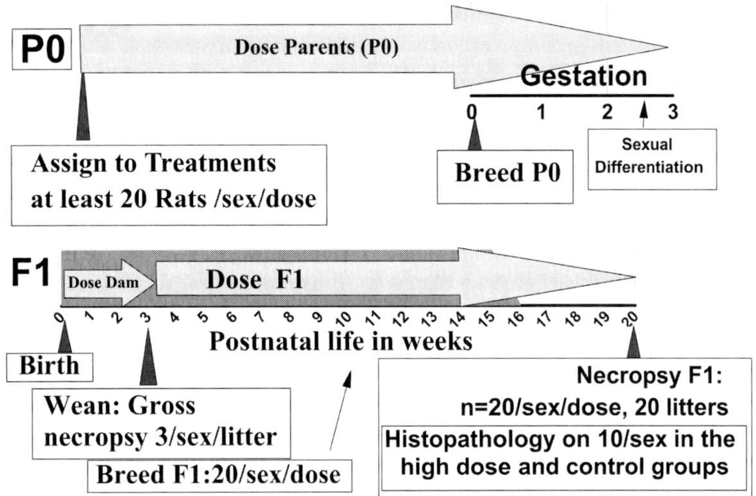

Fig. 5 A brief description of the standard USEPA 1996 multigenerational reproductive test guideline. This protocol uses more litters in the P0 generation than does the transgenerational study (Fig. 5), using at least 20 per sex per dose group. Litters can be standardized as neonates. At weaning three males and females per litter are examined for gross reproductive lesions, one animal per sex per litter is retained for further assessment (with dosing continued, usually oral, either gavage or dietary) and the rest of the pups are discarded. After generation of the F1a, and in some cases and F1b or F1c, the P0 animals are necropsied. Puberty, and fertility are assessed in the F1 and they too are necropsied (20 per sex per dose). Histology of the reproductive organs is required only on 10 animals per sex from the control and high-dose groups. If puberty was altered in the F1 generation, anogenital distance is measured in the F2. F2 pups are usually discarded at weaning.

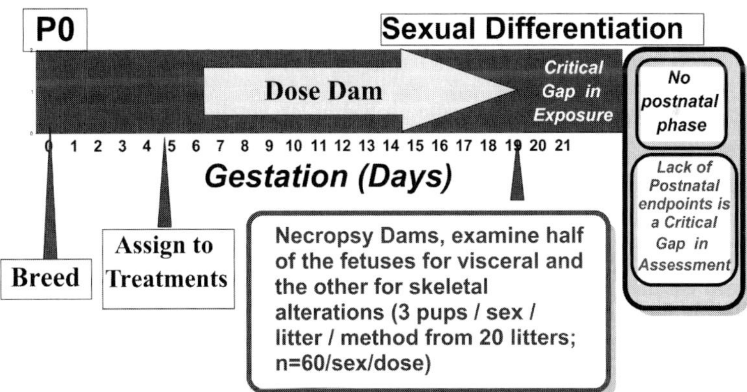

Fig. 6 General USEPA teratology/developmental toxicology test guideline study. Pregnant rats are dosed after implantation to a day or so before parturition. Previous to the current updated test guidelines, as found in most publications and pesticide studies submitted to the agency, dosing was terminated at day 15 of gestation. After dosing is terminated, the dams are euthanized, fetuses removed, and all the pups are examined for malformations, with one-half being used for visceral examination and the other half undergoing skeletal exam. These studies typically do not include any postnatal assessment of the exposed F1, while some studies examine the F1 animals during neonatal life.

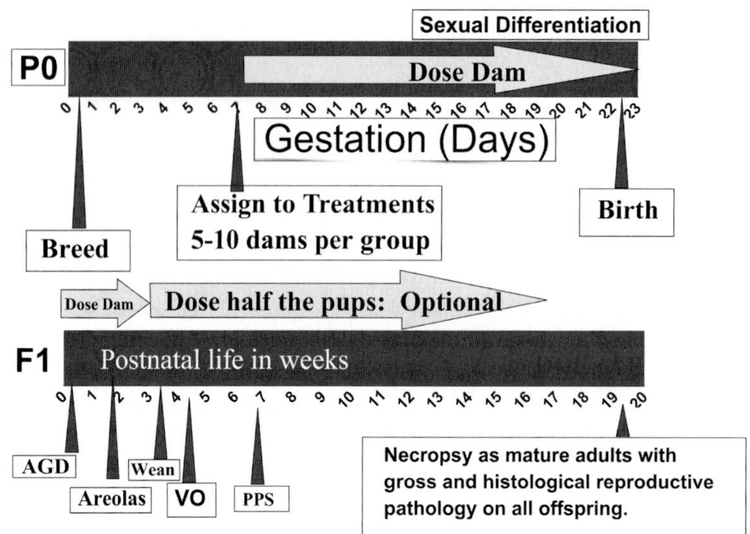

Fig. 7 General "transgenerational protocol". These protocols vary slightly from laboratory to laboratory and sometimes from study to study within a laboratory with regard to (i) the specific day of gestation that dosing is initiated, (ii) the day of gestation or lactation on which dosing is terminated, (iii) whether or not dosing is continued after weaning, with the F1 being directly exposed (rarely so), and (iv) the numbers of litters used in each dose group (5–10 is typical). However, in general the exposure is oral gavage and always includes the period of fetal sexual differentiation, the litters are not culled or standardized and all of the pups of the affected sex(es) are examined through puberty, young adulthood (5–9 months of age) and all individuals are examined thoroughly (gross morphology, histology, and weights of reproductive tissues) at necropsy. These tests are tailored to emphasize the endpoints known to be affected or suspected to be target tissues, based on other in vitro or in vivo screening studies.

TRANSGENERATIONAL TESTING FOR ANTIANDROGENS AND ANDROGENS

While the new USEPA multigenerational test provides for a comprehensive evaluation of the P0 or parental generation, too few F1 animals (offspring with developmental exposure) are examined after maturity to detect anything but the most profound reproductive teratogens [14,54]. While P0 animals within a dose group typically respond in a similar fashion to the chemical exposure, developmental events can be "all or none" with just a few animals displaying reproductive malformations in the lower dosage groups. Thus, a standard multigenerational protocol which examines one F1 animal per sex, per litter after maturity from 20 litters per dosage group can only detect statistically significant alterations if they are displayed by 25 % or more of the offspring. Histopathological alterations would have to be displayed by 50 % or more of the offspring since a histological examination of the reproductive tract is only required in 10 F1 animals per sex, per dose, far too few to detect anything but the most profound effects on reproductive development. As shown in Fig. 8, we have found that the antiandrogenic effects of linuron (epididymal and testicular hypoplasia) and vinclozolin (retained nipples) are displayed in the low-dose groups at rates well below these levels and, hence, would either go undetected at necropsy or would be described as nonsignificant. In a study of the phthalate ester di-*iso*-nonyl phthalate (DINP) [47], we found that a high dose administered to the pregnant dam induced a significant incidence of severe malformations in only 7 % of the male offspring (based upon effects consistent with the phthalate syndrome). In this regard, there is considerable uncertainty associated with LOAELs and NOAELS identified in multigenerational studies of phthalate esters that do not include either a complete assessment of the endpoints that constitute the phthalate syndrome or those included in the USEPA's new multigenerational test guidelines [55,56].

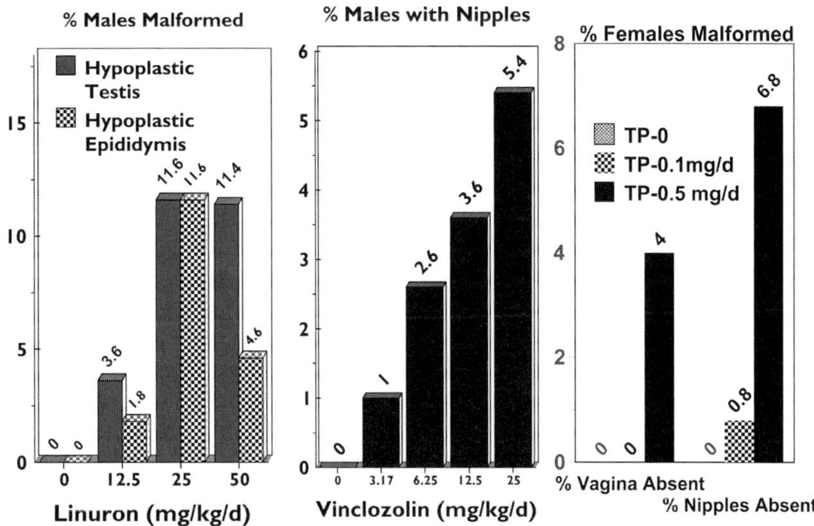

Fig. 8 In the standard multigenerational reproduction test, malformations induced by low-dosage levels of EDCs may go undetected (not observed and/or not being statistically significant) because too few F1 animals are thoroughly examined after reproductive maturity. Presented are examples of malformations that we have found in our studies that occur at a low incidence in lower-dosage groups. We are concerned that such effects will go undetected with a gross examination of only 20 F1 adult animals per sex per group and a histological examination of only 10 F1 adult animals per sex from the control and high-dose groups. Treatment with an androgen or antiandrogens induces low levels of malformations in the lower-dosage groups. [Data from refs. 14,40,61.]

When the female rat fetus is exposed to testosterone in utero, agenesis of the vagina and nipples also are seen at a low rate in the lower-dosage groups (Fig. 8). In fact, most of the low-dose effects of androgens in the female offspring (retained prostatic and vesicular tissues, nipple agenesis) are effects that are likely to be missed in a standard necropsy. There are additional factors besides detection of adverse effects at necropsy or during data analysis and interpretation that limit interpretation of data from the standard multigenerational reproduction test. The life-long exposure of both males and females in the F1 generation, which allows one to detect effects induced in utero, during lactation or from direct exposure after puberty, can confound the identification of when the effect was induced, i.e., during adulthood versus development or even the affected sex. For example, the current RED on linuron exemplifies how the registrant [57] and the USEPA (RED, 1995, TRED, 2002) have misinterpreted the stage of life and the mode of action for the effects of this pesticide on the testis and epididymis. This confusion arose from the fact that the F1 animals were dosed continuously and the reproductive effects, though actually induced in utero [6,14,40,58,59], were incorrectly attributed to a direct effect of linuron in the adult via increased serum LH and Leydig cell hyperplasia. In fact, linuron does not increase serum LH or induce Leydig cell hyperplasia in adult male rats [40] and the testicular effects result from pressure atrophy after puberty as the Sertoli cells form tight junctions and the fluid secreted into the lumen of the tubules is unable to flow normally out of the testis in animals presenting epididymal agenesis [58,59]. In our studies, the dosing period normally is terminated near birth, or at weaning which precludes misinterpretation of the developmental origin of reproductive effects. For DBP, the effects on fertility in P0 female rats is obscured by the more obvious effects on testis histology and sperm numbers in their breeding partners [40]. On the other hand, it is clear that our transgenerational protocols would not be appropriate for EDCs that induced low-dose alterations in the pubertal or adult animal. It is ironic that the teratology test examines all the animals carefully in a litter, but at an inappropriate stage of life, while the multigenerational test, which includes exposure during sexual differentiation and then examines these offspring as adults, discards most of the F1 animals before maturity.

In our "transgenerational" protocols [14,39,43–47,54,59], we typically use fewer litters (7–10 per dose group) but examine all the animals in each litter. A detailed methods paper describing how we execute this protocol is now available [60]. These protocols actually use fewer animals, but provide more statistical power to detect reproductive effects in the F1 generation. In addition, the measurement of anogenital distance (AGD) at 1–2 days of age and areolas/nipples at 12–13 days of age in the F1 generation is included. These endpoints are not only extremely useful "biomarkers" of effect, but the changes in AGD and nipples can be permanent. Moreover, these endpoints are highly correlated with malformations and other alterations in androgen-dependent tissues [59; Hotchkiss et al., in preparation]. Unfortunately, neither of these markers are measured in the F1 generation under the 1996 USEPA multigenerational test guidelines. In this regard, we propose that testing protocols should be tailored based upon the results of T1S such that more F1 male offspring are examined thoroughly (Table 1) as adults than is now required if the chemical appears to be antiandrogenic, while the examination of F1 females can be similarly enhanced (Table 2) for androgens.

It is important to reiterate that the above endpoints that are sensitive to antiandrogens or androgens in utero have not been shown to be sensitive to xenoestrogens or some other EDCs. We are not suggesting that all EDCs should be tested like the androgens or antiandrogens. Testing should be tailored based upon the pharmacological activity demonstrated in T1S. In addition, the developing fetus is not always the most sensitive life stage. Some EDCs, like fenarimol, ketoconazole, and other inhibitors of steroidogenesis, disrupt pregnancy by altering maternal maternal endocrine function in P0 dams at dosage levels that appear to be without direct effect on the offspring. In such cases, the standard USEPA multigenerational protocol, with minor enhancements, or a transgenerational protocol with exposure continued after weaning should be employed. For example, with xenoestrogens, the most consistent low-dose effect seen in rat multigenerational studies has been pseudoprecocious puberty in the female offspring. We believe that the transgenerational or in utero/lactational protocols fill a gap in the testing program for EDCs that should be used only on a case-by-case basis, as indicated by the results of T1S and any T1-Repeat. In cases where one is uncertain about how T2 should be tailored, a transgenerational study, including dosing of half the F1 after weaning, could be employed.

Table 1 Detecting antiandrogen syndromes.

Multitude of effects of antiandrogens in male rat offspring that should be evaluated in Tier 2 testing studies displaying this activity in Tier 1 screening or other assays.

Neonatal-infantile data

1. Anogenital distance at birth (1–3 days of age)
2. Areola/nipple retention in infant male rats at 13–14 days of age

External necropsy endpoints on all male rat offspring at maturity

1. Body weight any unusual malformations or anomalies, euthanize
2. Shave ventral surface from inguinal region to neck and count nipples and areolas (observer blind to treatment), record position of areolas and nipples
3. Check animals for hypospadias, epispadias, cleft phallus, and measure AGD
4. Note if testes obviously undescended
5. Note if inguinal region soiled with urine
6. Note if prepuce partially or entirely detached from glans penis, especially if a persistent thread of tissue is present along frenulum

Internal endpoints on all male rat offspring at maturity

1. Location of each testis (scrotal, abdominal, gubernaclum attached to abdominal wall)
2. Gubernacular cords, present or absent, and length in mm if abnormal
3. Note if present, cranial suspensory ligaments
4. Note if testes are small, absent, fluid filled, enlarged, appear infected or other
5. Note if epididymides are small, absent, or infected (record region of effects)
6. Note if ventral prostate is small, absent, or infected
7. Note if dorsolateral prostate is small, absent, or infected
8. Note if seminal vesicles are small, absent, infected, or one side larger than the other
9. Note if coagulating glands are small absent, infected, one side larger than the other or detached from seminal vesicles
10. Note if kidneys display hydronephrosis, calcium deposits
11. Note presence of hydroureter
12. Note presence of bladder stones or bloody in bladder

Weigh the following organs on all male rat offspring at maturity

1. Each testis individually (examine histology of each testis)
2. Each corpus plus caput epididymis (examine histology of each segment)
3. Each cauda epididymis (examine histology of each segment)
4. Entire seminal vesicle, plus coagulating glands with fluid as a unit, if possible
5. Entire ventral prostate, if possible
6. Each kidney
7. Paired adrenals
8. Liver
9. Levator ani plus bulbocavernosus
10. Cowper's glands as a pair, if possible
11. Glans penis
12. Pituitary
13. Brain

Histology on all male F1 offspring at maturity

1. Both testes
2. Both epididymides
3. Prostate glands
4. Any grossly abnormal reproductive tissues

Table 2 Detecting the androgenized female syndrome [61].

Multitude of effects of androgens in female rat offspring that should be evaluated in Tier 2 testing studies

Neonatal-infantile data in female rat offspring
1. Anogenital distance at birth (1–3 days of age)
2. Areola/nipple agenesis (complete or faint) in infant female rats at 13–14 days of age

External necropsy endpoints on all female rat offspring at maturity
1. Body weight any unusual malformations or anomalies, euthanize
2. Shave ventral surface from inguinal region to neck and count nipples and areolas (observer blind to treatment), record position of areolas and nipples
3. Check animals for cleft phallus and measure AGD and position of vaginal opening
4. Note if inguinal region soiled with urine

Internal endpoints on all female rat offspring at maturity
1. Location of ovaries in relationship to kidneys
2. Note if absent, cranial suspensory ligaments
3. Note if ovaries are small, cystic-fluid filled, enlarged, appear infected or other
4. Note if oviducts, uterus, or upper or lower vagina are small, absent, or infected (record region of effects) or fluid-filled
5. Note if ventral prostate tissue is present
6. Note if seminal vesicle tissue is present
7. Note if levator ani/bulbocavernosus muscle tissues are present
8. Note if other male tissues are present
9. Note if kidneys display hydronephrosis, calcium deposits
10. Note presence of hydroureter
11. Note presence of bladder stones or blood in bladder

Weigh the following organs in all female offspring at maturity
1. Paired ovaries (histology)
2. Uterus with fluid (histology)
3. Vagina
4. Each kidney
5. Paired adrenals
6. Liver
7. Pituitary
8. Brain

Histology on all female rat offspring at maturity
1. Both ovaries
2. Uterus
3. Vagina
4. Any suspected male reproductive tissues
5. Any grossly abnormal reproductive tissues

REFERENCES

1. T. Colborn and C. Clement. *Advances in Modern Environmental Toxicology. Vol XXI. Chemically-Induced Alterations in Sexual and Functional Development: The Wildlife/Human Connection*, Princeton Scientific Publishing, Princeton, NJ (1992).
2. T. Colborn, H. Bern, P. Blair et al. *Toxicol. Ind. Health* **14**, 1–8 (1998).
3. H. R. Andersen, A. M. Andersson, S. F. Arnold, H. Autrup, M. Barfoed, N. A. Beresford, P. Bjerregaard, L. B. Christiansen, B. Gissel, R. Hummel, E. B. Jorgensen, B. Korsgaard, R. Le Guevel, H. Leffers, J. McLachlan, A. Moller, J. B. Nielsen, N. Olea, A. Oles-Karasko, F. Pakdel, K. L. Pedersen, P. Perez, N. E. Skakkebaek, C. Sonnenschein, A. M. Soto et al. *Environ. Health Perspect.* **107** (Suppl. 1), 89–108 1999.
4. E. R. Bauer, A. Daxenberger, T. Petri, H. Sauerwein, H. H. Meyer. *APMIS.* **108** (12), 838–846 (2000).
5. C. Wong, W. R. Kelce, M. Sar, E. M. Wilson. *J. Biol. Chem.* **25; 270** (34), 19998–20003 (1995).
6. C. Lambright, J. Ostby, K. Bobseine, V. Wilson, A. K. Hotchkiss, P. C. Mann, L. E. Gray, Jr. *Toxicol. Sci.* **56** (2), 389–399 (2000).
7. V. S. Wilson, K. Bobseine, C. R. Lambright, L. E. Gray, Jr. *Toxicol. Sci.* **66** (1), 69–81 (2002).
8. C. Sultan, F. Paris, B. Terouanne, P. Balaguer, V. George, N. Poujol, S. Jeandel, J. Nicolas. *Horm. Reprod. Update* **7** (3), 314–322 (2001).
9. J. Legler, C. E. van den Brink, A. Brouwer, A. J. Murk, P. T. van der Saag, A. D. Vethaak, B. van der Burg. *Toxicol. Sci.* **48** (1), 55–66 (1999).
10. J. M. Rogers and M. S. Denison. *In Vitro Mol. Toxicol.* **13** (1), 67–82 (2000).
11. A. M. Vinggaard, V. Breinholt, J. C. Larsen. *Food Addit. Contam.* **16** (12), 533–542 (1999).
12. P. C. Hartig, K. L. Bobseine, B. H. Britt, M. C. Cardon, C. R. Lambright, V. S. Wilson, L. E. Gray, Jr. *Toxicol. Sci.* **66** (1), 82–90 (2002).
13. W. R. Kelce, E. Monosson, M. P. Gamcsik, S. C. Laws, L. E. Gray, Jr. *Toxicol. Appl. Pharmacol.* **126** (2), 276–285 (1994).
14. B. S. McIntyre, N. J. Barlow, D. G. Wallace, S. C. Maness, K. W. Gaido, P. M. D. Foster. *Toxicol. Appl. Pharmacol.* **167** (2), 87–99 (2000).
15. L. E. Gray, Jr., W. R. Kelce, T. Wiese, R. Tyl, K. Gaido, J. Cook, G. Klinefelter, D. Desaulniers, E. Wilson, T. Zacharewski, C. Waller, P. Foster, J. Laskey, J. Reel, J. Giesy, S. Laws, J. McLachlan, W. Breslin, R. Cooper, R. Di Giulio, R. Johnson, R. Purdy, E. Mihaich, S. Safe, T. Colborn et al. *Reprod. Toxicol.* **11** (5), 719–50 (1997).
16. P. Mak, F. D. Cruz, S. Chen. *Environ, Health Perspect.* **107** (11), 855–860 (1999).
17. L. E. Gray, Jr. *Toxicol. Lett.* **102–103**, 677–680 (1998).
18. L. E. Gray, Jr., J. Ostby, V. Wilson, C. Lambright, K. Bobseine, P. Hartig, A. Hotchkiss, C. Wolf, J. Furr, M. Price, L. Parks, R. L. Cooper, T. E. Stoker, S. C. Laws, S. J. Degitz, K M. Jensen, M. D. Kahl, J. J. Korte, E. A. Makynen, J. E. Tietge, G. T. Ankley. *Xenoendocrine Disrupters Tiered Screening and Testing: Filling Key Data Gaps Toxicology*, Dec 27, **181–182**, 371–382. Review (2002).
19. G. Ankley, E. Mahiach, R. Stahl, D. Tillitt, T. Colborn, S. McMaster, R. Miller, J. Bantle, P. Campell, R. Dickerson, M. Fry, J. Giesy, L. E. Gray, Jr. et al. *Environ. Toxicol. Chem.* **17**, 68–87 (1998).
20. G. T. Ankley, K. M. Jensen, M. D. Kahl, J. J. Korte, E. A. Makynen. *Environ. Toxicol. Chem.* **20**, 1276–1290 (2001).
21. G. T. Ankley, M. D. Kahl, K. M. Jensen, M. W. Hornung, J. J. Korte, E. A. Makynen, R. L. Leino. *Toxicol. Sci.* **67** (1), 121–130 (2002).
22. J. Kanno, L. Onyon, J. Haseman, F. Fenner-Crisp, J. Ashby, W. Owens, Organization for Economic Cooperation and Development. *Environ. Health Perspect.* **109** (8), 785–794 (2001).

23. J. Kanno, L. Onyon, S. Peddada, J. Ashby, E. Jacob, W. Owens. *Environ. Health Perspect.* **111** (12), 1550–1558 (2003).
24. J. Kanno, L. Onyon, S. Peddada, J. Ashby, E. Jacob, W. Owens. *Environ. Health Perspect.* **111** (12), 1530–1549 (2003).
25. L. Hershberger, E. Shipley, R. Meyer. *Proc. Soc. Exp. Biol. Med.* **83**, 175–180 (1953).
26. R. Dorfman. "Androgens and anabolic agents", in *Methods in Hormone Research, Vol. II, Bioassays*, R. I. Dorfman (Ed.), pp. 275–313, Academic Press, New York (1962).
27. J. C. O'Connor, S. R. Frame, L. G. Davis, J. C. Cook. *Toxicol. Sci.* **51** (1), 44–53 (1999).
28. J. M. Goldman, S. C. Laws, S. K. Balchak, R. L. Cooper, R. J. Kavlock. *Crit. Rev. Toxicol.* **30** (2), 135–196 (2000).
29. L. E. Gray Jr, J. Ostby, R. Sigmonn, J. Ferrell, G. Rehnberg, R. Linder, R. Cooper, J. Goldman, J. Laskey. *Reprod. Toxicol.* **2**, 281–287 (1988).
30. L. E. Gray, Jr., J. Ostby, J. Ferrell, G. Rehnberg, R. Linder, R. Cooper, J. Goldman, V. Slott, J. Laskey. *Fundam. Appl. Toxicol.* **12**, 92–108 (1989).
31. S. C. Laws, J. M. Ferrell, T. E. Stoker, J. Schmid, R. L. Cooper. *Toxicol. Sci.* **58** (2), 366–376 (2000).
32. M. S. Marty, J. W. Crissman, E. W. Carney. *Toxicol. Sci.* **52** (2), 269–277 (1999).
33. W. R. Kelce, C. R. Stone, S. C. Laws, L. E. Gray, J. A. Kemppainen, E. M. Wilson. *Nature* **375**, 581–585 (1995).
34. J. Ashby and P. A. Lefevre. *Regul. Toxicol. Pharmacol.* **26**, 330–337 (1997).
35. E. Monosson, W. R. Kelce, C. Lambright, J. Ostby, L. E. Gray, Jr. *Toxicol. Ind. Health* **15** (1–2), 65–79 (1999).
36. M. S. Marty, J. W. Crissman, E. W. Carney. *Toxicol. Sci.* **60** (1), 63–76 (2001).
37. T. E. Stoker, L. G. Parks, L. E. Gray, R. L. Cooper. *Crit. Rev. Toxicol.* **30** (2), 197–252 (2000).
38. W. Kelce, C. Lambright, L. Gray, K. Roberts. *Toxicol. Appl. Pharmacol.* **142**, 192–200 (1997).
39. L. E. Gray, Jr., J. S. Ostby, W. R. Kelce. *Toxicol. Appl. Pharmacol.* **129** (1), 46–52 (1994).
40. L. E. Gray, Jr., C. Wolf, C. Lambright, P. Mann, M. Price, R. L. Cooper, J. S. Ostby. *J. Toxicol. Ind. Health* **15** (1–2), 94–118 (1999).
41. P. M. D. Foster and B S. McIntyre. *Toxicol. Pathol.* **30** (1), 59–65 (2002).
42. L. E. Gray, J. Ostby, J. Furr, C. J. Wolf, C. Lambright, L. Parks, D. N. Veeramachaneni, V. Wilson, M. Price, A. Hotchkiss, E. Orlando, L. Guillette. *Hum. Reprod. Update.* **7** (3), 248–264 (2001).
43. E. Mylchreest, R. C. Cattley, P. M. D. Foster. *Toxicol. Sci.* **43** (1), 47–60 (1998).
44. E. Mylchreest, M. Sar, R. C. Cattley, P. M. D. Foster. *Toxicol. Appl. Pharmacol.* **15; 156** (2), 81–95 (1999).
45. E. Mylchreest, D. G. Wallace, R. C. Cattley, P. M. D Foster. *Toxicol. Sci.* **55** (1), 143–151 (2000).
46. E. Mylchreest, M. Sar, D. G. Wallace, P. M. D. Foster. *Reprod. Toxicol.* **16** (1), 19–28 (2002).
47. L. E. Gray, Jr., J. Ostby, J. Furr, M. Price, D. N. Veeramachaneni, L. Parks. *Toxicol. Sci.* **58** (2), 350–365 (2000).
48. V. D. Shultz, S. Phillips, M. Sar, P. M. D. Foster, K. W. Gaido. *Toxicol. Sci.* **64** (2), 233–242 (2001).
49. L. G. Parks, J. S. Ostby, C. R. Lambright, B. D. Abbott, G. R. Klinefelter, N. J. Barlow, L. E. Gray, Jr. *Toxicol. Sci.* **58** (2), 339–249 (2000).
50. B. S. McIntyre, N. J. Barlow, P. M. Foster. *Toxicol. Sci.* **62** (2), 236–249 (2001).
51. P. M. D. Foster, E. Mylchreest, K. W. Gaido, M. Sar. *Hum. Reprod. Update* **7** (3), 231–235 Review (2001).
52. J. Ostby, W. R. Kelce, C. Lambright, C. J. Wolf, P. Mann, L. E. Gray, Jr. *Toxicol. Ind. Health* **15** (1–2), 80–93 (1999).
53. V. S. Wilson, C. Lambright, J. Furr, J. Ostby, C. Wood, G. Held, L. E. Gray, Jr. *Toxicol. Lett.* (2003). In press.

54. P. M. D. Foster, R. C. Cattley, E. Mylchreest. *Food Chem. Toxicol.* **38** (1 Suppl.), S97–99, Review (2000).
55. R. H. McKee, R. T. Przygoda, M. A. Chirdon, G. Engelhardt, M. Stanley. *J. Appl. Toxicol.* **20**, 491–497 (2000).
56. L. J. Hushka, S. J. Waterman, L. H. Keller, G. W. Trimmer, J. J. Freeman, J. L. Ambroso, M. Nicolich, R. H. McKee. *Reprod. Toxicol.* **15** (2), 153–169 (2001).
57. J. C. Cook, L. S. Mullin, S. R. Frame, L. B. Biegel. *Toxicol. Appl. Pharmacol.* **119** (2), 195–204 (1993).
58. B. S. McIntyre, N. J. Barlow, M. Sar, D. G. Wallace, P. M. D. Foster. *Reprod. Toxicol.* **16** (2), 131–139 (2002).
59. B. S. McIntyre, N. J. Barlow, P. M. D. Foster. *Toxicol. Sci.* **65** (1), 62–70 (2002).
60. J. Ostby and L. E. Gray. *Curr. Protocol. Toxicol.* In press.
61. C. J. Wolf, A. Hotchkiss, J. S. Ostby, G. A. LeBlanc, L. E. Gray, Jr. *Toxicol. Sci.* **65** (1), 71–86 (2002).

Pure Appl. Chem., Vol. 75, Nos. 11–12, pp. 2143–2149, 2003.

Topic 3.12

Determination of acceptable exposure levels for humans for endocrine active substances: Use of animal models*,†

Penelope A. Fenner-Crisp‡

ILSI Risk Science Institute, One Thomas Circle, NW, Ninth Floor, Washington, DC 20005-5802, USA

Abstract: Regulatory authorities and other scientific organizations around the world have developed hazard/risk assessment practices that involve the derivation of numerical values said to represent "acceptable" or "safe" levels of human exposure to individual chemical substances. Human data would be preferable for this purpose since interspecies extrapolation would not be necessary. In most cases, however, these data are not available or are inadequate for this purpose. Therefore, results from studies conducted in non-human mammalian species are used as the principal or sole basis for hazard/risk assessment. Existing risk assessment frameworks developed for the evaluation of endpoints of toxicity for chemical substances are sufficiently flexible to incorporate knowledge concerning the mode(s)/mechanism(s) of action by which these endpoints occur, including those involving disturbance of normal endocrine status ("endocrine disruption"). Background on nomenclature, traditional practices using non-human animal models, and the nature and adequacy of data sets for deriving acceptable human exposures for chemicals, including endocrine active substances, are described, as is how/why the existing frameworks are adequate for the purpose of deriving numerical values for endocrine active substances. Projections on how assessment practices for these substances may evolve further in the future also are offered.

INTRODUCTION

Endocrine active substances that cause adverse effects often are characterized as "endocrine disruptors" when these effects can be linked by empirical evidence or inference to a disturbance of normal endocrine status. Endocrine disruption is not an adverse effect per se, but rather a mode or mechanism of action by which these substances may produce adverse effects [1]. Members of many different chemical classes and use categories (e.g., pesticides, human and veterinary drugs, and consumer products) may possess this capability. Up until the last few years, characterization or labeling of a chemical substance generally was only by the nature of the effect(s) it produced (e.g., "Chemical X is a developmental toxicant, a neurotoxicant, and a carcinogen") and not by the mode or mechanism of action by which the toxic effect is provoked. As the scientific community's understanding of the etiologies of toxic effects evolves, so does the practice of characterization of a chemical by effect *and/or* mode/mechanism of action. So, then, one might ask whether or not the long-standing hazard and risk assessment

---

*Report from a SCOPE/IUPAC project: Implication of Endocrine Active Substances for Human and Wildlife (J. Miyamoto and J. Burger, editors). Other reports are published in this issue, *Pure Appl. Chem.* **75**, 1617–2615 (2003).

†This paper represents the opinions only of its author and does not reflect opinions or policies of the International Life Sciences Institute or the ILSI Risk Science Institute.

‡Tel.: 202 659 3306; Fax: 202 659 3617; E-mail: pfennercrisp@ilsi.org

practices based primarily upon identification of an endpoint of toxicity require a major overhaul to accommodate the understanding of mode/mechanism of action. And, further, is "endocrine disruption" a special case? There is likely to be a range of opinions on both questions; however, this author would answer each question with a "no" in the belief that the existing assessment framework is flexible enough to accommodate knowledge of mode/mechanism and that "endocrine disruption" is just one of many modes/mechanisms that can be addressed within it.

NOMENCLATURE FOR "ACCEPTABLE" OR "SAFE" LEVELS OF HUMAN EXPOSURE

Over the past half-century or so, regulatory authorities and other scientific organizations around the world have developed procedures and practices that involve the application of mathematical techniques to representative studies from the body of available toxicity data on a chemical substance, with the intention of deriving numerical values that, as a matter of policy, are said to represent "acceptable" or "safe" levels of human exposure to that substance. This step of the risk assessment process occurs in the hazard characterization phase of the Risk Assessment Paradigm, with hazard characterization embracing both hazard identification and dose–response assessment. Exposure assessment and risk characterization round out the paradigm. Numerical values are known by a variety of names, depending, in part, upon the regulatory setting in which they are derived and, further, upon the organization developing the numbers. The key message with regard to nomenclature is that, even though different names may be given to these numerical values, they all represent essentially the same concept. Each represents, in the opinion of the originator of the number, an exposure dose that, when occurring over a specified time span, will be "acceptable" or "safe". The goal is to protect the human against adverse health consequences, should exposure to a chemical substance occur.

The longest-standing and most well-known numerical value developed to characterize "safety" or "acceptability" over a lifetime of oral exposure is the "acceptable daily intake" or ADI. This is a term historically associated with the hazard assessment that informs the setting of maximum residue levels for pesticides, veterinary drugs, and other additives in food (e.g., [2–4]). Some organizations use the term "tolerable daily intake" or "tolerable intake" when deriving numerical values for substances that are found inadvertently or unintentionally in food or elsewhere in the environment; that is, they have not been preapproved for a specific use (e.g., [5–7]). Examples of such substances are dichlorodiphenyltrichloroethane (DDT), dioxins, heavy metals, polychlorinated biphenyls (PCBs), or naturally occurring mycotoxins (e.g., aflatoxin on peanuts and grains). Alternatively, the U.S. Environmental Protection Agency (USEPA) prefers to use the term "reference dose" (RfD) for oral exposure scenarios and "reference concentration" (RfC) for inhalation exposure scenarios [8,9], and the U.S. Agency for Toxic Substances and Disease Registry (ATSDR) prefers to use the term "minimal risk level" to denote oral or inhalation exposure levels that agency deems to be without appreciable risk [10].

More recently, it has been realized that hazard and risk assessments were required for chemical exposure scenarios that were less-than-lifetime in duration. Perhaps the earliest implementation of such an approach is that of the drinking water program at the USEPA. While a number of formal drinking water standards were in place in the early 1980s, many additional substances were being found in drinking water supplies, and questions were being raised about their potential human health risks. Since standard-setting is a long and arduous process, and guidance was needed in the short term, the program developed the Health Advisory Program, which provided nonregulatory guidance to affected drinking water system and public health officials within hours or days of their initial inquiries. This guidance provided critical decision tools to determine if a "Do not drink" advisory should be issued, water from an outside source should be brought in, or no interventions were necessary. If the appropriate toxicological data were available, the program would develop Acute, Short-term, Longer-term, and Lifetime Health Advisories, each of which had the underpinnings of an RfD or its less-than-lifetime, exposure-duration-specific equivalent. In the late 1980s, the ATSDR began developing its series of toxicological profiles for substances found at hazardous waste sites. These profiles contain not only maximum residue levels (MRLs) for chronic (≥365 days) oral and/or inhalation exposures, but, where the data allow, MRLs for acute (from 1 to 14 days) and intermediate (from 14 to 364 days) exposure durations, as well [10]. In the early 1990s, the Pesticide Program at the USEPA began deriving acute RfDs for acute dietary exposure scenarios (single serving/meal or single day exposures) for pesticides with acute effects (e.g., cholinesterase-inhibiting insecticides and developmental toxicants) and intermediate RfDs for residential and occupational risk assessments. Within the past five years, the World Health Organization (WHO) Expert Panel of the (FAO)/WHO Joint Meeting on Pesticide Residues, an expert committee that provides technical advice to the Codex Committee on Pesticide Residues, has begun deriving acute RfDs for acutely toxic pesticides, as well [11].

Recently, a technical workgroup of the USEPA's Risk Assessment Forum, in its reevaluation of the agency's RfD/RfC development process, recommended that acute, intermediate, and long-term RfDs and RfCs be derived for all environmental substances within their regulatory purview [9]. This recommendation is tempered by the caveat that RfDs/RfCs for these multiple durations of exposure should be derived *only* if adequate and appropriate data are available to do so.

"TRADITIONAL" HAZARD/RISK ASSESSMENT FRAMEWORK

Traditionally, numerical values have been developed using toxicity data that provide two kinds of information. In the first instance, these data allow *hazard identification*; that is, the determination of the inherent toxic characteristics of the substance (e.g., whether it causes liver toxicity, infertility, or neurotoxicity). Secondly, data from studies that include multiple groups, including both exposed and unexposed subjects, allow *dose–response assessment*. Historically, data developed following chemical exposures in humans and in laboratory animal *mammalian* species have served as the basis for the development of numerical values. The desirability of using human data is obvious; no inter-species extrapolation need be done. However, adequate and relevant chemical-specific human data rarely are available for this purpose. Thus, we have come to depend upon a small number of non-human mammalian test species to provide the empirical base for hazard and risk assessment. These species have been chosen for two principal reasons: (1) assumed biological similarity to humans, and (2) economics and feasibility (e.g., relative ease of handling in a laboratory setting). The most commonly used species in regulatory testing paradigms are rats and mice, along with rabbits and dogs. Occasionally, guinea pigs, hamsters, and non-human primates also may be employed for testing. These seven species and/or others may be used in the research setting.

National and multinational regulatory agencies engaged in chemicals management generally have the authority to require the generation of a body of toxicological data for certain use categories prior to the introduction of those chemicals into commerce. This is particularly true for pesticides, human and veterinary drugs, and food additives. Some agencies also have the authority to require pre-market data on commodity chemicals destined for industrial, consumer, or other uses. Not all of these required data may be suitable for use in deriving numerical values. One criterion of suitability would be how well the study design can predict the potential for effects, or lack thereof (i.e., hazard characterization), in the exposure scenario for which the assessment is being conducted. A second criterion would be whether the study design includes multiple treatment groups and at least one control group so that *dose–response assessment* can be conducted. Historically, key assumptions involved when deriving tolerable daily intakes (TDIs)/ ADIs/RfDs/RfCs/MRLs are that the dose response for the parameter used as the basis for quantitation is monotonic and that a point of departure (POD) for quantitative assessment can be identified. This POD may be an observed no/lowest-observed-adverse-effect level (NOAEL or LOAEL) or a modeled benchmark dose representing some percentage of occurrence of effect (e.g., 1, 5, 10 %). Traditionally, safety or uncertainty factors then are applied to the selected POD to yield a "number". Selection of the factor(s) is informed by a modestly complex set of decision criteria. Detailed discussion of this component of hazard assessment can be found in Topics 3.13 and 3.14.

Pesticides are a use category of chemicals for which a robust set of toxicity data is generated in order for regulatory authorities to grant pre-market approval before allowing their introduction into commerce and the environment. As such, the resulting data are likely to be amenable to the development of "acceptable" or "safe" levels of human exposure for lifetime and less-than-lifetime exposure durations.

At the present time, the standard set of first-tier toxicology regulatory requirements focuses on identification of effects and studies are carried out in conformance with standardized consensus testing protocols. Roughly identical first-tier requirements exist for all Organization for Economic Cooperation and Development (OECD) member countries, including the members of the European Union. Many other non-OECD countries apply the OECD testing paradigm in their own regulatory programs.

An example of the standard first-tier set for food-use pesticides of conventional chemistry can be found in the U.S. Code of Federal Regulations [12,13]. Routinely required (i.e., first-tier) studies for food use pesticides of conventional chemistry include a battery of six tests to characterize acute toxicity (one each by the oral, dermal, and inhalation routes plus eye and skin irritation and dermal sensitization), subchronic feeding studies in rodent and non-rodent species, chronic feeding studies in rodent and non-rodent species, carcinogenicity studies in two rodent species, prenatal developmental toxicity studies in rodents and non-rodents, and a two-generation reproduction study in rodents. Second-tier studies may include dermal penetration, 21-day dermal, subchronic dermal, subchronic inhalation, and acute and subchronic neurotoxicity studies in rodents; acute and subchronic delayed neurotoxicity in hens; or a developmental neurotoxicity (DNT) study in rodents. These are triggered by some special characteristic of the pesticide (e.g., its chemical class) or by potential use and exposure patterns (e.g., residential uses) or by the results of routinely required studies. These study types have been found to have value in the derivation of numerical values, except for the six acute studies, which, by virtue of their study design and measurement parameters, are best suited only for hazard identification, but not for dose–response assessment. Embedded in all of the study protocols is the selection of specific parameters to be measured, which are thought to represent key manifestations of particular endpoint of toxicity. Over the years, as the state-of-the-science has evolved, so have the kind and number of endpoints measured. Many of these endpoints have come to be understood as events in endocrine disruptor mode/mechanisms of action.

Several key observations can be made about the traditional hazard/risk assessment framework for derivation of numerical values in human health assessment:

- Toxicological data development is focused primarily on toxic endpoint identification.
- The data used for derivation of numerical values come only from human exposure situations or studies conducted in other mammalian species.
- As a default, the dose responses for these endpoints are assumed to be monotonic and reflective of a threshold for the effect.

AN EVOLVING HAZARD/RISK ASSESSMENT FRAMEWORK FOR ENDOCRINE DISRUPTORS

Generally, in the evaluation of most chemical substances, pharmaceuticals being the principal exception, only after effects of concern have been identified are follow-up studies done to characterize the mode/mechanism of action by which these effects arise. Currently, there are no standardized consensus guidelines available for this purpose for any specific mode/mechanism, other than the mutagenicity/genotoxicity studies that help to inform the characteristics of a carcinogenic or heritable mutation effect. As discussed elsewhere in this conference, a significant effort is underway on a national (e.g., in the United States, in response to legislative mandates in the 1996 Food Quality Protection and Safe Drinking Water Acts [14]) and international scale (i.e., through the OECD [15]) to develop testing paradigms specifically to identify endocrine disruptors. These will be supported by the development and implementation of new and updated standardized and validated consensus protocols. Some of these study designs will define a toxicological profile, as do the traditionally conducted studies; others will assist in characterizing mode/mechanism of action. Worthy of note here is that, as the mode(s)/mechanism(s) of action (MOAs) of potential/identified endocrine active substances have been characterized, it has become apparent that many of the parameters that were measured originally only to identify effects of concern in the traditional studies also represent a key event or stage in the MOAs of the effect(s) of concern. Thus, as one component of a strategy to identify and characterize endocrine disruptors, it is reasonable and appropriate to build upon, and modify, traditional study designs for endpoint identification as the MOAs become better understood, rather than to start over.

How, then, might the traditional hazard/risk human health assessment framework used in deriving numerical values evolve to accommodate mode/mechanism of action information in the identification and assessment of endocrine active substances? In the near term, it should capitalize on the efforts underway to develop the screening and testing programs. In the first instance, continued improvements to traditional apical study designs in whole animal mammalian systems will contribute to more accurate understanding of endpoints of toxicity and also their MOAs. Other kinds of studies will be conducted in in vitro mammalian systems or in in vivo or in vitro nonmammalian systems. Some of the in vivo studies in nonmammalian systems include a relatively broad spectrum of apical endpoints, and, therefore, are structurally similar to the traditional apical mammalian studies currently used for number development. Others include few or no apical endpoints and better serve for elucidation of the MOA(s). Data from the non-apical mammalian and/or the nonmammalian studies still could have a role in both *hazard identification* and *dose–response assessment*, as follows:

Hazard identification

Currently, many endpoints are evaluated in the apical mammalian studies that are used for development of numerical values. Absent an understanding of the underlying MOA for the production of these endpoints, it is more difficult to sort out the effects of concern from the ones of lesser importance (i.e., the effects that are not adverse). Data from these other types of studies can help refine and focus the endpoint selection process. Instead of selecting the endpoint with the lowest NOAEL or lowest 1, 5, or 10 % benchmark dose (BMD) simply because it is the lowest, the important endpoints can be selected based upon their involvement in potentially adverse consequences. These important endpoints may or may not have the lowest NOAEL or BMD, but they would represent a more accurate assessment of potential hazard.

Dose–response assessment

As noted earlier, the default assumption for the shape of the dose–response curve for most toxic effects is monotonic, nonlinear. (There continue to be differences of opinion globally concerning the appropriate default assumption for carcinogenic responses.) The extrapolation techniques, then, were designed with the monotonic, nonlinear default in mind, but also could easily accommodate a linear response. However, as has been noted in many scientific papers in the literature, some of the dose responses for some effects observed following exposure to endocrine active substances exhibit neither a linear nor a monotonic, nonlinear dose response. Bimodal and "U"-shaped dose responses have been observed. Others may exist. This is not a problem if these effects are not selected as the critical ones for use in development of numerical values. It does pose a new challenge if they are. Obviously, then, extrapolation techniques that account for these other shapes must developed and applied. Defaults should be abandoned when credible data are available to accurately describe the biology. The data may validate the default or prove it incorrect. Empirical evidence should drive the assessment process.

Hazard characterization

As noted earlier, hazard characterization encompasses both hazard identification and dose–response assessment, but also is the phase of risk assessment in which the quantitative assessment, if done, is completed and described. Therefore, it is at this point in the process that the application of the procedure for deriving numerical values would be synthesized and summarized. In the future, it may be that MOA data play a role in further refinement of the numerical value, beyond those already described above in the hazard identification and dose–response assessment sections. No specific existing examples are known to the author and no hypothetical examples come to mind at this time. We, however, should remain alert to this possibility.

SUMMARY

The derivation of numerical values is often a key element in the assessment of human health hazard/risk from exposures to chemical substances. Traditional risk assessment practices typically employ toxicity data only from mammalian species (although this could be human data). These toxicity data generally identify only endpoints of effect, and inform little with regard to mode/mechanism of action. Absent empirical data, dose responses are generally assumed to be monotonic, nonlinear, although linear responses can be easily accommodated in the extrapolation process. An evolving framework for risk assessment of endocrine active substances/endocrine disruptors should incorporate relevant information on mode/mechanism of action, even if developed in nonmammalian systems. Empirical evidence indicates that sometimes the dose response for effects resulting as a consequence of disturbance of normal endocrine status is neither monotonic, nonlinear nor linear. If these effects are selected as critical to the development of a numerical value, extrapolation techniques that accommodate these different dose responses must be employed.

REFERENCES

1. U.S. EPA. *Special Report on Environmental Endocrine Disruption: An Effects Assessment and Analysis*, February, EPA/630/R-96/012, U.S. Environmental Protection Agency, Washington, DC (1997). <http://www.epa.gov/ORD/WebPubs/endocrine>
2. WHO JECFA. *Principles for the Safety Assessment of Food Additives and Contaminants in Food*, WHO Environmental Health Criteria No. 70, World Health Organization Expert Committee of the Joint FAO/WHO Expert Committee on Food Additives, World Health Organization, Geneva (1987). <http://www.inchem.org/documents/ehc/ehc/ehc70.htm>
3. WHO JMPR. *Principles for the Toxicological Assessment of Pesticide Residues in Food*, WHO Environmental Health Criteria, No. 104. Geneva, World Health Organization Expert Group on Pesticide Residues of the Joint FAO/WHO Meeting on Pesticide Residues, World Health Organization (1990). <http://www.inchem.org/documents/ehc/ehc/ehc104.htm>
4. U.S. FDA. *Food Contact Substance Notification Program's Cumulative Estimated Daily Intake/Acceptable Daily Intake Database*, U.S. Food and Drug Administration, Center for Food Safety & Applied Nutrition, Office of Food Additive Safety, August, Washington, DC (2002). <http://www.cfsan.fda.gov/~dms/opa-edi.html>
5. U.S. FDA. *Guidance Document for Cadmium in Shellfish*, U.S. Food and Drug Administration, Center for Food Safety & Applied Nutrition, Office of Food Additive Safety, August, Washington, DC (1993). <http://www.cfsan.fda.gov/~frf/guid-cd.html>
6. WHO IPCS. *Assessing Human Health Risks of Chemicals: Derivation of Guidance Values for Health-based Exposure Limits*, Environmental Health Criteria No. 170. World Health Organization International Program on Chemical Safety, World Health Organization, Geneva (1994). <http://www.inchem.org/documents/ehc/ehc/ehc170.htm>

7. European Commission Scientific Committee on Food. Opinion of the Scientific Committee on Food on Bisphenol A. European Commission Health and Consumer Protection Directorate-General, April, Brussels (2002). (http://europa.eu.int/foods/fs/sc/scf/index_en.html>
8. U.S. EPA. *Reference Dose (RfD): Description and Use in Health Assessments. Integrated Risk Information System (IRIS) Background Document 1A*, U.S. Environmental Protection Agency, Washington, DC (1994). <http://www.epa.gov.iris/rfd.htm>
9. U.S. EPA. *A Review of the Reference Dose and Reference Concentration Processes. External Review Draft*, May, EPA/630/P-02/002A, U.S. Environmental Protection Agency, Washington, DC (2002a). <http://www.epa.gov/ncea/raf/pdfs/RfDRfC/rfdrfcextrevdrft.pdf>
10. ATSDR. *Minimal Risk Levels for Hazardous Substances*, Agency for Toxic Substances and Disease Registry (2002). <http://www.atsdr.cdc.gov/mrls.html>
11. WHO JMPR. "Procedures for estimating acute reference dose", In *Pesticide Residues in Food–1997*, World Health Organization Expert Group on Pesticide Residues of the Joint FAO/WHO Meeting on Pesticide Residues, Rome (1998). <http://www.fao.org/waicent/faoinfo/agricult/agp/agpp/pesticid/jmpr/download/98/cont-intro.pdf>
12. 40 CFR Part 158. 2002. Code of Federal Regulations. Part 158.340 Toxicology data requirements. <http://www.access.gpo.gov/nara/cfr/index.html>
13. U.S. EPA. *Determination of the Sppropriate FQPA Safety Factor(s) in Tolerance Assessment*, U.S. Environmental Protection Agency, Office of Pesticide Programs, Washington, DC (2002b). <http://www.epa.gov/oppfead1/trac/science/determ.pdf>
14. U.S. EPA. Endocrine Disruptor Screening Program Web site, U.S. Environmental Protection Agency (2002c). <http://www.epa.gov/scipoly/oscpendo/index.htm>
15. OECD. Special Activity of the Test Guidelines Programme on Endocrine Disruptor Testing and Assessment, Organization of Economic Cooperation and Development (2002). <http://www1.oecd.org/ehs/endocrin/Subpage/Reports.htm>

Pure Appl. Chem., Vol. 75, Nos. 11–12, pp. 2151–2158, 2003.

Topic 3.13

Use of NOAEL, benchmark dose, and other models for human risk assessment of hormonally active substances*,†

R. Woodrow Setzer, Jr.[1,‡] and Carole A. Kimmel[2]

[1]USEPA, Office of Research and Development, National Health and Environmental Effects Research Laboratory, Experimental Toxicology Division, Pharmacokinetics Branch, MD B143-01 109 TW Alexander Drive, Research Triangle Park, NC 27711, USA; [2]USEPA Office of Research and Development, National Center for Environmental Assessment (8623D), Ariel Rios Building, 1200 Pennsylvania Ave. NW, Washington, DC 20460, USA

Abstract: The benchmark dose (BMD) is the dose of a substance that is expected to result in a prespecified level of effect, the benchmark response level or BMR. It is a general approach to characterizing dose response, applicable to any toxicant and endpoint. A BMD is conceptually superior to a "no observed adverse effect level" (NOAEL) for this purpose because of being less determined by experimental design, because it is a precisely defined entity, and because its precision can be estimated. Since a BMD is a single number, just as an NOAEL, it is tempting to use the BMD as a straightforward replacement for the NOAEL in the assessment process for calculating allowable daily intakes. However, the level of toxic response at an NOAEL is unknown, while that at a BMD is well defined. Use of the BMD approach potentially adds consistency and objectivity to the process of deriving reference values (RfDs, RfCs, or ADIs) for setting regulatory levels. To take advantage of this, BMRs need to be selected in a consistent way across studies and endpoints. This paper discusses some issues affecting the selection of BMRs, and presents an example of a BMD calculated for the effects of peripubertal exposure to the fungicide vinclozolin.

INTRODUCTION

The benchmark dose (BMD) was originally proposed in 1984 [1] as an alternative to the NOAEL (no observed adverse effect level) and LOAEL (lowest observed adverse effect level) for setting regulatory levels such as reference doses (RfDs), reference concentrations (RfCs), and acceptable daily intakes (ADIs). The RfD, RfC, or ADI approach is used for agents whose dose-responses are thought to be very nonlinear or threshold-like. In this methodology, the regulatory level is derived by first determining a point of departure (POD) based on the dose response for the most sensitive endpoint(s) relevant to humans, then dividing it by a series of uncertainty factors (UFs) [2]. Typically, these uncertainty factors include a factor for extrapolation from animal to human data (UF_A), a factor to account for uncertainty

*Report from a SCOPE/IUPAC project: Implication of Endocrine Active Substances for Human and Wildlife (J. Miyamoto and J. Burger, editors). Other reports are published in this issue, *Pure Appl. Chem.* **75**, 1617–2615 (2003).

†The opinions expressed in this paper are those of the authors and do not necessarily reflect the views or policies of the U.S. Environmental Protection Agency.

‡Corresponding author: E-mail: setzer.woodrow@epa.gov

about variability in the human population (UF_H), a factor for extrapolation from subchronic data to a chronic exposure scenario (UF_S), a factor for database deficiencies (UF_D), and a modifying factor (MF). Conventionally, NOAELs have been used as PODs, so when the lowest dose in the critical study is an LOAEL, an additional uncertainty factor has been used (UF_L). The default value for these uncertainty and modifying factors is 10, although an RfD or RfC with a total UF of >3000 is usually considered too uncertain to be reliable. Factors of 1, 3, or 10 are usually applied for the UFs depending on the available data, and chemical-specific pharmacokinetic and/or pharmacodynamic data can be used to adjust or replace these factors.

The BMD is used as an alternative to the NOAEL/LOAEL approach for a more quantitative way of deriving regulatory levels for health effects assumed to have a nonlinear (threshold-like) low dose–response relationship. Whereas NOAELs and LOAELs are discrete doses from a study, the BMD approach involves modeling the dose–response curve in the range of the observable data, and then using that model to interpolate an estimate of the dose that corresponds to a particular level of response, e.g., 5 or 10 % for quantal data, or some predefined change in response from controls for continuous data. A measure of uncertainty is also generally calculated, e.g., a confidence limit or Bayesian posterior [3], and the lower confidence limit on the dose used as the BMD is called the BMDL. The BMDL accounts for the uncertainty in the estimate of the dose response that is due to characteristics of the experimental design such as sample size. The BMDL is used as the basis for the point of departure (POD). In a recent health assessment done by the EPA for 1,3-butadiene [4], an additional factor for extrapolating from the BMDL to lower associated risk levels was applied (discussed further below).

Unlike NOAELs and LOAELs, BMDs are not constrained to be one of the experimental doses, and can thus be a more consistent basis for dose–response assessment. NOAELs do not correspond to a consistent response level and depend on sample size so that the NOAEL will be higher in studies with a smaller sample size, the opposite of what is desirable. In addition, NOAELs are usually associated with some definable level of risk, and are not threshold doses or "no effect levels". The slope of the dose–response curve is not usually considered in the NOAEL/LOAEL approach unless the slope is very steep or very shallow. If an NOAEL has not been defined in a particular study and only an LOAEL is available, an uncertainty factor is typically applied to account for the lack of an NOAEL (UF_L). The use of BMDLs as the POD has been the basis for several RfDs and RfCs in the IRIS database [12]. EPA's Draft Benchmark Dose Technical Guidance Document [6] outlines a number of considerations to be made in the derivation of BMDs and BMDLs.

The BMDL is also used as the basis for the POD for linear low-dose extrapolation, the dose–response assessment approach applied to most carcinogens [5]. In this case, once the POD is determined, risk is extrapolated linearly to a low dose corresponding to 10^{-5} to 10^{-6} risk.

The BMD can also be used to estimate relative potencies among different chemicals. Usually, in this case, the maximum likelihood estimate (BMD) rather than the lower confidence limit (BMDL) is used for the comparison.

The BMD approach can be used for dose–response modeling of all types of chemical and physical agents and associated endpoints, including endocrine active substances (EASs), regardless of the assumptions about low-dose linearity or nonlinearity. This is because dose–response modeling is done in the observable range and the BMD is typically related to a response rate near the lower end of the observable dose range. Whether the effect seen with EASs is a nonlinear or threshold-type response or is additive to background does not affect BMD calculation. Selection of the response level for deriving the BMD, i.e., the benchmark response, BMR, is the more difficult issue, especially for continuous endpoints.

ISSUES SURROUNDING SELECTION OF THE BMR

Type of data

Selection of the BMR depends on the kind of data being modeled, e.g., dichotomous (quantal) or continuous data. Other types of data are also encountered (e.g., categorical or graded responses), but are not dealt with specifically here. Such data are often converted to quantal data before modeling.

The approach to BMD development has been discussed most often in the literature for quantal data, primarily because a dichotomous response (i.e., whether a response is present or not) is somewhat easier to judge (e.g., tumor, malformation). BMRs have been expressed in terms of extra or additional risk, which are two ways of expressing the prevalence of adverse effects above background. EPA's draft BMD guidelines [6] recommend using extra risk as the more conservative approach. Thus, the BMD associated with an extra risk at a BMR above background is the dose where the following expression is true:

$$[P(d) - P(0)]/[1 - P(0)] \quad (1)$$

where P(d) is the risk at a dose = d and P(0) is the background risk at zero dose.

For continuous data, the BMR is expressed as a change in the mean from control values. The selection of the BMR is more difficult to determine for continuous data, because the goal is to base the BMR on a change that is biologically meaningful. However, for many endpoints, this degree of change has not been decided or agreed upon by relevant experts. The alternative, if no criteria have been developed for what degree of response is biologically meaningful, may be to use a change in mean response equal to one standard deviation of the control mean. Continuous data can be modeled as such or the data can be categorized (dichotomized) using the level of change considered meaningful or the 1 SD change, and modeled in the same way as for quantal data.

Selecting the benchmark response level

Once the dose response is established, the only factor that affects the magnitude of the benchmark dose is the selection of the BMR. How the BMR is selected can have a large effect on the way the resulting BMD is used in a subsequent risk analysis. For example, it is tempting to think of using the BMD like an NOAEL or LOAEL in a nonlinear risk assessment. With this interpretation, the BMD would be used along with the typical uncertainty and modifying factors appropriate for NOAELs or LOAELs. Several studies in the literature (e.g., [7–11]) have compared various approaches to setting the BMR by trying to ascertain the numerical relationship between BMDs calculated in different ways and NOAELs. However, this comparison was not meant to imply that BMDs should be direct substitutes for NOAELs. Rather, the comparison provided a reflection of the limit of detection and pointed out the differences in response rate at the NOAEL for different types of effects.

It is arguable that the approach to risk assessment based on NOAELs and LOAELs is weak because they are so strongly affected by the design of the bioassay from which they are generated. In addition, the level of effect actually present at the NOAEL and LOAEL is unknown and, therefore, they may not be as health-protective as desired. By contrast, the more objective BMD approach provides an opportunity to improve the consistency of risk assessments and their resulting health protectiveness.

One approach might be to select the BMR to reflect a constant level of toxicity, regardless of endpoint. This approach would certainly facilitate interpretation of the BMD, and would probably improve the consistency of risk assessments. However, there are problems with this proposal: for example, coming to common agreement on the amount of change in a continuous endpoint that is considered to be adverse; and, to some extent, the difficulty of equating changes in continuous endpoints with changes in the prevalence of discrete adverse effects. An alternative approach would be for experts to determine the degree of change that is considered biologically meaningful and adverse for each endpoint, thus removing the need to compare general levels of toxicity across endpoints. Taking the latter path would require a fair amount of effort to develop a consensus among regulatory toxicologists about adversity of responses for continuous data.

Given the difficulties that arise with the considered approaches to setting the BMR, it is tempting to simply fix upon a fixed default level of response, say, for example, 10 % increase in the prevalence of adverse effects, or a 1 SD change in the mean for continuous endpoints.

Use of BMDs in setting regulatory standards

Use of the BMD approach for nonlinear or threshold-type responses presents some challenges in terms of risk communication. This is because the BMD is associated with a particular level of response risk (5, 10 %), and the various uncertainty factors that are applied are not intended to reduce that risk, for the most part. On the other hand, BMD modeling can aid in determination of risk above the RfD and RfC values when exposures above those values occur.

Because UFs are applied multiplicatively they are acknowledged to overlap to some extent, so that the application of several UFs probably does effect some risk reduction. However, the intent of most of the UFs is not risk reduction. For example, the UF_A and UF_H deal, respectively, with the assumed differences in sensitivity between animals and humans, and the assumed variability in sensitivity among members of the human population. The UF_S and UF_D are intended to deal with various aspects of data deficiencies or limitations. Only the UF_L used for the LOAEL to NOAEL extrapolation is a true risk reduction factor. As indicated above, however, the NOAEL can be associated with a significant level of risk in animal bioassay studies (e.g., 5–40 %), and the UF_L may or may not adequately reduce the risk at the LOAEL to an acceptable level.

In a recent health assessment of 1,3-butadiene [4], a factor was applied to the BMDL for a quantal response in an attempt to reduce the risk associated with the POD and also to account for the slope at the BMD. This factor, which is a combination risk reduction factor and uncertainty factor, was termed the effect level extrapolation factor (ELF), and was applied to the POD. The UFs were then applied to this adjusted value. The ELF was determined as follows:

$$\mathrm{ELF} = X \times (\text{slope from } \mathrm{BMD}_x \text{ to } 0)/(\text{slope at the } \mathrm{BMD}_x) \tag{2}$$

where X is the % incidence at the BMR, and BMD_x is the benchmark dose for x level of response. To account for uncertainties about the level of risk at the POD and to insure adequate reduction, the minimum factor applied is intended to be greater than 1 (usually a minimum of 3) up to X. The minimum factor is determined by consideration of the level of response at X, the weight of the evidence, and the endpoint(s) used to determine the POD. Thus, using this approach, RfDs and RfCs are more likely to represent a negligible level of risk. Probabilistic approaches to determining the range of uncertainty around the RfD and RfC may be useful in estimating the range of risk above the RfD or RfC when exposures occur above those levels [14–16]. However, these approaches have not been adopted on a regular basis in risk assessment as yet.

High-dose effects

However the BMR is selected, it is important that the effects selected be relevant to the human situation. This is obvious when considering on which endpoints to base a risk assessment, but is just as important when considering the BMR for a particular endpoint. It is not uncommon for the dominant mechanism of toxicity to change as the dose level increases. Unless the BMR is chosen in the range of toxicity that is relevant to the human exposure range, the BMD that results will not be an appropriate summary of the dose response.

Consideration of model uncertainty

In the paper in which he introduced the idea of benchmark dose, Crump [1] specified that the BMR should be selected to be in the range of the data. This minimizes the effect of model choice on the value of the BMD. Formally, one can divide the uncertainty of a BMD estimate into that due to the data itself, and that due to the uncertainty about the "true" model. The farther a BMR is selected from the responses present in the data, the more the overall uncertainty about the BMD value is due to this latter model uncertainty. Unfortunately, conventional statistical methods do not capture model uncertainty when they quantify the uncertainty of a parameter estimate. Thus, it is important to select a BMR where the model uncertainty contribution to the overall uncertainty is minimal.

EXAMPLE: PERIPUBERTAL EXPOSURE TO THE ANTIANDROGENIC FUNGICIDE VINCLOZOLIN

The fungicide vinclozolin and its two metabolites M1 and M2 are androgen antagonists. They produce adverse effects when administered during sexual differentiation in the fetus or around the time of puberty, and alter sexual function in adult male rats. The study from which the data for this example were taken [17] examined the effects of exposure to vinclozolin around puberty on the male reproductive tract and serum hormone levels. In this example, benchmark doses were calculated for age at preputial separation, epididymal weight, seminal vesicle weight, ventral prostate weight, and serum concentrations of testosterone and luteinizing hormone. The data were provided by Dr. Gray as group means, standard deviations, and sample sizes.

Computing a benchmark dose requires that: (i) a BMR be selected; (ii) one or several appropriate dose–response models be fit to the data; (iii) one model be identified based on an assessment of the quality of the model fit to the data; (iv) the best-fitting model be used to calculate the benchmark dose; and (v) confidence limits or credible limits be computed for the estimate. Serum LH will be used to illustrate these steps, which were followed for all six endpoints.

For this example, the BMD is the dose at which the mean of the response variable is expected to change by an amount equal to the standard deviation of the control group. This level of change very roughly corresponds to the increase in prevalence of extreme individual observations (that is, more extreme than a few percent) by about 10 % [18]. The maximum likelihood fit of a linear, quadratic, power, and Hill model [19], was determined for each endpoint using EPA's BMDS software (see next section for source). For each endpoint, two separate models for the within-dose-group variance were entertained: (i) the variance is the same for all dose groups; and (ii) the variance for a dose group is proportional to the mean raised to a power (e.g., if the estimate of the power is 2.0, the coefficient of variation is constant). Thus, a total of eight models were fit to each of the six endpoints. The best-fitting model of the set of eight, as determined by the sample size corrected version of the Akaike Information Coefficient (AIC_c) [20], was used to calculate the BMD and BMDL (Table 1).

For this example, we show some of the details of fitting the Hill model to the serum LH data. Table 2 presents the original means and standard deviations as well as those predicted by the model, and the scaled residuals: differences between observed and predicted means scaled by their predicted standard errors. These latter quantities are useful adjuncts for assessing model fit, since values with large absolute values (say, greater than 2) indicate points that are not well described by the model. Graphical evaluation is an essential element in assessing overall model adequacy: Figure 1 shows mean serum LH with 95 % confidence intervals and the fitted curve. Both the figure and the tabulated scaled residuals demonstrate that the model fits the data quite well, especially the data for the lowest dose. This confirms that the Hill model is a good choice for the LH data.

The BMDs and BMDLs for all six endpoints are graphed in Fig. 2. While the remaining endpoints seem to form a cluster, with similar BMDs that overlap each other's confidence limits, serum LH stands apart with a substantially lower BMD.

Table 1 AIC_c values for the eight models considered for each of the six endpoints. The minimum AIC_c value for each endpoint is underlined. For five endpoints, the power parameter in the power model was estimated to be 1.0, so it and the linear models resulted in the same BMD, BMDL, and AIC_c values.

Model[a]	Endpoint					
	Age at preputial separation	Epididymal weight	Serum LH	Seminal vesicle weight	Serum testosterone	Ventral prostate weight
lin-hom	<u>106.59</u>	<u>339.77</u>	38.49	<u>430.45</u>	79.51	<u>317.69</u>
lin-het	108.45	341.64	18.96	432.62	<u>71.36</u>	319.08
quad-hom	108.97	341.05	40.64	432.92	81.94	320.16
quad-het	111.01	343.62	17.72	435.24	73.74	321.64
power-hom	<u>106.59</u>	<u>339.77</u>	38.49	<u>430.45</u>	79.51	<u>317.69</u>
power-het	111.08	341.58	18.96	432.45	<u>71.36</u>	321.69
Hill-hom	109.74	340.86	40.58	435.54	81.94	320.16
Hill-het	111.81	343.05	<u>15.7</u>	437.85	73.71	324.64

[a]The word before the hyphen indicates the model for the mean (lin = linear, quad = quadratic, power = power, Hill = Hill model); the word after the hyphen indicates the variance model, either constant (homogeneous) or modeled as a power of the mean (heterogeneous).

Table 2 Mean serum LH concentrations (ng/ml) with their standard deviations and sample sizes, along with the values predicted by the fitted model and differences between the observed and expected mean LH levels scaled by the predicted standard error ("scaled residuals").

		Mean		SD		Scaled
Dose	N	Observed	Expected	Observed	Expected	Residuals
0	10	0.62	0.61	0.23	0.23	0.017
10	10	1.10	1.07	0.67	0.54	0.055
30	10	1.33	1.54	0.73	0.93	–0.234
100	10	2.30	2.06	1.59	1.45	0.163

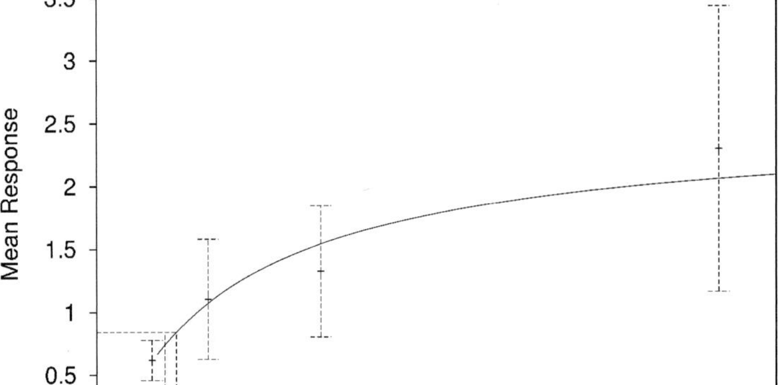

Fig. 1 Mean serum LH and 95 % confidence limits from [17]. The solid line is the fitted dose–response curve.

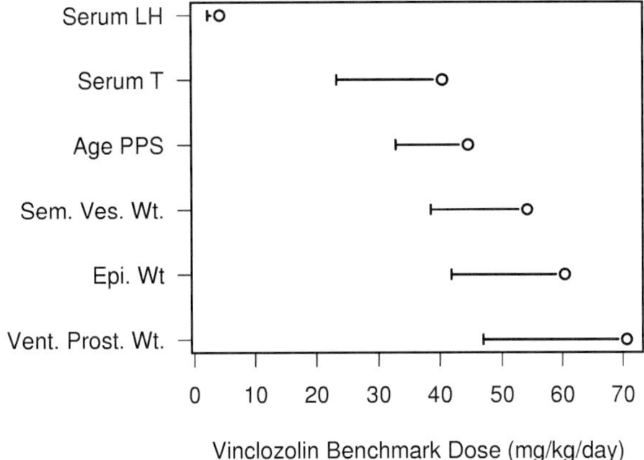

Fig. 2 BMDs and lower 95 % confidence limits for all six endpoints, based on the best-fitting model (determined by lowest AIC_c value) for each endpoint.

MODELING SOFTWARE FOR CALCULATING BENCHMARK DOSES

Many computer software packages allow nonlinear modeling of datasets, but most of these prove to be inadequate for benchmark dose modeling because they cannot calculate benchmark doses or their confidence limits. Currently (as of 1 October 2003), two packages have been created specifically for modeling toxicology data and calculating benchmark doses and their confidence limits:

BMDS: available for free download from the U.S. Environmental Protection Agency (<http://www.epa.gov/ncea>).

ToxTools: commercial software available from Cytel Software Corporation, Cambridge, MA (<http://www.cytel.com>).

In addition, BMD analysis can be carried out using general-purpose statistical software if it is flexible enough to allow the programming for calculation of the BMD and BMDL. This has the advantage that the analysis can be tailored to specific experimental designs and that there are no restrictions on the models that can be used. However, this approach requires substantially more statistical and programming skill than does using the special-purpose software.

RESEARCH PRIORITIES

There is much need for further research on the benchmark dose approach both in statistical methodology and in application to risk assessment. Methodologies need to be developed and applied to toxicology data for quantifying model uncertainty, which could allow extrapolation of the dose response to lower doses while tracking the uncertainty of doing so. Since health effects risk assessments are generally based on a review of effects on multiple endpoints based on multiple data sets, methods need to be developed for better modeling of multiple endpoints at a time, and for combining estimates across independent data sets.

On the applications side, more thought needs to be applied to the problem of extrapolating the BMD, usually derived from animal toxicology studies, to a safe human or ecological exposure level, especially for nonlinear effects.

REFERENCES

1. K. S. Crump. *Fundam. Appl. Toxicol.* **4**, 854–871 (1984).
2. D. G. Barnes and M. Dourson. *Regul. Toxicol. Pharmacol.* **8**, 471–486 (1988).
3. V. Hasselblad and A. M. Jarabek. In *Bayesian Biostatistics*, D. A. Berry and D. K. Stangl (Eds.), Marcel Dekker, New York (1995).
4. U.S. Environmental Protection Agency. National Center for Environmental Assessment-Washington Office, Office of Research and Development, U.S. Environmental Protection Agency, Washington, DC. EPA/600/P-98/001F, May (2002).
5. U.S. Environmental Protection Agency. National Center for Environmental Assessment, Office of Research and Development, U.S. Environmental Protection Agency, Washington, DC. NCEA-F-0644, July (1999). Available at <http://www.epa.gov/ncea/raf/cancer.htm>.
6. U.S. Environmental Protection Agency. National Center for Environmental Assessment, Office of Research and Development, U.S. Environmental Protection Agency, Washington, DC. EPA/630/R-00/001, October (2000). Available at <http://cfpub.epa.gov/ncea/>.
7. E. M. Faustman, B. C. Allen, R. J. Kavlock, C. A. Kimmel. *Fundam. Appl. Toxicol.* **23**, 478–486 (1994).
8. B. C. Allen, R. J. Kavlock, C. A. Kimmel, E. M. Faustman. *Fundam. Appl. Toxicol.* **23**, 487–495 (1994).
9. B. C. Allen, R. J. Kavlock, C. A. Kimmel, E. M. Faustman. *Fundam. Appl. Toxicol.* **23**, 496–509 (1994).
10. R. J. Kavlock, B. C. Allen, C. A. Kimmel, E. M. Faustman *Fundam. Appl. Toxicol.* **26**, 211–222 (1995).
11. J. R. Fowles, G. V. Alexeeff, D. Dodge. *Regul. Toxicol. Pharm.* **29**, 262–278 (1999).
12. U.S. Environmental Protection Agency. (2002). Online at <http://www.epa.gov/iris/index.html>.
13. U.S. Environmental Protection Agency. Vol. 1, EPA/600/R-99/001 (1997), and Vol. 2, EPA/600/R-98/155 (1998). Available at <http://www.epa.gov/nceawww1/colloquium.htm>.
14. S. J. S. Baird, J. T. Cohen, J. D. Graham, A. I. Shlyakhter, J. S. Evans. *Human Ecol. Risk Assess.* **2**, 79–102 (1996).
15. J. S. Evans, L. R. Rhomberg, P. L. Williams, A. M. Wilson, S. J. Baird. *Risk Anal.* **21**, 697–717 (2001).
16. J. C. Swartout, P. S. Price, M. L. Dourson, H. L. Carlson-Lynch, R. E. Keenan. *Risk Anal.* **18**, 271–82 (1998).
17. E. Monosson, W. R. Kelce, C. Lambright, J. Ostby, L. E. Gray, Jr. *Toxicol. Ind. Health* **15**, 65–79 (1999).
18. K. S. Crump. *Risk Anal.* **15**, 79–89 (1995).
19. U.S. Environmental Protection Agency. EPA/600/%-00/014F (2001).
20. K. P. Burnham and D. R. Anderson. *Model Selection and Inference*, p. 51, Springer Verlag, New York (1998).

Pure Appl. Chem., Vol. 75, Nos. 11–12, pp. 2159–2166, 2003.

Topic 3.14

Endocrine active substances and dose response for individuals and populations*

Hugh A. Barton‡

USEPA Office of Research and Development, National Health and Environmental, Research Laboratory, Experimental Toxicology Division, B143-01, Research Triangle Park, NC 27711, USA

Abstract: Dose–response characteristics for endocrine disruption have been major focuses in efforts to understand potential impacts on human and ecological health. Issues include assumptions of thresholds for developmental effects, effects at low doses with nonmonotonic (e.g., "U-shaped") behaviors, population vs. individual responses, and background exposures (e.g., dietary phytoestrogens). Dose–response analysis presents a challenge because it is multidisciplinary, involving biologists and mathematicians. Statistical analyses can be valuable for evaluating issues such as the reproducibility of data as illustrated for contradictory findings on low-dose effects. Mechanistically based modeling provides insights into how perturbations of biological systems by endocrine active substances can create different dose–response behaviors. These analyses have demonstrated that higher order behaviors resulting from the interaction of component parts may appear highly nonlinear, thresholded, low-dose linear, or nonmonotonic, or exhibit hysteresis. Some effects need to be evaluated as population impacts. For example, alterations in male:female ratio may be important at the population level even though not adverse for the individual. Descriptions of the contributions of background exposures to dose–response behaviors are essential. The challenge for improving dose–response analyses is to better understand how system characteristics create different dose–response behaviors. Such generalizations could then provide useful guidance for developing risk assessment approaches.

INTRODUCTION

Estimating the potential risks for human health or environmental toxicity from exposure to chemicals that can disrupt the endocrine system requires a quantitative understanding of the relationship between the dose of the endocrine active substance and the occurrence of effects [1–4]. The controversies about endocrine disruption have been somewhat unusual in environmental toxicology because issues of dose–response relationships have played a central role, somewhat similar to the continuing debates over whether or not cancer dose response is always low-dose linear [5–7]. This represents a challenge to the toxicology, endocrinology, and regulatory communities because dose–response assessment requires collaborations between those with biological and mathematical expertise. While physics, engineering, and population biology historically have required and utilized many forms of mathematics to progress, the introduction of mathematical analyses into many areas of biology, such as physiology and toxicol-

*Report from a SCOPE/IUPAC project: Implication of Endocrine Active Substances for Human and Wildlife (J. Miyamoto and J. Burger, editors). Other reports are published in this issue, *Pure Appl. Chem.* **75**, 1617–2615 (2003).

‡E-mail: habarton@alum.mit.edu

ogy, has been sporadic [8,9]. As reductionist biology is successfully giving scientists access to the pieces from which biological systems are composed, it is becoming increasingly apparent that understanding and fully describing behaviors of the intact biological systems will require sophisticated mathematical analyses. At the same time, the different disciplines need to understand each other's different perspectives and responsibilities. For example, pharmacologists are concerned with obtaining adequate doses to provide significant therapeutic benefits without causing unacceptable side effects, while public health professionals and risk assessors are attempting to ensure there will not be adverse impacts from relatively low doses in large populations. This review will describe some of the key issues that have been raised about the dose–response behaviors of endocrine-mediated toxicities and some of the approaches that have been used to begin to address them.

Dose–response issues for endocrine-mediated effects include a range of questions that reflect upon toxicity testing approaches and risk assessment methods.

- Is there a common dose–response shape characterized by threshold, nonlinearity, or linearity at low doses for endocrine-mediated effects? Do receptor-mediated effects exhibit a common dose–response relationship?
- Are effects during early development (i.e., in utero or postnatal through puberty) characterized by linear dose response?
- Are endocrine-mediated effects characterized by nonmonotonic (e.g. "U-" or "J-" shaped) dose–response relationships?
- Are effects commonly observed at low doses relative to the doses used in traditional toxicity studies, such that they are poorly predicted by such studies?
- When would alterations in the population distributions of characteristics (e.g., intelligence or numbers of males and females) be "adverse" even though the endpoint itself is not "adverse"?
- Are ecological species "sentinels" for potential endocrine-mediated effects in humans? How do their dose–response relationships compare?
- What are the contributions of normal dietary constituents to endocrine-mediated effects, particularly phytoestrogens?

Approaches to evaluating these issues have included experimental studies to provide additional data, statistical analyses of the data, and development of mathematical models for the biological systems and perturbations of those systems. These approaches will be discussed, with some perspectives provided on their abilities to provide answers. The paper begins with the interplay of statistical analyses and dose–response data, particularly with regard to issues of low-dose effects and reproducibility, summarizes the state of mechanistically based modeling, and describes the challenges of developing population approaches.

STATISTICAL ANALYSES OF DOSE–RESPONSE DATA

The major methods employed for analysis of dose–response data involve statistical fitting of curves and analyses using statistics to evaluate trends or group comparisons. These approaches form the underlying basis for most regulatory risk assessment analyses for both noncancer and cancer effects and are the subject of another discussion (Topic 3.13). Such analyses can provide useful insights about the behavior of data in the range of observation and the variability of dose–response data across labs or assay protocols. But, ultimately, these approaches are not powerful tools for providing a fundamental understanding of the complex and controversial issues centered on whether various effects exhibit thresholds, highly nonlinear behaviors, or low-dose linear responses.

It had been suggested that endocrine-mediated effects contrasted with other toxicities in that effects frequently occurred at unusually low doses. An extensive analysis of this issue has recently been completed using a definition of *low dose* as approximating human exposures or below those typically used in standard toxicity testing [10]. This peer review of data concluded that there were specific cases where low-dose effects apparently had been observed in experimental animals with estrogen receptor agonists, but that, in some cases, these findings had not been replicated, so there was no consistent or reproducible support for the existence of low-dose effects. No evidence for low-dose effects was reported for antiandrogens. The approximately 50 studies involved provide a good overview of the extensive experimental efforts to address this issue, including toxicity studies using five or six dose groups spanning several orders of magnitude in contrast to classic toxicity studies involving two or three dose groups often within a factor of 10 of each other. Clearly, there is a value to well-designed studies evaluating biological or adverse effects in the dose range to which humans are exposed, but, at the same time, it is necessary to keep in mind the significant limitations on the power of such studies due to the relatively small number of animals involved.

In addition, this peer review identified important issues for experimental design and statistical analysis of dose–response studies focusing on potential low-dose effects of endocrine active substances [11]. Many of the issues, such as handling body weight as a covariate versus calculating an organ/body weight ratio, apply much more broadly than to only endocrine-mediated effects.

Understanding the real variability of studies of endocrine-mediated responses with live animals is another difficult problem for dose–response analysis. Rarely are assays for toxicological effects repeated multiple times with the same chemical using the same exact methods, so data concerning the reproducibility of endpoints, especially their quantitative dose response, is limited. An analysis of published uterotrophic assay dose–response data following administration of estradiol found the dose–response characteristics to be highly variable [12]. The Hill equation (1)

$$\text{Response} = \text{Response}_0 + \text{Response}_{max} \times (\text{dose}^n/(\text{dose}^n + \text{ED}_{50}{}^n)) \quad (1)$$

was fitted to the data, where ED_{50} is the dose giving a response 50 % of maximal, and n is often referred to as the Hill coefficient. This equation, though loosely derived from biological underpinnings describing cooperative interactions in proteins [13], is frequently used empirically for fitting continuous response data exhibiting a maximum response. The Hill coefficient varied from 0.4 to 6.0 among the 12 uterotrophic datasets with estradiol; a coefficient of 1.0 is linear at low doses, while coefficients greater or less than 1.0 are sublinear and supralinear, respectively. The observed variability in dose response arises in part from the large number of variations in procedures for this bioassay (e.g., immature vs. ovariectomized adult females, injection vs. oral gavage). Clearly, this degree of variability and lack of assay standardization is unacceptable from the point of view of dose–response analysis, regardless of whether such data might be considered adequate for hazard characterization (i.e., qualitative characterization of the active form of a substance). To address issues of cumulative risk from compounds acting through the estrogen receptor, greater consistency is required [14].

MECHANISTICALLY BASED DOSE–RESPONSE ANALYSES

Mathematical descriptions of the biological processes and their perturbation by endocrine active substances can provide insights into how different dose–response behaviors are created. These analyses have demonstrated that quite varied dose–response behaviors can be obtained, depending upon the description of the system and the quantitative values for the parameters (e.g., affinities of ligand binding to receptors, Hill coefficients) [13,15].

A driving force behind the development of mechanistically based mathematical analyses is that statistical curve fitting approaches and qualitative evaluations do not appear able to provide sufficient insights and often reach contrasting conclusions that reflect the perspectives of the analysts [16–18]. Mechanistically based mathematical analyses provide a method to make an explicit description of biological processes (e.g., receptor–ligand binding, clearance processes, feedback regulation) that are hypothesized to be important and, thus, obtain understanding of the underlying basis for the observed dose–response behavior. For endocrine active substances, this requires describing the normal biological processes and their perturbation by the exogenous compounds. These models ultimately link models for the pharmacokinetics of the endogenous hormones, the pharmacokinetics of the exogenous chemical, and the pharmacodynamics of the endogenous system and its perturbation (e.g., Fig. 1).

The limitation of the mechanistically based analyses is that adult hormonal regulation is a complex process, and the changes occurring during fetal development and puberty are even more complex. Thus, strikingly different conclusions have been reached about the dose–response behaviors for dioxin-induced effects, by far the most extensively studied and mathematically modeled receptor-mediated processes in toxicology and risk assessment (though not, strictly speaking, hormonal responses). Some analyses have concluded that the effects, particularly induction of cytochromes P450, were ultimately hyperbolic (often described as Michelis–Menten, although this actually refers to enzyme kinetics, not receptor–ligand interactions) and, thus, low-dose linear [19–21]. In contrast, other analyses have focused upon the apparently "all or none" nature of induction in adjoining or nearby hepatocytes, which would have essentially the same free concentration of chemical, although the total concentration is affected by induction of P450s [22–26]. These analyses find the dose response for enzyme induction to be highly nonlinear, though hyperbolic when averaged over the entire liver. While this may appear to be the same position that is obtained from statistical curve fitting, the advantage is that the discussion is focused upon biological processes, which may be experimentally measurable so that, with time and resources, the questions of dose–response behavior could ultimately be resolved for that system. Thus, mechanistically based modeling may be able to provide insights into when and how threshold, highly nonlinear, or linear dose–response behaviors are created, so that the impact of chemicals could be evaluated, as appropriate, based upon these insights.

Mechanistically based models have been developed for a number of developmental and hormonal systems, as summarized in a workshop report on the topic [18]. This report also provides recommendations for risk assessment and research. While it has sometimes been assumed that linear systems of differential equations necessarily resulted in a low-dose linear behavior, this is not necessarily true for feedback systems. This is readily illustrated for positive feedback (sometimes referred to as feed-forward or autoinduction) systems in which the receptor concentration or the production of a high-affinity ligand (e.g., conversion of testosterone to dihydrotestosterone) is regulated by the ligand [15]. Not surprisingly, systems described with Hill equations (which are nonlinear if the exponent is significantly greater than 1.0) can give a range of dose–response behaviors that depend, in part, on the value of the

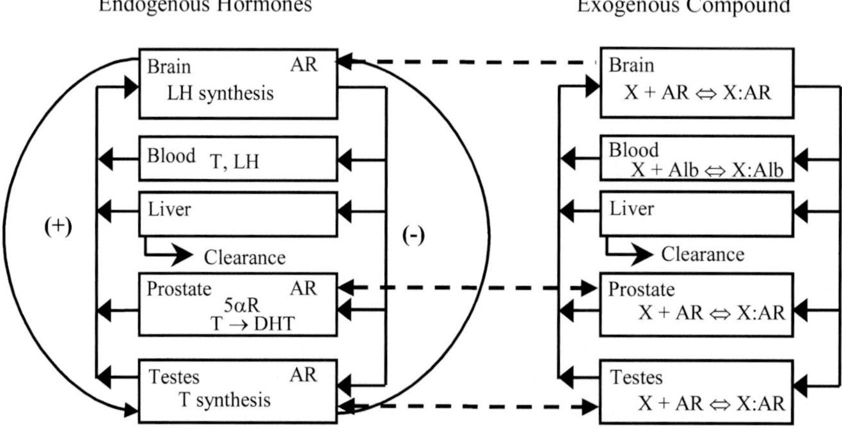

Fig. 1 *Pharmacokinetic and pharmacodynamic modeling for endocrine active substances.* Pharmacokinetic models for the endogenous hormones and exogenous compounds are needed to determine the appropriate tissue dosimetry. The endogenous hormone model incorporates feedback pharmacodynamic processes regulating its production (curved arrows). The exogenous compound, illustrated as an androgen receptor antagonist, interacts with the receptor and affects the pharmacodynamics processes. This affect will be in either the local tissue binding to the androgen receptor or, in the case of the brain, the feedback regulation of the endogenous hormone.

Hill coefficient or exponent [13]. This analysis found low-dose linear, nonmonotonic ("U-shaped"), and highly nonlinear dose response arising from a receptor-mediated system involving positive cooperativity, whether described with a Hill equation or two DNA binding sites for the receptor-ligand complex. Nonmonotonic behaviors can arise from other processes. For example, it has been hypothesized that the in vitro observation of a U-shaped dose response for androgen receptor-mediated response in the presence of dihydrotestosterone and hydroxyflutamide [27] might occur because the steroid receptor dimer bound to either the endogenous hormone, or an exogenous agonist was active for gene activation, but the dimer and one of each ligand were inactive (Rory Conolly, personal communication). Broader impacts on the animal resulting from altered endocrine function, such as caloric restriction due to reduced feeding, have the potential to create nonmonotonic behaviors (sometimes described as hormesis) and should also be considered [28]. Negative feedback is widely recognized to be involved in the creation of homeostatic conditions that are often associated with thresholds for toxic effects [29]. Finally, hysteresis (i.e., different dose–response behaviors depending upon progression from high-to-low dose or low-to-high dose) also was observed to result from positive feedback systems regulating receptor concentration and production of a high-affinity ligand [15].

Mathematical descriptions of the biology underlying the creation of varied dose–response behaviors are increasingly being explored for a wide range of biological systems, at levels of organization ranging from cells to whole organisms to populations of each of these [9]. Studies of embryonic development and cell cycling will clearly be highly relevant to gaining greater understanding of those situations where endocrine-mediated effects give different dose–response behaviors [30,31]. In particular, they will likely identify common types of processes used for creating those behaviors, even when factors such as the specific effector molecules or cell types are all different [15]. It is the interactions of these components in a higher-order system that generally creates the system dose–response behaviors.

POPULATION DOSE RESPONSE VS. INDIVIDUAL DOSE RESPONSE AND BACKGROUND EXPOSURES

Chemical risk assessments, particularly for noncarcinogenic effects, essentially evaluate the dose–response behavior for individuals representative of sensitive populations rather than estimate population risks. It has been suggested that endocrine-mediated effects may, in part, be evidenced as shifting distributions of a characteristic in the population rather than as directly causing an adverse effect. One example of this would be altered ratios of male and female offspring among turtles born from eggs exposed to estrogenic compounds [32–34]. It is interesting to note that temperature- and steroid-dependent turtle egg sex determination appears to be a highly nonlinear process at the individual level, involving positive feedback on both the steroid receptor and the high-affinity hormone synthesis enzymes in order to drive the turtle's development to be either male or female. However, the observation at the population level is that exogenous compounds can influence the chances that the positive feedback will proceed toward one sex or the other [34].

Another aspect involving populations is the widespread, but varied, exposures of populations to backgrounds of endocrine active substances, including persistent bioaccumulative compounds (e.g., dioxins/furans) and phytoestrogens. It has been suggested that the dose response could be low-dose linear if the endogenous hormones or the background exposures already resulted in some incidence of the effect in the population, and additional exposure worked through that common mechanism [35]. One complication with hormones and endocrine-mediated effects is that the same hormones (or exogenous compounds such as phytoestrogens) often appear to be responsible for both beneficial and adverse effects (e.g., associations of elevated estrogen status in women with increased risk of some cancers, decreased risks of others, decreased risks of osteoporosis). Thus, assessing the population impact of exposures to endocrine active substances is an area deserving effort, but requiring development of innovative tools and approaches. It should be particularly focused on bringing together knowledge about public health and toxicology.

RESEARCH PRIORITIES AND OTHER RECOMMENDATIONS

Research priorities in the area of dose–response assessment for endocrine active substances have been described previously [10,18,36]. Largely, the priorities previously outlined remain needs, as these were generally longer-term efforts. It has been recommended that prototype case studies implementing mechanistically based dose–response modeling be developed for endocrine-mediated effects [18]. Case studies have been suggested for impacts on adult prostatic function from in utero estradiol exposure or developmental antiandrogen exposures and the association of estrogen exposure and mammary/breast cancer, among others. This clearly continues to be one of the highest priority needs as this approach will build the understanding necessary to begin to discern how to appropriately generalize from these case examples to broader ranges of substances or broader ranges of effects.

A strong recommendation of one previous effort was that development of mechanistically based dose–response models should be a routine part of the risk assessment process [18]. Models for the pharmacokinetics of compounds are generally more readily developed than for their pharmacodynamic processes. Pharmacokinetic models can assist in addressing extrapolation issues that frequently arise in risk assessments, such as those across routes of exposure, among exposure regimens (e.g., continuous vs. episodic), and among species [3,37]. These models are also valuable for evaluating how internal doses would change with different exposures at different ages (e.g., children or the elderly). Selection of the appropriate internal dose-metric for such extrapolations is facilitated by knowledge of the pharmacodynamic processes, but doesn't require the same level of quantitative information as is often required to develop a mechanistically based pharmacodynamic model. Thus, implementation of mechanistically based modeling as a way of explicitly incorporating scientific data into risk assessment can and should proceed as an incremental process replacing default assumptions.

SUMMARY AND CONCLUSIONS

Issues of dose–response behavior are critically important to risk assessment for chemical-mediated toxicities [3]. The standard paradigm for chemical risk assessment in the United States involves an explicit dose–response assessment step. Continued progress toward better understanding dose–response behaviors is essential for moving risk assessments from default assumptions to more scientifically based approaches.

Theoretical mechanistically based modeling has clearly demonstrated that a range of dose–response behaviors can be obtained from receptor-based feedback-regulated systems, depending upon the specific characteristics of the system and the values of the parameters that describe it (e.g., affinity constants) [13,15,29]. These behaviors include: (1) highly nonlinear dose response, for example, created by positive feedback, (2) thresholds created by negative feedback regulation, (3) low-dose linear behavior when a system is essentially determined by the interaction of a ligand with a receptor, (4) nonmonotonic ("U-shaped) dose response arising from positive cooperativity in specific biological steps such as protein synthesis or from activity of dimers with the same ligand, but inactivity of mixed dimers with different ligands, and (5) hysteresis (dose–response behaviors that vary as dose increases or decreases) due to positive feedback on the receptor and synthesis of a high-affinity ligand. Thus, the challenge now is to better define when these different dose–response behaviors arise so that risk assessment approaches for endocrine active substances can be tailored appropriately.

ACKNOWLEDGMENTS

The critical review and input of Woody Setzer are acknowledged. The information in this article has been subjected to review by the National Health and Environmental Research Laboratory and approved for publication. Approval does not signify that the contents reflect the views of the Agency.

REFERENCES

1. H. A. Barton and M. E. Andersen. *Regul. Toxicol. Pharmacol.* **25**, 292–305 (1997).
2. H. A. Barton and M. E. Andersen. *Crit. Rev. Toxicol.* **28**, 363–423 (1998).
3. H. A. Barton, M. E. Andersen, H. J. Clewell, III. *Hum. Ecol. Risk Assess.* **4**, 75–115 (1998).
4. IUPAC. *Pure Appl. Chem.* **70** (9) (1998).
5. M. Crawford and R. Wilson. *Hum. Ecol. Risk Assess.* **2**, 305–330 (1996).
6. G. P. Daston. In *Issues and Reviews in Teratology*, Vol. 6, H. Kalter (Ed.), pp. 169–197, Plenum Press, New York (1993).
7. WHO/IPCS, Global Assessment of the State-of-the-Science of Endocrine Disruptors (2002).
8. D. E. Koshland, Jr. *Science* **280**, 852–853 (1998).
9. S. A. Levin, B. Grenfell, A. Hastings, A. S. Perelson. *Science* **275**, 334–343 (1997).
10. R. Melnick, G. Lucier, M. Wolfe, R. Hall, G. Stancel, G. Prins, M. Gallo, K. Reuhl, S. M. Ho, T. Brown, J. Moore, J. Leakey, J. Haseman, M. Kohn. *Environ. Health Perspect.* **110**, 427–331 (2002).
11. J. K. Haseman, A. J. Bailer, R. L. Kodell, R. Morris, K. Portier. *Toxicol. Sci.* **61**, 201–210 (2001).
12. H. A. Barton, M. E. Andersen, B. C. Allen. *Regul. Toxicol. Pharmacol.* **28**, 133–149 (1998).
13. M. C. Kohn and C. J. Portier. *Risk Anal.* **13**, 565–572 (1993).
14. S. H. Safe. *Environ. Health Perspect.* **103**, 346–351 (1995).
15. M. E. Andersen and H. A. Barton. *Toxicol. Sci.* **48**, 38–50 (1999).
16. R. M. Blair, H. Fang, D. Gaylor, D. M. Sheehan. *APMIS* **109**, 198–208 (2001).
17. L. E. Limbird and P. Taylor. *Cell* **93**, 157–163 (1998).
18. M. E. Andersen, R. B. Conolly, E. M. Faustman, R. J. Kavlock, C. J. Portier, D. M. Sheehan, P. J. Wier, L. Ziese. *Environ. Health Perspect.* **107** (Suppl 4), 631–638 (1999).
19. M. C. Kohn, G. W. Lucier, G. C. Clark, C. Sewall, A. Tritscher, C. J. Portier. *Toxicol. Appl. Pharmacol.* **120**, 138–154 (1993).
20. M. C. Kohn, C. H. Sewall, G. W. Lucier, C. J. Portier. *Toxicol. Appl. Pharmacol.* **165**, 29–48 (1996).
21. C. Portier, A. Tritscher, M. Kohn, C. Sewall, G. Clark, L. Edler, D. Hoel, G. Lucier. *Fundam. Appl. Toxicol.* **20**, 48–56 (1993).
22. M. E. Andersen, C. R. Eklund, J. J. Mills, H. A. Barton, L. S. Birnbaum. *Toxicol. Appl. Pharmacol.* **144**, 135–144 (1997).
23. M. E. Andersen, L. S. Birnbaum, H. A. Barton, C. Eklund. *Toxicol. Appl. Pharmacol.* **144**, 145–155 (1997).
24. M. E. Andersen and R. B. Conolly. *Hum. Exp. Toxicol.* **17**, 683–690 (1998).
25. M. E. Andersen and H. A. Barton. *Environ. Health Perspect.* **106**, 349–355 (1998).
26. M. J. Santostefano, V. M. Richardson, N. J. Walker, J. Blanton, K. O. Lindros, G. W. Lucier, S. K. Alcasey, L. S. Birnbaum. *Toxicol. Sci.* **52**, 9–19 (1999).
27. S. C. Maness, D. P. McDonnell, K. W. Gaido. *Toxicol. Appl. Pharmacol.* **151**, 135–142 (1998).
28. A. Turturro, B. S. Hass, R. W. Hart. *Hum. Exp. Toxicol.* **19**, 320–329 (2000).
29. J. D. Wilson. In *Receptor-Mediated Biological Processes: Implications for Evaluating Carcinogenesis. Proceedings of the Sixth International Conference on Carcinogenesis and Risk Assessment*, H. L. Pitzer, T. J. Slaga, W. F. Greenlee, M. McClain (Eds.), pp. 223–236, Wiley-Liss, Austin, TX (1992).
30. J. E. Ferrell, Jr. and E. M. Machleder. *Science* **280**, 895–853 (1998).
31. C. Thron. *Biophys. Chem.* **57**, 239–251 (1996).
32. D. Crews, A. R. Cantu, J. M. Bergeron. *J. Endocrinol.* **149**, 457–463 (1996).
33. D. Crews, J. M. Bergeron, J. J. Bull, D. Flores, A. Tousignant, J. K. Skipper, T. Wibbels. *Dev. Genet.* **15**, 297–312 (1994).

34. D. M. Sheehan, E. Willingham, D. Gaylor, J. M. Bergeron, D. Crews. *Environ. Health Perspect.* **107**, 155–159 (1999).
35. K. S. Crump, D. G. Hoel, C. H. Langley, R. Peto. *Cancer Res.* **36**, 2973–2979 (1976).
36. R. J. Kavlock, G. P. Daston, C. DeRosa, P. Fenner-Crisp, L. E. Gray, S. Kaattari, G. Lucier, M. Luster, M. J. Mac, C. Maczka, R. Miller, J. Moore, R. Rolland, G. Scott, D. M. Sheehan, T. Sinks, H. A. Tilson. *Environ. Health Perspect.* **104**, 715–740 (1996).
37. H. J. Clewell, 3rd, M. E. Andersen, H. A. Barton. *Environ. Health Perspect.* **110**, 85–93 (2002).

Pure Appl. Chem., Vol. 75, Nos. 11–12, pp. 2167–2179, 2003.

Topic 3.15

Endocrine disruption occurring at doses lower than those predicted by classical chemical toxicity evaluations: The case of bisphenol A*

John Ashby[‡]

Syngenta Central Toxicology Laboratory, Alderley Park, Cheshire, UK

Abstract: The meaning of the term "low dose" is discussed in relation to endocrine toxicity data for chemicals. Consideration is also given to experimental conditions likely to impinge on the interpretation and extrapolation of such low-dose effects, and the importance of gathering appropriate control data is emphasized. In the specific case of bisphenol A (BPA), it is concluded that despite the extensive endocrine disruptor (ED) database available for this chemical, it is still not possible to locate a single study that passes the most rudimentary scientific requirements—that the observations are capable of independent confirmation. Two possible explanations for this are considered. First, that BPA possesses subtle low-dose ED toxicities that only become evident under certain undefined experimental conditions. Until these conditions are defined and understood, it will be a matter of chance what individual investigators observe experimentally for BPA or any other chemical. Second, that the general failure of investigators to define and understand natural variability among control parameters monitored in ED studies allows artefactual positive results to be encountered for chemicals, especially in limited and nonreproduced studies. Whichever of these conclusions is correct, the positive low-dose data currently available for BPA cannot be extrapolated to humans with any confidence.

INTRODUCTION

Evaluation of the possibility that one or more environmental factors may be adversely affecting the reproductive capacity of humans and wildlife has been severely hampered by inadequate epidemiological evidence for induced effects, and by the absence of agreed and developed test methods and experimental data. Consequently, disparate experimental data have been published in the absence of a unified approach, and the field has been assailed by controversy, sinking, in some instances, to overt invective. The most controversial of issues, and the subject of this review, has been whether certain endocrine disruptors (EDs) are able to induce adverse effects at dose levels below those expected based upon the results of classical reproductive and developmental toxicity evaluations. Allied to this question is the suggestion that these "low-dose" toxicities follow an inverted-U shaped dose response, thus explaining the absence of effects at the "higher" doses used in the classical toxicity evaluations. The controversy that followed the initial report of such effects has centered on whether the effects reported can be independently reproduced, and more specifically, whether they are of general biological significance, or are limited to highly specific and undefined conditions of the experimental model used to establish them.

---

*Report from a SCOPE/IUPAC project: Implication of Endocrine Active Substances for Human and Wildlife (J. Miyamoto and J. Burger, editors). Other reports are published in this issue, *Pure Appl. Chem.* **75**, 1617–2615 (2003).

‡E-mail: John.Ashby@Syngenta.com

In attempting here to summarize the low-dose debate, two quotations from a recent book by Stephen Jay Gould [1] enable the discussion to be placed in a larger scientific context:

> "Many supposed debates in science arise from confusion engendered by differing uses of words, and not from deep conceptual muddles about the nature of things."
>
> "Replication 'with difference' builds the best case for a generality—for how can we prove a coordinating hypothesis unless we can apply it to multiple cases?"

Specific answers are not provided to the many questions posed in this article, but the rehearsing of these questions should aid their resolution. Primary attention is focused here on bisphenol A (BPA) as this chemical has the largest available low-dose ED database.

DISCUSSION

This review describes the underlying scientific issue awaiting resolution before a clear and scientifically justified position can be taken on low-dose endocrine toxicities. The issues raised are discussed in the context of a series of figures and tables.

Meaning of the terms "low-dose effects" and "inverted-U" dose response

The schematic shown in Fig. 1 demonstrates the paucity of human data on EDs. In contrast, there are well-established carcinogen databases for both humans and rodents, enabling firm correlations of activities to be established. The absence of a corresponding human database for EDs forces reliance on data derived from rodent studies. Most rodent ED data are derived from classical toxicology evaluations using dose levels related to the maximum tolerated dose (MTD) for the chemical. These studies are then used to define a no adverse effect level (NOAEL) for the chemical. There are three situations where one can consider the possibility of effects occurring below this NOAEL level. First, there is the possibility that a study with larger group sizes, and a higher resolving power, might have led to the observation of effects. This possibility applies to all toxicological evaluations. A more interesting possibility is that monitoring precursor events associated with the adverse effect in question might lower the NOAEL. This concern is also not unique to EDs. For example, it has yet to be considered whether the induction

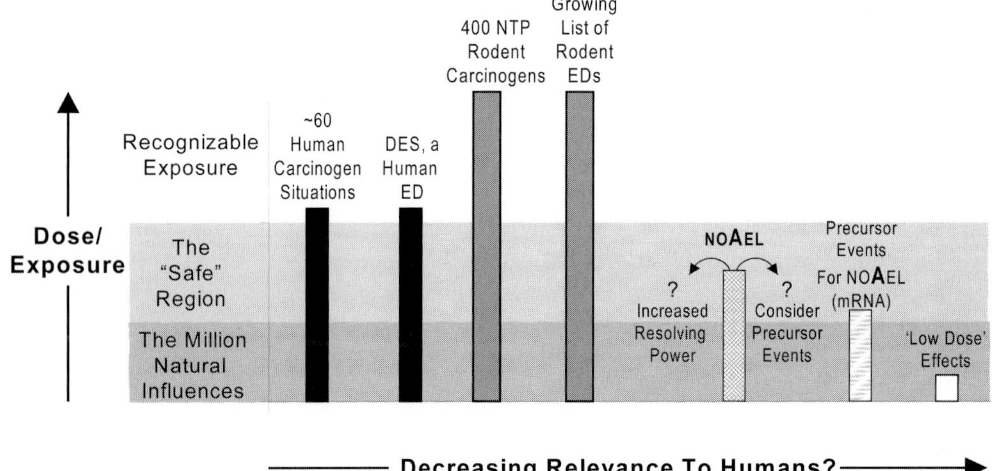

Fig. 1 Basis for considering an effect to be observed at a "low dose".

of subtle pathological changes in the rodent liver, such as the appearance of enzyme-altered foci, at sub-carcinogenic dose levels, should contribute to an assessment of the NOAEL for the liver carcinogen in question. Consideration of such questions hinge on the definition of an adverse effect, as opposed to an observed effect. The advent of genomic techniques will increase the need to answer such questions, as there one is faced with changes to the mRNA levels associated with a precursor to an adverse effect. General consideration by toxicologists of low-dose ED effects for chemicals was triggered by observations reported [2]. In that case, effects on the mouse prostate gland were reported at dose levels orders of magnitude lower (20 μg/kg) than would have been suggested by the standard toxicology data for this chemical (5 mg/kg; discussed later). Although these data [2] are often referred to as representing an inverted-U dose response there were, in fact, only two doses evaluated. These data, therefore, raise the prospect of a surge of ED activity for BPA within the region lying below its classical NOAEL and within the area of natural biological variability (for example, circadian rhythms, transitory changes after eating).

The concept of low-dose "inverted-U", and "U-shaped" dose–response relationships is not novel (reviewed in [3]), but to date there are few, if any, confirmed and agreed examples within the toxicology literature. Perhaps the most compelling example of a U-shaped dose response is provided by the data of Almstrup et al. [4]. These authors showed that certain phytoestrogens are able to inhibit the testosterone initiated proliferation of MCF7 cells at low doses, by inhibiting the conversion of testosterone to estradiol via aromatase enzymes, while stimulating proliferation at higher dose levels due to the intrinsic estrogenic activity of the phytoestrogens. The issue of relevance to this discussion is whether the effects reported for BPA are of general significance, because if they are, a new approach to establishing NOAEL levels for chemicals will be required. An assessment of the total available low-dose database for BPA is given later herein.

Integration of all relevant data

Consideration of low-dose chemical toxicities form only one possible component of current attempts to discern which environmental influences are associated with the adverse effects that are driving the science of endocrine disruption—effects such as the current incidences of human prostate, testicular, and breast cancer. To study chemical contaminants in isolation will delay resolution of these key underlying questions. The hypothesis that milk consumption may be associated with human prostate cancer (Fig. 2) [5] and the observation of reduced human sperm motility following the Kobe earthquake [6] provide two examples of the many complex issues underlying the questions being addressed in the field of ED.

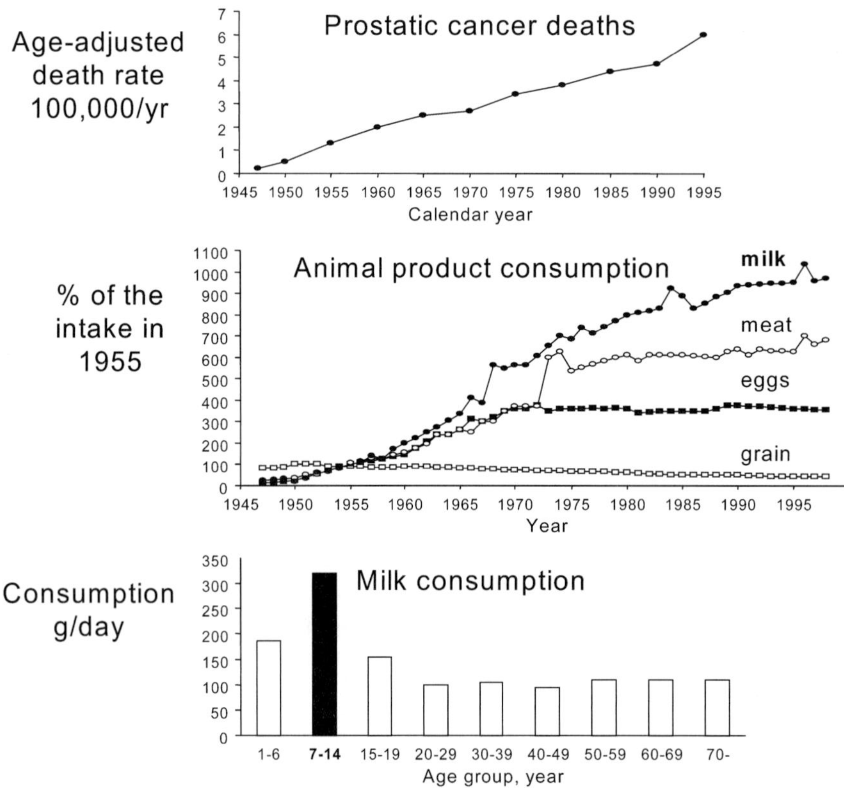

Fig. 2 Possible correlations between increases in prostate cancer in Japan and increases in milk consumption over the same period [5].

Consideration of the conditions under which experimental data are generated

Although the route of administration adopted in an ED study is rarely discussed by investigators, it has the power to affect seminally interpretation and extrapolation of the derived data. This can be illustrated by the data shown in Figs. 3 and 4. The data in Fig. 3 illustrate the highly significant observation that equally effective doses of diethylstilbestrol (DES), and the phytoestrogen genistein (GEN) in the immature mouse uterotrophic assay are also associated with equally effective carcinogenic responses in the mouse uterus [7]. The mouse data were extrapolated to infants consuming soy-based infant formula by the authors [7]. However, in terms of risk assessment, it is critical to note that the data shown in Fig. 3 were all generated using the subcutaneous route of administration, while infants are exposed to soy formula by the oral route. The uterotrophic activities of GEN in immature rodent uterotrophic assays shown in Fig. 4 illustrate the critical importance of the route of exposure selected; in particular, the relative insensitivity of the oral route. Thus, data generated for GEN using the subcutaneous injection route is of little value when estimating potential hazards posed by oral exposure to this agent. The general point intended here is that it is inappropriate to sum together data for a chemical generated using a variety of routes of exposure—but such is a common practice in ED risk assessment, as discussed later in relation to the low-dose ED effects reported for BPA.

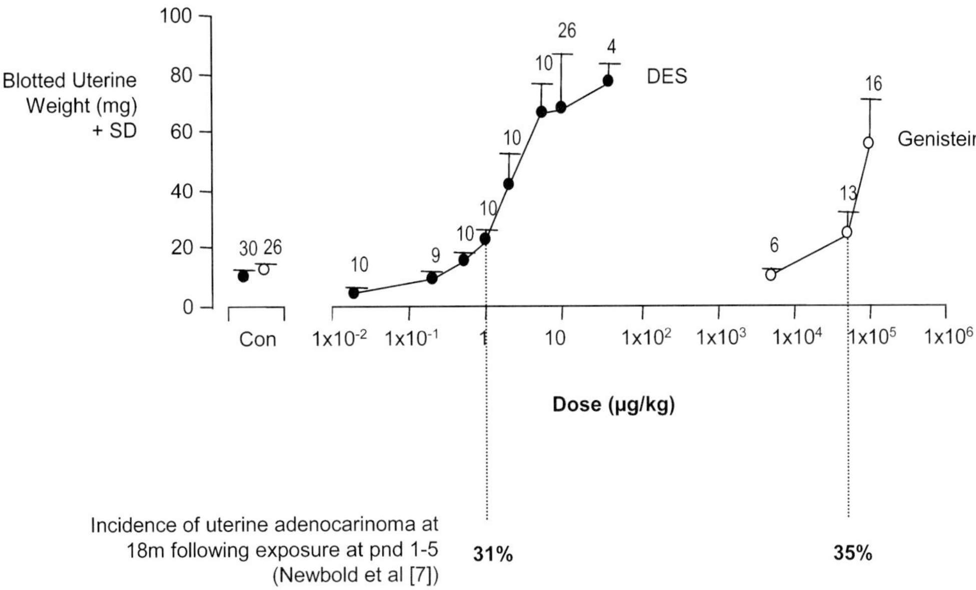

Fig. 3 Dose levels that yield equal levels of activity for DES and GEN in the immature mouse uterotrophic assay ([58] and Ashby et al., unpublished, 2002) are also associated with the induction of equal incidences of uterine adenoma in the mature mouse when exposed on postnatal days 1–5 [7].

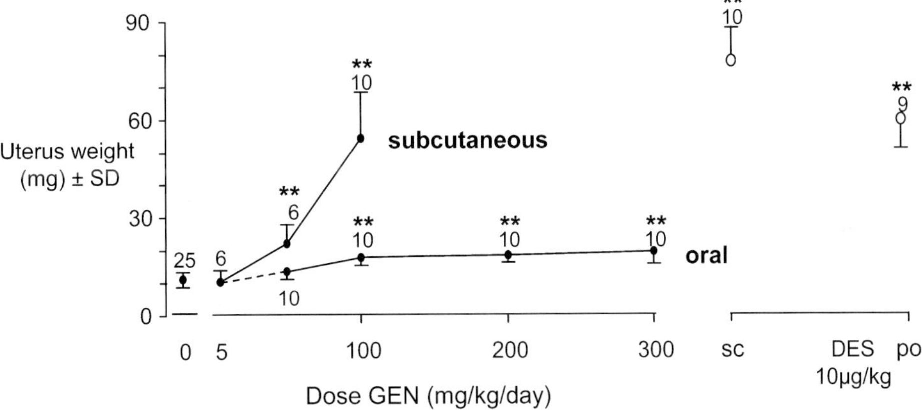

Fig. 4 Relative activity of GEN and DES in the immature mouse uterotrophic assay using either oral or subcutaneous routes of exposure [59].

Precursors of adverse effects

It is probable that precursor events to a toxic response will occur at dose levels below the NOAEL for the adverse toxic effect under study. This is illustrated by the data shown in Fig. 5. In the studies referred to therein, Gould et al. [8] confirmed the absence of uterotrophic activity for BPA in the immature rat at dose levels previously reported as inactive [9]. However, other effects were observed by Gould et al. in the treated uteri, despite the absence of uterine growth. Whether or not these changes are associated with the "beginnings" of an uterotrophic response, or with parallel phenomena, remains to be determined. Such questions will increasingly be posed by the trend to generate genomic data at dose levels below the NOAEL dose for adverse effects in the tissue under study. Such precursor/parallel ef-

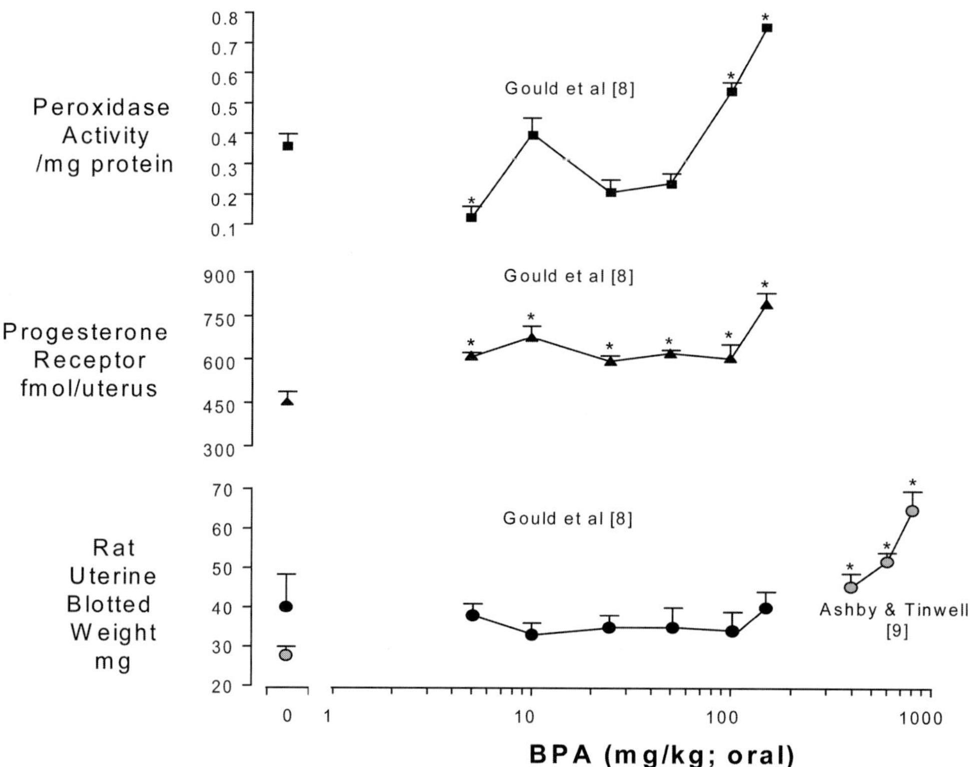

Fig. 5 Reported activities of BPA in the immature rat uterus.

fects are probably unconnected with the low-dose effects discussed for BPA in the next section, and the clear separation of these two low-dose effects will aid the interpretation of each of them in terms of human risk assessment.

"Low-dose" effects reported for BPA

The first report of low-dose effects induced by a chemical in the mouse prostate gland was for DES given orally [10]. Those data are shown in Fig. 6, together with mouse uterotrophic assay data for DES reported earlier ([11]; subcutaneous injection of DES and see Fig. 4). These increases in prostate weight were small and were not exceptional given that they were generated following exposure in utero, and that positive effects at similar dose levels were known for DES in the immature mouse uterotrophic assay (Fig. 4) [11]. Much more unexpected was the observation that BPA was as effective as DES at increasing mouse prostate gland weight (Fig. 7) [2] despite its essential inactivity in the mouse uterotrophic assay (oral or subcutaneous administration of BPA) [12]. Comparison of the effects shown in Figs. 6 and 7 reveals that BPA, a chemical generally considered to be a weak estrogen, is as active in the mouse prostate as is the reference and potent estrogen DES. It is this stark contrast of activities that has led to the many subsequent studies and discussions on BPA.

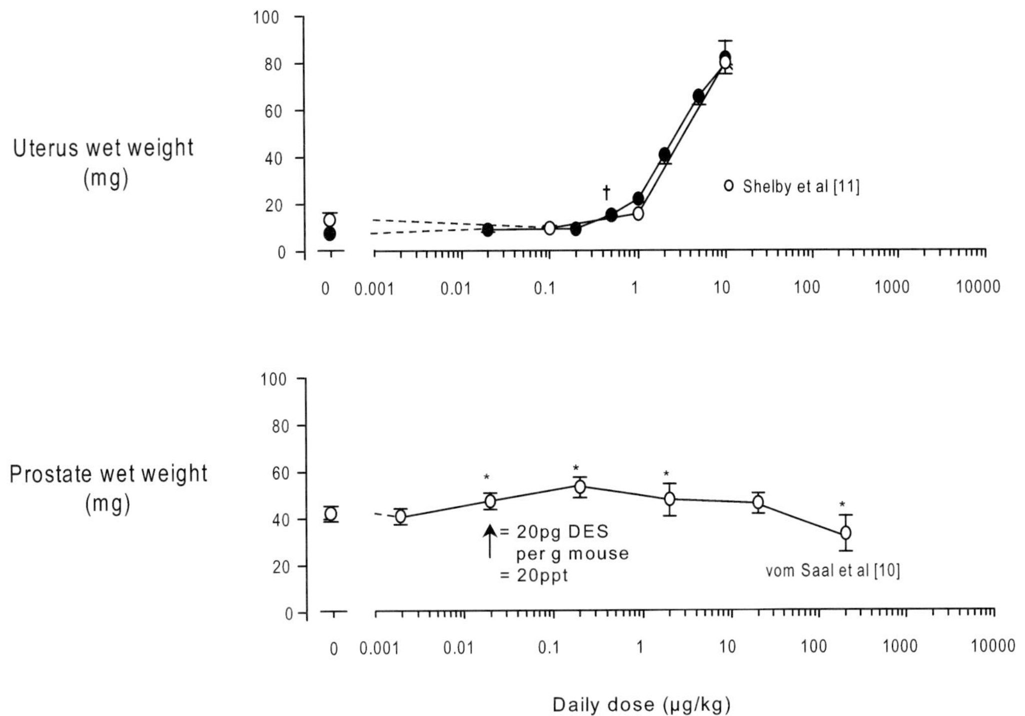

Fig. 6 Activity of DES in the immature mouse uterotrophic assay and comparison with changes in prostate weight induced following exposure in utero. *Reported as statistically significant.

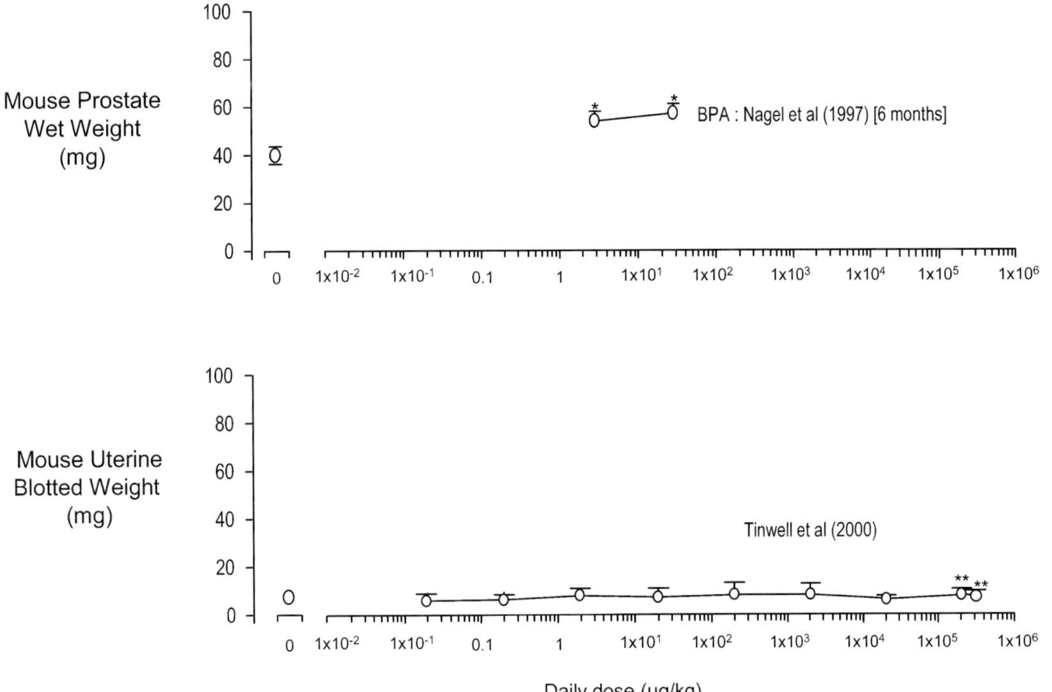

Fig. 7 Activity of BPA in immature mouse uterotrophic assay and comparison with changes in prostate weight induced following exposure in utero. *Reported as statistically significant.

Since those original studies on BPA many investigators have studied the ED effects of BPA in both mice (Table 1) and rats (Table 2). The rat multigeneration assay data for BPA [13,14] are generally considered by regulators to have defined its NOAEL at 5 mg/kg—against which value the low-dose effects reported for BPA can be assessed (unshaded areas of Tables 1 and 2).

Table 1 Reported activities of BPA in reproductive/developmental studies conducted in the mouse. Abstracts have been omitted. The shaded area of the table represents activities observed at or above the generally accepted NOAEL value for BPA in the rat [13,14].

Mouse strain	**Exposure period**	Route of admin.	Dose range	Min. positive dose	**Effects reported**	Ref.
CF_1	GD11–17	Oral	2 & 20 μg/kg	2 μg/kg	Increase in prostate & preputial gland wt. Decreased DSP & epididymal wt	[2,15]
CF_1	GD11–17	Oral	2.4 μg/kg	2.4 μg/kg	Increase in body wt; advance in age of 1st oestrus	[32]
CD-1	GD14–18 (in utero) ± GD14–18 (adult exposure during pregnancy)	Oral	10 μg/kg	10 μg/kg	Maternal behavior affected in mice exposed either in utero or during adulthood	[33]
ICR	GD11–17	s.c.	2 & 20 μg/kg	20 μg/kg	Advance in age at vaginal opening and at 1st estrus	[34]
CD-1	GD9–20	Implant	25 & 250 μg/kg	25 μg/kg	Increased mammary gland maturation	[35]
CD-1	GD16–18	Oral	50 μg/kg	50 μg/kg	Increased AGD & prostate wt; decreased epididymal wt	[36]
C57BL/6N	PND35–63 & PND35–91	DW	139 μg–12.7 mg/kg	139 μg/kg	Decreased testosterone; testicular morphology affected	[37]
ICR	GD11–17 PND1–5	s.c.	10 & 100 mg/kg	10 mg/kg 100 mg/kg	Decreased corpora lutea Increase vaginal stratification & polyovular follicles	[38]
CD-1	GD6–15	Oral	500–1250 mg/kg	500 mg/kg	Maternal & fetal toxicity	[39]
CD-1	Continuous breeding	Diet	300–1300 mg/kg	600 mg/kg	Reduction in testes wt and sperm motility	[40]
CF_1	GD11–17	Oral	0.2–200 μg/kg	–	No effects	[41]
CF_1	GD11–17	Oral	2 & 20 μg/kg	–	No effects	[24]
C57BL/6N	GD11–17; PND21–43; PND70–77	Oral	2–200 μg/kg	–	No effects	[42]

The top seven entries in Table 1 provide evidence of sub-NOAEL effects for BPA in the mouse. None of these data provide an independent confirmation of any of the other studies. Two of these studies are difficult to integrate with the others because they employed either subcutaneous injection or implantation of the BPA. The middle entries in Table 1 represent activities for BPA within its acknowledged active dose range. The final entries in Table 1 represent the results of three independent and unsuccessful attempts to confirm the original observations made for BPA [2,15]. One is therefore faced with a set of disparate observations made for low doses of BPA, and three unsuccessful attempts to confirm one of those observations. The strain of mouse used does not appear to be the critical determinant of activity for BPA (Table 1), nor does the diet employed (discussed further in [16]). Thus, the "generality of effect" mentioned earlier has yet to be established for low-dose effects of BPA in the mouse.

Table 2 Reported activities of BPA in reproductive/developmental studies conducted in the rat. Abstracts have been omitted except for the replicated studies by Welsch [51,52]. The shaded area of the table represents activities observed at or above the generally accepted NOAEL value for BPA in the rat [13,14].

Rat strain	**Exposure period**	Route of admin.	Dose range	Min. positive dose	**Effects reported**	Ref.
SD	GD6–21	Oral	20 µg–50 mg/kg	20 µg/kg	Multiple male and female parameters affect	[20]
SD	PND91–PND96	Oral	2 ng–200 mg/kg	20 µg/kg	Males only studied. Reduced daily sperm production at PND126	[21]
SD	GD1–PND21 GD14– PND6	Oral	40 µg/kg 400 µg/kg	40 µg/kg	Male and female play behavior affected	[18]
SD	Pregestation-PND21	Oral	40 µg/kg	40 µg/kg	Sexual activity impaired in males; sexual motivation and receptivity affected in females	[19]
SD	GD6–PND22	Oral	0.1 & 1.2 mg/kg	1.2 mg/kg	Altered patterns of estrous cyclicity	[43]
Wistar	Gestation & lactation	DW	~1.5 mg/kg	~1.5 mg/kg	Sexual differentiation of locus coeruleus and behavior affected; no effects on reproductive organs or sex hormones.	[44]
Wistar	GD8–birth	Implant	25 µg & 250 µg/kg	25 µg/kg	Differentiation pattern of periductal stromal cells of the ventral prostate affected	[17]
Wistar	PND2–12	s.c.	37 mg/kg	37 mg/kg	Reduction of epithelial cell height in efferent ducts only at PND18	[45]
Wistar	PND2–12	s.c.	37 mg/kg	37 mg/kg	Males only studied. Significant increase in testis wt & marked effects on pubertal spermatogenesis at PND18. Effects not evident at PND25. Increased testis wt at PND90–100	[46]
Wistar	PND2–12	s.c.	37 mg/kg	37 mg/kg	Increased plasma testosterone; increased germ cell volume/Sertoli cell at PND18	[47]
Wistar	PND22–32	s.c.	50 mg/kg	50 mg/kg	Males only studied. Increase in lateral prostate wt at PND120	[48]
F344	PND28–72	Diet	235–950 mg/kg	235 mg/kg	Significant decrease in seminalvesicle, dorso-lateral prostate, preputial glands, hypophysis and body wt; testicular toxicity at PND72	[49]
SD	2 generations	Oral	0.2 mg–200 mg/kg	–	Multiple parameters studied. AGD reduced in F1 generation, but considered not to be toxicologically significant by investigators.	[14]
SD	GD2–PND21	Oral	1 µg–10 mg/kg	–	Males studied. Significant increase in ventral prostate wt at PND177 but considered by investigators to be due to sampling design. Studied repeated by Welsch et al. 2001.	[50]
SD	GD2–PND21	Oral	1 µg–10 mg/kg	–	Females studies. No effects on developmental landmarks, estrous cyclicity, organ wts, fertility and fecundity observed (PND2–~10 months).	[51]
SD	GD2–PND21	Oral	1 µg–15 mg/kg	–	No effects observed	[52]

(*continues on next page*)

Table 2 (*Continued*).

Rat strain	**Exposure period**	Route of admin.	Dose range	Min. positive dose	**Effects reported**	Ref.
SD	3 generations	Diet	1 mg–500 mg/kg	–	Multiple parameters studied. Total live pups/litter and ovarian wt reduced, but considered by investigators to be due to toxicity. No effects at or below 5 mg/kg/day.	[13]
Wistar	Pregestation–PND22	Oral	4 µg–4 mg/kg	–	Males only studied. No effects at PND90 on tissue wts or developmental landmarks.	[53]
SD & AP	GD6–21	Oral	20 µg–50 mg/kg	–	No effects in male or female offspring	[22]
SD	PND91–PND96	Oral	20 µg–200 mg/kg	–	No effects on adult male sperm parameters	[23]
SD	PND1–5	s.c.	3 mg/kg	–	No effects in both sexes at PND21 and PND98	[54]
SD	GD11–PND20	Oral	3.2–32 mg/kg	–	No effects on both sexes (pubertal development in females only; male and female reproductive tissues PND180)	[55]
Wistar	PND2, 4, 6, 8, 10	s.c.	~10 mg/kg (0.5 mg/rat)	–	No effects of testicular wt; seminiferous tubule diameter; pituitary FSHb expression; inhibin expression in testes	[56]
AP	PND22–36; PND36–55	Oral	100–200 mg/kg	–	Males only studied. No effects on time of prepuce separation or wts of seminal vesicles, ventral prostate, testes or epididymides.	[57]
CD	GD6–15	Oral	160–640 mg/kg	–	No effects on fetus	[39]

The rat is the normal species used in chemical toxicity evaluation, and consequently, many ED studies on BPA in this species have been reported (Table 2). As is evident by the shaded area in Table 2, most of these studies gave negative results, including what are normally considered definitive rat two-generation [14] and three-generation [13] studies. Interest, therefore, centers on the four studies listed at the top of Table 2. These showed effects at dose levels well below the suggested NOAEL value of 5 mg/kg BPA and employed oral administration of the test chemical (one study is difficult to interpret due to use of an implant of BPA [17]). Two of the four studies [18,19] measured endpoints that are rarely encountered in toxicology (play behavior and sexual motivation, respectively) and are not assessed here. Thus, there are two "standard" ED toxicity evaluations of BPA in the rat [20,21]. Extensive studies aimed at confirming the first of these reports yielded uniformly negative results in two strains of rat, including the SD rats employed in the original publication [22].

Sakaue et al. [21] reported that exposure of 13-week-old SD rats to low doses of BPA for 6 days, with termination at 18 weeks, significantly reduced daily sperm production (DSP). Four separate attempts to confirm those observations yielded uniformly negative results [23]. Results from those four repeat studies, together with the data originally reported [21], are shown schematically in Fig. 8. The repeat studies initially employed RM3 diet, but two subsequent studies using Purina 5002, and a final study using CE2 diet (as used by Sakaue et al.) were also conducted. Discussions with Sakaue and coworkers ensured that every attempt was made to replicate the experimental conditions employed by Sakaue et al. Two points in Fig. 8 are of particular interest. First, Sakaue et al. [21] reported positive results from a repeat study. This is an unusual practice that adds weight to their observations. All of the test and control data reported by Ashby et al. [23] have DSP values within the range of the test data re-

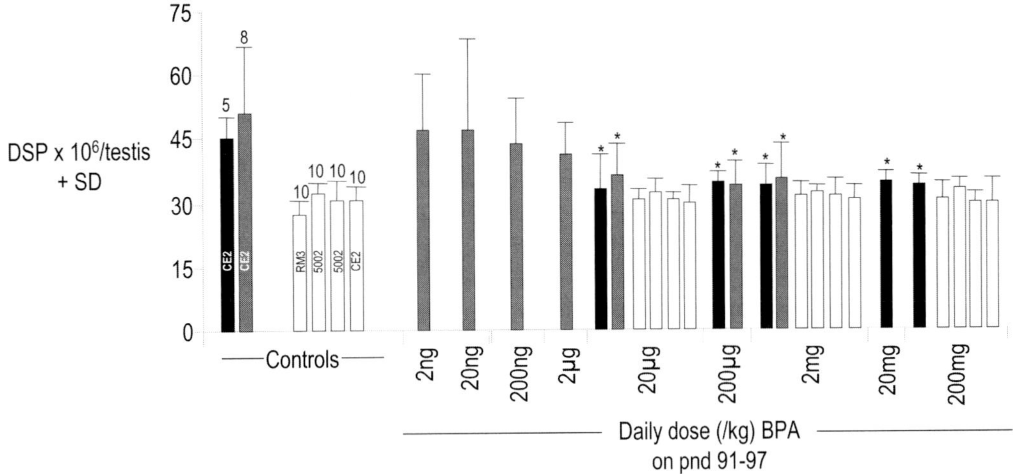

Fig. 8 Comparison of the daily sperm production (DSP) data published by Sakaue et al. [21] (black and gray columns) with those of Ashby et al. [23] (white columns). All data were generated in 13-week-old SD rats exposed orally to BPA for 6 days. Termination was at week 18. *Reported to be statistically significantly reduced by Sakaue et al. [21]. The control group sizes shown apply equally to all of the test.

ported by Sakaue et al. This suggests that an unaccountable difference in DSP control values between the two laboratories may be at the root of the divergence in test results for BPA.

Variation in ED parameter control values, both within and between laboratories, is emerging as one of the central issues in ED research. For example, similar arguments to those rehearsed above for the Sakaue repeat studies, have been made [16] to account for the failure [24] to confirm the original mouse prostate gland effects reported for BPA [2]. Variability in rat control testes weights has also been associated by Sharpe et al. [25] with the inability of several groups to confirm the reduction in rat testes weight induced by butyl benzyl phthalate [26]. In another case, similar problems with variability in control rat prostate gland weight were discussed [27] when attempting to confirm the data reported for nonylphenol (NP) [28]. A final example of this problem is the failure of Odum et al. [29,30] to confirm the reported activities [31] of NP and DES in the rat mammary gland. In that case, it was concluded that the use of a single control database over several independent studies in the initial report compromised the original observations. These several instances indicate that it would be profitable to understand, and then control, the several possible sources of control-variability for the major endpoints studied in ED. At present, this endeavor is being conducted ad hoc within a series of nonreplicated and often inconclusive chemical toxicity evaluations.

Despite the extensive ED database available for BPA it is still not possible to locate a single study that passes the most rudimentary scientific requirement—that the observations are capable of independent confirmation. Two explanations for this can be considered, as follows:

- BPA possesses subtle low-dose ED toxicities that only become evident under certain undefined experimental conditions. Until these conditions are defined and understood it will be a matter of chance what individual investigators observe experimentally for BPA or any other chemical.
- Failure to define and understand natural variability among control parameters monitored in ED studies allows artefactual positive results to be encountered for chemicals, especially in limited and unreproduced studies.

Which ever of these conclusions is correct, the positive low-dose data currently available for BPA cannot be extrapolated to humans with any confidence.

REFERENCES

1. S. J. Gould. In *I Have Landed*, pp. 43 and 246, Jonathan Cape, London (2002).
2. S. C. Nagel, F. vom Saal, K. A. Thayer, M. G. Dhar, M. Boechler, W. Welshons. *Environ. Health Perspect.* **105**, 70–76 (1997).
3. E. J. Calabrese. *Mutat. Res.* **51**, 181–189 (2002).
4. K. Almstrup, M. F. Fernandez, J. H. Petersen, N. Olea, N. E. Skakkebaek, H. Leffers. *Environ. Health Perspect.* **110**, 743–748 (2002).
5. D. Ganmaa, P. Y. Wang, L. Q. Qin, K. Hoshi, A. Sato. Medical Hypotheses **57**, 510–514 (2001).
6. M. Fukada, K. Fukada, T. Shimizu, W. Yomura, S. Shimizu. *Hum. Reprod.* **11**, 1244–1246 (1996).
7. R. R. Newbold, E. P. Banks, B. Bullock, W. N. Jefferson. *Cancer Res.* **61**, 4325–4328 (2001).
8. J. C. Gould, L. S. Leonard, S. C. Maness, B. L. Wagner, K. Conner, T. Zacharewski, S. Safe, D. P. McDonnell, K. W. Gaido. *Mol. Cell. Endocrinol.* **142**, 203–214 (1998).
9. J. Ashby and H. Tinwell. *Environ. Health Perspect.* **106**, 719–720 (1998).
10. F. S. vom Saal, B. G. Timms, M. M. Montano, P. Palanza, K. A. Thayer, S. C. Nagel, M. D. Dhar, V. K. Ganjam, S. Parmigiani, W. V. Welshons. *PNAS* **94**, 2056–2061 (1997).
11. M. D. Shelby, R. R. Newbold, D. B. Tully, K. Chae, V. L. Davis. *Environ. Health Perspect.* **104**, 1296–1300 (1996).
12. H. Tinwell, R. Joiner, I. Pate, J. Ashby. *Regul. Toxicol. Pharmacol.* **32**, 118–126 (2000).
13. R. W. Tyl, C. B. Myers, M. C. Marr, B. F. Thomas, A. R. Keimowitz, D. R. Brine, M. M. Veselica, P. A. Fail, T. Y. Chang, J. C. Seely, R. L. Joiner, J. H. Butala, S. S. Dimond, S. Z. Cagen, R. N. Shiotsuka, G. D. Stropp, J. M. Waechter. *Toxicol. Sci.* **68**, 121–146 (2002).
14. M. Ema, S. Fujii, M. Furukawa, M. Kiguchi, T. Ikka, A. Harazono. *Reprod. Toxicol.* **15**, 505–523 (2001).
15. F. S. vom Saal, P. S. Cooke, D. L. Buchanan, P. Palanza, K. A. Thayer, S. C. Nagel, S. Parmigiani, W. V. Welshons. *Toxicol. Ind. Health* **14** (1–2), 239–260 (1998).
16. J. Ashby. *Environ. Health Perspect.* **109**, A109–110 (2001).
17. J. G. Ramos, J. Varayoud, C. Sonnenschein, A. M. Soto, M. Monoz de Toro, E. H. Luque. *Biol. Reprod.* **65**, 1271–1277 (2001).
18. F. Dessi-Fulgheri, S. Porrini, F. Farabollini. *Environ. Health Perspect.* **110** (Suppl. 3), 403–407 (2002).
19. F. Farabollini, S. Porrini, D. Della Seta, F. Bianchi, F. Dessi-Fulgheri. *Environ. Health Perspect.* **110** (Suppl 3), 409–414 (2002).
20. C. Talsness, O. Fialkowski, C. Gericke, H-J. Merker, I. Chahoud. *Congen. Abnormal.* **40**, S94–S107 (2000).
21. M. Sakaue, S. Ohsako, R. Ishimura, S. Kurosawa, M. Kurohmaru, Y. Hayashi, Y. Aoki, J. Yonemoto, C. Tohyama. *J. Occup. Health* **43**, 185–190 (2001).
22. H. Tinwell, J. Haseman, P. A. Lefevre, N. Wallis, J. Ashby. *Toxicol. Sci.* **68**, 339–348 (2002).
23. J. Ashby, H. Tinwell, P. A. Lefevre, R. Joiner, J. Haseman. *Toxicol Sci.* **74**, 129–138 (2003).
24. J. Ashby, H. Tinwell, J. Haseman. *Regul. Toxicol. Pharmacol.* **30**, 156–166 (1999).
25. R. M. Sharpe, K. J. Turner, J. P. Sumpter. *Environ. Health Perspect.* **106**, A220–A221 (1998).
26. R. M. Sharpe, J. S. Fisher, M. R. Millar, S. Jobling, J. P. Sumpter. *Environ. Health Perspect.* **103**, 1136–1143 (1995).
27. J. Odum and J. Ashby. *Toxicol. Sci.* **56**, 400–404 (2000).
28. P. C. Lee. *Endocrine* **9**, 105–111 (1998).
29. J. Odum, I. T. G. Pyrah, J. R. Foster, J. Van Miller, R. L. Joiner, J. Ashby. *Regul. Toxicol. Pharmacol.* **29**, 184–195 (1999).
30. J. Odum, I. T. G. Pyrah, A. R. Soames, J. R. Foster, J. P. Van Miller, R. L. Joiner, J. Ashby. *J. Appl. Toxicol.* **19**, 367–378 (1999).
31. J. B. Colerangle and D. Roy. *Endocrine* **4**, 115–122 (1996).

32. K. L. Howedshell, A. K. Hotchkiss, K. A. Thayer, J. G. Vandenburgh, F. S. vom Saal. *Nature* **401** (6755), 763–764 (1999).
33. P. Palanza, K. L. Howdeshell, S. Parmigiani, F. vom Saal. *Environ. Health Perspect.* **110** (Suppl. 3), 415–422 (2002).
34. S. Honma, A. Suzuki, D. L. Buchanan, Y. Katsu, H. Watanabe, T. Iguchi. *Reprod. Toxicol.* **16** (2), 117–122 (2002).
35. C. M. Markey, E. H. Luque, M. Munoz de Toro, C. Sonnenschein, A. M. Soto. *Biol. Reprod.* **65** (4), 1215–1223 (2001).
36. C. Gupta. *Proc. Soc. Exp. Biol.* **224** (2), 61–68 (2000).
37. T. Takao, W. Nanamiya, I. Nagano, K. Asaba, K. Kawabata, K. Hashimoto. *Life Sci.* **65** (22), 2351–2357 (1999).
38. A. Suzuki, A. Sugihara, K. Uchida, T. Sato, Y. Ohta, Y. Katsu, H. Watanabe, T. Iguchi. *Reprod. Toxicol.* **16** (2), 107–116 (2002).
39. R. E. Morrissey, J. D. George, C. J. Price, R. W. Tyl, M. C. Marr, C. A. Kimmel. *Fundam. Appl. Toxicol.* **8** (4), 571–582 (1987).
40. NTP, NTIS PB86-103 207 (1985).
41. S. Z. Cagen, J. M. Waechter, S. S. Dimond, W. J. Breslin, J. H. Butala, F. W. Jekat, R. L. Joiner, R. N. Shiotsuka, G. E. Veenstra, L. R. Harris. *Toxicol. Sci.* **50**, 36–44 (1999).
42. T. Nagao, Y. Saito, K. Usumi, S. Yoshimura, H. Ono. *Reprod. Toxicol.* **16** (2), 123–130 (2002).
43. B. S. Rubin, M. K. Murray, D. A. Damassa, J. C. King, A. M. Soto. *Environ. Health Perspect.* **109**, 675–680 (2001).
44. K. Kubo, O. Arai, R. Ogata, M. Omura, T. Hori, S. Aou. *Neurosci. Lett.* **304** (1–2), 73–76 (2001).
45. J. S. Fisher, K. J. Turner, D. Brown, R. M. Sharpe. *Environ. Health Perspect.* **107**, 397–405 (1999).
46. N. Atanassova, C. McKinnell, K. J. Turner, M. Walker, J. S. Fisher, M. Morley, M. R. Millar, N. P. Groome, R. M. Sharpe. *Endocrinology* **141**, 3898–3907 (2000).
47. K. Williams, C. McKinnell, P. T. Saunders, M. Walker, J. S. Fisher, K. J. Turner, N. Atanassova, R. Sharpe. *Hum. Reprod. Update* **7**, 236–247 (2001).
48. T. E. Stoker, C. L. Robinette, B. H. Britt, S. C. Laws, R. L. Cooper. *Biol. Reprod.* **61**, 1636–1643 (1999).
49. O. Takahashi and S. Oishi. *Arch. Toxicol.* **75**, 42–51 (2001).
50. B. A. Elswick, D. B. Janszen, J. C. Gould, D. B. Stedman, F. Welsch. *Toxicol. Sci.* **54** (Suppl.), 256A (2000).
51. F. Welsch, B. A. Elswick, D. B. Stedman. *Toxicol. Sci.* **56** (Suppl.), 256A (2000).
52. F. Welsch, B. A. Elswick, D. B. Janszen, C. L. Robinette. *Toxicol. Sci.* **60** (Supplement), 73A (2001).
53. S. Z. Cagen, J. M. Waechter, Jr., S. S. Dimond, W. J. Breslin, J. H. Butala, F. W. Jekat, R. L. Joiner, R. N. Shiotsuka, G. E. Veenstra, L. R. Harris. *Regul. Toxicol. Pharmacol.* **30**, 130–139 (1999).
54. T. Nagao, Y. Saito, K. Usumi, M. Kuwagata, K. Imai. *Reprod. Toxicol.* **13**, 303–311 (1999).
55. S. Kwon, D. B. Stedman, B. A. Elswick, R. C. Cattley, F. Welsch. *Toxicol. Sci.* **55**, 399–406 (2000).
56. P. T. Saunders, G. Majdic, P. Parte, M. R. Millar, J. S. Fisher, K. J. Turner, R. M. Sharpe. *Adv. Exp. Med. Biol.* **424**, 99–110 (1997).
57. J. Ashby and P. A. Lefevre. *J. Appl. Toxicol.* **20**, 35–47 (2000).
58. J. Odum, P. A. Lefevre, H. Tinwell, J. P. Van Miller, R. L. Joiner, R. E. Chapin, N. T. Wallis, J. Ashby. *Toxicol. Sci.* **68**, 147–163 (2002).
59. J. Ashby. *APMIS* **108**, 805–813 (2000).

Pure Appl. Chem., Vol. 75, Nos. 11–12, pp. 2181–2193, 2003.

Topic 3.16
Environmental estrogens and sperm counts*

Harry Fisch[‡] and Robert Golden

Department of Urology, Columbia University, 944 Park Ave., New York, NY 10028, USA

Abstract: The term "environmental estrogen" refers to chemical substances that exhibit some degree of estrogen-like activity. The primary emphasis for potential adverse effects resulting from exposure to environmental estrogens is on in utero exposure because such exposure can occur during critical periods of organogenesis. Assessment of biological plausibility can be based, in part, on the extensive data on the effects of diethylstilbestrol (DES). The available evidence is too limited to judge with any confidence whether sperm counts have declined during the past 50 years. Based on both animal and human data with DES, it is biologically plausible that in utero exposures to exogenous estrogenic compounds are capable of reducing sperm production in adult men. However, the apparent existence of a maternal dose threshold for DES-induced effects on sperm counts undermines the likelihood that environmental estrogens, which are substantially less potent, are capable of causing similar effects.

INTRODUCTION

There is currently great uncertainty whether some synthetic chemicals, acting as "hormone-mimics" and released into the environment, might have the potential to disrupt the endocrine systems of humans. Although there is no question that hormones are potent modulators of biochemical and physiological function, the implication that in utero exposures to environmental endocrine modulators (e.g., xenoestrogens) have the capability to produce clinically detectable effects in humans is much less certain.

Although exogenous endocrine modulators can involve any hormonal system, the principle focus of this review is on estrogen-mediated effects. As used in this review, the term "environmental estrogens" refers to chemical substances that exhibit some degree of estrogen-like activity. In order to be accurate, this term must apply to both naturally occurring estrogenic compounds (e.g., phytoestrogens) and to synthetic estrogenic compounds. Naturally occurring estrogens in either humans or animals are referred to as endogenous estrogens, and all compounds with estrogenic properties entering the body from an outside source are referred to as exogenous estrogens.

The primary emphasis here for potential adverse effects resulting from exposure to environmental estrogens is on in utero exposure because such exposures can occur during critical periods of organogenesis. The concept of a critical window of exposure or sensitivity is an important factor that must be considered in animal studies and in human exposure situations. Abundant animal and human data demonstrate this phenomenon. For some adverse health outcomes (e.g., reproductive tract abnormalities) exposures to excessive amounts of estrogen during this critical period is essentially the only way that such effects could occur. For other adverse health endpoints, exposure to estrogen in utero may be a risk factor for subsequent adverse effects.

---

*Report from a SCOPE/IUPAC project: Implication of Endocrine Active Substances for Human and Wildlife (J. Miyamoto and J. Burger, editors). Other reports are published in this issue, *Pure Appl. Chem.* **75**, 1617–2615 (2003).

[‡]Corresponding author

POTENTIAL EFFECTS OF CONCERN

The following adverse health consequences to humans have been hypothesized to be associated with in utero exposure to environmental endocrine modulators (primarily environmental estrogens) [1,2]:

- adverse effects on male reproductive tract
- male and female fertility problems
- testicular cancer
- prostate cancer
- breast cancer
- endometriosis
- learning disability or delay
- alterations of sexual behavior
- immune system effects
- thyroid effects

It is biologically plausible to hypothesize that exposure (particularly in utero) to environmental estrogens could adversely affect humans since exposure to estrogen, whether in utero or in adulthood, can have biochemical, physiological, and specific target organ effects on development, reproduction, behavior, and metabolism [3,4]. Consequently, it is not unreasonable to hypothesize that exposure to estrogen-like compounds, whatever their source, could adversely affect human health. However, biological plausibility alone is not a sufficient basis for concluding that exposure to compounds identified as environmental estrogens has adversely affected humans.

CRITERIA FOR ASSESSING BIOLOGIC PLAUSIBILITY

DES paradigm

In assessing the likelihood that in utero exposure to environmental estrogens might be etiologically linked to any particular adverse effect, it is important to consider biological plausibility as part of the evaluation process. Assessment of biological plausibility can be based, in part, upon the extensive data on the effects of diethylstilbestrol (DES). There are considerable data, in both experimental animals and humans, on the effects of in utero exposure to DES on adult offspring. While the DES data can be used to judge the likelihood of adverse effects resulting from in utero exposure to environmental estrogens, because of the manner in which DES was used clinically, these data are not useful for assessing the likelihood of adverse effects that might result from postnatal exposure to environmental estrogens except in the mothers. Most of the adverse effects observed in both animals and humans following in utero exposure to DES are consistent with basic principles of dose response, as well as with the presence of maternal dose levels not associated with adverse effects in offspring. As reviewed in greater detail below, the clinical use of DES resulted in the massive exposure to 4 to 6 million women and their unborn children. Critical reliance upon the abundant DES database provides a useful tool for assessing many of the endpoints of concern according to the following scenarios:

- If an effect hypothesized as possibly resulting from in utero exposure to environmental estrogen is not reliably associated with in utero exposure to DES, it may be biologically implausible that such an effect (even if real) is mediated via an estrogen mechanism of action.
- If an effect hypothesized as possibly resulting from in utero exposure to environmental estrogen is similar to an effect reported following in utero exposure to DES, the likelihood that such an effect might be associated with exposure to environmental estrogen must be evaluated based upon comparative potency and dose–response considerations.

For some of the specific effects addressed in this review, it is imperative to note that the weight of the available evidence currently may be insufficient to reach any valid conclusion. The data may simply be inadequate to conclude one way or another whether in utero exposure to exogenous estrogen might be a risk factor.

DES as a model compound to judge potential effects of exogenous estrogen

DES is a synthetic estrogenic compound that is several times more potent than 17β-estradiol and hundreds to thousands of times more potent (based primarily on results from in vitro tests) than other compounds identified as exogenous environmental estrogens (not including oral contraceptives or estrogen replacement therapy), whether synthetic organochlorine compounds [e.g., certain metabolites of dichlorodiphenyltrichloroethane (DDT), selected polychlorinated biphenyl (PCB) congeners, aldrin, dieldrin, endrin] or naturally occurring phytoestrogens from plants [5–9]. The large difference in potency between DES and environmental estrogens may be important in assessing the likelihood that in utero exposure to environmental estrogens might be a risk factor for certain adverse health endpoints. Based upon substantial differences in estrogenic potency, the total maternal dose of DES required to produce adverse effects in offspring exposed in utero and maternal intakes of exogenous environmental estrogens that might be required to result in comparable estrogenic activity can be compared. While such comparisons may provide only a rough approximation, they can highlight potentially important qualitative (and perhaps even quantitative) differences between the dose–response characteristics of different estrogen receptor agonists.

It is critical to note that the approach outlined above is based on an estrogenic potency comparison between DES and various individual compounds identified as environmental estrogens. This, of course, is not representative of how real-world exposures to these compounds occurs. In reality, typical exposures, whether as a result of stored body burdens or from daily dietary intake, are far more likely to be to a complex mixture of weakly estrogenic compounds. Consequently, even though individual environmental estrogens may be less potent than natural estrogen or DES, there are two critical questions at the core of this issue that must be addressed. First, what are the *net* effects of simultaneous exposure to many such "weak" estrogens? Second, will the estrogenic sum of all such chemicals acting together at the level of the estrogen receptor result in effects that would not occur from exposure to just a few of these compounds? Speculation has centered around the idea that background adipose tissue concentrations and resulting serum levels of a number of persistent, bioaccumulating, weakly estrogenic compounds might have additive (or synergistic) net estrogenic activity capable of producing adverse effects. Several studies have attempted to address this issue. For example, when 10 different chemicals [i.e., endosulfan α and β, toxaphene, dieldrin, 2,3,4,5-tetrachlorobiphenyl, *p,p′*-DDT, 2,2′,3,3′,6,6′-hexachlorobiphenyl, *p,p′*-dichlorodiphenyldichloroethylene (*p,p′*-DDD), *p,p′*-dichlorodiphenyldichloroethylene (*p,p′*-DDE), and methoxychlor] already known to be weakly estrogenic in vitro in 7 (MCF-7) cells, were added to MCF-7 cells at one-tenth the concentration required for each to produce an estrogenic effect, the results indicated additive net estrogenic activity similar to 17β-estradiol [8].

In a provocative in vitro study using yeast cells containing human estrogen receptor, selected chemicals (i.e., endosulfan, dieldrin, toxaphene, and chlordane) were tested for estrogenic activity singly and in various combinations [10]. The results of this study suggested that endosulfan, dieldrin, and toxaphene had very low estrogenic potency, while chlordane was devoid of any estrogenic activity. However, various binary combinations of these chemicals demonstrated estrogenic potency from 160 to 1600 times greater than the activity of any single chemical. Chlordane, which had no estrogenic activity when tested alone, significantly enhanced the potency of the other chemicals tested. In addition, two PCB compounds were also tested in the same system (2′,4′,6′-trichloro-4-biphenylol and 2′,3′,4′,5′-tetrachloro-4-biphenylol) and shown to exhibit a similar pattern of synergistic estrogenic activity. For the two PCBs tested, the result in yeast cells was confirmed in Ishikawa cells, an endometrial cancer cell line transfected with human estrogen receptor. However, numerous attempts to replicate these findings have not been successful. Ten different estrogen-responsive assays (including induction of uterine wet weight, peroxidase activity, and progesterone receptor levels; induction of cell growth in MCF and human breast cancer cells; induction of reporter gene activity in two yeast-based assays; and competitive binding to human and mouse estrogen receptor) were used to test all of the binary mixtures noted above. The results confirmed additive activity, but no synergism was observed [11]. In two additional recent studies, one on a mixture of dieldrin and endosulfan tested in a yeast human estrogen transactivation assay and a rat uterotropic assay [12], and another on a dieldrin-toxaphene mixture tested in a mouse uterotropic assay, MCF-7 human breast cancer cells, and a yeast-based reporter gene assay, also failed to demonstrate any evidence of synergism [13]. Finally, the results of the original study reporting synergism [10] were recently withdrawn by the authors [14].

Effects of DES in humans

While estimates vary, DES was widely prescribed to several million pregnant women, with most use occurring between the late 1930s until 1971, in the mistaken belief that it was useful in the treatment of threatened or habitual abortion [15–17]. The use of DES during pregnancy was proscribed by the Food and Drug Administration (FDA) in 1971 with the discovery that a small number of women exposed in utero later developed vaginal clear cell adenocarcinoma [17,18]. As reviewed below, in utero exposure to sufficient maternal doses of DES is also associated with other adverse effects on the reproductive tracts of both males and females. In males, these effects include epididymal cysts, microphallus, cryptorchidism, testicular hypoplasia, decreased sperm count, and increased incidence of abnormal sperm [16,19–21]. In females, reported adverse effects include vaginal adenosis; clear cell adenocarcinoma; structural defects of the cervix, vagina, uterus, and fallopian tubes; infertility; and irregular menstrual cycles [16,19,22–24].

Because of the widespread clinical use of DES, there are some dose–response data on a number of adverse effects in both male and female offspring exposed in utero. During the peak years of DES use, different DES dosing regimens were used in different medical centers. Since the efficacy of DES had never been established by adequate clinical testing, there was no standard dosing regimen in use. This resulted in large groups of women receiving substantially different total DES doses during pregnancy. Such data permit rough approximations of the daily doses and total maternal doses of DES associated with some of the adverse effects on offspring exposed in utero. Table 1 summarizes the existing data on DES cohorts.

Table 1 Summary data on DES-exposed cohorts and estimated maternal exposures.

Cohort	Type of study	Estimated mean	Persons exposed total DES dose (g)	Non-DES-exposed controls
Mayo Clinic Sons [25]	Cohort	1.4	828	676
Connecticut Mothers [26]	Cohort	2.1	1531	1404
DESAD Study [27]	Cohort	4.2[a]	4014	24
Boston Collaborative Study [28][b]	Cohort	6.4	217	1033
DES Efficacy Trial [15,29,30][c]	Randomized clinical trial	11.6	840	
British Radomized Trial [31]	Randomized clinical trial	11.5	650	806
British Medical Research Council [32]	Randomized clinical trial	17.9	70	66

[a]Median dose based on 26 % of cohort with known DES doses.
[b]80 % of cohort from Boston Lying-In Hospital; DES mean total dose estimated for total cohort.
[c]University of Chicago.

Of central interest to the subject matter of this review is the fact that there are clinical studies that appear to identify no-effect total maternal dose levels and durations of exposure to DES for some adverse effects. These studies suggest that certain maternal DES dosing regimens were not sufficient to result in certain adverse effects to offspring exposed in utero. For example, a study conducted on a cohort from the Mayo Clinic, which appears to have employed the lowest DES dosing schedule of any clinical center in the United States, demonstrates a mean no-effect maternal dose level for effects attributable to in utero DES exposure [25]. This study was designed to determine whether males exposed in utero to DES had a higher frequency of urogenital abnormalities than an unexposed control group. Comparison of 265 DES-exposed males with 274 controls revealed that in utero exposure to DES did not increase the risk of any urogenital abnormalities, including penile length or diameter, testis length or width, epididymal cysts, or varicocele. There were also no adverse effects on sperm volume, density, motility, morphology, or fertility. The mean total administered maternal dose of DES was 1.4 g, with a median duration of exposure of 101 days. While it is possible that more subtle effects might have been present, only the effects noted above were studied. A potential problem with any such studies (whether in humans or animals) is that fetal exposure to DES may have occurred after the critical period of reproductive tract development. Records of DES doses and durations of exposure in this cohort reveal that the 25th and 50th percentiles of the first gestational day of exposure were on days 54 and 89, with exposure durations of 28 and 100 days, respectively [25]. Therefore, the timing of DES exposure is well within the window during which development of the reproductive tract is occurring. While the correlation between some endpoints (e.g., adenosis) and maternal doses is not perfect, the clinical data can provide an approximate benchmark level for investigating potential relationships between maternal exposure to DES (and by implication to other less potent environmental estrogens) and adverse effects in offspring exposed in utero.

ADVERSE EFFECTS ON MALE REPRODUCTIVE TRACT

Decreased sperm quality

Introduction

A great deal of attention has recently been directed to the allegation that sperm quality has been declining worldwide during the past 40 to 50 years [33–35]. Much of the attention in this debate has focused on the hypothesis that in utero exposure to environmental estrogens might be responsible. While certain chemicals have the ability to affect sperm quality if exposure to sufficient doses occurs over a sufficient period of time, such effects must be evaluated on a compound-by-compound basis. The questions raised with respect to the issue of declining sperm quality are extraordinarily complex, and much of the data collected to date are conflicting. Because of the profound implication of the claim that sperm quality (particularly sperm counts) worldwide has declined over the past 40 to 50 years, the evidence must be critically evaluated. Either adult or prenatal exposure to certain chemicals (both estrogenic and nonestrogenic chemicals) can affect sperm quality (i.e., counts, motility, or morphology). However, before it can be concluded that chemicals, whether estrogenic on nonestrogenic, have the potential to affect sperm quality, particularly worldwide, it is important to assess dose response and exposure data, potency, and possible interactive effects.

It is biologically plausible that in utero exposure to estrogen can affect sperm counts. In studies with mice, in utero exposure to DES is associated with decreased sperm production and abnormal sperm morphology in adult offspring [6,36]. There are also numerous studies of adult male humans prenatally exposed to DES; however, with respect to decreased sperm counts, the results of these studies are mixed. Decreased sperm count (115 million/ml in 87 controls vs. 91 million/ml in 134 DES-exposed) and abnormal Eliasson scores in 18 % of 134 DES-exposed men compared to 8 % in 87 placebo-exposed controls have been reported [29]. This study was conducted on a cohort of men exposed in utero to DES according to the dosing protocol in use at the University of Chicago where mean total maternal

DES dosages were, on average, 11 603 mg. Another small study reported lower average sperm counts in 20 DES-exposed men and pathologic Eliasson scores in 18 DES-exposed men, but no comparison controls were used and maternal DES dosages were unknown [20]. In contrast, in a cohort of men from the Mayo Clinic exposed in utero to DES in which total mean maternal DES doses were approximately 1.4 g, there were no differences between 110 DES-exposed men and unexposed men in sperm count, motility, or abnormal Eliasson scores [25].

While clinical data suggest that in utero exposure to DES at some maternal dose level can result in decreased sperm counts, the data also suggest the existence of an apparent maternal dose threshold for such effects. Even at the dose levels of DES that produced adverse effects on sperm following in utero exposure, the magnitude of such effects is not large, with average sperm counts reduced from 115 million/ml in men not exposed to DES to 91 million/ml in DES exposed men. While this shows a DES effect, it also demonstrates that an estrogen as potent as DES does not produce a decrease in mean sperm counts approaching the levels at which fertility might be affected. The failure of the maternal DES dosages used at the Mayo Clinic to cause any effects on sperm count or morphology underscores this point. As noted later in this review, fertility in the high-dose cohort of men (University of Chicago cohort) was not adversely affected, even after in utero exposure to DES at maternal doses sufficient to produce adverse effects on sperm counts and morphology. This raises the question of whether in utero exposure to environmental estrogens, which are significantly less potent than DES, could cause a decline in sperm counts.

Evidence for declining sperm quality

The critical evidence cited in support of the hypothesis that sperm quality has declined over the past 50 years is the Carlsen et al. [33]. Regression meta-analysis of 61 studies showing a steady decline in mean sperm counts from 1938 to 1990. The results of this study have been interpreted as demonstrating that in utero exposure to compounds identified as environmental estrogens (e.g., DDT, PCBs, and other hormonally active chemicals) might be responsible for this effect. Because of the biologic plausibility that prenatal exposure to estrogenic substances could affect adult sperm counts, it is important to determine whether decreased sperm counts are a real phenomenon, and if so, whether decreases are caused by exposure to environmental estrogens.

In general, the meta-analysis relied upon 61 studies selected from the years 1938 to 1990 that reported mean sperm counts; results were weighted by the number of subjects in each study [33]. While this approach appears valid, closer scrutiny of the methodology and data reveal a number of potential problems. It may not be reasonable to assume that methods and protocols for sperm collection and measurement from laboratories all over the world from 1938 to 1990 would be similar. Even in the United States, there is little standardization of methods for semen analysis, which suggests a substantial potential for error when studies spanning more than 50 years from all over the world are compared. There is also a wide range of biological variability in sperm counts, as well as numerous factors that may affect results, including temperature, season, period of abstinence, and others. Even though Carlsen et al. excluded studies that reported counting with computer-assisted or flow cytometric methods, the methodology relied upon (i.e., the use of various types of counting chambers) is fraught with numerous potential interpretive problems [33]. Depending solely on the semen counting chamber used (i.e., Cell VU, Microcell, Neubauer Hemacytometer, or Makler) mean sperm counts can vary from 35 to 51 million/ml based on calibration with glass beads of known size and concentration. Other documented technical sources of error include inter-individual counting differences, counter chamber variation, and poor pipetting technique [37]. Other studies demonstrate similar variation in sperm counts based on the counting chamber used. In particular, the use of the Makler chamber results in counts that can be as much as 65 % higher with significantly higher standard deviation, suggesting that counts made with this device are less reproducible [38]. There is also great variability in individual sperm counts from one ejaculate to the next, with counts varying from <20 million/ml to >100 million/ml [37]. Daily sperm production is also significantly reduced in aging men as is the number of Sertoli cells, which is also correlated with daily sperm production. Vascular degeneration, autoimmunity, or loss of Sertoli cells may contribute to the age-related decline in sperm production [39]. In evaluating the 61 studies in the meta-analysis, it is not clear to what extent age has been considered as a possible confounding factor. Finally, temperature has a profound effect on semen quality. In a pairwise study of 131 men who contributed semen specimens in summer and winter, there were significant decreases during the summer in sperm concentration, total sperm count per ejaculate, and concentration of motile sperm. Mean reductions for these parameters were 32, 24, and 28 %, respectively. These reductions correlated with lower birth rates during spring months [40]. These data demonstrate that seasonal variations in sperm counts may influence comparisons that fail to account for this phenomenon. The various factors noted above may be problems when sperm count results from different laboratories or clinics, in which different counting methods were used, are compared. In addition, certain methodologic practices may introduce systematic counting errors that would tend to produce results consistent with an apparent effect, when, in reality, no such effect was occurring. Significant intra- and interlaboratory variation also make it difficult to make meaningful comparisons of sperm morphology over time. These issues will require careful analysis before the results of comparative studies can be accepted as demonstrating a real effect.

Data not included in the 1994 meta-analysis

Additional studies that appear to refute the hypothesis

Additional studies, not included in the 1992 meta-analysis, appear to refute the hypothesis of declining sperm counts. The outcomes of these studies, which have not been integrated into the linear regression curve, illustrate the sensitivity of conclusions to data collection efforts. For example, for 1984, five studies were cited on a total of 290 men with mean sperm concentrations [reported in million/ml; (number of men)] of 83.9 (119), 72 (114), 58.9 (9), 59 (36), and 102 (12). Not cited was another 1984 study that reported mean sperm concentrations of 107 million/ml on 861 men [41]. This 1984 mean sperm concentration value, based on more subjects than 12 of the 13 pre-1970 studies relied upon, does not appear to support the hypothesis that sperm counts have declined. The meta-analysis precedes a 1992 study that reports mean sperm concentrations of 101.5 million/ml in 30 men [42], a level of sperm density similar to that in 1950. A longitudinal study of semen quality in Wisconsin men, also published in 1992, showed no decline in sperm concentration, motility, or ejaculate volume between 1978 and 1987. Mean sperm counts ranged from 55.6 million/ml to 105.7 million/ml [43]. A sharp rise in morphologically abnormal sperm between 1982 and 1983 coincided with a change in criteria used to identify abnormal sperm. A detailed historical analysis of sperm counts from 1951 to 1977 concluded that there was no significant change during this time period [44]. This interval, which encompasses about half of the 50-year period reviewed by Carlsen et al. [33], appears to refute the hypothesis that sperm counts have fallen over the years. Additional data are provided by a study of in utero exposure to maternal tobacco smoke and decreased semen quality in male offspring [45]. In this study, the mean sperm concentration over all exposure groups (n = 292) was approximately 106 million/ml. These data appear to have been collected in the United States in 1991. There is a marked contrast between the mean sperm count from this study and the mean sperm count of 54 million/ml (n = 54) reported by Carlsen et al. [33] for 1990, the last year considered in their review. A mean sperm count of 106 million/ml is more in line with the mean sperm count from the meta-analysis regression curve for about 1950 and does not support the hypothesis of declining sperm counts over the past 40 to 50 years. The issue of changes in semen quality over time was investigated in a retrospective analysis of 510 healthy men in the Seattle area who donated multiple semen samples between 1972 and 1993 [46]. The results of this study showed no decrease in sperm concentration, semen volume, total number of sperm per ejaculate, and percent normal sperm morphology over the 21-year period studied. There was actually a small, but statistically significant, increase in these parameters with time, although this finding was not considered to be biologically important. Multiple semen samples from each participant permitted the use of geometric mean values for each individual, providing more power to detect changes and reduce intrasubject variability due to different abstinence intervals.

In the most comprehensive study to date, Fisch et al. [47] investigated semen quality in 1283 men from the United States over a 25-year period between 1970 and 1994. This study included all men who banked sperm prior to vasectomy at sperm banks in New York, Minnesota, and California (i.e., no man was excluded from analysis with arbitrary rejection criteria). After controlling for age and duration of abstinence, there was a slight, but statistically significant increase in mean sperm concentration [77 million/ml in 1970 vs. 89 million/ml in 1994 ($p < 0.05$)], but no change in motility or semen volume during the 25-year period. A rather dramatic finding was the significant differences in mean sperm concentration and motilities between different sperm banks, with California the lowest (72.7 million/ml), Minnesota intermediate (100.8 million/ml), and New York the highest (131.5 million/ml). This striking geographical variation highlights a potentially significant confounder of studies such as Carlsen et al. [33], which are unable to account for this variable. As noted by Fisch et al. [47], the marked geographic variation found in their study had been reported previously [44,48]. After reviewing the papers relied upon by Carlsen et al. [33] in the meta-analysis, it was determined that the preponderance of earlier studies (which defined the upper end of the regression curve) were from New York, while, after 1970, only 50 % of men in the meta-analysis were from the United States, with only 25 % of these from New York. Based on the demonstrated geographic variability in semen quality across the U.S. and the possibility of similar variability around the world, the meta-analysis of Carlsen et al. [33] may be confounded by the inability to account for geographic variability in the parameters of concern. The likelihood that this has occurred is suggested by a reanalysis of all 61 studies included in the regression meta-analysis [49]. Of the studies comprising the meta-analysis, only 20 included more than 100 men, but these same studies represented 91 % of the total men studied. Of the studies prior to 1970, all were from the United States, with 80 % of these from New York, where sperm counts (then and now) are the highest. After 1970, only three studies were from the U.S., with many from third world countries, where sperm counts were low. Additional conformation of geographic variation is provided by the study reviewed above on 510 men from the Seattle area [46]. Mean sperm counts observed in this study (46.5 million/ml in 1972 and 52 million/ml in 1993) are similar to, but still lower than, mean sperm counts reported for California of 72.7 million/ml, suggesting not only an east–west gradation, but a north–south gradation, as well. Obviously, the significant influence of geographic variation needs to be considered when studies from all over the world are analyzed and compared for potential trends in sperm count over time.

It is beyond the scope of this review to resolve this complex issue. However, it is clear that additional study and analysis remain to be done in order to determine whether the trend reported by Carlsen et al. [33] as well as others is real or an artifact. While it may be tempting to hypothesize that exposure to compounds identified as environmental estrogens might be responsible for the observed effects, it is more likely premature to conclude that sperm counts are declining, much less to ascribe causes.

Alternative statistical analysis of the data

Of the 61 studies in the Carlsen et al. [33] meta-analysis, only 13 are from before 1970, and the results from these studies define the upper end of the regression curve. When only studies from 1970 to 1990 are considered, there is no downward trend in sperm counts. In a reanalysis of the 48 studies published since 1970, sperm counts actually increased between 1970 and 1990 [50]. Another reanalysis suggested that the reported decline in sperm concentration may have been accounted for entirely or in part by a change in lower reference values. This analysis concluded that the hypothesis that sperm counts have declined significantly between 1940 and 1990 was not supportable by the original data [51]. In the most comprehensive statistical reanalysis of the Carlsen et al. [33] study, in addition to reproducing the original linear regression, several other statistical models (quadratic, spline fit, and stairstep) were employed to determine the best model to analyze the data [52]. Part of the problem with the data concerns the highly nonuniform distribution of subjects and sperm measurements over the period of time under study (1938 to 1990; the majority of data were collected between 1970 and 1990). Of the four models used, only the linear model was consistent with the hypothesis that sperm counts have declined since 1940. The quadratic and spline fit models suggest that mean sperm counts have been increasing since 1940, while the stairstep model suggests a sharp drop in the mid-1960s, with levels constant since then. There is also an apparent inconsistency between the analysis of Carlsen et al. [33] and the data of MacLeod and Wang [44] noted above. Based on the linear regression model, the last 1000 men in the series studied by MacLeod and Wang [44] should have had a mean sperm count 20 % lower than what was actually observed. Overall, the meta-analysis illustrates the influence of methodologic issues on study results and conclusions. These include potential selection biases, variability in collection and analytical methods, uneven availability of data over the entire time period, and the questionable generalization of data obtained in different geographic areas. It is critical to note that the last 20 years of data, which contain 78.7 % of all the studies and 88.1 % of the total number of subjects, indicate no decrease in sperm counts; in fact, sperm counts were observed to have increased. Finally, a number of other mathematical models, which perform statistically better than the linear model when describing the recent data, suggest a much different explanation of the data. During the last 20 years, the interval with the most robust data, all the models, except the linear model, suggest constant or slightly increasing sperm counts.

Potential causes of observed effects on sperm production

As noted above, unless total in utero exposure to DES exceeded a certain maternal dose level, there do not appear to be adverse effects on sperm production (or any other related parameter) in adult offspring. In light of data suggesting a maternal DES dose threshold level for this effect, the dose–response characteristics and potency aspects of other potential chemical exposures must be considered in any attempts at explaining the observed declines in sperm production, if indeed such declines are real.

Dioxin [e.g., 2,3,7,8-tetrachloro-*p*-dibenzodioxin (TCDD)] is often hypothesized as a potential cause of wide-ranging environmental estrogen effects. With respect to decreased sperm counts, in utero exposure to a single dose of TCDD causes decreased sperm counts in adult male rats [53]. A recent study confirmed that in utero exposure to a single TCDD dose of 1 μg/kg reduces daily sperm production and epididymal sperm reserves [54]. In this study, the authors conclude that the effects of in utero TCDD exposure on sperm production are consistent with the results of Carlsen et al. [33] on declining sperm counts in humans. However, the relevance to humans of this effect in rats must be considered carefully. The single dose of TCDD used to produce a decline in daily sperm output was 1 μg/kg or 1 000 000 pg/kg, administered on day 18 of gestation, a dose level more than six orders of magnitude greater than typical daily intake levels of TCDD (0.3–0.6 pg/kg/day) [55]. Because of the high dose of TCDD used, this study is of questionable relevance for generalizing to human offspring exposed to TCDD in utero. A recent study also calls into question the potential relevance of the above findings. Administration to pregnant rats on day 15 of gestation of a single oral dose of TCDD (0.5, 1.0, 2.0 μg/kg) or indole-3-carbinol (I3C, 1.0 or 100 mg/kg), an Ah receptor agonist found in cruciferous vegetables (i.e., cauliflower, Brussels sprouts, broccoli) produced both common and different reproductive tract abnormalities in adult male offspring exposed in utero. These included decreases in the weights of seminal vesicles, prostate, and testicular parenchyma from one or more doses of TCDD and decreases in daily sperm production from one or more doses of I3C [56]. These findings underscore the complexity of assessing exposure and potential risks, and also (as reviewed in greater detail in the section on behavioral effects) calls into question the relevance of animal models for these kinds of effects.

In contrast to the in utero animal studies, extensive studies of the Air Force Ranch Hand population exposed to Agent Orange (i.e., TCDD) in Vietnam have not demonstrated any significant effects on sperm count, percent low sperm count, or abnormal sperm among adult men with elevated serum TCDD levels [initially 52–>292 parts per thousand (ppt; median 143 ppt for entire cohort) and currently >33 ppt median levels] [57]. However, studies on the Ranch Hand cohort do not address the issue of possible effects on sperm production following prenatal exposure to TCDD.

Although PCBs have also been hypothesized as potential causes of environmental estrogen effects based on the weak estrogenicity of some congeners, there is, as yet, no evidence suggesting that PCB exposure affects any male reproductive parameters, including sperm counts. A study of 37 currently exposed PCB workers [median serum PCB levels of 12 parts per billion (ppb)] and a comparison group (median PCB levels of 6 ppb), mean sperm concentrations were 65.5 million/ml and 67.2, respectively [58]. No other reproductive abnormalities were observed in this study. As with the Ranch Hand studies, this study also does not address the issue of possible effects subsequent to prenatal exposure to PCBs. Whether massive in utero exposure to PCBs and polychlorinated dibenzofurans (PCDFs) has adverse effects on sperm counts awaits appropriate follow-up studies of the male populations accidently exposed to PCBs and PCDFs at Yusho and Yu-Cheng.

Potential confounders of sperm count studies

Even assuming that sperm counts have declined, attributing any decline to environmental estrogens without consideration of other potential contributing factors may be unwarranted. For example, a significant increase in the use of marijuana worldwide, particularly evident in the 1960s, coincides with reports [33] of declining sperm counts. Several studies have demonstrated that the major active ingredient in marijuana can decrease sperm production and interfere with normal reproductive function and fertility in adult rats following in utero exposure [59–61]. Another possible factor that might play a role is cigarette smoking. There are conflicting reports concerning the potential contribution of smoking and subsequent effects on sperm quality. Heavy cigarette smokers (>19 cigarettes/day), compared to lighter smokers (<9 cigarettes/day), had a significantly lower total sperm count [62]. Heavy smokers also had a significantly lower total sperm count than non-smokers. However, heavy smoking did not appear to affect sperm density, motility, or morphology. The combination of heavy coffee drinking and heavy cigarette smoking inhibited sperm motility and increased the proportion of dead spermatozoa. Also potentially relevant in this regard is the effect of in utero exposure to maternal tobacco smoke and decreased semen quality in male offspring. An investigation of the possible role of maternal smoking on the reproductive system in adult male offspring found no significant effects of early exposure to maternal smoking on conventional measures of semen quality (i.e., volume, density, motility, and morphology). However, current smoking was associated with a significant decrease in the percentage of sperm with normal morphology [45]. Without consideration of such biologically plausible potential confounders, the possible influence of exposure to compounds identified as environmental estrogens on declining sperm counts is difficult to assess.

CONCLUSIONS

The available evidence is too limited to judge with any confidence whether sperm counts have declined during the past 50 years. Based on both animal and human data with DES, it is biologically plausible that in utero exposures to exogenous estrogenic compounds are capable of reducing sperm production in adult males. However, the apparent existence of a maternal dose threshold for DES-induced effects on sperm count undermines the likelihood that environmental estrogens, which are substantially less potent, are capable of causing similar effects. The meta-analysis by Carlsen et al. [33] is not an adequate basis from which to conclude that sperm counts have declined worldwide. The possible sources of error in that study are numerous and include methodological changes over time, selection bias, geographical variability, and statistical issues. While some additional data appear to support the central thesis advanced by Carlsen et al. [33], other data, in particular the large study by Fisch et al. [47], appear to refute the hypothesis that sperm quality has declined. This matter can be resolved only after all available data are comprehensively assessed, with adequate accounting of the potential sources of bias and error noted in this review. If the decline in sperm counts is real, careful evaluations of possible causes will include potential sources of estrogenic exposure, as well as other potential confounding factors (e.g., smoking, drugs, age). Finally, the documented lack of effects on sperm counts caused by elevated postnatal human body burdens of TCDD and PCBs suggests that sperm counts have not been adversely affected by exposure to these chemicals at typical environmental levels.

REFERENCES

1. W. R. Kelce, E. Monosson, M. P. Gamcsik. *Toxicol. Appl. Pharmacol.* **126**, 276 (1994).
2. National Wildlife Federation. *Hormone Copycats,* Great Lakes Natural Resource Center, Ann Arbor, MI (1994).
3. F. S. Greenspan and J. D. Baxter. *Basic and Clinical Endocrinology*, 4th ed., Appleton and Lange, Norwalk, CT (1994).
4. R. M. Berne and M. N. Levy. *Physiology*, 3rd ed., p. 1071, C. V. Mosby, St. Louis, MO (1993).
5. K. Verdeal and D. S. Ryan. *J. Food Protect.* **42**, 577 (1979).
6. J. A. McLachlan. "Rodent models for perinatal exposure to diethylstilbestrol and their relation to human disease in the male", in *Developmental Effects of Diethylstilbestrol in Pregnancy*, A. L. Herbst and H. A. Bern (Eds.), p. 148, Thieme-Stratton, New York (1981).
7. A. M. Soto, T.-M. Lin, H. Justicia, R. M. Silvia, C. Sonnenschein. "An 'in culture' bioassay to assess the estrogenicity of xenobiotics", in T. Colborn and C. Clement (Eds.), *Chemically Induced Alterations in Sexual Development: The Wildlife/Human Connection,* p. 295, Princeton Scientific Publishing, Princeton, NJ (1992).
8. A. M. Soto, K. L. Chung, C. Sonnenshein. *Environ. Health Perspect.* **102**, 380 (1994).
9. K. W. Gaido, L. S. Leonard, S. Lovell, J. C. Gould, D. Babai, C. J. Portier, D. P. McDonnell. *Toxicol. Appl. Pharmacol.* **143**, 205 (1997).
10. S. F. Arnold, D. M. Klotz, B. M. Collins, P. M. Vonier, L. J. Guillette, J. A. McLachlan. *Science* **272**, 1489 (1996).
11. K. Ramamoorthy, F. Wang, I.-C. Chen, S. Safe, J. D. Norris, D. P. McDonnell, K. W. Gaido, W. P. Bocchinfuso, K. S. Korach. *Science* **275**, 405 (1997a).
12. J. Ashby, P. A. Lefevre, J. Odum, C. A. Harris, E. J. Routledge, J. P. Sumpter. *Nature* **385**, 494 (1997).
13. K. Ramamoorthy, F. Wang, I.-C. Chen, J. D. Norris, D. P. McDonnell, L. S. Leonard, K. W. Gaido, W. P. Bocchinfuso, K. S. Korach, S. Safe. *Endocrinology* **138**, 1520 (1997b).
14. J. McLachlan. *Science* **277**, 462 (1997).
15. W. J. Dieckmann and M. E. Davis. *Am. J. Obstet. Gynecol.* **66**, 1062 (1953).
16. R. J. Stillman. *Am. J. Obstet. Gynecol.* **142**, 905 (1982).
17. R. M. Giusti, K. Iwamoto, E. E. Hatch. *Ann. Intern. Med.* **122**, 778 (1995).
18. A. L. Herbst, H. Ulfelden, D. C. Poskanzer. *N. Engl. J. Med.* **284**, 878 (1971).
19. M. Bibbo and W. B. Gill. *Obstet. Gynecol.* **49**, 1 (1977).
20. E. D. Whitehead and E. Leiter. *J. Urol.* **125**, 47 (1981).
21. B. E. Henderson, B. Benton, M. Cosgrove. *Pediatrics* **58**, 505 (1976).
22. A. L. Herbst, R. J. Kurman, R. E. Scully. *Obstet. Gynecol.* **40**, 287 (1972).
23. A. L. Herbst, D. C. Poskanzer, S. J. Robboy. *N. Engl. J. Med.* **292**, 334 (1975).
24. A. L. Herbst and M. M. Hubby. *Am. J. Obstet. Gynecol.* **141**, 1019 (1981).
25. F. J. Leary, L. J. Resseguie, L. T. Kurland, P. C. O'Brien, R. F. Emslander, K. L. Noller. *J. Am. Med. Assoc.* **252**, 2984 (1984).
26. O. C. Hadjimichael and J. W. Meigs. *J. Natl. Cancer Inst.* **73**, 831 (1984).

27. D. Labarthe, E. Adam, K. L. Noller, P. C. O'Brien, S. J. Robboy, B. C. Tilley, D. Townsend, A. B. Barnes, R. H. Kaufman, D. G. Decker, C. R. Fish, A. L. Herbst, J. Gundersen, L. T. Kurland. *Obstet. Gynecol.* **51**, 453 (1978).
28. O. P. Heinonen. *Cancer* **31**, 573 (1973).
29. W. B. Gill, G. F. B. Schumacher, M. Bibbo. *J. Urol.* **122**, 36 (1979).
30. A. J. Wilcox, D. D. Baird, C. R. Weinberg, P. P. Hornsby, A. L. Herbst. *N. Eng. J. Med.* **332**, 1411 (1995).
31. M. P. Vessey, D. V. Fairweather, B. Norman-Smith, J. Buckley. *Br. J. Obstet. Gynecol.* **90**, 1007 (1983).
32. Report to the Medical Research Council, "The use of hormones in the management of pregnancy in diabetics", *Lancet* **269**, 833 (1955).
33. E. Carlsen, A. Giwercman, N. Keiding, N. E. Skakkebaek. *Brit. Med. J.* **305**, 609 (1992).
34. R. M. Sharpe. *J. Endocrinol.* **136**, 357 (1993).
35. R. M. Sharpe and N. E. Skakkebaek. *Lancet* **341**, 1392 (1993).
36. J. McLachlan, R. Newbold, B. Bullock. *Science* **190**, 991 (1975).
37. E. Seaman, N. Bar-Chama, H. Fisch. *Mediguide Urol.* **7**, 1 (1994).
38. K. A. Ginsburg and D. R. Armat. *Fertil. Steril.* **53**, 882 (1990).
39. L. Johnson. *J. Androl.* **7**, 331 (1986).
40. R. J. Levine, R. M. Mathew, C. Brandon. *N. Engl. J. Med.* **323**, 12 (1990).
41. M. D. Whorton and C. R. Meyer. *Fertil. Steril.* **42**, 82 (1984).
42. D. Lerda. *Am. J. Ind. Med.* **22**, 567 (1992).
43. F. M. Wittmaack and S. S. Shapiro. *Wisconsin Med. J.* **91**, 477 (1992).
44. J. MacLeod and Y. Wang. *Fertil. Steril.* **31**, 103 (1979).
45. J. M. Ratcliffe and B. C. Gladen. *Reprod. Toxicol.* **6**, 297 (1992).
46. C. A. Paulsen, N. G. Berman, C. Wang. *Fertil. Steril.* **65**, 1015 (1996).
47. H. Fisch, J. Feldshuh, E. T. Goluboff, S. J. Broder, J. H. Olson, D. H. Barad. *Fertil. Steril.* **65**, 1009 (1996).
48. K. D. Smith and E. Steinberger. "What is digosperma?" in *The Testis in Normal and Infertile Men,* P. Troen and H. R. Nankin, (Eds.), p. 489, Raven Press, New York (1977).
49. H. Fisch and E. T. Goluboff. *Fertil. Steril.* **65**, 1044 (1996).
50. A. Brake and W. Krause. *Brit. Med. J.* **305**, 1495 (1992).
51. P. Bromwich, J. Cohen, I. Stewart, A. Walker. *Brit. Med. J.* **309**, 19 (1994).
52. G. W. Olsen, K. M. Bodner, J. M. Ramlow, C. E. Ross, L. I. Lipshultz. *Fertil. Steril.* **63**, 887 (1995).
53. T. A. Mably, D. L. Bjerke, R. W. Moore, A. Gendron-Fitzpatrick, R. E. Peterson. *Toxicol. Appl. Pharmacol.* **114**, 118 (1992).
54. D. L. Bjerke and R. E. Peterson. *Toxicol. Appl. Pharmacol.* **127**, 241 (1994).
55. United States Environmental Protection Agency (EPA), *Health Assessment Document for 2,3,7,8-Tetrachlorodibenzo*-p-*dioxin (TCDD) and Related Compounds*, Vols. I–III, EPA/600/BP-92/001c (1994).
56. C. Wilker, L. Johnson, S. Safe. *Fundam. Appl. Toxicol.* **141**, 68 (1996).
57. W. H. Wolfe, J. E. Michalek, J. C. Miner, A. J. Rahe. "An epidemiologic investigation of health effects in air force personnel following exposure to herbicides: Extract, reproductive outcomes, executive summary, introduction and conclusions", in *Air Force Health Study*, Epidemiology Research Division, Armstrong Laboratory, Brooks Air Force Base, TX (1992).
58. E. A. Emmett, M. Maroni, J. Jefferys, J. Schmith, B. K. Levin, A. Alvares. *Am. J. Ind. Med.* **14**, 47 (1988).
59. S. Dalterio, F. Badr, A. Bartke, D. Mayfield. *Science* **216**, 315 (1982).
60. S. Dalterio and A. Bartke. *Science* **205**, 1420 (1979).

61. B. S. Ahluwalia, S. U. Rajguru, G. H. Nolan. *J. Androl.* **6**, 386 (1985).
62. S. Osser, A. Beckman-Ramirez, P. Liedholm. *Acta. Obstet. Gynecol. Scand.* **71**, 215 (1992).

TOPIC 4

EFFECTS OF ENDOCRINE ACTIVE SUBSTANCES IN WILDLIFE SPECIES

Pure Appl. Chem., Vol. 75, Nos. 11–12, pp. 2197–2206, 2003.

Topic 4.1
Historical perspective on endocrine disruption in wildlife*

Peter Matthiessen

Centre for Ecology and Hydrology, Lancaster Environment Centre, Library Avenue, Bailrigg, Lancaster LA1 4AP, UK

Abstract: Endocrine disruption (ED) as a named field of research has been very active for over 10 years, but effects in wildlife that would now be labeled as ED have been studied since the 1940s. This paper briefly surveys the progress in wildlife studies that has been made to date and draws out the major themes and issues that have been identified. In particular, it discusses information concerning causative substances, modes of action, ubiquity of effects across taxa, individual- and population-level impacts, and the importance of low-dose and mixture effects. The main conclusion is that while most wildlife taxa are showing some ED effects at some locations, good evidence for population-level impacts is still limited to a few groups. In order to improve both the interpretation of field observations and the way in which environmental risk assessments are conducted, we need to develop an enhanced ability to predict effects on populations and communities from a knowledge of effects on individuals.

BEFORE "ENDOCRINE DISRUPTION"

It is a common misconception that the study of endocrine disruption (ED) began with the famous meeting at the Wingspread Conference Center, Racine, Wisconsin in July 1991 organized by Theo Colborn and coworkers [1]. Of course, there is no doubt that the current interest in this subject as a named research field dates from that time. For example, using the search term "endocrine disrup*" with the Science Citation Index (ISI Web of Science) in May 2002 brought up 1346 research article references, the first of which is Colborn et al. (1993) [2]. In fact, the first published mention of ED seems to be in Colborn and Clement (1992) [1]. Figure 1 shows how the field has exploded since then in terms of published papers, although it should be borne in mind that many relevant articles do not contain endocrine disruption in their titles, abstracts, or keywords. The true total of papers that have addressed the subject since 1993 is in excess of 2000 [26]. On the other hand, some of these relate to possible endocrine disruption in humans and mammalian models, and it is much harder to estimate the research activity on the wildlife side. Since 1981, for example, there have been 376 research papers alone that deal with tributyltin (TBT) and wildlife, and another 288 concerning alkylphenols and wildlife, so the total for all categories of endocrine active substances (EASs) is probably over 800. Of course, one of the almost unique features of this field of study, as exemplified by Colborn et al.'s first paper [2], is the close collaboration between wildlife scientists and clinicians or medical scientists, so perhaps we do not need to strive too hard to tease humans and other organisms apart.

*Report from a SCOPE/IUPAC project: Implication of Endocrine Active Substances for Human and Wildlife (J. Miyamoto and J. Burger, editors). Other reports are published in this issue, *Pure Appl. Chem.* **75**, 1617–2615 (2003).

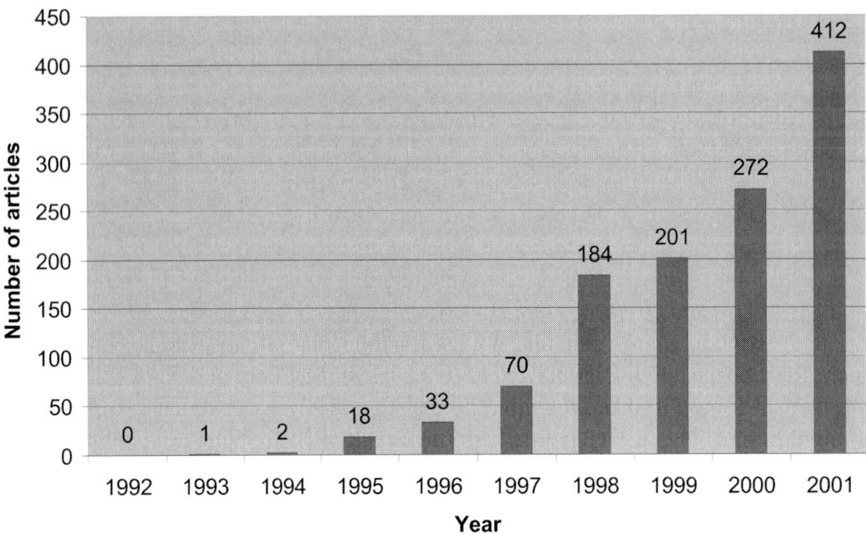

Fig. 1 Annual number of published scientific articles that refer to endocrine disruption. Note that articles which do not use this phrase are not included.

In reality, the study of ED has been in progress for much more than 10 years, although the full range of early work is not represented in Colborn and Clement [1]. Perhaps surprisingly, this book only contained 4 chapters (out of a total of 21) on evidence for ED in wildlife—Chap. 6 by W. P. Davis and S. A. Bortone on the effects of pulp-mill effluent on the sexuality of fish; Chap. 7 by J. F. Leatherland on thyroid abnormalities in Great Lakes salmon; Chap. 8 by G. A. Fox on abnormal sexual development in North American birds, but including some mention of TBT-related imposex in mollusks, and effects of pulp-mill effluents on fish; and Chap. 9 by P. J. H. Reijnders and S. M. J. M. Brasseur on hormonal disorders in marine organisms including mammals. Theo Colborn's major achievement in 1991 was to pull together information from very disparate fields of study (including both wildlife and humans) and come up with a new paradigm of toxicant action, but research in fields that would now be labeled as ED had in fact been published since the 1940s and 1950s. For example, work in the 1940s [3,4] revealed ED effects in livestock caused by natural and synthetic substances, respectively. Furthermore, synthetic substances with endocrine action had been known since the 1930s (e.g., estrogenic activity of diethylstilbestrol [5]). Even earlier, in the 1920s, it had been discovered that sexual condition could be induced in immature female rats by injecting small amounts of what was then known as ovarian follicular hormone from pigs [6], so the incredible potency of hormonally active substances is not a new idea.

As we now know, the new concept in 1991 was the attempt to explain many pathologies and other abnormalities in wildlife (and humans) in terms of endocrine interference through hormone mimicry and other mechanisms. Up until that time, the idea that synthetic substances with only slight or no resemblance to hormones or their antagonists, could be causing widespread biological effects in the natural world through interference with endocrine systems would have been dismissed as fanciful by all but a few. Furthermore, the effects on wildlife of discharges of naturally occurring hormonally active substances (e.g., phytoestrogens and androgens in pulp-mill effluent) had only been studied to a limited extent at that date.

Endocrine disruption is sometimes dismissed as an issue of primarily academic interest (or even self-interest), with little if any relevance for real ecosystems. This follows in a long tradition employed by critics with various axes to grind, who either deliberately or accidentally overlook the abundance or significance of evidence for particular phenomena. The present volume, for example, brings together an overwhelming body of knowledge which shows not only that ED in some wildlife is a widespread reality, but that it can cause population- or community-level damage given the right combination of conditions. There is certainly honest debate about the extent of such damage, but the quantity and quality of available data now make it clear that ED is a common mechanism whereby a range of pollutants is able to cause effects in wildlife. In earlier eras, similar "critical masses" of knowledge had to be assembled for toxicological phenomena caused, for example, by carcinogens, teratogens, and persistent organochlorine (OC) pesticides, before critics were forced to retreat. The best-known example of such an assembly of knowledge is probably Rachel Carson's final book *Silent Spring*, published in 1962, which blew the whistle about the impacts on wildlife being caused by OCs [7]. Interestingly, Carson drew attention to the ability of certain OCs to interfere with reproduction, although at that time the endocrine-disrupting mechanisms were unclear. Partly because she also attacked our unsustainable exploitation of nature, and partly because *Silent Spring* was written as a (highly literate) polemic rather than as a scientific treatise, Carson was heavily criticized by a variety of industrial interests, but time has of course proved her largely correct, both about OCs and about the need for sustainability.

In a real sense, these earlier discoveries prepared the ground for our study of ED. Crucially, they accustomed the world of ecotoxicology to the idea that certain substances in extremely small doses could have far-reaching biological effects which were completely unrelated to their acute toxicity. It is interesting to note, however, that while regulation of certain chemicals began to incorporate rodent-based tests for carcinogenic and teratogenic effects (essentially to protect humans), protection of wildlife against such substances was treated as a lower priority. Thus, even now, environmental risk assessment of most chemicals is based on acute toxicity data to which is applied some sort of empirical "safety factor", and it is doubtful whether such notorious pollutants as TBT would be caught even by the more sophisticated testing screens which are now applied to pesticides and biocides. It will probably be a few years yet before environmental risk assessment of chemicals for effects caused by ED becomes routine (see Topic 4.12 for information on the development of fish-based ED tests [8]).

DEVELOPMENTS SINCE 1991

Since 1991, several additional abnormalities in wildlife have been recognized as stemming from ED. These include the feminization of fish by natural and synthetic estrogens in sewage effluent (see Topics 4.3 and 4.5); abnormal interrenal function leading to weakening of the stress response in fish exposed to a range of contaminants [9]; and altered sex determination and differentiation in alligators and other reptiles which in some contaminated locations has been associated with population declines (Topic 4.7). There is also some evidence that limb deformities in amphibians may be caused by a form of ED related to environmental retinoids (e.g., Topic 4.6). Furthermore, laboratory studies have shown that amphibians can be feminized or demasculinized by EASs [10,11], although the available evidence suggests that the widespread declines in amphibian populations are probably related in complex ways to global warming rather than chemical pollution [12].

Much more information has become available since 1991 on the effects of organotins in mollusks which first appeared in about 1975 soon after TBT was introduced in antifouling paints [13]. TBT-induced masculinization in both bivalves and gastropods is now definitely established as a form of endocrine disruption which has led to invertebrate population and even community declines (e.g., Topic 4.2 [13,14]). On the other hand, although more information has also been accumulated on ED in birds (e.g., Topic 4.8 [15]), the data on effects in the field are still rather sparse, and it is even disputed whether ED occurs at all in wild birds [16]. DDE-induced eggshell thinning in raptors is often cited as an example of ED which had been observed as early as the 1960s [17], but the true mechanism has been elusive. The effect is probably a direct inhibition of prostaglandin synthesis in the shell gland mucosa, which in turn interferes with calcium metabolism [18], but it is a moot, though trivial, point whether this qualifies as an example of ED. Finally, it is perhaps surprising that more is not known about ED in mammalian wildlife given the extensive literature on experimentally induced ED in rodents, mustelids, and primates (several authors—present volume; [19]). The data on ED in some wild aquatic mammals are reasonably convincing (Topic 4.4), although experimental results are understandably sparse, but information from terrestrial mammals is still restricted to a few equivocal studies [20]. Whether this is due to genuine differences from aquatic mammals in exposure or susceptibility, or simply to an absence of suitable field investigations, is not clear.

In recent years, we have seen a plethora of books on ED and wildlife, none of which (in my opinion) quite match up to *Silent Spring* in terms of readability or impact. Nine of these were ably reviewed by David Peakall shortly before his untimely death [21]. They range from popular books in the *Silent Spring* tradition [22,23], to highly detailed scientific works that are similar in scope to the present volume [24]. Probably the most useful documents not reviewed by Peakall are Kime [25] which exhaustively covers ED in fish, and Damstra et al. [26] which reviews the effects of ED in both humans and wildlife for the International Programme on Chemical Safety. There are, of course, several others mainly concerned with possible ED in humans [27]. Finally, there are numerous web sites dedicated to ED in both wildlife and humans, including these useful ones: <http://endocrine.ei.jrc.it/index.html> maintained by the European Union, <http://www.epa.gov/endocrine> by the United States Environmental Protection Agency, <http://www.nihs.go.jp/hse/environ/endocrine.html> by the Japanese Institute of Health Sciences, and sites maintained by nongovernmental bodies such as Tulane University <http://www.som.tulane.edu/ecme/eehome/>.

MAJOR FEATURES OF ENDOCRINE DISRUPTION IN WILDLIFE

It is not the purpose of this article to review in detail the effects of EASs in wildlife—that job is done by other chapters in the present volume. Instead, with the benefit of the many years of research referred to above, the intention is to draw out some broad conclusions about how the field has progressed to date.

Causative substances

The list of potential EASs is now very long and diverse. For example, the European Union has published a candidate EAS list consisting of 553 man-made substances and 9 synthetic or natural hormones, which although very tentative (as it is based on patchy data and nonstandard testing methods), at least gives a crude picture of the possible scale of the problem [28]. This section focuses on those substances strongly suspected of causing effects in wildlife (as opposed to those that have been identified experimentally as EASs).

- Many examples of ED that have been reported in feral wildlife concern estrogens and their mimics causing receptor-mediated effects. For example, in the case of vitellogenesis and ovotestis in male fish, there is a strong association with estrogenic substances in treated sewage effluent [29,30]. Of these estrogens, natural and synthetic hormones appear to comprise the major part, although synthetic estrogen-mimics such as nonylphenol also contribute some effect. On the other hand, androgens in sewage do not appear to be a major issue for aquatic life [31], although natural substances in pulp-mill effluent can have androgenic effects in fish [32]. Pulp-mill effluent is actually a complex mixture of phytoestrogen derivatives and natural phenolics [33,34], both of which can cause a range of endocrine disturbances, not just at the receptor level, but also on various parts of the pituitary-gonadal axis [35].
- A ubiquitous group of contaminants, the polycyclic aromatic hydrocarbons (PAHs), is suspected of having a range of weak ED effects (depending on structure), including estrogenicity caused by agonistic action [36], antiestrogenicity via Ah-R and ER binding [37,38], androgenicity and anti-androgenicity—both AR-mediated [39]. PAHs do not biomagnify in vertebrates due to rapid metabolism, so they are unlikely to cause effects at a distance, but field studies in highly PAH-con-

taminated areas such as Puget Sound have revealed a range of possible ED effects in flatfish (e.g., delayed gonadal development, reduced reproductive success, and precocious female development) which have been attributed inter alia to PAHs [40–42]. It is difficult to attribute causality with precision, however, due to the frequent co-occurrence of PAHs in Puget Sound with various OCs.

- Of the other effects in wildlife which have been proven or strongly suspected to be caused by ED, the overwhelming majority are linked to the persistent organochlorines (OCs), especially the DDT, dioxin, and PCB families. Various isomers and congeners in these groups have been shown in the laboratory to affect steroid hormone levels or act agonistically/antagonistically at steroid and/or aryl hydrocarbon (Ah) receptors, and a host of field studies have implicated OCs in reproductive effects in various wildlife, from fish to mammals. One of the most well-studied phenomena concerns the impact of organochlorine contaminants on reproductive success of alligators in Florida (Topic 4.7 [43]), which have experienced altered sex hormone titres and reproductive tract anatomy that have in turn probably been the cause of population declines in some locations. Key features of many OCs are their persistence, long-distance transport, bioaccumulation and biomagnification. These properties imply that OC-related ED may occur in top-predators at locations remote from human activity, and this indeed could be happening in some wildlife groups ranging from swordfish to polar bears [44,45].
- Turning to synthetic nonestrogens/androgens, the major example identified to date is probably TBT, which interferes with testosterone metabolism in mollusks, causing imposition of male characteristics in females and reduced reproductive success [14,46]. TBT, although a biocide used in antifouling paints, was not designed to be an EAS. On the other hand, a range of substances such as ecdysteroid agonists and antagonists, and juvenile hormone analogs, have been deliberately designed as pesticides to control insects by ED, although there is little evidence that nontarget arthropods have been damaged by these pesticides under field conditions [47].
- Finally, it has been established that interrenal dysfunction can be caused in wild fish and other vertebrates by chronic exposure to a range of organic and inorganic pollutants, including heavy metals, PAHs, and PCBs (Topic 4.10 [9,48]). These effects have mainly been studied in North America, but in essence involve an impairment of the secretion of corticosteroid hormones such as cortisol. The affected organisms consequently develop a reduced ability to respond to stress, which in principle can cause a variety of deleterious outcomes. These effects deserve wider study because they are likely to result from virtually any kind of chronic pollutant exposure.

Modes of action

It will be apparent from the cases cited above that a wide range of modes of action has been found or strongly suspected to occur in wildlife. Of the receptor-mediated mechanisms, there is abundant evidence for estrogenic effects, mainly in connection with sewage effluent, and support also exists for activity in antiestrogenic, androgenic, and antiandrogenic pathways at some locations. Unsurprisingly, these result in various interferences with sex determination, sexual maturation, and/or successful reproduction. Agonistic and antagonistic actions at the ecdysteroid receptor also undoubtedly occur in insects in response to certain insecticides and phytochemicals [47,49], causing interference with molting, but it is unclear if other arthropods are also affected. Other suggested mechanisms (e.g., agonistic or antagonistic action by exogenous substances at retinoid receptors in amphibians, causing abnormal limb development—Topic 4.6) have not yet been widely established in the field. However, nonreceptor-mediated mechanisms are widespread, including interference with the synthesis, metabolic breakdown or excretion of hormones in a range of organs, and blockage of the controlling links between the pituitary and the hormone-synthesizing tissues. There are even examples of so-called "cross-talk" from one hormone system to another, such as the suspected interference with the thyroid system of fish (thyroxine build-up) due to pollutant-induced exhaustion of the cortisol response [50]. The reasons for other thyroid abnormalities, for example in Great Lakes salmonids, may not, however, be directly linked to contaminants [51,52]. Finally, there is evidence that larval exposure to some EASs can lead to heightened sensitivity to these substances in the resulting adults, presumably as a result of increased synthesis of hormone receptors [53], although this has not yet been proven to occur in the field. All these complex modes of action and interaction mean that the work of uncovering causative substances, or of predicting effects, can be very complicated. Indeed, one can only be confident of having arrived at the right explanation for abnormalities seen in the field if they can be satisfactorily replicated under laboratory conditions and then abolished or minimized by regulation of the suspect substance(s).

Ubiquity of effects across taxa

It is becoming clear that endocrine disruption can occur in a very wide range of animals. As well as having been observed (either in the field or the laboratory) in every class of vertebrate [24,54], it has also been recorded in insects, crustaceans, and mollusks [47]. It probably also occurs in echinoderms. By extension, it can therefore be expected in all animals which use some form of hormone signalling. This is not to say that the situation is well-studied in all the taxa referred to above. For example, although we now know rather a lot about ED in fish, our knowledge about other taxa tends to be sketchy. This particularly applies to most invertebrate phyla whose endocrine systems are poorly understood by comparison with vertebrates. However, vertebrate endocrinology is sufficiently similar across the phylum that it becomes feasible to use certain wildlife species as sentinels for effects in humans, always bearing in mind that exposure to EASs may differ greatly. It is noteworthy that ED seems to be particularly associated with aquatic wildlife [55], and this may be a consequence of the high pollutant exposure which water-breathers and their predators receive.

Individual- vs. population-level effects

The overwhelming majority of unequivocal ED effects seen in the field have been at the level of the individual. These range from subtle changes in biochemistry (e.g., VTG induction; altered hormone titres), through cellular changes (e.g., ovotestis) and gross morphological effects (e.g., imposition of secondary sexual characteristics; delayed gonadal development), to impacts on whole-body functions such as reproductive success (e.g., reduced egg-laying in TBT-exposed gastropods). On the other hand, there are few clear examples of population declines which have resulted from these processes, although the cases of TBT in mollusks and OCs in alligators, which have both been accompanied by reduced abundances in some locations, are the exceptions which prove that such effects are possible. DDE also caused population declines in some raptors due to eggshell-thinning, although it is still debated whether this is an example of ED. There are many other examples of population effects which have been tentatively attributed to ED (e.g., epizootics and population declines in some marine mammals; global declines in amphibians; reduced raptor populations; declines in Great Lakes salmonids), but causal relationships in these cases remain unclear [54]. This is, no doubt, partly due to the difficulty of proving cause and effect in species which are not amenable to experimentation, but it must also be recognized that not all impacts on individuals are translated through to the population level. A graphic illustration of this is provided by the phenomena of ovotestis and vitellogenesis in UK roach (*Rutilus rutilus*) populations where up to 100 % of males have been affected in severely estrogen-contaminated rivers without loss of populations [56]. Ovotestis in wild *R. rutilus* is accompanied by reduced gamete production [57] and probably by reduced reproductive success, but apparently not enough completely to overcome mitigating processes such as density dependence. This is not to say that population-level effects must necessarily be proven before ED is taken seriously, but they should at least be plausible in the light of the biology of the species in question.

Importance of low-dose effects and mixtures

Traditional ecotoxicology and environmental risk assessment are founded on the assumption that, below a certain dose (the threshold), no effects can occur due to the action of repair and excretory mechanisms, etc. However, as pointed out by Sheehan [58], if a dosed contaminant mimics an endogenous agonist, the threshold for effect will already have been exceeded at zero dose. This is supported by an analysis of 31 experimental mammalian ED datasets, 26 of which demonstrate modified Michaelis–Menten kinetics that have no threshold [59]. Similar dose–response relationships have been seen in temperature-dependent sex determination of estradiol-dosed turtle embryos [60], so it is not unreasonable to expect that such effects are occurring in wildlife populations. The situation is further complicated by the fact that some EAS dose–response curves are not monotonic. For example, it has been shown [61,62] that low estradiol doses in developing mouse embryos stimulate the growth of the prostate gland, while higher doses prevent it getting so large. These effects have not been shown to occur in free-living wildlife because it is usually impossible to have an adequate knowledge of dose under these conditions. However, the major implication is that environmental risk assessments which are often based on testing high concentrations in acute tests may tend to underestimate the effects of the low concentrations which usually occur in the field (Topic 4.9). Risk assessment becomes even more problematical when one considers that agonists or antagonists and their mimics can all act via the same receptor, so receptor-mediated ED in complex environmental mixtures is likely not to be predictable on the basis of the known concentration of a single EAS. Such EAS interactions have been shown in vivo to be less-than-additive, additive, or synergistic, depending on the organism, endpoint, and mixture [63–68], but much more work needs to be done to understand this complex field. At present, it is clear that the discovery of low dose and mixture effects of EASs will force a fundamental reappraisal of environmental risk assessment procedures for EASs.

CONCLUSIONS

The enormous amount of activity in the field of ED in wildlife over the last 10 years has born much fruit. It is now clear that almost all major wildlife groups are experiencing ED at some contaminated sites. Furthermore, such effects are being caused by a wider range of natural and synthetic substances than was anticipated in 1991, and by a wider range of mechanisms. It now seems likely that ED at many sites is caused by complex mixtures of substances, often acting at low concentrations which could not have been predicted on the basis of traditional toxicity tests using high concentrations of test substances. This has major implications for the environmental risk assessment of EASs. In a few cases, we now have convincing evidence that ED can lead to population declines and even disturbances at the community level, but our ability to predict such higher-order effects is still weak. This is a reflection of our imperfect knowledge about ecosystem structure and function, and is a substantial challenge for the future if we are to improve our interpretation of field observations.

REFERENCES

1. T. Colborn and C. Clement (Eds.). In *Chemically-Induced Alterations in Sexual and Functional Development: the Wildlife/Human Connection,* Advances in Modern Environmental Toxicology Vol. 21, p. 403, Princeton Scientific Publishing, Princeton, New Jersey (1992).
2. T. Colborn, F. S. V. Vom Saal, A. M. Soto. *Environ. Health Perspect.* **101**, 378–384 (1993).
3. H. W. Bennetts, E. J. Underwood, F. L. Shier. *Austr. Vet. J.* **22**, 2–12 (1946).
4. H. Burlington and V. F. Lindeman. *Proc. Soc. Exp. Biol. Med.* **74**, 48–51 (1950).
5. E. C. Dodds, L. Goldberg, W. Lawson, R. Robinson. *Nature* **141**, 247–248 (1938).
6. E. Allen and E. A. Doisy. *Am. J. Physiol.* **69**, 577–588 (1924).

7. R. Carson. *Silent Spring*. Houghton Mifflin, USA (1962).
8. S. Jobling. *Pure Appl. Chem.* **70**, 1805–1827 (1998).
9. A. Hontela. *Environ. Toxicol. Chem.* **17**, 44–48 (1998).
10. B. D. Palmer and S. K. Palmer. *Environ. Health Perspect.* **103** (Suppl. 4), 19–25 (1995).
11. T. B. Hayes, A. Collins, M. Lee, M. Mendoza, N. Noriega, A. A. Stuart, A. Vonk. *Proc. Nat. Acad. Sci.* **99**, 5476–5480 (2002).
12. J. M. Kiesecker, A. R. Blaustein, L. K. Belden. *Nature* **410**, 681–684 (2001).
13. C. Alzieu. *Ecotoxicology* **9**, 71–76 (2000).
14. P. Matthiessen and P. E. Gibbs. *Environ. Toxicol. Chem.* **17**, 37–43 (1998).
15. L. A. Feyk and J. P. Giesy. In: *Principles and Processes for Evaluating Endocrine Disruption in Wildlife*, R. J. Kendall, R. L. Dickerson, J. P. Giesy, W. P. Suk (Eds.), pp. 121–140, SETAC Press, Pensacola, Florida (1998).
16. A. Dawson. *Ecotoxicology* **9**, 59–69 (2000).
17. D. A. Ratcliffe. *Nature* **215**, 208–210 (1967).
18. C. E. Lundholm. *Comp. Biol. Physiol.* **118C**, 113–128 (1997).
19. L. E. Gray, J. Ostby, C. Wolf, C. Lambright, W. Kelce. *Environ. Toxicol. Chem.* **17**, 109–118 (1998).
20. C. F. Facemire, T. S. Gross, L. J. Guillette. *Environ. Health Perspect.* **103** (Suppl. 4), 79–86 (1995).
21. D. B. Peakall. *Ecotoxicology* **9**, 137–144 (2000).
22. T. Colborn, D. Dumanoski, J. P. Myers. *Our Stolen Future*, p. 306, Penguin, New York (1996).
23. D. Cadbury. *The Feminization of Nature. Our Future at Risk*, p. 303, Penguin, London (1997).
24. R. J. Kendall, R. L. Dickerson, J. P. Giesy, W. P. Suk (Eds.). In: *Principles and Processes for Evaluating Endocrine Disruption in Wildlife*, p. 491, SETAC Press, Pensacola, Florida (1998).
25. D. E. Kime. *Endocrine Disruption in Fish*, p. 396, Kluwer Academic Publishers, London (1998).
26. T. Damstra, S. Barlow, A. Bergman, R. Kavlock, G. Van Der Kraak (Eds.). In: *Global Assessment of the State-of-the-Science of Endocrine Disruptors*. WHO/ILO/UNEP International Programme on Chemical Safety, Geneva (available on-line: <http://endocrine.ei.jrc.it/final%20draft/who.html.>) (2002).
27. A.-M. Andersson, K. M. Grigor, E. Rajpert-De Meyts, H. Leffers, N. E. Skakkebaek. *Hormones and Endocrine Disrupters in Food and Water. Possible Impact on Human Health*, p. 571, Munksgaard, Copenhagen (2001).
28. European Commission. *Communication from the Commission to the Council and the European Parliament on the Implementation of the Community Strategy for Endocrine Disrupters– a range of substances suspected of interfering with the hormone systems of humans and wildlife, (COM(1999)706).* Commission of the European Communities, COM (2001) 262 final, p. 45 (2001).
29. C. Desbrow, E. J. Routledge, G. Brighty, J. P. Sumpter, M. Waldock. *Environ. Sci. Technol.* **32**, 1549–1558 (1998).
30. K. V. Thomas, M. Hurst, P. Matthiessen, M. J. Waldock. *Environ. Toxicol. Chem.* **20**, 2165–2170 (2001).
31. K. V. Thomas, M. R. Hurst, P. Matthiessen, M. McHugh, A. Smith, M. J. Waldock. *Environ. Toxicol. Chem.* **21**, 1456–1461 (2002).
32. I. Katsiadaki, A. P. Scott, M. R. Hurst, P. Matthiessen, I. Mayer. *Environ. Toxicol. Chem.* **21** (2002).
33. W. M. Howell and T. E. Denton. *Environ. Biol. Fish* **24**, 43–51 (1989).
34. M. G. Dubé and D. L. MacLatchy. *Environ. Toxicol. Chem.* **20**, 985–995 (2001).
35. G. Van Der Kraak, K. R. Munkittrick, M. E. McMaster, D. L. MacLatchy. In: *Principles and Processes for Evaluating Endocrine Disruption in Wildlife*, R. J. Kendall, R. L. Dickerson, J. P. Giesy, W. P. Suk (Eds.), p. 491, SETAC Press, Pensacola, Florida (1998).

36. J.-M. Nicolas. *Aquat. Toxicol.* **45**, 77–90 (1999).
37. J. M. Navas and H. Segner. *Environ. Sci. Pollut. Res.* **2**, 75–82 (1998).
38. J. Santodonato. *Chemosphere* **34**, 835–848 (1997).
39. A. M. Vinggaard, C. Hnida, J. C. Larsen. *Toxicology* **145**, 173–183 (2000).
40. T. K. Collier, L. L. Johnson, C. M. Stehr, M. S. Myers, J. E. Stein. *Mar. Environ. Res.* **46**, 243–247 (1998).
41. L. L. Johnson, S. Y. Sol, G. M. Ylitalo, T. Hom, B. French, O. P. Olson, T. K. Collier. Precocious sexual maturation and other reproductive anomalies in English sole from an urban waterway. International Council for the Exploration of the Sea, Copenhagen, ICES CM 1997/U:07, p. 15 + figs (1997).
42. L. L. Johnson, D. Misitano, S. Y. Sol, G. M. Nelson, B. French, G. M. Ylitalo, T. Hom. *Trans. Am. Fish. Soc.* **127**, 375–392 (1998), (2000).
43. L. J. Guillette, D. A. Crain, M. P. Gunderson, S. A. E. Kools, M. R. Milnes, E. F. Orlando, A. A. Rooney, A. R. Woodward. *Am. Zool.* **40**, 438–452 (2000).
44. M. C. Fossi, S. Casini, S. Ancora, A. Moscatelli, A. Ausili, G. Notarbartolo-di-Sciara. *Mar. Env. Res.* **52**, 477–483 (2001).
45. O. Wiig, A. E. Derocher, N. M. Cronin, J. U. Skaare. *J. Wildl. Dis.* **34**, 792–796 (1998).
46. E. Oberdörster and P. MClellan-Green. *Peptides* **21**, 1323–1330 (2000).
47. P. L. DeFur, M. Crane C. Ingersoll, L. Tattersfield. Endocrine disruption in invertebrates: endocrinology, testing, and assessment, p. 303, SETAC Press, Pensacola, Florida (1999).
48. A. Hontela. *Rev. Toxicol.* **1**, 1–46 (1997).
49. E. Oberdörster, M. A. Clay, D. M. Cottam, F. A. Wilmot, J. A. McLachlan, M. J. Milner. *J. Steroid Biochem. Mol. Biol.* **77**, 229–238 (2001).
50. T. Zhou, H. B. John-Alder, P. Weis, J. S. Weis. *Environ. Toxicol. Chem.* **18**, 2817–2823 (1999).
51. J. F. Leatherland. *J. Great Lakes Res.* **19**, 737–751 (1993).
52. J. F. Leatherland. *Guelph Ichthyol. Rev.* **2**, p. 67 (1994).
53. T. L. Metcalfe, C. D. Metcalfe, Y. Kiparissis, A. J. Nimi, C. M. Foran, W. H. Benson. *Environ. Toxicol. Chem.* **19**, 1893–1900 (2000).
54. Institute for Environment and Health. IEH Assessment on the Ecological Significance of Endocrine Disruption. Assessment A4. MRC Institute for Environment and Health, Leicester, UK, p. 262 (1999).
55. T. Colborn and K. Thayer. *Ecol. Applic.* **10**, 949–957 (2000).
56. S. Jobling, M. Nolan, C. R. Tyler, G. Brighty, J. P. Sumpter. *Environ. Sci. Technol.* **32**, 2498–2506 (1998).
57. S Jobling, N. Beresford, M. Nolan, T. Rodgers-Gray, G. C. Brighty, J. P. Sumpter, C. R. Tyler. *Biol. Reprod.* **66**, 272–281 (2002).
58. D. M. Sheehan. *Proc. Soc. Exp. Biol. Med.* **224**, 57–60 (2000).
59. R. M. Blair, H. Fang, D. Gaylor, D. M. Sheehan. *APMIS* **109**, 198–208 (2001).
60. D. M. Sheehan, E. Willingham, D. Gaylor, J. M. Bergeron, D. Crews. *Environ. Health Perspect.* **107**, 155–159 (1999).
61. F. S. Vom Saal, B. G. Timms, M. M. Montano, P. Palanza, K. A. Thayer, S. C. Nagel, M. D. Dhar, V. K. Ganjam, S. Parmigiani, W. V. Welshons. *Proc. Nat. Acad. Sci. USA* **94**, 2056–2061 (1997).
62. C. Gupta. *Proc. Soc. Exp. Biol. Med.* **224**, 61–68 (2000).
63. J. M. Bergeron, D. Crews, J. A. McLachlan. *Environ. Health Perspect.* **102**, 780–781 (1994).
64. J. M. Bergeron, E. J. Willingham, C. T. Osborn, T. Rhen, D. Crews. *Environ. Health Perspect.* **107**, 93–97 (1999).
65. J. Ashby, P. A. Lefevre, J. Odum, C. A. Harris, E. J. Routledge, J. P. Sumpter. *Nature* **385**, 494 (1997).
66. A. Kortenkamp and R. Altenburger. *Sci. Tot. Environ.* **221**, 59–73 (1998).

67. K. L. Thorpe, T. H. Hutchinson, M. J. Hetheridge, M. Scholze, J. P. Sumpter, C. R. Tyler. *Environ. Sci. Technol.* **35**, 2476–2481 (2001).
68. E. Silva, N. Rajapakse, A. Kortenkamp. *Environ. Sci. Technol.* **36**, 1751–1756 (2002).

Pure Appl. Chem., Vol. 75, Nos. 11–12, pp. 2207–2218, 2003.

Topic 4.2
Endocrine disruption in invertebrates*

J. Oehlmann[‡] and U. Schulte-Oehlmann

Department of Ecology and Evolution – Ecotoxicology, Johann Wolfgang Goethe University Frankfurt, D-60054, Frankfurt am Main, Germany

Abstract: Recent reports have shown that a number of xenobiotics in the environment are capable of interfering with the normal endocrine function in a variety of animals. The overwhelming majority of the studies on the effects of hormone-mimetic industrial chemicals were focused on findings in vertebrates. More detailed information about the effects on and mechanisms of action in invertebrates has only been obtained from a few cases, although invertebrates represent more than 95 % of the known species in the animal kingdom and are extremely important with regard to ecosystem structure and function. The limited number of examples for endocrine disruption (ED) in invertebrates is partially due to the fact that their hormonal systems are rather poorly understood in comparison with vertebrates. Deleterious endocrine changes following an exposure to certain compounds may easily be missed or simply be unmeasurable at present, even though a number of studies show that endocrine disruption has probably occurred. The well-documented case studies of tributyltin effects in mollusks and of insect growth regulators, the latter as purposely synthesized endocrine disruptors, are explained to support this view. According to our present knowledge, there is no reason to suppose that such far-reaching changes are in any sense unique. The additional existing evidence for ED in invertebrates from laboratory and field studies are summarized as an update and amendment of the EDIETA report from 1998. Finally, conclusions about the scale and implications of the observed effects are drawn and further research needs are defined.

INTRODUCTION

The hormonal regulation of biological functions is a common characteristic for all animal phyla, including invertebrates. While the basic endocrine strategy to regulate biological processes has been widely conserved [1], specific components of the endocrine system used in the various systematic groups have undergone significant evolutionary divergence resulting in distinct differences between the various biological taxa. This is especially true for invertebrates exhibiting a wide range of different chemical signaling systems, with some of them being unique to specific phyla. Other invertebrate groups seem to use at least partially (e.g., prosobranch mollusks) or totally (e.g., echinoderms) comparable hormones to vertebrates so that vertebrate-type sex steroids are produced in these groups and play a functional role [2–4]. Nevertheless, firm evidence of the role of these steroids in the endocrine system of invertebrates is still lacking for most phyla [5].

The endocrine systems of invertebrates generally regulate the same processes that are found in vertebrates such as development, growth, and reproduction. Because invertebrate species have devel-

---

*Report from a SCOPE/IUPAC project: Implication of Endocrine Active Substances for Human and Wildlife (J. Miyamoto and J. Burger, editors). Other reports are published in this issue, *Pure Appl. Chem.* **75**, 1617–2615 (2003).

‡Corresponding author

oped a huge diversity of life histories with characteristic events such as the formation of larval forms, often with a succession of different stages and/or pupation, metamorphosis, diapause or other types of resting stages, which do not occur in vertebrates, it is evident that endocrine systems of invertebrates are considerably more diverse than those found in vertebrates [2].

As it is virtually impossible to provide a complete overview of the various endocrine systems in invertebrates within the scope of this publication, we will provide a rough outline of generalities. For detailed information the reader is referred to a number of excellent reviews on invertebrate endocrinology [2,5–7]. The best-characterized invertebrate hormonal system is that of insects reflecting their economic and ecological significance and especially the need to control insect pests. Much less is known about a number of economically important aquatic groups such as crustaceans and mollusks, and knowledge on the remaining taxa is even more fragmentary.

In general, endocrine systems of invertebrates have not been documented in the same detail as vertebrates, nor have responses of invertebrate endocrine systems to suspected endocrine active substances (EASs) been studied with comparable intensity. Nevertheless, chemicals have been purposely synthesized to disrupt the endocrine system of a number of insects to aid their control. These so-called insect growth regulators (IGRs) were developed to intentionally interact with the hormonal system of these arthropods, acting as ecdysone agonists, antagonists, or juvenile hormone analogs. Perhaps one of the best-documented examples of the occurrence of EASs in the field is provided by tributyltin- (TBT-) induced imposex and intersex in gastropods [8]. The limited number of examples of endocrine disruption (ED) in invertebrates is partially due to the fact that their hormonal systems are poorly understood compared with vertebrates. Endocrine changes following an exposure to certain compounds may, therefore, be missed or simply be unmeasurable, even though some examples illustrate that invertebrates are susceptible to ED. Consequently, there is no reason to suppose that far-reaching changes as demonstrated by TBT and its effects on prosobranch populations are in any sense unique within the invertebrates [8]. Studies on ED in invertebrates are important because invertebrates represent not only more than 95 % of the known species in the animal kingdom, but also provide key species for ecosystem functioning and represent an insufficiently characterized although extremely important part of global biodiversity.

The article will summarize the existing evidence for ED in invertebrates from laboratory studies and field investigations. Because this objective was also covered by the EDIETA workshop some years ago [2], we will update and amend the information provided in the workshop proceedings.

INVERTEBRATE ENDOCRINE SYSTEMS AND THEIR SUSCEPTIBILITY TO ENDOCRINE-DISRUPTING COMPOUNDS

Any multicellular organism requires coordinated mechanisms that, besides direct cell contacts, involve chemical messengers. Consequently, all known invertebrate taxa make use of hormones to control biochemical, physiological, and behavioral processes in general as well as development, growth, and reproduction in particular. Because they are represented by more than 30 different phyla within the animal kingdom, it is not surprising that regulation of the above-mentioned processes by their endocrine systems is considerably more diverse than in vertebrates, which comprise only part of a single phylum, the Chordata. Despite the diversity in invertebrate endocrinology, some basic generalities can be made. Invertebrates use steroids, terpenoids, and peptide hormones, but the latter are by far the most common among these phyla [3,9]. While steroids are secreted in vertebrates from true glands, the secretory structures in invertebrates are often neuronal in origin and therefore referred to as neurosecretory organs or cells. Steroids, such as ecdysone and the vertebrate-type steroids, differ from terpenoid and especially peptide hormones in their physical and chemical properties, solubility, and resistance to degradation. A further issue that has to be emphasized is that certain compounds are likely to act as endocrine disruptors not only by a direct binding to receptors—acting as hormone-mimics (agonists) or as "antihormones" (antagonists)—but also indirectly by modulating endogenous hormone levels by interfering with biochemical processes associated with the production, availability, or metabolism of hormones or also by the modulation of receptors. Therefore, it is likely that the various endocrine systems in invertebrates are subject to modulation by an unforeseeable number of exogenous compounds.

Table 1 summarizes the occurrence of major hormone groups in various invertebrate taxa, based on a number of reviews on invertebrate endocrinology [2,5–7], but without intending to provide a complete list. The majority of invertebrates are not considered in the table because the endocrinology of these organisms remains largely unknown. It is obvious that the perhaps best-understood endocrine systems are those of insects, followed by crustaceans, echinoderms, and mollusks, although the latter are perhaps the most diverse of the invertebrate phyla, being second to the insects in number of identified species. The endocrine systems of the various classes of mollusks and even of major groups of gastropods—prosobranchs, opisthobranchs, and pulmonates—differ greatly, reflecting extreme differences in morphology and life histories. This can be exemplified by the vertebrate-type steroids, which do occur in prosobranchs and play a functional role. In contrast, there is no indication that opisthobranchs and pulmonates use steroids. Recent, still unpublished studies in Japan, Germany, and the United Kingdom demonstrated the occurrence of estrogen and androgen receptors in a number of marine and freshwater prosobranchs and characterized the receptors with regard to ligand-binding and structure (e.g., for *Thais clavigera*).

Table 1 Examples of reported hormones in different invertebrate taxa [2,5–7].

Taxon	Reported hormones (example, *controlled process*)
Coelenterata	Neuropeptides (glycine-leucine tryptophan amides = GLWamides, *metamorphosis*); thyroids (thyroxine, *strobilation*); retinoids (9-*cis*-retinoic acid, *strobilation*)
Nematoda	Ecdysteroids (reported but *functional role questionable*); terpenoids [juvenile hormone (JH) like hormones, *growth*]; neuropeptides (FMRFamide, *function unknown*)
Mollusca	Ecdysteroids (reported but *role questionable*); steroids (17ß-estradiol, testosterone, progesterone, *sexual differentiation, reproduction in prosobranchs*); terpenoids (JH reported but *role questionable*); neuropeptides [APGWamide, dorsal body hormone (DBH), *sexual differentiation, gonad maturation, spawning*; egg-laying hormone (ELH), *spawning*; FMRFamide, *neuromodulation*; molluscan insulin-like peptides (MIPs), *growth, development, energy metabolism*]
Annelida	Ecdysteroids (ecdysone, *role unknown*); neuropeptides (FMRFamide, *neuromodulation*)
Crustacea	Ecdysteroids (ecdysone, *molting, vitellogenesis*); steroids (17ß-estradiol, testosterone, progesterone, *role under debate*); terpenoids [methyl farnesoate (MF), *metamorphosis, reproduction*]; neuropeptides [androgenic hormone, *sexual differentiation, vitellogenesis inhibition*; crustacean hyperglycemic hormone family (CHH), *energy metabolism*; molt-inhibiting hormone (MIH), *ecdysteroid production*; vitellogenesis-inhibiting hormone (VIH), *vitellogenesis*]
Insecta	Ecdysteroids (ecdysone, *molting, egg maturation*); terpenoids (JH, *metamorphosis, reproduction*); neuropeptides [adipokinetic hormone (AKH), *energy metabolism*; allatostatin and allatotropin, *JH production*; bombyxin, *ecdysteroid production, energy metabolism;* bursicon, *cuticle tanning*; diapause hormone, *embryonic diapause*; diuretic hormone (DH), *water homeostasis*; ecdysis-triggering hormone (ETH) and eclosion hormone (EH), *ecdysis behavior*; FMRFamides, *neuromodulation*; prothoraciotrophic hormone (PTTH), *ecdysteroid production*]
Echinodermata	Steroids (progesterone, testosterone, 17ß-estradiol, estrone, *vitellogenesis, oogenesis, spermatogenesis, spawning*); neuropeptides (gonad-stimulating substance = GSS, *spawning*; maturation-promoting factor = MPF, *fertilization*)
Tunicata	Steroids (testosterone, 17ß-estradiol, *oogenesis, spermatogenesis, spawning*); neuropeptides (gonadotropin releasing hormone analogue, *gonad development*); thyroids (thyroxine, *probably tanning process during tunic formation*)

EVIDENCE FOR ENDOCRINE DISRUPTION IN INVERTEBRATES

The issue of ED in invertebrates has found an increasing scientific interest although only a limited number of confirmed cases have been reported. These are dominated by the antifouling biocide tributyltin (TBT) and its effects on prosobranch snails and by IGRs which were designed as EASs for use in insect pest control. The following sections will provide a summary of these confirmed examples of ED and an update of the detailed synopsis from the EDIETA workshop [2].

Organotin compounds and their effects in mollusks

The effects of TBT on prosobranch snails are one of the most complete examples of an EAS impact on aquatic invertebrates [8]. TBT compounds are mainly used as biocides in antifouling paints, but also in other formulations. They induce a variety of malformations in aquatic animals with mollusks as one of the most TBT-sensitive groups [10]. As the impact of TBT on nontarget organisms became apparent in the early 1980s, France drew up regulations to control TBT emission and banned the use of TBT antifoulings on small boats (length <25 m) in 1982, adopted later by other countries since 1987. Nevertheless, TBT pollution of coastal waters was found to have remained on a high level or even increased further in some regions. Consequently, the International Maritime Organization (IMO) decided in autumn 2001 to ban the application of TBT-based paints on all boats by January 2003 and the presence on ship hulls by January 2008.

The first adverse effects of TBT on mollusks were observed in *Crassostrea gigas* at the Bay of Arcachon, one of the European centers of oyster aquaculture, with ball-shaped shell deformations in adults and a decline of annual spatfall [11]. These effects led to a break-down of local oyster production with marked economic consequences. Laboratory and field analyses revealed that TBT was the causative agent with trace concentrations as low as 10 ng TBT/L in ambient water being effective [10]. Shell deformities in oysters, but also in other bivalves, were successfully applied as a biological marker of TBT effects in subsequent years. Another TBT effect in mollusks was first described in the early 1970s without identifying the organotin compound as the responsible cause at that time: A virilization of female prosobranchs, termed as imposex [12]. Imposex is characterized by the formation of a penis and/or vas deferens on females of gonochoristic prosobranch species and is induced at lower concentrations than all other described TBT effects. Furthermore, it is a specific response of organotin compounds under field conditions.

Imposex is known today in more than 150 prosobranch species, summarized in [2]. The gradual virilization of imposex affected females is described by a classification scheme with 6 stages, further divided in up into 3 different types (a–c) [13] having the advantage of being applicable for all affected species worldwide. Females are sterilized in the imposex stages 5 and 6 by a blockade of the pallial oviduct (stages 5a, b; Fig. 1) or by a split bursa copulatrix and capsule gland (stage 5c). The first possibility prevents the deposition of egg capsules, resulting in an accumulation of abortive capsular material in the pallial oviduct (stages 6a, b); the second mechanism prevents copulation and capsule formation. In young and sexual immature specimens of some muricid species, a protogyne sex-change can be induced by TBT concentrations, e.g., above 10 ng as Sn/L in *Nucella lapillus* [14] and above 2 ng as Sn/L in *Ocinebrina aciculata* [15].

The classification in 6 stages is the basis of the VDSI (vas deferens sequence index), calculated as the mean imposex stage of a population. It has been shown that imposex intensities, measured as the VDSI in a range of affected prosobranch species, show a highly significant correlation with TBT concentrations in ambient sea water, as demonstrated for the dog whelk in Fig. 2. Consequently, the degree of coastal TBT pollution can be assessed with high precision by a determination of imposex intensities in prosobranch populations. A further advantage of the VDSI is the possibility to perform comparisons of TBT sensitivities between different species and that the index is also a measure of the reproductive capability of a given population [16].

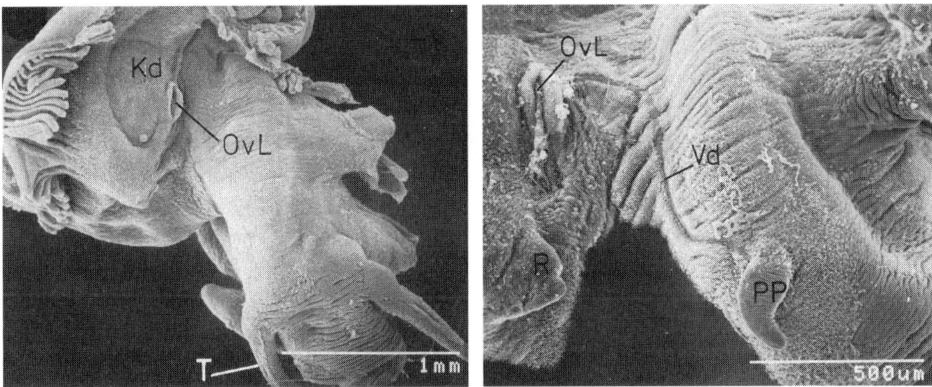

Fig. 1 *Hydrobia ulvae*. Scanning electron micrographs of females with opened mantle cavity. Left: female without imposex; right: sterilized imposex female with blocked oviduct. Abbreviations: Kd, capsule gland; OvL, ooparous oviduct opening (open left; blocked right); PP, penis; T, tentacle; Vd vas deferens.

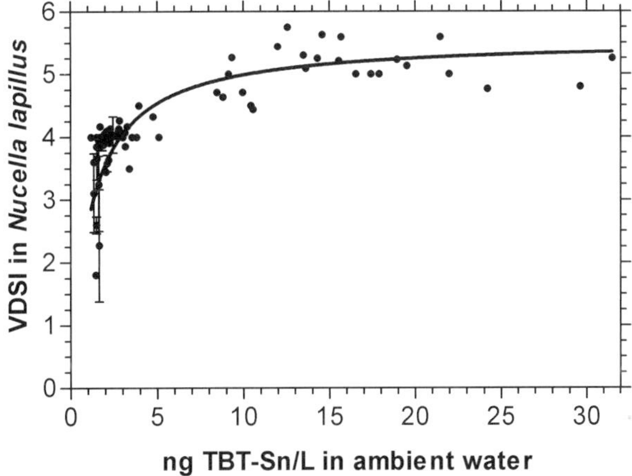

Fig. 2 *Nucella lapillus*. Relationship between aqueous TBT concentrations and imposex intensities: $y = (5.54\ x) / (1.12 + x)$; $n = 151$ population samples from 81 stations; $r = 0.688$; $p < 0.0005$.

The periwinkle *Littorina littorea* develops a closely related TBT-induced virilization phenomenon, termed as intersex [17]. Intersex-affected females exhibit male features on female pallial organs (inhibition of the ontogenetic closure of the pallial oviduct), or female sex organs are supplanted by the corresponding male formations. Comparably to imposex, the intersex response is a gradual transformation of the female pallial tract, which can be described by an evolutive scheme with four stages [17]. Intersex development restricts the reproductive capability of females. In stage 1, sperm are lost during copulation, and consequently the reproductive success is reduced. Females in stages 2–4 are definitively sterile because the capsular material is spilled into the mantle cavity (stage 2) or the glands responsible for the formation of egg capsules are missing (stages 3 and 4). Due to female sterility, periwinkle populations can be in decline, but are not likely to become extinct because of the planktonic veliger larvae, as long as aqueous TBT levels are not beyond mortality threshold concentrations for the larvae [18].

The assessment of intersex intensities in periwinkle populations is based on the same principle as described for the VDSI. The intersex index (ISI) is the mean intersex stage in a population. ISI values are highly significantly correlated to ambient TBT concentrations and can therefore be used together with or as an alternative to imposex assessments for the determination of the degree of coastal TBT pollution especially in regions with a relatively high level of contamination. In these areas, periwinkles are very common and can be sampled in sufficient numbers because *L. littorea* (a) is more tolerant of high TBT levels, (b) recruits from the plankton, and (c) can occur in areas where dog whelks have expired. Imposex in dog whelks and intersex in periwinkles have been used as combined biological markers for the convention-wide biological TBT effect monitoring of OSPAR (Oslo and Paris Commissions) [19].

Furthermore, TBT affects the community level and exhibits negative effects even beyond the scale of populations. Waldock et al. [20] analyzed the inter- and subtidal fauna of the River Crouch in a number of subsequent surveys between 1987 and 1992 and compared the results with older reports before the introduction of TBT-based antifouling paints. Overall, directional trends in community level attributed at a number of analyzed stations suggested a moderate improvement in environmental conditions over the sampling period, which was coincident with a marked decline in TBT concentrations at the stations. However, reference to historical data indicated that certain taxa that were previously frequent or common, especially snails, were only rarely recorded or still absent in the 1992 survey.

Much less attention has been paid to endocrine effects of TBT in freshwater ecosystems. The ramshorn snail Marisa cornuarietis and the hydrobiid *Potamopyrgus antipodarum* exhibit endocrine-mediated effects of TBT [21]. The latter and the netted whelk *Nassarius reticulatus* were used for a monitoring of androgenic activities in sediments of the River Elbe [22]. The majority of sediments exhibited marked androgenic activities and some of them, assigned to the ecological status classes IV and V according to the European Water Framework Directive, caused a maximum increase of imposex intensities in the netted whelk within four weeks.

At the molecular level, TBT interferes with hormone metabolism, most probably by an inhibition of the cytochrome P450-dependent aromatase, increasing the levels of androgens [8,23,24], while findings from the early 1980s showed that TBT inhibits the release of a neuroendocrine factor from the pleural ganglia, which is responsible for the suppression of penis formation in females, thus resulting in imposex development [25]. Although the factor has not been identified, recent results show that an administration of the neurohormone APGWamide can induce imposex in *Ilyanassa obsoleta* [26]. A possible explanation for these conflicting findings is the hypothesis of an even more pronounced analogy of vertebrate and prosobranch hormonal systems with neurohormones acting as releasing factors in both systematic groups, mediating steroid production and/or metabolism [27].

It has been accepted that imposex is induced almost typically by TBT [2], although at least in the marine rock shell *Thais clavigera* [28] and the freshwater ramshorn snail *Marisa cornuarietis* [29] not only TBT, but also triphenyltin (TPT) can promote imposex, while in other prosobranchs, such as *Nucella lapillus* and *Nassarius reticulatus*, TPT does not induce imposex [29].

One of the most important lessons to be learned from TBT and its effects in mollusks is that EASs may impact different levels of biological integration from molecules to communities affecting also the survival of populations in the field. Furthermore, the case history of TBT provides evidence that vertebrate-type steroids play an important functional role in a number of invertebrate groups, including prosobranchs.

Insect growth regulators (IGRs) as purposely synthesized EASs

The IGRs were developed to intentionally mimic, block, or otherwise interact with the hormonal system of insects, so that it is not too astonishing that they represent—next to TBT—a second group of xenobiotics which was rated by the EDIETA workshop as a confirmed case for ED. The workshop report [2] summarizes not less than 47 references, mainly laboratory studies but also a number of field investigations with a focus of IGR effects on nontarget species. Most of these studies were conducted with terrestrial species, while possible effects on aquatic insects, e.g., from IGR spray drift or run-off during or after agricultural application received only a little attention.

Ecdysteroid receptor agonists such as tebufenozide, methoxyfenozide, or RH 5849 induce symptoms of hyperecdysonism in terrestrial insects, delayed postembryonic development, and nymphal-adult intermediates. Tebufenozide interacts with the ecdysone receptor complex. Paradoxically, even though most insects and other arthropods use ecdysone as a molting hormone, tebufenozide is selectively toxic to the Lepidoptera [30]. Ecdysteroid antagonists such as azadirachtin or KK-42 prevent normal diapause induction, and induce an early termination of diapause or a precocious metamorphosis, while juvenile hormone (JH) analogs, such as fenoxycarb, methoprene, and pyriproxyfen, interfere with egg hatching, larval development, larval-pupal molts, and ecdysis and reduce the fertility and longevity of exposed specimens.

Despite these marked effects in insects, there are major terrestrial invertebrate classes for which there are no apparent data on the impact of IGRs, e.g., earthworms and mollusks. There are also no reported incidents of effects on terrestrial invertebrates of nonpesticidal endocrine disruptors that are present in the environment.

Occasionally, chitin synthesis inhibitors (CSIs), such as diflubenzuron, other benzylphenyl ureas, and related compounds [31] have been misreported as EASs. Although they interfere with molting, an endocrine-regulated process in arthropods, the mechanism of CSI action is purely nonendocrine. CSIs inhibit one of the steps of chitin synthesis selectively. Because this synthesis usually takes place at or during the time of molting, CSIs cause death during the molt, resembling the effects of endocrine-mimicking IGRs, albeit not via an endocrine pathway.

Further cases of endocrine disruption in aquatic and terrestrial invertebrates

Next to the effects of TBT in prosobranchs and of IGRs in insects, a number of further laboratory and field studies have been reported, where compounds exhibited effects on endocrine-regulated processes in marine, freshwater, and terrestrial invertebrates. For the period until late 1998, the EDIETA report [2] summarizes not less than 56 studies in which ED may have occurred although nonendocrine mechanisms are also possible for the observed effects (Table 2).

Table 2 Synopsis of studies, reported in [2], in which ED may have occurred, although nonendocrine mechanisms are also possible for the observed effects (f, freshwater; m, marine; t, terrestrial; tot, total).

Phylum	Contaminants	Effects	Cited studies
Mollusca	Cd, DDT, MCPA (2-methyl-4-chlorophenoxyacetic acid)	Fecundity alterations; spawning stimulation; inhibition of gonadial development	f: 1; m: 1; t: - **tot: 2**
Annelida	Volatile organic compounds from crude oil	Induction of spawning	f: -; m: 1; t: - **tot: 1**
Crustacea	Atrazine, Cd, DES (diethylstilbestrol), diazinon, dieldrine, diurone, endosulfane, Hg, lindane, methoprene, naphthalene, NP (nonylphenol), OP (octylphenol), PB, PCBs, PCP(pentachlorophenole), phthalate esters, Se, sewage outfall, simazine, TBT, Zn	Elevated ecdysteroids levels; interference with molt, growth, energy metabolism and fecundity; delayed maturity; mortality; increased intermolt duration; inhibition of larval development; abnormal coloration; disrupted testosterone metabolism; induction of cyprid major protein; intersexuality; retarded limb regeneration and limb abnormalities	f: 24; m: 11; t: - **tot: 35**
Insecta	NP, phthalate esters, metal containing effluents, PAHs, tannery and paper mill effluent, various organic and inorganic pollutants, Zn	Mortality; various mouthpart deformities; other pathomorphological changes; -interference with molt cycle and frequency	f: 6; m: -; t: 2 **tot: 8**
Echinodermata	Cd, estradiol, estrone, PCBs, PCP, Zn	Abnormal embryogenesis and development; low fertilization success; reduced ovarian growth; elevated or reduced steroids levels; increased oocyte growth	f: -; m: 10; t: - **tot: 10**

The crustaceans represent the systematic group providing the majority of cases of suspected ED shown in Table 2. While the examples for the aquatic environment are almost balanced between freshwater and marine species, only two studies report comparable effects in terrestrial arthropods.

Since the publication of the EDIETA report, not less than 25 new cases for ED in invertebrates have been published. These are summarized in Table 3 but without the intention of providing a complete list, because unpublished data from currently ongoing research programs for the identification of new suitable test species and sensitive endpoints could not be considered.

Table 3 Additional laboratory and field studies with evidence for ED, which have been published since the EDIETA workshop [2]. Abbreviations: LCT, life cycle test.

Phylum	Species (life stage)	Contaminant (conc. range)	Effects observed	Lab/ field	Ref.
Porifera	*Heteromyenia* sp., *Eunapius fragilis* (gemmulae - adults)	Bisphenol A (BPA) (0.16–160 mg/l), ethylbenzene (0.03–3 mg/l), NP (0.022–22 mg/l)	Reduced growth rates; developmental abnormalities with malformed water vascular systems (especially at lower concentrations)	Lab	[46]
Rotatoria	*Brachionus calyciflorus* (adults)	Flutamide, NP, testosterone (1–50 µg/l for all)	Fertilization of females inhibited	Lab	[55]
Mollusca	*Marisa cornuarietis* (adults, LCT), *Potamopyrgus antipodarum, Nucella lapillus Nassarius reticulatus* (adults)	BPA (0.05–100 µg/l), OP (1–100 µg/l)	Induction of "superfemales": stimulation of egg/embryo and spawning mass production, additional female sex organs (*Marisa*), oviduct malformations and increased female mortality (*Marisa*)	Lab	[32,52]
	M. cornuarietis, N. lapillus, N. reticulatus (adults)	TPT (0.005–0.5 µg as Sn/l)	Imposex development (*Marisa*); reduction of female sex glands; impairment of spermatogenesis and oogenesis	Lab	[29]
	M. cornuarietis, N. lapillus, N. reticulatus (adults)	Cyproterone acetate (1.25 mg/l), vinclozolin (0.03–1 µg/l)	Suppression of imposex development from TBT; reduction of male sex glands and penis; advancement of sexual repose	Lab	[33]
	Mya arenaria (adults)	Estradiol, NP, PCP, contaminated natural sea water	Induction of vitellogenin-like proteins by test chemicals; reduced levels in the field (due to (anti-)estrogens?)	Lab & field	[37]
	M. arenaria (adults)	Unknown (field survey) NP	Delayed gametogenesis; dysfunction of vitellogenesis	Field	[41]
	Crassostrea gigas (larvae)	(0.1–10000 µg/l)	Delayed development to D-stage; reduced survival; malformed D-larvae	Lab	[50]
Annelida	*Dinophilus gyrociliatus* (adults)	NP	Stimulation of egg production; reduced egg viability	Lab	[38]
Crustacea	*Daphnia magna* (LCT)	Ponasterone A (3.4–27 nM)	Reduced fecundity in F_2 generation; incomplete ecdysis; premature death	Lab	[34]
	D. magna (adults)	Cyproterone acetate (0.3–5 µM)	Molt-independent growth reduction; reduced offspring numbers	Lab	[48]
	D. magna (adults)	Androstenedione (6.2–25 µM), DES (0.75–3 µM), methoprene (0.08–0.32 µM)	Stimulation of abdominal process by DES and methoprene and of development of first antennae by andostenedione	Lab	[53]
	Gammarus pulex (adults)	Unknown (field survey)	Abnormal oocyte structure during vitellogenesis; reduced length and male/female size differences	Field	[42]

(*continues on next page*)

Table 3 (*Continued*).

Phylum	Species (life stage)	Contaminant (conc. range)	Effects observed	Lab/ field	Ref.
	Corophium volutator (juv. – adults)	NP (>10 μg/l)	Mortality; reduced growth; increased female fertility and antennae length	Lab	[39]
	Balanus amphitrite (larval stages)	NP (0.01–1 μg/l) estradiol (1 μg/l)	Induction of CMP (cypris major protein)	Lab	[35]
	Elminius modestus (larval stages)	NP (0.01–10 μg/l) estradiol (10 μg/l)	Altered timing of larval development; reduced growth	Lab	[36]
	Palaemonetes pugio (adults)	Pyrene	Induction of vitellin	Lab	[51]
Insecta	*Drosophila melanogaster* (cell line)	80 different test chemicals	Receptor-mediated ecdysteroid response as screening tool	Lab	[40]
	Lacanobia oleracea (larvae to adults)	Estradiol, methyltestosterone, thyroxine (1 mg/kg dietary dose)	Increased length and reduced weight of older larval stages; deformed pupae, reduced fecundity and egg vibility (only methyltestosterone)	Lab	[47]
	Chironomus riparius (LCT)	Tebufenozide (1–100 μg/l)	Mortality during pupation and emergence with sex-related differences	Lab	[43]
	C. riparius (LCT)	TBT (0.01–5 μg as Sn/l)	Ecdysteroid synthesis (males: increased females: decreased); development (males: faster, females: slower)	Lab	[44]
	C. riparius (LCT)	BPA (1–3000 μg/l), NP (1.9–2000 μg/l)	Alteration of vitellogenin/vitellin production in males	Lab	[45]
	C. riparius (larvae)	NP (10–100 μg/l)	Increased frequency of mouthpart deformities	Lab	[49]
Tunicata	*Ciona intestinalis* (larvae)	TBT (up to 10–5 M)	Reduced thyroxine production; block of metamorphosis	Lab	[54]

It is not possible to present and discuss the publications on ED in invertebrates, summarized in Table 3, in detail, but two general observations should be emphasized: (1) Endocrine-mediated effects of xenobiotics in terrestrial species still constitute less than 10 % of all reports, and (2) although single studies of ED in formerly ignored taxa are available now (e.g., for Porifera, Rotatoria, Tunicata), the overwhelming majority of publications focus on mollusks, crustaceans, and insects, thus continuing the main tendencies in the pre-1999 literature.

CONCLUSIONS ABOUT THE SCALE AND IMPLICATIONS OF EFFECTS

Despite the fact that an increasing number of research projects are investigating ED in invertebrates worldwide and that the published examples of potential effects of EASs in these groups has increased by almost 50 % in the last three years since the EDIETA workshop, the main conclusions are still valid [2]. With the exception of TBT effects in mollusks, which have been associated with a locally severe impact at the community level, and IGRs in terrestrial insects, there are only a few field examples of ED in invertebrates. Nevertheless, it is suspected that there are many more examples for ED affecting invertebrate populations and communities, though still undetected. This assumption is supported by the following indications: (1) The basic mechanisms of chemical signaling systems exhibit a considerable degree of conservatism throughout the animal kingdom [1] so that invertebrate endocrine function should be affected by the same or similar compounds as those of vertebrates [2,5]. (2) For purposes of pest control, a number of highly effective EASs have been intentionally developed to interfere with the hormonal systems of insects. There is no reason to suppose that such endocrine-mediating properties are unique for the IGRs, but rather reflect the fact that much less research has been undertaken for other invertebrate groups than insects. (3) ED in invertebrates has found far less attention than in vertebrates, probably because their hormonal systems are poorly understood, favoring investigations with vertebrates and especially fish as systematic groups for ecotoxicological research and routine analyses many scientists feel familiar with.

RESEARCH NEEDS

Our ignorance of invertebrate endocrinology is one of the main reasons for the unsatisfactory progress that has been made regarding ED in invertebrates. A further important point is that ED in vertebrates has attracted a higher degree of public and even scientific awareness, which is also reflected by funding resources and other economic circumstances making the general conditions for research with invertebrates less favorable. Nevertheless, the consideration of invertebrates in such research programs potentially offers a wealth of knowledge in understanding comparative and ecological aspects of ED [2]. For these reasons, invertebrates should have a high priority for further research, especially for the development of testing and monitoring techniques:

- More basic research on invertebrate endocrinology is needed, especially for groups that were almost totally neglected in the past (i.e., not considered in Table 1).
- Hormone receptors of invertebrates should be identified, cloned, and characterized, facilitating the identification of receptors that are shared by different groups. This would help to develop receptor-binding assays and other in vitro systems as a screening tool.
- Endocrine control of toxicological endpoints in tests should be characterized in more depth so that these endpoints can be used as valid measures for ED.
- New invertebrate tests with endocrine-regulated endpoints have to be developed or existing protocols amended. This will require a broad initiative with a variety of invertebrate assays. In a second step, these tests will have to be validated, including the use of reference compounds and positive controls known to have endocrine-disrupting properties in the systematic groups under investigation.
- It is important to emphasize that several sentinel species will be required since the endocrinology of invertebrates differs widely among taxa. Therefore, representatives of each phylum are needed. For the freshwater environment test and monitoring species should at least include Annelida, Mollusca, Crustacea, and Insecta, in the marine environment Coelenterata, Annelida, Mollusca, Crustacea, and Echinodermata, and an even more diverse list of taxa for terrestrial ecosystems.
- The current knowledge of valid endocrine-mediated endpoints in invertebrates is too incomplete to design specific monitoring programs for biological effects of EASs, perhaps with the exception of androgenic and estrogenic compounds and their effects in prosobranch snails. Nevertheless, it is obvious that carefully targeted monitoring programs are needed because effects in invertebrates are probably widespread but undetected.

REFERENCES

1. J. A. McLachlan. *Endocrine Rev.* **22**, 319–341 (2001).
2. P. L. deFur, M. Crane, C. Ingersoll, L. Tattersfield (Eds.). *Endocrine Disruption in Invertebrates: Endocrinology, Testing, and Assessment, Proceedings of the Workshops on Endocrine Disruption in Invertebrates, 12–15 December 1998, Noordwijkerhout, The Netherlands*, SETAC Press, Pensacola, FL (1999).
3. R. Lafont. *Ecotoxicology* **9**, 41–57 (2000).
4. G. A. Hines, S. A. Watts, S. A. Sower, C. W. Walker. *Gen. Comp. Endocrin.* **87**, 451–460 (1992).
5. L. C. V. Pinder, T. G. Pottinger, Z. Billinghurst, M. H. Depledge. *Endocrine Function in Aquatic Invertebrates and Evidence for Disruption by Environmental Pollutants*, Environment Agency, London. R&D Technical Report E67 (1999).

6. J. Hoffmann and M. Porchet (Eds.). *Biosynthesis, Metabolism and Mode of Action of Invertebrate Hormones*, Springer Verlag, New York (1984).
7. A. Dorn (Ed.). *Reproductive Biology of Invertebrates*, Vol. 10, Part A & B, Wiley, New York (2000f).
8. P. Matthiessen and P. E. Gibbs. *Environ. Toxicol. Chem.* **17**, 37–43 (1998).
9. A. Gorbman and K. Davey. In *Neureal and Integrative Animal Physiology*, C. L. Prosser (Ed.), pp. 693–754, Wiley-Liss, New York (1991).
10. G. W. Bryan and P. E. Gibbs. In *Metal Ecotoxicology: Concepts and Applications*, M. C. Newman and A. W. McIntosh (Eds.), pp. 323–361, Lewis, Ann Arbor (1991).
11. C. Alzieu, Y. Thibaud, M. Héral, B. Boutier. *Rev. Trav. Inst. Péch. Mar.* **44**, 306–348 (1980).
12. B. S. Smith. *Proc. Malacol. Soc. London* **39**, 377 (1971).
13. P. Fioroni, J. Oehlmann, E. Stroben. *Zool. Anz.* **226**, 1–26 (1991).
14. P. E. Gibbs, G. W. Bryan, P. L. Pascoe, G. R. Burt. *J. Mar. Biol. Assn. U.K.* **67**, 507–523 (1987).
15. J. Oehlmann, P. Fioroni, E. Stroben, B. Markert. *Sci. Total Environ.* **188**, 205–223 (1996).
16. J. Oehlmann, E. Stroben, U. Schulte-Oehlmann, B. Bauer, P. Fioroni, B. Markert. *Fres. J. Anal. Chem.* **354**, 540–545 (1996).
17. B. Bauer, P. Fioroni, I. Ide, S. Liebe, J. Oehlmann, E. Stroben, B. Watermann. *Hydrobiologia* **309**, 15–27 (1995).
18. P. Matthiessen, R. Waldock, J. E. Thain, M. E. Waite, S. Scrope-Howe. *Ecotoxicol. Environ. Saf.* **30**, 180–194 (1995).
19. Oslo and Paris Commissions (Ed.). *Agenda Item 9 of the Environmental Assessment and Monitoring Committee (ASMO).* OSPAR, Vila Franca (= ASMO 96/9/8-E) (1996).
20. R. Waldock, H. L. Rees, P. Matthiessen, M. A. Pendle. *J. Mar. Biol. Ass. U.K.* **79**, 225–232 (1999).
21. U. Schulte-Oehlmann, C. Bettin, P. Fioroni, J. Oehlmann, E. Stroben. *Ecotoxicology* **4**, 372–384 (1995).
22. U. Schulte-Oehlmann, M. Duft, M. Tillmann, B. Markert, B. Stachel, J. Oehlmann. *Biological effects monitoring with River Elbe sediments using* Potamopyrgus antipodarum *and* Hinia *(*Nassarius*)* reticulata *(Gastropoda: Prosobranchia)* (in German). ARGE Elbe, Hamburg (2001).
23. C. Bettin, J. Oehlmann, E. Stroben. *Helgoländer Meeresunters.* **50**, 299–317 (1996).
24. Y. Morcillo, A. Albalat, C. Porte. *Environ. Toxicol. Chem.* **18**, 1203–1208 (1999).
25. C. Féral and S. Le Gall. *C. R. Acad. Sc. Paris.* **295**, Série III, 627–630 (1982).
26. E. Oberdörster and A. O. Cheek. *Environ. Toxicol. Chem.* **20**, 23–36 (2001).
27. J. Oehlmann and U. Schulte-Oehlmann. In *Proceedings of the 21st Conference of the European Comparative Endocrinologists*, R. Keller (Ed.), pp. 63–69, Monduzzi Editore, Bologna (2002).
28. T. Horiguchi, H. Shiraishi, M. Shimizu, M. Morita. *Environ. Pollut.* **95**, 85–91 (1997).
29. U. Schulte-Oehlmann, B. Watermann, M. Tillmann, S. Scherf, B. Markert, J. Oehlmann. *Ecotoxicology* **9**, 399–412 (2000).
30. J. Oehlmann, U. Schulte-Oehlmann, M. Tillmann, B. Markert. *Ecotoxicology* **9**, 383–397 (2000).
31. M. Tillmann, U. Schulte-Oehlmann, M. Duft, B. Markert, J. Oehlmann. *Ecotoxicology* **10**, 373–388 (2001).
32. T. S. Dhadialla, R. R. Carlson, D. P. Le. *Ann. Rev. Entomol.* **43**, 545–569 (1998).
33. I. Ishaaya. In *Molecular Mechanism of Insecticide Resistance: Diversity Amongst Insects*, C. A. Mullin and J. G. Scott (Eds.). pp. 231–246, American Chemical Society, Washington, DC (1992).
34. W. S. Baldwin, R. Bailey, K. E. Long, S. Klaine. *Environ. Toxicol. Chem.* **20**, 1564–1569 (2001).
35. Z. Billinghurst, A. S. Clare, K. Matsumura, M. H. Depledge. *Aquat. Toxicol.* **47**, 203–212 (2000).
36. Z. Billinghurst, A. S. Clare, M. H. Depledge. *J. Exp. Mar. Biol. Ecol.* **257**, 255–268 (2001).
37. C. Blaise, F. Gagne, J. Pellerin, P. D. Hansen. *Environ. Toxicol.* **14**, 455–465 (1999).
38. M. H. Depledge and Z. Billinghurst. *Mar. Pollut. Bull.* **39**, 32–38 (1999).
39. R. J. Brown, M. Conradi, M. H. Depledge. *Sci. Total Environ.* **233**, 77–88 (1999).

40. L. Dinan, P. Bourne, P. Whiting, T. S. Dhadialla, T. H. Hutchinson. *Environ. Toxicol. Chem.* **20**, 2038–2046 (2001).
41. S. Gauthier-Clere, J. Pellerin, C. Blaise, F. Gagne. *Comp. Biochem. Physiol. C.* **131**, 457–467 (2002).
42. M. Y. Gross, D. S. Maycock, M. C. Thorndyke, D. Morritt, M. Crane. *Environ. Toxicol. Chem.* **20**, 1792–1797 (2001).
43. T. Hahn, M. Liess, R. Schulz. *Ecotox. Environ. Saf.* **49**, 171–178 (2001).
44. T. Hahn and R. Schulz. *Environ. Toxicol. Chem.* **21**, 1052–1057 (2002).
45. T. Hahn, K. Schenk, R. Schulz. *Environ. Pollut.* **120**, 525–528 (2002).
46. M. Hill, C. Stabile, L. K. Steffen, A. Hill. *Environ. Pollut.* **117**, 295–300 (2002).
47. A. E. Kirkbride-Smith, H. A. Bell, J. P. Edwards. *Environ. Toxicol. Chem.* **20**, 1838–1845 (2001).
48. G. A. LeBlanc and J. B. McLachlan. *Environ. Toxicol. Chem.* **18**, 1450–1455 (1999).
49. G. Meregalli, L. Pluymers, F. Ollevier. *Environ. Pollut.* **111**, 241–246 (2001).
50. H. E. Nice, M. C. Throndyke, D. Morritt, S. Steele, M. Crane. *Mar. Pollut. Bull.* **40**, 491–496 (2000).
51. E. Oberdörster, C. D. Rice, L. K. Irwin. *Comp. Biochem. Physiol. C.* **127**, 199–207 (2000).
52. U. Schulte-Oehlmann, M. Tillmann, D. Casey, M. Duft, B. Markert, J. Oehlmann. *UWSF – Z. Umweltchem. Ökotox.* **13**, 319–333 (2001).
53. A. W. Olmstead and G. A. LeBlanc. *Environ. Toxicol. Chem.* **19**, 2107–2113 (2000).
54. E. Patricolo, C. Mansueto, P. D'Agati, L. Pellerito. *Appl. Organomet. Chem.* **15**, 916–923 (2001).
55. B. L. Preston, T. W. Snell, T. L. Robertson, B. J. Dingmann. *Environ. Toxicol. Chem.* **19**, 2923–2928 (2000).

Pure Appl. Chem., Vol. 75, Nos. 11–12, pp. 2219–2234, 2003.

Topic 4.3
Endocrine disruption in wild freshwater fish*

Susan Jobling[1,‡] and Charles R. Tyler[2]

[1]Department of Biological Sciences, Brunel University, Uxbridge, Middlesex UB8 3PH, UK; [2]School of Biological Sciences, Hatherly Laboratory, Exeter University, Exeter, Devon, EX4 4PS, UK

Abstract: Endocrine disruption has been reported in freshwater fish populations around the world. This phenomenon ranges from subtle changes in the physiology and sexual behavior of fish to permanently altered sexual differentiation and impairment of fertility. Despite widespread reports of endocrine disruption in fish (and this is well characterized at the individual level), few studies have demonstrated population-level consequences as a result of exposure to endocrine-disrupting chemicals (EDCs). An exception to this is in Lake Ontario Lake trout where precipitous declines in the population have been linked with periods of high exposure to organochlorine chemicals (known EDCs). Recently, it has been established that roach (*Rutilus rutilus*) exposed to treated sewage effluent (that contains complex mixtures of EDCs) in UK rivers, have a reduced reproductive capacity. This, in turn, *may* have population-level consequences.

Evidence for a link between exposure to effluents from kraft mill (BKME) and sewage treatment works (STW) and altered reproductive function in freshwater fish is compelling. In most cases, however, a causal link between a specific chemical and a physiological effect has not been established. Indeed, identifying specific chemical(s) responsible for adverse effects observed in the wild is difficult, given that tens of thousands of man-made chemicals enter the aquatic environment and that mixtures of chemicals can have combination (e.g., additive) effects. Some EDCs are known to act at a number of different body targets to affect a variety of physiological processes, further complicating the identification of the causative agent(s).

Endocrine disruption appears to be particularly widespread in freshwater fish populations. There is little evidence, however, to suggest fish are more susceptible to EDCs relative to other wildlife. Notwithstanding this, there are some features of the endocrine physiology of fish that may be particularly susceptible to the effects of EDCs, including the processes of sex-determination and smoltification (in salmonids). Furthermore, their aquatic existence means that fish can be bathed constantly in a solution containing pollutants. In addition, uptake of chemicals readily occurs via the gills and skin, as well as via the diet (the major exposure route for most EDCs in terrestrial animals). The exposure of fish early life stages to the cocktail of EDCs present in some aquatic environments may be of particular concern, given that this is an especially vulnerable period in their development.

The challenge, from the point of view of ecological risk assessment, is to determine effects of EDCs on freshwater fish populations and freshwater ecosystems. In order to meet this challenge, high-quality data are required on the population biology of freshwater fish, on the effects of EDCs on their various life history characteristics, and comprehensive and ap-

*Report from a SCOPE/IUPAC project: Implication of Endocrine Active Substances for Human and Wildlife (J. Miyamoto and J. Burger, editors). Other reports are published in this issue, *Pure Appl. Chem.* **75**, 1617–2615 (2003).

‡Corresponding author

propriate population models. Basic information on the population biology of most species of wild freshwater fish is, however, extremely limited, and needs significant improvement for use in deriving a sound understanding of how EDCs affect fish population sustainability. Notwithstanding this, we need to start to undertake possible/probable predictions of population level effects of EDCs using data derived from the effects found in individual fish. Furthermore, information on the geographical extent of endocrine disruption in freshwater fish is vital for understanding the impact of EDCs in fish populations. This can be derived using published statistical associations between endocrine disruption in individual fish and pollutant concentration in receiving waters. Simplistic population models, based on the effects of EDCs on the reproductive success of individual fish can also used to model the likely population responses to EDCs. Wherever there is sufficient evidence for endocrine disruption in freshwater fish and the need for remediation has been established, then there is a need to focus on how these problems can be alleviated. Where industrial chemicals are identified as causative agents, a practical program of tighter regulation for their discharge and/or a switch to alternative chemicals (which do not act as EDCs) is needed. There are recent examples where such strategies have been adopted, and these have been successful in reducing the impacts of EDCs from point source discharges on freshwater fish. Where EDCs are of natural origin (e.g., sex steroid hormones from human and animal waste), however, remediation is a more difficult task. Regulation of the release of these chemicals can probably be achieved only by improvements in treatment processes and/or the implementation of systems that specifically remove and degrade them before their discharge into the aquatic environment.

INTRODUCTION

Endocrine disruption has been reported in freshwater fish populations in various parts of the world [reviewed in 1,2]. This phenomenon ranges from subtle changes in the physiology and sexual behavior of fish, to permanently altered sexual differentiation and impairment of fertility. Most of the data comes from studies in Europe and America, although evidence for endocrine disruption in freshwater fish has also been reported in Australia [3] and Japan [4]. Biological effects in wild freshwater fish that have been attributed to the effects of endocrine disruptors include the inappropriate production of the blood protein vitellogenin (VTG; the female-specific and estrogen-dependent egg yolk protein precursor) in male and juvenile fish, inhibited ovarian or testicular development, abnormal blood steroid concentrations, intersexuality and/or masculinization or feminization of the internal or external genitalia, impaired reproductive output, precocious male and/or female maturation, increased ovarian atresia (in female fish), reduced spawning success, reduced hatching success and/or larval survival, altered growth and development (thyroid hormone-like effects) and alterations in early development (altered rate or pattern) [2]. These effects may arise due to disruption of a range of endocrine-mediated mechanisms (including receptor-mediated processes, and/or interference with steroid metabolism and/or excretion), although nonendocrine toxicity could also explain some of these effects. Overall, current scientific evidence strongly suggests that certain effects observed in freshwater fish can be attributed to cocktails of chemicals that mimic and/or disrupt hormone function/balance. In most cases, however, the evidence of a causal link between a specific physiological disruptor and a specific effect is weak, largely due to the fact that freshwater fish and, indeed, all other wildlife, are exposed to a wide range of chemicals, that act at a number of different body targets, to affect a variety of physiological processes. Sewage treatment works (STWs), for example, (which often receive domestic, industrial and/or agricultural waste) release a complex (and ill-defined) mixture of natural and synthetic chemicals into the aquatic environment, following their partial or complete biodegradation during the treatment process. It is estimated that 60 000 man-made chemicals are in routine use worldwide and most of these enter the aquatic environment [5]. Identifying specific chemical(s) responsible for adverse effects observed in the wild is, thus, difficult and requires extensive laboratory studies to support the hypotheses drawn from field studies. Moreover, very recent studies investigating the interactive effects of mixtures of estrogenic chemicals in fish, using vitellogenin induction as an endpoint, have shown that combinations of steroid estrogens, alkylphenolic chemicals and a pesticide (methoxychlor) are additive in their effect [6]. This highlights the fact that even chemicals that are have slight effects on the endocrine system should be taken into consideration when assessing the effects of chemical mixtures in freshwater fish. A weak link in establishing whether observed adverse effects in freshwater fish are caused by exposure to EDCs is the lack of data documenting what freshwater fish are actually exposed to and what they take up into their bodies. Moreover, there is often a large discrepancy between the relatively high levels of pollutants generally used in laboratory studies and the low levels of these pollutants that actually occur in the aquatic environment. Exposures of fish to environmentally relevant concentrations of EDCs (and at the relevant life stages) are essential to adequately evaluate exposure/response relationships in field studies and produce credible risk assessments.

CURRENT EVIDENCE FOR ENDOCRINE DISRUPTION IN FRESHWATER FISH

Although there is a considerable amount of evidence for endocrine disruption in wild freshwater fish, only in a very few cases has a causal link between the presence of EDCs in freshwaters and altered endocrine function in exposed fish populations been demonstrated. In order to determine causality between an EDC and a particular perturbation, clearly, a relationship between exposure to the putative stressor and the effect of concern needs to be firmly established (e.g., decline in the population or reduced fertility). For a chemical to be designated an endocrine disrupter, exposure to the stressor has to result in an endocrine-mediated event (and at the relevant exposure concentration that occurs in the environment) that ultimately results in an effect of concern. In the following section of the review, documented examples of endocrine disruption in wild freshwater fish are described. Those examples for which there is considerable evidence for a link between exposure and effect are described first, followed by cases where the evidence is less convincing and/or where further research is much needed in order to provide a definitive association.

Reproductive abnormalities in freshwater fish living downstream of pulp- and paper-mill effluents

Over the last 10 years, a number of species of freshwater fish in Canada (white sucker, *Catostomus commersoni*; longnose sucker, *Catostomus catostomus*; lake whitefish, *Coregonus clupea formis* [7–15]) and Europe (perch, *Perca fluviatilis*; roach, *Rutilus rutilus*; [16,17]) living downstream of pulp- and paper-mill effluents have been found to exhibit an array of altered features in their reproductive development, including reductions in gonadal growth, inhibition of spermatogenesis, depressed sex steroids, reduced pituitary hormone concentrations, and delayed sexual maturity. In the studies on perch (but not for the suckers) viability of the developing larvae was also affected [18]. Lowered egg production and delayed reproduction have also been induced in fathead minnows in life-long exposures to bleached kraft mill effluents (BKMEs) [19]. Furthermore, the endocrine changes seen in wild fish are less severe during periods of reduced effluent discharge [20] and decrease with increasing distance from the effluent outfalls into the rivers. There is, thus, very strong evidence to suggest that something in the BKME is causing the adverse effects seen. The causative agents responsible for these reproductive effects in fish in Canada and Europe have, however, not been identified [21], although in a very recent in vivo study, using a toxicity identification and evaluation approach, Hewitt et al. [22] were able to provide the first evidence that at least one of the effects (the depression in steroid hormone concentrations) seen in wild fish in the vicinity of pulp mills may be due to products of the degradation of lignin. The authors showed that these chemicals were present in active fractions of the effluent that caused depressions in serum testosterone concentrations in mummichogs both in vitro and in vivo. Moreover, although not proven, other studies have suggested that the reproductive effects may (at least in part) be mediated through disruption of the process of steroidogenesis, by affecting the availability of cholesterol and pregnenolone and thus impairing steroid production by the gonads [23,24]. Still other in vitro studies suggest that mixtures of both estrogenic (e.g., β-sitosterol, lignans, stilbenes, and resin acids [25,26]) and androgenic chemicals (e.g., stigmastanol and a β-sitosterol degradation products [27]), together with Ah-receptor agonists (e.g., polychlorinated dibenzofurans and thianthrenes, dibenzothiphenes, and diphenyl sulfides), are found in these effluents, and these studies are supported by in vivo studies that show that white suckers, living in the vicinity of BKME discharges, rapidly accumulate chemicals that bind to the estrogen receptor, androgen receptor, and sex steroid binding protein [28]. Another study showed that during the spawning migration of white sucker in Jackfish Bay in Canada, returning fish were found to have altered pituitary function, as determined by depressed levels of luteinizing hormone (LH) in males and females compared with control fish from a reference location [29]. When taken together, the evidence shows clearly that the endocrine disrupters within BKME act at many targets in the hypothalamic-pituitary-gonad axis. Although it has not (thus far) been possible to link endocrine disruption (leading to deleterious effects on reproduction and development) in these various species of fish to a specific chemical or group of chemicals, it is clear that the endocrine effects are clearly linked to the constituents of pulp-mill effluents.

Interestingly, the multiplicity of androgenic-, estrogenic-, and steroidogenesis-inhibiting chemicals in paper-mill effluents reported for BKMEs in Canada has not been reported for BKMEs in Florida, USA. Instead, in Florida, only androgenic effects have been identified. In these studies, development of a male gonopodium was observed in female mosquito fish exposed to BKME (an androgenic effect [30,31]), but no apparent feminizing effects were seen in males. A recent in vitro study by Parks et al. [32] determined that the pulp-mill effluent from a Florida mill exhibited androgenic activity (determined by transcriptional activity of the androgen receptor) at levels sufficient to account for the masculinization of the female mosquitofish. It is not yet known whether the differences in effects of BKME on fish in Florida and Canada are due to differences in species sensitivities, or to different substances discharged into the BKME in Canada compared with that in Florida. Further characterization of the effluents is needed to more fully understand causation. The ecological significance of the physiological effects of BKME are not known, but could be hypothesized to result in the gradual impairment and eventual loss of reproductive function after continued BKME exposure. These seemingly intuitive population-level predictions have not, however, been observed directly in any wild population of fish exposed to BKME. Indeed, some recent evidence suggests the contrary, LeBlanc et al. [33], for example, recently observed a reduction in the intensity and duration of the spawning period in *Fundulus heteroclitus* exposed to BKME in the Mirichami Estuary, New Brunswick, Canada, but they also reported a simultaneous marked *increase* in reproductive investment and increased fecundity in these individuals.

Reproductive abnormalities in freshwater fish living downstream of sewage treatment works discharges

There is considerable (and increasing) evidence for endocrine disruption in freshwater fish populations living in stretches of river downstream of treated sewage effluent discharges in Europe [34–40], Canada and America [41–43] as well as more recent evidence of endocrine disruption in riverine carp in Osaka in Japan [4]. In the original work on freshwater fish, conducted in the United Kingdom, it was established that effluents from treated sewage effluents were estrogenic, inducing the production of vitellogenin, in male fish [44]. Vitellogenin is normally synthesized by the liver in female oviparous (egg-laying) vertebrates in response to estrogen and is sequestered by developing oocytes and stored as yolk to act as a nutrient reserve for the subsequent development of the embryo [45]. The production of VTG, therefore, is usually restricted to females. Male fish however, do contain the VTG gene(s), and exposure to both natural and synthetic estrogens can trigger its expression, resulting in the secretion of VTG in the blood plasma [46]. Vitellogenin is now one of the most widely used biomarkers for exposure to estrogen(s) in fish in freshwaters and it has been detected in the blood of both male and juvenile fish in rivers, lakes and streams contaminated by effluents from STWs and/or mixtures of estrogens [reviewed in 47]. Although almost all effluents tested in the United Kingdom have been shown to be estrogenic, causing induction of VTG in exposed male fish, there are some STW effluents in the United States that do not appear to be estrogenic to fish (they do not induce VTG [48,49]), probably due to the large dilution that occurs when the effluent reaches the receiving river and/or to the more extensive sewage treatment processes that are in place at these sites.

In addition to VTG production, exposure to treated sewage effluents has also been associated with deleterious effects on gonad differentiation and development [3,4,34–43,50] in various species of fish and with the abnormal development (feminization) of secondary sexual characteristics in male mosquitofish (*Gambusia affinis*) in Australia [3]. The most thoroughly studied effects are concerned with the widespread incidence of intersex reported in some species of freshwater fish in the United Kingdom, parts of Continental Europe, and the United States. Freshwater fish species in which an occurrence of intersex has been reported and deemed to be abnormal include the roach (*R. rutilus* [37]), bream (*Abramis abramis* [34,35]), the chub [38], gudgeon (*Gobio gobio* [36]), the barbel (*Barbus plebejus* [39]), the perch (*Perca fluviatilis*), the stickleback (*Gasterosteus aculeatus* [40]), and the shovelnose sturgeon (*Scaphirhynchus platyorynchus* [41]). Intersex as a consequence to exposure to effluent has been most intensively studied in the roach, a cyprinid fish common throughout lowland rivers in the United Kingdom and Europe. At some river sites downstream from large STW discharges in the United Kingdom, all of the "male" roach population has been reported to be intersex [37]. Intersex roach often have both male and female reproductive ducts, and many also have female germ cells (oocytes) within a predominantly male "testis". The number, pattern, and developmental stage of oocytes within testicular tissue in intersex roach vary greatly; the condition ranges from the presence of single primary oocytes scattered randomly throughout testicular tissue in a mosaic fashion, to a condition in the more severely feminized fish, where large areas of ovarian tissue occur that are clearly separated from testicular tissue [50]. Intersex roach also often have an altered endocrine status (altered plasma sex steroid hormone concentrations), and an elevated concentration of plasma VTG relative to normal male fish [51] and gonadal growth is often inhibited in severely intersex roach. More recent studies suggest that intersex roach (*R. rutilus*) also have impaired fertility relative to normal male fish from reference sites. Small numbers of wild roach in UK rivers were found that could not produce any gametes at all due to the presence of severely disrupted gonadal ducts [52]. Fertilization and hatchability studies have further shown that intersex roach (even with a low level of gonadal disruption—"mildly intersex") are compromised in their reproductive capacity and produce fewer offspring than fish from uncontaminated sites. In these studies, an inverse correlation was demonstrated between reproductive performance (defined as the ability to produce viable offspring) in intersex roach and severity of gonadal intersex. This, in turn suggests that the intersex condition is quite likely to have population level consequences, although further studies on wild populations are necessary to confirm or refute this.

In contrast to the effects observed in male roach, effects in female roach living in rivers contaminated by treated sewage effluents in UK rivers were less obvious [52]: There was a higher incidence of oocyte atresia and a slight, but statistically significant, lower fecundity in effluent-exposed fish compared with females from the reference sites. Interestingly, at some river sites, small proportions (up to 14 %) of the adult female fish (aged between 3 and 7 years) were sexually immature or sexually indifferent, and, although not proven, it is possible that these effects are also due to endocrine disruption.

There is substantive evidence (principally from lab-based studies, see below) to support that hypothesis that gonadal disruption in wild freshwater fish, inhabiting rivers that receive treated sewage effluents, is caused by estrogenic substances contained within these effluents. Moreover, the statistical associations between the various gonadal abnormalities that occur in wild freshwater fish and plasma VTG concentrations [34,37,51], adds further weight to the evidence that suggests these effects are all caused by estrogenic factors within the effluent.

Analyses of treated sewage effluents using toxicity, identification, and evaluation (TIE) approaches have shown that estrogens and their mimics are present in most, if not all, treated sewage effluents [53–55]. Studies in the United Kingdom have indicated that alkylphenolic compounds (e.g., nonylphenol, NP) and low levels of natural and synthetic steroidal estrogens (estradiol-17β estrone and 17α ethynylestradiol) are the primary estrogenic constituents of sewage effluents [53]. Moreover, laboratory studies have shown that the concentrations of the 17α ethynylestradiol, estradiol-17β, and estrone [6,44,55–58] or (in some industrial effluents) alkylphenolic chemicals [6,52], present in STW effluent in England are sufficient to explain the induction of vitellogenin synthesis in caged fish placed close to effluent discharges. Many rivers (in which the fish live) contain more dilute STW effluent, and thus the concentrations of estrogens in these rivers may not be high enough to induce plasma VTG that is seen in wild fish (based on short-term exposures). Longer-term exposures of freshwater fish to effluents have, however, been shown to reduce the threshold level for effect; a study by Rodgers-Gray et al. [59] found that exposure of roach to a STW effluent for 1 month induced a vitellogenic response at an effluent concentration of 37.9 +/– 2.3 %, but at an effluent concentration of only 9.4 +/– 0.9 %, after a 4-month exposure. The abnormal occurrence of VTG in wild freshwater fish is thus likely to occur, in many cases, as a result of long-term exposure to mixtures of estrogens present in effluents. It is probable that natural and synthetic estrogens, and in some instances, alkylphenolic chemicals found in STW effluents, also cause the effects on gonadal development and differentiation, and play a part in the evolution of intersexuality in wild fish; both groups of chemicals have been shown to do this under laboratory conditions. Concentrations of steroid estrogens and/or xenoestrogens required to induce these effect on the gonad, however, are higher than found in most effluents [e.g., 60–62], with the exception of some highly polluted rivers, and/or in times of drought (when river flow is low and the contribution made by effluent is high). Few studies have investigated whether environmentally relevant concentrations of estrogens within effluents, or indeed, the effluents themselves can cause the effects on the gonad duct seen in wild freshwater fish populations. In a study in which juvenile roach were exposed to a treated sewage effluent, it was proven that feminization of the development of the gonadal duct (prevalent in wild roach in UK and European rivers) occurs as a consequence of exposure to treated sewage effluent during the period of sexual differentiation [63]. Furthermore, a lab-based study has shown that gonad duct disruption can be induced in fish exposed to ethinyloestradiol at a concentration found in some treated effluents, when the exposure occurs during early life [64]. Although it is theoretically possible to produce an intersex or sex-reversed fish by exposure to sex steroid hormones or alkylphenols (usually during early life), in relatively short-term exposures, even higher concentrations are required to do so than for those inducing duct disruption. Furthermore, induction of altered sex cell development has not been shown in fish exposed to sewage effluents in controlled experiments. The reasons for this might be that the effluents used for the exposures [63] did not contain a sufficient concentration of the causative agent(s) and/or that the appropriate life stages have not been exposed and/or that the fish were not exposed long enough to cause this effect (the maximum duration of these exposures was 4 months). In our own unpublished studies on wild roach, we have found a positive correlation between the age of the fish (length of the exposure) and the severity of the intersex condition. This suggests that in real exposure scenarios (such as roach living in an effluent contaminated river in the United Kingdom), the longevity of the exposure might be of greater importance for disruptions in sex cell development (inducing oocytes in the testis), than the window in development during which the exposure occurs. In support of this hypothesis, NP has been shown to induce ovo-testes in the medaka, at a concentration of only 17 μg/l in the water when the exposure was life-long [65].

In summary, it seems that exposure of freshwater fish in the wild to natural steroidal and synthetic estrogens and, in some instances, alkylphenols cause inappropriate VTG induction and disruptions in the development of the reproductive ducts. Although not yet proven, it seems likely that these chemicals are also responsible for (or at least significantly contribute to) the occurrence of oocytes in the testes of male fish, for retarded testicular and ovarian development and delayed maturation.

Early life stage mortality syndrome and blue sac disease

There are very few studies that have demonstrated that freshwater fish are being impacted at the population level by exposure to a specific chemical (including EDCs). One such case, however, is for lake trout (*Salvelinus namaycush*) living in Lake Ontario where exposure to tetrachlorodibenzodifuran (TCDD) and coplanar polychlorinated biphenyls (PCBs) caused population declines because of negative impacts on reproductive success and early life survival [reviewed in 66]. The organochlorines induced a condition called blue sac disease, which is characterized by yolk sac edema, hemorrhaging, cranofacial abnormalities, and mortality in early larval development. Lake trout populations in Ontario declined precipitously during the 1950s when environmental concentrations of organochlorine chemicals were the at their highest. Subsequent, retrospective studies (based on measured PCB, PCDF, and PCDD residues in dated sediment cores) have established a strong relationship between the concentrations of TCDD, PCDDs, and PCDFs and the observed historical trends in lake trout reproduction, including the more recent signs of successful reproduction [67,68]. Laboratory studies have also shown that exposure to Ah (aryl-hydrocarbon) receptor agonists, including TCDD and coplanar PCBs, induces blue sac disease [69], but there is no evidence to show that these effects occur through an endocrine-mediated mechanism.

Reduced hatching success, low embryo survival, and slower rates of development in fry have also been reported in lake trout in the Great Lakes [70,71] and in Arctic char (*Salvelinus alpinus*) in Lake Geneva [72] and causally linked with exposure to coplanar PCBs, TCDDs, and PCDFs. Other conditions found in the Great Lakes fish during the 1960–1980s including early mortality syndrome and (in Baltic salmon) M74, resulting from thiamine deficiency, were thought to have a chemical etiology [73]. Like blue sac disease, M74 affects fry and is characterized by a loss of equilibrium, spiral swimming, lethargy, hemorrhaging, and death. There are data correlating incidences of M74 in Baltic salmon to elevated body burdens of PCDFs and coplanar PCBs and DDT [e.g., 74,75]. In none of these examples on fish in the Great Lakes, however, is there sufficient evidence to link the effects seen to a specific endocrine disruptor and/or their mixtures. Furthermore, the mechanisms via which these effects occur are generally unknown, and thus ascribing these effects to endocrine disruption at this time would be inappropriate.

Thyroid dysfunction in Great Lakes fish

Alterations in thyroid function have been reported in several wild populations of fish as a consequence of disruptions in their endocrine systems. Epizootics of thyroid hyperplasia and hypertrophy (affecting the whole population) have been reported in various species of salmonids in heavily polluted regions of the Great Lakes in the United States [76–79]. Although enlargement of the thyroid gland can occur as a result of iodine deficiency in the diet, this has been ruled out as a causative factor in the case of these salmonids. It was originally hypothesized that organochlorine contaminants, functioning as EDCs might be responsible for these effects [79]. Studies in the laboratory have shown that goiters and depressed thyroid hormone concentrations can be induced in rodents fed with contaminated fish from the Great Lakes, although fish fed with the same contaminated fish did not develop thyroid lesions [80,81]. Laboratory-based studies, however, have failed to identify the causative chemicals of these thyroid disruptions in the wild fish [reviewed in 82]. In summary, more than 40 years after the discovery of the thyroid dysfunctions in salmonids in the Great Lakes, although a chemical etiology has been established, the mechanism (endocrine, or otherwise) via which these effects occur is still uncertain. Very recently, thyroid abnormalities were also reported in mummichogs (*Fundulus heteroclitus*) from a polluted site (Piles Creek, New Jersey, USA) in the United States [83]. These effects have been loosely associated with exposure to a range of contaminants, especially mercury and petroleum hydrocarbons. When taken together, these studies suggest that thyroid function in fish appears to be sensitive to contaminant exposure generally.

Disruptions in adrenal physiology

There is a limited amount of evidence to suggest that environmental contaminants chronically stress fish, resulting in a compromised responsiveness of the HPI axis [84–89]. For example, Hontela et al. [84–86] demonstrated that yellow perch (*Perca flavescens*) and Northern pike (*Esox lucius*) from sites in Canada contaminated with heavy metals, PCBs, and PAHs were unable to produce cortisol in response to acute handling stress. Moreover, the adrenocorticotrophic hormone (ACTH)-producing cells (corticotrophs) in these fish were severely atrophied. Other studies by Hontela [87] have shown that both corticotrophs and the interrenal steroid producing cells undergo atrophy when fish are exposed to PAHs, PCBs, and heavy metals. It was speculated by the authors that the atrophy of the cells was a result of prolonged secretory hyperactivity of the cells. This hypothesis was later supported by studies on brown trout living in metal-contaminated waters that were shown to be hyper secreting ACTH and corticotrophin-releasing hormone [88,89]. More research is necessary to establish if the effects seen on the interrenal axis during exposure to specific contaminants have consequences to the health of affected fish populations.

ARE FRESHWATER FISH MORE SUSCEPTIBLE TO ENDOCRINE DISRUPTORS THAN OTHER ANIMALS?

Endocrine disruption appears to be particularly widespread in freshwater fish populations. There is little evidence, however, to suggest that fish are more susceptible to EDCs relative to other wildlife. Indeed, the evidence available on receptor binding affinities for chemicals that mimic sex steroid hormone, thyroid, and retinoic acid receptors suggests that vertebrates are likely to be similarly sensitive to environmental EDCs. Furthermore, there are many more similarities between the endocrine systems of fish and other higher vertebrates, notably with respect to the nature of the hormones, their receptors, and in the regulatory control of their endocrine system [7]. Notwithstanding this, there are more than 10 000 species of freshwater fish worldwide, displaying a high degree of heterogeneity in their physiology, anatomy, behavior, and ecology, and there are some features of the endocrine physiology of freshwater fish that may be particularly susceptible to the effects of EDCs, including those that determine sex (sex determination in fish has been shown to be especially sensitive to steroid hormones) and the process of smoltification in salmonids.

Living in the aquatic environment, fish can be bathed constantly in a solution of chemical pollutants. Furthermore, uptake of chemicals into fish can readily occur via the gills and skin, as well as via the diet (the major route of exposure to EDCs in terrestrial animals) [90]. Features of the gills including thin epithelial membranes and a large surface area coupled with the relatively high ventilation rates that occur in fish, facilitate the uptake of compounds from the water and their transfer into the blood stream. Some freshwater fish species are also top-predators and thus, are likely to bioconcentrate EDCs to a greater degree than other organisms at lower trophic levels. Freshwater fish are hypo-osmotic with their surroundings and thus a considerable movement of water into their bodies occurs down an osmotic gradient (taking chemicals with it). A major route of exposure to EDCs in fish during early life is from contaminants that have accumulated in lipid reserves within the egg as a consequence of maternal transfer during ovary development. These contaminants that have accumulated in the egg are mobilized when the lipid reserves are metabolized to fuel embryo development, exposing early life stages to especially high concentrations of EDCs at a time of greatest vulnerability to disruptions in their developing endocrine system. Furthermore, early life stages of fish have a limited capacity to metabolize and excrete contaminants, including EDCs. In situ exposures of fish have been used to assess both the bioavailability of EDCs, contained within complex mixtures, such as treated sewage effluents, and to determine non-point sources of pollution (agricultural run-off) and their biological effects. In such studies on rainbow trout, Larsson et al. [91] reported significant bioconcentration factors (in bile) for natural and synthetic sex steroid hormones (17β-estradiol, estrone, 17α ethinylestradiol) of up to 10 000-

fold after a 3-week exposure, whereas xenoestrogens (e.g., nonylphenol and bisphenol) bioconcentrate by several hundred to 1000-fold. Hewitt et al. [28] similarly obtained evidence for a very rapid uptake of EDCs in fish exposed to BKME, but here they also demonstrated that a rapid depuration of these chemicals occurs too. Apart from these two studies, however, there is very little information on the bioavailabilty of EDCs in wild fish or caged fish exposed to effluent discharges.

Another important issue that complicates determination of cause–effect relationships for EDCs, that is sometimes overlooked, is the possible time lag between the time of exposure and the biological response. Fish for example, living in the vicinity of sewage effluent outfalls will accumulate harmful contaminants in their tissues which may not cause any immediate deleterious effects, but which might affect the embryo development of their subsequent offspring. The biological responses in freshwater fish are very often especially influenced by physical environmental features, and concentrations of contaminants in the aquatic environment can vary widely temporally, and hence responses and effects may vary with season. All of these considerations are rarely taken into account in the analysis and interpretation of field-simulated exposures.

SCALE OF THE PROBLEM OF ENDOCRINE DISRUPTION IN FRESHWATER FISH

Endocrine disruption has only been studied in a small proportion of freshwater fish species, and data on cyprinids and salmonids dominate the literature. The differences in the sensitivities of different fish species to the effects of EDCs has not been comprehensively examined, although studies on the effects of pulp mill [92] and STW effluents [93], respectively, suggest that inter-species differences in sensitivity are likely to exist, between some fish species. Moreover, given the fact that endocrine disruption is commonly associated with exposure to effluents from domestic or industrial processes that enter rivers and streams, it seems likely that endocrine disruption in freshwater fish is more widespread than is currently documented. In the United Kingdom, for example, there are more than 70 000 consented discharges and 6500 of these are STW. Worldwide, each year, more than 5000 km^3 of water are used [94], and this figure is increasing every year. Furthermore, in some rivers, at times of low flow, up to 80 % of the river volume is made up of STW effluent discharge and this figure can be even higher in periods of drought. Using established statistical associations between endocrine disruption in freshwater fish and effluent dilution in receiving waters, theoretical predictions of the geographical extent of endocrine disruption can be estimated. In the United Kingdom, our own unpublished predictions, based on a statistical association between the concentration and dilution of the sewage effluent and the degree of feminization in wild fish exposed to the effluent, indicate that intersex fish are likely to exist at more than 50 % of 464 river sites that have effluent discharges with a population equivalent of more than 10 000. These predictions can be made using simple associative data from surveys of endocrine disruption in fish from a limited number of rivers (eight, in this case). Obviously, the more rivers from which one collects data, the better the predictions are likely to be. In our studies on roach in eight UK rivers, the following equation has been derived linking intersex in roach with effluent concentration in the rivers from where they were sampled: $\log (y + 1) = (0.000\,002\,88 * x) + 0.203$ where y = the intersex index and x = the concentration of effluent in the river (calculated by dividing the population equivalent of the effluent by the dilution factor for that effluent upon its entry into the river). The intersex index is a numerical index used to describe the degree of feminization of the gonads, based on their histological appearance [37]. The results from the UK analyses in roach for predicting intersexuality are illustrated in Fig. 1. This predictive map is now being validated, through determining the actual (observed) incidence of intersex at 46 of these study sites.

Additional studies in the roach have shown that the intersex index is correlated with fertility of intersex fish and hence, a predictive map of gamete quality can be constructed using the information on the intersex index and its relationship with fertility. Using this approach, we have estimated that at approximately 13 % of the 464 river sites selected for study in the United Kingdom (that receive effluent discharges with a PE of more than 10 000), the degree of intersexuality is estimated to be severe enough

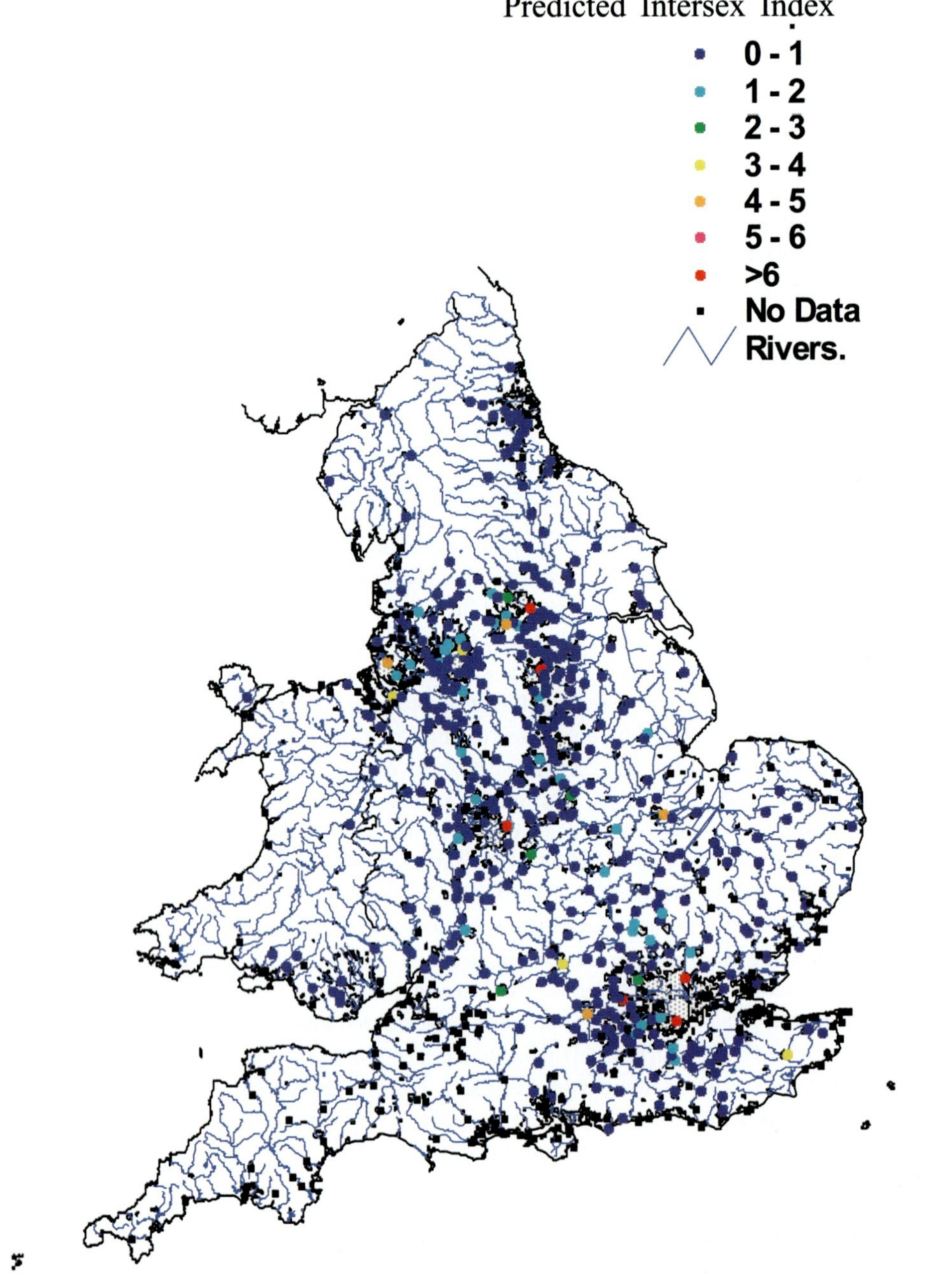

Fig. 1 Predicted intersex indices for populations of roach living downstream of sewage treatment works in selected rivers throughout the UK. The results were obtained using the following equation that links intersex in roach with effluent concentration in the rivers from where they were sampled: $\log(y + 1) = (0.000\,002\,88 * x) + 0.203$ where y = the intersex index and x = the concentration of effluent in the river (calculated by dividing the population equivalent of the effluent by the dilution factor for that effluent upon its entry into the river). The correlation between the intersex index and sperm quantity and quality predicts the following effects on fertilization success. Intersex index: 0–1 = little effect, 1–3 = slight effects, 3–4 = moderate effects, 4–>6: severe effects.

to have deleterious effects on gamete quality (fertilization success of intersex individuals would be predicted to be less than 60 %). At a further 16 % of the sites, the gamete quality of intersex and/or male fish would be predicted to be impaired, relative to male fish from reference sites. This system could be extended to other fish populations exposed to sewage and other types of effluent in order to provide a predictive map of the reproductive effects of endocrine disruption in freshwater fish populations worldwide.

Notwithstanding this, in any predictive study of ED in freshwater fish, it is important to establish the influence of age or longevity of exposure upon the effect that one is measuring. Our own investigations into the inter-relationships between age and intersexuality, for example, suggest that the intra-site variability in the degree of feminization that one observes in wild intersex roach is severely influenced by the ages of the fish collected, as older fish are more feminized than their younger counterparts (Jobling, unpublished data). Consequently, a perceived difference in the incidence and severity of intersex in fish collected from several different sites may be due to the differential age distributions of the fish sampled, rather than to differences in the endocrine-disrupting potencies of the various waters from whence the fish were collected.

POPULATION-LEVEL EFFECTS OF ENDOCRINE DISRUPTION

Much of the research on the effects of EDCs in freshwater fish has focused on effects at the individual level. A major challenge, from the point of view of ecological risk assessment, is to determine effects of endocrine disruption on populations and ecosystems. In order to meet this challenge, high-quality data on the population biology of freshwater fish, effects of EDCs on their various life history characteristics, and comprehensive and appropriate population models are needed. Basic information on the population biology of most species of wild freshwater fish is, however, extremely limited and it needs significant improvement for use in deriving a sound understanding of how EDCs affect fish population sustainability. Population growth rate in fish is determined by the balance between birth rate and the mortality rate. Collective fecundity and mortality thus predict the population's fate. In a large population with a stable age structure and sex ratio, future population size can be predicted from life table and fecundity data. This assumes, however, that each individual has an equal chance of contributing genes to the next generation, and this rarely happens in practice due to unequal sex ratios, differences in individual fertility, nonrandom mating, and variation in age structure. All of these factors influence the number of breeding individuals and hence, variation in the effective population size. In fish, juveniles are often not recruited to the adult population until they are 2–4 years old, and hence juvenile mortality and the rate of sexual maturity have a major bearing on the number of breeding individuals. Various external factors enhance population growth and others limit, and even prevent, population growth, and many of these are dependent on the density of the population. The most common density-dependent factors that limit population growth are food supply, space, predators, disease, and parasitism. Population-limiting factors in fish that are independent of population density include abiotic factors such as drought or floods.

Many of the parameters affecting fish population growth rate and sustainability are difficult to measure accurately in the field and are consequently, poorly understood. Whilst endocrine disrupters are known to affect factors such as individual fertility, and rate of sexual maturation and fecundity, a thorough assessment of these effects would require very extensive studies on the general life history and population biology of the exposed species compared with a reference population. Moreover, population declines are not usually caused by only one factor alone, but occur because of the effects of a multitude of factors.

Notwithstanding this, there are some examples in freshwater fish where there is substantial information on the consequences of endocrine disruption—on key reproductive parameters at the *individual* level, either from studies of wild populations, or from laboratory studies in which fish have been exposed to concentrations of EDCs known to be present in freshwaters. In the roach, for example, UK studies have provided sufficient evidence to show that widespread intersexuality, as a result of exposure to estrogenic sewage effluents, results in reduced fertility; there is a negative correlation between the degree of feminization of the intersex fish and their fertilization success ($r = -0.603$; $p < 0.001$). This information could be used albeit simplistically, to model the likely population responses to endocrine disruptors in the wild (in the absence of other factors that might also affect the population). Basic population models are already available, for this purpose. In a recent paper, Gleason and Nacci [95] attempted to model the effects of exposure to 17β-estradiol on populations of fathead minnow based (based on laboratory studies that showed egg production by females was negatively correlated with plasma VTG concentrations in the exposed males). The model predicted a negative linear correlation between the population growth rate of populations of fathead minnow and plasma vitellogenin production in males. Although these predictions are based on simple density-independent population models that require verification in realistic settings, they nevertheless provide a starting place for projecting population responses to EDCs from lab-based studies.

RECOMMENDATIONS

This account illustrates that there is very good evidence for endocrine disruption in freshwater fish. In order to more comprehensively assess the importance of endocrine disruption in freshwater fish, however, it will be necessary to put endocrine disruption into context with other environmental pressures that face freshwater fish populations. In our opinion, this will require research to define the global extent of the problem by expanding studies of endocrine disruption to other parts of the world and studies to extrapolate effects on individual fish to predict effects on populations and higher levels of biological organization.

Assessment of the extent of the problem of endocrine disruption in freshwater fish requires a more widespread sampling of a variety of wild populations of fish, ideally using nondestructive sampling methodologies and biomarkers. For biomarkers to be meaningful in this regard, efforts need to be directed at determining how they are related to the health of both individuals and populations. The biomarkers available for monitoring endocrine disruption are rather limited, and development of novel biomarkers should be encouraged to extend beyond those for estrogenic effects, with an emphasis on biomarkers that are indicative of reproductive and/or developmental effects and/or population responses. The presence of VTG in male fish, for example, is known to be negatively correlated with testicular growth and maturation [96]. In intersex fish, VTG is positively correlated with the degree of gonadal feminization [51], and hence also with their perceived reproductive success (which declines with increased degree of feminization [52]). A widespread assessment of VTG concentrations or of the degree of gonadal feminization in freshwater fish in a particular catchment could thus provide predictive information on the likely state of the testes or of their likely reproductive success, respectively. Vitellogenin concentration, however, could not (on its own) be used to directly predict the perceived fertility of a population of fish because even male fish, when exposed to estrogen for short periods of time, exhibit elevated plasma VTG concentrations, and little is known about the relationship between the timing and longevity of exposure to estrogen and the manifestation of gonadal feminization. Predictive maps and models thus need to be interrogated (by conducting both field and lab studies) to establish their validity. Information on what extent freshwater fish are exposed to EDCs in the wild is lacking, and hence more information is needed on what EDCs (and their concentrations) are present in the environment and to what extent they are absorbed and metabolized by freshwater fish. Moreover, the responses of fish to environmentally relevant mixtures of chemicals (containing EDCs) require further study and understanding.

Even with more widespread field data, research on endocrine disruption in wild freshwater fish is likely to be limited to those species in which large numbers of individuals are easily obtainable. A more global assessment of endocrine disruption in freshwater fish should ideally include the more rare and vulnerable species, although this is less practical. Current research on endocrine disruption in freshwater fish is limited to studies on a very few species, and there has been little or no attention given to the comparative sensitivities of different species of fish to EDCs, or indeed, of other animal species, on which freshwater fish may be dependent (e.g., many invertebrate species). Furthermore, current risk assessment strategies for endocrine disruption in freshwater fish are based on the responses of laboratory fish species and are unlikely to represent the full range of fish species that may be at risk in the wild. Laboratory studies will, therefore, need to be targeted at species with different life histories and different reproductive strategies, in order to compare the sensitivities of different fish species to EDCs and their mixtures. Moreover, an assessment of the implications of endocrine disruption on wild freshwater fish will require a comprehensive understanding of their physiology, endocrinology, and population biology, and thus a further recommendation for future research is to develop this information for sentinel species.

Wherever there is substantive evidence for endocrine disruption in freshwater fish, and the need for remediation has been established, there is a requirement to focus on how these problems can be alleviated. Where high quantities of industrial chemicals are used that are known to cause/contribute to endocrine disruption in freshwater fish, a program of tighter regulation for their discharge and/or a switch to alternative greener chemicals (which do not impact the endocrine system) is needed. The success of such programs can be illustrated by schemes implemented in both the United States and United Kingdom that have reduced the concentrations of EDCs discharged (either as a consequence of changes in industrial processes [97–98], or due to closure of a treatment plant [30]), which subsequently resulted in concomitant decreases in endocrine disruption in the exposed fish. Many known EDCs cannot, however, easily be eliminated at source, because they are of natural origin (e.g., sex steroid hormones from human and animal waste). For these types of contaminants, regulation of their release is likely to be achieved by improvements in treatment processes and/or the implementation of systems that specifically remove and degrade them. EDCs also enter the freshwater environment through non-point sources, but there has been very little study to assess the risk posed by these sources to freshwater fish. Studies of this nature are also needed.

ACKNOWLEDGMENTS

We gratefully acknowledge the U.K. Environment Agency for provision of data on effluent dilution and Richard Williams of CEH, Wallingford for his production of the predictive map in Fig. 1.

REFERENCES

1. D. E. Kime. *Endocrine Disruption in Fish*, p. 396, Kluwer Academic, Boston (1998).
2. G. Vos, E. Dybing, H. A. Greim, O. Ladefoged, C. Lambre, J. V. Tarazona, I. Brandt, A. D. Vethaak. *Crit. Rev. Toxicol.* **30**, 71–133 (2000).
3. J. Batty and R. Lim. *Arch. Environ. Contam. Toxicol.* **36**, 301–307 (1999).
4. A. Hassanin, S. Kuwatara, K. Ogawa, K. Hiramatsu, Nurhidayat, Y. Bukamoto, F. Sasaki. *J. Vet. Med. Sci.* **64**, 921–926 (2002).
5. B. S. Shane. *Introduction to Ecotoxicology*, CRC Press, Boca Raton, FL (1994).
6. K. Thorpe, M. Hetheridge, T. H. Hutchinson, M. Scholze, J. P. Sumpter, C. R. Tyler. *Environ. Sci. Technol.* **35**, 2476–2481 (2001b).
7. K. R. Munkittrick, M. E. McMaster, L. McCarthy, M. Servos, G. Van Der Kraak. *J. Toxicol. Environ. Health B* **1** (4), 347–371 (1998).
8. K. R. Munkittrick, C. B. Portt, G. J. Van Der Kraak, I. R. Smith, D. A. Rokosh. *Can. J. Fish Aquat. Sci.* **48**, 1371–1380 (1991).
9. K. R. Munkittrick, G. Van Der Kraak, M. McMaster, C. Portt. *Water Pollut. Res. J. Can.* **27**, 439–446 (1992b).

10. M. E. McMaster, G. J. Van Der Kraak, C. B. Portt, K. R. Munkittrick, P. K. Sibley, I. R. Smith, D. G. Dixon. *Aquat. Toxicol.* **21**, 199–218 (1991).
11. M. M. Gagnon, D. Bussieres, J. J. Dodson, P. V. Hodson. *Environ. Toxicol. Chem.* **14**, 317–327 (1995).
12. G. Van Der Kraak, K. R. Munkittrick, M. E. McMaster, C. B. Portt, J. P. Chang. *Toxicol. Appl. Pharmacol.* **115**, 224–233 (1992).
13. G. Van Der Kraak, T. Zacharewski, D. M. Janz, B. M. Sanders, J. W. Gooch. "Comparative endocrinology and mechanisms of endocrine modulation in fish and wildlife", In *Principles and Processes for Evaluating Endocrine Disruptions in Wildlife*, R. J. Kendall, R. L. Dickerson, J. P. Giesey, W. A. Suk (Eds.), SETAC, Pensacola, FL (1998).
14. R. L. Dickerson, J. P. Giesy, W. A. Suk (Eds.). *Principles and Processes for Evaluating Endocrine Disruption in Wildlife*, pp. 97–119, SETAC Technical Publication, Pensacola, FL (1998).
15. P. V. Hodson, M. McWhirter, K. Ralph, B. Gray, D. Thiverge, J. Carey, G. J. Van Der Kraak, D. McWhittle, M. Levesque. *Environ. Toxicol. Chem.* **11**, 1635–1651 (1992).
16. S. M. Adams, W. D. Crumby, M. S. Greely, L. R. Shugart, C. F. Saylor. *Ecotoxicol. Environ. Saf.* **24**, 347–360 (1992).
17. A. Karels, M Soimasuo, A. Oikari. *Water Sci. Technol.* **40**, 109–114 (1999).
18. A. Karels, E. Markkula, A. Oikari. *Environ. Toxicol. Chem.* **20**, 1517–1527 (2001).
19. M. E. McMaster, C. B. Portt, K. R. Munkittrick, D. G. Dixon. *Ecotoxicol. Environ. Saf.* **23**, 103–117 (1992).
20. T. G. Kovacs, J. S. Gibbons, L. A. Tremblay, B. I. O'Connor, P. H. Martel, R. I. Voss. *Ecotoxicol. Environ. Saf.* **31**, 7–22 (1995).
21. K. R. Munkittrick, G. J. Van Der Kraak, M. E. McMaster, C. B. Portt. *Environ. Toxicol. Chem.* **11**, 1427–1439 (1992a).
22. L. M. Hewitt, S. A. M. Smyth, M. G. Dube, Cl. Gilman, D. L. MacLatchy. *Environ. Toxicol. Chem.* **21**, 1359–1367 (2002).
23. D. L. MacLatchy and G. J. Van Der Kraak. *Toxicol. Appl. Pharmacol.* **134**, 305–312 (1995).
24. L. Tremblay and G. Van Der Kraak. *Environ. Toxicol. Chem.* **18**, 329–336 (1999).
25. P. Mellanen, T. Petänen, J. Lehtimäki, S. Mäkelä, G. Bylund, B. Holmbom, E. Mannila, A Oikari, R. Santti. *Toxicol. Appl. Pharmacol.* **136**, 381–388 (1996).
26. E. Rosa-Molinar and C. S. Williams. *Northeast Gulf Sci.* **7**, 121–125 (1984).
27. W. M. Howell and T. E. Denton. *Environ. Biol. Fish.* **24**, 43–51 (1989).
28. L. M. Hewitt, J. Parrott, K. Wells, M. K. Calp, S. Biddiscombe, M. McMaster, K. Munkittrick, G. Van Der Kraak. *Environ. Sci. Technol.* **34**, 4327–4334 (2000).
29. G. J. Van Der Kraak, K. R. Munkittrick, M. E. McMaster, C. B. Portt, J. P. Chang. *Toxicol. Appl. Pharmacol.* **115**, 224–233 (1992).
30. W. M. Howell, D. A. Black, S. A. Bortone. *Copeia* **980**, 676–681 (1980).
31. S. A. Bortone and R. P. Cody. *Bull. Environ. Contam. Toxicol.* **63**, 150–156 (1999).
32. L. G. Parks, C. S. Lambright, E. F. Orlando, L. J. Guillette, G. T. Ankley, L. E. Gray. *Toxicol. Sci.* **62**, 257–267 (2001).
33. J. LeBlanc, C. M. Couillard, J. C. F. Brethes. *Can. J. Fish. Aquat. Sci.* **54**, 2564–2573 (1997).
34. M. Hecker, C. R. Tyler, M. Hoffmann, S. Maddix, L. Karbe. *Environ. Sci. Technol.* **36**, 2311–2321 (2002).
35. C. Minier, G. Caltot, F. Leboulanger, E. M. Hill. *Analusis* **28**, 801–806 (2000).
36. R. van Aerle, M. Nolan, S. Jobling, L. B. Christiensen, J. P. Sumpter, C. R. Tyler. *Environ. Toxicol. Chem.* **20**, 2841–2847 (2001).
37. S. Jobling, M. Nolan, C. R. Tyler, G. Brighty, J. P. Sumpter. *Environ. Sci. Technol.* **32**, 2498–2506 (1998).
38. P. Flammarion, F. Brion, M. Babut, J. Barric, B. Migeon, P. Noury, E. Thybaud, C. R. Tyler, X. Palazzi. *Ecotoxicology* **9**, 127–135 (2000).

39. L. Vigano, A. Arillo, S. Bottero, A. Massari, A. Mandich. *Sci. Total. Environ.* **269**, 189–194 (2001).
40. J. Gercken and H. Sordyl. *Mar. Environ. Res.* **54**, 651–655 (2002).
41. L. C. Folmar, N. D. Denslow, V. Rao, M. Chow, D. A. Crain, J. Enblom, J. Marcino L. J. Guillette. *Environ. Health Perspect.* **104**, 1096–1101 (1996).
42. J. C. Harshbarger, M. J. Coffey, M. Y. Young. *Mar. Environ. Res.* **50**, 247–250 (2000).
43. L. C. Folmar, N. D. Denslow, K. Kroll, E. F. Orlando, J. Enblom, J. Marcino, C. Metcalf, L. J. Guilette. *Arch. Environ. Contam. Toxicol.* **40**, 392–398 (2001).
44. C. E. Purdom, P. A. Hardiman, V. J. Bye, N. C. Eno, C. R. Tyler, J. P. Sumpter. *Chem. Ecol.* **8**, 275–285 (1994).
45. C. R. Tyler and J. P. Sumpter. *Rev. Fish Biol. Fish.* **6**, 287–318 (1996).
46. J. P. Sumpter and S. Jobling. *Environ. Health Perspect.* **103** (Suppl. 7), 173–178 (1995).
47. D. E. Kime, J. P. Nash, A. P. Scott. *Aquacult.* **177**, 345–352 (1999).
48. K. M. Nichols, S. R. Miles-Richardson, E. M. Snyder, J. P. Giesy. *Environ. Toxicol. Chem.* **18**, 2001–2012 (1999).
49. R. A. Angus, S. A. Weaver, J. M. Grizzle, R. D. Watson. *Environ. Toxicol. Chem.* **21**, 1404–1409 (2002).
50. M. Nolan, S. Jobling, G. Brighty, J. P. Sumpter, C. R. Tyler. *J. Fish Biol.* **58**, 160–176 (2001).
51. S. Jobling, N. Beresford, M. Nolan, T. Rodgers-Gray, G. C. Brighty, J. P. Sumpter, C. R. Tyler. *Biol. Reprod.* **66**, 272–281 (2002).
52. S. Jobling, S. Coey, J. G. Whitmore, D. E. Kime, K. J. W. Van Look, B. G. McAllister, N. Beresford, A. C. Henshaw, G. Brighty, C. R. Tyler, J. P. Sumpter. *Biol. Reprod.* **67**, 515–524 (2002).
53. C. Desbrow, E. J. Routledge, G. C. Brighty, J. P. Sumpter, M. Waldock. *Environ. Sci. Technol.* **32**, 1549–1558 (1998).
54. W. Korner, P. Spengler, U. Bolz. *Environ. Toxicol. Chem.* **20**, 2142–2151 (2001).
55. S. A. Snyder, D. L. Villeneuve, E. M. Snyder. *Environ. Sci. Technol.* **35**, 3620–3625 (2001).
56. D. A. Sheahan, D. Bucke, P. Matthiessen, J. P. Sumpter, M. F. Kirby, P. Neall, M. Waldock. In *Sublethal and Chronic Effects of Pollutants on Freshwater Fish*, R. Muller and R. Lloyd (Eds.), Chap. 9, pp. 99–112, Cambridge, FAO, Fishing News Books, Blackwell Scientific (1994).
57. E. J. Routledge, D. Sheahan, C. Desbrow, G. C. Brighty, M. Waldock, J. P. Sumpter. *Environ. Sci. Technol.* **32**, 1559–1565 (1998).
58. G. H. Panter, R. S. Thompson, J. P. Sumpter. *Aquat. Toxicol.* **42**, 243–253 (1998).
59. T. P. Rodgers-Gray, S. Jobling, S. Morris, C. Kelly, S. Kirby, A. Janbakhsh, J. E. Harries, M. J. Waldock, J. P. Sumpter, C. R. Tyler. *Environ. Sci. Technol.* **34**, 1521–1528 (2000).
60. K. Kinnberg, B. Korsgaard, P. Bjerregaard, A. Jespersen. *J. Exp. Biol.* **203**, 171–181 (2000).
61. M. A. Gray, K. L. Teather, C. D. Metcalfe. *Environ. Toxicol. Chem.* **18**, 2587–2594 (1999).
62. T. J. Iwamatsu. *Exp. Zool.* **283**, 210–214 (1999).
63. T. P. Rodgers-Gray, S. Jobling, C. Kelly, S. Morris, G. Brighty, M. J. Waldock, J. P. Sumpter, C. R. Tyler. *Environ. Sci. Technol.* **35**, 462–470 (2001).
64. R. Van Aerle, N. Pounds, T. H. Hutchinson, S. Maddix, C. R. Tyler. *Ecotoxicology* **11**, 423–434 (2002).
65. H. Yokota, M. Seki, M. Maeda, Y. Oshima, H. Tadokoro, T. Honjo, K. Kobayashi. *Environ. Toxicol. Chem.* **20** (11), 2552–2560 (2001).
66. M. E. McMaster. *Water Qual. Res. J. Can.* **36** (2), 215–231 (2001).
67. S. Y. Huestis, M. R. Servos, D. M. Whittle, D. G. Dixon. *J. Great Lakes Res.* **22**, 310–330 (1996).
68. S. Y. Huestis, M. R. Servos, D. M. Whittle, M. van den Heuvel, D. G. Dixon. *Environ. Toxicol. Chem.* **16**, 154–164 (1997).
69. P. D. Guiney, P. M. Cook, J. M. Casselman et al. *Can. J. Fish Aquat. Sci.* **53**, 2080–2092 (1996).
70. M. J. Mac and C. C. Edsall. *J. Toxicol. Environ. Health* **33**, 375–394 (1991).

71. M. J. Mac, T. R. Schwartz, C. C. Edsall A. M. Frank. *J. Great Lakes Res.* **19**, 752–765 (1991).
72. G. Monod. *Bull. Environ. Contam. Toxicol.* **35**, 531–536 (1985).
73. M. Breitholtz, C. Hill, B. E. Bengtsson. *Ambio* **30**, 210–216 (2001).
74. B. E. Bengtsson, C. Hill, A. Bergman, I. Brandt, N. Johansson, C. Magnhagen, A. Sodergren, J. A. Thulin. *Ambio* **28**, 2–8 (1999).
75. P. J. Vuorinen, J. Passivirta, M. Keinanen, J. Koistinen, T. Rantio, T. Hyotylainen, L. Welling. **34**, 1151–1166 (1997).
76. J. F. Leatherland. *J. Great Lakes Res.* **19**, 737–751 (1993).
77. J. F. Leatherland and R. A. Sonstegard. *J. Fish Biol.* **16**, 539–562 (1980).
78. J. F. Leatherland, L. Lin, N. E. Down, E. M. Donaldson. *Can. J. Fish Aquat. Sci.* **46**, 2146–2152 (1989).
79. J. F. Leatherland and R. A. Sonstegard. *Comp. Biochem. Physiol.* **72C**, 91–100 (1982).
80. J. F. Leatherland. *J. Clean Technol. Environ. Toxicol. Occup. Med.* **6**, 381–395 (1997).
81. J. F. Leatherland and S. B. Barrett. *J. Great Lakes Res.* **19**, 149–159 (1993).
82. J. F. Leatherland. *Toxicol. Ind. Health* **14**, 41–57 (1998).
83. T. Zhou, H. B. John-Alder, J. S. Weis, P. Weis. *Mar. Environ. Res.* **50**, 393–397 (2000).
84. A. Hontela, J. B. Rasmussen, C. Audet, G. Chevalier. *Arch. Environ. Contam. Toxicol.* **22**, 278–283 (1992).
85. A. Hontela, P. Dumont, D. Duclos, R. Fortin. *Environ. Toxicol. Chem.* **14**, 725–731 (1995).
86. A. Hontela, C. Daniel, J. B. Rasmussen. *Ecotoxicology* **6**, 1–12 (1997).
87. A. Hontela. *Environ. Toxicol. Chem.* **17**, 44–48 (1998).
88. D. O. Norris, S. B. Felt, J. D. Woodling, R. M. Dores. *Gen. Comp. Endocrinol.* **108**, 343–351 (1997b).
89. D. O. Norris, S. Donahue, R. M. Dores, J. K. Lee, T. A. Maldonado, T. Ruth, J. D. Woodling. *Gen. Comp. Endocrinol.* **113**, 1–8 (1999).
90. G. Van Der Kraak, M. Hewitt, A. Lister, M. McMaster, K. Munkkittrick. *Hum. Ecol. Risk Assess.* **7**, 1017–1025 (2001).
91. D. G. J. Larsson, M. Adolfsson Erici, J. Parkkonen, M. Pettersson, A. H. Berg, P. E. Olsson, L. Forlin. *Aquat. Toxicol.* **45**, 91–97 (1999).
92. A. Karel, E. Markkula, A. Oikari. *Environ. Toxicol. Chem.* **20**, 1517–1527 (2001).
93. L. C. Sappington, F. L. Mayer, F. J. Dwyer, D. R. Buckler, J. R. Jones, M. R. Ellersieck. *Environ. Toxicol. Chem.* **20**, 2869–2876 (2001).
94. A. D. Shiklamanov. *Assessment of Water Resources and Water Availability in the World*, Stockholm, pp. 1–88, Stockholm Environmental Institute (1997).
95. T. R. Gleason and D. E. Nacci. *Hum. Ecol. Risk Assess.* **7**, 1027–1042 (2001).
96. S. Jobling, D. Sheahan, J. A. Osborne, P. Matthiessen, J. P. Sumpter. *Environ. Toxicol. Chem.* **15**, 194–202 (1996).
97. D. A. Sheahan, G. C. Brighty, M. Daniels, S. Jobling, J. E. Harries, M. R. Hurst, J. Kennedy, S. J. Kirby, S. Morris, E. J. Routledge, J. P. Sumpter, M. J. Waldock. *Environ. Toxicol. Chem.* **21**, 515–519 (2002).
98. D. A. Sheahan, G. C. Brighty, M. Daniels, S. J. Kirby, J. Kennedy, S. Morris, E. J. Routledge, J. P. Sumpter, M. J. Waldock. *Environ. Toxicol. Chem.* **21**, 507–514 (2002).

Pure Appl. Chem., Vol. 75, Nos. 11–12, pp. 2235–2247, 2003.

Topic 4.4

Effects of endocrine disruptors in aquatic mammals*

M. Cristina Fossi‡ and Letizia Marsili

Department of Environmental Sciences, Siena University, Via Mattioli 4, 53100 Siena, Italy

Abstract: In the last few decades, various studies have shown that aquatic mammals are sensitive to the toxicological effects of certain xenobiotic compounds, including the large class of endocrine-disrupting chemicals (EDCs). Since some EDCs, particularly organochlorines, tend to bioaccumulate and biomagnify in the aquatic food chain, various aquatic mammals, particularly those high in the food chain, such as pinnipeds, odontocete cetaceans, and polar bears, are potentially "at risk". The main aim of this chapter is to define the state of the art on effects of endocrine disruptors in aquatic mammals, both freshwater and marine. Another aim is to formulate recommendations for future research in this field and finally to define what can be done internationally for hazard/risk assessment and communication of the findings.

INTRODUCTION

Today there are over 4500 species of mammals on our planet. Aquatic mammals constitute a small percentage of this number, but have an extremely important ecological role in the marine and fresh water environments. Many aquatic mammal species, particularly marine mammals such as pinnipeds and odontocete cetaceans, as top-predators, exercise a dramatic regulatory control at community levels. Their decrease or disappearance in some areas could drastically alter community structure. In the last few decades, various studies have shown that several aquatic mammal species are sensitive to the toxicological effects of certain xenobiotic compounds [1–8], including the large class of endocrine-disrupting chemicals (EDCs). Fish-eating aquatic mammals may be extremely vulnerable to EDCs because of (a) their position in the food chain, (b) dependence on an aquatic/marine food web, (c) they live in areas influenced by industry and agriculture, and (d) their specific reproductive physiology [9]. Since several EDCs and particularly organochlorines, tend to bioaccumulate and biomagnify in the aquatic food chain [10,11], various aquatic mammals, particularly top-predators such as pinnipeds, odontocete cetaceans [12], and polar bears, are potentially "at risk" due to EDC contamination. Indeed, both reproductive and/or nonreproductive toxicities have been found in many fish-eating mammals that live in riverine or coastal areas where contaminant burdens are generally higher than in the open ocean.

EDCs are a structurally diverse group of compounds that may damage the health of humans, wildlife, fisheries, and their progeny, by interaction with the endocrine system [13–16]. They include chemicals used heavily in the past in industry and agriculture, such as polychlorinated biphenyls and organochlorine pesticides, and chemicals under current use, such as plasticizers and surfactants. Many

---

*Report from a SCOPE/IUPAC project: Implication of Endocrine Active Substances for Human and Wildlife (J. Miyamoto and J. Burger, editors). Other reports are published in this issue, *Pure Appl. Chem.* **75**, 1617–2615 (2003).

‡Corresponding author: Fax: ++39 0577 232930; E-mail: Fossi@unisi.it

known EDCs are estrogenic, affecting reproductive function. Because of the lipophilic persistent nature of most xenobiotic estrogens and their metabolites many bioaccumulate and biomagnify [10,11]; organochlorines are an example.

There are four types of organochlorine endocrine disruptors [17–35] commonly found in aquatic mammals [36–38] (Table 1): environmental estrogens, environmental androgens, antiestrogens, and antiandrogens. These endocrine disruptors act by mimicking steroid sex hormones, both estrogens and androgens, by binding to hormone receptors or influencing cell pathways (environmental estrogens and androgens), or by blocking and altering hormonal binding to hormone receptors (antiestrogens, anti-androgens).

Table 1 DDT metabolites and the PCB congeners with known estrogenic capacity commonly detected in aquatic mammals as measured in ER binding assays.

	Activity	Activity references	Potency[1]	aER binding IC_{50} (mM)[2]	ER binding IC_{50} (mM)[4]	ER binding RBA %[5]
DDTs	Estrogen	17				
p,p′-DDT	Estrogen	23		>50[3]	>1000	
	Antiandrogen	17,20,24				
	ER agonist	23				
o,p′-DDT	Estrogen	17,20,23,25,26,27	++	9.1	5	0.1
	Antiestrogen	22	+	>50[3]	>1000	
	Antiandrogen	22,23	+++			
	ER agonist	23				
p,p′-DDE	Estrogen	17,20,24,28	+			
	Antiestrogen	22,23	+			
	Androgen	22	+			
	Antiandrogen	22,23,29	++			
	ER agonist	23				
	AR agonist	23				
	AR antagonist	18,23				
o,p′-DDE	Estrogen	20,26		37.25		
	ER agonist	23				
p,p′-DDD	ER agonist	23			>1000	
o,p′-DDD	ER agonist	23		2.26		
PCBs		17,20,23,26,30,31				
Arochlor 1260	Estrogen	23,27,32				
	Effect on sexual differentiation	23,27,32				
	Gonadal abnormalities	23,27,32				
95	Estrogen	21,33	+			
99	Estrogen	21,33	++			
101	Estrogen	21,34				<0.001
118	Antiestrogen	21,34	++			
153	Estrogen	21,35	+++			0.004

[1]The most potent chemical for each activity was assigned a potency of four plus signs (++++), and the potency of all the chemicals expressed relative to this [21,22].
[2]Inhibitor concentrations necessary for 50 % inhibition (IC_{50}) of [^{3}H]17β-estradiol binding to estrogen receptor (aER) in the alligators. The aER binding IC_{50} value for 17β-estradiol was 0.0078 µM.
[3]Compounds that inhibited [^{3}H]17β-estradiol but were insoluble at concentrations necessary to achieve 50 % inhibition [19].
[4]Inhibitor concentrations necessary for 50 % inhibition (IC_{50}) of [^{3}H]17β-estradiol binding to ER in the rats. The ER binding IC_{50} value for 17β-estradiol was 0.002 µM [18].
[5]Relative estrogen receptor-binding affinities (RBAs). Competitive binding with estradiol in rat uterine ER preparations [21].

The main aim of this chapter is to define the state of the art on the potential effects of endocrine disruptors, with particular emphasis on organochlorines in aquatic mammals, both freshwater and marine. Another aim is to formulate recommendations for future research in this field and finally to define what can be done internationally for hazard/risk assessment and communication the findings.

STATE OF THE ART OF EDCs IN AQUATIC MAMMALS

High concentrations of polychlorinated biphenyls (PCBs) and DDTs known to be potential EDCs (Table 1) have been detected in different aquatic mammals. Moreover, polychlorinated dibenzo-*p*-dioxins (PCDDs) and dibenzofurans (PCDFs), that for their physicochemical properties are less transportable from terrestrial to marine environmental, are principally accumulated in terrestrial animals [39,40]. Both in vivo and in vitro studies show that these contaminants exhibit a broad spectrum of antiestrogenic responses [17]. The concentration-dependent effects of several PCDD and PCDF congeners as antiestrogens were determined in the aryl hydrocarbon (Ah)-responsive MCF-7 human breast cancer cell lines. For the PCDDs and PCDFs, the order of antiestrogenic potency was 2,3,7,8-tetrachlorodibenzo-*p*-dioxin > 2,3,7,8-tetrachlorodibenzofuran > 2,3,4,7,8-pentachlorodibenzofuran > 1,2,3,7,9-pentachlorodibenzofuran > 1,3,6,8-tetrachlorodibenzofuran [41]. Top predators often acquire large burdens of persistent pollutants through biomagnification of compounds received from contaminated prey [12,42]. Pinnipeds and cetaceans have relatively large amounts of blubber for insulation that readily retain considerable concentrations of highly persistent organochlorine insecticides, PCBs and dioxins (several thousand ppt in toxicity equivalents, TEQs), implying a higher risk from exposure of wildlife to dioxins and related compounds [40]. The estimated TEQ concentrations in the blubber of some cetacean species, such as northern right whale dolphin (*Lissodelphis borealis*) and Pacific white-sided dolphin (*Lagenorhynchus obliquidens*) from the northern North Pacific; Dall's porpoise (*Phocoenoides dalli*) from the Japan Sea; striped dolphin (*Stenella coeruleoalba*) off Sanriku and Fraser's dolphin (*Lagenodelphis hosei*) off Kii Peninsula, Japan; and hump-backed dolphin (*Sousa chinensis*) and finless porpoise (*Neophocaena phocaenoides*) from Hong Kong, exceeded the levels associated with immunosuppression in harbor seals (*Phoca vitulina*) [43]. Many reports have been published on PCDD and PCDF levels in aquatic mammals: Baikal seals [*Phoca (Pusa) siberica Gmel*] from Lake Baikal [44]; harp seals (*Pagophilus groenlandicus*) from Greenland Sea [45]; gray whale (*Eschrichtius robustus*); killer whale (*Orcinus orca*); false killer whale (*Pseudorca crassidens*); Risso's dolphin (*Grampus griseus*) and Dall's porpoise from British Columbia, and harbor porpoises (*Phocoena phocoena*) from British Columbia and central California [46]; beluga whales (*Delphinapterus leucas*) from the St. Lawrence River estuary [47]; dugong (*Dugong dugon*) from the Great Barrier Reef [48], and sea lions (*Otaria flavescens*) from Argentina [39]. The maximum levels were in the ringed seal from the Baltic Sea with 170 pg/g fat of TEQ. In any of these papers, relationships are investigated between PCDD and PCDF levels and endocrine disruptor effects. Some of the best evidence of the relationship between organochlorine chemicals (OCs) and reproductive toxicity comes from semi-field studies on seals [5] and mustelids [49].

FRESHWATER MAMMALS

Population of mustelids, including mink (*Mustela vison*) and Canadian otter (*Lutra canadiensis*), have declined in areas of the Great Lakes where these species have a high percentage of contaminated fish in the diet [50]. Following outbreaks of reproductive failure in commercial ranching operations, laboratory experiments showed that mink are extremely sensitive to OCs, particularly PCBs and dioxins. Wren [50] tried to test the hypothesis of reproductive dysfunction and population decline in wild mink feeding on Great Lakes fish. However, the evidence was not conclusive. A more recent study using mink demonstrated that exposure to low doses of PCBs over 18 months results in impaired reproduction [51]. Data from Ohio show that the mink harvest between 1982 and 1987 from contaminated counties bordering Lake Erie was consistently lower (380 animals per year) than those from counties far away from Lake Erie (850 animals per year), suggesting an effect of chemicals on mink populations [50]. Preliminary studies from Ontario also suggest that mink harvest is lower in potentially "high PCB exposure areas" compared with lower exposure areas.

Evidence is also available on the effect of EDCs on the harvest otters from four New York State counties adjacent to Lake Ontario and the St. Lawrence River. The harvest data from these four counties show that, between 1960 and early 1970 otter harvest remained stable and then increased. Increased harvest is consistent with improved water quality of Lake Ontario in the past 15 years. It is not possible to draw a causal link between the status of mink and otter populations and exposure to OCs from the Great Lakes without toxicological experiments, so a large amount of research and data analysis needs to be undertaken.

Harding et al. [52] assessed chlorinated hydrocarbon contamination of mink and river otters in Columbia and Fraser River systems of northwestern North America, in relation to their morphological condition. Livers were analyzed for residues of organochlorine pesticides, PCBs, dibenzo-*p*-dioxins, and dibenzofurans. Contaminant levels were relatively low compared to those documented in other North American populations, although they ranged higher than those detected during an earlier survey of these regional populations. Although a few individual animals with gross abnormalities of the reproductive system did not show high levels of contamination, there was a significant negative correlation between total PCB concentrations (such as Aroclor 1260) and baculum length in juvenile mink ($r = 0.707$; $p = 0.033$; $n = 8$).

In ranch-reared mink, used as a model in an experimental trial to investigate the potential effects of exposure to two petroleum products on sea otters (*Enhydra lutris*), females exposed to bunker C fuel oil in the diet had significantly reduced reproductive success (3.4 kits/female) although their offspring's only exposure to the petroleum products was in utero or during nursing. Sea otter populations consuming contaminated food or colonizing previously oiled habitats may therefore have reduced reproductive success [53]. In addition, the European otter (*Lutra lutra*) population in Europe declined dramatically in 1960s to 1980s, and exposure to PCBs leading to impaired reproduction has been considered a major cause of this decline [54]. A study of otter populations and levels of organochlorine pesticide residues and PCBs in otter feces (spraints) was made on rivers in East Anglia, England. Population and contaminant levels were compared against target values, the first based on the index value of a stable population of otters on the River Severn (Wales) and the second (4 mg/kg) taken as the "no effect level" for all individual contaminants [55]. Forty-four % of samples had concentrations of contaminants exceeding the "level of concern". Contamination, especially by PCBs, is therefore regarded as a factor affecting otter populations, which may not be viable in East Anglia without repeated releases of captive-bred animals [36].

MARINE MAMMALS

The marine environment is exposed to a variety of EDCs by an array of sources, for example, industrial effluent and agricultural run-off. The result is the potential exposure of marine mammals to EDCs, not only cetaceans and pinnipeds, but also mammals such as the polar bears that feed on marine organisms.

Polar bears

The polar bear (*Ursus maritimus*) is a top predator of the Arctic marine ecosystem. Polar bears prey primarily on ringed seals (*Phoca hispida*) and bearded seals (*Erignathus barbatus*), which live exclusively on the sea ice. They also consume larger prey such as walruses (*Odobenus rosmarus*) and white whales.

Pseudo-hermaphroditism observed in polar bears is thought to be an effect of EDCs. Two female polar bears at Svalbard (Norway) had both female and male genitals [56]. The two pseudo-hermaphrodites were genetically females but also had small ponise in front of their vagina. On different occasions in another two bears, aberrant genitalia morphology with a high degree of chloral hypertrophy has also been reported and classified as female pseudo-hermaphroditism. The observed rate of female pseudo-hermaphroditism in the Svalbard area was 1.5 % (4/269). The authors believe that the pseudo-hermaphroditism observed could be a result of endocrine disruption by OCs, especially PCBs, which concentrate in fat to very high levels.

Skaare et al. [57] found an association between organochlorine contaminants and retinol and thyroid hormones T_4 and triiodothyronine (T_3) in blood plasma of polar bears caught at Svalbard. Retinol concentration and the ratio of total T_4 (TT_4) to free T_4 (FT_4) decreased linearly with increasing concentrations of PCBs and HCB. The consequences of these associations for individuals and populations is unknown. Reduced plasma retinol concentration may be indicative of incipient retinol deficiency which manifests as impaired growth and development of epithelial tissue and reduced immune system and reproductive function. Low cub survival was reported in polar bears from Svalbard [58]. Immunoglobulin G (IgG) levels and OCs were determined in blood plasma of these polar bears [8]. IgG concentrations increased with age and were significantly higher in males than females. IgG was correlated negatively with PCB levels and with three PCB congeners (IUPAC numbers 99, 194, and 206). HCB was also correlated negatively with IgG, suggesting an immunotoxic effect.

A new compound, 4-hydroxyheptachlorostyrene (4-OH-HpCS) was identified as a major component in the chlorinated phenolic compound fraction of plasma of polar bears captured in the Nunavut Territory (Canada). Levels of 4-OH-HpCS ranged from 2.89 to 22.9 ng/g wet weight (w.w.) in polar bear plasma ($n = 30$) representing 2.89–24.8 % of total quantified chlorinated phenolic compounds [59]. Transthyretin (TTR) is assumed to be the main plasma protein responsible for specific binding of phenolic compounds in plasma, because most of the compounds, including 4-OH-HpCS, have a similar structure to the natural ligand, thyroxine (T_4). The study of Sandau et al. [59] indicated that the phenolic metabolites of relatively minor contaminants have the capacity to bind to circulating proteins, and their significance as potential endocrine-disrupting agents. The extent to which they disrupt thyroid hormone and retinol homeostasis cannot be underestimated.

Pinnipeds

Populations of harbor seals from the Dutch Wadden Sea had low reproductive success and declining population numbers that were attributed to the impact of PCBs [60]. Other studies showed that female harbor seals fed fish from the polluted Wadden Sea had a lower reproductive success (50 %) than seals fed less-contaminated fish. Implantation failure was found to be associated with lower levels of 17β-estradiol [61], induced by EDCs.

There is ample evidence that populations of Baltic ringed seals (*Phoca hispida bothnica*) and grey seals (*Halichoerus grypus*) have declined markedly over the past 100 years [62]. Overhunting and habitat destruction might have been the contributing factors, however, it is generally accepted that persistent pollutants (PCBs, DDTs), which adversely affect reproductive performance, was the major cause. Earlier studies showed that seals in the Baltic Sea (of which the Bay of Bothnia is the northernmost part) carry very high burdens of DDT and PCBs. Pregnant seals are reported [62] to have a much lower content of these compounds than nonpregnant animals. Females with over 70 mg PCB/kg of extractable fat are apparently unable to reproduce. Only 19 % of the grey seals in the Baltic proper have less than 70 mg PCB compared with 25 % in the Bay of Bothnia. In ringed seals, the values ranged from 39 to 54 % in the northernmost part of the Bay of Bothnia [63,64].

Most studies reporting concentrations of organochlorines in pinnipeds have investigated ringed, grey, and harbor seals [65]. Very few studies have been carried out on pinnipeds from the southern hemisphere. Before 1980, the highest mean wet-weight blubber concentrations of DDT and related metabolites (911 +/– 582 μg/g) were recorded in sea lions from California. The highest pre-1980 blubber concentrations of PCBs (1470 +/– 922 μg/g) were recorded in harbor seals from the Netherlands. Since 1980, the highest blubber concentrations of PCBs have been recorded in grey seals of the Dee estuary, UK [46.79 (10.17–116.68) μg/g]. The highest DDT concentrations measured in individual pinnipeds are in the 1–15 μg/g range recorded in ringed, grey, and harbor seals and Australian fur seals. Reported effects on pinnipeds suggested to result from organochlorine contamination, include skeletal deformities and impact on reproduction, such as uterine occlusion.

Cetaceans

Research indicates that some cetaceans, particularly odontocetes, have detectable and sometimes extremely high levels of substances known or suspected to be EDCs such as PCBs, DDTs, chlorinated pesticides, brominated flame retardants, and tributyltin (TBT). Biologists have reported a range of effects of these chemicals on cetaceans including immunosuppression, cancer, skin lesions, secondary infections and diseases, sporadic die-offs, and reduced reproductive success. There are several examples suggesting that exposure to OC insecticides and PCBs affected endocrine function and reproduction in marine mammals. For example, transformation of epididymal and testicular tissue has been observed in North Pacific minke whales (*Balaenoptera acutorostrata*) [66].

Another example is the endangered beluga whales of the St. Lawrence estuary (SLE), now amongst the most contaminated animals on earth, with tumors and reproductive problems [67,68]. A population of approximately 650 beluga inhabits a short segment of the SLE. Over 17 years (1983–1999), Martineau et al. [68] have examined 129 (or 49 %) of 263 SLE beluga carcasses reported stranded. De Guise et al. [69] reported in these animals a true hermaphrodite *Delphinapterus leucas*. This animal had two testicles, two separate ovaries, and the complete ducts of each sex; cervix, vagina, and vulva were absent. Mature spermatozoa were found in the lumen of seminiferous tubules in the testicles, and numerous involuted corpora lutea were recognized in the ovaries.

Increasing residue levels of PCBs and DDE in the blubber of Dall's porpoises were found to have a negative association with testosterone levels in blood. Testosterone levels decreased in a statistically significant way with increasing DDE concentrations. These results suggest that current levels of environmental contamination by persistent organochlorines can cause an imbalance in sex hormones and subsequent reproductive abnormalities in the wild. The other hormone measured, aldosterone, which has no sexual function, showed levels that were independent of the effects of PCBs and DDE [70].

In the endangered bowhead whale (*Balaena mysticetus*), not all sexual activity leads to conception. Pseudo-hermaphroditism has been reported in at least two males with testicular feminization [71], a relatively high incidence, as only 76 bowheads were closely examined between 1980 and 1989 [72]. Concentrations of most organochlorine contaminants in biological material of 20 bowhead whales from Barrow (Alaska, USA) were low compared to those in tissues of other cetaceans, especially odontocetes, but it is impossible to exclude the possibility that pseudo-hermaphroditism is independent of these contaminants.

Hot spots: The case study of Mediterranean cetaceans

Man-made EDCs range across all continents and oceans. Some geographic areas are potentially more threatened than others (Fig. 1); one of these is the Mediterranean Sea. This basin has limited exchange of water with the Atlantic Ocean, and is surrounded by some of the most heavily populated and industrialized countries in the world. Levels of some xenobiotics are therefore much higher here than in other seas and oceans [73,74]. In this peculiar environment, top predators (such as large pelagic fish and marine mammals) tend to accumulate large quantities of polyhalogenated aromatic hydrocarbons (PHAHs) and toxic metals [37,42]. Levels of PHAHs in a top predator of the Mediterranean, the striped dolphin, are 1–2 orders of magnitude higher than in Atlantic and Pacific individuals of the same species [75] (Fig. 1).

Nondestructive studies of Mediterranean cetaceans have revealed the hazard to which odontocete species are exposed in relation to EDCs in blubber [76]. Significant differences in total levels of

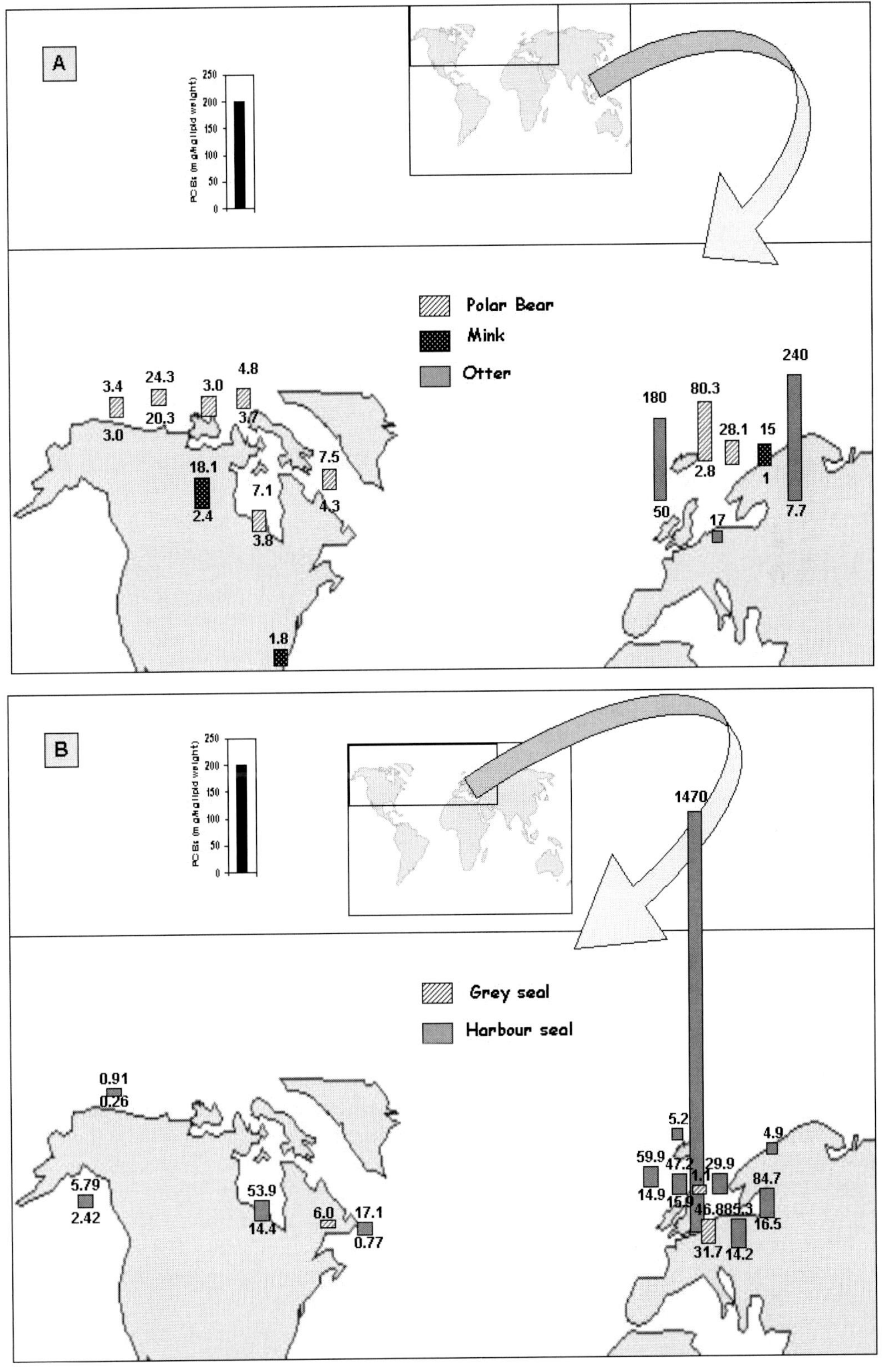

(continues on next page)

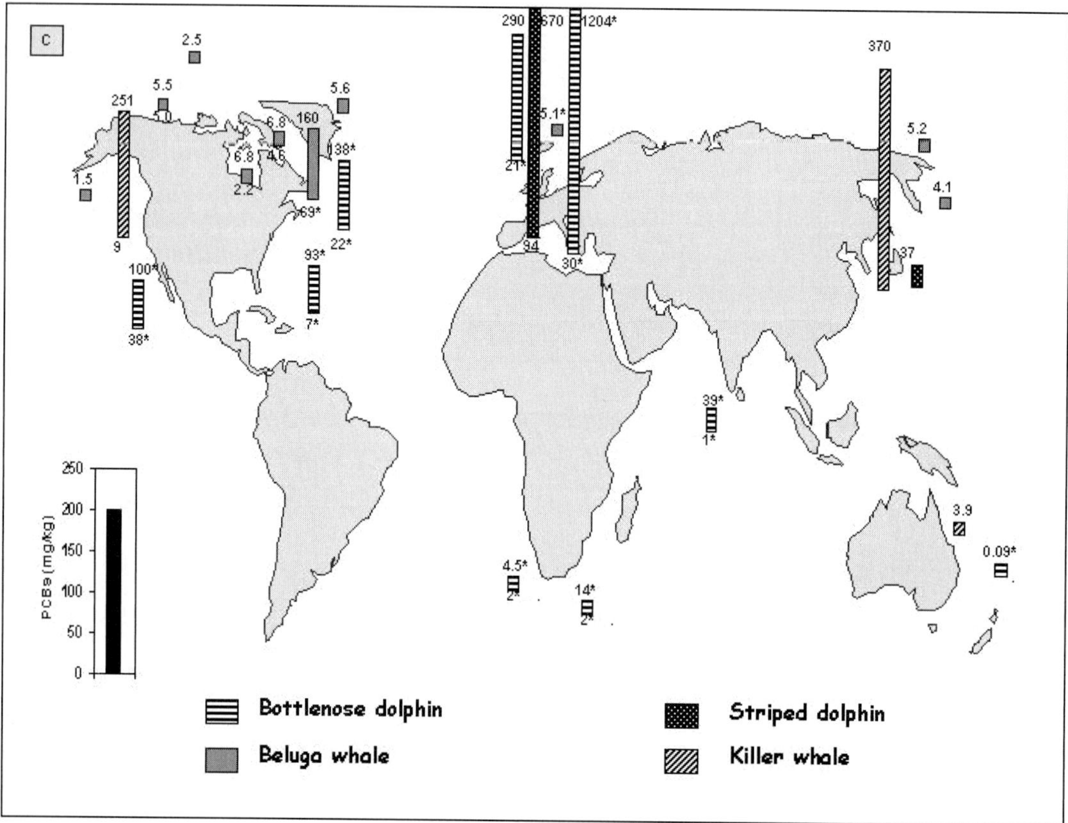

Fig. 1 Minimum and maximum mean levels of total PCBs (mg/kg lipid weight) in marine fish-eating mammals from various part of the world. **A** = freshwater mammals and polar bears; **B** = pinnipeds; **C** = cetaceans. Numbers with * in C are in mg/kg lipid weight, the others in mg/kg wet weight. Data in this figure were taken from the references listed in this paper.

organochlorine EDCs were found between odontocetes and mysticetes. Highest mean levels were found in striped dolphins [OC-EDCs = 40.0 μg/g fresh weight (f.w.)], followed by bottlenose dolphins (OC-EDCs = 24.3 μg/g f.w.) and common dolphins (OC-EDCs = 15.0 μg/g f.w.). Differences in organochlorine bioaccumulation and consequent potential risk due to endocrine disruptors, are primarily related to different positions in the marine food chain. Of the Mediterranean cetaceans investigated, the species with the highest levels of OCs and the highest biomarker response seems to be the striped dolphin. Interspecies differences in susceptibility to EDCs must be considered because high levels of contaminants and high biomarker responses do not necessarily mean high risk for the species.

Some cetaceans, such as the common dolphin, have almost completely disappeared from the Mediterranean Sea. Using skin biopsy as an ecotoxicological tool to study intraspecific susceptibility to contaminants, a high statistical correlation was found between benzo(a)pyrene monoxigenase (BPMO) activity and total DDTs, *p,p*′-DDE, *o,p*′-DDT, total PCBs and 153 congener (Pearson correlations significant at $p < 0.05$) in male specimens of common dolphin. This suggests that EDCs may be a major stress factor for common dolphin populations in the Mediterranean Sea. Similar correlation (Spearman rank correlations significant at $p < 0.05$) was obtained in fin whales sampled in the Ligurian Sea from 1992 to 1995 [77] between BPMO activity and organochlorine levels in skin biopsy specimens from males, but not females or males and females together. Future studies are needed to explore the role of detoxification enzymes and estrogen receptors (ERs) in interspecies susceptibility to EDC contaminants using fibroblast cell cultures of different species [78].

FUTURE RESEARCH NEEDS

The above review demonstrates that research is necessary to resolve uncertainties that remain in this field of study. Strengthening international collaboration in the following broad research areas is a step in this direction. Some specific recommendations for future research are described below.

Endocrine-mediated effects in aquatic mammals

Basic knowledge on endocrine systems of aquatic mammals and the range of mechanisms by which endocrine disruptors, especially organochlorines, interfere with reproductive success, immune function, and neurobehavioral parameters at key stages of life cycles, will provide the platform for understanding differences in the response of different species, both in a qualitative and quantitative sense. It is important to investigate interspecies susceptibility to EDCs, particularly in species living in hot spot areas (e.g., polar bears in Svalbard, Baltic ringed seals in the Baltic Sea, common dolphin in the Mediterranean Sea—see Fig. 1).

New methodology: Nondestructive techniques in the study of aquatic mammals

Development of a series of nondestructive techniques to evaluate residue levels and biomarker responses is recommended, in place of the lethal approach, for hazard assessment and conservation of endangered species of aquatic mammals [79]. In the last 10 years, much interest has been shown internationally [80] toward skin biopsies as a sensitive nonlethal technique for hazard assessment of free-ranging cetaceans exposed to EDCs [42,81,82].

The biopsy dart method has been used successfully on a range of cetacean species (at least 30) in oceans and the Mediterranean Sea. The response of fin whales to the dart was no change in behavior in 80 % of the cases. Skin biopsy is a powerful tool for toxicological studies for the following reasons: (a) it allows collection of a large number of samples across a wide geographic range; (b) it allows collection of sequential samples from the same animal if identified by photo identification or other techniques; (c) it is suitable for residue analysis of PHAHs, including dioxin group chemicals (suitable for calculation of TEQs), heavy metals, and polycyclic aromatic hydrocarbons (PAHs); (d) it is suitable for biomarker analysis: induction of MFO by enzyme assay (BPMO) and immunohistochemical assay (CYP1A1); DNA damage; fibroblast cell culture.

In conclusion, a number of successful studies show that cetacean skin biopsies are a powerful nonlethal tool for assessing ecotoxicological risk in marine mammals and aspects of feeding ecology and food preferences. In this context, it is also essential to develop more specific and sensitive biomarkers for detecting endocrine-mediated effects in individuals and populations.

Monitoring hot spots and species/populations at risk

Marine mammals are at the top of the food chain, and much monitoring data has been collected in the last few decades on tissue levels of persistent organic pollutants (POPs) from various species in different parts of the world (Fig. 1). It is essential to expand basic long-term monitoring programs of "sentinel" aquatic mammal species, using nonlethal techniques, particularly in "hot spots". This work can provide baseline data on population status in relation to EDC exposure. It is essential to improve international collaboration and cooperative research (e.g., Pollution 2000+) to assess exposure and effects of EDCs on a more global basis, particularly PCBs chosen as model compounds in selected species such as bottlenose dolphins and harbor porpoises. For conservation of biodiversity, it is important to focus on population/subgroups most likely to be susceptible to EDCs.

Future management

Management is essential at international level in order to define the main lines of hazard/risk assessment and communication to follow.

Causal criteria for assessing EDCs in aquatic mammals: A proposed framework

The UNEP volume entitled *Global Assessment of the State-of-the-Science of Endocrine Disruptors* [83] proposed a powerful framework to examine the hypothesis that chemicals with endocrine activity are causing adverse effects in wildlife populations. The framework that can be applied to hazard assessment of aquatic mammals begins with a clear statement of "hypothesis", which contains two distinct elements. First, the outcome of concern (e.g., decreased reproductive function in Baltic seals) is linked to a putative stressor (e.g., PCBs) that is affecting populations. Second, exposure to EDC stressors is postulated to result in endocrine-mediated events that ultimately lead to the outcome of concern. In order to evaluate the scientific evidence regarding their potential relationship we use scientific evidence on five aspects: (1) ***Temporality*** (e.g., high levels of PCBs are strongly correlated with reduced reproductive success in Baltic seal populations); (2) ***Strength of the association*** (e.g., the link between adverse reproductive outcomes and chemical etiology is weak); (3) ***Consistency of the observations*** (e.g., observation on a global scale provides evidence that the reproductive system of Baltic seals may be susceptible); (4) ***Biological plausibility*** (e.g., results of semi-field studies with Baltic seals provide general links with OCs exposure and reproductive outcome); and (5) ***Evidence of recovery*** (e.g., reproduction improves as levels of contamination decline).

MANAGEMENT, HAZARD/RISK ASSESSMENT, AND RISK COMMUNICATION

The proposed framework could enable identification of "aquatic mammal populations potentially at risk" and "hot spots" in which management of hazard/risk assessment and risk communication can be applied (see "hot spot areas" in Fig. 1). It is necessary to strengthen international collaboration in the following broad activities:

- Continuous monitoring of aquatic mammal populations in "hot spots" compared to the same species in control areas. Hazard assessment of "populations potentially at risk" in particular "hot spot" areas.
- International agreement for the reduction of input of new EDCs into the freshwater environment and coastal areas.
- International agreement for scientific research on a global scale focused on monitoring, conservation, and protection of endangered species of aquatic mammals.
- Identification of human populations "potentially at risk" due to consumption of cetacean-based whale products (identification of the "maximum daily input of EDCs").
- Transfer of scientific information to regulatory agencies at regional and international level for regulation and conservation.

CONCLUSION

Current knowledge makes it sufficiently evident that some aquatic mammal species, such as certain seal populations, have been adversely affected by environmental contaminants in the EDC category, but evidences are mounting to show at least indirectly, that the chemicals grouped as EDCs are affecting several aquatic mammals populations through some endocrine-dependent mechanisms. Research has largely focused on compounds that persist and bioaccumulate in organisms. Future research on endocrine-mediated effects in aquatic mammals, the development of new nondestructive monitoring methodologies, and validation of hazard/risk assessment procedures in endangered species/populations could enable better future protection and conservation of these ecologically essential animals.

REFERENCES

1. A. Bernhoft, J. U. Skaare, O. Wiig, A. E. Derocher, H. J. S. Larsen. *J. Toxicol. Environ. Health Part A* **59** (7), 561–574 (2000).
2. UNEP. *UNEP/POPS/CONF/4*, 22, Stockholm, Sweden (2001).
3. T. Colborn. *Environ. Toxicol. Chem.* **17** (1), 1–2 (1998).
4. A. Arukke, F. R. Knudsen, A. Goksoyr. *Environ. Health Perspect.* **105**, 418–422 (1997).
5. S. Tanabe, S. Watanabe, H. Kan, R. Tatsukawa. *Mar. Mamm. Sci.* **4**, 103–124 (1988).
6. T. Colborn, F. S. vom Saal, A. M. Soto. *Environ. Health Perspect.* **101**, 378–384 (1993).
7. T. Colborn, D. Dumanoski, J. P. Myers. *Our Stolen Future,* Dutton, Penguin Books, New York (1996).
8. T. Colborn, M. J. Smolen, R. Rolland. *Toxicol. Ind. Health* **14** (1/2), 9–23 (1998).
9. B. Gillesby and T. Zacharewski. *Environ. Toxicol. Chem.* **17** (1), 3–14 (1998).
10. H. O. Adami, L. Lipworth, L. Titus-Ernstoff, H. Chung-cheng, A. Hamberg, U. Anhlborg, J. Baron, D. Trichopoulos. *Cancer Cause Control* **6**, 551–566 (1995).
11. W. R. Kelce, C. R. Stone, S. C. Laws, L. E. Gray. *Nature* **375**, 581–586 (1995).
12. P. M. Vonier, D. A. Crain, J. A. McLachlan, L. J. Guillette Jr., F. A. Steven. *Environ. Health Perspect.* **104**, 1318–1322 (1996).
13. P. W. Wong and I. N. Pessah. *Mol. Pharmacol.* **49**, 740–751 (1996).
14. G. Hansen Larry. *Environ. Health Perspect.* **106** (1), 171–189 (1998).
15. P. Sohoni and J. P. Sumpter. *J. Endocrinol.* **158**, 327–339 (1998).
16. K. Hilscherova, M. Machala, K. Kannan, A. L. Blankenship, J. P. Giesy. *Environ. Sci. Pollut. Res.* **7** (3), 159–171 (2000).
17. K. Nesaretnam, D. Corcoran, R. R. Dils, P. Darbre. *Mol. Endocrinol.* **10**, 923–936 (1996).
18. J. A. Nelson. *Biochem. Pharmacol.* **23**, 447–451 (1974).
19. S. Safe, B. Astroff, M. Harris, T. Zacharewski, R. Dickerson, R. Romkes, I. Biegel. *Pharmacol. Toxicol.* **69**, 400–409 (1991).
20. A. M. Soto, C. Sonneschein, K. L. Chung, M. F. Fernandez, N. Olea, F. O. Serrano. *Environ. Health Perspect.* **103** (7), 113–122 (1995).
21. L. R. Cooper and R. J. Kavlock. *J. Endocrinol.* **152**, 159–166 (1997).
22. S. H. Safe. *Environ. Health Perspect.* **108**, 487–493 (2000).
23. M. Bergeron, J. Crews, J. A. McLachlan. *Environ. Health Perspect.* **102**, 780–781 (1994).
24. S. H. Safe. *Crit. Rev. Toxicol.* **24** (2), 87–149 (1994).
25. M. B. Matta, C. Cairncross, R. M. Kocan. *Environ. Toxicol. Chem.* **17**, 99–115 (1998).
26. T. M. Sajjd, Masters Thesis, University of Illinois at Urbana-Champaign, Urbana, IL (1996).
27. M. S. Wolf, D. Camann, M. Gammon, S. D. Stellman. *Environ. Health Perspect.* **105**, 13–14 (1997).
28. M. H. Li, Y. D. Zhao, L. G. Hansen. *Bull. Environ. Contam. Toxicol.* **53**, 583–590 (1994).
29. C. F. Mason and S. M. Macdonald. *Sci. Total Environ.* **138** (1–3), 147–160 (1993).
30. L. Marsili. *IJEP* **13**, 416–452 (2000).
31. A. Aguilar, A. Borrell, P. J. Reijnders. *Mar. Environ. Res.* **53** (5), 425–452 (2002).
32. B. Jimenez, M. J. Gonzalez, L. M. Hernandez, E. Eljarrat, J. Rivera, M. C. Fossi. *Chemosphere* **38** (3), 507–515 (1999).
33. S. Tanabe. *Mar. Poll. Bull.* **45**, 69–77 (2002).

34. V. Krishnan and S. Safe. *Toxicol. Appl. Pharmacol.* **120** (1), 55–61 (1993).
35. M. C. Fossi, L. Marsili, G. Neri, S. Casini, G. Bearzi, E. Politi, M. Zanardelli, S. Panigada. *Mar. Environ. Res.* **50** (1–5), 643–647 (2000).
36. T. B. Minh, H. Nakata, M. Watanabe, S. Tanabe, N. Miyazaki, T. A. Jefferson, M. Prudente, A. Subramanian. *Arch. Environ. Contam. Toxicol.* **39** (3), 398–410 (2000).
37. E. N. Tarasova, A. A. Mamontov, E. A. Mamontova, J. Klasmeier, M. S. McLachlan. *Chemosphere* **34** (11), 2419–2427 (1997).
38. M. Oehme, M. Schlabach, K. Hummert, B. Luckas, E. S. Nordoy. *Sci. Total Environ.* **162** (2–3), 75–91 (1995).
39. W. M. Jarman, R. J. Norstrom, D. C. G. Muir, B. Rosenberg, M. Simon, R. W. Baird. *Mar. Pollut. Bull.* **32** (5), 426–436 (1996).
40. D. C. G. Muir, C. A. Ford, B. Rosenberg, R. J. Norstrom, M. Simon, P. Béland. *Environ. Pollut.* **93** (2), 219–234 (1996).
41. D. Haynes, J. F. Muller, M. S. McLachlan. *Chemosphere* **38** (2), 255–262 (1999).
42. P. E. G. Leonards. Ph.D. thesis, Free University of Amsterdam (1997).
43. C. D. Wren. *J. Toxicol. Environ. Health* **33** (4), 549–85 (1991).
44. B. Brunstrom, B. O. Lund, A. Bergman, L. Asplund, I. Athanassiadis, M. Athanasiadou, S. Jensen, J. Ordberg. *Environ. Toxicol. Chem.* **20**, 2836–2840 (2001).
45. L. E. Harding, M. L. Harris, C. R. Stephen, J. E. Elliott. *Environ. Health Perspect.* **107** (2), 141–147 (1999).
46. J. A. Mazet, I. A. Gardner, D. A. Jessup, L. J. Lowenstine. *J. Wildl. Dis.* **37** (4), 686–692 (2001).
47. A. Roos, E. Greyerz, M. Olsson, F. Sandegren. *Environ. Pollut.* **111**, 457–469 (2001).
48. C. F. Mason and S. M. Macdonald. *Sci. Total Environ.* **138** (1–3), 127–145 (1993).
49. Ø. Wiig, A. E. Derocher, M. M. Cronin, J. U. Skaare. *J. Wildl. Dis.* **34** (4), 792–796 (1998).
50. J. U. Skaare, A. Bernhoft, Ø. Wiig, K. R. Norum, E. Haug, D. M. Eide, A. E Derocher. *J. Toxicol. Environ. Health Part A* **62** (4), 227–241 (2001).
51. Ø. Wiig. *Ursus.* **10**, 25–32 (1998).
52. C. D. Sandau, I. A. T. Meerts, R. J. Letcher, A. J. Mcalees, B. Chittim, A. Brouwer, R. J. Norstrom. *Environ. Sci. Technol.* **34** (18), 3871–3877 (2000).
53. P. J. H. Reijnders. *Neth. J. Sea Res.* **14**, 30–65 (1980).
54. P. J. H. Reijnders. *J. Reprod. Fert.* **90**, 403–409 (1990).
55. ICES. *ICES CM 1993/N:3* (1992).
56. E. Helle, M. Olsson, S. Jensen. *Fauna Flora (Stockholm)* **71** (2), 41–48 (1976).
57. J. G. Vos, E. Dybing, H. A. Greim, O. Ladefoged, C. Lambré, J. V. Tarazona, I. Brandt, A. D. Vethaak. *Crit. Rev. Toxicol.* **30** (1), 71–133 (2000).
58. J. D. Hutchinson and M. P. Simmonds. *Rev. Environ. Contam. Toxicol.* **136**, 123–167 (1994).
59. Y. Fujise, R. Zenati, H. Kato. *SC/50/RMP12 IWC Scientific Committee*, p. 14, Oman (1998).
60. S. De Guise, D. Martineau, P. Béland, M. Fournier. *Environ. Health Perspect.* **103** (4), 73–77 (1995).
61. D. Martineau, K. Lemberger, A. Dallaire, P. Labelle, T. P. Lipscomb, P. Michel, I. Mikaelian. *Environ. Health Perspect.* **110** (3), 285–292 (2002).
62. S. De Guise, A. Bisaillon, B. Seguin, A. Lagace. *Anat. Histol. Embryol.* **23** (3), 207–216 (1994).
63. A. Subramanian, S. Tanabe, R. Tatsukawa, S. Saito, N. Miyazaki. *Mar. Pollut. Bull.* **18** (12), 643–646 (1987).
64. R. J. Tarpley, G. H. Jarrell, J. C. George, J. Cubbage, G. G. Stott. *J. Mamm.* **76**, 1267–1275 (1995).
65. L. M. Philo, Jr., E. B. Shotts, J. C. George. In *The Bowhead Whale,* J. J. Burns, J. J. Montague, C. J. Cowles (Eds.), pp. 275–312, Society for Marine Mammalogy, Special Publication No. 2, Allen Press, Lawrence, KS (1993).

66. G. T. Ankley, E. Mihaich, R. Stahl, D. Tillitt, T. Colborn, S. McMaster, R. Miller, J. Bantle, P. Campbell, N. Denslow, R. Dickerson, L. Folmar, M. Fry, J. Geisy, L. E. Gray, P. Guiney, T. Hutchinson, S. Kennedy, V. Kramer, G. LeBlanc, M. Mayes, A. Nimrod, R. Patino, R. Peterson, R. Purdy, T. Ringer, L. Touart, G. Van Der Kraak, T. Zacharewski. *Environ. Toxicol. Chem.* **17**, 68–87 (1998).
67. A. Bernard, C. Hermans, F. Broeckaert, G. De Poorter, A. De Cock, G. Houins. *Nature* **401**, 231–232 (1999).
68. L. Marsili and S. Focardi. *Environ. Poll.* **91** (1), 1–9 (1996).
69. M. C. Fossi, S. Casini, L. Marsili. *Chemosphere* **39** (8), 1273–1285 (1999).
70. L. Marsili, M. C. Fossi, G. Notarbartolo Di Sciara, M. Zanardelli, B. Nani, S. Panigada, S. Focardi. *Chemosphere* **37** (8), 1501–1510 (1998).
71. L. Marsili, M. C. Fossi, G. Neri, S. Casini, C. Gardi, S. Palmeri, E. Tarquini, S. Panigada. *Mar. Environ. Res.* **50** (1–5), 649–652 (2000).
72. M. C. Fossi and L. Marsili. *Biomarkers* **2**, 205–216 (1997).
73. M. C. Fossi, G. Lauriano, M. Simmonds, F. Borsani. *J. Cet. Res. Manage.* **3**, 333–335 (2001).
74. M. C. Fossi, L. Marsili, C. Leonzio, G. Notarbartolo Di Sciara, M. Zanardelli, S. Focardi. *Mar. Poll. Bull.* **24** (9), 459–461 (1992).
75. M. C. Fossi, C. Savelli, L. Marsili, S. Casini, B. Jimenez, M. Junin, H. Castello, J. A. Lorenzani. *Chemosphere* **35** (8), 1623–1635 (1997).
76. WHO/UNEP/ILO. International Programme on Chemical Safety (IPCS) T. Damstra, S. Barlow, A. Bergman, R. Kavlock, G. Van Der Kraak (Eds.) (2001).

Pure Appl. Chem., Vol. 75, Nos. 11–12, pp. 2249–2261, 2003.

Topic 4.5
Endocrine disruption in marine fish*

Peter Matthiessen

Centre for Ecology and Hydrology, Lancaster Environment Centre, Library Avenue, Bailrigg, Lancaster LA1 4AP, UK

Abstract: This topic reviews the whole field of endocrine disruption (ED) in marine fish and compares this with our knowledge of the situation in freshwater species. In broad terms, similar types of ED have been observed in the two groups, although effects in the marine environment tend to be less marked, presumably due to dispersion and dilution. There are, however, some data which suggest that marine fish that are top-predators can experience ED due to biomagnification of organochlorines. Processes such as smoltification, metamorphosis, and hermaphroditism, which are common in some marine species, may be particularly susceptible to ED, but have as yet been scarcely studied. As with freshwater fish, firm links to population-level effects have not yet been demonstrated, although it is not unreasonable to suppose that they are occurring in some locations. The topic concludes with some recommendations for future research.

INTRODUCTION

Knowledge about the causes and effects of endocrine disruption (ED) is probably more comprehensive for fish than for any other animal group. A useful general review of the subject has been published by Kime [1], but additional material can be found in several publications [2–22]. Possible reasons for this focus on fish include their early identification as a group showing ED effects in the field, their sexual lability (exploited for many years by fish farmers), their marked exposure to contaminants via both oral and branchial routes, their relatively well-understood endocrinology, and their amenability to experimental manipulation. The breadth of such studies has meant that many different types of ED have been observed in fish, including interference with hormone receptor-mediated processes, with normal hormone titres via metabolic and excretory mechanisms, and with gonadal development.

The overwhelming preponderance of studies has been concerned with the sex steroids and their "downstream" actions, in particular the impacts of exogenous estrogens and their mimics. However, there is also more limited information available on disturbances of the thyroid and adrenocorticoid systems in fish, as well as a number of syndromes that may or may not be attributable to ED. A recent review of the subject that also highlights the many unique aspects of fish endocrinology, general biology, and evolution, has been produced by the International Program on Chemical Safety [23]. Despite their uniqueness and long evolutionary history (at least 400–500 million years), it is nevertheless important to bear in mind that fish possess endocrine systems which mirror in many respects those of more recently evolved vertebrates (all vertebrates have similar adrenal and sex steroid receptors, e.g., [24]). Fish, therefore, have the potential to act as sentinels for possible effects on other vertebrates, including

*Report from a SCOPE/IUPAC project: Implication of Endocrine Active Substances for Human and Wildlife (J. Miyamoto and J. Burger, editors). Other reports are published in this issue, *Pure Appl. Chem.* **75**, 1617–2615 (2003).

humans, although the differential exposure of various vertebrate groups is likely to be of major importance in determining the scale of such effects.

Although the literature is growing fast, the majority of it is related to freshwater fish, the group in which a variety of reproductive effects were first recognized in the field as types of ED [22,25–28]. There have been relatively few studies of marine fish, perhaps because contaminated discharges generally receive much greater dilution in salt waters than in rivers and lakes, a fact that probably discouraged an early search for effects in the sea. For a number of reasons, it is nevertheless timely to review progress in this field. For example, the diversity of marine fish is high by comparison with most freshwater assemblages (with some almost uniquely marine groups such as the 3000-strong elasmobranchs and chimaeras), and as we already know that there is considerable interspecific variation of response in fish to endocrine disruptors, this raises the possibility of new modes of action in marine species.

Secondly, marine fish other than the cyclostomes and elasmobranchs exhibit the unique ability to maintain their hypotonic body fluids by drinking, a behavior that probably exposes them to dissolved contaminants to a greater degree than freshwater fish. Their somewhat divergent physiology may also have a bearing on their responses to endocrine disruptors, although they generally possess similar endocrine systems to their freshwater relatives.

Thirdly, the well-known sexual plasticity of fish is particularly marked in some marine groups. Although the sexes are usually separate (gonochoristic), there are some functional hermaphrodites (e.g., some marine Serranidae and Sparidae), a feature which is rare in the vertebrates. Furthermore, in some other marine groups (e.g., the cyclostomes), the eggs and sperm can begin development in a single individual, with one gamete usually coming to dominate (although some individual hagfish may remain as sterile intersexes for life). In some coral reef fishes, individuals can change functional sex in response to social and environmental cues. There are examples of both protoandrous (sperm produced first) and protogynous (eggs first) genera. The ways in which sex reversal in fish is brought about are essentially unknown, although the involvement of steroid hormones is suspected [29]. It is interesting to note that there are usually germ cells of both sexes present in an individual fish of this type, although only one type of gamete is mature at a given time. Although these types of natural sex reversal are relatively rare among fish species, the ease with which fish farmers are able to change the phenotypic sex of some stocks through judicious application of exogenous steroids [30,31] is consequently not surprising. It should be noted that the performance of hormonally sex-reversed male and female fish is not comparable with normal ones [32]. The implications for endocrine disruption are clear.

The purpose of this paper is, therefore, to assess the nature and extent of ED in marine and estuarine fish, and to identify any substantive differences from the well-documented position in some fresh waters. The paper will focus on effects reported from the field, using laboratory studies where appropriate to help explain the observations.

VITELLOGENIN AND ZONA RADIATA PROTEIN INDUCTION AS BIOMARKERS FOR EXOGENOUS ESTROGENS, AND SPIGGIN INDUCTION AS A BIOMARKER OF ANDROGEN EXPOSURE

The processes of vitellogenesis and zonagenesis in normal female teleost livers are described in [33] and [34], respectively. They are both driven by endogenous estradiol (E2) secreted by the ovaries in response to environmental cues acting via the hypothalamus and pituitary. Vitellogenin (VTG) induction in male fish (which only contain very low endogenous E2, but whose livers nevertheless are able to synthesize VTG and zona radiata protein (ZRP) in response to exogenous E2) was the first effect of estrogen contamination to be reported from the freshwater environment [27], and a number of studies have now measured this biomarker in marine fish. It is worth bearing in mind that although VTG induction in males is an excellent biomarker of exposure to exogenous estrogens and their mimics acting via the hepatic estrogen receptors (demonstrated in many laboratory experiments, e.g., [35–37]), it can also be associated with liver enlargement and kidney damage (caused by its production and attempted excretion, respectively), and is generally accompanied by various degrees of reproductive interference at similar or lower ambient estrogen concentrations [38–43]. In other words, VTG induction in male fish is also a marker for a number of adverse effects.

Lye et al. [44,45] were the first to report VTG induction (and accompanying testicular abnormalities) in a male marine fish, the euryhaline flounder (*Platichthys flesus*). These were caught near a sewage treatment works (STW) discharge on the Tyne estuary (NE England) and later work suggested that alkylphenols may have been partly responsible [46]. More detailed surveys of *P. flesus* in 13 UK estuaries [35,36,47,48] showed that VTG induction is widespread in the males of this species, at titres of up to 20 mg VTG/ml plasma in the worst cases (Tees, Tyne, Clyde, and Mersey estuaries). Elevated VTG titres have even been observed in males migrating offshore to breed, although it is thought that this derives from earlier estuarine exposure. At some stations on the Mersey and Clyde, VTG titres appear to have declined over the period 1996–2001 during which the standard of sewage treatment in these areas has been improving, but the existence of real temporal trends remains to be confirmed. Males in other estuaries (e.g., Thames, Forth, Humber, and Dee) show medium or negligible amounts of VTG induction, despite large inputs of sewage effluent in some of these areas. Sediments in the Tees and Tyne are much more contaminated with estrogens than the overlying water [49], and there are indications that the flounder's benthic food organisms may be an important source of estrogen exposure [48]. However, the main causative substances have yet to be fully identified [49].

Male flatfish elsewhere have also been reported to show VTG induction, although not to the degree seen in UK fish. For example, Madsen et al. [50] report up to 86 μg VTG /ml in male *P. flesus* from 7 Danish coastal and estuarine stations, up to 2.2 μg VTG /ml was found in male *Pleuronectes yokohamae* from Tokyo Bay [51], and a maximum of only 0.06 μg VTG /ml was recorded in *P. yokohamae* from the Strait of Tsugaru [52]. In addition, VTG synthesis has been detected in more than 50 % of male English sole (*Pleuronectes vetulus*) from some organochlorine-contaminated areas of Puget Sound in the western United States [53], in male grey mullet (*Mugil cephalus*) from Osaka Bay in Japan [54], and in a few male *P. flesus* caught in a Dutch industrial harbor zone and in offshore spawning areas [55]. It is interesting to note that when larval sunshine bass (*Morone saxatilis* x *M. chrysops*) were experimentally exposed for 4 days to New York City sewage effluent, not only VTG but also estrogen receptor (ER) expression was induced [56]. This upregulation of ER has also been observed in the laboratory in trout exposed to octylphenol or E2 [57], and suggests that early exposure to estrogens may predispose fish to later hypersensitivity to estrogens. This has indeed been shown experimentally in male Japanese medaka, which gave a larger VTG response to E2 if they had been exposed as larvae to the estrogen mimic *o,p′*-DDT [58].

The studies reviewed above suggest that UK estuarine waters may be particularly contaminated with estrogens, and VTG mRNA has indeed been detected in males of another UK estuarine species, the viviparous blenny (*Zoarces viviparus*), caught at stations on the Forth, Tees, Clyde, and Tyne [48]. However, other studies have failed to find VTG in male eel (*Anguilla anguilla*) from some of the same UK estuaries (e.g., Tyne and Tees) in which male flounder are strongly vitellogenic [59,60], and VTG mRNA has been undetectable in male sand goby (*Pomatoschistus minutus*) from these estuaries [48,61]. In both these species, VTG can be induced by exposure in the laboratory to rather high estrogen doses, so it is assumed that the absence of VTG induction under estrogen-contaminated field conditions is due either to lower exposure than that experienced by flounder and blenny, or to lower sensitivity. Data discussed below suggest that wild UK gobies *are* exposed to estrogens, so it is possible that differential sensitivity is the main explanatory factor.

With the exception of some UK and Dutch flounder migrating offshore from contaminated estuaries to breed, none of the examples described above indicate that fish in the open ocean are experiencing significant estrogenic exposure. However, recent data on the Mediterranean swordfish (*Xiphias gladius*) a top-predator caught in the Straits of Messina near Sicily and in the open Atlantic around the Azores show that some males from the Straits are experiencing VTG and ZRP induction several times higher than Atlantic fish [62]. The data are based on rather few samples, but it appears that VTG and

ZRP induction increases with age, suggesting that it is related to the long-term bioaccumulation of lipophilic estrogenic substances. This is in contrast to UK flounder [47] where even immature fish show strong VTG induction at some sites, and known lipophilic estrogen-mimics are only bioaccumulated to a moderate extent in adults (e.g., maximum pooled hepatic ΣPAH = 0.36 mg/kg wet wt. in Tees and Mersey fish; max. pooled hepatic ΣPCB = 1.3 mg/kg wet wt. in Thames and Mersey fish; max. pooled hepatic ΣDDT = 0.49 mg/kg in Mersey fish). Note, however, that these levels of PCB contamination have been associated with reduced viable hatch of Baltic flounder fry [63]. It would, therefore, be worth investigating whether other predatory fish are experiencing significant estrogenic exposure under open-sea conditions.

It should be borne in mind that female fish can also be affected by exogenous estrogens. For example, Matthiessen et al. [47] reported elevated VTG titres in juvenile female flounder, but the already high natural levels in adult females were not further increased. The full implications of this require investigation, but it is possible that females may mature too early in estrogen-contaminated locations. It has been reported [64,65] that *P. flesus* exposed to harbor sediment showed premature vitellogenesis, although this could have been due to elevated endogenous estradiol titres rather than exogenous estrogens. Cadmium (not an estrogen mimic) has also been shown to cause increased vitellogenesis in female Atlantic croaker (*Micropogonias undulatus*), but this is thought to be caused by direct toxic action on the pituitary, leading to altered secretion of gonadotropin hormone (GtH) [66,67].

Finally, it is worth noting that *reduced* VTG titres have occasionally been reported in female fish [68–70] from contaminated marine harbors, etc., an effect which might be attributable to antiestrogen or androgen exposure, or to generalized stress [71–74], leading to lower VTG synthesis, which could have implications for normal egg development.

In summary, the available data show that demersal fish in some estuarine and coastal waters, and those pelagic top predators containing biomagnified contaminants, can be exposed to sufficient exogenous estrogen to cause VTG and/or ZRP induction in males, and/or premature induction in females. Unlike in rivers, where VTG induction seems to be associated primarily with natural and synthetic estrogenic hormones in STW discharges [75], the similar effects in estuaries (at least, those in the United Kingdom) appear to be linked to present or past industrialization, and to largely unknown contaminants adsorbed to sediments. Experimental data show that strong VTG induction in males can be accompanied by a range of other abnormalities [e.g., 40] and the following sections will demonstrate that some of these have also been observed in the field.

An analogous biomarker to VTG, but sensitive to androgens rather than estrogens, has been developed in the three-spine stickleback (*Gasterosteus aculeatus*) [76,77]. This is the protein spiggin, which is produced in the male kidney under the control of 11-ketotestosterone (11-KT), and is secreted via the bladder for use as a glue during nest construction. It has been shown [77] that spiggin is produced in females exposed to exogenous androgens, and that it is also induced during experimental exposure to an androgenic Baltic pulp-mill effluent. However, only a few STW discharges to UK estuaries appear to show androgenic activity in in vitro assays, the activity is solely attributable to natural androgens [78], and female sticklebacks caged in the vicinity of such discharges only show small spiggin responses [48,77]. Although these data are rather preliminary, it appears at present that androgens are relatively unimportant contaminants of sewage-contaminated estuaries by comparison with estrogens. The reverse, however, is probably true near pulp-mill discharges in the Baltic and elsewhere.

ALTERED SEX STEROID TITRES

Based on work in freshwater fish, impacts of contaminants on sex steroid titres might be expected in marine fish, but few have been reported to date. In principle, altered levels of sex steroids in plasma might be due to interferences with the control of steroid synthesis via the pituitary-gonadal axis, or to effects on steroid metabolism and excretion. For example, experimental exposure to pulp-mill effluent fractions containing low-molecular-weight phenolics (but not plant sterols) can cause depressions in testosterone (T) in the estuarine killifish (*Fundulus heteroclitus*) [79,80], probably via one or more independent mechanisms including action at the pituitary on GtH secretion, reductions in cholesterol availability, and inhibition of enzymes which catalyse sex steroid synthesis. Similar depression of plasma androgens has also been observed in perch (*Perca fluviatilis*) caught near a pulp-mill discharge in the Baltic Sea [81], and this effect probably explains the demasculinization of these fish (see below). In vitro laboratory experiments with *P. flesus* ovarian tissue exposed to high doses of polycyclic aromatic hydrocarbons (PAHs) have shown that they can inhibit the steroidogenic enzyme P450 17,20-lyase, leading to reduced secretion of androstenedione and E2 [82]. This may in turn explain the reduced VTG synthesis in some female marine fish living in contaminated environments (see above).

On the other hand, even less is known about the effects of estrogen exposure on steroid titres. One of the few observations made in marine fish [83] showed that male *P. flesus* from the estrogenically contaminated Tyne and Mersey estuaries contained up to 5 times more plasma E2 than fish caught in the open North Sea (but much less than seen in normal females). Furthermore, Dutch experiments in which female *P. flesus* were exposed to PAH- and PCB-contaminated harbor sediments for 3 years revealed elevated E2 and T titres, which were tentatively attributed to decreased steroid clearance via the cytochrome P450 system [65]. However, it has also been shown in the laboratory that maturing Atlantic croaker (*M. undulatus*) females exposed to E2 or the estrogen mimic *o,p′*-DDT experience increased GtH release and consequent ovarian growth, which may in turn lead to elevated E2 titres [84]. On the other hand, wild female kelp bass (*Paralabrax clathratus*) from the South California Bight showed *depressed* plasma GtH and E2, but elevated ovarian T production, and these effects were correlated with increased body-burdens of organochlorines [85]. Depressed plasma E2 has also been observed in flatfish (*Pleuronectes vetulus* and *P. bilineatus*) from areas of Puget Sound that are highly contaminated with organochlorines and PAHs [86,87].

The data described above show that endocrine disruptors can cause either increases or decreases in steroid titres of marine fish, depending on sex and contaminant type. These changes can in turn produce a cascade of effects, not the least of which are alterations in secondary sexual characteristics which are under sex steroid control.

ABNORMAL GONAD DEVELOPMENT OR SECONDARY SEXUAL CHARACTERISTICS

A number of instances of abnormal gonad development in marine fish have been attributed to pollutants, but these may of course not have been mediated by endocrine disrupting mechanisms. For example, the *Amoco Cadiz* oil spill off the coast of Brittany caused delayed ovarian development in plaice [88], and although this may have been an example of the antiestrogenic action of some PAHs, such effects could equally well be explained by generalized stress, or by reproductive toxins which do not act directly on the endocrine system.

The most frequently reported form of abnormal gonad development in wild marine fish which in some cases is almost certainly a form of endocrine disruption linked to estrogen exposure is the presence of intersex or ovotestis, i.e., primary or secondary oocytes in the testicular tissue of species which normally have completely separate sexes. The precise mechanism whereby this occurs is unknown, but the phenomenon can easily be induced experimentally through larval exposure (e.g., in sheepshead minnows [89]) but not when fish are already adult. In some of the very estrogenically contaminated UK estuaries (Mersey, Tyne, Clyde, and Forth), up to 15–20 % of male *P. flesus* and *Z. viviparus* at some stations show ovotestis (primary and occasionally secondary oocytes), and occasional examples of ovotestis have also been seen in *P. flesus* from less-contaminated situations [35,36,47,48,90]. However, ovotestis has never been observed in these species when caught from a relatively uncontaminated reference estuary (the Alde). It is noteworthy that the prevalence of ovotestis is rather low compared with that reported (up to 100 %) in freshwater roach from several UK rivers [22], but it is not known if this is due to interspecific differences in response, or to differential exposure.

There are clear interspecific differences, however, between *P. flesus* and *Z. viviparus* on the one hand, and the sand gobies *P. minutus/lozanoi* on the other [48]. Despite an intense sampling program in the United Kingdom, ovotestis has never been observed in the sand gobies, even though they inhabit similar areas to flounder and blenny. This mirrors the lack of VTG induction in gobies discussed above. However, recent data [48,61] show that many male gobies (up to 75 %) from estrogenic UK estuaries, while free of VTG and ovotestis, are exhibiting a condition known as morphologically intermediate papilla syndrome (MIPS) in which the urogenital papilla (used for depositing urine and sperm) takes on some of the characteristics of the equivalent female organ (used in oviposition). This condition has been induced in exposure experiments with E2 (10–1000 ng/l) in which juvenile fish were ambiently exposed for up to 32 weeks until adulthood, so it appears to be an effect of estrogen exposure on the expression of a secondary sexual characteristic. This suggests that the absence of VTG and ovotestis induction in wild gobies is more likely to be due to differential sensitivity than decreased exposure.

Examples of apparently abnormal, but relatively mild, ovotestis in marine fish have also been observed from a range of other locations, including the Seine estuary in France (*P. flesus* [91]), Tokyo Bay in Japan (*P. yokohamae* [51]), the southern Baltic in Germany (*Z. viviparus* [92]), and the Mediterranean (*Xiphias gladius* [115]). These cases generally consist of a few primary oocytes scattered throughout otherwise apparently normal testicular tissue. Indeed, other testicular changes such as abnormal spermatogenesis have not frequently been reported, although such abnormalities have been found in Tyne flounder [45], and the testicular tissue in some intersex flounder [47] appeared severely abnormal. Early exposure to estrogens may also cause precocious sexual development in females, and examples of this have been reported in marine flatfish [93,94], although it is not known whether the associated elevations in tissue concentrations of organochlorines were the causative factor.

As it is known that larval estrogen exposure of sufficient intensity can produce a completely female phenotype in genetically male fish [31], the cases of ovotestis cited above may well be incomplete examples of this phenomenon. This implies that some apparently normal females seen in contaminated estuaries etc. may be genetic males. This has not yet been tested in marine fish (although Nagler et al. [95] have observed this phenomenon in phenotypic female Chinook salmon from the Columbia River), due to an absence of suitable genetic sex markers, but it suggests that apparent sex ratios of fish populations exposed to estrogens may become biased toward females. Matthiessen et al. [47] did not observe sex ratios in UK flounder which differed substantially from 50:50, although it is extremely difficult to establish this with certainty under field conditions. More recently, it has been reported [48] that broods of *Z. viviparus* fry from the Forth, Clyde, and Tyne contain 52–60 % females, and although sufficient broods could not be caught from the reference estuary, other work [96,97] suggests that the sex ratio in uncontaminated broods of this species in the Baltic is 50:50. Furthermore, it has been shown [96,97] that broods from *Z. viviparus* females living near a large pulp-mill discharge on the Baltic are biased towards males (mean proportion of males up to 58 %), presumably in response to the known androgenic properties of pulp-mill effluent. It is interesting to note that this bias temporarily disappeared after a 17-day shutdown of the mill in 1999, which coincided with the period of embryonic gonadal differentiation [98].

REDUCED REPRODUCTIVE SUCCESS AND POPULATION-LEVEL EFFECTS

The observations discussed above raise the question of whether marine fish which have experienced disturbances including estrogen exposure, VTG induction, perturbed steroid titres, and ovotestis, etc. have been reproductively compromised, and whether populations are potentially at risk. This is not an easy question to answer because quantitative observations are hard to make in the field. Furthermore, any causative mechanisms for altered reproductive parameters or population declines in marine fish are likely to be multifactorial and include nonendocrine modes of action, as well as factors such as overfishing, habitat loss, and climate change [99]. Experimental work referred to above, however, certainly shows that the induction of VTG and ovotestis in males tends to be accompanied by a range of reproductive abnormalities in breeding groups, including reduced egg production, fertilization success, and fry survival. Whether any of these are likely to cause population declines in the field will depend on their magnitude, and on factors such as density-dependent compensatory effects, which tend to counteract population-level responses to relatively small changes in reproductive output.

There are some field data that suggest that high levels of lipophilic contaminants in marine fish can be associated with reduced reproductive success. For example, it has been shown [100] that a DDT-contaminated population of white croaker (*Genyonemus lineatus*) in California had decreased fecundity and/or spawning inducibility when individual levels of ovarian total DDT residues exceeded about 4000 ng/g wet wt. Similarly, a variety of flatfish species (e.g., *P. bilineatus* and *P. vetulus*) from contaminated areas of Puget Sound are showing precocious sexual maturation, retarded gonadal development, reduced egg weight, and reduced overall spawning success [87,94]. These fish are contaminated inter alia with a variety of aromatic hydrocarbons and PCBs, which are potentially implicated as causative agents, either as antiestrogens (some PAHs) or estrogen-mimics (some PCBs).

A number of marine field studies have linked reduced hatching success and fry survival to increased levels of lipophilic contaminants in eggs. For example, it has been shown [63] that elevated levels (>120 ng/g wet wt.) of PCBs in Baltic flounder (*P. flesus*) ovaries were correlated with impaired egg development and fry survival, and a similar study of Baltic herring (*Clupea harengus*) found that ovarian DDE residues of >18 ng/g wet wt. or PCB residues >120 ng/g wet wt. were significantly associated with reduced viable hatch [101]. High larval mortality and reduced hatching success in Baltic cod (*Gadus morhua*) have also been associated with organochlorines [102], but an early mortality syndrome known as M74 which is widespread in Baltic salmon (*Salmo salar*) seems to be caused by thiamine deficiency, although an indirect link to contaminants has not been ruled out [7,103].

In most of these species, there is massive redundancy in the numbers of fry produced, so it is unclear whether the effects described have significance for populations. One possible exception to this concerns the contamination of some Canadian rivers with the estrogen-mimic nonylphenol (sprayed as a pesticide coformulant during the control of forest pests). This was associated with subsequent major catch declines in returning salmon (*S. salar*) and blueback herring (*Alosa aestivalis*), and it has been speculated that the observations may have resulted from the known antagonism which gonadal steroids exert on the smoltification process [104]. In general, it remains to be established whether the causes of effects described above are indeed connected with endocrine disruption or some other effect. It should also be noted that the high levels of organochlorines and PAHs with which various observations have been associated are not necessarily causal because most study sites are generally contaminated with a complex of co-occurring substances. All these uncertainties imply the need for more laboratory-based reproductive studies with environmentally realistic dosing. Despite the difficulties, it is also necessary to conduct semifield experiments with free-breeding fish populations in order to establish true thresholds of environmental effect.

OTHER POSSIBLE EXAMPLES OF ENDOCRINE DISRUPTION IN MARINE FISH

At least two other major endocrine systems, the adrenal (located in fish in the interrenal and chromaffin tissue) and the thyroid, in addition to the estrogen signalling pathways, are subject to chemical interference in fish (see Topic 4.10 [8,105]), but such effects have not been widely studied in marine species. The adrenal system plays a vital part in responses to stress, while thyroid hormones influence a number of processes in fish, including neural development, metabolism, smoltification in salmonids, and metamorphosis in flatfish. It is worth noting, furthermore, that there is significant "cross-talk" between the different systems, so interfering with one is likely to cause knock-on effects in others. One example of this is the synergism between the thyroid hormones and the actions of gonadotropin hormone during early ovarian development [106]. Reciprocally, elevated plasma E2 can lead to depressed thyroid activity and reduced plasma levels of 3,5,3′-triiodo-L-thyronine (T3).

The only published example of interference with the thyroid system in marine fish concerns mummichogs (*Fundulus heteroclitus*) living in Piles Creek, New Jersey, an estuarine environment contaminated inter alia with heavy metals, PCBs and DDTs [107,108]. Compared with a clean reference site, these fish are sluggish with poor prey capture and avoidance of predators [109], and this has been attributed to interference with the thyroid system, perhaps through its involvement in neurological development. The fish from Piles Creek have larger thyroid follicles and follicular cell heights, and contain elevated titres of plasma thyroxine (T4), but not plasma or tissue T3. Furthermore, clean fish held in conditions simulating Piles Creek also develop elevated T4 titres. Interestingly, the Piles Creek fish do *not* show reproductive abnormalities, their gametes and embryos having developed some contaminant-resistance [110]. Mechanisms of action are still not fully understood, but "cross-talk" from the adrenal system may be partly involved [107]. Normally, the stress hormone cortisol stimulates conversion of T4 to T3, but chronic organic and metallic pollution can nonspecifically "exhaust" the cortisol-secreting interrenal cells [111,112], possibly leading in turn to a build-up of T4. However, there are several alternative (but not necessarily mutually exclusive) explanations, including heavy metal inhibition of the enzyme (5′-monodeiodinase) which catalyses T4 to T3 conversion. Pollution has been frequently shown to cause a reduced cortisol response to stress in freshwater fish, but the only known example at present from the marine environment concerns marbled sole (*Pleuronectes yokohamae*) caught in the contaminated Tokyo Bay [113]. This effect was associated with impaired cortisol release from the head kidney in response to adrenocorticotrophic hormone (ACTH) challenge in vitro.

The example of thyroid disturbance in *F. heteroclitus* serves well to illustrate the potential complexity of endocrine-disrupting effects in fish, and the extent of our ignorance of this subject outside the restricted (but important) field of estrogenic impacts.

CONCLUSIONS AND RECOMMENDATIONS

The main conclusion that arises from this brief review is that marine fish are no less susceptible to endocrine disruption than their freshwater relatives, and display as wide a range of effects as those seen in the latter group. Not surprisingly, most of the reported phenomena have been observed in estuarine and coastal fish, probably because they are more frequently studied, and generally exposed to higher levels of contaminants, than open-sea or oceanic species. Having said that, it is noteworthy that there may be a relationship between the bioaccumulation of lipophilic contaminants and endocrine disruption in large pelagic fish, which suggests that species living in the open sea should not be ignored in this respect. An analogy might be made with the large marine mammals, which are well known to accumulate lipophilic materials.

Marine fish have broadly similar endocrinology and physiology to freshwater species, so the lack of major differences is not unexpected. There are, however, some indications from this review that interference with the hormone systems controlling osmoregulation and smoltification may be an important factor in some euryhaline or marine species. In particular, the discovery that pollutant-induced disruption of the thyroid system can occur in marine fish opens up the possibility not only of impaired neural development, etc., but also of disrupted ionic balance in euryhaline species, and of potential interference with the metamorphosis of flatfish larvae. To the author's knowledge, these subjects have as yet received scant attention.

Protoandrous and protogynous species (which are relatively common in the sea) may also be at risk from endocrine disruptors that could trigger sex changes at inappropriate times. This possibility has not been investigated under natural conditions, but preliminary experimental work with the protogynous cuckoo wrasse (*Labrus bimaculatus/mixtus*) (Hylland in [74]) has shown that high doses of methyl testosterone (MT) were able to prevent the color changes associated with natural masculinization. This apparently paradoxical result was probably caused by the aromatization of MT to estrogen, but it suggests that further work on such species under polluted conditions might be worthwhile. Similar comments apply to marine protoandrous species such as the black porgy (*Acanthopagrus schlegeli*), which can be reversibly sex-changed by exposure to exogenous estrogens [114].

None of the available marine data have yet made a credible connection between the plethora of changes in individual fish caused by endocrine disruptors, and potential impacts on fish populations. If for no other reason than that marine fish form the basis of many important fisheries, this is a serious information gap. In particular, there is a need for experimental studies of free-breeding fish populations to establish the degree to which they are stable to a variety of endocrine-induced disturbances at the individual level. This endeavor would be assisted by the development of suitable population models, although their value would be limited by the availability of species-specific life history parameters.

In summary, the following research recommendations flow from this review:

- Experimental studies of free-breeding fish populations are required to investigate their stability to perturbation by endocrine disruptors.
- The potential endocrine-disrupting effects of bioaccumulated lipophilic substances, especially in long-lived pelagic fish species, need further research.
- More field studies of marine fish are needed to discover the importance of disruption of the adrenal and thyroid systems.
- The potential for interference with the endocrine control of osmoregulation and smoltification is worthy of further study.
- The likely impact of sex steroids and their mimics on naturally hermaphrodite, protogynous, and protoandrous marine fish species should be investigated in the field.

REFERENCES

1. D. E. Kime. *Endocrine Disruption in Fish*, Kluwer Academic, London (1998).
2. A. Hontela, J. B Rasmussen, G. Chevalier. *Water Pollut. Res. J. Can.* **28**, 767–780 (1993).
3. J. F. Leatherland. *J. Great Lakes Res.* **19**, 737–751 (1993).
4. J. F. Leatherland. *Guelph Ichthyol. Rev.* **2**, 67 (1994).
5. J. P. Sumpter. *Toxicol. Lett.* **82/83**, 737–742 (1995).
6. J. P. Sumpter, S. Jobling, C. R. Tyler. In *Toxicology of Aquatic Pollution: Physiological, Molecular and Cellular Approaches*, E. W. Taylor (Ed.), pp. 205–224, Cambridge University Press, Cambridge (1996).
7. H. Börjeson and L. Norrgren. In *Chemically Induced Alterations in Functional Development and Reproduction of Fishes*, R. M. Rolland, M. Gilbertson, R. E. Peterson (Eds.), pp. 153–166, SETAC Technical Publications Series, SETAC Press, Pensacola, FL (1997).
8. A. Hontela. *Rev. Toxicol.* **1**, 1–46 (1997).
9. S. V. Marcquenski and S. B. Brown. In *Chemically Induced Alterations in Functional Development and Reproduction of Fishes,* R. M. Rolland, M. Gilbertson, R. E. Peterson (Eds.), pp. 135–152, SETAC Technical Publications Series, SETAC Press, Pensacola, FL (1997).
10. P. Stahlschmidt-Allner, B. Allner, J. Römbke, T. Knacker. *Environ. Sci. Pollut. Res.* **4**, 155–162 (1997).
11. J. Schwaiger and R. D. Negele. *Acta Vet. Brno* **67**, 257–264 (1998).
12. A. Arukwe and A. Goksøyr. *Sarsia* **83**, 225–241 (1998).
13. J. P. Giesy and E. M. Snyder. In *Principles and Processes for Evaluating Endocrine Disruption in Wildlife,* R. J. Kendall, R. L. Dickerson, J. P. Giesy, W. A. Suk (Eds.), pp. 155–237, SETAC Technical Publication, SETAC Press, Pensacola, FL (1998).
14. P. Matthiessen. In *Principles and Processes for Evaluating Endocrine Disruption in Wildlife*, R. J. Kendall, R. L. Dickerson, J. P. Giesy, W. A. Suk (Eds.), pp. 239–247, SETAC Technical Publication, SETAC Press, Pensacola, FL (1998).

15. P. Matthiessen and J. P. Sumpter. In *Fish Ecotoxicology*, E T. Braunbeck, D. E. Hinton, B. Streit (Eds.), pp. 319–335, Birkhäuser Verlag, Basel (1998).
16. C. R. Tyler, S. Jobling, J. P. Sumpter. *Crit. Rev. Toxicol.* **28**, 319–361 (1998).
17. G. Van Der Kraak, K. R. Munkittrick, M. E. McMaster, D. L. MacLatchy. In *Principles and Processes for Evaluating Endocrine Disruption in Wildlife,* R. J. Kendall, R. L. Dickerson, J. P. Giesy, W. A. Suk (Eds.), pp. 249–265, SETAC Technical Publication, SETAC Press, Pensacola, FL (1998a).
18. G. Van Der Kraak, T. Zacharewski, D. M. Janz, B. M. Sanders, J. W. Gooch. In *Principles and Processes for Evaluating Endocrine Disruption in Wildlife,* R. J. Kendall, R. L. Dickerson, J. P. Giesy, W. A. Suk (Eds.), pp. 97–119, SETAC Technical Publication, SETAC Press, Pensacola, FL (1998b).
19. G. Van Der Kraak, M. Hewitt, A. Lister, M. E. McMaster, K. R. Munkittrick. *Human Ecol. Risk Assess.* **7**, 1017–1025 (2001).
20. M. E. McMaster. *Water Qual. Res. J. Can.* **36**, 215–231 (2001).
21. E. Oberdörster and A. O. Cheek. *Environ. Toxicol. Chem.* **20**, 23–36 (2000).
22. S. Jobling, M. Nolan, C. R. Tyler, G. Brighty, J. P. Sumpter. *Environ. Sci. Technol.* **32**, 2498–2506 (1998).
23. T. Damstra, S. Barlow, A. Bergman, R. Kavlock, G. Van Der Kraak (Eds.). In *Global Assessment of the State-of-the-Science of Endocrine Disruptors.* WHO/ILO/UNEP International Programme on Chemical Safety, Geneva (available on-line: <http://endocrine.ei.jrc.it/final%20draft/who.html>) (2002).
24. M. E. Baker. *Mol. Cell. Endocrinol.* **175**, 1–4 (2001).
25. W. M. Howell, D. A. Black, S. A. Bortone. *Copeia* 676–681 (1980).
26. G. J. Van Der Kraak, K. R. Munkittrick, M. E. McMaster, C. B. Portt, J. P. Chang. *Toxicol. Appl. Pharmacol.* **115**, 224–233 (1992).
27. C. E. Purdom, P. A. Hardiman, V. J. Bye, N. C. Eno, C. R. Tyler, J. P. Sumpter. *Chem. Ecol.* **8**, 275–285 (1994).
28. J. E. Harries, D. A. Sheahan, S. Jobling, P. Matthiessen, P. Neall, E. Routledge, R. Rycroft, J. P. Sumpter, T. Tylor. *Environ. Toxicol. Chem.* **15**, 1993–2002 (1996).
29. R. Reinboth. *Environ. Biol. Fish.* **22**, 249–259 (1988).
30. G. A. Hunter and E. M. Donaldson. In *Fish Physiology,* Vol. 9B, W. S. Hoar, D. J. Randall, E. M. Donaldson (Eds.), pp. 223–303, Academic Press, New York (1983).
31. F. Piferrer. *Aquaculture* **197**, 229–281 (2001).
32. T. J. Pandian and S. G. Sheela. *Aquaculture* **138**, 1–22 (1995).
33. S. J. Hyllner, D. O. Oppen-Berntsen, J. V. Helvik, B. T. Walther, C. Haux. *J. Endocrinol.* **131**, 229–236 (1991).
32. C. B. Lazier and M. E. MacKay. In *Biochemistry and Molecular Biology of Fishes* Vol. 2, P. W. Hochachka and T. P. Mommsen (Eds.), pp. 391–405, Elsevier Science, New York (1993).
35. Y. Allen, A. P. Scott, P. Matthiessen, S. Haworth, J. E. Thain, S. Feist. *Environ. Toxicol. Chem.* **18**, 1791–1800 (1999a).
36. Y. Allen, P. Matthiessen, A. P. Scott, S. Haworth, S. Feist, J. E. Thain. *Sci. Total Environ.* **233**, 5–20 (1999b).
37. L. J. Christensen, B. Korsgaard, P. Bjerregaard. *Aquat. Toxicol.* **46**, 211–219 (1999).
38. R. L. Herman and H. L. Kincaid. *Aquaculture* **72**, 165–172 (1988).
39. V. J. Kramer, S. Miles-Richardson, S. L. Pierens, J. P. Giesy. *Aquat. Toxicol.* **40**, 335–360 (1998).
40. L. C. Folmar, G. R. Gardner, M. P. Schreibman, L. Magliulo-Cepriano, L. J. Mills, G. Zaroogian, R. Gutjahr-Gobell, R. Haebler, D. B. Horowitz, N. D. Denslow. *Aquat. Toxicol.* **51**, 431–441 (2001).
41. A. O. Cheek, T. H. Brouwer, S. Carroll, S. Manning, J. A. McLachlan, M. Brouwer. *Environ. Health Perspect.* **109**, 681–690 (2001).

42. R. Länge, T. H. Hutchinson, C. P. Croudace, F. Siegmund. *Environ. Toxicol. Chem.* **20**, 1216–1227 (2001).
43. K. Van den Belt, R. Verheyen, H. Witters. *Arch. Environ. Contam. Toxicol.* **41**, 458–467 (2001).
44. C. M. Lye, C. L. J. Frid, M. E. Gill, D. McCormick. *Mar. Pollut. Bull.* **34**, 34–41 (1997).
45. C. M. Lye, C. L. J. Frid, M. E. Gill. *Mar. Ecol. Progr. Ser.* **170**, 249–260 (1998).
46. C. M. Lye, C. L. J. Frid, M. E. Gill, D. W. Cooper, D. M. Jones. *Environ. Sci. Technol.* **33**, 1009–1014 (1999).
47. P. Matthiessen, Y. T. Allen, C. R. Allchin, S. W. Feist, M. F. Kirby, R. J. Law, A. P. Scott, J. E. Thain, K. V. Thomas. *Science Series Technical Report* No. 107, Centre for Environment, Fisheries and Aquaculture Science, Lowestoft (1998).
48. P. Matthiessen, Y. Allen, S. Bamber, J. Craft, M. Hurst, T. Hutchinson, S. Feist, I. Katsiadaki, M. Kirby, C. Robinson, S. Scott, J. Thain, K. Thomas. *Mar. Environ. Res.* **54**, 645–649 (2002).
49. K. V. Thomas, M. Hurst, P. Matthiessen, M. J. Waldock. *Environ. Toxicol. Chem.* **20**, 2165–2170 (2001).
50. L. L. Madsen, T. Aagaard, B. Korsgaard, P. Bjerregaard. Poster paper presented at PRIMO Symposium, Plymouth (2001).
51. S. Hashimoto, H. Bessho, A. Hara, M. Nakamura, T. Iguchi, K. Fujita. *Mar. Environ. Res.* **49**, 37–53 (2000).
52. H. Bessho, S. Iwakami, N. Hiramatsu, A. Hara, S. Hashimoto. *Int. J. Environ. Anal. Chem.* **76**, 155–166 (2000).
53. D. P. Lomax, L. L. Johnson, W. T. Roubal, J. E. West, S. M. O'Neill, T. K. Collier. *Abnormal production of vitellogenin in marine fish from urban embayments in Puget Sound, Washington, USA.* Poster presented to 11th International Symposium on Pollutant Responses in Marine Organisms (PRIMO), Plymouth, UK, 10–13 July 2001 (2001).
54. K. Yoneyama, A. Hara, T. Matsubara, H. Ishibashi, K. Arizono, Y. Oshima, K. Fukudome, K. Kubo, K. Soyano. *Environ. Sci. (Tokyo)* **8**, 155 (2001).
55. A. D. Vethaak, J. Lahr, G. Grinwis, A. Gerritsen. *Toxicology* **164**, 12 (2001).
56. J. R. Todorov, A. A. Elskus, D. Schlenk, P. L. Ferguson, B. J. Brownawell, A. E. McElroy. *Mar. Environ. Res.* **54**, 691–695 (2002).
57. F. R. Knudsen, A. Arukwe, T. G. Pottinger. *Environ. Pollut.* **103**, 75–80 (1998).
58. T. L. Metcalfe, C. D. Metcalfe, Y. Kiparissis, A. J. Nimi, C. M. Foran, W. H. Benson. *Environ. Toxicol. Chem.* **19**, 1893–1900 (2000).
59. D. R. Livingstone, C. L. Mitchelmore, L. D. Peters, S. C. M. O'Hara, J. P. Shaw, B. S. Chesman, A. Doyotte, J. McEvoy, D. Ronisz, D. G. J. Larsson, L. Förlin. *Mar. Environ. Res.* **50**, 367–371 (2000).
60. L. D. Peters, A. Doyotte, C. L. Mitchelmore, J. McEvoy, D. R. Livingstone. *Sci. Total Environ.* **279**, 137–150 (2001).
61. M. F. Kirby, J. Bignell, E. Brown, J. A. Craft, I. Davies, R. A. Dyer, S. W. Feist, G. Jones, P. Matthiessen, C. Megginson, F. E. Robertson, C. Robinson. *Environ. Toxicol. Chem.* **22**, 239–251 (2003).
62. M. C. Fossi, S. Casini, S. Ancora, A. Moscatelli, A. Ausili, G. Notarbartolo-di-Sciara. *Mar. Environ. Res.* **52**, 477–483 (2001).
63. H. Von Westernhagen, H. Rosenthal, V. Dethlefsen, W. Ernst, U. Harms, P.-D. Hansen. *Aquat. Toxicol.* **1**, 85–99 (1981).
64. P. A. H. Janssen. Reproduction of the flounder, *Platichthys flesus* (L.) in relation to environmental pollution. Steroids and vitellogenesis. Ph.D. thesis, University of Utrecht, The Netherlands (1996).
65. P. A. H. Janssen, J. G. D. Lambert, A. D. Vethaak, H. J. T. Goos. *Aquat. Toxicol.* **39**, 195–214 (1997).
66. P. Thomas. *Mar. Environ. Res.* **28**, 499–503 (1989).

67. P. Thomas. *J. Exp. Zool.* **4**, 126–128 (1990).
68. R. B. Spies, J. J. Stegeman, D. W. Rice, B. Woodin, P. Thomas, J. E. Hose, J. N. Cross, M. Prieto. In *Biomarkers of Environmental Contamination,* J. F. McCarthy and L. R. Shugart (Eds.), pp. 87–121, CRC Press, Boca Raton, FL (1990).
69. E. Casillas, D. Misitano, L. L. Johnson, L. D. Rhodes, T. K. Collier, J. E. Stein, B. B. McCain, U. Varanasi. *Mar. Environ. Res.* **31**, 99–122 (1991).
70. J. J. Pereira, J. Ziskowski, R. Mercaldo-Allen, C. Kuropat, D. Luedke, E. Gould. *Estuaries* **15**, 289–297 (1992).
71. J. F. Carragher, J. P. Sumpter, T. G. Pottinger, A. D. Pickering. *Gen. Comp. Endocrinol.* **76**, 310–321 (1989).
72. S. M. Ruby, D. R. Idler, Y. P. So. *Aquat. Toxicol.* **26**, 91–102 (1993).
73. J.-M. Nicolas. *Aquat. Toxicol.* **45**, 77–90 (1999).
74. COMPREHEND. Final Report to the European Commission, Contract No. ENV4-CT98-0798 (2002).
75. C. Desbrow, E. J. Routledge, G. Brighty, J. P. Sumpter, M. Waldock. *Environ. Sci. Technol.* **32**, 1549–1558 (1998).
76. I. Katsiadaki, A. P. Scott, P. Matthiessen. In *Proceedings of the 6th International Symposium on the Reproductive Physiology of Fish,* B. Norberg, O. S. Kjesbu, G. L. Taranger, E. Andersson, S. O. Stefansson (Eds.), pp. 359–361, Institute of Marine Research and University of Bergen, 4–9 July 1999 (2000).
77. I. Katsiadaki, A. P. Scott, M. Hurst, P. Matthiessen, I. Mayer. *Environ. Toxicol. Chem.* **21**, 1946–1954 (2002).
78. K. V. Thomas, M. R. Hurst, A. Smith, M. McHugh, P. Matthiessen, M. Waldock. *Environ. Toxicol. Chem.* **21**, 1456–1461 (2002).
79. M. G. Dubé and D. L. MacLatchy. *Environ. Toxicol. Chem.* **19**, 2788–2796 (2000).
80. M. G. Dubé and D. L. MacLatchy. *Environ. Toxicol. Chem.* **20**, 985–995 (2001).
81. Å. Larsson, L. Förlin, E. Lindesjöö, O. Sandström. In *Proc. 3rd Int. Conf. on Environmental Fate and Effects of Pulp and Paper Mill Effluents, November 1997*, Rotorua, New Zealand (1997).
82. P. R. R. Monteiro, M. A. Reis-Henriques, J. Coimbra. *Aquat. Toxicol.* **48**, 549–559 (2000).
83. A. P. Scott, C. Stewart, Y. Allen, P. Matthiessen. In *Proceedings of the 6th International Symposium on the Reproductive Physiology of Fish*, B. Norberg, O. S. Kjesbu, G. L. Taranger, E. Andersson, S. O. Stefansson (Eds.), p. 382, Institute of Marine Research and University of Bergen, 4–9 July 1999 (2000).
84. I. A. Khan and P. Thomas. *Mar. Environ. Res.* **46**, 149–152 (1998).
85. R. B. Spies and P. Thomas. In *Chemically Induced Alterations in Functional Development and Reproduction of Fishes,* R. M. Rolland, M. Gilbertson, R. E. Peterson (Eds.), pp. 113–133, SETAC Press, Pensacola, FL (1997).
86. L. L. Johnson, E. Casillas, T. K. Collier, B. B. McCain, U. Varanasi. *Can. J. Fish. Aquat. Sci.* **45**, 2133–2146 (1988).
87. L. L. Johnson, D. Misitano, S. Y. Sol, G. M. Nelson, B. French, M. Ylitalo, T. Hom. *Trans. Am. Fish. Soc.* **127**, 375–392 (1998).
88. G. G. Stott, W. E. Haensly, J. M. Neff, J. R. Sharpe. *J. Fish Dis.* **6**, 429–437 (1983).
89. E. J. Zillioux, I. C. Johnson, Y. Kiparissis, C. D. Metcalfe, J. V. Wheat, S. G. Ward, H. Liu. *Environ. Toxicol. Chem.* **20**, 1968–1978 (2001).
90. M. G. Simpson, M. Parry, A. Kleinkauf, D. Swarbreck, P. Walker, R. T. Leah. *Mar. Environ. Res.* **50**, 283–287 (2000).
91. C. Minier, F. Levy, D. Rabel, G. Bocquené, D. Godefroy, T. Burgeot, F. Leboulenger. *Mar. Environ. Res.* **50**, 373–377 (2000).
92. J. Gercken and H. Sordyl. *Mar. Environ. Res.* **54**, 651–655 (2002).

93. L. L. Johnson, J. E. Stein, T. K. Collier, E. Casillas, B. McCain, U. Varanasi. U.S. Department of Commerce, National Oceanic and Atmospheric Administration, NOAA Technical Memorandum NMFS-NWFSC, pp. 1, 76 (1992).
94. T. K. Collier, L. L. Johnson, C. M. Stehr, M. S. Myers, J. E. Stein. *Mar. Environ. Res.* **46**, 243–247 (1998).
95. J. J. Nagler, J. Bouma, G. H. Thorgaard, D. D. Dauble. *Environ. Health Perspect.* **109**, 67–69 (2001).
96. D. G. J. Larsson, H. Hällman, L. Förlin. *Mar. Environ. Res.* **50**, 191–192 (2000a).
97. D. G. J. Larsson, H. Hällman, L. Förlin. *Environ. Toxicol. Chem.* **19**, 2911–2917 (2000b).
98. L. Förlin and D. G. J. Larsson. In *Male-biased sex ratios of fish embryos near a pulp mill: temporary recovery after a short-term shutdown.* Poster presented to the 11th International Symposium on Pollutant Responses in Marine Organisms, (PRIMO), 10–13 July 2001, Plymouth, UK (2001).
99. A. D. Rijnsdorp and A. D. Vethaak. In *Changes in reproductive parameters of North Sea plaice and sole between 1960 and 1995,* p. 9 + tables and figs., International Council for the Exploration of the Sea, Copenhagen, ICES C.M. 1997/U:14 (1997).
100. J. E. Hose, J. N. Cross, S. G. Smith, D. Diehl. *Environ. Pollut.* **57**, 139–148 (1989).
101. P.-D. Hansen, H. von Westernhagen, H. Rosenthal. *Mar. Environ. Res.* **15**, 59–76 (1985).
102. G. I. Petersen, J. Gerup, L. Nilsson, J. R. Larsen, R. Schneider. In *Body burdens of lipophilic xenobiotics and reproductive success in Baltic cod* (*Gadus morhua* L.), p. 22, International Council for the Exploration of the Sea, Copenhagen, ICES CM 1997/U:10 (1997).
103. G. Åkerman and L. Balk. *Am. Fish. Soc. Symp.* **21**, 41–61 (1998).
104. W. L. Fairchild, E. O. Swansburg, J. T. Arsenault, S. B. Brown. *Environ. Health Perspect.* **107**, 349–357 (1999).
105. R. M. Rolland. *J. Wildl. Dis.* **36**, 615–635 (2000).
106. D. G. Cyr and J. G. Eales. *Rev. Fish Biol. Fish* **6**, 165–200 (1996).
107. T. Zhou, H. B. John-Alder, P. Weis, J. S. Weis. *Environ. Toxicol. Chem.* **18**, 2817–2823 (1999).
108. T. Zhou, H. B. John-Alder, J. S. Weis, P. Weis. *Mar. Environ. Res.* **50**, 393–397 (2000).
109. G. M. Smith and J. S. Weis. *J. Exp. Mar. Biol. Ecol.* **209**, 75–87 (1997).
110. J. S. Weis and P. Weis. *BioScience* **39**, 89–95 (1989).
111. A. Hontela, D. Pierre, D. Dominick, R. Fortin. *Environ. Toxicol. Chem.* **14**, 725–731 (1995).
112. D. O. Norris, S. B. Felt, J. D. Woodling, R. M. Dores. *Gen. Comp. Endocrinol.* **108**, 343–351 (1997).
113. I. Kakuta. *Environ. Toxicol.* **17**, 1–6 (2002).
114. Y. H. Lee, J. L. Du, W. S. Yueh, B. Y. Lin, J. D. Huang, C. Y. Lee, M. F. Lee, E. L. Lau, F. Y. Lee, C. Morrey, Y. Nagahama, C. F. Chang. *J. Exp. Zool.* **290**, 715–726 (2001).
115. G. De Metrio, A. Corriero, S. Desantis, D. Zubani, F. Cirillo, M. Deflorio, C. R. Bridges, J. Eicker, J. M. de la Serna, P. Megalofonou, D. E. Kime. *Mar. Pollut. Bull.* **46**, 358–361 (2003).

Pure Appl. Chem., Vol. 75, Nos. 11–12, pp. 2263–2273, 2003.

Topic 4.6
Deformed frogs and environmental retinoids*

David Gardiner, Aristocle Ndayibagira, Felix Grün, and Bruce Blumberg[‡]

Department of Developmental and Cell Biology, University of California, Irvine, CA 92697-2300, USA

Abstract: Since the early 1990s, a substantial number of deformed frogs have been observed in North America, particularly in the upper Midwest and Canada. Attempts to understand the etiology of the deformed frog problem have met with limited success to date with nearly as many proposed explanations as research groups working on the problem. Models for the mechanism underlying the development of deformed frogs include parasite/predation, ultraviolet radiation, and chemical exposure. Each model has its strengths and weaknesses. Despite contentious debate among researchers, there is an overall consensus that the increasing prevalence of deformed frogs is the result of a water-borne contaminant that has recently appeared, or reached a critical concentration. Our detailed analysis of malformed frogs collected in Minnesota ponds and lakes suggested that limb patterning was being modified by the disruption of a retinoid-sensitive developmental signaling pathway. Accordingly, we focused in the identification and characterization of bioactive retinoids from lake water and showed that retinoid treatment of frog embryos at sensitive times of development could recapitulate the full spectrum of limb abnormalities observed in field specimens in the laboratory. These data have led to the conclusion that inappropriate modulation of retinoid signaling by environmental contaminants is the mechanism underlying the increased incidence of frog malformations.

INTRODUCTION

Development of the embryo is a very sensitive and vulnerable part of the life cycle of all animals. The molecular mechanisms controlling embryonic development are highly conserved among vertebrates, and components of these pathways are frequently reused in regulating adult physiology [1–3]. Hence, agents that disrupt physiology or development in animals are likely candidates to similarly impact human physiology and development. The amphibian model system is especially suitable for identifying potential risks to human development. This results both from strong conservation of developmental pathways and because amphibian development is accessible to environmental contaminants in the water. Early warnings provided by monitoring wild populations of amphibians are especially important in cases of non-point source contamination, where it is not possible to identify appropriate sites for direct chemical monitoring.

There has been a dramatic increase in the numbers of deformed amphibians found in North America and Japan since the early 1990s. At the same time, a worldwide decline in the numbers of

---

*Report from a SCOPE/IUPAC project: Implication of Endocrine Active Substances for Human and Wildlife (J. Miyamoto and J. Burger, editors). Other reports are published in this issue, *Pure Appl. Chem.* **75**, 1617–2615 (2003).

[‡]Corresponding author

many amphibian species, particularly the highly aquatic ones, has also been observed. This suggests that environmental modification is negatively impacting amphibian populations. While the cause of amphibian declines in relatively pristine environments remains unknown, there is an emerging consensus that the increasing prevalence of deformed frogs is the result of a water-borne contaminant that has appeared, or reached a critical concentration, in recent years. This conclusion is based on several lines of evidence. When it first became evident that the incidence of amphibian malformation was significantly greater than the historical rate [4,5], it was noted that the incidence of malformation was correlated with the life history of the frogs. Thus, highly aquatic species such as green frogs and mink frogs have high rates of malformations (>>50 %); whereas, primarily terrestrial species such as wood frogs have malformation rates closer to the historical rate (<1 %). Species that spend intermediate periods of embryonic and larval development in the water, such as leopard frogs, exhibit intermediate rates of malformations (5–50 %).

A number of models have been proposed to explain the occurrence of malformed frogs in North America. These include UV radiation, parasites, and environmental contamination of various sorts. The topic has been recently reviewed although it should be noted that the reviews contain significant errors of fact and interpretation as well as assumptions that are not clearly stated [6–8]. As Stocum correctly points out [8], the number of hypotheses, opinions and viewpoints about the causes of amphibian malformations currently exceeds the amount of data available with which to analyze the problem.

In the brief review that follows, we summarize the current state of knowledge regarding the deformed frog investigation. In contrast with other investigators who believe that there may be nearly as many different causes as sites under study [6], we conclude that there is evidence to support an hypothesis that a single causal mechanism can account for the vast majority of the observed phenotypes in wild amphibian populations.

ULTRAVIOLET RADIATION

Exposure to UV radiation was originally postulated to be a potential source of amphibian malformations. This model can potentially explain why deformed frog reports are highest in the Northern Midwest where UV radiation is abundant presumably due to ozone depletion in the upper atmosphere. But, as has been noted elsewhere [6–8], there is little cause to suspect that UV radiation is a significant source of amphibian malformations. UV treatment of embryos should lead to destruction of the apical ectodermal ridge. Since this structure is required for limb outgrowth, the expected result is bilateral, blunt-ended truncations. The types of malformations observed in field specimens are inconsistent with these effects in that blunt truncations are relatively rare.

There are data supporting two other potential roles for UV in amphibian malformations and declines. Blaustein and colleagues have demonstrated a positive association between UV exposure and amphibian decline [9,10]. A second possibility is that UV interacts with environmental chemicals leading to the production of new metabolites [11–13]. In this scenario, high levels of UV (as found in the upper Midwest) would interact locally with ubiquitous and otherwise innocuous chemicals to produce teratogens. This possibility could account for the geographic distribution of malformed frogs.

PARASITES

The parasite model for frog malformations is interesting and potentially significant. Considering that frog malformations have appeared sporadically in the literature for hundreds of years [14], it is seductive to consider that what we now observe is merely an expansion of a historical phenomenon and therefore, no cause for alarm.

What is commonly referred to as the parasite model [14–17] is really two conjoined models. Supernumerary limbs are explained by larval trematodes (*Ribeiroia ondatrae*) burrowing into the developing limb as first speculated by Sessions and Ruth [16]. Missing and reduced limbs are explained as predation by fishes and invertebrates [4,14,15,17]. Some authors believe that mechanical perturbations by the burrowing *Ribeiroia* larvae are responsible for limb duplications [14–16] although the available experimental evidence demonstrates only the production of duplicated digits by simulated mechanical perturbations in the laboratory [16]. Therefore, the conclusion that mechanical perturbation causes limb duplications is as yet unsupported by any published experimental data.

Johnson and colleagues performed laboratory experiments testing the association between *R. ondatrae* exposure and limb malformations [17,18]. These studies utilized *Ribeiroia* sp. metacercaria and wild-caught Pacific tree frog (*Hyla regilla*) [17] or Western toad (*Bufo boreas*) [18] tadpoles. Johnson and colleagues demonstrated that exposure to *Ribeiroia* metacercaria was associated with duplicated limbs, reduced limbs, missing limbs, and bony triangles in a dose-dependent manner [17,18]. The authors conclude that the exposure to *Ribeiroia* ondatrae metacercaria is associated with limb malformations and correctly observe that despite this correlation, no inferences were possible regarding the mechanism through which *Ribeiroia* act. One such mechanism could be mechanical perturbation, although one immediately wonders why only a single species of trematode appears to be responsible. It should be possible for many types of trematode and other parasite larvae to mechanically disrupt the limb field. Indeed, wild caught frogs are often heavily infested with parasites of all kinds with no correlation between parasite number or location and limb malformations. We consider it more likely that *R. ondatrae* either produce factors that mimic vertebrate signaling molecules or elicit the host to produce such factors at inappropriate times, locations, or levels during development. Other authors have reached similar conclusions [17,18], and this is a ripe area for future study.

It has been proposed that parasite infestation is the most parsimonious explanation for frog malformations [14]. However, proponents of this idea invoke two very different phenomena (parasites and predation) in a wholly nonparsimonious manner that explains only a subset of malformations. The majority of frog malformations in the field are limb reductions that are inconsistent with predation [5,7,19]. In addition, most affected animals die at metamorphosis with a variety of craniofacial and gastrointestinal defects in addition to limb deformities. Therefore, the parasite model, as currently articulated [14–16] is an unsatisfactory explanation for the majority of frog malformations found in the wild.

CHEMICALS

Studies by the NIEHS, based on an early amphibian development assay (FETAX: frog embryo teratogenesis assay xenopus) have detected agents in the water at sites with malformed frogs that induce developmental defects [20]. Field data regarding the occurrence of frog deformities indicate a correlation between intensive agriculture and deformed frogs [5]. Thus, agrochemical contamination of surface water is considered a source of the causative agent(s). In addition to malformations, the catastrophic declines and die-offs that have occurred globally appear in some cases to be a consequence of exposure to agrochemicals [21], although at this point the relationship between amphibian declines and malformations is unclear. It has been suggested that frog malformations are caused by different combinations of chemicals at each site through interactions between the chemical mixtures and components of the local water chemistry [20]. While it is unreasonable to exclude possible mechanisms, a priori, it seems equally unreasonable to speculate that different chemical mixtures at each site elicit the same or similar malformations.

RETINOIDS

Extensive research from many laboratories has established an essential role for retinoids in a large number of cellular processes (recently reviewed in [22–25]). Among these are development and patterning of limbs and the central nervous system. The precise quantities of retinoids present during early development are especially critical; serious developmental defects result from either too much or too little retinoid signaling. Retinoid signaling is primarily mediated through the activities of two classes of nuclear hormone receptors, the RARs and RXRs. The endogenous activators of RAR are thought to be all-trans retinoic acid (atRA) and 9-cis RA (9cRA). RXR is activated by 9cRA but not by atRA. RAR requires heterodimerization with RXR in order to bind DNA and activate transcription of RAR target genes. An interesting and significant feature of RXR is that, in addition to heterodimerizing with RAR, it is a common heterodimeric partner for ten other families of nonsteroidal nuclear receptors including the thyroid hormone and vitamin D3 receptors [26,27]. Therefore, inappropriate modulation of retinoid signaling would be expected to have pleiotropic effects on development and adult physiology.

We analyzed skeletal dysplasias observed in severely affected frogs from Minnesota and identified two classes of common limb abnormalities [19]. First, supernumerary or absent limbs were observed, suggesting that the process of limb initiation was being affected. Second, primary and supernumerary limbs both showed characteristic skeletal abnormalities, including truncated and phocomelic limbs, suggesting that limb growth and pattern formation are also being modified. In the phocomelic limbs, the skeletal elements within affected segments are folded back on themselves, such that the proximal and distal ends of the bone lie adjacent to one another and the mid-portion of the bone projects laterally, forming a "bony triangle" (BT) (Fig. 1).

Exposure of developing limb buds to teratogenic retinoids (e.g., retinoic acid) in frogs, mice, and chicks can induce the BT phenotype [28–31]. No analysis of the teratogenic mechanisms underlying this dysplasia has yet been carried out. We have not found evidence of any other teratogenic agent that induces this unique skeletal dysplasia, suggesting that BTs are diagnostic for retinoid exposure. It is possible that more than one agent is responsible for the whole spectrum of different observed abnormalities. However, we note that retinoic acid can induce supernumerary limbs in early mouse embryos [32,33] as well as developing limbs of chicks [31,34,35] and amphibians [36,37], and regenerating tails of amphibians [28,38,39]. Incomplete and missing limbs can also be induced in response to retinoid treatment, depending on the mode of exposure (reviewed in [40,41]). Thus, it is possible that a single agent, such as a retinoid, could be responsible for all of the observed malformation phenotypes.

The issue of the types of duplications induced by retinoid exposure has been raised previously [6,8,14,15]. In each case, the authors incorrectly state that retinoid exposure causes only proximodistal (PD) duplications in regenerating limbs and only limb reductions in developing limbs [42]. It has been shown that retinoid exposure at critical times in development can elicit complete limb duplications and even duplicated pelvic girdles in mammals [32] and frogs [36,37]. Retinoid exposure during tail regeneration also has the ability to reprogram the tails to regenerate into legs, which leads to the development of a cluster of satellite limbs in the metamorphosing frog [38,39]. Both duplicated pelvic girdles and satellite limbs have been observed in field specimens from deformed frog sites [43]. There are also significant and substantial differences between regeneration and development [40,41]. The observation that RA treatment causes PD duplications in regenerating limbs has no relevance when considering the ability of RA to induce complete limb duplications in developing animals. Lastly, it appears that con-

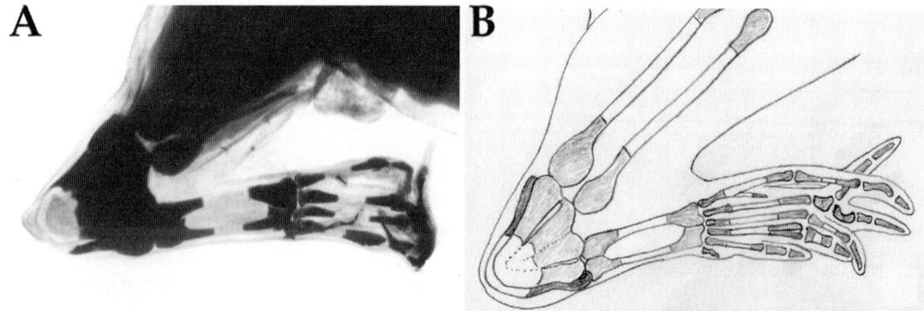

Fig. 1 Victoria blue stain to show a bony triangle in the zeugopodial (tibia/fibula) segment of the duplicated hindleg of a Minnesota deformed frog. A. Cartilage is darkly stained, the lighter area at apex of triangle is bone. Proximal and distal ends of the bones are next to one another. B. Diagram of skeletal elements in A.

clusions regarding mechanism may have been drawn without anyone actually doing the decisive experiments. A well-known hallmark of development is the existence of windows of sensitivity to perturbation. Treatments outside the window of sensitivity will not elicit effects. Therefore, conclusions regarding mechanisms of malformation drawn in the absence of a careful analysis of the windows of sensitivity to retinoid exposure are likely to be invalid.

We tested the effects of treating developing *Xenopus laevis* embryos with retinoids to evaluate the tenability of our hypothesis that altering retinoid signaling causes limb deformities. Pilot studies were undertaken to identify the stages of development where treatment with retinoids affected limb patterning [43]. A variety of effects on early development were observed, consistent with pervious studies in the literature on retinoid effects on amphibian development. The phenotypes include normal development, early embryonic lethality, duplicated limb buds, bony triangles and truncated limbs. These phenotypes could be elicited in response to treatment with retinol palmitate, atRA, and the synthetic RAR-specific activator TTNPB [43]. A notable result is that limb dysplasias were only observed when larvae were treated during stages of limb bud development (stages 48–55). Exposures at earlier developmental stages induced malformations in other organ systems (e.g., craniofacial, axial), but not in limbs [43]. In accord with our results (although they reach different conclusions) Degitz and colleagues have also shown that retinoid exposure causes multiple malformations ranging from bony triangles to limb reductions and lethality depending on the species and time of exposure [42].

The results from the pilot experiments led us to focus on treating limb bud stage tadpoles with TTNPB to ascertain the precise windows of sensitivity to retinoid exposure. We found that TTNPB induces all of the limb malformations observed in the field, including duplicated limbs, distal reductions, and bony triangles. TTNPB is very stable and induces malformations after relatively short exposures (3–24 h). Results from TTNPB treatment indicate that there are multiple developmental windows of sensitivity during limb bud development, and that these windows are remarkable short. Treatment at all stages is developmentally toxic at high doses and long exposures, although TTNPB treatment is not acutely toxic when embryos are treated after the gastrula stage. The surviving larvae typically die a few weeks after exposure.

Treatment at the very beginning of limb outgrowth (stages 49/50) with doses between 100–1000 nM TTNPB was lethal to virtually all of the tadpoles [43]. In contrast, less than 10 % mortality is observed at lower levels of TTNPB. No morphological changes were observed in limb buds 4 weeks after exposure at these lower doses.

Treatment of embryos beginning at stage 50 also led to very high mortality and nearly no malformations (Table 1). Treatment beginning one day later at stage 51 elicited a high incidence of defects (Table 1). Duplicated limbs were induced at low doses of TTNPB (40, 80 nM), whereas higher doses resulted in 100 % mortality. Treatment at stage 52 yielded much less mortality at low doses with substantial malformations at doses of 8 nM and higher. The predominant defects seen at this stage were bony triangles. Lastly, treatment at stages 53–55 induced hypomorphic limbs at high frequency [43].

Table 1 Treatment of *Xenopus laevis* tadpoles with TTNPB elicits stage-specific limb defects. *X. laevis* tadpoles were treated at the indicated stages with a concentration series of TTNPB for 24 h, then allowed to develop normally and scored for survival and the presence of malformed limbs.

	Stage 50		Stage 51		Stage 52	
TTNPB	% Mortality	% Malformed	% Mortality	% Malformed	% Mortality	% Malformed
None	12.5	0	40	0	10	0
8×10^{-9}	44.4	0	70	0	40	33.3
4×10^{-8}	25	0	41.6	50	10	66.6
8×10^{-8}	33.3	0	58.3	80	11.1	100
4×10^{-7}	50	0	100	NA	72.7	100
8×10^{-7}	50	0	100	NA	77	100

It is also notable that we observe both unilateral and bilateral defects in the treated animals, as has been observed in wild specimens from affected sites. Since TTNPB elicits these characteristic limb defects, we conclude that inappropriate regulation of RAR-mediated signaling during distinct developmental windows is the proximate cause of the limb defects observed in laboratory animals, and by inference those in the field. Interestingly, the developmental windows for the production of limb duplications, bony triangles and hypomorphic limbs appear to be separate and mostly nonoverlapping, within the precision of the experimental protocol.

Although *Xenopus* is a valuable laboratory model organism for experimental work, it is not a species that is native to North America and is not found in areas with high frequencies of amphibian malformations. For that reason, we and others conducted parallel experiments in native species. We showed that retinoids such as retinol palmitate, atRA and TTNBP can induce limb malformations, including BTs, in *Rana pipiens* (D. M. Gardiner, unpublished) and *Rana sylvatica* [43]. Degitz and colleagues have produced similar results in other native species [42].

The model that environmental retinoids is the cause of frog deformities predicts that retinoids will be found at sites where deformed frogs are found. Furthermore, it should be the case that the same compounds are not present, or are present at much lower levels at sites where deformed frogs do not occur. Therefore, we extracted hydrophobic substances from water samples, fractionated these by HPLC, and tested the fractions for their ability to activate RAR in transient transfection assays. A critical point and a fundamental difference between this approach and more commonly used approaches is that we are not making any assumptions about the nature of the compounds beyond that they are present in the water. Fractionation is guided simply by the ability of fractions to activate RAR. Active fractions were purified to homogeneity as judged by UV absorption spectra and then analyzed by electrospray (ES/MS) and electron impact mass spectroscopy (EI/GC/MS) for exact mass determination and fragmentation analysis to identify the causal agents.

We focused on two sites where deformed frogs are routinely found, the CWB site in Minnesota and a site in Mission Viejo, CA, almost 2000 miles away. We found fractions that activated RAR from both sites and showed that this activity could be recovered from the water samples by both solvent and solid phase extraction (Figs. 2, 3). The activity resolved into three closely spaced but distinct HPLC peaks, suggesting the presence of geometric isomers of the same compound. Notably, similar activity peaks were identified in the water samples from a vernal pond in Mission Viejo, CA as in the permanent lake in CWB Minnesota (Figs. 2, 3). These data suggest that the compound is unlikely to be natural in origin and provide an important confirmation of the laboratory results with retinoid exposure. Taken together, the laboratory and field observations strongly support our contention that an environmental retinoid(s) is present and responsible for frog deformities in the water samples from the sites tested. We infer that the same compounds will be found at other deformed frog sites and are currently establishing a testing program to test this hypothesis.

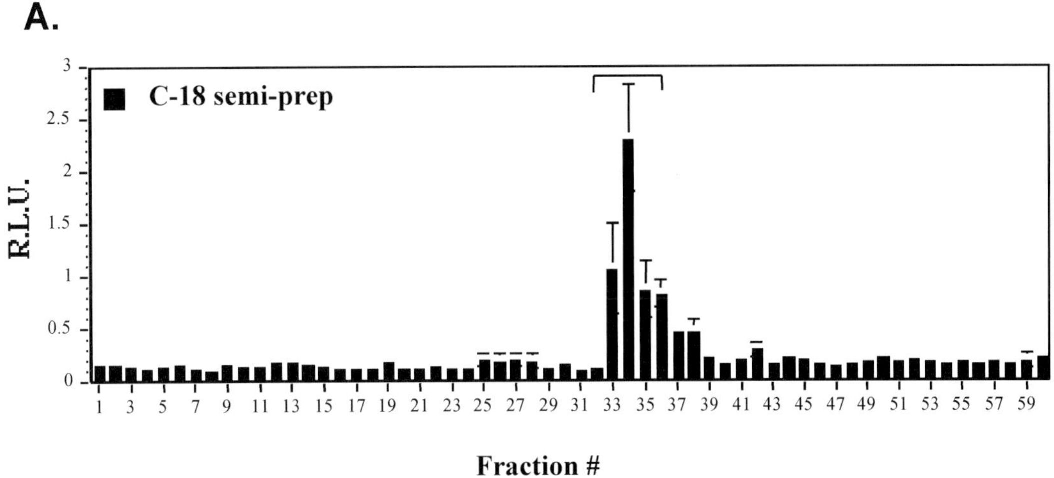

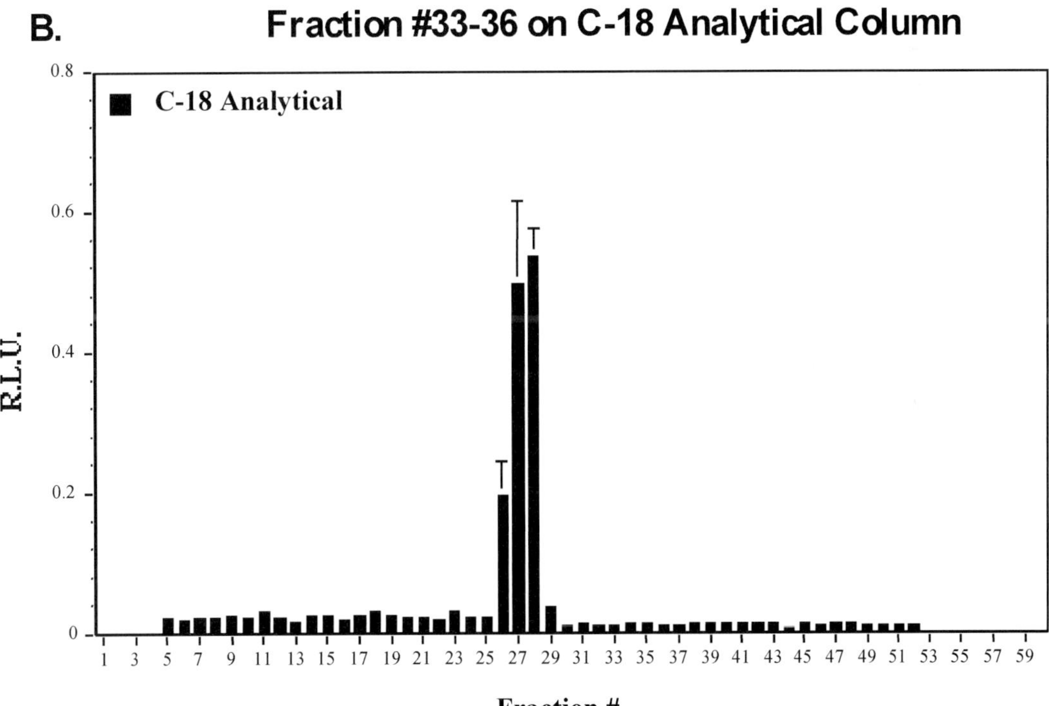

Fig. 2 Biologically active retinoids are found in a severely affected site. Solid phase extracts of lake water (2000 l) were fractionated by reversed-phase HPLC. Aliquots of column fractions were tested for their ability to activate RAR-dependent reporter gene transcription in transient transfection assays.

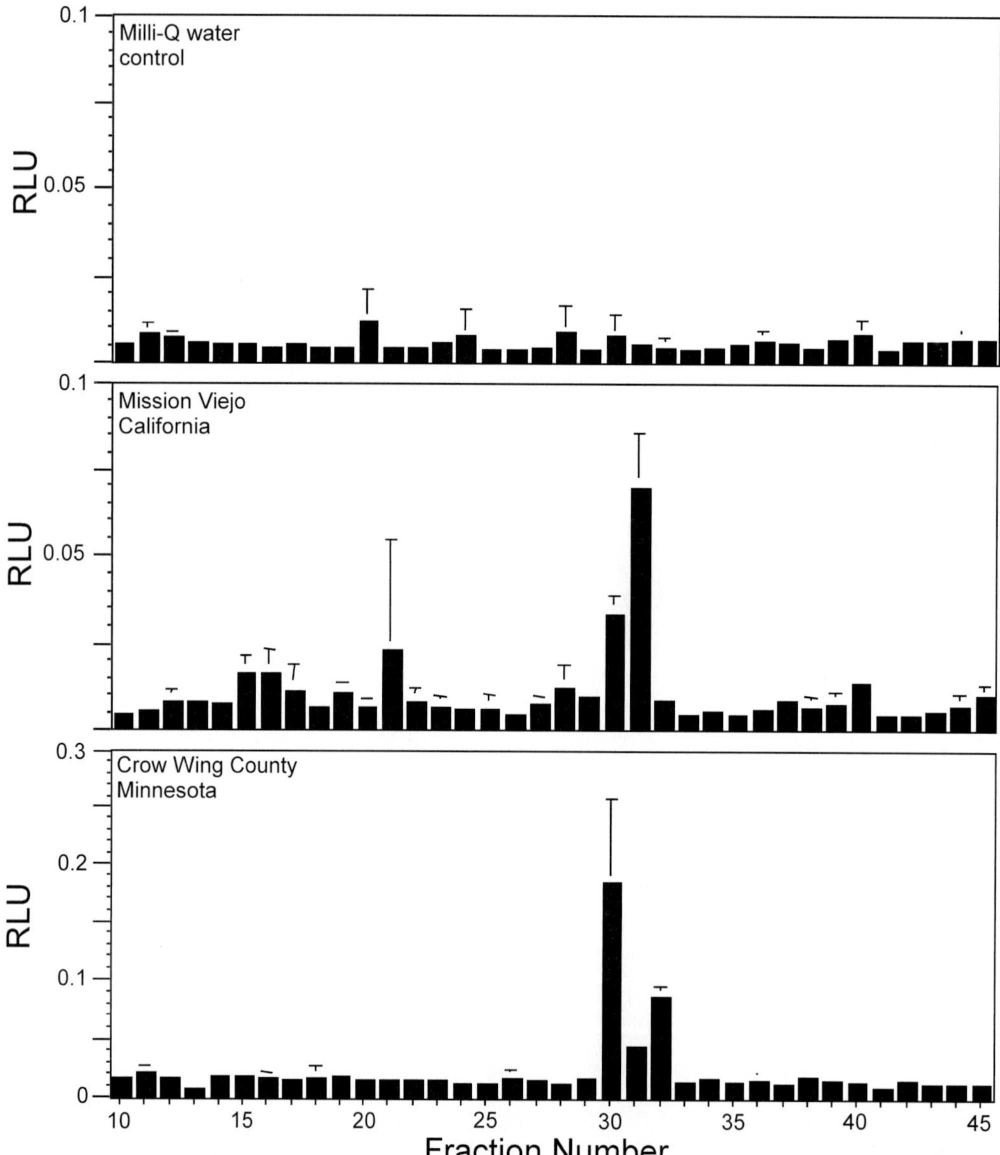

Fig. 3 Similar bioactive retinoid peaks are found in Mission Viejo, CA and Crow Wing County, MN. Organic extracts of lake water were prepared and fractionated by reversed-phase HPLC on an analytical C-18 column. Aliquots of column fractions were tested for their ability to activate RARα-dependent transcription of a luciferase reporter gene in transient transfection assays.

SUMMARY

The concept of parsimony dates back to the Middle Ages where its roots may be found in the writings of the English philosopher and theologian William of Ockham (1285–1349). His famous principle, often referred to as "Occam's razor", holds that in attributing an explanation to a set of observations, one should not make more assumptions than the minimum needed. The fundamental truth is that when considering a set of models to explain an observation, the **simplest** explanation that explains **all** of the observations is most often correct. Occam's razor metaphorically shaves off extraneous assumptions and variables that are not needed to explain the phenomenon in question. While it is formally possible that each and every site at which deformed frogs are found results from a different causative agent(s) or complex interactions between numerous agents and the local water chemistry, logic should tell us that the probability is exceedingly small. The most probable explanation for the vast majority of frog malformations will be the minimum set of causal agents that can explain all of the observed phenomena. As was described above, inappropriate modulation of retinoid signaling pathways during critical points in development can explain **ALL** of the observed frog deformities, even such bizarre phenotypes as an apparently normal frog with a cluster of ectopic limbs attached. Retinoids are capable of causing a wide variety of developmental abnormalities including craniofacial, gastrointestinal and neurological defects. Although little discussed, a common phenotype at affected sites is dead or dying frogs and retinoids can also account for these observations. We cannot completely exclude the possibility that frog malformations in the wild have multiple etiologies, but it is highly likely that the majority result from retinoid exposure.

Lastly, statements such as "With regard to an ecological enigma such as the malformed frogs, however, a focused light that illuminates the esoteric molecular or cellular processes of developmental biology poses the danger of blinding the investigator to the contextual clues shadowed in the surrounding environment" [6] should not be allowed to pass into the literature unchallenged. Although some authors may find the molecular and cellular processes of developmental biology to be "esoteric", cognoscenti who have followed the advances in the molecular genetics of development made in recent years will realize that the fundamental molecular and cellular processes underlying development cannot be ignored in the search for rational explanations. The vertebrate limb is not a black box that is acted on by a myriad of shadowy contextual clues in the environment, but rather the most intensively studied and well-understood model in vertebrate development. To treat it otherwise is reminiscent of the humoral theories of disease prevalent in past centuries.

RECOMMENDATIONS FOR FUTURE RESEARCH

Both parasites and retinoids can elicit frog malformations in the laboratory. Since only a single parasite species causes such malformations and it appears very unlikely that mechanical perturbation will be the cause, it is plausible that the parasites are providing or inducing a developmental signal in the affected animals. The application of Occam's razor leads to the inference that this agent will be a retinoid, alter the concentrations of retinoids in the animal, or otherwise modulate the expression of retinoid target genes. Thus, it would be interesting to identify the molecular mechanisms through which *Ribeiroia* act to cause malformations.

Since it seems beyond question that inappropriate retinoid signaling is playing an important role in the etiology of deformed frogs, the next step is to find out when and where such compounds are present in the environment. It is critical that a series of field surveys be undertaken that will reveal the presence, concentration, and stability of the candidate retinoids in bodies of water and determine whether drinking water supplies are affected. One would like to understand the route through which the compounds enter the animals (e.g., diet, transport across gills or skin through exposure to sediments or water) and whether human and animal populations are at risk from similar exposure. We believe that studying mechanisms can lead to important inferences about the nature and mode of action of environmental contaminants. This will lead to a deeper understanding of how animals interact with the environment and a rational basis for future risk assessment.

ACKNOWLEDGMENTS

Work in the authors' laboratory was supported by a grant from the Environmental Protection Agency (STAR G9D1 0090). We thank David Hoppe, William Souder, and Susan Bryant for stimulating discussions throughout the course of this work and Ronald M. Evans (Salk Institute for Biological Studies) for support in the early phases of this research.

REFERENCES

1. N. Dahmane, J. Lee, P. Robins, P. Heller, A. Ruiz i Altaba. *Nature* **389**, 876–881 (1997).
2. N. Dahmane, P. Sanchez, Y. Gitton, V. Palma, T. Sun, M. Beyna, H. Weiner, A. Ruiz i Altaba. *Development* **128**, 5201–5212 (2001).
3. J. Xie, M. Murone, S. M. Luoh, A. Ryan, Q. Gu, C. Zhang, J. M. Bonifas, C. W. Lam, M. Hynes, A. Goddard, A. Rosenthal, E. H. Epstein, Jr., F. J. de Sauvage. *Nature* **391**, 90–92 (1998).
4. D. M. Hoppe. In NAAMP III online: <http://www.mp1-pwrc.usgs.gov/naamp3> (1997).
5. M. Ouellet, J. Bonin, J. Rodriguez, J. L. DesGranges, S. Lair. *J. Wildl. Dis.* **33**, 95–104 (1997).
6. I. K. Loeffler, D. L. Stocum, J. F. Fallon, C. U. Meteyer. *Anat. Rec.* **265**, 228–245 (2001).
7. C. U. Meteyer, I. K. Loeffler, J. F. Fallon, K. A. Converse, E. Green, J. C. Helgen, S. Kersten, R. Levey, L. Eaton-Poole, J. G. Burkhart. *Teratology* **62**, 151–171 (2000).
8. D. L. Stocum. *Teratology* **62**, 147–150 (2000).
9. A. R. Blaustein, J. M. Kiesecker, D. P. Chivers, R. G. Anthony. *Proc. Natl. Acad. Sci. USA* **94**, 13735–13737 (1997).
10. J. M. Kiesecker, A. R. Blaustein, L. K. Belden. *Nature* **410**, 681–684 (2001).
11. R. C. Swartz, S. P. Ferraro, J. O. Lamerson, F. A. Cole, R. J. Ozretich, B. L. Boese, D. W. Schults, M. A. Behrenfeld, G. T. Ankley. *Environ. Toxicol. Chem.* **16**, 2151–2157 (1997).
12. G. T. Ankley, J. E. Tietge, D. L. DeFoe, K. M. Jensen, G. W. Holcombe, E. J. Durhan. *Environ. Toxicol. Chem.* **17**, 2530–2542 (1998).
13. J. J. LaClair, J. A. Bantle, J. Dumont. *Environ. Sci. Technol.* **32**, 1453–1461 (1998).
14. S. K. Sessions, R. A. Franssen, V. L. Horner. *Science* **284**, 800–802 (1999).
15. S. K. Sessions. *Science* **279**, 461–462 (1998).
16. S. K. Sessions and S. B. Ruth. *J. Exp. Zool.* **254**, 38–47 (1990).
17. P. T. Johnson, K. B. Lunde, E. G. Ritchie, A. E. Launer. *Science* **284**, 802–804 (1999).
18. P. T. Johnson, K. B. Lunde, R. W. Haight, J. Bowerman, A. R. Blaustein. *Can. J. Zool.* **79**, 370–379 (2001).
19. D. M. Gardiner and D. M. Hoppe. *J. Exp. Zool.* **284**, 207–216 (1999).
20. J. G. Burkhart, G. Ankley, H. Bell, H. Carpenter, D. Fort, D. Gardiner, H. Gardner, R. Hale, J. C. Helgen, P. Jepson, D. Johnson, M. Lannoo, D. Lee, J. Lary, R. Levey, J. Magner, C. Meteyer, M. D. Shelby, G. Lucier. *Environ. Health Perspect.* **108**, 83–90 (2000).
21. D. W. Sparling, G. M. Fellers, L. L. McConnell. *Environ. Toxicol. Chem.* **20**, 1591–1595 (2001).
22. B. Blumberg. *Seminars Cell Develop. Biol.* **8**, 417–428 (1997).
23. S. V. Bryant, T. Hayamizu, D. M. Gardiner. In *Experimental and Theorietcal Advances in Biological Pattern Formation*, H. G. Othmer (Ed.), pp. 37–43, Plenum Press, New York (1992).
24. P. Chambon. *FASEB J.* **10**, 940–954 (1996).
25. P. Kastner, M. Mark, P. Chambon. *Cell* **83**, 859–869 (1995).
26. B. Blumberg and R. M. Evans. *Genes Dev.* **12**, 3149–3155 (1998).
27. D. J. Mangelsdorf and R. M. Evans. *Cell* **83**, 841–850 (1995).
28. M. Maden. *Dev. Biol.* **159**, 379–391 (1993).

29. S. R. Scadding and M. Maden. *J. Embryol. Exp. Morphol.* **91**, 35–53 (1986).
30. S. R. Scadding and M. Maden. *J. Embryol. Exp. Morphol.* **91**, 55–63 (1986).
31. C. Tickle, J. Lee, G. Eichele. *Dev. Biol.* **109**, 82–95 (1985).
32. K. Niederreither, S. J. Ward, P. Dolle, P. Chambon. *Dev. Biol.* **176**, 185–198 (1996).
33. J. C. Rutledge, A. G. Shourbaji, L. A. Hughes, J. E. Polifka, Y. P. Cruz, J. B. Bishop, W. M. Generoso. *Proc. Natl. Acad. Sci. USA* **91**, 5436–5440 (1994).
34. C. Tickle, B. Alberts, L. Wolpert, J. Lee. *Nature* **296**, 564–566 (1982).
35. G. Eichele, C. Tickle, B. M. Alberts. *J. Cell Biol.* **101**, 1913–1920 (1985).
36. F. S. Liotti and G. A. Bruschelli. *Riv. Biol.* **62**, 99–133 (1969).
37. G. M. Bruschelli and G. Rosi. *Riv. Biol.* **64**, 271–283 (1971).
38. M. Maden and J. Corcoran. *Dev. Genet.* **19**, 85–93 (1996).
39. P. Mohanty-Hejmadi, S. K. Dutta, P. Mahapatra. *Nature* **355**, 352–353 (1992).
40. S. V. Bryant and D. M. Gardiner. *Dev. Biol.* **152**, 1–25 (1992).
41. D. M. Gardiner and S. V. Bryant. In *Cellular and Molecular Basis of Regeneration: From Invertebrates to Humans*, P. Ferretti and J. Gâeraudie (Eds.), Wiley, Chichester, New York (1997).
42. S. J. Degitz, P. A. Kosian, E. A. Makynen, K. M. Jensen, G. T. Ankley. *Toxicol. Sci.* **57**, 264–274 (2000).
43. A. Ndayibagira, F. Grün, B. Bumberg, D. M. Gardiner (2003). Submitted for publication.

Pure Appl. Chem., Vol. 75, Nos. 11–12, pp. 2275–2286, 2003.

Topic 4.7

Contaminant-induced endocrine and reproductive alterations in reptiles*

Louis J. Guillette, Jr.[1,‡] and Taisen Iguchi[2]

[1]Department of Zoology, 223 Bartram Hall, P.O. Box 118525, University of Florida, Gainesville, FL 32611, USA; [2]Department of Bioenvironmental Research, Center for Integrative Bioscience, National Institute of Basic Biology, Okazaki National Research Institutes, 38 Nishigonaka, Myodaiji, Okazaki 444-8585, Japan

Abstract: Many chemicals introduced into the environment by humans adversely affect embryonic development and the functioning of the vertebrate reproductive system. It has been hypothesized that many developmental alterations are due to the endocrine-disruptive effects of various environmental contaminants. The endocrine system exhibits an organizational effect on the developing embryo, altering gene expression and dosing. Thus, a disruption of the normal hormonal signals can permanently modify the organization and future functioning of the reproductive and endocrine system. We have worked extensively with contaminant-exposed and reference populations of the American alligator (*Alligator mississippiensis*) as well as performed a number of experimental studies exposing developing embryos to various persistent and nonpersistent pesticides. Using this species, we have described altered steroidogenesis, circulating hormone levels, and hepatic transformation of androgen and endocrine organ (gonad, thyroid) morphology in juvenile alligators living in polluted environments. Given the adverse observations reported to date, we recommend several important future needs:

1. Further development of "receptor zoos" and other molecular tools that include key reptiles from various major ecosystems, in addition to freshwater ecosystems.
2. Global studies extending the current knowledge base on crocodilians and freshwater turtles to comparable ecosystems on other continents, such as linked studies examining and extending current molecular to population-level studies in Florida (USA) to tropical and temperate regions of Africa, Australia, and South America.
3. Further studies of actual exposure, assimilation and excretion of contaminants by ectothermic vertebrates, especially reptiles that occupy high levels of the food chain.

INTRODUCTION

A central focus of comparative physiology and endocrinology has been the influence of environmental factors on the development and performance of various systems or whole organisms. Over the last century, it has been clearly established that such factors as temperature, pH, salinity, photoperiod, and gas tensions affect the endocrinology of vertebrates. As part of these studies, we have become aware of the influence of human activities on the biological performance of numerous species. The endocrine-dis-

*Report from a SCOPE/IUPAC project: Implication of Endocrine Active Substances for Human and Wildlife (J. Miyamoto and J. Burger, editors). Other reports are published in this issue, *Pure Appl. Chem.* **75**, 1617–2615 (2003).

‡Corresponding author

ruptive actions of various chemical contaminants have been a recent focus of much research [1–4]. This concern has become worldwide and goes beyond a focus on pesticides and industrial chemicals only. A number of pesticides, plasticizers or industrial chemicals and some naturally occurring plant or fungal compounds, have been shown to mimic naturally occurring steroids, act as hormone receptor antagonists or alter the enzymes responsible for hormone synthesis and degradation (Fig. 1). For example, hormones or chemicals with endocrine-like activity in food products and drinking water have become scientific and public concerns. Initial research from wildlife populations exposed to sewage or pesticides demonstrated endocrine disruptive activity from various classes of contaminants [2]. Recent studies have extended this concern to additional chemicals released from human activities such as antibiotics and a large number of pharmaceutical agents as well as chemicals listed as "inert" compounds in chemical formulations [5].

Contaminants can impact organisms in two fundamentally different ways, via organization or activation. First, embryonic exposure can cause "organizational" abnormalities by altering the chemical signals required for normal development and thus permanently changing the endocrine function and or response of an organ or organism [6]. Second, they can alter cellular signaling in mature systems, that is, disrupt "activational" signaling that would lead to altered organ system or organism performance. We will briefly overview our studies demonstrating both these effects in reptiles exposed chronically or acutely to various environmental contaminants or mixtures of contaminants.

Reptiles have received relatively little attention with regard to the effects of environmental contaminant exposure until recently [7], although they are valuable indicators of an ecosystem's condition and are exposed to environmental contaminants like other species [8]. Reptiles in general do not appear

- **Receptor Mediated**
  - **-Agonistic & Antagonistic**
    - *Estrogen receptor*
    - *Progesterone receptor*
- **Altered Enzyme Action**
  - *P450aromatase*
- **Altered Metabolism**
  - *Hepatic androgen*
- **Altered Hormone Availability**
- **Altered Gene Expression**
  - *P450aromatase*
  - *SF-1*

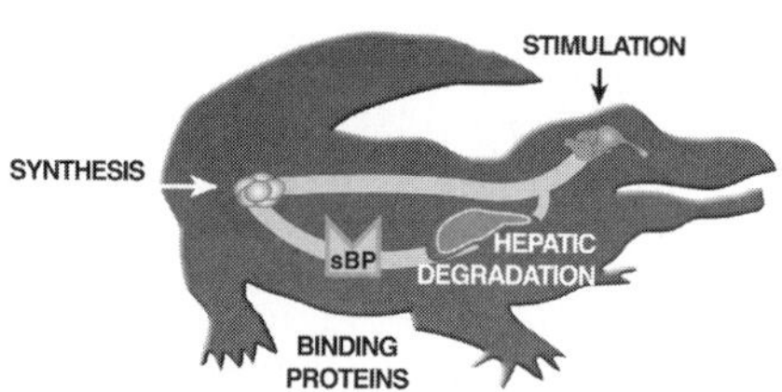

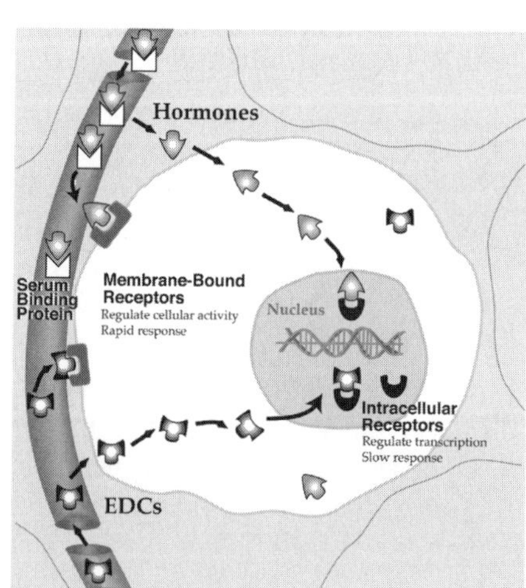

Fig. 1 Endocrine disruption in reptiles can occur at several levels of biological organization. To date, work has demonstrated that reptiles exposed to contaminants have the potential to alter the endocrine system by receptor-mediated agonistic or antagonistic actions. They can also alter enzyme action, such as alteration in P450aromatase activity or alterations in hepatic biotransformation of androgen. Recent studies have begun to demonstrate that alterations in gene expression also occur. See text for further explanations and references.

to be more or less affected by contaminants relative to other vertebrates, but given that the vast majority are predators or insectivores, they may be exposed to elevated levels of pesticides and other chemicals that bioaccumulate in food chains. Additionally, as with many other nonmammalian vertebrates, estrogens play an important and active role in the embryonic development of the ovary [9]; thus, reptiles could be susceptible to altered or abnormal gonadal development due to endocrine active contaminants that mimic or alter the estrogenic milieu of the developing embryo.

The American alligator is again common to many inland waters of the southeastern United States, following decades on the endangered species list. Recent studies from our laboratories have focused on this species because it is long-lived, a top-predator, easily obtained in large numbers (100s of neonates and juveniles can be obtained and sampled in a night), and individuals are large enough to provide adequate serum/urine samples for analysis of contaminants, hormones or other biomarkers. Further, an extensive history on population demography is available for this species in many lakes in Florida, USA. Further, in collaboration with the Florida Fish and Wildlife Conservation Commission, we have collected data on populations—recruitment and egg viability for the last 20 years—that continue to this day. Sex determination—temperature-induced sex—in alligators is different compared to that observed in mammals, but this mechanism of sex determination provides a very powerful tool for studying the role of contaminants capable of altering gender-specific steroid synthesis in wild animal populations. Additional studies have been performed on turtles, as they are an established model system for the study of environmental/temperature-induced sex determination [10].

STUDIES IN WILD POPULATIONS

We have examined seasonal and spatial (among lake) variation in plasma steroids, and have observed that sub-adult, juvenile animals (5–8 years of age) of both sexes exhibit seasonal (spring–fall) adult-like patterns of plasma sex steroids, even though they are not elevated to the same degree as in sexually mature individuals [11,12]. Sexual maturity occurs at approximately 12 years of age when they are almost 2 m in total length. Interestingly, this pattern suggests a multiyear period of puberty in alligators as is observed in humans. We have also studied pesticide interactions with the endocrine and reproductive systems in this species, allowing us to begin to understand the possible interactions between pesticide exposure and endocrine disruption.

Over the last decade, a common observation in laboratory and wildlife studies is an alteration in androgen synthesis and function in males [13,14]. For example, fish downstream from sewage treatment plants display elevated yolk protein—an estrogenic response—as well as exhibiting reduced androgen plasma concentrations, and increased estrogen concentrations [15–17]. Alligators living in several Florida lakes exhibit alterations of the reproductive and endocrine systems [18]. Many of these modifications are detectable at hatching, and persist throughout juvenile and sub-adult life. Examinations of the reproductive and endocrine systems of juvenile male alligators from Lake Apopka, have demonstrated elevations in plasma estradiol, reductions in androgen (T and DHT) and altered thyroxin concentrations (Fig. 2) as well as morphological abnormalities of the testis [11,19]. The alterations in plasma hormone concentrations persist to date, and also occur in females [18]. Altered endocrine parameters in juvenile alligators also occur in other Florida wetlands not associated with significant pesticide spills but that have a history of non-point source contamination, such as Lake Okeechobee and Lake Griffin, FL, USA [11,20,21].

Various factors have been shown to alter plasma androgen concentrations in alligators, the most common being seasonal variations due to temperature and photoperiod. However, depression in plasma androgen concentration has also been reported due to stress [22], and exposure to contaminated lakes could induce a stress response. We have examined basal and stress-induced (2 h capture stress) corticosterone levels in alligators from contaminated and reference lakes and found no differences among exposure groups in plasma corticosterone but significant differences in plasma testosterone concentrations [23]. This study, as well as a 20 h capture-stress study [12], suggests that the depressed testos-

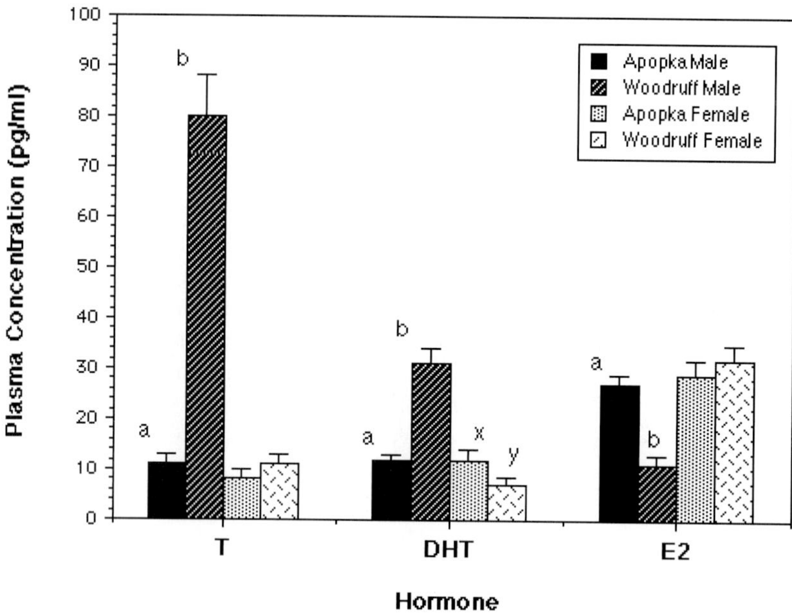

Fig. 2 Plasma concentration of three hormones, testosterone (T), dehydrotestosterone (DHT) and estradiol-17β (E2) in juvenile alligators from two Florida lakes, Lake Apopka and Lake Woodruff. Within a sex, differing superscripts indicate a significant difference. If no superscript is present, no significant difference was observed. [Data from refs. 11,19.]

terone concentrations reported in alligators from contaminated lakes are not due directly to stress, but does not rule it out as one of many factors influencing plasma sex steroid concentrations. Additional studies are required to examine ACTH as well as other parameters.

Altered plasma hormone concentrations can also result from alterations in hepatic biotransformation and metabolism of hormones [24]. We have recently documented alterations in hepatic testosterone biotransformation in juvenile alligators exposed to elevated organochlorines relative to lesser-exposed, reference individuals [25]. For example, hepatic oxido-reduction of testosterone displays a sexually dimorphic pattern—females higher than males—in alligators from Lake Woodruff, FL, a population with low organochlorine exposure, whereas populations with greater exposure, such as Lakes Apopka and Okeechobee, FL, show a lack of dimorphism (Fig. 3). Interestingly, the lack of dimorphism is due to females from the contaminated populations exhibiting a male-like pattern of androgen metabolism. Patterns in other enzyme activities are also altered. For example, sexual dimorphism is acquired in the activity of UDP-glucuronosyltransferase enzyme in one exposed population that is not observed in the reference population [25]. These data demonstrate that wildlife exposed to contaminants can display altered hepatic metabolism of hormones that appears to be a specific rather than a general response; that is, the animals exhibiting altered hepatic metabolism of testosterone do not have elevated EROD activities, indicative of a generalized heightened hepatic response to toxins.

Recent collaborative studies from our laboratories have begun to focus on alterations in the expression of genes essential for normal spermatogenesis or endocrine action, such as P450aromatase ($P450_{arom}$), steroidogenic factor-1 (SF-1) and the androgen and estrogen receptors. The genes for $P450_{arom}$ and SF-1 were previously cloned from alligator [26,27] and we have recently developed and validated a quantitative RT-PCR technique for assessing transcript number of the mRNA generated from these genes in alligators from Lakes Apopka, Woodruff, and Orange, FL (Bermudez, Katsu, Iguchi, and Guillette, unpublished data). Additionally, we have clones of alligator estrogen receptor alpha (aERα), aERβ and an androgen receptor (aAR), and recently validated quantitative RT-PCR tech-

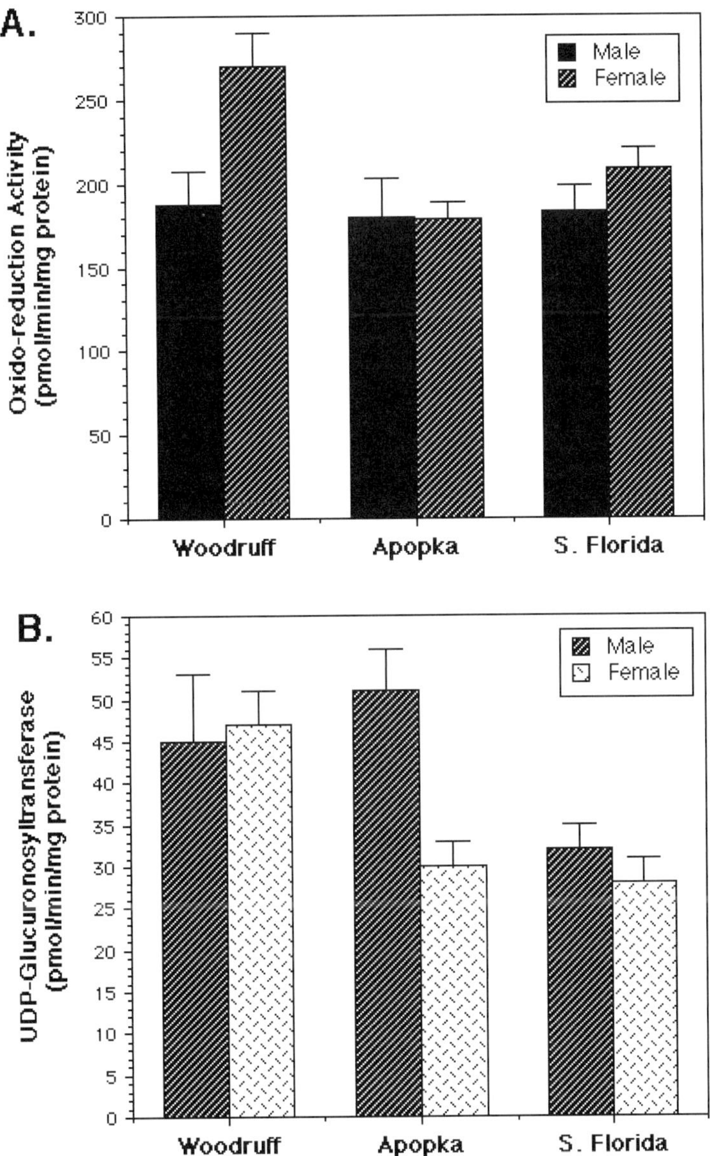

Fig. 3 Activity of hepatic enzymes on testosterone biotransformation in juvenile alligators from three regions in Florida, Lake Apopka, Lake Woodruff, and South Florida (the Lake Okeechobee watershed). Within a sex, differing superscripts indicate a significant difference. A star within a lake system indicates sexual dimorphism in testosterone biotransformation. [Data from ref. 25.]

niques to begin studies of gene expression in exposed and reference populations (Katsu, Guillette, Miyagawa, and Iguchi, unpublished data). Initial studies using quantitative RT-PCR suggest alterations in hepatic $P450_{arom}$ and SF-1 induction in animals from Lake Apopka (Bermudez and Katsu, unpublished data). Future studies will examine the relationship between gene induction/transcript number, cellular enzyme activity and plasma hormone concentrations in animals obtained from lakes having different pollution sources and histories.

LABORATORY-BASED STUDIES

Contaminants can impact organisms during embryonic development. Embryonic exposure can cause organizational abnormalities by altering the chemical signals associated with normal development [4,6]. Many of the abnormalities reported in the contaminant-exposed alligator populations appear to be organizational, in part, as they are present at birth and apparently continue throughout life. It has been proposed that many, but not all, of the actions of contaminants on embryos are a response to xenobiotic compounds interacting with hormone receptors. In order to support this hypothesis, chemicals that bind the receptor(s) must do so and must be present in concentrations high enough to induce effects.

A number of contaminants have been identified in alligator eggs [28], serum [29], and body tissues [30,31]. Concentrations of these compounds range from ppm concentrations of various organochlorine compounds in the eggs and or tissue of alligators to ppb concentrations of like compounds in the serum. Many of the compounds identified in the body exhibit an affinity for alligator estrogen (ER) and/or progesterone (PR) receptor proteins purified from the uterus [32,33]. Although the affinity of only a few of these compounds is relatively high (e.g., *o,p′*-DDT, DDOH, *trans*-nonachlor) for the aER, many compounds show an ability to displace estradiol from the aER, if present in high enough concentrations [32,33]. A recent study has also cloned an ERα from a crocodilian, *Caimen crocodylus* [34]. Studies using this cloned receptor transfected into mammalian HeLa cells demonstrated that at least one environmental estrogen, bisphenol A, was capable of binding to the receptor and inducing gene expression. We have recently cloned two estrogen receptors (ERα, ERβ), an androgen receptor and progesterone receptor from the American alligator (Katsu, Guillette, Miyagawa, and Iguchi, unpublished data). Using these recently cloned alligator receptors, we plan to transfect cells and test the ability of various chemicals, known to contaminate wild populations, to induce estrogenic, androgenic or progestogenic actions at the gene level.

In vitro system are useful as screens, but ultimately we need to address whether contaminants at known exposure levels in embryos alter development and the potential health of the individuals exposed. We have performed such studies on alligator embryos. Alligators have environmental sex determination in that temperature influences whether the gonad develops as an ovary or testis (Fig. 4). Recent studies examining alligator sex determination have begun to identify a number of genes associated with this phenomenon. Many of the genes commonly associated with sex determination in mammals and birds are also active during gonadal differentiation in alligators [35]. In other words, although differentiation is driven, in part, by temperature, sex determination in reptiles is not so unique as to be uninformative for

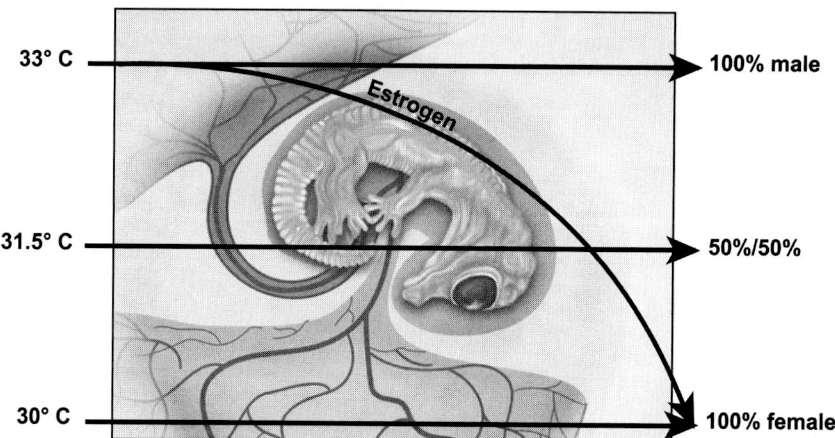

Fig. 4 Sex determination in alligators is temperature sensitive but estrogens can overcome male-producing temperatures and generate females. Several recent studies have demonstrated that environmental estrogens, such as *trans*-nonachlor and *p,p′*-DDD can act as estrogens in this system.

other vertebrates. In fact, many of the gene patterns seem to be identical to those observed during sex determination in birds, even though birds do not exhibit temperature-induced sex determination.

A number of studies using alligators and turtles have shown that various environmental contaminants are capable of altering (male to female) sex determination in reptilian embryos at concentrations as low as 100 ppt [12,36–38]. Using the average concentration of organochlorine pesticides or their metabolities (OCs) found in alligator eggs from Lake Apopka, FL, Willingham and Crews (1999) showed that sex reversal (male to female) could be induced in turtle embryos. They have also shown that steroidogenesis could be influenced by embryonic exposure to OCs. The concentrations used in these studies are clearly ecologically relevant as many studies examining contaminants in reptile eggs report ppb to ppm concentrations. The observation of altered steroidogenesis in OC-exposed embryos is important given recent data from our laboratories showing that alligators that were sex reversed by exogenous estrogens, in this case the naturally occurring estrogen estradiol-17β (E2), exhibited altered $P450_{arom}$ enzyme activity in the gonad and brain relative to reference animals developing under the influence of temperature alone [39]. That is, reference female embryos had elevated gonadal $P450_{arom}$ enzyme activity relative to females induced by E2 whereas E2-treated females had brain $P450_{arom}$ enzyme activity similar to males (Fig. 5). Previous studies with alligators have demonstrated that neonatal and

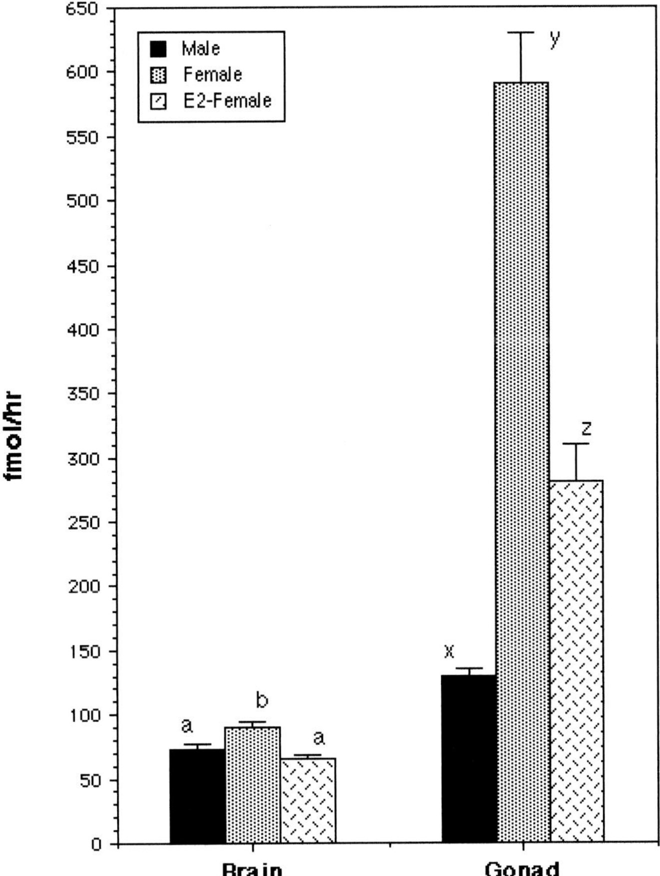

Fig. 5 Aromatase activity in the brain and gonad of stage 24 embryonic alligators. Stage 24 represents the stage immediately after the thermal sensitive period of sex determination. Differing superscript indicate a significant difference. E2 females where produced by exposing embryos to estradiol but incubating embryos at a male-producing temperature. Reference males and females were produced by incubating eggs and temperatures that produce 100 % males or females. [Data from ref. 39.]

juvenile alligators obtained from Lake Apopka had altered plasma sex steroid concentrations (reviewed in [18]) and that exposure to xenobiotics either experimentally or naturally during embryonic development can alter $P450_{arom}$ enzyme activity [40]. Thus, differentiation of a morphologically normal gonad can occur following exogenous hormone or xenobiotic exposure, but development of a normal endocrinology in that gonad may not occur.

Recent studies have begun to document the many mechanisms by which steroidogenesis can be altered in the gonad [41–43]. Any of these sites are potential targets for disruption. These can include alterations in the stimulus and inhibitory pathways involving gonadotropins, activins, and inhibins and as well as alterations in the molecular level activities involving receptor-specific actions, StAR protein, and the enzymes $P450_{scc}$ and 3β-HSD. Several recent papers clearly demonstrate that contaminants can alter the StAR-$P450_{scc}$-3β-HSD pathway [44,45]. Whether exposure of the embryo to endocrine active contaminants induces similar alterations in the StAR-$P450_{scc}$-3β-HSD pathway, causing permanent modifications in gonadal steroidogenesis, is unknown.

FUTURE NEEDS AND RECOMMENDATIONS

The current data on reptiles indicates that exposed populations, either due to polluted environments or in laboratory-controlled experiments, display altered reproductive, endocrine and immune systems (see [18,46]. Further, contamination need not be excessive, as low ppm or ppb concentrations are associated with alterations in the endocrine and reproductive systems. Further, even lower concentrations (parts per trillion) of various pesticides are capable of altering sex determination in reptiles, such as freshwater turtles and alligators. Much of the work to date has been performed on alligators from Florida, USA, but studies from the Lauretian Great Lakes of the United States and Canada, Australia, Africa, and Central America confirm that reptiles are contaminated with pollutants such as heavy metals and OC pesticides or their metabolites at levels seen in other vertebrates [8]. Their position in the food chain will influence bioaccumulation and biomagnification as it does with other vertebrates. Thus, reptiles are not unique in exposure or response. However, crocodilians may be very important ecological sentinels given their position at the top of most tropical and subtropical wetland ecosystems. Further, major advances in the endocrinological, reproductive and developmental biology of these animals at the molecular, organismal and population levels are providing important tools to further our understanding of the impacts of pollution on these important keystone species.

Given the above, we recommend several important future needs:

1. Further development of "receptor zoos" and molecular tools that include key reptiles from various major ecosystems, in addition to freshwater ecosystems.
2. Global studies extending the current knowledge base on crocodilians and freshwater turtles to comparable ecosystems on other continents, such as linked studies examining and extending current molecular to population level studies in Florida (USA) to tropical and temperate regions of Africa, Australia, and South America.
3. Further studies of actual exposure, assimilation and excretion of contaminants by ectothermic vertebrates, especially reptiles that occupy high levels of the food chain.

Receptor zoos

Although only one of several possible mechanisms by which endocrine disruption can occur, receptor-contaminant interactions are important as they can help identify chemicals that may be of major concern to a specific species of concern. Receptors from various species exhibit similar affinities for native hormones, such as that found between E2 and ERα from various species (see [47]). However, binding between the human ERα (hERα) or ERα from other species with environmental "estrogenic" chemicals can vary greatly—over several orders of magnitude. For example, a recent study examining the affinities between hERα and an ERα from a teleost fish, the medaka, indicates that affinities can be 100X or more different [48] suggesting that use of the receptor from one species, or even a few species, could greatly misrepresent the risk in other species. Several studies in reptiles have provided conflicting evidence of the estrogenicity of *p,p′*-DDE. For example, *p,p′*-DDE causes sex reversal in embryonic freshwater turtles and alligators in some studies [36,37], but has no effect at similar dose ranges in snapping turtles or green sea turtles [49,50]. Further, we have shown that *p,p′*-DDE shows some affinity for aER [32], but other studies have shown this chemical has no affinity for the hERα [51]. Given the relative power of modern molecular biology, it is relatively straightforward to develop transfected cell lines with receptors from a variety of species. We propose that reptiles be included in this approach because of their important positions as keystone species in many ecosystems and their position in the evolutionary biology of vertebrates. Several studies have begun this approach and these studies should be continued.

Further, it is important to note that this could also be a powerful approach in determining if endangered species, whether reptiles or other vertebrates, are at risk to endocrine disruption via receptor-based mechanisms. The difficulty, if not impossibility, of studying toxicology in whole organisms or even their tissues when that animal is in danger of extinction, suggests an approach of harvesting tissues from one individual and developing a receptor "zoo". Given that pollution is global and no ecosystem is pristine, this approach should be considered for all major endangered species, not just reptiles.

Global studies

One of the major weaknesses of current ecotoxicology is that many major studies are limited to specific geographical areas. For example, extensive field and laboratory-based studies have been performed on fish below sewage treatment works in the United Kingdom [17], on fish exposed to pulp mill effluents in Canadian waters [52,53], and on alligators living in Florida's freshwater lakes [18]. Other studies have been done on fish below sewage works or exposed to pulp mill effluent and other reptile studies are available. But linked studies, using similar approaches and interdisciplinary teams as well as international teams, are exceedingly rare. Many of the initial studies in endocrine disruption involved studies of wildlife at specific geographical locations. This generated a great deal of criticism suggesting these were isolated studies, based on isolated conditions and not representative of other populations. Today, available wildlife studies from diverse areas indicate that the original studies were not unique and do represent global problems. However, the same criticisms put forth against early wildlife studies of endocrine disruption are now leveled on studies examining human health effects. For example, our studies of alligators in central Florida were suggested to be unique and isolated to one lake. Studies have shown that criticism to be unwarranted as similar problems are seen on other Florida lakes [11,20]. Although a growing literature has documented endocrine disruption in wildlife from Europe, North America, and Japan, only a few studies have reported studies on wildlife from other areas. Given the large use of "older" chemicals, especially organochlorine pesticides in many tropical countries, their use throughout the year as well as the growing use of chemicals in every day activities in those nations, a greatly expanded effort to understand the extent of endocrine disruption in those countries is needed. Of particular concern, is the lack of a resource database on the fate, distribution and exposure of animals, including humans, in tropical areas. Current ecological risk and health policy is based largely on studies performed in northern temperate climates. Expanded studies in the tropics would add greatly to our understanding of ecotoxicology and endocrine disruption. Reptiles are important models given their abundance at all levels of the food chain. Further, reptiles are a protein source for many human populations in the tropics and thus serve as a source of exposure to contaminants.

Exposure, assimilation, and excretion

There are very few studies from even fewer ecosystems that document contaminant exposure in reptiles. Although the studies that do exist indicate that reptiles are not obviously different when compared to other vertebrates, there is little exposure data. Further, studies of hepatic degradation are relatively rare, although some important studies have been done (for examples, see [54]). In addition to basic studies of contaminant metabolism, work examining how contaminants alter hepatic biotransformation of hormones is essential. Work on a variety of species indicates that contaminant exposure can alter the endocrine system by modifying hepatic transformation of hormones [55,56]. We have recently reported that hepatic androgen metabolism was altered in contaminant-exposed alligators whereas enzymes responsible for generalized hepatic metabolism of toxins was not altered [25]. A real need is further studies, in reptiles and other vertebrates, on the role of hepatic biotransformation on hormone concentrations and the role of contaminants in altering these activities.

ACKNOWLEDGMENTS

The work cited here from our laboratories is made possible by the hard work of a large number of students and we thank them for their efforts. Work from the laboratory of LJG has been supported in part by grants from the U.S. Environmental Protection Agency (CR826357-01-1) and the U.S. Fish and Wildlife Service, whereas work in the laboratory of TI has been supported partly by a Grant-in-Aid for Scientific Research on Priority Areas (A) "The Environmental Risk of Endocrine Disruptors", the Ministry of Education, Culture, Sports, Science and Technology of Japan, and a grant from the Ministry of the Environment.

REFERENCES

1. G. T. Ankley, R. D. Johnson, N. E. Detenbeck, S. P. Bradbury, G. Toth, L. Folmar. *Rev. Toxicol.* **1**, 71–106 (1997).
2. L. J. Guillette, Jr. and D. A. Crain (Eds.). *Endocrine Disrupting Contaminants: An Evolutionary Perspective*, Taylor and Francis, Philadelphia (2000).
3. J. A. McLachlan. *Endocrine Rev.* **22**, 319–341 (2001).
4. T. Iguchi, H. Watanabe, Y. Katsu. *Horm. Behav.* **40**, 248–251 (2001).
5. D. W. Kolpin, E. T. Furlong, M. T. Meyer, E. M. Thurman, S. D. Zaugg, L. B. Barber, H. T. Buxton. *Environ. Sci. Technol.* **36**, 1202–1211 (2002).
6. L. J. Guillette, Jr., D. A. Crain, A. A. Rooney, D. B. Pickford. *Environ. Health Perspect.* **103** (Suppl. 7), 157–164 (1995).
7. D. A. Crain and L. J. Guillette, Jr. *Anim. Reprod. Sci.* **53**, 77–86 (1998).
8. D. W. Sparling, G. Linder, C. Bishop. *Ecotoxicology of Amphibians and Reptiles*, SETAC Pr., Pennsacola, FL (2000).
9. J. J. Bull, W. H. N. Gutzke, D. Crews. *Gen. Comp. Endocrinol.* **70**, 425–428 (1988).
10. D. Crews, J. M. Bergeron, J. A. McLachlan. *Environ. Health Perspect.* **103** (Suppl. 7), 73–77 (1995).
11. L. J. Guillette, Jr., A. R. Woodward, D. A. Crain, D. B. Pickford, A. A. Rooney, H. F. Percival. *Gen. Comp. Endocrinol.* **116**, 356–372 (1999).
12. A. A. Rooney. Ph.D. dissertation, University of Florida (1998).
13. L. E. Gray, Jr., J. Ostby, J. Furr, C. J. Wolf, C. Lambright, L. Parks, D. N. Veeramachaneni, V. Wilson, M. Price, A. Hotchkiss, E. F. Orlando, L. J. Guillette, Jr. *Human Reprod. Update* **7**, 248–264 (2001).
14. L. E. Gray, Jr., L. Parks, C. Lambright, E. F. Orlando, L. J. Guillette, Jr. In *Endocrine Disrupters in the Environment*, M. Metzler (Ed.), pp. 209–248, Springer-Verlag, Heidelberg (2002).

15. L. C. Folmar, N. D. Denslow, K. Kroll, E. F. Orlando, J. Enblom, J. Marcino, C. Metcalfe, L. J. Guillette, Jr. *Arch. Environ. Contam. Toxicol.* **40**, 392–398 (2001).
16. L. C. Folmar, N. D. Denslow, V. Rao, M. Chow, D. A. Crain, J. Enblom, J. Marcino, L. J. Guillette, Jr. *Environ. Health Perspect.* **104**, 1096–1101 (1996).
17. S. Jobling, M. Nolan, C. R. Tyler, G. Brighty, J. P. Sumpter. *Environ. Sci. Technol.* **32**, 2498–2506 (1998).
18. L. J. Guillette, Jr., D. A. Crain, A. Galle, M. Gunderson, S. Kools, M. R. Milnes, E. F. Orlando, A. A. Rooney, A. R. Woodward. *Am. Zool.* **40**, 438–452 (2000).
19. D. B. Pickford, L. J. Guillette, Jr., D. A. Crain, A. A. Rooney, A. R. Woodward. *J. Herpetol.* **34**, 233–239 (2000).
20. M. P. Gunderson, D. S. Bermudez, T. A. Bryan, D. A. Crain, S. Degala, T. M. Edwards, S. Kools, M. R. Milnes, A. R. Woodward, L. J. Guillette, Jr. *Environ. Toxicol. Chem.* **21**, 914–921 (2001).
21. E. A. Hewitt, D. A. Crain, M. P. Gunderson, L. J. Guillette, Jr. *Chemosphere* **47**, 1129–1135 (2002).
22. V. A. Lance and R. M. Elsey. *J. Exp. Zool.* **239**, 241–246 (1986).
23. L. J. Guillette, Jr., D. A. Crain, A. A. Rooney, A. R. Woodward. *J. Herpetol.* **31**, 347–353 (1997).
24. L. J. Guillette, Jr. and M. P. Gunderson. *Reproduction* **122**, 857–864 (2001).
25. M. P. Gunderson, G. A. LeBlanc, L. J. Guillette, Jr. *Environ. Health Perspect.* **109**, 1257–1264 (2001).
26. P. S. Western, J. L. Harry, J. A. M. Graves, A. H. Sinclair. *Gene* **241**, 223–232 (2000).
27. W. N. Gabriel, B. Blumberg, S. Sutton, A. R. Place, V. A. Lance. *J. Exp. Zool.* **290**, 439–448 (2001).
28. G. H. Heinz, H. F. Percival, M. L. Jennings. *Environ. Monit. Assess.* **16**, 277–285 (1991).
29. L. J. Guillette, Jr., J. W. Brock, A. A. Rooney, A. R. Woodward. *Arch. Environ. Contam. Toxicol.* **36**, 447–455 (1999).
30. J. Burger, M. Gochfeld, A. A. Rooney, E. F. Orlando, A. R. Woodward, L. J. Guillette, Jr. *Arch. Environ. Contam. Toxicol.* **38**, 501–508 (2000).
31. M. F. Delany, J. U. Bell, S. F. Sundlof. *J. Wildl. Dis.* **24**, 62–66 (1988).
32. P. M. Vonier, D. A. Crain, J. A. McLachlan, L. J. Guillette, Jr., S. F. Arnold. *Environ. Health Perspect.* **104**, 1318–1322 (1996).
33. L. J. Guillette, Jr., P. M. Vonier, J. A. McLachlan. *Toxicology* **181–182**, 151–154 (2002).
34. K. Sumida, N. Ooe, K. Saito, H. Kaneko. *Mol. Cell. Endocrinol.* **183**, 33–39 (2001).
35. P. S. Western and A. H. Sinclair. *J. Exp. Zool.* **290**, 624–631 (2001).
36. J. M. Matter, D. A. Crain, C. Sills-McMurry, D. B. Pickford, T. R. Rainwater, K. D. Reynolds, A. A. Rooney, R. L. Dickerson, L. J. Guillette, Jr. In *Principles and Processes for Evaluating Endocrine Disruption in Wildlife*, W. Suk (Ed.), pp. 267–289, SETAC Pr., Pensacola, FL (1998).
37. E. Willingham and D. Crews. *Gen. Comp. Endocrinol.* **113**, 429–435 (1999).
38. E. Willingham, T. Rhen, J. T. Sakata, D. Crew. *Environ. Health Perspect.* **108**, 329–332 (2000).
39. M. R. Milnes, R. Roberts, L. J. Guillette, Jr. *Environ. Health Perspect.* **110** (Suppl. 3), 393–396 (2002).
40. D. A. Crain, L. J. Guillette, Jr., A. A. Rooney, D. B. Pickford. *Environ. Health Perspect.* **105**, 528–533 (1997).
41. D. M. Stocco and B. Clark. *Endocrine Rev.* **17**, 221–244 (1996).
42. D. J. Bernard, S. C. Chapman, T. K. Woodruff. *Mol. Cell. Endocrinol.* **180**, 55–62 (2001).
43. D. M. de Kretser, K. L. Loveland, T. Meehan, M. K. O'Bryan, D. J. Phillips, N. G. Wreford. *Mol. Cell. Endocrinol.* **180**, 87–92 (2001).
44. L. P. Walsh, C. McCormick, C. Martin, D. M. Stocco. *Environ. Health Perspect.* **108**, 769–776 (2000).
45. L. P. Walsh and D. M. Stocco. *Biol. Reprod.* **63**, 1024–1033 (2000).

46. A. A. Rooney and L. J. Guillette, Jr. In *Crocodile Biology and Evolution*, G. Grigg, C. Franklin, F. Seebacher (Eds.), pp. 214–228, Surrey-Beatty & Sons, Chipping Norton, Australia (2001).
47. A. A. Rooney and L. J. Guillette, Jr. In *Endocrine Disrupting Contaminants: An Evolutionary Perspective*, L. J. Guillette, Jr. and D. A. Crain (Eds.), pp. 82–125, Francis and Taylor, Philadelphia (2000).
48. Ministry of Environment. *Report on the Test Results of Endocrine Disrupting Effects of Nonylphenol on Fish*, Ministry of Environment, Japan (2001).
49. M. J. Portelli, S. R. de Solla, R. J. Brooks, C. A. Bishop. *Ecotoxicol. Environ. Saf.* **43**, 284–291 (1999).
50. S. Podreka, A. Georges, B. Maher, C. J. Limpus. *Environ. Health Perspect.* **106**, 185–188 (1998).
51. W. R. Kelce, C. R. Stone, S. C. Laws, L. E. Gray, J. A. Kemppainen, E. M. Wilson. *Nature* **375**, 581–585 (1995).
52. M. E. McMaster. Ph.D. dissertation, University of Guelph (1995).
53. G. J. Van Der Kraak, K. R. Munkittrick, M. E. McMaster, C. B. Portt, J. P. Chang. *Toxicol. Appl. Pharmacol.* **115**, 224–233 (1992).
54. R. P. Ertl and G. W. Winston. *Comp. Biochem. Physiol.* **121C**, 85–105 (1998).
55. V. S. Wilson and G. A. LeBlanc. *Toxicol. Appl. Pharmacol.* **148**, 158–168 (1998).
56. V. S. Wilson, J. B. McLachlan, J. G. Falls, G. A. LeBlanc. *Environ. Health Perspect.* **107**, 377–84 (1999).

Pure Appl. Chem., Vol. 75, Nos. 11–12, pp. 2287–2303, 2003.

Topic 4.8

Review of the effects of endocrine-disrupting chemicals in birds*

John P. Giesy, Lori A. Feyk, Paul D. Jones‡, Kurunthachalam Kannan, and Thomas Sanderson

Michigan State University, Department of Zoology, National Food Safety and Toxicology Center, Institute for Environmental Toxicology, East Lansing, MI 48824, USA

Abstract: There have been several case studies of the impact of chemical contaminants on birds at the level of individuals or populations. While many of the chemicals involved in these incidents have been classified as endocrine-disrupting chemicals or endocrine active substances (EASs) the mechanisms by which these chemicals affect birds are not clearly or fully understood.

INTRODUCTION

There are a number of reasons that some bird populations have declined significantly since prehistoric times, including habitat loss, hunting, climate change, and population genetics, among others. However, some of the effects have been caused by exposure to anthropogenic chemicals [1]. Birds, particularly piscivorous birds, have been exposed to a number of toxic, synthetic compounds, particularly persistent bioaccumulative organic compounds [2]. Specifically, the persistent chlorinated hydrocarbons have accumulated to the greatest concentrations and have been associated with the most severe adverse effects [3]. Organochlorine pesticides (such as dieldrin) and organophosphorus insecticides that are acutely toxic have caused widespread mortalities of adult birds. Other chemicals have had more subtle effects in reproduction, such as deformities and embryo lethality [2,4,5]. These effects, have, in turn, caused declines in populations [6,7]. These effects that have been referred to as endocrine-disrupting effects [8].

The most dramatic chemical effect on reproductive performance in birds was the result of eggshell thinning, caused by DDE, a degradation product of DDT [9–13]. Since limitations have been placed on the manufacture and use of these persistent widespread contaminants, concentrations in fish and birds have decreased [14,15]. Currently, concentrations of DDE in birds from many areas of North America have decreased below concentrations associated with eggshell thinning. As a result, populations of some fish-eating, water birds have increased rapidly [16]. However, other adverse effects, such as localized impairment of reproductive performance [17] and anatomical defects [1,18] persist. While exposure to certain organochlorine compounds, such as DDT and its degradation products, is associated with adverse effects, the mechanism of action is still not fully understood. It was not known if the currently observed effects were due to existing contaminants and would abate as concentrations decreased further or if the effects were caused by some, as yet unidentified chemicals. It has been speculated that

*Report from a SCOPE/IUPAC project: Implication of Endocrine Active Substances for Human and Wildlife (J. Miyamoto and J. Burger, editors). Other reports are published in this issue, *Pure Appl. Chem.* **75**, 1617–2615 (2003).
‡Corresponding author

the observed effects acted through mechanisms that modulated the endocrine system [19]. It was further speculated that some of the observed effects, such as the deformities would have been observed during the 1950s through the 1970s at equal or greater rates of incidence, but were masked by the effects of eggshell thinning. It had been suggested that since concentrations of all of the known toxic chemicals in the food, tissues, and eggs of birds were declining the observed effects might not be due to synthetic, organic compounds. The suite of effects observed did not seem to be the types that could all be explained by disease or nutrition or a genetic "founder effect" since populations of birds had been greatly reduced due to the effects of DDE. While this might explain some of the effects, such as birth defects, it was not likely that all of the effects could be explained by genetic deficiencies. It was also suggested that the effects might be due to new, nonpersistent pesticides that could have effects at small concentrations, but would leave little or no trace in eggs of afflicted adults.

The suite of effects, observed in wild populations of birds, especially in the North American Great Lakes and the Baltic Sea have been compared to those caused by exposure to certain halogenated hydrocarbons in laboratory exposures [1,2]. Many of the observed field effects could be reproduced in controlled laboratory studies at tissue concentrations in the same range as those observed in field-collected organisms [20]. Therefore, Giesy and coworkers [2] compared concentrations of 2,3,7,8-tetrachlordibenzo-*p*-dioxin (TCDD) or TCDD equivalents (TCDD-EQ) in eggs of birds from the Great Lakes to concentrations known to cause effects. Based on concentrations observed in water birds and their eggs, it seemed unlikely that the PCDD and PCDF were responsible for the observed effects. These same effects were observed in birds exposed to polychlorinated biphenyls (PCBs) at concentrations similar to those observed [21]. However, a fairly poor correlation between the total concentrations of PCBs and egg lethality or birth defects was observed [17]. The weight of evidence demonstrated that there have been effects of these synthetic chemicals on birds. However, there is little evidence to support the conclusion that these effects are through modulation of the endocrine system. Furthermore, there is no evidence that the observed effects occur because any of these compounds act as estrogen agonist/antagonists.

The effects caused by synthetic chemicals on birds have been used to support what has been called the endocrine disruptor hypothesis [8]. Changes in reproduction, gender ratios, and eggshell thickness have been used to support a subhypothesis, the estrogen hypothesis. This hypothesis holds that the effects caused by exposure to persistent organic pollutants are caused by modulation of the endocrine system by these compounds acting as hormone agonists or mimics. In the case of estrogenic effects, it has been hypothesized that the adverse effects are caused by the xenoestrogen binding to the estrogen receptor (ER) and thus eliciting adverse effects.

To perform accurate ecological risk assessments it is always necessary to understand the mechanism of action. Only in this way can the effects of complex contaminant mixtures be interpreted and appropriate thresholds for protection be developed. It is unclear how classifying a compound as an "endocrine disruptor" assists in conducting an ecological risk assessment. What is required is an understanding of the mechanism of action, regardless of whether it involves direct or indirect action on the endocrine system or any endocrine-modulated effects at all. Here, we discuss some of the major effects observed in birds associated with exposure to environmental residues and discuss likely mechanisms of action. First, the processes of sexual differentiation and reproduction in birds are outlined with regard to their vulnerability to modulation by chemicals. Throughout this chapter, an attempt has been made to present and reconcile the results obtained from in vitro and in vivo laboratory studies, and field studies. The combined evidence from all these studies is crucial to understanding the complex phenomenon of endocrine modulation. The endocrine system of birds is described and the potential for xenoestrogens to affect these systems is examined. Some of the phenomena observed in birds that have been attributed to endocrine disruptors, especially chlorinated hydrocarbons or attributed to the effects of xenoestrogens are discussed. These include embryo-lethality, congenital deformities, skewed sex ratios, abnormal pairing behavior, and eggshell thinning.

EFFECTS OF SEX HORMONE MODULATORS ON SEXUAL DIFFERENTIATION

Gonadal differentiation in birds

In both birds and mammals, very early development is characterized by a sexually undifferentiated state [22]. Undifferentiated gonads are identical in "male" and "female" embryos/fetuses, and have the potential to develop into either state. The homogametic sex is the "default" sex, a differentiating hormone produced as a consequence of the "odd" chromosome causes the embryo/fetus to switch gender from the "default" sex. In mammals, the female is the homogametic sex (XX) "default" sex, the male is heterogametic (XY). The *Sry* region of the Y chromosome (the sex-determining region of the Y) encodes for maleness, its gene product is referred to as the H-Y antigen. The differentiating hormone for the mammalian testis is testosterone.

In birds, the homogametic default sex is male (ZZ), the heterogametic sex (ZW) is female. In birds the differentiating hormone for gonadal development is estrogen (c.f., testosterone in mammals). Female birds have an H-W antigen, analogous to the H-Y antigen in mammals, this gene product is the differentiation antigen for the heterogametic gonad (the avian ovary). Interestingly, this gene is not W-linked; it is present in both sexes and is activated by embryonic estrogen exposure [23]. It is, therefore, possible to cause avian sex-reversal from male to female, including expression of H-W antigen, by administration of estrogen to the genetically male egg during the critical period of embryonic development.

The early avian embryo has left and right Wolffian (preseminiferous tubules) and Müllerian ducts (preovarian ducts) which connect primordial gonads to the cloaca [24]. The Wolffian ducts serve as functional nephric ducts throughout embryonic development and normally regress and degenerate in females post-hatch. In males they are incorporated into the reproductive tract and are influenced by testicular hormones. Müllerian ducts regress in male embryos under the influence of testicular hormones. In females, the right Müllerian duct regresses such that only a small caudal remnant normally remains post-hatch, while the left duct develops into the primordial shell gland. In the absence of estrogen both gonads develop into testes, while during normal female development the left gonad develops into an ovary while the right gonad regresses [25].

Differentiation of brain and sexual behavior in birds

In mammal and bird behavior, neural substrates differentiate at the same time as, or after, the gonads [26], and are influenced by the hormonal secretions of the gonads. Adult behavior is influenced both by the organizational and activational effects of sex hormones. Organizational effects are permanent actions that establish the fundamental form (male or female) of an organ or behavior, they generally occur during critical periods of development. Activational effects are not permanent, are not limited to any critical period, and typically occur at or after puberty or sexual maturity. In both birds and mammals, estrogen is the primary differentiating hormone for behavior [27]. In male rats, the active differentiating hormone for copulatory behavior is estradiol, synthesized in the brain from testosterone supplied via the blood. In birds, ovarian estrogens directly feminize and demasculinize copulatory behaviors [28].

Early sex hormone administration can have profound and permanent sex-reversing effects on adult behavior in birds [29]. When male quail embryos (*Coturnix japonica*) are injected with 1 μg estradiol or 500 μg testosterone during a critical period of incubation, they can be completely behaviorally sex-reversed. As adults, they fail to mount, crow, or strut; they are completely demasculinized and are behaviorally indistinguishable from females [30]. This is due to a fundamental change in the neural substrate underlying behavior. The early organizational effect of estradiol confers a differential responsiveness to the activating effects of testosterone in adulthood: testosterone treatment restores copulation in castrated adult males, but is without effect in females. Female quail treated with an antiestrogen before hatching can be masculinized, and will mount other females as adults [29]. Therefore, during normal development of quail, an embryo develops as a behavioral male unless estrogen is present. Testosterone-induced organizational demasculinization requires the aromatization of testosterone to estradiol; the enzymatic aromatization activity in the embryonic quail brain is likely to be low, and thus protects male embryos from their endogenous secretions [31].

The sexual differentiation of behavior has also been extensively studied in the Zebra finch. This species exhibits sexual dimorphism of the brain and of behavior; normally only males sing, dance, and mount. The telencephalic nuclei (TN), higher vocal center (HVC), nucleus robustus archistriatalis (RA), nucleus magnocellularis of the anterior neostriatum (MAN), and area X of the lobus paraolfactorius (X) are larger and more extensively connected in males, which is essential for the learning and production of complex vocalizations [32]. The administration of exogenous estrogen to females during the first week post-hatch results in a profound organizational masculinization of brain and behavior [33–35]. This includes a neural masculinization of the TN, setting up a functional circuit in females similar to that in males enabling the learning and production of complex vocalizations [32]. When these females are stimulated with testosterone as adults, they engage in male behaviors, including singing and dancing [35]. Interestingly, males treated with estradiol during the first week post-hatch are demasculinized, and fail to mount as adults [35]. Thus, the pattern of behavioral sexual differentiation in the zebra finch is complex; estradiol treatment during the first week post-hatch masculinizes one behavior and demasculinizes another. A number of possible explanations for this have been proposed but for our purposes, it is sufficient to note that the process of sexual differentiation of behavior in this species is finely tuned, complex, and sensitive to estrogens.

The critical period for sexual differentiation of behavior varies with the biology of the bird species. The sex-reversing effects of early estrogen treatment on male Japanese quail behavior only occur if the treatment is before day 12 of the 18-day incubation [30], the estradiol-induced masculinization of the female zebra finch is only produced by treatment after hatching [35]. Quail are precocial (well-developed at hatching); zebra finches are altricial (poorly developed at hatching). The timing of sexual differentiation of behavior in these two species is consistent with the observation that precocial and altricial birds develop similarly, but hatch at different stages of development [29].

Susceptibility of birds to exogenous estrogens

There are several biological differences between birds and mammals that may confer differential sensitivity to estrogenic substances during sexual differentiation. In general, exposure of an avian embryo to exogenous estrogens during the critical period should have a greater adverse affect than would exposure of a mammalian fetus to estrogens during the same critical period. This is because in birds, estrogen is the differentiating hormone for both gonads and behavior. In contrast, estrogen is not the differentiating hormone for gonadal development in mammals. While exposure of the mammalian fetus to estrogen may cause changes in the development of sexual characteristics, the changes are not as profound as those that would be caused by testosterone exposure.

For the avian embryo to be exposed to exogenous estrogens during the critical period, these compounds must be passed from the female bird to her eggs. Some estrogenic pollutants, such as DDT, are readily transferred to lipid-rich yolk. It is possible for fish-eating birds to acquire substantial body burdens of such contaminants from their diet, and to transfer these to the egg yolk. However, if estrogenic substances do not pass from female to egg, the avian embryo is more protected from adverse effects than the mammalian fetus because avian embryos are isolated from the maternal blood stream during gonadal sexual differentiation.

Avian embryos may be more sensitive than mammalian fetuses to estrogenic chemicals, which require metabolic activation [25]. Water-soluble metabolic products of xenobiotics are not excreted from the avian egg, but remain in the blood circulation throughout incubation. The phase 1 metabolites of many organochlorine contaminants are considerably more estrogenic than the parent compounds [36,37]. It is possible that during the last half of incubation, the avian embryo may be capable of mixed-function oxygenase metabolism, including hydroxylation of organochlorines, and yet still be within the critical period for sexual differentiation of brain and behavior. This would be an interesting topic for future research.

Modulation of endocrine function during reproduction or development in birds by environmental chemicals: Laboratory evidence

Several persistent organochlorine pollutants exhibit estrogenic activity, and can influence reproduction and avian embryonic development. Kepone has a binding affinity for chicken estrogen receptor that is about 5000-fold less than that of 17β-estradiol [38]. Kepone stimulates growth and dramatic morphological changes in the oviduct of immature Japanese quail when administered at 160 μg/g in the diet [38]. Chronic ingestion of kepone at relatively great concentrations produces reproductive impairment, decreased eggshell quality, and decreased chick survival in several avian species [39].

o,p′-DDT and the di-demethylated metabolite of methoxychlor also bind to the estrogen receptor and produce estrogenic effects in vivo [40]. Altered gonadal development was observed following injection of *o,p′*-DDT or methoxychlor into gull eggs on day 1 of incubation [41,42]. The male gonads were feminized, right oviducts persisted in surviving females, the gross morphology of *o,p′*-DDT or methoxychlor injected eggs was intermediate between controls and estradiol injected eggs. It is difficult to judge the functional significance of these results since the criteria for feminization was a small histological change (i.e., a localization of primordial germ cells in a thickened cortex on the surface of the left testis). Ideally, the egg injection study with *o,p′*-DDT should be repeated, and hatchlings should be grown to maturity and their reproductive health assessed.

Organochlorine chemicals have also altered avian reproductive behavior in controlled studies. Adult ring doves fed a mixture of organochlorines exhibited alterations in hormone concentrations and reproductive behavior [43]. Consumption of contaminants led to a reduction and/or delay in behaviorally induced increases of sex hormones, contaminated females failed to respond to male courtship behaviors, and pairs receiving the highest dosage spent less time feeding the young. There was a marked dose-related decrease in fledging success, and the breeding cycle was greatly asynchronous in treated birds. The administration of a polychlorinated biphenyl (PCB) mixture to adult breeder doves resulted in aberrant incubation [44] and courtship [45] behaviors. PCB-dosed females were particularly affected in the later experiment, performing only a small number of courtship behaviors, resulting in a severe impairment of reproductive success.

Other environmental contaminants have also had a negative impact on avian reproduction due to endocrine disruption. Oral administration of crude oil caused elevated plasma corticosterone and thyroxine levels in herring gull and black guillemot nestlings, apparently related to depressed growth [46]. Exposure to the organophosphorus insecticide parathion may also impact avian incubation behavior and reproductive success [47,48], although such impacts are likely to be of a more acute and transient nature than organochlorine-induced reproductive impairment. Natural environmental estrogens, such as phytoestrogens in the diet, can also influence avian reproduction. California quail experienced a delayed onset of reproduction and decreased egg production when a plant extract containing isoflavones such as genistein was incorporated into their diet [49].

Sex ratio skew and female/female pairings in gull populations

An increase in female/female pairings in gull populations in DDT-contaminated regions is the most commonly cited evidence of organochlorine-induced estrogenic effects occurring in birds [42,50]. Due to difficulties encountered in accurately observing and sexing all birds in gull colonies, the incidence of female/female pairings is estimated by the number of nests containing abnormally large numbers of eggs ("supernormal clutches"). A single female gull typically lays 1 to 3 eggs. Situations by which 4-egg clutches arise are variable, unclear, and often not associated with female/female pairings [51].

Nests with five or more eggs are almost always the result of multiple-female associations. The number of nests with five or more eggs provides a reasonable assessment of the incidence of female/female pairing in a colony, although it probably under estimates the actual incidence.

Generally, female/female associations in gull colonies arise due to shortage of breeding males, as an increased incidence of supernormal clutches can be induced by removal of males [52]. While some supernormal clutches arise from female/female pairs, others are due to polygynous trios of two females and one male [53]. Females may associate with one another when males are unavailable to increase their probability of reproductive success, however slightly. A female may be able to achieve egg fertilization via copulation with a promiscuous male, but without a partner, will be unable to raise a chick as two nest attendants are necessary in gull colonies. Females may also associate to avoid aggression directed toward unmated birds in order to stay with the colony, and perhaps to increase their likelihood of obtaining a mate in future years [54]. Some studies demonstrate low reproductive success of supernormal clutches [55,56], others have demonstrated fledged offspring from supernormal clutches [57].

In gull colonies, a sex ratio skewed toward females can occur in several ways. It is quite common in rapidly expanding areas [58], as female gulls are less philopatric than males and are more common emigrants [59]. A skewed sex ratio can also occur if there is differential male mortality in a population, or in situations where there is a differential failure of male recruitment to the breeding population. Sex ratios skewed towards females were documented in several gull colonies during the 1960s and 1970s. The most dramatic and well-documented example occurred in the adult western gull population on Santa Barbara Island between 1972 and 1978 [56], the sex ratio was estimated to be 0.67 males/female. A supernormal clutch incidence of 0.6 to 1 % was observed in herring gulls inhabiting northeastern Lake Michigan during the period from1978 to 1981 [54,60]. Both the California and Great Lakes populations of gulls were exposed to relatively great concentrations of organochlorine compounds, including DDT, from the 1950s until the 1970s [41]. Several historical studies investigated the occurrence of supernormal clutches in the Laridae, utilizing literature sources and museum specimens, in order to determine whether incidences have actually changed in the pre- and post-DDT era. The incidence of supernormal clutches has actually decreased significantly for many species of terns throughout the United States [61]. Supernormal clutch incidence only increased significantly in three Laridae species since 1950: western gulls, Great Lakes herring gulls, and Caspian terns in the United States. Supernormal clutches were a regular occurrence in ring-billed and California gulls prior to the DDT era, and their occurrence has not changed over time [62]. Supernormal clutches were not found regularly in western or herring gulls until after 1950, since that time the sex ratio for their populations as a whole has changed dramatically towards an excess of females. Therefore, it appears that the shortage of males at breeding colonies resulted from a low male/female ratio in the population and not from a failure of feminized males to breed, since few of the museum specimens from which the sex ratios were calculated were collected from breeding colonies.

A decrease in the sex ratio of western and herring gulls could be due to a differential mortality, it is possible that male gulls are more susceptible to poisoning from persistent organochlorines. Male western gulls weigh about 25 % more than females, and feed higher up on the food chain [63]. Also, male gulls do not have the ability to excrete lipophilic contaminants by egg-laying. Therefore, males may accumulate greater body burdens of toxicants during their lifetimes than females. Alternatively, the cause of differential male mortality may not involve contaminants at all.

It has been suggested that the skewed sex ratios observed in western and herring gulls might have been caused by estrogenic environmental contaminants, either by differential male mortality or by feminization of male embryos resulting in chemical sterilization and failed recruitment into the breeding population [42]. While this is plausible, there is no direct evidence to support it. The hypothesis that females might have paired with the wrong sex due to chemical-induced masculinization is not supported by a detailed behavioral study that found no differences in behavior between females paired with females and those paired with males [64]. An attempt to correlate gonadal feminization with organochlorine contamination in Glaucous-winged gulls in Puget Sound was also inconclusive [42]. Thirty-one adult females from colonies spanning a range of contamination were trapped and sacrificed for gonadal inspection. Interestingly, the length of the right oviducts was correlated with the predicted degree of chemical contamination. However, the significance of the data is rather unclear, as all birds were successfully incubating clutches. Further, the most severe category was rated as greater than 10 mm length, while the literature indicates that a vestigial right oviduct of 9 to 10 mm is normal in the herring gull [65]. Unfortunately, although eggs were collected from the nests, no chemical data were presented. Finally, 4-egg clutches were considered "supernormal", but clutches of this size usually do not result from female-female associations [52] and are of dubious significance to the argument.

In conclusion, there is good evidence that there has been a fundamental change in the sex ratio of several North American gull populations in the post-DDT era, resulting in an overabundance of females in some colonies. The observations that the colonies most affected were in areas of great DDT contamination, and that a few DDT congeners have produced abnormal gonadal development in laboratory studies, support the hypothesis that environmental contaminants may have played a role in the sex ratio skew. However, direct evidence for that hypothesis is lacking, and any mechanism which could cause differential male mortality in gull populations could be a viable alternative explanation worthy of investigation.

MECHANISM OF DDE-INDUCED EGGSHELL THINNING

During the 1960s and 1970s, when DDT was in the North American environment at greater concentrations, populations of several sensitive bird species declined when individuals were unable to successfully incubate eggs because of abnormally thin shells [66]. Many of these species, such as the double-crested cormorant, experienced dramatic population increases after DDT was banned in the United States and environmental concentrations declined [67,68]. The eggshell-thinning effect of *o,p′*-DDT and its potent, stable metabolite *p,p′*-DDE in sensitive species is well known, even to the lay public. In fact, the contribution of DDT to population declines in bird species such as brown pelicans, peregrine falcons, and bald eagles is probably the most famous incident in wildlife ecotoxicology. It is, therefore, quite ironic that the mechanism by which DDTs cause eggshell thinning is still not understood.

Several possible mechanisms of DDE-induced eggshell thinning have been suggested however the mechanism appears to vary among species [66,69]. It is also difficult to study the mechanism as the most convenient laboratory species, including chicken and quail, are insensitive to this effect [70]. Research in this area within the last decade has been performed almost exclusively by one individual using one sensitive avian species [71–83].

An examination of the structure of DDE-induced thin eggshells can provide valuable clues about possible mechanisms of malformation. Some postulated mechanisms such as premature termination of shell formation, premature oviposition, effects on the protein matrix of the shell, effects on initiation sites of shell formation, or enhancement of shell growth inhibitors could be investigated by examining shell structure and composition. Indeed, abnormalities are observed in DDE-affected shells of several species. Species which produce eggs with a chalky valerite cover, including pelicans, cormorants, shags and gannets, produce eggs with a much reduced or absent cover following DDE exposure [69,84]. In these species, the shell-forming process is most impacted by DDE towards its termination. Other birds such as the great black-backed gull [69] and the grey heron [85] show a general reduction in all shell layers following DDE exposure. Changes in mineral composition of eggshells following DDE treatment have seldom been investigated [86].

Attempts to determine the mechanism of DDE-induced eggshell thinning have been based upon the a priori assumption that an alteration in Ca^{2+} metabolism of the shell gland is involved [81]. In two species of ducks, the effect of DDE administered in vivo has been localized to an interference of the translocation of Ca^{2+} from the shell gland mucosa cells to the shell gland cavity [75]. Calcium binding and transport have been studied, and the involvement of progesterone, prostaglandins, and calmodulin has been considered. There is substantial in vivo evidence that the shell gland actively transports Ca^{2+}

from the plasma into its lumen [87], and it has been suggested that this translocation of Ca^{2+} is driven by an ATP-dependent calcium "pump" which is inhibited by DDE [72]. Early reports indicated that calcium-dependent adenosine triphosphatase (Ca-ATPase) was reduced in the shell gland mucosa of DDE-treated birds which exhibited eggshell thinning [88,89]. Reduction of Ca-ATPase activity occurred only in DDE-sensitive species, and was considered a more likely candidate for the causation of eggshell thinning since such transport enzymes do not appear to be present in excess and a partial inhibition is generally accompanied by a parallel reduction in ion transport [88]. In ducks, the total content of calcium in the secreted lumen fluid was decreased following DDE treatment while the calcium content of the mucosal gland cells was increased, indicating that calcium translocation between the blood and shell gland mucosa was not impaired, but that translocation from the mucosa to the lumen was affected [75]. Detailed studies of the subcellular location and the timing of DDE-related effects on ATP-dependent Ca^{2+} uptake by and Ca^{2+}-Mg^{2+}-activated ATPase activity in the shell gland revealed complex relationships. DDE had no effect on the Ca^{2+} binding to mucosal homogenate of nonsecreting ducks (i.e., ducks with a fully calcified egg in the uterus), while DDE affected Ca^{2+} binding in actively "secreting" (i.e., in the process of forming a shell) ducks by decreasing mucosal microsomal binding, but increasing mucosal mitochondrial binding [75]. It was suggested that this effect of DDE might involve a decrease in the uptake of Ca^{2+} in secreting granules or in an interference with the stimulus-secretion coupling mechanism of calcium. Further studies revealed that DDE had an inhibitory effect on Ca^{2+} uptake of secreting shell gland mucosa, but there was generally no accompanying reduction of the Ca^{2+}-Mg^{2+}-ATPase activity [76]. While DDE markedly inhibited Ca^{2+}-Mg^{2+}-ATPase activity when added in vitro, when DDE was administered in vivo it often, but not always, resulted in an absolute or functional increase in Ca^{2+}-Mg^{2+}-ATPase activity despite a reduction in the rate of Ca^{2+} uptake. This indicates that DDE had an "uncoupling" effect on Ca^{2+} uptake and Ca^{2+}-Mg^{2+}-ATPase activity.

The current hypothesis regarding the mechanism of DDE-induced eggshell thinning involves an inhibition of prostaglandins (PGs) by the shell gland mucosa. Progesterone stimulated the synthesis of PGs in the eggshell gland mucosa of estrogen-primed chickens, and PGs play an important role in the control and regulation of reproduction in birds [82]. PG synthesis is decreased by *p,p′*-DDE in duck shell gland mucosa, both in in vitro experiments and following in vivo exposure [82]. PG synthesis was not inhibited by *p,p′*-DDT or *o,p′*-DDE in in vitro experiments, which indicates specificity between the proposed cause (PG synthesis inhibition) and the most potent eggshell thinning chemical (*p,p′*-DDE). However, in in vitro assays, a PCB Aroclor mixture (1242) was also an extremely potent PG synthesis inhibitor in shell gland mucosa. PCBs are not, however, potent eggshell thinning chemicals. It is unfortunate that the researchers did not conduct a parallel feeding trial with PCB 1242, in addition to their feeding trial with *p,p′*-DDE, in order to compare in vivo effects on PG synthesis. Also, the mechanism by which inhibition of PG synthesis might cause thin eggshells has not been definitively established. Some researchers have hypothesized that a furosemide-insensitive, PG-stimulated HCO_3^- transport could be inhibited in the shell gland mucosa of DDE-treated ducks, but further experiments have not supported that hypothesis [79].

Research into the mechanism of DDE-induced eggshell thinning has shown that the phenomenon is quite complex. Eggshell thinning is associated with a decreased quantity of calcium in affected eggs, and in the mallard duck the effect of DDE has been associated with a decreased transport of calcium from the eggshell gland mucosa to the lumen fluid. Treatment of birds with DDE has been associated with a variety of biochemical changes that could be related to changes in calcium transport. Many of these biochemical endpoints are interrelated, and it is difficult to determine which are the direct targets of DDE and which are merely coinfluenced by its action. The situation is further complicated by the fact that sensitivities to DDE-induced eggshell thinning vary among avian species, and in fact different mechanisms might be causing eggshell thinning in different species as evidenced by different gross eggshell defects.

Recent studies in New Zealand have highlighted the fact that eggshell thinning can be attributed to a variety of environmental factors. Investigations were begun into possible causes of eggshell thinning in Northern royal albatross nesting on one of New Zealand's sub-Antarctic islands. While the eggshell thinning observed was of the right type and magnitude to suggest a xenobiotic etiology the impact was limited to a single colony within the species range. Chemical analysis revealed background concentrations of persistent organochlorines including DDTs, but the concentrations measured were similar to those measured in other successful colonies [90]. Further investigation indicated that the cause of the shell thinning was crowding stress bought on by catastrophic habitat degradation, vegetation removal, and subsequent crowding stress [91]. This study illustrates both the complex nature of the eggshell thinning phenomenon and the degree of caution needed in investigating the relationship between xenobiotics and bird reproduction.

DEFORMITIES

Of the adverse effects observed to occur in colonial water birds of the North American Great Lakes region, the most obvious and that which can be most directly related to survival of individuals and populations are embryo lethality and developmental deformities. In the region, most of the embryos or chicks, which die during early development, have developmental deformities [92,93], particularly abnormalities which are of ectodermal origin [94]. One of the best-documented abnormalities, correlated with concentrations of planar chlorinated hydrocarbons (PCHs) in bird eggs, is the crossed-bill syndrome [93]. This suite of conditions found in Great Lakes wildlife has been named the Great Lakes embryo mortality, edema and deformity syndrome (GLEMEDS) [93]. This syndrome mimics chick edema disease, observed in the offspring of hens exposed to polychlorinated-dibenzodioxins and -dibenzofurans [95].

The few available case studies illustrating the effects of these chemicals are the reports of the Canadian Wildlife Service on studies performed on herring gulls living on Lake Ontario during the 1970s. In the late 1960s, anecdotal evidence circulated among field biologists of poor rates of hatching of Lake Ontario herring gull eggs [96]. Official Canadian Wildlife Service surveys began in 1971. Hatching rates of less than 20 % were found at some colonies in Lake Ontario. Initial examination of herring gull eggs and eggs of other species in Lake Ontario documented the presence of DDT and PCBs. However, analytical techniques at the time were incapable of measuring polychlorinated-dioxins (PCDDs), -furans (PCDFs), and the structurally similar PCB congeners. The characteristic symptoms in surviving chicks were similar to those of chick edema disease. Subsequent reanalysis in the late 1980s of a variety of eggs that had been collected from herring gull colonies in Lake Ontario in the 1970s were found to contain 1–3 μg/kg, wet wt. of TCDD. This chemically caused epizootic in Lake Ontario is probably the best-documented example of *dioxin-caused* effects on wildlife. During the last decade, the symptoms of chick edema disease and GLEMEDS have decreased significantly in the herring gulls of Lake Ontario [93], but more subtle, biochemical effects persist in all species of fish-eating birds inhabiting the Great Lakes [50].

Studies of nesting Forsters terns, conducted in Green Bay from 1983 to 1988 [4,97], and on double-crested cormorants and Caspian terns [18,92] demonstrated a similar suite of biological effects, but implicate PCHs other than TCDD. In Green Bay, Kubiak and coworkers found a variety of developmental deformities in the embryos and chicks of Forster's terns including growth deficiencies, deformities, and behavioral differences in parental care of eggs compared to an inland control colony where exposures were significantly less than on Green Bay [4]. Extrinsic adult behavioral abnormalities of inconsistent incubation led to a four-day-longer incubation period than the reference colony. Reciprocal transplant studies of eggs resulted in similar time delays ascribed to toxic substances in the eggs. That study suggested widespread, complex contaminant effects on the reproductive cycle, such as longer incubation times, smaller individuals, and wasting syndrome in those that did hatch. Similar, but less acute deformities, have been observed in double-crested cormorants and Caspian terns in upper Green

Bay where the total concentration of 2,3,7,8-TCDD equivalents (TCDD-EQ) ranged from 175 ng/kg wet wt. to greater than 440 ng/kg [18,92]. The majority (>90 %) of the TCDD-EQ in the eggs of cormorants and terns in the Great Lakes was due to non-*ortho*-substituted PCBs [98,99], rather than PCDD or PCDF, which accounted for between 2 and 9 % (12–22 ng/kg, wet wt.) of TCDD-EQ measured in water bird eggs in Lakes Superior, Huron, and Michigan. The primary contributions to the TCDD-EQ were due to the *dioxin-like* non-*ortho*-substituted PCBs, especially non-*ortho*-chlorinated congeners 126, 77, 169 and mono-*ortho*-chlorinated congeners 105 and 118. The understanding that dioxin-like effects in fish-eating, colonial water birds are due largely to non-*ortho*-substituted PCBs is an emerging consensus worldwide, except near TCDD point sources [100,101].

Available evidence supports the hypothesis that most, if not all, TCDD effects are mediated through the AhR [102,103], a cytosolic receptor protein that was first discovered by Poland and coworkers [104]. The AhR signaling transcription pathway is initiated by TCDD diffusion into the cell, where it binds with high affinity to the cytosolic AhR protein complex, which also includes heat shock protein 90 (Hsp90) and a 38-kDa, immunophilin-related protein [105,106]. The ligand binding activates AhR and stimulates the dissociation of AhR-associated proteins. The ligand-receptor complex is subsequently translocated into the nucleus, where it dimerizes with AhR nuclear translocator (ARNT) [103,107]. The heterodimers are capable of recognizing and binding DNA at the consensus sequence, GCGTG, of dioxin responsive elements (DREs) [108,109]. This action either increases or decreases the transcription of target genes [110,111], including cytochrome P450 (CYP1A1, CYP1A2) [112,113], NAD(P)H:quinone reductase [114], class 3 aldehyde dehydrogenase and glutathione S-transferase. The ARNT protein, however, does not function uniquely in the AhR signaling pathway, but also pairs with HIF-1α (hypoxia inducible factor 1α) to regulate genes active in response to low oxygen stress [115–117]. HIF-1α is continuously synthesized and degraded under normal oxygen tension. Hypoxic conditions inhibit the degradation of HIF-1α by the ubiquitin proteasome system, and triggers the nuclear localization of HIF-1α [118–120].

AhR, ARNT, and HIF-1α belong to bHLH (basic-helix-loop-helix)/PAS protein family. The bHLH motif is characteristic of a family of proteins that function as modulators of cell proliferation and differentiation. PAS proteins are found in representative organisms of all five kingdoms and may play a role in determining target gene specificity [121]. PAS proteins are involved in development and differentiation (Sim group, trachealess) [122,123], regulation of circadian clocks (Per, CLOCK) [124–125], sensing and responding to oxygen tension (HIF-1α, EPAS-1/HLF) [117,126], and steroid receptor signaling (SRC-1) [127]. The myogenic bHLH proteins autoregulate their own expression and cross-regulate the expression of other family members [128], since the secondary dimerization surface of the PAS domain can provide the specificity for dimerization among PAS/bHLH proteins [129]. PAS proteins, therefore, may behave in a way similar to the myogenic bHLH proteins and interact with each other through the PAS domain. Since ARNT has been found to dimerize with many PAS proteins such as AhR, Sim1, Sim2, HIF-1α, and EPAS-1 [126,130–133], it may act as a central regulator of PAS protein-dependent pathways.

Since many of the effects that have been observed in wildlife and used as support of the endocrine disruptor hypothesis are thought to be caused by TCDD and structurally similar, synthetic halogenated hydrocarbons, it is important to understand the mechanism of action, as it relates to the endocrine disruptor hypothesis. While the effects caused by TCDD can be classified as an effect on signal transduction, these effects are not covered by the narrow definition of receptor-mediated effects of steroid hormones. Thus, these effects would not be identified in a screening process so narrowly defined as to include direct receptor-mediated effects on the estrogen, androgen, or thyroid receptors.

Thyroid hormone and vitamin A

In addition to the primary steroid hormones, thyroxine and vitamin A are crucial to normal development of sexual and immune functions. Altered concentrations of thyroxin and vitamin A are frequently reported to co-occur with embryonic abnormalities in populations exposed to some synthetic halogenated hydrocarbons [93]. Individuals from these exposed populations have been observed to have altered sexual development [42], sexual dysfunction as adults [19], and immune system suppression [134,135]. The observations on adult sexual dysfunction are especially significant since young, which appear to be normal while raised by intoxicated parents, may become reproductively dysfunctional when mature [136,137]. Poor reproductive efficiencies and opportunistic diseases are characteristic of wild animals in exposed populations of the Great Lakes.

Vitamin A (retinol) has many functions in animals, such as in embryonic development, vision, maintenance of dermally derived tissues, immune competence, hematopoiesis, and reproduction. Vitamin A is necessary for normal embryonic development [138,139] and, thus, changes in the status of vitamin A in the plasma or liver may be responsible for the birth defects observed in birds, which have been exposed to PCHs. An inverse correlation has been observed between the concentration of vitamin A in serum and concentrations of PCHs in tissues of birds from Great Lakes. Both vitamin A and its storage form in the liver (retinal palmitate) were depleted in birds exposed to sublethal doses of the dioxin-like PCB congener 77 [140]. There are effects of PCH on concentrations of vitamin A in both the blood and liver. These effects are thought to be due to at least two processes. In blood, hydroxylated PCB metabolites bind to the carrier protein transthyretin [134,141,142]. Hydroxylated PCBs have been shown to be retained in the blood of birds [143]. In the liver, induction of hepatic enzymes involved in the degradative and mobilization pathways of vitamin A, such as, such as acyl-CoA:retinol acyltransferase and uridine diphosphate glucuronosyl transferase (UDPGT) are thought to alter the metabolic pathways involved in the storage and mobilization of vitamin A, which results in the observed depletion of retinols in the liver.

2,3,7,8-TCDD is known to have effects on both male and female steroid hormones. For instance, 2,3,7,8-TCDD has both estrogenic and antiestrogenic effects, in different tissues, depending on timing of exposures during development. Furthermore, 2,3,7,8-TCDD is a potent thyroxine agonist, which may account for its ability to cause wasting syndrome in homeotherms. The induction of the mixed function monooxygenase system can also reduce the concentrations of circulating steroid hormones, which can have adverse effects on the reproduction of wildlife [144].

Thyroid hormone, is an important regulator of development and metabolism and can be influenced by PCHs [134,139]. There are several possible mechanisms for the observed effects on circulating triiodo thyroxin (T_3) and thyroxin (T_4). First, hydroxy-substituted PCB congeners have been observed to displace T_4 from its carrier protein, transthyretin (TTR; prealbumin), which results in effects similar to thyroxin deficiency [141,142,145]. PCHs can induce UDPGT activity in the liver, which then decreases the concentration of TTR in the blood. Concentrations of TTR are not determined directly in the plasma, but rather, T_4 binding capacity is measured. Therefore, it is not possible to distinguish which of the two mechanisms may be causing the observed effects. TCDD also mimics the effects of thyroxine as a key metamorphosis signal during maturation [146]. 2,3,7,8-TCDD has also been shown to down-regulate the epidermal growth factor (EGF) receptor [147], which may result in disruption of the patterns of embryonic development at critical stages.

ALTERATIONS IN AVIAN THYROID HISTOLOGY BY GOITEROGENS

Goiter, an enlargement of the thyroid gland without hyperthyroidism, is endemic in Great Lakes wildlife [148]. Goiter is commonly caused by an iodine deficiency, although chemicals or defects that interfere with thyroid hormone synthesis or regulation can also cause the disease. Goiter can have a histological appearance which varies depending upon its cause. Avian goiter resulting from iodine deficiency is characterized by epithelial cell hyperplasia [149]. Epithelial cells are cuboidal to columnar, and the colloid is reduced and pale-staining. PCBs can cause goiter in animals by competitively binding to carrier proteins, such that thyroid hormones are not bound and are cleared more rapidly from the body. When there is a relative lack of thyroid hormones in the body, thyroid stimulating hormone (TSH) can stimulate thyroid hypertrophy or hyperplasia. While mammalian goiter caused by PCBs is a hyperplastic goiter which resembles that caused by iodine deficiency, the histological picture in birds is quite different. In birds, PCB treatment results in hypothyroidism, and a large-colloid goiter [140]. Thyroid weight, follicle size, and colloid are all increased, and epithelial cells are flattened. At great PCB doses, thyroid atrophy is observed [150]. The types of goiter induced by other organochlorine contaminants are more variable. DDT and DDE have produced a hypothyroidism in Japanese quail [151], but an epithelial cell hyperplasia in the pigeon [152]. Dieldrin produced an epithelial cell hyperplasia in the pigeon as well [152].

Herring gulls in the Great Lakes had enlarged thyroids when compared to a control colony in the Bay of Fundy [148]. The observed goiter was characterized by epithelial cell hyperplasia, few columnar epithelial cells, a diffuse microfollicular structure, and scant or absent luminal colloid. Since various etiologic agents produce different histological appearances in birds, histopathology can be useful in determining the cause of goiter in birds. It is clear that, in the case of the Great Lakes herring gulls, PCBs were not the cause of the observed goiter. While it is possible that DDE or dieldrin could have been a contributing factor, it is also possible that iodine deficiency or goiterogens in the forage could have been involved.

CONCLUSIONS

The stage in early life during which the sexual differentiation of gonads, brain, and behavior occurs is probably the most sensitive period to the influence of endocrine modulation by xenobiotics in birds. Exposure to relatively small quantities of exogenous hormones during the critical periods for sexual differentiation can result in severe and permanent alterations in gonadal and/or neural structure and in subsequent adult reproductive behavior. Alterations in gonadal structure, reproductive behaviors, and reproductive success have been experimentally induced by administration of exogenous hormones during critical developmental periods. In general, the administration of relatively great concentrations of organochlorine chemicals is required to elicit similar albeit less dramatic effects. While it is plausible that female/female pairings and low male/female ratios in gull populations could be related to environmental contaminants and endocrine modulation, any mechanism that could cause differential male mortality in gull populations may be a viable alternate explanation for these phenomena.

The mechanism by which DDE induces eggshell thinning in sensitive avian species continues to be investigated. It is quite possible that the mechanism may be different among various avian species. The quantity of calcium is decreased in thin eggshells, and a decreased transport of calcium from the eggshell gland mucosa to the lumen fluid has been implicated as the cause of eggshell thinning in at least one sensitive avian species. The modulation of calcium homeostasis is a mechanism by which xenobiotics could impact eggshell development. The possible interference of DDE with calcium transport enzymes or calcium regulators such as calmodulin, progesterone, or prostaglandins has been investigated, and the complex interplay of these regulatory factors have made it difficult to isolate the primary target of eggshell-thinning xenobiotics such as DDE.

Goiter is an endemic disease in some avian wildlife populations. It has been postulated that goiter may in some cases be caused by the modulation of thyroid hormone homeostasis by xenobiotic chemicals in the environment. Goiter has been experimentally induced in birds by the administration of organochlorine contaminants. The histological presentation of goiter can vary substantially depending upon the etiologic agent responsible. Histological evidence can be used to infer that PCBs were not the cause of goiter observed in Great Lakes herring gulls [148]. The observed goiter was characterized by epithelial cell hyperplasia with scant or absent luminal colloid, in marked contrast to the large colloid goiter commonly observed in birds experimentally treated with PCBs.

Regardless of the mechanism, it is clear that some synthetic organochlorine compounds have had an adverse effect on some species of birds in some locations. Some of these effects were due to the fact that these compounds are persistent and able to bioaccumulate and biomagnify into organisms at the top of the food chain. In some cases these were sensitive birds. There has been much discussion of the "endocrine disruptor hypothesis". As far as ecological risk assessments are concerned, this is somewhat of an artificial classification. While it is always important to know the mechanism of action of a compound, it is inappropriate to determine if a compound is or is not hazardous based on a single or a few mechanisms of action. The discussion of mechanisms of action presented here should make it evident that the simple classification of a compound as an endocrine disruptor or not is not very helpful in conducting the risk assessment. Specifically, classifying compounds on their ability to bind to specific steroid hormone receptors would not be very useful in screening compounds or providing the mechanistic information necessary to develop appropriate biomarkers or to interpret the potential effects of complex mixtures of compounds. There are many possible mechanisms through which persistent and bioaccumulative compounds can adversely affect organisms.

REFERENCES

1. J. P. Giesy, J. P. Ludwig, D. E. Tillitt. *Environ. Sci. Technol.* **28**, 128A–135A (1994).
2. J. P. Giesy, J. P. Ludwig, D. E. Tillitt. In *Dioxin and Health*, A. Schecter (Ed.), pp. 254–307, Plenum Press, New York (1994).
3. J. P. Ludwig, H. J. Auman, D. V. Weseloh, G. A. Fox, J. P. Giesy, M. E. Ludwig. *Colon. Water Birds* **18**, 60–69 (1995).
4. T. J. Kubiak, H. J. Harris, L. M. Smith, T. R. Schwartz, D. L. Stalling, J. A. Trick, L. Sileo, D. E. Docherty, T. C. Erdman. *Arch. Environ. Contam. Toxicol.* **18**, 706–727 (1989).
5. M. Gilbertson. *Chemosphere* **12**, 357–370 (1983).
6. D. B. Peakall. In *PCBs in the Environment,* Vol. II, J. S. Aaid (Ed.), Chap. 3, pp. 31–47, CRC Press, Boca Raton, FL (1986).
7. D. B. Peakall. In *Toxic Contamination in Large Lakes, Vol. II. Chronic Effects of Toxic Contaminants in Large Lakes,* N. W. Schmidtke (Ed.), pp. 39–54, Lewis Publishers, Chelsea, MI (1988).
8. T. Colborn, D. Dumanowski, M. J. Peterson. *Our Stolen Future. Are We Threatening our Fertility, Intelligence and Survival: A Scientific Detective Story*, Penguin Books, USA (1996).
9. D. V. Weseloh, S. M. Teeple, M. Gilbertson. *Can. J. Zool.* **61**, 427–436 (1983).
10. J. Struger and D. V. Weseloh. *Colon. Water Birds* **8**, 142–149 (1985).
11. J. Struger, D. V. Weseloh, D. J. Hallett, P. Mineau. *J. Great Lakes Res.* **11**, 223–230 (1985).
12. D. W. Anderson and J. J. Hickey. *Proc. XVth Int. Ornithol. Cong.* pp. 514–540 (1969).
13. J. E. Elliott, R. J. Norstrom, J. A. Keith. *Environ. Pollut.* **52**, 81–102 (1988).
14. P. C. Baumann and D. M. Whittle. *Aquat. Toxicol.* **11**, 241–257 (1988).
15. R. J. Allan, A. J. Ball, V. W. Cairns, G. A. Fox, A. P. Gilman, A. P. Peakall, D. A. Piekarz, J. C. Van Oosdam, D. C. Villeneuve, D. T. Williams. *Toxic Chemicals in the Great Lakes and Associated Effects*, Vols. I & II, Environment Canada, Department of Fisheries & Oceans, Health & Welfare, Canada (1991).
16. I. M. Price and D. V. Weseloh. *Can. Field Natural.* **100**, 474–482 (1986).
17. D. E. Tillitt, G. T. Ankley, J. P. Giesy, J. P. Ludwig, H. Kurita-Matsuba, D. V. Weseloh, P. S. Ross, C. Bishop, L. Sileo, K. L. Stromberg, J. Larson, T. J. Kubiak. *Environ. Toxicol. Chem.* **11**, 1281–1288 (1992).
18. N. Yamashita, S. Tanabe, J. P. Ludwig, H. Kurita, M. E. Ludwig, R. Tatsukawa. *Environ. Pollut.* **79**, 163–173 (1992).
19. T. Colborn and C. Clement. In *Advances in Modern Environmental Toxicology,* Vol. XXI, Princeton Scientific Publishing, Princeton (1992).
20. R. Eisler. USDI, USFWS Contaminant Hazard Review No. 8 (1986).
21. C. F. Tumasonis, B. Bush, F. D. Baker. *Arch. Environ. Contamn. Toxicol.* **1**, 312–324 (1973).

22. C. R. Austin, R. G. Edwards, U. Mittwoch. In *Mechanisms of Sex Differentiation in Animals and Man*, C. R. Austin and R. G. Edwards (Eds.), Academic Press, New York (1981).
23. U. Müller, M. T. Zenzes, U. Wolf, W. Engel, J. P. Weniger. *Nature* **280**, 142–144 (1979).
24. D. Price, J. J. P. Zaaijer, E. Ortiz, A. O. Brinkmann. In *Trends in Comparative Endocrinology*, E. J. W. Barrington (Ed.), *Amer. Zool.* ed., vol. **15** (Suppl. 1), Thomas J. Griffiths Sons, Utica, NY (1975).
25. D. M. Fry. *Environ. Health Perspect.* **103** (Suppl. 7), 165–171 (1995).
26. E. Adkins-Regan. In *Hormones and Reproduction in Fishes, Amphibians, and Reptiles*, D. O. Norris and R. E. Jones (Eds.), Plenum Press, New York (1987).
27. M. A. Ottinger, E. Adkins-Regan, J. Buntin, M. F. Cheng, T. DeVoogd, C. Harding, H. Opel. *J. Exp. Zool.* **232**, 605–616 (1984).
28. B. A. Schlinger and A. P. Arnold. *Endocrinology* **130**, 289–299 (1992).
29. E. Adkins-Reagan. *Trends Neurosci.* **10**, 517–522 (1987).
30. E. K. Adkins. *Neuroendocrinology* **29**, 178–185 (1979).
31. J. Balthazart and M. Schumacher. *J. Exp. Zool.* **232**, 595–604 (1984).
32. H. B. Simpson and D. S. Vicario. *J. Neurobiol.* **22**, 777–793 (1991).
33. M. E. Gurney and M. Konishi. *Science* **208**, 1380–1383 (1980).
34. H. B. Simpson and D. S. Vicario. *J. Neurobiol.* **22**, 755–776 (1991).
35. E. Adkins-Regan, V. Mansukhani, C. Seiwert, R. Thompson. *J. Neurobiol.* **25**, 865–877 (1994).
36. W. H. Bulger, R. M. Muccitelli, D. Kupfer. *Biochem. Pharmacol.* **27**, 2417–2423 (1978).
37. K. S. Korach, P. Sarver, K. Chae, J. A. McLachlan, J. D. McKinney. *Mol. Pharmacol.* **33**, 120–126 (1988).
38. V. P. Eroschenko and R. D. Palmiter. In *Estrogens in the Environment*, J. A. McLachlan (Ed.), pp. 305–324, Elsevier/North-Holland, New York (1980).
39. V. P. Eroschenko. *Toxicol. Environ. Health* **8**, 731–742 (1981).
40. D. Kupfer and W. H. Bulger. In *Estrogens in the Environment*, J. A. McLachlan (Ed.), pp. 239–262, Elsevier/North-Holland, New York (1980).
41. D. M. Fry and C. K. Toone. *Science* **213**, 922–924 (1981).
42. D. M. Fry, C. K. Toone, S. M. Speich, R. J. Peard. *Stud. Avian Biol.* **10**, 26–43 (1987).
43. M. L. B. McArthur, G. A. Fox, D. B. Peakall, B. J. R. Philogène. *Environ. Contam. Toxicol.* **12**, 343–353 (1983).
44. D. B. Peakall and M. L. Peakall. *J. Appl. Ecol.* **10**, 863–868 (1973).
45. G. M Tori and T. J. Peterle. *Bull. Env. Contam. Tox.* **30**, 44–49 (1983).
46. D. B. Peakall, J. Tremblay, W. B. Kinter, D. S. Miller. *Environ. Res.* **24**, 6–14 (1981).
47. D. H. White, C. A. Mitchell, E. F. Hill. *Bull. Environ. Contam. Toxicol.* **31**, 93–97 (1983).
48. R. S. Bennett, B. A. Williams, D. W. Schmedding, J. K. Bennett. *Environ. Toxicol. Chem.* **10**, 501–507 (1991).
49. A. S. Leopold, M. Erwin, J. Oh, B. Browning. *Science* **191**, 98–100 (1976).
50. G. A. Fox. In *Chemically-Induced Alterations in Sexual and Functional Development: The Wildlife/Human Connection*, T. Colborn and C. Clement (Eds.), pp. 147–158, Princeton Scientific Publishing, Princeton (1992).
51. M. R. Conover, D. E. Miller, G. L. J. Hunt. *Auk* **96**, 6–9 (1979).
52. M. R. Conover and G. L. J. Hunt. *Condor* **86**, 472–476 (1984).
53. M. R. Conover. *Condor* **86**, 467–471 (1984).
54. G. W. Shugart, M. A. Fitch, G. A. Fox. *Condor* **90**, 933–935 (1988).
55. R. W. Schreiber. *Condor* **72**, 133–140 (1970).
56. G. L. J. Hunt, J. C. Wingfield, A. Newman, D. S. Farner. *Auk* **97**, 473–479 (1980).
57. K. M. Kovacs and J. P. Ryder. *Auk* **100**, 658–669 (1983).
58. G. A. Fox and D. Boersma. *Wilson Bull.* **95**, 552–559 (1983).
59. G. Chabrzyk and J. C. Coulson. *J. Anim. Ecol.* **45**, 187–203 (1976).

60. M. A. Fitch and G. W. Shugart. *Behav. Ecol. Sociobiol.* **14**, 1–7 (1983).
61. M. R. Conover. *Wilson Bull.* **96**, 249–267 (1984).
62. M. R. Conover and G. L. J. Hunt. *Wilson Bull.* **96**, 619–625 (1984).
63. R. Pierotti. 1981. *Auk* **98**, 532–549 (1981).
64. G. L. J. Hunt, A. L. Newman, M. H. Warner, J. C. Wingfield, J. Kaiwi. *Condor* **86**, 157–162 (1984).
65. W. R. Boss and E. Witschi. *J. Exp. Zool.* **105**, 61–77 (1947).
66. A. S. Cooke. *Environ. Pollut.* **4**, 85–152 (1973).
67. J. P. Ludwig. *Jack-Pine Warbler* **62**, 91–102 (1984).
68. D. V. C. Weseloh and P. J. Ewins. *J. Great Lakes Res.* **20**, 443–456 (1994).
69. A. S. Cooke. *Environ. Pollut.* **19**, 47–65 (1979).
70. M. L. Scott, J. R. Zimmermann, S. Marinsky, P. A. Mullenhoff, G. L. Rumsey, R. W. Rice. *Poult. Sci.* **54**, 350–368 (1975).
71. C. E. Lundholm. *Acta Pharmacol. Toxicol.* **47**, 377–384 (1980).
72. C. E. Lundholm. *Acta Pharmacol. Toxicol.* **50**, 121–129 (1982).
73. C. E. Lundholm. *Comp. Biochem. Physiol.* **77B**, 655–663 (1984).
74. C. E. Lundholm. *Acta Pharmacol. Toxicol.* **54**, 400–407 (1984).
75. C. E. Lundholm. *Comp. Biochem. Physiol.* **78C**, 5–12 (1984).
76. C. E. Lundholm. *Comp. Biochem. Physiol.* **82C**, 1–16 (1985).
77. C. E. Lundholm. *Comp. Biochem. Physiol.* **89C**, 361–368 (1988).
78. C. E. Lundholm. *Comp. Biochem. Physiol.* **106C**, 389–394 (1993).
79. C. E. Lundholm. *Comp. Biochem. Physiol.* **109C**, 57–62 (1994).
80. C. E. Lundholm and K. Mathson. *Acta Pharmacol. Toxicol.* **52**, 390–394 (1983).
81. C. E. Lundholm. *Comp. Biochem. Physiol.* **88C**, 1–22 (1987).
82. C. E. Lundholm and M. Bartonek. *Arch. Toxicol.* **66**, 387–391 (1992).
83. C. E. Lundholm and M. Bartonek. *Arch. Toxicol.* **65**, 570–574 (1991).
84. R. W. Gould. *Bull. Environ. Contam. Toxicol.* **8**, 84–88 (1972).
85. A. S. Cooke, A. A. Bell, I. Prestt. *Environ. Pollut.* **11**, 59–84 (1976).
86. J. R. Longcore, F. B. Samson, J. F. Kreitzer, J. W. Spann. *Bull. Environ. Contam. Toxicol.* **6**, 345–350 (1971).
87. W. C. J. Eastin and E. Spaziani. *Biol. Reprod.* **19**, 505–518 (1978).
88. D. S. Miller, W. B. Kinter, D. B. Peakall. *Nature* **259**, 122–124 (1976).
89. G. J. Kolaja and D. E. Hinton. *Toxicol. Environ. Health* **3**, 699–704 (1977).
90. D. C. G. Muir, P. D. Jones, H. Karlsson, K. Koczansky, G. A. Stern, K. Kannan, J. P. Ludwig, H. Reid, C. J. R. Robertson, J. P. Giesy. *Environ. Toxicol. Chem.* **21**, 413–423 (2002).
91. C. J. R. Robertson, P. D. Jones. New Zealand: Science Publications, Department of Conservation. Wellington, New Zealand. *Science Poster #16* (1998).
92. J. P. Ludwig, J. P. Giesy, C. L. Summer, W. W. Bowerman, R. Aulerich, S. Bursian, H. J. Auman, P. D. Jones, L. L. Williams, D. E. Tillitt, M. Gilbertson. *J. Great Lakes Res.* **19**, 789–807 (1993).
93. M. Gilbertson, T. J. Kubiak, J. P. Ludwig, G. Fox. *J. Toxicol. Environ. Health* **33**, 455–520 (1991).
94. W. J. Rogan, B. C. Gladen, K. L. Hung, L. L. Koong, L. Y. Shih, J. S. Taylor, U. D. Yang, N. B. Ragan, C. C. Hsu. *Science* **241**, 334–336 (1988).
95. G. R. Higgenbotham, A. Huang, D. Firestone, J. Verrett, J. Reese, A. D. Campbell. *Nature* **220**, 702–703 (1968).
96. M. Gilbertson, R. D. Morris, R. A. Hunter. *Auk* **93**, 435–442 (1976).
97. D. J. Hoffman, B. A. Rattner, L. Sileo, D. Docherty, T. J. Kubiak. *Environ. Res.* **42**, 176–184 (1987).
98. D. E. Tillitt, G. T. Ankley, D. A. Verbrugge, J. P. Giesy, J. P. Ludwig, T. J. Kubiak. *Arch. Environ. Contam. Toxicol.* **21**, 91–101 (1991).

99. P. D. Jones, G. T. Ankley, D. A. Best, R. Crawford, N. Krishnan, J. P. Giesy, T. J. Kubiak, J. P. Ludwig, J. L. Newsted, D. E. Tillitt, D. A. Verbrugge. *Chemosphere* **26**, 1203–1212 (1993).
100. S. Tanabe, N. Kannan, A. Subramanian, S. Watanabe, R. Tatsukawa. *Environ. Pollut.* **47**, 147–163 (1987).
101. L. M. Smith, T. R. Schwartz, K. Feltz, T. J. Kubiak. *Chemosphere* **21**, 1063–1085 (1990).
102. A. B. Okey, D. S. Riddick, P. A. Harper. *Trends Pharmacol. Sci.* **15**, 226–232 (1994).
103. O. Hankinson. *Annu. Rev. Pharmacol. Toxicol.* **35**, 307–340 (1995).
104. A. Poland and J. C. Knutson. *Annu. Rev. Pharmacol. Toxicol.* **22**, 517–554 (1982).
105. Q. Ma and J. P. Whitlock, Jr. *J. Biol. Chem.* **272**, 8878–8884 (1997).
106. L. A. Carver and C. A. Bradfield. *J. Biol. Chem.* **272**, 11452–11456 (1997).
107. M. R. Probst, S. Reisz-Porszasz, R. V. Agbunag, M. S. Ong, O. Hankinson. *Mol. Pharmacol.* **44**, 511–518 (1993).
108. M. S. Denison, J. M. Fisher, J. P. Whitlock, Jr. *J. Biol. Chem.* **264**, 16478–16482 (1989).
109. L. Dong, Q. Ma, J. P. Whitlock, Jr. *J. Biol. Chem.* **271**, 7942–7948 (1996).
110. D. W. Nebert, A. Puga, V. Vasiliou. *Ann. NY Acad. Sci.* **685**, 624–640 (1993).
111. J. V. Schmidt and C. A. Bradfield. *Annu. Rev. Cell Dev. Biol.* **12**, 55–89 (1996).
112. J. P. Whitlock, Jr. *Annu. Rev. Pharmacol. Toxicol.* **30**, 251–277 (1990).
113. L. C. Quattrochi and R. H. Tukey. *Mol. Pharmacol.* **36**, 66–71 (1989).
114. L. V. Favreau and C. B. Pickett. *J. Biol. Chem.* **266**, 4556–4561 (1991).
115. K. Guillemin and M. A. Krasnow. *Cell* **89**, 9–12 (1997).
116. G. L. Semenza. *Hematol. Oncol. Clin. N. Am.* **8**, 863–884 (1994).
117. R. H. Wenger and M. Gassmann. *Biol. Chem.* **378**, 609–616 (1997).
118. L. E. Huang, J. Gu, M. Schau, H. F. Bunn. *Proc. Natl. Acad. Sci. USA* **95**, 7987–7992 (1998).
119. P. J. Kallio, I. Pongratz, K. Gradin, J. McGuire, L. Poellinger. *Proc. Natl. Acad. Sci. USA* **94**, 5667–5672 (1997).
120. S. Salceda and J. Caro. *J. Biol. Chem.* **272**, 22642–22647 (1997).
121. M. E. Hahn. *Comp. Biochem. Physiol.* **121C** (3), 23–53 (1998).
122. J. R. Nambu, J. O. Lewis, K. A. Wharton, Jr., S. T. Crews. *Cell* **67**, 1157–1167 (1991).
123. D. D. Isaac and D. J. Andrew. *Gene Dev.* **10**, 103–107 (1996).
124. Z. J. Huang, K. D. Curtin, M. Rosbash. *Drosophila. Sci.* **267**, 1169–1172 (1995).
125. D. P. King, Y. Zhao, A. M. Sangoram, L. D. Wilsbacher, M. Tanaka, M. P. Antoch, T. D. L. Steeves, M. H. Vitaterna, J. M. Kornhauser, P. L. Lowrey, F. W. Turek, J. S. Takahashi. *Cell* **89**, 641–653 (1997).
126. H. Tian, S. L. McKnight, D. W. Russell. *Genes Dev.* **11**, 72–82 (1997).
127. T.-P. Yao, G. Ku, N. Zhou, R. Scully, D. M. Livingston. *Proc. Natl. Acad. Sci. USA* **93**, 10626–10631 (1996).
128. E. N. Olson. *Genes Dev.* **4**, 1454–1461 (1990).
129. I. Pongratz, C. Antonsson, M. L. Whitelaw, L. Poellinger. *Mol. Cell. Biol.* **18**, 4079–4088 (1998).
130. J. B. Hogenesch, W. K. Chan, V. H. Jackiw, R. C. Brown, Y.-Z. Gu, M. Pray-Grant, G. Perdew, C. A. Bradfield. *J. Biol. Chem.* **272**, 8581–8593 (1997).
131. B.-H. Jiang, E. Rue, G. L. Wang, R. Roe, G. L. Semenza. *J. Biol. Chem.* **271**, 17771–17778 (1996).
132. P. Moffett, M. Reece, J. Pelletier. *Mol. Cell. Biol.* **17**, 4933–4947 (1997).
133. K. Sogawa, R. Nakano, A. Kobayashi, Y. Kikuchi, N. Ohe, N. Matsushita, Y. Fujii-Kuriyama. *Proc. Natl. Acad. Sci. USA* **92**, 1936–1940 (1995).
134. A. Brouwer. *Arch. Toxicol. Suppl.* **13**, 440–445 (1989).
135. D. W. Nebert. *Crit. Rev. Toxicol.* **20**, 153–174 (1990).
136. B. Brunström and J. Lund. *Comp. Biochem.* **91C**, 507–512 (1988).
137. B. Brunström and L. Andersson. *Arch. Toxicol.* **62**, 263–266 (1988).
138. M. H. Zile. *Proc. Soc. Environ. Biol. Med.* **201**, 141–153 (1992).

139. M. H. Zile, C. L. Summer, R. J. Aulerich, S. J. Bursian, D. E. Tillitt, J. P. Giesy, T. J. Kubiak. *Environ Toxicol. Pharmacol.* **3**, 277–288 (1997).
140. P. A. Spear and T. W. Moon. *Arch. Environ. Contam. Toxicol.* **14**, 547–553 (1985).
141. A. P. Brouwer. *Toxicol. Appl. Pharmacol.* **85**, 301–312 (1989).
142. A. P. Brouwer, J. H. Reijnders, J. H. Koeman. *Aquat. Toxicol.* **15**, 99–106 (1989).
143. E. Klasson-Wehler, Å. Bergman, Mathanasiadou, J. P. Ludwig, H. J. Auman, K. Kannan, M. Van den Berg, A. J. Murk, L. A. Feyk, J. P. Giesy. *Environ. Toxicol. Chem.* **17**, 1620–1625 (1998).
144. P. V. Hodson, M. McWhirther, K. Ralph, B. Gray, D. Thivierge, J. H. Carey, M. C. Levesque. *Environ. Toxicol. Chem.* **11**, 1635–1651 (1992).
145. T. Zacharewski, T. Harris, S. Safe, H. Thoma, O. Hutzinger. *Toxicology* **51**, 177–189 (1988).
146. J. D. McKinney, J. Fawkes, S. Jordan, K. Chae, S. Oatley, R. E. Coleman, W. Briner. *Environ. Health Perspect.* **61**, 41–53 (1985).
147. J. L. Newsted and J. P. Giesy. *Aquat. Toxicol.* **23**, 119–135 (1992).
148. R. D. Moccia, G. A. Fox, A. Britton. *J. Wildl. Dis.* **22**, 60–70 (1986).
149. D. K. Blackmore. *Vet. Rec.* **75**, 1068–1072 (1963).
150. D. J. Jefferies and J. L. F. Parslow. *Environ. Pollut.* **10**, 293–311 (1976).
151. E. P. Richert and K. V. Prahlad. *Poult. Sci.* **51**, 196–200 (1972).
152. D. J. Jefferies and M. C. French. *J. Wildl. Manage.* **36** (1), 24–30 (1972).

Pure Appl. Chem., Vol. 75, Nos. 11–12, pp. 2305–2320, 2003.

Topic 4.9

Wildlife as models for the study of how mixtures, low doses, and the embryonic environment modulate the action of endocrine-disrupting chemicals*

David Crews[1,‡], Oliver Putz[1], Peter Thomas[2], Tyrone Hayes[3], and Kembra Howdeshell[4]

[1]Institute of Behavioral Neuroendocrinology, University of Texas at Austin, Austin, TX 78712, USA; [2]Marine Science Institute, University of Texas at Austin, 750 Channel View Drive, Port Aransas, TX 78373, USA; [3]Department of Integrative Biology, University of California, Berkeley, 3060 Valley Life Sciences Building #3140, Berkeley, CA 94720-3140, USA; [4]Division of Biological Sciences, University of Missouri, 105 Lefevre Hall, Columbia, MO 65211-6190, USA

Abstract: This paper will review briefly the use of wildlife as models in the study of how mixtures, low doses, and the embryonic environment modulate the action of endocrine active substances (EASs). In so doing, it will show how the issue of low dosages must be considered within the context of mixtures present in the environment and the endocrine background of the exposed individual. That is, in nature, EASs usually are found in mixtures in which the constituent parts are in concentrations well below their NOAEL (no observed adverse effect level) as determined in single compound studies in the laboratory. In addition, exposure always occurs on organisms in various endocrine states. Thus, the issue of mixtures and dosages must always be considered within the context of the endocrine background. Finally, the effects of exposure are passed down through the generations. The question of exposure then at the level of the individual becomes very complicated, as it must take into account that at every life stage, the naturally occurring endocrine milieu of the organism (or tissue), any EAS burden inherited from the mother or built up over the individual's life, and the social environment in which the individual develops and interacts as an adult, will influence the response to acute exposure.

INTRODUCTION

At any given time, organisms in nature are exposed to constantly varying mixtures of natural or man-made chemicals that can act as hormones or antihormones. Generally, these endocrine active substances (EASs) are present in the environment only at very low concentrations, a circumstance that along with the low hormonal activities most of them display has caused many researchers to discount such contamination as a present danger [1]. However, EAS activities tend to be assessed in laboratory experiments that almost exclusively investigate effects of single agents and not combinations. It is therefore

*Report from a SCOPE/IUPAC project: Implication of Endocrine Active Substances for Human and Wildlife (J. Miyamoto and J. Burger, editors). Other reports are published in this issue, *Pure Appl. Chem.* **75**, 1617–2615 (2003).

‡Corresponding author

conceivable that the potential of some of these EASs to produce adverse health effects when acting in unison may have gone unnoticed. Few studies attended to the mixture effect problem, and in almost all cases the question of agent interaction was reduced to the exogenous contaminants alone, while interactions with endogenous hormones were by and large neglected [2]. Confusion in the field is further heightened by the inconsistency with which possible agent interaction is conceptualized; only too often are terms like synergism, summation, potentiation, or additivity used variably. To understand the environmental endocrine-disruption problem, we need to shift our main focus from assessing single compounds to determining if, and how, EASs interact with each other, and classifying their interaction using a well-defined and generally accepted terminology. It is also essential to establish how they interact with the endogenous milieu present in the organism at the time of exposure. In particular for wildlife studies, identifying potential animal models in which such mixture effects could best be determined is a mandatory objective for the field.

Recent research has challenged some assumptions traditionally presupposed by risk assessment protocols in toxicology. For example, a variety of investigations have shown that EASs do not necessarily have to reach a concentration threshold before adverse effects can be observed [3–5]. Also, dose–response profiles of certain EASs are not always monotonic, but can have a nonmonotonic shape [3,6]. While tissue differences in sensitivity are appreciated [7,8], the equally important issue of differential genetic predisposition to sensitivity is another phenomenon that is normally not considered in traditional risk assessment studies [8,9]; since inbred laboratory rodent strains are roughly equivalent to individuals of a natural species, such findings are directly relevant to extrapolation to and from the wild. Together, effect threshold, dose–response curve shape, and species/individual/tissue sensitivity can vary not only with the chemical tested, but also with the endpoint used for the test [8]. All three parameters are also closely connected to the problem of combination effects, since the proper prediction of adverse health effects caused by exposure to a mixture of chemicals depends on an accurate analysis of the dose–response relationship of each individual agent to the investigated endpoint [10–12].

The main difficulty of assessing combination effects lies within the complexity and variability of the tested model system, which is particularly true for in vivo experiments and explains why most approaches to identifying mixture effects employ in vitro assays [12–15]. However, understanding the effects of EASs requires an appreciation of the complexity of biological organization and how the different levels might interact. This requires a research strategy that not only continually revises concepts within each level, but also informs studies at higher (or lower) levels.

Animal studies clearly show that in some cases contaminant interactions occur [2,16], yet the nature of these interactions remains elusive [10]. It is interesting to note that in these studies particular organisms are found to be better model species than others [16]. One major difficulty in studying combination effects has been the formulation of a null hypothesis that can be tested without knowing every mechanistic detail by which the effect occurs [10,17]. Once mechanisms are taken into consideration for a chemical mixture model, the degree of complexity of the system, and thus of the model itself, increases to unworkable proportions and the resulting model quickly becomes impractical. Further, deciding on the nature of an observed chemical interaction depends entirely on our current understanding of the underlying mechanisms and can change as this knowledge evolves [17]; for example, a "synergistic" interaction of two agents could thus quickly turn into zero-interaction of the two compounds [18]. Since the measured effect does not change with our knowledge of mechanisms, it is warranted to have empirical models available that can predict chemical interactions solely based on the measured effects. In such an "input/output" (i.e., dose of chemical/effect) approach that basically treats mechanism as a black box, the null-hypothesis is zero-interaction and any deviation from the prediction marks an interaction [19].

Empirical approaches to assess combination effects require certain necessary conditions that have to be fulfilled or else the models fail. This is best exemplified in a paradox pointed out by

Berenbaum [18]. If a chemical A at a certain concentration d_i has an effect E_i, the effect it should have at twice the concentration, $2(d_i)$, should therefore also be twice as great, or $2(E_i)$. However, if the dose–response profile for chemical A is sigmoid, the effect at $2(d_i)$ will be greater than expected. Since a chemical cannot synergistically interact with itself, the observed effect is not summative and yet not synergistic, either. It was shown that effect summation requires linear dose–response relationships, which are rarely found or only at extremely low doses of a chemical [19]. This example shows how complex the problem of mixture effects is, and how difficult it can be to define the exact nature of an agent interaction.

This paper tries to conceptualize the current problem of studying mixture effects, in particular how to study it in wildlife species. We will show that low-dose effects, internal hormonal milieu, and the social context are subcomponents of this problem and how they could be taken into consideration in future experiments. In the following section, five animal groups will be reviewed as potential models for combination effects studies, followed by a discussion describing some newly emerging principles important for the field.

MODEL SYSTEMS

Teleosts

Some of the strongest evidence for endocrine disruption in wildlife has been obtained in fish populations exposed to effluents from sewage treatment plants and pulp mills containing xenoestrogens and antiandrogens, as well as in regions heavily contaminated with organochlorine pesticides such as the Great Lakes [20–23]. For example, measurement of estrogen-inducible proteins such as vitellogenin in male fish has been shown to be an especially sensitive biomarker of the presence of low concentrations of xenoestrogens in complex contaminant mixtures in freshwater and marine environments [20,24]. Van Der Kraak and coworkers demonstrated in white sucker (*Catostomus commersoni*) populations in Lake Superior that exposure to bleached kraft (pulp) mill effluent caused alterations in the reproductive fitness due to delayed age to sexual maturation, decreased gonadal size, a reduction in the expression of secondary sexual characteristics, reduced biosynthetic capacity of ovarian follicles, and elevated ovarian follicular apoptosis [25–29]. Thus, teleost fish have been used extensively as sentinels for detecting EASs in the environment and as models for investigating their mechanisms and sites of action on the hypothalamus-pituitary-gonadal (HPG) axis to disrupt reproduction. Knowledge gained from such field work along with substantial laboratory investigations on the underlying mechanisms by which EASs affect endocrine processes in teleosts provide an excellent basis from which the study of combination and low-dose effects can proceed.

A good example for how teleosts can play a pivotal role in elucidating EAS effects is an otherwise often ignored mechanism of endocrine disruption, that of neuroendocrine toxicity. Recent research in the Atlantic croaker (*Micropogonias undulatus*) by Peter Thomas and colleagues has shown that various EASs including organochlorine pesticides, PCB mixtures, and lead act at the hypothalamic level thereby impairing reproductive endocrine function [30,31]. In particular, it is changes in neurotransmitter concentrations in response to exposure to these neurotoxic chemicals that alters neuroendocrine function and ultimately disrupts reproduction. It could be demonstrated that the reproductive and neuroendocrine impairment in croaker after chronic exposure to environmentally relevant concentrations of the PCB mixture Aroclor 1254 (A1254) is associated with a reduction by 30–35 % of serotonin (5-HT) levels in hypothalamic regions involved in the neuroendocrine control of gonadotropin secretion [30]. Serotonin positively affects luteinizing hormone (LH) secretion in croaker by increasing gonadotropin releasing hormone (GnRH) production and synthesis of GnRH receptors in the pituitary. The decrease in hypothalamic 5-HT levels was due to inhibition of tryptophan hydroxylase, the rate-limiting enzyme in 5-HT synthesis, whereas the PCB mixture does not affect monoamine oxidase, the enzyme that converts 5-HT to its inactive metabolites. These studies suggest that the reproductive disruption occurs through a reduction of GnRH and its receptors in response to the lowered 5-HT activity. This was supported by the observation that the loss of reproductive function could be restored with an implant of GnRH or by treatment with 5-hydroxytryptophan, which bypasses the biosynthetic step catalyzed by tryptophan hydroxylase and restores 5-HT levels [32]. Also, treatment with the specific tryptophan hydroxylase inhibitor *p*-chlorophenylalanine mimicked all the neurotransmitter and neuroendocrine effects of the PCB mixture, giving further credibility to this hypothesis.

Although this work with Aroclor 1254 addresses the effects of a combination of EASs (i.e., PCBs), no conclusions about possible interaction of the individual constituents can be drawn. In order to establish whether or not the individual PCB congeners interact to cause the overall effect and, if so, what kind of an interaction they display, it is necessary to determine the dose–response profile for each of the major PCB congeners alone. Nonetheless, the work on the Atlantic croaker establishes useable endpoints that, along with the developed methodology, could be employed to assay the individual PCB congeners of Aroclor 1254 and thus determine whether a combination effect exists. It would then also be possible to characterize the type of interaction found in this particular PCB-mixture.

There is now convincing evidence that, in addition to the classic genomic mechanism of steroid action via binding and activation of nuclear steroid receptors, steroids also act at the cell surfaces of target tissues to initiate rapid, nongenomic responses, and that these actions are mediated by steroid membrane receptors. Several recent studies in teleost models have provided the first clear evidence that nongenomic steroid actions, like genomic ones, are susceptible to interference by xenoestrogens [33], and that this involves binding of these compounds to steroid membrane receptors [34]. Competition studies showed that the xenoestrogens Kepone and *o,p′*-DDD cause concentration-dependent displacement of progestin binding to the oocyte progestin membrane receptor over the range of 10^{-4} to 10^{-6} or 10^{-7} M, and also cause inhibition of progestin-induced oocyte maturation in vitro over the same concentration range (equivalent to 20–40 ppb, a tissue concentration frequently reported in fish from contaminated environments). Xenoestrogens can also interfere with the nongenomic actions of progestins to stimulate sperm motility as well as estrogens to inhibit testicular steroidogenesis in teleosts by binding to the membrane receptors thought to mediate these effects [35]. Taken together, these studies indicate that nongenomic steroid actions may be as susceptible to endocrine disruption by chemicals as are genomic actions and warrant further study. Possible interactive effects of individual chemicals in mixtures on this mechanism of endocrine disruption have not been investigated, although a complex mixture of polycylic aromatic hydrocarbons present in a water-soluble fraction of No. 2 fuel oil, has been shown to interfere with the nongenomic action of progestins to induce oocyte maturation in croaker [36].

Although there is now substantial evidence that a broad range of EASs can impair processes critical for the recruitment of fishes such as reproductive output and larval survival, comprehensive and reliable predictions of the impacts of environmental exposure to EASs on the size of fish populations are currently lacking. Density-independent mechanisms such as pollution often determine recruitment success at low spawning stock levels. In these cases, even small changes in fecundity and mortality rates of eggs and larvae may result in an eventual decline in population size. Currently, an interdisciplinary effort is underway to predict the population consequences of exposing croaker to environmentally realistic, low sublethal concentrations of EASs, both as mixtures and individual compounds. Integrated assessments of endpoints of critical stages of the reproductive cycle including sexual differentiation, production and maturation of the gametes, fertilization, and larval survival and ecological performance, are inputs into individual-based and matrix projection population models. Physiological models are being developed to integrate endocrinological, biochemical, and morphometric assessments of reproductive function at each critical adult stage. For example, plasma gonadotropin, estradiol-17β, vitellogenin, hepatic estrogen receptor concentrations, ovarian size, oocyte size distribution, and fecundity can be measured during the period of gamete production and gonadal growth in females to obtain an overall assessment of reproductive and endocrine functions at this stage. The endocrine and reproductive effects of EASs are often subtle and complex, involving multiple components of the reproductive endocrine system. This complicates the interpretation of both laboratory and field studies on EASs with the result that the potential long-term population consequences of environmental exposure to these compounds are difficult to predict. A variety of approaches, such as the modeling one described above, will be required to obtain a comprehensive understanding of risk of environmental contamination with EASs to fish populations and aquatic communities.

Amphibians

In light of current global declines in amphibian populations and their potential link with environmental contaminants, it is astonishing that effects of EASs on members of this taxon have not been examined more extensively. This lack of basic information is particularly surprising since their large clutch sizes, well-studied embryonic development, partially aquatic life style and complex life cycle of amphibians make them ideal model organisms for screening EASs, both singly and in combination.

In a comparative approach using multiple endpoints in a variety of species, Tyrone Hayes has found that exposure of amphibians to naturally occurring steroids like corticoids, estrogens, androgens and thyroid hormones causes dose-dependent differential responses that vary between species or even between developmental stages within a species [37–39]. For example, exogenous estradiol causes all young to develop as females in the South African clawed frog (*Xenopus laevis*), has no effect in the boreal toad (*Bufo boreas*) and, depending upon the dose, produces all males or all females in the leopard frog (*Rana pipiens*). Further, while it induces vitellogenin production in adult individuals, it does not affect vitellogenin synthesis in larval South African clawed frogs. In another species, the African reedfrog (*Hyperolius argus*), exogenous estrogens induce a color change (from a green dorsum to a reddish background with white spots), within androgens induce vocal sac development, and thyroid hormones induce tail reabsorption [40]. All of these changes can be produced by exposure over a six-day period. A survey of various compounds revealed that several steroidal estrogens induce color change as well as several synthetic estrogens (e.g., ethynyl estradiol, DES) and that tamoxifen blocks the color change when coadministered with estradiol. In addition, several phytoestrogens and various pesticides have estrogenic activity [41]. A number of androgens have also been examined for their ability to induce vocal sac development and the goitrogen, thiourea, has been shown to inhibit tail reabsorption in this species. In the South African clawed frog and the leopard frog, gonadal differentiation/sex ratio, laryngeal size, and time to metamorphosis are used as a measure of estrogen, androgen, and thyroid hormone activity, respectively [42].

In addition to documenting sexual chimeras in nature and relating their incidence with levels of the herbicide atrazine, Hayes has studied the effects of atrazine in the laboratory, finding that in the South African clawed frog it produces demasculinized and feminized gonads, hermaphrodites, and demasculinized secondary sex characters [42]. There is evidence suggesting that these effects are mediated by induction of the aromatase enzyme and not by direct interference of atrazine with the estrogen receptors [42]. Interestingly, the effects on the gonads in the South African clawed frog were produced at a dosage of 0.1 ppb, or more than 600 times lower than effective doses in in vitro mammalian studies and more than 30 million times lower than doses effective in in vivo mammalian studies. Similar effects are observed in the leopard frog, where larvae exposed to 0.1 or 25 ppb of atrazine result in hermaphrodites as well as males with testicular oocytes.

Large numbers of newly metamorphosed leopard frogs with malformed hind limbs discovered by middle school students on a field trip in rural Minnesota started a controversy about the possible link of such aberrations to EAS exposure. The frogs exhibited a variety of malformations of the hind limbs, including supernumery of bones, bone bridging and rotations, hypertrophy and hyperplasia of bones, but the majority of animals had truncated limbs, reduced bone segments and/or elements [43]. Based on their observations in a large integrative study, Loeffler et al. [44] concluded that "taken together [...] the data suggest that multiple causal agents converge in varying combinations, proportions, exposure times, and sequences upon different developmental stages of the limb bud to produce the range of observed malformation phenotypes". Such a synergism between trematode infection and pesticide exposure as cause of limb deformities has recently been suggested by Kiesecker [45]. In field and laboratory experiments, Kiesecker observed that parasite infection with the trematodes *Ribeiroia sp.* and *Telorchis sp.* was necessary for limb deformation in the wood frog (*Rana sylvatica*), but that previous exposure of tadpoles to low doses (based on EPA maximum contaminant levels for drinking water) of herbicides or insecticides increased the rate and degree of deformities. Kiesecker concluded that EAS exposure may decrease immunocompetency of exposed animals, thereby making them more susceptible to parasite infections, which may also explain why deformities have been reported for almost three centuries [46], and yet only recently dramatic increases in their occurrence have been observed.

On the other hand, the predominance of limb truncations has also been discussed as a possible effect of exposure to EASs that act similar to retinoic acid (RA). A natural derivative of vitamin A, RA is known to cause skeletal reduction most likely via the induction and maintenance of such genes as sonic hedgehog (Shh) and *Hox* genes in the developing or regenerating limb. Hence it has been suggested that certain EASs may cause limb truncations either by acting like RA binding to its receptors (RARs and RXRs) or by heterodimeric activation of the RA receptors. Bruce Blumberg (Topic 4.6) has isolated a compound from water of a Minnesota lake as a lead suspect potentially responsible for such an RA-like activity. Other possible causes that have been discussed (e.g., predation of limb buds in tadpoles, UV radiation) seem not to explain the complexity of the observed effects sufficiently.

Reptiles

Reptiles are particularly suitable models for studying the effects of environmental contaminants. A wide geographic distribution, presence in a variety of habitats, longevity, and carnivore lifestyle makes them reliable biomonitors. Furthermore, reptiles exhibit similar sensitivity to contaminants as have been reported for birds and mammals [47]. There is a substantial amount of information on the effects of EASs in reptiles available today, due largely to the seminal work by Louis Guillette and coworkers on the American alligator (*Alligator mississippiensis*) of Lake Apopka in Florida (Topic 4.7). The red-eared slider turtle (*Trachemys scripta elegans*) has also served as an excellent model system for endocrine disruption on several levels. Not only has it allowed for organismal (sex determination) and physiological (circulating steroid hormone titers) studies, it has also proven an invaluable model for studying EASs singly, in mixtures, and in low doses. A certain advantage of this species and the alligator over more traditional animal models is the plasticity of the sex determination process. In the red-eared slider incubation temperature, not sex chromosomes, determines gonadal sex during the midtrimester of embryonic development, a process known as temperature-dependent sex-determination (TSD). In the red-eared slider, eggs incubated at constant temperatures below 28.6 °C develop as males, eggs incubated at or above 29.6 °C develop as females [48].

The current working hypothesis concerning the mechanism of TSD in the red-eared slider is that sex steroid hormones are the physiological equivalent of incubation temperature with incubation temperature affecting the expression of genes involved in sex steroid synthesis and the sex steroid hormone receptors. A variety of evidence supports this interpretation. For example, incubation temperature effects can be overridden (i.e., an individual's putative gonadal sex can be reversed) by applying exogenous steroids to the egg during the thermosensitive period of sexual development. Importantly, the gonadal characteristics of these hormone-determined females are indistinguishable from those of temperature-determined females. Incubation at higher temperatures produces female hatchlings, but administration of aromatase inhibitor—which effectively halts the production of estrogens—to the eggshell will result in 100 % males. In sum, application of steroid hormones, steroidogenic enzyme inhibitors, and other chemicals onto the eggshell during incubation can redirect the effects of incubation temperature exogenous estrogens induce female sex determination, while nonaromatizable androgens induce male sex determination [48,49]. Further, both categories of hormones have strong dosage effects and also synergize with incubation temperature to induce gonadal differentiation.

The red-eared slider has proven to be an exceptional in vivo model system for mixtures and low doses of EAS exposure. Mixtures have been tested in two studies of the red-eared slider. Initially, Bergeron et al. [50] examined the effects of different PCB compounds on sexual differentiation and identified PCBs that alone cause a significant increase of female hatchlings at an all-male producing incubation temperature (26 °C). In combination, two of them synergized and increased ovarian development, whereas treatment with the individual PCBs required 10-fold higher concentrations. In a second study, eight compounds identified in the yolk of alligator eggs from Lake Apopka, Florida [51] were administered to red-eared slider eggs in the ecologically relevant concentrations identified in the alligator yolk. When all eight compounds were applied in a single-dose mixture, they significantly increased the ratio of females to males [51]. Results from single-compound exposures at the same dosages indicate that these compounds behave differently in combination than they do singly, again emphasizing the need for further studies using chemical mixtures reflecting proportions found in nature. Five of the compounds—the PCB mixture A1242, *trans*-nonachlor, *cis*-nonachlor, *p,p′*-DDE, and chlordane—altered sex ratio outcomes when applied to eggshells during development [51]. Aroclor 1242 produced the most powerful effects, shifting the ratio of females almost twofold, while chlordane had the greatest effect when combined with estradiol.

Research with the red-eared slider has challenged a central concept of traditional toxicology, namely the notion that a threshold (no observed adverse effect level, NOAEL) has to be reached below which adverse effects will not occur. Gaylor and coworkers [52] proposed the concept that a threshold dose may not exist when an exogenous molecule mimics an endogenous one by acting through the same mechanism. Given that a large number of turtle eggs can be manipulated at one time, it became possible to test this concept. First, in a retrospective analysis of published data on the effects of varying doses of estradiol at three different incubation temperatures on sex determination in the red-eared slider, Sheehan et al. [4] found that in each case the results fit the Michaelis–Menten model of a single protein-molecule interaction driving a reversible process. This in turn led to a larger study that determined that the Michaelis–Menten provided an ED_{50} of 5.0 with a 95 % confidence limit of ±2.0 ng (endogenous dose = 1.7 ± 1.3 ng; exogenous dose = 3.3 ± 1.7 ng) and an $r^2 = 0.90$ for fit of the modified equation. The lowest dose applied, 0.4 ng/10 g egg, increased the female fraction by 11.4 % beyond the temperature control. Considering that only 0.2 % of the estradiol applied to the eggshell ends up in the embryo [53], it becomes apparent that even very low dosages of steroid hormones or their mimics can have profound biological effects.

Other studies have determined the effects of low doses of xenobiotic compounds during development and after hatch [54]. Chlordane, a suspected antiandrogen in this species, does not affect aromatase activity in either the brain or the adrenal-kidney-gonad (AKG) complex of red-eared sliders. However, A1242 significantly increases aromatase activity levels in the red-eared slider brain—but not in the AKG—during a crucial developmental period. After this crucial period, A1242 causes an increase in aromatase activity in the AKG just prior to hatch. In complementary studies, basal steroid levels and steroid levels in response to follicle-stimulating hormone administration have been examined in hatchling males and females treated during embryogenesis with A1242, chlordane, or *trans*-nonachlor. Males treated with the A1242 or chlordane exhibit significantly lower testosterone levels than controls, while chlordane-treated females have significantly lower progesterone, testosterone, and 5α-dihydrotestosterone levels relative to controls. These results are similar to those reported by Guillette et al. [55] for juvenile alligators from Lake Apopka, Florida. Males treated with A1242 or *trans*-nonachlor display an elevated estradiol response to FSH administration vs. control males. Taken together, these results suggest that the effects EASs exert during embryonic development extend beyond birth. They also suggest the alterations in sex steroid hormone levels observed in animals from contaminated areas may result from EAS-induced alterations in the neuroendocrine axis controlling gonadal sex steroid hormone production. Whether the strong effects observed with A1242 are due to the interaction between the mixture's various PCB congeners remains unclear. However, as with the A1254 studies in fish, the results of the A1242 experiments with the red-eared slider provide a good starting point from which to further investigate possible effects of the individual mixture constituents, which then allows evaluating the nature of the interaction.

Recently, we extended these studies to the polyhalogenated aromatic hydrocarbons (PHHs) PCB-126 and dioxin (TCDD) [56]. While there was not a significant sex reversal at the dosages used (PCB-126 50, 500, and 5000 ng; TCDD 5, 50, 500 ng), it was possible to evaluate the delivery of these compounds to different compartments over time. Using the spotting method, eggs were treated at the middle of the sensitive period and analyzed 16 days after application. Results indicate that 90 % of PCB-126 and 96 % of TCDD was retained in the shell. For PCB-126, that portion of the total amount transferred across the shell (10 %, 705 ng) was distributed as follows: albumin - 14 %; yolk - 70 %; and embryo - 16 %. The total amount of TCDD transferred across the shell (4 %, 16 ng) was similarly distributed: albumin - 20 %; yolk - 55 %; and embryo - 25 %. Finally, this research revealed that the distributions of PCB-126 and TCDD in the interior of the egg were not proportional to lipid content. Rather, there was evidence of greater accumulation in both the albumin and the embryo relative to the yolk, suggestive of selective binding.

Finally, while there has been much attention paid to different forms of the estrogen receptor, it is equally important to keep in mind that there are multiple forms of estrogens and that each may have different binding affinities. For example, in the red-eared slider, estriol (E_3) is ten times more potent than is estrone (E_1) or estradiol (E_2) in overcoming the effects of a male-producing incubation temperature [16,49]. This difference is clearly evident when comparing either the dosage at which 50 % of the embryos are sex-reversed or the magnitude of the regression coefficients from statistical models. However, while E_3 is more potent than is E_2 and E_1 at reversing gonadal sex in the red-eared slider, it is less likely to synergize with temperature to reverse gonadal sex. These variations in dosage effects and synergy of different natural estrogens may be caused by differential affinity of the estrogen receptor for different ligands (i.e., higher affinity for more polar estrogens), cooperative binding in response to certain ligands (i.e., cooperative binding with more polar estrogens), and/or differential transactivation of downstream genes in the ovary determining cascade by certain ligands (i.e., more polar estrogens induce greater transcriptional activity). If any of these events are occurring, they could result in a synergistic increase of response at lower doses. These results have implications for the developmental mechanisms underlying sex determination and its sensitivity to the physical and chemical environment.

Birds

Some of the earliest observations on the endocrine impact of manmade environmental contaminants were made on birds, as may be best exemplified by Rachel Carson's milestone work *Silent Spring* [57]. In it, Carson suggested that the insecticide DDT might have caused the decline in passerine populations in the United States, a hypothesis that was soon thereafter confirmed. While Ratcliffe [58] in the late 1960s could only show that eggshell thinning in birds in England correlated with the onset of the local commercial use of DDT, others subsequently demonstrated that *o,p′*-DDT was indeed estrogenic and that it was responsible for the decline in shell thickness in many species [59]. It was also observed that in birds, exposure to *o,p′*-DDT caused feminization of embryos, alterations of sex ratios and breeding patterns, malformations, and abnormal sexual behavior [60,61]. Shortly thereafter, studies at the Great Lakes identified yet another source for endocrine disruption in piscivorous birds, PCBs. As apex predators, many waterfowls consume and accumulate PCBs that have made their way through the food chain into the adipose tissue of lake teleosts. The degree of bioaccumulation in birds is directly dependent on factors such as congener content and composition of prey, sex and age, and residence time of individual animals in PCB-contaminated areas [62]. With its long half-life and its presence in egg yolk [63], the PCB congeners were cause for great concern, which unfortunately was confirmed when studies showed that PCB exposures in birds caused the Great Lakes Embryo Mortality, Edema, and Deformity Syndrome (GLEMEDS) [64], as well as other effects such as altered liver enzyme activity [65].

Interestingly, PCB contamination has been demonstrated to cause aberrant behavior in birds, including egg-destroying behavior in captive mallards (*Anas platyrhynchos*) and gray herons (*Ardea cinerea*), decrease in nest defense behavior and nest attentiveness in herring gulls (*Larus argentatus*) and glaucous gulls (*L. hyperboreus*), merlins (*Falcon columbarius*), and prairie falcons (*F. mexicanus*), prolonged and increased aggressive behavior in male ring doves (*Streptopelia risoria*), and increased frequency of male courtship behavior leading to a delayed clutch initiation in the kestrel (*F. sparverius*) [64,66–72]. In a study testing the effects of in ovo exposure to PCBs in kestrels, Fernie et al. [73] reported that embryonic exposure suppressed ovipositioning in females, delayed clutch initiation, and reduced clutch sizes and fledgling success in both males and females. Post-hatch oral exposure of zebra finch chicks with estradiol benzoate (EB), but not 4-octylphenol (OP), was shown to cause sex-specific impairments of adult reproductive performance, which in some instances were additive when both sexes were treated [74]. These impairments included reduced egg production and increased egg breakage, reduced fertility, and reduced number of hatched chicks in pairs treated with 100 nmol EB/g body mass. In another study, Quaglino et al. [75] demonstrated that posthatch estrogen exposure increased the volume of brain nuclei involved in controlling singing, thus masculinizing the brain of female zebra finches. Interestingly, even high doses of OP, methoxychlor, or dicofol (100 nmol/g body mass) did not cause similar effects.

In spite of the important seminal work, research on the effects of EASs in birds has comparatively fewer current contributions to the field of EAS research than do teleost, amphibian, reptile, or mammalian studies.

Mammals

Mammals will continue to be the models of choice for application to human health issues. Rodent studies were instrumental in elucidating how DES results in severe endocrine disruption. There now have been a number of studies demonstrating how exposure early in life to environmentally relevant doses of EASs can cause alterations of reproductive organs. While one cannot consider laboratory rodents as wildlife, they are useful in illuminating an important problem that must be considered in endocrine disruption studies with wildlife. That is, it is important to understand that the internal hormonal milieu serves as a background against which exogenous hormones will act. Recent studies with laboratory rodents indicate that the endogenous hormonal milieu prior to or at the time of exposure can markedly alter the effect of EASs. For example, differences among individuals in responsiveness to estradiol have been associated with minute variations of hormone levels during critical periods in fetal development as a function of the individual's position in the uterus. This effect of fetal position within a uterine horn relative to the sex of neighboring fetuses is referred to as the intrauterine position effect. Research indicates that differences in background levels of estradiol among adjacent embryos during embryogenesis play a pivotal role in how EASs alter development.

The intrauterine position phenomenon has been described in mice, rats, gerbils, and pigs, as well as in twin fetus pregnancies in lambs and humans [76–79] and is due to the diffusion of sex steroid hormones (testosterone and estradiol) across the fetal amniotic and chorionic membranes of adjacent fetuses [80]. In mice, male fetuses developing between two females (2F fetuses) are exposed to an approximately 30 % higher concentration of estradiol than males developing between two male fetuses (2M fetuses). 2F female fetuses are exposed to approximately 35 % higher concentration of estradiol than are 2M fetuses of the same sex [78]. Fetuses positioned beside only one male fetus (1M fetuses) have more intermediate steroid hormone concentrations. These small differences in estradiol result in significant differences in the course of development and subsequent morphological, physiological, and behavioral characteristics [3,81–84].

The extreme sensitivity of the fetus to small differences in endogenous estradiol strongly suggested that humans and wildlife could be influenced by endocrine-disrupting chemicals even at the relatively low exposure levels typically encountered in most environments [85]. For example, prenatal exposure of male mice to environmentally relevant levels of bisphenol A [86–88] significantly alters the structure and function of the reproductive organs. Adult male offspring of CF-1 mice fed bisphenol A during their pregnancy have, relative to control males, significantly larger prostate and preputial glands, decreased weights of the seminal vesicles, epididymides and testes, as well as decreased sperm production [89,90]. Other reports of low-dose effects of bisphenol A in rodents include an accelerated rate of embryonic development, changes in the mammary gland, vagina, prostate, sperm production, epididymis, and pituitary response to estradiol (reviewed in Palanza et al. [91]).

Female young of pregnant mice fed 0 or 2.4 mg/kg bisphenol A in corn oil vehicle during days 11–17 of gestation are significantly heavier at weaning than control females [92]. Interestingly, the bisphenol A-exposed females with the highest background levels of estradiol in utero (2F females) are significantly heavier than the 2F control females, while bisphenol A-treated or control females exposed to the lowest levels of background estradiol in utero (2M females) show no difference in body weight at weaning. The body weight at weaning of females exposed to intermediate levels of background estradiol in utero (1M females) is also significantly heavier than their control 1M counterparts. Similarly, the body weight at weaning for bisphenol A-treated, 2F and 1M males are also significantly greater than control male pups from the same intrauterine positions [93].

Prenatal exposure to bisphenol A significantly accelerates the timing of puberty in female mice by decreasing the number of days between vaginal opening and first vaginal estrus [92]. An analysis according to intrauterine position indicates that 0M females show the greatest response to prenatal bisphenol A exposure, with the interval between vaginal opening and first vaginal estrus shortened by approximately 5 days relative to 0M controls. In contrast, bisphenol A-treated 2M females showed a similar timing of puberty as the 2M controls. The age at vaginal opening does not differ based on prenatal treatment or intrauterine position, which is expected since vaginal opening is not a marker of puberty in laboratory strains of mice [94]. The influence of background levels of sex steroid hormones due to intrauterine position has also been observed with regard to the effect of dioxin on prostate development in male rat fetuses. A single injection of dioxin (1 mg/kg) to pregnant rats on gestation day 15 resulted in a significant disruption of prostate development in 0M male fetuses, but no effect on prostate development in 2M males was observed [95].

GUIDING PRINCIPLES

From the discussion so far there emerge some principles that could guide future research on endocrine disruption in particular in wildlife populations. They can be divided into theoretical, experimental, and philosophical principles.

Theoretical principles

Combination effects have been characterized in various ways depending on different concepts. We believe that the vagueness with which terms like "synergism", "additivity", or "antagonism" have been used demands clarification and general agreement on definitions that could be used to guide future studies on mixtures. Obviously, definitions of different interaction types depend greatly on what type of model is used to determine agent interaction to begin with. As discussed above, empirical models that ignore underlying mechanisms have been suggested by some [18], while others give mechanistic approaches priority. Which of the two approaches is more promising and practical is a matter for debate; here we only want to emphasize the fact that without proper modeling of chemical interaction, combination effects cannot be determined reliably.

The complexity encountered in studies of EAS effects on the organismal level is reflected also on the cellular level, where actions via mechanisms other than nuclear steroid receptor binding of EASs can cause disruptions. For example, there is now convincing evidence that steroids also act at the cell surface of target tissues to initiate rapid, nongenomic responses and, further, that these actions are mediated by steroid membrane receptors. Such new findings have to be taken into consideration when EAS actions are being modeled and experimental data analyzed.

Although genetic predisposition, internal milieu, and molecular processes have been shown to be essential for an organism's responsiveness to EASs, epigenetic factors are equally demonstrable and need to be considered. This would include the aforementioned intrauterine position phenomenon, where very small increases in background estradiol, early in development, substantially increase the sensitivity of the female mouse to bisphenol A. It is likely that similar phenomena occur in other mammals, including humans. Even in species that characteristically have singleton births the fetus may experience similar epigenetic effects, as studies with humans have indicated. Endogenous levels of sex steroid hormones also vary among individual human fetuses due to a variety of factors such as the first vs. subsequent pregnancy [96], twin vs. singleton birth, [97,98], and race [99]. As a result, individuals with increased levels of background estradiol during development may form a sensitive subpopulation particularly susceptible to the effects of EASs on postnatal growth rate and puberty.

There also appears to be a tendency to regard the individual as separate from the social environment in which it develops. This is a mistake. For example, Adkins–Regan [100] has found that the hormones an individual is exposed to early in development as well as the social context in which it is reared are critical in the development of an individual's preference of a sex partner when adult. As others before her, she finds that if zebra finch nestling or embryos are treated with estradiol or fadrazole, an aromatase inhibitor, as adults they will be masculinized in their behavior and will prefer to pair with other females even if potential male partners are available. However, the zebra finch is a highly social species. Both parents rear the young, and they grow up in a mixed sex society. If adult males are removed from the breeding cages, so that young females are not exposed to male birds or to other male-female pairs during development, they also will no longer prefer males as sexual partners. What this suggests is that the developmental context can accentuate the effects of early endocrine disruption. Skewed sex ratios and feminized males thus can have impacts beyond the actual endocrine disruption of the individual.

Similarly, it is well known that the "society of the litter" can profoundly affect the development of adult sexuality and its underlying neural substrates in rodents. There is a long history of research in developmental psychobiology showing that adult sexual behavior can be altered either by manipulating hormones early in life or by stressing the mother during her pregnancy. Meaney [101] has shown that it is the quality and amount of care a pup receives from its mother that mediates these effects. Not only is the stress reactivity of the pup affected by the maternal care it receives but, as on reaching adulthood, these pups exhibit altered maternal behavior toward their own young, thereby perpetuating the effect. Another consideration is how the mother may interact with young that have been feminized or demasculinized by EAS exposure in utero. Moore [102] has demonstrated that mother rats lick the anogenital region of male pups more than they do female pups and, further, that this difference accentuates the copulatory behavior of the pup when it reaches adulthood. Thus, the alteration of the sexual differentiation process by early endocrine disruption might be magnified by the quality of care the young receive from mother.

Experimental principles

Although otherwise not congruent, most models of agent interaction agree that evaluation of the consequences of exposure to a mixture requires an adequate dose–response analysis of each component chemical [10,19]. But what constitutes an adequate dose–response study seems controversial. It has been suggested that at least three different concentrations have to be tested to get a sufficient profile [103]. However, the number of concentrations analyzed is only one aspect of the dose–response analysis; its dose range is another. It may be advisable to not only test EASs with the traditional standard safety margin of one or two orders of magnitude below the concentration that causes observable effects, but instead to go far below and above it. Testing over 6- to 7-log dose ranges seems sufficient to establish an adequate dose–response profile. It is worth mentioning that the concentration range to be tested may vary with the specific endpoint monitored.

Besides addressing the problem of how to define agent interactions properly, future mixture effect studies will also have to test environmentally relevant rather than experimentally convenient EAS combinations. Testing compounds that may interact with the same molecular site can help to test the accuracy with which empirical models predict EAS interaction [13], but in order to assess the risk posed by EAS contamination it is important to emulate the real-world situation in laboratory studies. Ideally, of course, mixture effects should be approached in wildlife studies, but these will have to go hand-in-hand with laboratory studies. The latter are indispensable since they alone can provide data with which the possible interaction of different EASs can be characterized.

One objective of this paper is to suggest what qualities are needed in good models for assessing endocrine disruption primarily in the wild. In general, a good model species for assessing combination effects, low-dose effects and the interaction of endogenous and exogenous hormones is characterized by (1) having one or more physiological endpoint(s) highly susceptible to EAS exposure, (2) having an endocrine physiology at different stages of their life history that is well studied, (3) have a general biology that makes them susceptible to environmental contamination with EASs, (4) occurring in sufficient abundance to enable replicate studies, and (5) have a sufficiently short generation time to make transgenerational studies feasible. The species discussed here are examples of models that fulfill all five requirements.

Philosophical principles

We know only what we study and, as a result, we tend to study only that which we know. Until the field biologist understands what is involved with laboratory studies, and the laboratory scientist has familiarity with the uncertainty of the field, research on endocrine disruption will continue to be dominated by subdisciplines that ignore one another.

The priority assigned to mechanistic studies is not only theoretically a problem, but also philosophically. Assuming that the existence of low-dose effects or EAS interactions can only be established reliably by identifying their molecular mechanisms presupposes a vertical cause-effect relationship where one can extrapolate from one level of organization to the next. However, as much as molecular mechanisms are crucial, reliance on this view is overly simplistic, for it ignores the phenomenon of emergence, where the properties of the whole cannot be predicted based on the properties of its parts. Even if more were known about cellular events underlying endocrine disruption, one would still have to study the effect on an organismal level. This is manifest in what we term the "Common Sense" Principle. Namely, if one sees an effect on the organismal level but fails to find molecular correlates, or simply can't think of a mechanism to explain it, that does not mean the effect is not real.

Finally, the notion of mechanistic studies as the ultimate judge of the ongoing controversies in the field also bears the philosophical dilemma that it is an inaccurate test of the hypothesis. The null-hypothesis, that low-dose exposure does not cause measurable effects, can only be falsified by finding low-dose effects. But since exactly that has been confirmed, mechanistic studies cannot do away with it. What they can do is to further our understanding of how low-dose effects and EAS interactions work on the cellular level. Put into context of the organism, this information is of immeasurable value and importance.

REFERENCES

1. S. Safe. *Environ. Health Perspect.* **108**, 487–493 (1995).
2. E. J. Clark, D. O. Norris, R. E. Jone. *Gen. Comp. Endocrinol.* **109**, 94–105 (1998).
3. F. S. vom Saal, B. G. Timms, M. M. Montano, P. Palanza, K. A. Thayer, S. C. Nagel, M. G. Dhar, V. K. Ganjam, S. Parmigiani, W. V. Welshons. *Proc. Natl. Acad. Sci. USA* **94**, 2056–2061 (1997).

4. D. M. Sheehan, E. Willingham, D. Gaylore, J. M. Bergeron, D. Crews. *Environ. Health Perspect.* **107**, 155–159 (1999).
5. R. M. Blair, H. Fang, D. Gaylor, D. M. Sheehan. *APMIS* **109**, 198–208 (2001).
6. O. Putz, C. Schwartz, S. Kim, G. A. LeBlanc, R. L. Cooper, G. S. Prins. *Biol. Reprod.* **65**, 1496–1505 (2001).
7. R. Steinmetz, N. A. Mitchner, A. Grant, D. L. Allen, R. M. Bigsby, N. Ben-Jonathan. *Endocrinology* **139**, 2741–2747 (1998).
8. O. Putz, C. Schwartz, G. A. LeBlanc, R. L. Cooper, G. S. Prins. *Biol. Reprod.* **65**, 1506–1517 (2001).
9. J. L. Spearow, P. Doemeny, R. Sera, R. Leffler, M. Barkley. *Science* **285**, 1259–1261 (1999).
10. A. Kortenkamp and R. Altenburger. *Sci. Total Environ.* **221**, 59–73 (1998).
11. A. Kortenkamp and R. Altenburger. *Sci. Total Environ.* **233**, 131–140 (1999).
12. E. Silva, N. Rajapakse, A. Kortenkamp. *Environ. Sci. Technol.* **36**, 1751–1756 (2002).
13. J. Payne, N. Rajapakse, M. Wilkins, A. Kortenkamp. *Environ. Health Perspect.* **108**, 983–987 (2000).
14. C.-Y. Chen and C.-L. Lu. *Sci. Total Environ.* **289**, 123–132 (2002).
15. A. M. Soto, K. L. Chung, C. Sonnenschein. *Environ. Health Perspect.* **102**, 380–383 (1994).
16. J. M. Bergeron, E. Willingham, C. T. I. Osborn, T. Rhen, D. Crews. *Environ. Health Perspect.* **107**, 93–97 (1999).
17. T. Colborn, F. S. vom Saal, A. M. Soto. *Environ. Health Perspect.* **101**, 378–384 (1993).
18. M. C. Berenbaum. *Pharmacol. Rev.* **41**, 93–141 (1989).
19. M. C. Berenbaum. *J. Theor. Biol.* **114**, 413–431 (1985).
20. C. E. Purdom, P. A. Hardiman, V. J. Bye, N. C. Eno, C. R. Tyler, J. P. Sumpter. *Chem. Ecol.* **8**, 275–285 (1994).
21. L. G. Parks, C. S. Lambright, E. F. Orlando, L. J. J. Guillette, G. T. Ankley, L. E. J. Gray. *Toxicol. Sci.* **62**, 257–267 (2001).
22. R. B. Spies and P. Thomas. In *Chemically Induced Alterations In Functional Development And Reproduction Of Fishes*, R. M. Rolland, M. Gilbertson, R. E. Peterson (Eds.), pp. 113–133 (1997).
23. P. Thomas, D. Breckenridge-Miller, C. Detweiler. *Mar. Environ. Res.* **46**, 136–167 (1998).
24. S. Hashimoto, H. Bessho, A. Hara, M. Nakamura, T. Iguchi, K. Fujita. *Aquatic Toxicol.* **49**, 37–53 (2000).
25. M. E. McMaster, G. J. Van Der Kraak, K. R. Munkittrick. *Comp. Biochem. Physiol. C Pharmacol. Tox. Endocrinol.* **112C**, 169–178 (1995).
26. M. E. McMaster, G. J. Van Der Kraak, C. B. Portt, K. R. Munkittrick, P. K. Sibley, I. R. Smith, D. G. Dixon. *Aquat. Toxicol.* **21**, 199–218 (1991).
27. M. E. McMaster, K. R. Munkittrick, G. J. Van Der Kraak, D. G. Dixon. *Excotox. Environ. Safe.* **23**, 103–117 (1992).
28. K. R. Munkittrick, C. B. Portt, G. J. Van Der Kraak, I. R. Smith, D. Rokosh. *Can. J. Fish Aquat. Sci.* **48** (8), 1371–1380 (1991).
29. D. M. Janz, M. E. McMaster, K. R. Munkittrick, G. J. Van Der Kraak. *Toxicol. Appl. Pharmacol.* **147**, 391–398 (1997).
30. I. A. Khan and P. Thomas. *Neurotoxicology* **18** (2), 553–560 (1997).
31. P. Thomas and I. A. Khan. "Mechanisms of chemical interference with reproductive endocrine function in sciaenid fishes", pp. 29–51 in *Chemically Induced Alterations in Functional Development and Reproduction of Fishes*, R. M. Rolland, M. Gilbertson, R. E. Peterson (Eds.), SETAC Technical Publications Series, Pensacola, FL (1997).
32. I. A. Khan and P. Thomas. *Biol. Reprod.* **64**, 955–964 (2001).
33. S. Ghosh and P. Thomas. *Mar. Environ. Res.* **39**, 159–163 (1995).

34. P. Thomas. In *Endocrine Disruptors: Effects on Male and Female Reproductive Systems*, R. K. Naz (Ed.), pp. 3–38, CRC Press, Boca Raton, Florida (1999).
35. A. K. Loomis and P. Thomas. *Biol. Reprod.* **62**, 995–1004 (2000).
36. P. Thomas and L. Budiantara. *Mar. Environ. Res.* **39**, 147–150 (1995).
37. T. B. Hayes. *Am. Zool.* **37**, 185–195 (1997).
38. T. B. Hayes. *Am. Zool.* **37**, 482–490 (1997).
39. T. B. Hayes. *J. Exp. Zool.* **281**, 373–399 (1998).
40. T. B. Hayes and K. P. Menendez. *Gen. Comp. Endocrinol.* **115**, 188–199 (1999).
41. N. N. Noriega and T. B. Hayes. *Comp. Biochem. Physiol. B* **126B**, 231–237 (2000).
42. T. B. Hayes, A. Collins, M. Lee, M. Mendoza, N. Noriega, A. A. Stuart, A. Vonk. *Proc. Natl. Acad. Sci. USA* **99**, 5476–5480 (2002).
43. C. U. Meteyer, I. K. Loeffler, J. F. Fallon, K. A. Converse, E. Green, J. C. Helgen, S. Kersten, R. Levey, L. Eaton-Poole, J. G. Burkhart. *Teratology* **62**, 151–171 (2000).
44. I. K. Loeffler, D. L. Stocum, J. F. Fallon, C. U. Meteyer. *Anat. Rec. (New Anat.)* **265**, 228–245 (2001).
45. J. M. Kiesecker. *Proc. Natl. Acad. Sci. USA* **99**, 9900–9904 (2002).
46. M. Ouellet. In *Ecotoxicology of Amphibians and Reptiles*, D. W. Sparling, G. Linder, C. A. Bisho (Eds.), pp. 617–661, SETAC, Pensacola, FL (2000).
47. R. J. Hall and D. R. J. Clark. *Enivron. Pollut. (Series A)* **28**, 45–52 (1982).
48. D. Crews, J. M. Bergeron, J. J. Bull, D. Flores, A. Tousignant, J. K. Skipper, T. Wibbels. *Devel. Gen.* **15**, 297–312 (1994).
49. D. Crews. *Zool. Sci.* **13**, 1–13 (1996).
50. J. M. Bergeron, D. Crews, J. A. McLachlan. *Environ. Health Perspect.* **102**, 780–781 (1994).
51. E. Willingham and D. Crews. *Gen. Comp. Endocrinol.* **113**, 429–435 (1999).
52. D. W. Gaylor, D. M. Sheehan, J. F. Young, D. R. Mattison. *Teratology* **38**, 389–391 (1988).
53. D. Crews, J. J. Bull, T. Wibbels. *Gen. Comp. Endocrinol.* **81**, 357–364 (1991).
54. E. Willingham and D. Crews. *Am. Zool.* **40**, 421–428 (2000).
55. L. J. Guillette, D. B. Pickford, D. A. Crain, A. A. Rooney, H. F. Percival. *Gen. Comp. Endocrinol.* **101**, 32–42 (1996).
56. R. W. Gale, J. M. Bergeron, E. Willingham, D. Crews. *Environ. Toxicol. Chem.* **21**, 2477–2482 (2002).
57. R. Carson. *Silent Spring,* Hougton Mifflin, Boston (1962).
58. D. A. Ratcliffe. *Nature* **215**, 208–210 (1967).
59. J. Bitman, H. C. Cecil, S. J. Harris, G. F. Fries. *Science* **162**, 371–372 (1968).
60. D. M. Fry and C. K. Toone. *Science* **231**, 919–924 (1981).
61. D. M. Fry, C. K. Toone, S. M. Speich, R. J. Peard. *Stud. Avian. Biol.* **10**, 26–43 (1987).
62. J. Struger and D. V. Weseloh. *Colonial Waterbirds* **8**, 142–149 (1985).
63. D. S. Frank, M. A. Mora, J. L. Sericano, A. L. Blankenship, K. Kannan, J. P. Giesy. *Environ. Toxicol. Chem.* **20**, 608–617 (2001).
64. D. S. Henshel, J. W. Martin, R. Norstrom, P. Whitehead, J. D. Steeves, K. M. Cheng. *Environ. Health Perspect.* **19**, 302–330 (1995).
65. M. G. Barron, H. Galbraith, D. Beltman. *Comp. Biochem. Physiol. C Pharmacol. Toxicol. Endocrinol.* **112C**, 1–14 (1995).
66. P. I. S. Milstein, I. Prestt, A. A. Bell. *Ardea* 171–257 (1970).
67. G. A. Fox, A. P. Gilman, D. B. Peakall, F. W. Anderka. *J. Wildl. Manage.* **42**, 477–483 (1978).
68. R. W. Fyfe, R. W. Risebrough, W. Walker II. *Can. Field Nat.* **90**, 346–355 (1976).
69. G. A. Fox and T. Donald. *Condor* **82**, 81–84 (1980).
70. M. L. B. McArthur, G. A. Fox, D. B. Peakall, B. J. R. Philogene. *Arch. Environ. Contam. Toxicol.* **12**, 343–353 (1983).

71. S. A. Fisher, G. R. Bortolotti, K. J. Fernie, J. E. Smits, T. A. Marchant, K. G. Drouillard, D. M. Bird. *Arch. Environ. Contam. Toxicol.* **41**, 215–220 (2001).
72. J. O. Boustnes, V. Bakken, K. E. Erikstad, F. Mehlum, J. U. Skaare. *J. Appl. Ecol.* **38**, 791–801 (2001).
73. K. J. Fernie, J. E. Smits, G. R. Bortolotti, D. M. Bird. *Arch. Environ. Contam. Toxicol.* **40**, 544–550 (2001).
74. J. R. Millam, C. B. Craig-Veit, A. E. Quaglino, A. L. Ericsen, T. R. Famula, D. M. Fry. *Horm. Behav.* **40**, 542–549 (2001).
75. A. E. Quaglino, C. B. Craig-Veit, M. R. Viant, A. L. Ericsen, D. M. Fry, J. R. Millam. *Horm. Behav.* **41**, 236–241 (2002).
76. L. C. Drickammer, A. S. Robinson, C. A. Mossman. *Ethology* **107**, 509–519 (2001).
77. J. W. Gill, B. J. Hosking, P. J. Holst, N. M. Fogarty, D. L. Hopkins, A. R. Egan. *Anim. Sci.* **66**, 375–382 (1998).
78. F. S. vom Saal. *J. Anim. Sci.* **67**, 1824–1840 (1989).
79. F. S. vom Saal, M. M. Clark, B. G. Galef, L. C. Drickamer, J. G. Vandenbergh. In *Encyclopedia of Reproduction*, E. Knobil and J. Neill (Eds.), pp. 893–900, Academic Press, New York (1999).
80. M. D. Even, M. G. Dhar, F. S. vom Saal. *J. Reprod. Fertil.* **96**, 709–716 (1992).
81. F. S. vom Saal and B. G. Timms. In *Endocrine Disruptors. Effects On Male And Female Reproductive Systems*, R. K. Naz (Ed.), pp. 307–327, CRC Press, Boca Raton (1999).
82. D. Nonneman, V. Ganjam, W. V. Welshons, F. S. vom Saal. *Biol. Reprod.* **47**, 723–729 (1992).
83. P. Palanza, S. Morley-Fletcher, G. Laviola. *Physiol. Behav.* **72**, 255–262 (2001).
84. B. G. Timms, R. E. Petersen, F. S. vom Saal. *J. Urol.* **161**, 1694–1701 (1999).
85. T. Colburn, F. S. vom Saal, A. M. Soto. *Environ. Health Perspect.* **101**, 378–384 (1993).
86. N. Olea, R. Pulgar, P. Perez, F. Olea-Serrano, A. Rivas, A. Novillo-Fertrell, V. Pedraza, A. M. Soto, C. Sonnenschein. *Environ. Health Perspect.* **104**, 298–305 (1996).
87. G. Schönfelder, W. Wittfoht, H. Hopp, C. E. Talness, M. Paul, I. Chahoud. *Environ. Health Perspect.* **110** (11), A703–A707 (2002).
88. Y. Takao, H. C. Lee, Y. Ishibashi, S. Kohra, N. Tominaga, K. Arizono. *J. Health Sci.* **45**, 39 (1999).
89. S. C. Nagel, F. S. vom Saal, K. A. Thayer, M. G. Dhar, M. Boechler, W. V. Welshons. *Environ. Health Perspect.* **105**, 70–76 (1997).
90. F. S. vom Saal, P. S. Cooke, D. L. Buchanan, P. Palanza, K. A. Thayer, S. C. Nagel, S. Parmigiani, W. V. Welshons. *Toxicol. Ind. Health* **14**, 239–260 (1998).
91. P. Palanza, K. L. Howdeshell, S. Parmagiana, F. S. vom Saal. *Environ. Health Perspect.* **110** (Suppl. 3), 415–422 (2002).
92. K. L. Howdeshell, A. K. Hotchkiss, K. A. Thayer, J. G. Vandenbergh, F. S. vom Saal. *Nature* **401**, 763–764 (1999).
93. K. L. Howdeshell and F. S. vom Saal. *Am. Zool.* **40**, 429–437 (2000).
94. R. L. Cooper, J. M. Goldman, J. G. Vandenbergh. In *Methods in Toxicology*, J. J. Heindel and R. E. Chapin (Eds.), pp. 45–56, Academic Press, New York (1993).
95. B. G. Timms, R. E. Petersen, F. S. vom Saal. *Toxicol. Sci.* **67** (2), 264–274 (2002).
96. K. Panagiotopoulou, K. Katsouyanni, E. Petridou, Y. Garas, A. Tzonou, D. Trichopoulos. *Cancer Causes Control* **1**, 119–124 (1990).
97. A. Ekbom, C. C. Hsieh, L. Lipworth, H. G. Adami, D. Trichopoulos. *J. Natl. Cancer Inst.* **89**, 71–76 (1997).
98. C. C. Hsieh, S. J. Lan, A. Ekbom, E. Petridou, H.-O. Adami, D. Trichopoulos. *Am. J. Epidemiol.* **136**, 1321–1326 (1992).
99. B. E. Henderson, R. Ross, L. Bernstein. *Cancer Res.* **48**, 246–253 (1988).
100. E. Adkins-Regan. *Arch. Sexual Behav.* **31**, 27–33 (2002).
101. M. J. Meaney. *Annu. Rev. Neurosci.* **24**, 161–192 (2001).

102. C. L. Moore. In *Advances in the Study of Behavior*, P. J. P. Slater, J. S. Rosenblatt, C. T. Snowdon, M. Milinski (Eds.), pp. 47–118, Academic Press, New York (1995).
103. D. Neubert. In *Approaches To Elucidate Mechanisms In Teratogenesis*, F. Welsch (Eds.), pp. 269–279, Hemisphere, Cambridge (1987).

Pure Appl. Chem., Vol. 75, Nos. 11–12, pp. 2321–2333, 2003.

Topic 4.10

Interactions of endocrine-disrupting chemicals with stress responses in wildlife*

T. G. Pottinger

NERC Centre for Ecology and Hydrology Windermere, The Ferry House, Far Sawrey, Ambleside, Cumbria LA22 0LP, UK

Abstract: The extent to which nonreproductive aspects of the endocrine system are affected by environmental contaminants is to a large extent unknown. However, an emerging body of data demonstrates that the neuroendocrine stress response is a sensitive target for disruption by a range of environmental contaminants, at a number of discrete loci. Several mechanisms are responsible for generating and sustaining the corticosteroid response to a stressor, including synthesis of the steroid, negative feedback at the pituitary and hypothalamus, and clearance via metabolism and conjugation in peripheral tissues and the liver. Laboratory and field studies provide evidence that these elements of the stress response are susceptible to interference by endocrine active substances. The functional significance to the individual of interference with this important adaptive mechanism remains to be established.

INTRODUCTION

Animals within the natural environment, whether aquatic, aerial, or terrestrial, are confronted with adverse and challenging conditions that can present a significant threat to their well-being or at worst, survival. The severity of any threat will be exacerbated if the ability of the animal to deploy its normal suite of adaptive or protective responses is compromised. It is the intention of this article to consider to what extent endocrine active substances (EASs) may alter the performance capacity of wildlife via modulation of the mechanisms that are employed in combating environmental stressors.

Despite considerable concern worldwide regarding the extent to which wildlife is being affected by exposure to low, but biologically active levels of EASs [1–3], there are as yet relatively few examples of endocrine disruption (ED) for which causality is established [4] and fewer still where population level impacts have been observed [5]. Indeed some skepticism has been expressed that ED is as significant a threat to wildlife and humans as has been supposed [6–8]. Most (though not all) documented or suspected cases of ED in wildlife concern effects within the reproductive axis, and some well-defined biomarkers of disturbance exist for a limited range of species [9,10]. For most vertebrate endocrine systems, clear markers of disruption are not yet established, and the extent to which nonreproductive endocrine systems, including those involved in adaptive stress responses, are at risk of disruption is not fully understood [11]. Because these limitations apply even more markedly to invertebrates [12,13] this article will consider vertebrates only.

The interrelationship between EAS and non-EAS stressors is potentially reciprocal:

*Report from a SCOPE/IUPAC project: Implication of Endocrine Active Substances for Human and Wildlife (J. Miyamoto and J. Burger, editors). Other reports are published in this issue, *Pure Appl. Chem.* **75**, 1617–2615 (2003).

- Exposure to EAS can modify the response of an animal to a non-EAS stressor.
- Exposure to a stressor may exacerbate effects arising from an existing EAS exposure.

Only the first of these possibilities will be considered in this article. This is because of the high level of uncertainty implicit in the second possibility. When considering ED in general, uncertainty lies in (i) making the link between measures of exposure to suspected EAS and physiological effects, (ii) interpreting biochemical and physiological effects in relation to whole organism fitness, and (iii) relating the relevance of responses in the individual to ecologically significant measures such as population sustainability [14]. These uncertainties are compounded and amplified where the interaction of two levels of disruption must be combined and interpreted. The primary response of vertebrates to most stressors is neuroendocrine in nature. Evaluating the likely interactions between alterations in the endocrine system that arise due to the stressor, and those that may be present due to exposure to EAS is a task that exceeds the scope of our existing knowledge base. Therefore, this article focuses on the mechanisms by which exposure to EAS can modify the response of an animal to a non-EAS stressor. That is, can EAS interfere with the adaptive capability of an animal?

What is a stressor?

The term stress is used in many different contexts leading to some confusion [15–17]. It is often employed in a generalized or nonspecific manner (e.g., ecosystem stress) or may be linked to a specific factor (e.g., pollutant stress). In a physiological, endocrine, or clinical context, stress is taken to have a very specific meaning (see below). A looser definition might include environmental factors that challenge the survival of the animal while not necessarily activating the neuroendocrine stress response. For example, exposure to organic or inorganic contaminants may evoke an adaptive (detoxification/elimination) response in the animal while not disturbing homeostasis sufficiently to activate the endocrine stress axis. For the purposes of this article, stressor will be employed to denote a destabilizing stimulus, the term stress response will define the primary response to the stressor, and exposure to a stressor with subsequent activation of the stress response will be considered to induce a state of stress in the animal. The destabilizing influence, or stressor, may be external or internal in origin.

Responses of wildlife to environmental stressors

Both aquatic and terrestrial animals are exposed to conditions that are potentially detrimental to their well-being. Adverse conditions can arise naturally, or through anthropogenic influences and may be rare, intermittent, or sustained. Four broad classes of stressor can be identified.

- Physical – (i) Abiotic: *temperature, winds, precipitation, flow regimes, suspended sediment, habitat alteration, UV exposure, e.g., [18–22].* (ii) Biotic: *conflict, predator damage, parasite damage e.g., [23].*
- Chemical – *alterations in dissolved oxygen or pH, direct or indirect exposure to pollutants, e.g., [24–26].*
- Physiological – *starvation, dehydration, salinity stress, disease (may arise as a consequence of exposure to physical or chemical stressors), e.g., [27–29].*
- Psychological – *threat of predation, intra- and inter-species conflict/competition, territoriality e.g., [30–33].*

Several constraints must be acknowledged when considering the interactions of physical, physiological, and psychological stressors with EASs in wildlife. The first and most frustrating of these is that we know relatively little about the physiological responses of wildlife to stressors in a natural environment. Because of the difficulties inherent in measuring physiological parameters that are altered by the process of measurement (capture, disturbance) there are few data that demonstrate the effects of environmental stressors on wildlife. Furthermore, very little is known about natural variation of factors that may be tertiary consequences of stress, such as disease patterns [34], among free-living animals. In contrast to this, for chemical stressors there is a sizeable section of ecotoxicological literature that concerns the measurement of biomarkers of exposure to a range of contaminants. Some responses are specific to a group of compounds (e.g., P450 induction, acetylcholinesterase inhibition) while others may be more general (heat shock proteins, DNA adducts, lysosomal stability, DNA strand breakage, SOD induction etc.).

By inference, the response of wildlife to stressors can be classified within one or more of three response types:

- Neuroendocrine stress response: *(adaptive/maladaptive) – nonspecific in nature, i.e., the response is common to a wide range of stressors, including all those listed above.*
- Behavioral response: *(avoidance/escape; modification of normal behavior; inappropriate behavior) – specific or nonspecific in nature, i.e., the response may be highly appropriate to the stimulus or may be entirely redundant. May be accompanied by a neuroendocrine response.*
- Physiological/biochemical response: *(induction of detoxification processes; direct toxic effects on biochemical processes; respiratory or osmoregulatory dysfunction; activation of nonspecific or specific immune system) – specific in nature, i.e., the response is directly dependent on the identity of the stressor. May be accompanied by a neuroendocrine and/or behavioral response.*

EASs interact with the endocrine system via a limited number of mechanisms. These may be direct (receptor-mediated augmentation, suppression, or modification) or indirect (nonreceptor-mediated interference). Given that all three classes of stress response listed above (neuroendocrine, behavioral, and physiological/biochemical) are either endocrine in nature, or are subject to control or modulation by the endocrine system, it is clear that all are potentially susceptible to the influence of EASs. The potential for interaction between EASs and the response of animals to stressors will be considered on the basis of the best evidence available from appropriate field and laboratory studies. At the present time, most of the available evidence concerns the influence of contaminants on the neuroendocrine stress response and this will therefore be the focus of this article.

INTERACTION OF ENDOCRINE-DISRUPTING CHEMICALS WITH STRESS RESPONSES

EASs and the neuroendocrine stress response

Neuroendocrine stress response

It is a central tenet of stress biology that the stress response is part of an adaptive strategy to cope with a perceived threat to homeostasis [35–39]. In response to a stressor, the animal alters its behavior and physiology to best serve the goal of reestablishing or preserving homeostasis. The stress response is highly conserved throughout the vertebrate taxa, emphasizing its adaptive value [see 40: nonhuman mammals; 41,42: fish; 43: reptiles; 44: birds]. The brief description below is drawn from these and the reviews of Chrousos and Gold [35], Johnson et al. [36], and Weissman [45].

The neuroendocrine stress response is initiated by perception of a threat. Identification of potentially threatening stimuli occurs within the higher centers of the central nervous system, and habituation or acclimation may occur to benign but initially stressful stimuli. The response comprises two elements; one originating within the sympathetic nervous system (rapid), the other a wholly endocrine response centered on the hypothalamic-pituitary-adrenal (HPA; or interrenal, HPI) axis (slow). Activation of the sympathetic nervous system results in the release of the catecholamines epinephrine (adrenaline) and norepinephrine (noradrenaline) into the circulation, primarily from the adrenal medulla (chromaffin tissue in lower vertebrates) and sympathetic nerve terminals. The catecholamines act at sites within the cardiorespiratory system resulting in optimization of the oxygen carrying and delivery capacity of the blood and of oxygen uptake from the environment. They also have positive effects on energy mobilization.

The endocrine response originates within the hypothalamus with the release of corticotropin-releasing hormone (CRH) that in turn stimulates the release of adrenocorticotropin (ACTH) from the pituitary. ACTH enters the general circulation and stimulates the synthesis and secretion of corticosteroids (cortisol or corticosterone, dependent on species) by the adrenal cortex (the interrenal in lower vertebrates) into the blood. These changes constitute the core or primary endocrine stress response, but stress also results in the increased or reduced secretion of several other pituitary hormones including prolactin, growth hormone, the gonadotropins, somatolactin, and the endorphins.

The stress response is conserved throughout the vertebrates, and must be assumed to offer significant adaptive value. It is therefore something of a paradox that activation of the stress response can also cause problems for the animal, that may be life-threatening in severity. Chronic or frequent activation of the response results in growth suppression, reproductive dysfunction, and immunosuppression. In higher vertebrates chronic or intermittent stress is linked to psychoneuroendocrine and emotional disruption, while in lower vertebrates normal behavior patterns can be disturbed.

Conventional (non-EAS) pollutants and the neuroendocrine stress response

It is well known that a neuroendocrine stress response may be initiated by exposure to toxic contaminants (e.g., in fish, see [46]) in addition to any specific detoxification mechanisms that might be activated. Apportioning the physiological consequences of this combined response is difficult because prolonged activation of the HPA axis is inherently harmful (as discussed above) and thus adverse effects of chronic stress may occur simultaneously with direct toxic effects of the contaminant. Although exposure to contaminants may elicit an endocrine response, this does not represent ED in the widely accepted sense; such a response can be considered to be a normal function of the intact HPA axis—a response to destabilizing influences (unless, of course, the endocrine response is itself damaged, in which case the non-EAS becomes an EAS). However, it would be difficult to discriminate between a stress response arising due to activation of the neuroendocrine cascade, and elevated corticosteroid levels that might arise due to direct stimulatory effects of contaminants on elements of the hypothalamic-pituitary-adrenal/interrenal cascade (see, e.g., [47,48]).

Endocrine-disrupting chemicals and the neuroendocrine stress response

In principle, all the elements of the neuroendocrine stress response are susceptible to interference by endocrine-disrupting contaminants, and therefore it must be assumed that EAS can affect the ability of an animal to mount an appropriate stress response. Although to date there have been no documented cases of direct interference with peptide signalling systems by xenobiotic hormone mimics it is likely that indirect effects occur, for example by interference in the biosynthesis of peptide hormones, or by alteration of regulatory mechanisms (e.g., steroid feedback on peptide secretion, a mechanism which operates extensively in the pituitary; see [49] for a possible example of disruption at this level). Certainly the distal portion of the HPA/I cascade is particularly susceptible to modulation because of its reliance on steroid hormone signalling. Several mechanisms are responsible for generating, sustaining and terminating the corticosteroid response to a stressor, including synthesis of the steroid, negative feedback at the pituitary and hypothalamus, and clearance via metabolism and conjugation in peripheral tissues and the liver. Laboratory and field studies have provided evidence that these elements of the stress response are susceptible to interference by EASs.

Effects of chemical contaminants on pituitary function

During the response to a stressor, the pituitary secretes ACTH following stimulation by CRH of hypothalamic origin. The ACTH-secreting cells (corticotropes) are also sensitive to negative feedback by corticosteroids. There are limited data that describe contaminant effects on pituitary function in this context. Exposure to various classes of organic contaminants results in the induction of those biotransforming enzymes that participate in both xenobiotic metabolism as well as endogenous substrate metabolism [50]. A cytochrome P450 monooxygenase system is known to exist in the vertebrate brain and induction of P450 isoenzymes has been reported to occur in the pituitary of rats and rainbow trout (*Oncorhynchus mykiss*) [51,52] treated with β-naphthoflavone and other organics [53]. In trout, the enzymes were found in gonadotropic cells and exposure to β-naphthoflavone resulted in alterations of gonadotrope function. It is likely that corticotropes could be similarly affected. Depending on the substrate specificity of the induced enzymes, this is a potential route by which steroid metabolism within the pituitary, and therefore feedback regulation of pituitary hormone secretion might be modulated.

Corticotrope dysfunction has been reported to occur in fish exposed to xenobiotics. In pike (*Esox lucius*) and perch (*Perca flavescens*) that were recovered from sites polluted by polycyclic aromatic hydrocarbons (PAHs), polychlorinated biphenyls (PCBs), and mercury, the corticotropes were atrophied compared to cells from fish at unpolluted reference sites. The fish that displayed abnormal corticotropes also failed to elevate blood cortisol levels in response to the stress associated with capture [54]. In this case, whether the atrophy was the result of direct toxic effects, or prolonged negative feedback suppression by corticosteroids, or the failure of some other regulatory mechanism, was unclear.

There are limited data that suggest that the bioactivity of key hormones associated with the stress response can be altered by exposure to organic contaminants. Rat anterior pituitary cells exposed to the aromatic hydrocarbon 2,3,7,8-tetrachlorodibenzo-*p*-dioxin (TCDD) in vitro secreted ACTH, both constitutively and in response to CRH stimulation, which was less potent in stimulating corticosterone secretion from cultured rat adrenal cells than was equimolar quantities of ACTH. The authors interpreted these results to indicate that TCDD was altering the bioactivity of the secreted ACTH [55]. Subsequent studies with TCDD have not produced identical results, although evidence has been found that TCDD disturbed HPA function in male rats [56] and that TCDD stimulated the synthesis and secretion of ACTH by the anterior pituitary under basal (unstressed) conditions, but reduced pituitary responsiveness to CRH stimulation [57].

Effects of chemical contaminants on plasma corticosteroids

Corticosteroids are synthesized de novo in the adrenal cortex (or interrenal tissue in lower vertebrates) following stimulation by ACTH. Some short-loop feedback by corticosteroids may occur at this level [58]. Peripheral conversion of corticosteroids occurs together with significant levels of biotransformation and conjugation in the liver [59]. Effects on circulating levels of corticosteroids have been reported for many fish species exposed to organic and inorganic contaminants.

Plasma levels of cortisol were lower in white sucker (*Catostomus commersoni*) exposed to bleached kraft pulp-mill effluent (BKME) than in fish from unpolluted sites [60] and in eels (*Anguilla anguilla*) exposed to 50 % secondary-treated BKME normal cortisol secretion was prevented and cortisol accumulated within the interrenal tissue [61]. Similarly, in schelly (*Coregonus lavaretus*) exposed to untreated or treated BKME, the increase in plasma cortisol following air exposure was attenuated [62]. Eels exposed to diesel oil water-soluble fraction also displayed an attenuated stress response [63], and the cortisol response to air exposure of marbled sole (*Pleuronectes yokohamae*) collected from Tokyo Bay was reduced in comparison with that of fish from less polluted reference sites [64]. Yellow perch (*Perca flavescens*) sampled from sites polluted by multiple organic and inorganic pollutants displayed an attenuated plasma cortisol response to capture and to ACTH challenge relative to the response of fish from a reference site [65,66], and the ability of inorganic contaminants to induce interrenal dysfunction was further indicated by the reduced interrenal responsiveness to stressors exhibited by yellow perch from lakes contaminated with Zn, Cu, and Cd [67] and by brown trout (*Salmo trutta*) from metal-contaminated sites [68]. Administration of β-naphthoflavone to rainbow trout resulted in a lower increase of plasma cortisol than was observed in control fish following a brief disturbance [69] or heat shock [70] and the administration of PCB 126 to tilapia (*Oreochromis mossambicus*) and rainbow trout via the diet resulted in reduced cortisol responsiveness to a confinement stressor [71,72].

Similar effects to those described for fish have been reported to occur in the mudpuppy, an aquatic salamander (*Necturus maculosus*), exposed to organic pollutants. Animals from impacted sites displayed a significantly lower corticosterone response to capture and confinement, and to ACTH challenge, than animals from reference sites [73].

Effects on corticosteroid levels have also been observed in contaminant-exposed birds. In a study of herring gull (*Larus argentatus*) embryos environmentally exposed to organochlorine contaminants in ovo at various Great Lakes sites, basal plasma corticosterone concentrations were inversely related to concentrations of organochlorine residues in the yolk sacs [74]. The authors interpreted these findings as indicating that organochlorine contamination was adversely affecting the HPA axis in contaminant-exposed gulls.

In marked contrast to these data for amphibians, birds, and fish, juvenile alligators (*Alligator mississippiensis*) sampled from a severely contaminated (agricultural runoff, sewage, pesticides) lake and an uncontaminated lake showed no difference in their ability to mount a corticosteroid response to capture and confinement stress [75]. This was despite there being significant differences in blood androgen levels in males from the two lakes, indicative of disruption of the reproductive endocrine system. The authors interpreted this as evidence that gonadal steroidogenic capacity may be modified without necessarily altering overall steroidogenic ability of the animal. Arguably, these findings simply emphasize that the site of lesions within the endocrine system is specific to the identity of the contaminants to which the animals are exposed or to the species concerned. It is also worth bearing in mind the methodological problems associated with measuring levels of hormones that are extremely sensitive to disturbance. If it is not possible to standardize methods of capture the resultant variability in corticosteroid levels can mask underlying differences.

Effects of chemical contaminants on interrenal responsiveness to ACTH

Alterations in the responsiveness of the adrenal/interrenal tissue to ACTH is evidently a primary factor underlying the modification of stress responsiveness by xenobiotics, presumably via interference in the detection and transduction of the ACTH signal or from a reduced steroidogenic capacity of the adrenal/interrenal tissue. In tilapia, the DDT metabolite *o,p′*-dichlorodiphenyldichloroethane (*o,p′*-DDD) suppressed the response of interrenal tissue to ACTH both in vitro [76] and in vivo [77]. In rainbow trout, *o,p′*-DDD was found to be adrenotoxic, disrupting the 3′,5′-monophosphate (cAMP) generation step [78]. The adrenotoxicity of *o,p′*-DDD is discussed further in the next section.

Interrenal tissue from β-naphthoflavone (BNF)-treated rainbow trout was less sensitive to ACTH in vitro than that from control fish. Treatment with BNF in vivo did not significantly affect either turnover or tissue distribution of cortisol, leading the authors to conclude that cortisol clearance mechanisms were not affected [69]. Exposure of interrenal cells from rainbow trout to the organochlorine pesticide endosulfan in vitro decreased the ACTH- or dbcAMP-stimulated cortisol secretion in a concentration-dependent pattern. The doses required to disrupt cortisol secretion were significantly lower than those that were lethal to the head kidney cells [79]. Reduced interrenal responsiveness to ACTH and dbcAMP was also exhibited by yellow perch from metal-contaminated lakes [67,80,81]. The suppressive effects of dietary PCB 126 on the stress response of tilapia and rainbow trout was also linked to impairment of interrenal steroidogenic capacity with both ACTH and cAMP-stimulated cortisol release from interrenal tissue in vitro being lower in PCB-treated fish [71,72]. In these studies, the attenuated response observed in response to cAMP stimulation confirmed that disruption occurred downstream of the ACTH receptor itself. Similarly, the reduced cortisol response to stress observed in marbled sole from Tokyo Bay was associated with impairment of cortisol release from the head kidney following in vitro ACTH treatment [64].

The impairment of the stress response in perch from polluted sites has been linked to the age of the fish. The plasma cortisol elevation in response to capture stress was reduced in perch of 4 years and older relative to fish from uncontaminated sites, but was not affected in fish younger than 4 years [81]. As in other studies, failure of the interrenal tissue of 4+ fish to respond fully to ACTH appeared to contribute to the disparity in responsiveness. The authors suggested that it is the cumulative effects of lifelong exposure to polluted conditions that accounts for the age effect. Age of the animal and duration of exposure are not factors that have been focused upon by other authors, and it may be that age is only of significance where the concentrations of contaminants to which animals are exposed are low, or exposure is intermittent.

Leopard frog (*Rana pipiens*) tadpoles exposed to a PCB congener 77-TCB showed decreased whole-body corticosterone content relative to controls both before and after injection with ACTH [82]. However, a study on toads (*Bufo terrestris*) exposed to coal combustion wastes provided results that contrast with most of the reports cited above. In exposed toads, corticosterone levels were high, and remained so after laboratory acclimation. Furthermore, injection of ACTH had no effect on circulating corticosterone levels in exposed toads, whereas an increase was observed in toads taken from reference sites [83]. The results suggest hyperactivation of the interrenal tissue, and given the considerable adverse effects of chronically elevated corticosteroid levels this response is likely to be more damaging to the individual than a dysfunctional stress response.

Effects of chemical contaminants on interrenal steroid synthesis and metabolism

The mechanisms by which steroidogenic capacity of the interrenal tissue might be altered by xenobiotics are not addressed in detail by most of the studies cited above. One mechanism may involve the P450 isoenzymes which are critical elements in the biosynthesis of steroids within the adrenal/interrenal tissue. There is potential for P450 enzymes, induced by the presence of a xenobiotic, to impact on the normal processing of steroid hormones of endogenous origin [84]. In rainbow trout treated with 3,3′,4,4′-tetrachlorobiphenyl (TCBP), hepatic uptake and catabolism of cortisol was increased, data interpreted by the authors to indicate that the fish may not be able to fully respond to a stressful stimulus [85]. Unfortunately, this was not investigated further (i.e., the fish were not subjected to a stressor) and the increased uptake and metabolism of cortisol was accompanied by plasma levels of cortisol in otherwise unstressed fish that were four-fold higher than those in control fish. Of course, it is almost impossible to obtain reliable estimates of plasma cortisol levels in unstressed free-living fish from a riverine or lacustrine environment, so we know nothing about baseline cortisol levels in fish from contaminant-exposed populations. There was no effect on cortisol production in the interrenal tissue of rainbow trout exposed to 3,4,5,3′,4′,5′-hexachlorobiphenyl (HCB) despite the conversion of progesterone to 17α-hydroxyprogesterone and 11-deoxycortisol being increased by HCB treatment [86]. The physiological implications of these observations are unclear. A further complication is presented by the fact that certain P450 isoenzymes in fish are modulated by estrogens and androgens [87,88], raising the possibility that estrogenic contaminants could exert effects on hepatic or interrenal steroid metabolism [89].

Several studies in mammals have suggested mechanisms by which contaminants may interfere with the function of the adrenal/interrenal tissue. In rats treated long-term with TCDD baseline corticosterone production by adrenal tissue was markedly reduced in vitro, although ACTH-stimulated release was comparable to that of controls, a result interpreted as indicating a reduction in the bioactivity of ACTH in the TCDD-treated animals [90]. In mouse adrenal tumor cells, cadmium chloride, acetate and sulfate inhibited basal steroid production and ACTH-stimulated steroid secretion in a dose-dependent fashion [91,92]. The authors postulated interference by Cd with both the transduction of the ACTH signal and elements of the steroidogenic pathway. Effects on basal secretion were clarified as being related to the inhibition of cholesterol and 25-hydroxycholesterol utilization [93], and both these and effects on ACTH-stimulated steroid secretion may be related to effects on adenyl cyclase activity [94]. Mercury treatment of rats caused dysfunction of adrenal steroid biosynthesis by inhibiting the activity of 21α-hydroxylase with the effect of lowering plasma levels of corticosterone [95]. A similar effect on plasma corticosterone levels was observed in rats exposed to Cu^{2+}. In this case, the reduction of corticosterone levels was attributed to effects on the cytochrome P450 enzymes responsible, in particular, for 11β-hydroxylation [96].

Perhaps the best characterized agents of adrenotoxicity are the DDT metabolites 3-methylsulfonyl-DDE and *o,p*′-DDD, both of which interfere with steroid synthesis in the adrenal cortex. The sulfone is a competitive inhibitor of 11β-hydroxylase, a key element of corticosteroid synthesis [97] and has been associated with adrenal dysfunction in wildlife [98,99] and in rodent [100] and human cell lines [101]. The metabolite *o,p′*-DDD reduces the elevation of plasma cortisol following stress in rainbow trout [102] and the corticosteroid-inhibiting properties of this metabolite have even been exploited therapeutically in the treatment of Cushing's syndrome [103]. In contrast to these data, a recent study on Arctic charr (*Salvelinus alpinus*) failed to detect any effects of a single (force-fed) oral administration of *o,p′*-DDD on the ACTH and cortisol response to a handling stressor [104]. Whether this result reflects the prolonged period between dosing and imposition of the stressor (28 days) or represents a genuine difference in species sensitivity is uncertain.

In mice, oral administration of the chlorinated insecticide lindane led to significant reductions in circulating corticosteroid levels [105]. This effect is evidently related to the adverse effects of lindane on the expression of steroidogenic acute regulatory (StAR) protein [106]. StAR mediates the intramitochondrial transfer of cholesterol to the P450 (side chain cleavage; SCC) enzyme—the rate-limiting and acutely regulated step in hormone-stimulated steroidogenesis. Lindane-induced reduction in second messenger production may also contribute to suppression of steroidogenesis [107].

Effects of chemical contaminants on the sensitivity of target tissues to corticosteroids

Corticosteroids exert their effects at target tissues by interaction with a specific receptor protein that may be either internal or external to the cell. Internal cytosolic receptors mediate effects of the steroid at the genomic level whereas membrane-bound cortisol receptors mediate more rapid responses via a coupled G-protein mechanism (e.g., depolarization) [108]. Corticosteroid receptors are "self-regulated", that is, prolonged elevation of the ligand induces a reduction in receptor abundance within the cell [109]. Therefore, there is scope for a corticosteroid mimic both to evoke a target tissue response by interaction with the receptor and also to cause a down-regulation of receptor and thereby reduce the sensitivity of the tissue to the endogenous ligand. Up-regulation of the estrogen receptor in rainbow trout liver has been observed in response to estrogen mimics [110,111], and there is no reason why a similar effect should not occur as a consequence of exposure to corticosteroid mimics.

There are few studies that have examined the interactions of xenobiotics with the corticosteroid receptor. Unsurprisingly, a range of known estrogen mimics displayed no affinity for the rainbow trout cortisol receptor [112]. However, Johansson et al. [113] found significant interaction of persistent PCB metabolites with the human glucocorticoid receptor using a response element and reporter enzyme construct. They demonstrated that a number of methylsulfonyl PCBs displayed IC50s of 2.7 μM or greater and interpreted these results to be of functional significance.

Effects of chemical contaminants on the magnitude of the stress response

An animal may modify the magnitude and/or duration of the stress response in order to address changing circumstances. This is particularly true during reproductively active periods, and attenuation of stress responsiveness has been reported in fish [114], birds [115], and mammals [116]. Changes in magnitude of the response appear to be controlled by gonadal steroids with androgens generally exerting a suppressive effect and estrogens a stimulatory effect on the response [117]. It is therefore conceivable that exposure to environmental androgens, estrogens, or agonists/antagonists of these compounds will also exert a modulating effect on the stress response.

The life-long characteristics of the stress response in an individual can also be altered as a consequence of prenatal or antenatal exposure to elevated levels of corticosteroids with both endocrine and behavioral consequences [118,119]. These changes arise from alterations in the differentiation of neurons that are involved in the negative feedback regulation of the HPA axis and in particular changes in glucocorticoid receptor gene expression in the hippocampus and frontal cortex [120]. The possibility therefore exists that parental exposure to environmental corticosteroid agonists/antagonists might result in the modification of important adaptive traits in their offspring, or that exposure of juveniles to corticosteroid agonists/antagonists will modify this important adaptive mechanism.

EASs and the behavioral response to stressors

The stress response itself may contain adaptive behavioral elements and exposure to stressors can also result in behavioral modification [121]. In general, stress exerts an inhibitory effect on "nonessential" activities such as reproductive behavior [122,123]. This aspect of the stress response of animals is less well studied than the underlying neuroendocrinology of stress and there are inadequate data available to support extensive speculation on the likely effects of EASs in this context. However, it is well established for vertebrates that much behavior is susceptible to influence, or is controlled, by steroid hormones, particularly reproductive behavior [124] and it is likely therefore that any stress-related behavioral activities that are steroid dependent will also be susceptible to disruption or modulation by EASs. This certainly appears to be the case for reproductive behavior in fish, which is altered by exposure to environmental ethinylestradiol [125].

EASs and the physiological/biochemical response to stress

The adaptive responses that are adopted by organisms exposed to potentially toxic contaminants are extremely well documented, and, as indicated above some of these mechanisms are regulated or modulated by the endocrine system. It is beyond the scope of this short article to address the complex issue of how these responses might be affected by EASs.

RESEARCH NEEDS

Research effort and public and governmental concern have understandably focused on the reproductive and developmental implications of ED, and even here matters are by no means resolved. The extent to which nonreproductive aspects of the endocrine system are affected by environmental contaminants is largely unknown. There is an increasing body of evidence that demonstrates that the neuroendocrine stress response is a sensitive target for disruption by a range of environmental contaminants, at a number of discrete loci. However, it remains to be established precisely how this affects the performance capacity of the individual. The benefit provided to the animal by possession of a functional neuroendocrine stress response is difficult to quantify, but is assumed to be substantial because of the conservation of this response throughout vertebrate taxa. If the response is attenuated, as much of the data presented here suggest is likely in polluted environments, what is the cost to the individual? Or is the cost of the attenuated stress response subsidiary to the presence of other effects (reduced growth, immunotoxicity, etc.) that may be more intrinsically harmful to both the individual animal and the population? It is clear that a considerable amount of research is required to address these issues that cannot be satisfactorily answered with the information currently at our disposal. The questions that require consideration include:

- How broad is the range of chemicals that interact with elements of the neuroendocrine stress response?
- Are the concentrations at which effects on the functioning of the stress response are observed lower than existing regulatory limits?
- Are these chemicals ubiquitous enough in the environment to cause widespread effects or are they restricted to localized hot spots?
- What is the functional significance to the animal of interference with the normal functioning of the stress response?

REFERENCES

1. P. M. Campbell and T. H. Hutchinson. *Environ. Toxicol. Chem.* **17**, 127–135 (1998).
2. T. Colborn and K. Thayer. *Ecol. Appl.* **10**, 949–957 (2000).
3. G. A. Fox. *Water Qual. Res. J. Can.* **36**, 233–251 (2001).
4. J. G. Vos, E. Dybing, H. A. Greim, O. Ladefoged, C. Lambre, J. V. Tarazona, I. Brandt, A. D. Vethaak. *Crit. Rev. Toxicol.* **30**, 71–133 (2000).
5. C. R. Tyler, S. Jobling, J. P. Sumpter. *Crit. Rev. Toxicol.* **28**, 319–361 (1998).
6. A. Dawson. *Ecotoxicology* **9**, 59–69 (2000).
7. R. W. Risebrough. *Hum. Ecol. Risk Assess.* **5**, 869–883 (1999).
8. R. J. Witorsch. *Regul. Toxicol. Pharm.* **36**, 118–130 (2002).
9. J. E. Harries, A. Janbakhsh, S. Jobling, P. Matthiessen, J. P. Sumpter, C. R. Tyler. *Environ. Toxicol. Chem.* **18**, 932–937 (1999).
10. I. M. Davies, A. Minchin, B. Bauer, M. J. H. Harding, D. E. Wells. *J. Environ. Monitor.* **1**, 233–238 (1999).
11. A. Dawson, T. G. Pottinger. Non-reproductive effects of endocrine disrupting chemicals. Report to the Department of the Environment, Transport and the Regions. Institute of Terrestrial Ecology: Huntingdon, p. 89 (1999).
12. P. L. DeFur, M. Crane, C. Ingersoll, L. Tattersfield (Eds.). *Endocrine Disruption in Invertebrates: Endocrinology, Testing, and Assessment.* Proceedings from the International Workshop on Endocrine Disruption in Invertebrates: Endocrinology, Testing, and Assessment.; 1998 Dec 12–15; Noordwijkerhout, The Netherlands, SETAC, Pensacola, Florida (1999).
13. L. C. V. Pinder, T. G. Pottinger, Z. Billinghurst, M. H. Depledge. *Endocrine function in aquatic invertebrates and evidence for disruption by environmental pollutants*, p. 150 Environment Agency R&D Technical Report E67 (1999).
14. G. Van der Kraak. *Pure Appl. Chem.* **70**, 1785–1794 (1998).
15. A. D. Pickering. In *Stress and Fish,* A. D. Pickering (Ed.), pp. 1–9, Academic Press, London (1981).
16. S. Levine. In *Animal Stress*, G. P. Moberg (Ed.). pp. 51–69, American Physiological Society, Bethesda, MD (1985).
17. F. Toates. *Stress. Conceptual and Biological Aspects.* Wiley & Sons: Chichester (1995).
18. R. Berghahn. *Mar. Ecol.- Prog. Ser.* **192**, 277–285 (2000).
19. R. G. Lake and S. G. Hinch. *Can. J. Fish. Aquat. Sci.* **56**, 862–867 (1999).
20. A. D. Lemly. *Ecotox. Environ. Safe.* **34**, 223–227 (1996).
21. L. M. Romero, J. M. Reed, J. C. Wingfield. *Gen. Comp. Endocr.* **118**, 113–122 (2000).
22. D. Saltz and G. C. White. *J. Wildl. Manage.* **55**, 1–16 (1991).
23. G. W. Schuett and M. S. Grober. *Physiol. Behav.* **71**, 335–341 (2000).
24. T. S. Gill, G. Leitner, S. Porta, A. Epple. *Comp. Biochem. Physiol.* C **104**, 489–495 (1993).
25. G. J. Vianen, G. E. E. J. M.Van den Thillart, M. Van Kampen, T. I. Van Heel, A. B. Steffens. *Neth. J. Zool.* **51**, 33–50 (2001).
26. T. Ytrestoyl, B. Finstad, R. S. McKinley. *J. Fish Biol.* **58**, 1025–1038 (2001).
27. J. E. Bly, S. M. A. Quiniou, L. W. Clem. *Dev. Biol. Stand.* **90**, 33–43 (1997).
28. K. D. Dunlap. *J. Herpetol.* **29**, 345–351 (1995).
29. J. Munro, C. Audet, M. Besner, J. D. Dutil. *Can. J. Fish. Aquat. Sci.* **51**, 2448–2456 (1994).
30. S. Creel. *Trends Ecol. Evol.* **16**, 491–497 (2001).
31. D. Eilam, T. Dayan, S. Ben-Eliyahu, I. Schulman, G. Shefer, C. A. Hendrie. *Anim. Behav.* **58**, 1085–1093 (1999).
32. K. Hirschenhauser, E. Mostl, B. Wallner, J. Dittami, K. Kotrschal. *Ethology* **106**, 63–77 (2000).
33. A. Scheuerlein, T. J. Van't Hof, E. Gwinner. *Proc. Roy. Soc. Lond. B Bio.* **268**, 1575–1582 (2001).

34. B. Wilson, H. Arnold, G. Bearzi, C. M. Fortuna, R. Gaspar, S. Ingram, C. Liret, S. Pribanic, A. J. Read, V. Ridoux, K. Schneider, K. W. Urian, R. S. Wells, C. Wood, P. M. Thompson, P. S. Hammond. *Proc. Roy. Soc. Lond. B Bio.* **266**, 1077–1083 (1999).
35. G. P. Chrousos. *JAMA* **267**, 1244–1252 (1992).
36. E. O. Johnson, T. C. Kamilaris, G. P. Chrousos, P. W. Gold. *Neurosci. Biobehav. R.* **16**, 115–130 (1992).
37. L. D. Dorn and G. P. Chrousos. *Endocrin. Metab. Clin.* **22**, 685–700 (1993).
38. W. Sutanto and E. R. de Kloet. *Lab. Anim.* **28**, 293–306 (1994).
39. C. Tsigos znd G. P. Chrousos. *Endocrin. Metab. Clin.* **23**, 451–466 (1994).
40. G. P. Moberg. *Animal Stress*, American Physiological Society, Bethesda, MD (1985).
41. S. E. Wendelaar Bonga. *Physiol. Rev.* **77**, 591–625 (1997).
42. G. K. Iwama, A. D. Pickering, J. P. Sumpter, C. B. Schreck. *Fish Stress and Health in Aquaculture*, Cambridge University Press, Cambridge (1997).
43. L. J. Guillette, A. Cree, A. A. Rooney. In *Health and Welfare of Captive Reptiles*, C. Warwick, F. L. Frye, J. B. Murphy (Eds.), pp. 32–81, Chapman and Hall, London (1995).
44. H. S. Siegel. *Brit. Poultry Sci.* **36**, 3–22 (1995).
45. C. Weissman. *Anesthesiology* **73**, 308–327 (1990).
46. J. A. Brown. In *Fish Ecophysiology,* J. C. Rankin and F. B. Jensen (Eds.), pp. 276–296, Chapman & Hall, London (1993).
47. L. L. Bestervelt, J. A. Pitt, W. N. Piper. *Toxicol. Sci.* **46**, 294–299 (1998).
48. S. Shridhar, A. Farley, R. L. Reid, W. G. Foster, D. A. Van Vugt. *Toxicol. Sci.* **63**, 181–188 (2001).
49. C. A. Harris, E. M. Santos, A. Janbakhsh, T. G. Pottinger, C. R. Tyler, J. P. Sumpter. *Environ. Sci. Technol.* **35**, 2909–2916 (2001).
50. J. B. Schenkman. *J. Steroid Biochem. Mol. Biol.* **43**, 1023–1030 (1992).
51. B. M. A. Näslund, G. Glauman, M. Warner, J.-A. Gustafsson, T. Hansson. *Mol. Pharmacol.* **33**, 31–37 (1988).
52. T. Andersson, L. Förlin, S. Olsen, A. Fostier, B. Breton. *Mol. Cell. Endocrinol.* **91**, 99–105 (1993).
53. D. J. Tom, L. E. J. Lee, J. Lew, N. C. Bols. *Comp. Biochem. Physiol. A* **128**, 185–198 (2001).
54. A. Hontela, J. B. Rasmussen, C. Audet, G. Chevalier. *Arch. Environ. Con. Tox.* **22**, 278–283 (1992).
55. L. L. Bestervelt, J. A. Nolan, W. N. Piper. *Neurotoxicol. Teratol.* **15**, 371–376 (1993).
56. J. A. Pitt, A. R. Buckalew, D. E. House, B. D. Abbott. *Toxicology* **151**, 25–35 (2000).
57. L. L. Bestervelt, J. A. Pitt, C. J. Nolan, Y. Cai, D. W. Piper, J. A. Dybowski, G. A. Dayharsh, W. N. Piper. *Toxicol. Sci.* **44**, 107–115 (1998).
58. C. S. Bradford, M. S. Fitzpatrick, C. B. Schreck. *Gen. Comp. Endocr.* **87**, 292–299 (1992).
59. T. G. Pottinger, T. A. Moran, P. A. Cranwell. *Fish Physiol. Biochem.* **10**, 55–66 (1991).
60. M. E. McMaster, K. R. Munkittrick, P. L. Luxon, G. J. Van Der Kraak. *Ecotoxicol. Environ. Safe.* **27**, 251–264 (1994).
61. M. A. Santos and M. Pacheco. *Ecotoxicol. Environ. Safe.* **35**, 96–100 (1996).
62. J. Lappivaara. *Arch. Environ. Con. Tox.* **41**, 55–64 (2001).
63. M. Pacheco and M. A. Santos. *Ecotoxicol. Environ. Safe.* **49**, 64–75 (2001).
64. I. Kakuta. *Environ. Toxicol.* **17**, 1–6 (2002).
65. A. Hontela, P. Dumont, D. Duclos, R. Fortin. *Environ. Toxicol. Chem.* **14**, 725–731 (1995).
66. C. Girard, J. C. Brodeur, A. Hontela. *Can. J. Fish. Aquat. Sci.* **55**, 438–450 (1998).
67. J. S. Laflamme, Y. Couillard, P. G. C. Campbell, A. Hontela. *Can. J. Fish. Aquat. Sci.* **57**, 1692–1700 (2000).
68. D. O. Norris, S. Donahue, R. M. Dores, J. K. Lee, T. A. Maldonado, T. Ruth, J. D. Woodling. *Gen. Comp. Endocr.* **113**, 1–8 (1999).

69. J. M. Wilson, M. M. Vijayan, C. J. Kennedy, G. K. Iwama, T. W. Moon. *J. Endocrinol.* **157**, 63–70 (1998).
70. N. Basu, C. J. Kennedy, P. V. Hodson, G. K. Iwama. *Fish Physiol. Biochem.* **25**, 131–140.
71. E. S. Quabius, P. H. M. Balm, S. E. Wendelaar Bonga. *Gen. Comp. Endocr.* **108**, 472–482 (1997).
72. E. S. Quabius, D. T. Nolan, C. J. Allin, S. E. W. Bonga. *Environ. Toxicol. Chem.* **19**, 2892–2899 (2000).
73. A. D. Gendron, C. A. Bishop, R. Fortin, A. Hontela. *Environ. Toxicol. Chem.* **16**, 1694–1706 (1997).
74. A. Lorenzen, T. W. Moon, S. W. Kennedy, G. A. Fox. *Environ. Health Persp.* **107**, 179–186 (1999).
75. L. J. Guillette, D. A. Crain, A. A. Rooney, A. R. Woodward. *J. Herpetol.* **31**, 347–353 (1997).
76. Z. Ilan and Z. Yaron. *J. Endocrinol.* **87**, 185–193 (1980).
77. Z. Ilan and Z. Yaron. *J. Fish Biol.* **22**, 657–669 (1983).
78. S. Benguira and A. Hontela. *Environ. Toxicol. Chem.* **19**, 842–847 (2000).
79 V. S. Leblond, M. Bisson, A. Hontela. *Gen. Comp. Endocr.* **121**, 48–56 (2001).
80. J. C. Brodeur, C. Girard, A. Hontela. *Environ. Toxicol. Chem.* **16**, 2171–2178 (1997).
81. J. C. Brodeur, G. Sherwood, J. B. Rasmussen, A. Hontela. *Can. J. Fish. Aquat. Sci.* **54**, 2752–2758 (1997).
82. K. A. Glennemeier and R. J. Denver. *Environ. Toxicol.* **16**, 287–297 (2001).
83. W. A. Hopkins, M. T. Mendonca, J. D. Congdon. *Comp. Biochem. Physiol. C* **122**, 191–196 (1999).
84. L. Förlin and C. Haux. *Aquat. Toxicol.* **6**, 197–208 (1985).
85. M. M. Vijayan, G. Feist, D. M. E. Otto, C. B. Schreck, T. W. Moon. *Aquat. Toxicol.* **37**, 87–98 (1997).
86. C. L. Miranda, M. C. Henderson, J.-L. Wang, H.-S. Chang, J. D. Hendricks, D. R. Buhler. *Comp. Biochem. Physiol. C* **103**, 153–157 (1992).
87. J. M. Navas and H. Segner. *Chem.-Biol. Interact.* **138**, 285–298 (2001).
88. D. R. Buhler, C. L. Miranda, M. C. Henderson, Y. H. Yang, S. J. Lee, J. L. Wang-Buhler. *Toxicol. Appl. Pharmacol.* **168**, 91–101 (2000).
89. A. Arukwe, L. Forlin, A. Goksoyr. *Environ. Toxicol. Chem.* **16**, 2576–2583 (1997).
90. L. L. Bestervelt, Y. Cai, D. W. Piper, C. J. Nolan, J. A. Pitt, W. N. Piper. *Neurotoxicol. Teratol.* **15**, 365–370 (1993).
91. O. P. Mgbonyebi, C. T. Smothers, J. J. Mrotek. *Cell Biol. Toxicol.* **9**, 223–234 (1993).
92. O. P. Mgbonyebi, C. T. Smothers, J. J. Mrotek. *Cell Biol. Toxicol.* **14**, 301–311 (1998).
93. O. P. Mgbonyebi, C. T. Smothers, J. J. Mrotek. *Cell Biol. Toxicol.* **10**, 23–33 (1994).
94. O. P. Mgbonyebi, C. T. Smothers, J. J. Mrotek. *Cell Biol. Toxicol.* **10**, 35–43 (1994).
95. J. C. Veltman and M. D. Maines. *Arch Biochem. Biophys.* **248**, 467–478 (1986).
96. J. C. Veltman and M. D. Maines. *Biochem. Pharmacol.* **35**, 2903–2910 (1986).
97. M. Johansson, C. Larsson, A. Bergman, B. O. Lund. *Pharmacol. Toxicol.* **83**, 225–230 (1998).
98. C.-J. Jönsson, B. O. Lund, B. Brunström, I. Brandt. *Environ. Toxicol. Chem.* **13**, 1303–1310 (1994).
99. B.-O. Lund. *Environ. Toxicol. Chem.* **13**, 911–917 (1994).
100. O. Lindhe, B.-O. Lund, A. Bergman, I. Brandt. *Environ. Health Perspect.* **109**, 105–110 (2001).
101. M. K. Johansson, J. T. Sanderson, B.-O. Lund. *Toxicol. In Vitro* **16**, 113–121 (2002).
102. S. Benguira, V. S. LeBlond, J. P. Weber, A. Hontela. *Environ. Toxicol. Chem.* **21**, 1753–1756 (2002).
103. H. Mosnierpudar, P. Thomopoulos, X. Bertagna, C. Fournier, D. Guiban, J. P. Luton. *Eur. J. Endocrinol.* **133**, 313–316 (1995).
104. E. H. Jorgensen, P. H. M. Balm, J. S. Christiansen, N. Plotitsyna, K. Ingebrigtsen. *Aquat. Toxicol.* **54**, 179–193 (2001).

105. P. Lahiri and S. Sircar. *Toxicology* **66**, 75–79 (1991).
106. L. P. Walsh and D. M. Stocco. *Biol. Reprod.* **63**, 1024–1033 (2000).
107. A. M. Ronco, K. Valdes, D. Marcus, M. Llanos. *Toxicology* **159**, 99–106 (2001).
108. D. W. Brann, L. B. Hendry, V. B. Mahesh. *J. Steroid Biochem. Mol. Biol.* **52**, 113–133 (1995).
109. K. L. Burnstein and J. A. Cidlowski. *Mol. Cell. Endocrinol.* **83**, C1–C8 (1992).
110. R. M. Donohoe and L. R. Curtis. *Aquat. Toxicol.* **36**, 31–52 (1996).
111. F. R. Knudsen, A. Arukwe, T. G. Pottinger. *Environ. Pollut.* **103**, 75–80 (1998).
112. F. R. Knudsen and T. G. Pottinger. *Aquat. Toxicol.* **44**, 159–170 (1999).
113. M. Johansson, S. Nilsson, B. O. Lund. *Environ. Health Perspect.* **106**, 769–772 (1998).
114. T. G. Pottinger, P. H. M. Balm, A. D. Pickering. *Gen. Comp. Endocr.* **98**, 311–320 (1995).
115. J. C. Wingfield and K. Hunt. *Comp. Biochem. Physiol.* **132**, 275–286 (2002).
116 N. C. Vamvakopoulos and G. P. Chrousos. *J. Clin. Invest*. **92**, 1896–1902 (1993).
117. T. G. Pottinger, T. R. Carrick, S. E. Hughes, P. H. M. Balm. *Gen. Comp. Endocr.* **104**, 284–295 (1996).
118. A. Catalani, P. Casolini, S. Scaccianoce, F. R. Patacchioli, P. Spinozzi, L. Angelucci. *Neuroscience* **100**, 319–325 (2000).
119. L. A. M. Welberg and J. R. Seckl. *J. Neuroendocrinol.* **13**, 113–128 (2001).
120. I. C. G. Weaver, P. La Plante, S. Weaver, A. Parent, S. Sharma, J. Diorio, K. E. Chapman, J. R. Seckl, M. Szyf, M. J. Meaney. *Mol. Cell. Endocrinol.* **185**, 205–218 (2001).
121. B. Silverin. *Anim. Behav.* **55**, 1411–1420 (1998).
122. I. T. Moore, M. P. Lemaster, R. T. Mason. *Anim. Behav.* **59**, 529–534 (2000).
123. T. G. Pottinger. In *Stress Physiology in Animals,* P. H. M. Balm (Ed.), pp. 130–177, Sheffield Academic Press, Sheffield (1999).
124. C. Fabre-Nys. *Rev. Reprod.* **3**, 31–41 (1998).
125. A. M. Bell. *Anim. Behav.* **62**, 775–780 (2001).

Pure Appl. Chem., Vol. 75, Nos. 11–12, pp. 2335–2341, 2003.

Topic 4.11

Effects of endocrine active substances in wildlife species: Genetic, biochemical, and physiological factors in variable susceptibility to endocrine disruptors*

Shin'ichiro Kawai[1,‡], Makito Kobayashi[2], and Hideo Kaneko[3]

[1]Department of Human Environmental Sciences, Kobe College, 4-1 Okadayama, Nishinomiya, Hyogo 662-8505, Japan; [2]Department of Biology, Division of Natural Sciences, International Christian University, 3-10-2 Osawa, Mitaka, Tokyo 181-8585, Japan; [3]Sumitomo Chemical Co. Ltd., 27-1, Shinkawa 2-chome, Chuo-ku, Tokyo 104-8260, Japan

Abstract: Responses to endocrine active substances (EASs) in animals are various, and differences between the responses among individuals, populations, and species are well known. These differences are observed not only in EASs but in most environmental chemicals including synthetic and naturally occurring ones. The basic differences in sensitivity to EASs are attributed to that of affinity or specificity of the receptors to EASs at the cellular level. Although the nucleotide sequences encoding for estrogen receptor proteins have been documented in several species and the functions of the receptors are the same, the ability to bind the natural hormones and the estrogenic xenobiotics is not necessarily identical. The reproductive endocrine system is basically common among vertebrates, but chemical types of hormones, physiological roles of hormones, and the basal blood levels of hormones differ among each species, especially in sex steroids. These differences cause various types of responses and sensitivity to EASs among animal species. Xenobiotic metabolism is important for the genetical, biochemical, and physiological factors concerning the influence of EASs. Some EASs directly inhibit cytochrome P450 (CYP) activity as was reported in tributyltin that inhibits CYP19 (aromatase) activity causing imposex in neogastropods. Some organochlorines including dioxins stimulate arylhydrocarbon (Ah) receptor-mediated xenobiotic metabolism, and result in the metabolic disruption of steroid hormones such as estrogen as were reported in eggshell thinning in birds of prey and uterus occlusion in seals. CYP activity greatly differs among wildlife species in both terrestrial and aquatic organisms, and these differences are significantly responsible for the multiple effects or toxicity of EASs. Sex and age differences also cause different responses to EASs and are largely due to the differences in xenobiotic metabolizing activities.

*Report from a SCOPE/IUPAC project: Implication of Endocrine Active Substances for Human and Wildlife (J. Miyamoto and J. Burger, editors). Other reports are published in this issue, *Pure Appl. Chem.* **75**, 1617–2615 (2003).

‡Corresponding author

INTRODUCTION

Endocrine disruption has been postulated as the cause of a large number of adverse effects on the health of various wild animal species. The majority of cases involve reproductive abnormalities that might be linked to population decline. Data supporting cause–effect relationships between the biological effects and exposure to certain specific chemical agents are extremely limited to a small number of cases. The impact of chemical pollution on the reproductive success and population sizes of wildlife species is often difficult to predict because of the ecological factors including geographic range, specific habitat, food habits, etc. Moreover, physiological, biochemical, and genetical factors are closely related to the differences of endocrine disruption between species, sex, and age of animals.

Responses to endocrine active substances (EASs) in animals are various, and the differences of responses among individuals, populations, and species are well-known facts. In this paper, the genetic, biochemical, and physiological factors influencing the variable susceptibility of wildlife species to EASs are described with special references to species, sex, and age differences. These differences are generally observed not only with EASs, but with most chemicals including synthetic and naturally occurring ones.

PHYSIOLOGICAL FACTORS

Sex, sexual status, age, endocrine system, and endogenous levels of sex steroids are important physiological factors influencing the susceptibility of aquatic and terrestrial wildlife species to EASs.

Sex differences and sexual status influences the levels of organochlorines in striped dolphins (*Stenella coeruleoalba*)

Endocrine disruption in marine mammals has been reported in seals since the 1970s, and uterus occlusion [1] and lesions of skull bones [2] were clearly noticed in the individuals that accumulated high levels of organochlorines such as PCBs. Organochlorines in these mammals are likely to cause endocrine disruption, especially those affecting estrogen metabolism, directly or indirectly through xenobiotic metabolism. Sex differences or sexual status affect the accumulation and excretion of organochlorines in wild animals. For example, PCB concentration and burdens in the blubber of striped dolphins migrating along the Pacific coast of Japan, were markedly low in mature females (lactating, pregnant, or resting) compared to mature males and immature individuals [3]. This is because of the transportation of PCB from mother dolphins to newborns through lactation. More than 60 % of organochlorines accumulated in lactating females are mobilized to transport during the lactation period of about 6 months. From these findings, physiological factors such as age, sex, and sexual status are concluded to be largely responsible for the uptake, accumulation, and excretion of organochlorines. Consequently, some adverse effects, including reproductive and immunological toxicity, might occur in marine mammals [4]. In addition, lipid content and lipid composition are responsible for the accumulation of organochlorines. Most organochlorines are accumulated in blubber, being a subcutaneous thick fatty tissue. Apart from blubber, high levels of organochlorines were observed in the tissues of high triglycerides content such as mammary gland, kidney, and pancreas. Though brain tissues show high lipid content, the lipids are mainly comprised of phospholipid and total cholesterol, and consequently the levels of organochlorines are low [5]. In this way, physiological characteristics such as lipid content and lipid composition of tissues or organs in certain animals drive the accumulation and fates of organochlorines in the body.

Reproductive cycles and developmental stages of carp related to the level of vitellogenin and 17β-estradiol (E2)

Reproductive stages and season are important for the understanding of sex steroid hormone levels and specific markers such as the egg protein precursor vitellogenin in fish. In the immature stages of male and female carp (*Cyprinus carpio*) (1 to 2 years), significant differences in blood E2 levels between the sexes are not observed. However, maturing male carp show rather higher levels of E2 in autumn as compared to females. After that, E2 levels in mature females clearly increase. Vitellogenin content in blood increases in accordance with the increase of E2 during winter to spring in female carp [6]. Though the blood vitellogenin levels in wild male fish are considered to be a good marker for the exposure to estrogenic substances in aquatic environments, endogenous sex steroid levels are significantly responsible for the low levels of vitellogenin in males and modify the effect of exogenous EASs.

Endogenous estrogens and androgens in male and female fish living in freshwater and marine environments

Chemical types of sex steroids and their roles differ among vertebrate species. E2 is a common female sex steroid, and its major role in oviparous animals is in the production of vitellogenin. In most vertebrates, E2 is involved in the occurrence of female sex behavior, but this is not the case in fish [7]. Testosterone is known as a major androgen in mammals, but this steroid is produced both in the ovary and testis in fish (Table 1), and blood levels of testosterone are higher in females than in males in some fish species.

The major androgens in amphibians and fish are dihydrotestosterone and 11-ketotestosterone, respectively [7]. Since these androgens are nonaromatizable, their mode of action and effects are considered to be different compared to those of testosterone or 17α-methyltestosterone, which are aromatized and bind to the estrogen receptor at some target organs. Although testosterone or 17α-methyltestosterone are often used as standard androgens in bioassays, the effects of EASs on nonaromatizable androgens should be also considered and examined depending on which chemical type of androgens the animal species possesses.

Endogenous hormone levels seem to be one of the factors that cause differences in sensitivity to EASs. When a certain type of hormone is exogenously administered to two animals in which blood concentrations of the endogenous hormone are different, effects of the administered hormone would be expected to be higher in the animal with low blood levels than in the one with high levels due to the competition between the exogenous and endogenous hormone in binding to the receptor. Large variations in blood levels of hormones have been reported in vertebrate species. Blood E2 concentrations of sexually matured female mammals are known to range from 10 to 200 pg/ml. Blood E2 concentrations in fish are much higher, and the variation among species is also large. Blood E2 levels in most female marine fish species examined are 10–2000 pg/ml while the levels range from 1 to 10 ng/ml in female cyprinid species during vitellogenesis. In salmonid species, blood E2 levels rise as high as 10 to 50 ng/ml. Blood E2 levels in male fishes have not been studied to the extent of those in females, but it

Table 1 Blood levels of sex steroid hormones in sexually mature fish.

	17β-Estradiol		Testosterone		11-Ketotestosterone	
	Male	Female	Male	Female	Male	Female
Salmonids	~0.5	10~40	100~200	100~200	100~200	–
Cyprinids	~0.5	5~10	5~100	5~100	1~20	1~10
Marine species	–	1~10	1~5	1~5	1~5	–

ng/ml

is reported that the levels rise up to 1000 pg/ml in common carp (*Cyprinus carpio*) during testicular development [6,8,9]. It is possible that large amounts of endogenous hormones compete with EASs in binding to the receptor, and the effects of EASs might therefore vary depending on the blood levels of endogenous hormones.

BIOCHEMICAL AND GENETICAL FACTORS

The following differences are considered to be important for species differences in the effects caused by EASs in wild animals.

Species differences in enzyme activity toward xenobiotics

Large species differences have been reported in the activity of several enzymes that play an important role in xenobiotic metabolism including microsomal oxidase, epoxide hydrolase, *N*-acetyltransferase, UDP-glucuronyltransferase, etc. [10]. For example, activity of microsomal oxidases was nearly five orders of magnitude higher in cow compared to cat, when benzo(a)pyrene was used as a substrate. There exists a multiplicity of enzymes in different species and this emphasizes the need for extreme caution in extrapolating metabolic information across species even with structurally similar chemicals. Clearly, these differences can lead to significant species differences in the action of xenobiotic EASs.

Species difference in the 17β-estradiol transactivation of the luciferase gene

Over the last decade, cloning studies of estrogen receptors have been performed with various vertebrates. Figure 1 shows the analysis of species differences in estrogen receptor (ER) dependent transactivation with E2 using reporter-gene assays [11]. Full-length ER cDNAs from human, rat, chicken, caiman (*Caiman crocodilus*), whiptail lizard (*Cnemidophorus uniparens*), African clawed frog (*Xenopus laevis*), and rainbow trout (*Oncorhynchus mykiss*) were prepared from hepatic mRNA by the RT-PCR method and inserted into expression plasmids. Both expression and reporter plasmids were transiently transfected into HeLa cells, and the estrogenic activity was analyzed in terms of induction of luciferase activity [11].

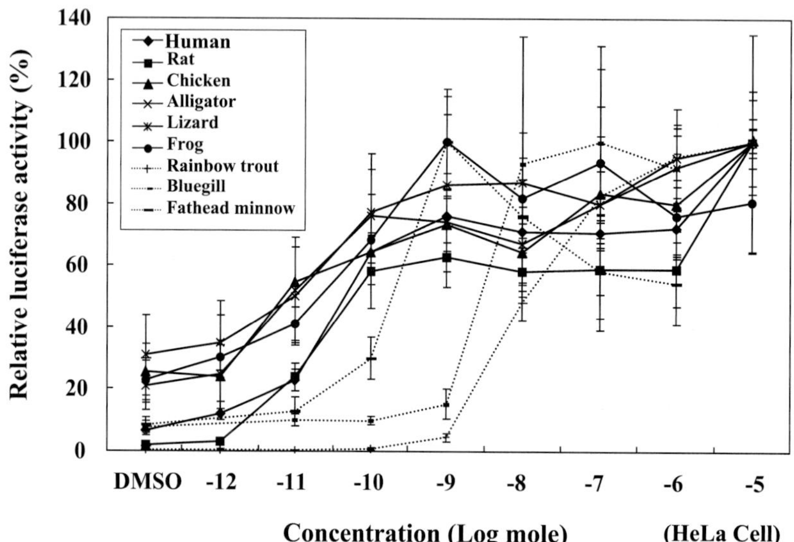

Fig. 1 Species differences in the transactivation of the luciferase gene (Receptor: ER, Ligand: 17β-estradiol).

No significant species differences in the transactivation by E2 were found among human, alligator, and lizard ER in Fig. 1. However, decreases in response to E2 were commonly observed in the fish species including bluegill, fathead minnow, and rainbow trout ERs in HeLa cells at 37 °C. The response of ER of rainbow trout to E2 was two-orders lower compared to that of human. This may be due to the temperature conditions of the reporter-gene assays, because thermo-dependent alteration in the affinity to ER has been reported in binding assays using ER of rainbow trout.

Figure 2 shows the elevation in response of rainbow trout ER to E2 in BF-2 cells derived and established from bluegill (*Lepomis macrochirus*) fry at the incubation temperature of 24 °C, compared to the result in HeLa cells at 37 °C [11]. This finding will be useful for assessing estrogenic effects of chemicals on wildlife species as well as humans.

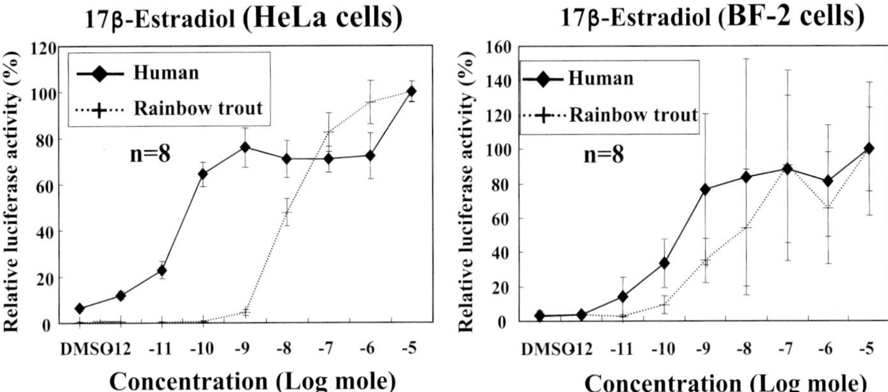

Fig. 2 Species differences in the transactivation of the luciferase gene: temperature effects (Receptor: ER, Ligand: 17β-estradiol).

Species differences in the amino acid sequences of cytochrome P450

Cytochrome P450 (CYP) and glutathione-*S*-transferase (GST) are important families of enzymes involved in the biotransformation of xenobiotics including EASs. A number of environmental pollutants are known as inducers of CYP and GST expression levels. Therefore, the induction of both enzymes by xenobiotics in wild life species is considered to be useful as a biomarker for monitoring levels of pollution [12]. The CYP family has been investigated in fish, amphibians, reptiles, gastropods, and crustaceans. Among these animals, studies of xenobiotic metabolism are highly advanced in fish, especially in rainbow trout (*O. mykiss*).

Species differences in the amino acid sequence of CYP has also been reported in several kinds of mammals relative to human [13], and the sequence of primates such as crab-eating macaque and marmoset are closely similar to human, while clear differences are reported between human and hamster or mouse.

Species differences in the amino acid sequences of ER-α

The ER is essential for the functional expression of estrogen. The ER is divided into A to F regions from the *N*-terminal, based on homology and function. The C-region being the DNA binding zone is important for the recognition of estrogen-responsive elements and for the binding of the receptor to DNA. As shown in Fig. 3, the amino acid sequences of the C-region are highly homologous (90 % more) among mammals (humans), birds (chickens), reptiles (lizards and alligators), amphibians (frogs), and fish (medaka). However, the amino acid sequence of the E-region which is important for the binding of es-

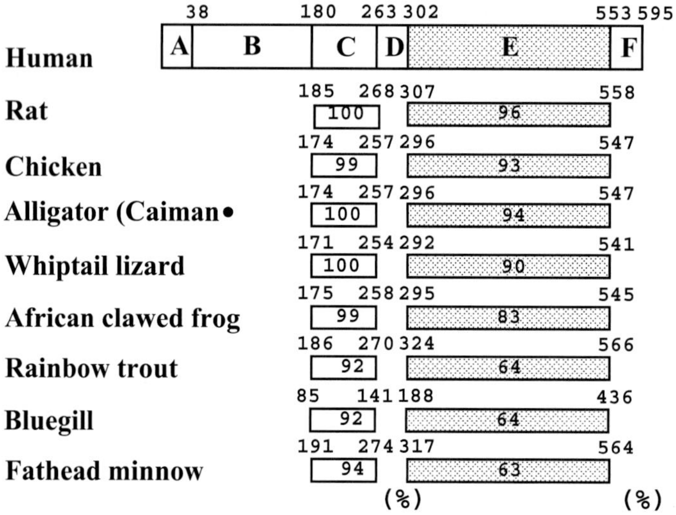

Human: 38 180 263 302 553 595 — A B C D E F
Rat: 185 268 307 558 — 100 96
Chicken: 174 257 296 547 — 99 93
Alligator (Caiman•: 174 257 296 547 — 100 94
Whiptail lizard: 171 254 292 541 — 100 90
African clawed frog: 175 258 295 545 — 99 83
Rainbow trout: 186 270 324 566 — 92 64
Bluegill: 85 141 188 436 — 92 64
Fathead minnow: 191 274 317 564 — 94 63
(%) (%)

Fig. 3 Comparison of the amino acid sequence similarities of the individual C and E domains for ER.

trogen to ER is not so homologous among species, and the homology compared to humans decreases in the following order [14].

Bird > Reptile > Amphibian > Fish

In fish, the homology is only 60 % to human E-region of ER. This evidence shows the species difference in the E-region of ER.

FUTURE STUDIES

Steroid hormones are lipophilic and do not dissolve easily in water. A very small percentage of the hormones are free in the blood, and most are carried bound to serum binding proteins such as sex hormone binding globulin (SHBG). Hormones bound to SHBG are protected from degradation by the liver enzymes, and are also considered to function as a reserve of available hormones. The concentration of SHBG, and the rate of degradation and excretion from organisms regulate the availability of steroid hormones to their target tissues. To date, little research has been conducted on the interaction of EASs with SHBG, although most EASs circulate freely in blood. Certain chemicals such as diethylstilbestrol (DES), octylphenol and *o,p*-DDT seem to have very low affinity for SHBG, because estrogenic activity of these chemicals was poorly inhibited by the presence of SHBG or albumin (nonspecific serum binding protein) in an in vitro assay when compared to the total inhibition of E2 activity [15]. Another way that EASs could alter the availability of hormones in an organism is to induce the production of SHBG, binding more free hormones and reducing their bioavailability to the cells. Some plant compounds, such as lignans and isoflavonoids, seem to stimulate SHBG synthesis in the liver in vivo and in vitro [16]. Consequently, information on SHBG in wildlife species is required.

Finally, the authors would like to emphasize that physiological factors are closely related and linked to ecological and ethological ones. Unfortunately, many findings on the adverse effects of various naturally occurring and synthetic chemicals including EASs are mostly obtained by laboratory experiments alone. Behavioral ecotoxicology will therefore be one of the most important disciplines in the near future.

SUMMARY AND CONCLUSIONS

Genetic or biochemical differences in ER influence gene expression and the synthesis of estrogen mediated or dependent functional proteins.

Differences of xenobiotic metabolism including CYP and conjugating activity are also important factors affecting the susceptibility of wild animals to EASs.

It is possible that endogenous sex steroids such as estrogens and androgens compete with EASs and cause differences in sensitivity to EASs depending on their titres in blood, especially in fish.

RECOMMENDATIONS

Obtaining information on normal levels of hormones and physiological response caused by endogenous hormones in each wild species is important. These data are essential to know whether the responses observed in animals are caused exogenously by EASs or physiologically by endogenous hormones.

Research on the interaction of EASs with SHBG should be conducted for the evaluation of availability of steroid hormones in their target tissues.

Information on CYP activity in wildlife species is very sparse, and studies on differences between species, age, and sex should be carried out.

REFERENCES

1. E. Helle, M. Olsson, S. Jensen. *Ambio* **5**, 261 (1976).
2. A. Bergman, M. Olsson, S. Reiland. *Ambio* **21**, 517 (1992).
3. S. Tanabe. *J. Oceanogr. Soc. Jpn.* **41**, 358 (1985).
4. K. Ohashi, N. Miyazaki, S. Tanabe, H. Nakata, R. Miura, K. Fujita, C. Wakasa, M. Uema, M. Shiotani, E. Takahashi, C. Kai. *Vet. Microbiol.* **82**, 203 (2001).
5. S. Kawai, M. Fukushima, N. Miyazaki, R. Tatsukawa. *Mar. Pollut. Bull.* **19**, 129 (1988).
6. F. Ito. "Freshwater fish", In *Problems of Endocrine Disruptors in Fisheries Environment*, S. Kawai and J. Koyama (Eds.), pp. 31-42, Kouseisha-Kouseikaku, Tokyo (2000) (in Japanese).
7. M. Kobayashi, N. E. Stacey, K. Aida, S. Watabe. "Sexual plasticity of behavior and gonadotropin secretion in goldfish and gynogenetic crucian carp". In *Reproductive Physiology of Fish*, B. Norberg, O. S. Kjesbu, E. Andersson, G. L. Taranger, S. O. Stefansson (Eds.), pp. 117–124, John Grieg AS, Bergen, Norway (2000).
8. S. Hashimoto. Marine Fish, In "*Problems of Endocrine Disruptors in Fisheries Environment* ", S. Kawai and J. Koyama, (Eds.), pp. 43–53, Kouseisha-Kouseikaku, Tokyo (2000) (in Japanese).
9. T. Matsumoto, M. Kobayashi, Y. Nihei, T. Kaneko, H. Fukuda, K. Hirano, A. Hara, S. Watabe. *Fish. Sci.* **68**, 1053 (2002).
10. J. B. Watkins III and C. D. Klassen. *J. Animal Sci.* **63**, 933 (1986).
11. K. Sumida, N. Ooe, K. Saito, H. Kaneko. *Mol. Cell Endocrinol.* **183**, 33 (2001).
12. K. Sumida, N. Ooe, K. Saito, H. Kaneko. *J. Steroid Biochem. Mol. Biol.* **84**, 33–40 (2003).
13. J. F. Payne, L. L. Fancey, A. D. Rahimtula, E. L. Porter. *Comp. Biochem. Physiol. C* **86**, 233 (1987).
14. T. Kamataki. *Ykugaku Zasshi* **115**, 370 (1999).
15. S. F. Arnold, M. K. Robinson, A. C. Notides, L. J. Guillette, Jr., J. A. McLachlan. *Environ. Health. Perspect.* **104**, 544 (1996).
16. H. Adlercreutz. *Environ. Health Perspect.* **103** (Suppl. 7), 103 (1995).

Pure Appl. Chem., Vol. 75, Nos. 11–12, pp. 2343–2353, 2003.

Topic 4.12

Development of fish tests for endocrine disruptors*

T. H. Hutchinson[1,‡], H. Yokota[2], S. Hagino[3], and K. Ozato[4]

[1]*AstraZeneca Global Safety Health & Environment, Brixham Environmental Laboratory, Brixham, Devon TQ5 8BA, UK;* [2]*Kurume Laboratory, Chemicals Evaluation and Research Institute (CERI), Japan;* [3]*Sumika Technoservice Corporation, Takarazuka-city, Japan;* [4]*Nagoya University Bioscience Center, Nagoya, Japan*

Abstract: International concern over endocrine active substances (EASs) has led to intensive research programs to establish fish reproductive and developmental toxicity tests for use in environmental (ecological) risk assessment. This chapter gives an overview of key themes of in vivo ecotoxicology research, including fish screening assays, partial life-cycle tests (the draft Organization for Economic Cooperation and Development (OECD) fish reproduction test and the new fish development test) and fish full life-cycle tests. In the context of the OECD test guidelines program, fish species of primary interest include fathead minnow, medaka, and zebrafish, while guppy, rainbow trout, sheepshead minnow, and three-spined stickleback are also of scientific importance. Critical factors for evaluation include baseline reproductive biology and definition of EAS sensitive life-stages. For regulatory applications, a critical review of existing fish EAS data suggests that apical adverse effect endpoints, namely development, growth and reproduction (e.g., fecundity, fertilization rates and hatching success) should be used to derive predicted no effect concentrations (PNECs) for the environmental risk assessment of EASs. In support of these apical adverse effect endpoints, biomarker responses (e.g., vitellogenin, gonadal-somatic index, and gonad histopathology) should be used to provide mechanistic data, compare species (e.g., cyprinids vs. salmonids) and allow extrapolation between laboratory and field studies.

INTRODUCTION

Over the past decade, a growing body of scientific information has highlighted the potential threats of chemicals to the reproductive health of fish [1]. Internationally, the Organization for Economic Cooperation and Development (OECD) is actively coordinating research efforts for EAS testing in fish and other animal species [2]. In this context, the OECD test guideline program seeks to ensure that proposed fish test guidelines measure biologically relevant endpoints and that these endpoints are reproducible between laboratories internationally. This OECD effort is supported by significant regional initiatives in Europe, Japan, and North America that collectively seek to identify a cost-effective battery of fish screening and testing assays for the regional ecological risk assessment of EASs. It is widely recognized that a range of fish tests for EASs are needed, including partial and life-cycle protocols with a range of freshwater and marine species.

*Report from a SCOPE/IUPAC project: Implication of Endocrine Active Substances for Human and Wildlife (J. Miyamoto and J. Burger, editors). Other reports are published in this issue, *Pure Appl. Chem.* **75**, 1617–2615 (2003).

‡Corresponding author

Importantly, different types of fish tests may be required according to different circumstances (e.g., persistent vs. unstable substances; constant release of the substance into surface waters or only limited episodic releases) [3]. Taking three illustrative scenarios: a fish full life-cycle test would be appropriate where an estrogenic chemical was being constantly discharged [4]; a fish development test or a fish reproduction test (both partial life-cycle tests) would be useful in addressing a nonbioaccumulative pesticide such as methoxychlor applied only seasonally [5,6]; while a short-term and inexpensive assay would be invaluable in rapidly screening potential EASs, complex mixtures, or effluents, and thereby help guide the experimental design of long-term studies [7–9].

In terms of test organisms, no single fish species monopolizes the useful attributes for use in environmental risk assessment protocols, consequently it is important to consider a suite of fish species when developing and validating test guidelines for EASs. Freshwater species of primary international interest include fathead minnow (*Pimephales promelas*) [6,9,10], medaka (*Oryzias latipes*) [5,11,12], and zebrafish (*Danio rerio*) [13,14]. Additionally, several other OECD fish species are also being used in EAS research, for example, guppy (*Poecelia reticulata*) [15], rainbow trout (*Oncorhynchus mykiss*) [8], sheepshead minnow (*Cyprinodon variegatus*) [16], and three-spined stickleback (*Gasterosteus aculeatus*) [17]. Against this background, this chapter summarizes key scientific developments in the establishment of new OECD test guidelines using fish, together with the use of such data in environmental risk assessments. Many of the principles are illustrated using the medaka, however, these examples apply equally to other OECD fish species.

SPECIES AND STRAIN SELECTION

Fathead minnow (*Pimephales promelas*)

This species is representative of the ecologically widespread Cyprinidae and has long been used to support water quality criteria through OECD and other test guidelines. This gonochoristic species (with no reports of natural hermaphrodites under natural conditions) is highly amenable to laboratory culture, and the life cycle is relatively rapid (4–5 months from embryo hatching to adult breeding at 25 °C). Fathead minnows are fractional spawners (typically, females produce batches of 50–100 transparent eggs every 3–5 days), the embryo-larvae are sensitive to EASs, and there is a growing body of information on the reproductive physiology and molecular endocrinology of this species [10,18–20]. To our knowledge there are no specific strains of fathead minnow available for EAS research, and wild-type populations are commonly used.

Medaka (*Oryzias latipes*)

The Japanese medaka is representative of the subtropical medaka (family Oryziatidae) and is widely used as small fish model in laboratory research and in ecotoxicology. The genetic sex determination, oogenesis, fertilization, and embryonic development of this species have been extensively studied [21,22]. In this gonochoristic species (showing no hermaphroditism under natural conditions), the generation time is short (typically 2–3 months), year-round spawning is daily under artificial conditions. Medaka sex is determined by XY chromosomes (i.e., female and male sex is determined by XX and XY chromosomes, respectively). It is known that some genes for pigmentation and DNA sequences identified by PCR are linked to sex chromosomes and very recently, a male determining gene on the Y chromosome (Dmy) was identified and cloned [23]. Using such markers, the genotypic sex of medaka can be accurately determined. Several medaka strains have been established in Japan for research into EASs (see <http://biol1.bio.nagoya-u.ac.jp:8000 for photographs>).

Orange-red variety or "Himedaka"
This strain with orange body color is commercially available and widely used for research and testing because it is the easiest to be maintained in laboratory conditions. This out-bred (wild-type) strain is well suited to testing EASs [24].

d-rR or S-rR (Yamamoto) strain
The genotypic sex of this strain can be identified using a color marker and the strain is therefore well suited to EAS testing [11]. The sex-specific body color is due to the location of the recessive gene (*r*) for the orange-red color in xanthophores on the X chromosome and the wild-type allele (*R*) on the Y chromosome (crossing-over rate between X and Y chromosomes is ca. 0.3 %). Expression of the body color occurs rather late, that is, some two weeks after hatching (depending on the laboratory conditions and feeding regime).

Qurt, FLF, and FLFII
In both the Qurt strain [26] and FLF strain [27] , the genotypic sex can be identified at early embryonic stages since leucophore differentiation occurs at the 2-day-old embryo stage in the male, but not in the female. This is due to the location of the recessive mutant gene (*lf*) for the white color in leucophores on the X chromosome and the wild-type allele (*LF*) in the Y chromosome (crossing-over rate of the color marker is rather high, ca. 3 to 4 %).

The Hd-rR.Y^{HNI} strain has the orange-red body and SL1 as markers for identification of the genotypic sex [28]. The genotypic sex can be identified accurately because the crossing-over rate of SL1 is almost zero. The FLFII strain was generated by crossing FLF with Hd-rR.Y^{HNI} . In the FLFII, the genotypic sex can be screened at early embryonic stages by the presence of leucophores, confirmed at larval stages by the presence of xanthophores, and finally reconfirmed by the PCR marker to eliminate errors in identification using pigment cells [29]. However, the *Dmy* sex marker may be used for all medaka strains, whereas the SL1 exists only in some specific strains. The FLF strain has been also used for studies of EASs [30].

See-through medaka
The see-through medaka is a fish model with a transparent body in the adult stage, as well as during embryonic stages [27]. The main internal organs, namely, heart, spleen, blood vessels, liver, gut, gonads, kidney, brain, spinal cord, lens, air bladder, and gills, in living adult fish are visible to the naked eye or with a simple stereoscopic microscope. Importantly, the see-through medaka will provide an opportunity for noninvasive studies of morphological and molecular events caused by exposure to EASs.

Zebrafish (*Danio rerio*)

This tropical species (family Cyprinidae) is native to the Indian subcontinent, is well suited to laboratory culture, and the life cycle is relatively rapid (3–4 months from embryo hatching to adult breeding at 25 ± 2 °C). Zebrafish are considered to be a gonochoristic-undifferentiated species (namely, both juvenile males and females pass through an ovary like stage before differentiating into the phenotypic sex). In males, this includes a period of juvenile hermaphroditism, followed by a histological intersex phase with both immature ovaries and testes prior to development of the mature male gonads [31]. This process of oocyte apoptosis in juvenile zebrafish has recently been reported by Uchida et al. [32]. Zebrafish are sensitive to EASs, and there is a growing body of information on the reproductive toxicology of this species [13,14].

SCREENING FOR POTENTIAL ENDOCRINE ACTIVITY

Based on current experience, we propose that the most promising approach to date is the establishment of a new screening protocol adapted from OECD test guidelines 204 and 215. The concept protocol is referred to as a "Non-spawning Fish Screening Assay" and is based upon the measurement of three core endpoints, namely vitellogenin, gonado-somatic index (GSI) and gonad histopathology [2]. For example, Panter et al. [33] reported a 21d adult fathead minnow assay measuring VTG and GSI response to estrogens. More recent work in Europe has extended this approach to juvenile fathead minnows using a variety of weak and potent antiestrogens [9]. Similarly, the Japanese Ministry of Environment [34] has successfully used a 14 to 21-day medaka screening assay to comprehensively detect benzophenone, di-cyclohexyl phthalate, di-ethyl phthalate, octylphenol, and triphenyltin chloride. Moreover, this approach has been successfully adopted for the rapid detection of xenoestrogens in sheepshead minnows [16] and zebrafish [13,14]. In addition to inducing VTG titres, xenoestrogen exposure has also been shown to reduce gonadosomatic index and cause histological changes in gonads of zebrafish after between 6 to 24 days [14]. In summary, available data indicate that the nonspawning fish screening test can be successfully in OECD fish species for the rapid detection (within 14–21 days) of a range of EASs. Other protocols that may have a future role in chemical testing programs are described below.

FISH DEVELOPMENT TEST (EXTENDED ELSs TEST)

It is now well established from aquaculture that fish early life-stages (ELSs) are sensitive to EASs. This fact, together with a reliance in regulatory ecotoxicology on the fish ELSs test for chronic hazard assessments [35] has led to the concept of a new "Fish Development Test" which can be seen as an extended ELSs test [36]. Depending on the suspected mode of action of an EAS (based on information gained in mammalian and fish screening assays), the assessment of developmental affects caused by EASs should include survival, growth, development, gonad histology, and VTG up to 90–100 days post-hatch. Demonstrations of aspects of this test concept have recently been reviewed [36]. Further details of the approach are illustrated for medaka.

Medaka development test

As medaka is a gonochoristic species (showing no hermaphroditism under natural conditions) a medaka development test could effectively address EAS-induced sex-reversal, one of the main endpoints of endocrine-disrupting effects. For example, the medaka sex-reversal assay is based on the d-rR and S-rR strains (see above). In the sex-reversal test, the sex-linked colors are unchanged by exposure to EASs but sex-reversal is identified by the gonad histology as well as the secondary sexual characteristics (dorsal and anal fins).

Fish are exposed to a chemical during their susceptible periods from the prelarva just after hatching to 28 days post-hatch and then reared for an additional 14 days in clean dilution water until their functional sexes become detectable. At the age of 42 days (at 24 °C), fish with a total length more than 20 mm are fixed and examined for secondary sexual characteristics on the dorsal fin (maximum length, and cleft depth between the last ray and preceding one) and anal fin [maximum length, length of second ray from the last, and appearance of small anal papillary processes (app) on the posterior region]. Thereafter, serial cross sections of gonads of the fish are made and observed microscopically.

The medaka sex-reversal test has been applied to a variety of chemicals, including 17α-ethinylestradiol (EE2), 17β-estradiol (E2), diethylstilbestrol (DES) 4-*t*-pentylphenol (4*t*PP), methyltestosterone (MT), and flutamide (Flu). The existence of a secondary sexual characteristic of males, small papillary processes on the anal fin, was confirmed in almost all genotypic males but not in control females (Fig. 1). In contrast, this character increased in genotypic females exposed to MT at 0.1 μg/l. No change was observed in either males or females exposed to Flu even at the highest concentration (1000 μg/l). The gonads of male and female control fish naturally differentiated to testis or ovary, respectively (Fig. 2). Gonads of genotypic males exposed to EE2, DES, 4tPP, and E2 differentiated into ovaries with LOECs of 0.032 μg/l, 0.032 μg/l, 10 μg/l, and 0.1 μg/l, respectively. On the con-

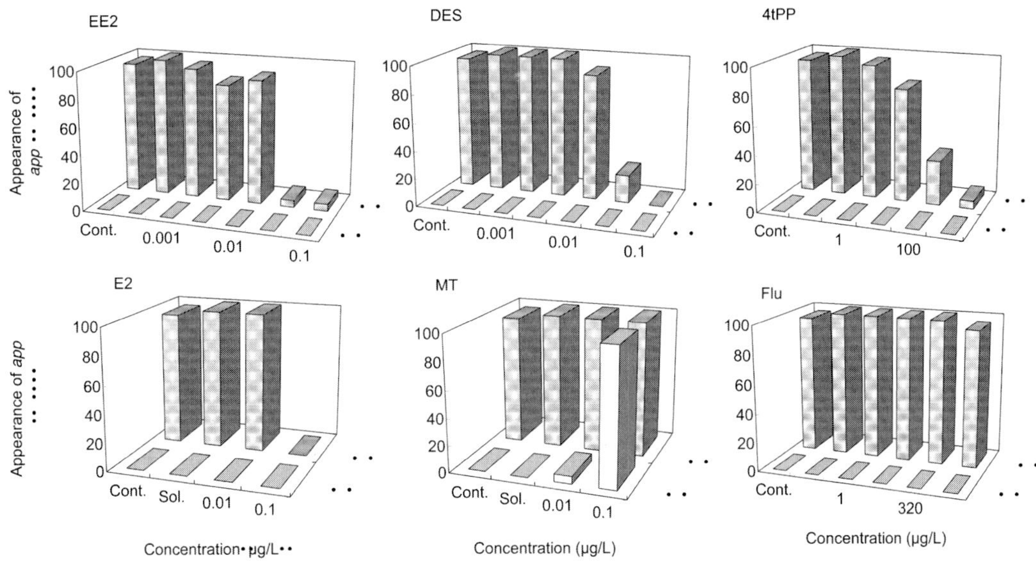

Fig. 1 Percentage of appearance of small papillary processes on anal fin of S-rR strain medaka exposed to ethynylestradiol (EE2), diethylstilbestrol (DES), 4-*t*-pentylphenol (4*t*PP), 17β-estradiol (E2), methyltestosterone (MT) and flutamide (Flu) (from Hagino et al., 2001).

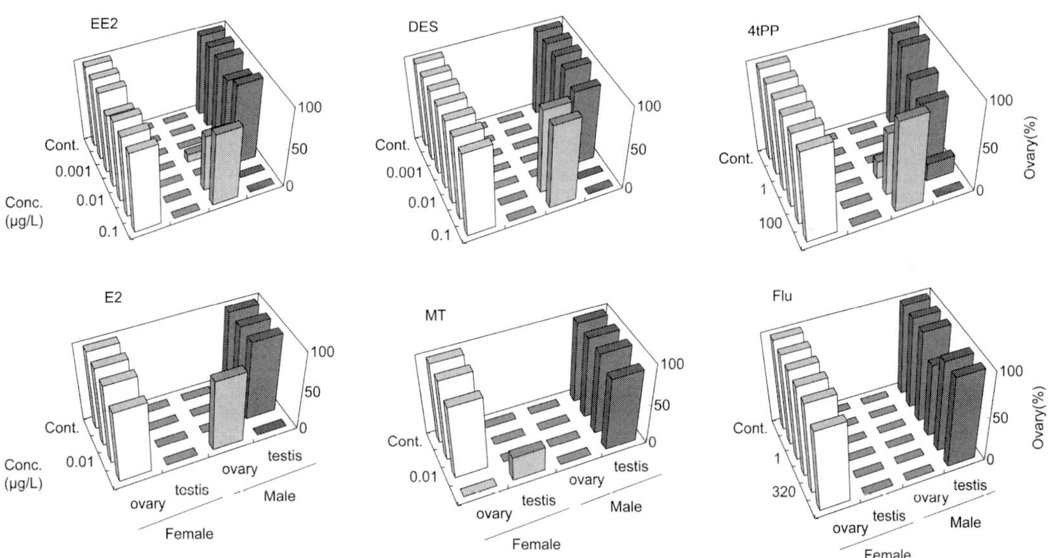

Fig. 2 Sex-reversal of gonads of S-rR strain medaka exposed to ethynylestradiol (EE2), diethylstilbestrol (DES), 4-*t*-pentylphenol (4*t*PP), 17α-estradiol (E2), methyltestosterone (MT), and flutamide (Flu) (from Hagino et al., 2001).

trary, gonads of genotypic females exposed to MT differentiated into testes at 0.1 μg/l. Gonads of both genotypic males and females exposed to Flu differentiated to testis and ovary, respectively.

Comparison between the no observable effect concentration (NOEC) or lowest observable effect concentration (LOEC) values for sex-reversal and the 96 h acute LC_{50} values of a chemical is useful to understand types of effect and mode of action. Potent (mammalian) estrogens and androgens caused sex-reversal on fish at extremely low levels, with 42d NOECs for sex-reversal being >100 000 times lower than 96 h LC_{50} values. The weak estrogenic chemicals produced slight effects on fish, with NOEC:LC_{50} ratios of up to 2600 (Table 1). Other chemicals suspected to have weak estrogenic effects had NOEC:LC_{50} ratios <100 [11]. Hagino et al. [37] reported that the androgenic action of MT was inhibited by Flu. When S-rR strain medaka were exposed to Flu at 1000 μg/l, methyltestosterone (MT) at 0.1 μg/l, or Flu at 1000 μg/l plus MT at 0.1μg/l, Flu alone induced neither estrogenic nor androgenic effects, MT alone induced sex-reversal from female to male, but no *app* were observed in any females simultaneously exposed to Flu and MT.

Table 1 Overall NOEC and LOEC (42-day values) of ethinylestradiol, diethylstilbestrol, 4-*t*-pentylphenol, 17ß-estradiol, methyltestosterone and flutamide on sex-reversal of S-rR strain medaka with acute toxicity/sex-reversal effect ratio.

Compound	Genotypic sex	96 hr-LC_{50} (μg/l)	NOEC on sex-reversal (μg/l)	LOEC on sex-reversal (μg/l)	LC_{50}/NOEC ratio	LC_{50}/LOEC ratio
Ethinylestradiol	male	1500	0.01	0.032	150 000	46 875
	female		0.1	>0.1	15 000	<15 000
Diethylstilbestrol	male	1400	0.01	0.032	140 000	43 750
	female		0.1	>0.1	14 000	<14 000
4-*t*-Pentylphenol	male	2600	1	10	2600	260
	female		1000	>1000	2.6	<2.6
17ß-Estradiol	male	3900	0.01	0.032	390 000	121 875
	female		0.1	>0.1	39 000	<39 000
Methyltestosterone	male	>10 000	0.1	>0.1	>100 000	100 000
	female		<0.01	0.01	>1 000 000	>1 000 000
Flutamide	male	3600	1000	>1000	3.6	<3.6
	female		1000	>1000	3.6	<3.6

FISH REPRODUCTION TEST

The concept of an adult fish reproduction test is seen by scientists who support the OECD work on EASs as a promising future test guideline [2]. This partial life-cycle protocol has been successfully demonstrated using the fathead minnow [6], medaka [7,38], and zebrafish [13,14]. For illustrative purposes, the approach using medaka is now outlined.

Recently, Japanese researchers have reported a 21-day medaka reproduction test adopted from the test design for fathead minnows and applied to estrogens. In this test protocol, mating pairs of reproductively mature medaka (3 to 6 month post-hatch) are exposed to EASs, and their reproductive performance (fecundity and fertility), mortality, behavior, and appearance are examined over a 21-day period. At the end of the exposure period, hepatic VTG is measured by ELISA [39] and gonads are examined microscopically.

Optionally, if potential trans-generational effects of EASs are to be assessed in medaka, the eggs spawned from females are collected and transferred to test chambers containing water with or without a test substance and then cultured until the early life stage or early mature stage (30 to 60 days post-hatch). Results from testing of estrogens by CERI suggests that the physiological and histological measures used were more sensitive to change than the fish's overt reproductive system; in other words, the elevated VTG levels and induction of testis-ova (mechanistic biomarkers) may not intrinsically be responsible for reproductive impairment (the adverse effect endpoint) [38]. Overall, these studies indicate that a 21-day medaka reproduction test can be used to detect the effects of estrogenic substances in terms of fecundity, gonadal histology, and VTG levels. More work is needed to extend this approach to a wider range of environmentally relevant EASs (including aromatase inhibitors, androgens, and antiestrogens), taking into account not only medaka but also other OECD fish species.

FISH FULL LIFE-CYCLE TEST

For EASs identified in mammalian and fish screening assays and which are expected to be continuously discharged into rivers or streams, the fish full life-cycle (FFLC) test represents the highest-tier single-species test [3]. Typically, fish are exposed to the test compound from newly fertilized (<24 h) embryo (F0 generation) to juvenile stage of the F1 generation offspring. At maturation of the F1 generation, breeding pairs are randomly selected in order to promote and record spawning activity. The endpoints analyzed in the existing FFLC study include spawning frequency, number of eggs produced, F0 fertility, viability of embryos, hatching success, growth and development (F0 and F1) [4,40,41].

To date, test substances evaluated in the medaka FFLC tests include bisphenol A (BPA), 4-nonylphenol (4-NP), E2 and EE2 (results summarized in Tables 2 and 3). In the BPA test, Yokota and coworkers did not evaluate the reproductive effects in the F0 fish. As described above, the FFLC test is proposed as a definitive test; therefore, this test must be able to quantitatively assess the concentrations of EASs at which there are developmental and reproductive effects that could lead to serious population impact. In CERI FFLC tests with these four estrogens, commonly observed effects related to their estrogenic properties were skewing of the sex ratio toward female and/or testis-ova development in the gonads, as well as decreased fecundity and/or fertility in the reproductive phase (Table 2). All of these effects appear to impair the reproduction ability of fish communities, indicating the capability of the medaka FFLC test to definitively evaluate the aquatic hazard of such EASs.

CONCLUSIONS AND RESEARCH RECOMMENDATIONS

It is clear that the numerous global observations of sexual disruption in wild fish populations have led to significant investment in fisheries research into EASs [1]. Given the wide range of testing scenarios envisaged (including single substances and complex effluents), it is proposed that the OECD effort toward validating fish test guidelines should continue to emphasize common scientific principles but remain inclusive of a modest range of important fish species: namely, fathead minnow, medaka, and zebrafish [2,36]. Four concept protocols seem useful: (1) nonspawning fish screening test; (2) fish development (extended ELSs) test; (3) fish reproduction test; and (4) fish full life-cycle test. Further priority research recommendations are outlined below:

Environmentally realistic exposures and integrated chemistry

Environmentally relevant routes of EAS exposure (primarily via the water or diet) should be used, with nominal concentrations verified by chemical analysis wherever feasible. Also, given reports of endocrine activity in commercial fish diets [42–44], these chemical analyses should not be restricted to the test substance per se but should also be applied to the fish diets. The purpose here is not to eliminate the use of such commercial fish diets but rather to gather critical baseline information on background levels of EASs (e.g., PCBs, organochlorine pesticides, phytoestrogens, steroids) that will help in the interlaboratory validation of test guidelines for OECD purposes.

Biologically relevant endpoints

Research needs to better define which endpoints can be used to directly measure the adverse effects of EASs in fish populations and therefore which endpoints should be used for setting water quality criteria. These adverse effect endpoints (suitable for the calculation of the PNEC) include fecundity, fertilization and hatching success, development, and growth. Further research is needed into the validation and interpretation of biomarkers (including vitellogenin, GSI, and gonadal histology) so as to help link data from both field and laboratory studies and help in pattern recognition across particular classes of chemicals.

Table 2 Summary of the effects observed in the FFLC test for medaka with several estrogens.

Test substance	Developmental stage										
	Parental generation (F0)						Progeny generation (F1)				
	Embryo	Larvae – Juvenile				Adult	Embryo	Larvae – Juvenile			
	Survival	Hatching	Survival	Growth	Sex diff.	Reprod.	Survival	Hatching	Survival	Growth	Sex diff.
Bisphenol A[1]	–	–	–	+ (Reduct.)	+ (Femin. and testis-ova induction)	n.t.	n.t.	n.t.	n.t.	n.t.	n.t.
17β-Estradiol[2]	–	–	+ (Reduct.)	+ (Reduct. and stimulation)	+ (Femin.)	+ (Low fecund.)	–	–	–	–	–
Ethinylestradiol[3]	–	–	–	–	+ (Femin. and testis-ova induction)	+ (Low fecund. and fertility)	+ (Reduct.)	–	–	–	–
4-Nonylphenol[4]	+ (Reduct.)	+ (Reduct.)	+ (Reduct.)	–	+ (Femin. and testis-ova induction)	+ (Low fertility)	–	–	–	–	+ (Femin. and testis-ova induction)

[1]Yokota et al. [12].
[2]Yokota [47].
[3]Yokota [48].
[4]Yokota et al. [41].
– : No adverse effects were observed.
+ : Adverse effects were observed.
n.t. : Not tested.

Table 3 Summary of LOEC determined for various developmental endpoints in the full life-cycle tests with several estrogens.

	LOEC (μg/l)			
Endpoint	Bisphenol A[1]	17β-Estradiol[2]	Ethynylestradiol[3]	4-Nonylphenol[4]
F0 generation				
Embryo survival	>1820	>0.891	>0.0306	183
Hatching	>1820	>0.891	>0.0306	183
Post-hatch survival	>1820	0.891	>0.0306	17.7
Growth	1820	0.0094	>0.0306	>51.5
Sex differentiation	1820	0.0927	0.0306	17.7
Fecundity	n.d.	0.0094	0.0101	>17.7
Fertility	n.d.	>0.0094	0.0101	17.7
Ratio of				
sex differentiation to post-hatch survival	<1	0.10	<1	1
fecundity or fertility to post-hatch survival	n.t.	0.011	<0.33	1
fecundity or fertility to sex differentiation	n.t.	0.10	0.33	1
F1 generation				
Embryo survival	n.t.	>0.0094	0.0101	>17.7
Hatching	n.t.	>0.0094	>0.0101	>17.7
Post-hatch survival	n.t.	>0.0094	>0.0101	>17.7
Growth	n.t.	>0.0094	>0.0101	>17.7
Sex differentiation	n.t.	>0.0094	>0.0101	8.2

[1]Yokota et al. [12]
[2]Yokota [47]
[3]Yokota [48]
[4]Yokota et al. [41]
n.t. : Not tested

Research beyond antiestrogens

While the focus of current debate is on the EASs affecting androgen, estrogen, and thyroid function in animals, it should be recognized that to date much of the published work on fish is limited to anti-estrogens. For example, there are relatively few data, supported by chemical analysis, quantifying the effects of antiandrogens or aromatase inhibitors on different life-stages of OECD fish species, although some aromatase data have been recently published [45,46]. The use of other fish species and novel endpoints (e.g., the spiggin biomarker for androgens) [17] suggests a potentially useful way forward.

REFERENCES

1. R. M. Rolland, M. Gilbertson, R. E. Peterson. *Chemically Induced Alterations In Functional Development and Reproduction of Fishes*, p. 194, Proceedings from the 1995 Wingspread Conference, SETAC Press, Pensacola, FL (1997).
2. OECD. *Report of the OECD Expert Consultation on Testing in Fish – EDF2*. Tokyo, March 2000. OECD, rue Andre Pascal, Paris (2000).
3. T. H. Hutchinson, R. Brown, K. E. Brugger, P. M. Campbell, M. Holt, R. Länge, P. McCahon, L. J. Tattersfield, R. van Egmond. *Environ. Health Perspect.* **108**, 1007–1014 (2000).
4. R. Länge, T. H. Hutchinson, C. P. Croudace, F. Siegmund, H. Schweinfurth, P. Hampe, G. H. Panter, J. P. Sumpter. *Environ. Toxicol. Chem.* **20**, 1216–1227 (2001).

5. A. C. Nimrod and W. H. Benson. *Aquat. Toxicol.* **44**, 141–156 (1998).
6. G. T. Ankley, K. M. Jensen, M. D. Kahl, J. J. Korte, E. A. Makynen. *Environ. Toxicol. Chem.* **20**, 1276–1290 (2001).
7. T. Shioda and M.Wakabayashi. *Water Sci. Technol.* **42**, 53–60 (2000).
8. K. L. Thorpe, T. H. Hutchinson, M. J. Hetheridge, J. P. Sumpter, C. R. Tyler. *Environ. Toxicol. Chem.* **19**, 2812–2820 (2000).
9. G. H. Panter, T. H. Hutchinson, R. Länge, C. Lye, J. P. Sumpter, M. Zerulla, C. R. Tyler. *Environ. Toxicol. Chem.* **21**, 319–326 (2002).
10. C. R. Tyler, R. van Aerle, T. H. Hutchinson, S. Maddix, H. Tripp. *Environ. Toxicol. Chem.* **18**, 337–347 (1999).
11. S. Hagino, M. Kagoshima, S. Ashida. *Environ. Sci.* **8**, 75–87 (2001).
12. H. Yokota, Y. Tsuruda, M. Maeda, Y. Oshima, H. Tadokoro, A. Nakazono, T. Honjo, K. Kobayashi. *Environ. Toxicol. Chem.* **19**, 1925–1930 (2000).
13. K. Van den Belt, R. Verheyen, H. Witters. *Arch. Env. Contam. Toxicol.* **41**, 458–467 (2001).
14. K. Van den Belt, P. W. Wester, L. Van de Ven, R. Verheyen, H. Witters. *Environ. Toxicol. Chem.* **21**, 767–775 (2002).
15. M. Bayley, M. Junge, E. Baatrup. *Aquat. Toxicol.* **56**, 227–239 (2002).
16. M. J. Hemmer, B. L. Hemmer, C. J. Bowman, K. J. Kroll, L. C. Folmar, D. Marcovich, M. D. Hoglund, N. D. Denslow. *Environ. Toxicol. Chem.* **20**, 336–343 (2001).
17. I. Katsiadaki, A. P. Scott, M. R. Hurst, P. M. Matthiessen, I. Mayer. *Environ. Toxicol. Chem.* **21**, 1946–1954 (2002).
18. S. Halm, M. Rand-Weaver, J. P. Sumpter, C. R. Tyler. *Fish. Physiol. Biochem.* **24**, 49–62 (2001).
19. K. M. Jensen, J. J. Korte, M. D. Kahl, M. S. Pasha, G. T. Ankley. *Comp. Biochem. Physiol. – Toxicol. Pharmacol.* **128**, 127–141 (2001).
20. M. D. Kahl, K. M. Jensen, J. J. Korte, G. T. Ankley. *J. Fish Biol.* **59**, 515–523 (2001).
21. K. Ozato and Y. Wakamatsu. *Develop. Growth Differ.* **36**, 437–443 (1994).
22. Y. Ishikawa. *Bioessays* **22**, 487–495 (2000).
23. M. Matsuda, Y. Nagahama, A. Shinomiya, T. Sato, C. Matsuda, T. Kobayashi, C. E. Morrey, N. Shibata, S. Asakawa, N. Shimizu, H. Hori, S. Hamaguchi, M. Sakaizumi. *Nature* **417**, 559–563 (2002).
24. M. A. Gray, K. L. Teather, C. D. Metcalfe. *Environ. Toxicol. Chem.* **18**, 2587–2594 (1999).
25. H. Wada, A. Shimada, S. Fukamachi, K. Naruse, A. Shima. *Zool. Sci.* **15**, 123–126 (1998).
26. Y. Wakamatsu. *Environ. Sci.* **9**, 419–426 (2002).
27. M. Matsuda, T. Kusama, T. Oshiro, Y. Kurihara, S. Hamaguchi, M. Sakaizumi. *Genes Genet. Syst.* **72**, 263–268 (1997).
28. Y. Wakamatsu, S. Pristyazhnyuk, M. Kinoshita, M. Tanaka, K. Ozato. *Proc. Natl. Acad. Sci. USA* **98**, 10046–10050 (2001).
29. S. Hatakeyama, Y. Sugaya, H. Takagi, E. Ishikawa, K. Ozato, Y. Wakamatsu. *Jpn. J. Environ. Toxicol.* **4**, 99–111 (2001).
30. H. Takahashi. *Bull. Fac. Fish. Hokkaido Univ.* **28**, 57–65 (1977).
31. D. Uchida, M. Yamashite, T. Kitano, T. Iguchi. *J. Exp. Biol.* **205**, 711–718 (2002).
32. G. H. Panter, R. S. Thompson, J. P. Sumpter. *Aquat. Toxicol.* **42**, 243–253 (1998).
33. Japanese Ministry of Environment. *Results of Assays and Tests In Evaluation of The Endocrine Disrupting Activities – Rodents and Fish.* Presented at the OECD EDTA6 meeting, Tokyo, 24–25 June 2002.
34. J. M. McKim. *J. Fish. Res. Bd. Can.* **34**, 1134–1154 (1977).
35. T. H. Hutchinson. Impacts of endocrine disrupters on fish development: opportunities for adapting OECD Test Guideline 210. *Environ. Sci.* **9**, 439-450 (2002).
36. S. Hagino, M. Kagoshima, S. Ashida, S. Hosokawa. *Environ. Sci.* **9**, 475–482 (2002).

37. I. J. Kang, H. Yokota, Y. Oshima, Y. Tsuruda, T. Yamaguchi, M. Maeda, N. Imada, H. Tadokoro, T. Honjo. *Chemosphere* **47**, 71–80 (2002).
38. H. Yokota, H. Morita, N. Nakano, I. J. Kang, H. Tadokoro, Y. Oshima, T. Honjo, K. Kobayashi. *Jpn. J. Environ. Toxicol.* **4**, 87–98 (2001).
39. H. Bresch, H. Beck, D. Ehlermann, H. Schlaszus, M. Urbanek. *Arch. Env. Contam. Toxicol.* **19**, 419–427 (1990).
40. H. Yokota, M. Seki, M. Maeda, Y. Oshima, H. Tadokoro, T. Honjo, K. Kobayashi. *Environ. Toxicol. Chem.* **20**, 2552–2560 (2001).
41. C. Pelissero and J. P. Sumpter. *Aquaculture* **107**, 283–301 (1992).
42. M. Kobayashi, H. Ishibashi, T. Moriwaki, T. Koshiishi, S. Ogawa, T. Matsumoto, K. Arizono, S. Watabe. *Environ. Sci.* 157 (2002).
43. T. Matsumoto, M. Kobayashi, S. Kawai, T. Moriwaki, S. Watabe. Estrogenic activity in fish diet determined by yeast estrogen-screen assay. *Environ. Sci.* 157 (2002).
44. G. T. Ankley, M. D. Kahl, K. M. Jensen, M. W. Hornung, J. J. Korte, E. A. Makynen, R. L. Leino. *Toxicol. Sci.* **67**, 121–130 (2002).
45. M. Zerulla, R. Länge, T. Steger-Hartmann, G. H. Panter, T. H. Hutchinson, D. R. Dietrich. *Toxic. Lett.* **131**, 51–63 (2002)
46. H. Yokota. *Proceedings of the International Symposium Endocrine-Disrupting Substance Testing in Medaka*, Nagoya, pp. 111–112, Japan, 17–20 March 2000.
47. H. Yokota. *Report of the International Symposium on Environmental Endocrine Disrupters 2000*, pp. 323–330, Yokohama, Japan, 16–18 December 2000.

Pure Appl. Chem., Vol. 75, Nos. 11–12, pp. 2355–2360, 2003.

Topic 4.13
Endocrine disruption in wildlife: The future?*

John P. Sumpter

Department of Biological Sciences, Brunel University, Uxbridge, Middlesex, UB8 3PH, UK

Abstract: Probably the only thing that can be said with certainty about the future of this field of ecotoxicology is that predicting it is foolish; the chances of being right are very slim. Instead, it seems to me likely that unexpected discoveries will probably have more influence on the field of endocrine disruption than the outcomes of all the planned experiments. It is certainly true that chance discoveries, such as masculinized fish in rivers receiving paper-mill effluent, imposex in molluscs due to exposure to tributyltin, and feminized fish in rivers receiving effluent from sewage-treatment works, have been pivotal in the development of the field of endocrine disruption in wildlife. I consider that further such discoveries are likely, but I do not know which species will be affected, what effects will be found, what chemical(s) will be the cause, or what endocrine mechanism(s) will underlie the effects. The recent realization that many pharmaceuticals are present in the aquatic environment only underscores the range of effects that could, in theory at least, occur in exposed wildlife. What is somewhat easier to predict is the research that will be conducted in the immediate future, which will build upon what is known already. For example, it is clear that wildlife is rarely, if ever, exposed to single chemicals, but instead is exposed to highly complex, ill-defined mixtures of chemicals, including many that are endocrine active in various ways. We need to understand much better how chemicals interact, and what overall effects will occur upon exposure to such mixtures. We also need to move from assessing effects at the individual organism level, to understanding the consequences of these effects at the population level. Then, we need to determine the significance of any population-level effects due to endocrine disruption in comparison with the impact of many other significant stressors (e.g., over-exploitation, habitat loss, climate change) that also negatively impact wildlife. Such research will be difficult, and time-consuming, and will probably produce many surprises. All I can be fairly certain about is that the next few years are likely to be as interesting and exciting as the last few have been.

INTRODUCTION

While thinking about this chapter, it became apparent to me that I was finding it much more difficult trying to predict the future than I do writing research papers based on results. Many years ago I read that predicting the future was foolish, because unanticipated events usually have a great deal of impact, and influence things in ways that cannot be predicted. At best, knowledge of the past perhaps makes the future less surprising than it otherwise might have been. It is certainly the case that, ten years ago, when I first became involved in the field of endocrine disruption, I had no idea that it would grow as it has,

*Report from a SCOPE/IUPAC project: Implication of Endocrine Active Substances for Human and Wildlife (J. Miyamoto and J. Burger, editors). Other reports are published in this issue, *Pure Appl. Chem.* **75**, 1617–2615 (2003).

to become a major field of biological sciences in its own right. This failure of mine to predict the enormous growth in activity in the field over the last ten years [1] does not bode well for my attempts in this chapter to predict the next ten years. Nevertheless, I will try and use my knowledge and experiences gained over the last ten years to, if not accurately predict future developments in the field of endocrine disruption (the chances of being correct are, I think, very slim), perhaps provide some pointers as to where I think major advances might be made.

Given the history of the development of the field of endocrine disruption, it seems to me likely that unexpected discoveries will probably have more influence on the field than the outcomes of all carefully planned experiments. It is certainly true that chance discoveries, such as the presence of masculinized fish in rivers receiving effluents from paper and pulp mills [2], imposex in many species of molluscs due to exposure to tribulyltin (TBT) emanating from antifouling paints used on ships [3], and feminized fish in rivers receiving effluent from sewage-treatment works (STW) [4,5] have been pivotal in the development of the field of endocrine disruption in wildlife. I consider that further such discoveries are likely, but I do not know what species will be affected, what effects will be found, what chemical(s) will be the cause, or what disrupted endocrine mechanisms will underlie the effects. The recent realization that many pharmaceuticals are present in the aquatic environment [6], many of which are intended to alter the endocrine status of the recipient (intended or otherwise), only underscores the range of effects that could, at least in theory, occur in exposed wildlife.

It is probably easiest to predict (or at least try to predict) the immediate future—say the next year or two—less easy to predict reliably the intermediate future—say the next five to ten years, and probably impossible to predict the distant future. Hence, I will begin by speculating on possible developments in the near future, and finish with some more, very speculative, ideas about the more distant future.

THE NEXT FEW YEARS

One thing it is possible to say with a fair degree of certainty is that we will learn a lot more in the next few years about issues that we already know something about, but presently lack detail and depth of knowledge. One of my colleagues recently described this phenomenon as the field getting fatter! We need to do this if we are to put endocrine disruption in wildlife onto a solid foundation.

Let me provide an example. It is now well established that intersex roach (a small member of the carp family of fishes native to much of northern Europe) are widespread in British rivers [5]. In some rivers all of the "males" are, in fact, intersex. Further, intersexuality seems to be associated with the presence of STW effluents. Recent research has demonstrated that intersex fish can also be found in polluted areas of the marine environment [7,8], and that they are not confined to the United Kingdom; intersex fish occur in Spain, [9], Germany, [10], and Italy [11]. Hopefully, research presently underway will better define exactly where intersex fish are found (in what countries, and at which sites in these countries), and how badly affected the fish are. Close scrutiny of the "testes" of these fish, perhaps using modern techniques such as immunohistochemistry and gene arrays, should help not only to define the problem, but also provide information about the mechanisms underlying the intersexuality. Thus, within the foreseeable future I would hope that we will have a much clearer picture concerning the worldwide prevalence of this example of endocrine disruption in wildlife, and perhaps also have a better idea as to exactly what developmental processes are affected, how, and when. This building on our rather flimsy present state of knowledge will provide the more solid foundations required to enable us to progress with more confidence, and in unity. All new areas of research develop in this manner; initially surprising results are built upon (or refuted) by subsequent research, which provides the solid foundation to progress with confidence.

A second example of how the field may progress in the next few years concerns the identification of the causes of the examples of endocrine disruption reported in wildlife. However, I should say that I feel less confident about trying to predict the future of this aspect of endocrine disruption, because I consider it a very difficult one. Generally speaking, it is very difficult to confidently identify the chemical, or chemicals, causing a particular effect on wildlife. This is because a very large number (around 100 000) man-made chemicals are in everyday use, most of which will reach the environment one way or another. Here they will degrade, to varying extents, and at different speeds, to a wide range of intermediate chemicals. We know very little about these processes. Hence, we know little about the exposure of wildlife to chemicals; ignorance outweighs knowledge at every stage of the process.

Occasionally, a specific effect can be closely associated with the use of a particular chemical, greatly aiding establishing cause and effect. Such was the case with imposex in molluscs and TBT, although it still took ten years to link the two. More often, however, circumstances do not allow such links to be established, in which case other approaches need to be undertaken if the causative chemicals are to be identified; such was the case when a toxicity identification and evaluation (TIE) approach was used to show that the main estrogenic chemicals in STW effluents of domestic origin were natural and synthetic sex steroids [12]. However, in other well known examples of endocrine disruption in wildlife, such as the "feminized" alligators in some lakes in Florida, it has proved much more difficult to identify the causative chemical(s). Much more research is needed in this area, if we are to be able to find solutions to the problems that have been identified in wildlife (see later). I predict, and I hope I'm not wrong with this one, that significant advances will be made in this area in the forseeable future. Perhaps different approaches, such as utilizing bile as a "sink" of chemicals to which the organism has been exposed [13], will aid progress. Chemical analysis of adipose tissue (fat), which as an approach hasn't been used much yet in studies of endocrine disruption in wildlife, might also pay dividends. Despite the undoubted difficulties, future research must address the issue of causality, because without it little progress in resolving the problem can be achieved.

So, to summarize my views on the immediate future, I think most research (much of which will already be underway) will build upon existing issues within endocrine disruption. Hence, we should learn much more about these existing issues. By this I do not mean to imply that this "follow-up" research will be in any way inferior to that which produced the original findings. It may not be so novel, but it should (and will, I hope) be of high quality. Without it endocrine disruption (with few exceptions, such as imposex in molluscs) will remain on shaky ground, and be open to criticism.

THE INTERMEDIATE FUTURE

Where will we be ten years from now? Of course I don't know. However, I do feel confident enough to say that technological advances will occur, will be applied to the field of endocrine disruption, and the results gained will shed new light on the field. I will attempt to predict how two or three of these advances (technological and theoretical) will influence the field.

It is very clear to everyone that this is the era of molecular biology. However, molecular approaches have yet to make much of an impression in the field of endocrine disruption (especially of wildlife, partly because of our lack of knowledge of their molecular physiology), although there are recent signs that this position is changing. For example, Nancy Denslow's group in the United States has begun to study gene expression in fish exposed to endocrine-disrupting chemicals [14]. The rapid spread of microarray technology into comparative physiology (using laboratory animals such as the zebrafish) will certainly allow a quantum leap forward to be made in our understanding of how chemicals affect gene expression. It is likely that tens, if not hundreds, of genes will have their expression altered when an organism is exposed to an endocrine-disrupting chemical. The problem will then be to interpret the multitude of data in a meaningful way, in order to improve our understanding of what adverse effects are caused by chemicals, and what the consequences of these effects are. This area of research, now called toxicogenomics, may well be a major area of interest in the next few years. Presently it is difficult to see how toxicogenomics can be applied to wildlife, when we know little or nothing about the genome of the organisms of interest. However, I may well be surprised by how rapidly advances in technology and our understanding of the genomics of wildlife allow microarray technology to be applied to wildlife, to address fundamental issues about how chemicals in the environment cause changes in gene expression: Will different chemicals produce distinct "footprints"? It may even become possible to ascertain if persistent changes in the environment (such as the presence of estrogenic chemicals due to input of STW effluents) create a selective pressure strong enough to lead to altered genotypes, a type of environmental genomics. Put another way, is it possible that fish exposed to estrogenic effluent for many generations evolve the ability to be less adversely affected by the effluents, and hence reproduce more successfully, than would naive (unadapted) fish? Such questions may well become answerable, and hopefully answered, in the foreseeable future.

Besides technological advances, there are likely to be theoretical advances that lead to significant advances in our understanding of endocrine disruption in wildlife. One example will probably be improved predictability of the effects of mixtures of endocrine active substances (EASs). As already stated, the environment usually contains not one or two, but many, EASs. Thus, wildlife are almost always exposed to complex (presently usually ill-defined) mixtures of chemicals, including many EASs with various types of activity (estrogenic, antiandrogenic, etc.). How will organisms respond to these complex mixtures, and will it be possible, based on sound scientific principles, to understand these responses of organisms well enough to be able to accurately predict them? If this could be achieved, regulators could introduce combination effects into legislation, and hence better protect wildlife from the effects of EASs.

Recently, biomathematical concepts initially developed decades ago have been applied to the field of endocrine disruption, in collaboration between ecotoxicologists and biomathematicians/biostatisticians. Relatively simple experiments, using binary mixtures in vivo [15], and somewhat more complex mixtures in vitro [16], have shown that it is possible to accurately predict the effects of these (admittedly fairly simple) mixtures of EASs. Future research needs to expand this work, to include more complex (and environmentally realistic) mixtures, both in terms of the number of chemicals in the mixtures tested, and the variety of endocrine activities. This research cannot proceed without the collaboration of scientists from different disciplines, especially ecotoxicologists, physiologists, and biomathematicians. It will advance not only our knowledge of the effects of mixtures of EASs, but also lead to major advances in the mathematical and statistical principles underpinning the effects and their predictability. Major gains in risk assessment procedures should follow.

The examples above illustrate that, in the future, major advances in our understanding of endocrine disruption may well require the collaboration of scientists with quite distinct areas of expertise. Research in the area of biological sciences has become progressively more multidisciplinary, involving larger numbers of researchers (witness the genome sequencing projects of the last few years), and I think this movement will also be required in endocrine-disruption research if it is to progress significantly. For example, one of the key questions to address is whether the effects observed in individual organisms (intersexuality, impaired or inappropriate reproductive behavior, elevated vitellogenin concentrations, etc.) lead to adverse population-level effects. Probably only in the case of TBT and imposex in molluscs has it been shown that endocrine disruption can lead to population-level effects (local extinctions, in that particular case). In all other cases, we know much less about the consequences of the endocrine disruption. Most often we do not even know if the effects reported are adverse, in the sense that they reduce the fitness of the affected individual (for example, reproduction could be impaired or even prevented). Only very recently [17] has it been possible to demonstrate that intersexuality in fish does lead to adverse effects at the individual level, in this case to reduced fertility. But, this does not help us in assessing whether populations of fish are threatened, which is what we really want to know. We could, of course, sit back and wait to find out, but this would not be a very responsible strategy. An alternative approach is to model the possible consequences at the population level, which would require fish physiologists, population geneticists, chemists, and population modellers to work together to develop and test the predictive models. I hope that in the future such groups of diverse scientists will form, to tackle some of the key outstanding questions. Up until now, only scientists with similar expertise tend to work together, which usually means that only specific, narrow issues within the wider field are addressed.

Funding agencies have recently become aware of this problem of scientists working in isolation and are beginning to use their clout to "encourage" scientists to work together more. I support this movement and predict that it will become much more common for groups (sometimes large groups) of research scientists, incorporating wide ranges of expertise, to form to address a question or hypothesis that cannot be tackled by small projects. Many of the most important presently unanswered questions in the field of endocrine disruption (are there population-level effects; can the effects of complex mixtures be reliably predicted; what are wildlife exposed to, etc.?) will require broadly based approaches if they are to be addressed successfully.

THE DISTANT FUTURE

I said at the beginning of this chapter that it was foolish to try and predict the future, and that the further one projects into the future, the more unreliable (and plain wrong) these predictions are likely to be. Therefore, I do not propose to try and guess where the field of endocrine disruption will be in ten plus years, other than to say that I think the field will not have gone away (as some suggested when it expanded rapidly, sometimes on not much more than hype, in the 1990s). I hope that the field will mature into a major area of research within biological sciences, incorporating such diverse disciplines as ecology, animal physiology, hydrology, molecular genetics, and chemistry. In many respects, endocrine disruption can be seen as the issue that opened our eyes to the unexpected effects that chemicals released into the environment can have on wildlife. It taught us how little is known about what chemicals are present in the environment (especially the aquatic environment), how they behave in the environment, and what effects (if any, of course) they have on our fauna. We are in a state of ignorance, which is why we shouldn't be surprised when unexpected discoveries, such as the worldwide effects of TBT on molluscs, are reported. More will surely follow. Once one realizes that around 100 000 man-made chemicals are in everyday use, and entering our environment, where thousands of species of animals live, most of which have hardly been studied, if studied at all, it is difficult not to conclude that some surprises await discovery. As we learn more about the endocrine systems of organisms (new hormones are still regularly discovered, even in that most studied of species, *Homo sapiens*), it becomes apparent that chemicals could perturb it in many ways. Perhaps in the future the focus will shift from chemicals adversely affecting reproduction to those affecting other physiological processes—most are, after all, controlled by hormones.

Perhaps I can conclude the serious side of this chapter with a note of hope and optimism (after being rather negative so far). As pointed out in Peter Matthiessen's historical perspective chapter (Topic 4.1), many thousands of research papers concerned with endocrine disruption in wildlife have already been published, and many more will surely follow. The vast majority (probably 99 % or more) are concerned, in a multitude of different ways, with describing the problem. Of course this is probably appropriate: we need to understand the problem first and foremost. However, as researchers charged with sustaining the quality of our environment, we have a duty to provide potential solutions to problems, not only just describe one problem after another. I would like to see, in the future (but soon!) more emphasis being placed on finding solutions to the problems associated with endocrine disruption in wildlife. There are encouraging signs that this is occurring already, albeit not to a great extent. For example, improvements in treatment processes at STW would aid the removal of endocrine-disrupting chemicals, and hence reduce the concentrations of these chemicals to which aquatic wildlife were exposed [18]. In addition, other researchers are engaged in developing and applying new technologies to improve the performance of STW, such as the use of titanium oxide catalysts to remove endocrine-disrupting chemicals from effluents [19].

Finally, please do not judge me in ten years time on whether or not I was successful in predicting the future of this field of research. I am not a clairvoyant, and do not believe people can predict the future. However, I look forward to the next ten years of research in this still very young field of science, and to the many surprises it will undoubtedly bring. I also look forward to that research improving the quality of our environment for the myriad species that live there, so that their lives can also be improved.

REFERENCES

1. P. Matthiessen. *Pure Appl. Chem.* **75** (11/12), 2197–2206 (2003).
2. W. M. Howell, D. A. Black, S. A. Bortone. *Copeia* 676–681 (1980).
3. B. S. Smith. *Proc. Malacol. Soc. London* **39**, 377 (1971).
4. C. E. Purdom, P. A. Hardiman, V. J. Bye, N. C. Eno, C. R. Tyler, J. P. Sumpter. *Chem. Ecol.* **8**, 275–285 (1994).
5. S. Jobling, M. Nolan, C. R. Tyler, G. Brighty, J. P. Sumpter. *Environ. Sci. Technol.* **32**, 2498–2506 (1998).
6. T. Ternes. *Water Res.* **32**, 3245–3260 (1998).
7. Y. Allen, A. P. Scott, P. Matthiessen, S. Haworth, J. E. Thain, S. Feist. *Environ. Toxicol.Chem.* **18**, 1791–1800 (1999).
8. S. Hashimoto, H. Bessho, A. Hara, M. Nakamura, T. Iguchi, K. Fujita. *Mar. Environ. Res.* **48**, 1–17 (1999).
9. M. Petrovic, M. Solé, M. López de Aldo, D. Barceló. *Environ. Toxicol. Chem.* **21**, 2146–2156 (2002).
10. J. Gerken and H. Sordyl. *Mar. Environ. Res.* **54**, 651–655 (2002).
11. L. Viganò, A. Arillo, S. Bottero, A. Massari, A. Mandich. *Sci. Total Environ.* **269**, 189–194 (2001).
12. C. Desbrow, E. J. Routledge, G. Brighty, J. P. Sumpter, M. Waldock. *Environ. Sci. Technol.* **32**, 1549–1558 (1998).
13. D. G. J. Larsson, M. Adolfsson-Eric, J. Parkkonen, M. Petterson, A. H. Berg, P. E. Olsson, L. Forlin. *Aquat. Toxicol.* **45**, 91–97 (1999).
14. N. D. Denslow, C. J. Bowman, R. J. Ferguson, H. S. Lee, M. J. Hemmer, L. C. Folmar. *Gen. Comp. Endocrinol.* **121**, 250–260 (2001).
15. K. L. Thorpe, T. H. Hutchinson, M. J. Hetheridge, M. Scholze, J. P. Sumpter, C. R. Tyler. *Environ. Sci. Technol.* **35**, 2476–2481 (2001).
16. E. Silva, N. Rajapakse, A. Kortenkamp. *Environ. Sci. Technol.* **36**, 1751–1756 (2002).
17. S. Jobling, S. Coey, J. G. Whitmore, D. E. Kime, K. J. W. Van Look, B. G. McAllister, N. Beresford, A. C. Henshaw, G. C. Brighty, C. R. Tyler, J. P. Sumpter. *Biol. Reprod.* **67**, 515–524 (2002).
18. A. C. Johnson and J. P. Sumpter. *Environ. Sci. Technol.* **35**, 4697–4703 (2001).
19. Y. Ohko, K.-I. Iuchi, C. Niwa, T. Tatsuma, T. Nakashima, T. Iguchi, Y. Kubota, A. Fujishima. *Environ. Sci. Technol.* **36**, 4175–4181 (2002).

WORKSHOP 1

EFFECTIVENESS OF QSAR FOR PRESCREENING OF ENDOCRINE DISRUPTOR HAZARD

Pure Appl. Chem., Vol. 75, Nos. 11–12, pp. 2363–2373, 2003.

Workshop 1.1

Mechanism-based QSAR approach to the study of the toxicity of endocrine active substances*

C. Dias Selassie‡, R. Garg, and S. Mekapati

Chemistry Department, Pomona College, Claremont, CA 91711, USA

Abstract: Mechanism-based QSAR models for interactions between various ligands and the estrogenic receptor are developed by using well-developed physicochemical parameters. Common features of these QSARs are identified, and deficiencies in some datasets are highlighted. The relative binding affinities of various substituted hexestrols to estrogen receptors are examined in terms of their steric, electronic, and hydrophobic attributes. Different QSARs for hexestrols and tamoxifens reveal that steric effects are of overriding importance in variations of binding affinity. In the few cases where a large number of diverse substituents are located on aromatic rings, electronic effects emerge and suggest that electron-donating groups enhance binding to the receptor while hydrophobicity plays a marginal role in decreasing binding affinity. With substituted phenols bearing alkyl-type substituents and substituted hydroxy-biphenyls, the binding is strictly dependent on hydrophobicity and size. These QSAR models are described in detail and examined together to illustrate the utility of lateral validation in mechanistic interpretation.

INTRODUCTION

The realization that agents from the environment of both synthetic and natural origin can produce reversible or irreversible biological effects in individuals or populations by interfering with the synthesis, transport, distribution, and binding of endogenous hormones has led to extensive investigations of these endocrine active substances (EASs) [1]. Many EASs are estrogenic in nature (xenoestrogens) and exert agonist and antagonist effects via the estrogen receptor (ER), which functions as a ligand-activated transcriptional regulator [2]. Those that are synthetic include pesticides, food antioxidants, and metabolites of nonionic surfactants; and naturally occurring ones are plant secondary metabolites and mold metabolites [3]. Mechanistic toxicological studies of environmental pollutants have sparked a great deal of scientific interest in the chemico-biological interactions between the ER and various EASs [4].

The general mechanisms for receptor-mediated responses are numerous and include binding to the receptor at the cell surface, cytoplasm, or nucleus. The interaction with a specific DNA sequence (estrogen response elements) and other coregulatory proteins results in alterations in gene transcription and ultimately changes in regulation of cell proliferation and differentiation. Thus, the mechanism of action of EASs is complex and multipronged. It could involve inhibition of hormone synthesis, metabolism, and gene activation.

The ER, which is the target of a large number of EASs, is a member of the nuclear hormone receptor gene superfamily. It is characterized by modular structural organization with distinct domains

*Report from a SCOPE/IUPAC project: Implication of Endocrine Active Substances for Human and Wildlife (J. Miyamoto and J. Burger, editors). Other reports are published in this issue, *Pure Appl. Chem.* **75**, 1617–2615 (2003).

‡Corresponding author

that are associated with DNA binding, hormone binding, and *trans*-activation [2]. There are two ER isoforms; ERα and ERβ, which exhibit some similar, but mostly distinct, tissue distribution patterns and vary in their ligand-binding ability [5]. Compounds with distinct and diverse structures bind to ER with varying degrees of affinity and potency [5,6]. 17β-Estradiol acts as a pure agonist, while other compounds such as Faslodex (ICI182780) act as antagonists [7], and still others, termed selective ER modulators, demonstrate both agonist and antagonist activity. Raloxifene and tamoxifen fall into the latter category. The diversity in structure and physicochemical properties of these various ER-binding ligands warrants a thorough examination of their binding affinities in terms of their electronic, steric, and hydrophobic attributes. Utilization of the quantitative structure–activity relationship (QSAR) paradigm allows for delineation of the structural attributes that determines affinity in binding [8].

Our mechanism-based approach to the study of the interactions of various ligands in cellular systems and with isolated estrogen receptors utilizes well-developed physicochemical parameters to develop appropriate models, which take on added significance when subjected to the process of lateral validation. Thus, an individual QSAR is considered to be weak when on its own and significant when bolstered by other similar QSAR. The availability of a sophisticated database, which allows for lateral validation of a biological QSAR from a physical organic viewpoint, as well as added biological perspective, makes this approach useful and amenable to mechanistic interpretation [9]. Specific examples of binding interactions between different types of estrogen receptors (mouse, rat, lamb, calf, human) and their nonsteroidal ligands will now be described. Common features of these QSARs will be pinpointed, and inadequacies in some data sets will be identified.

METHODOLOGY

All physicochemical descriptors are auto-loaded, and multiple regression analysis is used to derive the QSARs that are executed with the C-QSAR program [9]. While comparing different QSARs, however, it must be borne in mind that variations in quality of testing in different laboratories will have an effect that cannot be estimated.

The parameters used in this article have been discussed in detail along with their applications [8]. In brief, Hammett σ, σ^+, and σ^- constants are electronic parameters that apply to substituent effects on aromatic system. Normal σ values are utilized for substituents on aromatic systems where resonance between substituent and reaction center is minimal. σ^+ and σ^- are employed where there is a strong resonance interaction between substituent and reaction center.

Clog *P* represents the calculated partition coefficient in octanol/water and is a measure of hydrophobicity of the whole molecules. π is the hydrophobic parameter for substituents attached to benzene. CMR is the calculated molar refractivity for the whole molecule. MR is calculated using the Lorentz–Lorenz equation and is described as follows: $[(n^2 - 1)/(n^2 + 2)](MW/d)$, where n is the refractive index, MW is the molecular weight, and d is the density of a substance. Since there is a little variation in n, MR is largely a measure of volume with a small correction for polarizability. MR values are generally scaled by 0.1. MR can be used for a substituent or for the whole molecule. Clog *P* and CMR are calculated values for the neutral form of partially ionized molecules. B1, B5, and L are Verloop's STERIMOL parameters for substituents. B1 is a measure of the width of the first atom of a substituent, B5 is an attempt to define the overall volume, and L is the substituent length. All the biological data have been collected from the literature (see individual QSAR for respective references).

RESULTS

Metahexestrol and hexestrol derivatives

Metahexestrol and hexestrol derivatives were synthesized by Hartmann et al. [10,11], and their relative binding affinities (RBA) to estrogen receptors from calf uterine cytosol at 4 °C were determined. QSARs 1 and 2 were derived for di-X-metahexestrol **1** and di-X-hexestrol **2** derivatives, respectively.

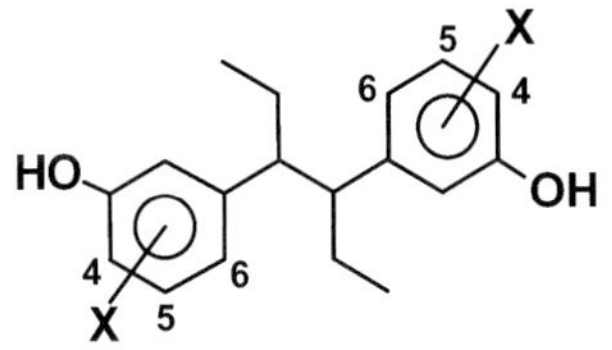

*RBA of di-X-metahexestrols **1** to calf uterine cytosol ER at 4 °C/16 h*

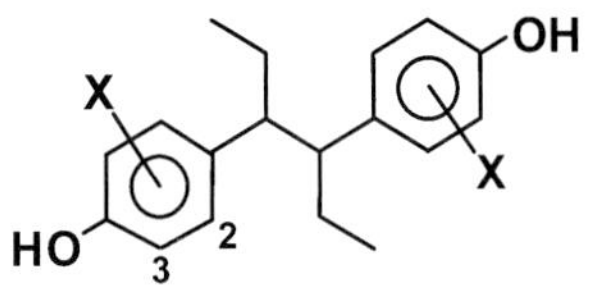

$$\text{Log RBA} = -0.90(\pm 0.32)\,\sigma_X^{+} - 1.26(\pm 0.12)\,L_{4X} - 2.73(\pm 0.33)\,B5_{5X} + 6.13(\pm 0.61) \qquad (1)$$

$n = 13$; $r^2 = 0.989$; $s = 0.141$; $q^2 = 0.969$; *outliers*: X = 4-CH_2OMe and 4-NH_2.

*RBA of di-X-hexestrols **2** to calf uterine cytosol ER at 4 °C/16 h*

$$\text{Log RBA} = -0.65(\pm 0.28)\,\sigma_X - 1.40(\pm 0.27)\,MR_{3X} - 1.01(\pm 0.54)\,MR_{2X} + 1.41(\pm 0.26) \qquad (2)$$

$n = 20$; $r^2 = 0.909$; $s = 0.281$; $q^2 = 0.864$; *outlier:* none

In these and following QSARs, n is the number of compounds included in the analyses, r is the correlation coefficient, s is the standard deviation, and q is the predictive (leave-one-out) correlation coefficient. The figures in parentheses are the 95 % confidence interval of regression coefficients and the intercept. The subscripts attached to parameter symbols represent substituents and/or their position. Substituents examined in the above two series are on the phenolic rings. In QSAR 1, X varies from H to such substituents as 4-F, -Cl, -Br, -Me, -Et, -NO_2, -CH_2OH, 5-OH, -Cl, 6-F, -Cl, and -Me; log RBA varies in a range of –2.00 (X = 5-OH) to 1.18 (6-Me). In QSAR 2, X is one of substituents such as H, 3-OH, -F, -Cl, -Br, -I, -Me, -Et, -CH_2NMe_2, -CH_2OR (R = H, Me, Et), -NO_2, -NH_2, 2-OH, -F, -Cl, -Br, -Me, and Et; the log RBA variations are from –1.40 (X = 3-CH_2NMe_2) to 1.51 (X = 2-OH).

These QSARs contain only electronic and steric terms. The presence of Hammett constants indicates that electron-donating substituents enhance binding by either increasing electron density on the ring or decreasing the acidity of the phenolic group. Increased electron density on the phenolic ring would enhance dipolar interactions between the ring and electron-deficient amino acid residues in the binding domain. The presence of steric descriptors such as MR and Verloop's L and B5 terms and the negative dependence of RBA on them emphasize the constrained locale of the binding site and the deleterious effects of substituent bulk, particularly in the meta positions of both metahexestrols and hexestrols. Hydrophobicity is not significant in variations of these ligand–ER interactions.

Further modifications of the side chains of hexestrol to compound series **3** and **4** were undertaken by the Katzenellenbogen group [12,13].

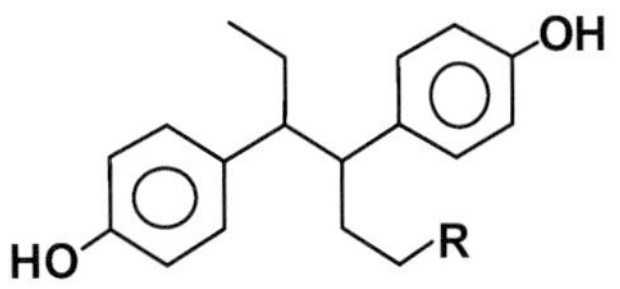

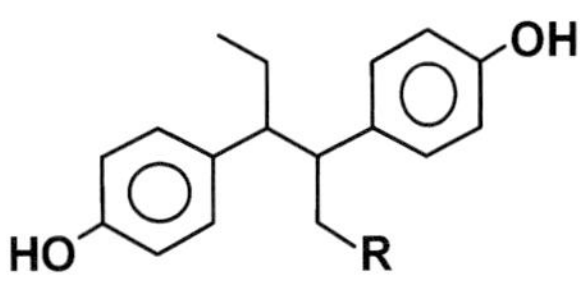

*RBA of halogenated side chain hexestrols **3** to lamb uterine cytosol ER at 0 °C/16 h*

$$\text{Log RBA} = -0.56(\pm 0.30)\ \text{B1}_R + 2.98(\pm 0.52) \quad (3)$$

$n = 5$; $r^2 = 0.920$; $s = 0.090$; $q^2 = 0.754$

*RBA of functionalized side chain hexestrols **4** to lamb uterine cytosol ER at 0 °C/16 h*

$$\text{Log RBA} = 0.67(\pm 0.25)\ \pi_R - 1.04(\pm 0.24)\ \text{MR}_R + 2.75(\pm 0.42) \quad (4)$$

$n = 11$; $r^2 = 0.928$; $s = 0.274$; $q^2 = 0.845$; *outlier:* R = $CONH_2$

In QSAR 3, R varies among H, F, Cl, Br, and I; log RBA from 1.78 (R = I) to 2.48 (R = H). QSAR 4 includes R = CH_2R' (R′ = H, F, Br, I, or OH) and COR″ (R″ = Me, *n*-Bu, OMe, O-*n*-Pent, CHN_2, or NH-*n*-Pent); log RBA varies from –0.82 (R = CONH-*n*-Pent) to 2.48 (R = Me). In both QSARs 3 and 4, a significant negative dependence on steric parameters indicates that the binding site has limited bulk tolerance. QSAR 4 pinpoints the importance of hydrophobic side chains in enhancing RBA.

Heiman et al. synthesized and evaluated the binding affinity of dihalogenated hexestrols **5** [14]. Utilization of this data resulted in the formulation of QSAR 5, while a similar study of 4-X-deoxyhexestrol derivatives **6** [15] led to the formulation of QSAR 6.

OH Y X HO

5

X HO

6

*RBA of aromatic ring-halogen hexestrols **5** to lamb uterine cytosol ER at 0 °C/16 h*

$$\text{Log RBA} = -1.25(\pm 0.61)\ \text{B1}_Y - 0.71(\pm 0.56)\ \text{B1}_X + 4.49(\pm 0.99) \quad (5)$$

$n = 7$; $r^2 = 0.953$; $s = 0.218$; $q^2 = 0.842$

*RBA of 4-X-deoxyhexestrols **6** to immature rat uterine cytosol ER at 0°C/20~24 h*

$$\text{Log RBA} = -0.64(\pm 0.17)\ \text{MR}_X - 1.24(\pm 0.31)\ \sigma_X + 1.67(\pm 0.26) \quad (6)$$

$n = 27$; $r^2 = 0.822$; $s = 0.286$; $q^2 = 0.760$; *outliers:* X = SO_2N_3, NO_2, OCH_2COOH, OCH_2COCH_2Cl, and OCH_2COCH_2Br

QSAR 5 includes compounds **5** with combinations of X (H, F, Br, or I) and Y (H, F, Br, or I). X and Y substituents are defined so that $\text{B1}_Y \geq \text{B1}_X$. The log RBA value varies between 0.42 (X = Y = Br) and 2.48 (X = Y = H). QSAR 6 is for compounds **6** in which X = H, OH, OCH_2R [R = H, CH_2OH, $CH_2O(CH_2)_2Cl$, -$CHCH_2O$- (oxiranyl), or CH=CH_2], OCH2CH(OH)R′ (R′ = Me, CH2OH, CH2Cl, or CH2Br), OCH2COR″ (R″ = Me, CH=N_2, or OEt), NH_2, N_3, SO_2F, COOH, CH_2R''' (R‴ = OH, Cl, Br, or OEt), COCH2R⁗ (R⁗ = H, Cl, or Br), CN, or C(=NH)OEt. The log RBA is in a range from –0.48 (X = SO_2F) to 2.48 (X = OH). The loss of a phenolic OH group in 4-X-deoxyhexestrol derivatives **6** clearly leads to a substantial decrease in binding affinity.

A similar study of binding of DES derivatives **7** to rat cytosolic ER [16] led to the formulation of QSAR 7, in which R = Me, COOPh, COOPh(4′-Cl), $CONEt_2$, or CH_2X [X = $OCONH(CH_2)_2Cl$, OEt, OH, I, OSO_2Me, or Cl] and the log RBA varies from –0.52 ($COOC_6Cl_5$; $COOC_6H_4$-4′-Cl) to 1.0 (Me).

*RBA of DES derivatives **7** to rat uterine cytosol at 18°C/30 min*

$$\text{Log RBA} = -0.24(\pm 0.12)\ \text{MR}_{\text{R}} + 0.69(\pm 0.34) \tag{7}$$

$n = 10$; $r^2 = 0.725$; $s = 0.248$; $q^2 = 0.457$; *outliers:* R = COOH and $(CH_2)_4Me$

Alkyl and phenyl phenols

Alkyl-phenolic compounds have also been implicated as weak ER ligands [14,17]. Using a recombinant yeast cell assay expressing human ER, the estrogenic potential of alkyl phenols **8** was determined in terms of the concentration (M) required for inducing half the maximal gene activation response produced by 17β-estradiol [17]. This data is now used to formulate QSAR 8. Substituents included are 2-*t*-Bu, -*s*-Bu, -CHMe$(CH_2)_7$Me, and -$CMe_2CH_2CMe_3$, 3-*t*-Bu, 4-Me, -Et, -*n*-Pr, -*t*-Bu, -*n*-Pent, -*n*-Hept, -*n*-Non, -CH(Me)R (R = Et, Hex, Oct, or Decyl), and -$C(Me)_2R'$ (R′ = Et, Pr, CH_2-*i*-Pr, or CH_2-*t*-Bu), and 2-Me-4-Non, the activity varies from 1.00 (4-Me) to 6.98 (4-CMe_2CH_2-*t*-Bu).

*Induction of half maximal gene activation by alkyl phenols **8***

$$\text{Log } 1/\text{EC}_{50} = 1.39(+/-\ 0.48)\text{Clog } P - 3.69(\pm 0.90)(\beta \cdot 10^{\text{Clog } P} + 1) + 1.05(\pm 0.24)\ \text{MR}_{4\text{X}} - 3.00(\pm 1.65) \tag{8}$$

$n = 21$; $r^2 = 0.927$; $s = 0.633$; $q^2 = 0.874$; Clog $P_o = 4.8$; *outliers*: X = 2,4-*t*-Bu_2, 2-*n*-Non-3-Me, and 2-*n*-Non-4-Me.

QSAR 8 shows a strong dependence of estrogenicity on hydrophobicity and size. The gene activation response increases as hydrophobicity increases up to a Clog *P* of 4.8 and then sharply decreases as hydrophobicity continues to increase. The extensive hydrophobic pocket may be constricted by some bulky amino acid residues at the binding site. The bulk of the substituent in the para position may help to anchor/lodge the substituent firmly within the confines of the hydrophobic pocket. The large number of diverse alkyl phenols in the set allows for a thorough evaluation of the role played by hydrophobicity.

In a similar study of the inhibition of ER binding of a smaller series of alkyl phenols **8** [18], the following QSAR 9 was developed. Despite the adequate range in hydrophobcity, a nonlinear model in Clog *P* could not be assessed because of the limited number of data points and lack of diversity in substituent type including 2-, 3-, 4-*t*-Bu, 4-*n*-Non, and 4-C(Me_2)R [R = *n*-Pent, and -Ph(4′-OH)]. The activity ranges from –2.60 (3-*t*-Bu) to 0.08 (4-*n*-Non). The positive dependence on bulk of substituents persists in this QSAR 9.

*Inhibition of mouse uterine ER binding by X-phenols **8***

$$\text{Log } 1/\text{K}_i = 0.63(\pm 0.26)\ \text{MR}_{\text{X}} - 2.67(\pm 0.75) \tag{9}$$

$n = 6$; $r^2 = 0.920$; $s = 0.405$; $q^2 = 0.858$; *outlier:* none

For the binding of a more diverse set of 4-hydroxybiphenyls **9** to mouse uterine cytosol ER [19], QSAR 10 indicates the bulk of the substituents in the phenyl ring (X) helps enhance binding to the ER, but those on the phenolic ring (Y) fail to do so.

*Half-maximal specific binding of 4-hydroxybiphenyls **9** relative to estradiol to mouse uterine ER*

$$\text{Log } 1/IC_{50} = 2.56(\pm 0.82)\ MR_{2X} + 1.13(\pm 0.95)\ MR_{4X} + 4.84(\pm 0.46) \quad (10)$$

$n = 11$; $r^2 = 0.883$; $s = 0.294$; $q^2 = 0.778$; *outlier:* [X = 2,3,5,6-Cl_4-4-OH, Y = H].

In the hydroxybiphenyls **9**, X is either one of H, 4-OH, 4-Cl, 2,5-Cl_2, 2,6-Cl_2 3,4,6-Cl_3, 2,3,4,5-Cl_4, 2-Cl-4-OH, and 3,5-Cl_2-4-OH, while Y is mostly H except for 3-Cl and 2,6-Cl_2. The activity range is from 5.00 (X = 4-OH, Y = H) to 7.38 (X = 2,4,6-Cl_3, Y = H).

Tamoxifen derivatives

Tamoxifen derivatives exert their antiestrogenic action via an ER mediated process [20]. From a study by Gilbert et al. on the RBA of a novel series of basic diphenylethylenes **10**, the following QSAR 11 was developed. The combinations of substituents (R_1, R_2, and R_3) are from R_1 = *i*-Pr, *n*-Bu, *n*-Pent, *n*-Hex, *n*-Oct, and CH_2Ph, R_2 = OH, $O(CH_2)_2NMe_2$, $O(CH_2)_2NEt_2$, and $O(CH_2)_2N(CH_2)_4$, and R_3 = OH, $O(CH_2)_2NMe_2$, and $O(CH_2)_2NEt_2$; the activity range is between –1.70 [*n*-Hex, $O(CH_2)_2NEt_2$, and OH] and 0.00 [*i*-Pr, OH, and $O(CH_2)_2NEt_2$].

*RBA of diphenyl ethylenes **10** to lamb uterine cytosol ER at 25 °C/5 h*

$$\text{Log RBA} = 0.63(\pm 0.26)I_Z - 0.64(\pm 0.21)\ MR_{R1} + 0.36(\pm 0.52) \quad (11)$$

$n = 13$; $r^2 = 0.855$; $s = 0.204$; $q^2 = 0.744$; *outliers*: [$R_1 = CH_2Ph$, $R_2 = O(CH_2)_2N(Et)_2$, R_3 = OH]; and [R_1 = *n*-Oct, $R_2 = O(CH_2)_2N(Et)_2$, R_3 = OH].

Overall, the binding of these series of compounds to lamb uterine cytosol ER was very low as the intercept indicates. I_Z is an indicator variable, which is unity when $R_3 = O(CH_2)_2NMe_2$ or $O(CH_2)_2NEt_2$, and R_2 = OH but zero when R_3 = OH and $R_2 = O(CH_2)_2NMe_2$, $O(CH_2)_2NEt_2$, or $O(CH_2)_2N(CH_2)_4$. An increase in the bulk of the R_1 group and "*E*-isomery" in the geometry are detrimental to binding to the ER.

In an extensive study of tamoxifen analogs, Hardcastle et al. [21,22] focused on the following series **11** and **12** in which $NR_2 = NMe_2$ or $N(CH_2)_4$ (a pyrrolidine ring). With the iodine derivatives **11** (n = 2~10), QSAR 12 was formulated. The log RBA varies from –0.40 (n = 10, $NR_2 = NMe_2$) to 1.40 (n = 4, $NR_2 = NMe_2$). Note that the indicator variable I takes a value of 1 when NR_2 is $N(CH_2)_4$ and 0 when NR_2 is NMe_2.

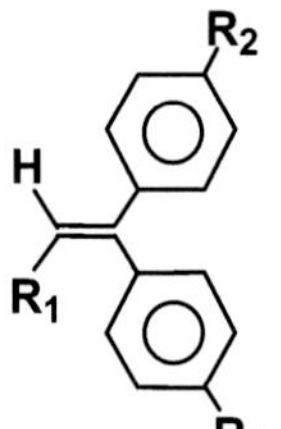

10 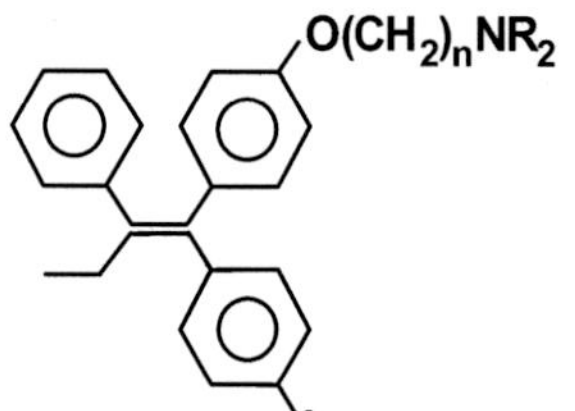 **11** 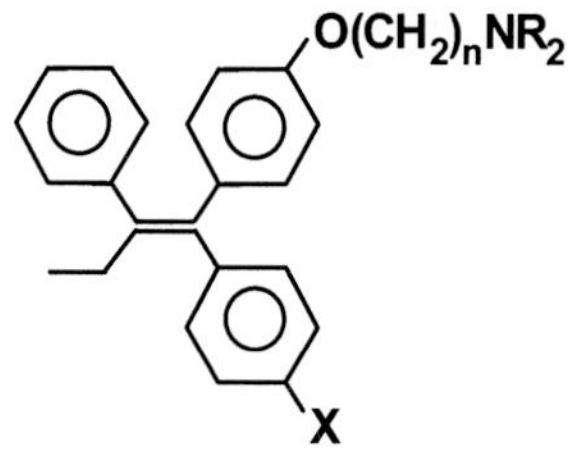 **12**

*RBA of 4-iodo-tamoxifen analogs **11** to immature rat uterine cytosol at 4 °C/16 h*

$$\text{Log RBA} = -0.40(\pm 0.10)\ \pi_4 + 0.36(\pm 0.25)\ I + 1.18(\pm 0.21) \quad (12)$$

$n = 16$, $r^2 = 0.860$, $s = 0.212$, $q^2 = 0.778$, *outlier:* none

RBA of tamoxifen analogs **12** *to immature rat uterine cytosol ER at 4 °C /16h*

$$\text{Log RBA} = -0.48(\pm 0.14)\ L_{4X} + 3.18(\pm 0.63) \qquad (13)$$

$n = 11$; $r^2 = 0.874$; $s = 0.183$; $q^2 = 0.827$; *outlier:* [X = H, $n = 2$, $NR_2 = NMe_2$]

In QSAR 12, the strong dependence on the hydrophilicity of the basic side chains is to be expected since there is a lack of variation in electronic character. The pyrrolidine ring on the side chain enhances RBA. For the 4-X-tamoxifen derivatives **12**, in which $n = 2 \sim 4$ and X = I, OH, *n*-Bu, or C≡CH with the activity varying from 0.30 ($n = 2$, $NR_2 = NMe_2$, X = *n*-Bu; $n = 2$, $NR_2 = N(CH_2)_4$, X = *n*-Bu) to 2.00 ($n = 2$, $NR_2 = NMe_2$, X = OH), a decrease in the length of the substituent leads to enhanced RBA. QSARs 12 and 13 suggest that substituents on the phenyl rings bind to restricted, polar space on the ER.

Ruenitz et al. examined the RBA of a small series of triarylethylenes **13** [23], in which X = H, F, or Br and Y = H or Cl, and log RBA ranges from 1.74 (X = Br, Y = H) to 2.25 (X = Y = H).

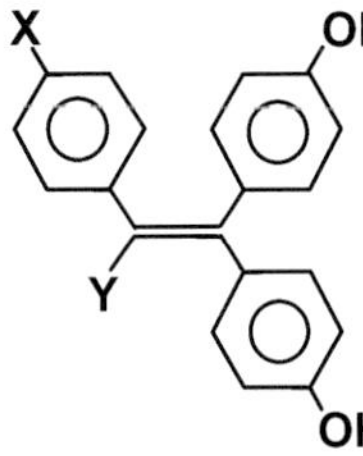

13 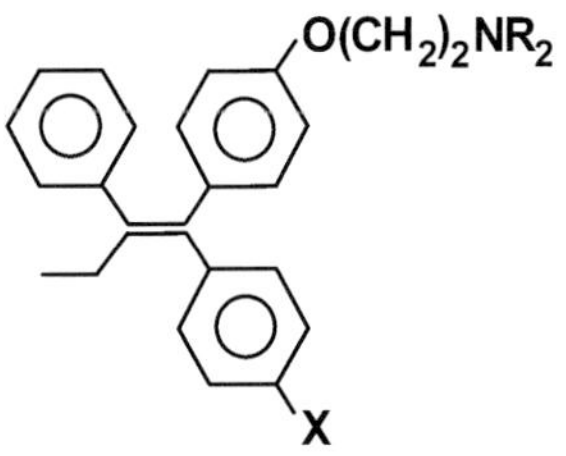 **14-*trans*** 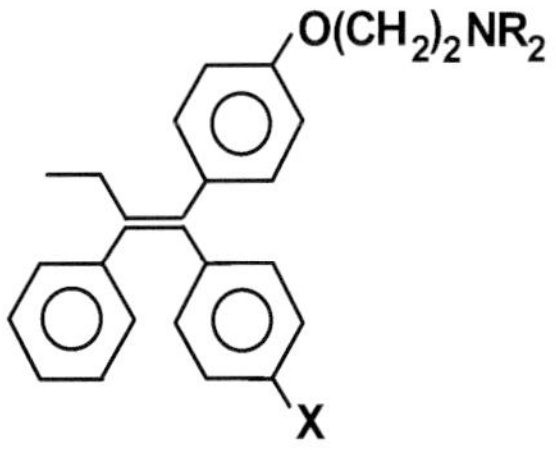 **14-*cis***

RBA of triarylethylenes **13** *to rat uterine ER at 4 °C/4 h*

$$\text{Log RBA} = -2.61(\pm 2.24)\sigma_X^{+} + 2.15(\pm 0.19) \qquad (14)$$

$n = 4$; $r^2 = 0.926$; $s = 0.084$; $q^2 = 0.660$; *outlier*: [X = I ,Y = H]

The iodine derivative (X = I, Y = H) was not well predicted, and a size effect might be implicated. Unfortunately, the small number of analogs in the data set precludes further exploration with other parameters.

The analysis of a series of 4-substituted tamoxifen derivatives **14** (trans and cis) synthesized and evaluated by McCague et al. [24] led to the formulation of QSAR 15. The substituent X varies from H to Cl, Br, I, SH, $S(O)_nMe$ ($n = 0 \sim 2$), CHO, and $CH(O)CH_2$ (oxiranyl). The activity changes from –1.00 (*cis*-SMe) to 1.30 (*trans*-CHO). The data set includes both trans and cis geometric isomers. Trans and cis designate the relative positions of the ethyl group and the phenyl substituted with the basic side chain. Thus, an indicator variable I_{cis} is assigned a value of 1 for *cis*-isomers.

RBA of 4-X-tamoxifens **14** *to calf uterine cytosol at 18 °C/30 min*

$$\text{Log RBA} = -1.20(\pm 0.33)\ I_{cis} + 0.93(\pm 0.68)\ \sigma_X + 0.25(\pm 0.26) \qquad (15)$$

$n = 16$; $r^2 = 0.835$; $s = 0.303$; $q^2 = 0.779$; *outlier: trans*-CH_2OH

The positive Hammett ρ value (+0.93) indicates that electron-withdrawing groups on the aromatic ring enhances RBA. They would decrease the basicity of the basic side chain. The detrimental effects of cis configuration in binding are noted; thus the *trans*-isomers bind more effectively to the ER.

DISCUSSION

Careful comparison of the various models developed for the binding of hexestrols to various mammalian ERs reveals a strong dependence on steric effects. Steric hindrance, whether represented by MR, which describes mostly volume, or Verloop's B1 (or B5 and L), which delineates width (length) of substituents, is of overriding importance in QSARs 1~7. Surprisingly variations in hydrophobicity do not play a significant role in these ligand–ER interactions except in QSAR 4 where modification is on the hexestrol side chain. A negative Hammett ρ value shows up when a diverse set of substituents is located on the aromatic rings. Merely extending an alkyl chain on an aromatic ring does not bring electronic diversity to a set of compounds. These features of hexestrol binding do suggest that although the ER may be overwhelmingly hydrophobic, the surface of the binding site is "craggy" and has lots of nooks and crannies, which make it "sterically challenged".

The negative correlation with bulk and size (as delineated by an indicator variable and Verloop's parameters) persists in QSARs 11, 13, and 15 of the binding of tamoxifen derivatives to the estrogen receptor. Electronic effects are only seen when the substituents on the phenyl ring vary in electronic character, otherwise hydrophilicity accounts for the variance in the data of a limited number of data sets. These results suggest that there must be patches of polar residues that are accessible by hydrophilic substituents on the aromatic rings of tamoxifen. QSAR 15 does not fit the normal pattern and reasons for its anomalous behavior are not apparent, although the incubation time is unusually short and thus equilibrium may have not been attained.

Phenolic compounds are rather interesting and their behavior is contradictory to what is seen with the substituted hexestrols and tamoxifens. Various phenols bearing alkyl type substituents have been assessed for their ability to interact with the receptor via a direct binding or yeast cell assay. In the two cases that were examined, the lack of diversity in substituent type precluded the inclusion of electronic parameters in QSARs 8 and 9. Thus, ER binding is strictly dependent on substantial hydrophobicity and minimal steric hindrance. A set of hydroxybiphenyls also behaves in a similar fashion as shown in QSAR 10. Phenols are smaller in size as compared to the hexestrols and tamoxifens and thus they have more flexibility. It has been assumed that the phenolic OH group generally superimposes itself on the OH group of the "A" ring of estradiol. But the strong dependence on hydrophobicity suggests that the binding mode of the X-phenols could be altered and the phenol could maneuver into the "D" ring region where the OH group could bind in the region of the 17-OH group of estradiol. The presence of an anchoring, polar amino acid residue could ground the phenols. The positive dependence on MR terms in QSARs 8, 9, and 10 suggests that a type of interaction toward an optimum could be occurring. Modeling studies would be invaluable in shedding light on this unusual behavior of substituted phenols binding to the ER.

Two comparative studies are worth mentioning—one, which examines different assay conditions and one which evaluates differential binding to α and β isoforms of the ER. Bignon et al. [25] examined the reactivity of a series of cyano-tamoxifen derivatives **15** under two experimental conditions from which QSARs 16 and 17 were formulated. Each of the substituents R, R_1, and R_2 in compounds **15** is either one of H, Me, OH, OMe, $OCHMe_2$, $O(CH_2)_2NEt_2$, and NMe_2.

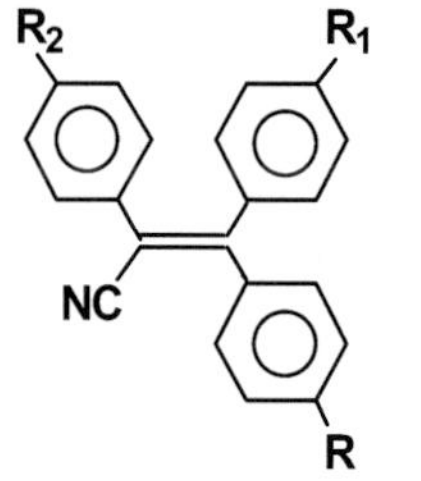

R_2 R_1 NC R

15

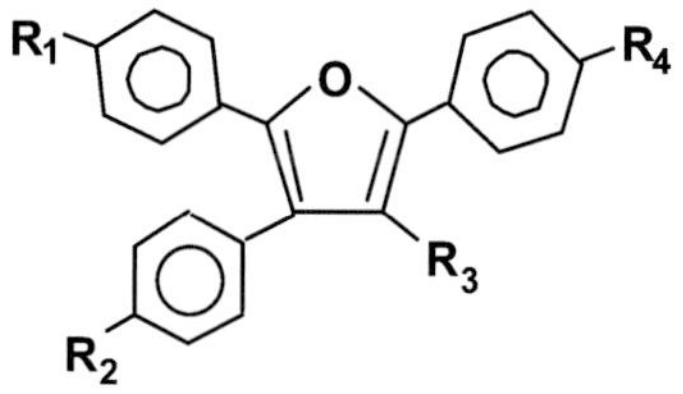

R_1 O R_4 R_3 R_2

16

RBA of cyano-tamoxifen derivatives ***15*** *to rat uterine cytosol ER at 25 °C/5 h*

$$\text{Log RBA} = 2.28(\pm 0.58)\ I_{OH} - 1.11(\pm 0.49)\ \pi_{4R} - 1.45(\pm 0.49) \tag{16}$$

$n = 23$; $r^2 = 0.861$; $s = 0.522$; $q^2 = 0.815$; *outlier:* [R = R_1 = $O(CH_2)_2NEt_2$, R_2 = H]

RBA of cyano-tamoxifen derivatives ***15*** *to rat uterine cytosol ER at 0 °C/2 h*

$$\text{Log RBA} = 2.08(\pm 0.47)\ I_{OH} - 0.46(\pm 0.39)\pi_{4R} - 0.94(\pm 0.38) \tag{17}$$

$n = 24$; $r^2 = 0.861$; $s = 0.437$; $q^2 = 0.805$; *outlier:* none

In QSAR 16, log RBA covers a range of –2.00 [R = R_1 = $OCHMe_2$, R_2= H and R = R_1 = H, R_2 = 4-$OCH_2CH_2N(Et)_2$] to 2.22 (R = R_1 = R_2 = OH). In QSAR 17, the lowest activity was also –2.00 (R = R_1 = $OCHMe_2$, R_2 = H), while the highest was 2.10 [R = OH, R_1 = $O(CH_2)_2NEt_2$, R_2 = H]. I_{OH} is an indicator variable that denotes the positive impact of the presence of an OH moiety at any of R, R_1, and R_2 positions. A comparison of QSARs 16 and 17 notes strong similarities in the presence of the key descriptors: I_{OH} and π_4. However, the magnitude of the sensitivity to hydrophilic effects varies considerably. It practically doubles at higher temperatures and longer reaction times. In both cases, RBA is enhanced by the presence of hydrophilic OH substituents, in particular, at the R position on the ring cis to the cyano group. The more negative π_{4R} term in QSAR 16 than that in QSAR 17 would indicate that the extent of desolvation of the substituent at 25 °C exceeds that at 0 °C. This excellent study points out the danger of combining data from different assays which may have been run under different conditions.

Mortensen et al. compared the RBA of a series of triphenyl-furans **16** to isoforms of lamb uterine cytosol ER [26]. From their data, the following QSAR models 18 and 19 were formulated for the RBA to ERα and ERβ receptors, respectively.

RBA of triphenylfurans ***16*** *to ERα receptor of lamb uterine cytoxol at 0 °C/1~24 h*

$$\text{Log RBA} = -0.80(\pm 0.43)\ MR_{R3} + 1.15(\pm 0.46)\ I_{R2} + 1.75(\pm 0.51) \tag{18}$$

$n = 9$; $r^2 = 0.868$, $s = 0.227$; $q^2 = 0.729$; *outliers*: [R_3 = Me, R_1 = R_2 = R_4 = OH], [R_3 = Et, R_1 = R_2 = OH], [R_3 = Et, R_1 = R_4 = H, R_2 = OH]

RBA of triphenylfurans ***16*** *to ERβ receptor of lamb uterine cytoxol at 0 °C/18~24 h*

$$\text{Log RBA} = 2.46(\pm 0.99)\ MR_{R3} - 6.36(\pm 0.23)\ \log(\beta \cdot 10^{MR}{}_{R3} + 1) - 1.23(\pm 0.77) \tag{19}$$

$n = 10$; $r^2 = 0.899$; $s = 0.197$; $q^2 = 0.819$; MRO = 1.45, *outliers:* [R_3 = Pr, R_1 = R_2 = R_4 = OH], [R_3 = Et, R_1 = R_4 = H, R_2 = OH]

In the triphenyl-furans **16**, R_1, R_2, and R_4 = H or OH, and R_3 = Me, Et, Pr, Bu, or Ph(4-OH). The ERα binding activity ranges from 0.83 (R_1 = OH, R_2 = R_4 = H, R_3 = Et) to 2.15 (R_1 = R_2 = R_4 = H, R_3 = Et), while the ERβ binding activity ranges from –0.60 [R_1 = R_2 = R_4 = OH, R_3 = Ph(4-OH)] to 1.18 (R_1 = H, R_2 = R_4 = OH, R_3 = Pr).

QSAR 18 and 19 are significantly different, which suggests that the binding sites in the α and β receptors have marked differences in their topography. In QSAR 18, I_{R2} is an indicator variable that acquires a value of 1 when R_2 = OH. Its presence suggests that a hydrogen bond formation could occur between the OH moiety and perhaps a polar residue at the binding site. The negative dependence on the bulk/polarizability parameter MR_{R3} suggests that the binding site of R_3 substituents may be highly constrained and lined with residues of intermediate to marked polarity. The β receptor appears to be more extensive with less constraints and a moderate size cavity as implied by the fact that the optimum MR value of R_3 is 1.45. The lack of the I_{R2} indicator variable in QSAR 19 suggests that the R_2 substituent on the phenyl ring at the 2-position may not be in close proximity to a H-bond acceptor or donor in the ERβ binding site.

CONCLUDING REMARKS

The utility of the present type of "2D QSAR" in delineating receptor–ligand interactions in mostly congeneric series has been well documented while the process of lateral validation via comparative QSAR has helped bringing consistency to the various models [8]. Nevertheless, despite the proliferation of in vitro and in vivo assays to describe binding to steroid hormone receptors, the development of extensive 2D QSAR models has been limited.

The CoMFA (comparative molecular field analysis) is often viewed as an extension of QSAR and addressed as three-dimensional (3D). CoMFA has been utilized to develop several QSAR models for binding to various steroid hormone receptors [27]. However, despite the development of regression equations, the results are best mapped out and visualized as 3D color-coded contour plots. Semi-quantitative generalizations are drawn from these graphics but clear-cut quantitative aspects of the models are minimized or lost. Thus, rigorous and direct comparisons or extrapolations cannot be made between CoMFA models for different receptors, unlike "2D QSAR".

One of the attributes of CoMFA is its ability to combine diverse data sets and examine them as a whole. This operation can also be a weakness since it implicitly suggests that all the members of that set bind to a receptor at the "same" binding site and in the "same" mode. In the presence of definitive knowledge of the alignment and conformation of the individual molecules, it becomes necessary to examine numerous alignments and conformations and select the best combination that generates a CoMFA model with the highest predictive power (high q^2) [28]. With the 2D QSAR, the lateral validation of the individual QSARs allows for direct comparison of the descriptor terms, their contributions, and other statistical terms.

The action of endocrine-active substances is not limited to receptor binding and other non-receptor binding events should be considered. Our experimental results with simple and complex phenols (including eight estrogenic phenols) in a fast growing murine leukemia cell line (L1210), which acts as a model for fast growing cells (e.g., in embryogenesis), provide an underlying basis for our concern. The strong dependence of cytotoxicity of phenols on the OH bond-dissociation energy suggested that free-radical forming ability (coupled with weak hydrophobic contributions) was critical to inducing the cellular toxicity. The estrogenic phenols were included and were well predicted by this model [29]. The low coefficient with the log P term suggests that interaction with a hydrophilic nucleotide sequence may be operative [29]. Thus, the ability of estrogenic phenols to interact with DNA at higher doses suggests that the effects of these endocrine-active substances could be diverse and of major concern.

Comparisons between our results and those previously published [4] on the binding of hexestrol and tamoxifen derivatives to the ERα, establish a consistency in models, which emphasize the steric intolerance and moderate electron density requirements of the receptor. Hydrophobicity as represented either by Clog P or π is practically nonexistent and appears to not play a significant role in receptor binding. These results are also reflective of what emerged from an extensive analysis of estradiol derivatives [4]. The lack of hydrophobic terms in QSARs delineating ER–ligand interactions is perplexing. It may well be that the basic estradiol footprint (ABCD rings) present in estradiol, hexestrol, and tamoxifen derivatives occupy the sterically challenged, hydrophobic space only allowing limited mobility and subsequently variability to the substituents. Thus only small molecules like X-phenols will have options in binding site choices (either the A ring area or D ring environs).

Classical QSAR is invaluable in identifying structural features or properties that can affect binding to a receptor at the molecular level and thus impact biological activity. It can also pinpoint molecules that behave in an anomalous fashion. QSAR has great utility in mechanistic interpretation as well as predictability within a congeneric series, particularly with regard to binding affinities. QSAR models can be utilized effectively to predict ER-binding affinity within spanned space and thus have potential as priority setting tools for EASs.

REFERENCES

1. J. A. Katzenellenbogen. *Environ. Health Perspect.* **103**, 99–101 (1995).
2. M. J. Tsai and B. W. O'Malley. *Annu. Rev. Biochem.* **63**, 451–486 (1994).
3. B. Gutendorf and J. Westendorf. *Toxicology* **166**, 79–89 (2001).
4. H. Gao, J. A. Katzenellenbogen, R. Garg, C. Hansch. *Chem. Rev.* **99**, 723–744 (1999).
5. G. G. J. M. Kuiper, B. Carlsson, J. Grandien, E. Enmark, J. Haggblad, S. Nilsson, J. A. Gustafsson. *Endocrinology* 863–870 (1997).
6. T. Barkhem, B. Carlsson, Y. Nilsson, E. Enmark, J. A. Gustafsson, S. Nilsson. *Mol. Pharmacol.* **54**, 105–112 (1998).
7. A. Howell, S. Downey, E. Anderson. *Eur. J. Cancer* **32A**, 516–588 (1996).
8. C. Hansch and A. Leo. *Exploring QSAR: Fundamentals and Applications in Chemistry and Biology*, American Chemical Society, Washington DC (1995).
9. C. Hansch, D. Hoekman, A. Leo, D. Weininger, C. D. Selassie. *Chem. Rev.* **102**, 783–812 (2002).
10. R. W. Hartmann, A. Heindl, H. Schonenberg. *J. Med. Chem.* **27**, 577–585 (1984).
11. R.W. Hartmann, W. Schwarz, H. Schonenberger. *J. Med. Chem.* **26**, 1137–1144 (1983).
12. R. Goswami, S. G. Harsy, D. F. Heiman, J. A. Katzenellenbogen. *J. Med. Chem.* **23**, 1002–1008 (1980).
13. S. W. Landvatter and J. A. Katzenellenbogen. *J. Med. Chem.* **25**, 1300–1307 (1982).
14. D. F. Heiman, S. G. Senderoff, J. A. Katzenellenbogen, R. J. Neeley. *J. Med. Chem.* **23**, 994–1002 (1980).
15. J. A. Katzenellenbogen, R. J. McGorrin, T. Tatee, R. J. Kempton, K. E. Carlson, D. H. Kinder. *J. Med. Chem.* **24**, 435–450 (1981).
16. K. Krohn, K. Kulikowski, G. Leclercq. *J. Med. Chem.* **32**, 1532–1538 (1989).
17. E. J. Routledge and J. P. Sumpter. *Environ. Toxicol. Chem.* **15**, 241–248 (1996).
18. C. L. Waller, T. I. Oprea, K. Chae, H. K. Park, K. S. Korach, S. C. Laws, T. E. Wiese, W. R. Kelce, L. E. Gray, Jr. *Chem. Res. Toxicol.* **9**, 1240–1248 (1996).
19. K. S. Korach, P. Sarver, K. Chae, J. A. McLachlan, J. D. McKinney. *Mol. Pharmacol.* **33**, 120–126 (1988).
20. J. Gilbert, M. Fuentes, T. Ojasoo, J.-C. Dore, M. Pons. *J. Med. Chem.* **40**, 1104–1111 (1997).
21. I. R. Hardcastle, M. G. Rowlands, R. M. Grimshaw, J. Houghton, M. Jarman, A. Sharff, S. Neidle. *J. Med. Chem.* **39**, 999–1004 (1996).
22. I. R. Hardcastle, M. G. Rowlands, J. Houghton, I. B. Parr, G. A. Potter, M. Jarman, K. J. Edwards, C. A. Laughton, J. O. Trent, S. Neidle. *J. Med. Chem.* **38**, 241–248 (1995).
23. P. C. Ruenitz, J. R. Bagley, N. T. Nanavati. *J. Med. Chem.* **31**, 1471–1475 (1988).
24. R. McCague, G. Leclercq, N. Legros, J. Goodman, G. M. Blackburn, M. Jarman, A. B. Foster. *J. Med. Chem.* **32**, 2527–2533 (1989).
25. E. Bignon, M. Pons, A. C. Crastes de Paulet, J. C. Dore, J. Gilbert, J. Abecassis, J. G. Miquel, T. Ojasoo, J. P. Raynaud. *J. Med. Chem.* **32**, 2902–2103 (1989).
26. D. S. Mortensen, A. L. Rodriguez, K. E. Carlson, J. Sun, B. S. Katzenellenbogen, J. A. Katzenellenbogen. *J. Med. Chem.* **44**, 3838–3848 (2001).
27. R. D. Cramer III, D. E. Patterson, J. D. Bunce. *J. Am. Chem. Soc.* **110**, 5959–5967 (1988).
28. R. A. Hughes, T. Hams, E. Altmann, D. McAllister, R. Vlahos, A. Robertson, M. Cushman, Z. Wang, A. G. Stewart. *Mol. Pharmacol.* **61**, 1053–1069 (2002).
29. C. D. Selassie, A. J. Shusterman, S. Kapur, R. P. Verma, L. Zhang, C. Hansch. *Perkin Trans. 2* 2729–2733 (1999).

Pure Appl. Chem., Vol. 75, Nos. 11–12, pp. 2375–2388, 2003.

Workshop 1.2

Regulatory application of SAR/QSAR for priority setting of endocrine disruptors: A perspective*

Weida Tong[1,‡], Hong Fang[2], Huixiao Hong[2], Qian Xie[2], Roger Perkins[2], Jeanne Anson[1], and Daniel M. Sheehan[1]

[1]*FDA's National Center for Toxicological Research (NCTR), Jefferson, AR 72079, USA;* [2]*Logicon ROW Sciences, Jefferson, AR 72079, USA*

Abstract: Some seven years have passed since the U.S. legislature mandated the Environmental Protection Agency (EPA) to develop and implement a screening and testing program for chemicals that may disrupt the delicate endocrine system. The envisioned EPA program has evolved to incorporate a tiered scheme of in vitro and in vivo assays, and considered QSAR as a viable method to set testing priorities. At the U.S. Food and Drug Administration's (FDA) National Center for Toxicological Research (NCTR), the Endocrine Disruptor Knowledge Base Project has developed models to predict estrogen and androgen receptor binding. Our approach rationally integrates various QSAR models into a sequential "Four-Phase" scheme according to the strength of each type of model. In four hierarchical phases, models predict the inactive chemicals that are then eliminated from the pool of chemicals to which increasingly precise but more time-consuming models are subsequently applied. Each phase employs different models selected to work complementarily in representing key activity-determining structure features in order to absolutely minimize the rate of false negatives, an outcome we view as paramount for regulatory use. In this paper, the QSAR models developed at NCTR, and particularly how we integrated these models into the "Four-Phase" system will be discussed for a number of datasets, including 58 000 chemicals identified by the U.S. EPA.

INTRODUCTION

A large number of environmental chemicals are suspected of disrupting endocrine function by mimicking or antagonizing natural hormones in experimental animals, wildlife, and humans. There is growing concern among the scientific community, government regulators, and the public that these endocrine-disrupting chemicals (EDCs) in the environment are adversely affecting human and wildlife health [1,2]. Adverse outcomes have been observed in experimental animals and wildlife; potential effects on humans include reproductive and developmental toxicity, carcinogenesis, immunotoxicity, and neurotoxicity, among others [3]. EDCs may exert adverse effects through a variety of mechanisms, such as estrogen receptor (ER)-mediated mechanisms of toxicity.

The scientific debate surrounding EDCs has grown contentiously, in part owing to the fact that some suspected EDCs are produced in high volume, and many chemicals are economically important. These public and regulatory concerns led to government regulatory actions [4] and expanded research across Europe, Japan, and North America. The U.S. Congress in 1996 mandated that the Environmental

*Report from a SCOPE/IUPAC project: Implication of Endocrine Active Substances for Human and Wildlife (J. Miyamoto and J. Burger, editors). Other reports are published in this issue, *Pure Appl. Chem.* **75**, 1617–2615 (2003).

‡Corresponding author

Protection Agency (EPA) should develop a strategy for screening and testing a large number of chemicals found in drinking water and food additives [4] for their endocrine disruption potential. In response to Congressional action, the EPA established the Endocrine Disruptor Screening and Testing Advisory Committee (EDSTAC), which includes scientific expertise from government, academia, and industry. EDSTAC recommended a two-tier (Tier 1: screening and Tier 2: testing) strategy to screen and test estrogenic, androgenic, and thyroidal activities for a large number of chemicals. To accomplish this, chemicals will be screened (Tier 1) using a multiple endpoint strategy that includes more than 20 different in vitro and in vivo assays recommended by EDSTAC [5]. Although more than ~87 000 chemicals were initially selected for evaluation, many are polymers, leaving about ~58 000 chemicals for evaluation in Tier 1. The number that will progress to the testing step (Tier 2) [6], is not known. Processing chemicals through both tiers will require many years and extensive resources. Hence, the EPA has adopted an approach requiring priority setting before Tier 1 (<www.epa.gov/scipoly/oscpendo/>).

Among the types of hormonal activities, estrogenic activities have been most widely studied. Estrogenic endocrine disruption can result from a variety of biological mechanisms. We found a strong linear correlation among a diverse group of chemicals between binding affinities with the ER from the rat uterine cytosol and those with the human ER of the α-subtype [7]. Furthermore, the rat ER binding data also correlates strongly with the results from assays measuring estrogenicity using downstream events, i.e., the yeast-based reporter gene and the MCF-7 cell proliferation assays. Importantly, chemicals positive in uterotrophic responses (in vivo estrogenic activity) are also positive in the ER binding assay, indicating that binding affinity is a good predictor of in vivo activity with few false negatives observed [8]. These findings demonstrate that ER binding is the major determinant for estrogenic EDCs, and therefore the prediction of the rat ER binding affinity provides an important piece of information for priority setting.

Structure–activity relationship (SAR) and/or quantitative structure–activity relationship (QSAR) models have proven their utility, from both the pharmaceutical and toxicological perspectives, for identification of chemicals that might interact with ER. While QSAR models (QSAR is used hereafter to encompass models that predict activity on either an ordinal or categorical scale rather than only on a quantitative scale) in pharmaceuticals identify high-affinity ligands, they are particularly effective in toxicology in separating active and inactive chemicals, and in rank-ordering chemicals according to potency. Developing a useful QSAR model heavily depends on many factors, particularly, including the quality of biological data, the descriptor selection, and the choice of statistical approaches. Since any QSAR approach has strengths and weaknesses, the careful selection of a specific model or a combination of models needs to be done in accordance with the intended application.

This review first summarizes our motives and efforts to set up a robust training set (the NCTR dataset) for developing ER QSAR models. The development and validation of the QSAR models to predict ER binding is then described, as well as the rationale for integrating models into a hierarchical scheme for use in priority setting of potential estrogenic EDCs. The review will conclude with a discussion of some key issues for applying QSAR models for regulatory purposes.

NCTR ER DATASET: A ROBUST TRAINING SET FOR QSAR DEVELOPMENT

Although an effective QSAR model depends on a number of factors, the most critical is a training set with high-quality biological data. It is desirable that the training set has the biological data coming from the same assay protocol, and contains a sufficiently large number of chemicals with diverse structure. Most importantly, both the biological activity/potency data and structural characteristics in terms of various descriptors are evenly distributed as far as possible in a range reflecting the domain in which chemicals to be predicted are located.

A number of QSAR models for ER binding developed several years ago [9–13], including some of our early works [14–17], were based on datasets available in the literature. These datasets were too small and/or lacked structural diversity [11,12,14]. Although these models yield good statistical results in the training and cross-validation steps and explain some structural determinants for ER binding, they had limited applicability in predicting the ER-ligand binding affinity of chemicals that, in fact, cover a wide range of structural diversity.

In order to obtain an adequate training set to develop more robust QSAR models, we developed and validated a rat ER binding assay and assayed 232 chemicals [18,19]. The ER binding activity is represented by relative binding affinity (RBA), where the RBA value for the endogenous ER ligand, 17β-estradiol (E2), was set to 100. This NCTR dataset contains chemicals that were selected to cover the structural diversity that bind to ER with an activity distribution ranging over six orders of magnitude. Figure 1 compares the chemistry space/domain of the NCTR dataset with other literature datasets used for early ER QSAR models.

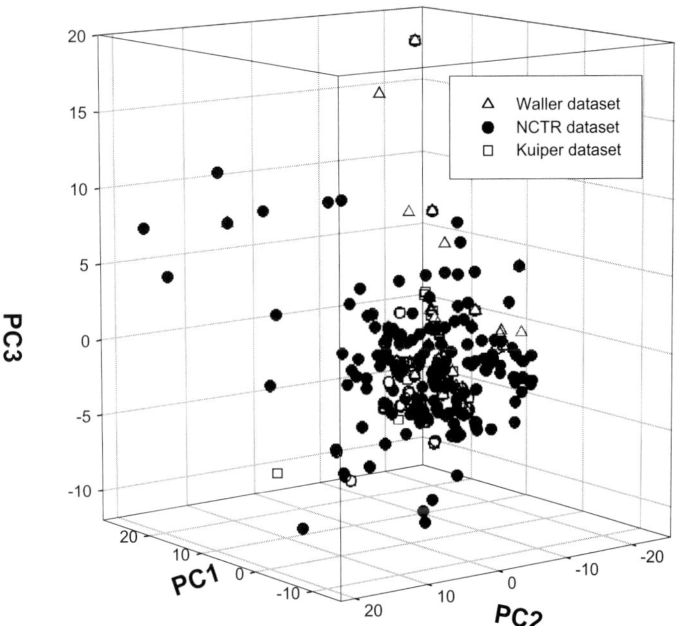

Fig. 1 Comparison of structural diversity of NCTR, Waller, and Kuiper datasets in a chemistry space defined by three principal components (PC) of structural descriptors.

DEVELOPMENT OF QSAR MODELS FOR ER BINDING

We evaluated a number of QSAR approaches useful for predicting ER binding affinity ranging from simple rejection filters often used for drug-like chemical identification to more sophisticated QSAR models often used in the lead optimization. The methods selected for the four-phase approach are discussed below.

Rejection filters

Rejection filters are useful to quickly exclude chemicals from further evaluation. We investigated various physicochemical parameters to use as rejection filters. The ideal filter to be used in the earliest stage of modeling should (1) not generate any false negatives and (2) be able to significantly reduce the number of chemicals for further evaluation. Two rejection filters, molecular weight range and lack of a ring motif, were found to satisfy the two criteria. Chemicals matching any one of these two filters were excluded from subsequent models. The first rejection filter is a molecular weight range, set to <94 and

>1000. The molecular weight of phenol, 94, was considered as the lowest limit for a chemical to bind to ER, whereas a molecular weight of 1000 was considered as the upper limit. The second rejection filter requires that an ER ligand contains at least one ring structure of any size based on the finding from a large literature survey that no known estrogens were found to lack a ring [20].

Structural alerts

Structural alerts are key 2D structural fragments associated with ER binding. Figure 2 depicts the three structural alerts, i.e., the steroid, DES, and phenolic skeletons, selected to identify potential ER binders. Each alert independently characterizes a unique structural feature important for the ER binding. Any chemicals containing any one of these structural alerts are considered to be a potential ER binder.

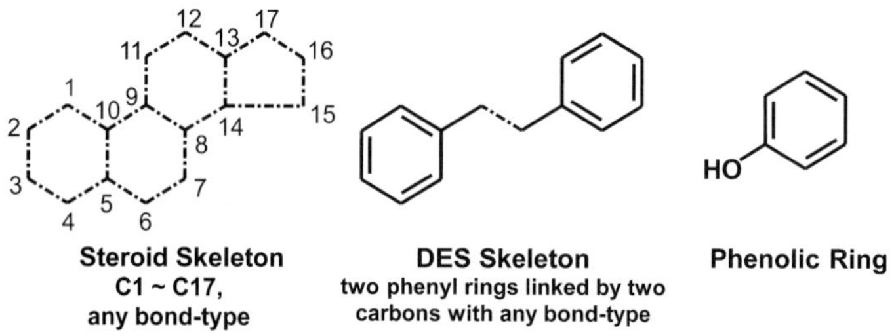

Steroid Skeleton
C1 ~ C17,
any bond-type

DES Skeleton
two phenyl rings linked by two carbons with any bond-type

Phenolic Ring

Fig. 2 Three structural alerts defined by 2D substructural features commonly observed for ER ligands.

These three substructures were selected as structural alerts because the length and width of both steroid and DES skeletons conform well to the dimension of the ER binding pocket. In addition, while most endogenous hormones contain the steroid skeleton, most strong estrogens have two benzene rings separated by two carbon atoms [20]. It has been long understood that the phenolic ring is often associated with estrogenic activity [21]. In fact, the contribution of the phenolic ring in ER binding is much more significant than any other structural feature [20]. By overlaying the crystal structure of ER complexes of E2, 4-hydroxytamoxifen, raloxifene, and DES based on their common protein residues at the binding site, we found that the phenolic rings of all four of these ligands are closely positioned at the same location as shown in Fig. 3, allowing hydrogen bond interactions with Glu 353 and Arg 394 of the receptor and a water molecule [22].

Pharmacophore queries

The pharmacophore is represented as a 3D arrangement, in which such molecular features as H-donor, H-acceptor, and hydrophobic center are positioned and combined with a certain geometry, that is supposedly needed for a molecule to exhibit a certain type of biological activity [23]. The pharmacophoric arrangements are used as the queries. A query-matched chemical is considered positive and segregated for further evaluation. One of the advantages of the 3D pharmacophore searching is that it can identify chemicals whose 3D structures are similar to the template structure (normally, a highly active chemical) that may not be discernable by chemists in 2D.

The bound ligand-ER crystal structures [24,25] guided our selection of pharmacophore queries. Using 3D structures of E2, raloxifene, 4-hydroxytamoxifen, and DES in the ER-bound conformation as templates, all possible molecular features as well as molecular shapes were delineated. For each template, any of three to six features were combined to form pharmacophore queries. These queries were sorted according to their discriminatory power to separate active from inactive chemicals in the NCTR dataset. A chemical with any of multiple 3D conformations (up to 100) [26–28] matching the query was considered to be active. Queries with high discriminatory power were further evaluated for their bio-

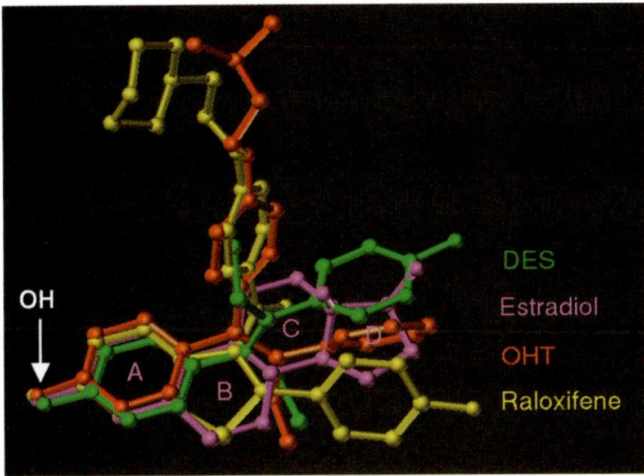

Fig. 3 Relative positions of estradiol (E2), DES, raloxifene, and 4-hydroxytamoxifen (OHT) in the complex with ER.

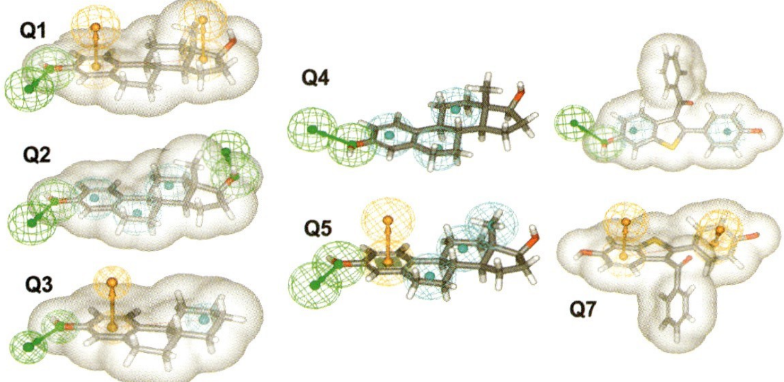

Fig. 4 Seven pharmacophore queries where green mesh balls represent H-bond acceptor sites, blue mesh balls represent hydrophobic centers and yellow mesh balls represent aromatic centers. Solid white surfaces represent shape constraint.

logical relevance based on careful SAR examinations of a large number of chemicals for their binding affinities to ER [20] in conjunction with analysis of the ligand-ER crystal structures [24,25]. Finally, the Tanimoto similarity score was used to determine the uniqueness of each query. Through this process, we identified seven queries that provide unique pharmacophoric signatures for the ER binding (Fig. 4). A chemical could match none, a few, or many of the seven separate queries. We generally found that the number of matches increased in direct proportion to measured activity among the training-set chemicals. Thus, the number of pharmacophore matches could be used to rank-order chemicals in accordance with the potential activity.

CLASSIFICATION MODELS

Classification is a supervised learning technique that provides categorical prediction (e.g., active/inactive). A number of classification methods were evaluated to categorize chemicals as ER binders or non-binders. While the methods differ in a number of ways, they generally produce similar results [29]. We found that the nature of the descriptors used, and more specifically the effectiveness in which descriptors encode the structural features of the molecules related to the ER binding activity, is far more important than the specific method employed. The selection of biologically relevant descriptors is the critical step to develop a robust model. We found the Genetic Algorithm to be the preferred method to identify the most biologically relevant descriptors from a large set of descriptors.

For example, using the best 10 descriptors selected by the Genetic Function Approximation approach [30,31] from among 153 descriptors, we were able to construct a Decision Tree model for the classification. The model displays a series of YES/NO rules to classify chemicals into active and inactive categories based on five meaningful descriptors: the Phenolic Ring Index (0 or 1), log P, two types of parameters related to the charged partial molecular surface area, and a geometric descriptor connected with the breadth of the molecule. The model identified the Phenolic Ring Index as the most important descriptor for ER binding. If chemicals contained a phenolic moiety but also had log P values larger than 1.49, they were more likely to be ER binders. In contrast, chemicals without a phenolic moiety were less likely to be ER binders unless they had relatively large hydrophobicity and charged surface area, and breadth of the structure.

3D-QSAR/CoMFA model

We evaluated three different techniques for the QSAR modeling—CoMFA, CODESSA (COmprehensive DEscriptors for Structural and Statistical Analysis), and HQSAR (Hologram QSAR)—for their utility (predictivity, speed, accuracy, and reproducibility) to quantitatively predict the ER binding activity [16,22]. For three relatively small datasets under investigation, the CoMFA and HQSAR procedures were of a comparable high performance [16]. These two procedures were further investigated and compared, particularly for their predictivity, by using the NCTR dataset and two other test sets [22]. We found that CoMFA performed better for the training set as well as for predicting two different test sets.

To develop a CoMFA model, the molecules of interest must first be aligned to maximize superposition of their steric and electrostatic fields in 3D. The critical and difficult aspect of the CoMFA procedure is choosing the most appropriate set of alignment rules for a structurally diverse training set. Fortunately, crystal structures of the ER complex of four ligands were published [24,25] that aided the derivation of rational CoMFA alignment rules. The CoMFA model based on the crystal structure-guided alignment is statistically robust. With the conventional $r^2 = 0.91$ (r: the correlation coefficient) and the cross-validated q^2 (leave-one-out) = 0.66, (q: the predictive correlation coefficient) indicating that it is both internally consistent and highly predictive.

Model validation

Concordance, specificity, and sensitivity [32] are commonly used to assess the quality of a classification model, while a quantitative regression model is assessed using r^2 [14]. The current challenge in QSARs is no longer in constructing a model that is statistically sound to predict the activity within the training set, but in developing a model with the capability to accurately predict the activity of untested chemicals. Most experts in the QSAR field, as well as the present authors, assert that a model's predictive capability at least needs to be demonstrated using some sort of cross-validation procedure. All models developed in our laboratory are validated using the leave-one-out technique for quantitative models and leave-10-out for qualitative models. It is worthwhile to point out that cross-validation methods only assess the interpolation within the training set, and have a limited ability to validate the prediction of untested chemicals that are structurally different from the training-set chemicals.

When additional data are available, the models are validated by predicting chemicals, not used in the training set but with the known activity data (the test set). The major difference between the cross-validation and external validation is that the chemicals selected in the latter case are in a sense of randomness, providing a more rigorous evaluation of the model's predictive capability for untested chemicals. We strongly believe that the confidence in a model's predictive capability can be tested and validated when predictability has been demonstrated on an external test set.

It is usual that the QSAR training set barely contains enough chemicals to develop a statistically robust model such that setting aside chemicals for use in external validation (10~20 % of the dataset is recommended) is not possible. One approach for selecting a test set is to identify datasets in the literature with the same type of activity. In such cases, care must be taken to avoid interlaboratory and assay variability among different data sources. It is desirable that the potential test set has activity data measured in an assay protocol as similar as possible to that for the training set to keep the variability at minimum. Another important consideration in selecting the test set is to ensure that chemicals in the dataset relate to the real problem in question. Based on these considerations, the datasets reported by Kuiper et al. [33], Waller et al. [10], and Nishihara et al. [34] were selected as test sets. In Kuiper's study, the pure human ERα was used, whereas the mouse uterine cytosol that primarily contains ERα was used in Waller's data. The Nishihara dataset contains 517 chemicals tested with the yeast two-hybrid assay, of which over 86 % are pesticides and industrial chemicals.

NCTR "FOUR-PHASE" SYSTEM AS A PRIORITY SETTING TOOL FOR REGULATORY APPLICATION

The objective of priority setting is to rank order, from most to least importance, a large number of chemicals to undergo more resource-intensive experimental evaluations. The strategy of QSAR is highly dependent on the application domain and goals. The QSAR used in drug discovery is to increase the chance of finding active chemicals or "hits" that may become "lead" chemicals, and false positives are of great concern. In contrast, a good priority setting method for the regulatory application should have minimal false negatives. False negatives constitute a crucial error, because they will be assigned a lower priority for subsequent evaluation. In addition, the methods should provide reasonable quantitative accuracy for true positives, as those with higher affinities will generally be of higher priority.

The QSAR procedures described in the previous section have strengths as well as weaknesses, and they all have a degree of prediction error. All procedures and particularly those that only provide active/inactive predictions can be optimized to minimize either the overall prediction error or the false negative or positive rate. Decreasing false negatives is achieved at a cost of increasing false positives and vice versa. Because selecting a single procedure is problematic, we adopted an approach of rationally combining different QSAR procedures into a sequential "Four-Phase" scheme according to the strength of each type of procedures. A progressive phase paradigm is used to screen out chemicals and thus reduce the number of chemicals to be considered in each subsequent phase. The four phases work in a hierarchical manner, incrementally reducing the size of the dataset while increasing precision of the prediction during each phase. Within each phase, different models are selected to work complementarily in representing key activity-determining structure features in order to minimize the rate of false negatives. The overall architecture of the NCTR "Four-Phase" system for identification of ER ligands is illustrated below and in Fig. 5.

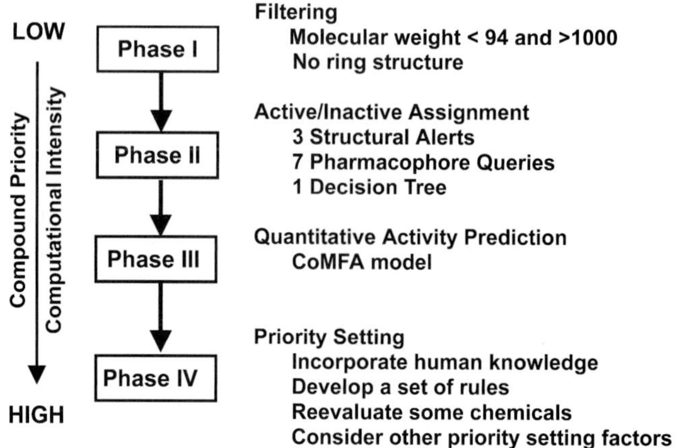

Fig. 5 NCTR "Four-Phase" system as a priority-setting tool.

Phase I: Filtering – Two filters, the molecular weight range and ring-structure indicator described above, were selected to efficiently eliminate chemicals very unlikely to have ER binding activity. As shown in Table 1, the two rejection filters correctly eliminated 6 and 98 inactive chemicals from the NCTR and Nishihara datasets, respectively. No false negative was introduced with these two rejection filters. For the Nishihara dataset, the data size was reduced by some 21 %. This suggests that, for real-world applications, the rejection filters significantly reduce the number of chemicals for further evaluation with minimum risk of introducing false negatives.

Table 1 Results of two rejection filters for the NCTR, Nishihara and Walker datasets*.

Datasets		Eliminated by				Number (%) of eliminated chemicals
		MW range				
		MW		Ring		
Sets	Size	Active	Inactive	Active	Inactive	
NCTR	232	0	0	0	6	6 (2.6 %)
Nishihara	463	0	28	0	89	98 (21.2 %)
Walker	558 230	16 048		1495		16 689 (28.7 %)

*Table lists the number of chemicals eliminated by either MW range or lack of ring criteria as well as their combination. No active chemical was rejected by these two filters.

Phase II: Active/Inactive Assignment – This phase categorizes chemicals from Phase I as either active or inactive. The three structural alerts, seven pharmacophore queries, and the Decision Tree classification model (11 models in total) discussed above were used in parallel to discriminate between active and inactive chemicals. To ensure the lowest possible false negative rate in Phase II, a chemical predicted to be active by any of the 11 models is presumed active and subsequently evaluated in Phase III, while only those predicted to be inactive by all these models are deemed inactive and eliminated from further evaluation. Since structural alert, pharmacophore, and Decision Tree methods incorporate and weigh differently the various structural features that endow a chemical with the ability to bind the ER, the combined outputs derived from the three approaches are complementary in minimizing false negatives.

All active chemicals in the NCTR, Waller [10], Kuiper [33], and Nishihara [35] datasets were identified by combining the 11 models.

Phase III: Quantitative Predictions – In Phase III, the CoMFA model described above [22] was used to make a quantitative activity prediction for chemicals categorized as active in Phase II.

Chemicals with higher predicted binding affinity are ranked at higher priority for further evaluation in Phase IV.

Phase IV: Rule-Based Decision-Making – In this final stage of the integrated system, we believe that a set of rules needs to be developed as a knowledge-based or expert system to make a priority setting decision. The system is useful only after incorporating accumulated human knowledge and expertise (i.e., rules). This system can make decisions on individual chemicals based on the rules in its knowledge base. Computational chemists, toxicologists, and regulatory reviewers should jointly develop and define the rules. The following are suggestions for such rules:

1. Special attention needs to be placed on the following chemicals, which may need to be reevaluated by assaying or modeling according to the "IF-THEN" scheme depicted in Fig. 6 [20]:
   1) The chemical is predicted to be inactive, but its structure information has been modified during structural preprocessing, e.g., by "correcting" such chemicals as mixtures and organic salts into separate entities.
   2) The chemical whose structure is dissimilar to all those that have been used to train and test the models.
   3) The chemical is active in Phase II, but inactive in Phase III.

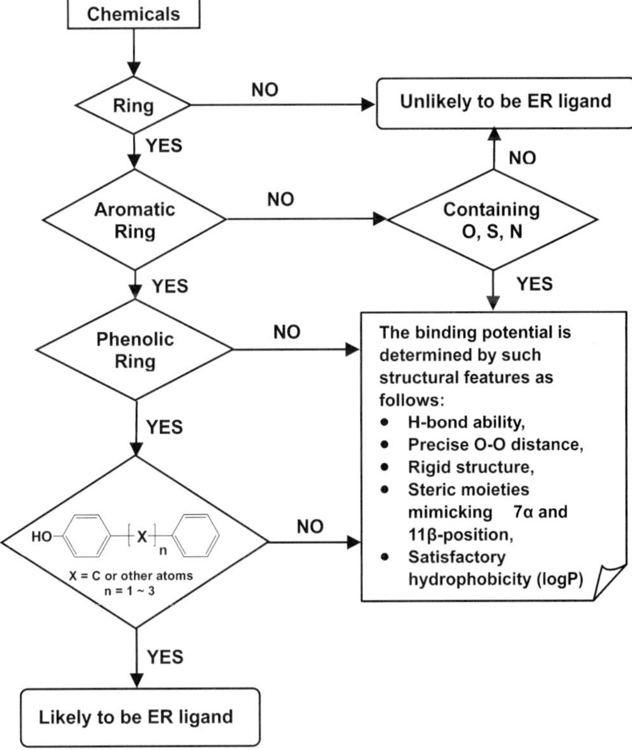

Fig. 6 Identification of ER ligands using a set of "IF-THEN" rules: (a) IF non-cyclic structure, THEN it is unlikely to be an ER ligand; (b) IF a chemical has non-aromatic ring structure, THEN unlikely to be an ER ligand if does not contain an O, S, N or other heteroatoms for H-bonding. Otherwise, binding potential is dependent on the occurrence of key structural features; (c) IF chemical has a non-OH aromatic structure, THEN binding potential is dependent on the occurrence of the key structural features; and (d) IF chemical contains a phenolic ring, THEN it tends to be an ER ligand if it contains any additional key structural features. For the chemicals containing a phenolic ring separated from another benzene ring with the bridge atoms ranging from none to three, it will most likely be an ER ligand.

2. Information on the level of human exposure and production, environmental fate, and other public health related parameters can be used independently or jointly incorporated for priority setting.

The NCTR "Four-Phase" system has been validated using a number of existing datasets, including the E-SCREEN assay data [36], the yeast two-hybrid reporter-gene assay data [35], and other datasets [10,37–41]. The system has produced no false negative for these test datasets. The system was recently applied to two environmental datasets recognized by EPA as important chemical subsets of potential EDCs:

1. HPV-Inerts dataset, containing 623 high-production-volume inerts (HPV-Inerts), which is a subset of the Toxic Substances Control Act (TSCA) Inventory. The EPA is including HPV-Inerts in version 2 of the Endocrine Disruption Priority Setting Database (EDPSD2), and there was a need to prioritize HPV-Inerts for further experimental evaluation [42]. Of 623 chemicals, 166 chemicals were either mixtures or their structures were not available, leaving 457 chemicals for prediction.

2. Walker dataset [43], containing a large and diverse collection of known pesticides and industrial chemicals as well as some food additives and drugs. The database contains 92 964 chemicals, of which the Chemical Abstract Service (CAS) Registry number is available, that will probably have to be evaluated for their potential endocrine disruption. After eliminating chemicals for which structures were not available [43] and/or 3D structures could not be generated [32], a final dataset of 58 391 chemicals were predicted.

Table 2 summarizes the priority setting results for these two datasets using the NCTR "Four-Phase" system. Even when only the Phase I and II protocols are used, the system dramatically reduces the number of potential estrogens by some 80~85 %, demonstrating effectiveness in eliminating the most unlikely ER binders from further expensive experimentation. The Phase III CoMFA model further reduces the data size by about 5~10 %. Importantly, the quantitative binding affinity prediction from Phase III also provides an important rank-ordering value for priority setting.

Table 2 Size reduction of two environmental datasets processed by the NCTR "Four-Phase" system.

	HPV-Inerts	Walker
Original data size	457	58 391
After Phases I and II	15.7 %	12.0 %
After Phases III	9.8 %	–

QSAR APPLICATION IN PERSPECTIVE

The QSAR procedures are applied extensively to a wide range of scientific disciplines including chemistry, biology, and toxicology [44,45]. In both drug discovery and environmental toxicology [46], QSAR models are now regarded as a scientifically credible tool for predicting and classifying the biological activities of untested chemicals. However, analogous to misinterpretation of experimental results, QSAR results can also be misleading when limitations of the procedures are not well understood. Presented below are our experientially based suggestions of the salient aspects in applying QSARs for toxicology and regulation.

1. **Limitation in the fundamental principle of the QSAR** – Any QSAR model will produce some degree of error. This is partially due to the inherent limitation to predict a biological activity solely based on the chemical structure. One can argue from the principles of chemistry that the molecular structure of a chemical is key to understanding its physicochemical properties and ultimately its biological activity and the influence on organisms. Since both molecular structure and physico-

chemical properties are associated with the chemical itself, the relationship between structure and physicochemical properties should be apparent and, therefore, more accessible using the QSARs. In contrast, the biological activity of a chemical is an induced response that is influenced by numerous factors dictated by the levels of biological complexity of the system under investigation. The relationship between structure and activity is thus more implicit and thereby poses a more challenging problem in QSAR applications.

2. **Limitation in extrapolation of QSAR models** – A chemical can be represented in three distinct, but also related, structural representations such as 2D substructures, 3D pharmacophores, and physicochemical properties. If a biological mechanism is mainly related to the chemical structure (probably in the case for receptor binding), QSARs become meaningful using the aforementioned structural features. However, we often find that, even for a simple mechanism such as ER-binding, some features may well represent binding dependencies for one structural class, whereas other features will better represent binding dependencies for a different structural class [20]. In such cases, caution is warranted in interpreting QSAR results for the chemical classes that are not well represented in the training set. In other words, no matter how rigorous is the validation procedure used, the model may give incorrect predictions for some chemicals since the entire chemistry space of active chemicals is unknown. This realization is especially important when a QSAR model is used for regulatory application since it could lead to false negatives.
3. **Limitation in processing chemical structure** – Most computational chemistry programs accept only discrete organic chemicals for the descriptor calculation and QSAR modeling. Unfortunately, most toxicological databases contain chemicals that are not necessarily discrete organic chemicals, but sometimes are mixtures. Thus, it is necessary to process molecular structures of a toxicological dataset by "correcting" chemicals with separate entities (e.g., mixtures, organic salts, the presence of H_2O and HCl, etc.), and eliminating those chemicals whose descriptors cannot be calculated. Such procedures might lead to a prediction that might not reflect the real activity of the preprocessed chemical.
4. **Limitation in quality and transformation of biological data** – Predictions from any model are intrinsically no better than the experimental data employed to calibrate the model. Any limitations of the assay used to generate the training data apply equally to the model's predictions. Furthermore, additional limitations could be introduced by transforming the biological data for use in the QSAR modeling. When developing quantitative models, inactive chemicals are usually either not used because their exact values cannot be determined or included by assigning an arbitrary value. Either case will tend to decrease the model sensitivity in prediction of chemicals with low activity. For qualitative models, false negatives and false positives depend on the defined cut-off value to distinguish active from inactive. As the cut-off value is lowered, it is likely that the error will increase even for a well-designed and executed assay. The increased experimental error in close proximity to the cut-off value will be transferred to the QSAR model, which in turn will increase false prediction for chemicals with activity in this region. For example, as defined by EPA, the cut-off log RBA value to distinguish ER binders from non-ER binders is set to –3.0. This cut-off was used to develop the models in the NCTR "Four-Phase" system. There are 31 chemicals exhibiting the binding affinity in our assay that were assigned to be inactive because their RBA is below this cut-off value. Among these 31, 14 chemicals have the RBA within 0.3 log units of the cut-off value ("within 0.5 log units" is expected for a good assay). Since the real activity classification for these chemicals are unknown, assigning them arbitrarily as inactive to train QSAR models introduces errors in prediction for chemicals with similar structures. Thus, caution must be taken in interpreting QSAR prediction results for these chemicals. Apparently, a high confidence in the prediction can only be expected for chemicals with the log RBA more positive than –2.5.

While the results presented in this article clearly show both the feasibility and utility of using QSARs for various applications, it is important to realize that any QSAR model has to be considered as a *living* model that will be improved whenever new data is available. In other words, the model development should be a recursive process that alternates between incorporating new data in the model and using the model to choose new chemicals for assay [47,48]. As depicted in Fig. 7, the process starts with an initial set of chemicals from the literature for QSAR modeling [14–16]. Next, the preliminary QSAR models are used prospectively to define and rationalize a set of chemicals that may further improve the model's robustness and predictive capability. These new chemicals are assayed, and the data are then used to challenge and refine the QSAR model. Thus, the process results in the iterative increase of the chemistry space of the training set.

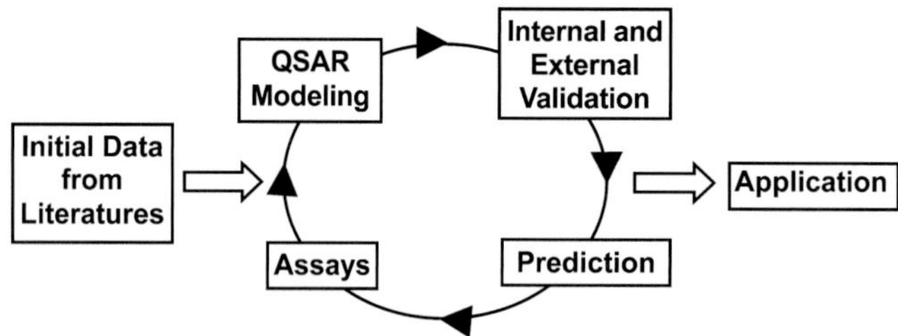

Fig. 7 The recursive process to develop QSAR models.

Several benefits accrue from the integration of the experimental and modeling efforts. Immediate feedback can be given to the experimentalists so that suspected assay problems can be rapidly investigated. Also, as the models evolve, the modelers can select the chemicals for subsequent testing, based on considerations of structural diversity and activity range. Each new assay data point coming from the lab becomes a challenge to the evolving model; the result is either further confirmation of its validity, identification of a limitation, or an outlier prediction. Failure of the model also provides important information, such as identification of the need for new data based on a rational understanding of the dependence of activity on structure. Regardless of the cause of model failure, a research hypothesis is spawned during each iteration process that should lead to new data and/or an improved training set, and, ultimately, an improvement to the living model.

ACKNOWLEDGMENTS

The authors gratefully acknowledge the U.S. Environmental Protection Agency (EPA) and the American Chemistry Council for financial support.

REFERENCES

1. B. Hileman. *Chem. Eng. News* **72**, 19–23 (1994).
2. B. Hileman. *Chem. Eng. News* **75**, 24–25 (1997).
3. R. J. Kavlock, G. P. Daston, C. DeRosa, P. Fenner-Crisp, L. E. Gray, S. Kaattari, G. Lucier, M. Luster, M. J. Mac, C. Maczka, R. Miller, J. Moore, R. Rolland, G. Scott, D. M. Sheehan, T. Sinks, H. A. Tilson. *Environ. Health Perspect.* **104** (Suppl. 4), 715–740 (1996).
4. US-Congress, The Food Quality Protection Act (FQPA) and the Safe Drinking Water Act (SDWA) (1996).
5. L. E. Gray, Jr. *Toxicol. Lett.* **102–103**, 677–680 (1998).
6. M. Patlak. *Environ. Sci. Technol.* **30**, 540A–544A (1996).
7. H. Fang, W. Tong, R. Perkins, A. Soto, N. Prechtl, D. M. Sheehan. *Environ. Health Perspect.* **108**, 723–729 (2000).

8. T. Zacharewski. *Environ. Health Perspect.* **106** (Suppl. 2), 577–582 (1998).
9. S. Bradbury, O. Mekenyan, A. GT. *Environ. Toxicol. Chem.* **15**, 1945–1954 (1996).
10. C. L. Waller, T. I. Oprea, K. Chae, H. K. Park, K. S. Korach, S. C. Laws, T. E. Wiese, W. R. Kelce, L. E. Gray, Jr. *Chem. Res. Toxicol.* **9**, 1240–1248 (1996).
11. T. E. Wiese, L. A. Polin, E. Palomino, S. C. Brooks. *J. Med. Chem.* **40**, 3659–3669 (1997).
12. B. R. Sadler, S. J. Cho, K. S. Ishaq, K. Chae, K. S. Korach. *J. Med. Chem.* **41**, 2261–2267 (1998).
13. W. Zheng and A. Tropsha. *J. Chem. Inf. Comput. Sci.* **40**, 185–194 (2000).
14. W. Tong, R. Perkins, R. Strelitz, E. R. Collantes, S. Keenan, W. J. Welsh, W. S. Branham, D. M. Sheehan. *Environ. Health Perspect.* **105**, 1116–1124 (1997).
15. W. Tong, R. Perkins, L. Xing, W. J. Welsh, D. M. Sheehan. *Endocrin.* **138**, 4022–4025 (1997).
16. W. Tong, D. R. Lowis, R. Perkins, Y. Chen, W. J. Welsh, D. W. Goddette, T. W. Heritage, D. M. Sheehan. *J. Chem. Inf. Comput. Sci.* **38**, 669–677 (1998).
17. L. Xing, W. J. Welsh, W. Tong, R. Perkins, D. M. Sheehan. *SAR QSAR Environ. Res.* **10**, 215–237 (1999).
18. R. Blair, H. Fang, W. S. Branham, B. Hass, S. L. Dial, C. L. Moland, W. Tong, L. Shi, R. Perkins, D. M. Sheehan. *Toxicol. Sci.* **54**, 138–153 (2000).
19. W. S. Branham, S. L. Dial, C. L. Moland, B. Hass, R. Blair, H. Fang, L. Shi, W. Tong, R. Perkins, D. M. Sheehan. *J. Nutr.* **132**, 658–664 (2002).
20. H. Fang, W. Tong, L. Shi, R. Blair, R. Perkins, W. S. Branham, S. L. Dial, C. L. Moland, D. M. Sheehan. *Chem. Res. Toxicol.* **14**, 280–294 (2001).
21. G. M. Anstead, K. E. Carlson, J. A. Katzenellenbogen. *Steroids* **62**, 268–303 (1997).
22. L. M. Shi, H. Fang, W. Tong, J. Wu, R. Perkins, R. Blair, W. Branham, D. Sheehan. *J. Chem. Inf. Comput. Sci.* **41**, 186–195 (2001).
23. H. Hong, N. Neamati, S. Wang, M. C. Nicklaus, A. Mazumder, H. Zhao, T. R. Burke, Y. Pommier, G. W. A. Milne. *J. Med. Chem.* **40**, 930–936 (1997).
24. A. M. Brzozowski, A. C. Pike, Z. Dauter, R. E. Hubbard, T. Bonn, O. Engstrom, L. Ohman, G. L. Greene, J. A. Gustafsson, M. Carlquist. *Nature* **389**, 753–758 (1997).
25. A. K. Shiau, D. Barstad, P. M. Loria, L. Cheng, P. J. Kushner, D. A. Agard, G. L. Greene. *Cell* **95**, 927–937 (1998).
26. A. Smellie, S. D. Kahn, S. Teig. *J. Chem. Inf. Comput. Sci.* **35**, 285–294 (1995).
27. A. Smellie, S. D. Kahn, S. Teig. *J. Chem. Inf. Comput. Sci.* **35**, 295–304 (1995).
28. A. Smellie, S. L. Teig, P. Towbin. *J. Comput. Chem.* **16**, 171–187 (1995).
29. L. M. Shi, W. Tong, H. Fang, R. Perkins, J. Wu, M. Tu, R. Blair, W. Branham, J. Walker, C. Waller, D. Sheehan. *SAR QSAR Environ. Res.* **13**, 69–88 (2002).
30. D. E. Clark and D. R. Westhead. *J. Comput. Aided Mol. Des.* **10**, 337–358 (1996).
31. S. Forrest. *Science* **261**, 872–878 (1993).
32. H. Hong, W. Tong, H. Fang, L. M. Shi, Q. Xie, J. Wu, R. Perkins, J. Walker, W. Branham, D. Sheehan. *Environ. Health Perspect.* **110**, 29–36 (2002).
33. G. G. Kuiper, J. G. Lemmen, B. Carlsson, J. C. Corton, S. H. Safe, P. T. van der Saag, B. van der Burg, J. A. Gustafsson. *Endocrin.* **139**, 4252–4263 (1998).
34. T. Nishihara, J. Nishikawa, T. Kanayama, F. Dakeyama, K. Saito, M. Imagawa, S. Takatori, Y. Kitagawa, S. Hori, H. Utsumi. *J. Health Sci.* **46**, 282–298 (2000).
35. T. Nishirara, J. Nishikawa, T. Kanayama, F. Dakeyama, K. Saito, S. Imagawa, S. Takatori, Y. Kitagawa, S. Hori, H. Utsumi. *J. Health Sci.* **46**, 282–298 (2000).
36. A. M. Soto, C. Sonnenschein, K. L. Chung, M. F. Fernandez, N. Olea, F. O. Serrano. *Environ. Health Perspect.* **103** (Suppl. 7), 113–122 (1995).
37. K. W. Gaido, L. S. Leonard, S. Lovell, J. C. Gould, D. Babai, C. J. Portier, D. P. McDonnell, *Toxicol. Appl. Pharmacol.* **143**, 205–212 (1997).
38. N. G. Coldham, M. Dave, S. Sivapathasundaram, D. P. McDonnell, C. Connor, M. J. Sauer. *Environ. Health Perspect.* **105**, 734–742 (1997).

39. E. J. Routledge and J. P. Sumpter. *J. Biol. Chem.* **272**, 3280–3288 (1997).
40. E. J. Routledge, J. Parker, J. Odum, J. Ashby, J. P. Sumpter. *Toxicol. Appl. Pharmacol.* **153**, 12–19 (1998).
41. C. A. Harris, P. Henttu, M. G. Parker, J. P. Sumpter. *Environ. Health Perspect.* **105**, 802–811 (1997).
42. J. D. Walker, H. Fang, R. Perkins, W. Tong. *QSAR Comb. Sci.* **22**, 89–105 (2003).
43. J. D. Walker, C. W. Waller, S. Kane. "The Endocrine Disruption Priority Setting Database (EDPSD): A Tool to Rapidly Sort and Prioritize Chemicals for Endocrine Disruption Screening and Testing". In: *Handbook on Quantitative Structure Activity Relationships (QSARs) for Predicting Chemical Endocrine Disruption Potentials*, J. D. Walker (Ed.), SETAC Press, Pensacola, FL (2001).
44. C. Hansch and A. Leo. *Exploring QSAR - Fundamentals and applications in chemistry and biology*, American Chemical Society, Washington, DC (1995).
45. C. Hansch, B. R. Telzer, L. Zhang. *Crit. Rev. Toxicol.* **25**, 67–89 (1995).
46. S. Bradbury. *Toxicol. Lett.* **79**, 229–237 (1995).
47. W. Tong, W. J. Welsh, L. Shi, H. Fang, R. Perkins. *Environ. Toxicol. Chem.* **22**, 1666–1679 (2003).
48. R. Perkins, H. Fang, W. Tong, W. J. Welsh. *Environ. Toxicol. Chem.* **22**, 1680–1695 (2003).

Pure Appl. Chem., Vol. 75, Nos. 11–12, pp. 2389–2396, 2003.

Workshop 1.3

QSAR prioritization of chemical inventories for endocrine disruptor testing*

Patricia Schmieder[1,‡], Ovanes Mekenyan[2], Steven Bradbury[3], and Gilman Veith[1]

[1]U.S. Environmental Protection Agency, Office of Research and Development, National Health and Environmental Effects Research Laboratory, Mid-Continent Ecology Division, 6201 Congdon Blvd., Duluth, MN 55804, USA; [2]Laboratory of Mathematical Chemistry, University "Prof. As. Zlatarov", 8010 Bourgas, Bulgaria; [3]U.S. Environmental Protection Agency, Office of Pesticide Programs, Environmental Fate and Effects Division, Washington, DC 20460, USA

Abstract: Binding affinity between chemicals and the estrogen receptor (ER) serves as an indicator of the potential to cause endocrine disruption through this receptor-mediated endocrine pathway. Estimating ER-binding affinity is, therefore, one strategic approach to reducing the costs of screening chemicals for potential risks of endocrine disruption. While measuring ER binding with in vitro assays may be the first choice in prioritizing chemicals for additional in vitro or in vivo estrogenicity testing, the time and costs associated with screening thousands of chemicals is prohibitive. Recent advances in 3D modeling of the reactivity of flexible structures make in silico methods for estimating ER binding possible. One technique, the common reactivity pattern (COREPA) approach, was applied to development of reactivity patterns for ER relative binding affinity based on global nucleophilicity, interatomic distances between nucleophilic sites, and local electron donor capability of the nucleophilic sites. The reactivity patterns provided descriptor profiles for order-of-magnitude RBA ranges of training set chemicals. An exploratory expert system was subsequently developed to predict RBA and rank chemicals with respect to potential estrogenicity. A strategy is presented for extending initial exploratory 3D QSAR models beyond current training sets to increase applicability to more diverse structures in large chemical inventories.

INTRODUCTION

The concern over potential hazards posed by environmental chemicals that can mimic endogenous estrogens is evidenced by screening and testing programs under development by governments of the United States, European Union, and Japan. Assessing the risk of potential endocrine-disrupting chemicals (EDCs) is particularly challenging due to the diverse nature of chemical structures believed to initiate activity through a common toxic pathway, i.e., direct chemical interaction with the estrogen receptor (ER). When attempting to institute an efficient approach to predict the potential hazard of EDCs, or any chemical hazard, the following question arises. How are efforts best focused on those chemicals most likely to cause adverse effects without empirically testing all chemicals of regulatory concern?

*Report from a SCOPE/IUPAC project: Implication of Endocrine Active Substances for Human and Wildlife (J. Miyamoto and J. Burger, editors). Other reports are published in this issue, *Pure Appl. Chem.* **75**, 1617–2615 (2003).

‡Corresponding author

The use of quantitative structure–activity relationships (QSARs) is a powerful in silico technique useful for prioritizing chemicals for subsequent empirical assessments. Specific challenges faced in developing and applying QSAR to rank and prioritize within large chemical inventories are addressed herein, and include the proper selection of methods that reflect the state-of-the-science, both in regard to what is understood of the underlying biology and chemistry, as well as what computational approaches are possible and applicable.

A critical consideration when applying predictive models is establishing the applicability of any model to a defined universe of chemicals of concern in a particular risk assessment. Thus, the most difficult step may not be QSAR development, per se, but determining when a model is sufficiently improved to satisfy criteria defined for model acceptance. Recently, an international effort has been undertaken to define criteria for acceptance of QSAR in risk assessment [1]. These criteria include the ability to provide predictions for a specified chemical domain of concern within acceptable limits of error. The users (e.g., the regulatory community applying the models) specify the chemical domain of concern and the limits of acceptance. Acceptable error is dependent upon the risk assessment scenario being employed. For instance, predictions used to prioritize chemicals for further testing can have larger uncertainty bounds than those upon which final regulatory determinations are to be made.

A scheme for iterative QSAR model development, as presented in Fig. 1, allows for model development, assessment, and improvement through a series of steps, following rigid scientific criteria, until the model is determined acceptable for use in risk assessment. The scheme is conceptually based upon earlier QSAR approaches used to predict acute toxicity to aquatic organisms for thousands of industrial chemicals [2–7]. Lessons learned from these earlier efforts emphasize the importance of defining the chemical structure space to which models will be applied, and employing approaches which are designed to obtain the maximum amount of new information from any empirical testing conducted for purposes of model improvement.

The scheme in Fig. 1 starts with identification of an adverse effect of concern in a risk assessment. Knowledge, or hypothesis, of the underlying toxicity pathway likely producing the adverse effect is needed to identify a well-defined biological endpoint that is plausibly related to the adverse outcome. An endpoint is also well defined if it can be demonstrated that the chemical concentration and form applied in a study are directly linked to the observed alteration in the biological endpoint. In the case of

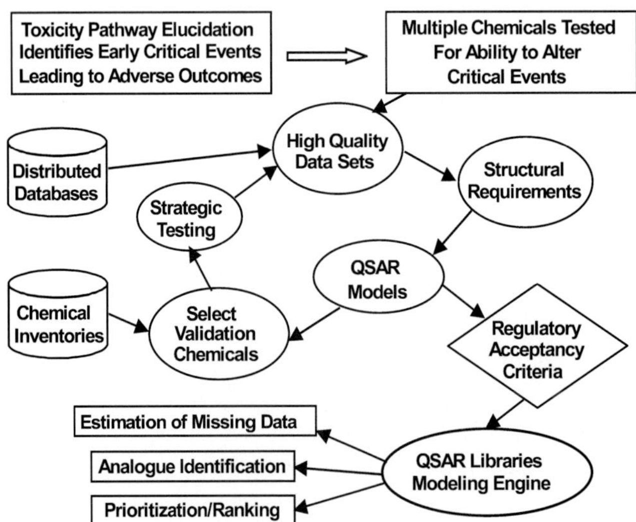

Fig. 1 Iterative QSAR model development scheme: A generalized scheme for development of predictive models to meet regulatory acceptance criteria for applicability to large chemical inventories.

receptor binding, this would require, for instance, assurance that the chemical applied (and not a metabolite of the chemical) is responsible for the apparent affinity, and that displacement of the endogenous ligand is due to competitive binding at the active site and not to some other alteration in protein structure. The next step described in Fig. 1 involves compiling a high-quality data set for the well-defined endpoint, obtained through laboratory determinations or by accessing well-documented databases. It is then possible to identify chemical structural parameters associated with the biological activity, followed by, or simultaneous to, development of a preliminary QSAR model using a training set of high-quality data. The model is evaluated against predetermined criteria (defined in consultation with intended model users). A first iteration through the scheme will likely require proceeding with the selection of validation chemicals. Chemicals used to evaluate and improve the model are selected from the chemical universe of concern, i.e., to which the model will be applied. These strategically selected chemicals are subsequently tested, the results compared to model predictions, and the new information added to expand the training set and improve the model.

Strategic selection of test chemicals intended for model improvement is greatly facilitated by first determining the extent to which a training set covers the larger chemical universe (the domain of applicability) and identifying areas of the chemical universe not well represented in the initial training set. The model is most efficiently improved when strategic chemical selection and testing are focused on the most dissimilar chemicals in the domain of applicability, those chemicals most distinct from each other and most dissimilar from training set chemicals used to develop the model. The results obtained using a strategic testing approach also provide the means to assess the relevance of molecular descriptors from the original model to the newly tested chemicals, as well as allowing an assessment of the model's predictive power. Incorporation of new test data to expand the training set also provides an incremental improvement in predictive power with an increase in the number of chemicals relative to the number of descriptors. The QSAR model is advanced using the expanded training set and reevaluated using criteria established for regulatory acceptance. If criteria are met, the model will be considered for the intended use, e.g., priority setting, hazard identification, or chemical classification. The application of this iterative scheme, as presented in Fig. 1, is essential to the successful development of QSAR for use in risk assessments where such models are applied to large numbers of chemicals showing wide structural variation. The following sections address specific considerations important to the development of robust predictive models for: (1) application in chemical risk assessment (specifically, prioritization of chemicals for further testing); (2) to provide defensible coverage of a large and diverse chemical universe; and (3) based upon a well-defined endpoint for which plausible linkage to an in vivo effect of concern has been established.

RECEPTOR BIOLOGY AND CHEMICAL INTERACTIONS

The endocrine system is a multifaceted and complex set of interconnected pathways that are controlled at multiple points through numerous feedback loops. Determining key points of vulnerability along a pathway, which could be altered by xenobiotic chemical interaction, is necessary to assess, and subsequently predict, the potential for chemical disruption of endocrine systems. Significant advancements have been made in the understanding of actions and interactions of components of the nuclear hormone receptor superfamily, such as, requirements for activation factors, cross-talk with other signal transduction pathways, and critical regulatory enzymes associated with endocrine pathways. Perhaps the most studied means of chemical disruption of endocrine systems, to date, is that initiated by direct chemical–ER interaction. Even in the absence of complete pathway elucidation, enough is known to reliably assert that the binding of xenobiotics to the ER (of both mammalian and nonmammalian species) can potentially alter normal reproduction. It is also now appreciated that a multistep ligand–receptor–effector process is involved in gene activation through this transcription factor, with the induction of protein conformational change (and potentially conformational change of flexible ligands) identified as an important part of this interaction [8].

The role of ligand conformation in the binding interaction has been especially troublesome. With respect to screening chemicals, active chemicals can be missed because the minimum energy conformation did not meet steric requirements for receptor binding or because the electronic descriptors of the minimum energy conformation did not indicate significant binding affinity. Through evaluation of both thermodynamic and kinetic considerations [9] we have shown that multiple stable conformational states can be found within a few kcal/mole of the lowest energy conformer (e.g., ±20 kcal/mol). Activation energies between these states are on the order of 3–4 kcal/mol [9–10], thus establishing the potential for multiple conformers to impart sufficient energy for binding. This suggests that all energetically reasonable conformers of a chemical, not solely a single lowest energy conformer, should be considered. Further, there can be considerable variability in the value of important stereoelectronic parameters among the many conformations [11]. Therefore, failure to assess the possible influence of conformational flexibility of ligands on their potential activity could lead to unreliable screening methods.

COREPA: QSAR evolution for flexible ligands

Until recently, applications of QSAR methods to prediction of adverse ecotoxicological effects for large chemical data sets were limited to 2D depiction of chemicals for classification and model development [2,3,5–7,12]. This early reliance on 2D depiction was driven largely by computing time associated with 3D structures as well as with computing stereoelectronic parameters for a given structure. With greater emphasis on applying QSAR to predict toxic effects initiated by chemical binding to biological receptors, screening methods were required to detect not only those chemicals that fit the structural requirements for a given receptor but also whether each chemical had a plausible conformation and flexibility which permitted binding to the receptor. Recent advances in algorithms for rapidly converting 2D structures to an array of plausible 3D structures [13] now make it possible to develop 3D QSAR screening methods for libraries of thousands to tens of thousands of chemicals. Therefore, the use of 3D approaches is not only warranted, but is feasible. Various receptor-free and receptor-docking 3D QSAR approaches have been recently reviewed [14].

Departing from the minimum-energy presumption for structure gave rise to the significant 3-D QSAR challenges of quantifying the flexibility of chemicals (conformers) when attempting to predict chemical interaction with macromolecules. Software to systematically generate plausible conformations for chemicals and substitute distributions of molecular descriptors for point estimates in developing QSARs is centered around algorithms to identify **CO**mmon **RE**activity **PA**tterns (COREPA) through probabilistic classification [11,15–18]. In this present research on binding affinity to the ER, COREPA was used to solve a number of combinatorial problems. First of all, the experimental data indicate that there are multiple binding mechanisms at the receptor. Secondly, the training set of chemicals contains many flexible chemicals with numerous conformations. Thirdly, each of the projected conformations produce a complete set of stereoelectronic and other molecular descriptors to form a distribution of each descriptor for a given test chemical. COREPA is designed to manage the combinatorial explosion of parameters for each test chemical and identify patterns between molecular descriptors and binding affinity.

The complexity of modeling conformer flexibility at receptors arises from the major computational factors that include rotation around each acyclic single bond, flipping of free corners in saturated rings, and creating mirror images of saturated ring systems. The result is that a single flexible chemical may have hundreds of energetically reasonable conformers and COREPA applies a "genetic" algorithm [22,23] to select representative conformers for stereoelectronic parameter estimations [20,21]. After a 2D structure is automatically converted to an extensive list of plausible 3D conformers, parameters such as E_{HOMO} and E_{LUMO} (energies of the highest occupied, and lowest unoccupied molecular orbitals, respectively), distance between and charges on electronegative atoms, and related stereoelectronic parameters are computed for each conformer and added to the topological indices for the chemical. The COREPA data sets are then used with training databases to explore possible relationships using multivariate probabilistic clustering algorithms to identify the structural requirements for different classes of binding affinity and formulation of QSARs.

The COREPA can also be used in a screening mode for large inventories of heterogeneous chemicals by matching the 3D structural requirements for ER binding with each chemical in the inventory. This screening process not only checks if the structure requirements of a minimum energy structure fits the receptor but also whether each chemical is flexible enough to conform to the receptor. Of course, to screen thousands of chemicals in large inventories, the computational time needed can be substantial. While the combinatorial explosion may still be a limiting factor in screening thousands of chemicals, it is feasible to compute permanent conformation files for each chemical in the inventory just once and use these data in screening assessments for many different QSAR models. Advanced statistical approaches for simultaneous multidimensional parameter estimation and comparison to classify biological activity of chemicals are also under development. These approaches, which are currently being developed for classification of large chemical databases are described elsewhere [19].

Prioritizing within large chemical inventories: A European Union case study

The development of a preliminary COREPA model to predict the ER-binding affinity was presented by Bradbury et al. [11], with extension to more diverse structures (from 46 to 115 chemicals) and species (from hER to rat and mouse ER) [24]. The model was experimentally tested within the Endocrine Disruption Activity of Environmental Pollutants project administrated by the European Union (EU). It was used to predict the rank order of chemicals in EU inventories, which might potentially disrupt endocrine systems through the ER signaling pathway. A subset of chemicals predicted to have activity at varying affinity levels was tested in a series of receptor binding, gene activation, and in vivo assays [20].

Consistent with a predicted extremely small occurrence rate of ER binders among all industrial chemicals, the in silico screening did not identify any chemicals that would bind strongly (RBA > 10 %) within EU high production volume chemicals. Of 907 high production volume EU chemicals assessed, seven compounds were predicted to have some, but low, ER relative binding affinity (RBA ~ 1 %). Four of these were randomly selected for empirical evaluation of hER binding and gene expression and found to elicit weak estrogenic activity. An additional 14 chemicals of the 900 predicted to be inactive were randomly sampled and found to have no activity, as predicted. The model was further used to predict potential for ER binding for an additional 63 000 lower production volume chemicals from the EU IUCLID database. Ten chemicals, of 200 predicted to have RBA > 10 %, were randomly selected for testing and all of them were found to be strongly estrogenic in the two assays used previously. These results were confirmed by additional in vitro assays.

The EU exercise represents the progression through the first half of the scheme in Fig. 1. While results were promising, completion of this scheme would require an evaluation of the extent of coverage provided by the initial model training set, in relation to a chemical inventory for which predictions are sought, e.g., EU high-production-volume chemicals. Techniques are available to visualize training sets within the chemical structure space of the universe of applicability (Fig. 2), using stereoelectronic parameters associated with ER binding [11]. As mentioned previously, the next step in application of the iterative model development scheme is to strategically choose chemicals to test that would give maximize the coverage of unknown structure space. The goal is to increase the size of the more certain "interpolation space" of the model, and reduce the size of the less certain "extrapolation space". For instance, the hatched points in Fig. 2 show chemicals in a chemical domain of concern that are all outside of the training set region but "evenly" distributed across the outer edges of the extrapolation region. Algorithms are currently under development to systematically and iteratively select chemicals furthest from any tested chemical. Upon completion of empirical testing, the new data is assessed and, if necessary, added to the training set to improve its applicability. Theoretically the incremental gain in model improvement should diminish with each iteration through the entire scheme shown in Fig. 1. The num-

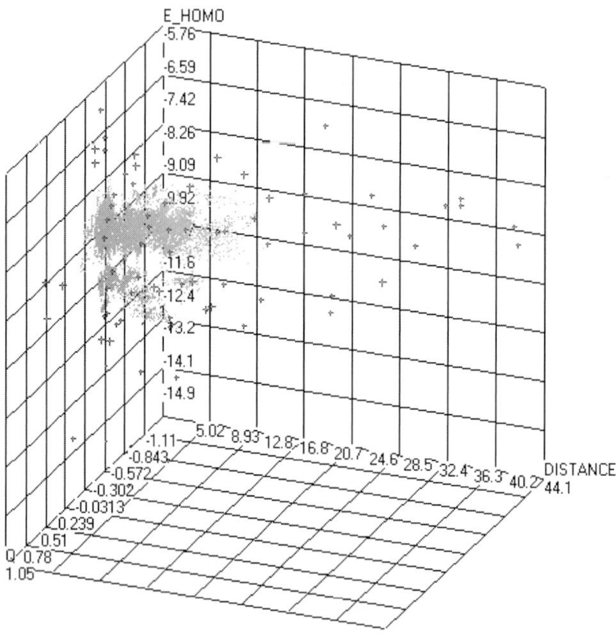

Fig. 2 Chemical selection to cover structure space: An illustration of the chemicals chosen to evenly represent chemical structure space (universe of applicability) of a large industrial chemical inventory, plotted using parameters found associated with ER-binding affinity in the COREPA-QSAR model: E_{HOMO}: energy of the highest occupied molecular orbital; DISTANCE: interatomic distance between nucleophilic sites; and Q: charge on nucleophilic sites. Filled points represent conformationally multiplied training set chemicals used to develop the initial QSAR model. Cross-hatched points represent those chemicals in the universe of applicability that are furthest (most dissimilar) from training set chemicals. Selection of these chemicals for empirical testing serves to expand the structural diversity of the training set to cover that encountered in the universe, thus increasing the applicability of the model.

ber of additional chemicals to test to improve the model will depend on many factors as previously described, and deserves further exploration.

The discussion so far has addressed how to assess and incrementally improve the domain of applicability of a QSAR model. Additional focus should also be given to improving the mechanistic transparency of predictive models. This requires verification of hypothesized chemical/biological interactions linked to the toxicological endpoint, in conjunction with more mechanistic rationale for chemical descriptors used to explain the biological/toxicological action predicted by a QSAR. Attempts to improve model transparency and assess model applicability are considered essential for broader international acceptance of a role for predictive models in human health and environmental risk assessments [1].

SUMMARY

A scheme was presented for development of predictive models for regulatory risk assessment. The prioritization of chemicals, within large inventories, for further testing to identify potential endocrine disruptors was used as an example. Successful application of the scheme requires the integration of empirical testing and further model refinements until regulatory acceptance criteria are met. Current knowledge of the biology of receptor-mediated pathways indicates that consideration of chemical conformational flexibility will improve predictive models for these biological endpoints. Considerable advances in computational techniques now allow the rapid generation and assessment of energetically reasonable conformers of chemicals. The COREPA approach was used to illustrate how recent advances have been applied to prioritization for further toxicological testing. However, additional advances are needed to achieve efficient application of these techniques to extremely large chemical data sets of regulatory concern. The successful application of the QSAR development scheme, as presented, will facilitate a close collaboration between modelers, data collectors (toxicologists), and users (risk assessors) to iteratively develop and improve models for regulatory acceptance.

DISCLAIMER

This paper has been subjected to review by the USEPA National Health and Environmental Effects Research Laboratory and approved for publication. Approval does not signify that the contents reflect the views of the agency, nor does mention of trade names or commercial products constitute endorsement or recommendation for use. Research associated with this paper was funded in part through an EPA cooperative research agreement (CR828826).

REFERENCES

1. J. Jaworska, M. Comber, C. Auer, C. J. van Leeuwen. International Workshop of the Regulatory Acceptance of (Q)SARs for Human Health and Environmental Endpoints. Proceedings of CEFIC-LRI and ECETOC Workshop, Setabul, Portugal, 4–6 March (2002).
2. G. D. Veith, D. J. Call, L. T. Brooke. *Can. J. Fish. Aquat. Sci.* **40**, 743–48 (1983).
3. G. D. Veith, D. L. DeFoe, M. L. Knuth. *Drug Metab. Rev.* **15**, 1295–1303 (1985).
4. C. M. Auer, J. V. Nabholz, K. P. Baetcke. *Environ. Health Perspect.* **87**, 183–187 (1990).
5. G. D. Veith and S. J. Broderius. *Environ. Health Perspect.* **87**, 207–11 (1990).
6. S. P. Bradbury. *Toxicol. Lett.* **79**, 229–237 (1995).
7. C. L. Russom, S. P. Bradbury, S. J. Broderius, D. E. Hammermeister, R. A. Drummond. *Environ. Toxicol. Chem.* **16**, 948–967 (1997).
8. J. A. Katzenellenbogen, B. W. O'Malley, B. S. Katzenellenbogen. *Mol. Endocrin.* **10**, 119–131 (1996).
9. J. M. Ivanov, O. G. Mekenyan, S. P. Bradbury, G. Schüürmann. *Quant. Struct.–Act. Relat.* **17**, 437–449 (1998).
10. T. E. Wiese and S. C. Brooks. *J. Steroid Biochem. Mol. Biol.* **50**, 61–73 (1994).
11. S. Bradbury, V. Kamenska, P. Schmieder, G. Ankley, O. Mekenyan. *Toxicol. Sci.* **58**, 253–269 (2000).
12. H. J. M. Verhaar, C. J. van Leeuwen, J. L. M. Hermens. *Chemosphere* **25**, 471–491 (1992).
13. J. M. Ivanov, S. H. Karabunarliev, O. G. Mekenyan. *J. Chem. Inf. Comput. Sci.* **34**, 234 (1994).
14. P. Schmieder, G. Ankley, O. Mekenyan, J. Walker, S. Bradbury. *Environ. Toxicol. Chem.* - QSAR Annual Review Issue **22** (8), 1844–1854 (2003).
15. O. Mekenyan, J. Ivanov, G. Veith, S. Bradbury. *Quant. Struct.–Act. Relat.* **13**, 302–307 (1994).
16. O. Mekenyan, J. Ivanov, S. Karabunarliev, S. Bradbury, G. Ankley, W. Karcher. *Environ. Sci. Technol.* **31**, 3702–3711 (1997).
17. O. Mekenyan, N. Nikolova, S. Karabunarliev, S. Bradbury, G. Ankley, B. Hansen. *Quant. Struct.–Act. Relat.* **18**, 139–153 (1999).
18. S. P. Bradbury, O. G. Mekenyan, G. T. Ankley. *Environ. Toxicol. Chem.* **17**, 15–25 (1998).
19. O. Mekenyan, N. Nikolova, P. Schmieder. *J. Mol. Struct. (TheoChem)* **622**, 147–165 (2003).
20. T. Hurst. *J. Chem. Inf. Comput. Sci.* **34**, 90 (1994).
21. O. Mekenyan, V. Kamenska, R. Serafimova, L. Poellinger, A. Brower, J. Walker. *SAR QSAR Environ. Res.* **13**, 579–595 (2002).

22. D. E. Goldeberg. *Genetic Algorithms in Search Optimization and Machine Learning*, Addison-Wesley, Reading, MA (1989).
23. A. W. R. Payne and R. C. Glent. *J. Mol. Graph.* **11**, 74 (1993).
24. O. Mekenyan, V. Kamenska, P. Schmieder, G. Ankley, S. Bradbury. *Toxicol. Sci.* **58**, 270–281 (2000).

Pure Appl. Chem., Vol. 75, Nos. 11–12, pp. 2397–2403, 2003.

Workshop 1.4

Nature of the ligand-binding pocket of estrogen receptor α and β: The search for subtype-selective ligands and implications for the prediction of estrogenic activity*

John A. Katzenellenbogen[‡], Rajeev Muthyala, and Benita S. Katzenellenbogen

Departments of Chemistry, Molecular and Integrative Physiology, University of Illinois, Urbana, 61801 IL, USA

Abstract: The ligand-binding pockets of estrogen receptor alpha and beta (ERα and ERβ) appear to have subpockets of different size and flexibility. To find ligands that will discriminate between the two ER subtypes on the basis of affinity or efficacy, we have prepared compounds of varying size, shape, and structure. We have evaluated the binding affinity of these compounds and their potency and efficacy as transcriptional activators through ERα and ERβ. In this manner, we have identified a number of ligands that show pronounced ER subtype selectivity. These studies also highlight the eclectic structure–activity relationships of estrogens and the challenges inherent in developing computational methods for the prediction of estrogenic activity.

INTRODUCTION

The actions of estrogens are mediated by the estrogen receptor (ER), an intracellular protein that functions as a ligand-regulated transcription factor [1]. There are two subtypes of the estrogen receptor, ERα and ERβ; these subtypes have different tissue distributions and are thought to regulate different estrogen responses [2]. Therefore, ligands that are selective in activating or inhibiting these two subtypes could have a desirable pattern of pharmacological activity that might be useful in regulating fertility, in treating menopausal symptoms, and in preventing and treating breast cancer [1].

LIGAND-BINDING POCKET OF THE ESTROGEN RECEPTOR EXHIBITS DYNAMIC AND PLASTIC CHARACTER

Before the structure of ER ligand-binding domains (LBDs) was elucidated by X-ray crystallography, pharmacophore analysis suggested that the hormonal steroid was surrounded by the protein, but that there was not a continuous, intimate contact between the protein and the ligand in the ligand-binding pocket [3]. The fact that strategically placed lipophilic substituents on steroidal estrogens, notably at positions 7α and 11β, can result in large increases in binding affinity suggested that there were some preformed pockets in the receptor at these points around the ligand (Fig. 1). In addition, the fact that even

*Report from a SCOPE/IUPAC project: Implication of Endocrine Active Substances for Human and Wildlife (J. Miyamoto and J. Burger, editors). Other reports are published in this issue, *Pure Appl. Chem.* **75**, 1617–2615 (2003).

[‡]Corresponding author

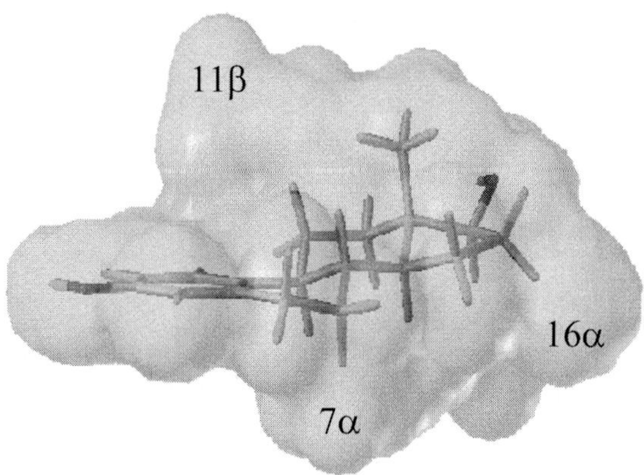

Fig. 1 Preformed pockets that were predicted to surround estradiol in the ligand-binding pocket of the estrogen receptor, according to pharmacophore analysis based on the affinity of substituted estradiol derivatives. The continuous surface represents the protein surface surrounding the ligand, which is shown as a skeleton representation [3].

larger substituents were tolerated at these and other sites on the steroid skeleton without major losses of binding affinity further suggested that there were regions of the ligand-binding pocket that were substantially deformable. On the other hand, it was clear that even small substituents at other ligand sites caused dramatic losses of binding affinity. Thus, *the ligand-binding pocket of ER seems to be characterized by regions that had extra space, regions that were readily deformable, and regions that lacked such space or plasticity* [3].

The crystal structures of various ER ligand complexes that have appeared over the past five years have confirmed this general view of the ER ligand-binding pocket and have refined it further to a great degree [4–6]. Overall, all of these structures show the ER LBD to be constituted almost completely of α-helical elements that are oriented largely in an antiparallel manner and to have a "triple sandwich" structure [7]. There are, however, substantive differences between the ER complexes with ligands of different size and shape.

For example, ER complexes with the agonists estradiol (E2) and diethylstilbestrol (DES) show the ligand engulfed by the LBD, but residing in pockets having substantially different shape (Fig. 2) [5,6]. The ER structure that surrounds E2 showed, as predicted, preformed pockets below the 7α and above the 11β positions that are not filled by the ligand. By contrast, in the ER-DES structure, these pockets are filled by the two ethyl groups that extend upward and downward from the ligand. In this structure, however, the portions of the ER that surround the periphery of the ligand in the E2 structure have moved inward to compensate for the lack of bulk in the middle region of the "thinner" DES ligand.

In substantial contrast to the two ER-agonist bound structures, ER complexes with the antiestrogens raloxifene and hydroxytamoxifen have an altered helical topology at the C-terminal region of the LBD [5,6]. Because the large basic side chain of these ligands, which is the signature structural element of their antiestrogenic character, does not fit into a fully closed ligand binding pocket, the C-terminal helix-12 of the LBD becomes repositioned (Fig. 2). By this movement, the ligand-binding pocket is opened up to accommodate the basic side chains, and helix-12 occludes a hydrophobic groove on the surface of the LBD, which in the ER agonist structure functions as a docking site for helical sequences present in certain coactivator proteins.

Thus, the ER-ligand crystal structures highlight the fact that the region of the ER LBD that surrounds the ligand is, in fact, not in uniform, intimate contact with the ligand. Furthermore, it can be reshaped and even reorganized by changes in ligand size and shape. The fact that the ER interacts with its ligands in a dynamic and plastic manner presents special challenges for methods that might be used

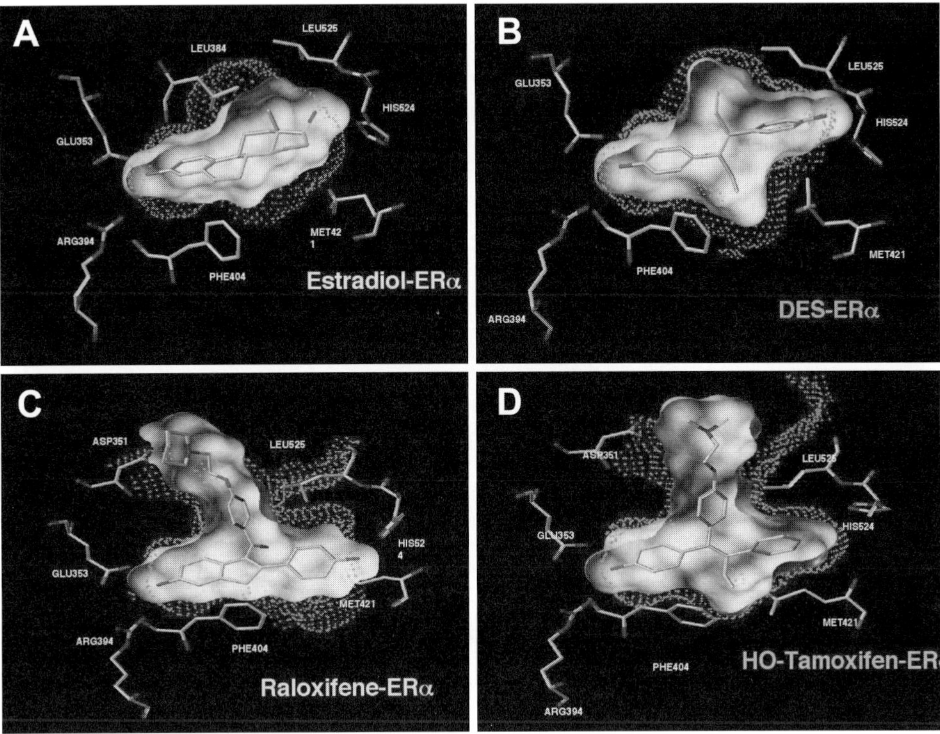

Fig. 2 X-ray crystal structures showing the fit of the ligands, estradiol (**A**), diethylstilbestrol (DES, **B**), raloxifene (RAL, **C**), and hydroxytamoxifen (TOT, **D**), in the LBD of the ERα. The continuous surface represents the protein surface surrounding the ligands, which are shown as skeleton representations [5,6].

to predict the estrogenic potency of new chemical entities, as well as for the development of various types of selective ligands for this receptor. Nevertheless, methods such as structure-based modeling of ER–ligand interaction and the development of structure/affinity/activity-based pharmacophore analysis provide the best basis for appreciating the "eclectic" nature of structure–activity relationships of estrogens.

At this point, it is not clear whether other members of the nuclear receptor family have a similar degree of dynamic and plastic character as does ER. The retinoic acid receptors appear to bind their ligands more tightly and show less variation of receptor structure as ligand structures change, but it is not clear how widespread this apparent reduced plasticity is [8–11]. In contrast, other members of the nuclear receptor superfamily, notably the orphan receptors, seem to bind their ligands even more loosely than does ER, with large preformed and unfilled pockets being noted in the crystal structures [12].

DEVELOPING LIGANDS THAT ARE SELECTIVE FOR THE ER SUBTYPES, ERα AND ERβ

The dynamic, plastic nature of the ER LBD also plays a role in ligand design, especially in the design of ligands that discriminate between the two ER subtypes, ERα and ERβ. The LBDs of ERα and ERβ have only 60 % sequence identity. However, the interior of the ligand-binding pockets of the two subtypes is much more similar, with 22 or 24 contact residues being identical [4]. Nevertheless, the two subpockets seem to have somewhat different size and flexibility, and this seems to be a critical distinction that is useful in the development of subtype-selective ligands.

To find ligands that will discriminate between the two ER subtypes on the basis of affinity or efficacy, we have prepared compounds of varying size, shape, and structure. We have evaluated the binding affinity of these compounds and their potency and efficacy as transcriptional activators through ERα and ERβ. Ligands have been designed in part through a consideration of the structure of the ER LBD and the universe of known ER ligands, and in part through a consideration of their expedited synthesis by combinatorial chemistry means [13]. In this manner, we have identified a number of ligands that show pronounced ER subtype selectivity (Fig. 3).

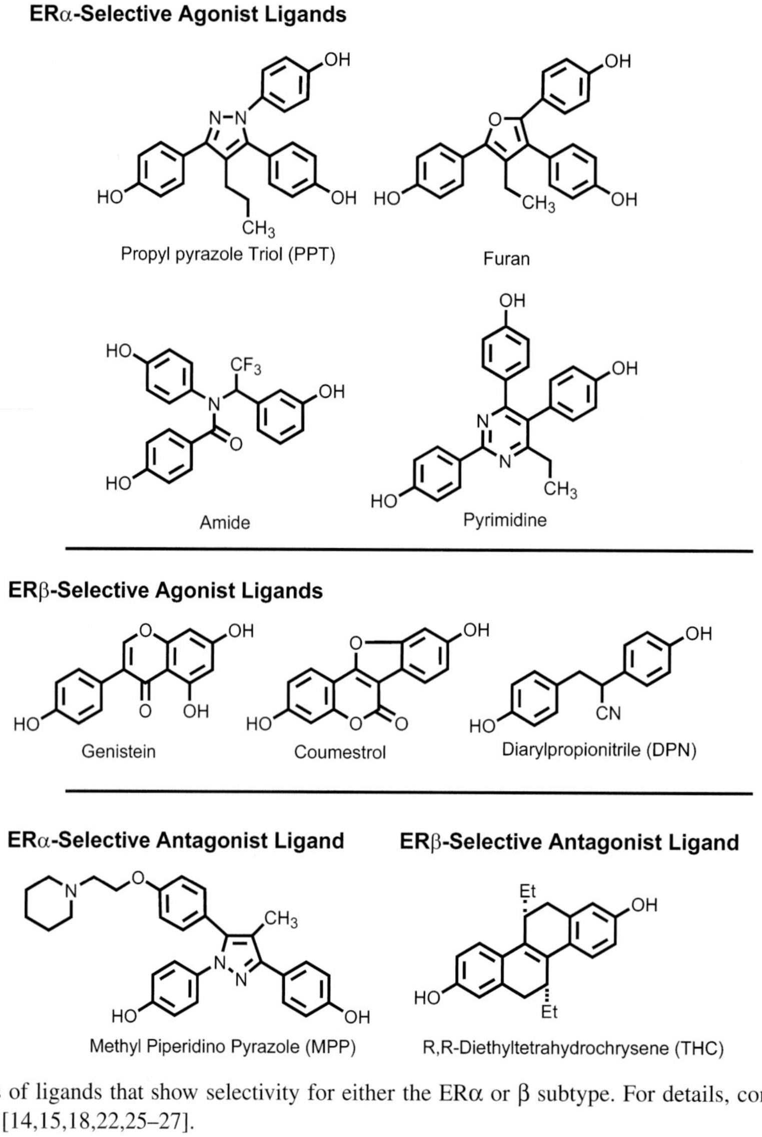

ERα-Selective Agonist Ligands

Propyl pyrazole Triol (PPT)

Furan

Amide

Pyrimidine

ERβ-Selective Agonist Ligands

Genistein

Coumestrol

Diarylpropionitrile (DPN)

ERα-Selective Antagonist Ligand

Methyl Piperidino Pyrazole (MPP)

ERβ-Selective Antagonist Ligand

R,R-Diethyltetrahydrochrysene (THC)

Fig. 3 Structures of ligands that show selectivity for either the ERα or β subtype. For details, consult references noted in the text [14,15,18,22,25–27].

Ligands that are selective for ERα

Certain triaryl amides show potency preferences as agonists for ERα that can be as great as 500-fold; they function as agonists on both ERα and ERβ, but in cell-based assays of gene transcription, they activate ERα at much lower concentrations [14].

Several triaryl-substituted five-membered heterocycles show exceptionally large potency and efficacy preferences for ERα [15–18]. The best of these are triaryl-alkyl-substituted pyrazoles and furans that function as complete ERα agonists, but are almost completely inactive on ERβ. The very best heterocycle we have found so far is called propyl pyrazole triol (PPT), a compound that is ca. 10 000-fold more potent on ERα than on ERβ and showed ERα-selective effects in vivo [19], but the molecular basis for its ERα selectivity is not fully understood [15]. Other larger ring heterocycles, such as tetrasubstituted pyrimidines and pyrazines, also retain greater potency and efficacy on ERα than on ERβ [20]. Ligands in these heterocycle classes can easily be synthesized in a combinatorial fashion [21].

Substituted tetrahydrochrysene ligands, in some cases, function as potent agonists on ERα, but are potent antagonists on ERβ [17,22]. This character is, however, a delicate function of substituent size and stereochemistry, and the best compound is the *R,R*-enantiomer of a diethyl-tetrahydrochrysene called *R,R*-THC. The difference in efficacy of *R,R*-THC on the two ER subtypes appears to arise from its optimal fit in the ERα ligand-binding pocket and its suboptimal fit in the slightly smaller ERβ pocket [23,24].

Ligands that are selective for ERβ

It has proved to be a greater challenge to discover ligands that are highly selective for ERβ. While certain phytoestrogens such as genistein and coumestrol have higher affinity for ERβ than ERα [24], in cell-based transcription assays, they do not show much difference in potency. Some simple diarylethane systems do show considerable affinity and potency preference as agonists on ERβ [25]. The best of these, called diarylpropionitrile (DPN), will activate ERβ at 100-fold lower concentrations than ERα. The ERβ selectivity of this ligand seems to result from its preferential interaction with a key methionine residue that is present only in the ERβ LBD [28]. In general, ERβ selective ligands seem to be smaller and more polar than ERα-selective ligands.

ER subtype-selective antagonists

Antagonists based on the highly ERα-selective pyrazoles have been developed that retain a large degree of ERα preference as antagonists [26,27,29,30]. The best of these, a compound called methyl piperidino pyrazole (MPP), can block estrogen action through ERα at a concentration that has insignificant inhibitory effect on ERβ [27]. So far, no compound has been reported that is more a potent antagonist of ERβ than of ERα. It should be noted that the THC compound, noted above, is a complete ERβ antagonist, but has substantial agonist activity on ERα [22].

GENERAL SCHEME FOR CONSIDERING DIMENSIONS OF ER SUBTYPE SELECTIVITY

Selectivity for the ER subtypes ERα and ERβ can be based on differences in ligand-binding affinity, ligand potency, or ligand efficacy. In general, one might expect that differences in ligand-binding affinity would be directly reflective of differences in potency; however, this is not universally true, and there can be surprising deviations in both directions. There are compounds that show substantial ER subtype differences in affinity, such as the phytoestrogens, yet do not show major differences in potency; by contrast, certain triarylamides that have only modest affinity preference for ERα show large potency differences in favor of ERα [14]. These disjunctions between affinity selectivity and potency selectivity probably have several origins, but most likely reflect preferential interactions the ER subtype ligand complexes have for the various coregulatory proteins that mediate their effects on transcription [31].

Ligands that show subtype differences in efficacy or degree of agonist vs. antagonist character between the two ER subtypes seem, at least in all reported cases, to show higher levels of agonism on ERα than on ERβ. The structural basis for this trend is not yet clear [23], but it might indicate that the most difficult pharmacological class of ER subtype-selective ligands to develop will be those that are more agonistic on ERβ than on ERα.

CONCLUSIONS AND IMPLICATIONS FOR MODELING OF ESTROGEN ACTIVITY

Our studies on the development of ER subtype-selective ligands, as well as the crystallographic determination of the structure of ER ligand complexes, illustrate the complexity of the structural response of the ER to ligands of different shape and size. In the process, these results also highlight the eclectic structure–activity relationships of estrogens. In the face of clear evidence for the dynamic and plastic character of the estrogen receptor, it can be anticipated that the most successful approaches to structure-based modeling of ER ligands or of structure–activity based pharmacophore modeling will be those that incorporate, in some appropriate fashion, consideration of the dynamic and plastic elements involved in ER–ligand interaction. It is hoped that such considerations will enable these models to be raised to a higher degree of reliability and utility for both pharmaceutical development and regulatory consideration.

ACKNOWLEDGMENTS

We are grateful for support of this research through grants from the National Institutes of Health, the Department of Energy, and the U.S. Army Medical Research Command. The contributions of numerous coworkers from the laboratories of John and Benita Katzenellenbogen are gratefully acknowledged.

REFERENCES

1. B. S. Katzenellenbogen and J. A. Katzenellenbogen. *Breast Cancer Res.* **2**, 335–344 (2000).
2. K. Pettersson and J. A. Gustafsson. *Annu. Rev. Physiol.* **63**, 165–192 (2001).
3. G. M. Anstead, K. E. Carlson, J. A. Katzenellenbogen. *Steroids* **62**, 268–303 (1997).
4. A. C. Pike, A. M. Brzozowski, R. E. Hubbard, T. Bonn, A. G. Thorsell, O. Engstrom, J. Ljunggren, J. Gustafsson, M. Carlquist. *EMBO J.* **18**, 4608–4618 (1999).
5. A. K. Shiau, D. Barstad, P. M. Loria, L. Cheng, P. J. Kushner, D. A. Agard, G. L. Greene. *Cell* **95**, 927–937 (1998).
6. A. M. Brzozowski, A. C. Pike, Z. Dauter, R. E. Hubbard, T. Bonn, O. Engström, L. Öhman, G. L. Greene, J.-A. Gustafsson, M. Carlquist. *Nature* **389**, 753–758 (1997).
7. J.-M. Wurtz, W. Bourguet, J.-P. Renaud, V. Vivat, P. Chambon, D. Moras, H. Gronemeyer. *Nat. Struct. Biol.* **3**, 87–94 (1996).
8. J.-P. Renaud, N. Rochel, M. Ruff, V. Vivat, P. Chambon, H. Gronemeyer, D. Moras. *Nature* **378**, 681–689 (1995).
9. P. F. Egea, N. Rochel, C. Birck, P. Vachette, P. A. Timmins, D. Moras. *J. Mol. Biol.* **307**, 557–576 (2001).
10. B. P. Klaholz, A. Mitschler, D. Moras. *J. Mol. Biol.* **302**, 155–170 (2000).
11. P. F. Egea, A. Mitschler, N. Rochel, M. Ruff, P. Chambon, D. Moras. *EMBO J.* **19**, 2592–2601 (2000).
12. H. E. Xu, M. H. Lambert, V. G. Montana, K. D. Plunket, L. B. Moore, J. L. Collins, J. A. Oplinger, S. A. Kliewer, R. T. Gampe, Jr., D. D. McKee, J. T. Moore, T. M. Willson. *Proc. Natl. Acad. Sci. USA* **98**, 13919–13924 (2001).
13. B. E. Fink, D. S. Mortensen, S. R. Stauffer, Z. D. Aron, J. A. Katzenellenbogen. *Chem. Biol.* **6**, 205–219 (1999).
14. S. R. Stauffer and J. A. Katzenellenbogen. *Bio. Med. Chem.* **8**, 1293–1316 (2000).
15. S. R. Stauffer, C. J. Coletta, R. Tedesco, J. Sun, B. S. Katzenellenbogen, J. A. Katzenellenbogen. *J. Med. Chem.* **43**, 4934–4947 (2000).

16. S. R. Stauffer, Y. R. Huang, C. J. Coletta, R. Tedesco, J. A. Katzenellenbogen. *Bio. Med. Chem.* **9**, 141–150 (2001).
17. J. Sun, M. J. Meyers, B. E. Fink, R. Rajendran, J. A. Katzenellenbogen, B. S. Katzenellenbogen. *Endocrinology* **140**, 800–804 (1999).
18. D. S. Mortensen, A. L. Rodriguez, K. E. Carlson, J. Sun, B. S. Katzenellenbogen, J. A. Katzenellenbogen. *J. Med. Chem.* **44**, 3838–3848 (2001).
19. H. A. Harris, J. A. Katzenellenbogen, B. S. Katzenellenbogen. *Endocrinology* **143**, 4172–4177 (2002).
20. U. Ghosh, D. Ganessunker, V. J. Sattigeri, K. E. Carlson, D. J. Mortensen, B. S. Katzenellenbogen, J. A. Katzenellenbogen. *Bioorg. Med. Chem.* **11**, 629–657 (2003).
21. S. R. Stauffer and J. A. Katzenellenbogen. *J. Comb. Chem.* **2**, 318–329 (2000).
22. M. J. Meyers, J. Sun, K. E. Carlson, B. S. Katzenellenbogen, J. A. Katzenellenbogen. *J. Med. Chem.* **42**, 2456–2468 (1999).
23. A. K. Shiau, D. Barstad, J. L. Radek, M. J. Meyers, K. W. Nettles, B. S. Katzenellenbogen, J. A. Katzenellenbogen, D. A. Agard, G. L. Greene. *Nature Struct. Biol.* **9**, 360–364 (2002).
24. G. G. J. M. Kuiper, B. Carlsson, K. Grandien, E. Enmark, J. Häggblad, S. Nilsson, J.-Å. Gustafsson. *Endocrinology* **138**, 863–870 (1997).
25. M. J. Meyers, J. Sun, K. E. Carlson, G. A. Marriner, B. S. Katzenellenbogen, J. A. Katzenellenbogen. *J. Med. Chem.* **44**, 4230–4251 (2001).
26. D. S. Mortensen, A. L. Rodriguez, J. Sun, B. S. Katzenellenbogen, J. A. Katzenellenbogen. *Bioorg. Med. Chem. Lett.* (2001).
27. J. Sun, Y. R. Huang, W. R. Harrington, S. Sheng, J. A. Katzenellenbogen, B. S. Katzenellenbogen. *Endocrinology* **143**, 941–947 (2002).
28. J. Sun, J. Baudry, J. A. Katzenellenbogen, B. S. Katzenellenbogen. *Mol. Endocrinol.* **17**, 247–258 (2003).
29. Y. R. Huang and J. A. Katzenellenbogen. *Org. Lett.* **2**, 2833–2836 (2000).
30. S. R. Stauffer, Y. R. Huang, Z. D. Aron, C. J. Coletta, J. Sun, B. S. Katzenellenbogen, J. A. Katzenellenbogen. *Bio. Med. Chem.* **9**, 151–161 (2001).
31. J. A. Katzenellenbogen, B. W. O'Malley, B. S. Katzenellenbogen. *Mol. Endocrinol.* **10**, 119–131 (1996).

Pure Appl. Chem., Vol. 75, Nos. 11–12, pp. 2405–2410, 2003.

Workshop 1.5

Fragment molecular orbital study of the binding energy of ligands to the estrogen receptor*

Kaori Fukuzawa[1], Kazuo Kitaura[2], Kotoko Nakata[3], Tsuguchika Kaminuma[4], and Tatsuya Nakano[3,‡]

[1]Fuji Research Institute Corporation, 2-3 Kanda Nishiki-cho, Chiyoda-ku, Tokyo 101-8443, Japan; [2]National Institute of Advanced Industrial Science and Technology, 1-1-1 Umezono, Tsukuba, Ibaraki 305-8568, Japan; [3]National Institute of Health Sciences, 1-18-1 Kamiyoga, Setagaya-ku, Tokyo 158-8501, Japan; [4]Chem-Bio Informatics Society, 4-3-16 Yoga #301, Setagaya-ku, Tokyo 158-0097, Japan

Abstract: We examined the published data for the binding affinity of typical ligands to the α-subtype of the human estrogen receptor with use of an approximate molecular orbital method applicable to interacting molecular clusters. An ab initio procedure for "molecular fragments" proposed recently to deal with such macromolecules as proteins was applied to the molecular orbital calculations. The receptor protein was primarily modeled using 50 amino acid residues surrounding the ligand. For a few ligand-receptor complexes, the binding energy was also calculated with use of 241 amino acid residues contained in the entire binding domain. No significant difference was found in the calculated binding energy between the complex modeled with ligand-surrounding 50 amino acids and that with residues of the entire domain. The calculated binding energy was correlated very well with the published relative binding affinity for typical ligands.

INTRODUCTION

The effect of estrogenic ligands is induced by their binding to the estrogen receptors (ERs) [1–3]. Since a variety of unknown compounds could bind to the ligand-binding domain (LBD) of the ER and exert hormone-like effects on human and wildlife health, the ER is an important research target for the development of therapeutic agents [3,4] as well as the screening of endocrine disruptors [5]. A number of experimental and theoretical efforts have been carried out for the mechanism of the interaction of ligands with the ER LBD. Most of the theoretical works, however, have stood on empirical force field approximations [6–8]. Although they are suited for calculating macromolecules in terms of the computational time, empirical approaches may not be accurate enough theoretically. Hoping to establish a time-saving and versatile computational procedure for biomacromolecules, we recently proposed the fragment molecular orbital (FMO) method [9]. Here, we report the result of our FMO study for the interaction of ligands with the α-subtype of ER carried out to elucidate its submolecular mechanism theoretically and accurately.

*Report from a SCOPE/IUPAC project: Implication of Endocrine Active Substances for Human and Wildlife (J. Miyamoto and J. Burger, editors). Other reports are published in this issue, *Pure Appl. Chem.* **75**, 1617–2615 (2003).

‡Corresponding author

METHODS

In the FMO method [9], a single molecule or a molecular cluster (a group of molecules interacting to each other noncovalently) is dealt with after being divided into fragments to which electron pairs are assigned according to certain rules. The molecular orbitals (MOs) for fragments and fragment pairs (combinations of two fragments) are calculated under conditions under which the orbitals are forced to localize as the closed shell within the corresponding region. For fragments to which no electron pair is allocated from the bond when detached in the fragmentation, the MO is built from usual atomic basis functions of the constituent atoms according to the conventional linear combination of atomic orbitals to yield molecular orbitals (LCAO-MO) framework. For fragments in which bonding electron pair is left, the atomic valence basis function of the partner atom, with which the fragment is connected originally, is used additionally in the LCAO-MO model. The initial calculation for each fragment MO yields the initial electron density distribution.

The Hamiltonian for each fragment is composed to include the terms for the electrostatic potential governed by electrons in the surrounding fragments and all nuclei in the molecule. Since the electrostatic potential of each fragment depends on the electron distribution of surrounding fragments, the electron density distribution of each fragment is calculated first using the initial electron distribution calculated in a manner described above. A set of "Schrödinger" equations for every fragment with the initial electron density is solved iteratively until the electron density distribution for all fragments converges self-consistently. Likewise, the Hamiltonian of each fragment pair has the terms for the potential arising from electrons in the surrounding fragments and the terms from every nuclear charge in the molecule. The set of equations for fragment pairs is solved using the electrostatic potential from the converged electron density distribution of the surrounding fragments. The potential energy of fragments and fragment pairs at the HF/STO-3G level is calculated to estimate the energy of the total system.

The ligand molecules examined here are shown in Fig. 1. The coordinates of heavy atoms in the ER complex of EST, RAL, DES, and OHT were fixed as being equivalent to those of the PDB files, entries 3ERE, 1ERR, 3ERD, and 3ERT, of the Research Collaboratory for Structural Bioinformatics (RCSB) Protein Data Bank (PDB), respectively [10–12]. For ligands such as ESTA, GEN, TAM, BISA, BISF, CLO, and OHC, the PDB files for the ERα complex are not available. Thus, the binding geometry of the first two ligands was approximated first by superimposing the "phenoxy" substructure of the phenol moiety on that of EST in the 3ERE, while that of the others was by superimposing their "phenoxy" substructure or corresponding phenyl group on that of OHT in the 3ERT file. Then, the geometry of GEN was approximated by that in the ERβ-GEN complex taken from the PDB 1QKM file. TAM, CLO, and OHC were modeled with the Insight II system [13] based on the geometry of OHT, and the others were optimized using the HF/6-31G(d) method. The geometry of hydrogen atoms was modeled with the Insight II system [13] and the CHARMm force field calculations [14].

Hydrogen bonds, occurring between the ligand and surrounding residues directly as well as through the mediation of a single water molecule, have been shown to stabilize the ER ligand binding [15]. In this study, the most stable geometry of the hydrogen bond network was calculated at the HF/6-31G(d) level [16] with use of a model molecular cluster consisting of such hydrogen-bonding residues in the LBD as Glu 353, Leu 387, Arg 394, and His 524, each of the ligands and the single water molecule (Model 3).

The entire LBD of the receptor protein containing 241 amino acid residues (Model 1) was used for the calculation only for some ligands. The binding domain was, however, primarily modeled with use of 50 amino acid residues "directly" surrounding the ligand (Model 2) as displayed in Fig. 2. To make the fragmentation of the receptor protein, the peptide chain was divided at the Cα atom into blocks of every two residues in a manner as shown in Fig. 3. The ligand as well as the hydrogen-bonding water molecule was treated as a single fragment.

All the FMO calculations were carried out with an FMO program package, ABINIT-MP [17], mostly on dual Pentium III 1-GHz clusters equipped with 32 processor units. The time required for cal-

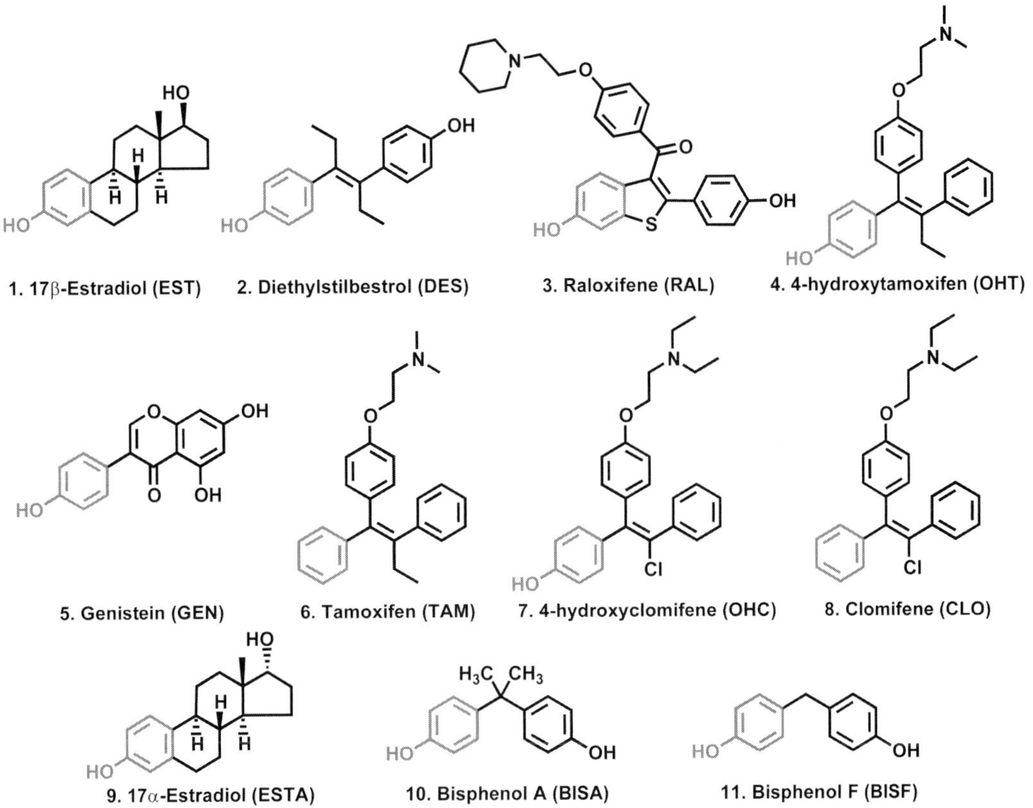

Fig. 1 Ligands used for the calculation of the binding energy. Light black substructures represent the moiety to be superimposed with the corresponding moiety in reference compounds.

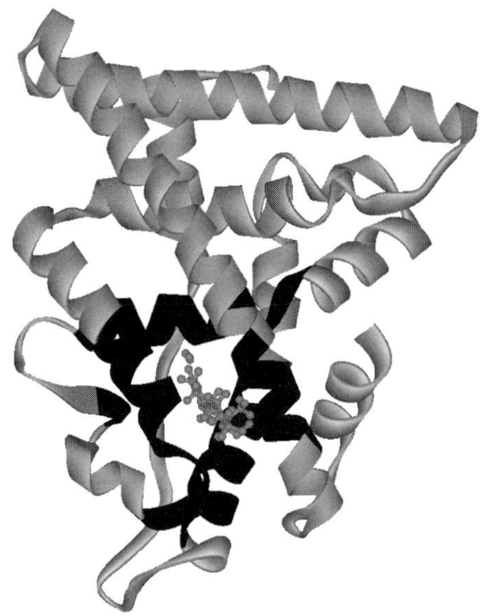

Fig. 2 The ribbon display of the ERα LBD complexed with 17β-estradiol (1, EST). Model 1 including 241 residues is shown as the entire picture. Fifty residues surrounding "directly" the ligand for Model 2 are dark-colored. The ligand and the water molecule are displayed inside the matrix using ball and stick.

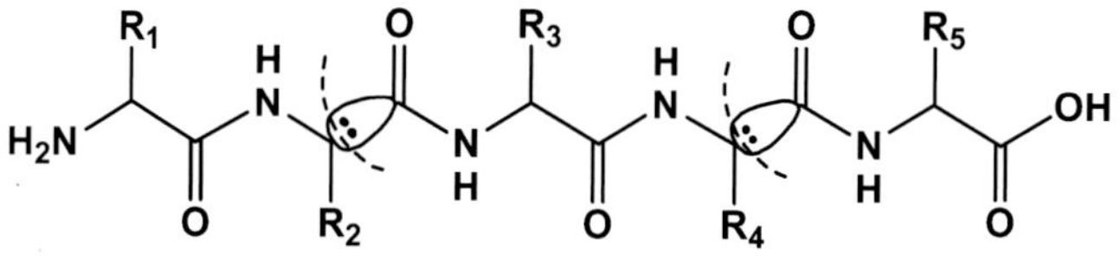

Fig. 3 Fragmentation of peptides indicated as broken arcs.

culating entire ERα LBD containing 241 residues with ca. 4000 atoms was about 14 h. The accuracy of the FMO method has been examined using crambin, a protein series with 46 residues [9]. The ab initio total energy values calculated at the HF/STO-3G level for [Pro^{22}, Leu^{25}]crambin with and without the FMO approximation are –17779.5030 and –17779.5024 a.u., respectively, corresponding to a difference below 0.5 kcal/mol. The computational time is "drastically" reduced with the FMO procedure compared to that without the FMO approximation.

RESULTS AND DISCUSSION

The energy of each of the three systems, i.e., the receptor, $E_{receptor}$, ligand E_{ligand}, and the ER ligand complex, $E_{complex}$, can be calculated from the sum of energy values of fragments and the counterpart for fragment pairs within each system under certain conditions [9]. In the calculation of the $E_{receptor}$ value, the hydrogen-bonding water molecule was included as a fragment along with "dipeptide" fragments. The binding energy for a given ligand (ΔE_{ligand}) can be expressed in eq. 1 as the difference in the energy between complex and components.

$$\Delta E_{ligand} = E_{complex} - (E_{receptor} + E_{ligand}) \quad (1)$$

The binding energy relative to that of 17β-estradiol (EST), $\Delta\Delta E_{ligand}$, in eq. 2 is the value to be compared with the experimental relative binding affinity (RBA) value. The RBA value of 17β-estradiol is defined as 100.

$$\Delta\Delta E_{ligand} = -(\Delta E_{ligand} - \Delta E_{EST}) \quad (2)$$

The $\Delta\Delta E_{ligand}$ values estimated using Model 2 are plotted against the published values of log (RBA/100) in Fig. 4.

The ligands 1–6, 9, and 10, of which the experimental RBA value is known, are shown as a circle in Fig. 4. For these 8 compounds, the correlation between $\Delta\Delta E$ and log (RBA/100) seems to be promising, the correlation coefficient r being 0.837. In particular, there is a very good correlation (r = 0.931) for the 7 ligands omitting TAM (6). From the correlation equation (n = 8), the log (RBA/100) value of ligands 7, 8, and 11, of which the RBA value is unknown, can be estimated with use of the calculated $\Delta\Delta E$ value. These 3 compounds are shown as a square in the plot.

The $\Delta\Delta E$ value was also calculated according to Model 1 for the complex of ligands 1~4. The result was almost identical with that calculated with Model 2. The difference in the $\Delta\Delta E$ value between two models was mostly below 3 kcal/mol, suggesting that the binding between ER and ligand is local. Another interesting finding was a difference in the charge distribution between complexed and individual component molecules. The total charge of ligands was changed to be negative with the values –0.00 ~ –0.18 when complexed with ER. The greatest negative charge influx occurs from Glu 353 to ligands, and a slight efflux is observed into Arg 394 and His 524. Such charge transfer is highly related with the binding energy. In fact, the ΔE tends to be greater with the increase in the difference of the charge distribution. Thus, most of the stabilization in the ER–ligand docking arises from the ligand-Glu 353 interaction. This observation seems to indicate that the charge is variable in the ER–ligand interaction, and therefore atomic charges should be calculated dynamically instead of using fixed charges as in classical calculations.

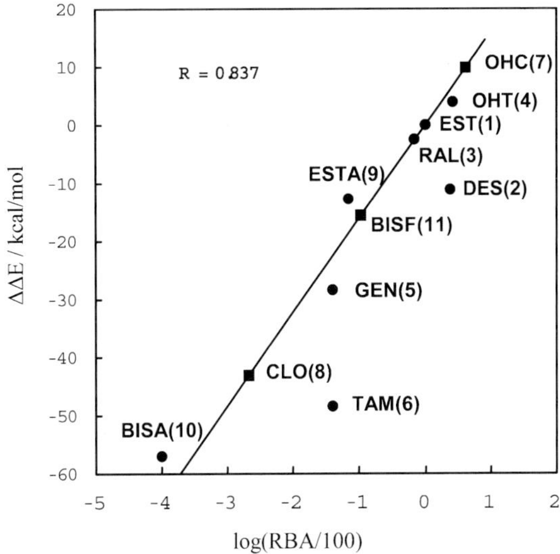

Fig. 4 Relationship between calculated relative binding energy ($\Delta\Delta E$) and experimental relative binding affinity [log (RBA/100)] of eight ligands (●), and the estimation of log (RBA/100) for three ligands (■). The regression line is drawn so that it is forced to pass the origin of coordinates.

To summarize, we have applied the ab initio FMO method to ER ligand binding which allows us to accurately predict the relative binding energy of xenoestrogenic ligand molecules from a "single" energy calculation. Given a variety of compounds, some of which could bind to the ER, such methods as we have proposed may provide a powerful tool for assessing the affinity of putative xenoestrogens in silico prior to biological studies. For further improvements, it is necessary to optimize not only the hydrogen bond, but also the geometry of the ligand and surrounding residues to estimate possible effects, in particular, those according to induced-fit in the ER ligand binding. Such functions are under development in our group.

REFERENCES

1. K. Paech, P. Webb, G. G. J. M Kuiper, S. Nilsson, J.-Å. Gustafsson, P. J. Kushner, T. S. Scanlan. *Science* **277**, 1508–1510 (1997).
2. G. G. J. M. Kuiper, J. G. Lemmen, B. Carlsson, J. C. Corton, S. H. Safe, P. T. van der Saag, B. van der Burg, J.-Å. Gustafsson. *Endocrinology* **139**, 4252–4263 (1998).
3. T. Barkhem, B. Carlsson, Y. Nilsson, E. Enmark, J.-Å. Gustafsson S. Nilsson. *Mol. Pharmacol.* **54**, 105–112 (1998).
4. S. Nilsson, G. Kuiper, J.-Å. Gustafsson. *Trends Endocrinol. Metab.* **9**, 387–395 (1998).
5. C. Sonnenschein and A. M. J. Soto. *J. Steroid Biochem. Molec. Biol.* **65**, 143–150 (1998).
6. S. P. Bradbury, O. G. Mekenyan, G. T. Ankley. *Environ. Toxicol. Chem.* **17**, 15–25 (1998).
7. B. C. Oostenbrink, J. W. Pitera, M. M. H. van Lipzig, J. H. N. Meerman, W. F. van Gunsteren. *J. Med. Chem.* **43**, 4594–4605 (2000).
8. P. D. Kirchhoff, R. Brown, S. Kahn, M. Waldman, C. M. Venkatachalam. *J. Comput. Chem.* **22**, 993–1003 (2001).
9. T. Nakano, T. Kaminuma, T. Sato, K. Fukuzawa, Y. Akiyama, M. Uebayasi, K. Kitaura. *Chem. Phys. Lett.* **351**, 475–480 (2002).
10. H. M. Berman, J. Westbrook, Z. Feng, G. Gilliland, T. N. Bhat, H. Weissig, I. N. Shindyalov, P. E. Bourne. *Nucleic Acids Res.* **28**, 235–242 (2000); <http://www.rcsb.org/pdb/>.

11. A. M. Brzozowski, A. C. W. Pike, Z. Dauter, R. E. Hubbard, T. Bonn, O. Engström, L. Öhman, G. L. Greene, J.-Å. Gustafsson, M. Carlquist. *Nature* **389**, 753–758 (1997).
12. A. K. Shiau, D. Barstad, P. M. Loria, L. Cheng, P. J. Kushner, D. A. Agard, G. L. Greene. *Cell* **95**, 927–937 (1998).
13. InsightII Version 98.0, Molecular Simulations Inc., San Diego, CA (1998).
14. CHARMm, Version 25.2, Revision: 98.0731.
15. D. M. Tanenbaum, Y. Wang, S. P. Williams, P. B. Sigler. *Proc. Nat. Acad. Sci. USA* **95**, 5998–6003 (1998).
16. Gaussian 98, Revision A.7, M. J. Frisch et.al., Gaussian, Inc., Pittsburgh PA (1998).
17. ABINIT-MP: <http://moldb.nihs.go.jp/abinitmp/>.

WORKSHOP 2

TOXICOGENOMICS AS A RATIONAL APPROACH TO ENDOCRINE DISRUPTOR RESEARCH

Pure Appl. Chem., Vol. 75, Nos. 11–12, pp. 2413–2414, 2003.

Workshop 2.1
Toxicogenomics: Impact on human health*

James K. Selkirk‡ and R. W. Tennant

National Institute of Environmental Health Sciences, National Institutes of Health, Research Triangle Park, NC 27709, USA

Abstract: Toxicology is the science of adverse effects of chemicals, drugs, environmental agents, and stressors. Genomics defines the structure, sequence (code), and function of the entire DNA complement of organisms. The interface of these diverse disciplines is called toxicogenomics and is based upon the application of genomic technologies to define globally the changes in gene expression (both mRNA and proteins) as a consequence of exposures. DNA microarray technology enables the simultaneous measurement of transcription of thousands of genes using microchips containing thousands of probes of complementary DNA (cDNA) immobilized in a predetermined array. The ultimate application of this technology to toxicology holds great promise but faces several formidable problems. With the solution to these problems, it will be possible to develop a substantial database of Chemical Effects in Biological Systems (CEBS). Such a database will provide the capacity to relate specific changes in gene expression to specific adverse effects and to look for similar pathways in different organisms. Such data will provide an objective way of assessing surrogate systems for reporting or predicting potential adverse effects in humans. While the potential for toxicogenomics is thus very high for studying active substances, the task must be approached in a deliberate, incremental manner to insure that only high-quality data are compiled and analyzed. This workshop will present the latest results of research conducted in leading international laboratories studying endocrine-mediated toxicity.

Toxicogenomics can be described as a new approach to understanding the genetic mechanisms and biochemical pathways to disease by environmental toxicants via the simultaneous analysis of gene and protein expression. For the first time, by careful comparison of perturbations between normal and diseased metabolic pathways and variance in expressed genes, investigators can decipher the complete mechanism of how and where toxicants disrupt normal processes that lead to disease. Over the last decade, the genomic revolution has been fueled by the meteoric rise in analytical technology, and the science of toxicogenomics represents a logical extension of the knowledge gained from the fields of gene and protein discovery. Publication of the human genome sequence draft five years earlier than anticipated was greatly dependent on the increased rate of sequence acquisition made possible by high-throughput technology and more sophisticated algorithms for gene discovery and identification.

Toxicologists and environmental health scientists have studied the effects of the environment on human health for many years. Adverse environmental effects have been identified, and important progress has been made in mitigating exposure to harmful agents such as X-radiation, UV-light, lead, pesticides, and dioxins. Toxicological research has attempted to develop an efficient, cost-effective, and comprehensive strategy for predicting and preventing toxic responses in humans. However, progress to-

*Report from a SCOPE/IUPAC project: Implication of Endocrine Active Substances for Human and Wildlife (J. Miyamoto and J. Burger, editors). Other reports are published in this issue, *Pure Appl. Chem.* **75**, 1617–2615 (2003).

‡Corresponding author

ward this goal has been proportionate to the existing technologies and level of scientific knowledge. The field of toxicology could not have risen to this current challenge by using only the less efficient technologies of the past several decades.

The genomic knowledge and technological developments surrounding the human genome program have yielded a great legacy to toxicogenomic research because the road to understanding the genetic and biochemical pathways to disease from environmental toxicants has been significantly widened by this wealth of genetic information. Furthermore, the accelerated discovery of genetic knowledge of both human and nonhuman genomes has empowered toxicology to reach beyond its classical boundaries of pathology and clinical chemistry to include all the genes and proteins in the biochemical pathways toward the manifestation of disease. Thousands of genes and proteins can now be analyzed simultaneously; it is now possible and practical to carefully map the effects of a toxic chemical as a function of dose and time against perturbations in networks of expressed genes and proteins. In the next decade, the culmination of all these measurements will begin to elucidate all the gene and protein modulations in an organ, tissue, or cell as it proceeds through its defense against the toxicant to reach recovery or toward a disease outcome or cell death.

The National Institute of Environmental Health Sciences is uniquely positioned with the National Center for Toxicogenomics (NCT) to provide leadership for the development of a unified strategy for toxicogenomics studies and a public knowledge base with the informatics infrastructure to allow scientists worldwide to share equally in its benefits and products. By providing a focus for technological coordination and basic research, a centralized public knowledge base and a center for coordination among all the partners in academia and the pharmaceutical and chemical industries, the NCT will facilitate this diverse national effort. The NCT seeks to achieve not only economies of time, cost, and effort, but also contributions to the successful development of a broad scientific consensus on the toxicogenomic applications to the improvement of human health.

The NCT has established five goals for its toxicogenomics program: (1) facilitate further development of gene expression and proteomic methodology; (2) create a public database relating environmental stressors to biological responses; (3) collect information relating environmental exposures to disease; (4) develop an improved paradigm for use of computational mathematics for understanding responses to environmental stressors; and (5) identify biomarkers of disease or exposure to enhance environmental health. It is important to realize that accomplishing these goals is a long-term effort and that the magnitude of this task is far greater than the physical resources and intellectual capacity of any individual institution. Therefore, the NCT is being established as a unique fusion of intramural laboratories, extramural grants, cooperative research agreements, and resource contracts.

Although the field of toxicogenomics is in its early stages and will be a long-term effort, it is already clear that the benefits of this methodical and comprehensive approach will be both extensive and exciting. The breath and scope of knowledge derived from this effort will be used to understand the underlying mechanism of disease by toxic chemicals, as well as for drug discovery, and also will be a critical platform for developing interventional and remedial strategies to interrupt the disease process.

The program presented at the SCOPE/IUPAC conference clearly showed a portion of the spectrum of research that will be embraced by the field of toxicogenomics. The following extended abstracts cover the presentations at the conference in greater detail to allow the reader a more complete appreciation for the excitement toxicogenomics is generating throughout the research community.

Pure Appl. Chem., Vol. 75, Nos. 11–12, pp. 2415–2418, 2003.

Workshop 2.2
Temporal responses to estrogen in the uterus*

Kirsten C. Fertuck, on behalf of Tim R. Zacharewski

Biochemistry and Molecular Biology, National Food Safety and Toxicology Center, and Institute for Environmental Toxicology, Michigan State University, East Lansing, MI 48824-1319, USA

Abstract: The estrogen receptor acts primarily as an estrogen-inducible transcription factor in target organs, however, until recently the identification of individual responsive transcripts has been cumbersome. In the present study, oligonucleotide GeneChip microarrays were used to describe and analyze the levels of approximately 6500 transcripts in the uteri of immature, ovariectomized mice at various times following oral exposure to 100 µg/kg of ethynyl estradiol. The most reproducible responses were identified and subjected to K-means clustering, and functional annotations were obtained for the transcripts within each cluster. Connections were made between the observed transcriptional responses and the known physiological responses to estrogen exposure, focusing here on effects on uterine cell cycle progression leading to estrogen-stimulated uterine growth.

INTRODUCTION

Efforts to identify compounds that act as estrogenic endocrine disruptors in vivo often use the rodent uterus as a model, because the rapid and robust stimulation of uterine tissue proliferation is a very characteristic response to estrogen exposure [1]. A dominant role of the estrogen receptor is as an estrogen-inducible transcription factor, and the modulation of transcript levels of a number of genes both prior to and during this proliferative response have been described. Some of these have then been used as indicators of the estrogenic action of exogenous compounds, with lactoferrin and complement component C3 being particularly well characterized estrogen-inducible markers [2]. However, more detailed examinations have indicated that estrogens may vary in the specific transcript profiles that they induce, which may be reflective of differences in their mechanism of action [3,4]. More study is therefore needed in order to identify the responses that are truly indicative of an estrogenic response, with the hope of eventually identifying a reliable subset that signals adverse responses to estrogenic exposure.

While the rat is a very commonly used species in uterine studies, mouse was used as the animal model of choice in the present study for several reasons. First, results from these studies can complement and be compared to results of other murine uterotrophic and microarray studies within the laboratory. As well, maintenance and dosing is relatively cost-effective, public databases of sequence and functional information are quite well annotated, and results can be compared with those of possible future studies using mice with targeted disruptions in genes of interest. However, a major drawback is the small tissue size and consequently low RNA yield from the mouse uteri compared with larger species.

*Report from a SCOPE/IUPAC project: Implication of Endocrine Active Substances for Human and Wildlife (J. Miyamoto and J. Burger, editors). Other reports are published in this issue, *Pure Appl. Chem.* **75**, 1617–2615 (2003).

APPROACH

In the present study, duplicate immature, ovariectomized C57BL/6 mice were treated with 100 μg/kg 17α-ethynyl estradiol (EE), an orally active pharmaceutical estrogen, or with vehicle for 2, 8, 12, or 24 h. Additional groups were administered three daily doses of EE or vehicle before sacrifice, which induced the approximately 5-fold increase in uterine weight characteristic of the widely used uterotrophic assay for estrogenic effect [1,5]. RNA was extracted from individual uteri, and samples were separately hybridized to Affymetrix Mu11KSubA GeneChip arrays (Santa Clara, CA) using the methods provided by the manufacturer. The Affymetrix MicroArray Suite 5.0 software was used to obtain a Signal intensity value for each probe set, which is the set of sequences that report the relative abundance of one transcript.

The five-step approach that was used to analyze the resulting microarray data is described below. Using this approach, the most reproducible, treatment-induced responses were identified, the temporal response patterns were characterized, and the associated gene functions were identified and discussed.

Screening for active responses

In a method that will be published separately, a modification of the empirical Bayes approach [6] was employed to analyze all data at all time points, yielding one summary statistic per probe set indicative of the degree and reproducibility of the response over time relative to that of the time-matched controls. A cut-off value of 0.99 was arbitrarily assigned, and all probe sets with associated p1z values greater than the cut-off value, which consisted of approximately 15 % of all measured responses, were retained for further analysis.

Screening active responses for treatment effect

Probe set responses that were identified as being active in the first screening step were then further screened using an analysis of variance. In this method, responses that were significant at the arbitrarily chosen $p < 0.01$ level in either Treatment or in the Treatment*Time interaction were considered to have a significant treatment effect, with the approximately 400 responses representing 45 % of those retained by the first screening step. The remaining responses with solely a significant Time effect were rejected since this is suggestive of either vehicle or circadian responses.

Clustering

For the 400 responses that passed the second stage of the screening procedure, the averaged duplicate Signal responses at each time point, normalized by the corresponding time-matched control values, were clustered by the K-means method in GeneSpring (Redwood City, CA). Though the 3 × 24 h treatment group was the only group to receive multiple doses, this group was included with all others for clustering purposes since the relationship of probe set responses before and during estrogen-induced proliferation is of interest.

In K-means clustering the number of clusters must be specified by the user. By examining results of between 2 and 15 clusters, it was judged that seven K-means clusters appeared to most fully describe the 400 responses. These seven clusters could be generally described as follows: upregulated only at 2 h; upregulated only at 8 h; upregulated at 8 and 3 × 24 h; upregulated at 12 h; upregulated only at 24 h; upregulated at 24 and 3 × 24 h; and upregulated at 3 × 24 h only.

Annotation

Following the assignment of the 400 responses to the seven clusters, the identities of the responses were determined and functions were ascribed where possible, using public repositories of gene information and customized Perl scripts that return hyperlinked Excel output files. First, the GenBank accession ID associated with each probe set was used to query the National Center for Biotechnology Information (NCBI; <http://www.ncbi.nlm.nih.gov>) UniGene database. In approximately 60 % of cases a gene name, abbreviation, and UniGene ID were returned, while in most other cases the accession ID corresponded to an unnamed expressed sequence tag (EST). In 23 cases, the accession ID was not contained within the UniGene database, but a BLAST search of the probe set sequence in the GenBank databases allowed identification of a putative gene name in 18 cases.

Additional information for probe sets with associated gene names was then obtained from the NCBI LocusLink database, including mRNA and protein RefSeq IDs, chromosomal location, NCBI PubMed references, and Gene Ontology names and IDs. In addition, the LocusLink ID for the human homolog was obtained where available, which then provided links to human databases such as the NCBI Online Mendelian Inheritance in Man (OMIM) database and the Weizmann Institute GeneCards database (<http://bioinfo.weizmann.ac.il/cards>).

These scripts were also used to annotate hand-curated published information for uterine estrogen-responsive genes, and UniGene ID was then used as the identifier with which to allow comparison of the present experimental results with published results.

A future interest is to expand the capabilities of dbZach, the in-house database used for microarray data analysis and interpretation, to support GeneChip data. This database currently supports only cDNA microarray data, but its features are also applicable to other array types. dbZach stores LocusLink, UniGene, and Gene Ontology information, as well as toxicology data of interest, and links it automatically to stored microarray data and sample annotation information to facilitate analysis.

Interpretation

For simplicity, we focus here on effects of estrogen on the transcription of genes related to cell cycle progression. A well studied effect of estrogen exposure in the uterus is the stimulation of synchronized cell division which results in greatly increased uterine tissue mass within a relatively short period of time. This cell cycle progression, in which shortly after estrogen exposure quiescent cells are recruited from G_0 to G_1, followed by the initiation of DNA synthesis in S phase at approximately 15 h and the progression to G_2/M beginning at approximately 24 h [7,8], requires the coordinated regulation of expression of specific genes. Some of these specific responses were confirmed from previous studies, while others have not been previously reported in this system but based on their known function can be reconciled with the temporal pattern of gene expression observed.

For example GADD45α, which was highly induced at 2–8 h, is known to be critical for the DNA damage and repair checkpoint prior to S phase and to be induced only during G_1. Cyclin D2, which is also required for the G_1/S transition, was also upregulated during this time, while a number of other transcripts for proteins involved in the start of DNA synthesis in S phase were found in clusters that were upregulated at approximately 8–12 h. Other genes required for S phase, as well as some required for M phase progression, were upregulated later (24 h). No genes associated with cell cycle progression were identified in the cluster that was upregulated at 3×24 h, the time at which maximal growth has been attained.

In some cases, the response obtained did not confirm results reported in the literature, or two different probe sets interrogating different regions of the same transcript yielded highly different results. In these cases, the specific sequences being probed, and the characteristics and performance of the probes composing the probe set were examined in more detail. The Affymetrix database NetAffx (<http://www.affymetrix.com>) was used to obtain specific probe sequence information, and to examine the number and overlap of the probes within the set, while dChip (<http://biosun1.harvard.edu/complab/dchip>) was used to examine the performance of the different probes within the set. These sources of information, in addition to the suffix flag codes provided with the probe set IDs, was helpful in explaining unexpected or discordant results.

CONCLUSIONS

In the present study, a statistically rigorous method of reducing temporal microarray responses to estrogen in the mouse uterus to the most active and reproducible responses was presented. Automated scripts were used to obtain data from local copies of public database information, both for mouse genes and for the human homologs where available. These scripts were also used to annotate literature search information for estrogen-induced responses in the uterus, and all of the assembled information was used in the subsequent physiological interpretation. This interpretation, which is ongoing, aims to link the observed temporal transcriptional responses to the tissue-level changes that lead to the rapid, coordinated, and extensive uterine proliferation following estrogen exposure. It is anticipated that the large-scale characterization of transcriptional responses to estrogen in the uterus will be valuable in better understanding the physiological effects of estrogen exposure, which in turn can be used to evaluate the presence and toxicological significance of estrogenic effects of exogenous compounds.

ACKNOWLEDGMENTS

Assistance with scripting by Ms. Shraddha Pai and Ms. Raeka Aiyar is gratefully acknowledged. Financial support from ES 011271, as well from fellowships from Michigan State University and the Society of Toxicology, have also been gratefully received.

REFERENCES

1. E. Padilla-Banks, W. N. Jefferson, R. R. Newbold. *Environ. Health Perspect.* **109**, 821–826 (2001).
2. W. N. Jefferson, E. Padilla-Banks, R. R. Newbold. *Reprod. Toxicol.* **14**, 103–110 (2000).
3. D. A. Zajchowski, K. Kauser, D. Zhu, L. Webster, S. Aberle, F. A. White, H. L. Liu, R. Humm, J. MacRobbie, P. Ponte, C. Hegele-Hartung, R. Knauthe, K. H. Fritzemeier, R. Vergona, G. M. Rubanyi. *J. Biol. Chem.* **275**, 15885–15894 (2000).
4. H. K. Hamadeh, P. R. Bushel, S. Jayadev, K. Martin, O. DiSorbo, S. Sieber, L. Bennett, R. Tennant, R. Stoll, J. C. Barrett, K. Blanchard, R. S. Paules, C. A. Afshari. *Toxicol. Sci.* **67**, 219–231 (2002).
5. H. D. Lauson, C. G. Heller, J. B. Golden, E. L. Sevringhaus. *Endocrinology* **24**, 35–44 (1939).
6. B. Efron, R. Tibshirani, J. D. Storey, V. Tusher. *J. Am. Stat. Assoc.* **96**, 1151–1160 (2001).
7. S. V. Cheng, B. S. MacDonald, B. F. Clark, J. W. Pollard. *Exp. Cell Res.* **160**, 459–470 (1985).
8. J. Z. Guo and J. Gorski. *Mol. Endocrinol.* **2**, 693–700 (1988).

Pure Appl. Chem., Vol. 75, Nos. 11–12, pp. 2419–2422, 2003.

Workshop 2.3

Application of toxicogenomics to the endocrine disruption issue*

Tomoyuki Shirai‡ and Makoto Asamoto

Department of Experimental Pathology and Tumor Biology, Nagoya City University Graduate School of Medical Sciences, 1-Kawasumi, Nagoya 467-8601, Japan

Abstract: Toxicogenomics can be expected to be a useful method for detecting the carcinogenic potential of endocrine active substances (EASs) in the short term with the generation of understanding of mode-of-action and mechanisms when a reliable database with information about proteomics and informatics is established. At present, there are no concrete epidemiological data supporting any exogenous EAS contribution to hormone-related organ carcinogenesis in humans. However, with the establishment of appropriate animal models and analysis of genomic-scale gene expression, risk identification and evaluation should be facilitated within a relatively short period, and this approach eventually promises to contribute a great deal of risk management regarding EASs.

INTRODUCTION

Over the past several years, great concern has been raised as to possible disruption of hormonal systems by environmental chemical substances. The issue of endocrine disruption (ED) has attracted the attention of scientists as well as the media. Although such chemicals were defined as "exogenous substances that alter function(s) of the endocrine system and consequently cause adverse health effects, in an intact organism or its progeny, of (sub)populations" at the Weybridge Workshop, 1996, no firm criteria were adopted for assessment of adverse health effects. As far as endocrine systems are concerned, the issue is basically confined to the hypothalamus-pituitary-gonads and thyroid axis. Diverse phenotypic changes that arise from disturbance of the actions of female, male, and/or thyroid hormones through agonistic and antagonistic mechanisms or alteration of the homeostatic control balance are clearly concerns for health. In addition to their potential influence on reproductive parameters such as spermatogenesis, the sexual cycle, and development of sexual organs, they may affect processes underlying endometriosis, carcinogenesis in hormone-related organs, and neurotoxicity leading to intelligence impairment and emotional instability (Fig. 1).

The simultaneous analysis of expression of thousands of genes as endpoints using cDNA chips or microarrays should allow toxicologists a new comprehension of toxicological issues. Toxicogenomics, the combined field of toxicology and genomics, thus has become a focus for the research community and regulatory authorities as a new approach to understanding mode-of-action. It can provide us with very helpful data relevant to difficult areas such as dose–response relationships, species-to-species extrapolation, and exposure assessment that cannot be resolved with traditional toxicological techniques

*Report from a SCOPE/IUPAC project: Implication of Endocrine Active Substances for Human and Wildlife (J. Miyamoto and J. Burger, editors). Other reports are published in this issue, *Pure Appl. Chem.* **75**, 1617–2615 (2003).

‡Corresponding author

- Reproductive dysfunction (spermatogenesis, sexual cycle, endometriosis)
- Enhancement of carcinogenesis (breast, uterus, prostate, testis)
- Immune toxicity (hyporesistance to infection)
- Developmental disorders of sexual organs (reproductive organ malformation, hypoplasia)
- Elevated carcinogenic potential in the 2nd generation (vagina, breast, uterus, prostate, testis)
- Neurotoxicity in the 2nd generation (growth retardation, intelligence impairment, emotional instability)

Fig. 1 Presumed adverse effects of EDCs.

(Fig. 2). Application of genomic technology to the ED issue can thus be expected to overcome limitations of conventional methods. One possibility is that toxicogenomics will facilitate differentiation of gene responses specific to endocrine activity from those associated with nonspecific general stress. For example, sequential analysis of gene expression profiles in endometrium with histological alterations after treatment with 17β-estradiol or related estrogenic compounds may give us estrogen-specific gene alterations, as presented by Orphanides in Workshop 2.5. As noted above, ED may lead to diverse and complex phenotypic outcomes. Since toxicogenomics is defined as elucidation of phenotypic alteration in gene expression, it causes a very broad spectrum of possibilities.

- Establish signatures of specific chemically induced altered gene expression
- Identify biomarkers of exposure and toxicity
- Predict toxicity of unknown agents
- Classify and predict phenotypes of toxicity
- Delineate models/mechanisms of action
- Allow extrapolation from one species to another (from animals to human beings)

Fig. 2 Toxicogenomics-expected promise.

Therefore, in order to make toxicogenomic approaches to the ED issue efficient, it is necessary to focus on one particular aspect for analysis. Among diverse potential effects exerted by endocrine active substances (EASs), elevation of cancer risk in endocrine-related organs is one with most concern. Thus, concentration of effort on hormone-related carcinogenesis is worthy of particular attention, because hormone-related carcinogenesis is assumed to occur by nongenotoxic mechanisms, with a long periods between exposure and appearance of neoplasms. Disadvantages with long-term animal experimentation to detect EAS-associated cancer risk may be overcome by application of toxicogenomics accompanied by relevant elucidation of mode-of-action. Predictive toxicology, i.e., prediction of toxicological adverse effects in terms of a pattern or set of genes is one important task of toxicogenomics. Whether any signature of gene alteration specific to carcinogenesis can be established remains for future elucidation.

HORMONE-RELATED CARCINOGENESIS AS A TARGET OF TOXICOGENOMICS

Carcinogenic risk associated with an EAS can be divided into three categories: (1) potential carcinogenicity of the EAS itself as seen in cases of vaginal adenocarcinomas in young girls exposed to diestilbesterol in their uteri; (2) induction of a condition predisposing to cancer development, so that it is accelerated; and (3) promotion or enhancement of cancer development by prolonged action. In line with these three modes-of-action, we must construct a database of expression profiles for gene alteration using cDNA microarray or chip technology.

There are several organs in which carcinogenic responses could be modified by an EAS. These include the breast, endometrium, prostate, testis, and ovary. For elucidation of gene expression profiles related to carcinogenic processes in individual organs/tissues, appropriate animal models are essential (Fig. 3). Carcinogenic potential for suspicious chemicals could be explored by administration of DES

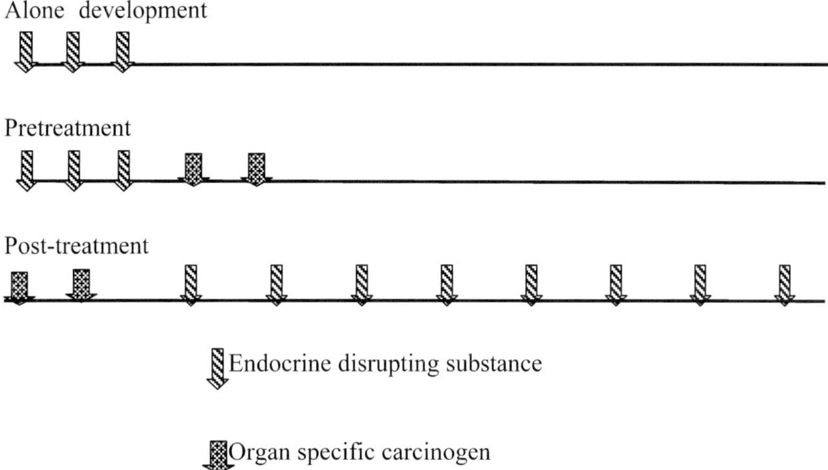

Fig. 3 Possible reference animal models.

to rats or mice [1], with tissue samples taken from the uterus, mammary glands, or vagina at certain time points for analysis of changes in alterations of expression at the genomic level. Promoting/enhancing effects of EAS could be evaluated by application of the two-stage carcinogenesis concept to the organs of interest. For mammary glands, 7,12-dimethylbenz[a]anthracene (DMBA)- or *N*-methyl-*N*-nitrosourea (MNU)-initiated rat models have been widely utilized. As a reference chemical, E2 can be given after the initiation until a sufficient yield of mammary tumors is attained [2]. During the tumor induction, changing in gene expression can be analyzed sequentially to determine whether a tumor promotion specific gene profiles exists. Similar approaches could be applied using established animal models for endometrial and prostate carcinogenesis. For the former there is a rat model with intrauterine administration of ENNG and then E2 [3], and for the latter, DMAB- or PhIP-initiated rat models are good candidates [4]. Unfortunately, at the present there are no appropriate animal models for testicular or ovarian carcinogenesis. The possibility of increased likelihood of cancer development in offspring if they were pre- or postnatally exposed to EASs is of great concern. Animal models could be envisaged with application of estrogen during gestation and lactation to mothers and then exposure of offspring to potential carcinogen at a certain point or for an appropriate period after weaning. Boylan and Calhoon [5] have already shown that prenatal exposure to DES potentiates mammary tumorigenesis in rats given DMBA. However, in spite of several attempts, in utero and lactational exposure to bisphenol A or 4-nonylphenol did not produce a predisposition for cancer development in the prostate or uterus [6]. It can be expected that a database of gene profiles associated with EASs will become available as and only when appropriate and reproducible animal models are established.

Confidence in results might be elevated by combination with traditional toxicological and toxicopathological findings. Evaluation of any influence on cancer development in hormone-related organs after early life exposure to endocrine disruptors is likely to be very difficult, because there is little knowledge about susceptibility. If cancers do develop due to exposure to environmental chemicals, genes associated with chemical activation or detoxification, DNA repair genes, and genes associated with cellular replication might be targets. In addition, genes enconding hormone metabolic enzymes such as 5α-reductase and aromatase, expression of hormone receptor-related and coactivator genes are obvious candidates for assessment.

In Japan, a consortium project on toxicogenomics for carcinogen detection has just commenced. This project was initiated by the Ministry of Economy, Trade and Industry and is supported by a grant from the New Energy and Industrial Technology Development Organization. As reference compounds, quaternary categorized chemical carcinogens in the liver such as mutagenic or nonmutagenic hepatocarcinogens and mutagenic nonhepatocarcinogens and noncarcinogens are administered to rats for short periods and database of gene expression profiles is being constructed with tissues sampled at various time points using in-house cDNA microarrays. Proteomic studies are also carried out at the same time. Once carcinogenesis-related genes are selected from nonspecific altered genes, the information should contribute to understanding at mechanisms of EAS-associated carcinogenesis.

Promotion or antipromotion effects of possible EASs could be assessed by evaluation of expression profiles of genes involved in cell replication such as cyclines, cycline-dependent kinases (CDKs), and CDK inhibitors, as well as their counterparts associated with cell death pathways and cell differentiation. It can be envisaged that efforts with specific organs like the mammary glands, endometrium, prostate glands, and thyroid, which depend upon hormonal activities, will be most rewarding.

CONCLUSION

Toxicogenomics can be expected to be a useful method to detect the carcinogenic potential of EASs in the short term with generation of understanding of mode-of-action and mechanisms when a reliable database with information about proteomics and informatics is established. At present, there are no concrete epidemiological data supporting any exogenous EAS contribution to hormone-related organ carcinogenesis in human beings. However, with establishment of appropriate animal models and analysis of genomic-scale gene expression, risk identification and evaluation should be facilitated with a relatively short period, and this approach eventually promises to contribute a great deal of risk management regarding EASs.

ACKNOWLEDGMENT

This paper was supported by the Society for Promotion of Toxicologic Pathology, Nagoya.

REFERENCES

1. IARC Monographs on the Evaluation of Carcinogenic Risks to Humans. *Overall Evaluations of Carcinogenicity: An updating of IARC Monographs*, Vols. 1 to 42, suppl. 7, p. 278, Lyon, France (1987).
2. IARC Monographs on the Evaluation of Carcinogenic Risks to Humans. *Sex Hormones (II)* Lyon, France (1979).
3. T. Ichihara, H. Yoshino, N. Imai, T. Tsutsumi, M. Kawabe, S. Tamano, T. Shirai. *Proc. Ann. Meeting of Jpn. Cancer Assoc.* **64**, 206 (2002).
4. T. Shirai, S. Takahashi, C. Lin, M. Futakuchi, K. Kato, S. Tamano, K. Imada. *Mutat. Res.* **462**, 219 (2000).
5. E. S. Boylan and R. E. Calhoon. *J. Toxicol. Environ. Health* **5**, 1059 (1979).
6. K. Niwa, T. Tanaka, H. Mori, Y. Yokoyama, T. Furui, H. Mori, T. Tamaya. *Jpn. J. Cancer Res.* **82**, 1391 (1991).

Pure Appl. Chem., Vol. 75, Nos. 11–12, pp. 2423–2427, 2003.

Workshop 2.4

Transcript profiles elicited by developmental exposure to endocrine-mediated toxicants*

George P. Daston‡ and Jorge M. Naciff

Miami Valley Laboratories, Procter & Gamble, Cincinnati, OH 45253, USA

Abstract: Genomics can be applied in toxicology as a means of identifying modes of action, for generating hypotheses on the relationships of gene activity and toxicity, and for better characterizing the nature of dose–response relationships at low doses. This paper illustrates each of these applications with examples from our research on endocrine active compounds. We have determined that agents that bind estrogen receptors produce a characteristic transcript profile in estrogen-responsive tissues of the fetal and juvenile rat. The transcript profile is diagnostic of the mechanism of action. The transcripts that are up- or down-regulated by estrogens belong to a number of functional groups, such as growth factors, pro-apoptotic factors, transcription factors, and steroid metabolizing enzymes. There are a number of testable hypotheses that can be generated from these data regarding the relationships between changes in these genes and developmental or physiological responses to estrogens. We have determined that the sensitivity of gene expression changes is high, making it possible to define the shape of the dose–response curve at dose levels of estrogen several orders of magnitude below those that cause a physiological response (in this case, a uterotrophic response). The shape of the dose response is monotonic: both the number of genes changed and the intensity of the up- or down-regulation decreases with decreasing dose.

Our lab is using toxicogenomics to identify molecular fingerprints, which we call transcript profiles, that are specific for steroid hormone receptor-mediated mechanisms of action. The main reason for conducting this research is to develop screening methods based on these profiles. However, the work also has general utility in understanding the life-stage specificity of gene expression; characterizing dose–response relationships, particularly at the low end of the dose–response curve; and in providing the basis for testing hypotheses on the role of individual genes and combinations of genes in development and physiological processes.

Regulatory agencies around the globe are engaged in validating screening systems for endocrine-mediated toxicants. The screening batteries are extensive, but few of the screens use developing organisms and none use developing mammal embryos. This is despite the general belief that early development is the most sensitive to endocrine-mediated toxicity, and the fact that manifestations after developmental exposure tend to be permanent. One reason why screening assays have not used mammalian embryos is that the adverse effects produced by endocrine mechanisms are often latent and cannot be detected using traditional toxicology methods until well after exposure occurs, often not until the animal becomes sexually mature. The long lag time between exposure and outcome has made development-based screening infeasible.

Although the ultimate manifestations of toxicity, which occur at the tissue and organ level, take a long time to be manifested, it is likely that there are much more rapid, and perhaps persistent, changes

*Report from a SCOPE/IUPAC project: Implication of Endocrine Active Substances for Human and Wildlife (J. Miyamoto and J. Burger, editors). Other reports are published in this issue, *Pure Appl. Chem.* **75**, 1617–2615 (2003).

‡Corresponding author

in gene expression that precede and are responsible for the biological response. In fact, gene expression is an integral part of the signal transduction pathway for steroid hormones. Therefore, we have stated the hypothesis that toxicants that act via interference with normal steroid hormone receptor function will produce a characteristic pattern of gene expression in sensitive tissues. If true, then it should be possible to use these profiles as a basis for screening for endocrine active agents in developing tissues, because the gene expression response will be rapid and specific to a particular mechanism of action. The general support for this hypothesis is the knowledge that gene expression is an integral step of the steroid hormone signal transduction process. Empirically, it has been reported that prenatal exposure to potent estrogens [e.g., diethylstilbestrol (DES)] elicits expression of a few genes in the mouse reproductive system that are known to be estrogen-responsive.

We tested the hypothesis by examining the gene expression changes induced by transplacental exposure of rat fetuses to agents with estrogenic activity. (Similar work is underway for antiandrogens and thyroid toxicants, but these will not be discussed here.) The agents we used were ethinyl estradiol, a potent estrogen; bisphenol A, a chemical with weak estrogenic activity; and genistein, a phytoestrogen with potency between that of ethinyl estradiol and bisphenol A (dosages are listed in Table 1). For each compound, the highest dosage was selected because it has been reported to have estrogenic activity in a uterotrophic assay, and the dosages were adjusted for potency such that the pharmacological activity of the compounds was roughly comparable across compounds, even though on a mg/kg basis the dosages were different.

Table 1 Dosages used to elicit estrogen-specific transcript profiles in fetal rats.

17α–EE	0.5	1	10	(μg/kg/day)
Bisphenol A	5	50	400	(mg/kg/day)
Genistein	1	10	100	(mg/kg/day)

Pregnant Sprague–Dawley rats were assigned to treatment groups randomly and administered the test compound daily from gestation day 11 to gestation day 20. The rationale for this treatment interval is that the reproductive system primordia arise on gestation day 12 and continue to develop and mature during this period. Gestation day 20 is the day of gestation when most segment II developmental toxicity studies are ended. In selecting this treatment regimen, it was our thought that, if successful, transcript profiling could be added to regular developmental toxicity studies to enhance their ability to detect endocrine-mediated toxicity. The rats were sacrificed 2 h after the final treatment on gestation day 20. Fetuses were evaluated for external abnormalities, sexed, and the uterus and ovaries of females, and testes and epididymides of males, were removed and immediately placed in RNALater. Subsequently, mRNA was isolated from the tissues and gene expression evaluated using Affymetrix U34A gene chips. These microarrays contain sequences for about 8000 annotated genes and about 1000 expressed sequence tags (ESTs), which represent mRNA sequences that have been isolated from a rat library, but for which the sequence (and function) of the gene have not been determined. Tissues from five fetuses within a litter were pooled to provide sufficient RNA. RNA from at least five litters per treatment group was hybridized to the microarrays, which provided us with sufficient statistical power to support our conclusions about which genes were induced or repressed by treatment.

We observed some apparent toxicity in the high-dose ethinyl estradiol group, in the form of vaginal bleeding in a number of the dams, and early parturition in one. Prominent nipples and areolae were observed in male and female fetuses exposed to the highest dosages of bisphenol A or ethinyl estradiol, and in female fetuses at lower dosages of these compounds. There were no effects on uterine or ovarian morphology or histology of the fetuses.

Each estrogenic compound significantly altered the pattern of gene expression in uterus and ovaries. The expression level of around 5 % of the genes was changed according to at least one statistical criterion that we used. Applying extremely rigorous statistical criteria, we identified a subset of 66 genes that were changed by all three estrogenic compounds, in a dose-related manner. We consider this panel of 66 genes to be the transcript profile for estrogens in the fetal rat uterus and ovaries.

A partial list of genes that are upregulated (Table 2) include many that have already been reported in the literature to be estrogen-responsive at some life-stage. This includes the progesterone receptor and intestinal calcium-binding protein. There were also a number of genes that are involved in steroid metabolism, as well as growth factors and other regulators of cell differentiation and function. In addition to the genes previously reported to be estrogen-responsive, we have discovered a number of other

Table 2 Partial list of genes regulated by estrogens in fetal rat uterus and ovaries.

Upregulated:

- Intestinal Ca-binding protein
- 11-β-OH steroid dehydrogenase
- Progesterone receptor
- Vascular a-actin
- Uterus-ovary-specific transmembrane protein
- FSH-regulated protein
- Asp aminotransferase
- Low-density lipoprotein R/LDLR
- Dermo-1 protein
- *trk* B (receptor Tyr-kinase)
- Growth potentiating factor
- Growth hormone receptor
- Interleukin 4 receptor/IL-4R
- Insulin-like growth factor 1
- Non-neuronal enolase
- C-CAM4
- Glucocorticoid-regulated kinase
- NGF-induced factor A
- Glutathione-*S*-transferase
- Na-K-ATPase-β-3 subunit
- NTAK α-2-1p (activator of Erb-kinases)
- Protein Tyr-phosphatase

Downregulated:

- Retinol-binding protein
- Glutathione-*S*-transferase M5
- Phosphodiesterase I
- Apolipoprotein CI
- Calmodulin-binding protein
- K-channel regulatory protein
- Distal-less 3 homeobox protein
- VLDL receptor
- Phosphofructokinase-C
- *Ssecks* 322, mitogenic regulatory gene
- *c-ret*, Tyr-kinase receptor
- Polymerase alpha
- Precursor interleukin 18
- Carboxypeptidase E

Data from J. M. Naciff, M. L. Jump, S. M. Torontali, G. J. Carr, J. P. Tiesman, G. J. Overmann, G. P. Daston. *Toxicol. Sci.* **68**, 184–199 (2002).

genes, including ESTs, that have not previously been reported to be estrogen-responsive. An analysis of the genomic sequence of the annotated genes indicates that not all of these contain consensus estrogen response element sequences, suggesting that their expression is secondary to changes in the expression of genes that are immediately activated by the receptor complex, or that they contain other response elements that bind the receptor.

There are also a number of genes that are downregulated by treatment (Table 2). Again, the genes identified represent a mixture of functions.

We have verified the results of the microarrays using an independent technique, quantitative reverse transcriptase-polymerase chain reaction (RT-PCR), for a subset of the genes in the transcript profile. The microarray results are reproducible, and the fold-changes observed by either technique are comparable.

We have compared the fetal transcript profile to the profile that is induced in the uterus and ovaries or rats that are mature enough to produce a uterotrophic response to estrogen exposure. For this experiment, we used a typical uterotrophic assay treatment regimen in which rats were treated daily on postnatal days 21–24, then sacrificed. We used ethinyl estradiol (0.1, 1, or 10 µg/kg/day) as the estrogen for this set of experiments. The dosages of ethinyl estradiol that we used produced a dose-related increase in uterine weight and uterine epithelial cell height, as well as accumulation of fluid in the uterine lumen at the highest dose level. The expression level of over 600 genes was changed according to some statistical criterion, and the transcript profile, composed of those genes that gave a consistent, dose-related response, consisted of 227 genes, much larger than the fetal profile for estrogens.

As with the fetal profiles, there were a number of genes identified in the juvenile profiles that have been previously reported to be estrogen-responsive, such as complement component 3. One impressive difference between the fetal and juvenile profiles was the extent to which gene expression could be changed by treatment. In the juvenile rat, the most robust increases in gene expression were on the order of 300–400-fold, whereas the most marked increases in the fetus were less than 20-fold. It is not clear at this point whether this represents some form of generalized repression of estrogen responsiveness in the fetus, or that the difference is attributable to a few genes in the juvenile profile that have particularly robust responses.

There is considerable, but not complete overlap between the fetal and juvenile uterine transcript profiles. Of the 80 most highly expressed genes in the juvenile profile, 87 % were also expressed in the fetus. On the other hand, only 45 % of the genes present in the fetal profile were also present in the juvenile profile. It is possible that the genes that are unique to the fetus are developmentally important, or that the genes unique to the juvenile are critical for the uterotrophic response (which is not mounted in the fetus), or both. The differences in the profile will serve as the basis for generating and testing hypotheses about the role of specific genes in physiological or developmental events. An example of one such hypothesis will be given later.

We have begun to use toxicogenomics to better characterize the nature of the dose–response curve, particularly at the low end of the curve. This is particularly important in that some labs have reported that estrogens at very low dosages produce qualitatively different changes in responsive tissues than are observed after exposure to exaggerated dose levels. Other laboratories have been unable to repeat these studies, and the weight of evidence is strongly on their side. However, a panel of scientists convened by the U.S. National Institute of Environmental Health Sciences (NIEHS) was unable to conclude, for a few examples of this low-dose phenomenon, which results were correct. They cited the many sources of experimental variability that could be confounding the results, thereby making it impossible at this point to come to a definitive conclusion.

One possible way forward is to evaluate responses at a more fundamental level of biological organization (i.e., gene expression), the thought being that changes in gene expression, being a more immediate response to the toxicant–receptor interaction, may be more sensitive than more traditional endpoints of organ weight and histology. If low-dose phenomena are real, then one would expect it to be reflected in the dose–response behavior of gene expression.

To test this hypothesis, we again used the uterotrophic assay protocol and ethinyl estradiol. In this instance, we used two experimental blocks, the first consisting of the relatively high dosages of ethinyl estradiol used in the previous experiment, and the second consisting of a block in which the high dose (0.1 µg/kg/day) was the same as the low dose in the first block, and the lowest dose being two orders of magnitude lower. We used different diets for the two blocks, standard alfalfa- and soy-based lab chow in the first instance, and a casein-based, phytoestrogen-free diet in the second. Comparisons of the two control groups, and the 0.1 µg/kg/day ethinyl estradiol groups that were in common between the two blocks were made to determine whether there was any effect of diet. We also reviewed data from the controls in our fetal transcript profile experiments, in which the phytoestrogen-free diet was used for the genistein experiment and standard lab chow for the other two compounds.

There were some differences in the profiles that were attributable to diet; however, they were subtle and the majority of the changes appear to be attributable to some aspect of the diet unrelated to estrogen content.

As for the dose–response curve for gene expression in the uterotrophic protocol, change in gene expression was a more sensitive endpoint than uterine weight or morphometry, in that a statistically significant number of genes was changed at a lower dose level than that for which a uterotrophic or histological change could be detected. Furthermore, a few elements of the transcript profile were detectable even at the lowest dosage of ethinyl estradiol used. However, both the number of genes and their level of expression decreased in a monotonic manner with decreasing dose, supporting the conclusion that, at least for this tissue and treatment paradigm, there is not a low-dose phenomenon.

One of the strong potential uses of toxicogenomic analysis is its use in formulating and testing hypotheses about the role of genes in particular physiological processes. One such example, for the uterotrophic response, is given here.

One of the genes that is strongly upregulated by estrogen in the juvenile uterus is matrilysin, a matrix metalloproteinase. One of the substrates for matrilysin is osteopontin, a secreted protein that has been reported to induce apoptosis. Not only is osteopontin degraded by matrilysin, but its expression is downregulated by estrogen; therefore, this signal that would promote an increase in apoptosis, manifested at the tissue level as a decrease in uterine weight, is doubly inhibited. Another gene that is upregulated by estrogen is CD44, a membrane protein that functions to bring matrilysin and the precursor of an epidermal growth factor, HB–EGF, into proximity. Matrilysin activated the growth factor, enabling it to interact with the receptor tyrosine kinase Erb B4, which functions to promote cellular proliferation and tissue changes such as angiogenesis. These changes promote tissue growth, and in this case may be partially responsible for the uterotrophic response. This hypothetical pathway is testable through the use of knock-out animals or other techniques that allow the temporary manipulation of mRNA levels. It is only one of the many hypotheses that can be formulated from the data we have generated thus far, and only serves to illustrate the power of toxicogenomic technology.

In conclusion, we have provided evidence that transcript profiles can be used to identify a mechanism of action, in this case, estrogenicity. We have shown that the response can be detected in the fetus, and that transcript profiles could serve as the basis for the development of improved screens for endocrine active compounds. One particular advantage is that transcript profiling provides a practical way to incorporate developing mammalian systems into a screening approach. We have also shown that transcript profiles can be generated in more mature life-stages. The profiles from fetus and juvenile rat uterus are both recognizable as being estrogen-induced, but their differences suggest different roles for gene expression in the fetus and adult (or near-adult, in the case of the juvenile model). We have used toxicogenomics to evaluate dose response and to formulate hypotheses as to how changes in gene expression may control tissue responses. These hypotheses are testable, and, together with dose–response information, may provide the basis for a quantitative, biologically based approach for understanding the relationship of changes in gene expression to toxicity, and for using this information to improve our predictions about human risk from endocrine-mediated toxicants.

Pure Appl. Chem., Vol. 75, Nos. 11–12, pp. 2429–2432, 2003.

Workshop 2.5

Use of gene expression profiling to understand the transcriptional program associated with estrogen-induced uterine growth*

Jonathan G. Moggs, Damian G. Deavall, and George Orphanides[‡]

Syngenta Central Toxicology Laboratory, Alderley Park, Macclesfield, Cheshire, SK10 4TJ, UK

Abstract: The use of gene expression data in predictive and mechanistic toxicology is hindered by a lack of information on the relationships between transcriptional events and physiological and pathological changes. We discuss the analysis of these relationships using the rodent uterotrophic response as a model experimental system for estrogen-induced uterine growth.

INTRODUCTION

The successful survival and growth of an organism depends upon highly regulated cellular and molecular interactions that are orchestrated by coordinated patterns of, and changes in, gene expression. Not unexpectedly therefore, the adverse health effects that may result from exposure to xenobiotics are invariably associated, either directly or indirectly, with alterations in gene expression. There are now available a variety of technology platforms that permit measurement of simultaneous changes in the expression of many hundreds or many thousands of genes, and in the context of toxicology the application of such methods for multiple transcript profiling is termed "toxicogenomics". The judicious application of such approaches will facilitate the more detailed interrogation of the mechanisms through which adverse effects are induced by chemicals and drugs, with the promise in the future of informing hazard identification and risk assessment paradigms.

With the advent of new genomic technologies, it is now possible to identify, rapidly and holistically, the molecular alterations associated with exposure to toxicants. However, the increase in the rate at which these data can be generated has not been matched by corresponding advances in the ability to interpret them into biologically meaningful information. The use of gene expression data in mechanistic and predictive toxicology is hindered by a lack of information on the relationships between transcriptional events and physiological and pathological changes. We have begun to analyze these relationships in studies designed to improve our understanding of how estrogenic chemicals induce uterine growth and cellular differentiation. Using the rodent uterotrophic response to estrogenic compounds as a model experimental system [1], we are applying transcript profiling to elucidate the molecular events that lead to uterine growth induced by a reference estrogen (17β-estradiol).

---

*Report from a SCOPE/IUPAC project: Implication of Endocrine Active Substances for Human and Wildlife (J. Miyamoto and J. Burger, editors). Other reports are published in this issue, *Pure Appl. Chem.* **75**, 1617–2615 (2003).

‡Corresponding author

MOLECULAR MECHANISMS OF ESTROGEN SIGNALING

The cellular effects of estrogenic compounds are mediated by two subtypes of estrogen receptor, ERα and ERβ. These receptors belong to a superfamily of nuclear receptors that act as ligand-activated transcription factors [2–4] and show tissue-specific distribution patterns, with both ER subtypes being expressed in the rodent uterus. In the classical model of estrogen receptor action, estrogenic compounds diffuse across the plasma membrane and bind to ER, inducing the receptor to dimerize and bind to specific estrogen response elements (EREs) in the promoter regions of target genes to regulate gene expression. However, it is now established that the mode of action of ER is much more complex than this classical model.

Both ER subtypes possess two transactivation domains, termed AF-1 and AF-2. In addition to ligand-dependent activation via the AF-2 region, ER may influence gene expression in a ligand-independent manner via the AF-1 domain. Activation of AF-1 is due, at least in part, to phosphorylation of conserved serine residues. As AF-1 and AF-2 may interact in a synergistic fashion to influence promoter activity [5], there is considerable potential for interplay between ER ligand binding and the activation of kinase signaling cascades to influence gene expression. ER modulates gene expression in the nucleus by forming protein complexes with cofactors [6,7]. These ER-associated cofactors facilitate and augment the recruitment of the basal transcriptional machinery, and thus mediate ER-dependent transcription. The tissue-specific expression of these cofactor proteins may provide another level at which specificity in the effects of estrogens in different tissues can be achieved. Indeed, it has been shown that tissue-determined differences in the action of selective estrogen receptor modulators (SERMs), such as tamoxifen, which is used to treat hormone-dependent breast cancer, may be due to differences in the tissue-specific expression profile of ER cofactors [8]. In addition to ERE-bound ER forming complexes with cofactors, ER may itself be capable of exhibiting cofactor activity, as it has been shown to interact with DNA-bound AP-1 transcription factor and determine transcriptional responses in a manner that does not depend on its own direct interaction with DNA [9,10].

In addition to the nuclear activity of ER, rapid extranuclear actions have been demonstrated, including activation of mitogen-activated protein kinases that have pleiotropic cellular effects at both genomic and nongenomic levels [11,12]. The realization that ERs employ diverse molecular mechanisms for regulating gene expression has generated a need for more holistic studies of alterations in gene expression in response to estrogenic chemicals, for whereas a relatively small number of genes have been described as being directly ER-responsive, estrogens may yet influence the expression of many more genes via the activation of alternative signaling cascades.

GENOMIC APPROACHES FOR UNDERSTANDING THE MOLECULAR MECHANISMS OF ESTROGEN ACTION

Transcript profiling represents a powerful tool for measuring the expression levels of thousands of genes in response to given toxicants [13]. It may thus be used to implicate certain genes in the mode of action of a toxicant, and these leads can be further studied to link them to more classical toxicity endpoints. Gene expression profiles identified using established toxicants may subsequently be compared to those seen with novel compounds to formulate predictive hypotheses on the mode of action of the novel compound. For example, the expression profile in response to estrogen may be readily compared to that obtained with other estrogenic (or potentially estrogenic) compounds which may differ in their relative affinities for ER and thus differ in potency and potential to induce toxicity. These approaches may be further enhanced by combining transcript profiling with transgenic technologies, using either transgenic animals or adapted cell lines. By applying transcript profiling to animal models in which a component of a pathway (e.g., ERα and/or ERβ) relevant to a given mode of action is perturbed, the direct target genes can be separated from those whose regulation is not directly governed by that pathway. Cell lines facilitate advanced mechanistic studies, as they may be engineered to knock in (or out) a particular pathway of interest in isolation, which can be more readily studied in a homogeneous cell type in which interference from other pathways may be minimized.

GENOMIC ANALYSIS OF THE TRANSCRIPTIONAL PROGRAM ASSOCIATED WITH ESTROGEN-INDUCED UTERINE GROWTH

One consequence of ER being capable of mediating transcription of its target genes in a variety of ways is that estrogenic compounds may induce a broad spectrum of cellular responses. However, the molecular mechanisms leading to pleiotropic estrogen-induced cellular responses in vivo have not yet been well described. Recent studies employing microarray analyses of gene expression changes at a single time point in the rodent uterus have revealed a range of novel genes that are responsive to estrogenic chemicals [14,15]. These genes encode proteins involved in a wide range of functions and biological pathways. Additional kinetic studies, in which the temporal pattern of gene expression in response to estrogen is compared to gravimetric and histopathological measurements of uterine growth, have great potential for enhancing our understanding of the relationships between estrogen-induced transcriptional networks and physiological responses. This type of approach is exemplified by the elucidation of the transcriptional program associated with the response of human fibroblasts to serum [16]. Hierarchical clustering of genes into groups on the basis of their temporal patterns of expression provided a detailed view of the molecular events involved in the control of the transition from G_0 to a proliferating state. This study also revealed an unexpected relationship between many components of the transcriptional program and the physiology of wound repair. A similar kinetic approach was used to reveal the co-expression of DNA replication fork genes during estrogen-induced mitogenesis in a human breast cancer cell line [17]. It is noteworthy that transcript profiling experiments can generate enormous data sets, the interpretation of which relies heavily on computational methods for classifying and displaying gene expression patterns [18]. Identification of common molecular or biological functions amongst large numbers of temporally coregulated genes requires the use of additional bioinformatic tools, such as functional annotation databases (e.g., Affymetrix NetAffx™ Analysis Center [19]; <http://www.affymetrix.com/analysis/index.affx>) and biological pathway maps (e.g., KEGG Encyclopedia; <http://www.genome.ad.jp/kegg/kegg2.html>). One additional consideration in designing experiments for a genomic analysis of the transcriptional program associated with estrogen-induced uterine growth is the choice of dosing regime for a given estrogenic chemical. Use of a single dose of estrogen may be necessary in order to avoid the complexity inherent with overlapping waves of gene expression that may be activated following the three successive daily administrations of E_2 that are usually employed using immature animals in the standard rodent uterotrophic assay [1]. Additional controls, comparing changes in gene expression occurring naturally during increased estrogen production at the onset of puberty during normal mouse development with the transcriptional programs associated with uterine-growth in immature mice exposed to an exogenous estrogen (i.e., a chemically induced uterotrophic response), should facilitate our understanding of the consequences of short- vs. long-term changes in hormonally regulated gene expression.

SUMMARY AND PERSPECTIVES

The genomic analysis of transcriptional programs associated with estrogen-induced uterine growth has the potential to reveal a wealth of novel information on the molecular events initiated by exposure of the rodent uterus to exogenous estrogen. This approach should also allow a comprehensive analysis of the molecular mechanisms of xenoestrogen (e.g., genistein and diethylstilbesterol) action. Furthermore, the identification of coregulated clusters of estrogen-responsive genes offers the possibility of gaining novel insights into the molecular mechanisms that regulate their expression. This could be achieved through extensive bioinformatic analyses of their regulatory regions [13]. The recent completion of the first draft rodent genome sequence [20] should greatly facilitate this phase of analysis. Characterizing the biological signaling pathways that regulate transcriptional responses to estrogenic compounds will lead to a better understanding of their molecular mechanisms of action.

REFERENCES

1. J. Odum, P. A. Lefevre, S. Tittensor, D. Paton, C. A. Harris, E. D. Routledge, J. P. Sumpter, J. Ashby. *Regul. Toxicol. Pharmacol.* **26**, 176–188 (1997).
2. K. Dechering, C. Boersma, S. Mosselman. *Curr. Med. Chem.* **7**, 561–576 (2000).
3. L. L. Hart and J. R. Davie. *Biochem. Cell Biol.* **80**, 335–341 (2002).
4. J. G. Moggs and G. Orphanides. *EMBO Reports* **2**, 775–781 (2001).
5. R. Metivier, G. Penot, G. Flouriot, F. Pakdel. *Mol. Endo.* **15**, 1953–1970 (2001).
6. Y. Shang, X. Hu, J. DiRenzo, M. Lazar, M. Brown. *Cell* **103**, 843–852 (2000).
7. N. J. McKenna and B. W. O'Malley. *Cell* **108**, 465–474 (2002).
8. Y. Shang and M. Brown. *Science* **295**, 2465–2468 (2002).
9. M. P Gaub, M. Bellard, I. Scheuer, P. Chambon, P. Sassone-Corsi. *Cell* **63**, 1267–1276 (1990).
10. K. Paech, P. Webb, G. G. Kuiper, S. Nilsson, J. A. Gustafsson, P. J. Kushner, T. S. Scanlan. *Science* **277**, 1508–1510 (1997).
11. E. Falkenstein, H-C. Tillmann , M. Christ, M. Feuring, M. Wehling. *Pharmacol. Rev*. **52**, 513–555 (2000).
12. J. G. Moggs, D. G. Deavall, G. Orphanides. *Bioessays* **25**, 1–5 (2003).
13. G. Orphanides, J. G. Moggs, T. C. Murphy, J. W. Edmunds, W. D. Pennie. *Toxicogenomics*, T. Inoue and W. D. Pennie (Eds.), pp. 20–28, Springer, Tokyo (2003).
14. J. M. Naciff, M. L. Jump, S. M. Torontali, G. J. Carr, J. P. Tiesman, G. J. Overmann, G. P. Daston. *Toxicol. Sci.* **68**, 184–199 (2002).
15. H. Watanabe, A. Suzuki, T. Mizutani, S. Khono, D. B. Lubahn, H. Handa, T. Iguchi. *Genes Cells* **7**, 497–507 (2002).
16. V. R. Iyer, M. B. Eisen, D. T. Ross, G. Schuler, T. Moore, J. C. F. Lee, J. M. Trent, L. M. Staudt, J. J. Hudson, M. S. Boguski, D. Lashkari, D. Shalon, D. Botstein, P. O. Brown. *Science* **283**, 83–87 (1999).
17. E. K. Lobenhofer, L. Bennett, P. L. Cable, L. Li, P. R. Bushel, C. A. Ashfari. *Mol. Endocrinol.* **16**, 1215–1229 (2002).
18. R. A. Young. *Cell* **102**, 9–15 (2000).
19. G. Liu, A. E. Loraine, R. Shigeta, M. Cline, J. Cheng, S. Chervitz, D. Kulp, M. A. Siani-Rose. *ACM SAC Bioinformatics Track* (2002).
20. Mouse Genome Sequencing Consortium. *Nature* **420**, 520–562 (2003).

WORKSHOP 3

THE NEED FOR ESTABLISHING INTEGRATED MONITORING PROGRAMS

Pure Appl. Chem., Vol. 75, Nos. 11–12, pp. 2435–2444, 2003.

Workshop 3.1

Need for establishing integrated programs to monitor endocrine active compounds*

Helmut Segner

Centre for Fish and Wildlife Health, University of Bern, Post Box, CH-3001 Bern, Switzerland

Abstract: Environmental monitoring programs on endocrine active compounds (EACs) have been used to document the level of exposure and to assess the possible association to the occurrence of developmental and reproductive disorders in wildlife. The establishment of causal links between exposure and effect data, however, was found to be difficult due to, for example, the presence of confounding factors or limited understanding of EAC mechanisms and interactions, but also because of conceptual and methodological limitations of current monitoring strategies. In order to provide plausibility of an EAC etiology for a developmental or reproductive alteration in a wildlife population, integrated monitoring programs are needed that will use a combination of complementary approaches: methods for a targeted search for suspected EACs in an environmental mixture, analysis of internal EAC doses instead of external EAC concentrations, utilization of mechanism-based endpoints in bioanalytical and effect monitoring, investigation of the basic biology and physiology of wildlife sentinel species, laboratory replication of field effects, as well as consideration of epidemiological and weight-of-evidence criteria in the design and data evaluation of monitoring programs.

INTRODUCTION

A number of synthetic as well as natural chemicals are capable of interfering with the endocrine system of wildlife, including fish, amphibians, reptiles, birds, and mammals. These so-called endocrine active compounds (EACs) may lead to activational or organizational alterations of endocrine homeostasis, development, and reproduction in exposed organisms. The consequences of EAC exposure have attracted much attention due to the potential impact on survival and sustainability of wildlife populations. In fact, monitoring studies on a range of wildlife species have shown the vulnerability of developmental and reproductive processes, and exposure to chemicals has been suspected to be the causative factor in many studies [e.g., 1–3]. However, to date a chemical etiology could be unequivocally demonstrated only in a few cases. In a review of field monitoring studies, Monosson [4] identified eight cases where scientists felt confident on a chemical etiology of the observed developmental and reproductive disturbances, while in the majority of the studies there was only some suspicion that contaminants are causative for the adverse effects. Thus, the proof of causative links between environmental chemical exposure and adverse biological alterations in endocrine-related parameters of wildlife continues to be a subject of debate.

*Report from a SCOPE/IUPAC project: Implication of Endocrine Active Substances for Human and Wildlife (J. Miyamoto and J. Burger, editors). Other reports are published in this issue, *Pure Appl. Chem.* **75**, 1617–2615 (2003).

EXAMPLES OF MONITORING STUDIES ON DEVELOPMENTAL AND REPRODUCTIVE DISTURBANCES IN WILDLIFE WITH SUSPECTED CHEMICAL ETIOLOGY

In most of the field research, wildlife populations dependent upon aquatic food webs such as fish and fish-eating birds have been in the focus. The aquatic environment is a sink for many chemicals that can be bioaccumulated from water or sediment or can be bioconcentrated along the food webs. The release of EACs into the aquatic environment occurs, e.g., via sewage treatment plants (STPs), and the observation that STP effluents possess endocrine activity [5], was one of the hallmarks that led to public concern on EACs.

One of the best studied examples of endocrine disruption in aquatic organisms is the phenomenon of imposex in prosobranch mollusks, particularly neogastropods. The term imposex has been coined to describe the superimposition of male characteristics such as penis and vas deferens onto female mollusks. Available evidence shows that worldwide more than 100 species of prosobranchs are affected by imposex [6]. Abundant field data indicate a correlation between this irreversible sexual abnormality and environmental tributyltin (TBT) contamination [7–9], pointing to TBT as the causative agent of imposex. Further support for a TBT etiology comes from the observation that with declining environmental concentrations of TBT, subsequent to TBT bans in many countries, a recovery of affected field populations of mollusks did occur [9]. Finally, laboratory experiments demonstrated that TBT is able to induce imposex in mollusks at environmentally realistic concentrations and in a concentration-dependent way [10,11]. The TBT example satisfies most of the criteria of Gray et al. [12], for a cause–effect relationship between an EAC and observed effects in wildlife, in particular correlation between effect and exposure, presence of the same effect in related species, induction of a specific pathognomic effect, disappearance of the effect with chemical cleanup, and ability to replicate the effect in the laboratory using relevant concentrations of the suspected EAC. One criterion that is not fulfilled in the TBT example is the identification of the mechanism of action. To date, the precise mechanism by which TBT leads to imposex formation has not been proven unequivocally.

The importance of understanding the mode of action in the diagnostic assessments of EAC-suspected field cases may be illustrated by the studies on the impact of pulp- and paper-mill effluents on fish populations. A number of studies have shown that pulp- and paper-mill effluents can induce alterations in endocrine homeostasis and reproductive fitness in exposed feral fish [13]. Perhaps the most extensive of these studies deal with white sucker (*Catostomus commersoni*) exposed to bleached kraft pulp-mill effluent (BKME) in Jackfish Bay, Lake Superior, USA. These fish displayed an array of altered physiological and reproductive parameters, including cytochrome P4501A induction, elevated apoptotic DNA fragmentation in ovarian follicular cells, alterations of circulating sex steroid levels, reduced gonad size, reduced fecundity, and delayed sexual maturity [13–15]. Examination of white sucker collected during their spawning migration demonstrated that a number of functions within the pituitary-gonad axis were affected by exposure to BKME [16]. Laboratory studies with fathead minnows confirmed it is the effluent exposure that induces the reproductive changes [17]. However, identification of the chemicals responsible for the BKME effects has been not achieved yet. Plant-derived compounds such as β-sitosterol have been discussed as causative agents for some of the effects associated with BKME exposure. Treatment of fish with β-sitosterol showed that this compound binds to the estrogen receptor and has estrogenic properties, however, it appears that the β-sitosterol effects are mediated not only through the estrogen-receptor pathway but involves other mechanisms as well [18,19]. Further, these studies indicated that it is unlikely that β-sitosterol is solely responsible for the endocrine and reproductive changes. However, the full identification of the responsible chemicals requires that the mechanisms underlying the BKME-induced responses are known, so that a directed search for those chemicals in the BKME which interfere with the relevant molecular targets and processes in exposed organisms will be possible.

The double-crested cormorants (*Phalacrocorax auritus*) at the U.S. Great Lakes may illustrate the difficulty in assessing the role of EACs relative to other factors in determining recruitment and dynamics in widlife populations. Great Lake populations of this fish-eating bird fell dramatically throughout the 1950s and 1960s, and by 1970, the double-crested cormorant was nearly extirpated from the Great Lakes [20]. DDT-induced eggshell thinning is believed to have been the primary cause of this cormorant decline in the Great Lakes. With the banning of DDT in 1972, cormorant populations in the Great Lakes began to recover and increased from approximately 100 breeding pairs in 1973 to more than 10 000 breeding pairs in the 1990s [20]. However, in the post-DDT era, cormorant populations continued to suffer from embryo mortalities and birth defects [21,22]. In contrast to the DDT-induced syndrome, which is characterized by normal embryonic development with mortality induced by eggshell breakage, the post-DDT syndrome is characterized by in ovo mortality, edema, and deformities (GLEMEDS: Great Lakes embryo mortality, edema, and deformities syndrome; [23]). Epidemiological investigations, findings from laboratory studies with other avian species, as well as analytical studies on the correlation between embryo deformities and their contaminant burdens indicated that GLEMEDS is caused by the accumulation of dioxin-like compounds, i.e., compounds that act through the arylhydrocarbon-receptor pathway [21,22,24,25]. Possibly, the GLEMEDS effects have been present already during the DDT era, but have been masked by the eggshell thinning. The case of GLEMEDS in double-crested cormorants provides an example of a field study where the mechanism of action, the causative chemicals, and the adverse effects at the level of the individual are known. However, the ecological relevance of the dioxin-induced adverse effects is questionable since cormorant populations strongly increased during the 1990s, despite the continued contamination with dioxin-like chemicals. It appears that chemical contamination is just one of the cumulative set of factors that contributed to the previous population decline of cormorants; other factors involved could be changes in human persecution of cormorants, alterations in the abundance of the main forage fish of cormorants, the alewife (*Alosa pseudoharengus*), or altered interspecific competition, for instance, because of different sensitivities of Larids and cormorants to developmental toxicants [22].

CONFOUNDING FACTORS TO THE ESTABLISHMENT OF CAUSE–EFFECT RELATIONSHIPS BETWEEN EAC EXPOSURE AND RESPONSES IN WILDLIFE

Monitoring provides an approach to assess the association between the scale and magnitude of environmental EAC contamination, and the biological responses in wildlife organisms and populations. The principal questions that have to be answered by impact monitoring are whether an impact does occur, by which factor, and how "large" the impact is [26]. As exemplified by the case studies above, the most compelling challenge hereby is perhaps not finding out what is wrong, but in finding out what is causing the effects. The difficulty in linking cause and effect at either the population or individual level in monitoring programs on endocrine disruption may stem from a number of factors, some of them indicated in the following:

- Exposure assessment of wildlife to EACs is a critical component in evaluating cause–effect relationships. While some compounds that interfere with the endocrine system, e.g., PCBs, are persistent and bioaccumulative, other EACs such as alkylphenols or estrogens are biodegradable and/or metabolizable. Very little is known about the fate of natural and synthetic EACs in the environment [27]. For the natural estrogen, 17β-estradiol (E2), it has been shown that the steroid was degraded by river microorganisms with half-lives between 0.2 to 9 days at 20 °C, while the synthetic estrogen, ethynylestradiol (EE2), was found to be more recalcitrant to catabolism [28]. Larsson et al. [29], observed in a study on a Swedish sewage treatment work, that both natural estrogens and EE2 were present in unconjugated form although humans primarily excrete the conjugates. This suggests that deconjugation occurred within the sewage system. The ratio between EE2 and natural estrogens was higher than the theoretical value based on human secretion ratios of natural and synthetic estrogens, indicating a faster degradation of the natural estrogens. The findings from chemical analytical investigations on environmental concentrations of EE2 are variable, with some studies reporting no detectable concentrations of EE2 in the environment, while other studies found EE2 levels in the low ng/l-range [27,29,30]. These concentrations are sufficient to induce biological effects [31,32], but are close to the current limits of chemical analytical detection. Considering the lipophilicity of EE2, bioaccumulation is to be expected, however, to date few studies have addressed this question. Larsson et al. [29] found that bile of fish caged downstream to the sewage plant contained estrogenic substances at concentrations 10^4–10^6 times higher than water levels, indicating a strong bioaccumulation of these substances. Lai et al. [33] calculated EE2 bioaccumulation factors for a range of aquatic organisms. The calculations predicted that bioaccumulation would occur in all organisms, with the bioaccumulation factors ranging from 1.8 to 332. EE2 was predicted to exhibit a higher bioaccumulation than the natural estrogens. Compared to actually measured data as provided in the literature, the predicted values for fish were approximately 1000 times less than the values observed in laboratory tests, while for invertebrates, the modeled values were less than two orders of magnitude below laboratory results.

- The time- and concentration-dependency of EAC-induced biological responses may differ from that of other hazardous substances. There are discussions ongoing whether EACs evoke unusual concentration–response relationships such as U-shaped and inverted U-shaped curves [34]. Further, sensitivity of organisms to EACs can vary for specific periods during their life cycle ("sensitive window" concept). Oviparous species possess complex life cycles that contain transitional periods in which new functions emerge. The transitional life stages are often the critical windows for exogenous perturbation. In mammals, with an early determination of sexual differentiation, the embryonic and fetal period may be the critical period for EAC action, while this may be different in many teleost species with a high degree of plasticity in phenotypic sex throughout their life span [35–37].
- Species differences of endocrine physiology and reproductive strategies are of particular importance for the assessment of EAC effects, since they may lead to different responses under identical exposure situations [27,36]. For instance, sexual differentiation in mammals is primarily controlled by androgens, in reptiles by androgens and estrogens, and by estrogens in birds; hence, it is expected that androgen analogs would be more likely to have an effect on the sexual differentiation of mammal, rather than avian, species [36,37]. The specific reproductive strategy of oviparous species offers several targets for EAC action that do not exist in mammals, including external fertilization, eggshell production, egg yolk production/utilization, hatching, and metamorphosis [3].
- Natural fluctuations of developmental and reproductive rates in wildlife populations can be high, which makes the detection of an—additional—impact of EACs on population dynamics difficult. Further complications arise from possible compensatory responses to EAC impact, for instance, an EAC-induced increase of embryo mortalities may be compensated by reduced mortalities at later stages of development. Not only population-level fluctuations, but also changes at the organism level may be difficult to judge with respect to the role of EACs. For instance, for roach, (*Rutilus rutilus*), it has been shown that intersex, i.e., the simultaneous occurrence of both male and female gametes in the gonad, is inducible by exposure to environmental estrogens [38,39]. This gonadal condition appears to occur also spontaneously, but it is not known at what frequency. Jobling et al. [38] reported that in roach from a laboratory control, 4 % of the fish examined showed intersex condition, and at field control sites, incidence of intersexuality ranged from 4 to 18 %. According to the authors, a "low level of intersexuality can be considered as natural", however, it can be difficult to exactly define what a "low level" is.
- Developmental and reproductive alterations at the organism level as well as alterations of population dynamics in wildlife can be induced by a number of factors other than EACs. In the environment, wildlife is not only exposed to EACs, but to other stressors such as overfishing of populations, habitat loss or change, predation, food availability, water temperature, etc. The

combination of multiple stressors and inherent attributes of wildlife biology makes it difficult to tease out the relative contribution of exposure to EACs in endocrine-disruptive effects. Frequently, locations showing elevated levels of EACs may be also contaminated by other anthropogenic compounds. Further, environmental factors other than chemical contamination, such as pathogens or thermal stress can impair reproduction, growth, and development. A well-known example from fish is the influence of temperature on sexual differentiation, also parasite infection has been shown to modify sexuality in fish [40]. For birds, the Seychelles warbler provides an interesting example as this species can manipulate the sex ratio of their offspring such that a breeding pair produces predominantly males or females, depending on the availability of food [41]. These few examples may illustrate the environmental plasticity of endocrine-regulated processes in many vertebrate species. Therefore, disturbances in endocrine parameters of wildlife species, e.g., skewed sex ratios, are not necessarily an indication of endocrine disruption.

- A similar problem as the multiple stressor issue discussed above is the fact that under real world conditions, organisms typically are not exposed to single but to mixtures of EACs. These can include mixtures of chemicals with identical action, e.g., estrogenic activity, but also mixtures of compounds with differing activity. It has been suggested that the possible adverse effects of estrogenic EACs may be neutralized by the presence of antiestrogenic compounds, but conclusive data are missing. Currently, we have no thorough understanding of the behavior of EACs in mixtures in order to be able to explain and understand endocrine-disrupting effects that may arise from exposure to EAC mixtures.

INTEGRATED CHEMICAL–BIOLOGICAL MONITORING

From the previous sections, the problems of monitoring studies in linking observed biological responses to exposure to suspected EACs are apparent. These difficulties come in part from the lack of current understanding of many important interactions, but they are also due to methodological and conceptual limitations in monitoring programs. Clearly, monitoring has to be more than a "just measuring" activity. Effective monitoring requires understanding of the nature, as well as temporal and spatial scales of both, the disturbance and the responses. Monitoring programs that are limited in their methodological or conceptual approach will be not sufficient to provide comprehensive answers. Also, the purely correlative use of chemical and biological monitoring measures is restricted in its ability to establish causative links. What we actually need are monitoring approaches that can identify the relevant chemicals in a targeted search (i.e., to identify those chemicals having an endocrine potency among the many chemicals present in an environmental sample), and, at the same time, the monitoring approach has to be able to sort out the relevant factors among the many confounding factors which confuse the cause–effect relationship. This asks for the use of modern methodologies of exposure and effects assessment, for instance, combined chemical–bioanalytical techniques or the use of a broad array of biological endpoints. In addition, however, a weight-of-evidence approach will have to be used to reduce the level of uncertainty. Several investigators have proposed criteria which can help to make a case for causal inference and to evaluate whether wildlife populations have been affected by EACs [4,12,57,58]. An example of such criteria is provided by the work of Ankley and Giesy [60], who suggested the following criteria for establishing cause–effect relationships: (1) documentation of effects in individuals, (2) documentation of adverse effects in populations, (3) coherence between effects observed in populations vs. those in individuals, (4) identification of a plausible mechanism of action consistent with effects in individuals (possibly through laboratory studies—see above), (5) positive identification of specific contaminants operating through this mode of action (possibly by combined bioanalytics/chemical analytics—see above), (6) reasonableness of dose–response relationships, (7) evidence of the recovery of populations or individuals upon removal of the stressor. Hill [60], suggested nine epidemiological criteria to be used when a correlation is observed between an exposure and an effect, to help decide whether there is a case for inferring that exposure causes the effect: strength of association, consistency of association, specificity of association, time order, biological gradient/dose–response relation, biological plausibility, coherence, experimental evidence, and analogy. The usefulness and applicability of the weight-of-evidence approach and epidemiological criteria in an ecological context is demonstrated in the review of Rollands [61], upon the effects of pollution on reproduction and survival of early life stages in teleosts.

An important step for the success of a monitoring study is the definition of which questions the study is actually supposed to answer and in what detail. The precise formulation of the monitoring objective has important implications for the selection of the monitoring design and of the variables and measurement parameters to be monitored. For instance, if the monitoring study is intended as an initial screen to identify potentially impacted sites, the approach has to be different than when the aim is to identify the EACs at an impacted site. This may be exemplified by comparing the approach taken by the European project COMPREHEND, which aimed to examine the distribution of endocrine active STP effluents across European countries [42], or the approach taken by Routledge [43], who aimed to identify the compounds responsible for the estrogenic activity of STP effluents in the United Kingdom. Another question to be considered in the design of a monitoring program is whether we deal with point sources or non-point sources. Critical parameters for the success of a monitoring study are the temporal and spatial scale, i.e., the recognition of cause–effect linkages by comparison the "before-and-after-impact" situation, or by comparing a control and an impact site [26]. To find suitable control sites, i.e., locations that are as similar as possible in all respects to the impact location, except for the presence of the putative impact (EAC contamination), can be difficult. In densely populated areas such as Europe, a location not impacted by anthropogenic activity hardly exists. In riverine systems, often upstream locations are taken as controls, since they are usually less impacted, at least they may have not yet received STP effluents and may be not surrounded by urbanized or agricultural areas. However, upstream sites are often different to downstream site in many aspects other than the chemical impact, for instance, flow and temperature regimes. Therefore, downstream/upstream differences may be due to background spatial changes that are unrelated to EAC impact. In addition, since both the control and the impact site are located in the same stream, it cannot be excluded that the dynamics of a variable at one point influence the dynamics of this variable at the other site, i.e., strictly speaking, the two locations are statistically not independent [26]. Alternatively, no control sites may be used, but the monitoring study may focus on the comparison of different sites with differing levels of EAC impact.

A methodological aspect critical for the success of EAC monitoring is exposure assessment. Usually, this is done by the analytical determination of preselected compounds. However, this approach may easily miss relevant chemicals, which makes the establishment of cause–effect relationships problematic, if not impossible. Integrated monitoring programs should take advantage of recently developed advanced chemical analytical as well as bioanalytical technologies, which can be valuable complements to the classical analytical approach. Mechanistic understanding of the modes of EAC action has allowed the construction of rapid, cost-effective bioassays, e.g., assays such as the yeast estrogen-receptor assay which detects substances acting through estrogen-receptor binding [44], or the enzyme-linked receptor assay [45]. These assays enable to screen large numbers of samples for endocrine potencies that are mediated through the estrogen receptor. The more laborious and costly chemical analysis can then be restricted to samples that were positively identified in the screening bioassay.

Bioanalytical methods can only indicate that an endocrine activity is present in a complex environmental sample or in a tissue extract, however, they cannot identify the individual compounds being responsible for the biological activity. This requires the combination of the bioanalytical method with chemical analytical techniques. The targeted search for EACs exerting a specific mode of action, for instance, binding to the estrogen-receptor, among the many chemicals present in an environmental sample is possible, for instance, by the bioresponse-linked instrumental analysis [45]. Biomolecular components—in this example, estrogen receptors—are used to selectively extract and subsequently analyze estrogen receptor-binding substances from a complex sample. A different approach is the bioassay-directed fractionation [43,46]. In this method, the environmental sample is subjected to chemical fractionation, for instance, according to lipophilicity, and the resulting fractions are assessed for endocrine activity by means of a specific bioassay. Positive fractions are further separated until a fraction is left which contains one or only few substances that can then be determined by means of chemical analytics. An important part of bioassay-directed fractionation is to confirm that the identified substance, at the concentration present in the environmental sample, is able to induce the observed biological response at a magnitude comparable to that of the sample (confirmation step). An example of the application of bioassay-directed fractionation to identify EACs in environmental samples is the British study on estrogenic STP effluents. Caging experiments with trout had shown that a number of STP effluents were estrogen-active [5]. As causative agents, alkylphenol derivatives were suspected, however, since effluents contain highly complex mixtures of substances, it was difficult to prove this hypothesis. Desbrow et al. [47], therefore applied the bioassay-directed fractionation to search for the estrogenic compounds in the STP effluents. Despite the complexity of the composition of the effluents, only a small number of estrogenic compounds were identified using this approach; namely, the synthetic estrogen EE2, the natural estrogens 17β-estradiol and estrone, and alkylphenolic compounds.

When addressing EAC exposure in monitoring studies, to date the focus has been mainly on the assessment of external exposure, however, more attention should be given to internal exposure and organ dosimetry, i.e., accumulation and disposition of EACs in target tissues. With respect to oviparous vertebrates, xenobiotic residues in endocrine organs, gametes and early life stages may be particularly critical [48]. While there is a good database available concerning the bioaccumulation of highly lipophilic compounds such as PCBs in reproductive organs and gametes, much less is known about the accumulation of more polar EACs. A well-known fact is that maternal steroids and thyroids are deposited in the eggs of oviparous vertebrates, however, the possible additional deposition of hormonally active exogenous substances and the short- and long-term biological consequences of this additional burden have hardly been investigated [49].

In addition to improved exposure assessment, improved effects assessment is essential for future EAC monitoring. The selection of appropriate effect measures depends on the purpose of the monitoring study and can be decisive for its success. One of the more frequently used effect measures is VTG induction in male fish [5,50,51]. This parameter offers a number of advantages, as it is based on a known mechanism, specific for estrogens, sensitive, and it shows a concentration-dependent graded response. The relevance of vitellogenin induction with respect to adverse alterations at the organism or population level, however, remains questionable. What would be important to include, in addition to the molecular response parameter, vitellogenin, are other more individual- or population-related endpoints in order to provide evidence that induction of vitellogenin correlates to alterations at the individual or population level. The concomitant observation of, e.g., reproductive dysfunction and induction of vitellogenin would give a hint to a possible causative role of environmental estrogens in the observed reproductive disturbance.

In the face of multiple exposures, and in the face of multiple modes of endocrine disruption, effects monitoring has to go beyond the measurement of single parameters such as vitellogenin, but should apply multiple response measures. For this purpose, candidate measures can be selected according to their specificity, mechanistic basis, threshold values, effect sizes, and effect quantification. The advantage of using a broader set of mechanistically based effect measures is the capability to differentiate between the impact of EACs and other toxicants, since EACs are often found at complexly contaminated sites with many chemicals showing different modes of action being present [52,53]. Multiple response measurements may be facilitated by recently developed techniques such as gene arrays. Initial attempts to utilize such techniques for EAC assessment have been made in laboratory studies [54,55], however, their potential for assessing EAC impact in field studies still needs to be demonstrated.

Laboratory studies can be an essential component of monitoring programs. Through laboratory studies, it can be confirmed whether biological responses observed in the field are in fact caused by water-borne factors, i.e., when controlled laboratory exposure to environmental matrices is able to induce the biological response [39]. Laboratory studies provide plausibility on observed associations of

EAC exposure and effect in that they reveal underlying mechanisms of action. These experiments establish effect threshold values, and they determine the rate of increase of the response in relation to EAC concentration. Further, laboratory studies are important to establish basic biological traits and response patterns to EAC exposure of the monitored species, e.g., knowledge of developmental patterns of endocrine systems, or recognition of sensitive periods during development [12]. Finally, laboratory studies are essential in studying the mixture issue, i.e., the combined effect of various substances with identical or different modes of endocrine action. Since it will be not feasible to test all possible combinations of EACs as they may occur in the environment, the function of laboratory studies on EAC mixtures is to develop a rational basis to understand EAC interactions in environmental matrices [56].

Monitoring the association between EAC exposure and biological effects will always include some degree of reasonable speculation [62]. It is clear that there will be not the one and only monitoring strategy for all problems, but for each individual case or purpose, different pieces of the puzzle will have to be assembled and utilized to provide plausibility for an EAC-related etiology. Depending on the specific aims of the study, integrated monitoring will have to include detailed demographic information where adverse effects are occurring in wildlife individuals or populations, improved assessment of extent and pattern of EAC exposure and accumulation, consideration of epidemiological criteria, realistic replication of environmental exposure in the laboratory, knowledge upon mechanisms of action, or evaluation of the role of confounders. As such, integrated monitoring is not a new instrument, but the consequent and, importantly, the purpose-driven combination of existing and newly emerging techniques and concepts.

REFERENCES

1. T. Colborn, F. S. vom Saal, A. M. Soto. *Environ. Health Perspect.* **101**, 378–384 (1993).
2. C. R. Tyler, S. Jobling, J. P. Sumpter. *Crit. Rev. Toxicol.* **28**, 319–361 (1998)
3. A. Fairbrother, G. T. Ankley, L. S. Birnbaum, S. P. Bradbury, B. Francis, L. E. Gray, D. Hinton, L. L. Johnson, R. E. Peterson, G. van der Kraak. In *Reproductive and Developmental Effects of Contaminants in Oviparous Vertebrates.* R. T. Di Giulio and D. E. Tillitt (Eds.), pp. 283–362, SETAC Press, Pensacola, FL (1999).
4. E. Monosson. In *Chemically Induced Alterations in Functional Development and Reproduction of Fishes.* R. M. Rolland, M. Gilbertson, R. E. Peterson (Eds.), pp. 177–194, SETAC Press, Pensacola, FL (1997).
5. C. E. Purdom, P. A. Hardiman, V. J. Bye, N. C. Eno, C. R. Tyler, J. P. Sumpter. *Chem. Ecol.* **8**, 275–285 (1994).
6. P. Fiorini, J. Oehlmann, E. Stroben. *Zool. Anz.* **226**, 1–26 (1991).
7. G. W. Bryan, P. E. Gibbs, L. G. Hummerstone, G. R. Burt. *J. Mar. Biol. Assoc. UK* **66**, 611–640 (1986).
8. P. Matthiessen and P. E. Gibbs. *Environ. Toxicol. Chem.* **17**, 37–43 (1998).
9. S. M. Evans and G. J. Nicholson. *Sci. Total Environ.* **258**, 73–80 (2000).
10. P. E. Gibbs, G. W. Bryan, P. L. Pascoe, G. R. Burt. *J. Mar. Biol. Assoc. UK* **70**, 639–656 (1988).
11. M. Brick, U. Deutsch, P. Fiorini. *Helgol. Meeresunters.* **50**, 319–325 (1996).
12. L. E.Gray, J. Ostby, C. Wolf, C. Lambright, W. Kelce. *Environ. Toxicol. Chem.* **17**, 109–118 (1998).
13. M. E. McMaster, G. van der Kraak, K. R. Munkittrick. *J. Great Lakes Res.* **22**, 153–171 (1996).
14. K. R. Munkittrick, G. van der Kraak, M. E. McMaster, C. B. Portt. *Water Res. Can.* **27**, 439–446 (1992).
15. D. M. Janz, M. E. McMaster, K. R. Munkittrick, G. van der Kraak. *Toxicol. Appl. Pharmacol.* **147**, 391–398 (1997).
16. G. van der Kraak, K. R. Munkittrick, M. E. McMaster, C. B. Port, J. P. Chang. *Toxicol. Appl. Pharmacol.* **115**, 224–233 (1992).

17. T. G. Kovacs, J. S. Gibbons, L. A. Tremblay, B. I. O'Connor, P. H. Martel, R. I. Voss. *Ecotoxicol. Environ. Safe.* **31**, 7–22 (1995).
18. D. L. MacLatchy and G. van der Kraak. *Toxicol. Appl. Pharmacol.* **134**, 305–312 (1995).
19. L. Tremblay and G. van der Kraak. *Environ. Toxicol. Chem.* **18**, 329–336 (1999).
20. G. H. Heinz. In *Principles and Processes for Evaluating Endocrine Disruption in Wildlife.* R. Kendall, R. Dickerson, J. Giesy, W. Suk (Eds.), pp. 141–154, SETAC Press, Pensacola, FL (1998).
21. G. A. Fox, B. Collins, E. Hayakawa, D. V. Weseloh, J. P. Ludwig, T. J. Kubiak, T. C. Erdman. *J. Great Lakes Res.* **17**, 158–167 (1991).
22. J. P. Ludwig, H. J. Auman, D. V. Weseloh, G. A. Fox, M. E. Ludwig. *Colonial Waterbirds* **18**, 60–69 (1995).
23. M. Gilbertson, T. Kubiak, J. Ludwig, G. Fox. *J. Toxicol. Environ. Health* **33**, 455–520 (1991).
24. J. P. Ludwig, H. Kurita–Matsuba, H. J. Auman, M. E. Ludwig, C. L. Summer, J. P. Giesy, D. E. Tillitt, P. D. Jones. *J. Great Lakes Res.* **22**, 172–197 (1996).
25. D. P. Ryckman, D. V. Weseloh, P. Hamr, G. A. Fox, B. Collins, P. J. Ewins, R. J. Norstrom. *Environ. Monitor. Assess.* **53**, 169–175 (1998).
26. B. J. Downes, L. A. Barmuta, P. G. Fairweather, D. P. Faith, M. J. Keough, P. S. Lake, B. D. Mapstone, G. P. Quinn. *Monitoring Ecological Impacts. Concepts and Practice in Flowing Waters*, p. 433, Cambridge University Press, Cambridge (2002).
27. J. Miyamoto and W. Klein. *Pure Appl. Chem.* **70**, 1829–1845 (1998).
28. M. D. Jürgens, K. I. E. Holthaus, A. C. Johnson, J. J. L. Smith, M. Hetheridge, R. J. Williams. *Environ. Toxicol. Chem.* **21**, 480–488 (2002).
29. D. G. J. Larsson, M. Adolfsson–Erici, J. Parkkonen, M. Petterson, A. H. Berg, P. E. Olson, L. Förlin. *Aquat. Toxicol.* **45**, 91–97 (1999).
30. S. A. Snyder, T. L. Keith, D. A. Verbrugge, E. M. Snyder, T. S. Gross, K. Kannan, J. P. Giesy. *Environ. Sci. Technol.* **33**, 2814–2820 (1999).
31. R. Länge, T. H. Hutchinson, C. P. Croudace, F. Siegmund, H. Schweinfurth, P. Hampe, G. Panter, J. P. Sumpter. *Environ. Toxicol. Chem.* **20**, 1216–1227 (2000).
32. H. Segner. K. Caroll, M. Fenske, C. R. Janssen, G. Maack, D. Pascoe, S. Schäfers, G. F. Vandenbergh, M. Watts, A. Wenzel. *Ecotox. Environ. Safe.* **54**, 302–314 (2003).
33. K. M. Lai, M. D. Scrimshaw, J. N. Lester. *Sci. Total Environ.* **289**, 159–168 (2002).
34. R. L. Dickerson, A. Brouwer, L. E. Gray, D. R. Grothe, R. E. Peterson, D. M. Sheehan, C. Sills–McMurry, M. A. Wiedow. In *Principles and Processes for Evaluating Endocrine Disruption in Wildlife,* R. Kendall, R. Dickerson, J. Giesy, W. Suk (Eds.), pp. 69–96, SETAC Press, Pensacola, FL (1998).
35. L. D. Arcand–Hoy and W. H. Benson. *Environ. Toxicol. Chem.* **17**, 49–57 (1998).
36. P. M. Campbell and T. H. Hutchinson. *Environ. Toxicol. Chem.* **17**, 127–135 (1998).
37. D. M. Fry. *Environ. Health Persp.* **103** (Suppl. 7), 165–171 (1995).
38. S. Jobling, M. Nolan, C. R. Tyler, G. Brighty, J. P. Sumpter. *Environ. Sci. Technol.* **32**, 2498–2506 (1998).
39. T. P. Rodgers–Gray, S. Jobling, C. Kelly, S. Morris, G. Brighty, M. Waldock, J. P. Sumpter, C. R. Tyler. *Environ. Sci. Technol.* **35**, 462–470 (2000).
40. T. Wicklund, L. Lounasheimo, J. Lom, G. Bylund. *Dis. Aquat. Org.* **26**, 163–171 (1996).
41. J. Komdeur, S. Daan, J. Tinbergen, C. Mateman. *Nature* **285**, 522–525 (1997).
42. R. I. L. Eggen, B. E. Bengtsson, C. T. Bowner, M. Gibert, A. Gerritsen, K. Hylland, A. C. Johnson, P. Leonards, T. Nakari, L. Norrgren, J. P. Sumpter, M. F. Suter, A. Svenson, A. D. Pickering. *Pure Appl. Chem.* **75** (11/12), 2445–2450 (2003).
43. E. J. Routledge. *Pure Appl. Chem.* **75** (11/12), 2461–2466 (2003).
44. E. J. Routledge and J. P. Sumpter. *Environ. Toxicol. Chem.* **15**, 241–248 (1996).

45. M. Seifert, L. Wen, M. Alberti, U. Kausch, B. Hock. *Pure Appl. Chem.* **75** (11/12), 2451–2459 (2003).
46. D. A. Sheahan, G. F. Brighty, M. Daniel, S. J. Kirby, M. R. Hurst, J. Kennedy, S. Morris, E. J. Routledge, J. P. Sumpter, M. J. Waldock. *Environ. Toxicol. Chem.* **21**, 507–514 (2002).
47. C. Desbrow, E. J. Routledge, G. C. Brighty, J. P. Sumpter. *Environ. Sci. Technol.* **32**, 1549–1558 (1998).
48. K. Kleinow, J. Baker, J. Nichols, F. Gobas, T. Parkerton, D. Muir, G. Monteverdi, P. Mastrodone. In *Reproductive and Developmental Effects of Contaminants in Oviparous Vertebrates.* R. T. Di Giulio and D. E. Tillitt (Eds.), pp. 9–111, SETAC Press, Pensacola, FL (1999).
49. T. Madigou, P. Le Goff, G. Salbert, J. P. Cravedi, H. Segner, F. Pakdel, Y. Valotaire. *Aquat. Toxicol.* **53**, 173–186 (2001).
50. Y. Allen, A. P. Scott, P. Matthiessen, S. Haworth, J. E. Thain, S. Feist. *Environ. Toxicol. Chem.* **18**, 1791–1800 (1999).
51. M. Petrovic, M. Sole, M. J. Lopez de Alda, D. Barcelo. *Environ. Toxicol. Chem.* **21**, 2146–2156 (2002).
52. E. Noaksson, U. Tjärnlund, A. T. C. Bosveld, L. Balk. *Toxicol. Appl. Pharmacol.* **174**, 160–176 (2001).
53. M. S. Sepuvelda, W. E. Johnson, J. C. Higman, N. D. Denslow, T. R. Schoeb, T. S. Gross. *Sci. Total Environ.* **289**, 133–144 (2002).
54. B. P. Bradley, E. A. Shrader, D. G. Kimmel, J. A. Meiller. *Mar. Environ. Res.* **54**, 373–377 (2002).
55. P. Larkin, L. C. Folmar, M. D. Hemmer, A. J. Poston, H. S. Lee, N. D. Denslow. *Mar. Environ. Res.* **54**, 395–399 (2002).
56. E. Silva, N. Rajapakse, A. Kortenkamp. *Environ. Sci. Technol.* **36**, 1751–1756 (2002).
57. J. C. Lamb, R. Balcomb, C. M. Bens, R. L. Cooper, J. W. Gorsuch, P. Matthiessen, M. M. Peden-Adams, E. O. Voit. In *Principles and Processes for Evaluating Endocrine Disruption in Wildlife.* R. Kendall, R. Dickerson, J. Giesy, W. Suk (Eds.), pp. 17–37, SETAC Press, Pensacola, FL (1998).
58. G. W. Suter, S. B. Norton, S. M. Cormier. *Environ. Toxicol. Chem.* **21**, 1101–1111 (2002).
59. G. T. Ankley, J. P. Giesy. In *Principles and Processes for Evaluating Endocrine Disruption in Wildlife,* R. Kendall, R. Dickerson, J. Giesy, W. Suk (Eds.), pp. 349–367, SETAC Press, Pensacola, FL (1998).
60. A. B. Hill. *Proc. Royal Soc. Med.* **58**, 295–300 (1965).
61. R. M. Rolland. *Fish Fisheries* **1**, 41–72 (2000).
62. G. van der Kraak. *Pure Appl. Chem.* **70**, 1785–1794 (1998).

Pure Appl. Chem., Vol. 75, Nos. 11–12, pp. 2445–2450, 2003.

Workshop 3.2

Search for the evidence of endocrine disruption in the aquatic environment: Lessons to be learned from joint biological and chemical monitoring in the European Project COMPREHEND*

Rik I. L. Eggen[1,‡], Bengt-Erik Bengtsson[2], C. T. Bowmer[3], Anton A. M. Gerritsen[4], Michel Gibert[5], Kjetil Hylland[6], Andrew C. Johnson[7], Pim Leonards[8], Tarja Nakari[9], Leif Norrgren[10], John P. Sumpter[11], Marc J.-F. Suter[1], Anders Svenson[12], and Alan D. Pickering[13]

[1]Swiss Federal Institute for Environmental Science and Technology (EAWAG), Überlandstrasse 133, CH-8600, Dübendorf, Switzerland; [2]Stockholm University, Sweden; [3]TNO, Nutrition and Food Research Institute, The Netherlands; [4]Institute for Inland Water Management and Waste Water Treatment (RIZA), Lelystad, The Netherlands; [5]Anjou Recherche/Vivendi Water, Paris, France; [6]Norwegian Institute for Water Research (NIVA), Oslo, Norway; [7]NERC Centre for Ecology and Hydrology, Oxfordshire, UK; [8]Institute for Fisheries Research (RIVA-DLO), IJmuiden, The Netherlands; [9]Finnish Environment Institute (FEI), Helsinki, Finland; [10]Swedish University of Agricultural Sciences (SLU), Oslo, Sweden; [11]Brunel University, Middlesex, UK; [12]Swedish Environmental Research Institute Ltd (IVL), Sweden; [13]NERC Centre for Ecology and Hydrology, Cumbria, UK

Abstract: Between January 1999 and December 2001, the European Community project COMPREHEND was performed. The overall aim of COMPREHEND was to assess endocrine disruption in the aquatic environment in Europe, consequent to effluent discharge, with emphasis on estrogenic activity. COMPREHEND demonstrated the widespread occurrence of estrogenic effluents across Europe and presented evidence of impacts on a range of wild fish species. Using a variety of bioassays in combination with chemical analytical methods, estrogenic steroids of human origin from domestic wastewater effluents were identified as the most pervasive problem, although alkylphenols may be important estrogenic components of some industrial effluents. New tools have been developed for the identification of estrogenic effluents, and recommendations are made for the improvement of existing techniques. We have shown that individual fish within natural populations may be feminized to varying degrees, but it has not been possible to show, using traditional fish population parameters, that the survival of fish populations is threatened. However, laboratory-based fish life-cycle studies demonstrate the sensitivity of fish to estrogen (and androgen) exposure and

---

*Report from a SCOPE/IUPAC project: Implication of Endocrine Active Substances for Human and Wildlife (J. Miyamoto and J. Burger, editors). Other reports are published in this issue, *Pure Appl. Chem.* **75**, 1617–2615 (2003).

‡Corresponding author: Tel.: +41 (0) 1 823 5320; E-mail: eggen@eawag.ch

how this might lead to complex (and potentially damaging) genetic changes at the population level. New approaches to this problem, utilizing recent advances made in the field of molecular and population genetics, are recommended. Finally, a study of estrogenic and androgenic activity of waste waters during the treatment process has shown that some of the existing wastewater treatment technologies have the potential to eliminate or minimize the hormonal activity of the final effluent.

In the period between January 1999 and December 2001, the community program of research on endocrine disruptors and environmental hormones (COMPREHEND), a European Community-financed project (ENV4-CT98-0798), was performed. In COMPREHEND, coordinated by Prof. Alan D. Pickering, 12 partners from 7 European countries were involved. COMPREHEND was designed with the overall aim of assessing the evidence for endocrine disruption in the aquatic environment in Europe, consequent to effluent discharge. Particular emphasis was given to the estrogenic activity of both domestic and industrial effluents and their impacts on fish (both freshwater and marine). The overall objectives of the COMPREHEND program were to: (1) examine the occurrence and distribution of endocrine-disrupting effluents across a range of European countries using existing fish-exposure techniques for the detection of reproductive interference, (2) analyze those effluents which are shown to be capable of interfering with the fish endocrine system for the principal causative agents, (3) investigate the influence of partitioning within the water column of known endocrine disrupters in relation to their impact on fish, (4) assess available evidence and collect new information on the impacts of endocrine disruptors on aquatic wildlife, (5) develop and improve tools (both in vivo and in vitro) for the rapid detection of endocrine disruptors and investigate the application of existing in vitro techniques for the direct screening of complex effluents. In this manuscript, the lessons learned and suggestions for future activities in this field are described. Many more details can be found in published data or upcoming publications from the laboratories involved.

COMPREHEND has shown that estrogenic effluents are widespread across Europe. Approximately one-third of the municipal sewage treatment works (STW) effluents examined during the 3-year program were found to be strongly estrogenic and capable of stimulating vitellogenesis in juvenile or male fish (from a range of species) after a 2–3 week exposure period. Several industrial wastewater effluents (principally those involved with chemical/pharmaceutical manufacture) were also found to be significantly estrogenic to exposed fish. However, other municipal STW effluents and industrial waste waters were apparently nonestrogenic or only weakly estrogenic. This variability in estrogenic activity was detected by both in vivo and in vitro techniques and was characteristic of effluents from most countries examined during the survey. We conclude that strongly estrogenic municipal effluents will be reasonably commonplace across mainland Europe (as they are in the United Kingdom) and, therefore, there is the potential for estrogenic endocrine disruption of aquatic wildlife in all countries (the situation with industrial effluents will depend very much upon the nature of the industrial processes). National sewage effluent surveys have already been undertaken in the United Kingdom, the Netherlands (as part of the LOES/COMPREHEND program), and in Sweden. The Swedish survey was initiated as a direct result of the Swedish involvement (three laboratories) in the COMPREHEND program and demonstrates how EU funding may act as a lever to secure additional national funding for environmental research and monitoring programs. We are also aware of additional effluent survey work in France (indeed, we invited environmental scientists from Cemagref to all the COMPREHEND meetings and they were able to participate in all but one of the six planning/reporting meetings), in Austria and, most recently, in Switzerland. Several of the COMPREHEND senior scientists have been involved in the peer-review process of these new initiatives and one of the COMPREHEND partners (EAWAG) is heavily involved in the Swiss national program. Thus, an increasing number of European countries are now investing significantly in studies on aquatic endocrine disruption.

The in vivo approach to measuring effluent estrogenicity, by measuring induced vitellogenin (VTG) levels in the blood of immature fish (rainbow trout, carp, and cod) exposed in situ to the effluent was difficult to control and standardize. This approach, whilst being a useful indicator of estrogenicity, is not suitable for the routine assay of effluents within any statutory framework. Inconsistencies were found in the current methods for blood plasma vitellogenin analyses. Enzyme-linked immunoabsorbent assay (ELISA) techniques are preferable to radioimmunoassay, in terms of the facilities required and their ease of operation, but most existing ELISA techniques lack the sensitivity of radioimmunoassay (RIA). However, the only homologous rainbow trout ELISA tested during COMPREHEND did have the required sensitivity, and we recommend that future assays use homologous systems wherever possible. A further problem was the lack of an agreed VTG standard for the rainbow trout (or for any other species), and an interlaboratory comparison of existing "purified" VTG preparations is urgently required.

Several in vitro assays (human, yeast, and fish cell-based) for the direct measurement of estrogenic activity were tested for use on complex effluents. Sample preparation, in most systems, consisted of solid-phase extraction and elution, but did not include further fractionation. There were marked differences in sensitivity of the in vitro assays, but specificity to known estrogens was reasonably consistent, irrespective of the nature of the assay. Thus, in vitro systems based upon fish estrogen receptors produced similar estimates of estrogenic activity in effluents to those based on human estrogen receptors. Assays based on genetically modified yeast cells (the YES and YAS assays) were sufficiently robust and reliable for most purposes and were considerably simpler to perform than those assays requiring sterilized cell-culture techniques. If insensitivity is a problem, we recommend that effort is spent increasing the sensitivity of the yeast cell systems. The YES assay does have the potential for standardization in routine effluent testing, but COMPREHEND identified two sources of possible interference. Toxicity of the effluent to the yeast cells can, in some cases, be overcome by dilution but suppression of the estrogenic response at the level of the receptor requires that the assay protocol is modified, in order to detect such effects. It was demonstrated that alkylphenol acetates suppress the activity of estrogenic steroids and of alkylphenols. The point must be made, however, that in vitro studies are not a substitute for in vivo studies. We have shown, for example, that EE2 is approximately ten times more estrogenic to zebrafish than is E2, whereas most in vitro assays failed to separate their relative potencies.

The techniques for chemical analysis of alkylphenols and related compounds and of bisphenol A were reliable and repeatable (at the $\mu g\ l^{-1}$ concentration), but problems were experienced with the measurement of estrogenic steroids (at the $ng\ l^{-1}$ concentration) in such complex matrices. Bisphenol A was detectable in municipal effluents, but at concentrations less than 5 $\mu g\ l^{-1}$ and nonylphenol, the most abundant of the alkyl phenols, was generally below 2 $\mu g\ l^{-1}$. Estrone (E1) measurements were the most consistent in terms of recovery, and a good correlation was obtained in a comparison of the techniques in two laboratories for measurements of the same set of wastewater samples. There was poor agreement with estradiol (E2) measurement, and both laboratories experienced very poor recoveries with estriol (E3) and low sensitivity with ethinylestradiol (EE2). Estrone measurements in STW effluents showed a good degree of correlation with estrogenic activity (as measured with in vitro assays) and estrogenic steroids E1 and E2 were generally in the 0 to 10 $ng\ l^{-1}$ range. EE2, however, was often at or below the limit of detection (approximately 1 $ng\ l^{-1}$). We recognize the need for an interlaboratory comparison of analytical techniques for steroids in complex effluents. It may be necessary to develop new analytical approaches, and the role(s) of RIA needs to be considered alongside more traditional analytical chemistry. Estrogenic steroids were generally below the limits of detection for most industrial waste waters (unless there was a significant component of the effluent originating from domestic/human sources within the industrial plant). The strong estrogenicity of two industrial effluents (specialty chemicals manufacture and textile) correlated with relatively high levels (up to 5 $\mu g\ l^{-1}$) of nonyphenol (NP) and nonylphenol ethoxylates (NPEs) and, in the effluent from another chemical manufacturing plant, the toxicity identification and evaluation (TIE) approach identified a hydroxyphenyl hexanoic acid as the principal estrogenic contaminant. The highest level of bisphenol A (BPA) (1.14 $\mu g\ l^{-1}$) was found in the effluent from a pharmaceutical plant.

TIE identified E2, E1, and EE2 as the principal estrogenic components of domestic raw sewage, with EE2 and E1 dominating the estrogenic activity of the final effluent. Taking into consideration the potencies of the various estrogenic compounds measured in municipal STW effluents, we conclude that natural and synthetic steroids, of human origin, are by far the most important estrogenic components and are responsible for most of the estrogenic effects seen in vivo and in vitro. EE2 may be particularly important in this respect, but the limitations of our current analytical techniques are a major constraint to confirming the importance of this component of the contraceptive pill. TIE also provided evidence of "cooperative" effects between the different steroids, making the measured activity (YES assay) approximately three times greater than the sum of the activity of the individual components. Currently, there is no consensus on the best approach for experimental design and data analysis to measure potential interaction between estrogenic compounds in complex mixtures. This is clearly an important area and one which requires more research. In this context, we are pleased that two new, EU-funded research programs (ACE and EDEN) will focus primarily on the effects of mixtures of endocrine disruptors. Several of the original COMPREHEND consortium are participants in the EDEN program (under Framework V) and will address the issue of mixtures in the natural environment.

Laboratory studies demonstrated that natural suspended sediments in the water do not modulate, to any significant extent, the estrogenicity of octylphenol to fish and it is concluded that the most important route of exposure for both alkylphenols and steroids is directly from the water, rather than indirectly from contaminated suspended sediments or via the food chain. However, a Dutch survey of the distribution of potential xenoestrogens in the aquatic environment found alkylphenols and their ethoxylates at particularly high concentrations in suspended, particulate matter in fresh water. Whether such material is biologically available to fish requires further study, as does the role of bed sediments as a potential route of exposure for truly benthic/burrowing animals.

Strong androgenic activity (YAS assay) was found in the influent to domestic STWs and was presumed to be of human origin, but most of the activity disappeared during the wastewater treatment process. Androgenicity (and some estrogenicity) was also detected in some pulp-mill effluents when zebrafish were exposed to effluents dilutions or to wood sterols in the laboratory, but the environmental consequences of this are unknown.

Thus, COMPREHEND has shown that hormonally active substances (primarily steroids) are being discharged into the aquatic environment at concentrations initially high enough to induce estrogenic responses in experimentally exposed fish. However, in most circumstances the effluents are rapidly diluted following discharge and, therefore, the question arise as to whether native fish in the receiving waters are adversely impacted. Samples of wild fish, taken in the vicinity of some of the known estrogenic municipal effluents were found to have abnormally high levels of VTG (a female egg protein) in the blood of juvenile or male fish. Moreover, increased levels of intersexuality (presumed males with oocytes present in the testis) were found in the same vicinity. The species of wild fish showing such evidence of estrogen exposure were the common bream, the common carp, the roach, and the gudgeon, all belonging to the carp family (the Cyprinidae). There was no evidence that the gudgeon (a benthic species) was significantly more affected than the roach (a more pelagic species). Exposure of fertilized brown trout eggs to rivers impacted by sewage effluents resulted in significant impacts on embryonic development, but it is too early to say whether these were mediated in any way by endocrine disruption. Nevertheless, it is interesting that some of the novel, estrogen-sensitive gene products identified as part of COMPREHEND had strong homologies with growth and transcription factors. The role of such proteins in (and possible estrogenic influences on) normal cellular development is an area that merits further research.

All the evidence obtained thus far is indicative of localized areas of estrogenic impact on individual fish, but it tells us little about the longer-term consequences at the population level. An analysis of one of the largest freshwater fish population data sets (for bream in the Netherlands) available to science revealed only limited circumstantial evidence of possible impacts of endocrine disruption at the population level. Thus, any significant deviation of sex ratios in bream populations away from a presumed normality of 50/50 was always in favor of females. In addition, evidence was found in one population of a significant decline over 30 years in the testis size of sexually mature male fish, whereas the female fish in the same population showed no change in gonad weight. However, we recognize that many factors might influence sex ratios in fish populations (and, indeed, relative gonad size) and accept that the analytical approach to fish population parameters adopted during COMPREHEND can never show cause and effect. Moreover, the size of cyprinid fish populations in particular are often determined by stochastic factors, such as food availability during the early life stages, and these may mask any effects of endocrine disruption on population structure. We are of the opinion that a new approach to this problem is required and recommend the development of reliable tools for the genetic sex determination of fish and the use of population genetics to investigate the potential impacts of endocrine disruption. If, for example, a population is being maintained by just a small percentage of the sexually mature adults, it might be possible to recognize this in terms of a reduction in genetic diversity within the population. Such a situation might not have an immediate impact on fish numbers, but it might make the population more susceptible to other environmental stresses, by limiting the genetic diversity available for future natural selection.

In view of the difficulties inherent in any study of natural fish populations exposed to endocrine disrupting influences, we also examined impacts of known estrogens and of wastewater effluents on fish (primarily zebrafish, with some work on the stickleback) held under laboratory conditions, including studies of chronic exposure over two complete generations. Chronic exposure to as little as 0.6 ng l^{-1} EE2 (below the limits of chemical detection for most effluents) was sufficient to sex-reverse male fish, and 1.5 ng l^{-1} stimulated vitellogenesis in juvenile fish. The stickleback studies indicated a particularly sensitive window of exposure to estrogens during the first two weeks post-hatch. Sex reversal in the opposite direction could be induced by low-level androgen (methyltestosterone, MT) exposure but higher concentrations of MT caused feminization—a clear case of a U-shaped dose–response curve. The feminizing effects of MT were also observed in a study of natural sex reversal in the cuckoo wrasse. One of the more unusual results from chronic exposure studies was the effect of wood sterols (and pulp-mill effluent) on the sex ratios of successive populations of zebrafish. The first progeny (F1) of the chronically exposed fish were predominantly male but then reversed in the F2 generation to predominantly female. It is possible that this effect may be caused by the successful spawning of sex-reversed, F1 genetic females. Whatever the explanation, more work is now required into determination of the genetic sex of individual fish and the reproductive competence of sex-reversed fish.

COMPREHEND was unable to find any evidence of endocrine disruption in a brackish-water crustacean *Nitocra spinipes* exposed to estrogenic substances, and we question whether this species has a functional estrogen receptor. Some evidence of ecdysteroid receptor antagonism was found for a waste water effluent from a chemical manufacturing plant but we are not aware of any evidence of problems in aquatic invertebrates consequent to such endocrine disruption.

Whenever the estrogenic activity of the raw sewage (influent) was measured (using in vitro techniques) it was found to be strongly estrogenic. The fact that some of the final effluents were nonestrogenic implies that some existing STWs have the capacity to remove most, if not all, of the estrogenic activity. Similar observations also apply to androgens in municipal STWs. Chemical analyses and TIE indicated that the estrogenic activity in the raw sewage was caused by E2, E1, and EE2. E1 to E2 ratios were much higher in the final effluent, indicating conversion of E2 to E1 during the treatment process, but EE2 may be the dominant estrogenic component of the effluent (difficult to confirm in the unfractionated effluent because of insufficient sensitivity in the EE2 analysis). A comparison of different types of STWs and a study of estrogenicity during the various stages of treatment confirmed that the majority of the activity was lost during secondary biological treatment. Indeed, in systems with primary treatment only, androgenic and estrogenic activity could actually increase, presumably because of deconjugation. A major factor determining the estrogenicity of the final effluent was the residence time during the treatment process, with effluents from treatment plants with residence times over 15 h having minimal activity. Thus, the solution to this potential environmental problem may already exist within the existing technology. However, alternative technologies, aimed to avoid the release of estrogenic compounds in the waste streams already at the source, must be seriously analyzed as well.

In summary, the following conclusions may be drawn from the COMPREHEND program.

- Estrogenic wastewater effluents (both domestic and industrial) are widespread across Europe.
- Steroids are the principal estrogenic compounds in municipal STW effluents.
- Improved analytical techniques for the measurement of estrogenic steroids in complex matrices are urgently needed.
- Nonylphenol may play a significant estrogenic role in some industrial effluents.
- For in vivo screening, fish vitellogenin assays need to be improved and standardized.
- Current in vitro screening techniques can be used to assess the estrogen receptor-mediated estrogenicity and the androgen receptor-mediated androgenicity of unfractionated effluents.
- We need increased molecular mechanistic understanding of endocrine disruption. This as a basis for proper risk assessment and the development of novel bioanalytical tools.
- The principal xenoestrogen exposure route for freshwater fish is directly from the water.
- There is evidence of endocrine disruption in individual fish from wild stocks in the vicinity of estrogenic effluents.
- We need a new approach (using molecular genetic techniques) for the study of impacts at the population level.
- Sex reversal in either direction is a possible consequence of exposure to hormonally active compounds.
- Chronic exposure to some effluents can cause masculinization in the F1 generation, followed by feminization in the F2 generation.
- We could not find evidence of crustacean sensitivity to xenoestrogens.
- Existing wastewater treatment technologies have the *potential* to eliminate most of the estrogenic and androgenic activity before discharge to the environment.
- Technologies aimed at avoiding the release of endocrine active compounds in the waste stream might be an alternative.

ACKNOWLEDGMENTS

This work was financed by the European Community (ENV4-CT98-0798) and the Swiss federal office for education and science (BBW, grant 98.0090).

Pure Appl. Chem., Vol. 75, Nos. 11–12, pp. 2451–2459, 2003.

Workshop 3.3

Biomonitoring: Integration of biological endpoints into chemical monitoring*

M. Seifert‡, Li Wen, M. Alberti, U. Kausch, and B. Hock

Technische Universitaet Muenchen, Center of Life Sciences Weihenstephan, Chair of Cell Biology, Alte Akademie12, D-85350 Freising, Germany

Abstract: Biomonitoring is currently performed at two levels, assessing exposure to pollutants and effects monitoring by bioassays. As an example for the first approach, vitellogenin (VTG) in male fish of *Abramis brama* as an endpoint for estrogen exposure is discussed. However, similar changes of VTG or VTG-like proteins in the hemolymph of mussels could not be detected. Enzyme-linked receptor assays for monitoring estrogenic effects at the molecular level serve as an example for the second category. Applications of the enzyme-linked receptor assay (ELRA) developed in our laboratory are presented. Detection limits of 0.02 μg/l 17β-estradiol were recently achieved with the chemiluminescent format. Although effect monitoring provides information in terms of toxicity equivalents, it is not possible to relate the signals to specific pollutants and their concentrations. For this purpose, chemical analysis is required. New approaches are reported for the direct coupling of bioassays and chemical analysis. This concept is defined as bioresponse-linked instrumental analysis. It combines biomolecular recognition, initiating a biological effect, and chemical analysis. In addition to the classical bioanalytical approaches, new strategies in genomics and proteomics have been developed. This may lead to multimarker approaches opening this area to environmental analytics.

INTRODUCTION

Biological responses can be exploited for environmental monitoring. Especially programs in the assessment of water quality monitoring take advantage of this approach. Biomonitoring is carried out at several levels of complexity: the level of populations and species as well as the suborganismic and molecular level. Two different strategies have been used in the past: the application of (1) bioindicators for assessing exposure to pollutants and (2) biotests, biosensors, and reporter organisms for monitoring the effects of pollutants. Endocrine disruptors (EDCs) are used in this chapter for highlighting current approaches to biomonitoring at the suborganismic level. Although test systems with whole organisms currently provide the only means to measure integral effects, their disadvantages are due to the high variability of the results resulting from the heterogeneity of organisms, high costs, and the difficulty to relate damage to individual pollutants. In contrast, suborganismic tests can provide more specific and faster responses. But it is clear that structure and concentration of pollutants are only revealed by chemical analysis. Therefore, new approaches are required for a tighter coupling of bioassays and chemical analysis. Bioresponse-linked instrumental analysis combines biomolecular recognition, initiating a bi-

*Report from a SCOPE/IUPAC project: Implication of Endocrine Active Substances for Human and Wildlife (J. Miyamoto and J. Burger, editors). Other reports are published in this issue, *Pure Appl. Chem.* **75**, 1617–2615 (2003).

‡Corresponding author

ological effect, and chemical analysis. In addition, it is necessary to learn more about effects and mechanisms of EDCs in living organisms. One of the main problems of bioanalytics is due to the fact that test systems with whole organisms do not provide specific information on the mechanism of the effect. In contrast, suborganismic tests can only cover a limited range of toxicity classes. This gap can be bridged by the application of multi-endpoint approaches, which can be realized in the form of DNA microarrays and 2D gel electrophoresis.

VITELLOGENIN AS A BIOINDICATOR FOR ENDOCRINE DISRUPTION

Biomarkers are valuable tools for estimating the exposure of organisms to environmental pollutants. A well-known example is the use of vitellogenin for assessing estrogenic effects in male fish [14]. VTG is an ancient transport and storage protein that serves as a ligand for the delivery of nutrients to the egg yolk. In male fish, the levels are very low, but they rise dramatically in the presence of estrogenic compounds. For instance, plasma levels in the mg/ml range are detected in male as well as immature fish after injection of 17ß-estradiol [13,15]. Figure 1 illustrates this effect in bream *(Abramis brama)* where an exposure of male fish was carried out with ethinylestradiol (EE2). The SDS gel shows an additional band of 120 kDa that is present in untreated females, but missing in male controls.

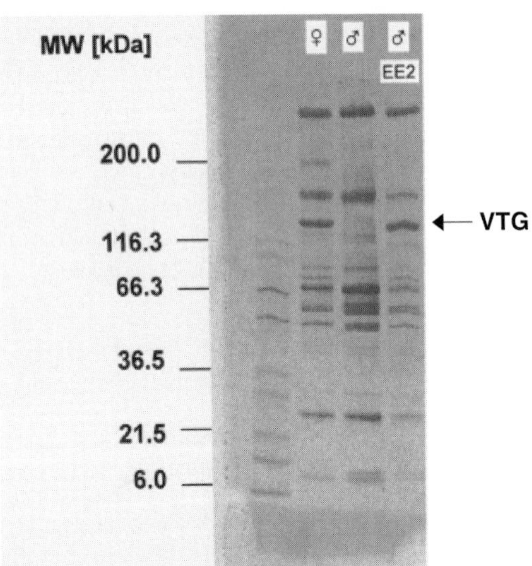

Fig. 1 Effect of ethinylestradiol on male fish *(Abramis brama)*. The animals were exposed to EE2 and compared to untreated female and male fish. After separation of serum proteins by SDS gel electrophoresis, an additional band at 120 kDa is seen.

There is considerable interest in the use of biomarkers from bivalve species for estrogens. Mussels are sessile and filter huge amounts of water. They should therefore be main targets for estrogenic contaminants. However, it is still under debate whether there is a dose–response relationship for estrogens and VTG in mussels. If the situation was similar to fish, VTG or VTG-like proteins in mussel hemolymph would be an ideal endpoint for an exposure to estrogenic compounds and allow the monitoring of a large number of samples, for instance by enzyme-linked immunosorbent assay (ELISA). Therefore, experiments were carried out to examine whether hemolymph is a carrier of VTG or VTG-like proteins or other proteins under estrogenic control that could be targets for indicator assays. For this purpose, 1D and 2D polyacrylamide gel electrophoresis (PAGE) of mussel plasma were performed. First, it was found that the protein concentration of mussel plasma prepared from the hemolymph is very low compared to other animals (0.36 mg/ml for *Anodonta* and 0.40 mg/ml for *Mytilus* plasma).

This means that mussel plasma contains 50–200 times less soluble protein than fish. Even when compared to other mollusks (snails and octopus), mussel plasma has a very low protein concentration.

Mussels were either injected or exposed to water (hormone concentrations up to 2 µg/l 17ß-estradiol, E2). Control experiments with male fish showed clearly that the parameters chosen were suited to strongly induce VTG in plasma. However, no significant change of proteins in response to E2 in the plasma of treated mussels could be detected. Figure 2 shows the results for the freshwater mussel (*Anodonta cygnea*).

Even 2D electrophoresis did not reveal any protein, which was induced by estrogen [10].

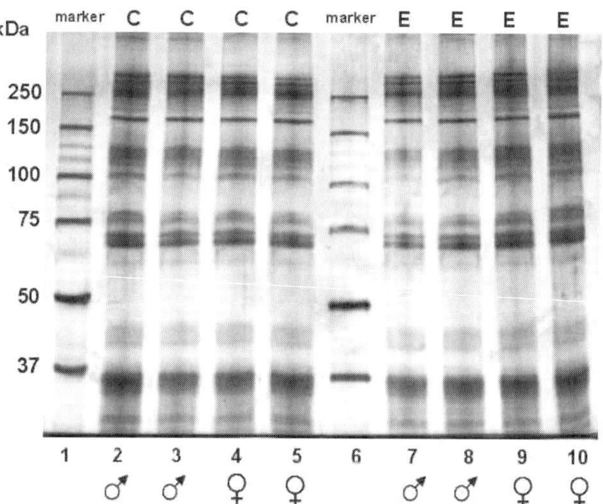

Fig. 2 SDS-PAGE (7.5%) under reducing conditions of the pooled plasma ($n = 5$) from male and female *A. cygnea* after exposure to 2 µg/l E2 for three weeks (E) and compared to control mussels (C).

USE OF GENE EXPRESSION PATTERNS FOR THE MEASURING OF MULTIPLE ENDPOINTS

Biomarkers like VTG are suitable tools for the detection of estrogenic compounds in environmental compartments. However, the mechanisms of action in target organisms are not entirely clear. For example, it is open whether different estrogenic compounds cause the same effect or trigger a variety of effects.

For this reason other approaches have to be applied, which provide information on effect mechanisms. As estrogenic compounds act by altering the gene expression in different tissues, one option is to examine gene expression patterns. This can be done in different ways. For the mRNA level, rt-PCR and DNA microarray techniques can be applied. For the protein level, 2D gel electrophoresis combined with mass spectrometry is used. To carry out gene expression experiments, zebrafish as a common model organism were exposed to xenobiotics, and the resulting alterations of the gene expression patterns in different tissues are mapped. DNA microarray techniques are based on the hybridization of two complementary nucleic acid strands. The probe consists of an oligonucleotide with the sequence of a specific gene and is immobilized on a suitable surface such as a glass slide. The mRNA from a target tissue is reverse transcribed to cDNA. In this process, a fluorescent dye is incorporated into the cDNA. The cDNA is then hybridized to the probe. After readout with a fluorometer, the gene expression pattern can be analyzed.

On the protein level, gene expression patterns can be determined by 2D gel electrophoresis giving information on the proteins expressed at the time the sample is collected. After picking differentially expressed spots, an identification of the protein can be carried out by mass spectrometry, e.g.,

MALDI-TOF. Using massive parallel approaches of DNA microarrays and/or 2D gel electrophoresis, a global view of the gene response in an organism can be gained.

For the detection of estrogenic effects in zebrafish (*Brachidanio rerio*) reverse transcriptase-polymerase chain reaction (RT-PCR) and DNA microarray approaches are developed in our group. Fish exposures, were carried out in 20 l glass tanks for five days. A flow-through system with a flow-rate of 14.4 ml/min fresh water in the tank provided stable conditions for the fish. Plants, gravel, or plastic (e.g., filtering unit) was omitted in the tank. For the exposures the concentration was kept at 1µg/l 17ß-estradiol. To control the endocrine activity in the water, samples were taken three times a day and measured by the ELRA (cf. below).

The RT-PCR method starts with a phenol-extraction of RNA from gonads and liver of female and male fish. Then mRNA was transcribed to cDNA by reverse transcription (RT). Specific sequences of VTG gene 3 (vg3), aromatase (ovary form), and estrogen receptor α (ER) were used as primers for the PCR. The products are 304 bp (vg3), 410 bp (ER), and 610 bp (aromatase) and were electrophoresed in an 1 % agarose gel (Fig. 3).

In the control fish, vg3 was found mainly in liver tissue of the female fish. Also in the gonads of female fish vg3 was expressed, but weaker than in the liver. In the male fish no vg3 expression could

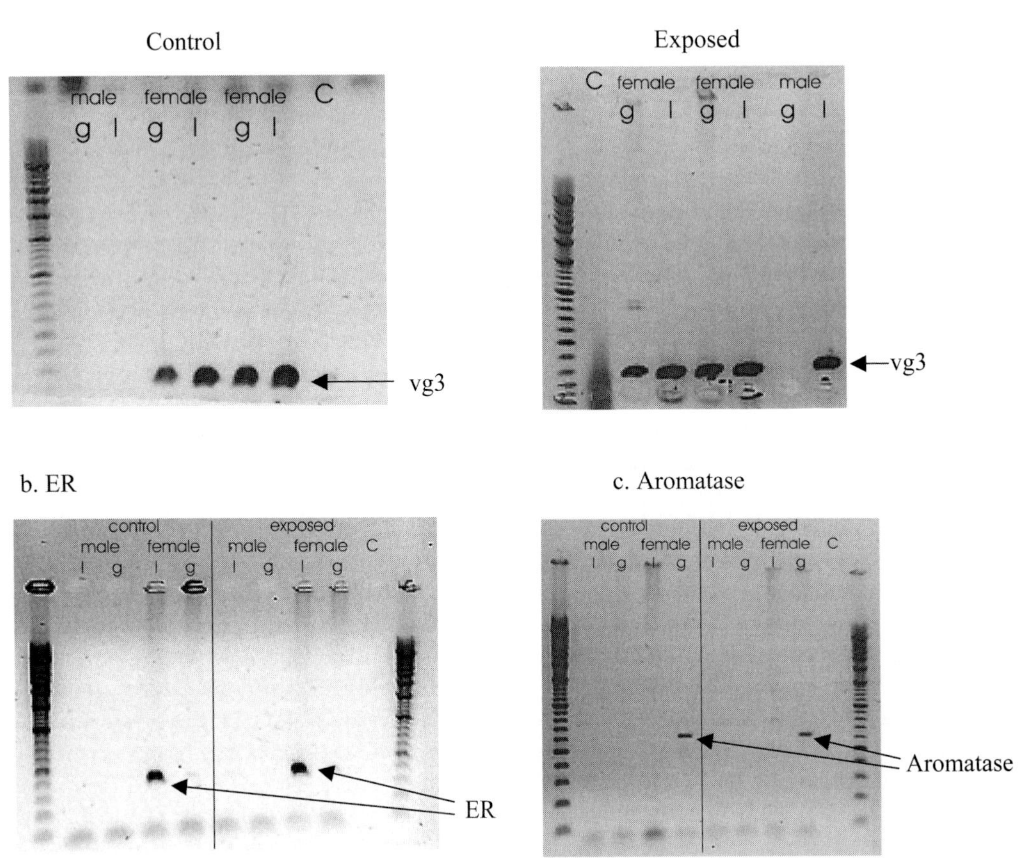

Fig. 3 RT-PCR of zebrafish mRNA of exposed fish and controls (l: liver; g: gonads; C: controls without template).

be found. Since VTG is a vertebrate egg yolk protein, only female zebrafish should express the vitellogenin gene under normal conditions.

In the exposed group, female fish show vg3 mRNA in gonad and liver tissue. Due to 17ß-estradiol in the water, also male fish show a strong expression of VTG in the liver, but not in the gonads.

The second gene we looked at codes for the ERα. We found expression of ER mainly in the liver and much weaker in the gonads of female fish. No expression could be detected in the male individuals. After exposure, no change in the expression pattern could be detected. The amount of ER expressed in females seems to be slightly increased. This is currently under investigation.

The third primer pair we used was complementary to aromatase (ovary form), an enzyme that catalyzes the conversion of androgens to estrogens. Again, no mRNA could be found in male fish, no matter whether they were exposed or not. Therefore, the expression of aromatase does not seem to be under positive control of 17ß-estradiol in male zebrafish.

The exposed and nonexposed females show aromatase expression in the gonad tissues, but not in liver, which illustrates the importance of the gonads for the conversion of androgens.

Further investigations on gene expression patterns are currently carried out in our lab using mainly DNA microarray techniques with a largely expanded number of different zebrafish genes.

USE OF ESTROGEN RECEPTORS FOR THE DETERMINATION OF ESTROGENIC COMPOUNDS

In the environment, estrogenic compounds not only occur as residues of natural sexual hormones of humans and animals, but also as constituents of plants and fungi (phyto- and mycoestrogens). Also, pharmaceuticals and xenobiotics can act as estrogens. The wide variety of potentially estrogenic compounds makes it difficult to predict the estrogenicity of xenobiotics on a structural basis [9]. Thus, monitoring strategies cannot be based exclusively on chemical analysis of environmental samples for known or suspected structures. In addition, an analytical system should be capable of detecting estrogenic compounds, even if they are yet unknown. There are different ways of measuring estrogenic compounds in the environment. The most common ones are cell proliferation assays (E-screen assay with hormone-dependent growth of MCF-7 breast cancer cells), reporter-gene assays and receptor-ligand interaction assays. For receptor-ligand interaction assays, hormone receptors have to be available. This can be achieved by laborious purification from tissues containing the receptor or, more elegantly, by recombinant techniques.

As an example, estrogen receptor α (ERα) can be produced in a recombinant yeast expression system. Briefly, yeast (strain BJ3503) was transformed with an expression plasmid (kindly provided by Prof. D. P. McDonnell, Duke University, NC) containing the gene for the human ERα. After induction with Cu^{2+}, the receptor was expressed as an ubiquitin fusion protein. After removing the cell walls by enzymatic digestion, the cells were lysed by hypotonic shock to liberate the receptor. The cell extract was centrifuged (at 22 000 × g) and the supernatant was used for further purification. The receptor was partially purified by affinity chromatography with heparin agarose. The recombinant receptor was undegraded, soluble, and biologically active. After affinity purification, receptor yields in the range of 10 µg were obtained routinely from 5 l yeast batch cultures.

The receptor properties, especially the affinity characteristics, were determined with a surface plasmon resonance sensor (Biacore). The affinity data obtained for the produced receptors is well comparable to literature data (Table 1).

Table 1 Comparison of K_D values reported for the human estrogen receptor α.

Receptor	K_D	Method	Ref.
Recombinant human estrogen receptor α	2.3×10^{-10}	SPR-sensor	[11]
Human estrogen receptor α isolated from mammary tumor cytosol	4.2×10^{-10}	Radioreceptor assay	[4]
Human estrogen receptor α isolated from MCF-7	2.0×10^{-10}	Radioreceptor assay	[8]

A simple and cost-effective test format to measure biomolecular interactions is the enzyme–linked receptor assay (ELRA). The ELRA is a microwell-based receptor-binding assay using the human ERα. Measurements are carried out in 96-well microwell-plates. In the first incubation step, an estradiol-BSA conjugate is absorbed to the walls of the microwells. In the second (competition) step, an estradiol solution of defined concentration is added together with the ERα. After the receptor binding reaction, a biotinylated mouse antiestrogen-receptor antibody is added. A streptavidin-peroxidase-biotin enhancement system, is applied. By using the luminescent substrate luminol, a significant improvement of the detection limit of estradiol (0.02 μg/l) could be achieved recently (Fig. 4), compared to the standard assay (0.1 μg/l, [12]).

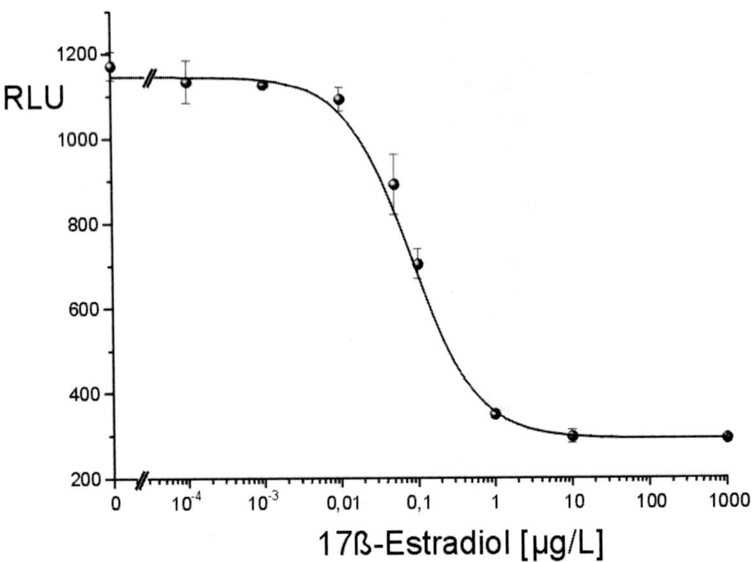

Fig. 4 ELRA calibration curve, using the luminescent substrate luminol. The detection limit is 0.02 μg/l.

The ELRA approach has proven to be robust and highly suited for the detection of natural and synthetic estrogens as well as xenoestrogens in field studies. Figure 5 shows the comparison of spiked lake water measurements, which were carried out with the ELRA and HPLC as reference method. A close correlation ($r = 0.987$) could be achieved.

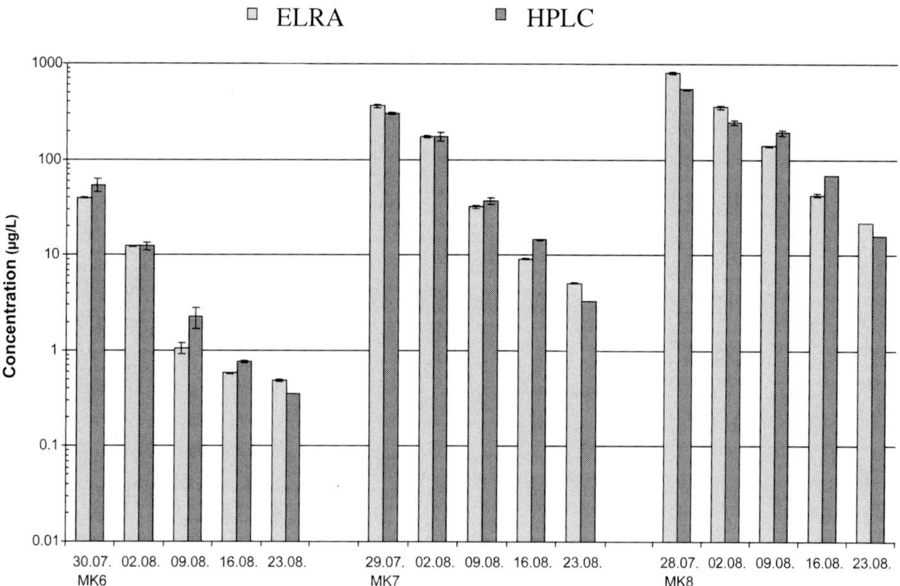

Fig. 5 ELRA applied for real water samples. The results show the measurement of spiked lake water from a mesocosm (MK) experiment. Reference analyses were carried out by HPLC. The mesocosm experiment and HPLC measurements were performed by the Institute of Ecological Chemistry, GSF (Neuherberg, Germany).

BIORESPONSE-LINKED INSTRUMENTAL ANALYSIS

A comparison between biological and chemical approaches to environmental analysis reveals that they are clearly complementary. Bioanalysis provides information on the (potential) biological activity of a sample, chemical analysis on its composition. Therefore, it would be advantageous to combine both principles—the goal of bioresponse-linked instrumental analysis (BLIA). The proper application of this strategy reduces chemical analysis to those samples or fractions of samples that contain bioeffective compounds. The concept has been put forward by Bilitewski et al. [1] and Hock [5].

The biomolecular components serve as targets for bioactive substances. Although binding assays can already be performed on this basis providing toxicological or pharmacological equivalents, only chemical analysis yields information on the responsible substances. Chemical analysis is therefore performed with those substances that are bound by the biological recognition element and are therefore bioactive.

There are principally two strategies for BLIA (Fig. 6): (1) Sample separation followed by bioassays; biological active samples are automatically analyzed by instrumental analysis. An excellent example is given by Schobel et al. [11] who report the identification of estrogenic compounds by bioresponse-linked instrumental analysis (BLIA). (2) Binding assays providing toxicity equivalents are followed by the separation of bound and free binding proteins. Bound ligands are analyzed by MS. Obst and Brenner-Weiß [7] report the further development of this approach using EDCs as an example.

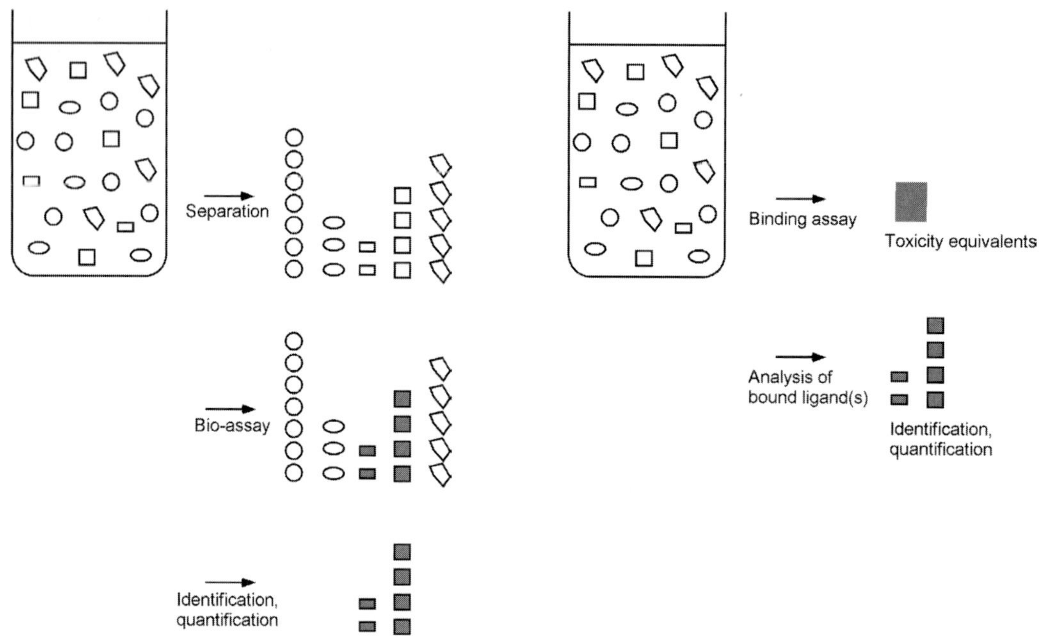

Fig. 6 Two approaches to BLIA. Left: Sample separation, e.g., by LC, followed by bioassay and MS. Right: Binding assay followed by analysis of bound analyte, e.g., by MS.

DISCUSSION

Approximately 5 million chemicals are presently known and 80 000 in use; 500–1000 are added per year [3]. This creates a problem for environmental analysis when relevant pollutants are to be identified. Since usually mixtures of substances are present that can attenuate or enhance individual effects, chemical analysis is not sufficient to estimate a potential damage. For this purpose, bioanalysis is additionally required. BLIA is considered a breakthrough as it links bioanalysis with chemical analysis. Advances in instrumental analysis as well as a better access to biological target structures have contributed to realize this concept. BLIA in parallel obtains chemical and toxicological data and focuses analytics on relevant samples.

A disadvantage of subcellular and related tests is due to the fact that they cover only a small section of the total spectrum of possible bioeffects. This creates a dilemma between tests with whole organisms, which register integral effects, but usually do not link them to specific classes of toxicity, and suborganismic tests, which detect selective effects but do not provide information on effects on the whole organisms. A compromise is offered by two approaches, the analysis of gene expression patterns as well as multifunctional tests.

A change of gene expression patterns not only indicates general stress situations, but also specific effects of pollutants on the organism. In other words, multiple effects can be detected. An important task for the forthcoming years is to expand the list of suitable marker genes, whose expression indicates toxic effects. Available genes include the gene for vitellogenin, aromatase, cytochrome P456. This strategy may be particularly useful in exposition studies, especially in ecosystems. A most attractive aspect is the possibility to detect chromic effects.

Multifunctional tests are the opposite of multianalyte methods, as they are used in immunochemistry [2,6]. Multifunctional tests combine several subcellular bioassays. However, it is not clear, yet, how powerful these measuring systems have to be in order to mirror effects on the whole organism. This question is related to the classical philosophical dispute between reductionism and holism and deals with the problem, to which extent analytical methods provide information on complex processes in organisms. It is expected that bioinformatics and systems biology will considerably promote this field.

Progress in science in the last decades has increasingly shed light on risks for public health as well as the environment involved in endocrine disruptors. Unfortunately, scientific ignorance for the unintended consequences will always be behind the development and application of new substances that might have an influence on the hormonal system of animals and humans. It is therefore essential to implement a durable and adequate risk assessment and to ensure that reassessment of the issue will always take place when new knowledge that might influence the previous risk estimation becomes available.

ACKNOWLEDGMENTS

We gratefully acknowledge the financial support by the BMBF (grant 02WU9647/0 and 02WU9993/8). We thank Dr. Karl-Werner Schramm for providing the HPLC data. We acknowledge Michaela Krötz and Stefanie Haindl for their excellent technical assistance. Dr. Martin Riffeser has contributed the mussel data.

REFERENCES

1. U. Bilitewski, G. Brenner-Weiß, P. D. Hansen, B. Hock, E. Meulenberg, G. Müller, U. Obst, H. Sauerwein, F. W. Scheller, R. Schmid, H. Schnabl, F. Spener. *TRAC.* **19**, 428–433 (2000).
2. R. P. Ekins, R. Chu, E. Biggart. *Ann. Biol. Clin.* **48**, 655–666 (1990).
3. K. Fent. *Ökotoxikologie*, Georg Thieme Verlag, Stuttgart (1998).
4. H. J. Grill, B. Manz, O. Belovsky, B. Krawielitzki, K. Pollow. *J. Clin. Chem. Clin. Biochem.* **21**, 175–179 (1983).
5. B. Hock. *Bioresponse-linked Instrumental Analysis*, Teubner-Verlag, Stuttgart-Leipzig-Wiesbaden (2001).
6. H. Kido, A. Maquieira, B. D. Hammock. *Anal. Chim. Acta* **411**, 1–11 (2000).
7. U. Obst and G. Brenner-Weiß. *Chem. Unser Zeit* **36**, 156–162 (2002).
8. N. Olea, N. Devleeschower, G. Leclercq. *Rev. Esp. Fisiol.* **41**, 373–380 (1985).
9. Oosterkamp, B. Hock, M. Seifert, H. Irth. *TRAC.* **16**, 544–549 (1997).
10. M. Riffeser and B. Hock. *Comp. Biochem. Physiol.* **132**, 75–84 (2002).
11. U. Schobel, M. Frenay, D. A. Van Elswijk, J. M. McAndrews, K. R. Long, L. M. Olson, S. C. Bobzin, K. Irth. *J. Biomol. Screen* **6**, 291–303 (2001).
12. M. Seifert, S. Haindl, B. Hock. *Anal. Chim. Acta* **386**, 191–199 (1999).
13. C. Silversand, S. J. Hyllner, C. Haux. *J. Exp. Zool.* **267**, 587–597 (1993).
14. J. P. Sumpter and S. Jobling. *Environ. Health Perspect.* **103**, 173–178 (1995).
15. Z. Yao and L. W. Crim. *Comp. Biochem. Physiol.* **113B**, 247–253 (1996).

Pure Appl. Chem., Vol. 75, Nos. 11–12, pp. 2461–2466, 2003.

Workshop 3.4

Identifying the causative agents: The use of combined chemical and biological strategies in monitoring programs*

Edwin John Routledge

Department of Biological Sciences, Brunel University, Uxbridge, Middlesex, UB8 3PH, UK

Abstract: The toxicity identification and evaluation (TIE) approach that identified unconjugated natural and synthetic steroid estrogens to be the major estrogenic component in domestic sewage treatment work effluent is described. Given the increasing popularity of TIE-based approaches in ecotoxicology research, a number of important considerations, limitations, and possible future directions of TIE studies are also discussed.

BACKGROUND

In 1994, Purdom and colleagues reported that all UK effluents were estrogenic to fish [1]. These effluents were mainly domestic (rather than industrial) in source, indicating that the estrogenic component(s) were likely to be domestic in origin, and were probably common to most of them. At the time that this research was initiated, there were no specific examples where an estrogenic effect on wild fish had been conclusively linked to a particular chemical emanating from sewage-treatment works (STW) effluents, although many chemicals (including ethynylestradiol and alkylphenolic chemicals) had been implicated. This project was therefore commissioned by the UK Environment Agency in order to identify and quantify the estrogenic substances present in domestic sewage effluent and, once identified, to assess their effects on fish at environmentally relevant concentrations [2].

GENERAL APPROACH

STW effluent composed of domestic and industrial waste release a complex and ill-defined mixture of chemicals, their metabolites, and biotransformation products into the aquatic environment. Although effluents have been tested for their toxicity to aquatic organisms, few chemicals within an effluent have been tested, on an individual basis, for their toxicity or endocrine-disrupting activity either in vivo or in vitro. As it was clearly not practical to identify, quantify, and test all the individual substances present in effluent, a toxicity identification and evaluation (TIE) approach was employed [3]. Using this bioassay-directed fractionation procedure, STW effluent was chemically separated into fractions of decreasing complexity, which were then assessed for estrogenic activity in vitro. Fractions identified as active in the bioassay were separated further until the causal compounds could be identified by GC/MS. This procedure, through a series of steps involving separation and resolution, simultaneously eliminates inactive compounds in the mixture, and isolates chemicals that are biologically active, without any preconceived ideas about their identity.

*Report from a SCOPE/IUPAC project: Implication of Endocrine Active Substances for Human and Wildlife (J. Miyamoto and J. Burger, editors). Other reports are published in this issue, *Pure Appl. Chem.* **75**, 1617–2615 (2003).

SELECTION OF A SUITABLE BIOASSAY

Selection of a suitable in vitro bioassay for detecting estrogenic chemicals was important if the effluent fractionation scheme was to succeed. It was recognized from an early stage that the bioassay employed would have to be robust, reproducible, sensitive, and specific, with a clear endpoint. Moreover, given the possibility that a large number of fractions would need to be tested (we screened 500 fractions per effluent in triplicate), the bioassay would also need to be rapid, inexpensive to run, and high throughput. This meant that many of the established tests for estrogenicity were unsuitable, particularly those using extended in vivo techniques. At the commencement of the fractionation project we acquired the recombinant yeast estrogen screen [4]. Following validation, it became apparent that the yeast-based system was rapid (four days), sensitive (could consistently detect 2 ng 17β-estradiol/l in the assay medium), specific, reproducible, robust, and high-throughput (32 fractions could be tested in triplicate on a single 96-well microtitre plate), making it an ideal test system to screen large numbers of effluent fractions throughout the project. Moreover, the bioassay detects natural estrogens as well as known xenoestrogens; thus, we expected the bioassay to produce a quantitative measurement of estrogenic compounds regardless of the identity of the chemical (or chemicals) responsible for the estrogenic activity of STW effluent. An important consideration when selecting a bioassay is the amount of sample required to perform the assay, as the more sample that is used for biological testing, the less sample that is subsequently available for further analytical chemistry.

CRITERIA FOR SITE SELECTION AND SAMPLING

Seven domestic STW effluents were selected based on convenience of sample collection, continuity with previous sampling (caged fish) studies and to provide a range of treatment processes. Southend-on-Sea STW was selected for method development as it received only primary treatment, and therefore was expected to contain high concentrations of the estrogenic chemicals under investigation, and would provide a readily measurable response to enable isolation of the estrogenic fraction(s). Discrete samples were taken from each of the STWs, in preference to a 24-h bulked composite sample, as the stabilities of the estrogenic compounds at this stage were unknown. However, the consistency of the estrogenic activity in each effluent sample was determined on three separate occasions.

COARSE FRACTIONATION OF SEWAGE EFFLUENT

Neat ("whole" sample) effluent was collected and tested for estrogenicity, after which it was crudely fractionated to determine the contribution of the particulate ("filtered" sample) and volatile phase ("purged" sample) to the overall estrogenicity of the effluent. The effluent was also tested after C18 extraction in order to evaluate the contribution of organic compounds in the sample to the estrogenic activity. For the assay controls, the same procedure was carried out using distilled, deionized water to provide a "procedural blank", which was analyzed alongside the effluent fraction at each stage. At the five-fold dilution, the "whole" effluent sample from Southend-on-Sea contained estrogenic activity equivalent to approximately 10 ng 17β-estradiol/l. Our results also indicated that the estrogenic activity was mainly present in the dissolved phase, as extraction of the effluent liquid phase using a C18 SPE cartridge caused a >80 % reduction in estrogenic activity compared with the "whole" effluent.

SOLID-PHASE EXTRACTION USING C18 COLUMNS

Solid-phase extraction (SPE) columns containing C18 reversibly bind a broad range of organic compounds, which can be eluted with a range of different solvents according to their polarity, enabling the selective removal, and concentration of organic compounds from complex mixtures. By controlling the elution sequence of the compounds retained on the C18 matrix, the biologically active components can be isolated within discrete fractions appropriate for higher-resolution procedures. Since most of the activity was retained on the C18 column, it appeared that the active compounds were not inorganic substances (which would not bind), but were organic substances. The organic compounds were eluted from the C18 column using a series of volumes of methanol/water mixtures (0, 25, 50, 75, 80, 85, 90, 95, and 100 % methanol), which were collected in separate vials. The estrogenic activity was found to elute within the 50 to 85 % methanol fractions. Elution of the SPE cartridges with solvents of low polarity to nonpolar did not yield any further activity, indicating that the compounds were mid-polar organics. However, as effluent contains numerous mid-polar organic constituents, which would elute in a similar manner on the C18 column, we could not ascertain whether the chemical(s) responsible for the estrogenic activity were similar at each site at this stage.

FINE FRACTIONATION USING HIGH-PERFORMANCE LIQUID CHROMATOGRAPHY

Following their assessment for estrogenic activity, the active fractions were combined and reconcentrated prior to the next fractionation stage using high-performance liquid chromatography (HPLC). Gradient elution (40–100 % methanol) was employed to separate organic chemicals of medium polarity, so that the most polar (weakly retained) compounds were eluted before those of low polarity (more strongly retained). As the sample components were eluted from the separating column by the carrying solvent, they were monitored through an ultra-violet (UV) detector, and discrete fractions were collected at 1.5 or 1 min intervals over 45 min into separate vials which were then assessed for estrogenic activity. All the effluent samples analyzed produced a single highly estrogenic fraction, which eluted approximately 25 min after the start of the HPLC run, which suggested that the activity was due to a single compound, or a set of closely related compounds, that were eluted concurrently using the HPLC conditions employed. The UV profile of the chromatographic run indicated that the sample was still a highly complex mixture. Reversed-phase HPLC, at this stage, was unable to resolve the mixture into fractions composed of a few identifiable compounds. It was therefore impossible to determine whether the activity within the fraction was due to a highly abundant component, or whether it was due to minor components concealed within the UV profile. Attempts to identify the components present in the fraction by HPLC/MS were inconclusive, due to the complex nature of the mixture.

ANALYSIS OF ESTROGENIC FRACTION USING GC/MS

The active components within the effluent sample were transferable, by liquid/liquid extraction from the methanol/water HPLC fractions into DCM (nonpolar solvent), enabling their analysis by gas chromatography/mass spectrometry (GC/MS). The major components of the extracts were found to be a mixture of isomers of α-terpineol (a terpenoid alcohol widely used in detergents and cosmetics), which had been previously identified in sewage effluent. There were also around 20 unidentified minor components that occurred in all three extracts. Standards of the identified terpineol, of both pure and technical (an isomeric mixture) grade were purchased, and solutions spanning the concentration range present in the fraction were tested for estrogenicity in the yeast screen. None of the terpineols tested were found to be active in the yeast screen. Following the elimination of terpineols as the estrogenic components in the active fraction, it was considered necessary to focus on the large number of potentially active trace compounds. An additional fractionation step was developed using a "shallower" (55–60 % methanol) gradient in order to reduce the sample complexity further. Using the shallow HPLC elution gradient, the active fraction from the first HPLC run was resolved into two separate peaks of estrogenic activity, eluting between 29–31 and 31–33 min, respectively. The total estrogenic activity in the combined active samples was equivalent to around 100 ng 17β-E2/l in the neat effluent, indicating that the minor components in the fraction were potent estrogenic compounds. At this stage, it was also noted that the active fraction behaved in the yeast screen in a manner similar to steroid estrogens, in which assay color development occurred rapidly relative to the xenoestrogens that we had tested. Based on this observation, it was postulated that the estrogenic component might be a steroidal compound.

BEHAVIOR OF STEROIDS IN THE FRACTIONATION SYSTEM

The behavior of steroids in the fractionation system was evaluated using the synthetic estrogen, ethynylestradiol (EE2). If steroids (or steroid-like compounds) were responsible for the activity in the effluent, we would expect their chromatographic retention time to coincide with that of the active fraction. Therefore, in separate chromatographic runs, EE2, procedural blanks and the active effluent fraction were injected onto the HPLC column, and fractions were collected at 1 min intervals using both gradients. The EE2 standard co-eluted with the estrogenic activity present in the effluent samples using the 40–100 % methanol gradient. In the 55–60 % gradient run, the EE2 standard coincided with the second peak of activity in the effluent sample. This result was confirmed using a standard solution containing estrone (E1), 17β-estradiol (E2) and EE2, which was analyzed under the same GC/MS conditions. The elution time of the standard compounds coincided with retention time of the peaks identified in the effluent extract. These results indicated that estrogenic steroids could be responsible for the estrogenic activity in domestic STW effluent. In both cases, the blanks were inactive, as expected.

IDENTIFICATION AND QUANTIFICATION OF ESTROGENIC COMPOUNDS BY GC/MS

Working on the premise that trace amounts of steroidal compounds were responsible for the estrogenicity observed in the effluent fraction, it was necessary to work with larger volumes of effluent (20 l) in order to concentrate the effluent samples further, and obtain sufficient quantities of substance to enable identification by GC/MS. The standard GC/MS library searching routine identified potent natural and synthetic steroidal estrogens in the extract, namely E1, E2, and EE2; all present in an unconjugated (free) biologically active form. Potential improvements to the GC/MS method, such as creating silyl derivatives of the sterols (less polar compounds) which were more amenable to gas chromatography, resulted in an improved peak shape for the steroids. However, this advantage was largely offset by a loss of sample during the additional manipulation. The results indicated that estrone and 17β-estradiol were present in the effluents at concentrations ranging from 1 ng/l up to 80 ng/l and 50 ng/l, respectively. Ethynylestradiol (up to 7 ng/l) was detected in one third of the samples collected, and was associated with effluents containing the highest steroid loads.

CONCLUSIONS

In this project, a TIE approach, modified for our specific needs, identified natural and synthetic steroidal estrogens as candidate compounds responsible for the estrogenic activity observed in domestic STW effluent [5]. Proficient chemical manipulation of the effluent sample, together with the use of a rapid, robust and high-throughput bioassay were key factors responsible for the success of the project. Despite the complexity of the composition of domestic effluents, only a small number of highly estrogenic compounds were identified. Three estrogenic steroids (E2, E1, and EE2) were positively identified in the active effluent fractions tested in the ng/l range. None of the activity identified was found to be associated with alkylphenolic chemicals. This finding was supported by direct measurements, indicating that alkylphenols, when present, were below the threshold for detection in the yeast screen.

The results suggested that the concentrations of EE2 detected in most of the samples were too low to account for the magnitude of the vitellogenic response observed in male fish exposed to domestic STW effluents. There was also little, if any, data to address the issue of sensitivity of fish exposed to natural estrogens via the water at the time. Hence, it was not possible to conclude whether the concentrations of E2 and E1, reported in this study, would, or would not, be estrogenic to fish. To address some of these issues, and thus put the results obtained from the fractionation studies into an environmental context, in vivo laboratory tank trials were conducted in which rainbow trout and roach were exposed to low concentrations of steroid estrogens via the water. The results of these exposures demonstrated that the levels of steroidal estrogens present in STW effluent were estrogenic to fish, and that their effects were additive [6]. The previously reported observations of estrogenic activity in domestic effluents and some rivers could therefore be attributed to these hormones.

LIMITATIONS OF THE TIE APPROACH

The fractionation procedures described here are time-consuming and have a number of drawbacks. Toxicity was a problem associated with certain effluent fractions, which may have prevented the detection of estrogenic components within the samples. For example, a tentative identification of an estrogenic response at one concentration in the bioassay, may have been followed by a toxic effect at a higher concentration. Therefore, all fractions were tested using a range of volumes in an attempt to avoid complications associated with toxicity, and to observe dose-response effects. Changes in toxicity could result with each chemical manipulation of the sample, due to alterations in the bioavailability of toxins. Conversely, interactive effects may be reduced, as the mixtures were resolved into single components.

The results from the domestic effluent fractionation procedure illustrated that the majority of the activity in vitro occurred within a single fraction associated with natural steroidal estrogens. However, further increases in the concentration of the HPLC fractions tested in the yeast screen resulted in the appearance of other more weakly active fractions, which were also eluted in the mid-polar range. Most of the synthetic estrogens identified to date are reported to be lipophilic compounds which bioaccumulate in whole animals. Therefore, the potential contribution of these weakly active fractions to the estrogenic activity in vivo must not be underestimated, particularly in view of the fact that biological effects can be augmented by bioaccumulation, biomagnification, metabolism, and pharmacokinetics. Thus, the biological activity of a compound in vivo may be two or more orders or magnitude greater than they appear in vitro. It is therefore possible that in vitro assays may inherently underestimate the true environmental relevance of the "weaker" active components in the sample.

TIEs—THE WAY FORWARD

Given the power of TIE approaches to identify causal compound in complex mixtures, it is likely that future TIE studies will concentrate on samples with very different compositions (including landfill leachates, industrial effluents, agricultural runoffs, and pulp-mill effluents). New developments in bioassays will also broaden the range of measurable endpoints; many of which may be determined simultaneously using microarrays. In silico (QSAR) technology may assist in the identification of unknown metabolites and biotransformation products with inadvertent biological activity. Small-scale, short-term, in vivo assays (employing early life stage) may be developed for TIE studies using existing or novel model species, including invertebrates.

REFERENCES

1. C. E. Purdom, P. A. Hardiman, V. J. Bye, N. C. Eno, C. R. Tyler, J. P. Sumpter. *Chem. Ecol.* **8**, 275–285 (1994).
2. Environment Agency. R&D Project 490, ISBN 0 11 310124 4, The Stationery Office, London (1997).
3. D. I. Mount and L. Anderson-Carnahan. EPA Report EPA/600/3-88/034. EPA Deluth, MN 55804 (1998).
4. E. J. Routledge and J. P Sumpter. *Environ. Toxicol. Chem.* **15**, 241–248 (1996).
5. C. Desbrow, E. J. Routledge, G. C. Brighty, J. P. Sumpter, M. Waldock. *Environ. Sci. Technol.* **32**, 1549–1558 (1998).

6. E. J. Routledge, D. Sheahan, C. Desbrow, G. C. Brighty, M. Waldock, J. P. Sumpter. *Environ. Sci. Technol.* **32**, 1559–1565 (1998).

Pure Appl. Chem., Vol. 75, Nos. 11–12, pp. 2467–2475, 2003.

Workshop 3.5
Closing the gap between exposure and effects in monitoring studies*

Donald E. Tillitt[‡] and Diana M. Papoulias

Columbia Environmental Research Center, U.S. Department of Interior, Geological Survey, 4200 New Haven Road, Columbia, MO 65201, USA

Abstract: A major challenge to contaminant monitoring programs is the selection of an appropriate suite of measurements for assessing exposure and effects. Early monitoring programs relied solely on residue analysis to detect the organochlorine compounds that were in use at that time. A shift to the use of more transient, less persistent chemicals required that a new set of tools be developed to determine if an organism had been exposed. This led to the development of cellular and biochemical assays that could indicate the presence of these types of chemicals in biota and the environment. However, it was recognized that measures of contaminant presence alone were insufficient to assess the health of biota. As a result, considerable research began to be directed toward development of diagnostic tools for measuring chemical effects in fish and wildlife. Today, contaminant monitoring programs follow a paradigm for study design that emphasizes not only the use of measures of exposure, but also measures of effect.

Using data from our monitoring and research studies for hormonally active substances, we discuss a variety of metrics of exposure and effects and their application to specific chemicals, and the current information gaps. We conclude that although several bioindicators of exposure and effect have been promoted and used, to date there continues to be a poor association between cause and effect for endocrine active substances. In part, this is due to the limited number of diagnostic tools that are available and to a lack of basic toxicological information concerning toxicokinetics and mechanisms of action of hormonally active chemicals in fish and wildlife species. In the foreseeable future, both tissue and environmental residue data, despite the many limitations, will continue to be an important component of monitoring programs for hormonally active chemicals as we continue to develop and validate more specific bioindicators of exposure and effects.

INTRODUCTION

Monitoring for effects of chemicals in the environment has become more important over the past decade. This is due to changes that have occurred in the types and patterns of chemicals used in agriculture, industry, and household products. The shift in chemicals used in commerce and industry has been from persistent, hydrophobic chemicals such as polychlorinated biphenyls (PCBs) and 2,2-bis(*p*-chlorphenyl)-1,1,1-trichloroethane (DDT) to less persistent chemicals, such as organophosphate pesticides. More stringent regulations that control the use and release of chemicals into the envi-

*Report from a SCOPE/IUPAC project: Implication of Endocrine Active Substances for Human and Wildlife (J. Miyamoto and J. Burger, editors). Other reports are published in this issue, *Pure Appl. Chem.* **75**, 1617–2615 (2003).

[‡]Corresponding author

ronment are responsible for these changes in use patterns from persistent, bioaccumulative chemicals to more water-soluble, less persistent chemicals. These "new generation" chemicals are not detected in tissues, and therefore exposure must be measured in the environment. This creates a gap between the exposure metric, environmental (e.g., water) concentration, and the potential for effects in fish and wildlife. Exposure can no longer be measured directly in the organism, as it could for the organochlorine compounds, and other persistent, bioaccumulative chemicals.

Compounding this difficulty in measuring the exposure to fish and wildlife for these "new generation" chemicals is the fact that complete models for exposure and toxicity are lacking. Exposure models that estimate the dose an aquatic organism receives at a given water concentration are inadequate for many of these chemicals. Moreover, the toxicity dose–response relationships are unknown for a vast number of these new chemicals as well as many of the not-so-new chemicals. The necessary toxicity testing in standard laboratory organisms has simply not been conducted for many of these chemicals. Additionally, the relative sensitivity of many fish and wildlife species of concern to natural resource managers are not known. As such, even when dose–response relationships have been evaluated in surrogate species, the evaluation of environmental effects remains a challenge. Last, there is little understanding of the effects of multiple chemical exposures on aquatic organisms. We simply do not understand how chemicals interact once they reach the target organ or tissue.

As a result of this lack of knowledge and the associated uncertainty, it is nearly impossible to predict toxicity or potential toxicity of the less persistent chemicals alone or in combination with the older bioaccumulated compounds. This is particularly true for that group of chemicals reported to cause endocrine-related effects in fish and wildlife species. Biological monitoring of health effects in fish and wildlife species is required to begin to evaluate any potential effects that may be caused by these chemicals. Therefore, the objectives of this presentation were to address two issues in monitoring for effects of endocrine active compounds (EACs):

Do we still need to measure the "old" organochlorine pesticides and industrial chemicals in large-scale monitoring programs?

How do we close the gap between metrics of exposure and estimates of effects in a time of more chemicals with less persistence?

IS THERE A NEED FOR CONTINUED MONITORING OF THE ENVIRONMENT FOR ORGANOCHLORINE PESTICIDES AND INDUSTRIAL COMPOUNDS?

Chemical monitoring efforts in the United States, Europe, and elsewhere in the world have been designed to measure persistent chemicals in biotic and abiotic matricies. The concentrations of chemicals measured in this fashion are compared with known threshold toxicity values to estimate or determine relative risk. The results of most of these monitoring efforts have been consistent with one another over the past three decades. Most environmental monitoring of chemicals has demonstrated the same temporal trends in concentrations of the persistent bioaccumulative compounds. There have been declines, often exponential decreases, in the concentrations of organochlorine chemicals in most of the more contaminated environments. Examples of these declines include concentrations of PCBs in Great Lakes fish and wildlife [1–3], DDT and PCB declines in the biota of the Baltic Sea [4,5]; and fish in the National Contaminants Biomonitoring Program (NCBP) [6,7]. Regulations were instituted in the late 1970s that banned or restricted the use of persistent organochlorine chemicals. The declines in concentrations of these chemicals in biota and the environment have been dramatic, more than originally estimated in some cases, and underscore the importance of regulatory actions. Globally, monitoring of the environment for organochlorine pesticides and industrial chemicals has demonstrated those declines with the exception of remote areas, such as the Arctic [8,9]. Evidence is controvertible as to whether concentrations of organochlorine pesticides and industrial compounds are on the decline or increase in these areas.

Even with the almost universal declines in concentrations of the persistent, bioaccumulative chemicals in the environment, there are still strong reasons and rationale for resource management agencies to continue to monitor the environment for these chemicals. First, areas of concern still exist. While declines have been the rule, there are still areas of environmental contamination that are well above a threshold for toxicity in fish and wildlife species. Second, environments that have experienced declines in concentrations of many persistent, bioaccumulative chemicals are in some cases still experiencing untoward effects, which may be linked to these chemical contaminants [10,11]. There are populations of fish and wildlife that continue to have symptoms of toxicity consistent with some of the organochlorine pesticide or industrial compounds. Although concentrations of these compounds have declined, these populations may be indicating that thresholds for effects need further refinement. Additionally, there may be other stressors impinging on these populations, or there may be joint effects of these chemicals in combinations that are not predicted by our laboratory-based assessments with single compounds. Until we understand the true causes for the observed population effects in some of these contaminated environments, we must continue to monitor the potential causes, which include organochlorine pesticides and industrial chemicals. Third, there remains the continued threat of new releases of some of the banned or restricted organochlorine pesticides and industrial compounds. In some cases, large amounts of these chemicals continue to be found either in landstocks or still in use. For example, even though PCBs have not been manufactured in the United States for a quarter of a century, the majority of the PCBs produced are still found in use in electrical devices [12].

The efforts to monitor persistent organochlorine pesticides and industrial compounds need to be redirected and evaluated, but not eliminated, at this point in time. There are a number of steps that could be taken to reduce the costs associated with organochlorine chemical monitoring. Steps to reduce the cost associated with persistent organochlorine pesticides and industrial compounds should include: reduced replication of analyses at a given site; composite of samples; increased temporal periods between sampling events; and increased use of new technologies. Alternative forms of analysis, such as immunoassays or bioassay screening techniques, can provide the exposure information at substantially reduced costs as compared with routine analytical chemistry analysis. Immunoassay procedures for PCBs in biota [13–15] have proven to be quantitative, and are about 20 % of the costs of the most simple PCB analysis. Similarly, the H4IIE bioassay to screen for dioxin-like chemicals in fish, wildlife, and other environmental samples is now routinely and cost-effectively used [16].

CLOSING THE GAP BETWEEN EXPOSURE AND EFFECTS IN CHEMICAL MONITORING

The veterinary products, human antibiotics, pharmaceutical drugs, among countless other chemicals that are found in our sewage treatment plant wastewater effluents present new challenges to environmental scientists and resource management agencies. The number of such chemicals that are released into the environment increases each year, yet the sources and amounts of these releases are generally not well understood. However, some specific efforts have been directed toward detection of these chemicals in streams [17]. In the United States for example, the National Water-Quality Assessment (NAWQA) program has developed methods for many of these less persistent compounds [17]. The challenge that then arises is to bridge the gap between the environmental concentrations that predict exposure to fish and wildlife species and the evaluation of the effects that may occur as a result of those exposures. Threshold values for effects of many of these chemicals are simply not known. This gap is further widened by the fact that the responses of organisms to endocrine active agents, such as steroid hormones, can have nonmonotonic responses to increases in exposure. These nontypical dose–response curves may have an "inverted U" or "J-shape", as opposed to the more classic sigmoid curve [18,19]. A consequence of these more complex dose–response relationships is that our ability to interpret measures of chemical exposure is diminished. Limitations in our ability to estimate individual chemical exposures and interpret those exposures with respect to potential effects are further exacerbated by the complex interactions of multiple chemicals. The effects of multiple chemicals with similar modes of action or the action of chemicals with dissimilar modes of action on the same target organs and tissues are far from being understood. These complex interactions may follow simple, recognized relationships, but in most cases it is not readily apparent what type of interaction occurs.

Biological monitoring is the most direct and prudent action to address the uncertainties of chemical exposure in the environment. The U.S. Geological Survey Biomonitoring of Environmental Status and Trends (BEST) program is an example of a biological monitoring approach that is designed to develop multiple lines of evidence of contaminant exposure and effects in the aquatic environment [20]. The approach includes four major categories of methods: (1) biomarkers, (2) toxicity tests and bioassays, (3) community health, and (4) residue analysis. The use of these methods is intended to provide results that constitute a "weight-of-evidence" approach for identifying effects from a wide variety of contaminants including hormonally active chemicals. The BEST program is unique among national monitoring programs with its emphasis on characterizing the effects of environmental contaminants on the health of the biota and their supporting habitat. BEST accomplishes this through the application of both chemical analysis and biological response measurements. In addition to applying measures of biochemical, physiological, morphological, and histopathological responses, BEST assesses and synthesizes existing information from other monitoring efforts to incorporate multiple lines of evidence into an ecological risk assessment approach. An important component of the BEST program is its emphasis on integrating established monitoring tools with research to develop new methods for addressing emerging issues and problems.

BIOMONITORING AS AN INDICATOR OF ECOSYSTEM HEALTH

The BEST program's study of environmental contaminant effects on fish in the Mississippi River Basin (MRB) was a demonstration project for the BEST program [20]. The MRB is the largest basin in North America. More than 72 million people live in the basin where agriculture, mining, and industry are the main activities generating contaminants of concern. The BEST MRB project served to illustrate how well a selected set of fish health biomarkers performs, the limits on interpretation of the results from fish health biomarkers, and how monitoring studies can serve as a platform for focused investigations.

Among the objectives of the MRB study were documentation and comparison of reproductive biomarker response, and identification of sites where fish reproductive health may be affected by chemical exposure [20]. Forty-eight stations were sampled for this study at which 10 males and 10 females of carp (*Cyprinus carpio*) or bass (*Micropterus sp.*) were collected and evaluated. The bioindicators of endocrine disruptors included: gonad pathologies, sex steroid and vitellogenin levels, and GSI. Tissue residue data were also obtained for organochlorines and elemental contaminants.

An analysis of the tissue data and biomarker responses at a station and overall comparison among stations provides some idea of the association between biomarker response and exposure to potential endocrine disruptors. Of the 48 stations evaluated, 31 stations were identified as having contaminant levels of concern based on either tissue concentrations or biomarker response. Twenty-three stations were identified as having tissue residue levels sufficiently high to potentially cause endocrine disruption. Likewise, 13 of the stations were identified as having contaminant levels of concern based on biomarker response. However, biomarker and tissue data coincided for only 8 of the 31 stations. These data suggest that (1) contaminants not analyzed for in tissues caused the biomarker responses in at least 5 of the 31 stations; and (2) this suite of biomarkers did not reveal an endocrine effect of the chemicals found at 15 of the 31 stations, although those same chemicals have been shown to cause endocrine-related effects in laboratory tests.

The finding that biomarker and tissue residue data corresponded at just 8 of 31 stations with possible endocrine-disrupting chemical contamination suggests that there may be only a weak relationship between biomarkers measured and the chemicals detected. Correlation analysis on the full MRB data set was used to evaluate the relationship between the fish biomarkers and the chemicals found in their tissues and showed reasonable correspondence between the organochlorines (cyclodiene pesticides and PCBs) and biomarkers based on extant toxicological information. *p′p′*-DDE, predominantly an anti-estrogen, correlated negatively with steroid and vitellogenin levels and positively with atresia. The results for mercury were ambiguous as it correlated positively with atresia and estradiol levels although laboratory studies indicate this may only be true for atresia; estradiol levels have been shown to correlate negatively with mercury.

Despite a general concordance between biomarkers and tissue residue data, the relationships between contaminants measured and biomarker responses at specific stations were difficult to reconcile. Station 111 was one of the eight stations at which results from the biomarker and tissue data indicated exposure to possible endocrine disruptors. Bass captured at this station had PCB concentrations averaging 1 µg/g and a high incidence of ovotestes. However, the existing toxicological data does not indicate that intersex results from exposure to PCBs. Furthermore, sex steroid levels in males were >2-fold higher than those in their counterparts from a reference location. That PCBs tend to have a depressive effect on sex steroid levels further suggests that there is no link between the high levels of PCBs and the biomarker responses at this station. The presence of ovotestes has been attributed to estrogen agonists such as those found in sewage effluents, yet exposure to estrogens typically depresses sex steroid levels in fish. The data are perhaps better explained by the presence of an aromatase enhancer, such as atrazine, which would increase endogenous estradiol levels and which in turn could also induce the ovotestes.

Monitoring data alone cannot provide define cause and effect relationships between environmental exposures and effects in fish and wildlife. Biological monitoring programs are typically designed to provide generalized information on spatial and temporal trends. Large-scale monitoring programs are usually designed to provide a large amount of information, but the depth of the studies does not allow determination of etiologies for any observed effects. The BEST program uses fish health indicators to identify areas of concern based on organismal responses and exposures. While the combined biological and chemical indicators of exposure to endocrine disruptors used in the BEST program may suggest cause and effect linkages, their greatest value is in identifying sites of potential concern. Once effects are observed in biological monitoring efforts, such as the BEST program, focused investigations are required to understand the origin of these effects. Quite often these focused investigations take the form of field-laboratory combined investigations that afford some of the complexities of the field, yet some of the control of laboratory investigations. Combined these are the studies that further our understanding of the effects of complex mixtures on endocrine systems.

FOCUSED INVESTIGATIONS CAN ORIGINATE FROM BIOLOGICAL EFFECTS MONITORING

Results from monitoring studies, such as BEST, also serve as a platform on which focused research questions can be formulated. For example, the results from Station 111, described above, led to an integrated lab–field study to understand whether fish with clinical signs of endocrine disruption (i.e., hormone imbalances, ovotestes) and tissue concentrations of known endocrine disrupting chemicals exhibit reproductive impairment. The establishment of a connection between the indicators of endocrine disruption and reproductive performance is necessary to better understand the potential significance of this condition to fish populations. The goal of this focused investigation was to capture wild fish in spawning condition to determine whether fish with signs of endocrine disruption can successfully spawn and produce healthy progeny. Fish from the contaminated site (Station 111) and a reference site were transported to laboratory ponds and maintained throughout spawning. Various observations and measurements were made on adults and offspring during the spawning, hatching, and rearing phases. Results from this study indicated that relative to the reference fish, the overall health of the adult bass from Station 111 was poor. Conclusions regarding reproductive success were confounded however, due to the unexpected appearance of clinical signs of endocrine disruption in the reference fish and the fact that the spawning behavior of fish from Station 111 differed unpredictably from that of the reference fish.

The exposure of model organisms to extracts of environmental mixtures can be a powerful tool for interpreting the reproductive significance of chemical and biomarker results, when evaluated in conjunction with controlled reproduction studies [21,22]. These combined studies with standard test chemicals and mixtures of chemicals from the environment can be an important set of tools for elucidation of cause and effect linkages [23]. The d-rR strain of medaka is a useful teleost model because this strain bears a sex-linked color marker allowing easy determination of genetic sex even when phenotypic sex has been altered. Medaka are amenable to several exposure methods. Egg injection is one method that has been perfected for use with this species. Egg injection mimics the maternally derived exposure to the egg of the persistent bioaccumulative compounds [24]. It has also been useful in studying the effects of stage specific early embryonic exposure to the more water soluble chemicals. The eggs, hatchlings, and older stages are easily exposed in traditional water baths as well. In some instances, a combination of injection and water-borne exposures may be useful as it has been shown that while an early exposure may not have obvious immediate effects, it may potentiate the effects of exposures later in life.

Laboratory-based studies using medaka have helped us characterize a number of indicators of endocrine disruption and assess the cause and effects links. For example, because we cannot easily determine genetic sex of the monitored organisms, generally sex reversal cannot be determined from field collections. Although indirect indicators such as skewed sex ratios from field collections may suggest sex reversal or sex-specific mortality. The d-rR medaka model can elucidate the effects of environmental conditions on phenotypic sex.

Secondary sexual characteristics are quite sensitive to estrogens and androgens and many fish are sexually dimorphic. Medaka have distinct sex-specific fin shapes, and the males develop breeding tubercles on the anal fin that have been shown to be very sensitive to androgens. Also, the urogenital duct systems are distinct between the genders. Established early in development, the ducts can be altered depending on exposure to estrogens or androgens during the period of sexual differentiation. Additionally, because the secondary sex characteristics are hormone-defined, they are not seen in normal immature juveniles, but can be induced through exposure. Therefore, precocious appearance of secondary sex characteristics or quantifiable changes in the characters can be an indicator of exposure.

Gonad and gamete pathologies are important indicators of exposure to and effects of endocrine disruptors, but are difficult to interpret because of the many variables in addition to hormonally active substances that can alter normal endogenous hormonal homeostasis. The medaka model is useful for understanding this complexity through characterizing the effects of multiple biotic and abiotic variables in part because much of the normal reproductive and endocrine biology is already understood. For example, the pattern of ovotestes observed in the MRB study has also been observed in both juvenile and adult male medaka exposed to aqueous solutions of natural and synthetic estrogens. However, it is also known that abrupt temperature and pH changes among other factors can induce ovotestes. Using the medaka model, we can examine the development of ovotestes in vivo or use ex vivo cultures of germ cells to look at effects of chemicals on development of such abnormalities.

Vitellogenin has been a keystone biomarker of estrogen exposure. But there remain difficulties in interpreting field results. Often, as in the MRB study, the vitellogenin levels don't coincide with endogenous estradiol levels or stage of development and individual variation seems to be great. Replacing or complimenting this measurement with measurement of the zona radiata protein, choriogenin, which has been shown to be more sensitive to estrogen than vitellogenin in medaka and other species, may be warranted. And for both measures, using juveniles that should have normally very low levels of endogenous sex hormones would be preferable.

Studies with medaka and other species have shown that estrogen and androgen levels are affected by exposure to certain chemicals including natural and synthetic sex steroids. The underlying mechanisms however are still unknown and this contributes to the difficulty in interpreting their significance in monitoring studies. In the MRB example, males tended to have higher than normal levels of 11-ketotestosterone and estradiol for their stage of development, time of year, and geographic location. Sensitive steroid test kits designed to use small volumes of plasma facilitate the use of models such as the medaka in designing studies to interpret such monitoring results.

SUMMARY

A shift from the manufacture and usage of persistent bioaccumulative chemicals to chemicals that are water soluble or are readily metabolized and eliminated from an organism has necessitated a concomitant paradigm shift in our approach to monitoring chemicals in the environment. No longer can we assess exposure and predict effects based solely on tissue residue information. Contemporary monitoring programs must be modified so as to maintain some aspects of traditional residue-based evaluations, yet also incorporate biological measures of exposure and effect consistent with the toxicological properties of the new generation of chemicals. Using data from our monitoring and research studies for hormonally active substances, we have discussed a variety of metrics of exposure and effects, their application to specific chemicals, and the current information gaps. We conclude that although several bioindicators of exposure and effect have been promoted and used, to date there continues to be a less robust association between cause and effect than could formerly be made based on tissue residue information. In part, this is due to the limited number of diagnostic tools that are currently available, to a lack of basic toxicological information concerning toxicokinetics and mechanisms of action of chemicals (e.g., endocrine disruptors) in fish and wildlife species, and to a lack of understanding of the effects of environmental mixtures. In the foreseeable future, measurement of chemical exposure in tissues and in the appropriate environmental compartment (i.e., water), will continue to be an important component of monitoring programs despite the many limitations. We must continue to develop and validate more specific bioindicators that will help close the gap between exposure and effects of endocrine-disrupting chemicals.

Research needs for development and implementation of successful monitoring programs for endocrine active compounds

1. **Understand basic endocrine function in representative fish and wildlife species.** The species important to resource managers often do not have some of the basic information on endocrine function. Basic endocrine and life history information such as seasonal steroid hormone cycles, gonad development patterns, signals and patterns of sexual maturation, and cues for spawning and seasonal migrations. This information is need for species of fish and wildlife that are important to resource managers.
2. **Develop dose–response models for EACs.** The complexity of endocrine function and the vast numbers of new chemicals introduced into commerce requires that model dose–response relationships continue to be developed for EACs.
3. **Develop new technologies for monitoring EACs (ELISA, bioassay).** Screening systems for both the quantification of EACs in the environment and evaluation of effects (biomarkers) in fish and wildlife need to be developed or validated for environmental use. As a better understanding of basic endocrine function and models of response to EACs are developed, our knowledge of key control points of reproductive processes will aid in the development of assays, biomarkers, and other tools for environmental monitoring for EACs.
4. **Develop estrogen/androgen responsive genetic markers for screening with microarrays.** Toxicogenomic responses of fish and wildlife species toward EACs have the potential to provide a better understanding of initial events in responses to EACs. Abnormal or altered changes in gene expression are often some of the primary events in an organism's response to EACs. The tools being developed in genomics, such as microarrays, have the potential to allow screening of hundreds or thousands of genes simultaneously and nonlethally. However, the utility of these tools is

dependent on extensive characterization of genes and the dose and temporal response patterns of those genes to EACs.

5. **Target indicators that are significant relative to higher-order functional effects need to be validated for EACs.** Metrics of population- and community-level effects of EACs in fish and wildlife with linkages to behavioral, physiological, and biochemical measures need to be developed and validated for environmental monitoring. Apical endpoints, those that integrate the combined responses to multiple chemicals or multiple stressors, need continued evaluation for incorporation into environmental monitoring for EACs. Most measures used in contaminant monitoring are organismal level of biological organization. These are important and tend to be the best defined, yet we need to continue to strive towards measures of population-level indicators of reproductive health.

ACKNOWLEDGMENTS

The authors would like to acknowledge the BEST program, directed by Timothy Bartish, and the collaborating scientists: C. J. Schmitt, J. E. Hinck, J. J. Whyte, V. Blazer, T. Gross, J. Coyle, and P. Anderson. The results of the BEST MRB study used as the example in this presentation are a result of the combined efforts of these scientists and their support staff. Funding for this work and the BEST program was provided by the U.S. Geological Survey, Department of Interior.

REFERENCES

1. P. C. Baumann and D. M. Whittle. *Aquat. Toxicol.* **11**, 241–257 (1988).
2. D. V. Weseloh, T. W. Custer, B. M. Braune. *Environ. Pollut.* **59**, 141–160 (1989).
3. D. E. Tillitt, C. J. Schmitt, D. V. Weseloh, C. M. Bunck. In *Ecotoxicological Risk Assessment of the Chlorinated Organic Chemicals,* J. Carey, P. Cook, J. Giesy, P. Hodson, D. Muir, W. Owens, K. Solomon (Eds.), pp. 259–288, SETAC Press, Pensacola, FL (1998).
4. M. Olsson. In *PCBs and the Environment. Vol. III*, J. S. Waid (Ed.), pp. 181–208, CRC Press, Boca Raton, FL (1986).
5. M. Olsson and L. Reutergårdh. *Ambio* **15**, 103–109 (1986).
6. C. J. Schmitt and W. G. Brumbaugh. *Arch. Environ. Contam. Toxicol.* **19**, 731–747 (1990).
7. C. J. Schmitt, J. L. Zajicek, P. H. Peterman. *Arch. Environ. Contam. Toxicol.* **19**, 748–782 (1990).
8. R. J. Norstrom and D. C. G. Muir. *Sci. Total Environ.* **154**, 107–128 (1994).
9. D. J. Gregor, A. J. Peters, C. Teixeira, N. Jones, C. Spencer. *Sci. Tot. Environ.* **160/161**, 117–126 (1995).
10. A. Fairbrother, G. T. Ankley, L. S. Birnbaum, S. P. Bradbury, B. Francis, L. E. Gray, D. Hinton, L. L. Johnosn, R. E. Peterson, G. van der Kraak. In *Reproductive and Developmental Effects of Contaminants in Oviparous Vertebrates,* R. T. Di Giulio and D. E. Tillitt (Eds.), pp. 283–362, SETAC Press, Pensacola, FL (1999).
11. D. E. Tillitt, G. T. Ankley, J. P. Giesy, J. P. Ludwig, H. Kurita, D. V. Weseloh, P. S. Ross, C. A. Bishop, L. Sileo, K. L. Stromborg, J. Larson and T. J. Kubiak. *Environ. Toxicol. Chem.* **11**, 1281–1288 (1992).
12. S. Tanabe. *Environ. Pollut.* **50**, 5–28 (1988).
13. J. L. Zajicek, D. E. Tillitt, J. N. Huckins, J. D. Petty, M. E. Potts, D. A. Nardone. In *Environmental Immunochemical Methods*, J. Van Emon, C. Johnson, Gerlach (Eds.), Chap. 26, pp. 307–325, ACS Symposium Series 646, American Chemical Society, Washington, DC (1996).
14. J. L. Zajicek, D. E. Tillitt, T. R. Schwartz, C. J. Schmitt. *Chemosphere* **40**, 539–548 (2000).
15. J. A. Lasrado, C. R. Santere, J. L. Zajicek, J. R. Stahl, D. E. Tillitt, D. Deardorff. *J. Food Sci.* **68** (1), 133–7 (2003).
16. J. J. Whyte, C. J. Schmitt, D. E. Tillitt. *Crit. Rev. Toxicol.* **34** (1), (2004).

17. D. W. Kolpin, E. T. Furlong, M. T. Meyer, E. M. Thurman, S. D. Zaugg, L. B. Barber, H. T. Buxton. *Environ. Sci. Technol.* **36** (6), 1202–12 (2002).
18. E. J. Calebrese and L. A. Baldwin. *Hum. Experiment. Toxicol.* **19**, 2–31 (2000).
19. H. A. Tilson. *Environ. Health Perspect.* **106** (Suppl. 3), 807–811 (1998).
20. C. J. Schmitt (Ed.). *Biomonitoring of Environmental Status and Trends (BEST) Program: Environmental Contaminants and their Effects on Fish in the Mississippi River Basin*, p. 241, U.S. Geological Survey, Biological Science Report 2002-0004 (2002).
21. M. K. Walker, P. M. Cook, B. C. Butterworth, E. W. Zabel, R. E. Peterson. *Fundam. Appl. Toxicol.* **30**, 178–186 (1996).
22. P. M. Cook, E. W. Zabel, R. E. Peterson. In *Chemically-Induced Alterations in the Functional Development and Reproduction of Fishes*, R. Rolland, M. Gilbertson, R. Peterson (Eds.), Chap. 2, pp. 9–27, SETAC Press, Pensacola, FL (1997).
23. P. J. Wright and D. E. Tillitt. *Aquat. Toxicol.* **47** (2), 77–92 (1999).
24. M. K. Walker, E. W. Zabel, G. Åkerman, L. Balk, P. J. Wright, D. E. Tillitt. In *Techniques in Aquatic Toxicology*, G. Ostrander (Ed.), Chap. 4, pp. 41–72, CRC Press/Lewis Publishers, Boca Raton, FL (1996).

WORKSHOP 4

SIMPLE, RAPID ASSAY FOR CONVENTIONAL DEFINITIVE TESTINGS OF ENDOCRINE DISRUPTOR HAZARD

Pure Appl. Chem., Vol. 75, Nos. 11–12, pp. 2479–2482, 2003.

Workshop 4.1

Simple, rapid assays for conventional definite testing of endocrine disruptor hazard: Summary and recommendations*

Shoji Fukushima[1] and Alexius Freyberger[2,‡]

[1]*Department of Pathology, Osaka City University Medical School, Osaka, Japan;*
[2]*BAYER AG, PH PD P Toxicology, Wuppertal, Germany*

Abstract: Study protocols for the characterization of endocrine active compounds presented in Workshop 4 included the enhanced Organization for Economic Cooperation and Development (OECD) test guideline (TG) 407, the medium-term rat liver and rat multi-organ carcinogenicity assays, and an enhanced one-generation reproduction study.

The outcome of rat studies on flutamide and ethinylestradiol indicated that these strongly active compounds can readily be detected even with a low animal number using the enhanced OECD TG 407. Both newly added (such as male accessory sex organ weights, histology of pituitary, vagina and male mammary gland) and already included parameters contributed to the detection of endocrine effects. Thorough evaluation of the results of 20 studies conducted with 10 compounds thought to interfere with the endocrine system by different mechanisms will identify the most appropriate enhancements to the current OECD TG 407.

Medium-term rat liver and rat multi-organ carcinogenicity assays are well recognized in the International Conferences on Harmonization for Pharmaceutical Chemicals. They have been successfully used to detect carcinogenic and modifying potentials of new chemicals within a relatively short time and can be applied to endocrine active compounds. Dose–response studies on nonylphenol, bisphenol A, and styrene using the rat liver carcinogenicity assay did not reveal effects of any of these compounds on the development of preneoplastic lesions in rat liver.

The enhanced one-generation reproduction study protocol included treatment of pregnant female rats from gestation day 0 through to lactation day 21, and examination of all offspring. Half of the animals were necropsied at weaning, the remaining animals were examined for vaginal opening, preputial separation, estrous cyclicity, and sperm characteristics and were necropsied at adulthood. In a pilot study ethinylestradiol inhibited maternal fertility at dose levels similar to those effective in the uterotrophic assay.

It is recommended to rapidly evaluate the conducted enhanced OECD TG 407 studies and to enhance the current OECD TG 407 appropriately. Further compounds with different mechanisms of action should be studied in the one-generation reproduction study to further investigate the usefulness of this protocol. The established medium-term carcinogenicity assays can be used to study carcinogenic potential rapidly. Use of female animals and inclusion of carcinogens targeting at breast and uterus should be considered in order to explore further the predictibility of this model.

*Report from a SCOPE/IUPAC project: Implication of Endocrine Active Substances for Human and Wildlife (J. Miyamoto and J. Burger, editors). Other reports are published in this issue, *Pure Appl. Chem.* **75**, 1617–2615 (2003).

‡Corresponding author

In the introduction to the workshop, Shoji Fukushima pointed to a number of areas that require particular emphasis when considering rapid assays for endocrine active substances (EASs) and the effects of these compounds on disease processes like reproductive lesions or neoplasia. First of all, and fundamental to the understanding of the mechanism of action of EASs, is the question of how an agent interacts with nuclear receptors or hormone synthesis, and which are the downstream events that then ensue. Related to this are metabolism of both EAS and endogenous hormones and the fate and potential activity of corresponding EAS metabolites and their environmental impact. In addition to studies on rodents, providing comprehensive information on a great diversity of endpoints, for human risk assessment due attention must be paid to any epidemiological evidence that may be available. It may also be helpful to take into account the effects of EASs in wildlife mammalian species should such information be available.

Of particular importance are the dose response and the appropriate definition of threshold doses. The possibility of homeostasis playing a role also deserves special consideration. Clearly, appropriate choice of representative test compounds and screening is essential, and given the likelihood of species specificity mammals must be the animals of choice for testing. Furthermore, the complex influence of hormones and EASs, acting simultaneously on many organs and tissues, means that in vivo approaches must be given high priority if a comprehensive understanding is to be obtained. Attention must also be concentrated on the stage in life when sensitivity would be expected to be highest on the basis of mechanistic and developmental considerations. This underlies the selection of models for presentation in our workshop.

In Workshop 4.1, by Freyberger et al., the enhanced OECD test guideline (TG) 407, was introduced. This subacute protocol for detecting endocrine modulation was discussed at the international level under the umbrella of the OECD, and the initial version was the basis for the reported feasibility study on flutamide (FLU). Furthermore, data on ethinyl estradiol (EE) using a modified protocol with fewer enhancements that is presently under investigation were presented for comparison. The enhancements considered were additional determination of: (a) thyroid-related hormones; (b) estrus cyclicity from vaginal smears starting in exposure week 4 to ensure necropsy of all females in the diestrus stage of the estrus cycle in week 5; (c) the number and morphology of cauda epididymal spermatozoa; (d) organ weights of ovaries, uterus, thyroid, male accessory reproductive organs, and pituitary; and (e) histopathological investigation of pituitary, male and female mammary gland, epididymes, pancreas. and vagina. For the study, groups of 5 male and 5 female rats were orally gavaged with FLU at 0, 1, 10, or 100 mg/kg body weight (b.w.)/day or with 0, 0.01, 0.05, or 0.2 mg/kg EE/day for at least 28 days. Two studies (A and B) were run in parallel to assess intra-laboratory variation and the potential increase in sensitivity with doubling of the animal numbers.

Endocrine activity of both FLU and EE was readily demonstrated with the enhanced OECD TG 407 protocol. Relative weights of the male accessory sex organs (MASO) were consistently decreased by the highest doses of FLU and EE. Treatment-related increase of relative uterine weights in EE-treated animals was only observed with the combined data at the high dose, indicating that uterine weight in young adult animals is not a sensitive measure of estrogenicity.

Consistent histopathological changes in studies A and B were atrophy of MASO and decreased tubular size of the epidiymus at 100 mg/kg FLU, along with increased numbers of PAS-positive cells from 10 mg/kg, hypertrophic basophilic cells at 100 mg/kg, and Leydig cell hypertrophy from 10 mg/kg, the latter indicating activation of the hypothalamic-pituitary-gonadal axis. Furthermore, microvesicular cytoplasmic vacuoles were observed in the zona fasciculata of the adrenals at the highest dose. With EE, similar atrophy of the MASO was recognized at the high dose, degeneration of the germinal epithelium and Leydig cell atrophy was apparent at 0.2 mg/kg, and feminization of the mammary gland was observed from 0.05 mg/kg. In contrast to the FLU study, reduced vacuolation in the zona fasciculata of the male adrenal at the high dose was seen with EE. In FLU-treated females, no endocrine-mediated histological changes were observed, while EE caused a striking discrepancy between the diagnosis of the stage of the female cycle by vaginal smear cytology (diestrus) and vaginal and uterine morphology (estrogenized tissue) that was consistently observed in all treatment groups. Furthermore, an increase of early stage follicles was detected at the high dose. This findings underline the important role of histology in the detection of EASs.

Asssessment of thyroid-related hormones did not contribute to the detection of the endocrine activity of FLU, but the data from the combined analysis revealed an increase in thyroid stimulating hormone (TSH) levels in both sexes treated with EE and increased thyroxine levels in females. However, there was a lack of any clear dose relationship, and the data from the paired studies differed and changes could not be linked to estrogenic activity and thus did not contribute to sensitivity.

Treatment with EE did not affect sperm, whereas FLU at the highest dose increased the frequency of abnormal spermatozoa, especially in study B. A decreased epididymal sperm count was also noted on combined evaluation of studies A and B.

In general, the doubling of group size with analysis of combined data from the paired studies elevated the sensitivity although the additional significant alterations revealed had been already tentatively identified using the conventional group of five animals. This, however, was not the case with histopathology. Furthermore, doubling the animal number did not increase the sensitivity of detection of endocrine-mediated effects above the level already obtained by histopathological examination of groups of five animals.

From the present results for the strongly active compounds FLU and EE, inclusion of spermatology and determination of thyroid hormones and TSH in the final guideline cannot be recommended. Furthermore, use of vaginal smear cytology as a measure to determine the female cycle appears to be problematic in estrogen-treated animals. We must now await the evaluation of a phase 2 investigation that was performed in 13 laboratories in different countries using 10 compounds known or suspected to interact with the endocrine system through different mechanisms to finally decide which parameters should be included in the guideline and whether weakly active compounds can also sensitively be detected.

In Workshop 4.2, given by Imaida, medium-term in vivo tests for carcinogenicity or modifying potential of exogenous in rats were described, along with the body of results so far obtained. In the medium-term liver bioassay, animals are initiated with a single dose of diethylnitrosamine (DEN) then administered test compound starting two weeks thereafter with performance of partial hepatectomy in week 3. The effects of six weeks exposure to a compound on development of glutathione-*S*-transferase-positive putative preneoplastic foci are then assessed at the end of week 8. Distinction can be made between initating and promoting effects by inclusion of a group without DEN initiation. A total of 313 chemicals have already been analyzed, and the efficacy of the system for detection of hepatocarcinogens has thereby been well established. Comparison of results with this test and the two-year long-term rat assay confirmed similar dose-dependence and validated use of GST-P-positive foci as surrogate endpoint lesions. Dose–response studies of nonylphenol (doses of 25, 250, and 2000 ppm in the diet), bisphenol (40 and 160 mg/kg b.w., ig, six times per week) and styrene (250 and 1000 mg/kg b.w., ig, six times per week) did not reveal effects of any of these EASs on development of preneoplastic lesions, in terms of either number or area.

In the rat medium-term multiorgan bioassay described, male F344 rats are treated sequentially with five carcinogens, DEN, *N*-methyl-*N*-nitrosourea, dihydroxy-di-*n*-propylnitrosamine, *N*-butyl-*N*-(4-hydroxy-butyl)nitrosamine and dimethylhydrazine, primarily targeting the liver, lung, urinary bladder, and colon, respectively, then exposed to test compound for 24 weeks. A total of 63 chemicals have so far been tested in this model, all of 17 hepatocarcinogens and 19/22 carcinogens targeting other organs were positive. It is strongly recommended that representative EASs be now examined using this approach. Use of female animals and inclusion of carcinogens targeting the breast and uterus would be particularly beneficial in order to further explore the predictability of this model.

With both of these medium-term tests, particular advantages are the relatively small number of animals necessary, the short period of exposure, and the possibility of using surrogate markers for detecting carcinogenicity or modifying potential of EASs at low doses or in combination.

In Workshop 4.3, Aoyama described an enhanced one-generation reproductive study in rats for detecting endocrine-disrupting effects of chemicals, along with data obtained from pilot studies using EE. The presently used two-generation study protocol requires a relatively long period and only a small number of offspring is followed to adulthood. For the proposed new assay, pregnant females are treated with the test substance from gestation day 0 through to lactation day 21, the F1 offspring from half of the litters in each dose group then being killed for necropsy at weaning. The remaining half are examined for sexual maturation, estrous cyclicity, and/or sperm production. The parameters include fertility and gestation indices of dams, numbers of implants and pups delivered, sex ration and AGD of pups, the viability of pups during the lactation period, gross pathology of dams, and observations of F1 offspring such as organ weights at weaning and after maturation, sexual maturation (vaginal opening/preputial separation), estrous cyclicity, sperm characteristics, and gross and histopathological findings in all animals. Other additional endpoints that can be flexibly added include mRNA expression of appropriate genes. Following subcutaneous injection of EE effects on maternal fertility (inhibition of pregnancy and reduction of litter size) could be detected at dose levels corresponding to those effective in the uterotrophic assay. The same treatment of dams limited to the organogenetic or perinatal period did not disclose this effect. Thus, the days 0–5 of gestation are of prime importance. Findings observed in F1 offspring in the pilot study were slightly increased anogenital distance in female newborns and down- or upregulation of mRNAs of certain genes. Genes for androgen receptor in the prostate and IGF-1 in the uterus may be particularly sensitive markers for monitoring potential estrogenic effects.

Clearly, the different properties of the proposed single and two-generational approaches will require further comparative studies using the same representative test compounds at different doses, with appropriate consideration of data from other models.

Overall, the workshop pointed to potential variabilities of detecting EASs, especially for weakly active agents. It is recommended that several different approaches are evaluated in parallel to provide the most comprehensive profile possible before any conclusions are drawn.

Pure Appl. Chem., Vol. 75, Nos. 11–12, pp. 2483–2489, 2003.

Workshop 4.2

Testing of endocrine active substances using an enhanced OECD test guideline 407: Experiences from studies on flutamide and ethinylestradiol*

Alexius Freyberger‡, Peter Andrews, Elke Hartmann, Rolf Eiben, Ingo Loof, Ulrich Schmidt, Michael Temerowski, Andree Folkerts, Michael Becka, Bernhard Stahl, and Martin Kayser

BAYER AG, PH PD P Toxicology and PH PD Biometry and Pharmacometry, D-42096 Wuppertal, Germany

Abstract: Groups of five male and five female Wistar rats were treated by gavage with 0, 1, 10, and 100 mg/kg body weight (b.w.) flutamide (FLU) or 0.01, 0.05, and 0.2 mg/kg b.w. of ethinylestradiol (EE2) for at least 28 days according to an enhanced Organization for Economic Cooperation and Development (OECD) test guideline (TG) 407 to investigate which of the current and/or additional parameters would detect effects on the endocrine system and to provide data on intralaboratory variability. Two identical studies were performed in parallel on each compound. Common enhancements were determination of thyroid hormones (T3, T4) and thyroid stimulating hormone (TSH), of the stage of the estric cycle to ensure necropsy of females in diestrus, of the number and morphology of epididymal sperm, and of additional organ weights (e.g., male accessory sex organs, MASO) and histopathology of additional organs (e.g., pituitary, vagina). Endocrine-mediated findings consistently observed in these studies were decreased relative weights of MASO at 100 mg/kg FLU and at 0.2 mg/kg EE2, histological changes in pituitary and testes at ≥10 mg/kg and in MASO, epididymis and adrenals at 100 mg/kg in FLU-treated males, histological changes in the mammary gland at ≥0.05 mg/kg and in testes, MASO and adrenals at 0.2 mg/kg in EE2-treated males, estrogenization of uterus and vagina (despite necropsy in diestrus) at ≥0.01 mg/kg EE2, and changes in the ovary at 0.2 mg/kg EE2. Spermatology was insensitive (EE2) or revealed changes only at the maximum tolerated dose (MTD). Determination of T3, T4, and TSH did not contribute to the detection of the endocrine effects (FLU) or provided equivocal results. Doubling the group size to 10 animals by combining the studies run in parallel did not increase the sensitivity of detection of endocrine-mediated effects above the level obtained by histopathological examination of groups of five animals. Only some of the proposed enhancements evaluated were helpful in detecting the endocrine-mediated effects of FLU and EE2. Evaluation of studies according to an enhanced TG 407 on 10 compounds with different endocrine activities will identify the most appropriate enhancements.

*Report from a SCOPE/IUPAC project: Implication of Endocrine Active Substances for Human and Wildlife (J. Miyamoto and J. Burger, editors). Other reports are published in this issue, *Pure Appl. Chem.* **75**, 1617–2615 (2003).

‡Corresponding author

INTRODUCTION

In response to concerns that current toxicological test guidelines (TGs) may not address the detection of interactions of chemicals with the endocrine system sufficiently, EMSG had proposed enhancements to the current subacute rodent Organization for Economic Cooperation and Development (OECD) TG 407 by parameters thought to be effective in detecting endocrine modulation. The enhancements had been discussed at the international level under the umbrella of OECD, and the resulting final version of the enhanced protocol [1] was the basis for a feasibility study on FLU [2]. Experiences from this and other studies resulted in a reduction of the enhancements. This streamlined protocol [3] was then used for the testing of a broad range of endocrine active substances. For each compound, two studies (designated A and B) were run in parallel with the conventional animal number of five rats per dose and sex in order to assess intralaboratory variability. Moreover, individual (n = 5) and combined (n = 10, designated as Combined) evaluation of the results was performed in order to investigate a potential increase in sensitivity when doubling animal numbers. Our laboratory was entrusted with the testing of the flutamide (FLU, feasibility study) and ethinylestradiol (EE2). In the following, we discuss the efficiency of current and additional parameters of the enhanced TG 407 protocol for the detection of endocrine-mediated changes induced by FLU and EE2. Only enhancements common to both studies will be considered, namely the additional determination of (a) thyroid-related hormones; (b) estrus cyclicity from vaginal smears starting in exposure week 4 to ensure necropsy of all females in the diestrus stage of the estrus cycle in week 5; (c) the number and morphology of cauda epididymal spermatozoa; (d) organ weights of ovaries, uterus, thyroid, male accessory reproductive organs and pituitary; and (e) histopathological investigation of pituitary, mammary gland, epididymides, pancreas, and vagina.

MATERIAL AND METHODS

SPF-bred Wistar rats (Hsd Cpb:WU) obtained from Harlan–Winkelmann (Borchen, Germany) were 7 weeks old at initiation of treatment. The study design followed the enhanced protocol [1,3]. The high dose was chosen to represent a mean therapeutic dose (MTD), the low dose was chosen to result in a no observed adverse effects level (NOAEL). Groups of 5 male and 5 female rats were orally gavaged with 0, 1, 10, or 100 mg/kg body weight (b.w.) FLU per day or with 0, 0.01, 0.05, or 0.2 mg/kg b.w. EE2 per day for at least 28 days. Two identical studies were run concurrently for each compound. Females were sacrificed beginning at day 28, provided they were in the diestrus stage. Animals which did not come into diestrus were sacrificed on day 32. In the FLU study, one male rat per group and study was necropsied per day from day 28 to 32, whereas in the EE2 study males were necropsied on days 28 and 29. Appropriate statistical tests were performed. Results of manipulative behavioral tests and incidences of abnormal spermatozoa and of histopathological findings were not subjected to statistical analysis [for details see refs. 2,4].

RESULTS

General findings

All animals survived to necropsy. Body weight gains were biologically significantly and consistently reduced by more than 30 % in high-dose males treated with FLU or EE2, whereas females were less affected. Treatment-related clinical signs (FLU only) and alterations in hematological and clinical chemical parameters were moderate and mainly restricted to the high dose [for details see refs. 2,4].

Effects on estrus cyclicity

All FLU-treated females were sacrificed in the diestrus stage of the female cycle according to vaginal smear cytology and subsequent histopathology of the vaginal epithelium. Two females (0.05 mg/kg, study B) of the investigation on EE2 did not enter into diestrus during the observation period and were sacrificed in estrus, whereas all other females were classified as diestric by vaginal smear cytology when necropsied.

Effects on organ weights

FLU strongly increased liver weight in both sexes at the high dose. Other changes in organ weights were restriced to male rats (Table 1). Effects of EE2 treatment on organ weights in male animals are shown in Table 2. Remarkable changes in organ weights in EE2-treated female rats were a dose-dependent, strong increase of liver weights and an increased uterine weight observed only at the high dose and following combined analysis of studies A and B.

Table 1 Effects of FLU treatment on terminal body weights and relative organ weights in male rats.

		Control	1 mg/kg	10 mg/kg	100 mg/kg
Terminal body wt (g)					
	Study A	352 ± 15	318 ± 32	338 ± 38	308 ± 17
	Study B	327 ± 19	311 ± 36	313 ± 32	288 ± 20
	Combined studies	339 ± 21	315 ± 32	326 ± 36	298 ± 20**
Organ wt (mg/100 g)					
<u>Pituitary</u>	Study A	3 ± 0.2	3 ± 0.1	3 ± 0.3*	4 ± 0.3**
	Study B	3 ± 0.4	3 ± 0.5	3 ± 0.4	3 ± 0.8
	Combined studies	3 ± 0.3	3 ± 0.4	3 ± 0.3	3 ± 0.7
Adrenals	Study A	20 ± 3	22 ± 1	21 ± 2	24 ± 3
	Study B	19 ± 2	21 ± 3	24 ± 5	24 ± 3*
	Combined studies	19 ± 2	21 ± 2	22 ± 4*	24 ± 3**
Testis, left	Study A	469 ± 12	494 ± 41	462 ± 39	490 ± 5
	Study B	480 ± 24	421 ± 96	499 ± 42	496 ± 39
	Combined studies	475 ± 19	457 ± 80	480 ± 43	493 ± 26
Epididymis, left	Study A	185 ± 20	186 ± 18	152 ± 20	133 ± 35**
	Study B	198 ± 12	179 ± 29	166 ± 19	106 ± 18**
	Combined studies	191 ± 17	183 ± 23	159 ± 20*	120 ± 30**
<u>Ventral prostate</u>	Study A	124 ± 16	92 ± 14	92 ± 21	32 ± 14**
	Study B	121 ± 26	90 ± 11	93 ± 10	30 ± 21**
	Combined studies	123 ± 20	91 ± 12	93 ± 15	31 ± 15**
<u>Seminal vesicles</u>	Study A	473 ± 104	433 ± 86	333 ± 38*	103 ± 14**
<u>plus dorso-lateral</u>	Study B	467 ± 37	462 ± 102	391 ± 108	101 ± 21**
<u>prostate</u>	Combined studies	470 ± 73	448 ± 90	362 ± 82**	102 ± 17**

Tissues were weighed fresh or after fixation (pituitary). Enhancements are underlined. Asterisks indicate a significant difference from the corresponding controls at the $p \leq 0.05$ (*) and $p \leq 0.001$ (**) level.

Table 2 Effects of EE2 treatment on terminal body weights and relative organ weights in male rats.

		Control	0.01 mg/kg	0.05 mg/kg	0.2 mg/kg
Terminal body wt (g)					
	Study A	321 ± 25	311 ± 21	293 ± 11	277 ± 22**
	Study B	304 ± 25	336 ± 34	301 ± 42	269 ± 28
	Combined studies	312 ± 25	323 ± 30	297 ± 29	273 ± 24**
Organ wt (mg/100g)					
Pituitary	Study A	3 ± 0.4	3 ± 0.7	3 ± 0.6	4 ± 0.4
	Study B	3 ± 0.8	3 ± 0.7	4 ± 0.7	5 ± 1.2
	Combined studies	3 ± 0.7	3 ± 0.7	3 ± 0.7	4 ± 1.0*
Thyroid	Study A	5 ± 0.9	5 ± 0.8	6 ± 0.4	6 ± 0.4*
	Study B	5 ± 1.2	5 ± 1.0	5 ± 0.7	6 ± 0.9
	Combined studies	5 ± 1.0	5 ± 0.8	5 ± 0.7	6 ± 0.7*
Adrenals	Study A	15 ± 3	16 ± 2	19 ± 3*	24 ± 2*
	Study B	16 ± 2§	17 ± 1§	20 ± 4	25 ± 3
	Combined studies	16 ± 2	17 ± 2	20 ± 3**	25 ± 3**
Testes	Study A	1051 ± 123	1091 ± 196	1101 ± 57	1037 ± 155
	Study B	955 ± 90	959 ± 137	917 ± 299	1110 ± 50
	Combined studies	1003 ± 113	1025 ± 174	1009 ± 225	1073 ± 115
Epididymides	Study A	346 ± 16	336 ± 43	341 ± 22	310 ± 101
	Study B	324 ± 31	344 ± 45	283 ± 96	335 ± 32
	Combined studies	335 ± 26	340 ± 42	312 ± 72	323 ± 72
Ventral prostate	Study A	126 ± 12	144 ± 27	113 ± 26	67 ± 31*
	Study B	141 ± 29	143 ± 16	113 ± 24§	68 ± 14**
	Combined studies	133 ± 22	144 ± 21	113 ± 23	68 ± 23**
Dorso-lat. prostate	Study A	151 ± 17	159 ± 27	136 ± 42	99 ± 41
	Study B	163 ± 26	143 ± 24	144 ± 53§	85 ± 17**
	Combined studies	157 ± 22	151 ± 25	140 ± 44	92 ± 30**
Seminal vesicles	Study A	230 ± 39	234 ± 42	204 ± 26	102 ± 38**
	Study B	273 ± 58	253 ± 15	225 ± 60§	122 ± 33**
	Combined studies	252 ± 51	244 ± 31	213 ± 42	112 ± 35**
Coagulating glands	Study A	46 ± 9	40 ± 8	37 ± 5	17 ± 11*
	Study B	49 ± 13	50 ± 10	40 ± 16§	19 ± 9**
	Combined studies	48 ± 11	45 ± 10	39 ± 10	18 ± 9**

§, $n = 4$. No multiple comparisons were calculated. Tissues were weighed fresh or after fixation (pituitary, thyroid, male accessory sex organs). Enhancements are underlined. Asterisks indicate a significant difference from the corresponding controls at the $p \leq 0.05$ (*) and $p \leq 0.001$ (**) level.

Histological findings

In the liver, centrilobular hepatocellular hypertrophy and cytoplasmic change was induced by FLU in males and females at the high dose. Effects of FLU on endocrine organs and hormone sensitive tissues in males are shown in Table 3. In females, only a slightly increased incidence of increased interstitial glands in the ovaries was observed at the high dose. Effects of EE2 are shown in Table 4. Most striking was a discrepancy between the diagnosis of the stage of the female cycle by vaginal smear cytology (diestrus) and vaginal and uterine morphology (estrogenized tissue).

Table 3 Important histological findings in endocrine tissues of FLU-treated male rats.

Dose (mg/kg body wt)		1			10			100		
Study		A	B	A & B	A	B	A & B	A	B	A & B
Pituitary	Increased number of PAS-positive cells				⊕	⊕	⊕	⊕	⊕	⊕
	Intracytoplasmic inclusions and hypertrophic basophilic cells							⊕	⊕	⊕
Testis	Leydig cell hypertrophy				⊕	⊕	⊕	⊕	⊕	⊕
Epididymis	Decreased tubular size and increased interstitial tissue							⊕	⊕	⊕
Accessory sex organs	Atrophy							⊕	⊕	⊕
Adrenals	Microvesicular cytoplasmic vacuoles in zona fasciculata							⊕	⊕	⊕

A & B, combined evaluation of studies A and B; ⊕, changes detected. Enhancements are underlined. The asterisk indicates a significant difference from the corresponding controls at the $p \leq 0.05$ (*) level.

Table 4 Important histological findings in endocrine tissues of EE2-treated rats.

Dose (mg/kg body wt)		0.01			0.05			0.2		
Study		A	B	A & B	A	B	A & B	A	B	A & B
Changes in male rats										
Testes	Degeneration of germinal epithelium, Leydig cell atrophy							⊕	⊕	⊕
Prostate	Atrophy							⊕	⊕	⊕
Seminal vesicles	Atrophy					⊕	⊕	⊕	⊕	⊕
Coagulating gland	Atrophy				⊕			⊕	⊕	⊕
Mammary gland	Feminization	⊕			⊕	⊕	⊕	⊕	⊕	⊕
Adrenals	Reduced vacuolation in zona fasciculata							⊕	⊕	⊕
Changes in female rats										
Ovaries	Increase of early stage follicles							⊕	⊕	⊕
Uterus	Increased epithelial height and other correlates of estrogenic action	⊕	⊕	⊕	⊕	⊕	⊕	⊕	⊕	⊕
Vagina	Reduced number of diestric animals	⊕	⊕	⊕	⊕	⊕	⊕	⊕	⊕	⊕
	Keratinization and other correlates of estrogenic action	⊕	⊕	⊕	⊕	⊕	⊕	⊕	⊕	⊕
Adrenals	Cytoplasmic eosinophilia								⊕	⊕

A & B, combined evaluation of studies A and B; ⊕, changes detected. Enhancements are underlined.

Hormone determinations

Determination of thyroid related hormones did not contribute to the detection of the endocrine activity of FLU [2]. In EE2-treated rats mainly combined analysis suggested increased thyroid stimulating hormone levels in both sexes and increased thyroxine levels in females. However, these findings were not clearly dose-related, and hormone levels of the individual studies greatly differed from one another [4].

Spermatological findings

Effects of FLU and EE2 treatment are given in Table 5.

Table 5 Effects of FLU and EE2 on epididymidal sperm count and morphology.

		Control	Low dose	Mid dose	High dose
Right epididymidal spermatozoa counts (10^3 spermatozoa/mg)					
Flutamide	Study A	602 ± 170	705 ± 286	780 ± 157	498 ± 221
	Study B	687 ± 210	701 ± 318	901 ± 269	270 ± 109
	Combined studies	645 ± 186	703 ± 286	840 ± 217	384 ± 203*
Ethinylestradiol	Study A	715 ± 160	664 ± 88	618 ± 210	743 ± 239[§]
	Study B	553 ± 153	496 ± 189	733 ± 214[§]	799 ± 249
	Combined studies	634 ± 170	580 ± 165	669 ± 207	774 ± 231
Abnorm spermatozoa (%)					
Flutamide	Study A	1.3	1.0	1.6	3.1
	Study B	1.3	1.3	1.4	7.9
	Combined studies	1.3	1.2	1.5	5.4
Ethinylestradiol	Study A	0.6	1.6	0.1	1.3[§]
	Study B	0.2	1.4	0.8[§]	0.4
	Combined studies	0.4	1.5	0.4	0.8

[§]One male with aspermia not taken into account. Enhancements are underlined.

CONCLUSIONS

It was feasible to include all enhancements into the testing routine. The endocrine activity of FLU and ethinylestradiol (EE2) was readily demonstrated using the enhanced OECD TG 407 protocol. Both already existing and newly added parameters contributed to the detection of endocrine-mediated changes and individual studies A and B corresponded well to one another. Additional organ weights (mainly male accessory sex organs) and additional histological investigation of pituitary, epididymis, male mammary gland, and vagina were helpful in detecting endocrine mediated changes. Thyroid related hormones showed high variability and did not contribute to the detection of the endocrine activity of FLU. Changes of these hormones in EE2-treated animals could not be clearly related to estrogenic activity and thus did not contribute to sensitivity. Spermatology was insensitive at all or revealed changes only at the MTD. The use of vaginal smear cytology as a measure to determine the female cycle appears to be problematic in estrogen treated animals. Increasing the animal number from five to ten animals per sex and per dose level did not increase the sensitivity of detection of flutamide or EE2 induced endocrine-mediated effects above the level already obtained by histological examination of groups of five animals.

Care must be taken to include in the future version of the OECD TG 407 only those additional endpoints that have been proven to contribute to a reliable and sensitive detection of endocrine-mediated effects. Thorough evaluation of the results of 20 studies on 10 different compounds known or suspected to interact with the endocrine system through different mechanisms that have been tested in the validation phase 2 will provide a sound basis for the identification of the most appropriate enhancements.

ACKNOWLEDGMENT

The authors wish to thank EMSG for substantial financial support of this investigation.

REFERENCES

1. OECD. *OECD Protocol for Investigating the Efficacy of the Enhanced TG 407 Test Guideline, Rationale for the Investigation and Description of the Protocol*, Paris (1999).

2. P. Andrews, A. Freyberger, E. Hartmann, R. Eiben, I. Loof, U. Schmidt, M. Temerowski, M. Becka. *Arch. Toxicol.* **75**, 65–73 (2001).
3. OECD. *OECD Protocol for Investigating the Efficacy of the Enhanced TG407 Test Guideline (phase 2), Rationale for the Investigation and Description of the Protocol*, Paris (2000).
4. P. Andrews, A. Freyberger, E. Hartmann, R. Eiben, I. Loof, U. Schmidt, M. Temerowski, A. Folkerts, B. Stahl, M. Kayser. *Arch. Toxicol.* **76**, 194–202 (2002).

Pure Appl. Chem., Vol. 75, Nos. 11–12, pp. 2491–2495, 2003.

Workshop 4.3

Application of rat medium-term bioassays for detecting carcinogenic and modifying potentials of endocrine active substances*

Katsumi Imaida[1,‡], Seiko Tamano[2], Akihiro Hagiwara[2], Shoji Fukushima[3], Tomoyuki Shirai[4], and Nobuyuki Ito[5]

[1]*Faculty of Medicine, Kagawa Medical University, 1750-1 Ikenobe, Miki-cho, Kita-gun, Kagawa 761-0793, Japan;* [2]*Daiyu-kai Institute of Medical Sciences, 64 Gura, Nishiazai, Azai-cho, Ichinomiya 491-0113, Japan;* [3]*Osaka City University Medical School, 1-4-3 Asahi-machi, Abeno-ku, Osaka 545-8585, Japan;* [4]*Nagoya City University Graduate School of Medical Sciences, 1 Kawasumi, Mizuho-cho, Mizuho-ku, Nagoya, 467-8601, Japan;* [5]*Nagoya City University Medical School, 1 Kawasumi, Mizuho-cho, Mizuho-ku, Nagoya, 467-8601, Japan*

Abstract: Two in vivo bioassay methods, a rat medium-term liver bioassay and a rat multi-organ bioassay, can be used for detecting carcinogenic or modifying potentials of endocrine active substances (EASs) on endocrine disruption (ED). The first bioassay, the rat medium-term liver bioassay, is fundamentally based on the two-step hypothesis of liver carcinogenesis; initiation with diethylnitrosamine (DEN, 200 mg/kg b.w., ip) is followed by test chemical administrations during the second stage, in combination with 2/3 partial hepatectomy. It requires only eight weeks for animal experimental treatment and a further few weeks for quantitative analysis of immunohistochemically demonstrated gluthathione-*S*-transeferase placental form positive hepatic foci. A total of 313 chemicals/substances have already been analyzed, and the efficacy of the system for hepatocarcinogenesis has thereby been well established. This bioassay also provides information concerning dose responses and inhibitory potentials of test chemicals. Several possible EASs, most of them categorized as pesticides, have already been examined in this bioassay, and dose–response studies of nonylphenol, bisphenol A, and styrene have also been tested. Another bioassay, a medium-term, multi-organ bioassay system, using five different chemical carcinogens—DEN, MNU, BBN, DMH, and DHPN—has also been established for rapid detection of not only hepatocarcinogens, but also other organ-targeted carcinogens. These medium-term bioassays are particularly useful and reliable methods for detecting carcinogenic or modifying potentials of low doses of test chemicals, such as EASs, and these methods can be used for the effects of chemical mixtures of EASs.

INTRODUCTION

The risk of carcinogenic or modifying potentials of endocrine active substances (EASs) on endocrine disruption (ED) to humans is of great concern. It is urgently necessary to establish practical and reli-

*Report from a SCOPE/IUPAC project: Implication of Endocrine Active Substances for Human and Wildlife (J. Miyamoto and J. Burger, editors). Other reports are published in this issue, *Pure Appl. Chem.* **75**, 1617–2615 (2003).

‡Corresponding author

able bioassay methods for detecting carcinogenic and modifying potentials of EASs, which are effective at very low doses. Two in vivo bioassay methods, a rat medium-term liver bioassay and a rat multi-organ bioassay, can be used for detecting those risks from EASs.

RAT MEDIUM-TERM LIVER BIOASSAY FOR CARCINOGENS

Male 6-week-old F344 rats were initially given a single ip injection of diethylnitrosamine (DEN, 200 mg/kg body weight, b.w.) dissolved in saline to initiate hepatocarcinogenesis (Fig. 1). Two weeks later, animals received test chemicals and were subjected to two-thirds partial hepatectomy (PH) at week 3. The animals were sacrificed for quantitative analysis of glutathione-*S*-transferase placental form (GST-P) positive liver foci at week 8. Carcinogenic or modifying potentials of test chemicals are scored by comparing the numbers and areas per cm^2 of induced GST-P-positive foci in the livers of test chemical-treated groups with those of corresponding control groups given DEN alone. A positive response is defined as a single increase in the quantitative values of GST-P-positive foci, a negative response is defined as no change or a decrease. The results obtained have been compared with reported *Salmonella*/microsome and long-term carcinogenicity test findings for the same compounds.

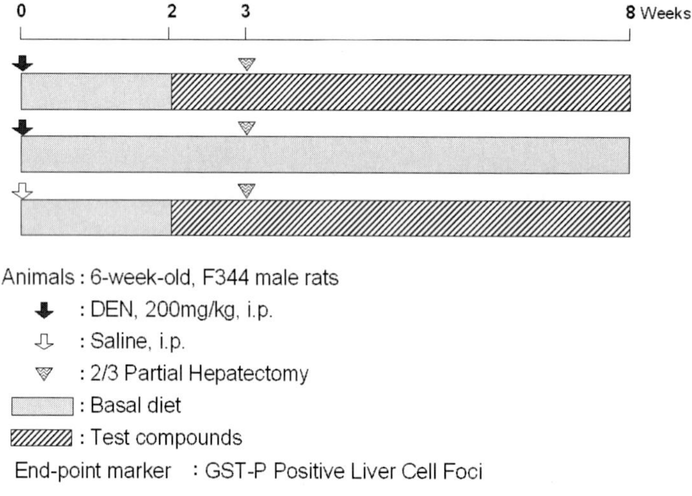

Fig. 1 Rat medium-term liver bioassay for carcinogens.

Results of 313 chemicals in the medium-term liver bioassay

Of a total of 313 chemicals examined, 60 out of 65 known hepatocarcinogens (92 %) gave positive results (Table 1). Five hepatocarcinogens that proved negative all belonged to the peroxizome proliferator group that depresses GST-P expression. Carcinogens targeting organs other than liver gave fewer positive results (10 out of 43, 23 %). One of the 48 chemicals reported as noncarcinogenic was found to be positive in this assay, but this might suggest that the chemical is a liver tumor promoter rather than the finding being a false-positive.

Table 1 Positive rates for 313 compounds of different categories (%).

Test compound	Ames test			Total
	+	–	Unknown	
Hepatocarcinogen	30/31 (97)	29/33 (88)	1/1 (1000)	60/65[a] (92)
Nonhepatocarcinogen	7/26 (27)	2/15 (13)	1/2 (50)	10/43 (23)
Noncarcinogen	0/6 (0)	1/40 (3)	0/2 (0)	1/48[b] (2)
Unknown	4/14 (29)	30/86 (36)	14/57 (24)	48/157 (31)
Total	41/77 (53)	62/174 (35)	16/62 (25)	119/313 (38)

[a]Negative; 5 peroxisomal proliferators, such as clofibrate, and DEHP, etc.
[b]Positive; malathione.

Results of possible EASs, which have been tested in the rat medium-term bioassay

Table 2 showed the results of possible EAS compounds, which have already been examined in the bioassay. Most of all chemicals are categorized as pesticides, but some other chemicals, such as hormone-related medicines, have also been examined. Eight chemicals have been showed positive, and two chemicals showed negative in this bioassay.

Table 2 Results of possible EASs analyzed in rat medium-term liver bioassay.

EAS	Dose (ppm)	Route	GST-P positive foci Number	(cm^2)	Area (mm^2/cm^2)		Results
Alachlor	2000	D	10.49***	(6.98)	1.33**	(0.69)	↑
Aldrin	50	D	8.46***	(4.35)	0.83***	(0.33)	↑
Atrazine	500	D	7.72	(6.98)	0.79	(0.67)	–
Benomyl	5000	D	6.86	(6.46)	0.46	(0.44)	–
Chlordane	500	D	3.59	(3.25)	0.49**	(0.25)	↑
DDT	10	D	12.17***	(7.09)	0.88**	(0.52)	↑
Dieldrin	100	D	14.41***	(9.09)	1.41***	(0.77)	↑
Permethrin	4000	D	11.51**	(8.86)	1.03	(0.80)	↑
Trifluralin	5000	D	14.36***	(6.82)	1.26*	(0.52)	↑
Vinclozolin	2000	D	16.08***	(8.32)	1.48***	(0.72)	↑

*$P < 0.05$
**$P < 0.01$
*** $P < 0.001$
( ): Respective control values

Dose–response study of nonylphenol in the medium-term liver bioassay

Dose–response studies of nonylphenol have been completed in this medium-term liver bioassay. Male F344 rats were given a single ip injection of DEN at a dose of 200 mg/kg b.w., and starting two weeks later, received nonylphenol at doses of 2000, 250, and 25 ppm in the basal diet. Quantitative data (numbers and areas) for GST-P-positive liver foci in each treated group and the control group were 4.7, 4.2, 6.1, and 6.1/cm^2; and 0.31, 0.30, 0.41, and 0.44 mm^2/cm^2, respectively. These results indicate that nonylphenol did not show clear dose response on induction of GST-P-positive foci, preneoplastic liver lesions, at the doses examined.

Dose–response study of bisphenol A and styrene in the medium-term liver bioassay

Male F344 rats were given DEN, ip, and then animals received bisphenol A at doses of 40 and 160 mg/kg b.w., ig, or styrene at doses of 250 and 1000 mg/kg b.w., ig, six times per week, for six weeks. Quantitative values of GST-P-positive liver foci in bisphenol A created groups and in the control group were 4.9, 5.2, and 6.0/cm^2 and 0.39, 0.42, and 0.42 mm^2/cm^2, respectively. Those values are not statistically different between groups. The GST-P values in styrene-treated and the control groups were 3.3, 1.5, and 6.0/cm^2 and 0.37, 0.10, and 0.42 mm^2/cm^2, respectively, and these values are statistically different between groups. Therefore, the results indicate that styrene showed dose–response inhibitory effects on induction of GST-P-positive liver foci at doses examined.

Since styrene showed inhibitory effects in the previous study, further low-dose studies were conducted in the same bioassay. After injection of DEN, animals received styrene at doses of 0.0006, 0.006, and 0.6 mg/kg b.w., ig, six times per week, for six weeks. The lowest dose of styrene examined was the almost same level found in our environment. The GST-P values in styrene-treated and the control groups were 8.3, 8.7, 9.0, and 8.4/cm^2 and 0.80, 0.67, 0.84, and 0.74 mm^2/cm^2, respectively. These values were not statistically different, and indicate that at the levels found in our environment, styrene did not show any modifying effect on the induction of GST-P liver foci.

RAT MEDIUM-TERM MULTIORGAN BIOASSAY

Male 6-week-old F344 rats were treated sequentially with five carcinogens (DEN, 100 mg/kg b.w. in saline, ip, single dose at the commencement; *N*-methyl-*N*-nitrosourea, 20 mg/kg b.w. in citrated-buffered solution, ip, four doses on days 2, 5, 8, 11; dihydroxy-di-*N*-propylnitrosamine, 0.1 % in drinking water during weeks 3 and 4, *N*-butyl-*N*-(4-hydroxybutyl)nitrosamine, 0.05 % in drinking water during weeks 1 and 2; 1,2-dimethylhydrazine, 40 mg/kg b.w. in saline, sc, four doses on days 14, 17, 20, 23 (DMBDD treatment) (Fig. 2). After those treatment, the animals were given test substances for 24 weeks from week 5. All animals were sacrificed and subjected to complete necropsy, and all organs/tissues were histopathologically and immunohistochemically examined.

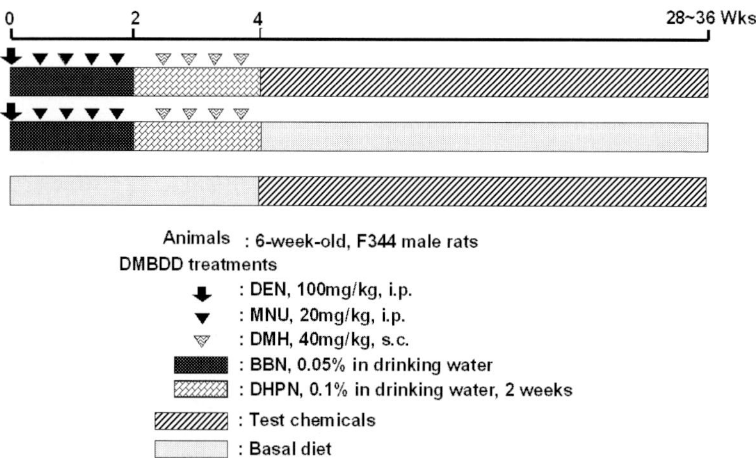

Fig. 2 Rat multi-organ bioassay (DMBDD methods).

Results of 63 chemicals in the medium-term multiorgan bioassay

Sixty-three chemicals have been tested in this bioassay (Table 3). All 17 hepatocarcinogens (100 %) and 19/22 (86 %) of the nonhepatocarcinogens were positive in the bioassay. Five noncarcinogens were negative. For chemicals with unknown carcinogenicity, the positive rate was 9/19 (47 %). This bioassay is useful for analysis of carcinogenic or modifying potential of test chemicals when their target organs are other than liver. This bioassay system can also be useful for dose–response studies, including at very low doses, and can be used for analyzing risk of carcinogenic potentials of EDCs at low doses.

Table 3 Results of 63 test compounds in a rat multiorgan bioassay for carcinogens.

Category of chemicals	Ames' test (%)			
	Positive	Negative	Unknown	Total
Hepatocarcinogen	12/12 (100)	5/5 (100)	0/0 (0)	17/17 (100)
Nonhepatocarcinogen	10/11 (91)	8/10 (80)	1/1 (100)	19/22 (86)
Noncarcinogen	0/1 (0)	0/4 (0)	0/0 (0)	0/5 (0)
Unknown	0/1 (0)	6/11 (55)	3/7 (43)	9/19 (47)
Total	22/25 (88)	19/30 (63)	4/8 (50)	45/63 (71)

CONCLUSIONS

These medium-term liver and multiorgan bioassay systems are very useful tools for detection of not only genotoxic but also nongenotoxic carcinogens. Positive results obtained in a relatively short period closely correlate with the long-term carcinogenicity test. The bioassays are particularly useful and reliable methods for detecting carcinogenic or modifying potentials of low doses of test chemicals, such as EASs, and these methods can also be used for the effects of chemical mixtures of EASs.

REFERENCES

1. N. Ito, K. Imaida, M. Asamoto, T. Shirai. *Mutat. Res.* **462**, 209–217 (2000).
2. M. A. Moore, H. Tsuda, S. Tamano, A. Hagiwara, K. Imaida, T. Shirai, N. Ito. *Toxicol. Pathol.* **27**, 237–242 (1999).
3. T. Shirai, M. Hirose, N. Ito. *IARC Sci. Publ.* **146**, 251–272 (1999).
4. N. Ito, R. Hasegawa, K. Imaida, M. Hirose, T. Shirai. *Exp. Toxicol. Pathol.* **48**, 113–119 1996 (1996).
5. N. Ito, R. Hasegawa, K. Imaida, M. Hirose, M. Asamoto, T. Shirai. *Crit. Rev. Oncol. Hematol.* **21**, 105–133 (1995).
6. N. Ito, R. Hasegawa, K. Imaida, S. Takahashi, T. Shirai. *Drug Metab. Rev.* **26**, 431–442 (1994).
7. R. Hasegawa and N. Ito. *Food Chem. Toxicol.* **30**, 979–992 (1992).
8. N. Ito, T. Shirai, R. Hasegawa. *IARC Sci. Publ.* **116**, 353–388 (1992).
9. N. Ito, K. Imaida, R. Hasegawa, H. Tsuda. *Crit. Rev. Toxicol.* 19, 385–415 (1989).
10. N. Ito, H. Tsuda, M. Tatematsu, T. Inoue, Y. Tagawa, T. Aoki, S. Uwagawa, M. Kagawa, T. Ogiso, T. Masui, et al. *Carcinogenesis* **9**, 387–394 (1988).

Pure Appl. Chem., Vol. 75, Nos. 11–12, pp. 2497–2501, 2003.

Workshop 4.4

Enhanced one-generation reproductive toxicity study in rats for detecting endocrine-disrupting effects of chemicals*

Hiroaki Aoyama[1,‡] and Katsushi Suzuki[2]

[1]*Laboratory of Reproductive Toxicology, Institute of Environmental Toxicology, Japan;* [2]*Department of Veterinary Physiology, Nippon Veterinary and Animal Science University, Japan*

Abstract: An enhanced one-generation reproductive toxicity study in rats without adjusting a litter size during the lactation period is proposed as a rapid and reliable bioassay for providing the data concerning adverse and/or low-dose effects of suspected endocrine disruptors. In this study, pregnant females are treated with the test substance from gestation day 0 through lactation day 21, in principle. F1 offspring from one-half of the litters in each dose group are killed and necropsied at weaning, while those from the remaining litters are examined for sexual maturation, estrous cyclicity, and/or sperm production. A series of pilot studies with ethynylestradiol as a reference chemical have suggested that the exposure of estrogenic chemicals during the early gestation period is critical for detecting effects on fertilization and/or implantation of eggs and survival of implants, and that expression of some genes including AR in the prostate and IGF-1 in the uterus of F1 offspring may be sensitive markers for monitoring potential estrogenic effects of the test compound.

INTRODUCTION

Not a few chemicals are suspected of having endocrine-disrupting effects on living organisms. These include pesticides and/or their active metabolites (*o,p′*-DDT, *p,p′*-DDE, methoxychlor, and vinclozolin), phenols (bisphenol A and nonylphenol), and phthalates (DEHP and DBP). Some of these chemicals have been confirmed to interact with estrogen and/or androgen receptors by in vitro and in vivo screening assays [1–4]. Otherwise, the compounds are suggested to modulate the activity of key enzymes to synthesize steroid hormones [5–8]. As for many endocrine active compounds and/or suspected endocrine disruptors, however, adversity of their endocrine effects remains unclear at present. This paper focuses on the rapid and reliable bioassay for predicting adverse reproductive effects of chemicals on human health based on endocrine and other mechanisms for the future risk assessment.

AVAILABILITY OF A CURRENT TWO-GENERATION REPRODUCTION STUDY FOR DETECTING ADVERSE REPRODUCTIVE EFFECTS OF ENDOCRINE DISRUPTORS

The most common way to detect adverse effects of suspected endocrine disruptors may be conducting a two-generation reproduction study in rats according to the authorized guidelines [9–11]. Although the

---

*Report from a SCOPE/IUPAC project: Implication of Endocrine Active Substances for Human and Wildlife (J. Miyamoto and J. Burger, editors). Other reports are published in this issue, *Pure Appl. Chem.* **75**, 1617–2615 (2003).

‡Corresponding author

assay is made to examine reproductive toxicities of chemicals regardless of their underlying mechanism(s), this assay is expected to be able to detect adverse effects of endocrine disruptors. We have conducted a series of in vitro and in vivo experiments including a two-generation reproduction study to confirm the availability of these assays for use in evaluating endocrine-disrupting effects of pesticides [12]. Our results demonstrated that adverse reproductive effects of an estrogenic pesticide, methoxychlor, could be sufficiently detected by a current two-generation reproduction study, in which reproductive parameters in parental females such as estrous cyclicity, pregnancy rates, and numbers of implants and pups delivered as well as male reproductive endpoints of sperm counts were clearly affected dose-dependently (Table 1). Our results also suggest that the measurement of uterine weights of F1 and F2 female weanlings is useful to evaluate the potential estrogenicity of the test compounds. In addition, adverse effects of antiandrogenic pesticide, vinclozolin, have also been successfully evaluated by the two-generation reproduction study [13].

Table 1 Results of a two-generation reproductive toxicity study in rats with methoxychlor.

Findings	Dose levels (ppm)		
	10	500	1500
General toxicity on parental animals			
Reduced body weights/body weight gains	–	++	+++
Reduced food consumption	–	++	+++
Reduced accessory sex organ weights in males	–	–	+
Reproductive toxicity on parental animals			
Prolonged estrous cycle in females	–	+	++
Decreased sperm counts in males	–	–	+
Reduced fertility (pregnancy rate) in females	–	–	+
Decreased number of implantation sites	–	+	++
Decreased number of pups delivered	–	+	++
Toxicity on offspring			
Reduced body weights in pups	–	+	++
Reduced thymus weights in weanlings	–	+	++
Additional endpoints			
Decreased estradiol concentration in parental females	–	+	+
Increased uterine weights in female weanlings	–	++	+++

The observations described above suggest that the current two-generation reproduction study can be used for evaluating the adversity of treatment-related effects that are caused by endocrine active compounds and/or suspected endocrine disruptors. However, this assay may not always be the best assay to screen the adverse effect(s) of suspected endocrine disruptors. One of the disadvantages of a current two-generation reproduction study is the fact that the assay takes relatively a long time (approximately 36 weeks) to obtain the result, so that only the limited number of studies can be conducted. In other words, we can evaluate only a limited number of suspected endocrine disruptors closely. Secondary, a small number of offspring (usually 1/sex/litter) being examined at adulthood may cause (1) missing low incidence finding(s) and/or (2) overestimation of chemical effect(s) due to unexpectedly deviated data from an "odd fellow" in a certain litter of a certain group. For example, we experienced malformed offspring due to spontaneous mutation occurring in the breeding colony of the supplier during the course of conducting developmental and reproductive toxicity studies [14,15]. Matsumoto et al. [16] also reported the presence of spontaneous abnormalities in sperm production in a certain strain of rats. If the mutant characters and/or strain-specific spontaneous abnormalities became evident after maturation and if scientists might select the pups carrying the concealing abnormalities for post weaning examination as a representative of the litter without knowing the fact, these may lead scientists to misinterpret the toxicity of test compound. Contrary, if the test substance actually induced a low incidence of abnormalities that became evident after weaning and if scientists might conduct the postweaning examination by using a normal pup as a representative of the litter, this may again cause a misinterpretation of the result.

AVAILABILITY OF AN ENHANCED ONE-GENERATION REPRODUCTIVE TOXICITY STUDY FOR DETECTING ADVERSE REPRODUCTIVE EFFECTS OF ENDOCRINE DISRUPTORS

Studies have shown that adverse reproductive outcome can be elucidated by the treatment of animals with endocrine disruptors during the period of sexual differentiation regardless of the underlying mechanisms [17–20]. Based on these facts and those described above, we propose an enhanced one-generation reproductive toxicity study in rats without adjusting a litter size during the lactation period as a candidate for rapid and reliable bioassay for providing precise adverse effect data. Animals in our enhanced one-generation study are treated with a suspected endocrine disruptor during the entire period of gestation and lactation (from gestation day 0 through lactation day 21), which is followed by postmortem examination of dams after weaning of pups and observation of F1 offspring at weaning, during sexual maturation, and at terminal sacrifice after sexual maturation (Fig. 1). The exposure period for the offspring can be extended if necessary. Essential endpoints include fertility and gestation indices of dams, numbers of implants and pups delivered, sex ratio and AGD of pups, viability index of pups during the lactation period, gross pathology of dams, and observations of F1 offspring such as organ weights at weaning and after maturation, sexual maturation (vaginal opening/preputial separation), estrous cyclicity, sperm analysis and gross- and histopathological examinations. Optional endpoints such as histopathological examination of dams and hormone measurements and/or quantitative analysis of mRNA expression in the target organs of offspring may be added flexibly according to the suspected endocrine mechanism(s) of the test compound.

Pilot studies have been conducted to confirm the availability of this assay by using ethynylestradiol (EE) as a reference chemical. We conducted a series of experiments in which the basic protocol (animal husbandry, dose levels of EE, administration route, endpoints, and the timing of each observation) was kept unchanged while only the treatment period was variable among the three. An administration route (subcutaneous injection) and dose levels (0, 0.01, 0.03, 0.10, 0.30, and 1.00 µg/kg/day) were determined according to the protocol used in OECD (Organization for Economic Cooperation and Development) validation exercises of rat uterotrophic assay [21]. The results demonstrated that the ef-

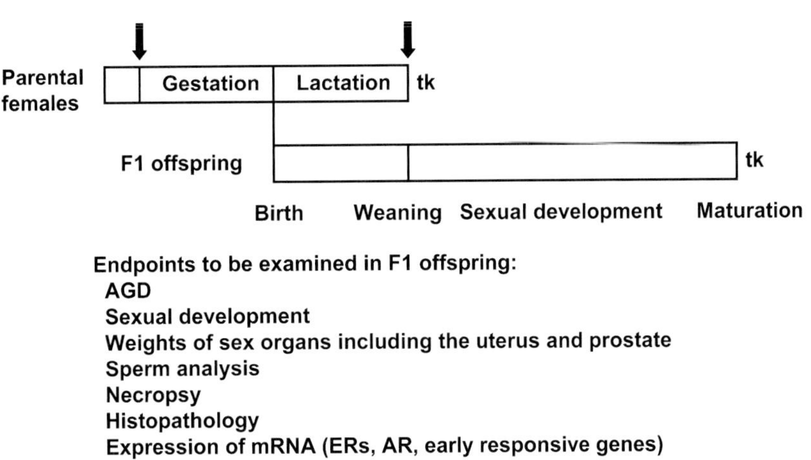

Fig. 1 Schematic explanation of a trans-generation assay.

fect of EE on maternal fertility (inhibition of pregnancy and reduction of litter size) could be detected when the chemical was administered to dams throughout the gestation and lactation period (Table 2), while the same treatment of dams during the organogenetic or perinatal period never disclosed this effect. The sensitivity of this endpoint in the gestational and lactational exposure study was almost equivalent to that of uterine weights in the uterotrophic assay. These results suggest that the exposure of estrogenic chemicals during the early gestation period (day 0 through day 5 of gestation) is critical for detecting effects on fertilization and/or implantation of eggs or survival of implants, and that animals should be treated throughout the gestation and lactation period in the future one-generation reproductive toxicity study for evaluating adverse effects of suspected endocrine disruptors with uncertain mechanism(s) of action. In the present studies, no clear adverse effects were observed in F1 offspring in terms of sexual differentiation, organ weights, female estrous cyclicity, and male sperm production. However, anogenital distances in female newborns were slightly increased, and expressions of mRNA in certain genes were up- or downregulated in all treated groups. Although toxicological meanings of these effects remain to be elucidated at present, the present results suggest that the expression of some genes including AR in the prostate and IGF-1 in the uterus may be sensitive markers for monitoring potential estrogenic effects of the test compound.

Table 2 Results of an enhanced one-generation reproductive toxicity study in rats with ethynylestradiol.

	Dose levels (μg/kg/day)				
Findings	0.01	0.03	0.10	0.30	1.00
Effects on dams					
Reduced fertility (pregnancy rate)	–	–	–	–	+++
Decreased number of implants	–	–	–	–	+++
Decreased number of pups delivered	–	–	–	++	a
Effects on F1 offspring					
Increased AGD in female newborns	+	+	+	b	
Down- or up-regulated mRNA expression[c]	+	+	+		

[a]Data not available because no dam in this group became pregnant.
[b]Excluded from evaluation because only a few pups were obtained in this group.
[c]Expression of mRNA was quantitatively analyzed in the prostate (ER-α, ER-β, AR and IGF-1) and the uterus (ER-α, ER-β, IGF-1 and IL-6) of F1 offspring at weaning and after sexual maturation.

ACKNOWLEDGMENTS

The works presented here were conducted at the Institute of Environmental Toxicology, Hatano Research Institute of Food and Drug Safety Center, and Safety Research Institute for Chemical Compounds, Co. Ltd. under the support from the Ministry of Agriculture, Forestry and Fishery, Japan and the Ministry of Environment, Japan. The authors are grateful to all persons in charge of the project.

REFERENCES

1. J. Nishikawa, K. Saito, J. Goto, F. Dakeyama, M. Matsuo, T. Nishihara. *Toxicol. Appl. Pharmacol.* **154**, 76–83 (1999).
2. M. D. Shelby, R. R. Newbold, D. B. Tully, K. Chae, V. L. Davis. *Emviron. Health Perspect.* **104**, 1296–1300 (1996).
3. W. R. Kelce, C. R. Stone, S. C. Laws, L. E. Gray, J. A. Kemppainen, E. M. Wilson. *Nature* **375**, 581–585 (1995).
4. W. R. Kelce, L. E. Gray, E. M. Wilson. *Reprod. Fertil. Dev.* **10**, 105–111 (1998).

5. B. J. Davis, R. R. Maronpot, J. J. Heindel. *Toxicol. Appl. Pharmacol.* **128**, 216–223 (1994).
6. B. J. Davis, R. Weaver, L. J. Gaines, J. J. Heindel. *Toxicol. Appl. Pharmacol.* **128**, 224–228 (1994).
7. M. Ema, E. Miyawaki, K. Kawashima. *Toxicol. Lett.* **111**, 271–278 (2000).
8. J. C. Corton, C. Bocos, E. S. Moreno, A. Merritt, R. C. Cattley, J. A. Gustafsson. *Biochimie* **79**, 151–162 (1997).
9. MAFF in Japan. Test guidelines for agricultural chemicals, 12-Nousan-No. 8147 (Japanese) (2000).
10. OECD. OECD Guidelines for Testing of Chemicals, Section 4, No. 416 (2001).
11. USEPA. Health Effects Test Guidelines, OPPTS 870,3800 (1998).
12. H. Aoyama, R. E. Chapin, K. Ebino, S. Teramoto. *Cong. Anom.* **39**, 156 (1999).
13. I. Matsuura, T. Saitoh, M. Ashina, Y. Wako, H. Iwata, M. Namiki, N. Hoshino, Y. Ishizuka, O. Katsuta, K. Takahashi, M. Tsuchitani, Y. Ikeda, M. Miura. Abstracts from annual meeting of the Japan Society of Endocrine Disruptors Research, p. 243 (2002).
14. H. Aoyama, S. Teramoto, Y. Shirasu. *Teratology* **37**, 159–166 (1988).
15. M. Kaneda, S. Teramoto, Y. Shirasu. *Teratology* **40**, 77–84 (1989).
16. K. Matsumoto, S. Matsumoto, T. Yoshida, Y. Ohshima. *J. Toxicol. Sci.* **24**, 63–68 (1999).
17. K. S. Hirsch, E. R. Adams, D. G. Hoffman, J. K. Markham, N. V. Owen. *Toxicol. Appl. Pharmacol.* **86**, 391–399 (1986).
18. R. E. Chapin, M. W. Harris, B. J. Davis, S. M. Word, R. E. Wilson, M. A. Mauney, A. C. Lockhart, R. J. Smialowicz, V. C. Moser, L. T. Burka, B. J. Collins. *Fundam. Appl. Toxicol.* **40**, 138–157 (1997).
19. L. E. Gray, C. Wolf, C. Lambright, P. Mann, M. Price, R. L. Cooper, J. Ostby. *Toxicol. Ind. Health* **15**, 94–118 (1999).
20. L. E. Gray, J. Ostby, J. Furr, M. Price, D. N. Veeramachaneni, L. Parks. *Toxicol. Sci.* **58**, 350–365 (2000).
21. J. Kanno, L. Onyon, J. Haseman, P. Fenner-Crisp, J. Ashby, W. Owens. *Environ. Health Perspect.* **109**, 785–794 (2001).

WORKSHOP 5

PRECAUTIONARY PRINCIPLE/APPROACH AND WEIGHT OF EVIDENCE IN ENDOCRINE DISRUPTOR ISSUES

Pure Appl. Chem., Vol. 75, Nos. 11–12, pp. 2505–2513, 2003.

Workshop 5.1

Making decisions in the 21st century: Scientific data, weight of evidence, and the precautionary principle*

Joanna Burger‡

Environmental and Occupational Health Sciences Institute, and Consortium for Risk Evaluation with Stakeholder Participation, Division of Life Sciences, Cell Biology and Neurosciences, Rutgers University, Piscataway, NJ 08854, USA

Abstract: Traditionally, science has progressed by slow steps involving the accumulation of studies showing particular effects, leading eventually to a general consensus. However, with increasing development and industrialization, environmental problems have escalated faster than the ability to collect sufficient data to form clear consensus among scientists. Since managers require scientific information to make decisions about management, regulation, and public policy, the gap has been partially filled by two approaches: weight of evidence and the precautionary principle. I suggest that both are useful for making decisions about endocrine active substances, although few papers in the refereed literature link the precautionary principle with endocrine active substances. As with most public policy decisions, these involve an iterative process whereby scientific inquiry must continue to fill data gaps, and to determine if the decisions made by these processes are still appropriate and protective of human and ecological health. The precautionary principle is most useful when it continues to inform and help direct research to fill data gaps in our understanding of environmental problems, such as the effect of endocrine active substances on endocrine disruption.

INTRODUCTION

There is worldwide concern for global climate change, yet there are many global changes that involve increases in human populations, shifts in their distribution, concentration of people along coasts, shifts in land use, increases in the temporary movement of people, massive industrial and suburban development, and increases in technology with environmental consequences. Increasing industrialization, technology, and human populations clearly place people and their ecosystems at risk, particularly with respect to environmental degradation and contamination. Although the width of the dense population band along coasts may vary, increasing development places demands on fragile land–ocean margins and associated ecosystems, partly because it is in this region that transfer and transportation of products occur, allowing for the possibility of environmental contamination. In the coming years, the human dimensions of environmental health sciences, and conservation and protection of biodiversity will gain even more importance as global changes in population size and distribution, land use, and increased use of chemicals occur in many societal domains (agriculture, medicine, industry). While the intrinsic value

*Report from a SCOPE/IUPAC project: Implication of Endocrine Active Substances for Human and Wildlife (J. Miyamoto and J. Burger, editors). Other reports are published in this issue, *Pure Appl. Chem.* **75**, 1617–2615 (2003).

‡E-mail: burger@biology.rutgers.edu.

of ecosystem protection and biodiversity are clear to some [1,2], the human health consequences of increased reliance on chemicals, petrochemicals, and pharmaceuticals are clear to all [3].

In the past, we have relied on sound scientific data to reach solutions for environmental problems, and to restore ecosystems. It is no longer enough to decry the continual loss and degradation of habitat and biodiversity. We must seek creative solutions that are advantageous both for protecting human health and the environment. However, the pace of technological development, the increase in the number and widespread use of chemicals, and the time required to adequately test these chemicals, petrochemicals, and pharmaceuticals increases, hence decisions must often be made before the data are all available. In the past, environmental agencies have relied on a cause–effect evidence, derived from evidence, hopefully leading to certainty. However, often this is no longer the case. Since managers require scientific information to make decisions about management, regulation, and public policy, the gap has been filled by two approaches: weight of evidence and precautionary principle. These two approaches should involve an iterative process whereby regulatory actions are taken, while scientific inquiry must continue to fill data gaps, and to determine if the decisions made by these processes are still appropriate and protective of human and ecological health.

In this paper, I explore weight of evidence and the precautionary principle as they are generally used, contrast them, and propose that both are useful for making decisions about endocrine active substances, with the caveat that scientific inquiry must continue, both to increase our knowledge base and to determine whether regulatory actions were protective of human and ecological health. Below, I will discuss briefly why making decisions about endocrine active substances requires these two approaches, define and describe some of the ways the two approaches have been used generally, and finally return to discuss briefly how they will be particularly useful for making decisions about endocrine active substances.

SPECIAL CASE OF ENDOCRINE ACTIVE SUBSTANCES

Governmental agencies, the private sector, and the public are increasingly interested in assessing the well-being of both humans, and other species within their ecosystems. Human health risk assessment and ecological risk assessment have emerged as separate paradigms embodying the disciplines of toxicology and exposure assessment [4,5]. In many cases, the two risk assessors either examine human health [6,7] or ecological health [8–14], although a few volumes have included both human and ecological risk assessment [15]. There have been some attempts to show the interconnections between human and ecosystem health [16,17], but these usually have not dealt with methodology. An important development, however, has been the development of conceptual models for exposure in food chains that include pathways for human exposure [18].

Endocrine active substances provide an excellent example of an environmental issue that has important consequences for both human and ecological health. The question of whether environmental contaminants are inducing adverse health effects in humans and wildlife because of disruptions to the endocrine system is one of the more important questions of our age, and is, for example, one of the highest research priorities of the U.S. Environmental Protection Agency [19]. Endpoints for assessment of the effects of endocrine active substances include multigenerational measures. Endocrine active substances in the broadest sense include not only man-made chemicals, but natural agents that occur in the diet [20,21], however, it is often interpreted as referring only to man-made chemicals. A major concern is that persistent bioaccumulating chemicals affect fetal development by acting like estrogens or antiestrogens [22].

Initially, concerns arose because of the observation that some synthetic chemicals in the environment were associated with adverse developmental and reproductive effects in wildlife [23]. The effects of exposure (in utero) to the potent estrogen diethylstilbestrol (DES) in children of treated women (and later tests with treated mice) further led to concern. The pesticide DDT interfered with female reproduction in birds. There is information on associations between synthetic chemical and adverse biological outcomes that are normally mediated by the endocrine system [23], although the mechanisms are not always clear [21]. While it is always difficult, costly, and time-consuming to determine the cause and effect of a chemical in the environment, the task for endocrine active substances is made more difficult by the mixture of chemicals humans and wildlife are normally exposed to [21]. It is not the purpose of this paper to discuss the scientific data, but to examine the current state of our knowledge and the consequences of the state of the science.

The main difficulty is the conflict between the need on the part of governmental agencies and the public for concrete data demonstrating the links between specific chemicals and outcomes and the state of our knowledge regarding endocrine active substances. We need or want to know more than science seems able to produce. However, the potential consequences of waiting to make a decision until every aspect is scientifically proven beyond a shadow of doubt may be severe if endocrine disruption is occurring in humans and wildlife, and if the effects are intergenerational. Thus, there is a need to adopt methods that will allow reasonable decisions in the face of uncertainties and knowledge gaps.

Below, I discuss two methods of dealing with knowledge gaps and uncertainties regarding the effects, mechanisms, and magnitude of effects of chemicals that act as endocrine active substances to cause endocrine disruption. While quantitative risk assessment provides another approach [24], often the data necessary for such assessments are not available.

WEIGHT OF EVIDENCE

The phrase "weight of evidence" is fashionable, ambiguous, and difficult to define. It suggests a quantitative ranking of evidence, or the qualitative appraisal of many different forms of evidence to arrive at a conclusion [25]. Sometimes, much of the evidence is qualitative or otherwise not suitable for statistical treatment. Evidence derives from epidemiological and clinical studies, long-term laboratory assays, and predictive short-term tests, the latter two with animal models [25–27]. Epidemiology can provide useful information on human populations, but it has the disadvantages of lacking rigorous controls, has difficulties determining exposure, and is usually reactive, rather than preventive [27]. A preponderance of data used in weight-of-evidence approaches are derived from animals tests, often with mammals.

Weight-of-evidence approaches to consensus often comes from scientific committees that are empowered by organizations such as the United Nations, International Agency for Research on Cancer, International Program on Chemical Safety, International Labor Organization, World Health Organization, and the Scientific Committee for Problems of the Environment, among others. Within countries, such committees are organized by academies of science, such as the U.S. National Academy of Sciences [5,9,21]. Such committees function by convening a diverse group of scientists to evaluate the evidence and reach consensus views—they weigh the evidence [27].

In the case of endocrine disruption, much of the weight of evidence comes from epidemiology studies of wildlife in nature [23]. While such studies can provide epidemiological evidence, they do not shed light on the mechanisms [21]; this must await controlled laboratory experiments. Further, a weight-of-evidence approach, with endocrine active chemicals or any other, can only be used when there is evidence or studies to consider. Where such evidence exists, scientific committees can pull the data together to support a weight-of-evidence conclusion, which can be used for regulatory purposes [28].

PRECAUTIONARY PRINCIPLE

"In order to protect the environment, the precautionary approach shall be widely applied by states according to their capabilities. Where there are threats of serious or irreversible damage, lack of full scientific certainty shall not be used as a reason for postponing cost-effective measures to prevent environmental degradation." (Rio Declaration of the United Nations Conference on Environmental Development, 1992, see Applegate [29]).

This definition partly reflected an articulation by Bergen [30]. The precautionary principle is invoked when there is reason to assume that chemicals or technology introduced into the environment may create hazards, either directly or indirectly, to humans or other receptors. There does not have to be evidence of a causal relationship [31]. The hazard has to be more than plausible and usually severe and irreversible. The precautionary principle is a way of dealing with uncertainties or where there is a clear knowledge deficit [32–35]. It deals with uncertainty by "staying on the safe side" [36].

The precautionary principle is a leading principle in some environmental law, such as in Germany [37] and in much of Europe [38], where it is sometimes explicitly referred to. It is embodied in some U.S. legislation (e.g., Toxic Substances Control Act), but not by name. In the United States, it does not offer legal standing [38]. It was adopted by the countries in the charter of the European Union (EU Treaty, 1993 [29]). They wish to use the precautionary principle to reduce the use of toxic chemicals by developing safe industrial technologies and invoking criminal sanctions for suppression and manipulation of information about chemicals and their effects [39]. The precautionary principle only recently has entered environmental policy debates in the United States [29]. Applegate [29] argued that while U.S. law contains some elements of the precautionary principle, precaution is usually balanced against cost.

There are four main elements to the precautionary principle: (1) trigger for regulatory action, (2) timing of the regulatory action, (3) nature of the regulatory response, and (4) a regulatory strategy [29]. While Applegate and others have focused on the regulatory nature of the precautionary principle, others have focused more on environmental management. Recently, Kriebel et al. [40] suggested a different set of guidelines suggested by the precautionary principle: (1) taking preventive action in the face of uncertainly, (2) shifting the burden of proof to those who want to do an action or use a chemical, (3) exploring a range of alternatives, and (4) increasing public participation in decisions and management. In other words, the principle suggests that regulators, policy-makers, and law-makers should err on the side of caution.

The precautionary principle has been invoked in such diverse areas as conservation [41], salmon farming and other fisheries [42,43], marine ecosystems [44], occupational medicine standards [45], and breast cancer reduction [46]. In the latter case, Davis et al. [46] suggested that prudent precautionary principles suggest that reducing exposure to avoidable risk factors should receive high priority. To some extent, it is this aspect that has been applied to the issue of endocrine active substances. The precautionary principle can provide guidance for regulatory initiatives to reduce the risks caused by chemicals [47], particularly to children [48].

Remarkably, in the refereed literature the precautionary principle has not been linked with endocrine active substances. For example, a search of papers on MedLine for the last five years yielded 252 papers on the precautionary principle, but none linked either weight of evidence or precautionary principle with the term. Both terms, however, have been used extensively, although mention of the precautionary principle began only in the mid-1980s (Fig. 1, MedLine). An environmental science and pollution search indicated only four papers linking endocrine disruption with the precautionary principle, and seven linking endocrine disruption with weight of evidence. The two terms have been used extensively mainly over the last ten years in other contexts (Table 1). Similarly, weight of evidence has been mentioned less over the last five years, compared to the precautionary principle (Fig. 2, Cambridge Scienctific Abstracts, environmental science and pollution search).

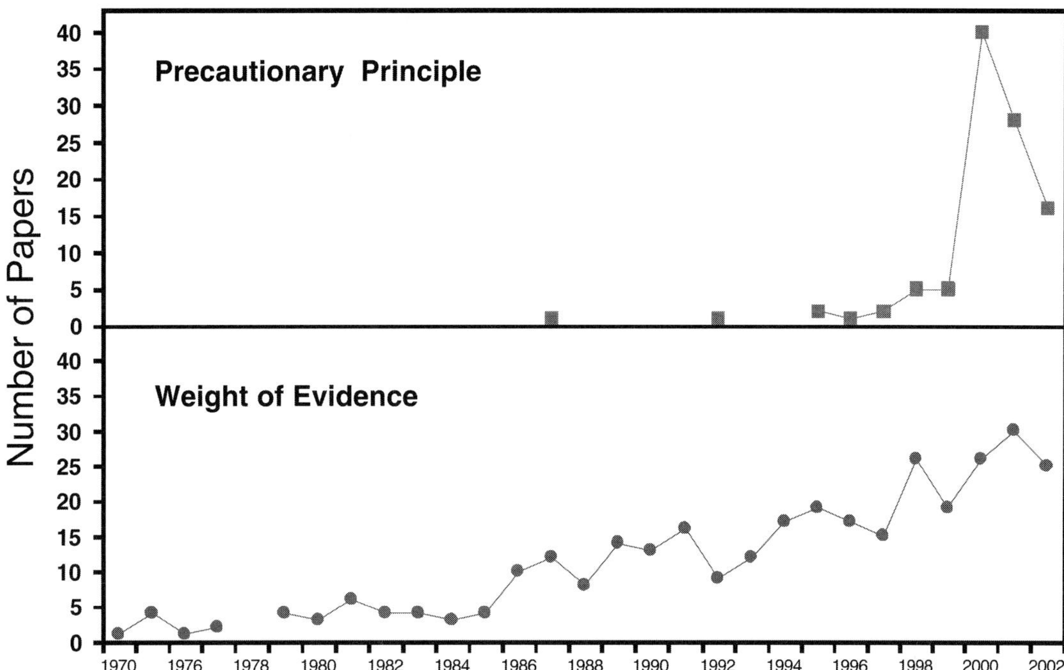

Fig. 1 Number of papers mentioning the precautionary principle and weight-of-evidence approaches, from a MedLine search.

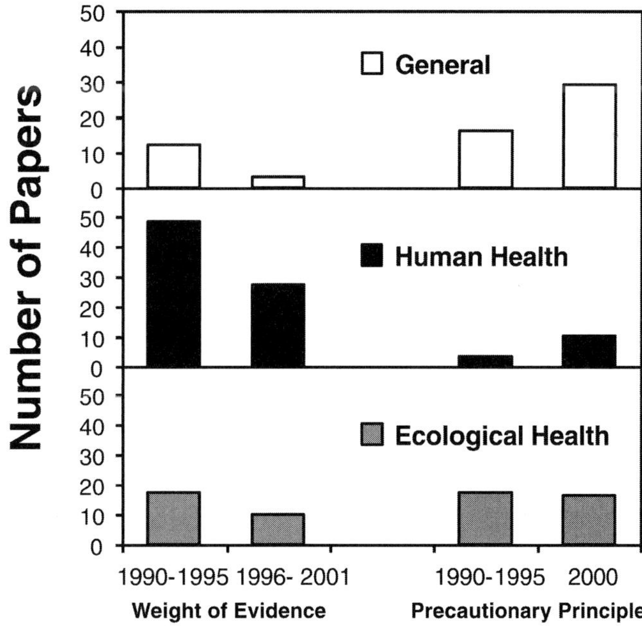

Fig. 2 Number of papers mentioning the precautionary principle and weight-of-evidence approaches, from a Cambridge Scientific Abstracts search.

Table 1 References to weight of evidence and precautionary principle in the CSA–Environmental Science and Pollution database.

	Weight of evidence		Precautionary principle	
Number of papers	1990–1995	1996–2001	1990–1995	1996–2001
General				
Laws, regulations, and guidelines	2	1	3	6
Risk assessment	5	1	3	8
General theory	2	1	6	13
Global warming	3		4	2
Total	12	3	16	29
Human health				
Endocrine disruptors	3	4		2
Chemicals	28	18	2	5
Physical agents	2			
Carcinogenicity	5	1	1	3
Neurotoxicity	3			
Occupational health	1			
Social (seat belts, EMF)	2	1		
Epidemiology	3	2		
Surgery/medicine	3			
Toxicity testing	1	1		
Total	48	27	3	10
Ecological health				
Endocrine disruptors				2
Marine ecosystems and pollution	2	3	7	7
Marine ecosystem function	1	2	2	1
Freshwater and pollution	4	4	1	3
Freshwater and chemicals/pollutants	2		1	
Invertebrates	3			1
Fish	3			1
Other vertebrates	1		1	
Sustainability/biodiversity	1	1	5	1
Total	17	10	17	16

CONCLUSIONS

Recently, Goldstein [49] argued that the precautionary principle is really the modern formulation of the Hippocratic principle that says "above all do no harm". Since human health is tied to global health, there are, he argues, reasons to act cautiously. Yet, the imprecise definition of the principle restricts its usefulness as a goal. Further, he argues persuasively that the use of this principle should not preclude further scientific inquiry to further our knowledge of the initial problem, and to ascertain whether our precautionary actions were warranted [49]. The "principle" and scientific research are not antithetical [50], as Holm and Harris [51] have lamented. This seems to be an important aspect to bear in mind, particularly for endocrine active substances where the scientific evidence is often contradictory.

Goldstein's [49] arguments, and those of others, suggest that while the precautionary principle may be useful in formulating our current public policy decisions regarding endocrine active substances, caution would also argue that we continue to conduct research to determine cause-and-effect and mechanisms [52]. While European and American environmental agencies are acting on the precautionary principle with respect to possible endocrine disruption [19], it is clear that on-going research on endocrine disruption is critical to our future actions. Uncertainties in themselves should not stifle future research, but instead should inform and stimulate research.

There are some criticisms of the precautionary principle, including that current regulations are already precautionary, it is not scientifically sound because it advocates making decisions without scientific justification, and it might stifle new technologies [40]. While these are not unfounded, there are situations where it might be difficult to obtain sufficient scientific justification until it is too late to prevent disaster, particularly in the case of ecological receptors (such as endangered species). The precautionary principle is useful in public policy when the failure to act may cause potentially serious of irreversible threats to health or the environment [35].

In summary, there is a continuum of three approaches that are used for regulation: traditional scientific data leading to regulation, weight of evidence, and the precautionary principle. I would argue that all three approaches are useful in the realm of management, regulation, and policy-making (Fig. 3). The problem is determining when to use each of the approaches. Clearly, scientific certainty and the degree of consequences influence when each is used.

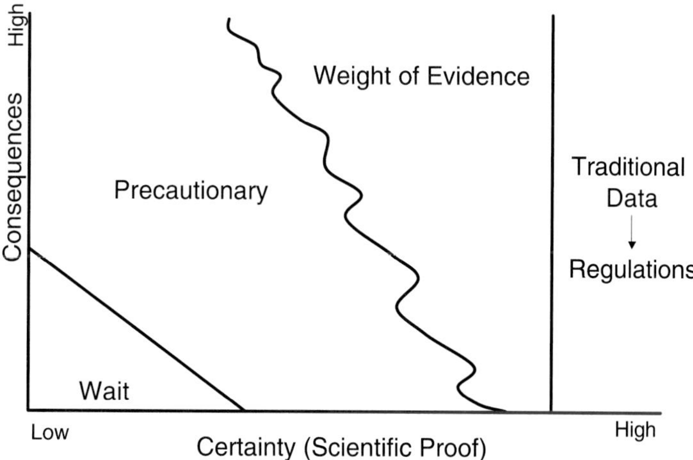

Fig. 3 Schematic of the relationship between differing approaches to risk management based on traditional scientific data, weight of evidence, and precautionary principle as a function of certainty and consequences.

ACKNOWLEDGMENTS

I thank the many people who have discussed these topics with me over the years, including M. Gochfeld, B. D. Goldstein, B. Friendlander, C. Powers, and C. Safina. This research was funded by the International Union of Pure and Applied Chemistry, the Japanese government, National Institute of Environmental Health Sciences (ESO 5022), the Consortium for Risk Evaluation with Stakeholder Participation (CRESP) through the Department of Energy (AI # DE-FC01-95EW55084, DE-FG 26-00NT 40938), and the Environmental and Occupational Health Sciences Institute.

LITERATURE CITED

1. C. R. Tracy and P. F. Brussard. *Ecol. Appl.* **4**, 205–207 (1994).
2. J. Bengtsson, H. Jones, H. Setala. *TREE* **12**, 334–336 (1997).
3. M. Gochfeld. In *Environmental Medicine*, S. Brooks, M. Gochfeld. J. Herzstein, M. Schenker, R. Jackson (Eds.), pp. 3–8, Mosby, New York (1995).
4. National Research Council (NRC). *Risk Assessment in the Federal Government*, National Academy Press, Washington, DC (1983).

5. National Research Council. *Issues in Risk Assessment*, National Academy Press, Washington, DC (1993).
6. National Research Council. *Science and Judgement in Risk Assessment*, National Academy Press, Washington, DC (1994).
7. M. L. Mendelsohn, J. P. Peeters, M. J. Normandy. *Biomarkers and Occupational Health*, Joseph Henry Press, Washington, DC (1995).
8. P. J. Sheehan, D. R. Miller, G. C. Butler, P. Bourdeau (Eds.). *Effects of Pollutants at the Ecosystem Level*, Wiley, Chichester, UK (1984).
9. National Research Council. *Ecological Knowledge and Environmental Problem Solving*, National Academy Press, Washington, DC (1986).
10. S. M. Bartell, R. H. Gardner, R. V. O'Neill. *Ecological Risk Estimation*, Lewis Publishers, Boca Raton, FL (1992).
11. J. Cairns, Jr., B. R. Niederlehner, D. R. Orvos. *Predicting Ecosystem Risk*, Princeton Scientific, Princeton, NJ (1992).
12. G. W. Suter II (Ed.). *Ecological Risk Assessment*, Lewis Publishers, Boca Raton, FL (1993).
13. D. Peakall. *Animal Biomarkers as Pollution Indicators*, Chapman & Hall, London (1992).
14. R. A. Linthurst, P. Bourdeau, R. G. Tardiff. *Methods to Assess the Effects of Chemicals on Ecosystems*, Wiley, Chichester, UK (1995).
15. V. B. Vouk, G. C. Butler, D. G. Hoel, D. B. Peakall (Eds.). *Methods for Estimating Risk of Chemical Injury: Human and Non-human Biota and Ecosystems*, Wiley, Chichester, UK (1985).
16. J. Burger and M. Gochfeld. In *Interconnections Between Human and Ecosystem Health*, R. T. DiGiulio and E. Monosson (Eds.), pp. 127–148, Chapman & Hall, London (1996).
17. R. T. DiGiulio and E. Monosson. *Interconnections Between Human and Ecosystem Health*, Chapman & Hall, London (1996).
18. G. W. Suter II. *Environ. Health Perspect.* **105**, 1282–1283 (1997).
19. R. J. Kavlock. *Chemosphere* **39**, 1227–1236 (1999).
20. G. E. Dunaif, S. S. Olin, J. Scimeca, J. A. Thomas. *Human Diet and Endocrine Modulation: Estrogenic and Androgenic Effects*, ILSI Press, Washington DC (1998).
21. National Research Council. *Hormonally Active Agents in the Environment*, National Academy Press, Washington, DC (1999).
22. S. Ringvold and J. A. Rottingen. *Tidsskr Nor. Laegeforen* **117**, 66–70 (1997).
23. T. Colburn and C. Clement (Eds.) 1992. *Chemically-induced Alterations in Sexual and Functional Development: The Wildlife/human Connection*, Princeton Scientific Publishing, Princeton, NJ (1992).
24. A. Tukker. *Risk Anal.* **22**, 821–832 (2002).
25. R. L. Carter. *Hum. Toxicol.* **7**, 411–418 (1988).
26. E. Somers. *Regul. Toxicol. Pharmacol.* **4**, 99–106 (1984).
27. E. Somers. *Regul. Toxicol. Pharmacol.* **6**, 388–398 (1986).
28. W. E. Wagner. *Hum. Ecol. Risk Assess.* **6**, 459–477 (2000).
29. J. S. Applegate. *Hum. Ecol. Risk Assess.* **6**, 413–443 (2000).
30. J. Bergen. *Intl. Environ. L.* **1**, 429 (1990).
31. E. Hagenah. *Environ. Pollut.* **100**, 13–18 (1999).
32. A. Blowers. *Urban Stud.* **30**, 775–796 (1992).
33. P. Healey and T. Shaw. The treatment of environment by planners: evolving concepts and policies in development plans. Univ. of Newcastle upon Tyne, Department of Town and Country Planning, Newcastle upon Tyne (1993).
34. European Commission (EC). Communication from the Commission on the precautionary principle. COM, Brussels (2000).
35. European Environment Agency (EEA). Late lessons from early warnings: the precautionary principle 1896-2000. European Communities, Luxembourg (2001).

36. D. Counsell. *Environ. Plan. Manage.* 42, 45–61 (1999).
37. F. H. Lesner. In *Handworterbuch des Umweltrechts*, O. Kimminich, H. F. Von Lersner, P. Ch. Storm (Eds.), pp. 2701–2710, Erich Schmidt, Berlin, Germany (1999).
38. P. Sand. *Int. J. Global Legal Stud.* **1**, 293–323 (1994).
39. S. S. Epstein. *Int. J. Health Serv.* **30**, 353–371 (2000).
40. D. Kriebel, J. Tickner, P. Epstein, J. Lemons, R. Levins, E. L. Loechler, M. Quinn, R. Rudel, T. Schettler, M. Stoto. *Environ. Health Perspect.* **109**, 871–876 (2001)
41. W. Kuhlmann. *Wild Earth Fall* 67–71 (1997).
42. A. Ross. *Mar. Pollut. Bull.* **20**, 372–374 (1989).
43. R. M. Peterman and M. M'Gonigle. *Mar. Pollut. Bull.* **24**, 231–234 (1992).
44. P. Johnston, D. Santillo, R. Stringer. *Nat. Res. For.* **23**, 157–167 (1999).
45. A. Strijkel and L. Reijnders. *Occup. Environ. Med.* **52**, 304–312 (1995).
46. D. L. Davis, D. Axelrod, L. Bailey, M. Gaynor, A. J. Sasco. *Environ. Health Perspect.* **106**, 523–529 (1998).
47. L. G. Ahrens and E. Salter-Green. *Int. J. Occup. Environ. Health* **6**, 289–295 (2000).
48. J. A. Tickner and P. Hoppin. *Int. J. Occup. Environ. Health* **6**, 281–288 (2000).
49. B. D. Goldstein. *Risk Policy Rep.* March, 39–40 (2000).
50. B. D. Goldstein. *Environ. Health Perspect.* **107**, 594–595 (1999)
51. S. Holm and J. Harris. *Nature* **400**, 398 (1999).
52. M. Giampietro. *Ambio* **31**, 466–470 (2002).

Pure Appl. Chem., Vol. 75, Nos. 11–12, pp. 2515–2519, 2003.

Workshop 5.2

Precautionary principle and endocrine active substances*

Bernard D. Goldstein‡

University of Pittsburgh, Graduate School of Public Health, 103 DeSoto Street, Pittsburgh, PA 15261, USA

Abstract: The precautionary principle has been central to many of the debates concerning the appropriate approach to the threat posed by endocrine active substances (EASs). This newly emerging principle has been applied to issues as diverse as persistent organic pollutants and the European trade barrier on beef from hormone-treated cattle.

INTRODUCTION

There is growing interest in the precautionary principle as both a rationale and a legal mechanism for control of agents potentially harmful to the environment and to public health. Described below is an overview of certain of the aspects of the precautionary principle pertinent to the difficult issues posed by compounds with effects on the endocrine system [1].

One of the earliest major international statements of the precautionary principle was in the 1992 Rio Declaration: "Nations shall use the precautionary approach to protect the environment. Where there are threats of serious or irreversible damage, scientific uncertainty shall not be used to postpone cost-effective measures to prevent environmental degradation" [2].

The focus on protective action despite scientific uncertainty is central to the many formulations of the precautionary principle. More recent definitions have extended the precautionary principle to include protection of public health as well as the environment. In some cases, the definitions seem to lower the bar for action under the precautionary principle by not requiring the "serious or irreversible damage" or "cost-effectiveness" found in the Rio definition, e.g., the Wingspread statement [3].

Of note is that the European Community (EC) has recently published a major document on the precautionary principle in which it does not provide a definition, instead in essence calling for regulatory actions to be judged by the extent to which they have a precautionary impact [4]. This failure to define a term being advocated for use in international treaties has led to some cynicism among U.S. officials who tend to see the precautionary principle as an excuse for European trade protection [5], a point recognized by European Union Environmental Commissioner Wallstrom, who stated, "We do not spend our days in Brussels—as some might think—in Machiavellian plotting to apply precaution to the detriment of U.S. businesses" [6].

THE PRECAUTIONARY PRINCIPLE AND THE SAFETY OF BEEF FROM HORMONE-TREATED CATTLE

Part of the American cynicism about the precautionary principle stems from a trade dispute about beef from hormone-treated cattle in which endocrine effects are a central issue. The World Trade

*Report from a SCOPE/IUPAC project: Implication of Endocrine Active Substances for Human and Wildlife (J. Miyamoto and J. Burger, editors). Other reports are published in this issue, *Pure Appl. Chem.* **75**, 1617–2615 (2003).

‡Tel.: 412 624 3001; E-mail: bdgold@pitt.edu

Organization (WTO) responded to a complaint by the United States and Canada that the EC was unfairly excluding beef derived from animals treated with growth hormone. The EC argued in part that the precautionary principle provided an adequate justification for their exclusion, that under this principle there was sufficient evidence of the possibility of an adverse health effect, including cancer, to ban importation of beef from hormone-treated animals without a formal risk assessment. This led the WTO to consider whether the precautionary principle was an established principle of law applicable to adjudication of international trade disputes. The WTO ruling against the EC's ban on beef from hormone-treated animals was also upheld by the WTO Appellate Body [7,8]. The WTO appeared to support standard risk assessment approaches and to reject the argument that the precautionary principle was a reasonable basis for trade barriers, at least at that time. But the ruling was complicated by a number of considerations. Among the other U.S. and Canadian arguments most pertinent to the present paper were that the EC was inconsistent in not having limits on the levels of residues of similar hormones present naturally in foods, that an EC scientific body previously evaluated the hormone residue issue and had not found significant evidence of risk, and that the Codex Alimentarius Commission failed to support the EC position on beef from hormone-treated cattle as being a consumer risk. In addition, the WTO seemed to agree with the U.S. and Canadian argument that a major reason for the EC allowing potentially carcinogenic growth-promoting antibiotics in swine but not potentially carcinogenic growth-promoting hormones in cattle was its surplus of beef but not pork products, i.e., it was a trade barrier, not a health issue.

THE PRECAUTIONARY PRINCIPLE, DISTRUST OF SCIENCE AND GOVERNMENT, RISK PERCEPTION, AND EASs

Increasing distrust of science and of government is cited as a reason for the acceptance of the precautionary principle. This is said to be particularly true for Europe where a series of incidents such as "mad cow disease" in Britain and unnecessary HIV in the hemophilia population in France have undermined public confidence in governmental science and health pronouncements. The preamble to the 2000 EC Communication on the Precautionary Principle states "The Communication recalls that a number of recent events have undermined the confidence of public opinion and consumers..." [4]. In the United States, frustration with the slow pace of science and risk-based regulation of hazardous air pollutants fostered changes in the regulation of hazardous air pollutants in the 1990 Clean Air Act Amendments that fit well under the precautionary principle. These include a shift in the burden of proof so that it is now necessary to demonstrate the safety of a chemical rather than its risk of harm, and a primary reliance on maximum available control technology rather than risk assessment as a regulatory control device. This distrust and frustration is understandable. All too often, industry and government have used scientific uncertainty as a means to delay unwanted but necessary regulation [9,10].

In the case concerning beef from hormone-treated animals, the European Community argued that the public's perception of the risk due to hormone residues in their food, separately from scientific factors, is a reasonable basis for banning this product. While in this case, the WTO appellate body appeared to implicitly reject the proposition that risk perception can substitute for or overcome scientific evidence of risk, a subsequent WTO appellate body ruling appears to have brought risk perception back into acceptability [11,12].

Environmental groups have been particularly active and effective in expressing concern about the potential harmfulness of EASs and have couched their concerns in the context of the precautionary principle. This perception of hormonal interactions as being a particularly dangerous problem worthy of precautionary action might well account for the rapid and widespread public acceptance of the purported marked synergistic effects of hormones [13], a finding that needed to be withdrawn after it could not be replicated. In essence, the public has been primed to expect that the endocrine system is particularly susceptible to environmental agents and that subtle effects in this system could have wide ranging health and environmental consequences. The extent to which this is true is the subject of much other discussion within this volume. Whatever the case, the obvious danger in regulating on the basis of perception is that manipulation of perception is the stock in trade of both politicians and advertising managers, and such manipulation counters the potential value of scientific enquiry into the truth of these important issues.

THE PRECAUTIONARY PRINCIPLE, RISK ASSESSMENT, RISK MANAGEMENT, AND SCIENTIFIC RESEARCH

As a simplification, two types of actions can be discussed as falling under the precautionary principle: those that add additional levels of protection within standard risk assessment and risk management approaches (e.g., additional prudence in default assumptions, more stringent safety factors, and acceptable risk levels); and those that go beyond usual risk-based approaches (e.g., shifts in the burden of proof, actions based upon hazard without sufficient information to assign risk) [12]. Within the standard risk paradigm, a key issue for EASs that cause cancer at high doses through hormonal effects is whether they should be held to the standard conservative assumption that any single molecule of a carcinogen can cause cancer. This supposition is pertinent to issues concerning the mechanism by which hormonal agents produce cancer—a tumor-promoting effect would, at least to some, argue that a one-hit cancer causation model is inappropriate.

One of the concerns about the precautionary principle is that it will downgrade the value of research that in the long run could lead to definitive answers concerning the risk of EASs [14,15]. There are two reasons for this concern. Firstly, some advocates of the precautionary principle view risk science as antithetical to postmodern democracy and risk assessors as a technocracy at the beck and call of industrial interests [16]. Secondly, once an action has been taken under the precautionary principle, there appears to be a tendency for the research funding agency to look for new problems to study, rather than persist to find out if the precautionary action is justified. Invoking the precautionary principle by definition means that there is a finite probability that an erroneous action with significant societal and/or economic cost has been taken—if there were certainty, or the cost was minimal, there would be no need to invoke the precautionary principle. In fact, it is a truism that the more precautionary a society, the more likely it is to make costly mistakes [14]. Accordingly, the precautionary principle would seem to provide a major justification for basic mechanistic research and for research to discover if the precautionary action was in fact justified. It would be particularly problematic to erroneously take a precautionary action for a concern such as the disappearance of amphibian species, which has been ascribed to EASs, but not do the follow up research needed to discover if the action was erroneous as the effect on amphibians was due to some other factor. Unfortunately, the experience of a marked decline in support for hazardous air pollutant research following the passage of the 1990 U.S. Clean Air Act Amendments suggests that regulating on the basis of the precautionary principle may well interfere with obtaining the needed understanding to make appropriate decisions.

Another potential weakness of the use of the precautionary principle is also exemplified by EASs—the difficulty in assessing the potential for net positive trade-offs between the valuable and negative aspects of a situation. This has been central to the argument about genetically modified foods and malnutrition [17]. Similarly, agents capable of modulating hormonal action can have a suite of beneficial and nonbeneficial effects. This is particularly seen for hormone replacement therapy and for agents capable of treating or preventing breast cancer through hormonal effects. Davis et al. presented a framework for viewing the impact of complex hormonal interactions on breast cancer risks through the lens of the precautionary principle while pointing out that certain phytoestrogens appear to be protective [18]. And a recent *New York Times* editorial argues for the continued use of DDT for the prevention of malaria despite the acknowledged harmful effects of DDT [19].

It is unclear whether the current interest in the precautionary principle reflects a move toward heightened concern about health and safety in developed countries in keeping with increased longevity and a feeling of health entitlement. An alternate explanation may be a world view that prefers natural over synthetic products, that is willing to accept phytoestrogens with little or no concern, but wishes no possible exposure to "unnatural" sources of hormonal agents. The latter view is supported, at least in the United States, by the fact that in the case of herbal agents, many of which contain high levels of estrogenic substances to the point of toxicity [20], the burden of proof was actually shifted against precaution. The 1994 Dietary Supplement Act now requires the U.S. Food and Drug Administration prove harm before regulating these products.

One scientific approach that should come into increased use in a world in which the precautionary principle is a major regulatory approach is that of surveillance. Intervention before there are adverse effects can best be accomplished if there is sufficient investment in developing and measuring early indicators of exposure and effect, coupled with an understanding of human and ecosystem susceptibility. In the case of EASs, the decrease in most locales of body and ecosystem burdens of dioxins and PCBs is reassuring to the extent it suggests that the worst has been seen, but the seeming increase in burdens of polybrominated agents raises reasons for concern in keeping with an argument for precautionary action.

PERSISTANT ORGANIC POLLUTANTS

There are a number of facets of the EAS issue as it relates to persistent organic pollutants (POPs) that appear to be particularly pertinent to the precautionary principle. Notable about many of the agents that are under this heading is that their toxicity to humans remains less than fully convincing. The Seveso accident led to sufficient dioxin exposure to cause full-blown cases of chloracne, a skin condition that is diagnostic of significant body burdens of dioxins and related compounds. Yet there have been no clearly demonstrable long-term effects in the more than two decades of follow-up. One controversial finding has been a difference in birth ratio consistent with an EAS effect, but this is still unconfirmed and the issue of the effect of persistant organic pollutants (POPs) on sperm counts and birth ratios remains controversial [21,22]. While in some cases POPs are unquestionably animal carcinogens, and are capable of causing endocrine effects at high doses in laboratory tests, the evidence for adverse effects in humans remains controversial. For others, such as PCBs, the effects in humans at high doses are unquestionable and, at least in part, are expressed through endocrine disruption [1,23].

To some advocates of the precautionary principle, debates as to the human health effects of POPs are almost meaningless. The key issue is that these compounds are both persistent in the environment and harmful to ecosystems. Under the precautionary principle, such compounds should be banned without any further debate as to their effect in humans.

On the other hand, those arguing against the need for the precautionary principle as an additional regulatory approach can also use the history of the control of POPs. Traditional POPs such as PCBs are banned and no longer acceptable for use in OECD (Organization for Economic Cooperation and Development) countries. Most importantly, research has led to predictive approaches and assays that permit recognition of persistence as a characteristic of a new chemical before it is marketed. The availability of these predictive approaches, coupled with the significant legal and financial penalties that are now imposed on a chemical company that markets a persistent organic compound, make it highly unlikely that any reputable chemical company will move forward on developing or marketing such a compound.

REFERENCES

1. Y. Aoki. *Environ. Res.* **86**, 2–11 (2001).
2. United Nations Conference on Environment and Development. Final Declaration, Principle 15, Rio De Janeiro (1992).
3. C. A. Raffensberger and J. Tichner (Eds.). *Protecting Public Health and the Environment: Implementing the Precautionary Principle*, p. 353–354, Island Press, Washington, DC (1999).

4. Commission of the European Communities. *COM* 2000:1 (2000).
5. Codex Alimentarius Commission, FAO/WHO. Risk Analysis: 1) Working Principles for Risk Analysis. Additional Comments of the United States. CX/GP 00/3. Rome (2000).
6. M. Wallstrom. EU and US Approaches to Environment Policy – Are We Converging or Diverging? European Institute, Washington, DC, 25 April 2002 (2002). <www.eurunion.org/index.htm>.
7. EC Measures Concerning Meat and Meat Products (Hormones), WT/DS26//R/USA and, WT/DS48//R/CAN, Panel Report, 18 August 1997.
8. EC Measures Concerning Meat and Meat Products (Hormones), WT/DS26/AB/R, 1998.
9. S. Rampton and J. Stauber. *Trust Us, We're Experts: How Industry Manipulates Science and Gambles with your Future*, Penguin Putnam, New York (2002).
10. European Environmental Agency. Environmental issue report No. 22, Late Lessons from Early Warnings: the Precautionary Principle 1896–2000, Luxembourg, Office for Official Publications of the European Communities (2001).
11. European Communities – Measures Affecting Asbestos and Asbestos-Containing Products, WT/DS135/AB/R, Appellate Body Report, 12 March 2001.
12. Goldstein and Carruth, *Risk Analysis*. In press.
13. S. F. Arnold, D. M. Klotz, B. M. Collins, P. M. Vonier, L. J. Guillette Jr., J. A. McLachlan. *Science* **272**, 1489–1492 (1996). [Retracted publication].
14. B. D. Goldstein. *Environ. Health Perspect.* **107**, A594–A595 (1999).
15. B. D. Goldstein and R. S. Carruth. In *The Precautionary Principle: Implications for Research and Prevention in Environmental and Occupational Health*, proceedings of the Collegium Ramazzini, p. 60, Bologno, Italy (2002).
16. M. O'Brien. *Making Better Environmental Decisions*. MIT Press, Cambridge, MA (2000).
17. H. E. Cauvin. *New York Times* **52,231**, A5 (2002).
18. D. L. Davis, D. Axelrod, L. Bailey, M. Gaynor, A. J. Sasco. *Environ. Health Perspect.* **106**, 523–529 (1998).
19. Anonymous Editorial. *New York Times* **52,341**, A26 (2002).
20. R. S. DiPaola, H. Zhang, G. H. Lambert, R. Meeke, E. Licitra, M. M. Rafi, B. T. Zhu, H. Spaulding, S. Goodin, M. B. Toledano, W. N. Hait, M. A. Gallo. *N. Engl. J. Med.* **339**, 785–791 (1998).
21. S. H. Safe. *Environ. Health Perspect.* **108**, 487–493 (2000).
22. W. H. James. *Environ. Health Perspect.* **109**, A250 (2001).
23. Y. L. Guo, G. H. Lambert, C. C. Hsu. *Environ. Health Perspect.* **103** (Suppl. 6), 117–122 (1995).

Pure Appl. Chem., Vol. 75, Nos. 11–12, pp. 2521–2529, 2003.

Workshop 5.3

Why epidemiology of endocrine disruptors warrants the precautionary principle*

Michael Gochfeld‡

Environmental and Community Medicine, UMDNJ-Robert Wood Johnson Medical School and Environmental and Occupational Health Sciences Institute and Consortium for Risk Evaluation with Stakeholder Participation, 170 Frelinghuysen Road, Piscataway, NJ 08854, USA

Abstract: The precautionary principle is controversial, and critics invoke the need to wait for "sound science" before taking "costly" regulatory action. For human health effects, epidemiologic results are often considered more valuable than toxicologic studies in animals. Direct human evidence on the endocrine effects of environmental chemicals has been slow to accumulate because of inherent sample size limitations of exposed populations and over-conservative hypothesis testing approaches. Moreover, human health outcomes may take decades to emerge. Indeed, even huge population-based studies of hormone replacement therapy have been inconclusive regarding both benefits and risks. This paper argues that certain intrinsic standard epidemiologic methods are stacked to avoid making a type I error. Moreover, these combine with extrinsic limitations (long latency, high cost), leading me to conclude that reliance solely on epidemiology to provide definitive answers, will almost inevitably delay the discovery of meaningful associations warranting timely action for protection of public health. There are several ways in which the inherent conservatism of epidemiology is arrayed against preventative regulatory steps, hence a precautionary approach is warranted while awaiting the results of additional studies which for various reasons may be very long delayed or even impossible.

INTRODUCTION

Although formally established in the literature only during the 1990s, the precautionary principle is firmly entrenched in society, dating back to antiquity. Proverbs such as "better safe than sorry" and " an ounce of prevention is worth a pound of cure" are articulations of the precautionary principle. The precautionary principle emerged mainly in the context of new technologies such as bovine growth hormone and genetically modified organisms, but can also be extended to regulation of environmental pollutants including endocrine active substances. Applied to new technologies with uncertain adverse outcomes, it behooves those who will profit from a proposed technology to demonstrate conclusively its safety or that its benefits will outweigh its harm. In its most frequent incarnation, it means that the existence of uncertainties, unresolved facts, or the need for more research shall not be invoked to delay controls or regulatory action, particularly when the consequences of inaction or delay are serious or irreversible.

*Report from a SCOPE/IUPAC project: Implication of Endocrine Active Substances for Human and Wildlife (J. Miyamoto and J. Burger, editors). Other reports are published in this issue, *Pure Appl. Chem.* **75**, 1617–2615 (2003).

‡Tel.: 732 445 0123, ext. 627; Fax: 732-445-0130; E-mail: <gochfeld@eohsi.rutgers.edu>

The most frequent definition (Rio definition of 1992) is "Nations shall use the precautionary approach to protect the environment…. Where there are threats of serious or irreversible damage, scientific uncertainty shall not be used to postpone cost-effective measures to prevent environmental degradation." Or in simplest terms a potential hazard is not innocent until proven guilty. It is commonly stated that the precautionary principle has been adopted in Europe but abjured in the United States. Actually, a number of U.S. environmental statutes embody precautionary language. For example, the Toxic Substances Control Act (1977): "among the many chemical substances and mixtures which are constantly being developed and produced, there are some whose manufacture, processing, distribution in commerce, use, or disposal may present an unreasonable risk of injury to health or the environment…"

Agencies responsible for regulating environmental pollution on both the national and international scale set emissions standards or contaminant levels based, in part, on the results of scientific studies published in the peer-reviewed literature. Yet, even science-based standard setting is often affected by economic and technical considerations. Most agencies, the general public, and particularly scientists, like to think that standards are based primarily on science and that as new scientific information becomes available, standards will be revised. This, however, is not always (or even often) the case. In the United States, for example, the Occupational Safety and Health Standards are mostly based on pre-1970s science—attempts to update them having been stayed in court. Moreover, for some kinds of contaminants, particularly where effects are subtle or occur at very low levels, or where funding is not directed, scientific data are slow to accumulate.

Therefore, the precautionary principle is increasingly invoked to take regulatory action in the absence of incontrovertible scientific data demonstrating harm. The precautionary principle is considered controversial, and indeed those who oppose or caution about its application usually refer to it in the extreme sense of complete banning of a new technology or a toxic material. However, in practice precaution represents a spectrum of actions ranging from those that can be clearly and confidently based on adequate and sound data to interim actions taken while additional data are pending. Moreover, precautionary actions are themselves not irreversible. Taking preliminary remediation steps or delaying the introduction of new technologies may incur costs, but can be reversed when information on safety or risk become available. To extricate "precaution" from controversy, advocates now refer to "precautionary approaches" rather than a "principle".

Ideally, environmental health policy ought to reflect scientific consensus and risk assessment and management. Risk assessment is never completely free of uncertainties, and often it is necessary to articulate policies when science and risk assessment have yet to provide clear answers. I view risk assessment and precaution as complementary, rather than opposite approaches. There are many environmental health issues where uncertainties abound, and the role of endocrine active substances (EASs) and their consequences are fraught with uncertainties. The consequences of uncontrolled exposure can be significant, at least to individuals, yet the difficulty of demonstrating or validating connections between chemical and consequences, remains challenging, particularly in epidemiologic studies of humans. Since Colburn et al. [1] popularized the issue of "endocrine disruptors", opponents have argued that the much higher dose effects of intrinsic hormones such as estradiol (E2) or pharmacologic hormones, would mask or vitiate any possible effects of the environmental estrogens. This very point, however, strengthens the argument for a precautionary approach. The controversy over epidemiologic studies of hormone replacement therapy—where studies have involved many thousands of women—highlight the difficulty of arriving at scientific consensus. The main issue involves resolving the uncertain benefits vs. the uncertain harms [2,3].

EPIDEMIOLOGIC STUDIES OF ENVIRONMENTAL EASs

Although there are abundant epidemiologic studies of endogenous hormones and pharmacologic use of hormones (e.g., diethylstilbestrol and hormone replacement therapy), epidemiologic studies of endocrine affects in humans from environmental contaminants are relatively limited [4]. The organochlorine pesticides were probably the first environmental EASs recognized through their effects on avian reproduction. In humans, exposure to the pesticide DDT appeared to impair lactation in both U.S. and Mexican women [5]. Exposures to polychlorinated diphenyls and dibenzofurans in the Yu-Cheng (Taiwan 1978–79) disaster, was associated with a variety of developmental deficits including shortened penises [6], a finding reminiscent of Florida alligators exposed to organochlorines [7]. Bertazzi et al. [8] reported a low rate of breast and endometrial cancer in women exposed to dioxin after the Seveso (Italy) explosion, raising the question of an antiestrogenic effect of dioxin. The putative thyroid effects following an accidental exposure to polybrominated biphenyl compounds proved elusive [9]. Recently, Goh et al. [10] reported a complex relationship between occupational exposure to trichloroethylene and insulin levels, with no pattern for steroid levels. Probably the greatest recent controversy has been over the proposed relation of organochlorines to breast cancer, summarized by Arab (Topic 3.2).

PREMISE

In the face of controversy over the meaning of animal and human studies, the delay in reaching consensus, and the possibility of irreversible harm, the precautionary approach is viewed as a way of reducing exposure to EASs. The alternative wait-and-see approach, anticipates reassuring results when epidemiologic studies on human health impacts are finally completed decades from now. This paper examines the limits of epidemiology and why even apparently reassuring results (i.e., statistically negative or inconclusive) cannot be the sole factor influencing risk management. I arrive at the same conclusion as Tukker (2002), that epidemiology has "a limited role in leading to preventive action". I identify almost a dozen reasons for this limitation, only a few of which will be discussed here.

For the purpose of this paper I will assume that in the absence of a precautionary approach, no regulatory action would be taken without convincing epidemiologic evidence of causation, usually in the form of multiple, statistically significant studies in relevant populations.

Table 1 Aspects of the epidemiology literature which delay consensus on causation OR "Why epidemiology stacks the deck".

INTRINSIC TO METHODOLOGY
Alpha of 0.05 is too restrictive (avoiding type I error seems more important than avoiding type II error).
Focus on confidence limits retains the 0.05 level.
Willingness to question statistically significant results if sample size small.
Misclassification biases toward the null hypothesis, making type II error more likely.
Participation bias reduces power.
Multiple comparison corrections, over-correct, making type II error more likely.
Tendency to use two-tailed test, when one-tailed test is warranted makes type II error more likely.
Confounders are always available to take the blame and vitiate an association or delay its acceptance.
EXTRINSIC TO METHODOLOGY
Population size imposes a limit on power.
Opposition to meta-analysis.
Human health effects may have long latency.
Epidemiologic studies take several to many years to plan, execute, and publish.
Large-scale epidemiologic studies are prohibitively expensive.

EPIDEMIOLOGY "TO THE RESCUE?"

Before embarking on a critique of epidemiology from one who teaches the subject, it is important to recognize its value. In many aspects of environmental health, data obtained in animals are viewed with suspicion or skepticism when applied to humans. In the area of carcinogenesis, an agent that causes cancer in two test species, in both sexes, and in more than one organ, can be considered a probable human carcinogen. But only epidemiologic studies can validate definition as a "known" human carcinogen. In risk assessment, for example, there is a large and growing literature on the means and limitations of extrapolating from animal data to humans. And interspecies extrapolation is one of the main areas of uncertainty [11].

Epidemiology affords the opportunity to eliminate this uncertainty. Studies on asbestos and cancer, for example, have mainly been accomplished in humans. Likewise, arsenic and cancer are based on human data, since this is one carcinogen for which no animal model exists [12]. But the expectation that epidemiology can eliminate uncertainty and provide convincing answers for all audiences for all problems is an illusion. Epidemiology has severe limitations imposed by costs, methodology, and available populations [13].

INTRINSIC METHODOLOGIC ISSUES

Hypothesis testing in epidemiology and toxicology

In the mid-1900s, hypothesis testing emerged as a central paradigm in science. Scientists developed an idea or hypothesis, framed it in terms of a null hypothesis (Ho), gathered data observationally or experimentally, analyzed the data, and rejected or failed to reject the null hypothesis. Since the early 20th century, the 0.05 level was arbitrarily used as the criterion for rejecting the null hypothesis, so that rejecting the Ho at the alpha level of 0.05, meant that there was only a 5 % chance or one chance in 20 that the Ho was actually true (or should not have been rejected) [14]. This amounts to a 5 % probability of making a type I error (a false positive or rejecting the Ho when it is true and should not be rejected).

There are many fields where being 95 % percent confident of an experimental result is desirable or even necessary, but it is not a universal necessity. Moreover, this form of hypothesis testing does not allow an a priori definition of the converse probability of making a type II error, failing to reject the null hypothesis when it is false. This is contingent on the power of a study, power being the probability that if the Ho is false, the study will generate data that rejects the Ho. Rarely is power addressed in scientific publications, and often it cannot easily be calculated retrospectively. However, a power of 80 % is often considered desirable, meaning that even in a well-designed study, one has a 20 % chance of making a type II error. Thus, in the evolution of hypothesis testing, avoiding a type I error became more important—approximately four times more important—than avoiding a type II error. This means, in effect, that the deck is stacked against detecting an effect that is real.

Consider an example where drug A is being developed as a cure for disease D. A cautious investor might insist on being 95 % confident that A is indeed effective against D, but a consumer suffering from D would be content with a much lower level of confidence, such as 80 % or perhaps even 51 %. Thus, the traditional alpha of 0.05 is merely a statistical convention, not an absolute threshold for significance.

Hypothesis testing applies equally to toxicology and epidemiology, but in the former, at least, the experimenter has the capability—though not always the resources—to design studies of sufficient power.

Hypothesis testing has dominated the thinking of most scientists alive today. But simple hypothesis testing vis-à-vis the Ho, is already being replaced. Computerized techniques yielding absolute estimates of probability, afford new ways of looking at scientific data and providing confidence intervals around point estimates.

What does statistically significant mean

When we require a study to be statistically significant we look to see whether Ho is rejected at some a priori alpha level, usually the 0.05 level. Where effects are subtle or power is low, lack of statistical significance at the 0.05 level is not reassuring. Knowing the power of a study in advance is often valuable, but sobering. Far too many epidemiologic studies lack the power to reject the Ho of no association. Negative studies abound, with p values in the range of 0.06 to 0.20 which are interpreted as "negative" or "not significant", but which are actually potentially positive studies which simply lacked power. Indeed, performing a study that lacks power may be unethical, unless one can clearly indicate the lack of power.

Alternative alphas

Perhaps the easiest way to deal with the inherent statistical bias favoring type II errors in epidemiology is to set alternative alpha values or ignore alpha altogether. With small sample sizes and low power, an alpha of 0.05 is an invitation to a false-negative study. Setting alpha at 0.10 has been shown as a reasonable basis when a one-tailed test should have been invoked. It is also reasonable when a study lacks power to reject at the 0.05 level. In any case, readers should look with suspicion on "negative" studies particularly where p values are in the range of 0.05 to 0.10. Dispensing with alpha altogether is also reasonable, since modern statistical programs can calculate an absolute P value for any statistic. It is simple enough to publish the result and the P value and let the readers bring their own interpretation to the data.

CONFIDENCE LIMITS

Confidence limits have been extensively used in biometrics for nearly 20 years, with almost all epidemiologic publications on relative risks or odds ratios, including the 95 % confidence limits. If the confidence limits include 1 (i.e., the lower 95th % CL < 1), then this is interpreted as not statistically significant at the 0.05 level. Taking a hypothetical point estimate of 2.0 (95 % CL 0.5, 8.0), it is equally likely that the true value is 8.0 as it is that it is 0.5, and it is more likely than not that the true point estimate is above 1.0. Therefore, slavish reliance on confidence limits can be just as deceiving as simply rejecting or failing to reject the Ho.

Possible decision rule for those who insist on 0.05

- If Ho is rejected at the 0.05 level, then regulate or prevent exposure.
- If Ho is not rejected at the 0.05 level but the power is not adequate, then regulate or prevent exposure.
- If Ho is not rejected at the 0.05 level and power is adequate, then no regulation or prevention needed at this time [15].

Biological significance vs. statistical significance

One commonly hears that a particular study yields a statistically significant result, "but is it biologically significant?" That is, even in the face of statistical significance, readers may question the meaning of a result. It is important that the converse question gets equal billing—"even if it isn't statistically significant, does that mean that it's not biologically significant?" One of the commonest reasons for not achieving statistical significance is lack of power in the study design, usually because of small sample size. If a study achieves statistical significance (rejects the Ho), then it cannot, by definition lack power, even though the results may remain open to question and confirmation by additional studies.

Multiple comparisons

Not only does an alpha of 0.05 carry with it an unacceptably high probability of making a type II error, but there is another barrier imposed. Once one undertakes a study, particularly an expensive epidemiologic study, it is desirable to investigate as many endpoints or as many exposures as possible. Thus, if one is studying the effect of soy milk formulas on infant development comparing soy- to cow-based or breast-fed babies, a series of endpoints might be considered including height attainment, weight for height, bone ossification, cognitive development, and gonadal and genital development. If for example, only the latter were statistically significant at the 0.05 level, one might interpret this as evidence for a phytoestrogen effect. But skeptics would argue, "no", when you have six endpoints that are being tested, that increases the chance that any one of them might emerge as "significant" by chance alone, therefore you have to correct for multiple comparisons. Statisticians apply post hoc tests for multiple comparisons, for example, the Bonferroni test.

However, if you consider that each of the six hypotheses is an independent null hypothesis, any one of which might be biologically as well as statistically significant, it is apparent, that a multiple comparison test merely increases the probability of a type II error and is inherently fallacious. Rothman [16] provides a detailed statistical treatise on why one should not correct for multiple comparisons in most epidemiologic studies.

Confounders

Unlike multiple comparisons, confounding is a real complication in interpreting epidemiologic studies. There are many sources of confounding. When the study population is large and resources unlimited, it is often possible to stratify samples and correct for confounding. In reality, however, confounding often cannot be eliminated in a single study, and critics are quick to seize upon this. Confounding arises when both the independent variable (e.g., soy milk) and the dependent variable (growth rate) are correlated with a third variable that is not tested or testable. Socioeconomic status (SES) is a frequent confounder. If people with higher SES are more likely to go to pediatricians who recommend soy milk (more expensive) and also enjoy better nutrition and medical care, then a relation between soy and growth rate would be confounded by SES, and this might obscure a negative relationship between soy and growth, if SES had been corrected for.

Authors, editors, referees, or agencies, invoke confounding as a way of explaining or undermining positive results. In its famous study of dioxin exposure and cancer in 10 industries, the National Institute for Occupational Safety and Health qualified its positive study by noting that "we cannot exclude the possible contribution of factors such as smoking and occupational exposure to other chemicals," even though the nature of the study design made confounding highly unlikely [17]. More recent papers [18] attempt to quantify the possible impact that confounders might have on a relative risk (RR) or odds ratio (OR).

Misclassification bias

In any epidemiologic study where an exposed and referent population are compared, it is assumed that most or all of those in the exposed population were actually exposed while most or all of the referent population were unexposed. It is apparent that this is rarely likely to be the case. In an industrial-based study, not all workers have the same exposure. In a study of fish-eating populations exposed to mercury in the Seychelles and Faroes, not all families ate the same large quantities of fish. Where possible epidemiologists try to rank exposures, creating at least a crude dose–response relationship. In addition to exposure misclassification, there can be outcome misclassification. Some people with the disease may be overlooked. Other people may be ascribed a disease that they did not have. This is particularly a problem with death certificate studies of mortality causes. Any random misclassification in exposure or outcome biases toward the null hypothesis [19], making a type II error more likely. Since no study can absolutely ascertain exposure, this is a nearly universal conservative bias.

Subject protection and institutional review boards

Some historical epidemiologic studies, most notably the Tuskeegee syphillis study, showed that subjects in epidemiologic studies might be placed in jeopardy by the study itself. Even in controlled clinical trials, subjects on placebos may be denied promising therapies. Therefore, following the World Medical Association's Declaration of Helsinki [20], human subjects protection and institutional review boards (IRBs) emerged, and their review is now required for every study involving human subjects. Increasingly, IRBs are tightening the standards for epidemiologic studies. In some cases, full disclosure of a study's purpose may create additional biases by influencing whether subjects participate or how they respond. There is an increasing concern in the public health community that IRBs are jeopardizing the future of important areas of epidemiologic research, and it may become impossible to expect epidemiology to ever address the questions now posed of it.

Participation bias

One consequence of increased requirements posed by IRBs is a reduction in participation rates. In toxicology, experimental rodents are not given a choice about participating in studies. Most epidemiologic studies require consent of the subjects, at least a willingness to answer questions or provide blood samples. This is now being extended to review of records of past illnesses, where no physical harm to subjects occurs. Some subjects, particularly African Americans, are suspicious of any epidemiologic studies and refuse to participate, thereby leading to the under-representation in all studies and the lack of ethnic-specific information. Participation bias stacks the deck in two ways. It reduces the sample size, thereby reducing the power of the study. It is always a source of suspicion in interpreting the results (whether negative or positive). Thus, a positive study with a high refusal rate may be dismissed on the grounds that the results were influenced by selection or participation bias.

EXTRINSIC EPIDEMIOLOGY ISSUES

Population limitations

A major limitation of epidemiology is that for any give type of exposure, the population that can be studied is not infinite. For example, only a few thousand people are occupationally exposed to putative xenoestrogens. The tobacco industry took delight in claiming that the cancers attributable to smoking were probably due to alcohol consumption since the two exposures were highly correlated. Finding smokers who consumed little or no alcohol was challenging, and finding populations with neither, even more so. If there are only a few hundred people exposed to a particular chemical, then power is limited by the available universe, even if all participated.

Uneven receptor populations

Risk assessment and standard setting often take into account that among humans a subset (usually ill-defined) may be at substantially greater than average risk. A common default value of 10 is used, assuming that susceptible people are 10 times more vulnerable than average. Epidemiologic studies may or may not incorporate susceptible individuals, and the proportion of susceptibles in a study may vary greatly. In the absence of specific investigations (such as genotyping) this proportion is unknown. If the proportion is low, the impact of an exposure will be underestimated; if the proportion of susceptibles is high the impact may be overestimated. This is another type of misclassification.

Even when exposure is correctly classified, the inclusion of a high percentage of nonsusceptibles in the exposed population militates against detecting an effect, even when the same percentage is present in the controls. For example, a study of breast cancer that includes a large number of Asian women would be biased toward the null [18].

Meta-analysis and the precautionary principle

Meta-analysis offers several ways of combining studies, each of which alone lacks power, but which in combination may generate sufficient power to allow confidence in the outcome. Despite initial skepticism [22], meta-analyses are performed with increasing frequency and can be used to demonstrate or refute a proposed association. Often, however, meta-analysis does not provide a definitive answer as Hebert-Croteau [23] concluded in an analysis of estrogen and colon cancer: "Inadequate assessment of exposure, poor control of confounding factors, and changing patterns of use over time might have contributed to the slow emergence of this association postulated almost two decades ago. Additional large studies are needed to replicate this finding and explain the exact mechanism of this putative protective effect."

CONCLUSION

It is this last conclusion that highlights the need for precautionary approaches to many environmental health issues, including endocrine active substances. Epidemiologic studies are slow, often weak, and seldom conclusive, and decades may be required to reach consensus. Several standard defaults in epidemiology stack the deck against detecting associations, particularly weak associations, making it much more likely that a false Ho will be accepted (type II error) than that a true Ho will be rejected (type I error). The failure to report power, the setting of alpha at the artificially high default of 0.05, correcting unnecessarily for multiple comparisons, or blaming positive effects on hidden confounders are all ways of making detection of biologically significant associations quite unlikely. Faced with this problem, it is not only reasonable, but desirable to take effective precautionary actions to mitigate exposures to putatively hazardous materials, while awaiting additional scientific evidence, particularly when the consequences of inaction or laissez-faire, may be serious or irreversible.

ACKNOWLEDGMENTS

Joanna Burger, Russellyn Carruth, Bernard Goldstein, and many other colleagues have provided valuable insights on the precautionary principle. This project was partially funded by National Institute of Environmental Health Sciences (ESO 5022), the Department of Energy (AI # DE-FC01-95EW55084, DE-FG 26-00NT 40938) grant to the Consortium for Risk Evaluation with Stakeholder Participation (CRESP), and the conference hosts SCOPE and IUPAC.

REFERENCES

1. T. Colburn, D. Dumanoski, J. P. Myers. *Our Stolen Future*, Dutton, New York (1996).
2. C. Schairer, H.-O. Adami, R. Hoover, I. Persson. *Epidemiology* **8**, 59–65 (1997).
3. U.S. Preventive Services Task Force. *Ann. Intern Med.* **137**, 834–839 (2002).
4. National Research Council. *Hormonally Active Agents in the Environment,* National Academy Press, Washington DC (1999).
5. B. C. Gladden and W. J. Rogan. *J. Pediatrics* **119** (Part 1), 58–563 (1991).
6. Y. L. Guo, T. J. Lai, S. H. Ju, Y. C. Chen, C. C. Hsu. *Organohalogen Compd.* **14**, 235–238 (1995).
7. L. J. Guillette, Jr., D. B. Pickford, D. A. Crain, A. A. Rooney, H. F. Percival. *Gen. Comp. Endocrinol.* **101**, 32–34 (1996).

8. P. A. Bertazzi, A. C. Pesatori, D. Consonni, A. Tironi, M. T. Landi, C. Zocchetti. *Epidemiology* **4**, 398–406 (1993).
9. K. Kreiss, C. Robert, H. E. Humphreys. *Arch. Environ. Health* **37**, 141–147 (1982).
10. V. H.-H. Goh, S.-E. Chia, C.-N. Ong. *Environ. Health Perspect.* **106**, 41–44 (1998).
11. H. R. Pohl and H. G. Abadin. *Regul. Toxicol. Pharmacol.* **22**, 180–188 (1995).
12. National Research Council. *Arsenic in Drinking Water*, National Academy Press, Washington DC (1999).
13. A. Tukker. *J. Epidemiol. Commun. Health* **56**, 883–884 (2002).
14. R. A. Fisher. *Statistical Methods for Research Workers,* Oliver & Boyd, Edinburgh (1925).
15. R. M. Peterman and M. M'Gonigle. *Mar. Pollut. Bull.* **35**, 231–234 (1992).
16. K. J. Rothman. *Epidemiology* **1**, 43–46 (1990).
17. M. A. Fingerhut, W. E. Halperin, D. A. Marlow, L. A. Piacitelli, P. A. Honchar, M. H. Sweeney, A. L. Greife, P. A. Dill, K. Steenland, A. J. Suruda. *N. Engl. J. Med.* **324**, 212–218 (1991).
18. M. Goodman, M. Kelsh, K. Ebi, J. Ianuzzi, B. Langholz. *Epidemiology* **13**, 50–58 (2002).
19. A. H. Wu. *Environ. Health Perspect.* **107** (Suppl. 6), 873–877 (1999).
20. World Medical Association Declaration of Helsinki. World Medical Association (1964) (as amended) see <http://www.wma.net/e/policy/17-c_e.html>.
21. S. L. Chow. *J. Psychol.* **121**, 259–271 (1987).
22. N. Hebert-Croteau. *Cancer Epidemiol. Biomarkers Prev.* **7**, 653–659 (1998).

Pure Appl. Chem., Vol. 75, Nos. 11–12, pp. 2531–2534, 2003.

Workshop 5.4

General process for the risk assessment of pesticides that interact with or affect the endocrine system*

Karen Hamernik

Office of Pesticide Programs, U.S. Environmental Protection Agency, 1200 Pennsylvania Avenue NW (7509C), Washington, DC 20460-0001, USA

Abstract: The U.S. Environmental Protection Agency's Office of Pesticide Programs evaluates human health risk associated with exposure to pesticide chemicals. Chemical hazard and exposure assessment are components of the risk assessment process. For the risk assessment of single chemical conventional-type pesticides, there may be multiple exposure scenarios depending on the use pattern. Examples include acute and chronic dietary, and short-, intermediate-, and long-term occupational/residential exposures. For hazard assessment, available toxicity data and a weight-of the-evidence approach are used in the process of selecting appropriate toxicity endpoints for relevant exposure scenarios. The pesticide registration process requires that certain types of supporting toxicity data be submitted by the registrant depending in part on the chemical use pattern (e.g., food use). Types of toxicity data that might be submitted and used in hazard assessment include acute, subchronic, chronic, carcinogenicity, mutagenicity, metabolism, reproduction, developmental, neurotoxicity, and mechanistic studies. There may be data from multiple exposure routes (e.g., oral, dermal, inhalation) and from the scientific literature to consider. Dose–response information is also taken into account. In endpoint selection for a chemical, endocrine system-related effect(s) and dose–response relationship(s) are assessed in context of other types of effects, toxicities, and dose–response relationships noted. Endocrine system-related endpoints may include frank effects (e.g., endocrine organ hyperplasia or cancer) or precursor events (blood hormone level elevations). Endocrine system-related endpoints are generally treated like other cancer or non-cancer toxicity endpoints (e.g., hepatic cancer, neurotoxicity) in the risk assessment process. For chemicals with evidence of endocrine system interaction(s), an endocrine system-related effect may or may not be the most sensitive or relevant endpoint for a particular risk assessment exposure scenario. Some chemical examples will be presented. In the final risk assessment, hazard assessment information is integrated with exposure information. The assessment may be adjusted, at some point, for uncertainties in hazard or exposure data. An aggregate risk assessment, in which multiple sources or routes of exposure are considered, is typically performed for occupational and residential exposure scenarios. A cumulative risk assessment may be considered for groups of chemicals with a common mechanism of toxicity.

The U.S. Environmental Protection Agency (EPA) Office of Pesticide Programs (OPP) evaluates risks associated with exposure to pesticide chemicals. This paper presents examples of some of the consid-

*Report from a SCOPE/IUPAC project: Implication of Endocrine Active Substances for Human and Wildlife (J. Miyamoto and J. Burger, editors). Other reports are published in this issue, *Pure Appl. Chem.* **75**, 1617–2615 (2003).

erations that might be used in the process of assessing risks of pesticide exposure to human health, including risks from chemicals that interact with, perturb, or affect the endocrine system. In order to further narrow down a broad topic for purposes of the symposium, the perspective of the presentation is taken primarily (although not exclusively) from general experience with the single chemical risk assessment of conventional pesticides such as insecticides, fungicides, rodenticides, and herbicides. This paper should not be considered as a comprehensive review of the subject and is not intended to convey what has been, should be, or will be done in the risk assessment of any particular substance. It should be noted that the agency conducts risk assessments on other types of materials and exposure scenarios. In addition, modifications in the current approaches to the assessment of chemical risk are possible with advances in information or technology or changes in policy.

Chemical hazard and exposure assessments are components of the risk assessment process. There may be numerous factors and conditions to be considered in these steps. For instance, the type of chemical to be evaluated (e.g., conventional pesticide) and the type of risk assessment to be performed (e.g., single chemical) are identified. In order to assess hazard for particular exposure scenarios, the use pattern of the chemical is ascertained. Examples of use patterns include food uses in which a chemical is used agriculturally and residues might be found in the diet and non-food uses in which a chemical might be used in swimming pools or on turf or roadside weeds. Many different exposure scenarios are possible. For example, dietary exposure to pesticides may be acute and occur singly or within one day or may be chronic and occur repeatedly over periods of time up to a lifetime. Other possible situations are short, intermediate, or long-term occupational exposures to workers that mix, load, or apply pesticides or pick crops in the field and residential exposures through household uses of chemicals. Consideration is also given to which population groups might be susceptible to certain pesticide exposure scenarios. One possible group is the general population. Another is females, age 13 years and older, representing females that might experience pesticide exposure during the child-bearing years.

An important part of hazard assessment is the selection of appropriate toxicity endpoints for relevant chemical exposure scenarios, such as acute and chronic dietary exposures. In the endpoint selection process, the available toxicity data and dose–response information for a chemical is evaluated in a weight-of-the-evidence (WOE) approach to determine pertinent effects and dose levels (e.g., no observable adverse effect levels, NOAELs; points of departure) for use in performing risk assessments. With the WOE approach, toxicity endpoints are not necessarily chosen from the toxicity study with the lowest NOAEL. Rather, the WOE approach involves evaluation and consideration of all of the available relevant data for a chemical and aids in providing an overall picture of a chemical's effects and dose–response relationships. Therefore, individual studies are looked at in the context of other available information for the chemical and not in isolation. The WOE approach can be used in assessing both noncancer effects and the carcinogenicity potential for a chemical. Sound science and rationally based scientific judgments should be used in implementing the WOE approach to select toxicity endpoints. Some points to consider in the WOE assessment include evaluating whether a selected endpoint makes sense in the context of the all of the available data, whether an endpoint has human relevance, whether the species in which a potential endpoint is observed is an appropriate model for human disease, and whether the potential endpoint has toxicological significance. Other examples of information that might be considered in a WOE analysis are the shape of the dose–response curve, dose-spacing issues and evidence for progression of events leading to a lesion or particular type of toxicity.

There may be several sources of toxicity data available for the evaluation of a given chemical using the WOE approach. First, the pesticide registration process has requirements for certain types of toxicology data that are submitted to the agency for analysis. Requirements depend in part on the use pattern (e.g., food use, non-food use) for a particular chemical. Examples include acute, subchronic, chronic, carcinogenicity, mutagenicity, metabolism, dermal absorption, reproduction, developmental, and neurotoxicity studies. Mechanistic or special studies to focus in on a particular issue may also be performed (although not necessarily required). Toxicity data may be obtained from studies using different routes of exposure (e.g., oral, dermal, inhalation) and from different species. Data from the open literature may also be used in the WOE evaluation. In addition, toxicity data may be available from structurally related compounds or compounds with other types of similarities to the chemical under review.

Some pesticides affect or interact with the endocrine system. Available toxicity data from the types of toxicology data sources noted above may provide evidence of direct effects on endocrine organs, tissues or cells or of indirect effects on the endocrine system, including such phenomena as perturbation of endocrine system homeostasis leading to hypertrophy and/or tumor formation in an endocrine tissue. In general, endocrine system-related effects and dose–response relationships are assessed in context of other types of effects, toxicities, and dose–response relationships noted for a chemical. Endocrine system-related effects considered for use as endpoints in risk assessment may include frank effects (e.g., endocrine organ hyperplasia; cancer), or precursor events (e.g., sustained blood hormone level elevations). Sometimes, observed effects are relatively nonspecific in nature such that it is difficult to judge whether the effects are really endocrine system-related.

For use in hazard assessment incorporating a WOE approach, endocrine system-related endpoints, cancer or non-cancer, are generally treated like other cancer (e.g., hepatic cancer) or non-cancer (e.g., splenic necrosis) endpoints. However, for Food Quality Protection Act (FQPA) considerations, it is possible that some kinds of endocrine system-related endpoints as well as other types of endpoints (such as some types of reproductive and neurotoxicity findings) might have particular relevance in evaluating the relative susceptibility of infants and children vs. adults to an agent. It should be noted that an endocrine system-related effect may or may not be the most sensitive or relevant endpoint for a particular risk assessment exposure scenario. For example, a chemical may have more than one toxic effect (e.g., testicular atrophy and neurotoxicity) with the neurotoxic effect perhaps occurring at a dose lower than that of the testicular atrophy.

Next, two simplified hypothetical applications of the WOE approach to toxicity endpoint selection will be presented using toxicology data that include indications of possible endocrine system-related interactions. Both examples are artificial in nature and are composites drawn from experience with toxicology data on numerous chemicals over the years. The examples are for illustrative purposes only, over-simplify a more complex process, and do not represent an evaluation by the agency of any particular pesticide nor of how the agency would definitely use such data in a hazard or risk assessment.

The first hypothetical example concerns a fungicide with food uses. Five developmental toxicity studies were performed by the oral route with the chemical. In a rabbit study, no effects were observed in offspring. Four rat studies were available in which similar dose levels of the test material were administered. Although effects on sexual differentiation in male offspring only were observed in all of these studies, they occurred at lower doses in two of the studies that had been conducted using a modified dosing regimen. Maternal effects were observed only at doses higher than those at which effects in offspring were noted. Findings in a rat reproductive toxicity feeding study included endocrine organ weight changes and developmental delays and feminization of male offspring at doses at which parental toxicity was observed but which were much higher than those at which effects were seen in the rat developmental toxicity studies. The effects on male development in the rat studies were consistent with proposed antiandrogen activity of the chemical. This interpretation was supported by other submitted toxicity data on the fungicide (e.g., subchronic, chronic, cancer, metabolism, and mechanistic studies) and information from the literature on this and similar chemicals (e.g., antiandrogen pharmaceuticals, other pesticides with antiandrogen activity). One possible use of these data might be to establish a toxicity endpoint and NOAEL based on the in utero effects noted in male offspring in the developmental toxicity studies. This type of endpoint has relevance for a risk assessment conducted for the population group of human females of child-bearing age (13 years and older). Since the developmental effects noted could possibly occur after a single dose of the chemical, the endpoint could be appropriate for an acute exposure scenario.

The second hypothetical example concerns a herbicide with food uses. It is structurally related to other chemicals known to induce thyroid tumors in rats. Two rat chronic/carcinogenicity feeding studies were available. Thyroid hyperplasia was observed in one and thyroid tumors were noted in the other but only at doses considered to be excessive. Thyroid effects were not found in other species. No developmental or reproductive effects were observed in the developmental and reproductive toxicity studies. Although no obvious endocrine-related effects were noted in a dog chronic feeding study, hemosiderosis and anemia were observed at the lowest dose tested. Ovarian atrophy was found in one long-term mouse feeding study, along with uterine tumors, and was also noted in one long-term rat feeding study, along with prostate atrophy. The results of both of these rodent studies were confounded by the presence of a chemical additive and the effects were not seen in other long-term feeding studies in these species conducted without the additive. However, two other nonendocrine organ tumor types were observed in both rat long-term feeding studies performed with and without the additive present. From these data, an endpoint for chronic toxicity (for possible use in a chronic dietary exposure scenario and risk assessment) could be supported based on nonendocrine system-related effects (e.g., on the hemosiderosis and anemia observed in the dog study) rather than on endocrine system-related findings. Ovarian and prostate atrophy were only seen in the presence of the chemical additive and thyroid hyperplasia was observed only at excessive doses of the chemical. The two nonendocrine system-related tumor types could be considered as more feasible candidates for the cancer risk assessment. Uterine tumors were only noted in the presence of the chemical additive and thyroid tumors occurred only at an excessive dose in the rat and there were no signs of thyroid effects in other species.

In preparing the final risk assessment for an individual chemical, hazard assessment information is integrated with exposure information. The assessment may be adjusted, at some point, for uncertainties in hazard or exposure data. FQPA considerations are factored in, as necessary. Examples of other types of risk assessments conducted in OPP are aggregate and cumulative risk assessments. In aggregate risk assessments, multiple sources or routes of exposure may be combined. Cumulative risk assessment may be considered for groups of chemicals determined to have a common mechanism of toxicity. Ideally, final risk assessments will provide useful information for chemical risk management.

Pure Appl. Chem., Vol. 75, Nos. 11–12, pp. 2535–2541, 2003.

Workshop 5.5

Role of the precautionary principle in the EU risk assessment process on industrial chemicals*

Jack de Bruijn[‡], Bjorn Hansen, and Sharon Munn

European Chemicals Bureau, Joint Research Centre, 21021 Ispra (VA), Italy

Abstract: This paper discusses the practical implementation of the precautionary principle in the area of management of industrial chemicals in the European Union. An analysis of a number of recent cases where the precautionary principle was invoked shows that the main reason for doing so were the uncertainties in the risk assessment (or the underlying effects or exposure data), which were, according to the scientific experts, so high that the "normal" level of certainty could not be obtained. The challenge for the future is to try to develop general guidance or rules that will support the policymakers in their decision as to whether this uncertainty is so large that action is warranted or whether it is acceptable to wait until further information has become available.

INTRODUCTION

Taking regulatory action on the basis of the precautionary principle (PP) is often presented as an alternative to taking action based on an assessment of risks. In practice, however, many references in international law to the precautionary principle, such as the 1992 Rio Declaration, refer to the use of this approach when there are threats of serious irreversible damage, but there is lack of full scientific evidence. This implies that some sort of identification of the potential hazards and or risks has to be present in order to allow decision-makers to decide on any action to be taken. In February 2000, the European Commission published a Commission Communication on the precautionary principle providing a general framework for its use in EU policy [1]. This paper addresses specifically the practical implementation of the approach in the area of management of industrial chemicals.

COMMISSION COMMUNICATION ON THE PRECAUTIONARY PRINCIPLE

The aim of the Commission Communication was to outline the Commission's approach to using the precautionary principle, to establish guidelines for it, to build a common understanding of how to assess, manage and communicate risks that science is not yet able to evaluate fully, and to avoid unwarranted recourse to the precautionary principle, as a disguised form of protectionism [1]. According to this communication, the implementation of an approach based on the precautionary principle should start with a scientific evaluation, as complete as possible, and where possible, identifying at each stage the degree of scientific uncertainty. Once the scientific evaluation has been performed as best as possible and the conclusions of this evaluation show that the desired level of protection for the environment

---

*Report from a SCOPE/IUPAC project: Implication of Endocrine Active Substances for Human and Wildlife (J. Miyamoto and J. Burger, editors). Other reports are published in this issue, *Pure Appl. Chem.* **75**, 1617–2615 (2003).

‡Corresponding author: Fax: 0039-0332-785862; E-mail: jack.de-bruijn@jrc.it

or a population group could be jeopardized, this may provide a basis for triggering a decision to invoke the precautionary principle. The conclusions should also include an assessment of the scientific uncertainties and a description of the hypotheses used to compensate for the lack of the scientific or statistical data. An assessment of the potential consequences of inaction should be considered and may be used as a trigger for action by the decision-makers. Clearly, the potential application of the precautionary principle is relevant to the risk management phase.

It is important to recognize that the communication indicates that the application of the precautionary principle should not be confounded with the prudential approach used in the context of assessing the risks. In assessing risks, a wide range of uncertainties have to be dealt with. Examples are the extrapolation of effects seen in laboratory test systems to real world conditions, the extrapolation from single species test results to effects on complex multispecies ecosystems, the extrapolation from effects seen in animal studies to potential effects in humans, and, last but not least, the extrapolation from short or medium-term exposure to lifetime actual exposure conditions. All of these uncertainties are usually taken care of in the risk assessment process, for instance by applying uncertainty or safety factors to the results of single species animal studies. It is the job of the risk assessment experts to provide their ultimate view on the reliability of their assessments, as well as on the remaining uncertainties. The overall level of uncertainty will affect the foundation for protective or preventive action in the risk management phase.

The Commission Communication on the precautionary principle was endorsed both by the European Parliament and the Council of Ministers who in their resolutions on this subject called upon the Commission to actively apply the guidelines and incorporate the principle in drawing up its legislative proposals [2,3].

RISK MANAGEMENT OF INDUSTRIAL CHEMICALS IN THE EUROPEAN UNION

In Europe, the potential risks of industrial chemicals with high-production volumes are assessed under Council Regulation (EEC) No. 793/93 [4]. This regulation introduced a comprehensive framework for the evaluation and control of existing chemicals. It foresees that this evaluation and control of the risks should be carried out in four steps: (1) data collection, (2) priority setting, (3) risk assessment, and (4) risk reduction. In short, the process is as follows: draft risk assessment reports are developed by a Member State, mainly on the basis of information provided by the producing companies. These draft reports are discussed in what is called technical meetings (TMs) of experts from Member States, industry, and NGO's. Once agreed by the TM, the risk assessments are sent to the Scientific Committee for Toxicity, Ecotoxicity and the Environment (CSTEE), which is asked whether it can support the conclusions of the assessments. Subsequently, the conclusions are forwarded to the so-called competent authorities (CAs) of the Member States, which is the policy body responsible for agreeing the conclusions and, where relevant, discussing and proposing the necessary risk reduction strategy. In the implementation, a separation of the risk assessment and risk management processes has been carried through as was also recommended by the Council in their resolution on the precautionary principle [3].

One should realize that this regulation deals with the assessment of the risks of high-production volume chemicals produced in volumes >1000 tonnes per year. The deliberate use of such large volumes implies that human activities almost inevitably lead to pollution.

Risk assessment is then the tool used to evaluate if these inevitable emissions lead to "unacceptable risk". Different judgments on the need to take action can be made by policy makers compared to scientists when there is not sufficient scientific evidence to demonstrate that an "unacceptable risk" exists, though also not enough to demonstrate that the risks are "acceptable". The case studies below will describe some of these situations.

CASE STUDIES

Since the implementation of Regulation 793/93, risk assessment reports have been prepared and discussed for approximately 113 out of 141 priority substances [5]. For approximately 70 of these, the conclusions have been agreed and sent forward to the risk managers to discuss the necessary development and implementation of risk management activities. The four cases described below provide useful information on how the relationship between the risk management process and the actual implementation of the precautionary principle has worked in practice.

Nonylphenol(ethoxylates)

The risk assessment for nonylphenol and nonylphenol(ethoxylates) was finalized in 1999 [6]. The conclusion on the environmental part of the assessment was that there is concern for the aquatic compartment due to the fact that for a range of uses of the substance it was estimated that the predicted environmental concentrations (PECs) would exceed the predicted no-effect concentration (PNEC). This PNEC was calculated using all the aquatic toxicity data present on nonylphenol, including all available data on estrogenic effects. Since these effects generally occurred at higher concentrations, it was concluded that the calculated PNEC should be protective for estrogenic effects in fish as well. The conclusions of the report were supported by the CSTEE [7] and adopted by the CAs. Based on this risk assessment, the Commission is currently discussing the possible implementation of restrictions for marketing and use of a range of uses of these substances in the context of Directive 76/769 on Restrictions on the marketing and use of certain dangerous substances and preparations.

In conclusion the available information allowed the development of science-based conclusions on the risks of the substance that triggered the necessary risk reduction. In this case, there was no need to instigate the precautionary principle.

Pentabromodiphenylether

The risk assessment for pentabromodiphenylether (pentaBDPE) was also finalized in 1999 [8]. The conclusions of the substance were reached after intensive discussions in the TM. The human health section of the assessment concluded that there is a need for further information to clarify the potential risks to infants that are potentially exposed to breast milk or infant formulae prepared from cow's milk. The experts, however, were particularly concerned by the studies that indicated rapidly increasing concentrations of pentaBDPE in women's breast milk in Sweden. The report was reviewed by the CSTEE who agreed that further information is needed to better identify the potential risks of the substance [9].

The conclusions of the risk assessment, including a discussion on the uncertainty around them, were presented to the CAs. The CAs decided that in light of the properties of and available information on pentaBDPE, and in the absence of adequate scientific knowledge, the time it would take to gather the information needed to enable an adequate scientific evaluation, was considered to be unacceptable. Therefore, risk reduction measures should be taken immediately. This conclusion was subsequently forwarded to the relevant bodies within the Commission who developed a ban on the marketing and use of the substance, which has recently been implemented through Directive 2003/11/EC.

Bisphenol A

Both the sections on environment and human health of the risk assessment of bisphenol A have recently been finalized [8]. Interestingly, the conclusions on both parts are very similar in the sense that the TM was of the opinion that further information on the toxicity and ecotoxicity of this substance is needed. On the environment, the discussion focused on the effects to be taken into account in setting the PNEC for water. A preliminary PNEC was derived using the "traditional" endpoint hatchability, but the experts also recognized the present studies showing potential effects on spermatogenesis in fish and the observations of superfeminization in snails, both occurring at much lower concentrations. However, the TM agreed that regarding these studies there were a number of scientific questions unanswered and that further investigations were needed.

The discussion on the human health aspects of the substance focused on the developmental toxicity. The TM agreed that overall, in standard developmental studies in rodents, there is no convincing evidence that bisphenol A is a developmental toxicant. However, the available and apparently conflicting data from studies conducted using low doses (in the μg/kg range) do raise uncertainties and the majority of the experts felt that these studies could not be disregarded. The data though could also not be used to conclude with sufficient certainty that an *unacceptable* risk exists. It was considered that further testing could reduce the uncertainties. The CSTEE reviewed the report and agreed on the need for more information although they wondered whether in the area of the potential low-dose effects more studies would actually help in resolving the issue since many high-quality studies are already available [10,11].

This issue was referred to the CAs who agreed unanimously that further work was required to resolve the uncertainties surrounding the potential for bisphenol A to produce adverse effects on development at low doses. In addition, it was agreed that a provisional NOAEL (no observed adverse effect level) for developmental effects, derived from the rat multigeneration study, should be used in the risk characterization in the interim, while awaiting the outcome of further testing, with the aim of identifying those scenarios which were clearly of concern irrespective of the outcome of the further testing.

In conclusion, on both parts of the risk assessment uncertainties were identified in the data which precluded drawing definitive conclusions regarding the risks of the substance. The experts, however, identified areas for improvement of the database and the risk managers agreed that this route should be followed rather than implementing temporary risk reduction measures. Hence, this is an example where after careful consideration of the results of the risk assessment as well as of the remaining uncertainties, it was decided not to invoke the precautionary principle yet.

Octa- and decabromodiphenylether

The risk assessment reports for these two flame retardants were recently finalized after intensive debate between the risk assessment experts which lasted more than five years and entailed the delivery of substantial information by the producing industry [12,13]. The TM recognized that there are significant uncertainties surrounding the scientific evidence relating to the evaluation of the environmental risks of these substances. These uncertainties are based on their very high persistence, the possible debromination and the uptake of the substance and its debromination products in biota via the food chains. Especially the fact that recent studies indicate that these substances are taken up through the food chain and may end up in eggs of predatory birds caused the TM experts to express their concern about the potential risks for secondary poisoning. The TM agreed that the uncertainties might be reduced if substantial additional information was collected and/or generated but recognized also that it would take considerable time to collect and/or generate this information and that it might not be sufficiently comprehensive to remove all of the uncertainty. Other experts, however, were of the opinion that the information currently provided in the reports and the above-mentioned uncertainties about the actual or potential risks for secondary poisoning should lead to a conclusion that there is a need for risk reduction measures.

The CSTEE also reviewed the results of the assessments and in the case of octaBDPE concluded by using historical evidence from substances with comparable persistence and bioaccumulating properties that there might be a risk to top-predators, including humans [14]. On decaBDPE their conclusion was that long-term effects of this persistent and, at least in some organisms, bioaccumulating substance cannot be excluded. They concluded that further research is needed but also that the uncertainty warrants risk reduction measures directly [15].

The conclusions of the RARs were discussed by the CAs who came to a somewhat similar conclusion as with pentaBDPE that in light of the properties of and available information on the two substances, the time it would take to gather the information needed to enable an adequate scientific evaluation, was considered to be unacceptable. Therefore, immediate risk reduction measures should be developed.

The Commission has recently implemented risk reduction measures for octabromodiphenylether through Directive 2003/11/EC. For decabromodiphenylether, the discussion is still ongoing. This is another example where, due to the high uncertainties in the outcome of the quantitative risk assessment, it was decided to invoke the precautionary principle.

Dealing with uncertainties

If we closely analyze these examples for the reasons why in certain cases the precautionary principle was invoked whereas in others was not, it basically boils down to situations where the uncertainties in the risk assessment (or the underlying effects or exposure data) are so high that the "normal" level of certainty cannot be obtained. As stated in the Commission communication, this judgment is crucial for the risk managers that face the dilemma of having to act or not to act. Since general rules have until now not been developed the next question is whether we can identify substances or groups of substances for which due to their inherent properties the acceptable risk concept fails due to the high uncertainties is estimating these risks.

The first example of a group of substances for which science cannot provide acceptable risk levels are nonthreshold carcinogenic substances. Since for these substances no safe level can be established the policy reaction is one of risk avoidance by exposure minimization. As a result of that, it is in the EU forbidden to use these substances in consumer products.

A second category for which there can be very high uncertainty in predicting the no-effects level and/or the long-term exposure concentration in the environment and in biota are persistent, bioacummulating and toxic (PBT) substances. For such substances, EU experts have recently concluded that the risks cannot be quantified with the appropriate level of certainty [16]. However, as a result of a benchmarking process using historical evidence from known dangerous substances that have caused long-term effects in the environment, it has been possible to develop criteria for the identification of PBTs and vPvBs.

The question whether endocrine active substances should be treated similarly is less easy to answer. EU experts on reproductive toxicity decided recently that in the context of the risk assessment of chemicals they recognize endocrine disruption more as a mechanism than as an adverse health effect [17]. The Commission in its Community strategy for endocrine disruptors also indicated that the scientific knowledge in this area is limited, with regard to test methods for identification and regarding the actual magnitude of effects that such substances may cause [18]. These substances need not be treated differently from other chemicals, as they will eventually be addressed in the normal assessment, in particular under the reproductive toxicity evaluation, but the Commission urged Member States to give priority to testing, classification and risk assessment of known or suspected endocrine active substances.

CONCLUSION

The Commission in its Communication on the precautionary principle clearly argues for a separation between the scientific assessment of risks and the subsequent application, if necessary, of the precautionary principle based, among others, on the results of this assessment. This separation has worked well in the context of the implementation of the existing substances regulation. The key question to be answered is when and under what conditions the uncertainties in the risk assessment are so high that the "normal" level of certainty is not obtained. The challenge for the future is to try to develop general guidance or rules that will support the policymakers in their decision whether this uncertainty is so large that action is warranted or whether it is acceptable to wait until further information has become available.

REFERENCES

1. CEC 2000. Communication from the Commission on the precautionary principle, COM/2000/0001/final.
2. EP (European Commission). European Parliament resolution on the Commission communication on the precautionary principle (COM(2000)1-C5-0143/2000 – 2000/2086 (COS) (2000).
3. EC (European Commission). Council Resolution on the Precautionary principle. Annex III to the Presidency Conclusions of the Nice European Council Meeting, 7–9 December 2000, SN 400/00 ADD 1 (2000).
4. CEC 1993. Council Regulation (EEC) n° 793/93 of 23 March 1993 on the evaluation and control of the risks of existing substances.
5. European Chemical Bureau. Existing chemicals, <http://ecb.jrc.it/existing-chemicals/>.
6. EC (European Commission). 4-nonylphenol (branched) and nonylphenol. European Union Risk Assessment Report, Vol. 10, Publication No. EUR 20837 (2002). Internet publication at <http://ecb.jrc.it>.
7. Opinion of the CSTEE on the results of the Environmental Risk Assessment of Nonylphenol, straight chain (CAS N° 84852-15-3) and branched chain (CAS N° 25154-52-3), 4 February 2000, <http://europa.eu.int/comm/food/fs/sc/sct/outcome_en.html>.
8. EC (European Commission). Diphenyl ether, pentabromo derivative. European Union Risk Assessment Report, Vol. 5. Publication No. EUR 19730. Office for Official Publications of the EC, Luxembourg (2001). Internet publication at <http://ecb.jrc.it>.
9. Opinion of the CSTEE on the results of the Human Risk Assessment of Pentabromodiphenyl ether (CAS N° 32534-81-9), Adopted on 19 June 2000, <http://europa.eu.int/comm/food/fs/sc/sct/outcome_en.html>.
10. EC (European Commission) (2002). Bisphenol-A. Draft European Union Risk Assessment Report, Internet publication at <http://ecb.jrc.it>.
11. Opinion of the CSTEE on the risk assessment of Bisphenol A - Human Health part (Cas no: 80-05-7). Adopted on 22 May 2002, <http://europa.eu.int/comm/food/fs/sc/sct/outcome_en.html>.
12. Opinion of the CSTEE on the results of the risk assessment of Bisphenol A - Environmental part (CAS no: 80-05-7). Adopted on 27 June 2002, <http://europa.eu.int/comm/food/fs/sc/sct/outcome_en.html>.
13. EC (European Commission). Diphenyl ether, octatabromo derivative. European Union Risk Assessment Report, Vol. 16, Publication No. EUR 20403 (2002). Internet publication at <http://ecb.jrc.it>.
14. EC (European Commission). Bis(pentabromophenyl) ether. European Union Risk Assessment Report, Vol. 17, Publication No. EUR 20402 (2002). Internet publication at <http://ecb.jrc.it>.
15. Opinion of the CSTEE on the results of the Risk Assessment of: Diphenyl ether, octabromo derivative - Environmental and Human Health Part (CAS No.: 32536-52-0). Adopted on 31 October 2002, <http://europa.eu.int/comm/food/fs/sc/sct/outcome_en.html>.
16. Opinion of the CSTEE on the results of the Risk Assessment of: Bis(pentabromophenyl)ether - Environmental and Human Health Part (CAS No.: 1163-19-5). Adopted on 31 October 2002 <http://europa.eu.int/comm/food/fs/sc/sct/outcome_en.html>.
17. EC (European Commission). Technical Guidance Document on Risk Assessment in support of Commission Directive 93/67/EEC on Risk Assessment for new notified substances and Commission Regulation (EC) No 1488/94 on Risk Assessment for existing substances, and Directive 98/8/EC of the European Parliament and of the Council concerning the placing of biocidal products on the market, Parts 1–4, 2nd ed. Publication No. EUR 20418/EN/1, EUR 20418/EN/2, EUR 20418/EN/3, EUR 20418/4. Office for Official Publications of the EC, Luxembourg (2002). Internet publication at <http://ecb.jrc.it/>.

18. CEC 1999. Communication from the Commission to the Council and the European Parliament - Community strategy for endocrine disrupters - A range of substances suspected of interfering with the hormone systems of humans and wildlife, COM/1999/0706 final.

Pure Appl. Chem., Vol. 75, Nos. 11–12, pp. 2543–2545, 2003.

Workshop 5.6

Differing perspectives on the use of scientific evidence and the precautionary principle*

Joanna Burger[‡]

Environmental and Occupational Health Sciences Institute, and Consortium for Risk Evaluation with Stakeholder Participation, Division of Life Sciences, Cell Biology and Neurosciences, Rutgers University, Piscataway, NJ 08854, USA

Abstract: Different governments and agencies are approaching the use of scientific evidence, weight of evidence, and the precautionary principle in different ways. The European Community has used the precautionary principle in situations where the consequences are great, data are unavailable or will be costly (in terms of money and time) to obtain, or data are difficult or impossible to obtain. Other countries, such as the United States, have a risk assessment process that has built-in safety or uncertainty factors which are themselves precautionary. Risk management decisions can be made on the basis of adequate studies, risk assessment, weight-of-evidence approaches, and the application of the precautionary principle. While weight of evidence has been used in the United States for increased research funding and regulator action with respect to some chemicals that are hormonally active, the European Community has applied the precautionary principle.

INTRODUCTION

There are several approaches to the regulation of man-made environmental insults, such as anthropogenic chemicals, including using traditional scientific studies, weight of evidence, and the precautionary principle. All of these can be used as a basis for risk assessment, leading to risk management. All scientific studies have uncertainties, which must be accounted for in both risk assessment and risk management. While weight-of-evidence approaches can identify serious impacts, the precautionary principle can lead to interim management, allowing the necessary research to continue. When data gaps are filled, they can be used to inform both risk assessment and risk management.

This workshop dealt with the issues surrounding the history of weight of evidence and the precautionary principle, the use of these approaches in risk assessment and risk management, the kinds of precautionary actions societies can take to protect themselves and the environment, and lessons learned from early warnings. Each of these will be discussed briefly below, followed by general recommendations from the workshop.

While there is information and many studies on endocrine active substances (often called endocrine disruptors, or endocrine-disrupting chemicals), there is still much to be learned about fate and effects, and mechanisms of action. Particular needs for more studies include: the effects of low doses and of mixtures, identification of target or sensitive species, and the identification of subpopulations at risk. Endocrine active substances clearly offer a fertile field for precaution.

*Report from a SCOPE/IUPAC project: Implication of Endocrine Active Substances for Human and Wildlife (J. Miyamoto and J. Burger, editors). Other reports are published in this issue, *Pure Appl. Chem.* **75**, 1617–2615 (2003).

[‡]E-mail: burger@biology.rutgers.edu

ENDOCRINE ACTIVE SUBSTANCES AND RISK MANAGEMENT APPROACHES

Risk management requires some level of understanding of the hazard and the risks that humans and other components of ecosystems face. The difficulty is in determining the level of risk and the attendant effects. The discussions regarding the use of traditional scientific studies, weight-of-evidence approaches, and the precautionary principle have largely focused on the differences as if these were discrete approaches. As Burger suggests in the first paper in this series, these are approaches on a continuum, and they are not mutually exclusive. The precautionary principle may provide for interim management, while research continues to address some of the uncertainties. "Precautionary actions" are not irreversible if new data suggests another risk management action. The precautionary principle is a management tool, not a risk assessment tool. Precautionary actions can include additional research, public warnings, emission reductions, substitution of alternatives, or partial or total bans. Such precautionary actions can be based on traditional studies, weight-of-evidence approaches, or the precautionary principle.

All three approaches (traditional studies, weight of evidence, precautionary principle) have been used extensively in risk management (often involving regulation), and references to them appear most often in government documents or the gray literature. However, for a full scientific dialogue, all approaches would benefit from a full discussion in the peer-reviewed literature. Overall, references to a weight-of-evidence approach go back to the 1970s, while the precautionary principle is mentioned in the peer-reviewed literature mainly since the 1990s (MedLine and Cambridge Scientific Abstracts, see Workshop 5.1). Even so, weight of evidence has been invoked much more than the precautionary principle in discussions of effects. Weight of evidence has been invoked more for human health studies than for ecological studies, and the precautionary principle has been invoked more for ecological studies than for human health studies. As Goldstein points out (see Workshop 5.2), the apparent conflict between the European use of the precautionary principle and the American cynicism about it stems from trade disputes, such as growth hormones in beef. Goldstein notes that endocrine active substances are persistent and harmful to ecosystems, and thus require some regulation, regardless of the debate about their effect on humans (see Workshop 5.2 for a further discussion).

Evaluating the effects of endocrine active substances on humans is difficult because the main line of evidence comes from epidemiological studies. Direct human evidence on the endocrine effects of chemicals has been slow to accumulate because of sample size limitations in exposed populations, and over-conservative hypothesis-testing approaches (see Workshop 5.3).

U.S. ENVIRONMENTAL PROTECTION AGENCY APPROACHES TO RISK ASSESSMENT

The U.S. Environmental Protection Agency (EPA) uses risk assessment methods to examine and regulate the effects of chemicals, and endocrine active substances are treated the same way (see Workshop 5.4). The EPA methodology embodies a conservative approach, which it considers precautionary. Assumptions and uncertainty factors are precautionary in themselves. However, the nature of the risk assessment means that the most sensitive, nontrivial endpoint is used as a basis for regulation. Thus, endocrine effects of a chemical may not be the basis for regulation, even though such effects may exist.

A EUROPEAN APPROACH

The European Environmental Agency (EEA) reviewed over 100 years of environmental management, including the examination of retrospective case studies of endocrine active substances. The EEA concluded that there were many lessons that should have been apparent much earlier. If precautionary actions were taken when problems first became evident, long-term consequences and costs would have been avoided.

The European Union established guidelines for the precautionary principle (see Workshop 5.5), which urged unwarranted recourse of the precautionary principle. The precautionary principle should be applied in the standardized process of risk assessment, risk management, and risk communication. There are uncertainties in all risk assessment processes, and these should be assessed in the application of the precautionary principle as is possible.

RECOMMENDATIONS AND CONCLUSIONS

1. Precautionary principle and weight of evidence are complementary approaches, particularly when risk assessment points to high severity, irreversibility, and high uncertainty (especially when time for collecting adequate data is long).
2. Precaution *now* can be reversed as new information becomes available; failure to take precaution when warranted may lead to irreversible consequences that cannot be undone when new information becomes available.
3. Interpretation of epidemiological and toxicological studies on endocrine active substances should take into account type II errors (false negatives) and lack of power, as well as type I errors (false positives). Need for more scientifically positive studies should not delay interim (precautionary) action for endocrine active substances.
4. There is a need to assess situations in which the precautionary principle was used for endocrine active substances. Was it warranted? Did the precautionary principle provide early action before catastrophic effects? Were there unanticipated consequences or were new hazards introduced inadvertently? These should inform our actions for endocrine active substances.
5. There is a need to examine the relationship between the precautionary principle and merely being cautious in various regulatory regimes, in several different countries.
6. There is a need to examine as a case study how weight of evidence and the precautionary principle have been specifically applied to endocrine active substances, in a variety of organisms, species guilds, ecosystems, and countries.
7. There is a need to assess situations in which regulatory actions were not taken soon enough. How could the precautionary principle have helped? Can the costs of delaying action be compared to the costs of taking action possibly prematurely?
8. There is a need to examine the use of weight of evidence and the precautionary principle across nations and continents for efficacy, usefulness, timeliness, and suitability of regulatory or management action.
9. Examples of the application of weight of evidence and the precautionary principle should be examined for the trade-offs between severity of adverse effects, irreversibility, degree of uncertainty, and costs deriving from each process, particularly for endocrine active substances.
10. Scientists engaged in traditional scientific inquiry on endocrine active substances should be encouraged to consider the broader aspects of use of their data for regulatory and management actions.

WORKSHOP 6

RISK MANAGEMENT OPTIONS FOR ENDOCRINE DISRUPTORS IN NATIONAL AND INTERNATIONAL PROGRAMS

Pure Appl. Chem., Vol. 75, Nos. 11–12, pp. 2549–2553, 2003.

Workshop 6.1

Risk management options for endocrine disruptors in national and international programs*

James C. Lamb, IV[1], Herman B. W. M. Koëter[2], Richard Becker[3], Andreas Gies[4], Les Davies[5], Tohru Inoue[6], Abigail Jacobs[7], Gwynne Lyons[8], Mitsuo Matsumoto[9,‡], and Gary Timm[10]

[1]BBL Sciences, BBL, Inc., Reston, VA 20190, USA; [2]Organization for Economic Cooperation and Development, Paris, France; [3]American Chemical Council, USA [4]UBA, Germany; [5]Department of Health and Ageing, Australia; [6]NIHS, Japan; [7]U.S. Food and Drug Administration, USA; [8]World Wildlife Fund, UK; [9]Chemical Management Policy Division, Ministry of Economy, Trade and Industry, 1-3-1 Kasumigaseki, Chiyoda-ku, Tokyo 100-8901, Japan; [10]U.S. Environmental Protection Agency, USA

Abstract: This workshop was convened to address common issues and concerns associated with risk management of endocrine-disrupting chemicals (EDCs). The talks described the tools and policies for key Japanese, Australian, German, and U.S. regulatory agencies. The agencies participating in the workshop were responsible for the regulation of various substances including: chemicals, pesticides, environmental contamination, pharmaceuticals, and food additives. The panel also described the role of the Organization for Economic Cooperation and Development (OECD) in standardizing the tools and validation of testing and screening methods. The panel also included nongovernmental organizations presenting the views of the World Wildlife Fund, and the chemical industry from industrialized nations; each organization described its concerns and proposed approaches to risk management of EDCs. This summary highlights the most important areas of common points of view of government, industry, and environmentalists. We also try to identify issues upon which viewpoints diverge.

GOVERNMENT AGENCY VIEWPOINTS

The presentations in this workshop discussed governmental authority to regulate chemicals, pesticides, food additives, drugs, and contaminants in the air and water. Concerns were expressed for the potential risks to both wildlife and humans of exposure to endocrine active substances (EASs).

---

*Report from a SCOPE/IUPAC project: Implication of Endocrine Active Substances for Human and Wildlife (J. Miyamoto and J. Burger, editors). Other reports are published in this issue, *Pure Appl. Chem.* **75**, 1617–2615 (2003).

‡Corresponding author

Japan

In his presentation on certain aspects of the Japanese government efforts, Dr. Inoue explained that little is known about the experimental evidence for human effects of EDCs [1]. Data gaps are most pertinent in exposure monitoring and understanding of the dose response relationship. He explained that the relevance of the low-dose effects issue, if any, must be better understood. He noted that this may also lead to a better understanding of issues relating to additivity and synergism. He further informed the meeting attendees of the Japanese Intergovernmental Agencies cooperation on the EAS issue, and the extensive research efforts being undertaken by the Japanese government. He acknowledged that the evidence of human health effects for EASs is uncertain at this time, and that considering the biological plausibility of such effects, human data should be a major focus of future research.

In his talk and paper [2], Dr. Matsumoto gave a full overview of all Japanese government-sponsored research activities in the endocrine disruptor (ED) area. The Ministry of Economy, Trade and Industry (METI) facilitates trade activities, while the Ministry of Health, Labour, and Welfare (MHLW) promotes human health and worker safety. The Ministry of the Environment (MoE) is responsible for environmental protection, the Ministry of Agriculture, Forestry and Fisheries (MAFF) covers natural resources and food safety, and the Ministry of Education, Culture, Sports, Science and Technology (MEXT) promotes basic studies and research. Japan has formed an inter-ministry collaboration body that brings the efforts of these ministries to promote efficient use of resources and appropriate risk assessment and risk management.

Germany

Dr. Gies provided his view of the German and European perspective on EAS regulation. He explained in his talk and manuscript [3] the plans for the new European Directive for new and existing chemicals. Chemical production is growing faster than the population. He described evidence on imposex in snails and endocrine disruption in fish as evidence of adverse effects in wildlife from EDCs. Effects are reported worldwide and include population effects. In his opinion, environmental effects could be considered as early indicators of human effects. He reported to the meeting that there is a candidate list of more than 500 suspect EDs. He described changes in chemical regulation under the European Union's New Chemical Policy that shifts the burden of proof to the manufacturer. He called for intensified research, increased coordinated monitoring programs, facilitated flow of information, and promoted improvements to quantitative risk assessment, especially the use of wildlife data as a model for humans and vice versa.

Australia

Dr. Les Davies, of the Office of Chemical Safety, TGA Department of Health and Ageing, explained the current legislation in Australia [unpublished presentation] and focused on a few cases including the exposures of sheep to contaminated clover, resulting in infertility in sheep. Many of the issues that have concerned the Office of Chemical Safety have included contamination of veterinary feed, foods, and industrial chemicals. Australian chemical regulatory agencies support the improved screening and testing for potential EDCs. However, endocrine activity is one of many chemical or biological properties of concern, and other mechanisms can also be affected by excessive exposure. He mentioned that an extensive package of tests required for drugs, pesticides, new HPV chemicals, and food additives is unlikely to miss any ED effect. The Australian government is concerned about the protection of the environment and of public health as it is intimately related to the environment.

United States

The United States government's approach to regulating EASs was described for two regulatory agencies, the U.S. Environmental Protection Agency (USEPA) and the U.S. Food and Drug Administration (USFDA).

Dr. Gary Timm of the USEPA ran through U.S. laws that deal with chemicals [unpublished presentation] and informed the meeting that risk management of EDCs as a unique category is not yet done in the United States. There are 12 major laws that apply to the regulation of chemicals in the United States. Some of these laws (e.g., Toxic Substances Control Act, Federal Insecticide Fungicide and Rodenticide Act, and Federal Food Drug and Cosmetic Act) focus on the manufacture, distribution, and use of substances or products. Other laws (e.g., Clean Air Act and Clean Water Act) limit or control emissions or releases into the environment, and some laws (e.g., Resource Conservation and Recovery Act) address the damage already caused by releases. Each law has unique criteria for determining acceptable risk, and for managing that risk.

Risk is the driving element in the U.S. legislation, and this does not leave much room for the precautionary principle. Risk concepts in the United States in some cases include the consideration of benefits in a risk-benefit balancing formula. Such a method is applied to the registration of pesticides under FIFRA or management of chemicals under TSCA. In other cases, risk is the only criterion, and reasonable certainty of no harm is the standard applied under most of the Federal Food Drug and Cosmetic Act (FFDCA). The USEPA has found that the current state of the science does not support a special consideration of the low-dose hypothesis in the risk assessment or risk management process. While encouraging further research, the risk management process takes no specific account of this controversial area.

The current issues related to drug research and approval in the United States were presented by Dr. Abigail Jacobs and are described in her manuscript [4]. For drugs, there is no standard nonclinical battery of tests. She noted that in vitro tests can be useful in understanding the mechanism of action, but that even for EASs, in vitro studies were not necessarily of predictive value. Dr. Jacobs discussed many of the technical issues, such as the choice of test tissue, choice of receptor, and details of the test system. If there are risk issues, they can be solved in various ways. For drugs, warnings and precautions may be sufficient risk management techniques, because of the voluntary and relatively controlled nature of the exposure. In conclusion, an endocrine signal is not considered equal to a risk, and each case is evaluated on its own merits. The risk–benefit assessment is the paramount consideration for the USFDA and drug evaluation.

ROLE OF THE ORGANIZATION FOR ECONOMIC COOPERATION AND DEVELOPMENT (OECD)

Dr. Koëter of the OECD gave an overview of the EDTA (Task Force on Endocrine Disrupters Testing and Assessment) work and explained in detail the recently accepted conceptual framework for testing and assessment, comprising five levels [5]:

1. sorting and prioritization
2. in vitro assays providing mechanistic data
3. in vivo assays about single endocrine mechanisms
4. in vivo assays providing data about multiple endocrine mechanisms
5. in vivo assays providing adverse effects data from endocrine and other mechanisms

He described several of the test methods being evaluated for potential value in assessing EDCs. He also presented a list of reference chemicals for estrogenic agonism and antagonism, androgenic agonism and antagonism, thyroid agonism and toxicity, and aromatase and 5 α-reductase inhibitors.

He further mentioned the "Share the Burden" program that would allow countries to use data and assessments developed by other countries. He explained that risk management is driven by risk assessment, science, economic factors, social factors, and cultural factors.

The OECD is managing several complicated and aggressive programs designed to improve and standardize testing and assessment methods. Many nationalities are participating actively in these exercises.

ENVIRONMENTALIST ORGANIZATION VIEWPOINT

Dr. Lyons of the World Wildlife Fund (WWF), in her presentation and paper [6], put emphasis on the European legislation. She argued that the precautionary principle should be applied to EDCs. In the current proposal, the decision-making process does not really apply to this principle. She made a plea for enforcement of the use of substitute chemicals for endocrine disruptors, when available. The availability of substitutes should be grounds to refuse authorization. She criticized the European Community (EC) by moving away from minimal risks to acceptable risks. She made the following conclusions:

- The existing regulatory policy does not protect adequately health and the environment.
- It is too onerous a burden for regulators to prove unacceptable risk.
- Only 12 persistent organic pollutants (POPs) and a few other chemicals have been phased out in most countries.
- There is a need for a new paradigm toward comparative risk assessment, in order to find safer alternatives to EDCs.

She also presented views on the European Union (EU) Authorization Process and presented reasons to change the current conditions of the process.

INDUSTRIAL VIEWPOINT

Dr. Rick Becker's presentation and paper [7] on behalf of the American Chemistry Council, the European Chemical Industry Council (CEFIC), and Japan Chemical Industry Association, emphasized that endocrine disruption should not be considered as an endpoint for classification and, in fact, as such not as an adverse effect but as a mechanism. He stressed that the relative potency of EDCs would be an important element to consider in risk assessment. He presented the view that the risk management process should be based upon risks and benefits, considering certain alternatives and societal concerns. The risk assessment process is based upon collecting data and reviewing and assessing hazard potential, then reviewing and assessing exposures. Risk assessment includes evaluating individual and population effects. He noted that work needed to be done to identify representative taxonomic groups for the environmental assessment.

Dr. Becker described a holistic process for looking at EASs and potential adverse effects. The guiding principles and efforts of the international chemical industry-sponsored Long-Range Research Initiative (LRI) were described. Dr. Becker also proposed a process for dealing with nonvalidated, new, and novel methods or data.

CONCLUSIONS

Though not unanimous on all points, this diverse workshop demonstrated areas of considerable agreement and focus. Divergent viewpoints have also been highlighted by the papers, presentations, and this review. No one should expect such complicated and important issues to be quickly or easily resolved. Significant political or cultural differences were evident by the various organizations, and the needs of different regulatory systems. Research efforts have been extensive and successful in many areas. That research is being increasingly applied in the risk assessment and risk management processes.

REFERENCES

1. T. Inoue. *Pure Appl. Chem.* **75** (11/12), 2555–2561 (2003).
2. S. Oikawa and M. Matsumoto. **75** (11/12), 2609–2611 (2003).
3. A. Gies. **75** (11/12), 2563–2574 (2003).
4. A. Jacobs, P. Brown, J. Farrelly, J. E. Fisher, D. Morse. **75** (11/12), 2605–2607 (2003).
5. H. B. W. M. Koëter. **75** (11/12), 2613–2615 (2003).
6. G. Lyons. **75** (11/12), 2593–2604 (2003).
7. S. Webb, R. Taalman, R. Becker, K. Onuma, K. Igarashi. **75** (11/12), 2575–2591 (2003).

Pure Appl. Chem., Vol. 75, Nos. 11–12, pp. 2555–2561, 2003.

Workshop 6.2

Hormonally active agents and plausible relationships to adverse effects on human health*

Tohru Inoue[‡]

Center for Biological Safety and Research, National Institute of Health Sciences, 1-18-1 Kamiyohga, Setagaya-ku, Tokyo 158-8501, Japan

Abstract: A hormonally active compound was first identified in the book *Silent Spring* by Rachel Carson in 1962, implicating the effect of pesticides such as DDT and the derivatives. Nearly four decades later, the book *Our Stolen Future* by Theo Colborn et al., and other pertinent publications have revisited and broadened the issue regarding a variety of possible chemicals and the area exposed. Translation and publication became available in Japan within the last four years. Since then, Japan joined the member countries involved in the global issue of endocrine disruptors, the "environmental hormone".

Although a significant number of chemicals possessing a hormone-like action have been recognized for many years, and the action of their biological plausibility related to the receptor-mediated effects strongly suggests possible human effects comparable to hormonal changes in wildlife, little is known about evidences or adversities in experimental animals and humans. The most essential key to resolving these dilemmas may be to understand the mechanism of actions (i.e., a possible low-dose issue). In other words, the mechanism at the low-dose effect may be resolved simultaneously by the mechanism of three major questions linked to the low-dose issue; namely, threshold, possible oscillation, and additive and/or synergistic action.

INTRODUCTION

The objective of this paper is to summarize all currently available information on hormonally active agents and plausible relationships to adverse effects on human health from the standpoints of the mechanisms of action of these chemicals.

It is not uncommon to come across agrochemicals and industrial chemicals that have hormone-mimic effects. These chemicals, the so-called "environmental hormones", often accumulate at detectable levels in the environment, and it has been feared that they may have adverse effects not only on wildlife but also on human beings. Following reports of feminization and decreased colony size of wild creatures, and reports suggesting a possible association of these chemicals with abnormalities of reproductive organs and oncogenesis in humans, attention has focused on the possibility that these occurrences may be associated with exposure to endocrine-disrupting chemicals (EDCs). In this connection, we would like to draw the attention of the reader to a Japanese translation of the book *Our Stolen Future*, written by Theo Colborn et al.

*Report from a SCOPE/IUPAC project: Implication of Endocrine Active Substances for Human and Wildlife (J. Miyamoto and J. Burger, editors). Other reports are published in this issue, *Pure Appl. Chem.* **75**, 1617–2615 (2003).
‡Tel.: +81-3-3700-1564; Fax: +81-3-3700-1622; E-mail: tohru@nihs.go.jp

This paper will review the subjects related to EDCs, the courses of arguments regarding the possible hazards of these chemicals, and current medical subjects pertaining to them.

CHEMICALS WITH HORMONE-MIMIC ACTIONS

Substances with hormone-mimic effects can be divided into four groups:

- hormones found in vivo;
- medicines with hormone-mimic actions manufactured for use in hormonal therapy, etc.;
- plant hormones known to exert phytoestrogen-like actions; and
- chemicals found in environments that can interact with hormone receptors.

In addition, substances that do not interact with hormone receptors but exert effects on gonads by their modifying effects on steroid metabolism may be deemed as hormone-mimics in the broader sense of the term. In this paper, however, emphasis shall be placed on the hormone-mimic actions mediated by receptors that play essential roles in the mechanism of actions of hormone-mimics.

CHARACTERISTICS OF THE RECEPTOR-MEDIATED ACTIONS OF HORMONE-MIMICS

The receptor-mediated actions of hormone-mimics are fundamentally characterized by the similarity in structures of the receptors involved, crossing the barrier of animal species. These characteristics allow us to speculate the possibility that the actions of these chemicals exerted in nature may also occur in humans.

Second, since similarities in the structure of various sex steroids and hormones are also known, it is possible that each individual hormone-mimic exerts diverse effects by acting on male hormone receptors, female hormone receptors, and nuclear receptors (including many orphan receptors), etc.

Third, many of these chemicals are excreted from the living body in the form of conjugated inactive substances instead of as degraded metabolites. They may also be eliminated in the unchanged form. Therefore, if feces and urine containing these substances are eliminated into river water, it is plausible to imagine that even inactivated hormones can sometimes become active and exert hormone-mimic actions in the environment. This is one of the characteristics unique to this class of chemicals.

Receptor-mediated responses involve many unresolved questions. Various undefined elements may be involved, including the relationship between receptor binding and signals, the relationship between receptor-ligand binding (ligand: substances that can bind to receptors) and the dissociation of ligands from receptors, signal cross-talks, involvement of unknown nuclear receptors, etc.

The actions of these chemicals add to the effects of intrinsic hormones. For this reason, these chemicals may exert their actions in a way different from that known for other chemicals that do not have structural or functional counterparts in vivo. For example, stimulation of hormone receptors by these extrinsic chemicals may modify homeostasis in vivo, leading to down-modulation of the physiological stimulation of these receptors by the intrinsic ligands. Therefore, the influence of the continued effects of environmental hormones needs special study.

PITFALL IN THE EFFECTS OF HORMONE-MIMICS

We must distinguish the interactions of endocrine hormone-mimics with hormone receptors from the hazards caused to endocrine tissue. Bearing this in mind, let us now summarize the problems related to the effects of hormone-mimics.

Antagonistic effects maintaining homeostasis

The endocrine system is regulated by homeostatic mechanisms. It is not uncommon for the effects of small amounts of hormone-mimics to interfere slightly with these mechanisms, often with no adverse influence; this is well known. However, this is not always the case. There seems to be a group of genes that act antagonistically to each other in the maintenance of homeostasis.

With the uterotrophic assay, which is used to check for estrogenic activity, the ovary is removed in advance and the blood level of the intrinsic female hormone is reduced to the minimum. Under the thus-created extremely shrinking state of the uterus, the test substance (a chemical or hormone) is administered to evaluate for its effects on the inflation of the uterus. This test (checking for growth of the uterus in ovariectomized animals) is designed to evaluate the hormone activity and effects of hormone-mimics under conditions of blockade of homeostasis.

This test method itself is valid. However, there is no sufficient rational evidence that indicates that the responses observed under such indirect control conditions of the living body can serve as an indicator of the health hazards of hormone-mimics. Although the ovo-testes seen in lower vertebrates may be used if the effects observed were to be valid as such an indicator, there is no consensus on what is valid as an indicator of the health hazards of EDCs when mammals are used as experimental animals.

Down-regulation of the expression of receptors

It is known that the expression of gene-encoding receptors is down-regulated by continuous stimuli, leading to reduced receptor activity. This can lead to a paradoxical outcome wherein the effects observed in the presence of low levels of a substance are not seen at high levels of the same substance. If this phenomenon occurred in individual organisms, the dose–response relationship will be nonlinear.

This means that extrapolation of results obtained at high levels of the chemicals, to conditions where low levels of the same substance are present, would be difficult. It is needed to test the validity of this hypothesis; analysis of the mechanisms underlying this phenomenon if the hypothesis were indeed valid, are thus important. Studies to resolve these questions are now under way.

Data gap concerning the effects of female hormones

In mature women, there are high levels of physiological hormones in vivo, and these are subject to cyclic control. It has been proposed that girls with inadequate physical growth begin menstruation at lower ages and undergo sexual maturation earlier than usual, and that hormone-mimics in these subjects can precipitate breast cancer.

The weak links in this hypothesis have been pointed out, and it has been shown experimentally that estrogen by itself may be teratogenic, although this tendency has been shown to be weak. It is known that organisms are programmed such that excessive exposure to estrogens during the intrauterine period or other developmental stages is avoided.

There are many open questions as to the mechanism by which mature females remain physiologically stable, even when exposed daily to high levels of estrogen (400 pM/l). Some additional dramatic effects may be needed to disturb this homeostatic physiology.

Multigeneration tests and effects on fetuses

It has been shown that exposure to hormones or hormone-mimics during intrauterine or early neonatal periods can lead to irreversible changes in the pattern of development. This susceptibility period is short, extending from the 13th gestational day to about one week after birth. These effects are the so-called "intrauterine window effects."

In animal studies involving observation of experimental animals for two or more generations, no effects of EDCs have been demonstrated. The question therefore arises as to why window effects are observed during the short period mentioned above. It is unknown whether or not these effects really do occur, and if they do, how they are produced.

Delayed growth of the thalamic nucleus specific to males (called sexual dimorphic nucleus) is seen in male rats treated with female hormones. We may say that under conditions of homeostasis of the physiological hormones in mature individuals, exposure to dose levels that usually cause only reversible changes can lead to irreversible changes, if the exposure occurs during genesis, morphogenesis, or functional development. However, there are no ample data endorsing this view in humans.

Considering the biological plausibility inferred from the experimental data accumulated to date*, we may say that there are no sufficient data that clearly rule out this view. Close attention has therefore been paid to these effects in children.

New theories of methodology, focusing on effects in fetuses and children, are now being developed, primarily in the United States, or the World Health Organization, within the framework of children's program, etc.

HEALTH HAZARDS AT LOW LEVELS OF EXPOSURE

Chemicals used for agriculture or industrial purposes are marketed, in general, only after their effects on living beings have been investigated. We may therefore understand that they are used on the premise that the possibility of these chemicals exerting hazardous effects on health at relatively high-dose levels has been almost ruled out. Nevertheless, problems with EDCs have begun to be highlighted. These problems may not be confined to those related to the accumulation of these substances through food chains in the ecosystem, but also to the additional possibility that these chemicals may exert effects at low-dose levels even if they have been declared safe at high-dose levels. The latter possibility may apply, however, only to some cases and not to others.

We may say that a major issue pertaining to EDCs that must be resolved urgently is whether or not they pose health hazards at low-dose levels. This issue can be summarized into the following three questions:

- presence/absence of threshold level;
- presence/absence of synergistic or additive effects; and
- possibility of extrapolation of high-dose effects to low-dose levels (i.e., presence/absence of a linear dose–response relationship).

No clear-cut answers have as yet emerged to these questions. Considering the above-mentioned characteristics of the effects of hormones, it is plausible to imagine how difficult it may be to resolve these questions.

To determine if these chemicals exerted hazardous effects on health at low-dose levels, the following basic questions may need to be considered; their biological plausibility is hardly denied.

- Regarding the presence or absence of threshold levels, it seems likely that many chemicals suspected of being EDCs can easily permeate across the cell membrane, which is composed of phospholipids. Therefore, assuming that one receptor molecule reacts with one chemical molecule, the lower limit of the dose level exerting the chemical's effects would be extremely low.

  Of course, since the probability of the binding of a ligand to the receptor will be low if the dose level is low, we cannot say that there is no threshold level for the effects seen in the low-

*Biological plausibility: Likelihood of a phenomenon as judged by considering the difference or similarity of elements of reactions in individual organisms, on the basis of the results of a series of related biological experiments. (Cf. probability)

dose-level range. In fact, for bisphenol A (which has been attracting close attention because of its hazardous effects on health at low-dose levels), the presence/absence of a threshold level has not yet been reported. It seems rational, therefore, to assume that these health hazards occur in a very low-dose-level range.

- If we consider not only the affinity of each substance for the receptor, but also the nonlinearity of responses (e.g., waveform responses as a result of reduced receptor expression following an increase in dose level), it is possible to assume that there are U-shaped or reverse U-shaped reactions, or oscillational dose–response curves. Interim data endorsing such a view are being accumulated.
- Regarding the possibility of synergistic or additive effects, the observation of additive effects among different nuclear receptors has been reported. Data yielded by analysis of interactions between receptor signals also suggest such a possibility. In fact, the dose–response curves for some composite materials were reported to be additive, but not synergistic.

Thus, the questions on health hazards at low-dose levels have several aspects:

- type of receptor-mediated actions of the hormone mimics;
- diverse reactive characteristics on the part of the receptors;
- diverse modification during expression of intracellular signals; and
- factors involved in irreversible changes related to morphogenesis and functional development.

Resolution of all these aspects of the question will lead to clarification of the mechanism of actions of the substances from each of the aforementioned standpoints. While these questions are among the hottest research themes at present, they are certainly unlikely to be resolved easily.

At a workshop held in North Carolina, USA, in October 2000, health hazards of chemicals at low-dose levels were discussed. Investigators for and against the possibility of these substances posing health hazards at low-dose levels gave detailed accounts of their studies, and no definitive conclusions could be reached, as the arguments of both sides appeared to be tenable.

This means that reports affirming the plausibility of these substances posing health hazards at low-dose levels in animal experiments cannot be immediately rejected. The workshop concluded by pointing out the necessity of paying attention to the possible hazards on fetuses and neonates.

HEALTH HAZARDS OF HORMONE-MIMICS TO HUMANS

The possibility of health hazards of hormone-mimics to human beings have not been supported by adequate epidemiological data, and the number of cases for which the data clearly endorse such effects is quite small. The U.S. National Research Council (NRC) emphasizes the necessity of conducting further epidemiological studies on this topic (NRC, 1999).

In conclusion, this paper summarizes the current knowledge concerning the health hazards of hormone-mimics to humans. Reports dealing with the effects of these substances on humans are confined to those pertaining to the effects of dioxins and polychlorinated biphenyls (PCBs); the validity and usefulness of these results have not yet been established.

The following information is based on case studies conducted to date.

Health hazards of dioxins

Regarding health hazards of dioxins, two-year dosing studies revealed weight loss and liver damage, and three-generation reproductive studies in rats disclosed intrauterine death and a decrease in litter size. Onset of endometriosis in rhesus monkeys has also been reported.

A causal relationship of EDCs to the following episodes in humans has been suggested: biased male-to-female ratio in children born in the dioxin-exposed Seveso area of Italy, and increased incidence of cleft palate in the Diemerzeedijk district of the Netherlands, probably due to steroids. In both of these cases, the U.S. Environmental Protection Agency (USEPA) did not affirm a causal relationship, and classified them as cases requiring special attention.

No consensus has been reached concerning the relationship of hypothyroidism observed in the inhabitants along Lake Michigan to the ingestion of PBB- (polybrominated biphenyls-) contaminated fish.

Effects on mature females (e.g., increased incidence of breast cancer)

No reports affirm the effects of dioxins on mature human females (e.g., effects on breast cancer or endometriosis as discussed below). There are many unresolved questions on this topic. However, none of the studies conducted in mature experimental animals revealed data endorsing the plausibility of occurrence of such effects. On the other hand, it is known that the age at menarche is lower and the incidence of breast cancer higher in females exposed to dioxins. Some investigators cite these data when discussing the health hazards of dioxins. It is also known that females exposed to dioxins are often taller.

In European countries, a height increase of about 3.5 mm per year and an approximately one-year decrease in the age at menarche have been reported during the past 30 years. It is difficult to identify the influence of extrinsic endocrine factors on these changes, and no studies addressing this issue have been reported to date. Although several studies have been published concerning the effects of female hormone preparations, including pills used for contraception and hormone replacement therapy in postmenopausal women, no studies have provided data that establish the effects of EDCs.

Endometriosis

Endometriosis is a disease of unexplained origin that is seen in primates with sexual cycles. It has been pointed out that this disease tends to be more severe in individuals exposed to dioxins (2,3,7,8-tetrachlorodibenzo-*p*-dioxin [TCDD] and to PCBs). Data yielded from experiments in rhesus monkeys are used as evidence to corroborate the causal relationship between dioxins and endometriosis. Thus, we cannot rule out the biological plausibility of these effects. However, no reports affirming the causal relationship in humans have been published.

Possibility of other effects on humans

Biological plausibility has also been considered for the following effects of hormone-mimics on humans: qualitative dysfunction of human sperm, effects on neurobehavior of neonates, and immune functions. The effects on immune functions have been suggested by reports of cases with Yu-sho (PCB intoxication).

CONCLUSION

The International Program of Chemical Safety (IPCS), a section of the World Health Organization, has released a Web site publication "Global Assessment of the State-of-the Science of Endocrine Disruptors" (GAED), June 2002 (URL: <http://ehp.niehs.hih.gov/who/>). WHO/IPCS started the GAED program in March 1998 after the publication of *Our Stolen Future* (Theo Colbone et al., 1996). The publication took three years to edit; covering a policy to document all the published pertinent literatures, to summarize them as descriptive manner solely based on those published literatures. Twenty-seven expert scientists and 20 independent peer-reviewers participated in editing the GAED.

Other reports on nonylphenol and octylphenol, released by the Japanese Ministry of Environment (MoE), revealed an "ovotestes" formation that was observed in the assay of the laboratory experimental fish (*Medaka*) exposed to doses close to those recorded in the monitoring fields in the MoE surveillance. Further, phthalates, such as di-(2-ethylhexyl)phthalate, di-cyclohexylphthalate, and butylbenzylphthalate, as selected and prioritized chemicals by the MoE, showed some unique data in different endpoints, including mRNA expression, in dose ranges lower than those no observed effect levels (NOELs) and/or no observed adverse effect levels (NOAELs) reported previously.

The effects of EDCs on human health are unknown at this moment. However, due to the biologically plausible data currently accumulated, the existence of endocrine disruptions under certain circumstances seems to be a reality. Thus, by the time of the SCOPE/IUPAC symposium, the EDC research for the next stage may shift from plausibility to possibility, and put forward further mechanistic research.

Pure Appl. Chem., Vol. 75, Nos. 11–12, pp. 2563–2574, 2003.

Workshop 6.3

Government view of endocrine disruption in wildlife*

Andreas Gies

German Federal Environmental Agency (UBA) P.O. Box 330022, 14191 Berlin, Germany

Abstract: Like hardly any other issue in ecotoxicology, endocrine disruption has given rise to public concern. Reproductive, behavioral, and immunological effects in wildlife were publicly not only understood as possible threats to wildlife populations, but also as early warning signals that human health could be at risk. Above all, the public has been concerned about negative outcomes in reproductive health, and effects like feminization in fish were regarded as evidence for the biological plausibility of the hypothesis that environmental levels of hormonally active chemicals are high enough to affect human reproductive health.

Public concern has been mirrored by several parliamentary and governmental decisions emphasizing the need for extensive research and rapid measures to reduce the risk associated with endocrine-disrupting substances.

Endocrine disruption in wildlife is clearly a priority issue. At least in densely populated areas like Europe, symptoms of endocrine disruption in wildlife cannot only be detected in areas with abnormally high levels of pollution, but have also occurred in main river systems, estuaries, and even in the open sea. Imposex in mollusks and feminization in fish that were clearly related to disturbances in the hormonal system of these organisms by exogenous substances have been used as markers in monitoring programs. Though symptoms of endocrine disruption can be clearly identified, it is much more difficult to link these outcomes to causative chemicals or mixtures of substances. Natural and pharmaceutical hormones, phytoestrogens, pesticides, and industrial chemicals may all play a role to a different degree depending on the site under study. This means that several different risk-reduction strategies have to be applied, including bans of substances, use restrictions, and installation and optimization of sewage treatment works embedded in a strategy for the overall reduction of chemical input into the environment.

It should be noted that, in addition to national and international regulatory actions taken by state authorities, a considerable reduction of the environmental input could be achieved in several countries by voluntary actions taken by industry.

Regulatory bodies are still facing major problems in the field of risk assessment and risk reduction. Association between effects and causative agents or mixtures are in many cases weak. Important tools for risk assessment such as dose–response relationships or the existence of thresholds are not yet agreed on. These uncertainties are the reason that many national governments and the European Commission have identified precaution as the main element in chemicals policy for the management of endocrine disruptors.

*Report from a SCOPE/IUPAC project: Implication of Endocrine Active Substances for Human and Wildlife (J. Miyamoto and J. Burger, editors). Other reports are published in this issue, *Pure Appl. Chem.* **75**, 1617–2615 (2003).

This paper is based on documents of the German Federal Environmental Agency, but solely represents the view of the author from a regulatory perspective and emphasizes the wildlife aspects of endocrine disruption.

INTRODUCTION

For more than 60 years, scientists have known that a large number of environmental chemicals can have hormonal activity though their structure does not resemble those of known hormones [1]. As early as 1936, Dodds and Lawson described the estrogenic activity of bisphenol A [2] and as early as the 1950s and 1960s the hormonal activities of phytoestrogens like genistein [3] and common pesticides such as methoxychlor [4] and DDT [5] were already known. Nonetheless, it was not until the early 1990s that the issue of disruption of the endocrine system from chemical exposure became the subject of public and political interest [6].

Publications of various conference documents and scientific articles in the early 1990s totally changed the character of the debate on endocrine disruptors. With these articles a widespread public interest was induced on possible effects of hormonally active environmental chemicals.

In 1991, a group of scientists met in Wingspread, Wisconsin. At the end of their meeting they worded a declaration [6] pointing to developmental impairments in wildlife and laboratory animals that were obviously caused by chemicals interfering with the hormonal system. The effects observed became visible in different species such as birds, fish, and shellfish, and were different in nature, probably affecting different hormonal systems such as the system of sexual hormones and thyroid hormones. Despite all the differences observed in the character of the effects they had several common characteristics:

- The chemicals may have effects in embryos, fetuses, and newborns that differ from those in adult organisms.
- In many cases, the effects may not (or not only) be seen in the parent generation exposed to them, but rather in its progeny.
- There are particular times when organisms are sensitive to these chemicals. Outside these windows of sensitivity, the chemicals may have no effect on the developing organisms.
- The effects may be delayed. Though the organism is exposed in very early life stages, negative effects may not become visible until adulthood.

The participants of the Wingspread meeting saw these effects in wildlife not only as indicators that wildlife populations were possibly at risk, but also warned that the developing human fetus may be negatively influenced by environmental chemicals disturbing the hormonal system.

The debate on the possibility of human health being harmed by substances with effects on the endocrine system was further sparked by a supposed reduction in men's sperm quality in industrialized countries. A meta-analysis published in 1992 (and since then frequently reanalyzed and controversially discussed) originally concluded that the sperm count in ejaculates from test subjects had fallen by approximately 50 % between 1938 and 1990 [7]. It was hypothesized that exogenous chemicals with estrogenic activity may have caused these changes [8,9] (for review see Topic 3.4 [10]).

Not only toxicologists and ecotoxicologists had to critically review their paradigms in the ensuing discussion on endocrine disruption. People working in the government or in governmental agencies were also challenged by this debate that quickly had political dimensions. Those responsible for the safe use of chemicals in society were confronted with several problems and open questions.

First, there were two different groups of experts in the agencies: those responsible for human toxicology and those for ecotoxicology. In most countries, both groups are differently educated, have a different professional history, and work in different ministries or agencies. The lack of communication between these two subgroups of regulators may be one reason for the rather slow reaction of regulatory bodies on the new challenges posed by endocrine disruptors.

Second, no one working in the field of chemical safety was prepared to debate how far effects in wildlife could serve as indicators for human health risks. No one really knew whether, for example, feminization of fish populations in surface waters by estrogenic substances could be interpreted as an early warning signal that similar processes may be induced in humans.

In addition, the regulatory agencies did not know how many out of the approximately 50 000 chemicals that are marketed today are interfering with the hormonal control of life processes.

It was also unclear to what extent wildlife populations were endangered by these types of substances, whether phenomena like morphological and physiological feminization in fish, masculinization in snails, or alteration of behavioral patterns in birds were common throughout the environment and would subsequently need rapid and radical regulatory measures or whether they were restricted to some hot spots, heavily polluted parts of the landscape, and could be easily tackled with measures at a local scale.

Last, several questions arose concerning how to test and assess chemicals in the routine processes of chemical notification. Those questions were associated with the unique features of endocrine disruptors. Do standard tests sufficiently cover transgenerational effects? Were safety factors used for normal chemicals also sufficient for endocrine disruptors? Are there concentration thresholds below which a risk for wildlife is unlikely?

Ten years after these questions were posed to officials, it is time to review whether we have succeeded in answering them and what degree of uncertainty we still face.

IS ENDOCRINE DISRUPTION WIDESPREAD IN THE ENVIRONMENT?

To analyze the size of the problem posed by endocrine disruption, it is useful to view this problem in relation to the overall endangerment of natural biological systems. Ever since industrialized societies changed the world dramatically, biological diversity on earth has been at risk. Many species have vanished or are endangered. In an industrialized country like Germany, 3 % of 15 850 animal species for which data are available have died out and a further 35 % are endangered [11]. In most cases, it is impossible to understand the reasons why these organisms are in danger, but in many cases drastic changes in the structure of the landscapes and surface waters created unfavorable conditions for the survival of vulnerable species.

Additionally, toxic contamination may play its part in endangering species and populations, though its part in total hazards can hardly be quantified. For macroinvertebrate communities in rivers, for example, toxic contamination is likely to be of the same importance as unfavorable ecological conditions [12]. But even if general toxic effects can be observed in wildlife, in most cases the mechanism of toxicity is not known. Adverse toxic effects like imposex that can be correlated directly to endocrine disruption have been described only rarely. To answer the question whether endocrine disruption is a widespread phenomenon or only restricted to special places with extremely high contamination, one is dependent on data from the aquatic environment. The majority of information as to chemical disruption of the endocrine systems of free-living organisms has been derived from aquatic ecosystems. This is due to the fact that most research projects have concentrated on this medium. But there is no reason to suppose that there is no impact on terrestrial ecosystems.

Two phenomena related to endocrine disruption have been intensively investigated during recent years: masculinization of mollusks and feminization of fish. Therefore, these two examples are suitable to look at regional patterns and the worldwide severity of changes due to the disturbance of the hormonal system by exogenous substances.

Organotins and invertebrates

The incidence of certain disruptions in the fertility and development of marine and limnetic mollusks, such as imposex (female snails are expressing male sex organs additionally to female organs), prosobranch gastropods, or shell malformation and disruption of larval development in oysters, are considered to be direct effects of aquatic pollution by organic tin compounds, especially tributyltin and triphenyltin (TBT and TPhT) [13]. TBT is a nonsteroidal compound, which, for example, is used in antifouling paint for ships. Direct correlations between the occurrence of imposex and TBT pollution were found above all in the vicinity of harbors [14,15], and the degree of imposex in prosobranch snails is related to the degree of TBT exposure [16]. This effect can be shown at TBT concentrations as low as 5 ng/l (as Sn) [17]. More than 150 species of prosobranch snails have been found to be affected by organotin compounds [18].

Occurrence of imposex has been described in many parts of the world: Japan [19], Spain [20], Northern Europe [21], Malta [22], Taiwan [23], Korea [24], Portugal [25], Ireland [26], Argentina [27], Thailand [28], and Australia [29].

In the case of imposex induction by TBT and TPhT, the threat for biodiversity is clearly worldwide due to the global nature of the shipping industry. Population decreases and local extinction have been observed in many places around the world at concentrations that are extremely low and frequently found in the environment. This is clearly one example for the necessity to develop global strategies for chemical control.

In the case of imposex, environmental monitoring revealed that a class of animals was seriously threatened that is not represented in the battery of test methods used routinely to assess chemicals. The threat that TBT and TPhT pose to prosobranch snails was unpredictable by our current assessment methods. This should not be surprising as for decades specific toxicity for some animal classes is one of the selection criteria applied in the development of pesticides. Some invertebrate classes in particular have unique hormonal systems [30]. Effects on these endocrine systems may hardly become evident if a chemical passes routine testing procedures.

Feminization of fish

Male and immature fish in contaminated inland, coastal, and marine waters have been found to show incidence of high concentrations of vitellogenin in their blood. This precursor of yolk proteins is usually not found in male and juvenile fish as only females produce yolk protein. The production of vitellogenin is under the control of the estrogen receptor. Vitellogenin is an indicator for contamination by estrogens and substances with estrogenic effects [31–33].

Organisms that are induced to synthesize vitellogenin by external estrogenic stimuli may also exhibit other more or less pronounced negative effects, including changes in steroid metabolism, liver atrophy, delayed testicle growth, and occurrence of eggs in male testicles (ovotestis) in juvenile and adult males, as well as disruption of gamete production associated with reduced reproductive success [34,35]. Induction of vitellogenin production in male fish and feminization to different degrees are common in many parts of the world. These phenomena have been reported in Japan [36], Europe including Finland [37], Italy [38], Switzerland [39], Germany [40,41], the United Kingdom 42–45], and the United States [46,47].

The environmental inputs associated with feminization phenomena in fish are highly variable, including: paper-mill effluents [37], effluents of sewage treatment works [39,43], chemical spills and deliberate chemical applications [48], or complex pollution from human activities [36,41]. Even in places without visible input of polluted waters, like the Wahnbach drinking-water reservoir in Germany, slight elevation of the estrogenic biomarkers have been observed [40], giving room to speculations that natural phytoestrogens or deposition of airborne pollution may be the cause. In large rivers like the Elbe, vitellogenin in male fish is elevated at different sites. Wild bream show different patterns of biomarker response in different stretches of this river with different profiles of chemical pollution [41], suggesting that there was no single chemical that alone could explain the observed inhibitory effects on sexual development.

Both examples—masculinization in mollusks and feminization in fish—may illustrate that endocrine disruption in wildlife is a problem warranting worldwide attention and is not restricted to hot spots with uniquely high levels of environmental contamination. In the first case, the syndrome is clearly associated with a small group of chemicals with TBT as the most prominent member. In the latter case, different polluting chemicals and mixtures have been identified as putative causative substances, including industrial chemicals, natural hormones, and pharmaceutical estrogen analogs. Though these two examples indicate that endocrine disruption is not limited to only a few hot spots, very little is known about the geographic patterns of the occurrence of endocrine disruption. This is particularly true for those areas outside the industrialized agglomerations of Europe and North America.

REDUCING RISKS

The example of fish feminization in surface water clearly shows that the problem of endocrine disruption in wildlife is much more complex than commonly assumed. Several different chemicals are probably involved, and consequently different measures will be needed to reduce the adverse effect of endocrine disrupters in the environment depending on the chemical characteristics of the substance, its use patterns, and its mobility and degradability within the environment. These measures will include the ban of persistent endocrine-disrupting substances. This will be necessary if the substances are widely used in products, and their input into the environment is diffuse so that they hardly can be controlled at the source. TBT and polychlorinated biphenyls (PCBs) are examples for such chemicals where worldwide bans are under way. In other cases, bans are no feasible solution for the problems posed by the substances, and technical solutions have to be sought. The modification of sewage treatment works may be the strategy of choice to remove natural and synthetic hormones more efficiently from sewage and to protect aquatic species. Additionally, several different measures may be useful to reduce the free use of these substances, ranging from voluntary commitments by industry to cease or reduce use of certain chemicals or legally binding use restrictions, to eco-labeling products that do not contain endocrine disruptors (EDs). In the last few years, both for persistent and nonpersistent EDs, substantial progress has been made to minimize their impact on the environment.

Difficulties assessing the risk of endocrine disruptors

For tracing the chemical contamination of wildlife species, the maintenance of an environmental specimen bank in Germany has proven useful. For this specimen bank, funded by the Ministry for the Environment, tissue samples of free-living animals are taken and stored at ultra-low temperature. Tissue concentrations of environmental contaminants can be analyzed retrospectively. Several persistent and nonpersistent EDs have been analyzed in samples, which were taken in 1985 and later [49]. The results revealed time trends and consecutively reviewed the success of regulatory actions that were taken [49].

Nevertheless, it should be remembered that we are still not able to quantify the risk these substances pose to wildlife. Monitoring wildlife populations is only conducted at a very limited number of sites worldwide, and monitoring programs hardly ever include observations of endpoints that may be related to the disturbance of the hormonal system of free-living animals. This lack of coordinated monitoring programs is a factor obstructing targeted environmental policy in the field of endocrine disruption. Our picture of ongoing processes in wildlife is far from being complete. Incidental reports of observations made in the environment, like hermaphrodite polar bears [50] or feminized fish, are worrying the public and causing the impression that public services are not able to characterize the real situation in the environment.

As for many invertebrates (but also some vertebrates), only little is known about the hormonal regulation of life processes, making it virtually impossible to detect any disturbances of these processes [30]. It has only recently been discovered that mollusks are sensitive to estrogenic chemicals like nonylphenol and bisphenol A [51], leading to the characterization of an estrogen receptor in this class of animals in which estrogenic regulation had not been suspected.

SUSPECT CHEMICALS

Besides the question of how widespread endocrine disruption may be in the environment, the most urgent question is how many chemicals possess endocrine-disrupting properties. As only a small minority of all chemicals has been tested for endocrine-disrupting properties, no answer to this question will be comprehensive. According to the definition of EDs agreed upon at the Weybridge workshop, in vivo testing is necessary in all cases to decide whether a chemical is an endocrine disruptor or not, as an endocrine disruptor has to show adverse effects in an intact organism or its progeny. Gülden et al. have published a list of substances in surface waters suspected of being EDs [52]. This list of more than 200 chemicals contains many substances for which a definitive judgement is impossible, due to a lack of valid in vivo studies. A candidate list of 564 substances suspected of being EDs was compiled in 2000 by BKH Consulting Engineers for the EU Commission (DG ENV). Subsequently, this list has been condensed to less than 300 substances, which have given rise to concern.

It should be emphasized that even substances definitively classed as EDs are more or less a random selection, as systematic, large-scale testing programs have not yet been conducted (also a consequence of there being no standardized methods). Furthermore, the EDs do not possess a limited set of clearly describable structural characteristics, making a prognosis as to the total number of disruptors impossible at this time. Nevertheless, comprehensive screening of available literature has shown that we have to face the likelihood that some 100 chemicals with endocrine-disrupting properties are relevant to this problem. Considering that in the European Union (independent of the problem of endocrine disruption), less than 100 chemicals were finally assessed during the last 10 years, it seems unlikely that the traditional forms of chemical risk assessment and risk reduction are able to manage the problem of endocrine disruption in an acceptable time frame. To avoid delays, a targeted risk assessment, focusing on endocrine effects, should be initiated in cases where a comprehensive assessment is not expected in the foreseeable future. The aim of the EU Commission is presently to achieve agreement on a list of substances that should be subject to such a targeted risk assessment.

TESTING FOR ENDOCRINE DISRUPTORS

The aim of every strategy in the field of chemicals policy is to identify risks associated with chemicals before human health or ecosystems are harmed. In this attempt, biological testing of chemicals plays a key role to identify the characteristic risks a chemical may pose. Many of these tests using laboratory animals as a surrogate for wildlife individuals are internationally harmonized today. Particularly in the case of chemicals interfering with the hormonal system of wildlife species, little is known about their natural hormone system. This makes it difficult to develop internationally harmonized tests for endocrine disruption in animal classes, other than those commonly used in chemical and pesticide testing. Hormonal regulation differs so widely among different taxa that the strategy to test only few animal species as representatives for the whole animal kingdom may be obsolete in this case.

Internationally harmonized test procedures for EDs have been developed under the framework of the OECD in the Endocrine Disruptor Testing and Screening Workgroup (see Topic 4.11 for information on the development of fish-based tests [53]). For three years, this group has focused on developing and validating a short-term test for young and adult fish, a test with fish early life-stages (based on the OECD 210 test), a fish reproduction test, a fish full life-cycle test, and two short-term tests for identifying estrogenic and androgenic effects in rats (Uterotrophic and Hershberger assay), as well as an extended 28-day test on oral toxicity to rats exposed to repeated doses (enhanced TG 407). Although the group conducted very intensive work, and even if their present work is continued, it will only cover a small part of the hormonal systems present in the animal kingdom and the possible mechanisms involved in endocrine disruption.

Today we have to face the situation that comprehensive testing of chemicals for their hormone system disturbing properties is a goal that cannot be achieved in the near future. Comprehensive testing is desirable to exclude delayed transgenerational effects and low-dose effects due to unusual dose–response curves that have been observed for some of these chemicals. Nevertheless, due to the multitude of hormone systems existing in wildlife organisms and the even greater multitude of possible adverse effects, a comprehensive test strategy seems not to be practicable. This makes a precautionary approach for the regulation of EDs even more important than in other fields of chemicals policy.

ASSESSING THE RISK OF ENDOCRINE DISRUPTORS

Even if the hazard assessment of a chemical concludes that a chemical is an endocrine disruptor, regulatory bodies have to face the problem of how to conduct a quantitative risk assessment [54]. Many elements of a quantitative risk assessment both for humans and wildlife are still debated controversially.

In traditional risk assessment, a threshold for effects of a chemical agent is assumed. Below that threshold, the body is able to compensate for the exposure, above it, effects become visible. There are some types of effects that do not show thresholds, like mutagenic and carcinogenic effects where theoretically one single molecule may lead to an adverse effect like cancer induction. It has been doubted that such a threshold exists for EDs [55,56]. The absence of a threshold would make it impossible to define a safe concentration.

Risk assessment for wildlife species aims to protect the population rather than the individual. In many cases, adverse effects on the individual level can be detected (e.g., lower sperm production) but the consequences for wildlife populations remain unclear [57].

The largest problem for environmental risk assessment of EDs is the extrapolation of high concentrations to low ones. Some studies suggest that effects of low concentrations of EDs may be more pronounced than of high concentrations [58]. Though this issue is currently subject to heated discussions and intense research, it seems not to be justified to apply linear extrapolation models to data derived from studies with EDs. As a consequence, regulatory bodies do not have a valid model to undertake quantitative risk assessment to date. The German Federal Environmental Agency has recently proposed to perform risk assessment according to procedures agreed for other substance classes, but to apply additionally an extra safety factor of 3 to 5 in human health and environmental risk assessments until the major problems of the methodology of risk assessment for endocrine-disrupting substances are solved.

REGULATION OF ENDOCRINE DISRUPTORS—WHAT HAS HAPPENED DURING THE LAST 10 YEARS?

Some of the most prominent persistent EDs like dioxins, PCB, DDT, toxaphene, aldrine, and dieldrine will be banned worldwide by the Stockholm convention on persistent organic pollutants (POPs), which has been signed in 2001 and will probably enter into force in 2004. This worldwide ban provides that environmental concentrations of these substances will continue to fall, as is observed in most industrialized countries. Humans and other organisms on the end of the food chain will profit most from the further worldwide reduction of the POPs. These substances are probably responsible for some wildlife effects, like reproductive disorders in polar bears [50] and immune suppression in seals [59] (that may be one of the causes of the seal die-offs in the 1980s and very recently in the Baltic and the North Seas). The hypothesis that disturbances of the hormonal systems of these vertebrate species may be part of the mechanisms causing these adverse effects has been discussed.

Because POPs are subject to long-range transport, the highest concentrations in animals were observed in northern and arctic regions to which these substances are transported. Though the POP treaty is open for the incorporation of further substances, work with the aim of internationally prohibiting other persistent pollutants is only progressing slowly, so that the hope to phase out other persistent EDs (such as pentachlorophenol) globally under the framework of this treaty will probably be disappointed at least for the next few years.

The use of TBT is already heavily restricted in many industrialized countries. The application in antifouling paints on vessels less than 25 m in length is prohibited in Canada, Australia, and Europe, while Japan totally banned the use of TBT. The major problem internationally, and for the protection of the sea, is the use of TBT as an antifouling agent on the hulls of ocean-going ships. In October 2001, the International Maritime Organization (IMO) adopted a convention on the control of harmful antifouling systems on ships (AFS-Convention) agreeing that the application of all antifoulings containing TBT should be globally terminated by 2003 and that a complete ban on the presence of TBT antifoulings on ship's hulls should be in place by 2008. The agreed timetable of this treaty will probably be delayed, as the process of ratification in the signing countries is slower than anticipated. Nevertheless, with this agreement in the framework of the IMO, the main source for the input of TBT into the environment will be stopped.

Due to consequent reduction of TBT (and in some countries TPhT) in the environment, invertebrate populations are recovering in many parts of the world [50,60,62] though in others the imposex is still widespread despite a 10-year partial ban of TBT [61]. This may be due to continuous use in large ship paintings or to high contents of TBT in the sediments that may be subject to remobilization. It was reported from the United Kingdom that not only prosobranch snails, but also many taxa of the macrobenthic fauna recovered after TBT was banned for use as hull paint on small boats, and TBT concentrations dropped in an estuary [62,63].

The remaining source of TBT in the environment is its use in many products as a biocide. TBT is still quite frequently found in textiles and other consumer products, though in Europe the TBT industry does not support any other use than in antifoulings. TBT has been found in effluents of Bavarian sewage treatment works in concentrations up to 180 ng/l, which is well above the concentration of 1 ng/l, causing adverse effects in invertebrates [51].

In the European Union, risk assessments for some EDs of major concern have been finalized recently or are close to finalization. These substances include nonylphenol, bisphenol A, and some phthalate plasticisers. For some uses, risk-reduction measures are necessary and risk-reduction strategies are currently being drafted.

It should be noted that, in addition to national and international regulatory actions taken by state authorities, a considerable reduction of environmental inputs could be achieved by voluntary actions taken by industry. The German washing and cleansing industry voluntarily committed to phase out the use of alkylphenolethoxylates (APEs) in household detergents by 1992 and in industrial detergents by 1999. Due to this commitment, the consumption of APEO by the washing and cleansing industry has fallen in Germany by approximately 90 % relative to the 1980s, marking a great success. Nevertheless, considerable remaining quantities (more than 100 t/a) continue to be traded, especially by foreign companies or those who are not members of the German industrial associations, and further measures are required to guarantee the success of the agreement in the long term.

In the case of bisphenol A, European industry has agreed to terminate use of this substance as a stabilizer for polyvinylchloride (PVC), thus closing down one of the two major sources of bisphenol A for the environment. The environmental input caused by recycling thermal paper, the other major source, needs further regulatory action. It should be noted that relatively high concentrations of bisphenol A found in the environment could not be attributed to any identified source and is likely due to diffuse input from bisphenol A-containing products or from degradation products of plastics manufactured using bisphenol A. However, in the absence of a product register, quantifying the emissions at this level is not possible.

POLITICAL IMPLICATIONS

Even more than most other issues in the field of environmental protection, the debate on endocrine disruption was never only a scientific one. Maybe due to the fact that impairments in wildlife species were understood as early warning signals for possible threats to human health, this issue has always been important in political discussions throughout Europe, Asia, and America.

In Europe, the European Parliament, European Commission, and many national governments in the European Union have come to recognize the importance of the issue of EDs and have been asking for more activity with a greater integration into the chemicals safety program. Key documents are the commission's communication on a community strategy for EDs [64] and the document on its implementation [65], which set out the necessity of further research, informing the public, and taking political action. Short-term proposals are prioritizing the various substances, primarily with respect to the risks associated with their hormonal effects, applying existing legal regulations (e.g., assessing high priority substances according to the Existing Substances Regulation), and deciding on monitoring programs, international coordination, and information for the general public. In the medium term, xenobiotic EDs should be determined and assessed, and impetus given to research and development of an improved evaluation of the consequences. This has already begun. In the long term, the EU legal framework on chemicals, crop protection agents, and biocides may require adjustments.

In two decisions in 1999 and 2000, the European Parliament encouraged the commission in its aim to compile a list of EDs and advised the commission to supply sufficient resources for endocrine disruptor research in its research framework programs. The EU parliament called upon the commission in the strongest terms to take rapid action to reduce the risks from EDs, rather than waiting for further tests.

In Germany in August 1999, a decision of the German Bundestag called for a staged but drastic reduction in discharges of proven EDs, drawing on the similar decision by the European parliament on January 26, 1999. Furthermore, it asked that those chemicals that can also reach groundwater and drinking supplies, and that can regularly be shown to have done so should be banned, and that limits for drinking water should be determined. The use of environmental chemicals should also be reduced where there is reason to suspect that they are EDs. Domestically, special measures should be taken for alkylphenol (ethoxylate)s, phthalates, and TBT compounds.

Similarly, as early as 1996 the U.S. Congress directed the U.S. Environmental Protection Agency to develop a screening program to determine whether certain substances may have hormonal effects in humans. This led to the development of an Endocrine Disruptor Screening Program, aiming to test all chemicals being marketed in the United States systematically for their endocrine-disrupting properties.

These interventions from political bodies helped to speed up hazard and risk assessment for endocrine-disrupting chemicals, and to allocate research funds in this field. The parliament and government decisions picked up public concern and emphasized application of the precautionary principle to achieve risk reduction quickly. Nevertheless, in many cases the time frames set for risk assessment and risk reduction were exceeded by the regulatory bodies due to complexity of the legal framework underlying chemicals policy.

WHAT HAS TO BE DONE?

From a governmental view, the following are main priorities for further work:

- Set up a program for coordinated monitoring combining chemical analysis and effect monitoring. This program should include not only fish and mammals, but also invertebrate species, birds, reptiles, and amphibians.
- Promote research on problems associated with quantitative risk assessment, such as extrapolation of high doses to low doses and existence of thresholds.

- Promote research on integrated risk assessment, particularly whether and how wildlife species can serve as models for endocrine disruption in humans and vice versa.
- Continue the work on a list of potential EDs and seek international cooperation in this field.

REFERENCES

1. H. Seibert (Ed.). In *Endocrinically Active Chemicals in the Environment. Umweltbundesamt,* pp. 10–14, UBA-Texte 3/96, Berlin (1996).
2. E. C. Dodds and W. Lawson. *Nature* **137**, 996 (1936).
3. R. B. Bradbury and D. E. White. *Vit. Horm.* **12**, 207–233 (1954).
4. W. W. Tullner. *Science* **133**, 647–648 (1961).
5. J. Bitman, H. C. Cecil, S. J. Harris, G. F. Fries. *Science* **162**, 371–372 (1968).
6. A. E. Fischli, T. Godfrain, I. F. H. Purchase. *Pure Appl. Chem.* **70**, 1863–1865 (1998).
7. H. Bern et al. In *Chemically-Induced Alterations in Sexual and Functional Development: The Wildlife/Human Connection*, T. Colborn and C. Clement (Eds.), pp. 1–8, Princeton Scientific Publishing, Princeton, NJ (1992).
8. E. Carlsen, A. Giwercman, N. Keiding, N. E. Skakkebaek. *Brit. Med. J.* **305**, 609–613 (1992).
9. R. M. Sharpe and N. E. Skakkebaek. *Lancet* **341**, 1392–1395 (1993).
10. N. E. Skakkebaek and R. M. Sharpe. *Pure Appl. Chem.* **75** (11/12), 2023–2038 (2003).
11. M. Binot, R. Bless, P. Boye, H. Gruttke, P. Pretscher. *Rote Liste gefährdeter Tiere Deutschlands. Schr. R. f. Landschaftspfl. u. Natursch.* **55** (1998).
12. E. T. Peeters, A. Dewitte, A. A. Koelmans, J. A. van der Velden, P. J. den Besten. *Environ. Toxicol. Chem.* **20**, 2883–2891 (2001).
13. J. Oehlmann and U. Schulte Oehlmann. *Pure Appl. Chem.* **75** (11/12), 2207–2218 (2003).
14. J. Oehlmann, E. Stroben, U. Schulte-Oehlmann, B. Bauer, P. Fioroni, B. Markert. *Fresenius J. Anal. Chem.* **354**, 540–545 (1996).
15. S. Takahashi, S. Tanabe, I. I. Takeutchi, N. Miyazaki. *Arch. Environ. Contam. Toxicol.* **37**, 50–61 (1999).
16. G. W. Bryan, P. E. Gibbs, L. G. Hummerstone, G. R. Burt. *J. Marine Biol. Assoc. UK* **68**, 611–640 (1986).
17. C. Bettin, J. Oehlmann, E. Stroben. *Helgol. Meeresunters.* **50**, 299–317 (1996).
18. P. Matthiessen, T. Reynoldson, Z. Billinghurst, D. Brassard, P. Cameron, G. Chandler, I. Davies, T. Horiguchi, D. Mount, J. Oehlmann, T. Pottinger, P. Sibley, H. Thompson, A. Vethaak. In *Endocrine Disruption in Invertebrates: Endocrinology, Testing, and Assessment*, P. deFur, C. Crane, C. Ingersoll, C. L. Tattersfield (Eds.), pp. 199–270, SETAC Press, Pensacola, FL (1999).
19. T. Horiguchi, C. Hyeon-Seo, H. Shirashi, Y. Shibata, M. Soma, M. Morita, M. Shimizu. *Sci. Total Environ.* **214**, 65–70 (1998).
20. Y. Morcillo and C. Porte. *Environ. Res.* **81**, 349–354 (1999).
21. J. G. Vos, E. Dybing, H. A. Greim, O. Lagefoged, C. Lambre, J. V. Tarazona, I. Brandt, A. D. Vethaak. *Crit. Rev. Toxicol.* **30**, 71–133 (2000).
22 V. Axiak, A. J. Vella, D. Agius, P. Bonnici, G. Cassar, R. Casson, P. Chircop, D. Micallef, B. Mintoff, M. Sammut. *Sci. Total Environ.* **258**, 89–97 (2000).
23. T. C. Hung, W. K. Hsu, P. J. Mang, A. Chuang. *Environ. Pollut.* **112**, 145–152 (2001).
24. W. J. Shim, S. H. Kahng, S. H. Hong, N. S. Kim, S. K. Kim, J. H. Shim. *Mar. Environ. Res.* **49**, 435–451 (2000).
25. M. F. Pessoa, A. Fernando, J. S. Oliveira. *Environ. Toxicol.* **16**, 234–241 (2001).
26. A. J. Power and B. F. Keegan. *Mar. Pollut. Bull.* **42**, 761–772 (2001).
27. P. E. Penchaszadeh, A. Averbuj, M. Cledon. *Mar. Pollut. Bull.* **42**, 790–791 (2001).
28. M. Bech. *Environ. Pollut.* **117**, 421–429 (2002).

29. T. J. Reitsema, J. A. Thompson, P. Scholtens, J. T. Spickett. *Mar. Pollut. Bull.* **44**, 257–261 (2002).
30. P. L. deFur, M. Crane, C. G. Ingersoll, L. Tattersfield. *Endocrine Disruption in Invertebrates: Endocrinology, Testing and Assessment*, SETAC-Publications, Pensacola, FL (1999).
31. J. Sherry, A. Gamble, P. Hodson, K. Solomon, B. Hock, A. Marx, P. Hansen. In *Impact Assessment of Hazardous Aquatic Contaminants; Concepts and Approaches*, S. S. Roa (Ed.), Lewis Publishers, Boca Raton, FL (1999).
32. P. Matthiessen. *Pure Appl. Chem.* **75** (11/12), 2249–2261 (2003).
33. S. Jobling and C. R. Tyler. *Pure Appl. Chem.* **75** (11/12), 2219–2234 (2003).
34. D. E. Kime. *Endocrine Disruption in Fish*, Kluwer Academic, Boston (1998).
35. A. Arukwe and A. Goksøyr. *Sarsia* **83**, 225–241 (1998).
36. S. Hashimoto, H. Bessho, A. Hara, N. Nakamura, T. Iguchi, K. Fujita. *Mar. Environ. Res.* **49**, 37–53 (2000).
37. A. Karels, E. Markkula, A. Oikari. *Environ. Toxicol. Chem.* **20**, 1517–1527 (2001).
38. M. C. Fossi, S. Casini, S. Ancora, A. Moscatelli, A. Ausili, G. Notarbartolo-di-Sciara. *Mar. Environ. Res.* **52**, 477–483 (2001).
39. T. Wahli, W. Meier, H. Segner, P. Burkhardt-Holm. *Histochem. J.* **30**, 753–758 (1998).
40. J. Lehmann, F. Paris, F.-J. Stürenberg, V. Blüm. In: LÖBF Jahresbericht 1999, Recklinghausen; Germany (2000).
41. M. Hecker, C. R. Tyler, M. Hoffmann, S. Maddix, L. Karbe. *Environ. Sci. Technol.* **36**, 2311–2321 (2002).
42. Y. Allen, P. Matthiessen, A. P. Scott, S. Haworth, S. Feist, J. E. Thain. *Sci. Total Environ.* **233**, 5–20 (1999).
43. S. Jobling, N. Beresford, M. Nolan, T. Rodgers-Gray, G. C. Brighty, J. P. Sumpter, C. R. Tyler. *Biol. Reprod.* **66**, 272–281 (2002).
44. R. van Aerle, T. M. Nolan, S. Jobling, L. B. Christiansen, J. P. Sumpter, C. R. Tyler. *Environ. Toxicol. Chem.* **20**, 2841–2847 (2001).
45. M. G. Simpson, M. Parry, A. Kleinkauf, D. Swarbreck, P. Walker, R. T. Leah. *Mar. Environ. Res.* **50**, 283–287 (2000).
46. M. S. Sepulveda, W. E. Johnson, J. C. Higman, N. D. Denslow, T. R. Schoeb, T. S. Gross. *Sci. Total Environ.* **289**, 133–144 (2002).
47. L. C. Folmar, N. D. Denslow, K. Kroll, E. F. Orlando, J. Enblom, J. Marcino, C. Metcalfe, L. J. Guilette. *Arch. Environ. Contam. Toxicol.* **40**, 392–398 (2001).
48. P. Matthiessen and J. P. Sumpter. *EXS* **86**, 319–335 (1998).
49. A. Gies, C. Gottschalk, P. Greiner, W. Heger, M. Kolossa, B. Rechenberg, E. Rosskamp, C. Schroeter-Kermani, K. Steinhaeuser, C. Throl. In *Sustainable and Precautionary Risk Assessment and Risk Management of Chemicals*. UBA-Texte 30/01, pp. 1–54, Berlin (2001).
50. J. U. Skaare, A. Bernhoft, A. Derocher, G. W. Gabrielsen, A. Goksoyr, E. Henriksen, H. J. Larsen, E. Lie, O. Wiig. *Toxicol. Lett. Mar.* **15**, 103–109 (2000).
51. J. Oehlmann, U. Schulte-Oehlmann, M. Tillmann, B. Markert. *Ecotoxicology* **9**, 383–397 (2000).
52. M. Gülden, A. Turan, H. Seibert. *Substanzen mit endokriner Wirkung in Oberflächengewässern.* UBA-Texte 46/97, p. 362, Berlin (1997).
53. T. Hutchinson. *Pure Appl. Chem.* **75** (11/12), 2343–2353 (2003).
54. S. M. Barlow. *Chemosphere* **39**, 1287–1292 (1999).
55. J. J. Amaral Mendes and E. Pluygers. In: VI Cogliano, *Perspectives on Biologically Based Cancer Risk Assessment*, NATO series on Challenges of Modern Society, E. G. Luebeck and G. A. Zapponi (Eds.), Vol. 25, pp. 81–182, Kluwer Academic/Plenum, New York (1999).
56. D. M. Sheehan, E. Willingham, D. Gaylor, J. M. Bergeron, D. Crews. *Environ. Health. Perspect.* **107**, 155–159 (1999).
57. J. P. Sumpter. *Toxicol Lett.* **102–103**, 337–342 (1998).
58. E. J. Calabrese and L. A. Baldwin. *Toxicol. Sci.* **71**, 246–250 (2003).

59. H. Van Loveren, P. S. Ross, A. D. Osterhaus, J. G. Vos. *Toxicol. Lett.* **15**, 319–324 (2000).
60. P. Matthiessen and P. E. Gibbs. *Environ. Toxicol. Chem.* **17**, 37–34 (1998).
61. C. P. Gibson and S. P. Wilson. *Mar. Environ. Res.* **55**, 101–112 (2003).
62. H. L. Rees, R. Waldock, P. Matthiessen, M. A. Pendle. *Mar. Pollut. Bull.* **42**, 137–144 (2001).
63. P. Matthiessen, R. Waldock, J. E. Thain, M. E. Waite, S. Scrope-Howe. *Ecotoxicol. Environ. Saf.* **30**, 180–194 (1995).
64. European Commission. Communication from the Commission to the Council and the European Parliament. Community Strategy for Endocrine Disrupters. COM (99)706 (1999).
65. European Commission. Communication from the Commission to the Council and the European Parliament on the implementation of the Community Strategy for Endocrine Disrupters—a range of substances suspected of interfering with the hormone systems of humans and wildlife. COM (2001) 262 (2001).

Pure Appl. Chem., Vol. 75, Nos. 11–12, pp. 2575–2591, 2003.

Workshop 6.4

Risk perception: A chemical industry view of endocrine disruption in wildlife*

Simon Webb[1,‡], Rob Taalman[1], Richard Becker[2], Kaz Onuma[3], and Koichi Igarashi[3]

[1]CEFIC European Chemical Industry Council, Ave. E. Van Nieuwenhuyse 4, B-1160 Brussels, Belgium; [2]American Chemical Council, 1300 Wilson Blvd., Arlington, VA 22209, USA; [3]Japanese Chemical Industry Association, Kazan Building 3-2-4 Kasumigaseki Chiyoda-Ku, Tokyo 100-0013, Japan

Abstract: Manufactured chemicals are essential to the vast array of goods and services that contribute to modern life. Their benefits are innumerable, and society is entirely dependent upon them. At the same time, there is an increasing awareness of the concept of environmental impacts. The challenge is to achieve the appropriate balance between the benefits and risks from chemicals, so that we all may enjoy the benefits of chemicals without significant detriment to current and future human and wildlife health. Ecological risk assessment is the mechanism that allows potential environmental chemical exposure to be benchmarked against hazardous properties so that risk is acceptable and environmental health is not impaired. Chemical management decisions based on such assessments are said to be risk-based. Within the context of environmental risk assessment practice for endocrine disruption, industry would support a position that:

- endorses the risk assessment process;
- recognizes that endocrine disruption is not an adverse effect per se, but rather a potential mechanism of action;
- gives precedent to population-level effects instead of individual-level effects;
- employs a tiered approach to hazard assessment;
- emphasizes, standardizes, and validates effects testing methodologies;
- recognizes that exposure per se does not necessarily constitute a risk;
- considers relative potency (i.e., evaluation of the dose levels and mechanisms producing toxic adverse effects and determining whether the critical effect arises via an endocrine mechanism or another mechanism);
- benchmarks risk against loss of benefits; and
- evaluates risk within the context of overall risk from both natural and anthropogenic substances with common modes of action.

To help address uncertainty surrounding the risk from endocrine active substances (EASs) to wildlife, the chemical industry—via the Long-Range Research Initiative (LRI)—has implemented a research program aimed at identifying and addressing knowledge gaps and establishing internationally harmonized testing methodologies in cooperation with other stakeholders. Details of individual projects within the current LRI research program are presented.

*Report from a SCOPE/IUPAC project: Implication of Endocrine Active Substances for Human and Wildlife (J. Miyamoto and J. Burger, editors). Other reports are published in this issue, *Pure Appl. Chem.* **75**, 1617–2615 (2003).

‡Corresponding author

INTRODUCTION

Manufactured chemicals are essential to the vast array of goods and services that contribute to modern life. Their benefits are innumerable, and society is entirely dependent upon them. At the same time, there is an increasing awareness of the potential environmental impacts of chemicals. Consequently, there are increasing stakeholder demands for public information about chemicals in consumer products and about emissions to the environment. The implicit aim of these stakeholders is one of encouraging less hazardous production and reduced use of chemicals in general. The challenge to civil society is to achieve the appropriate balance between the benefits and risks from chemicals, so that we all may enjoy the benefits of chemicals without significant detriment to current and future human and wildlife health. However, it is inevitable—given the variety in volume and scope of physicochemical characteristics of chemicals in commerce—that some environmental exposure of both human populations and wildlife in the ecosystem will take place. However, exposure per se does not constitute risk. Likewise, a compound may be potentially hazardous based on intrinsic properties, but only via exposure will it be able to exert such potential and result in adverse effects. It will not pose a risk where exposure is limited. This was recognized by Paracelsus in his oft-quoted dictum "*Sola dosis facit venenum—the dose makes the poison*". Risk is therefore a function of both exposure and potential effects.

Ecological risk assessment is the mechanism that allows potential environmental chemical exposure to be benchmarked against hazardous properties so that risk is acceptable and environmental health is not impaired or otherwise impacted. Chemical management decisions based on such assessments are said to be risk-based. Risk-based approaches differ from exposure- and hazard-based approaches. Exposure-based decisions relate to pollution prevention or minimization of exposure irrespective of chemical properties. Hazard-based decisions employ risk avoidance by avoiding any exposure to chemicals with specific hazards (e.g., carcinogens, teratogens). When decisions are hazard- rather than risk-based, the implication is that any exposure to a chemical with the specific inherent properties under consideration is unacceptable, regardless of actual risk.

With the advances of risk and exposure sciences during the last 20 years, decisions to control, limit, or otherwise manage the use of chemicals are increasingly risk-based. Risks can and should be scientifically assessed using actual data when available. To garner the confidence and acceptance of the scientific community, public, and stakeholders, risk assessments should be conducted in a scientifically rigorous manner, using the best available/scientifically credible hazard and exposure information, and risk assessment techniques and reasonable safety factors. All relevant information should be made public. The resulting risk assessment should be scientifically defensible and therefore open to review by independent scientific experts. These basic principles are included in the International Council of Chemical Associations' (ICCA's) "Principles for Risk-Based Decision Making", which describes a reasoned approach to protecting the environment and public health [1].

During the last decade, considerable research has focused on evaluating potential hazards, exposures and impacts from exposures to environmentally relevant levels of EASs on human health and ecological well-being. The "endocrine disruptor hypothesis"* postulated that certain substances could act like natural hormones and, even at low levels typically found in the environment, could cause adverse effects by interfering with the endocrine systems of wildlife and humans. To help address uncertainties, it was recognized that considerable research would be necessary to increase scientific understanding and determine whether adverse effects in humans and wildlife were arising from hormone-mediated

*As defined during the 1996 Weybridge Workshop [5], "An endocrine disruptor is an exogenous substance that causes adverse health effects in an intact organism, or its progeny, secondary to changes in endocrine function." It is generally understood that the definition of "endocrine disruption" encompasses both the endocrine mechanism of action and adverse health effects. Such a definition is consistent with our understanding of the science in that it implicitly recognizes that while substances may interact with the endocrine system, they may not adversely affect health or the ecosystem. For example, natural variations in hormone levels and reversible or transient effects associated with many natural or synthetic substances are well known.

processes from exposure to chemicals in the environment. The chemical industry—via the LRI—implemented a research program aimed at identifying and addressing knowledge gaps and establishing internationally harmonized testing methodologies in cooperation with other stakeholders. Working with academic and research institutions and governments, members of the chemical industry in Europe, North America, and Asia under the aegis of the ICCA sponsor fundamental research on the potential of chemicals to interact with and affect the hormone system and cause adverse effects. ICCA members have pledged to conduct research through an open and transparent process at institutions selected through a competitive peer-reviewed process. The results of this research will be made available to the public and acted upon by industry in a timely manner. An overview of individual projects within the current LRI research program is presented below within the context of specific research recommendations from the Scientific Committee for Toxicity, Ecotoxicity, and the Environment (CSTEE) [2] and International Program on Chemical Safety (IPCS) [3]. Full details of individual research projects can be obtained from the LRI Web sites (<www.cefic.org/lri>, <www.endocrineresearch.com>, and <http://www.nikkakyo.org>).

ENDOCRINE DISRUPTION IN WILDLIFE

IPCS [3] evaluates the extent to which wildlife—both vertebrate and invertebrate—have plausibly been affected by environmental exposures to EASs. Case studies involving mammals, birds, reptiles, amphibians, and invertebrates were all considered (see Table 1). That there are examples of effects observed in wildlife associated with EASs is apparent, particularly in areas "which have received extensive chemical contamination". Many of the case studies have previously been highlighted in other reviews such as Van Der Kraak [4] and CSTEE [2]. The case studies also demonstrate the problems regarding the establishment of cause and effect relationships between chemical exposures and physiological dysfunction across the various taxa considered—particularly when compounded by a wide variety of other potential influential factors such as habitat, food availability, etc.

Table 1 Wildlife case studies from the IPCS-GAED report (from IPCS [3]).

Case studies and associated contaminants	Conclusions
Mammals:	
• Reproductive dysfunction in mustelids (PCBs, dioxins, furans, OC insecticides) • Reproductive dysfunction in marine mammals (PCBs, DDE/DDT) • Reproductive dysfunction in feral rodents (PCBs, cadmium) • Defects in reproductive, endocrine, and immune systems of Florida panthers • Pseudohermaphroditism in bears (teratogenic herbicides, androgenic plant alkaloids, PCBs) • Pathological lesions in marine mammals (PCBs, DDT/DDE/DDD, PHAHs)	• *"Sufficient evidence that feral mammals have been adversely impacted by environmental contaminants."* • *"Limited evidence....that these effects are mediated through endocrine dependent mechanisms."* • *"Difficult to assess the mode of action of environmental chemicals in feral mammals" due to a "lack of knowledge regarding their endocrinology & reproductive biology and how other environmental stressors affect these processes."*

(*continues on next page*)

Table 1 (*Continued*).

Case studies and associated contaminants	Conclusions
Birds:	
• Alterations in behavior/reproductive success of colonial waterbirds (DDE, PCBs, mirex/photomirex) • Abnormal reproductive morphology (dioxins, PCBs, mirex) • Sex ratio skew/female–female pairings in gulls (DDT) • Eggshell thinning (DDT/DDE) • Great Lakes Embryo Mortality, Edema and Deformity Syndrome – GLEMEDS (PCHs, PCBs, DDT, dioxins/furans)	• *"Oviparous reproductive strategy and certain life history traits create avenues of exposure that makes these species more vulnerable to EDCs than traditional animal models or humans."* • *"Although exposure to environmental contaminants can have dramatic effects on endocrine regulated process and overall population fitness, the mechanism need not involve endocrine disruption."* • *"Individuals may experience endocrine disruption, which may or may not be linked to effects on reproduction and population fitness."*
Reptiles:	
• Developmental abnormalities in Lake Apopka alligators (dicofol, DDD, DDE, chloro-DDT, PCBs, dieldrin, endrin, mirex, oxychlorodane) • Developmental abnormalities in Great Lakes snapping turtles (PCBs, dioxins/furans)	• *"Some developmental processes in reptiles…are susceptible to endocrine disruption."* • *"Although some reptile populations have been impacted by environmental contaminants with endocrine-disrupting properties, it is unclear how widespread the phenomenon is."* • *"There is insufficient data to evaluate whether aquatic reptiles are at greater risk of endocrine disruption compared to terrestrial reptile species for which we have limited data."*
Amphibians:	
• Amphibian population changes (DDT, DDE, DDT, dieldrin) • Deformities/malformations ("retinoid mimics," PCB 126)	• *"There are insufficient data to implicate EDCs as causative agents in amphibian declines."* • *"There is not enough conclusive evidence to state that environmental contaminants are responsible for the observed malformations."*
Fish:	
• Vitellogenin induction in juvenile male fish/altered gonadal development (estrogenic compounds in sewage effluent, 17β estradiol, estrone, 17α ethinyl estradiol, alkylphenols, alkylphenol ethoxylates, phytoestrogens in pulp/paper-mill effluents) • Reproductive abnormalities/altered sex steroid levels induced by pulp/paper-mill effluents (β sitosterol, sterols, lignans, stilbenes, resin acids) • Reduced reproductive success and population-level effects (PAH, DDT/DDE, PCBs, dioxins) • Altered adrenal physiology (PAHs, PCBs Scientific Committee for Toxicity, Ecotoxicity and the Environment, heavy metals) • Early life stage mortality (dioxins, furans, PCBs) • Thyroid dysfunction (PCBs, perchlordecone, mercury petroleum hydrocarbons)	• *"Endocrine disruption is undoubtedly occurring in wild fish populations."* • *"In most cases the precise modes of action are poorly understood."* • *"The compounds responsible for the observed effects may be due to both synthetic and natural compounds."* • *"There is limited understanding of how the existing endocrine disruption affects population fitness."*

(*continues on next page*)

Table 1 (*Continued*).

Case studies and associated contaminants	Conclusions
Invertebrates: • Imposex in gastropods (TBT) • Disruption of ecdysteroid-regulated processes in crustaceans (insecticides) • Disruption of juvenoid-regulated processes in crustaceans (methoprene, atrazine, nonylphenol) • Molting disturbances and deformities in insects (fenoxycarb, tebufenozide, phytoecosteroids, " DDE, heavy metals, pulp/paper-mill effluents)	• *"The diversity of the invertebrate phyla creates numerous challenges in determining the potential risks of EDCs" to invertebrate health.* • *"Compounding these challenges is the poor understanding of the endocrinology of most invertebrates.* • *"The effects of EDCs in vertebrates will not necessarily be similar to those observed in invertebrates."* • *"Conversely invertebrates are susceptible to endocrine-disrupting properties of compounds that are not problematic in vertebrates."*

EPISTEMOLOGICAL FRAMEWORKS—ASSOCIATION AND CAUSALITY

An endocrine disruptor is defined within the context of an adverse effect upon a complete or intact organism [5]. In contrast, the goal of ecological risk assessment is to protect communities and the functioning of ecosystems from the effects of chemical pollutants. Historically, it has proved problematic to categorically demonstrate that adverse effects at the ecosystem or population level have resulted from exposure to an EAS. Van Der Kraak [4] recognized that defining the extent to which environmental chemicals affect the functioning of the endocrine system and consequently contribute to adverse health effects in wildlife is a "controversial and highly charged issue." This was attributed to a lack of defined criteria for establishing causality vis-à-vis chemical exposure and any observed adverse effects. Identifying a possible association between chemical exposure and an observed adverse effect typically involves an element of speculation. The presence of a compound with a known mode of action in the environment offers the possibility that such exposure is responsible for the observed effect. Such associations of an observed effect with a possible cause are relatively easy to identify and examples abound. Demonstrating that an actual causal relationship exists is much more difficult. Various guidelines and criteria have been applied to aid assessments of causality [3,4,6]. The application of such frameworks allows expert judgement as to whether an association is causal or not. These are qualitative assessments and not quantitative risk assessments that ascribe probability of adverse effects. They can also be used for gap analysis to guide future research needs for individual cases. The starting point is a hypothesis linking an adverse effect to a particular stressor via an endocrine-mediated mode of action. Evaluation is typically on the basis of the five aspects listed below (Table 2), with an overall evaluation of the strength of evidence regarding the linkage between the adverse effect and exposure.

Examples of the use of the framework relating to wildlife effects as dealt with in the IPCS-GAED report [3] are summarized in Table 3. Strong evidence for an EAS-mediated adverse effect was noted for TBT-induced imposex in gastropods, vitellogenin induction in male fish exposed to sewage effluent, and reproductive alterations in fish exposed to bleached kraft-mill effluent (BKME). Moderate evidence for an EAS-mediated adverse effect was noted for decreased reproductive function in Baltic seals, eggshell thinning in birds, and reproductive abnormalities in Lake Apopka alligators. Overall, the evidence that wildlife has been impacted adversely following exposure to EASs is deemed "extensive" —particularly in areas where levels of environmental contamination are known to be high. This is despite the problems inherent in determining the extent of potential effects of EASs on wildlife (i.e., exposures to multiple substances and a high number of potential species with differing life histories, physiology, and endocrine systems). The key unresolved question identified in the report is whether areas with lower (background) levels of contamination likewise constitute a risk to wildlife. The examples cited relate to aquatic ecosystems that are highlighted as sinks of environmental chemicals, particularly those that bioaccumulate (i.e., PCBs, dioxins/furans, and DDT/DDE). Many of the compounds highlighted have essentially already been eliminated from production (e.g., PCBs, DDT) or otherwise restricted (e.g., TBT), although some remain widely distributed in the environment.

Table 2 Framework for evaluation of causality (from IPCS [3]).

Element for evaluation	Aspect
Temporality	Does exposure precede adverse impacts upon health?
Strength of the association	Examination of the incidence rate in a population, the extent to which other factors may have contributed, the risk attributable to exposure, and shape of the dose–response curve.
Consistency of the observations	How frequently are similar or dissimilar observations made (across regions, across species, across exposure)?
Biological plausibility	Evaluation of the mechanism of action of the compounds of concern. Is the substance an endocrine disruptor?
Evidence of recovery	Examines whether the adverse effects are reversible if exposure is reduced/ removed.

CENTRAL TENETS FOR RISK ASSESSMENT OF ENDOCRINE ACTIVE SUBSTANCES

The examples of effects of EASs on wildlife highlighted above amply illustrate the need to establish appropriate tools with which to assess risk in both a prospective fashion for new chemical entities and a retrospective fashion for existing compounds in commerce. In proposing the use of risk assessment in this context, there are several central tenets or guiding principles to promote a scientifically rigorous approach for assessing and managing potential risk.

Risk assessment concept

Hutchinson et al. [7] detail a tiered approach to ecological risk assessment for suspected EASs. This framework draws from the previous suggestions by Stevens et al. [8,9] that, without evidence to the contrary, it is scientifically sound to assess and evaluate endocrine-related effects in the same manner as adverse effects caused by other modes of action. This stance was reiterated by the CSTEE [2] who confirmed that the underlying concept of risk assessment is scientifically sound and applicable to the issue of endocrine disruption. Due to regulatory considerations, this focus on ecosystem structure and function is modified somewhat and ecotoxicological assessments are based on a limited number of laboratory species as a consequence of technical and economic considerations. However, this concept is still appropriate if the ecotoxicity assessment is conducted with representative taxonomic groups, ecologically relevant endpoints (i.e., reproduction), and acceptable extrapolation from laboratory to ecosystem. The concept of ecological risk assessment is therefore general and not related to mode of action [2]. The ecotoxicological consequences of endocrine disruptors should therefore be assessed by general nonspecific endpoints relevant for the detection of population–community effects, but the tests must cover potential consequences of hormonal alterations. The approach is questionable only if ecotoxicity testing is unable to detect toxicity/endpoints related to endocrine disruption or when the uncertainty factors/probability cut-offs employed in extrapolation from the laboratory to the ecosystem are not appropriate (i.e., acute/chronic, chronic/multispecies ratios).

Table 3 Illustrative examples of the evaluation of the association between environmental exposure to EASs and adverse wildlife effects (from IPCS [3]). Each element was ranked from weak (*) to strong (****), and the overall strength to support the hypothesis and EAS mechanism was evaluated as either weak, moderate, or strong.

Statement of hypothesis		Evaluation factor					Overall strength of evidence	
Outcome	Stressor	Temporality	Strength of association	Consistency	Biological plausibility	Recovery	For hypothesis	For EAS mechanism
Imposex in marine gastropods	Tributyltin (TBT)	****	****	****	***	****	Strong	Strong
Decreased reproductive function in Baltic seals	PCBs	***	**	***	***	****	Strong	Moderate
Great Lakes embryo mortality and edema syndrome in birds	PCHs, PCBs	****	****	****	****	****	Strong	Weak
Eggshell thinning in colonial waterbirds	DDE and other DDT metabolites	****	****	****	***	****	Strong	Moderate
Reproductive abnormalities in Lake Apopka alligators	Dicofol and agricultural pesticides	****	***	***	***	**	Moderate	Moderate
Developmental abnormalities and reproductive failure in Lake Ontario trout	Dioxins and co-planar PCBs	****	****	***	****	****	Strong	Weak
Vitellogenin induction in fish exposed to sewage treatment plants in England	Estrogenic contaminants	****	****	***	****	**	Strong	Strong
Reproductive alterations in fish exposed to bleached kraft mill effluent in Ontario	Bleached kraft mill effluent (BKME)	****	****	***	****	***	Strong	Strong

Adverse effect vs. mechanism of action

Endocrine disruption is not an adverse effect per se, but rather a potential mechanism of action or descriptor of functional change that may lead to adverse health effects as measured by well-established endpoints such as reproductive toxicity or impaired development [2,10]. The Weybridge definition of "endocrine disruption" encompasses both the endocrine mechanism of action and adverse health effects (i.e., "an endocrine disruptor is an exogenous substance that causes adverse health effects in an intact organism, or its progeny, secondary to changes in endocrine function" [5]). This definition implicitly recognizes that, while substances may have the potential to interact with the endocrine system, they do not necessarily adversely affect wildlife or ecosystem health. Evidence that a substance interacts with a component of the endocrine system through a particular mechanism does not provide any information on whether that substance causes other biological changes, particularly adverse health effects. By adopting a definition of endocrine disruption that includes evidence of adverse effects and not just evidence of a potential for interaction with the endocrine system, the focus is on understanding what is most important to protect the environment.

Structure and function of ecosystems

In ecological risk assessment, population-level effects take precedent over individual-level effects [11,12]. Within the context of endocrine disruption, the CSTEE [5] has similarly emphasized that ecological risk assessment is intended to evaluate risk of effects on the structure and function of ecosystems. Endocrine alterations at the individual level may or may not result in consequences on populations, communities, or ecosystems. A suitable strategy must therefore focus on endpoints relevant to the detection of population- or community-level effects. Understanding the mechanism is therefore secondary to ecological effects that can elucidate effects on populations, communities, and ecosystems [13]. Ecological effects are those that manifest themselves at the population level or higher. The most common unit of study in the population and endpoints of concern are those related to population size or reproductive capacity. Higher levels of organization include the community (an assemblage of species) and the ecosystem (communities within their physical and biological environment). Population-level effects are more difficult to detect than effects on individuals due to background variability. Population-level measures include size, age structure, sex ratios, recruitment, and biomass. Other measures are used for communities (i.e., keystone species, sentinel species, predator number, and predator–prey ratio) or ecosystems (i.e., species diversity, species richness, species abundance, species assemblage, primary production, biomass change, nutrient cycling, and energy flow). These require even greater interpretation before cause–effect relationships can be demonstrated. A variety of potential pyramid effects at the community or ecosystem level have been identified [13]. They include bioaccumulation, biomagnification, cascading effects (effects following from one effect on a component of the ecosystem), keystone effects (changes in population size of species linked with a community such as predator-prey), and matrix effects (those occurring in adjacent communities or ecosystems by virtue of proximity).

Tiered hazard assessment

Screening assays should be short-term, rapid, and inexpensive assays designed to detect specific hormonal activity. They should be based on a discrete mode or mechanism of interaction with the endocrine system. Screening assays are designed to detect substances that have the potential to interact with one or more components of the endocrine system. They do not detect adverse effects per se and cannot be interpreted as short-term and predictive tests for hazard assessment purposes. Positive results in a screening assay/battery (based upon a weight of evidence evaluation) indicate a high priority for evaluation in a definitive test. Hazard assessment for risk characterization must rely primarily on the results of definitive tests. This is not to say that screening assays are used only to prioritize substances for further testing, but their use in hazard identification and risk characterization should be constrained to providing mechanistic clarification regarding adverse effects demonstrated by definitive tests. Longer-term in vivo tests (e.g., mammalian reproduction studies, fish life-cycle tests) should be employed to identify adverse effects using relevant, atypical endpoints. Dose–response characterization should also be a goal of such tests. Hazard identification for risk characterization should rely primarily on the results of definitive tests. Some sources have implied that "endocrine disruption" is a new and previously unknown type of toxicity, and that current toxicological testing cannot address such potential effects. This is clearly not the case. Work is in progress via the Organization for Economic Cooperation and Development (OECD) Endocrine Disruptor Testing and Assessment (EDTA) Work Group to review and, where appropriate, change/introduce screening and testing protocols. Indeed, this is an explicit recommendation of the CSTEE [5].

Hutchinson et al. [7] provide a clear line of reasoning that the potential for exposure should drive the selection of appropriate test organisms for hazard assessment. Part of the international OECD effort is to nominate, discuss, and reach consensus on a practical number of representative taxonomic groups and specific species within each group to employ in screens and tests to investigate the potential hazard of EASs. Hazard identification is an assessment of the qualitative toxicity of a chemical, and constitutes an initial step in the risk characterization process. A defensible hazard characterization for hormonally active chemicals requires not only summarizing toxicological screening and testing data (hazard identification), but also an objective evaluation of whether the effects produced are adverse and whether adverse effects are due to a hormonal activity of the chemical. A "weight of evidence" (WoE) evaluation is an objective and balanced interpretation of the totality of scientific evidence regarding hormonal activity and adverse effects that might result from an endocrine mechanism. In conducting a WoE evaluation, a tiered, hierarchical approach is recommended to guide development and use of screens and tests for hormonally active agents. In such a tiered approach, results from definitive tests must outweigh or supersede results from screening assays. An example of such a WoE framework is provided in the Endocrine Modulation Steering Group (EMSG), [14]. The principles presented in this approach are based on "sound evaluation practices and are in line with CSTEE approaches to chemical hazard and risk assessment" [10]. A concrete example of the application of a WoE approach to wildlife with the context of endocrine disruption is that of Johnson [15].

The Weybridge [5] workshop recommended the development of a hazard identification and characterization strategy for endocrine modulation that involved three broad stages or tiers:

- **Initial assessment.** Assemble and use available information (e.g., data on toxicity) to gather evidence on whether a chemical may or may not lead to adverse effects associated with endocrine modulation or use other available information (e.g., structure–activity relationships [SARs]) to gather evidence on whether a chemical may or may not take part in modes of action underlying endocrine modulation (e.g., receptor binding). For the purpose of prioritization.
- **Screening.** Use in vitro and in vivo short-term assays to assess whether a chemical could act via a specific mode or mechanism of action that may lead to endocrine modulation. It was envisioned that in vitro assays could precede in vivo assays. However, the results of in vivo screening assays were deemed essential before proceeding to the testing tier to consider absorption, metabolism, etc.
- **Testing.** Enhanced short-term (subchronic) and long-term (chronic) test guidelines would be employed to determine whether a chemical elicited a particular hazard or adverse effect(s) and would characterize the dose–response relationship for any adverse effect(s) that were observed.

A tiered hierarchical framework provides the most efficient, and therefore protective, mechanism to obtain, organize, and interpret necessary data. The tiered approach uses available information and/or less resource-intensive screening to determine the extent to which substances warrant more resource-intensive testing. Such an approach allows for optimal use of finite resources (which includes the use of laboratory animals, laboratory capacity, and financial resources) and is necessary to expeditiously evaluate chemicals of greatest concern [7].

Test methodologies

The CSTEE [2] emphasized that ecological risk assessment is intended to evaluate risks on the structure and function of ecosystems. Ecotoxicity testing must therefore focus on relevant endpoints for the detection of population–community effects (i.e., reproduction and development). All laboratory toxicity tests should be standardized and validated in advance of their use in a regulatory testing program to assess endocrine activity. Use of standardized and validated toxicity methods facilitates scientific interpretation of study results, promotes clear and consistent risk assessment analyses, and enhances confidence in the use of test results for protection of wildlife. When new toxicity testing is planned, the study protocols should follow globally harmonized guidelines to promote mutual acceptance of data, and thus preclude additional and repetitive testing at a later date. That said, some concerns have been expressed regarding the ability of low-tier-level tests in the detection of ecological risk of endocrine disruptors [2]. Concerns relate to the suitability of test species (i.e., use of parthenogenic organisms such as *Daphnia*) and epistemological uncertainties when extrapolating from acute lethality to long-term effects [2]. This is problematic, as risk assessment follows a tiered approach. Higher-tier tests are only employed when low-tier evaluations indicate potentially unacceptable risks. Hence, the concern regarding the capability of low-level tiers to detect ecological risk from endocrine disruptors. The CSTEE [2] concluded that existing test guidelines cannot detect all endocrine-disrupting effects, and that current test guidelines have to be enhanced or new guidelines developed, preferably via international cooperation to avoid duplication (e.g., OECD). Reliance on in vitro assays for predicting in vivo, EAS-mediated effects was not recommended as they were deemed to be prone to false negatives and false positives. Major emphasis was placed on in vivo assays. Priority was placed on establishing long-term tests to detect ecologically relevant effects related to endocrine disruption. This includes the enhancement of the OECD 407 repeated oral toxicity and OECD 416 reproduction toxicity mammalian tests, development of enhanced early-life stage (ELS) and partial fish life-cycle studies, enhancement of existing avian tests, and development of appropriate invertebrate test systems. In many instances, industry has provided laboratory resources to assist in the development of assay validation data [16–18].

The foundation of the OECD framework for evaluating potential EASs will make use of standardized, validated, and internationally harmonized test methods. However, as basic research on endocrine mechanisms advance, new and novel scientific methods have and will continue to be reported. These new and novel types of studies are significantly different from laboratory studies using standardized and validated techniques. Research laboratory studies reporting novel test methods, nonstandardized and nonvalidated methods, or nonstandard test species may lack quality criteria for regulatory purposes (i.e., reliability of test methods, relevance/significance of endpoints). Under such circumstances, such studies should not be used for regulatory action per se, but rather trigger further testing and/or method validation efforts. Recognizing that new and novel methods and studies with nonstandard species may provide important scientific information, it is clear that they cannot be ignored. The preferred action would be to evaluate the substance of concern in one of the wide variety of existing validated test methods using standardized OECD methods and species (or similarly validated scientific methods, for example, those promulgated by the International Standards Organization [ISO], ASTM International [ASTM], or the U. S. Environmental Protection Agency [USEPA]). Alternatively, the new or novel test method could be subjected to standardization and validation within the OECD EDTA program or within a similar formal program if the method is viewed as necessary to augment or replace (an) existing test method(s) in the screening and testing battery. For a new or revised test method to be considered validated for regulatory purposes, it should meet criteria such as those specified by the Interagency Coordinating Committee on the Validation of Alternative Methods (ICCVAM; see <http://iccvam.niehs.nih.gov/docs/guidelines/criteria.htm>).

Exposure assessment

It is important to keep in mind that hazard identification and characterization alone are still insufficient to characterize risk. Likewise, exposure per se does not constitute an adverse effect. Risk characterization requires integration of scientific data and knowledge relating to both hazard and exposure. It is therefore important to evaluate if a compound is present in the environment. If the compound is not present or the ecological systems of interest are not exposed, then it can be concluded that the agent is not a factor in the system. There can be no risk without exposure. Risk only occurs when a threshold corresponding to a no-effect concentration (at the population level) is exceeded. The corollary is that for all substances there are exposures below which harm will not occur. Acceptance of this fact is important as advances in analytical techniques mean that an ever-increasing number of natural and anthropogenic compounds are being detected in the environment via an expanding array of increasingly sensitive techniques. For example, within the aquatic environment contaminants are now frequently reported in the ng/l range or even lower. It will be impractical, and in many cases undesirable, to ban all such compounds purely on the basis of identified presence in the environment. There is also a question of scale of risk to be considered. Is exposure localized or widespread? This has obvious implications for any potential chemical management strategies. Within the context of endocrine disruption, it is possible that potential effects are separated in time from critical exposures. Impacts upon adult development may theoretically be dependent upon exposures in an earlier (more sensitive) life phase. This gives rise to the concept of critical windows of exposure and the need to consider temporal as well as spatial aspects in exposure assessment.

Relative potency

An important consideration is whether the critical toxic mechanism for a particular substance is via endocrine disruption (i.e., what is the probability of achieving an effective exposure that elicits an endocrine-mediated effect relative to the probability of inducing prior major systemic toxicity via an alternative mechanism) [19]. From a pragmatic perspective, substances should not be regulated on the basis of potential endocrine effects that only occur at extreme doses above those incurring other critical toxicity. This approach has similarly been proposed by the CSTEE [10]. They state that only in situations where the endocrine-disrupting effects are critical (i.e., the most potent) in comparison to other toxic hazards, should the endocrine disruption be considered for hazard and risk assessments: "Little importance should be assigned to situations where the no-observed effect level for endocrine disruption is substantially higher than that for other adverse effects caused by a chemical".

Risk/benefit

Any decision to manage compounds on the basis of purported endocrine hazard should examine potential benefits, and costs of and lack of action to human or wildlife health. The scope of such considerations should not be restricted to economic considerations alone. Cost/benefit analyses feature as one of the general principles of application in the European Commission (EC) communication on the precautionary principle [20]. There may be circumstances where zero emissions may be technically infeasible, and perhaps not even desirable, as it may entail loss of critical benefits or imply consumption of alternative economic or material resources that will also carry societal or environmental costs. For example, natural and synthetic estrogens are consistently believed responsible for the majority of the estrogenic effects observed in fish populations near sewage treatment plants [21,22]. Yet, the loss of benefit associated with a ban on ethinyl estradiol used in the contraceptive pill is an example of a critical benefit that society is not prepared to countenance. Risk management of effects in fish populations from steroids in sewage effluent would therefore be limited to the imposition of improved sewage treatment technologies rather than an outright ban on use per se. This relative balance of risk and loss of benefit needs to be considered in any risk management deliberations.

Effects benchmarking

Several compounds (e.g., nonylphenol) have been identified as being estrogenic in fish. Such observations have led to calls for such compounds to be banned on the basis of their potential hazard to fish populations in watercourses impacted by domestic wastewater treatment plant effluent discharges. This ignores the likely fact that the large majority of the estrogenic potential of sewage influent is undoubtedly attributable to natural and synthetic sex steroids such as estradiol, estrone, and ethinyl estradiol [21,22]. Under such circumstances, management actions on nonylphenol—which has impacts on a local level—would not have addressed the causal factor for the large majority of impacted watercourses where widespread exposure to steroids is more critical. The problem would have remained, as causality would not have been affected. Within the environment, the risk from substances with common modes of action should therefore be considered collectively and compared with likely background concentrations of naturally occurring but similarly acting compounds. Any distinction between the two is artificial [10]. Likewise, the impacts of exogenous agents should be benchmarked against endogenously occurring compounds with a similar mode of action (e.g., exposure to xenoestrogens via the environment should be benchmarked against endogenous steroid levels) or other exogenous agents of natural origin such as dietary phytoestrogens. Consistent with this stance is the fact that one of the general principles of application in the EC communication on the precautionary principle [20] is that of *nondiscrimination* (i.e., "comparable situations should not be treated differently").

FUTURE RESEARCH ADDRESSING RISK FROM ENDOCRINE DISRUPTORS

Specific research recommendations identified in the IPCS-GAED report (IPCS, 2002) are highlighted below (Table 4). These demonstrate the current incomplete understanding of the issue and the need to address uncertainties. Many of the points identified reflect earlier similar recommendations by the likes of the CSTEE [2] or the National Research Council (NRC) [13]. Also presented are details of the current LRI-funded/cofunded research program that are relevant to the specific research recommendations. These projects reflect attempts by industry to reduce uncertainty surrounding the risk from EASs to wildlife and include consideration of issues such as mixtures. Full details, including lists of publications that have arisen from this research, are provided on the LRI Web sites.

Table 4 IPCS research areas and LRI activities.

Biology underlying endocrine-mediated effects
• Expand basic knowledge about endocrine systems in wildlife. • Elucidate the range of mechanism by which endocrine disruption may interfere with reproductive/population success, immune function, neurobehavioral, and development of cancer, at all levels of biological organization and at key stages of life cycles. **Relevant LRI-sponsored research (lead researcher and institute):** - Genetic sex determination of fish (R. Devlin, Fisheries and Oceans, Canada) - Estrogens and neuro-endocrine regulation of reproduction in fish (H. Goos, University of Utrecht, Netherlands) - Aquatic invertebrates (L. Pinder, NERC Institute of Freshwater Ecology, UK) - Effects of DES on the fetus considering its placental transport (N. Utoguchi, Teikyo University, Japan) - Molecular and cellular approach on the reproductive tract abnormalities induced by prenatal exposure to DES in mice (T. Nagao, Hatano Research Institute, Japan)

(*continues on next page*)

Table 4 (*Continued*).

- Relationship between the effects to reproduction and vitellogenin induction of fish by EDCs (J. Koyama, Kagosima University, Japan)
- Toxicological significance of vitellogenin synthesis induced by estrogens exogenously administered in the inbred and wild medaka (S. Hamaguchi, Niigata University, Japan)
- A whole lake experiment to examine the effects of a synthetic estrogen on aquatic populations (K. Kidd, Environment Canada)

Methodology

- Develop improved methodologies to assess dose-response relationships at environmentally relevant concentrations.
- Develop more specific and sensitive biomarkers for detecting endocrine-mediated effects in individuals and populations.

Relevant LRI-sponsored research (lead researcher and institute):

- Fish 14-day Vitellogenin Screening Assay TG Screen (T. Hutchinson, Astra-Zeneca Brixham Environmental Laboratory, UK)
- Fish extended early life stage and reproductive chronic assays (T. Hutchinson, Astra-Zeneca Brixham Environmental Laboratory, UK and R. Bogers, NOTOX, NL)
- Endocrine disruption in the marine environment – EDMAR (P. Matthiessen, CEFAS, UK)
- Endocrine disruption in the aquatic environment: Laboratory investigations of endocrine active chemicals (J. Sumpter, Brunel University, UK)
- Development, validation, and application of in vitro and in vivo test systems for nonestrogenic EDCs in wildlife—especially amphibians (A. Murk, University of Wageningen, Netherlands)
- Avian two-generation toxicity test development (I. Chahoud, Free University of Berlin, Germany)
- Validation of OECD test guidelines for the uterotrophic, Hershberger, and enhanced TG 407 assays (EMSG industry consortium)
- Field-deployable methods for measuring endocrine-mediated modulation of songbird reproduction (L. Brewer, Springborn Laboratories, USA and A. Fairbrother, Parametrix, USA)
- Development, application, and validation of a sheepshead minnow estrogen-responsive cDNA microarray (M. Hemmer and L. Folmar, NHEERL, USEPA)
- Evaluation of eastern fence lizards (*Sceloporus undulatus*) and western fence lizards (*S. occidentalis*) as reptile models for assessment of endocrine mediated toxicity (L. Talent and D. Janz, Oklahoma State University, USA)
- Avian one-generation reproductive study with Japanese quail incorporating endocrine endpoints (L. Brewer, Spingborn Laboratories, USA)
- Effect of EDCs on the development of the cerebral nerve system (R. Kishi, Hokkaido University, Japan)

Monitoring

- Increase long-term monitoring of "sentinel" wildlife species to provide baseline data on population status.
- Improve international collaboration and cooperative research to assess the exposure and effects of EDCs on wildlife populations on a more global basis.

Relevant LRI-sponsored research (lead researcher and institute):

- Environmental effects on uterine tissues of Baltic Seals with special emphasis on organochlorins and uterine Leimyomas (M. Olssen, Swedish Natural History Museum, Sweden)
- Examination of bull sperm as a sentinel species for male reproductive health (J. Van Os, Global Institute for Study of Natural Resources, Netherlands)

(*continues on next page*)

Table 4 (*Continued*).

Identifying Endocrine Disruptors

- Continue to identify chemicals (persistent and nonpersistent, naturally occurring, and anthropogenic) that are the most likely candidates for high-impact effects in populations at environmentally relevant concentrations.
- Identify "hotspots" for exposure or effects that warrant particular concern.
- Focus work on populations/subgroups most likely to be susceptible to endocrine disruptors.
- Assess the role of endocrine disruptors relative to other environmental stressors on the fitness of the population.

  Relevant LRI-sponsored research (lead researcher and institute):
  - Endocrine-modulating effects in fish along the Elbe River and in reference areas. Assessment of risks related to habitat conditions and the natural variability of endocrine functions (L. Karbe, University of Hamburg, Germany)
  - Ecological hazard assessment for endocrine toxicity of chemical mixtures (C. Borgert and T. Gross, USGS/University of Florida, USA)

Database Development

- Develop better global data, especially in countries outside North America and Europe, on status and trends of environmental contamination, exposure, and health outcomes.
- Improve international coordination for sharing information on effects caused by endocrine disruption.

  Relevant LRI-sponsored research (lead researcher and institute):
  - Studies on actual conditions in relation to the influences of chemical substances in Japanese wild animals (M. Uchiyama, Toyama University, Japan)

DISCUSSION

Most legal frameworks require regulators to demonstrate that under certain defined conditions, a substance poses an unacceptable risk. Implicit in the application of the precautionary principle is a shift of the so-called "burden of proof" from regulatory agencies to demonstrate risk to manufacturers or formulators to demonstrate that no harm to the environment will occur. This reversal of the burden of proof ignores the scientific fact that it is always impossible to prove a negative; thus, it can never be proven that any given activity or substance will pose no harm to the environment [23]. The claim that a particular chemical causes adverse effects in the environment is feasible in principle to evaluate. One example is sufficient to prove the claim and supply the burden of proof. The opposing claim that the chemical does not cause adverse effects in the environment can never be lifted. Situations will exist that cannot be adequately covered by investigations. Such a claim can therefore only be evaluated by induction. The better and more comprehensive the investigation, the more likely the claim, but there can never be complete certainty. Only example would falsify the claim. This restriction on lifting the burden of proof applies to all chemicals regardless of intrinsic hazardous properties. Environmental acceptance of any given compound can therefore only be demonstrated by some degree of induction and reasonable likelihood according to acceptable criteria [23]. Zero-risk and absolute safety can never be achieved or proven for any issue or substance. It is consequently a question of confidence or acceptability of whether uncertainty has been reduced sufficiently. The maximum that can be done is to provide evidence that the activity or substance in question has been adequately tested and evaluated against potential known effects. This has always been the stance of the chemical industry. It is also reflected somewhat in the EC communication on the precautionary principle [20]. Within the context of the so-called general principles of application under *proportionality*, there is an explicit statement that "measures based on the precautionary principle...must not aim at zero risk, something which rarely exists." In other words, there must be a proportional response to a potential risk.

Environmental nonacceptance implies some form of risk management. Substitution of a compound requires a full evaluation of all candidate substitute compounds. All will carry economic, societal, and environmental costs as well as benefits. Perception of the relative balance of these will be very different for each stakeholder. The societal process of the appreciation and perception of risk does not necessarily make use of a rational and technically informed procedure. Judgment is based on personal impression founded on a limited and unstructured subset of information with qualitative rather than quantitative attributes. The scientific process of risk assessment is more formally structured and subject to standardized, rational, and scientifically accepted rules and procedures. In contrast, risk perception is not dependent upon a formal evaluation scheme, but on a large number of subjective variables (e.g., education, natural/social environment, ethical, political and religious beliefs, physical and mental health). Herein lies a major problem with the use of the precautionary principle in this context. All regulatory decisions need to take place within a transparent and predictable framework, otherwise all actions such as substitution are precluded, as the acceptability of any alternative outcome to all stakeholders will be unpredictable and not apparent. Without such predictability and transparency, any decision framework may be seen as an arbitrary black-box process. Given these restrictions, risk assessment must provide the foundation for effective regulation. Other approaches potentially lead to arbitrary regulatory decisions.

Where risk assessment cannot be conducted with appropriate certainty, the precautionary principle may be invoked, particularly where effects are perceived as being potentially serious and/or irreversible. This implicit primacy of risk assessment is reflected in the EC communication on the precautionary principle [20]. Use of the precautionary principle as a management strategy is always preceded by an initial evaluation of risk. The application of the precautionary principle in this way therefore assumes that potentially adverse effects have been identified and that scientific evaluation does not allow the determination of risk with sufficient certainty. Given that scientific knowledge is not static, the degree of certainty may be subject to future revision. Industry would therefore also suggest that any use of the precautionary principle is potentially transitory pending other data (i.e., subject to review). This is one of the general principles of application in the EC communication [20]. It states that precautionary "measures should be maintained as long as the scientific data are inadequate, imprecise, or inconclusive." Measures are therefore provisional pending the availability of more reliable data.

One of the problems with ecological risk assessment is that the focus is on the effects on the structure and function of the ecosystem, a difficult concept to relate. The assessment process itself is complex, requires sophisticated tools, and is difficult to communicate. As well as communicating risks, it is also important to describe benefits so that there can be a fair and factually oriented debate between stakeholders. In the first instance, the provision of factual scientific information to stakeholders is important. Explaining assessment tools and results will aid understanding of risks. Secondly, participation is important. All stakeholders should be empowered to express their views to allow productive debate. This is essential to establish trust and social responsibility. The public will judge all scientists from regulators, NGOs, and industry on their combined capacity to manage issues together in a timely fashion.

Regardless of the decision basis (i.e., hazard or risk), there are certain common research needs. Within the context of endocrine disruption, the LRI has attempted to address gaps in knowledge and data via collaborative and cooperative projects. Much work has been sponsored dealing with the development and validation of mammalian and environmental effects testing methodologies that can be used for comparative hazard or risk assessment. Exposure data is similarly required when assessing risk or causal relationships between exposure and wildlife health effects. This includes sources, fate and transport processes, transformation, and degradation. This is also a focus of LRI-sponsored research. Great emphasis has been placed on publication and sharing of data with all stakeholders for informed policy and regulatory decision making based on best available scientific knowledge. Any instances where there is an overestimation of risk will be due to an ignorance or lack of the technical facts leading to uncertainty. The solution to such conflict is to reduce ignorance through the provision of data on toxicity, exposure routes, and effects of environmental contaminants, thereby contributing to an informed decision-making process. This is the central aim of the LRI.

REFERENCES

1. ICCA. Policy paper on "Principles for Risk Based Decision Making". International Council of Chemical Associations. <http://www.icca-chem.org/issues.htm> (1995).
2. CSTEE. CSTEE opinion on human and wildlife health effects of endocrine-disrupting chemicals with emphasis on wildlife and Ecotoxicology test methods. Scientific Committee for Toxicity, Ecotoxicity and the Environment, European Commission. <http://europa.eu.int/comm/food/fs/sc/sct/out37_en.pdf> (1999).
3. IPCS. Global Assessment of the State of the Science of Endocrine Disruptors, T. Damstra, S. Barlow, A. Bergman, R. Kavlock, G. Van Der Kraak (Eds.), International Programme on Chemical Safety (WHO, ILO, UNEP), Geneva, Switzerland (2002).
4. G. Van Der Kraak. *Pure Appl. Chem.* **70** (9), 1785–1794 (1998).
5. European Commission. European workshop on the impact of endocrine disruptors on human health and wildlife, 2–4 December 1996, Weybridge, UK. Report of the Proceedings (Report EU 17549). European Environment Agency, European Commission DG XII: Copenhagen, Denmark, (1997).
6. G. T. Ankley and J. P. Giesy. In *Principles and Processes for Evaluating Endocrine Disruption in Wildlife*, R. J. Kendall, R. L. Dickerson, J. P. Giesy, W. A. Suks (Eds.), pp. 349–367, Society of Environmental Toxicology and Chemistry Press, Pensacola, FL (1998).
7. T. H. Hutchinson, R. Brown, K. E. Brugger, P. M. Campbell, M. Holt, R. Lange, P. McCahon, L. J. Tattersfield, R. van Egmond. *Environ. Health Perspect.* **108** (11), 1007–1014 (2000).
8. J. J. Stevens, W. Gfeller, L. Machemer, K.-H. Leist. *J. Toxicol. Environ. Health, Part B* **1**, 59–79 (1997a).
9. J. J. Stevens, A. Tobia, J. C. Lamb, C. C. Tellone, F. and O'Neal. *J. Toxicol. Environ. Health* **50**, 415–431 (1997b).
10. CSTEE. CSTEE opinion on BKH Consulting Engineers report "Towards the establishment of a priority list of substances for further evaluation of their role in endocrine disruption". Scientific Committee for Toxicity, Ecotoxicity and the Environment, European Commission, <http://europa.eu.int/comm/food/fs/sc/sct/out73_en.html> (2000).
11. G. T. Ankley, R. D. Johnson, N. T. Detenbeck, S. P. Bradbury, G. Toth, L. C. Folmar. *Rev. Toxicol. B. Environ. Toxicol.* **1**, 231–267 (1997).
12. M. R. Taylor, P. Holmes, R. Duarte-Davidson, C. D. N. Humfrey, P. T. C. Harrison. *Sci. Total Environ.* **233** (1–3), 181–191 (1999).
13. NRC. Hormonally Active Agents in the Environment. Committee on Hormonally Active Agents in the Environment, Board on Environmental Studies and Toxicology, Commission on Life Sciences, National Research Council, US National Academy of Sciences. National Academy Press: Washington DC (2000).
14. EMSG. Towards the establishment of a weight of evidence approach to prioritizing action in relation to endocrine disruption. Endocrine Modulation Steering Group, LRI-CEFIC: Brussels. <http://www.cefic.be/lri/emsg/infopack/22.pdf> (2000).
15. I. Johnson. Study on the scientific evaluation of 12 substances in the context of endocrine disruptor priority lists of action. Report to the European Commission (contract No.: 12834-0) by WRc-NSF, UK (2002).
16. P. Andrews, A. Freyberger, E. Hartmann, R. Eiben, I. Loof, U. Schmidt, M. Temerowski, M. Becka. *Arch. Toxicol.* **75**, 67–73 (2001).
17. P. Andrews, A. Freyberger, E. Hartmann, R. Eiben, I. Loof, U. Schmidt, M. Temerowski, A. Folkerts, B. Stahl, M. Kayser. *Arch. Toxicol.* **76**, 194–202 (2002).
18. J. Kanno, L. Oyon, J. Haseman, P. Fenner-Crisp, J. Ashby, W. Owens. *Environ. Health Perspect.* **109** (8), 785–794 (2001).
19. P. W. Harvey and I. Johnson. *J. Appl. Toxicol.* **22**, 241–247 (2002).

20. European Commission. Communication from the Commission on the Precautionary Principle COM(2000) 1. <http://europa.eu.int/comm/dgs/health_consumer/library/pub/pub07_en.pdf> (2000).
21. C. R. Tyler and E. J. Routledge. *Pure Appl. Chem.* **70** (9), 1795–1804 (1998).
22. NOAA/NCCOS. Endocrine Disruption in Fish: An Assessment of Recent Research and Results. Technical Memorandum NOS NCCOS CCMA 149. National Oceanic Atmospheric Administration's National Centers for Coastal Ocean Science (NOAA/NCCOS). <http://www.nccos.noaa.gov/documents/endocrinedisruptor.pdf> (2002).
23. P. Harremoës. Precaution and incertitude: Lessons from scientific surprises. Paper presented at "*Bridging the Gap – Sustainability Research and Sectoral Integration*", Stockholm, Sweden 9–11 May 2001. <http://www.bridging.environ.se/bridgdok/harrepap.pdf> (2001).

Pure Appl. Chem., Vol. 75, Nos. 11–12, pp. 2593–2604, 2003.

Workshop 6.5

Endocrine active substances and the need to improve environmental protection: An environmentalist's perspective*

Gwynne Lyons

World Wildlife Fund European Program, 17 The Avenues, Norwich NR2 3PH, UK

Abstract: This paper reviews the existing concerns and presents conclusions and recommendations for action.

The first section outlines an environmentalist's perspective and predicts that endocrine active substances (EASs) pose a high risk for wildlife populations. The second section explains that risk assessment, as currently practiced for individual chemicals, does not adequately protect the environment, and suggests several improvements that should be made. However, it is argued that the properties of endocrine disruption lead to increased uncertainty in the risk assessment, such that countries wishing to achieve a high level of protection would be justified in implementing precautionary controls.

The third section outlines some conclusions and additionally provides an appraisal of the responses of governments and industry. Furthermore, several recommendations are made, particularly the imperative for research funding to be substantially increased and guaranteed for many years to come. Also highlighted is the need to hasten the development of screens and tests for ecotoxicity, and for governments to commit to a comprehensive sorting, screening, and toxicity testing program for all chemicals to which significant exposure occurs.

AN ENVIRONMENTALIST'S PERSPECTIVE OF RISKS OF ENDOCRINE ACTIVE SUBSTANCES (EASs) TO WILDLIFE

The term "risk assessment" is typically used to describe "the *estimation* of the incidence and severity of the adverse effects likely to occur in an environmental compartment due to actual or predicted exposure to *a substance*" [1]. Thus, to many scientists working on chemicals, the term "risk assessment" suggests a detailed assessment of the studies that have been undertaken on an individual chemical. This is a very narrow "reductionist" focus, which merely seeks to determine whether a chemical is individually found in the environment at levels in excess of its predicted no-effect concentration (PNEC).

In attempting to establish the potential scope of the threat that EASs pose to wildlife, it is important to evaluate the bigger picture. Despite the lack of understanding regarding the full consequences of the phenomenon of endocrine disruption (ED), this section argues that in order to identify appropriate policy responses to EASs, likely environmental risks must be assessed both at the micro or single-substance level, and at the macro level. The term "macro-level risk assessment for endocrine disruptors" is

*Report from a SCOPE/IUPAC project: Implication of Endocrine Active Substances for Human and Wildlife (J. Miyamoto and J. Burger, editors). Other reports are published in this issue, *Pure Appl. Chem.* **75**, 1617–2615 (2003).

coined to refer to an estimation of whether adverse effects are likely to occur, and the possible severity of those effects, due to current exposures to EASs in general, some of which may not yet be identified.*

As noted above, the evaluation of risk is not an exact science, but rather it is "an estimation". Thus, risk assessment should evaluate and use available scientific data, but it should also be able to consider expert judgement or "best guesses" where scientific data are not available.

At the macro level, several questions are pertinent, including the following:

a) Is the environment exposed to EASs?
b) Could critical effects in wildlife populations be passing unnoticed?
c) Are suspected EASs adequately controlled, and are more EASs likely to be identified in the future?
d) Are many species likely to be under threat?

Finally, in light of the responses to the above questions, is it likely that endocrine EASs, either singly or in combination, are causing effects in wildlife that might impact at the population level?** These questions are answered below.

a) Chemical monitoring data confirm that EASs are found in the environment [3,4], and studies show that local wildlife populations have already been adversely affected. In many countries, sublethal effects due to EASs have been reported in fish [see 5,6].

   However, although certain chemicals with endocrine active properties have been found in the environment, there is insufficient environmental monitoring to determine the range of chemicals that are actually present. Global chemical production has escalated in recent decades from around 1 million tonnes per year in 1930 to around 400 million tonnes by the year 2000 [7]. This will no doubt have led to a dramatic increase in the exposure of the environment to a myriad of chemicals, including EASs.

b) Critical effects in wildlife populations could certainly be passing unnoticed for many reasons, not least the paucity of environmental monitoring. Trends in many endocrine endpoints, such as the occurrence of intersex, or even population trends, cannot be easily ascertained for most species. The diversity of ways in which chemicals can influence endocrine systems also makes it difficult to identify such effects in the wild. Even in different fish species exposed to the same estrogenic compounds, it appears that the effects may vary. Thus, several fish species show intersex characteristics and vitellogenin production on exposure to estrogens, while male sand gobies exhibit feminization of the urogenital papilla [8].

   Numerous wildlife populations have declined without the precise causes for these declines being known in many cases. The living planet index, which integrates data on the abundance of 282 forest species, 195 freshwater species, and 217 marine species, shows an overall decline of about 37 % between 1970 and 2000 [9]. Globally, species are being lost at an alarming rate, and many are insecure. In the United States, for example, it is suggested that one-third of the native flora and fauna is at risk [10]. This lends weight to the need for a more cautious approach to pollution control. Habitat destruction is often implicated in the decline of wildlife, but this paper conjectures that pollutants such as EASs *might* be partly responsible in some cases.

   Neurological alterations, leading to behavioral effects, often represent the earliest observable manifestation of toxicity. However, effects on behavior may be subtle and difficult to recog-

*Theo Colborn could be said to have pioneered the technique of macro-level risk assessment, in that in 1991 she brought together experts to agree to the now famous Wingspread consensus statement on EASs [2]. By getting consensus amongst recognized experts, she gave legitimacy to the technique, which by its nature is even more subjective than micro risk assessment.

**Unlike human health risk assessors, ecotoxicologists in the EU risk assessment process do not strive to protect every single fish, but rather the fish *population*. Thus, the ultimate aim of the risk assessment for the environment is to protect the structure and function of ecosystems, and so the protection goal is at the population or community level [11,12].

nize, and some species are particularly difficult to study. It is speculated that effects on behavior might be responsible for some wildlife species being less able to escape predation, rear offspring, or cope with the stress of a changing environment. Such speculation is not unreasonable, as the developing nervous system is especially vulnerable to toxicants. Furthermore, in humans, in utero exposure to PCBs (polychlorinated biphenyls), which have been shown to disrupt thyroid hormones in laboratory experiments [13], has been linked to intellectual impairment in children [14–16]. Indeed, chemicals with endocrine active properties have already been associated with behavioral effects in some wildlife species. Lack of parental attentiveness, for example, has contributed to reproductive problems in Forster's tern [17]. However, although behavioral effects attributed to pollutants have been reported in birds nesting in colonies [18], it is more difficult to pick up effects in birds that nest individually. Nevertheless, recent studies suggest that pollutants may be affecting behavior and egg hatchability in male kestrel (*Falco sparverius*) [19]. Other effects on reproduction and sex-specific behaviors may be missed in the wild. In the laboratory, for example, estrogen exposure has been shown to affect songbirds, causing females to sing like males [20].

Critical effects in wildlife populations may also be passing unnoticed because delays can occur between exposure and the effects becoming evident. For example, adverse effects on an animal's ability to reproduce may not be evident until the animal reaches maturity. Transient or even "one-off" exposure to certain pollutants during developmental periods can result in irreversible effects in later life [21]. Effects may also be delayed because it may take some time before pollutants build up in biota to reach levels able to cause harm. EASs are certainly being passed on in the food chain. For example, experiments with male flounder showed that they did not produce vitellogenin (the female egg yolk protein) when caged in estrogen-contaminated estuaries in the United Kingdom. However, when fed on mussels (*Mytilus edulis*) that had been kept in the industrially contaminated Tees, they did show vitellogenin induction [22]. Attention should not just be focused on lipophilic compounds, because other chemicals may be building up to harmful levels in certain organs, or in blood. For example, PFOS (perfluorooctane sulfonate), a chemical with thyroid-disrupting properties, has been found bound in the blood of numerous wildlife species throughout the world [23].

Moreover, the full effects of even well-studied EASs may take years to come to light, even in species for which data would be expected to be well scrutinized. For example, DDT was shown to affect reproduction in birds back in the 1960s, but it was not until 2001 that new research was published linking this pollutant to low infant birth weight and premature births in the United States [24]. The effects of EASs may also pass unrecognized because effects and even deaths may be solely attributed to natural causes like disease. For example, researchers have suggested that impaired immune function due to pollutants may have been at least partly to blame for the mass mortality in some cetacean populations [25–27]. The endocrine system plays an important role in the function and development of the immune system, and so disruption might result in increased susceptibility to disease.

c) Chemicals that are already suspected to be EASs are not adequately controlled. This is because in the past, toxicity tests did not evaluate endocrine endpoints adequately, and potential low-dose effects were not fully investigated. Also, many toxicity assessments were not conducted on sensitive life stages (e.g., early life stages and during times of rapid cell replication). Thus, the potency of many substances may be significantly underestimated. Furthermore, for many chemicals, even basic toxicity data are lacking, such that it is a fair assumption that more EASs will be identified in future.

d) Many species are likely to be under threat, as the way in which the endocrine system works shows many similarities between species. For example, some steroid hormones found in mammals are also found in other vertebrates, such as fish and amphibia, as well as in some invertebrate groups, such as mollusks. Unfortunately, relatively few species have been examined for their responses to man-made hormone-disrupting chemicals, and therefore it is not possible to predict the most sensitive species, nor is it possible to predict the most critical endpoints in most species. Effects may not even be limited to animals, as estrogen-mimicking chemicals may affect plant growth by disrupting the process of nitrogen fixation [28].

Macro risk assessment conclusions

The endocrine system includes a wide range of organs and hormones, such that there is a vast number of potential ways in which the system can be disrupted. Man-made EASs are certainly found in the environment, and it is likely that critical effects in wildlife populations are passing unnoticed. Also, given the lack of toxicity data on the bulk of chemicals in commerce today, many currently unsuspected chemicals may in the future be found to contribute to endocrine-disrupting effects. EASs may act together in an additive or more than additive manner. Even currently suspected chemicals may not be adequately controlled, because low-dose effects have not yet been properly investigated, and current toxicity testing regimes may not pick up endocrine disruption. Many species could potentially be under threat, because the endocrine system has many similarities in living creatures from different phyla. Finally, given all these concerns, it can be concluded that there is a very real likelihood that EASs will, in the future, be found to be causing population-level effects in many more wildlife species. Thus, the macro-level risk assessment predicts that EASs pose a high risk for wildlife populations. It is certainly reasonable to suggest that the scale of the phenomenon of endocrine disruption is likely to be considerably greater than currently known, and that many wildlife populations may be impacted.

AN ENVIRONMENTALIST'S PERSPECTIVE ON IMPROVEMENTS THAT NEED TO BE IMPLEMENTED IN FRAMEWORKS TO ASSESS THE RISKS OF INDIVIDUAL CHEMICALS

This section focuses on risk assessment, as it is the central pillar of chemical-based legislation, which provides controls over the registration, marketing, and use of substances. However, it must be recognized that controls over the marketing and use of certain chemicals do not preclude the need for extensive environmental monitoring and surveillance. This must operate in tandem to identify and prevent effects both from substances that are not marketed (e.g., natural hormones found in sewage effluents, degradation products, and unwanted by-products), and from substances that have either not undergone hazard and risk assessment, or are being used in a harmful manner.

Formal frameworks for assessing the risks from individual chemicals are laid down in various national and regional legislative regimes, but unfortunately these tend to ignore important concerns. Such omissions undermine the validity of these assessments. For example, because it is difficult to find a scientifically acceptable way of dealing with the complexity, risk assessment tends to ignore the likelihood of possible additive effects. Also, the likely ramifications for wildlife populations of subtle effects from low-level exposures are often largely ignored, as it is difficult to show whether or not these effects will affect the population level. Scientists performing risk assessments seem loath to make predictions about these concerns, because few robust scientific data are available.

Several of the key improvements that need to be implemented in the risk assessment of individual chemicals are listed below:

- Substances should be tested at low doses. Hormone-disrupting chemicals modulate a system that is physiologically active. Therefore, the traditional assumption that there is a threshold dose that must be exceeded before a response is seen may not apply. Receptor-mediated responses can first increase and then decrease as the dose increases, due to receptor down-regulation and other compensatory responses. Thus, it should not be assumed that dose–response relationships are always monotonic, because this is contradicted by many examples from the literature [29,30]. The potential low-dose effects on behavior should be fully evaluated.

- There is a need to try to minimize animal testing, but it may not be sufficient to look for effects in just one or two species, because it is widely accepted that there may be remarkable species variation in the sensitivity to chemicals. In particular, there is a need to develop test methods for amphibia and invertebrates, and sentinel species need to be identified. Mollusks, for example, seem particularly sensitive to certain EASs. In addition, if exposure is likely to occur, tests may need to be undertaken to determine the potential effects on birds and on plants. A battery of tests may therefore be necessary, particularly given the diverse physiological mechanisms in different animals. As an example, eggshell thinning in birds would not be predicted by testing on mammals, fish, or invertebrates.
- In the risk assessment of an individual chemical, the possibility for additive or interactive effects due to combinations of substances should be considered, where exposure need not be simultaneous or via the same route.* There is already evidence to suggest that the exposure to several EASs may result in a combined response more than the threshold for effects, even though individually each chemical is below its effect level [31,32]. Risk assessment as currently practiced usually only evaluates the effects of the individual chemical on a previously unexposed laboratory animal, but in the real world an animal may be exposed to a variety of substances, via several exposure routes, including air, water, and food, as well as from the remobilization of contaminants already stored in the organism's adipose tissue. It will be exceedingly difficult to accurately predict the interactive effects of exposure to numerous hormonally active substances at varying concentrations. Possible options for dealing with this might include the imposition of further "safety" or "assessment factors" in the risk assessment, or the development of a toxic equivalent factor approach. However, some researchers have already identified that a toxic equivalent factor approach, as used for dioxin and dioxin-like PCB mixtures, is unlikely to be a useful approach for these compounds, because EASs may behave in an additive manner for some effects, but antagonistic for others [35]. Risk assessment of chemicals should also always be based on aggregate risk, such that all exposure routes to the substance should be considered.**
- A more prudent approach (that is, a more cautious approach) should be brought to bear when deciding whether subtle low-dose effects are likely to affect human health or to impact the population level of wildlife. The burden of proof should be lowered such that regulations should deliver "reasonable certainty of no harm". This is explained in more detail below, where the European Union (EU) situation is used as an example.

The need for a more prudent approach in risk assessment

Much of the recent heated debate on EASs has focused on conflicting studies relating to whether certain EASs, such as bisphenol A, can cause effects at very low-dose levels. In the future, science will hopefully resolve this issue. By 2002, low-dose effects due to bisphenol A had been reported in several species, including fish [36], mollusks [37], and mammals [38–43].

However, this debate may obscure more intractable reasons for differences in the perception of the risks from EASs. That is, what are the implications of some of the effects that can be attributed to

*As an example of such an omission, the potential additive effects due to concurrent exposure to di(2-ethylhexyl)phthalate (DEHP) and di(*n*-butyl)phthalate (DBP) are unable to be considered within the EU risk assessment of DEHP under the Existing Substances Regulation, despite the fact that exposure to several phthalates is likely [33], and that additivity has been shown to occur [34].

**The U.S. Food Quality Protection Act, for example, does consider the cumulative effects of pesticide residues and other substances that have common mechanisms of action, and also considers aggregate risk from non-food exposure.

low doses of EASs in laboratory experiments, and which effects should be taken forward in risk assessment to drive the regulation of these chemicals? As an Organization for Economic Cooperation and Development (OECD) document noted, "judgements are unclear and certainly not unanimous on issues such as whether or not hormonal effects without obvious toxic effects should be considered adverse" [44].

In the EU, the debate has been aired at the Technical Meetings for the Existing Substances Regulation. In these meetings, representatives from the Member States assess the risks of priority industrial chemicals, which were placed on the EU market prior to 1981. In the meetings discussing the risk assessment for humans, the implications of a positive result in the uterotrophic assay has been an area of disagreement. Representatives of the United Kingdom have argued that the dose of bisphenol A causing an increase in uterine weight in laboratory animals should not be taken forward in risk assessment. They consider that the uterotrophic assay is only a screening test, which while it can be used to indicate potential estrogenic activity, should not be used for risk characterization purposes because it does not measure functional changes in reproductive parameters. However, representatives from other Member States, such as Denmark and Sweden, disagree and have argued that an increase in uterine weight should be considered adverse.

The World Wildlife Fund (WWF) supports the Swedish and Danish viewpoint that effects on uterus weight should be taken forward in risk assessment. Exposure to xenoestrogens at levels causing biological effects should be considered undesirable. Indeed, an increased uterine weight is probably not the most sensitive endpoint for estrogenicity [45], such that even lower dose levels need to be taken forward in the risk assessment. Estrogenic effects should be considered undesirable because it is well known that increased exposure to endogenous estrogen increases the risk of breast cancer. Moreover, our knowledge of the roles played by hormones in unborn and newborn infants are still being unravelled, such that it would be better to be cautious about the possible long-term effects of perturbing normal function. It is, for example, interesting to note that during the first three months of life, male babies have high levels of male hormones (around 50 % of adult levels) [46,47]. It is not known exactly why this is, but it is believed that the subsequent behavior of the individual is imprinted at this time [48]. Therefore, interference in hormonal processes at this age could have significant consequences to development.

Similarly, in the EU Technical Meetings discussing the environmental risk assessment, there has been vigorous debate as to whether some of the effects that have been noted in laboratory experiments are liable to translate into an effect on fish population numbers. For example, in an experiment on fish exposed to bisphenol A, an effect on spermatogenesis was noted at a dose level lower than that causing effects on egg hatchability [36]. At the EU Technical Meeting in March 2001, the UK representative argued that only the dose level causing effects on egg hatchability should be used for risk assessment, since the lower dose causing effects on sperm could be disregarded, as it had been explicitly shown not to affect hatchability. However, other Member States were not satisfied, and did not want to dismiss the effects on spermatogenesis as unimportant for the risk assessment. This was because although effects on spermatogenesis may not affect the population level in the laboratory, this may not hold true in the wild, where a small difference in sperm count might well have an impact on reproductive performance.

Stuart Dobson of the UK Centre for Ecology and Hydrology has noted that the distinctions that are made between findings that are likely to affect the population level, and those that are not, may be unrealistic and somewhat artificial, since there is no accepted way of reliably determining which is which [49]. Even traditional endpoints, such as egg hatchability, have no definite population effect, since survival to adulthood and breeding is density-dependent. Thus, the distinction is deductive rather than scientific because there is no population dynamics component in the risk assessment process.

Opinions of EU technical experts have also been divided as to how to interpret other effects recorded in laboratory experiments. For example, low doses of bisphenol A have been shown to affect sword length in swordtail fish (*Xiphophorus helleri*) [50]. This is considered to be a secondary sexual characteristic in males, but UK experts did not consider that a reduction in sword length would impact on the population level. The draft Risk Assessment Report, written by the UK rapporteur, stated, "The significance of the changes in sword length is not understood. It is thought that the length of the sword has an influence on mating success, with female fish preferring males with longer swords, but it is not clear what degree of change should be considered to be significant." However, it could be argued that altered pairing might make the population less able to cope with other stresses. The United Kingdom appears to want to be assured about the levels causing significant or population-level effects, but this approach is fundamentally flawed because it would be very difficult, if not impossible, to prove the ecological relevance of all the effects recorded in the laboratory. The UK approach is likely to reduce or potentially eliminate safety margins that *might* exist due to the imposition of assessment factors designed to take account of intra- and inter-species variations. Nevertheless, this is the situation that prevails.

WWF suggests that dose levels giving rise to effects that can be *reasonably predicted* to give rise to a population-level effect, should be taken forward in risk assessment. Thus, effects on spermatogenesis, sword length, or vitellogenin production should all be considered potentially adverse for the population, and these dose levels taken forward in risk assessment. The complexities of ecosystems, and the external threats they face, are such that the environment should be given the benefit of any doubt. If the population level of just one species is directly adversely affected, then this may have as yet unknown consequences for many other species. For example, tributyltin (TBT) caused imposex and decline in numbers of mollusks, but several other species were also affected [51–53]. Therefore, it is argued that there is a need for a more cautious or more prudent approach to be taken in risk assessment.

Similarly, in the United States there is a need to lower the hurdle that regulators have to climb before they can impose control measures. For example, the Toxic Substances Control Act (TSCA) requires that an "unreasonable risk" must be shown. However, environmentalists would argue that the burden of proof should be reversed and lowered, such that industry should have to provide data and regulations should deliver "reasonable certainty of no harm".

JUSTIFICATION FOR INVOKING THE PRECAUTIONARY PRINCIPLE

Each member of the World Trade Organization has the independent right to determine the level of environmental or health protection that they consider appropriate. For countries that want to promote a high level of protection, making modifications to the risk assessment methodology may not be adequate. Instead, recourse to the precautionary principle may be justified for EASs.

Where the risk assessment indicates that there are reasonable grounds for concern that the desired level of protection for the environment (or future generations) could be jeopardized, then precautionary risk management can be taken, even though the scientific evidence is insufficient, inconclusive, or uncertain.

EASs have properties that specifically lead to increased uncertainty in the estimation of whether they pose an acceptable risk. Such increased uncertainty arises due to the greater potential for interactive effects, because EASs frequently act via a common mechanism of action, or via pathways that converge on a common mechanism of action. EASs also give rise to increased uncertainty in relation to potential low-dose effects (including behavioral effects), and what these might mean for the organism. Thus, EASs are likely to cause adverse effects in ways that challenge traditional risk assessment models. Given this, it is argued that precautionary action can be justified. In the EU, there have already been several notable political statements on the need to take precautionary action on EASs [54,55], although little concrete action to reduce the risks has yet been seen.

Opponents of the precautionary principle ridicule it as being a tool that can be used arbitrarily even in the absence of evidence that a hazard exists. This is not the case. The intent is that precaution can be invoked when the risk is unproven but nevertheless plausible in the light of existing scientific knowledge. The European Commission's communication on the precautionary principle states that recourse to the precautionary principle presupposes (a) the identification of potentially negative effects resulting from a phenomenon, product, or procedure, and (b) a scientific evaluation of the risk, which because of the insufficiency of the data, or their inconclusive or imprecise nature, makes it impossible to determine with sufficient certainty the risk in question [56].

Recourse to the precautionary principle might result in any one or more of a range of options, from requiring more research to implementing a ban on the chemical. Therefore, for example, a possible precautionary risk management option would be the imposition of a requirement to get permission before the EAS could be used, and to permit its use only in certain applications, considering its socio-economic benefits, the lack of safer chemicals, and measures to minimize exposure. Indeed, this is the approach that the EU may take to control the risks posed by these chemicals, an approach that is actively supported by WWF. Thus, in 2001, the Environment Ministers of the EU suggested that when scientifically valid test methods and criteria are available for EASs, then new EU legislation should require EASs to be subject to a prior authorization procedure [57]. However, as of 2003, the outcome of the EU review of chemicals legislation was still a matter of debate.

CONCLUSIONS AND RECOMMENDATIONS

Conclusions and appraisal of some government and industry responses

- To identify appropriate policy responses to EASs, there is a need to assess the likely environmental risks both at the *micro (*or single-substance level), and at the *macro level.* The term "macro level risk assessment for endocrine disruptors" is coined to refer to an estimation of whether adverse effects are likely to occur, and the possible severity of those effects, due to current exposures to EASs.
- The macro-level risk assessment predicts that EASs pose a high risk for wildlife populations. It is certainly reasonable to suggest that the phenomenon of endocrine disruption is considerably greater than currently known, because critical effects in wildlife populations are likely to be passing unnoticed.
- Risk assessment as currently practiced for individual chemicals does not adequately protect the environment, and improvements could be made.
- However, given the increased uncertainty in determining the risks posed by EASs, countries wishing to implement a high level of protection would be justified in taking precautionary action.
- Regulatory processes are clearly inadequate, in that many chemicals are currently marketed without even basic toxicity data being available [58–61]. Undoubtedly, industry has a legacy of neglect, and could do more to help develop new test methods, and to identify EASs using the available test methods.
- The U.S. government's response to the issue of endocrine disruption has been praiseworthy, in that it has a mandatory screening and testing program. These commitments need to be met.
- Japan is also to be congratulated, in that Japanese scientists are very active in developing screens and tests, including DNA micro-array systems [62]. Moreover, Japan is reportedly to implement a tiered approach to chemical testing consisting of initial sorting, followed (where appropriate) by prescreening, screening, and definitive testing [63].
- Industry groups overseeing research programs tend to "down-play" or miscommunicate the risks. For example, the European Chemical Industry Council (CEFIC) 1999 Newsletter focusing on endocrine disruption concluded the following, "Even though CEFIC's ED research program is yielding results on the potential effects of endocrine disrupters on human health and wildlife, we're still a long way from finding out if or how this is happening" [64]. In contrast to this view, there are many reports showing that endocrine disruption is certainly occurring in wildlife. Indeed, an eminent scientist in the field, Dr. Peter Matthiessen (Centre for Ecology and Hydrology, UK) has said that there is abundant evidence that wildlife is being affected and that endocrine disruption in wildlife is no longer just a hypothesis [65].

- "Public health professionals need to be aware that the 'sound science' movement is not an indigenous effort from within the profession to improve the quality of scientific discourse, but reflects sophisticated public relations campaigns controlled by industry executives and lawyers whose aim is to manipulate the standards of scientific proof to serve the corporate interests of their clients" [66]. Industry should, of course, be entitled to air their views honestly, and such views should be given due consideration, but industry funding should not be hidden.
- The CEFIC has estimated that between 1996 and 2005, it will have allocated about 17 million Euros *in total* to research into endocrine disruption [Taalman, personal communication, 29.01.02]. This is a remarkably small amount, which equates to just 0.004 % of the *annual* value of EU chemicals production.
- There are grounds for criticizing industry's actions to date and for suggesting a certain amount of duplicity. On the one hand, industry wants to be seen to be responsible and tries to show that it has invested a significant amount of money in research, while on the other hand it actually allocates a relatively small proportion of money to research, and seeks to undermine the concerns. That said, it is always difficult to generalize, and some companies are obviously better than others.

Recommendations

- Government and industry research funding into endocrine disruption should be substantially increased and guaranteed for many years to come, because we are still a long way from understanding the extent of the problem.
- Long-term monitoring of the environment is needed, and this should include both the aquatic and terrestrial environment, and species from all trophic levels.
- More effort is particularly needed to hasten the development of screens and tests for ecotoxicity. Developing OECD test guidelines is always a lengthy process, but there is a need for far more commitment from governments.
- Consideration should be given to reviewing the OECD screening information data sets (SIDS), and enhancing some of the OECD test guidelines with respect to ED effects and/or adding new screens.
- Governments should commit to a comprehensive sorting, screening, and testing program for chemicals to which significant exposure occurs. International coordination is needed to prevent unnecessary duplication of testing.
- In particular, given the concern about endocrine disruption in the EU, the Member States of the EU should agree to a screening and testing strategy for EASs, which should be implemented as far as possible in the forthcoming new EU chemicals legislation.
- More resources and international coordination and effort are needed to speed the development of sophisticated non-animal test methods for screening and testing chemicals for endocrine disruption. However, governments should test chemicals to the extent necessary to protect human health and the environment.
- Substances should be tested at low doses, because there may not be a threshold dose, and the dose–response relationship may not be monotonic.
- In risk assessment, exposure to combinations of substances (cumulative and aggregate risks) must be considered.
- In environmental risk assessment, a more prudent approach (that is, a more cautious approach) should be taken when deciding whether a particular effect might have population-level repercussions. Regulatory frameworks should require "reasonable certainty of no harm".
- Given the increased uncertainty in the risk assessment of EASs, WWF considers that precautionary action to try to eliminate or minimize exposures to EASs is warranted.

REFERENCES

1. L. Tattersfield, P. Matthiessen, P. Campbell, N. Grandy, R. Lange. SETAC-Europe/OECD/EC Expert Workshop on Endocrine Modulators and Wildlife: Assessment and Testing (EMWAT), SETAC Europe, Brussels (1997).
2. T. Colborn and C. Clement. *Advances in Modern Environmental Toxicology, Vol. XXI, Chemical-Induced Alterations in Sexual and Functional Development: The Wildlife/Human Connection*, Princeton Scientific Publishing, Princeton, NJ (1992).
3. Environment Agency of England and Wales. Endocrine disrupting substances in wildlife: A review of the scientific evidence and strategic response. Environment Agency, Bristol (1998).
4. M. Hewitt and M. Servos. *Water Qual. Res. J. Canada* **36** (2), 191–213 (2001).
5. EU Committee on Toxicity Ecotoxicity and the Environment. *CSTEE Opinion on Human and Wildlife Health Effects of Endocrine-Disrupting Chemicals, with Emphasis on Wildlife and on Ecotoxicology Test Methods*. DGXXIV, Consumer Policy and Consumer Health Protection, Brussels (1999).
6. T. Damstra, S. Barlow, A. Bergman, R. Kavlock, G. Van Der Kraak. Global assessment of the state-of-the-science on endocrine disruptors. International Programme on Chemical Safety (IPCS), WHO (2002).
7. Commission of the European Communities (CEC). White Paper: Strategy for a future Chemicals Policy, 2001 (COM92001) 88 final, CEC, Brussels (2001).
8. P. Matthiessen, Y. Allen, S. Bamber, J. Craft, M. Hurst, T. Hutchinson, S. Feist, I. Katsiadaki, M. Kirby, C. Robinson, S. Scott, J. Thain, K. Thomas. *Mar. Environ. Res.* **54** (3–5), 645–649 (2002).
9. J. Loh. Living Planet Report, WWF International, Gland (2002).
10. B. A. Stein. *Environment* **43** (7), 12–22 (2001).
11. EU Scientific Committee on Toxicity, Ecotoxicity and the Environment (CSTEE). Opinion expressed at the 19th CSTEE plenary meeting, C2/JCD/csteeop/Ter9100/D(0), 9 November, Brussels (2000).
12. Technical Guidance Documents in support of the Commission directive 93/67/EEC on risk assessment for new notified substances and the Commission regulation (EC) 1488/94 on risk assessment for existing substances (2002).
13. D. C. Morse, A. Plug, W. Wesseling, K. J. can den Berg, A. Brouwer. *Toxicol. Appl. Pharmacol.* **139**, 252–261 (1996).
14. J. L. Jacobson and S. W. Jacobson. *N. Engl. J. Med.* **335** (11), 783–78 (1996).
15. S. Patandin, C. I. Lanting, P. G. Mulder, E. R. Boersma, P. J. Sauer, N. Weisglas-Kuperus. *J. Pediatr.* **134** (1), 33–41 (1999).
16. T. Darvill, E. Lonkey, J. Reihman, P. Stewart, P. Pagano. *Neurotoxicology* (6), 1029–1038 (2000).
17. T. J. Kubiak, H. J. Harris, L. M. Smith, T. R. Schwartz, D. L. Stalling, J. A. Trick, L. Sileo, D. E. Docherty, T. C. Erdman. *Arch. Environ. Contam. Toxicol.* **18** (5), 706–727 (1989).
18. G. L. Hunt and M. W. Hunt. *Science* **196**, 1466–1467 (1977).
19. S. A. Fisher, G. R. Bortolotti, K. J. Fernie, J. E. Smits, T. A. Marchant, K. G. Drouillard, D. M. Bird. *Arch. Environ. Contam. Toxicol.* **41** (2), 215–220 (2001).
20. A. E. Quaglino, C. B. Craig-Veit, M. R. Viant, A. L. Erichsen, D. M. Fry, J. R. Millam. *Horm. Behav.* **41** (2), 236–241 (2002).
21. P. Eriksson, E. Jakobsson, A. Fredriksson. *Dioxin 98, Vol. 35, Organohalogen Compounds*, Stockholm (1998).
22. P. Matthiessen, J. McLachlan, P. Myers, I. Callard. *APMIS Supplement* **109** (103), S187–S188 (2001).
23. J. P. Geisy and K. Kannan. *Environ. Sci. Technol.* **35** (7), 1339–1342 (2001).
24. M. P. Longnecker, M. A. Klebanoff, H. Zhou, J. W. Brock. *Lancet* **358** (9276), 110–114 (2001).
25. R. Dietz, M. P. Heide-Jorgensen, T. Harkonen. *Ambio* **18**, 258–264 (1989).

26. H. Van Loveren, P. S. Ross, A. D. Osterhaus, J. G. Vos. *Toxicol. Lett.* **15** (112–113), 319–24 (2000).
27. EU Committee on Toxicity Ecotoxicity and the Environment. *CSTEE Opinion on Human and Wildlife Health Effects of Endocrine Disrupting Chemicals, with Emphasis on Wildlife and on Ecotoxicology Test Methods.* DGXXIV, Consumer Policy and Consumer Health Protection, Brussels (1999).
28. J. E. Fox, M. Starcevic, K. Y. Kow, M. E. Burow, J. A. McLachlan. *Nature* **413**, 128–129 (2001).
29. National Toxicology Program. *Report of the Endocrine Disruptors Low Dose Peer Review,* USA (2001).
30. F. S. vom Saal, B. G. Timms, M. M. Montano, P. Palanza, K. A. Thayer, S. C. Nagel, M. D. Dhar, V. K. Ganjam, S. Parmigiani, W. V. Welshons. *Proc. Natl. Acad. Sci.* **94** (5), 2056–2061 (1997).
31. Institute for Environment and Health and Environment Agency National Centre for Ecotoxicity and Hazardous Substances. The Identification of oestrogenic effects in wild fish – phase II, R & D Technical Report W2-014/TR, principal investigators J. Sumpter, S. Jobling, C. Tyler, Environment Agency, Bristol (2002).
32. E. Silva, N. Rajapakse, A. Kortenkamp. *Environ. Sci. Technol.* **36** (8) (2002).
33. Centers for Disease Control and Prevention. *National Report on Human Exposure to Environmental Chemicals*, Atlanta, USA (2001).
34. P. M. Foster, K. J. Turner, N. J. Barlow. *Toxicologist* **66** (1-S), 233 (2002).
35. K. Gaido, L. You, S. Safe. SCOPE/IUPAC International Symposium on Endocrine Active Substances, program and collective abstracts, November 17–21, Yokohama (2002).
36. P. Sohoni, C. R. Tyler, K. Hurd, J. Caunter, M. Hetheridge, T. Williams, C. Woods, M. Evans, R. Toy, M. Gargas, J. P. Sumpter. *Environ. Sci. Technol.* **35** (14), 17–25 (2001).
37. J. Oehlmann, U. Schulte-Oehlmann, M. Tillmann, B. Markert. *Ecotoxicology* **9**, 383–397 (2000).
38. S. C. Nagel, F. S. vom Saal, K. A. Thayer, M. G. Dhar, M. Boechler, W. V. Welshons. *Environ. Health Perspect.* **105** (1), 70–76 (1997).
39. F. S. Vom Saal, P. S. Cooke, D. L. Buchanan, P. Palanza, K. A. Thayer, S. C. Nagel, S. Parmigiani, W. V. Welshons. *Toxicol. Ind. Health* **14** (1/2), 239–260 (1998).
40. K. L. Howdeshell, A. K. Hotchkiss, K. A. Thayer, J. G. Vandenbergh, F. S. vom Saal. *Nature* **401**, 763–764 (1999).
41. C. Gupta. *Proc. Soc. Exp. Biol. Med.* **224**, 61–68 (2000).
42. M. Sakaue, S. Ohsako, R. Ishimura, S. Kurosawa, J. Yonemoto, C. Tohyama. *J. Occup. Health* **43**, 185–190 (2001).
43. A. S. Al-Hiyasat, H. Darmani, A. M. Elbetieha. *Eur. J. Oral Sci. Apr.* **110** (2), 163–7 (2002).
44. OECD (Organization for Economic Cooperation and Development). *Initiatives to share the burden of the testing and assessment of endocrine disrupting chemicals*, ENV/JM(2001)20, 11 April, OECD, Paris (2001).
45. C. M. Markey, C. L. Michaelson, E. C. Veson, C. Sonnenschein, A. M. Soto. *Environ. Health Perspect.* **109** (1), 55–60 (2001).
46. M. G. Forest, P. C. Sizonenko, A. M. Cathiard, J. Bertrand. *J. Clin. Invest.* **53**, 819–828 (1974).
47. N. J. Bolton, J. Tapanainen, M. Koivisto, R. Vihko. *Clin. Endocrinol.* **31**, 201–207 (1989).
48. B. D. Sachs, R. L. Meisel. In *The Physiology of Reproduction* 1st ed., Vol. 2, E. Knobil and J. Neill (Eds.), pp. 1444–1447, Raven Press, New York (1988).
49. Risk Policy Analysts (RPA). Report from the UK Environment Agency's National Centre for Ecotoxicology and Hazardous Substances Peer Review Group, Consultation on BPA, RPA, Norfolk (2001).
50. H. I. Kwak, M. O. Bae, M. H. Lee, Y. S. Lee, B. J. Lee, K. S. Kang, C. H. Chae, H. J. Sung, J. S. Shin, J. H. Kim, W. C. Mar, Y. Y. Sheen, M. H. Cho. *Environ. Toxicol. Chem.* **20** (4),787–95 (2001).

51. Health Council of the Netherlands. *Hormone Disruptors in Ecosystems.* 1999/13E, Health Council of the Netherlands, The Hague (1999).
52. R. Waldock, H. L. Rees, P. Matthiessen, M. A. Pendle. *J. Mar. Biol. Assoc. UK* **79**, 225–232 (1999).
53. H. L. Rees, R. Waldock, P. Matthiessen, M. A. Pendle. *Mar. Pollut. Bull.* **42**, 137–144 (2001).
54. The Environment Council of the EU, Council conclusions on the Communication from the Commission to the Council and the European Parliament: Community strategy for endocrine disrupters – a range of substances suspected of interfering with the hormone systems of humans and wildlife, 7526/00, Brussels, 7th April 2000. Conclusion 5 stresses "that the precautionary principle must be applied in order to respond quickly and effectively to the problem…". Conclusion 6 considers "that for endocrine disrupters, there is a need to develop quick and effective risk management strategies for substances which may, on the basis of a preliminary scientific evaluation, have potential adverse effects on human health or the environment…".
55. The European Parliament, Resolution A5-0197/2000 on the Commission communication to the Council and the European Parliament on a Community strategy for endocrine disrupters – a range of substances suspected of interfering with the hormone systems of humans and wildlife (COM(1999) 706-C5-0107/2000 – 2000/2071(COS)), October 2000. Paragraph C states "whereas limitations and uncertainties in the available scientific data and, consequently, discrepancies in the observed effects – linked with stated disagreements between scientists as to the importance and/or interpretation of data – must all lead to the application of the precautionary principle".
56. Commission of the European Communities. Communication from the Commission on the precautionary principle, COM(2000)1, Brussels (2000).
57. Environment Council of the European Union. Council Conclusions from their meeting of 7 June on the Commission of the European Communities (CEC). *White Paper: Strategy for a Future Chemicals Policy, 2001 (COM92001) 88 final,* CEC, Brussels (2001).
58. National Academy of Sciences (NAS), National Research Council. *Toxicity Testing, Strategies to Determine Needs and Priorities,* National Academy Press, Washington DC (1984).
59. Environmental Defence Fund (EDF). *Toxic Ignorance*, EDF, Washington DC (1997).
60. US EPA, Office of Pollution Prevention and Toxics. *Chemical Hazard Data Availability Study*, US EPA, Washington DC (1998).
61. R. Allanou, B. G. Hansen, Y. van der Bilt. *Public Availability of Data on EU High Production Volume Chemicals. EUR 18996 EN.* European Commission Joint Research Centre, Italy (1999).
62. OECD. Thought starter paper: Sharing the work on testing and assessment of endocrine disruptors. 15 May 2002. ENV/JM/TG/EDTA(2002)4.
63. OECD. Thought starter paper: Framework for the testing and assessment of endocrine disruptors, 15 May 2002. ENV/JM/TG/EDTA(2002)2.
64. CEFIC. The Long Range Research Initiative Update. Issue 2. The chemical industry's research program on endocrine disruption.
65. The ENDS Report. May, **280**, Environmental Data Services, London (1998).
66. E. K. Ong and S. A. Glantz. *Am. J. Pub. Health* **91** (11), 1749–1757 (2001).

Pure Appl. Chem., Vol. 75, Nos. 11–12, pp. 2605–2607, 2003.

Workshop 6.6

Endocrine disruption and the USFDA's Center for Drug Evaluation and Research*

Abigail Jacobs‡, Paul Brown, James Farrelly, J. Edward Fisher, and David Morse

U.S. Food and Drug Administration, HFD-540 5600 Fishers Lane, Rockville, MD 20857, USA

Abstract: Drugs may have intended or unintended endocrine effects. Drug evaluation may include both in vitro and in vivo evaluations of toxicity and developmental/reproductive effects. After a signal is identified, human relevance is of utmost concern. An integration "tool" that formalizes a weight-of-evidence approach has been developed to assess concern about reproductive/developmental toxicity to humans. This approach can be used to assess concern about an endocrine disruption signal. A signal alone does not mean a concern for humans. An effect needs to have biologic relevance, and exposure thresholds for effects may exist. Risk/benefit for a particular drug is a clinical decision and may vary by the drug indication. Risk management for an identified concern could include wording in patient communications, tracking distribution or limited distribution, and patient or pregnancy registries.

INTRODUCTION

Drugs may have intended or unintended endocrine effects. Several drug products have potent intended endocrine effects, such as oral contraceptives, selective estrogen receptor modulators, estrogen receptor antagonists, luteinizing hormone agonists, and some drugs with effects on the thyroid gland or with 5α-reductase inhibitory activity or aromatase inhibitory activity. Other drug products may have adverse effects on the thyroid gland, unintended luteinizing hormone agonist activity, or unintended inhibition of 5α-reductase or aromatase activity. The intended effects are the reason for developing particular drug products, whereas the unintended effects may only become apparent during drug development.

IDENTIFICATION OF A HAZARD

Drug evaluations that might reveal endocrine effects include both in vitro and in vivo evaluations of toxicity and developmental/reproductive effects. Nonclinical in vitro evaluations of drugs could include receptor binding assays, transcriptional activation assays, gene expression assays, or enzyme inhibition assays. Specific in vivo endocrine evaluations could include uterotrophic assays for estrogenic activity and the Hershberger assay for androgenic activity. Other in vivo studies that could reveal endocrine effects include repeated dose toxicity studies that include clinical chemical analyses, organ weight measurements, histologic evaluation, and various reproductive and developmental toxicity studies. If an endocrine effect is suspected, additional endpoints evaluating reproductive performance or focused on

*Report from a SCOPE/IUPAC project: Implication of Endocrine Active Substances for Human and Wildlife (J. Miyamoto and J. Burger, editors). Other reports are published in this issue, *Pure Appl. Chem.* **75**, 1617–2615 (2003).

‡Corresponding author

specific aspects of the male and female reproductive systems can be added to standard studies to enhance the probability of detecting an effect. Carcinogenicity studies may reveal effects on the thyroid gland or on levels of luteinizing hormone or prolactin.

A positive endocrine finding in an in vitro nonclinical study must be carefully evaluated to determine its relevance to in vivo conditions. An in vitro effect without an in vivo adverse effect may not be considered to be a signal. Binding to a receptor or any other in vitro biologic activity does not necessarily translate into adverse effects for humans under clinical conditions of use. In vitro effects may depend on the tissue or species tested. A false-positive from an in vitro study may be detrimental to development of a drug product and may detract from a drug's therapeutic benefit. The value of in vitro studies may be greatest when they help clarify a mechanism of an observed in vivo effect.

EVALUATION OF CONCERN FOR ADVERSE HUMAN ENDOCRINE EFFECTS, AN INTEGRATED ASSESSMENT

After an in vivo endocrine or reproductive developmental signal is identified, human relevance is of utmost concern. An integration "tool" that formalizes a weight-of-evidence approach has been developed in the Center for Drug Evaluation and Research (CDER), U.S. Food and Drug Administration (USFDA), to assess concern about reproductive/developmental toxicity to humans. The draft document is described in the federal register and is available on the Center for Drug Evaluation, USFDA, Web site [1]. Ordinarily, the integration process would be based on an evaluation of a complete set of the expected general toxicology, reproductive toxicology, and pharmacokinetics studies. This evaluation would include an assessment of the ability of the drug to produce a positive finding in the relevant animal studies, or whether doses used were large enough to induce toxicity of some kind. The evaluation would also compare animal and human pharmacodynamic effects, animal and human metabolism and disposition, animal and human pharmacologic and toxic effects, and drug exposures in animal studies in relation to the highest proposed dose in humans. Three classes of reproductive toxicity are evaluated for degree of concern for human risk: fertility, parturition, and lactation. Four classes of developmental toxicity category are evaluated: mortality, dysmorphogenesis (structural alterations), alterations to growth, and functional toxicities. In this formalized approach to a weight-of-evidence evaluation, six factors are considered: signal strength part 1; signal strength part 2; pharmacodynamics; concordance between test species and humans; relative exposures in animals and humans; and class alerts in humans. Each factor has several elements. Signal strength part 1 includes cross-species concordance, multiplicity of effects, and adverse effects at different stages of development. Signal strength part 2 includes considerations of parental toxicity, the dose response, and rarity of the event. For each contributory element within a factor, the quality and type of data are considered. An assignment of increased (+1), decreased (–1), or no change (0) in the level of concern is made for each factor in an endpoint for an identified signal. The values for the factors are summed to arrive at an overall conclusion for an endpoint.

Sometimes, concern for an adverse endocrine effect can be estimated more simply by considering the biologic relevance for an effect, existence of a threshold for the effect, exposure in animals at which an effect occurred or did not occur, and availability of monitorable biomarkers in humans.

RISK/BENEFIT OF ENDOCRINE EFFECTS OF DRUG PRODUCTS

For a drug product, risk is never considered in isolation from the benefit of the drug product. Risk/benefit for a particular drug product is a clinical decision and will vary with the drug indication. An endocrine effect could be desirable or undesirable. The more serious the indication, the greater the tolerance for adverse effects. Risk management for an identified concern could include monitoring patients, wording in patient communications regarding the drug product, tracking distribution or limited distribution of the drug product, and patient or pregnancy registries. Risk management programs are developed in cooperation with the drug sponsors. The Center for Drug Evaluation and Research (of the

USFDA), and the Center for Biologics Evaluation and Research (of the USFDA), are jointly working to develop final guidance documents that will address good risk assessment, risk management, and pharmacovigilance practices.

REFERENCES

1. <www.fda.gov/cder/guidance/4625dft.pdf>.

Pure Appl. Chem., Vol. 75, Nos. 11–12, pp. 2609–2611, 2003.

Workshop 6.7

Relevant activities for risk management of endocrine disruptors in Japanese government agencies*

Shinichi Oikawa and Mitsuo Matsumoto‡

Chemical Management Policy Division, Ministry of Economy, Trade, and Industry, 1-3-1 Kasumigaseki, Chiyoda-ku, Tokyo 100-8901, Japan

Abstract: Similar to other countries, issues concerning endocrine disruptors have been attracting people's attention, as well as concerns, in Japan. Consequently, relevant ministries have been conducting several activities and projects. A summary of the current activities related to endocrine disruptors being conducted by Japanese government agencies is provided in this paper.

INTRODUCTION

There are six ministries in Japan that are conducting or coordinating activities and projects related to endocrine disruptors. Each ministry deals with the issues in the same field from different viewpoints. For example, the Ministry of Economy, Trade, and Industry (METI) deals with the issue mainly from the viewpoint of facilitating industrial activities, while the Ministry of Health, Labor, and Welfare (MHLW) works to promote human health as well as safety of workers. The Ministry of Environment (MoE) works to protect the environment, and the Ministry of Land, Infrastructure, and Transport (MLIT) works to preserve the aquatic environment such as rivers, sewage, water systems, marinas, and the sea. The Ministry of Agriculture, Forestry, and Fisheries (MAFF) works to maintain the safety and quality of food, and the Ministry of Education, Culture, Sports, Science, and Technology (MEXT) promotes basic studies and research on endocrine disruptors.

INTERMINISTRY COLLABORATION BODY

To achieve the maximum use of limited resources among Japanese government agencies and to conduct activities efficiently, a collaboration body called Interministry Meeting for Collaboration on Endocrine Disruptors was established in 1998. The meetings, held at least twice a year or whenever considered necessary, include information exchanges and discussions about taking comprehensive measures for endocrine disruptors.

By combining several activities of different ministries, the comprehensive measures can be materialized. The following are objectives of these comprehensive measures:

- prevent possible effects on human health and/or environmental system;
- control emissions to environment;

*Report from a SCOPE/IUPAC project: Implication of Endocrine Active Substances for Human and Wildlife (J. Miyamoto and J. Burger, editors). Other reports are published in this issue, *Pure Appl. Chem.* **75**, 1617–2615 (2003).

‡Corresponding author

- promote safety of products such as household goods, foods, packaging, and pesticides, as well as safety of workplaces;
- provide technical guidance to industries and develop alternative materials and products; and
- establish the basis for appropriate risk management measures of endocrine disruptors in Japan.

The activities and projects of the major three ministries (METI, MHLW, and MoE) are explained in the following paragraphs.

ACTIVITIES AND PROJECTS OF THE MINISTRY OF ECONOMY, TRADE, AND INDUSTRY

The METI has been developing the 3-D QSAR system since fiscal year (FY) 2000. With the input of in vitro test data for approximately 1000 chemical's, a prototype QSAR system will be established early in 2003. Both in vitro and in vivo screening methods have also been developed and conducted for certain numbers of chemicals. The METI has conducted some in vivo definitive tests, and development activities for improved definitive testing methods have been started recently. The METI also has a project to collect and analyze information and data concerning endocrine disruptors as a basic tool.

The METI has been developing the 3-D QSAR system since FY 2000 and expects that the prototype system will be established early in 2003. The whole development activity is planned to be finalized by the end of FY 2004 (the end of March 2005). In addition, in vitro test data for 500 chemicals will be collected and used to improve the prototype system during the following two years. The QSAR system will be used to prescreen thousands of industrial chemicals for additional in vivo and in vitro screening assays, as well as definitive testing.

Concerning the development of in vitro screening assays, several projects are ongoing. The reporter gene assay using the "high-throughput prescreening (HTPS)" system and the receptor binding assays have been developed for four to five years. As a new project of in vitro assays, the METI has recently started to develop a thyroid hormone effects assay and a receptor nonmediated endocrine-disrupting assay.

The METI is also developing in vivo screening methods such as uterotrophic assay, Hershberger assay, and a repeated dose 28-day oral toxicity study. Data collected while developing these screening methods are also used to improve the above-mentioned 3-D QSAR system.

Concerning the in vivo definitive testing, the METI has conducted two-generation reproduction toxicity testing. Seven chemicals have been tested, and results are open to the public. As a new project, the METI has recently started to develop in utero and through-lactational exposure tests as the improved in vivo definitive test methods.

To support those activities or projects, the METI has been collecting information including scientific data relevant to endocrine disruptors, such as testing and evaluation methods, hazard and exposure assessment data, and information on risk assessment activities throughout the world. Based on the data collected through this project and data taken while developing relevant test methods, the METI has so far issued hazard assessment reports for 15 chemicals.

The METI has one other planned new project. Under the new project, industrial chemicals with the production or import volume of more than 100 tons per year (approximately 2500 chemicals) will be screened to assess their endocrine-disrupting activity. For this screening purpose, a provisional testing and assessment scheme will be set up according to the outcome of the METI's technical development projects described so far. During FY 2003 to 2004, approximately 1000 chemicals are planned to be screened under the scheme, and obtained results will be used to further improve relevant test methods including the QSAR system.

ACTIVITIES AND PROJECTS OF THE MINISTRY OF HEALTH, LABOR, AND WELFARE

The activities of the MHLW are, in general, based on the recommendations by the Study Group on Health Effects of Endocrine-Disrupting Chemicals. The mission of the study group is to comprehensively review the latest findings on endocrine-disrupting chemicals in light of human health, and to recommend further actions.

Their activities and projects are categorized into the following seven fields:

- identify target chemical substances through the HTPS system;
- perform studies on the Organization for Economic Cooperation and Development (OECD) screening methods;
- develop full testing methods to identify/confirm endocrine disruptors;
- establish methods for sampling;
- clarify the inverted U effect (low-dose effect);
- collect and analyze exposure and epidemiological data; and
- promote risk communication.

Under these seven fields of recommendations, the ministry has set 18 action items for further activities. Some of them have been finalized, but most of them are on-going projects with plans to be finalized by FY 2005.

ACTIVITIES AND PROJECTS OF THE MINISTRY OF ENVIRONMENT

The activities of the MoE are based on the report, "SPEED '98, Strategic Programs on Environmental Endocrine Disruptors", which was published in 1998. Following the report, the ministry has been conducting the following activities:

- environmental monitoring and investigation of effects on wildlife;
- research and development of testing methods;
- environmental risk assessment/management; and
- sharing information and establishing an international network.

The MoE identified 65 chemical substances as "suspected chemicals" in the "SPEED '98" report, and have been testing and assessing these chemical substances. They have assessed the environmental risk of the "suspected chemicals", including 12 chemicals in FY 2000, 8 chemicals in FY 2001, and another 8 chemicals in FY 2002.

CONCLUSION

Further information can be obtained from the following URLs:

METI: <www.meti.go.jp/english/report/data/g020205ae.html>
MHLW: <www.nihs.go.jp/hse/endocrine-e/index.html>
MoE: <www.env.go.jp/en/topic/edcs.html>

Activities and projects related to endocrine disruptors are in the beginning stage in Japan. Consequently, measures for risk management of endocrine disruptors have not been developed or established. On the other hand, dialogues between governments and Japanese industries have been held to exchange information as well as views and opinions. As a result, with the view of social responsibility, Japanese companies have voluntarily developed and started risk management programs for some suspected chemical substances such as nonylphenol and bisphenol A. This kind of voluntary risk management activity with the cooperation from industries is believed to be one of the most efficient options under the current situation.

Pure Appl. Chem., Vol. 75, Nos. 11–12, pp. 2613–2615, 2003.

Workshop 6.8

Advancing the testing and assessment of chemical substances for endocrine disruption: OECD activities*

Herman B. W. M. Koëter

Environment, Health and Safety Division, Organization for Economic Cooperation and Development, Paris, France

Abstract: This paper describes the establishment of a number of expert groups in the Organization for Economic Cooperation and Development (OECD) to manage the work related to the development and validation of test methods to identify and characterize the human health and aquatic environmental hazards of endocrine-disrupting chemicals. In addition, the concept of testing and assessment, using the most adequate testing tools without being prescriptive, is explained, and a coordinated approach for international work sharing under the OECD umbrella is described.

INTRODUCTION

The most important aspect of the Organization for Economic Cooperation and Development (OECD) work of the Environment, Health, and Safety Division focuses on the regulatory management of chemicals, taking into account the possible effects of chemicals on human health and the environment. In the context of the Chemicals Program, the work covers chemicals, and includes, e.g., industrial chemicals and pesticides. In 1996, the OECD launched a Special Activity on Endocrine Disruptor Testing and Assessment with the following objectives of:

- providing information and coordinating activities;
- developing new and updated Test Guidelines to detect endocrine disruptors; and
- harmonizing hazard and risk characterization approaches.

This activity was initiated at the request of the Member countries and the Business and Industry Advisory Committee (BIAC) to the OECD to promote that testing and assessment approaches for endocrine disruptors would not substantially differ among countries. In order to manage the work of the Special Activity, the Task Force on Endocrine Disruptors Testing and Assessment (EDTA) was established in the same year.

DEVELOPMENT AND VALIDATION OF TEST METHODS

In the first years of the Special Activity, the focus was largely on developing new and updating existing test methods. The development of selected tests as OECD Test Guidelines involved formal validation of these tests. To that end, validation management groups (VMGs) were established to validate mammalian tests (VMG-mammalian) and ecotoxicity tests (VMG-eco), respectively. In June 2002, the

*Report from a SCOPE/IUPAC project: Implication of Endocrine Active Substances for Human and Wildlife (J. Miyamoto and J. Burger, editors). Other reports are published in this issue, *Pure Appl. Chem.* **75**, 1617–2615 (2003).

EDTA agreed that a third VMG was needed to validate in vitro and other nonanimal tests (VMG-in vitro). Current test method development and validation activities include:

- uterotrophic assay [to detect mammalian (anti)estrogen-mimicking substances];
- Hershberger assay [to detect mammalian (anti)androgen-mimicking substances];
- enhancement of Test Guideline 407 [to detect (anti)estrogen, (anti)androgen, and thyroid hormone-related effects in mammals];
- fish screening and full life-cycle tests and reproduction test(s) in birds (to screen and test antiestrogens and antiandrogen effects;
- amphibian metamorphosis test (for thyroid hormone function disruption); and
- invertebrate tests.

CONCEPTUAL FRAMEWORK FOR TESTING AND ASSESSMENT APPROACHES

To provide a framework for the testing and assessment of potential endocrine disruptors in which the various newly developed tests would fit, a conceptual framework was developed. The conceptual framework is intended to apply to both new and existing substances as well as for different chemical sectors such as pharmaceuticals, industrial chemicals, and pesticides. It was developed taking into account:

- The views of Member countries as expressed through answers to a questionaire and the OECD's Appraisal of Test Methods for Sex Hormone and Disrupting Chemicals (OECD Monograph No. 21).
- Proposed testing schemes such as those developed at relevant workshops, notably the European Workshop on the Impact of Endocrine Disruptors on Human Health and Wildlife (the Weybridge Workshop), and the Joint Society of Environmental Toxicologists and Chemists (SETAC) and OECD/EC Expert Workshop on Endocrine Modulators and Wildlife: Assessment and Testing (the EMWAT Workshop).
- The work of national activities such as the U.S. Environmental Protection Agency's (USEPA's) Endocrine Disruptors Screening and Testing Advisory Committee (EDSTAC) and research activities in Japan.
- Industry initiatives such as those being undertaken by the European Chemical Industry Council (CEFIC).

The conceptual framework has been reconsidered and substantially revised by the EDTA in its June 2002 meeting in Tokyo. There it was confirmed that the conceptual framework is not a testing scheme, but rather a toolbox to place the various tests that can contribute information for the detection of the hazards of endocrine disruption. The toolbox is organized into several compartments or levels, each corresponding to a different level of biological complexity (for both toxicological and ecotoxicological areas). Even though the toolbox may be full of testing tools, this does not imply that they all will be needed for assessment purposes.

SHARING THE WORK OF TESTING AND ASSESSMENT

As this ongoing cooperative work on test method development is starting to bear fruit and the tools for testing and assessment of possible endocrine disruptors are taking shape, the next step in the process was to start using these tools. Considering (i) the vast amount of chemicals currently in use that need to be considered, (ii) the time pressure to identify and assess endocrine-disrupting chemicals as expeditiously as possible, (iii) the number of studies necessary for screening and, as appropriate, full hazard assessment, and (iv) the lack of resources needed for this work, it was considered appropriate to find ways to share (at least some of) the work internationally. In June 2001, Member countries agreed to share the actual testing of existing substances for possible endocrine-disrupting activities and to find ways to share assessments of tested chemicals. Member countries further agreed that the work-sharing activity should start as a small activity that could gradually increase if proven successful. First priority would include grouping chemicals of interest, sharing information and test results of high-throughput screens, and exchanging information on ongoing animal tests. Sharing assessment reports of specific substances was also considered extremely useful. This work has now begun and will be expanded in coming years.

Author Index

Subject Index

F

G